The ENVIRONMENT ENCYCLOPEDIA *and* DIRECTORY

The ENVIRONMENT ENCYCLOPEDIA *and* DIRECTORY

EUROPA PUBLICATIONS LIMITED

First published 1994

18 Bedford Square, London WC1B 3JN, England

Australia and New Zealand
James Bennett (Collaroy) Pty Ltd, 4 Collaroy Street,
Collaroy, NSW 2097, Australia

Japan
Maruzen Co Ltd, POB 5050, Tokyo International 100-31

British Library Cataloguing in Publication Data
A catalogue record for this book is available from
the British Library
ISBN 0-946653 94 1
ISSN 1352-6480

Printed on Reprise 100 paper
(no virgin pulp used,
100% post-consumer waste,
Totally Chlorine Free—TCF)

Printed in England by
Staples Printers Rochester Limited,
Rochester, Kent

Bound by
Hartnolls Ltd,
Bodmin, Cornwall

FOREWORD

The ENVIRONMENT ENCYCLOPEDIA AND DIRECTORY has been produced to meet the growing demand for information about environmental affairs. It brings together factual, cartographic, directory, bibliographic and biographic data concerning the environment in a single volume. The significance of environmentalism, both for the present day and for the future, is examined in the introductory essay.

The book is divided into five main sections. Firstly, there is a series of 20 maps, both global and regional, dealing with climate, vegetation, population, food, desertification, rainforests, the atmosphere, energy, protected areas, fishing, marine resources, pollution, and other subjects of relevance to the environment.

The A–Z of the Environment provides almost 1,000 definitions of terms relating to the environment, such as **acid rain, Bhopal,** the **Common Agricultural Policy,** the **Great Barrier Reef, Lake Baikal,** the **nitrogen cycle** and so on (cross-references within the definitions are given in bold type).

The Directory of Environmental Organizations lists the leading international environmental organizations, followed by the appropriate government bodies and non-governmental organizations in every country of the world. There are approximately 3,000 different entries, giving contact names, addresses, telephone, telex and fax numbers, and a brief outline of each organization's activities.

The Bibliography of Environmental Periodicals alphabetically lists nearly 1,000 relevant titles, giving addresses, telephone, telex and fax numbers, names of editors, foundation dates, and details about frequency, subject matter, language and circulation.

Finally, the Who's Who in the Environment provides biographical details, addresses and telephone, telex and fax numbers for approximately 600 leading personalities involved in environmental affairs around the world.

December 1993.

ACKNOWLEDGEMENTS

The Editors would like to thank all the organizations, periodicals and individuals listed in *The Environment Encyclopedia and Directory* who completed questionnaires and returned them to us.

Special acknowledgements are due to the following: David Nicholson-Lord, for his introductory essay; Edward Oliver, for the maps; Pearce Wright, for his expertise, good humour and perseverance in compiling the A–Z of the Environment; and Alex Lipinski, for his valuable advice during the compilation of the Who's Who section.

With regard to the maps on pages 7–26 of this book, we are particularly grateful for the permission to make use of material from the following sources: *Atlas of the Environment*, by Geoffrey Lean and Don Hinrichsen (Helicon, 1992); *World Environment Atlas* (mapping © Bartholomew, 1991; an imprint of HarperCollins Publishers; extracts taken from pages 18–39 of Bartholomew World Environment Atlas; reproduced by kind permission of Bartholomew); *The Environment Atlas*, by David Wright (George Philip, 1992); and *5000 Days to Save the Planet*, by Edward Goldsmith (Paul Hamlyn, 1990).

CONTENTS

LIST OF ABBREVIATIONS

AAA	Agricultural Adjustment Administration
AAAS	American Association for the Advancement of Science
AAF	Army Air Force
AASA	Associate of the Australian Society of Accountants
AB	Aktiebolag; Alberta; Bachelor of Arts
ABA	American Bar Association
AC	Companion of the Order of Australia
ACA	Associate of the Institute of Chartered Accountants
ACCA	Associate of the Association of Certified Accountants
Acad	Academy, Académie
Accad	Accademia
Accred	Accredited
ACIS	Associate of the Chartered Institute of Secretaries
ACP	African, Caribbean and Pacific Group of States; American College of Physicians
ACS	American Chemical Society
ACT	Australian Capital Territory
ADC	Aide-de-camp
Adm	Admiral
Admin	Administration, Administrative, Administrator
AERE	Atomic Energy Research Establishment
AF	Air Force
AFC	Air Force Cross
AFESD	Arab Fund for Economic and Social Development
affil	affiliated
AFL	American Federation of Labor
AG	Aktiengesellschaft (Joint Stock Company)
Agric	Agriculture
Ags	Aguascalientes
ai	ad interim
AIA	American Institute of Architects; Associate of Institute of Actuaries
AIAA	American Institute of Aeronautics and Astronautics
AIB	Associate of the Institute of Bankers
AICE	Associate of the Institute of Civil Engineers
AIChE	American Institute of Chemical Engineers
AID	(US) Agency for International Development
AIDS	Acquired Immune Deficiency Syndrome
AIEE	American Institute of Electrical Engineers
AIME	American Institute of Mining Engineers; Associate of the Institution of Mining Engineers
AIMechE	Associate of the Institution of Mechanical Engineers
AK	Alaska; Knight of the Order of Australia
Akad	Akademie
AL	Alabama
ALS	Associate of the Linnaean Society
Alt	Alternate
AM	Albert Medal; Alpes Maritimes; Master of Arts; Member of the Order of Australia
Amb	Ambassador
AMICE	Associate Member of the Institution of Civil Engineers
AMIEE	Associate Member of the Institution of Electrical Engineers
AMIMechE	Associate Member of the Institution of Mechanical Engineers
ANU	Australian National University
AO	Officer of the Order of Australia
Apdo	Apartado
approx	approximately
Apptd	Appointed
AR	Arkansas
ARA	Associate of the Royal Academy
ARC	Agriculture Research Council
ARCS	Associate of the Royal College of Science
ARIBA	Associate of the Royal Institute of British Architects
ARSA	Associate of the Royal Scottish Academy; Associate of the Royal Society of Arts
ASEAN	Association of South East Asian Nations
ASLIB	Association of Special Libraries and Information Bureaux
ASME	American Society of Mechanical Engineers
Asoc	Asociación
Ass	Assembly
Asscn	Association
Assoc	Associate
ASSR	Autonomous Soviet Socialist Republic
Asst	Assistant
avda	avenida
Ave	Avenue
Aug	August
AZ	Arizona
b	born
B AGR	Bachelor of Agriculture
B AGR SC	Bachelor of Agricultural Science
B ARCH	Bachelor of Architecture
B CH, B CHIR	Bachelor of Surgery
B COM(M)	Bachelor of Commerce
B ECONS	Bachelor of Economics
B ED	Bachelor of Education
B ENG	Bachelor of Engineering
B LIT(T)	Bachelor of Letters
B LL	Bachelor of Laws
B MUS	Bachelor of Music
B PHIL	Bachelor of Philosophy
B SC	Bachelor of Science
BA	Bachelor of Arts; British Airways
BAAS	British Association for the Advancement of Science
BAFTA	British Academy of Film and Television Awards
BAO	Bachelor of Obstetrics
BAS	Bachelor in Agricultural Science
BBC	British Broadcasting Corporation
BC	British Columbia
BCE	Bachelor of Civil Engineering
BCL	Bachelor of Canon Law; Bachelor of Civil Law
BCN	Baja California Norte
BCS	Bachelor of Commercial Sciences; Baja California Sur
BD	Bachelor of Divinity
Bd	Board
BDS	Bachelor of Dental Surgery
BE	Bachelor of Education; Bachelor of Engineering
Beds	Bedfordshire
BEE	Bachelor of Electrical Engineering
Berks	Berkshire
BFA	Bachelor of Fine Arts
BFI	British Film Institute
BIM	British Institute of Management
biog	biography
BIPM	Bureau International des Poids et Mesures
BIS	Bank for International Settlements
BL	Bachelor of Laws
Bldg	Building
BLS	Bachelor in Library Science
blvd	boulevard
BM	Bachelor of Medicine
BMA	British Medical Association
Bn	Battalion
BP	Boîte postale (Post Office Box); British Petroleum
BPA	Bachelor of Public Administration
Brig	Brigadier
BS	Bachelor of Science; Bachelor of Surgery
BSA	Bachelor of Scientific Agriculture

LIST OF ABBREVIATIONS

BT	Baronet
Bucks	Buckinghamshire
c	child, children
c	circa
C CHEM	Chartered Chemist
C ENG	Chartered Engineer
C LIT	Companion of Literature
C PHYS	Chartered Physicist
C SC	Candidate of Sciences
CA	California; Chartered Accountant
CABI	Commonwealth Agricultural Bureaux International
Cambs	Cambridgeshire
Camp	Campeche
Cand	Candidate, Candidature
Capt	Captain
Cards	Cardiganshire
CARICOM	Caribbean Community and Common Market
CB	Companion of the (Order of the) Bath
CBC	Canadian Broadcasting Corporation
CBE	Commander of (the Order of) the British Empire
CBI	Confederation of British Industry
CBIM	Companion of British Institute of Management
CBS	Columbia Broadcasting System
CC	Companion of Order of Canada
CCP	Chinese Communist Party
Cdre	Commodore
CE	Chartered Engineer, Civil Engineer
CEAO	Communauté Économique de l'Afrique de l'Ouest
Cen	Central
CEO	Chief Executive Officer
CERN	Conseil (*now* Organisation) Européen(ne) pour la Recherche Nucléaire
CFC	chlorofluorocarbon
CGT	Confédération Général du Travail
CH B	Bachelor of Surgery
CH M	Master of Surgery
Chair	Chairman, Chairperson, Chairwoman
Chem	Chemistry
Chih	Chihuahua
Chis	Chiapas
CIA	Central Intelligence Agency
Cia	Compagnia (Company)
Cía	Compañía (Company)
Cie	Compagnie (Company)
CIEE	Companion of the Institution of Electrical Engineers
C-in-C	Commander-in-Chief
circ	circulation
CIS	Commonwealth of Independent States
CLD	Doctor of Civil Law (USA)
CM	Master in Surgery
CMG	Companion of (the Order of) St Michael and St George
CNAA	Council for National Academic Awards
CNRS	Centre National de la Recherche Scientifique
Co	Company; County
CO	Colorado; Commanding Officer
Coah	Coahuila
Col	Colima; Colonel
Coll	College
Comm	Commission
Commd	Commissioned
Commdg	Commanding
Commdr	Commander, Commandeur
Commdt	Commandant
Commr	Commissioner
Conf	Conference
Confed	Confederation
Cons	Conservative; Conservative and Unionist Party
Contrib	Contribution, Contributor
Corp	Corporate
Corpn	Corporation
Corresp	Correspondent, Corresponding
CP	Communist Party; Caixa Postal/Case Postale (Post Office Box)
CPA	Certified Public Accountant; Commonwealth Parliamentary Association
CPSU	Communist Party of the Soviet Union
cr	created
CSCE	Conference on Security and Co-operation in Europe
CSIRO	Commonwealth Scientific and Industrial Research Organization
CT	Connecticut
CVO	Commander of the Royal Victorian Order
Cttee	Committee
d	daughter(s)
D ARCH	Doctor of Architecture
D CN L	Doctor of Canon Law
D COMM	Doctor of Commerce
D ECON	Doctor of Economics
D EN D	Docteur en Droit
D EN MED	Docteur en Medicine
D ENG	Doctor of Engineering
D ÈS L	Docteur ès Lettres
D ÈS SC	Docteur ès Sciences
D HIST	Doctor of History
D HUM LITT	Doctor of Humane Letters
D IUR	Doctor of Law
D IUR UTR	Doctor of both Civil and Canon Law
D JUR	Doctor of Law
D LIT(T)	Doctor of Letters; Doctor of Literature
D(R) MED	Doctor of Medicine
D MIN SCI	Doctor of Municipal Science
D MUS	Doctor of Music
D PHIL	Doctor of Philosophy
D SC	Doctor of Science
D SC S	Doctor of Social Science
D TECH	Doctor of Technology
D THEOL	Doctor of Theology
DB	Bachelor of Divinity
DBA	Doctor of Business Administration
DC	District of Columbia
DCL	Doctor of Civil Law
DCS	Doctor of Commercial Sciences
DCT	Doctor of Christian Theology
DD	Doctor of Divinity
DDS	Doctor of Dental Surgery
DE	Delaware
Dec	December
Del	Delegate, delegation
Denbighs	Denbighshire
Dept	Department
Desig	Designate
Devt	Development
DF	Distrito Federal
DFA	Doctor of Fine Arts
Dgo	Durango
DH	Doctor of Humanities
DHL	Doctor of Hebrew Literature
DIC	Diploma of Imperial College
DIP AD	Diploma in Art and Design
DIP AGR	Diploma in Agriculture
DIP ED	Diploma in Education
DIP(L) ENG	Diploma in Engineering
Dir	Director
Dist	District
Div	Division; Divisional
DLS	Doctor of Library Science
DM	Doctor of Medicine (Oxford)
DMD	Doctor of Dental Medicine
DMV	Doctor of Veterinary Medicine
DN	Distrito Nacional
DO	Doctor of Ophthalmology
DPH	Diploma in Public Health
DPM	Diploma in Psychological Medicine
Dr	Doctor
DR AGR	Doctor of Agriculture
DR ING	Doctor of Engineering
DR IUR	Doctor of Laws
DR JUR	Doctor of Laws
DR OEC (PUBL)	Doctor of (Public) Economy
DR RER NAT	Doctor of Natural Sciences
DR RER POL	Doctor of Political Science
DR SC	Doctor of Sciences
DR SC NAT	Doctor of Natural Sciences
DS	Doctor of Science
DST	Doctor of Sacred Theology
DTM(& H)	Diploma in Tropical Medicine (and Hygiene)
DUP	Diploma of the University of Paris
E	East, Eastern
EC	European Communities
ECA	Economic Co-operation Administration; Economic Commission for Africa

ECAFE	Economic Commission for Asia and the Far East
ECE	Economic Commission for Europe
ECLAC	Economic Commission for Latin America and the Caribbean
Econ(s)	Economic(s)
ECOSOC	Economic and Social Council
ed	edited; editor
Ed	Editor
ED	Doctor of Engineering (USA)
ED D	Doctor of Education
ED M	Master of Education
Edn	Edition
Educ	Education
EEC	European Economic Community
EFTA	European Free Trade Association
EIA	Environmental Impact Assessment
EIB	European Investment Bank
EM	Master of Engineering (USA)
Emer	Emeritus
Eng	Engineering
ENG D	Doctor of Engineering
EP	European Parliament
EPA	Environmental Protection Agency
ESCAP	Economic and Social Commission for Asia and the Pacific
ESCWA	Economic and Social Commission for Western Asia
esq	esquina (corner)
ETH	Eidgenössische Technische Hochschule (Swiss Federal Institute of Technology)
EURATOM	European Atomic Energy Community
Exec	Executive
Exhbn	Exhibition
Ext	Extension
f	founded, founder
F ENG	Fellow, Fellowship of Engineering
FAA	Fellow of Australian Academy of Science
FAATS	Fellow Australian Academy of Technological Sciences
FACC	Fellow of the American College of Cardiology
FACCA	Fellow of the Association of Certified and Corporate Accountants
FACE	Fellow of the Australian College of Education
FACP	Fellow of American College of Physicians
FACS	Fellow of the American College of Surgeons
FAHA	Fellow Australian Academy of the Humanities
FAIA	Fellow of the American Institute of Architects
FAIAS	Fellow of the Australian Institute of Agricultural Science
FAIM	Fellow of the Australian Institute of Management
FAO	Food and Agriculture Organization
FASE	Fellow of Antiquarian Society, Edinburgh
FASSA	Fellow Academy of Social Sciences of Australia
Fax	Facsimile
FBA	Fellow of the British Academy
FBIM	Fellow of the British Institute of Management
FBIP	Fellow of the British Institute of Physics
FCA	Fellow of the Institute of Chartered Accountants
FCAE	Fellow Canadian Academy of Engineering
FCGI	Fellow of the City and Guilds of London Institute
FCIA	Fellow Chartered Institute of Arbitrators
FCIB	Fellow Chartered Institute of Bankers
FCIC	Fellow of the Chemical Institute of Canada
FCIS	Fellow of the Chartered Institute of Secretaries
FCO	Foreign and Commonwealth Office
FCSD	Fellow Chartered Society of Designers
FCT	Federal Capital Territory
FCWA	Fellow of the Institute of Cost and Works Accountants
Feb	February
Fed	Federal, Federation
FFCM	Fellow of Faculty of Community Medicine
FGS	Fellow of the Geological Society
FGSM	Fellow of the Guildhall School of Music
FIA	Fellow of the Institute of Actuaries
FIAL	Fellow of the International Institute of Arts and Letters
FIAM	Fellow of the International Academy of Management
FIAMS	Fellow of the Indian Academy of Medical Sciences
FIARB	Fellow of the Institute of Arbitrators
FIB	Fellow of the Institute of Bankers
FIBA	Fellow of the Institute of Banking Associations
FICE	Fellow of the Institution of Civil Engineers
FICHEME	Fellow of the Institute of Chemical Engineers
FID	Fellow of the Institute of Directors
FIE	Fellow of the Institute of Engineers
FIEE	Fellow of the Institution of Electrical Engineers
FIEEE	Fellow of the Institute of Electrical and Electronics Engineers
FIJ	Fellow of the Institute of Journalists
FIL LIC	Licentiate in Philosophy
FIM	Fellow of the Institute of Metallurgists
FIME	Fellow of the Institute of Mining Engineers
FIMECHE	Fellow of the Institute of Mechanical Engineers
FIMI	Fellow of the Institute of the Motor Industry
FINSTF	Fellow of the Institute of Fuel
FINSTM	Fellow of the Institute of Marketing
FINSTP	Fellow of the Institute of Physics
FINSTPET	Fellow of the Institute of Petroleum
FIPM	Fellow of the Institute of Personnel Management
FIRE	Fellow of the Institution of Radio Engineers
FITD	Fellow of the Institute of Training and Development
FL	Florida
FLA	Fellow of the Library Association
FLS	Fellow of the Linnaean Society
fmr	former
fmrly	formerly
FNI	Fellow of the National Institute of Sciences of India
FNZIA	Fellow of the New Zealand Institute of Architects
FoE	Friends of the Earth
FRACO	Fellow of the Royal Australian College of Ophthalmologists
FRACP	Fellow of the Royal Australasian College of Physicians
FRACS	Fellow of the Royal Australasian College of Surgeons
FRAES	Fellow of the Royal Aeronautical Society
FRAI	Fellow of the Royal Anthropological Institute
FRAIA	Fellow of the Royal Australian Institute of Architects
FRAIC	Fellow of the Royal Architectural Institute of Canada
FRAM	Fellow of the Royal Academy of Music
FRAS	Fellow of the Royal Asiatic Society; Fellow of the Royal Astronomical Society
FRBS	Fellow of the Royal Society of British Sculptors
FRCM	Fellow of the Royal College of Music
FRCO	Fellow of the Royal College of Organists
FRCOG	Fellow of the Royal College of Obstetricians and Gynaecologists
FRCP(E)	Fellow of the Royal College of Physicians (Edinburgh)
FRCPI	Fellow of the Royal College of Physicians of Ireland
FRCP(UK)	Fellow of the Royal College of Physicians (United Kingdom)
FRCS(E)	Fellow of the Royal College of Surgeons (Edinburgh)
FRECONS	Fellow of the Royal Economic Society
FRES	Fellow of the Royal Entomological Society
FRFPS	Fellow of the Royal Faculty of Physicians and Surgeons
FRGS	Fellow of the Royal Geographical Society
FRHISTS	Fellow of the Royal Historical Society
FRHORTS	Fellow of the Royal Horticultural Society
FRIBA	Fellow of the Royal Institute of British Architects
FRIC	Fellow of the Royal Institute of Chemists
FRICS	Fellow of the Royal Institute of Chartered Surveyors
FRMETSOC	Fellow of the Royal Meteorological Society
FRPS	Fellow of the Royal Photographic Society
FRS	Fellow of the Royal Society
FRSA	Fellow of the Royal Society of Arts
FRSAMD	Fellow of the Royal Scottish Academy of Music and Drama
FRSC	Fellow of the Royal Society of Canada; Fellow of the Royal Society of Chemistry
FRSE	Fellow of the Royal Society of Edinburgh
FRSL	Fellow of the Royal Society of Literature
FRSM	Fellow of the Royal Society of Medicine
FRSNZ	Fellow of the Royal Society of New Zealand
FRSS	Fellow of the Royal Statistical Society
FRSSA	Fellow of the Royal Society of South Africa
FSA	Fellow of the Society of Antiquaries

FSIAD Fellow of the Society of Industrial Artists and Designers
FTI Fellow of the Textile Institute
FTS Fellow of Technological Sciences
FZS Fellow of the Zoological Society

GA Georgia
GATT General Agreement on Tariffs and Trade
GB Great Britain
Gen General
GIS Geographic Information System
Glam Glamorganshire
Glos Gloucestershire
GmbH Gesellschaft mit beschränkter Haftung (Limited Liability Company)
GOC (in C) General Officer Commanding (in Chief)
Gov Governor
Govt Government
GPO General Post Office
Grad Graduate
Gro Guerrero
GRSM Graduate of the Royal School of Music
GSO General Staff Officer
Gto Guanajuato

HABITAT United Nations Centre for Human Settlements
Hants Hampshire
hc honoris causa
HE His Eminence; His (or Her) Excellency
Herefords Herefordshire
Herts Hertfordshire
Hgo Hidalgo
HH His (or Her) Highness
HI Hawaii
HLD Doctor of Humane Letters
HM His (or Her) Majesty
Hon Honorary; Honourable
Hosp Hospital
HQ Headquarters
HRH His (or Her) Royal Highness
Hunts Huntingdonshire

IA Iowa
IAEA International Atomic Energy Agency
IATA International Air Transport Association
IBRD International Bank for Reconstruction and Development (World Bank)
ICAO International Civil Aviation Organization
ICC International Chamber of Commerce; International Computing Centre
ICE Institution of Civil Engineers
ICFTU International Confederation of Free Trade Unions
ICI Imperial Chemical Industries
ICJ International Court of Justice
ICOM International Council of Museums
ICSC International Civil Service Commission
ICSU International Council of Scientific Unions
ID Idaho
IDA International Development Association
IDB Inter-American Development Bank
IEE Institution of Electrical Engineers
IEEE Institution of Electrical and Electronic Engineers
IFAD International Fund for Agricultural Development
IFC International Finance Corporation
IFRB International Frequency Registration Board
IL Illinois
ILO International Labour Organization
IMCO Inter-Governmental Maritime Consultative Organization
IMECHE Institution of Mechanical Engineers
IMF International Monetary Fund
IMO International Maritime Organization
IN Indiana
Inc Incorporated
Ind Independent
Insp Inspector
Inst Institut; Institute; Institution
INSTRAW International Research and Training Institute for the Advancement of Women
Int International
INTELSAT International Telecommunications Satellite Organization
IOC International Olympic Committee
IPU Inter-Parliamentary Union
ITU International Telecommunications Union
IUCN World Conservation Union
IUPAC International Union of Pure and Applied Chemistry
IUPAP International Union of Pure and Applied Physics

Jal Jalisco
Jan January
JCB Bachelor of Canon Law
JCD Doctor of Canon Law
JD Doctor of Jurisprudence
JP Justice of the Peace
Jr Junior
Jt Joint
JU D Doctor of Law
JU DR Doctor of Law, Juris utriusque Doctor (Doctor of both Civil and Canon Law)
JUD Juris utriusque Doctor (Doctor of both Civil and Canon Law)

KCB Knight Commander of (the Order of) the Bath
KS Kansas
KY Kentucky

L EN D Licencié en Droit
L ÈS L Licencié ès Lettres
L ÈS SC Licencié ès Sciences
L PH Licentiate of Philosophy
L TH Licentiate in Theology
LA Los Angeles; Louisiana
Lab Laboratory; Labour; Labour Party
Lancs Lancashire
LDS Licentiate in Dental Surgery
Legis Legislative
Leics Leicestershire
LHD Doctor of Humane Letters
LIC EN DER Licenciado en Derecho
LIC EN FIL Licenciado en Filosofía
LIC MED Licentiate in Medicine
Lieut Lieutenant
Lincs Lincolnshire
LITT D Doctor of Letters
LL B Bachelor of Laws
LL D Doctor of Laws
LL L Licentiate of Laws
LL M Master of Laws
LM Licentiate of Medicine, or Midwifery
LORCS League of Red Cross and Crescent Societies
LRAM Licentiate of the Royal Academy of Music
LRCP Licentiate of the Royal College of Physicians
LSE London School of Economics
Ltd(a) Limited; Limitada

m marriage, married; metre(s)
m. million
M AGR Master of Agriculture (USA)
M ARCH Master of Architecture
M CH Master of Surgery
M CH D Master of Dental Surgery
M COM(M) Master of Commerce
M DIV Master of Divinity
M ED Master of Education
M ENG Master of Engineering (Dublin)
M PH Master of Philosophy (USA)
M SC Master of Science
M SC S Master of Social Science
MA Massachusetts; Master of Arts
Maj Major
Man Management, Manager, Managing
Math Mathematical, Mathematics
MB Bachelor of Medicine; Manitoba
MBA Master of Business Administration
MBE Member of (the Order of) the British Empire
MCE Master of Civil Engineering
MCL Master of Civil Law
MD Doctor of Medicine; Maryland
MDS Master of Dental Surgery
ME Maine
mem member
MEP Member of the European Parliament
Méx México
MFA Master of Fine Arts
Mfg Manufacturing
Mfrs Manufacturers

MI	Michigan
MIA	Master of International Affairs
MICE	Member of the Institution of Civil Engineers
Mich	Michoacán
MICHEME	Member of the Institution of Chemical Engineers
Middx	Middlesex
MIEE	Member of the Institution of Electrical Engineers
MIGA	Multilateral Investment Guarantee Agency
Mil	Military
MIMARE	Member of the Institute of Marine Engineers
MIMECHE	Member of the Institution of Mechanical Engineers
MIMINE	Member of the Institution of Mining Engineers
MINSTT	Member of the Institute of Transport
MISTRUCTE	Member of the Institution of Structural Engineers
MIT	Massachusetts Institute of Technology
MJ	Master of Jurisprudence
MN	Minnesota
MO	Missouri
Mon	Monmouthshire
Mor	Morelos
Movt	Movement
MP	Member of Parliament
MPA	Master of Public Administration (Harvard)
MRAS	Member of the Royal Asiatic Society
MRC	Medical Research Council
MRCP(E)	Member of the Royal College of Physicians (Edinburgh)
MRCP(UK)	Member of the Royal College of Physicians (United Kingdom)
MRCS(E)	Member of the Royal College of Surgeons (Edinburgh)
MRCVS	Member of the Royal College of Veterinary Surgeons
MRI	Member of the Royal Institution
MRIA	Member of the Royal Irish Academy
MRIC	Member of the Royal Institute of Chemistry
MS	Master of Science; Master of Surgery; Mississippi
MT	Montana
MTS	Master of Theological Studies
MU DR	Doctor of Medicine
MUS BAC or B	Bachelor of Music
MUS DOC or D	Doctor of Music
MUS M	Master of Music (Cambridge)
MVD	Master of Veterinary Medicine
N	North, Northern
NAS	National Academy of Sciences (USA)
NASA	National Aeronautical and Space Administration
Nat	National
NATO	North Atlantic Treaty Organization
Nay	Nayarit
NB	New Brunswick
NBC	National Broadcasting Corporation
NC	North Carolina
ND	North Dakota
NE	Nebraska; North East
NF	Newfoundland
NH	New Hampshire
NJ	New Jersey
NL	Nuevo León
NM	New Mexico
no	number, numero
Northants	Northamptonshire
Notts	Nottinghamshire
Nov	November
nr	near
NS	Nova Scotia
NSW	New South Wales
NT	Northern Territory
NV	Nevada
NW	North West
NWT	North West Territories
NY	New York
NZ	New Zealand
NZIC	New Zealand Institute of Chemistry
OAPEC	Organization of Arab Petroleum Exporting Countries
OAS	Organization of American States
OAU	Organization of African Unity
Oax	Oaxaca
OBE	Officer (of the Order of) the British Empire
Oct	October
OECD	Organization for Economic Co-operation and Development
OH	Ohio
OK	Oklahoma
ON	Ontario
OPEC	Organization of the Petroleum Exporting Countries
OR	Oregon
Org	Organization
Oxon	Oxfordshire
PA	Pennsylvania
Parl	Parliament, Parliamentary
PC	Privy Councillor
PD B	Bachelor of Pedagogy
PD D	Doctor of Pedagogy
PD M	Master of Pedagogy
PE	Prince Edward Island
Pembs	Pembrokeshire
Perm	Permanent
PH B	Bachelor of Philosophy
PH D	Doctor of Philosophy
PH DR	Doctor of Philosophy
PH L	Licentiate of Philosophy
PHARM D	Docteur en Pharmacie
pl	place, platz, ploschad (square)
PLC	Public Limited Company
PMB	Private Mail Bag
PO(B)	Post Office (Box)
POW	Prisoner of War
PQ	Province of Quebec
Pres	President
Prin	Principal
Prof	Professor
Prov	Province, Provincial
Pty	Proprietary
Publ(s)	Publication(s)
Publr	Publisher
Pue	Puebla
Pvt	Private
Q Roo	Quintana Roo
QC	Queen's Counsel
Qro	Querétaro
RA	Royal Academy
RACP	Royal Australasian College of Physicians
RAF	Royal Air Force
RAM	Royal Academy of Music
RCA	Radio Corporation of America; Royal Canadian Academy; Royal College of Art
RCRA	Resource Conservation and Recovery Act
Rd	Road
Regt	Regiment
Rep	Representative, Represented
Repub	Republic
retd	retired
Rev	Reverend
RI	Rhode Island
RIBA	Royal Institute of British Architects
RNZAF	Royal New Zealand Air Force
RS DR	Doctor of Social Sciences
RSA	Royal Scottish Academy; Royal Society of Arts
RSL	Royal Society of Literature
RSPB	Royal Society for the Protection of Birds
Rt Hon	Right Honourable
Rt Rev	Right Reverend
s	son(s)
S	South, Southern
SA	Sociedad Anónima, Société Anonyme; South Africa
SAE	Society of Aeronautical Engineers
Salop	Shropshire
SALT	Strategic Arms Limitation Treaty
SB	Bachelor of Science (USA)
SC	South Carolina
SC B	Bachelor of Science
SC D	Doctor of Science
SD	South Dakota
SE	South East
SEATO	South East Asia Treaty Organization
Sec	Secretary
Secr	Secretariat

LIST OF ABBREVIATIONS

Sept	September
Sin	Sinaloa
SIPRI	Stockholm International Peace Research Institute
SJD	Doctor of Juristic Science
SK	Saskatchewan
SLP	San Luis Potosí
SM	Master of Science
Soc	Société, Society
Son	Sonora
SpA	Società per Azioni
Sq	Square
Sr	Senior
St	Saint; Street
Staffs	Staffordshire
STB	Bachelor of Sacred Theology
STD	Doctor of Sacred Theology
STL	Licentiate of Sacred Theology
STM	Master of Sacred Theology
str	strasse
Supt	Superintendent
SW	South West
Tab	Tabasco
Tamps	Tamaulipas
Tech	Technical, Technology
Tel	Telephone
Temp	Temporary
TH B	Bachelor of Theology
TH D	Doctor of Theology
TH DR	Doctor of Theology
TH M	Master of Theology
Tlax	Tlaxcala
TN	Tennessee
Trans	Translation, Translator
Treas	Treasurer
TU(C)	Trades Union (Congress)
TV	Television
TX	Texas
UAE	United Arab Emirates
UED	University Education Diploma
UK	United Kingdom (of Great Britain and Northern Ireland)
UKAEA	United Kingdom Atomic Energy Authority
UMIST	University of Manchester Institute of Science and Technology
UN(O)	United Nations (Organization)
UNA	United Nations Association
UNCED	United Nations Conference on Environment and Development
UNCITRAL	United Nations Commission on International Trade Law
UNCTAD	United Nations Conference on Trade and Development
UNDP	United Nations Development Programme
UNDRO	United Nations Disaster Relief Office
UNEP	United Nations Environment Programme
UNESCO	United Nations Educational, Scientific and Cultural Organisation
UNFPA	United Nations Fund for Population Activities
UNHCR	United Nations High Commissioner for Refugees
UNICEF	United Nations International Children's Emergency Fund
UNIDIR	United Nations Institute for Disarmament Research
UNIDO	United Nations Industrial Development Organization
UNIFEM	United Nations Development Fund for Women
UNITAR	United Nations Institute for Training and Research
Univ	University
UNPF	United Nations Population Fund
UNRISD	United Nations Research Institute for Social Development
UNRRA	United Nations Relief and Rehabilitation Administration
UNRWA	United Nations Relief and Works Agency
UNSCEAR	Secretariat of the United Nations Scientific Committee on the Effects of Atomic Radiation
UNU	United Nations University
UNV	United Nations Volunteers
UPU	Universal Postal Union
US(A)	United States (of America)
USAF	United States Air Force
USSR	Union of Soviet Socialist Republics
UT	Utah
VA	Virginia
Ver	Veracruz
Vic	Victoria
Vol(s)	Volume(s)
VT	Vermont
vul	street (Ukraine)
W	West, Western
WA	Washington (State); Western Australia
Warwicks	Warwickshire
WEU	Western European Union
WFC	World Food Council
WFP	World Food Programme
WFTU	World Federation of Trade Unions
WHO	World Health Organization
WI	Wisconsin
WIDER	World Institute for Development Economics Research
Wilts	Wiltshire
WIPO	World Intellectual Property Organization
WMO	World Meteorological Organization
Worcs	Worcestershire
WTO	World Tourism Organization
WV	West Virginia
WWF	World Wide Fund for Nature
WY	Wyoming
Yorks	Yorkshire
YT	Yukon Territory
Yuc	Yucatán
Zac	Zacatecas

INTERNATIONAL TELEPHONE CODES

To use this table:

Select the International Code for the country from which the call is to be made (column 2), then the Country Code for the country to which the call is going (column 1). To make the call, dial the International Code plus the Country Code plus the number required (adding area codes provided in the entries where necessary). For example, to call a person in Brussels from the USA, the following numbers would have to be dialled: 011 (International Code from the USA) + 32 (Country Code for Belgium) + (2) (area code) + the number required.

	Column 1 Country Code To	Column 2 International Code From
Afghanistan	93	*
Albania	42	*
Algeria	213	00
Andorra	33 628	0
Angola	244	*
Antigua and Barbuda	1 809	011
Argentina	54	*
Armenia	7	810
Australia	61	0011
Australian External Territories:		
Christmas Island	672 4	*
Cocos (Keeling) Islands	672 2	*
Norfolk Island	672 3	*
Austria	43	00
Azerbaijan	994	810
The Bahamas	1 809	011
Bahrain	973	*
Bangladesh	880	*
Barbados	1 809	011
Belarus	7	810
Belgium	32	00
Belize	501	*
Benin	229	*
Bhutan	975	*
Bolivia	591	*
Bosnia and Herzegovina	38	99
Botswana	267	*
Brazil	55	00
Brunei	673	*
Bulgaria	359	00
Burkina Faso	226	*
Burundi	257	*
Cambodia	855	*
Cameroon	237	*
Canada	1	011
Cape Verde	238	*
The Central African Republic	236	*
Chad	235	*
Chile	56	00
China, People's Republic	86	00
China (Taiwan)	886	*
Colombia	57	90
The Comoros	269	*
The Congo	242	*
Costa Rica	506	*
Côte d'Ivoire	225	*
Croatia	38	99
Cuba	53	*
Cyprus	357	00
'Turkish Republic of Northern Cyprus'	905	99
Czech Republic	42	00
Denmark	45	009
Danish External Territories:		
Faeroe Islands	298	*
Greenland	299	*
Djibouti	253	*

	Column 1 Country Code To	Column 2 International Code From
Dominica	1 809 44	0
The Dominican Republic	1 809	*
Ecuador	593	01
Egypt	20	*
El Salvador	503	0
Equatorial Guinea	240	*
Eritrea	251	*
Estonia	372	810
Ethiopia	251	*
Fiji	679	*
Finland	358	990
Finnish External Territory:		
Åland Islands	358 28	990
France	33	19
French Overseas Departments:		
French Guiana	594	*
Guadeloupe	590	*
Martinique	596	*
Réunion	262	*
French Overseas Collectivités Territoriales:		
Saint Pierre and Miquelon	508	*
French Overseas Territories:		
French Polynesia	689	*
New Caledonia	687	*
Wallis and Futuna Islands	*	*
Gabon	241	*
The Gambia	220	*
Georgia	7	810
Germany	49	00
Ghana	233	*
Greece	30	00
Grenada	1 809	011
Guatemala	502	00
Guinea	224	*
Guinea-Bissau	245	*
Guyana	592	*
Haiti	509	*
Honduras	504	00
Hungary	36	00
Iceland	354	90
India	91	00
Indonesia	62	*
Iran	98	*
Iraq	964	*
Ireland	353	16
Israel	972	00
Italy	39	00
Jamaica	1 809	011
Japan	81	001
Jordan	962	*
Kazakhstan	7	810
Kenya	254	*
Kiribati	686	*
Korea, Democratic People's Republic (North Korea)	850	*
Korea, Republic (South Korea)	82	001

	Column 1 Country Code To	Column 2 International Code From
Kuwait	965	*
Kyrgyzstan	7	810
Laos	856	*
Latvia	371	810
Lebanon	961	*
Lesotho	266	*
Liberia	231	*
Libya	218	*
Liechtenstein	41 75	*
Lithuania	370	810
Luxembourg	352	00
Macedonia	38	99
Madagascar	261	*
Malawi	265	*
Malaysia	60	*
Maldives	960	*
Mali.	223	*
Malta	356	0
Marshall Islands	692	*
Mauritania	222	*
Mauritius	230	*
Mexico	52	98
Micronesia	691	*
Moldova	373	810
Monaco	33 93	19
Mongolia	*	*
Morocco	212	00
Mozambique	258	*
Myanmar	95	*
Namibia	264	*
Nauru	674	*
Nepal	977	*
Netherlands	31	09
Netherlands Dependencies:		
Aruba	297	*
Netherlands Antilles	599	*
New Zealand	64	00
New Zealand's Associated Territories:		
Cook Islands	682	*
Niue	683	*
Nicaragua	505	*
Niger	227	*
Nigeria	234	*
Norway	47	095
Norwegian External Territories:		
Svalbard	47 80	095
Oman	968	*
Pakistan	92	*
Panama	507	*
Papua New Guinea	675	*
Paraguay	595	*
Peru	51	*
The Philippines	63	*
Poland	48	00
Portugal	351	00 (07 Porto)
(Madeira	351 91)	
Portuguese Overseas Territory:		
Macau	853	*
Qatar	974	*
Romania	40	*
Russian Federation	7	810
Rwanda	250	*
Saint Christopher and Nevis	1 809	011
Saint Lucia	1 809 45	0
Saint Vincent and the Grenadines	1 809 45	0
San Marino	39 549	0
São Tomé and Príncipe	239	*
Saudi Arabia	966	*
Senegal	221	*

	Column 1 Country Code To	Column 2 International Code From
Seychelles	248	*
Sierra Leone	232	*
Singapore	65	005
Slovakia	42	00
Slovenia	38	99
Solomon Islands	677	*
Somalia	252	*
South Africa	27	*
Spain	34	07
Sri Lanka	94	*
Sudan	249	*
Suriname	597	*
Swaziland	268	*
Sweden	46	009
Switzerland	41	00
Syria	963	*
Tajikistan	7	810
Tanzania	255	*
Thailand	66	001
Togo	228	*
Tonga	676	*
Trinidad and Tobago	1 809	01
Tunisia	216	00
Turkey	90	99
Turkmenistan	7	810
Tuvalu	688	*
Uganda	256	*
Ukraine	7	810
United Arab Emirates	971	*
United Kingdom	44	010
United Kingdom Crown Dependencies	44	010
British Dependent Territories:		
Anguilla	1 809 497	011
Ascension Island	247	*
Bermuda	1 809	011
British Virgin Islands	1 809	011
Cayman Islands	1 809 94	0
Falkland Islands	500	*
Gibraltar	350	*
Hong Kong	852	001
Montserrat	1 809 491	011
Pitcairn Islands	*	*
Saint Helena	290	*
Turks and Caicos Islands	1 809 946	0
United States of America	1	011
United States Commonwealth Territories:		
Northern Mariana Islands	670	*
Puerto Rico	1 809	135
United States External Territories:		
American Samoa	684	*
Guam	671	*
Palau	680	*
United States Virgin Islands	1 809	011
Uruguay	598	*
Uzbekistan	7	810
Vanuatu	678	*
Vatican City	39	00
Venezuela	58	00
Vietnam	84	*
Western Samoa	685	*
Yemen	967	*
Yugoslavia	38	99
Zaire	243	*
Zambia	260	*
Zimbabwe	263	*

* No Information/Contact International Operator

INTRODUCTORY ESSAY

David Nicholson-Lord writes on environmental issues for *The Independent* and *The Independent on Sunday*. He is the author of *The Greening of the Cities* (Routledge, 1993).

ENVIRONMENTALISM: FAD OR FACT OF LIFE?

David Nicholson-Lord

It has become fashionable in the 1990s to talk of the fading of environmentalism. For all the hyperbole surrounding the Earth Summit in Rio de Janeiro in 1992, the idea has gained ground that 'greenery' has lost its gloss, that recession and unemployment have turned the minds of politicians, industry and the public to different issues—most notably economic survival. The bubble of green consumerism has burst, proclaim the critics, scarcely concealing their delight. Never more than skin deep, it has proved itself a temporary fad, part of the cosmetic packaging of the 1980s, when people could afford a conscience. In the 1990s a conscience, particularly a green one, has become a luxury, and one we are decreasingly inclined to indulge. What matters is the state of one's mortgage, one's job, one's bank balance.

To the extent that such views express anything more than the casual (but accurate) perception that the environment is no longer 'in the news' as much as it was, they owe their provenance to a widespread backlash against environmentalism. This has been more pronounced in the USA than in Western Europe, but it is proving itself almost as complex a phenomenon as the movement it denigrates. There are elements in it of right-wing ideology—hostility to new laws and regulations, to the 'nanny state', to the so-called 'eco-fascists' who are alleged to be planning a new green authoritarianism. There is the widely held belief that conservationists are killjoys, delighting in finding things to ban. There is the much more widespread reaction against what are often perceived as scare tactics. There is resentment at perpetually being made to feel guilty or gloomy—the green version of 'compassion fatigue'.

There is also a sense of hopelessness and powerlessness. What can we do about the destruction of the rainforests on the other side of the world? What difference will our own small efforts make to the ozone layer? Better—less stressful, certainly—to ignore it. It might just go away.

Regrettably the environment, as an issue, will not go away. We are already set on a course that, without drastic alteration, will bring a warming climate and a thinning ozone layer at least until the second half of the next century, and probably for a long time afterwards. World food supply is at or near its limits, yet world population continues to increase. Less than a century from now, when our babies are great-grandparents, there may be 11,000m. to 12,000m. people on the planet—roughly twice present numbers—heralding the onset of what some specialists have called the 'demographic winter', lasting perhaps a thousand years, in which human beings and their paraphernalia will crowd into every semi-habitable corner of the planet, banishing animals to zoos and nature to parks and ticket-only wilderness areas.

Who is to say whether the heaviest price we pay will be physical or psychological? We have some idea of the costs of global warming: climatic instability, severe agricultural disruption, rising sea-levels, hundreds of millions of 'environmental refugees'. We also know that ozone depletion will cause hundreds of thousands more deaths from skin cancer as well as serious but as yet incalculable effects on the productivity of land and sea—increased radiation impairs photosynthesis, stunting the growth of plants and marine organisms. But what of the psychological effects of turning city living into a planetary life style? What will the costs be in neurosis and stress? What will a world be like in which overcrowding and congestion have become a normal daily experience? We may 'adapt' to it but, to borrow from Lord Byron, how will we adapt to the adaptation?

Global warming and ozone depletion are the 'big' issues that put the environment high on the political agenda in the late 1980s—and which remain there, albeit somewhat lower down, in the 1990s. Yet consider some of the other ingredients of what we term environmentalism. Many, while not overtly global in their impact, carry clear implications for health: the quality of food, of drinking water and bathing water, for example. Some, such as transport, noise, air pollution, we encounter every day: they directly affect our well-being. Others—the loss of habitats and species, the preservation of green space—bear more on what we regard as the quality of life. Is there any reason to suppose these concerns will disappear? Isn't it more likely, in fact, that we will become more, rather than less, concerned about health, well-being and quality of life?

One's answer to this question will depend, in part, on one's views on how and why environmentalism began. For many people, the decisive year was 1988, when thousands of seals died in the North Sea, toxic waste ships ranged coasts seeking a port of entry, drought and storms affected large areas of the planet—causing the first failure in decades of the American grain harvest—and leading climatologists announced that global warming had turned from forecast into fact. In the UK Margaret Thatcher announced her startling conversion to the green cause and work hurriedly began on an environmental White Paper, conceived as a corner-stone of government policy. In the USA George Bush, campaigning for the White House, promised he would be the 'environmental President': the greenhouse effect, he declared, would be as nothing compared with the White House effect.

In fact, the origins of modern environmentalism go back to the early 1960s—specifically to 1962, when Rachel Carson published her devastating indictment of pesticide use, *Silent Spring*. The green cause was rapidly taken up by the 'baby boom' generation, the generation of student protest, flower power, Woodstock and Earth Day—the first Earth Day was in 1970. In 1972 the first major international conference on the environment was held in Stockholm, and a year later the oil crisis generated widespread alarm in the West over forecast shortages of energy and other primary resources. Tracts such as the Club of Rome's *Limits to Growth* and *The Ecologist's Blueprint for Survival*, both published in the same year as the Stockholm conference, were among many warnings that time was running out. A new breed of campaigning environmental organization was born, Greenpeace and Friends of the Earth the best known, specializing in high-profile protest and direct action.

Time, however, has not run out—or perhaps it is safer to say that it does not appear to have run out. We still have our oil, relatively cheap and in relative abundance. The threatened shortages have failed, apparently, to materialize. Moreover, the recession of the late 1970s and early 1980s appeared to dissipate environmental concern just as its successor did more than a decade later.

Yet although apocalypse failed to arrive, environmentalism did not go away. By the mid-1980s membership of green groups in the UK was estimated at between 3m. and 3.5m. In the early 1990s, for all the talk of green fatigue, the figure is put at over 5m. Environmental groups now dominate the voluntary sector. The biggest, the National Trust, has 2.4m. members, compared with 300,000 in 1971. Campaigning organizations such as the Royal Society for the Protection of Birds, the World Wide Fund for Nature, Friends of the Earth, Greenpeace and the Royal Society for Nature Conservation all saw steady gains throughout the 1970s and 1980s. The RSPB, for example, with 98,000 members in 1971 and 441,000 in 1981, now has around 850,000. Compare these figures with the membership of political parties, churches and other voluntary groups and their significance is unmistakable. The three main political parties may muster a million members between them. On English church census day in 1989 3.7m. adults attended services.

The polling organization MORI has a different method of measuring environmental concern. According to MORI, which has been tracking attitudes to the environment since 1969, anyone who has taken at least five environmental initiatives, such as writing to an MP, is classed as a 'green activist' and in 1992 23% of the adult population—about 10m. people—came into this category. Eighteen million people—four in ten of the British population aged 14 and over—were 'green consumers': they had bought a product for environmental reasons.

Although these figures have declined from their peak in 1991, they are still at historically high levels. Green issues, for example, took up 3% of MPs' post in 1987, but in 1992 the figure was 34%. According to MORI, any attempt to 'write off' environmentalism is fundamentally misconceived, as green consciousness and behaviour have become a part of people's lives.

The longer perspective, in fact, shows environmental concern slowly rising over three decades—an apparently irresistible green tide which ebbs and flows but each time leaves the watermark a little higher. In part this reflects the move of the 'baby boom' generation into positions of power and influence. The 'baby boomers', through film, television and the media, have then transmitted their concerns to the next generation. But there is clearly something more complex and profound involved. The former Archbishop of Canterbury, Dr Robert Runcie, is one of many who have compared environmentalism to a religious movement. Given the decline of organized religion and of Utopian political creeds such as Socialism, is that such a fanciful analogy? Certainly the green movement draws much of its strength from a powerful sense of kinship with nature, rediscovered in the late 20th century, at a time—and this is surely no coincidence—when the very existence of nature was suddenly perceived to be under threat.

What is probably distinctive about the latest phase of environmentalism, however, is that the green tide has found its way into vast new areas of social and political life, previously largely untouched by environmental concern. Of these, the most significant is industry. Not only is there money in 'greenery' (the OECD estimates that the global market for environmental goods will increase from £200,000m. to £300,000m. per year over the next decade), there is also market advantage (as exemplified by the Body Shop, which uses public relations as its only method of advertising). Business analysts already speak of 'green chip' companies. There is now a British Standard—BS7750—for environmental management. Polls of business leaders carry a similar message—that under pressure from legislation, consumers and its own growing sense of responsibility, industry expects to spend more and more of its time on environmental issues. To quote Peter Morgan, Director-General of the Institute of Directors, the environment 'can no longer be seen to be a fad'.

One can go further and argue that many of those doom-laden predictions of the early 1970s have been proved right. Secure in their affluent and well protected corner of the planet, Westerners may feel that life has continued more or less as normal. But suppose you are among the millions in the Third World who, because of population pressures and deforestation, have to trek increasing distances for firewood? What about the growing prevalence of 'man-made' disasters—for example, floods in Bangladesh caused by loss of tree cover hundreds of miles upstream in the Himalayas? What of the growth in absolute global poverty in the 1980s, or the 200m. people who, according to the American population biologist Paul Ehrlich, have died 'needlessly' through hunger or hunger-related diseases since 1968?

The truth is that disaster is not 'around the corner', it is here and now. But it is a slow-burning disaster, distant and to a degree invisible. It becomes ever harder, however, to disguise it from ourselves. Was the Gulf War not just such a resource conflict as the 1970s doom-sayers predicted? Is not the Scottish fishing industry now threatened with quotas—and economic collapse—precisely because of over-exploitation of the seas (in other words, we are in danger of 'running out' of fish)?

The greatest illusion of all, one might conclude, is that environmentalism is a passing fancy. Whatever its current placing on the political agenda, it is anchored both in human consciousness and in global reality. Much of the 'greening' of the late 1980s may well have been skin-deep—in which case the 1990s can be said to represent a new 'post-cosmetic' phase of 'greenery', taking it beyond the outer layers and towards the central, vital organs of the body politic.

London
1993

MAPS

1 Climate

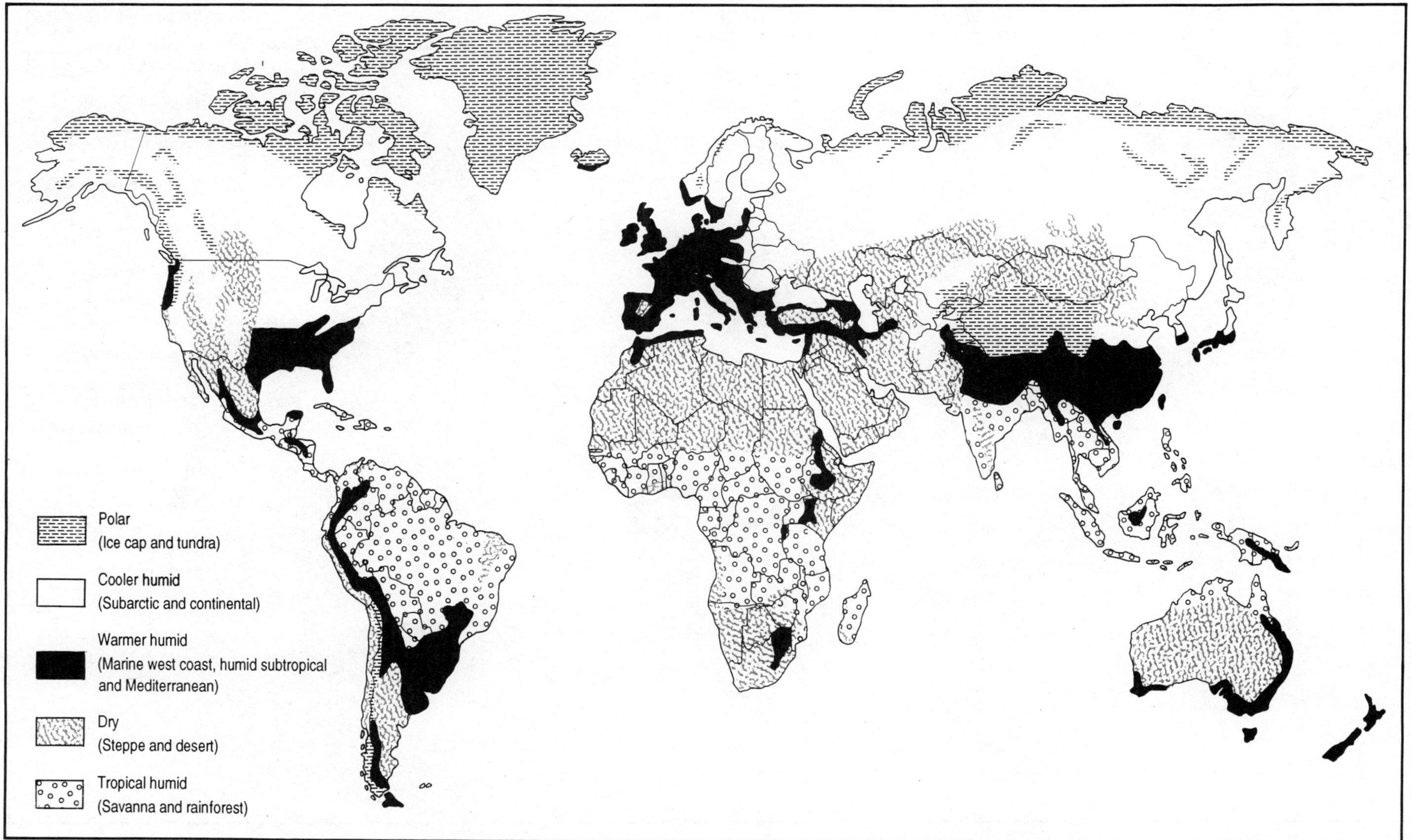

Source: *Atlas of the Environment* , Helicon, 1992

2 Vegetation

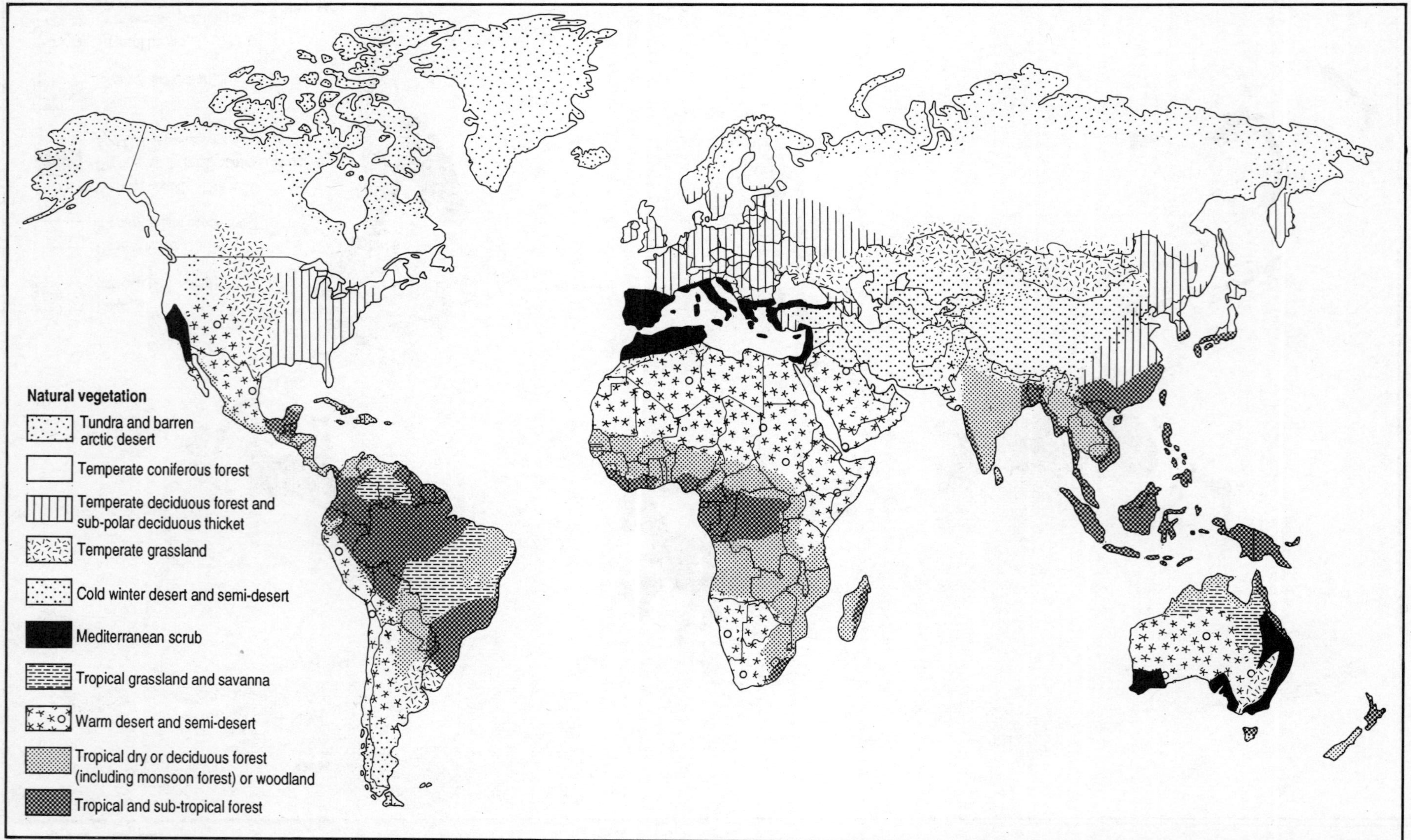

Sources: *Atlas of the Environment*, Helicon, 1992; *Environment Atlas*, George Philip, 1992

3 Population Density

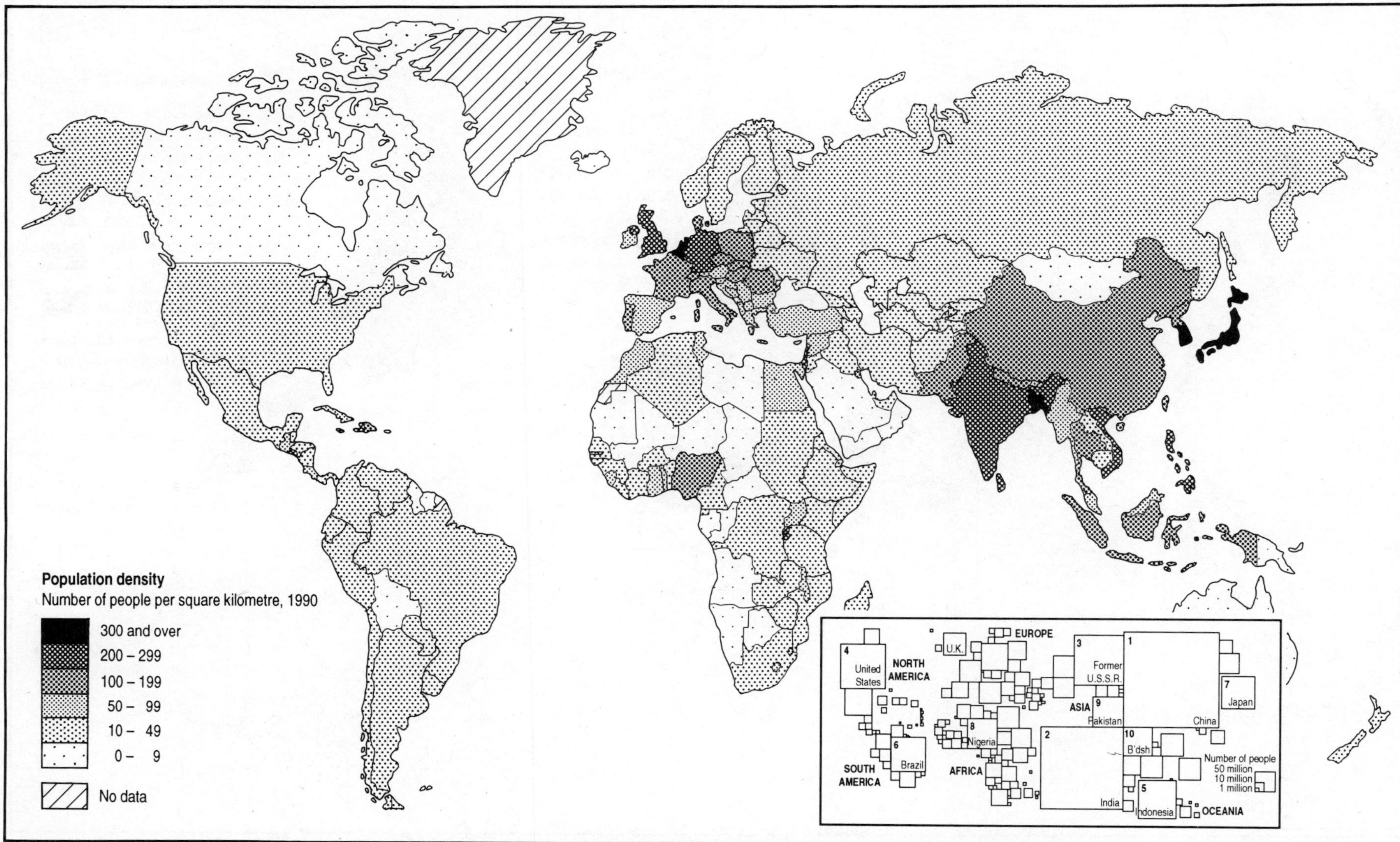

Sources: *Atlas of the Environment*, Helicon, 1992; *Environment Atlas*, George Philip, 1992

4 Population Growth

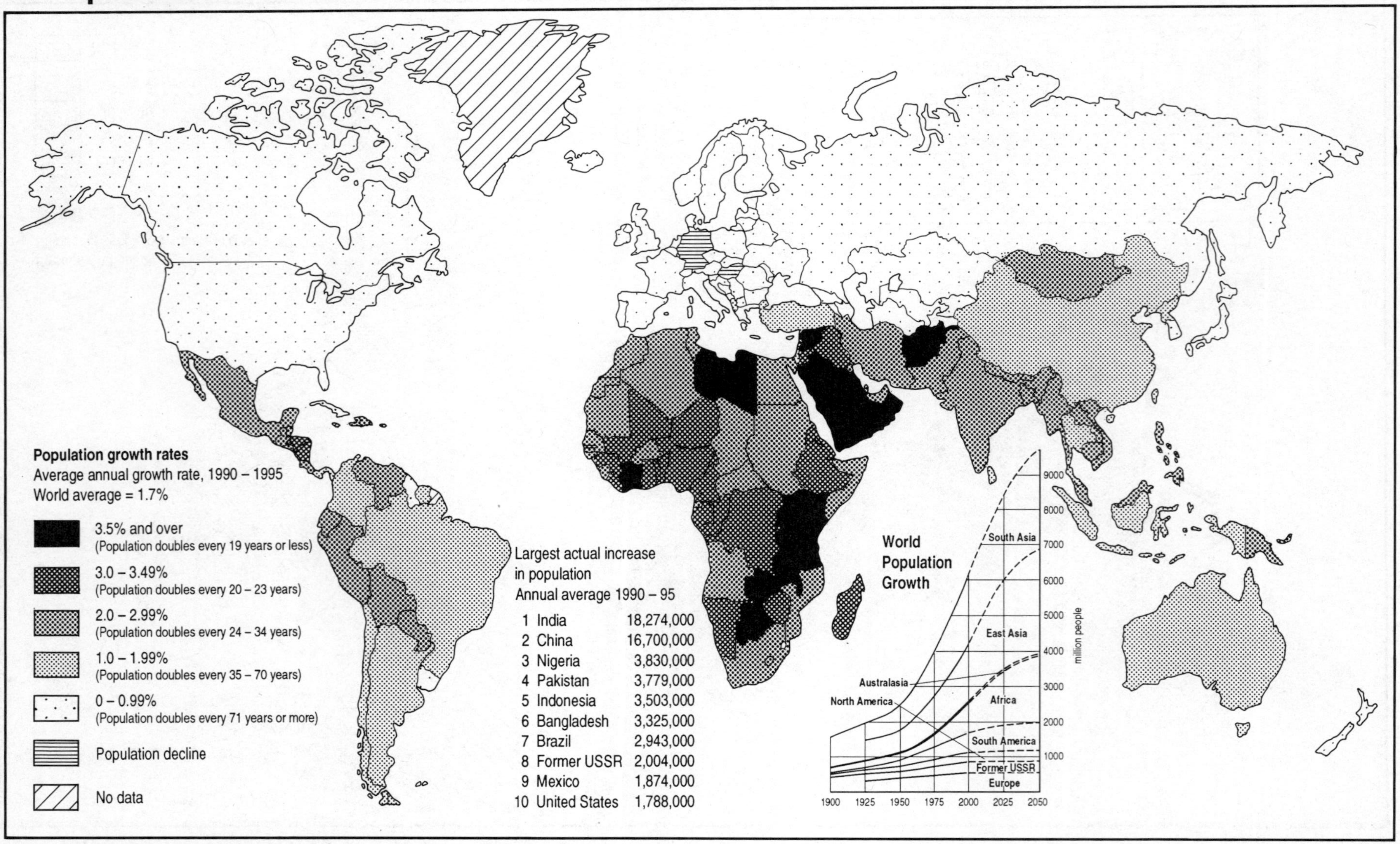

Largest actual increase in population
Annual average 1990 – 95

1	India	18,274,000
2	China	16,700,000
3	Nigeria	3,830,000
4	Pakistan	3,779,000
5	Indonesia	3,503,000
6	Bangladesh	3,325,000
7	Brazil	2,943,000
8	Former USSR	2,004,000
9	Mexico	1,874,000
10	United States	1,788,000

Sources: *Atlas of the Environment*, Helicon, 1992; *Environment Atlas*, George Philip, 1992

5 Food Distribution

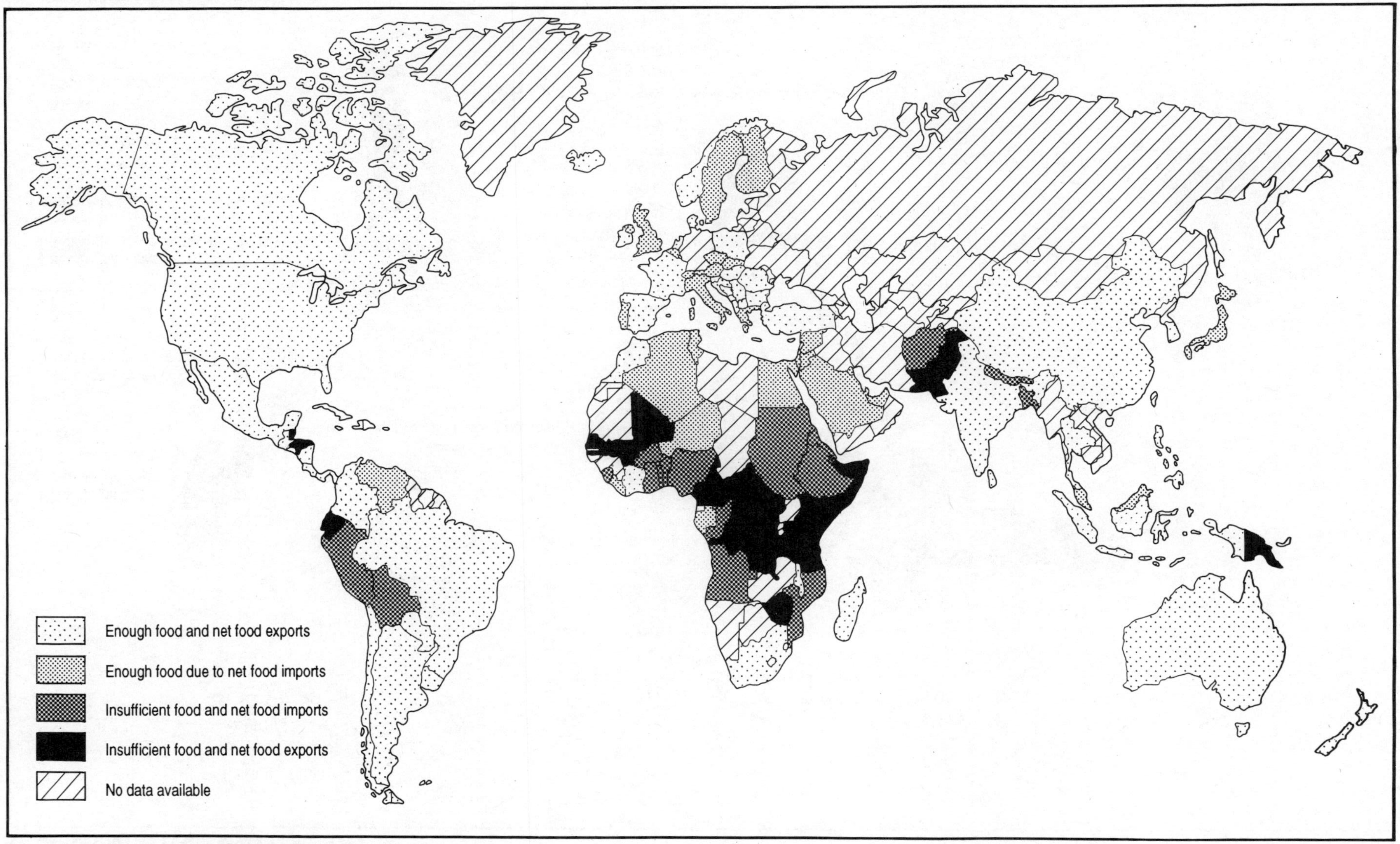

Source: *World Environment Atlas*, Bartholomew, 1991

6 Desertification

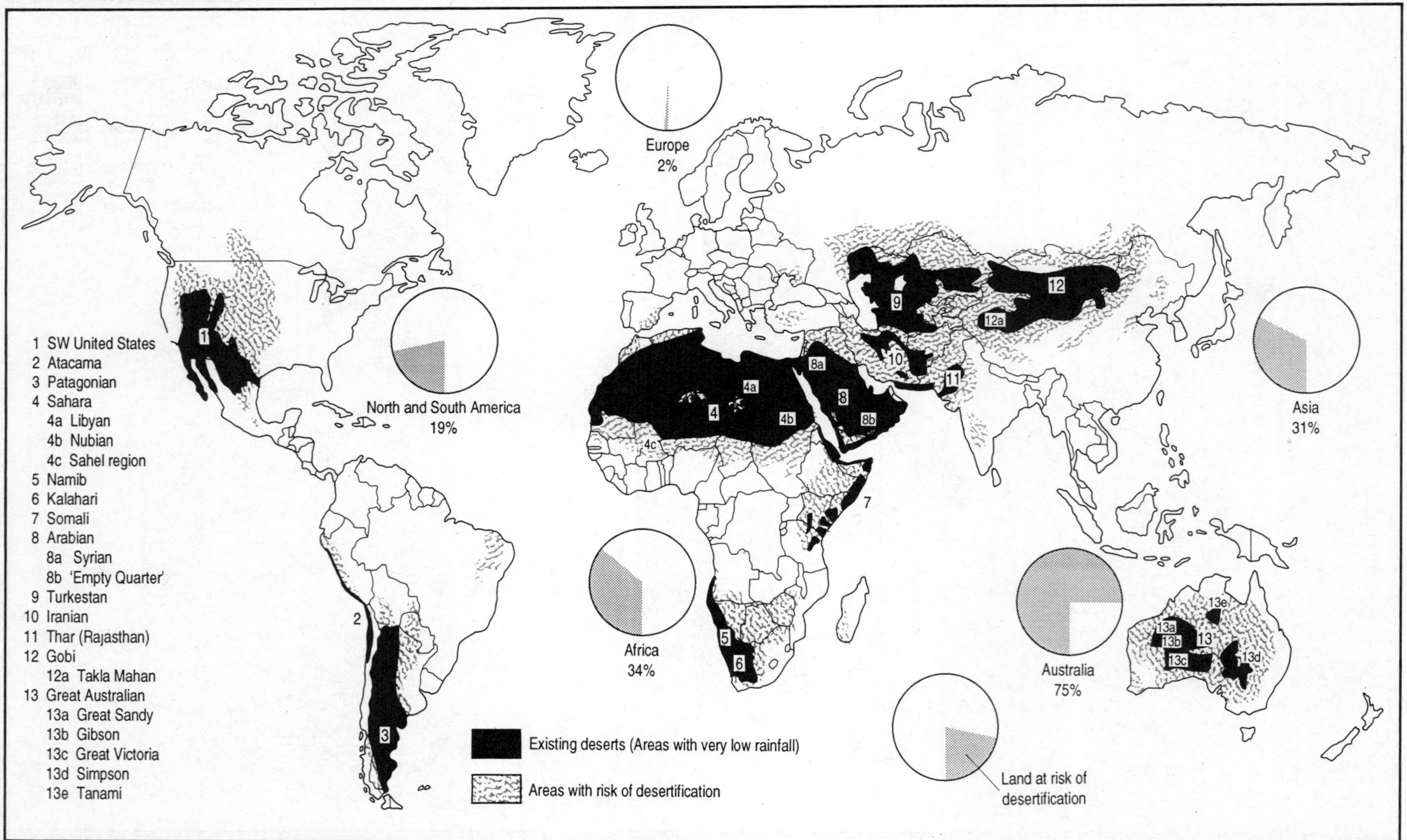

Sources: *Atlas of the Environment* , Helicon, 1992; *World Environment Atlas*, Bartholomew, 1991

7 Rainforest Destruction

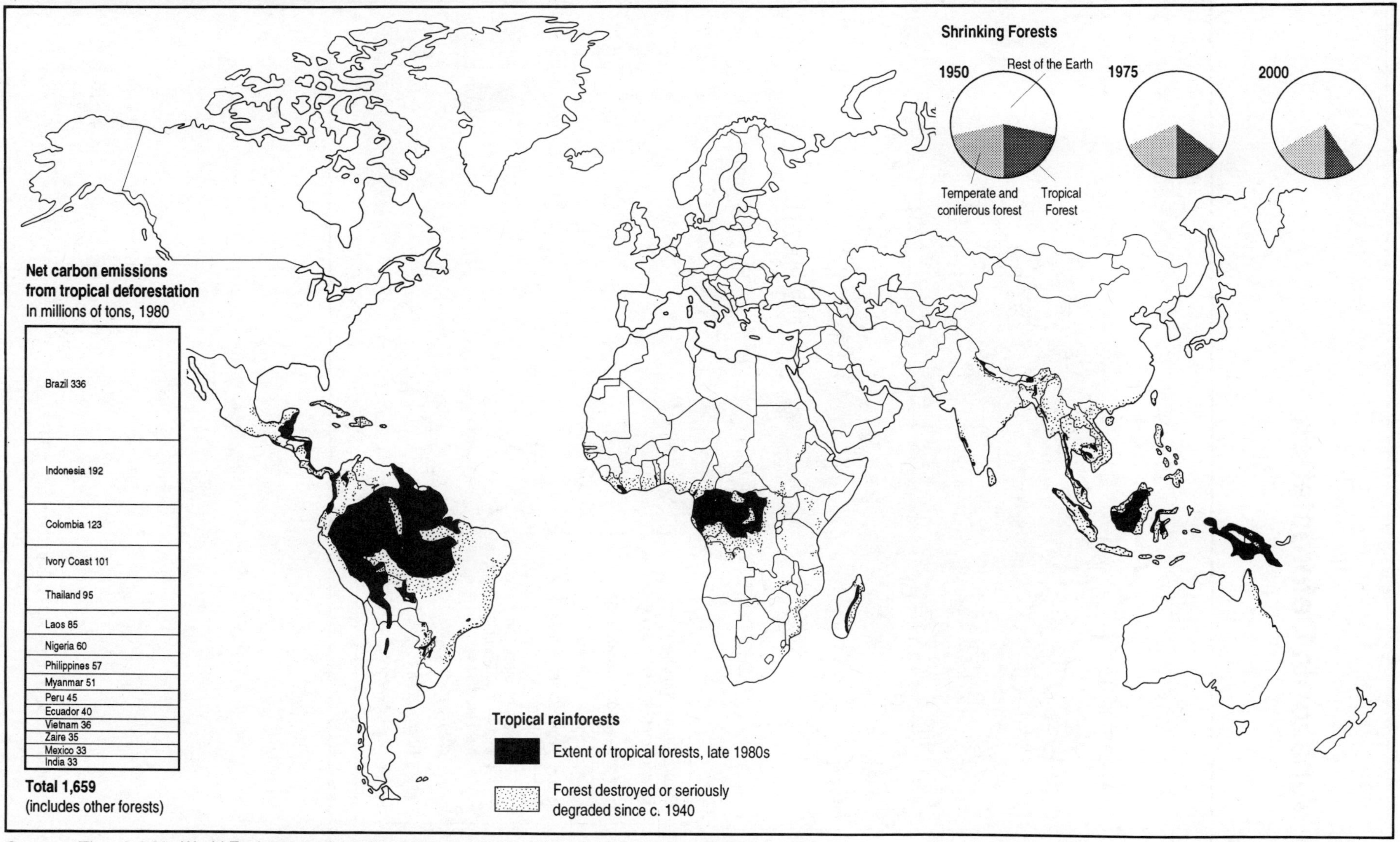

Sources: *Time*, 2.1.89; *World Environment Atlas*, Bartholomew, 1991; *Atlas of the Environment*, Helicon, 1992

8 Amazonia Deforestation

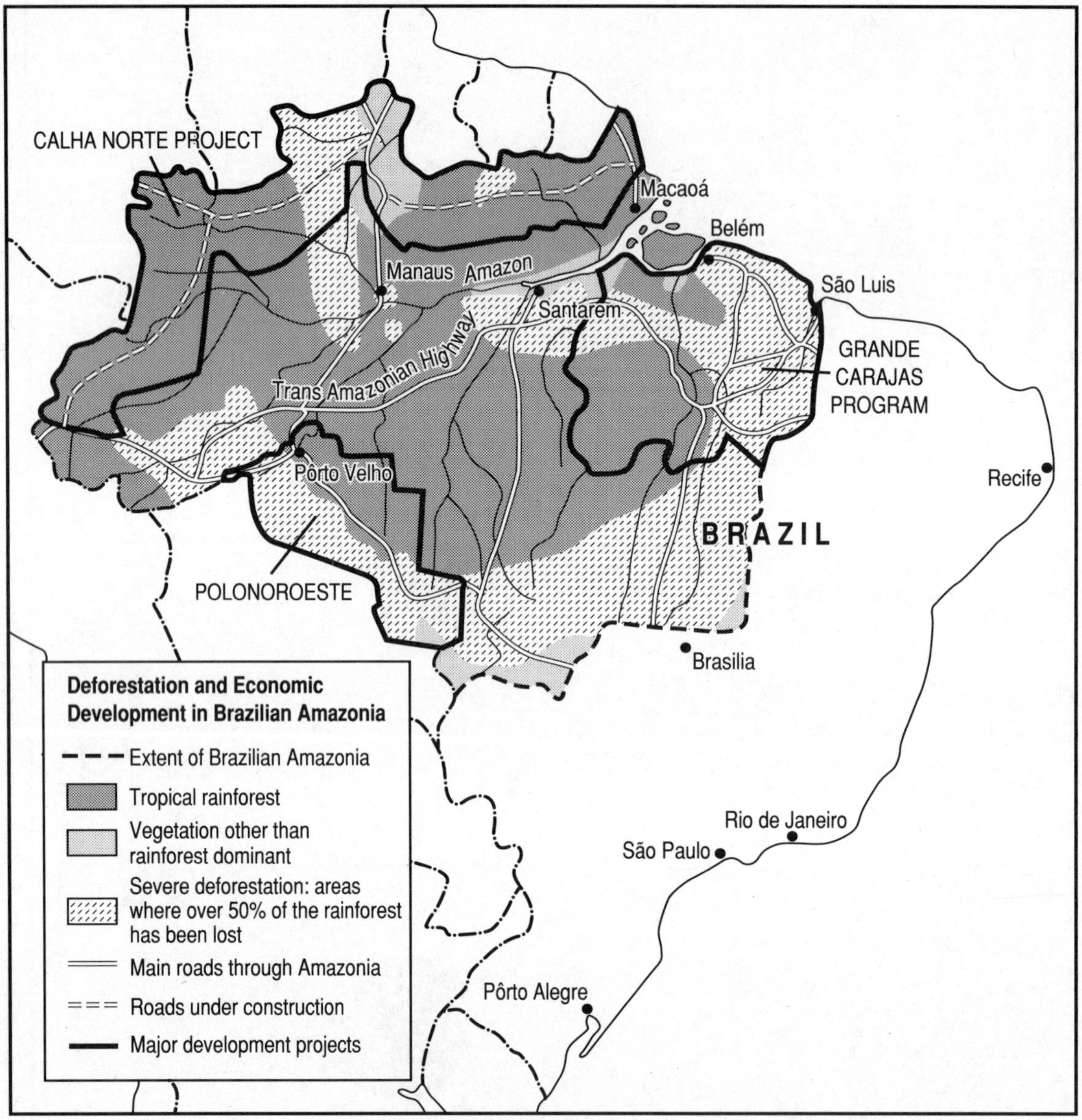

Source: *World Environment Atlas*, Bartholomew, 1991

9 Atmospheric Pollution and Acid Rain

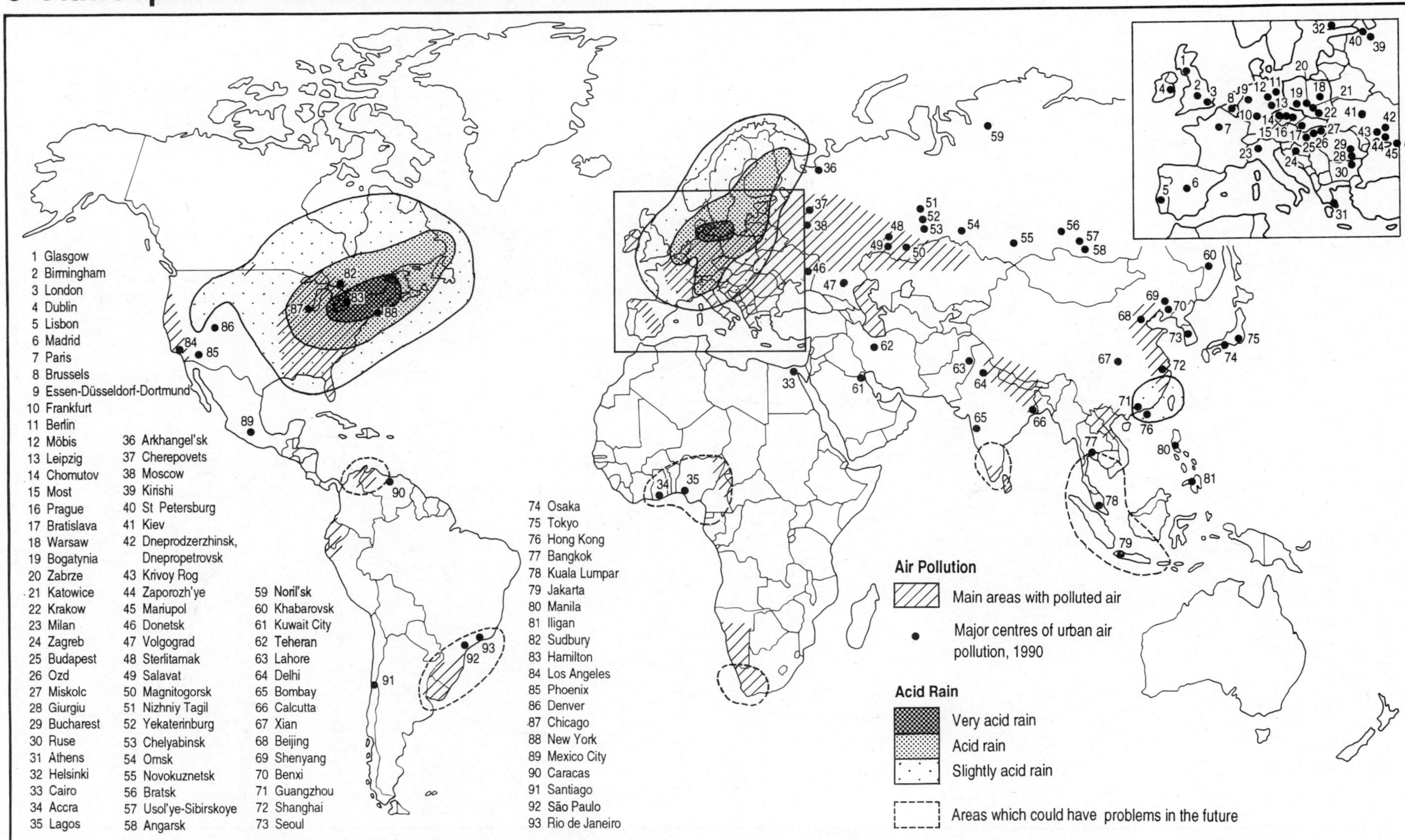

Sources: *Environment Atlas*, George Philip, 1992; *Atlas of the Environment*, Helicon, 1992

10 Greenhouse Effect

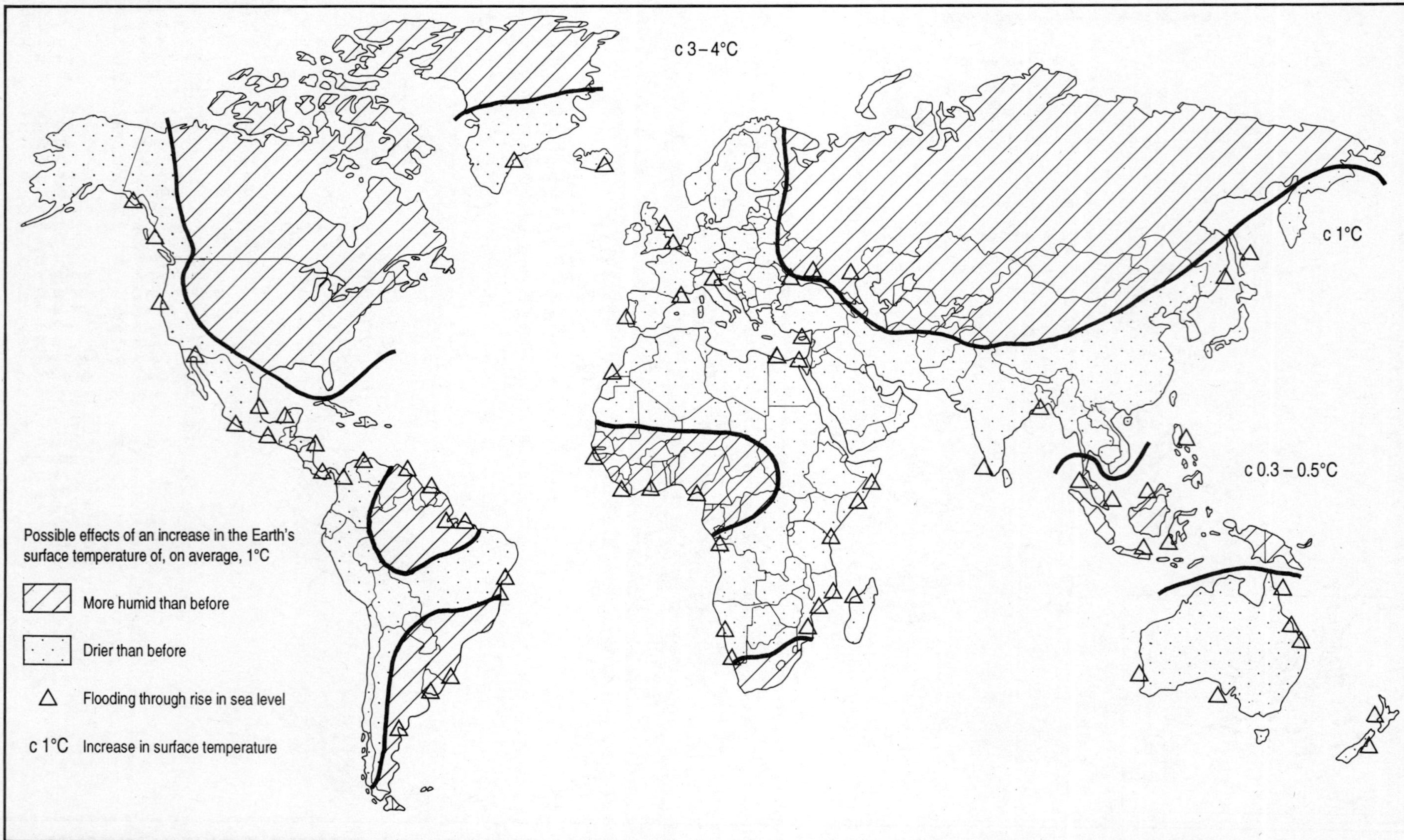

Source: *World Environment Atlas*, Bartholomew, 1991

11 Energy Production

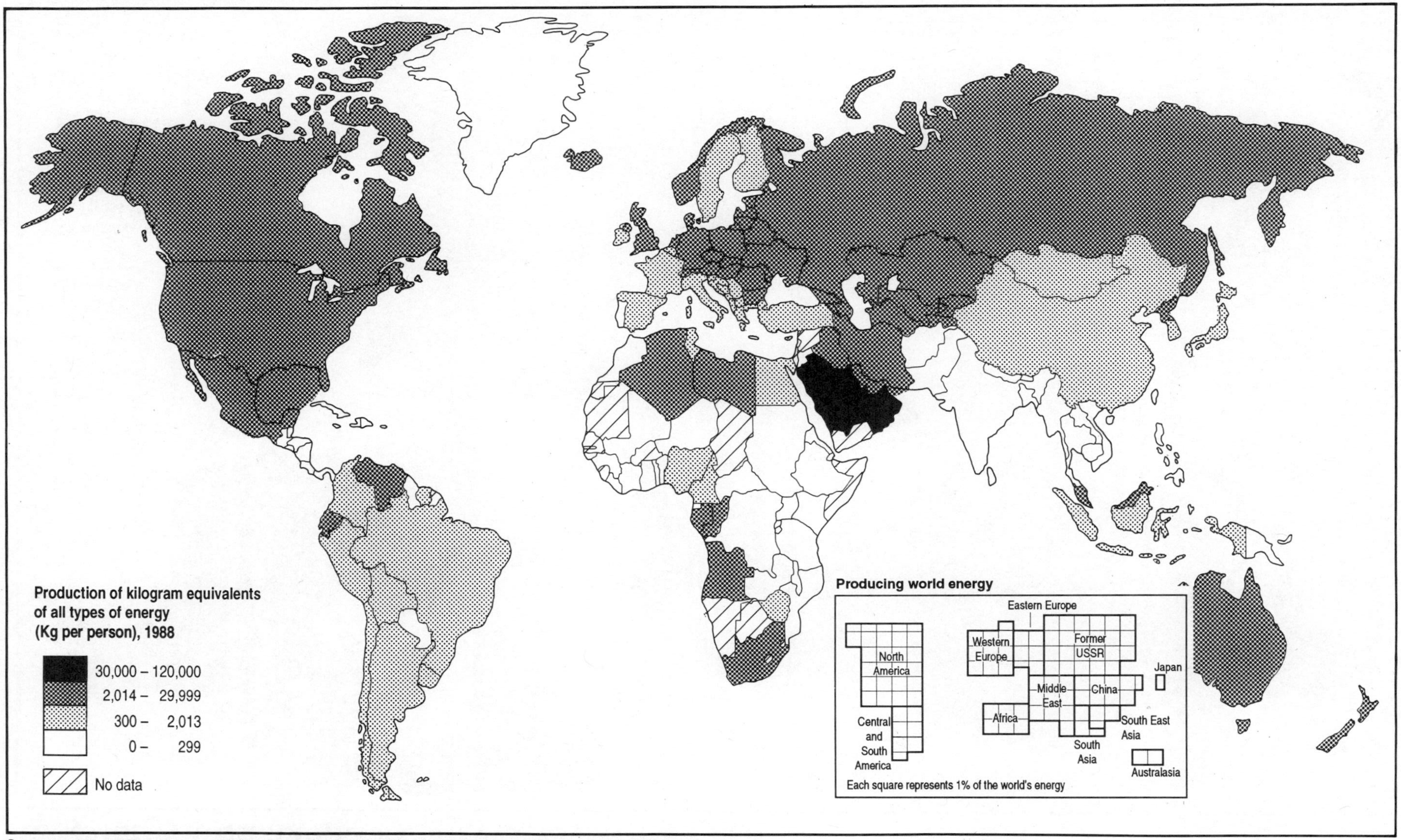

Sources: *World Environment Atlas*, Bartholomew, 1991; *Environment Atlas*, George Philip, 1992

12 Energy Consumption

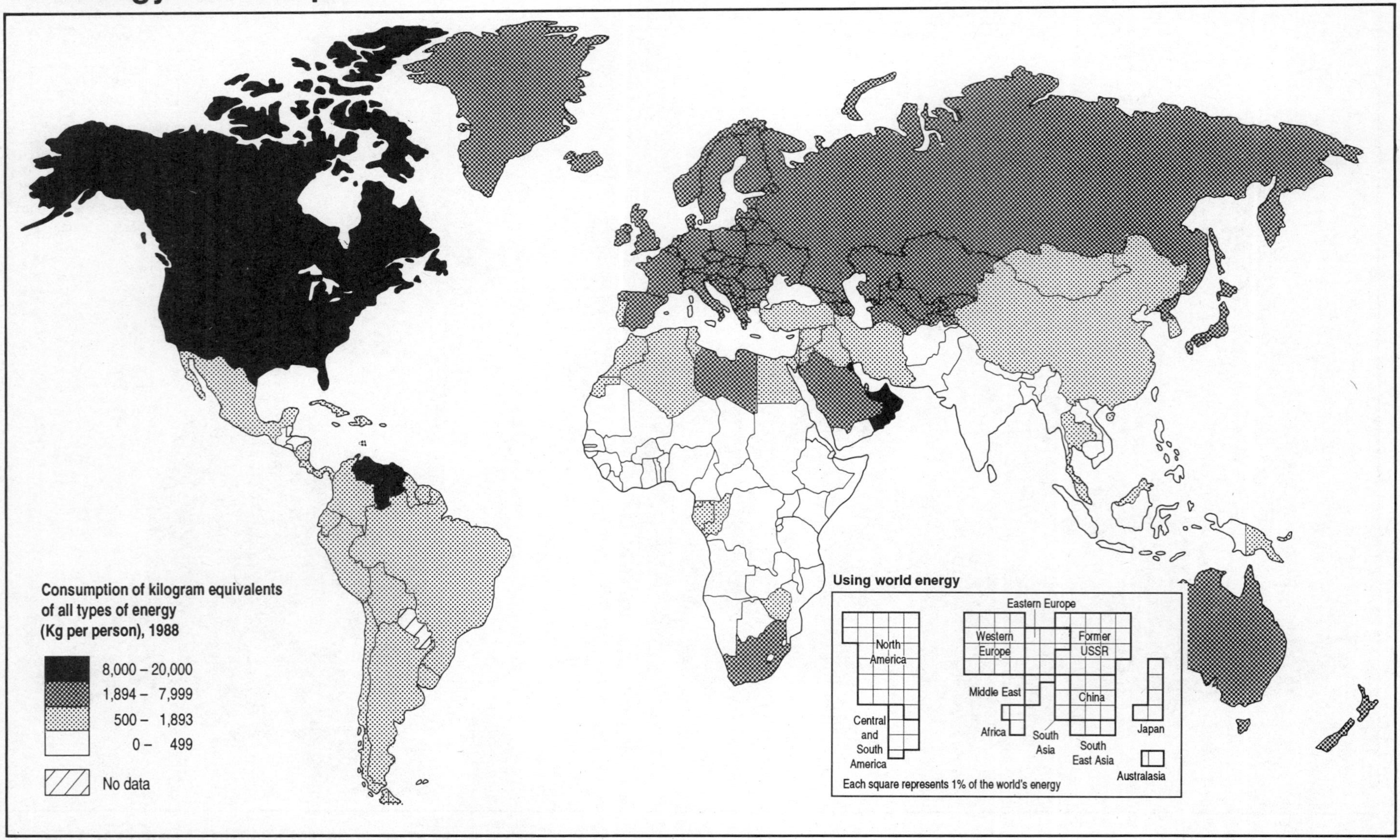

Sources: *World Environment Atlas*, Bartholomew, 1991; *Environment Atlas*, George Philip, 1992

13 Protected Areas and World Heritage Sites

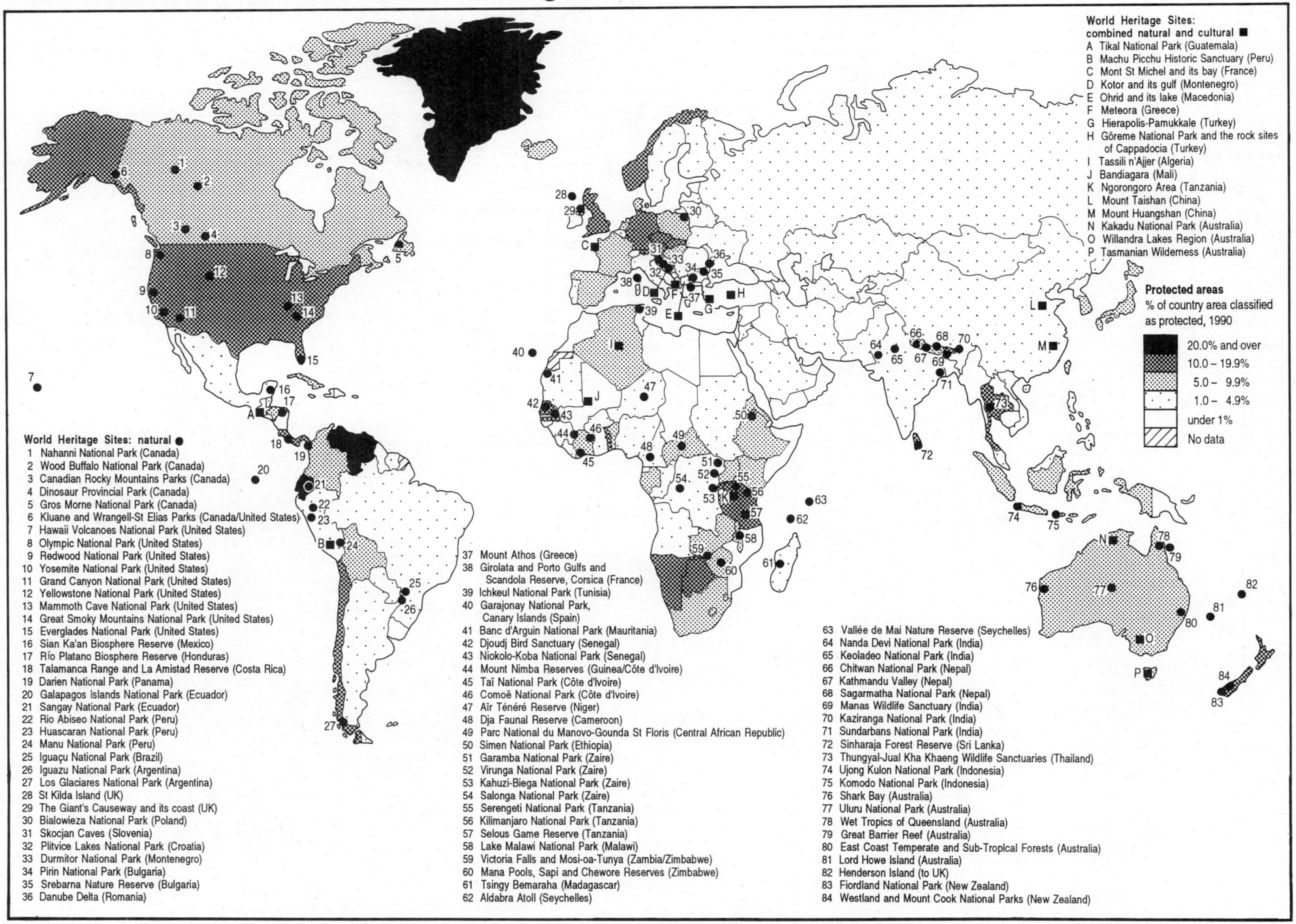

Source: *Atlas of the Environment*, Helicon, 1992

14 Fishing

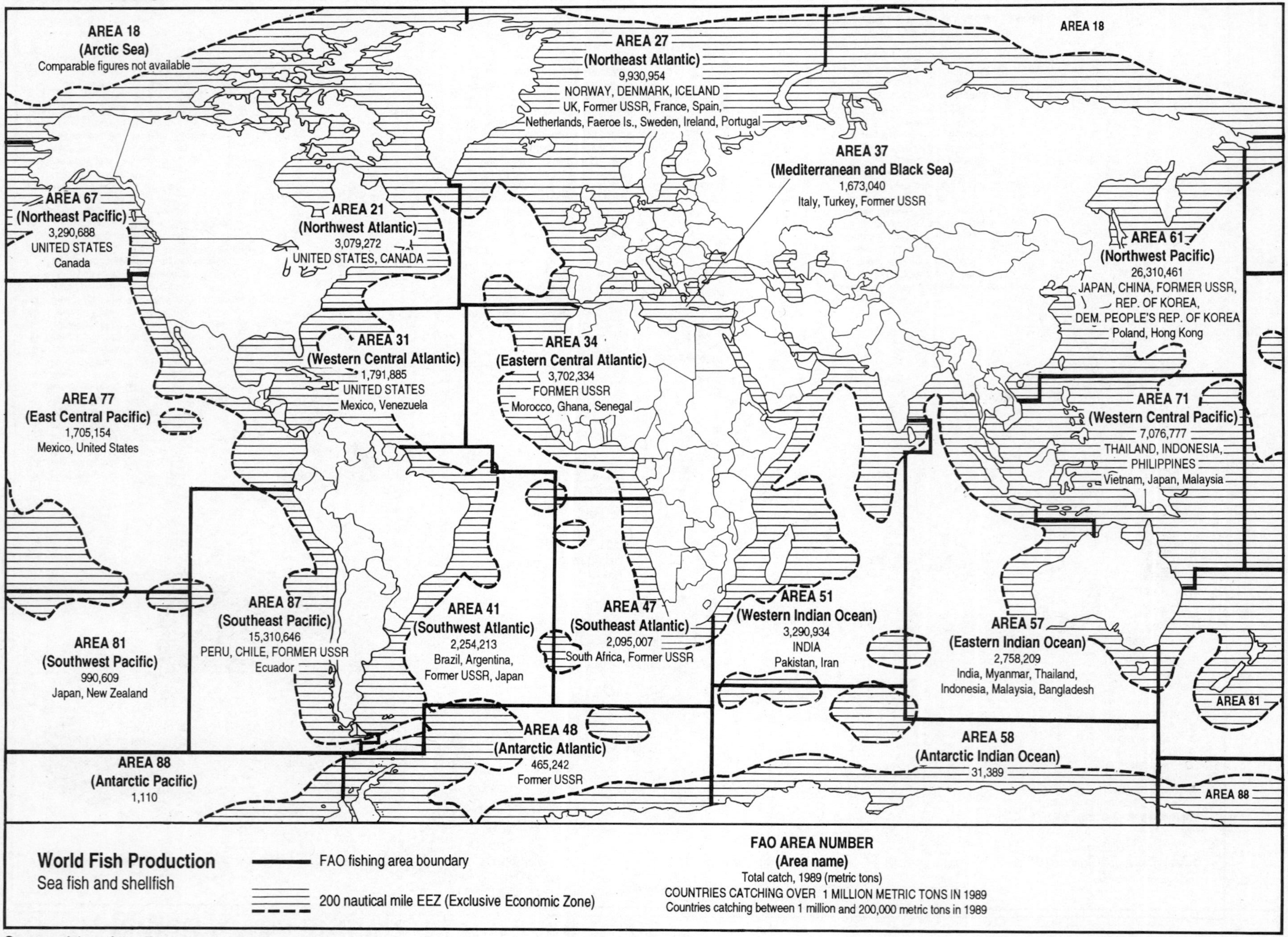

Source: *Atlas of the Environment*, Helicon, 1992

15 Sea Mineral Resources

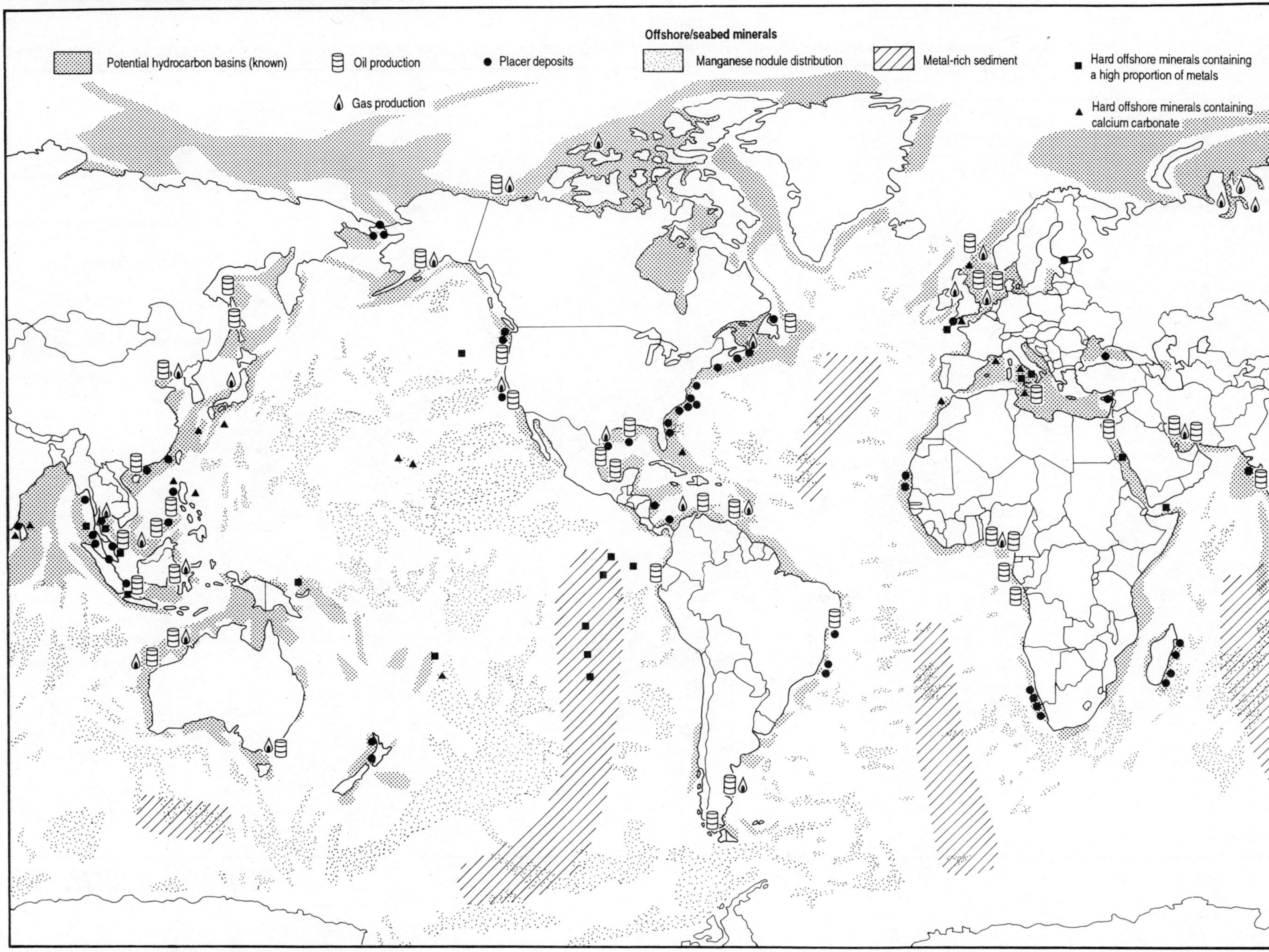

Source: *Atlas of the Environment*, Helicon, 1992

16 Water Pollution

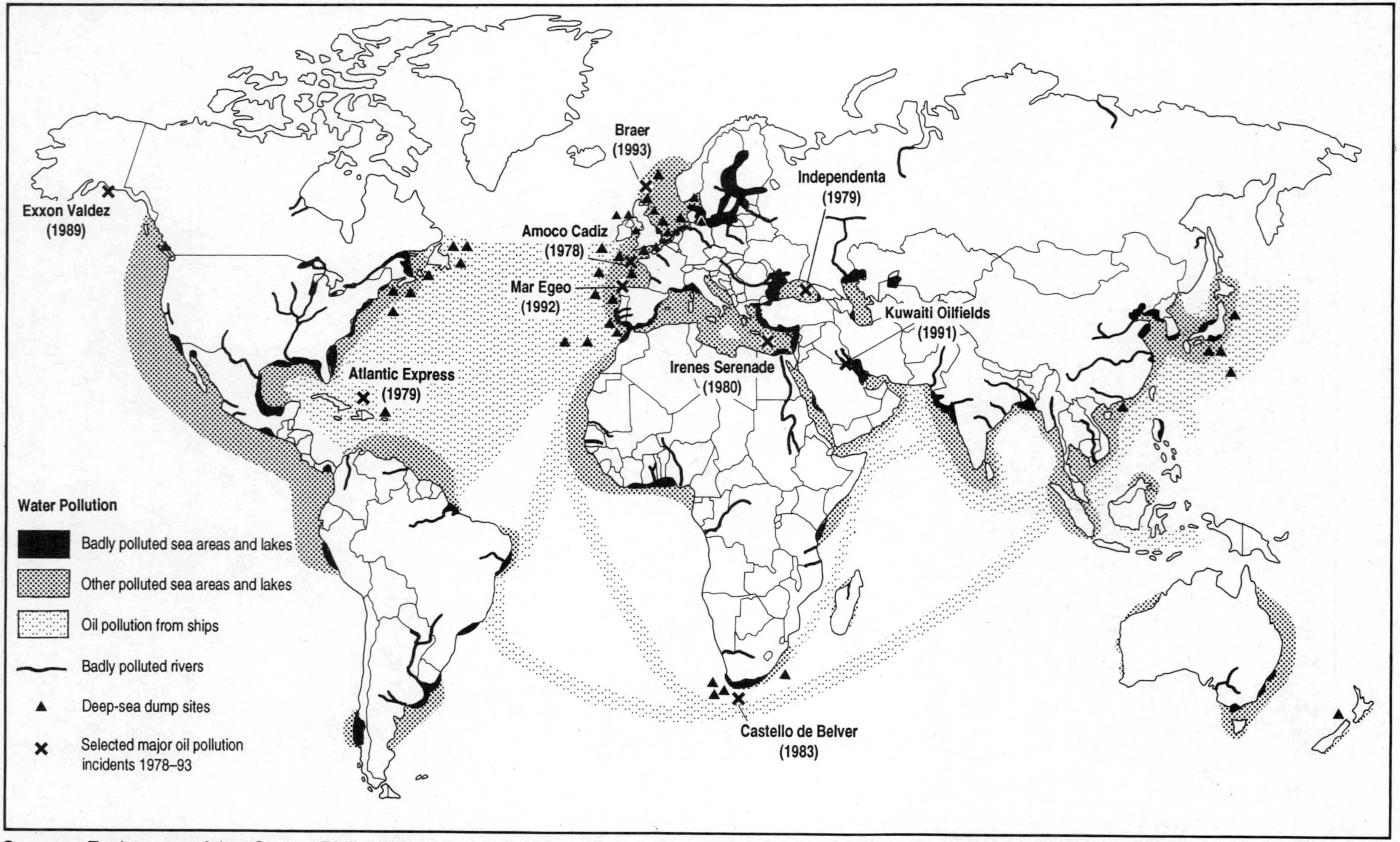

Sources: *Environment Atlas*, George Philip, 1992; *World Environment Atlas*, Bartholomew, 1991; *Atlas of the Environment*, Helicon, 1992

17 Antarctica Resources and Territorial Claims

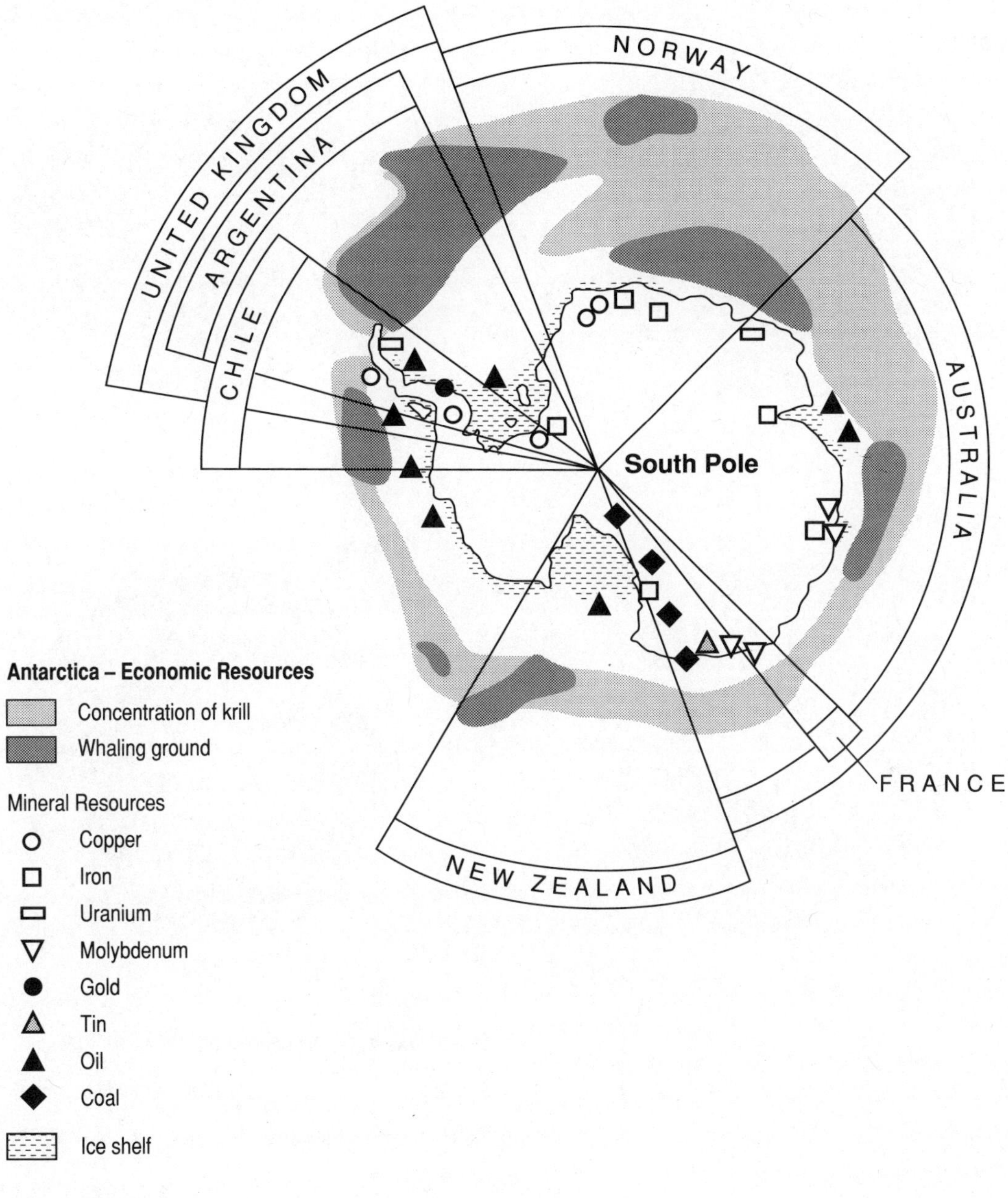

Source: *World Environment Atlas*, Bartholomew, 1991

18 Antarctic Ozone Hole

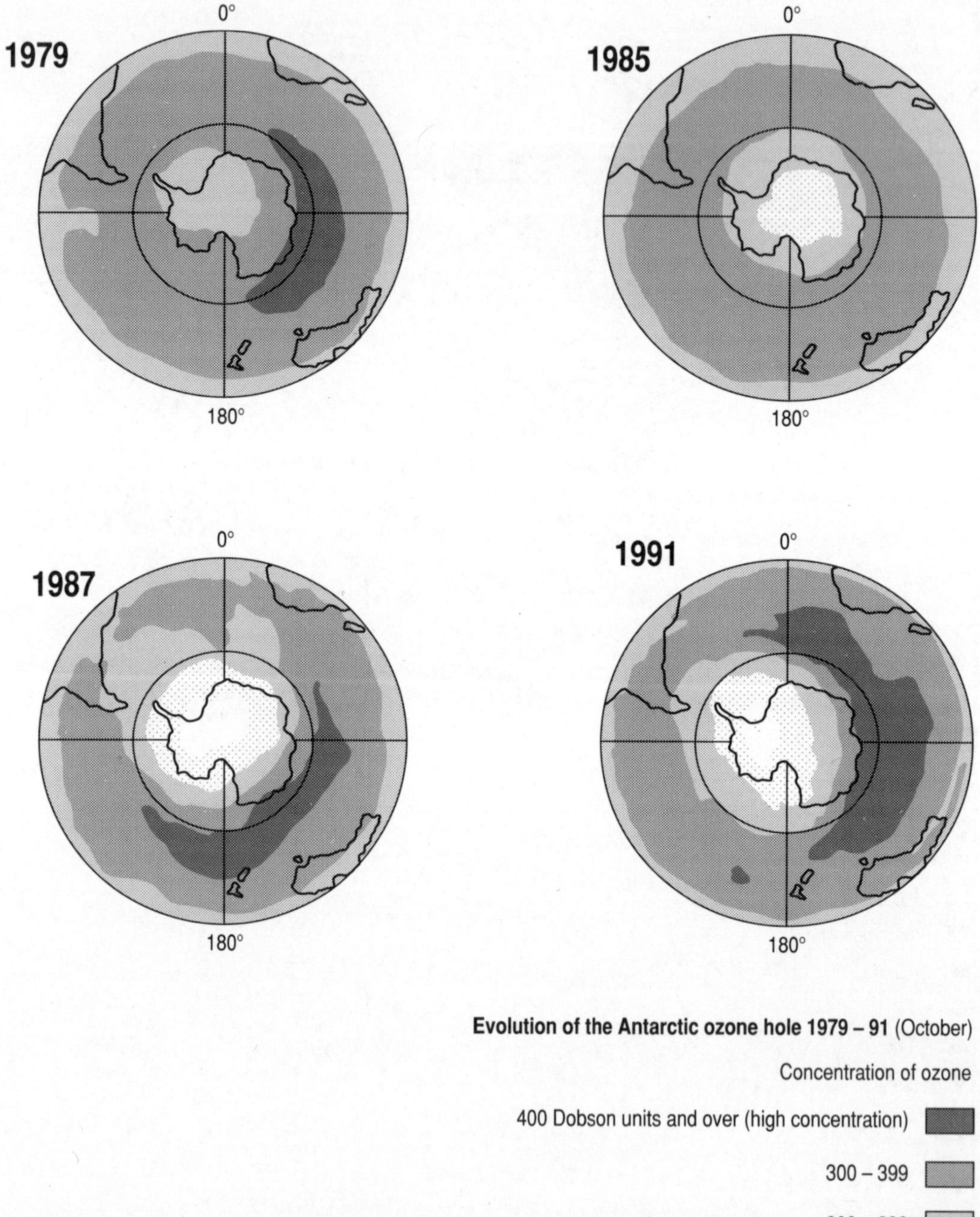

Evolution of the Antarctic ozone hole 1979 – 91 (October)

Concentration of ozone

400 Dobson units and over (high concentration)

300 – 399

200 – 299

150 – 199

under 150 (low concentration)

Source: *Atlas of the Environment*, Helicon, 1992

19 Chernobyl Radiation Pollution

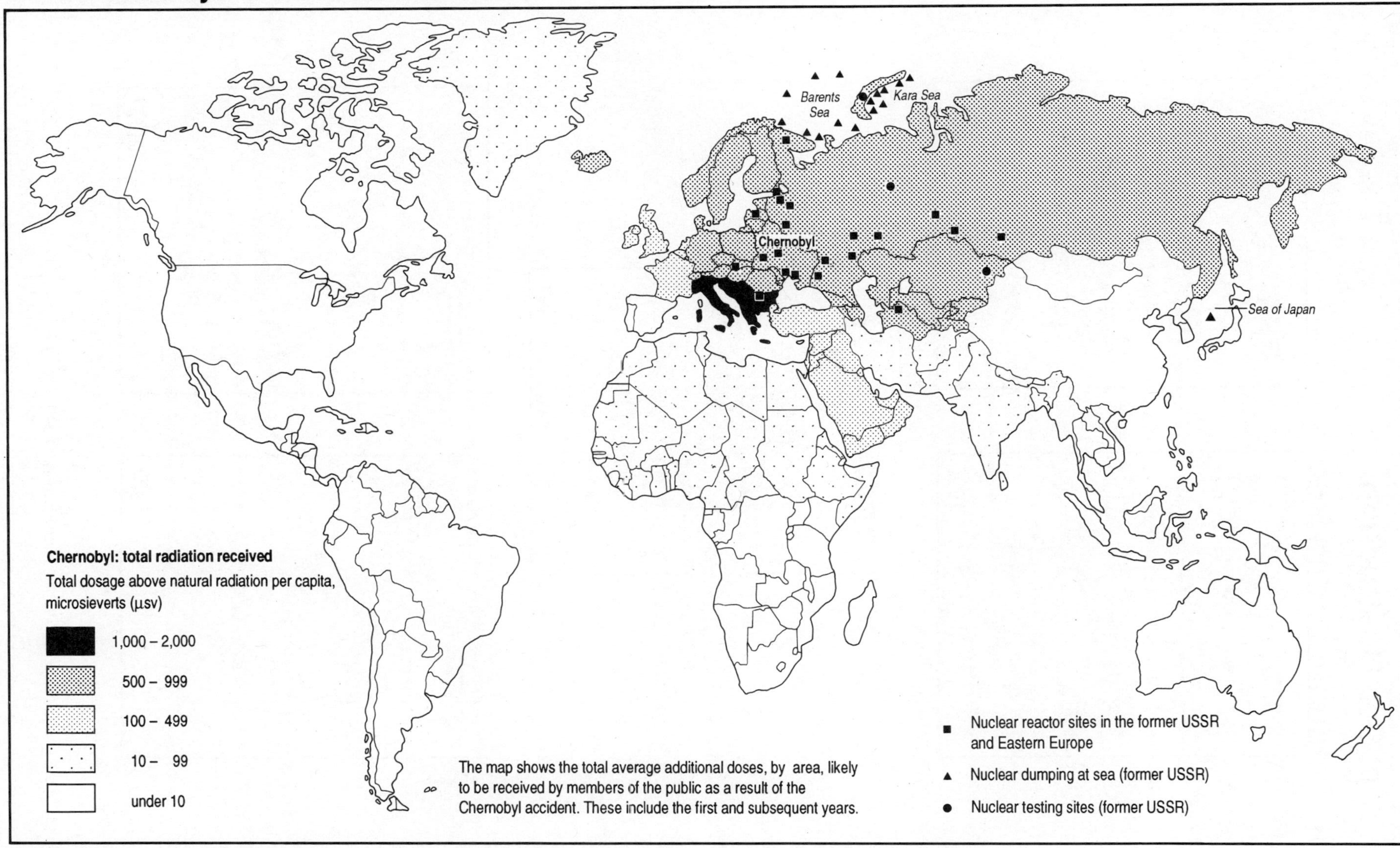

Sources: *Atlas of the Environment*, Helicon, 1992; *The Times*, 8.4.93; *The Guardian*, 20.3.93; *Le Monde*, 21.4.93; *Asiaweek*, 14.4.93

20 Aral Sea Destruction

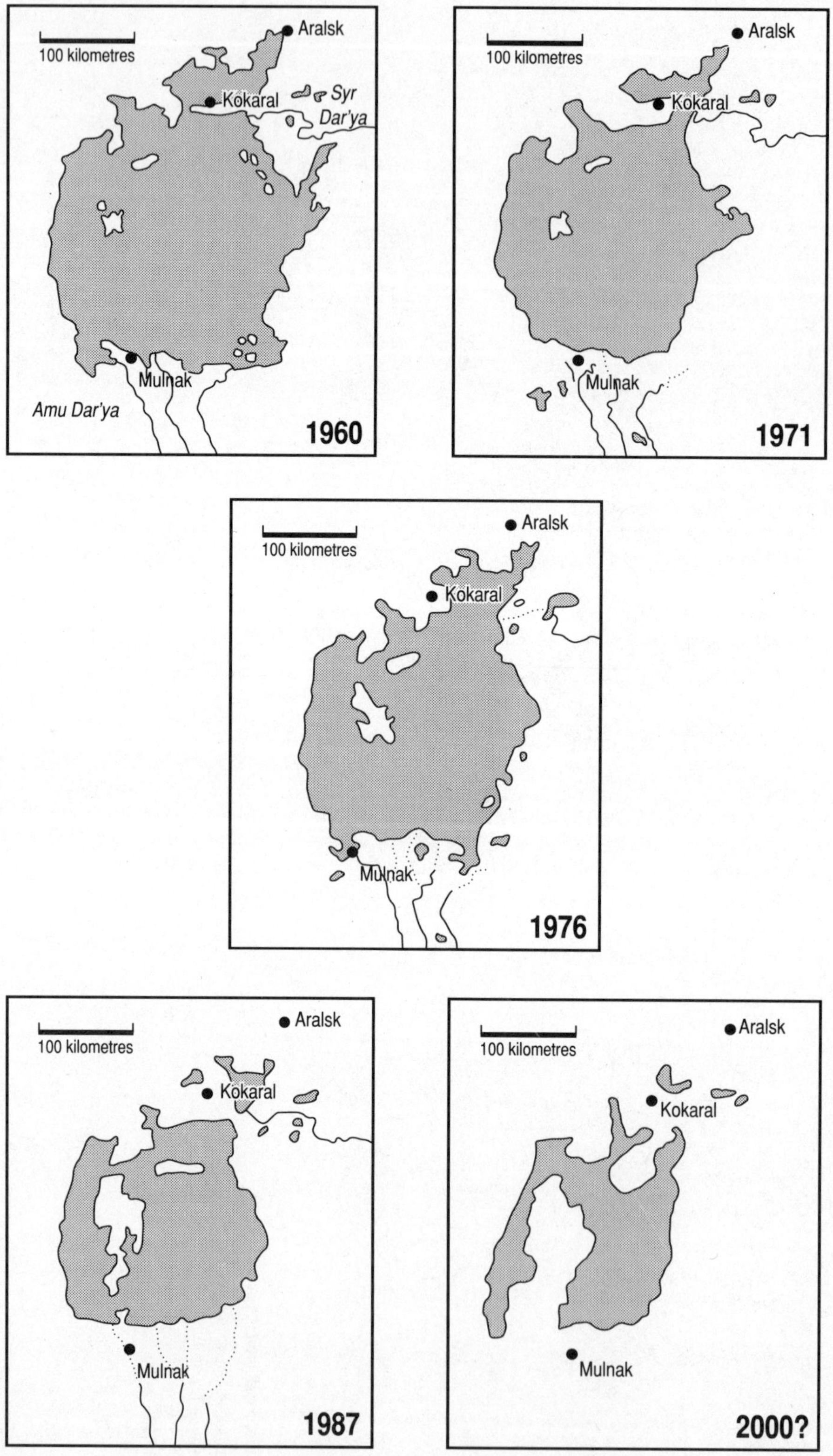

Source: *5000 Days to Save the Planet,* Paul Hamlyn, 1990

A–Z
of the
ENVIRONMENT

Pearce Wright

Pearce Wright was formerly Science Editor of *The Times,* and is now a freelance writer in science and medicine.

Note: Words printed in **bold** type indicate that an entry on the subject so marked is to be found elsewhere in the A–Z of the Environment.

A

ABIOTIC refers to the non-living components of an **ecosystem**—the rocks, minerals and water. They are essential for living plants and animals (the **biotic** components) of an ecosystem, providing the essential elements and nutrients that are necessary for growth. The abiotic elements also include the climatic and **pedologic** components of the ecosystem. They are characterized by low-energy, high-**entropy** materials.

ABSCISIC ACID is one of the five major plant hormones; each has powerful properties that come into play at the appropriate parts of the life-cycle. Abscisic acid inhibits growth. Synthetic **agrochemicals** are used to mimic the effect of the natural inhibitors and growth retardants in order to prevent sprouting of stored onions, potatoes and other root crops. They are used to control flower blooms and to regulate the maturing of cereal crops.

ABSOLUTE DROUGHT is the British definition of a period of at least 15 consecutive days in which less than 0.01 inches of rainfall is recorded. In the USA **dry spell** has a similar meaning. See also **drought.**

ABSORBED DOSE is a term employed when measuring the effect on the body of exposure to nuclear radiation. The unit of **activity** of a radioactive substance says how many atomic disintegrations are happening each second as it decays. However, it says very little about the biological effects of the substance, because it does not indicate whether the radiation it produces, when its atoms disintegrate, is in the form of **alpha, beta** or **gamma radiation.** Each of these behaves differently on the body. One measure of the effect of radiation is the amount of energy it deposits in a piece of tissue. This quantity is the absorbed dose, and the unit of absorbed dose is the gray (Gy).

ABSORPTION is the process by which a substance retains radiant energy, heat and light waves, instead of reflecting, refracting or transmitting them. Minor gases, such as **ozone,** nitrous oxide, **water vapour** and **carbon dioxide** (see greenhouse gases), have a significant effect on atmospheric absorption. Oxygen and nitrogen, the main atmosphere gases, do not absorb very much radiation. Another definition of absorption is the process in which the energy of **electromagnetic radiation** is taken up by a molecule and transformed into a different form of energy. See also **albedo.**

ABYSSAL DEPTHS are the very deep parts of the ocean beyond the **continental shelf,** but not as deep as the ocean trenches. The abyssal plains are extensive flat areas of seabed in the abyssal depths. Although very deep, there is still marine life on the abyssal sea floor, but of a rather exotic type. The communities form strange, sometimes grotesque and often exquisite, samples of fish, crustaceans, sponges, sea-lilies, sea-squirts, sea-pens and segmented worms. These parts of the ocean have become environmentally threatened as they are areas which have been used for the disposal of radioactive waste, but that are now subject to the **London Convention** and other agreements on the control of **waste disposal.**

ACCELERATED EROSION describes the speeding up, either directly or indirectly by human activity, of erosive processes. For example, **deforestation** or **overgrazing** can increase surface run-off and result in **soil erosion. Urbanization** also causes accelerated erosion, because concrete and tarmac increase surface run-off by obstructing the surface water which flows through cracks and pores in soil and rocks.

ACCEPTABLE DAILY INTAKE (ADI) is the measurement of the amount of any chemical substance that can be safely consumed by a human being in a day. Calculations are usually based on the maximum level of a substance that can be fed to animals without producing any harmful effects. This is divided by a 'safety factor' to allow for the differences between animals and humans and to take account of the variation in human diets. The ADI is set by international agencies: the Joint Expert Committee on Food Additives, a body of scientists from the **Food and Agriculture Organization (FAO)** and the **World Health Organization (WHO);** the European Community's Scientific Committee for Food; and the US Food and Drug Administration (FDA). The levels are set and discussed independently by each of the agencies.

ACCLIMATIZATION refers to the process by which plants and animals become accustomed to unusual environmental conditions, so that they are able to live and reproduce. Until recently, the word described the natural adjustment by plants and animals to changes in climate. For example, plants can undergo changes that mean they have to synthesize new types of protein and membranes that will be more tolerant of cold weather. Now, acclimatization is also used to describe ideas for **genetic engineering** of plants to create modified strains that could withstand climate changes due to global warming. The most remarkable illustration is the ice-minus project.

An **ACID** is a substance that contains hydrogen and splits up in water, liberating hydrogen ions and creating conditions that are chemically reactive.

ACID RAIN describes a type of pollution which washes out of the atmosphere as dilute sulphuric and nitric acids. It tends to be a regional rather than a global phenomenon, linked to particular industrial activities and meteorological conditions. It is not just rain; it includes more than normally acidic snow, mist, sleet, fog, gas and dry particles. Acid rain upsets the balance of nature, disrupting **ecosystems,** and destroys forests and woodlands, plants and crops; kills aquatic life by altering the chemical balance of lakes and rivers; and corrodes building materials and fabrics. The pollutants are caused principally by discharges from power station chimneys of **sulphur dioxide** and nitrogen oxides released by burning fossil fuels, coal and oil. Other by-products emitted by industry to the air and by vehicles can add to the mixture. The airborne pollutants interact chemically with **water vapour.** Some drifts back to earth within a few miles of the source of emission. Some is carried high into the atmosphere, where it finds prevailing winds that carry it hundreds of miles before it falls back to earth as dust, rain, snow or mist. Discharges from very tall chimneys of modern fossil fuel power stations in Britain have been detected in Scandinavia. A small part of acid rain in Europe is believed to come from the USA. After fog killed 4,000 people in London in 1952, a major campaign for clean air began. The UK Clean Air Act was adopted in 1956. City-based power stations were closed and larger replacements, with tall chimneys, were built in more remote areas—but close to coal-mining regions. After a debate on acid rain at the first **UN Conference on the Human Environment,** held in

Stockholm in 1972, international pressure grew for controls. Thirty-three countries signed the Geneva Convention, Long Range Transport of Air Pollutants, deploring acid rain. The **World Wide Fund for Nature** identified the most seriously affected forest and lakeland areas of Europe in a report in 1986. Pollutants can be removed by gas scrubbers on power stations and **catalytic converters** on cars.

ACOUSTICS are sounds that are both pleasant, like music, and distressing, like **noise pollution** from traffic, aircraft or rowdy neighbours. Acoustics includes not only the reception of sounds by the ear, but also their production, transmission and use. Musical acoustics is concerned with the design and making of musical instruments. Engineering acoustics deals with the production of the more technical aspects of sound production, such as microphones, amplifiers and loudspeakers. Architectural acoustics concentrates on obtaining the best reception of sound in enclosed spaces, especially auditoriums and concert halls. Environmental acoustics refers to the technical means of controlling noise, such as reducing sound levels by insulating heavy machinery and lessening vibrations in vehicles.

ACTINIDES are a group of 15 radioactive elements produced when uranium fuel is irradiated in a nuclear reactor. They include **plutonium, americium** and **neptunium.** All of the group have atomic numbers higher than uranium, and so are also referred to as the **transuranic elements.** The health hazard presented by the actinides, if they are released into the environment, comes from the potency of their radioactive characteristics. They are **alpha-emitters,** and therefore can cause intense localized damage in tissues if absorbed into the body.

ACTINOMETER is the general name for any instrument which measures radiant energy, particularly **solar energy.**

ACTION PLAN FOR THE HUMAN ENVIRONMENT, see **United Nations Conference on the Human Environment.**

ACTIVATION PRODUCTS are produced when normal metal and concrete construction materials and components employed in building a nuclear power station come under constant irradiation by neutrons. The materials absorb neutrons and become radioactive, and strong emitters of **gamma-radiation,** creating difficulties in dismantling nuclear reactors and disposal of the pieces.

ACTIVE LAYER is the top layer of soil above **permafrost,** or permanently frozen ground. The active layer is subject to seasonal freezing and thawing, and during the melting period it becomes very mobile and can lead to mudslides.

The **ACTIVITY** of a radioactive substance refers to the number of nuclei which **decay** per second. Until recently the unit of activity was the **curie** (Ci). This is equal to 37,000 million decays per second. Since it is too large a unit to be useful in many applications, scientists use the measurements millicurie (mCi, one thousandth), microcurie (uCi, a millionth), or nanocurie (nCi, a thousand-millionth of a curie). For convenience, the curie is being replaced by the **becquerel** (Bq). One becquerel is equal to just one decay a second, and 1 nCi equals 37 Bq, 1 uCi equals 37 kBq (kilobecquerels) and 1 mCi equals 37 MBq (megabecquerels).

ADAPTABILITY refers to the ability of an organism to change its mode of behaviour and even its physiology, by **acclimatization,** in order to survive under new conditions that would otherwise be too stressful. In ecological terms, when something is adapted, it means that its structure is well suited to its function.

An **ADDITIVE** is a substance mixed in small quantities with another product to modify the chemical or physical state of the commodity. Additives are used to make food look visually more attractive, in the case of colouring agents, as well as to preserve and extend the life of the product. Additives may also become environmental pollutants, such as tetra-ethyl **lead,** introduced into petrol as an **anti-knock agent.** It was developed as a cheaper way of blending high octane fuels for use in high compression automobile engines. It improves the process of combustion as well as eliminating 'knocking' or 'pinking'. Unfortunately, tetra-ethyl lead is a very toxic form of lead when discharged into the air from car exhausts. It is especially dangerous for young children in whom it causes brain damage. For regulation on additives to food, see **E-numbers.**

ADENOSINE TRIPHOSPHATE (ATP), is one of the two fundamental power units in animals and plants. ATP is the principal energy-carrying compound in the cells of all living organisms. It is found wherever critical energy conversion processes are taking place, such as in the conversion of sunlight by **photosynthesis** to plant tissue or in extracting the energy from sugar at the instant it is demanded by the muscle of an animal. It is used to bring about such biological processes as muscle contraction, the transport of ions and molecules across cell membranes, and the synthesis of biomolecules. ATP is regenerated using the chemical energy obtained from the oxidization of food.

ADOBE is the building material used in some developing countries to build houses. It is made by mixing clay-based soil with water and adding straw or animal dung to bind it together. Made into bricks and dried in the sun it provides cheap house-building material. Adobe is comparatively simple to produce, it is fire-resistant, absorbs humidity and is a good heat insulator. Sometimes bamboo, or something similar, is added to the structure of houses, schools and hospitals. This strengthens the building and protects them from storms or even small earth tremors.

ADVANCED GAS-COOLED REACTOR (AGR) was the name of the British-designed nuclear reactor the UK Government chose for its second nuclear power programme. The AGR followed the first generation of **Magnox** nuclear power stations. The plan was to increase Britain's nuclear electricity capacity from the 5,500 megawatts provided by Magnox stations to 11,000 megawatts. The new plan was announced in 1964, just three years after the first two Magnox power stations had been switched on. The aim of the AGR was to achieve higher efficiency by reaching higher temperatures. The AGR, with a nuclear core of 520 tons of fuel, is more compact than the Magnox. However, it has a slightly higher percentage, two per cent, of enriched **uranium-235.** The AGR programme encountered many design difficulties, delays and excessive costs. It was the subject of successive parliamentary and government inquiries. However, the first two AGRs eventually came into operation in February 1976, at **Hinkley Point,** in Somerset, and Hunterston, in Scotland.

AEDES AEGYPTI is the most important of the aedes family of **mosquitoes** found in the tropics and subtropics because it transmits serious illnesses such as **dengue fever** and yellow fever. It is found in estuaries and river areas, usually near human settlements. The female will lay her eggs in any available stagnant water: puddles, discarded tin cans, old tyres, coconut shells, even holes in trees. Other species of Aedes are carriers of **filariasis** and other viruses that cause various forms of encephalitis. Much of this problem is caused by poverty which produces an unsanitary environment. Attempts to eliminate pesticide sprays have only added to the environmental depreciation.

AEROBIC RESPIRATION, see **respiration.**

AEROLOGY is the scientific study of all the free **atmosphere.**

AERONOMY is the scientific study of the upper **atmosphere,** where the **ionization** of gas molecules occurs.

AEROSOLS are small particles in the atmosphere that can affect climate, depending on how they reflect or absorb radiation. Natural events like volcanic eruptions can produce aerosol clouds. Human-produced aerosols include the hairsprays, deodorants, polishes and other consumer goods that used **chlorofluorocarbons (CFCs)** as propellants.

AFFORESTATION is the planting of trees on land which was previously used for other uses than forestry. See **forests and woodlands** and **reafforestation.**

AFLATOXINS are a group of poisonous substances produced by the fungus mould, *Aspergillus flavus,* which grows on seeds and grain. They were first identified in the 1960s when a large flock of birds died on a poultry farm in the UK in strange circumstances. It was found that they had been fed with meal contaminated with a poisonous mould. Aflatoxins can damage the immune system and cause liver cancer. Most of the reported cases in people have come from the Indian subcontinent and East Africa, where environmental conditions suit aflatoxin growth. Undernourished children are at most risk. Aflatoxins are a major problem in the tropics and subtropics because they thrive in moist and warm conditions and also because stored cereals and nuts, especially **groundnuts,** are often contaminated. The use of chemical **fertilizers** is thought to play a major part in spread of aflatoxin contamination because they tend to increase the water content of the crop and encourage the growth of mould.

AFRICAN ELEPHANTS have been hunted for their valuable ivory which is used to make piano keys, jewellery and ornaments. They are the largest land mammal and live in rainforests or **savannah.** In recent years, in most of the countries of Africa, the elephant population has rapidly declined, sometimes because of illegal poaching, but mostly because its habitat has disappeared as a result of human encroachment. The **Convention on International Trade in Endangered Species (CITES)** placed the African elephant on Appendix 1, the highest protection designated under the Convention in 1989. In 1985, CITES had set up the ivory quota system in an attempt to maintain the ivory trade within realistic limits. Most of the 103 member countries agreed not to allow the sale of ivory. In recent years, several countries in southern Africa have reported that their elephant populations have increased. South Africa and Zimbabwe have carried out culling programmes. Given adequate space, the African elephant makes a valuable contribution to a stable ecosystem. Because they only digest about a half of the vast amount of leaves, grass, fruits and bark they eat, their dung provides essential nutrients to restore the soil of the savannah. As humans have taken over parts of the elephants' range, elephants have eaten crops and become regarded as a pest. The Asian elephant is also considered to be an **endangered species. Vanishing habitats,** rather than poaching, are their greatest threat. The Asian elephant has been largely domesticated and is widely used in the logging industry.

An **AFTERBURNER** is a gadget fitted to the exhaust flues of furnaces and also to the exhaust systems of motor vehicles. They remove polluting gases and particles, which are the result of incompletely combusted fuel, by incineration and break down other chemical molecules associated with **combustion** into inert chemicals. The **catalytic converter,** which is fitted to vehicles to remove unburned hydrocarbons, is an example of an afterburner that neutralizes a complicated mixture of molecules. The catalytic converter was first introduced in Los Angeles to overcome the pollution caused by **photochemical smog,** the type of atmospheric pollution that is created by the action of sunlight on the unburned hydrocarbons and nitrogen oxides released from automobiles.

AGENDA 21, the Rio Declaration on Environment and Development, is a comprehensive programme of action to be achieved by governments, development agencies, United Nations organizations and independent sector groups in every area where human activity affects the environment. Agenda 21 was adopted by the United Nations Conference on Environment and Development, the **Earth Summit.** Agenda 21 is the response to the UN General Assembly's call, in December 1989, for a global meeting to devise strategies to halt and reverse the effects of **environmental degradation** 'in the context of increased national and international efforts to promote sustainable and environmentally sound development in all countries'. Behind Agenda 21 is the belief that humanity has reached a defining moment in its history. Present policies deepen the economic divisions within and between countries and this increases poverty, hunger, sickness and illiteracy world-wide and is causing the continued deterioration of the **ecosystem** on which we depend for life on earth. Agenda 21 concentrates on ways to change course and how to improve living standards for those in need with the implementation of a global partnership for **sustainable development.**

AGENT ORANGE was the herbicide which the US military sprayed in Vietnam to defoliate vast tracts of jungle that provided cover for the Viet Cong, whom they were fighting. The ecological impact of the action caused concern among many scientists. However, there was even greater concern about the health of the rural populations in Vietnam. Agent Orange is an organochlorine **herbicide,** containing traces of an impurity known by the initials TCDD, which is a member of the **dioxin** family of chemicals, regarded as among the most poisonous ever made.

AGRICULTURAL HABITATS are those areas created by man to cultivate crops and graze domestic animals in order to provide food. Domestication of plants and animals brought great changes in the environment. In creating agricultural habitats man removed almost everything within each habitat that had a detrimental effect on his activities, such as **pests,** weeds and indigenous plants and animals, and left only those organisms which were helpful to his environment. The result was that the diversity of plant and animal species was greatly reduced. A large increase in the human population has led to an increase in **intensive agriculture** and the mechanization of farming methods, and more and more land throughout the world being turned over to farming. Industrialization in

Europe, from the end of the 18th century, had a major impact on agriculture, both in the provision of new machinery that could improve the efficiency of agriculture and in the production of synthetic fertilizers to improve the fertility of the soil. Modern agricultural methods are energy intensive and require extensive use of **pesticides** and **fungicides.**

AGROCHEMICALS are the man-made **fertilizers** and **pesticides** that were increasingly produced in industrialized countries following the development of **intensive agricultural** practices in the 1950s. They are the products of the modern chemical and petrochemical industries. Manufactured **nitrate** fertilizers have produced huge rises in the yields per acre of cereals, vegetables, grazing pasture and fruit. However, the high crop yields have become dependent on very heavy doses of agrochemicals. World fertilizer use rose from 15 million tons in 1950 to over 150 million tons in 1990. One estimate suggests that if agrochemical use stopped overnight, the world's harvest would be halved. On the debit side, high use of agrochemicals has been followed by the **leaching** of fertilizers into waterways where they cause **eutrophication** and **algal blooms.** Nitrates from fertilizers are also leaking into **groundwater** aquifers, which supply drinking water. The worrying problem for water engineers is that it may take decades for nitrates to work their way through the overlying rocks before reaching the groundwater's natural reservoirs. Pesticide poisoning comes from direct contamination or from residues that have accumulated in the **food chain,** causing extensive deaths among wildlife and raising anxiety about human health. An estimate by the **World Health Organization** suggests that nearly three million people suffer from acute pesticide poisoning every year; of these, 200,000 are workers in the developing countries who die because they are not trained properly in the use of protective masks and coveralls when spraying fields.

AIR POLLUTION can be caused by natural and man-made activities. It usually refers to the release of ash and soot or radioactive gases from factories and power stations, or to automobile emissions. It can also be caused in rural areas by **pesticide** sprays from fields, or dust from mines or quarrying. Natural sources, such as dust blowing from deserts, the ash and dust from volcanic eruptions and salt from the sea being blown inland by the wind, are other contributory factors. The gases rising from bogs, marshes and other sources of decomposing material can also cause air pollution. The large particles fall to the ground near the source. Smaller particles are carried long distances by the wind until they are washed out by the rain, resulting in **acid rain.** Air pollution has a damaging effect on health. Apart from irritated eyes and breathing problems, studies have shown that high **lead** content in the air in cities and towns probably results in brain damage and a reduction in the intelligence of children. Air pollution causes severe **environmental degradation:** in the towns it causes the fabric of buildings to deteriorate and metals to tarnish, because of the acidity of the pollution. Throughout the world air pollution is a serious cause for concern and great efforts are being made to reduce it through implementation of clean air legislation, the use of smokeless fuels and by fitting **catalytic converters** to cars. The emission of **carbon dioxide** and gases, such as **CFCs** and **methane,** can cause an enhanced **greenhouse effect.** The build up of **chlorofluorocarbon (CFC)** gases has made the international community act to restrict their use.

AIR QUALITY refers to the type and maximum concentration of man-produced **pollutants** that should be permitted in the atmosphere. Before the recent development of environmental awareness and the formation of the Green movement, there had been an outcry provoked by thousands of deaths caused by chest disease as a result of the devastating London smogs of the 1950s. The tragedy led to the first clean air legislation. The immediate problem was cured by creating 'smokeless zones' in major cities. However, there were new 'invisible' sources of pollution developing, such as **sulphur dioxide** from power stations and lead in car exhausts. The next stage depended on agreeing a European Community policy on air pollution. It was slow in coming. In 1984 measures were agreed in an EC Directive regulating industrial emissions into the air. The delay was partly due to the effect of the 1973 oil crisis, when the industrialized countries were compelled to rethink their energy policies. The new prices and production quotas adopted by the member countries of the OPEC cartel produced a scarcity of light crude oil with a low sulphur content and then forced its price up. The need to restrict consumption could not be agreed between EC countries. One consequence, for example, was that a proposal adopted in 1975 to limit the sulphur content of heavy fuel oil burnt in power stations was withdrawn.

After 1980 the results of many scientific studies into problems like **transboundary pollution** were being published. Atmospheric pollution moved to the top of the political agenda in June 1983, at the Stuttgart meeting of the European Council which called for immediate action to save Europe's forests. The call followed the publication of research results confirming that damage to forests in northern Europe was caused by acid rain exported across the North Sea from British power stations and from industry in central Europe. In 1984 an EC Air 'Framework' Directive issued regulations limiting discharges from industrial plants, and especially power stations, of sulphur dioxide, nitrogen oxides and dust. Continuing concern about acid rain led to more stringent legislation, limiting vehicle exhaust emissions, the phasing in of compulsory fitting of **catalytic converters** on cars and the use of lead-free petrol. In 1985 the EC joined the **Vienna Convention for the Protection of the Ozone Layer,** which came into force in 1988. It included a programme for phased reduction in the output of **CFCs** by EC manufacturers. In seeking to control pollution the EC has a diverse approach. The overriding regulation states precisely what ambient air quality standards must be maintained in terms of maximum allowed levels of sulphur dioxide, dust and other solid particles, lead and nitrogen oxides. Environmental standards are set for polluting products, such as the limit on the sulphur content of gas and oil, lead in petrol, the content of vehicle exhausts and restrictions on power stations.

AIR QUALITY STANDARD is the amount of atmospheric pollution allowed during a defined time in a specific area. It is set and monitored by public health authorities. The standards vary from area to area. In fact, the standards are almost impossible to enforce because polluting gases are mostly invisible, and also because the gases are quickly dispersed. The first standards were set in the UK after London was seriously affected by smog in 1952 and, after a government inquiry, the **Clean Air Act** was passed. In the USA the **photochemical smog** experienced in California led to the passing of the Air Quality Act in 1970, which set national emission standards for hazardous air pollutants from factories, automobiles and aircraft. This Act also allowed a private citizen to sue any person who violates an emission standard.

AIRBORNE DISEASES describes those infections, including the common cold, influenza and **tuberculosis,** in which **bac-**

teria are transmitted through the air from an infected person, usually by coughs and sneezes. Other allergic ailments, such as **hay fever,** are caused by airborne substances, pollen, rusts and dust in the atmosphere. There is mounting evidence that pollution, particularly from automobile emissions, is contributing to the rising number of people who suffer allergic reactions to substances in the environment.

AIRLINE WASTE RECYCLING measures have been introduced by airlines in order to reduce waste by collecting articles, such as seat covers and cushions, plastic coverings and carpet, magazines, cans, plastic glasses and cups, for scrap reprocessors. When Northwest Airlines pioneered an on-board recycling scheme for materials that would have otherwise been dumped on a **landfill** site, its employees collected 1,812 tons of scrap during the first year. Savings in landfill costs alone were over £150,000. By the beginning of 1993, the American Green Consumer Newsletter calculated that Northwest flights were recycling more than 2,600 tons of in-flight waste a year, representing a saving of about 15,500 trees and 3,532 cubic yards of landfill waste a year.

ALAR is the trade name for the chemical daminozide, which is used on vegetables and fruit, and particularly apples, to increase crop yields while, at the same time, stunting the growth of the plants. Evidence in the 1980s suggested that the chemical was a **carcinogen,** and in 1989 it was banned in the USA when a direct association was found between its use and cancerous growths in animals. The main manufacturer, Uniroyal, withdrew Alar from world markets because it was no longer thought to be safe. In the UK it was still used because government scientists felt the risk was low. However, farmers voluntarily stopped using it.

ALBEDO is the proportion of sunlight reflected from a surface. It is an important factor in the heating of the **atmosphere** and its general circulation around the planet. The albedo of the Earth, and hence the climate, may be changed by alterations in forest cover, the size of deserts and the amount covered by ice. Observations by satellites have provided more precise measurements about the scope albedo provides for calculating the possibility of **global warming.** The value of any albedo varies, depending on the nature and properties of the terrestrial surface. The incident angle of the sun's radiation is also very important. Water has a low albedo of around five per cent when the sun is high in the sky, but acts like a mirror and reflects about 70% of solar radiation when the sun's rays strike the surface obliquely.

ALDRIN is a toxic chlorinated hydrocarbon **insecticide.** Because it very quickly disappears from the soil it was thought to be a 'safe' insecticide, after it was first produced in 1948. However, once in the atmosphere it is converted into dieldrin, which accumulates in the bodies of animals and enters the **food chain.** The numbers of birds of prey, in particular, were dramatically reduced. Since strict regulations on its use were introduced in many countries during the 1960s and 1970s, the numbers of most afflicted species have gradually improved. In 1979 the EC issued a directive to all member states banning the marketing and sale of aldrin in plant protection products, but permitting its use in agriculture, horticulture and food storage.

ALGAE are simple, green, aquatic plants without stems, roots or leaves. They are among the microscopic organisms that form the start of the **food chain.** Algae are found floating in the sea and fresh water, but they also grow on the surface of damp walls, rocks, the bark of trees and on soil. They contain chlorophyll and other pigments that let them grow by **photosynthesis,** which is the process that uses sunlight to convert **carbon dioxide,** water and other salts into a source of food. During the summer algae flourish near the surface of the oceans and they are referred to as a **carbon sink.** On land, algae can be useful in improving the fertility of soil by **nitrogen** fixation, the process by which nitrogen in the atmosphere is converted into a type of **nitrate** that is essential for crop plants. Algae can also be troublesome in streams and lakes polluted by fertilizer run-off or sewage wastes. Nitrates in the effluent cause algae to grow in thick blankets, and far too quickly to be handled by the natural cycle of growth and decomposition, under the process known as **eutrophication.** There is not enough oxygen in the water needed to carry out decomposition and the stream or lake becomes lifeless.

ALGAL BLOOM refers to excessive and rapid growth of **algae** and other aquatic plants when they are stimulated to grow too quickly by pollution. It takes place when there are too many **nutrients** in the water and is aggravated when accompanied by a rise in temperature. Although the algae grow quickly they soon die because they have swallowed up all the water's nutrients. As they decompose they tend to rise to the surface and form a green slime. The natural processes that decompose decaying plants depend on the oxygen available in the water, and when the algae becomes too prolific the process of decay is said to impose too great a biochemical oxygen demand (**BOD**). Algal blooms have increased because higher levels of **nitrogen** and **phosphates** from agricultural areas have leached from the fields into water courses.

The **ALKALI INSPECTORATE** was the first agency established in the world to control emissions from the chemical industry. It was established by the British Government over 100 years ago. In the 1960s it was given wider duties for monitoring of industrial **air pollution** in the UK. However, environmentalists were dissatisfied on several counts. Firstly, the inspectorate's reports on individual companies were kept private. In fact, they were covered by the Official Secrets Act. Secondly, the criteria for accepting emissions were not based on absolute pollution control standards, but on the principle of the so-called **best practical means.** In other words, if an industrialist persuaded the alkali inspector that a chemical works was employing the best practical means of operation, the activities of the works would continue. The inspectorate is now part of **Her Majesty's Inspectorate of Pollution,** an organization which has a far wider remit and is answerable to the Department of the Environment.

ALP ACTION was established in 1990 by the Bellerive Foundation. The Foundation, which has Prince Sadruddin Aga Khan as its president, is a group of environmentalists, scientists, industrialists and businessmen, whose aim is to raise funds to co-ordinate grass roots **conservation** efforts, preserve unique **habitats,** encourage **reafforestation,** revive traditional agriculture, and control the excesses of tourism in the Alps. Alp Action also has a programme of initiatives designed to increase public awareness of these problems.

ALPHA RADIATION is one of the three principal forms of **ionizing radiation** to which people on earth are exposed. The radiation is only produced by some radioactive elements as they go through their process of **decay,** but it applies particularly to the radiation emitted by the **actinides.** Alpha radiation consists of relatively large fragments of an atomic nucleus with two protons and two neutrons, which is the same number found in the nucleus of a helium atom. Consequently, alpha particles are relatively cumbersome compared with **beta** and **gamma rays.** However, alpha particles carry a lot of energy which can easily be absorbed by a wafer thin layer of tissue. If alpha particles are being emitted by a radioactive element that has entered the lungs or the bone marrow, the damage done to the tissue can trigger a radiation-induced cancer. The damage is caused by the ionizing radiation effect within the cells of tissue, producing changes in the chemistry of the substance that can bring about the death of a cell or its transformation into a malignant one.

ALPHA-EMITTERS are radioactive elements which in the process of **decay** release **alpha radiation.**

ALPINE refers to any high mountain. Alpine is also used to describe a mountainous terrain with significant glacial erosion; a type of mountain plant, related to a cold but not dry mountain environment; a type of mountain climate; a mountain glacier; and a mountain soil occurring above the tree line.

The **ALPS** link seven European countries: France, Switzerland, Germany, Liechtenstein, Italy, Austria and the former Yugoslavia. They are being damaged by pollution, and the blanket of smog that comes from European industrial centres and drifts across the mountains. The Alps are a valuable **ecosystem.** They are the natural habitat of a wide variety of flora and fauna, some of it unique to the area. The Alpine forests protect houses and winter sports centres against rock falls and landslides, and also stabilize the banks of its rivers. However they have been badly affected by **acid rain** and **air pollution.** Large dams have destroyed the ecological balance of rivers and **wetlands,** and many rivers are polluted to such an extent that the fish are contaminated. Several governments, especially the Swiss, have brought in strict legislation designed to control the emission of pollutants and **greenhouse gases.** Some animal species, including the **otter,** lynx, brown bear and wolf, have virtually disappeared from the Alps. Bearded vultures, bred in captivity, have been returned to the area. **Conservation** groups have been very active in the Alps, particularly on issues like acid rain and the pressure of tourism. **Alp Action** has been established by the Bellerive Foundation to co-ordinate conservation programmes.

ALTERNATIVE FARMING SYSTEMS are the methods of agriculture developed in response to the growing awareness of the environmental damage caused by **intensive agriculture** and farming practices relying on heavy doses of **agrochemicals.** In Europe, the development of intensive farming was encouraged by the inducements and subsidies offered under the **Common Agricultural Policy (CAP)** of the European Community. There is now evidence that many farmers are turning to alternative systems. It does not mean that they do not use chemical **fertilizers** or **pesticides,** but rather that their use has been reduced. Nor does it mean that farmers are returning to old-fashioned farming methods. These farmers still use modern machinery and hybrid seeds. They concentrate on practices that conserve resources and make for a healthier soil. They use animal manures that add nutrients to the soil, and use new methods to reduce plant diseases and increase crop yields.

ALUMINIUM became implicated as an **environmental health** hazard in the 1980s on two counts. Biomedical scientists looking for possible causes of **Alzheimer's** disease, the premature senility indicated by loss of memory and confusion, found a circumstantial link with aluminium. The theory is a controversial one. Post-mortem examinations compared two groups of people. One group, with a higher incidence of Alzheimer's, had lived in an area of Britain where public drinking water supplies were dosed heavily with aluminium sulphate to remove peaty materials. The analyses of brain cells showed microscopic particles of aluminium, together with other changes in tissue used as the definitive diagnosis of the disease. The second event occurred at **Camelford,** in Cornwall, when a bulk delivery of **aluminium sulphate** was tipped directly into the local drinking water supply instead of into storage tanks. The pollution caused serious intestinal and dermatological illnesses in the local population.

ALZHEIMER'S is a disease of premature ageing, characterized by a loss of memory and confusion. Doctors are uncertain about the primary cause, but they know from post-mortem analyses that it damages cells in a specific part of the brain associated with memory and learning. However, there are indications that environmental factors, such as **aluminium** contamination, might be implicated in the onset of the disorder.

AMBERGRIS is a waxy substance, produced by sperm **whales,** which is found floating in the ocean. It is thought to protect the whale's digestive system against the hard beaks of squid which form the major part of its diet. Ambergris is a black, greasy and smelly substance, and, when dried, it turns greyish-black, smelling faintly of musk. It was used at one time as a fixative in perfume manufacture, trapping the volatile aromatic materials that are dissolved in alcohol. Ambergris has now been largely replaced by synthetic materials.

AMERICIUM (Am) is one of the radioactive elements that did not exist before 1942 and which were first created by nuclear **fission** in the first nuclear reactor. Americium is one of the **high-level waste** products from the nuclear industry that have caused controversy because there is no way to dispose of them. Other highly active wastes include **plutonium** and **neptunium,** which remain in the waste effluents produced after reprocessing spent **nuclear fuel** (see **nuclear reprocessing**). Americium belongs to a group of elements that are also called the **actinides** or **transuranic.** They get their group name from the fact that they have **atomic numbers** greater than 92, and so lie beyond uranium in the periodic table. Before nuclear fission and its by-products were discovered, **uranium** was the heaviest natural element. Americium is particularly hazardous because it is an **alpha-emitter** and has a **half-life** of over 450 years.

AMINO ACIDS are the building blocks of life. They are the basic components from which proteins are assembled. Proteins are formed by various combinations and proportions of the 20 most common amino acids. Plants and many micro-organisms synthesize amino acids from simple inorganic compounds. Animals have to rely on their diet to obtain theirs. They can also be produced deliberately in **genetically modified organisms (GMOs)**—strains of bacteria and fungi—in fermentation tanks. All living organisms, from the simplest of life forms, the viruses, to the human animal, are composed mainly of proteins. Amino acids themselves consist of many combinations of just four elements: carbon, hydrogen, oxygen and nitrogen. There are also amino acids that never occur in protein and which are essential nutrients.

AMMONIA plays a critical part in the **nitrogen cycle,** which is a natural process through which plants get nitrogen from the atmosphere. Nitrogen-fixing bacteria are able to achieve a similar reaction and were used by scientists before the **Haber process** was invented. If the temperature and pressure are normal, these bacteria release ammonium ions which are converted by nitrifying bacteria into nitrite and nitrate ions. Ammonia is a colourless gas with a strong smell and is very soluble in water. It is also a principal ingredient of many chemical **fertilizers.** It is produced by the Haber process, and more than 80 million metric tons are used every year, in the manufacture of nitric acid, fertilizers, explosives, dyes and resins.

AMOCO CADIZ was the oil tanker that was wrecked off the northern French coast in 1978 spilling more than 200,000 metric tons of crude oil. More than 100 miles of the Brittany coast was contaminated. An estimated 30,000 seabirds were killed, along with some 240,000 metric tons of fish and shellfish.

ANGIOSPERMS are a highly evolved group of vascular plants, characterized by the production of true flowers which, after pollination, form a well protected seed. Angiosperms differ from the simpler plants because their cells have undergone complex specialization to form distinct tissue types. The mature plant has four parts: stem, root, leaf and flower. There are two groups of angiosperms: monocotyledons, which include sedges, rushes and grasses and have been bred and used to produce all of our cereal crops; and dicotyledons, which include all the colourful flowering herbs and hardwood trees, such as oak, beech, teak and rosewood.

ANIMAL SOCIETIES refers to the particular way different species of animals group together, if only for sexual reproduction. When animals live in tight-knit groups for all or part of their lives, the groups are those commonly known as a herd, shoal, school, pack, swarm, etc. The value of such groups may be mutual protection, or for defence of a feeding area in order to restrict the numbers of animals to a level that a habitat can support. Some groupings, such as ants, bees and termites, have rigid caste systems. In these cases members of the group are kept together by passing food from one to another (trophallaxis) and by **pheromones,** which give a common odour to the colony. Thus, a strange bee attempting to enter a hive is stopped, investigated by guard bees and repulsed. Of course, bees play a crucial role in many ecosystems as one of the main agents of pollination of plants.

ANOPHELES is a type of mosquito that seriously affects human health. The malarial parasite, *Anopheles plasmodium,* is transmitted into the body by the bite of female anopheles. There are estimated to be more than 350 species of anopheles and some transmit the parasites which cause Bancroftian **filariasis,** which can result in blindness.

The **ANTARCTIC ICE SHEET** is the huge blanket of ice that covers the Antarctic continent. It forms the southern ice cap, which is kept in a constant balance by snow accumulation and the release of icebergs into the Antarctic Ocean. If temperatures were to increase slightly, perhaps by the 1.5°–2.5° Celsius predicted as a result of the **greenhouse effect,** the Antarctic ice cap would be likely to get thicker, because of a subsequent increase in snowfall. In principle that would lower the mean sea level by 20–90 cm by the year 2100. Some glaciologists estimate the effect would be counterbalanced by the melting of the Greenland ice sheet in the Northern hemisphere, which would leave a net rise in the sea level of 20 cm.

The **ANTARCTIC OZONE HOLE** was the name given to the first report of changes in the amount of ozone in the upper layers of the **atmosphere** over the South Pole. It was produced in 1985, by a team led by Joseph Farman of the **British Antarctic Survey.** The team reported in the scientific journal *Nature* that they had detected, over a number of seasons, a 40% drop in the usual springtime formation of **ozone** over **Antarctica.** The team, using equipment developed in the 1920s, discovered the lowest concentration of ozone ever recorded. Eleven years before, in 1974, the chemists F. Sherwood Rowland and Mario J. Molina, of the University of California, had already predicted a decline in global ozone. They had demonstrated in the laboratory mechanisms involving man-made chemicals which, if repeated in the environment, would destroy ozone. The international scientific community was sceptical, particularly as state-of-the-art monitoring equipment on sophisticated satellites had not detected ozone depletion in the upper atmosphere. It was later found that measurements made by satellites had in fact observed the depletion of ozone, but the scientists had failed to interpret the data when processing the extensive records. Faced with the evidence of Farman's group, scientists began work to determine whether the depletion would spread beyond Antarctica. Arguments continued about whether the observation of the Antarctic team could have been caused by natural forces or by synthetic chemicals. Tests to confirm the potential sources of chemical destruction were inconclusive, but stated that 'a chemical mechanism is fundamentally responsible for the hole'. A larger expedition to the Antarctic, with 150 scientists and back-up teams, representing 19 organizations and four nations, was launched, using satellite, aircraft, balloon and ground measurements. It was discovered that, compared with long-term records, the average ozone concentration in a region twice as large as the USA declined by nearly half from August to October and in some areas had vanished completely.

The **ANTARCTIC WHALE SANCTUARY** was proposed by the French Government and the **International Whaling Commission** during a meeting of conservation groups called by **Greenpeace,** which was held in Stockholm in December 1992, and at which measures for the protection of marine mammals were discussed.

ANTARCTICA, the Earth's last great wilderness, has a strong relationship with the global **ecosystem** and an important role in the global climate system. The continent acts as a 'refrigerator' which affects the global atmosphere and **ocean circulation.** The 1961 Antarctic Treaty was designed to ensure that Antartica would be used exclusively for peaceful purposes. Nuclear explosions and the disposal of radioactive wastes are forbidden. Several nations with an interest in Antarctica (the USA, the former Soviet Union, the UK, Australia, New Zealand, Chile and Argentina) signed the treaty, agreeing to the establishment of joint scientific research projects. The treaty forbids the exploitation of living resources in Antarctica or any activity that would cause deterioration of those resources. The treaty has been respected and wild animals and the sparse plant life have been protected. However, there is concern about the **environmental degradation** of the continent. This has been brought about by depletion of the **ozone layer,** by industrial pollutants originating at lower altitudes and by damage caused by humans, in the form of construction, waste, disturbance to plants and animals, and fuel spillages. The global effects of ozone depletion are not yet fully understood, but as the ozone layer shields the Earth from hazardous solar radiation its depletion may have adverse effects on the Antarctic environment, particularly on marine life. For example, increased radiation may affect plankton, essential for the Antarctic **food chain.** In October 1991 a draft agreement was signed by 24 countries in Madrid. This would ban mineral and oil exploration in Antarctica for at least 50 years. The agreement, the Protocol on Environmental Protection, calls for the Antarctic to be declared a natural reserve devoted to peace and science. It includes new regulations for wildlife protection, waste disposal, marine pollution and continued monitoring.

ANTHRACITE is a type of **coal** with a very high carbon content and a small amount of volatile matter. Anthracite is a hard, shiny coal, which burns with no smoke or flame and generates a great deal of heat.

ANTHRAX is an infectious disease of wild and domesticated animals caused by the organism *Bacillus anthracis.* Under certain conditions this organism forms highly resistant spores which persist and keep their virulence in contaminated soil, or other material, for many years. Almost all animals, especially herbivores, are vulnerable to the disease, but usually it occurs in cattle, sheep, goats, and horses, who are all infected as they graze on contaminated land. Domestic pets or animals kept in captivity usually catch it by eating contaminated food. Anthrax can be transmitted to man by direct contact with, inhalation and ingestion of infected material, such as wool, hair, hides, bones or the carcasses of infected animals. It is also occasionally transmitted by wearing furs or leather goods or using brushes which are contaminated by spores. Anthrax is one of the oldest recorded diseases in the world, dating back to a reference by Moses in the book of Exodus. In the 18th and 19th centuries anthrax reached epidemic proportions in southern Europe. In the USA, an epidemic in Texas and Washington State in 1974 was traced to saddle bags contaminated with goat hair which had been imported from Pakistan. In 1863, anthrax became the first disease in which a specific micro-organism was identified as the cause. It was also the first infectious disease against which a bacterial vaccine, developed by Louis Pasteur in 1881, was found to be effective, and which subsequently led to the development of bacteriology and immunology. To control the disease, strict quarantine regulations are used during outbreaks of the disease, including the burning of diseased carcasses.

ANTHROPOGENIC describes an object or disturbance to the environment which is man-made or which follows from the emissions of harmful substances as a result of human activities.

ANTI-KNOCK AGENTS are substances that are added to petrol to make sure the fuel burns more efficiently. Engine knock, a metallic knocking sound and loss of power caused by inefficient **combustion** of fuel, is produced by the petrol-air mixture burning too quickly. It results in lots of explosions in too many parts of the cylinder at once instead of a single smooth detonation which pushes the piston down. The **diesel engine,** invented in 1892 by Rudolf Diesel, avoided engine knock by putting the fuel-air mixture under high pressure, so the heat of compression alone was enough to ignite it. Petrol is a complex mixture of hydrocarbons, some of which burn more quickly than others. The amount of knock produced by a given petrol is measured and known as its **octane rating.** One of the principal functions of oil refining is to produce a hydrocarbon mixture with a high octane rating. The object is to meet the demands of car manufacturers, who through the years have produced new engines with an increasingly high **compression ratio.** This means the petrol-air mixture is compressed to greater densities before ignition. The task of producing high octane ratings was made easier by the use of chemicals that, when added in small quantities, reduce knock. The most effective of these was tetra-ethyl **lead.** Leaded petrol was introduced in 1925.

ANTIMONY has some medical applications in the form of antimony tartrate, and some other special formulations are used for treating **schistosomiasis** and some fungal infections. However, if it is present in the air as an **aerosol,** concentrations of one part in 10,000 form a lethal nerve poison.

ANTS are found throughout the world, but are most common in hot climates. The ant is a member of the insect family formicidae, of which there are some 8,000 species. Some species are regarded as pests when they invade homes and gardens, or are those which inflict painful stings, or which protect the **aphids** that cause great damage to plants. Ants live together in organized colonies and their social behaviour is among the most complex in the insect world. Most live in nests, either in the ground, under rocks, or above the ground in nests built out of twigs, sand or gravel. Others live in old logs and timbers, trees or the hollow stems of weeds. They feed on both plants and insects. The fire ant was introduced into Alabama from South America and spread throughout the southern states of the USA in the mid-1970s. It has a painful sting and is a serious nuisance because it builds large mounds as nests. The honey ants feed on honeydew, a digestive secretion from some aphids, by gently stroking the aphid's abdomen with their antennae. In the African tropics, tailor, or weaver, ants make their nests out of leaves glued together with silk secreted by their larvae. The most notorious species of ant, army ants, are nomadic and set out on organized raiding parties, destroying plant and animal life as they go. They rest from time to time while the queen lays her eggs, then continue their march, carrying the growing larvae with them.

The **APHELION** is the point at which the orbit of a planet is at its farthest point from the sun. In the case of the Earth, it occurs early in July. It has relevance in the theory of **climatic change** proposed by **Milankovitch.**

APHIDS are part of a world-wide family of over 3,600 species, which includes the greenfly. They are about the size of a pinhead and are serious plant pests. They stunt the growth of plants, transmit plant viruses and also deform leaves, buds and flowers. They are very prolific. Wingless females produce living young without fertilization during the summer. When the plant containing mother and offspring dies, some aphids grow two pairs of wings and fly to another plant. They are kept under control with chemical sprays and by natural predators, including ladybird beetles.

AQUACULTURE is the cultivation and harvest of freshwater or marine animals and plants, in ponds, tanks, cages or on protected beds. This is usually done in inland waters, estuaries or coastal waters. It is estimated that commercial fish farming accounts for more than 10% of the world's fish needs. Fish farming usually concentrates on molluscs, including oysters, mussels and clams, because they are usually immobile and fetch high prices. Shrimps and salmon are also farmed, but the stock have to be caught in the wild first, so that they can be brought up to commercial standards in pens. Aquaculture is not new. In Asia freshwater fish have been farmed for some 4,000 years, usually on small farms. Today, more than half the fish eaten in Israel, and a quarter in China and India, comes from fish farms. In other parts of the world new forms of aquaculture are being developed and encouraged as a result of fears of a world shortage of fish in the next century. For example, **seaweed** is being grown as food for humans and livestock, and for industrial materials.

AQUIFERS are layers of rock, sand or gravel that can absorb water and allow it to flow. An aquifer acts as a **groundwater** reservoir when the underlying rock is impermeable. This may be tapped by wells for domestic, agricultural or industrial use. A serious environmental problem arises when the aquifer is contaminated by the seepage of sewage or toxins from waste dumps (see **leachate**). If the groundwater in **coastal areas** is over-used salt water can seep into the aquifer.

The **ARAL SEA** was once one of the world's largest inland bodies of water. Over a period of 25 years it has shrunk dramatically and, today, it has lost 70% of its water; more than 26,250 square miles have been turned into salt flats and desert. A scheme designed to irrigate a large area of a desert in the former Soviet Union diverted water from the lake's major tributaries, the River Amu and the River Syr Darya. The environment of the surrounding area has suffered severe damage. There are violent dust storms along the shores and the sea's waters are now too saline for fish and too shallow for ships. In the irrigated desert area, water in the irrigation channels is badly polluted, drinking water is of a very low quality, and **soil erosion** has increased rapidly.

The **ARCTIC** generally refers to the north polar regions, but it has also become used to denote the territory lying within the Arctic Circle and to the landscape, climate and plant and animal life found in those regions. Both the Arctic and Antarctic play a very large part in maintaining the Earth's **heat balance.** Since a large part of the Earth's water is frozen in the ice fields, snow and glaciers of the polar regions, the possible melting caused by **global warming** could have world-wide repercussions. Melting could cause catastrophic **sea-level** rises and **climatic change.** Both polar areas only absorb energy during their respective summer period and radiate heat throughout the year. Each year the outflow of energy must be balanced, more or less, by the inflow, otherwise the polar regions would become progressively colder. The vastly different geography of the two polar regions means that there are equally large differences in the way they influence the global **climate.**

Unlike Antarctica, which is a huge land mass with bordering oceans and covers about three times the area of the Arctic, the northern region is an ocean, with the land of three continents at its perimeter: North America, Asia and Europe. However, the continental shelves that extend into the Arctic ocean make it one of the world's most nutrient-rich seas and most fertile fishing grounds. The Earth is also nearest the sun during the Antarctic summer, so its **atmosphere** receives seven per cent more solar radiation than reaches the northern atmosphere during the Arctic summer. The southern limits of the Arctic are more difficult to define and include areas of Scandinavia warm enough to support reptiles. Some ecologists set the limit as the boundary where the sparse scrub of the **polar tundra** and the conifer-dominated **taiga** forest meet. The taiga covers more than half of Canada and much of Alaska and Scandinavia. Only the top 50 cms of the tundra thaw each summer, but many species of migratory birds use it as their breeding ground. The protection of flora and fauna and the exploitation of the mineral resources of the region are the subject of the **Arctic Treaty.** Many scientists believe global warming could melt the tundra's **permafrost.** Aside from destroying the tundra's ecosystem, they fear melting would release vast amounts of **methane** which would accelerate the **greenhouse effect** on a world-wide scale.

The **ARCTIC TREATY** was an agreement made between the former Soviet Union, Canada, China and Japan in 1990. The four countries pledged to collaborate on the protection of the environment and scientific research in the **Arctic.** The treaty covers the monitoring of pollution, exchanges of technical information and funding for environmental protection programmes. The countries also agreed to co-ordinate research into timber harvesting, wildlife management and offshore oil exploration.

ARSENIC was used in many household and garden pesticides in the form of arsenous oxide, copper acetoarsenite (Paris green), calcium, or lead arsenite. Fruits sprayed with them may contain enough arsenic to be toxic.

ARTESIAN WELLS are bore holes sunk into **aquifers** to extract the **groundwater.** The term is applied loosely to any well in which water rises under pressure, but not necessarily to the surface. If the water level is higher than the outlet of the well then the water will flow upwards under pressure. If the well-head is above the water level in the aquifer the water has to be pumped up and this is known as a sub-artesian well.

ASBESTOS is the generic name for a group of fibrous mineral silicates. It includes blue asbestos (crocidolite), white asbestos (chrisotite) and brown asbestos (amosite). After they are mined the asbestos fibres are separated from the rock and are spun into a cloth. When inhaled the fibres penetrate the lungs and the tissues of the bronchial tubes, resulting in asbestosis, a crippling lung disease. Asbestos also causes cancer of the lung and the gastro-intestinal tract, and mesothelioma, a malignant cancer of the inner lining of the chest. However, because it is a poor conductor of electricity and highly resistant to heat it has been widely used over the years in fire-fighting suits, and building and insulating materials. In 1968 blue and brown asbestos were banned in Britain, but white asbestos continues to be sold.

The **ATMOSPHERE** is the envelope of colourless gases surrounding the planet in a 500 km-thick layer. It consists primarily of **nitrogen** (over 79%), **oxygen** (20.9%) and **carbon dioxide** (0.03%), as well as traces of argon, krypton, xenon, neon and helium, plus **water vapour,** traces of **ammonia,** organic matter, **ozone** and various salts. The proportion of most gases in the **lower atmosphere** is relatively stable, although the level of water vapour, which is a major **greenhouse gas,** can vary greatly. Weather and climate conditions are contained in the lowest zone of the atmosphere, the **troposphere.** The atmosphere is particularly efficient at filtering potentially harmful radiation, particularly **ultraviolet radiation.** It also acts as a shield against solid matter from space, by burning up any particles with heat generated by air friction.

ATMOSPHERIC CHEMISTRY is the study of a particularly sensitive indicator of chemical processes. Discoveries of the **ozone hole** in the **Antarctic** showed how little is known about basic atmospheric chemistry and interactions of the natural and man-made molecules in the atmosphere, and their subsequent effects on the environment.

ATMOSPHERIC CIRCULATION describes the general movement and circulation of air, which transfers energy between different levels of the **atmosphere.** The mechanisms of circulation are very complicated. They involve the transfer of energy between the oceans and the atmosphere, the land and the atmosphere as well as the different levels of the atmosphere. Some atmospheric movements occur frequently enough to be recognized as the trade winds and the all-important **jet stream.**

ATMOSPHERIC DUST is the quantity of fine particles discharged into the atmosphere which can cause a dust veil and **climatic change.** The particles include emission from volcanic activity and **aerosol** clouds from agricultural and industrial activities. The major sources of natural aerosols are dust from drought-stricken soils, sea-salt spray, volcanic eruptions and soot from natural forest and brushwood fires. Major sources of **anthropogenic** aerosols are the burning of **fossil fuels** and **slash-and-burn agriculture** and intensive land development in the developing world. Although the heaviest particles are soon washed out of the atmosphere by rainfall, or simply by gravity, they still affect the **radiation** balance in the atmosphere. Depending on the nature of the materials creating the dust veil, the effect may add to or counteract warming trends attributed to **global warming.** The effect on warming depends on the quantity and types of the particles and the nature of the land or ocean below. A method of calculating the severity of volcanic dust in terms of the mass of material ejected, and the duration and maximum extent of the spread, was devised by Hubert H Lamb by way of a **dust veil index.** Regional effects of particular atmospheric dusts may be more significant than their global effect. **Sulphur** aerosols in the Northern Hemisphere may be causing more low cloud cover, because the aerosols increase the condensation rates in clouds.

ATMOSPHERIC ENERGY is the capacity of the Earth's atmosphere to continue to transport heat that is the source of energy for the winds. Atmospheric energy depends on several kinds of energy: **solar energy,** which is transferred by radiation from the sun; latent energy, which is released by the condensation of **water vapour;** geopotential energy, which is related to gravity and height above the Earth's surface; and kinetic energy, which is continually being converted from heat energy by movement of air molecules. Because there is twice as much annual solar radiant energy at the equator as at the poles, the atmosphere acts as a huge heat engine. This continuous heat transfer provides the energy to maintain **atmospheric circulation** and the planetary winds.

ATMOSPHERIC GASES refer to the natural and man-induced vapours and gases in the atmosphere like **water vapour, carbon dioxide, methane** and **CFCs,** which are responsible for **global warming.** The composition of dry air is 78% nitrogen and 21% oxygen. Among the smaller quantities of other gases there is about 0.03% of carbon dioxide, which is one of the **greenhouse gases.** The reference to dry air is critical to calculations about global warming, because water vapour is the principal greenhouse gas. However, air composition varies enormously by season and region.

ATMOSPHERIC POLLUTION, see **air quality.**

The **ATOMIC NUMBER** defines an element by the number of protons it contains in the nucleus of a single atom. Elements are classified by the number of protons they have. The first element in the periodic table, **hydrogen,** has one proton. Helium has two, and **carbon** has six protons. Among natural elements, **uranium** has the most protons, with 92.

ATOMIC PILE is the term that was used to describe early types of nuclear reactors. It is now regarded as old-fashioned and inappropriate for the types of reactors used in nuclear power stations.

ATTRACTANTS are substances that lure insects to traps or poison-bait stations.

AUDUBON is a name inextricably linked to the origins and rise of the modern conservation movement. John Audubon was a naturalist who in the early 19th century painted all known species of North American birds. His books, illustrated with coloured plates, remain classic works on ornithology. Audubon's publications were among a growing number of writings and lectures by naturalists on wildlife and conservation that led to the development of the modern environmental movement, which was marked by the foundation of the **Sierra Club** in 1892. The fact that the modern **conservation** movement had its beginnings in the New World has been attributed to the fact that within the memory of a single generation there had been extreme man-made changes in the landscape, as well as disruption and destruction of wildlife.

AVAILABLE NUTRIENTS are the elements in a soil solution that can be readily taken up by plants.

B

BACKGROUND RADIATION is the radiation resulting from natural sources, as opposed to man-made sources, and to which people are exposed in everyday, normal life. It comes from the sun and terrestrial origins, such as rocks and soil. On average, every person receives two millisieverts (mSv) (see **sievert**) of background radiation each year. Amounts vary depending on location. At high altitudes there is more cosmic radiation than at sea-level. Other parts of the world have rocks containing small amounts of uranium and thorium that release **radon** gas into the air.

BACTERIA are a group of single-cell micro-organisms, the smallest of the living organisms. Some are vital to sustain life, while others are responsible for causing highly dangerous human diseases, such as **anthrax, tetanus** and **tuberculosis.** Bacteria are found everywhere, in the soil, water and air. For instance, fleas on black rats in the holds of Italian ships brought a bacterium to Messina in 1347 which caused the Black Death (bubonic plague) and ravaged Europe's population. In the first scientific use of **biological control** for pest management, a beetle was introduced from Australia to California to destroy cottony cushion scale, a pest that was destroying the state's citrus groves.

A **BACULOVIRUS** is a class of insect virus which has been used as a **genetically modified organism (GMO)** to create a **biopesticide,** an agent that kills insect pests by **biological control.** The baculovirus is used as a carrier, or vector, replacing healthy genes with another carrying the virus which will kill an insect pest.

The **BALTIC SEA** is the partially landlocked, fairly shallow sea that is bordered by Norway, Sweden, Finland, Estonia, Latvia, Lithuania, Poland, Germany and Denmark. It is the largest expanse of **brackish** water in the world and stretches from southern Denmark almost to the **Arctic** circle, a total of 160,000 square miles. The Baltic is a haven for wildlife. It is also tideless, sluggish and has been a victim of pollution from the countries on its shores over many years, from a variety of sources: toxic wastes, **fertilizer** run-offs, oil spills and discharges, and domestic wastes. Fish are contaminated and the Baltic seal is practically extinct. In 1974 the Helsinki Convention was the first regional agreement drawn up to protect an enclosed sea from land-based pollutants. In recent years pollution from polychlorinated biphenyls (PCBs), **DDT** and **mercury** has shown signs of decreasing.

BANGLADESH could be one of the countries most vulnerable to flooding from rises in sea levels caused by **global warming.** The country is periodically inundated by monsoon floods, which, on average, cover 20% of the territory every other year and 37% one year in every ten. In 1987 and 1988 40% and 60% of the territory, respectively, was flooded. As they are used to repetitive flooding, the Bangladeshi people have always lived and coped with the problem. However, the difference now is that the country has a growing **population,** which has been doubling every 30 years, putting great pressure on limited land resources and the demand for food. The potential impact of **climatic change** from global warming could be disastrous. Bangladesh has set up a Flood Plan Co-ordination Organization to conduct studies to find a lasting solution. It has also asked the **World Bank** for help in developing a five-year Flood Action Plan.

BARNACLES are found in almost all marine environments and cause considerable damage by fouling ships, piers and offshore installations. There are more than 800 species. Some have shells, while others are exposed. Some live as parasites on corals or other **crustaceans,** while others have a **commensal relationship,** living in communities attached to **whales,** sea turtles or coral reefs.

The **BASEL CONVENTION ON THE CONTROL OF TRANSBOUNDARY MOVEMENTS OF HAZARDOUS WASTES AND THEIR DISPOSAL** was an agreement drawn up under the auspices of the **United Nations Environment Programme** (UNEP). It came into effect in May 1992 and was adopted by 116 governments and the European Community. Its main aim was to impose strict controls on the trans-border transport of hazardous wastes, and on their disposal. It gives developing countries the right to turn away shipments of hazardous waste from their shores. It means that a country exporting substances that it considers hazardous has to get permission for their delivery from the proposed importer, who then has 60 days to issue a 'prior informed consent'. The ultimate goal of the convention is to cut the generation of hazardous wastes to a minimum. The **recycling** industry in the developed countries considers that the convention is too strict. Before it came into force, the **Organization for Economic Co-operation and Development** (OECD) approved new rules to protect the recycling industry which threaten to undermine the convention. The rules place hazardous waste substances into three categories. Those on the red list, dangerous hazardous wastes, will still remain under the convention's control. Amber substances on the OECD's list of rules give the importer 30 days, not the convention's 60 days, to accept the load and once permission is given it is valid for a year. The European Community included the lists in its waste regulation, which was officially adopted in 1993. Critics are alarmed because some substances on the OECD's green list, including **lead, cadmium,** and plastics, are not considered dangerous by member states and could, therefore, be exported anywhere in the world.

BATS are the only mammals which are capable of true flight. They also have a highly developed 'navigation' system of acoustic orientation. There are some 900 species of bat throughout the world, mostly insectivorous. They eat vast quantities of insects and are important in balancing, and perhaps controlling, insect populations. They are also important to the environment because of their role in pollination and seed dispersal, and because their **guano** droppings are valuable as a fertilizer. Some bats feed on fruit, pollen and nectar. The vampire bats of tropical America are the only bats which cause man a serious problem. They feed on the blood of mammals and large birds, and transmit cattle diseases. They are sometimes suspected as carriers of **rabies.** In 1981 the Wildlife and Countryside Act gave protection to all 15 species of bats normally found in Britain. The EC Agreement on the Conservation of Bats in Europe was opened for signature in London in December 1991.

BECQUEREL is the modern unit for expressing the radioactivity of a substance. It is superseding the **curie.** One becquerel is equal to the decay, or disintegration, of one atomic nucleus a second.

BEES are members of a huge and widely distributed group of insects. Many of them play a major role in the environment as pollinators of plants. They are important, too, as producers of honey and beeswax. See also **bumblebee.**

BENTHOS is the highly varied collection of plant and animal life that inhabits the sea bottom, the Benthic zone. Most are found in relatively shallow seas less than 200 metres deep. The type of life found in the zone depends on the proportion of sand and rock, the amount of light, and the temperature of the water. Organisms have been able to colonize inhospitable regions of the zone because of extreme specialization. Benthos, an important part of the **food chain,** occurs in many forms: some are fixed to the sea floor, while others are capable of active movement on or within the sediment on the bottom. In shallow waters, the Benthic zone is also home to molluscs, such as oysters and mussels, crustaceans, such as crabs and lobsters, and worms, starfish and bottom-dwelling fish, including plaice and cod, which are important human food sources.

BENZENE forms 3%–5% of petrol, depending on the national regulations that limit its use. Benzene is a colourless, liquid hydrocarbon, discovered in 1825 by Faraday as an extract from coal tar. However, in 1865, in a flash of inspiration, the German chemist Frederick Kekule recognized that the benzene molecule was a ring of six carbon atoms, and it was this that made benzene so industrially important. The benzene ring is the building block of a vast family of organic chemical compounds known as the aromatics. It was on this discovery that the creation of the modern petrochemical and organic chemical industries was based. Benzene is also well known as a **carcinogen,** or cancer-causing agent. New limits on industrial exposure of 100 parts per 1,000 million, proposed in the USA in 1993, mean the air in American factories would have benzene levels less than those commonly found in European cars and in congested urban areas.

The **BERNE CONVENTION** is the name given to the Convention on the Conservation of European Wildlife and Natural Habitats. It aims to protect all forms of endangered wildlife in specified parts of Europe.

BEST PRACTICAL MEANS was a term first used in Britain by the **Alkali Inspectorate** (now HM Inspectorate of Pollution) to describe a compromise by which industrial sites are permitted to emit higher levels of polluting gases than are usually accepted because of special local conditions. This meant that equipment and processes which did not meet new pollution standards, but had not reached their economic expiry time, could still be used. It could also apply to a factory if it was the only source of employment in an area and would have to close if forced to make alterations in order to bring it in line with new anti-pollution measures. The use of best practical means has caused a great deal of confusion. Because there is no precise definition for 'best practical means', it has to be interpreted according to local conditions by local health inspectors. However, stricter pollution laws and more efficient equipment to monitor pollution mean that 'best practical means' is now relied upon less and less.

BETA RADIATION is the name given to the **ionizing radiation** which is produced as a stream of high speed electrons emitted by certain types of radioactive substance when they **decay.** The intensity of **radiation** energy produced in human tissue by a beta particle is a hundred times less than that produced by an **alpha radiation** particle, but it travels slightly deeper into tissue.

BHOPAL is the capital of Madhya Pradesh in central India and site of one of the worst environmental disasters in world history. In the early hours of 3 December 1984 an estimated 45 tons of the gas methyl isocyanate, a highly volatile and toxic chemical used in the manufacture of **pesticides,** escaped from a tank at a plant owned by Union Carbide India, a subsidiary of Union Carbide, an international company based in the USA. The gas drifted over the densely populated area surrounding the plant, killing many inhabitants instantly. There was widespread panic as thousands of others tried to escape from the city. The official death toll was 2,500. As many as 50,000 other people were treated for respiratory problems, temporary blindness and eye irritations, lung damage and severe vomiting, which were caused by exposure to the gas. The health of an estimated 200,000 people was affected and a year after the leak thousands were still being treated for side-effects. A report on the disaster concluded that it was caused by a combination of badly designed equipment and sub-standard operating and safety procedures. By 1987 more than 500,000 local residents, more than two-thirds of Bhopal's population, had put in claims for compensation. In 1989 the Indian Supreme Court awarded damages of US $470m. against Union Carbide.

BIOACCUMULATION is a process that can be both environmentally useful and environmentally hazardous. It can also occur naturally or be formed as part of a planned biotechnology process. Bioaccumulation is the accumulation of materials by an organism that are not an essential component or nutrient of that organism. Usually it refers to the accumulation of metals, but it can apply to bioaccumulation of persistent synthetic substances such as organochlorine compounds. Many organisms, such as plants, fungi and bacteria, will accumulate metals when grown in solutions containing them. The process can be employed usefully as a purification process to remove toxic **heavy metals** from waste water and contaminated land. The hazard arises when crops are grown unwittingly on polluted land, and the plants become the route for poisons to enter the **food chain.**

The **BIOCHEMICAL ENERGY CYCLE** involves the absorption of **solar energy** by plants and its conversion into **carbohydrate** compounds, which are then available as food for animals.

BIODEGRADABLE refers to waste substances that can be decomposed by bacteria or other biological means. Commonly, this suggests that the decomposed remains will be non-toxic and will not accumulate in the **food chain.** Paper, wood, wool and leather are biodegradable.

BIOFUELS are the bulk biological materials that can be converted into liquid fuels, such as ethanol and methanol, either for use as fuel or as the raw material feedstocks of the chemical industry. The process may begin with crops being deliberately grown for this purpose, and these range from wood pulp and sugar cane, to seeds like maize and oil-seed rape that can be converted into methanol. However, the initial **biomass** can be wet materials such as starch, sugar, sewage waste water and other effluent used in **fermentation.** Processing of sugar cane, by fermentation and distillation, was developed in Brazil in response to the 1973 energy crisis in order to produce ethanol. Brazil developed a major fuel called Proalcool, producing over 3,000 million gallons of it by 1990. In the USA an ethanol-based fuel called Gasohol was developed. Methanol is less easy to make. Another gaseous biofuel, or **biogas,** is hydrogen, which is made by **photolysis** of water, which almost mimics the process of **photosynthesis.** However, in photosynthesis in plants the hydrogen is not created in the form of a gas, but is used to make sugars for growth of the plant.

BIOGAS is a gas, rich in **methane,** which is produced by the **fermentation** of animal dung, human sewage or crop residues in an air-tight container. It is used as a fuel, to heat stoves, lamps, run small machines and to generate electricity. The residues of biogas production are used as a low-grade organic fertilizer. Biogas fuels do not usually cause any pollution to the atmosphere, and because they come from **renewable energy** resources they have great potential for future use.

The **BIOGEOCHEMICAL CYCLE** is the movement of chemical elements in a circular pathway, from organisms to physical environment, back to organisms. The process is termed a nutrient cycle if the elements concerned are **trace elements,** which are essential to life. A biogeochemical cycle occurs when vegetation decomposes and minerals are incorporated naturally in the **humus** for future plant growth.

BIOLOGICAL CONTROL is a form of pest control that is an alternative to using synthetic chemicals. It is the control of one species by another. The most famous example was the introduction of myxomatosis into Australia in order to control rabbits. Another organism used for years as a biological control agent has been the bacteria *Bacillus thuringiensis* which produces a protein that kills caterpillars. *Bacillus thuringiensis* has been used as a spray. Recently scientists isolated the protein and have used it to make a **biopesticide.**

BIOLOGICAL DIVERSITY describes the incalculable, natural variety of the world's flora and fauna and their habitats.

Although the record of all living species is still incomplete, the concern about a reduction in biological diversity rests on reports produced by the **World Wide Fund for Nature** and other scientific groups, which point to the threat of extinction faced by plants and animals as a result of man-made phenomena, such as **acid rain** and **deforestation.** Biodiversity is also a possible casualty of **global warming,** which could also cause further loss of forests, and the disruption of **wetlands** and the **polar tundra,** complex **ecosystems** that have existed for millennia. Many species might be unable to migrate rapidly enough to cope with **climatic change** in order to find a suitable habitat. Since species of flora and fauna are tied together in a web of interdependence, there is concern that the whole ecosystem could unravel as a result of the destruction of a vital link. Forests, for example, have adapted geographically to a narrow temperature and moisture range. In northern temperate zones there are profound changes in the belts of vegetation, coinciding with a shift of four degrees Celsius in the average temperature.

The **BIOLOGICAL DIVERSITY CONVENTION** was agreed in May 1992 after four years of negotiations under the sponsorship of the **United Nations Environment Programme (UNEP).** It was presented to the United Nations Conference on Environment and Development (the **Earth Summit**) and signed by 154 states. It contains provisions intended to ensure effective national action to curb the destruction of biological species, **habitats** and **ecosystems.** Among them are: that countries regulate to conserve their biological resources; the imposition of legal responsibility upon nations for the environmental impact of their private companies in other countries; technology transfer on preferential and concessional terms, where it does not prejudice property rights or patents; the regulation of biotechnology firms; access to and ownership of genetic material; and compensation of developing countries for extraction of their genetic materials. The Convention also embraces the idea that industrialized countries must help developing countries financially and with know-how. It also accepts that the responsibility for setting up a network of protected areas rests first with each country, and that the first beneficiaries of the conservation and sustainable use of wild plant and animal species should be the rural communities and indigenous peoples whose traditional knowledge and respect for those resources has preserved them for centuries. The Convention does not specifically refer to **genetically modified organisms** and the ecosystems of which they are a part, but such organisms are understood to be included in the broader form referring to living modified organisms resulting from **biotechnology.** The Convention does not include the idea of drawing up a global list of protected areas. Instead, every country which is a party to it will put together its own list of protected areas in the expectation that, eventually, they will become part of a global list.

The **BIOLOGICAL EFFECTS of IONIZING RADIATION (BEIR)** is a special committee, established by the US National Academy of Sciences, to investigate the way **ionizing radiation** affects the human body. As a result of its findings, it also proposes safety limits. The committee said, in a report published in 1972, that there is a risk in any dose of radiation, but contradicted this in another report eight years later. The world's scientific community is uncertain about the effects of exposure to low-level radiation over a long period because there is little information on this issue. Most observations are based on information gained from observing survivors of the atomic bomb explosions in Japan at the end of the Second World War.

BIOLOGICAL ENVIRONMENT is a phrase with a special context in environmental medicine. It considers the individual human body as a biological environment and a normal, rich habitat for parasites and other organisms which present no threat. It is the means by which disease-producing organisms from other environments invade the human environment that is at the centre of environmental medicine.

BIOLOGICAL MAGNIFICATION, or biomagnification, relates to the behaviour of certain contaminants when absorbed into the body, especially **chlorinated hydrocarbons.** These chemicals get into the food chain and are hazardous because they are very stable and highly soluble in the fatty tissue. They resist the processes of **metabolism** and accumulate in the body. Substances that possess the characteristics of biomagnification include the pesticides **DDT,** dieldrin, **aldrin,** endrin, heptachlor and mirex. They also include the polychlorinated biphenyls (PCBs) a family of chemicals used extensively in industry since the mid-1930s.

BIOLOGICAL POLLUTION is the disturbance of the ecological balance by the accidental or deliberate introduction of a foreign organism, animal or plant species into an environment. The introduction of Japanese knotweed into Britain, where it is now on the rampage, and the escape of mink are examples. In the USA the classic example is the sea lamprey fish, which has been devastatingly destructive for commercial fisheries. It gained access to the **Great Lakes** through the Welland Ship Canal built in 1829 to bypass Niagara Falls. Lamprey prey on trout and other stocks.

BIOMASS refers strictly speaking to the total weight of all the living things in an ecosystem. However, it has come to refer to the amount of plant and crop material that could be produced in an ecosystem for making **biofuels** and other raw materials used in industry, for example. Biomass can be divided into several areas of activity. Algal biomass refers to single-celled plants that can be grown in ponds, by **aquaculture,** to make food materials. Plant biomass refers to crops like sugar cane.

A **BIOME** is a large geographic area, and more precisely one of the 14 major ecological regions into which the world can be divided. Each biome comprises a very complex **biotic** community of a myriad of plants and animals. Biomes are defined in terms of their animals and plant life, which in turn reflect the major climatic regime of the region: **polar tundra, savannah** and **forest, grassland, steppelands** and **desert, mountains, oceans** and **ice sheet.**

BIOPESTICIDES are naturally occurring biological agents used to kill pests by causing specific biological effects rather than by inducing chemical poisoning. The idea is based on mimicking processes that arise naturally, such as protection of the coffee bean by its caffeine content, or use of a bacterial substance, known as the BTK protein, that interferes with the absorption of food by some insects. The argument for biopesticides, and against conventional chemical pesticides, is that they are more easily **biodegradable** and more target specific.

BIOREMEDIATION describes the use of biological processes, using micro-organisms, to clean up pollutants. It ranges from sewage treatment to methods of destroying or removing toxic materials in polluted soil and waste tips. Contaminated sites can be inoculated with appropriate microbes.

The **BIOSPHERE** is the part of the Earth that supports life and consists of the soil, oceans and waterways and the atmosphere up to 15 km above the Earth's surface. It has three main functions for plants and animals: providing a safe habitat so that individual organisms can complete their life cycle; providing a stable habitat to allow evolution of species; and providing a system in which energy is provided by the sun and the materials essential to life are recycled. The biosphere depends on a complex series of inter-relationships between the inorganic (soil, water and air) and living organisms. The biosphere supports a multitude of different **ecosystems,** each one interacting with its neighbour. Hence, a change in one ecosystem triggers a knock-on effect, changing neighbouring systems.

BIOSPHERE RESERVES are protected land and coastal areas that are approved under the **Man and the Biosphere** programme (MAB) in conjunction with the **Convention on International Trade in Endangered Species (CITES).** Each reserve has to have an **ecosystem** that is recognized for its diversity and usefulness as a **conservation** unit. The reserves have at least one core area where there can be no interference with the natural ecosystem. A transition zone surrounds this and within it scientific research is allowed. Beyond this is a buffer zone which protects the whole reserve from agricultural, industrial and urban development. Biosphere reserves and buffer zones were regarded as examples of a new generation of conservation techniques. A 1992 **World Resources Institute** report, published in collaboration with the **United Nations Environment Programme** and the **United Nations Development Programme,** said that many of the reserves were created simply by giving the designation to existing **national parks** or **nature reserves,** with no new land or separate regulations to go along with the designation. Agencies have also not received enough money to make the complicated zoning system work. The report also said that buffer zones were meant to enhance the park or reserve's conservation aims and to do this there needs to be control over human activities within it, while at the same time encouraging economic incentives to people living near the park, so that the area of the natural habitat in question can be widened. In practice this only works if the management has the authority and ability to carry out the regulations.

BIOTA is the general description of all the flora and fauna of the planet. It covers all living things from the tiniest submicroscopic organism to the elephant and giant redwood tree, all of which are supported by the **biosphere.**

BIOTECH is a combination of biology and technology. It is used to describe developments in the application of biological organisms for commercial and scientific purposes. So 'bio' stands for biology and the science of life, and 'tech' stands for technology, or the tools and techniques that the biotechnologists have in their workbox. Those tools and techniques include micro-organisms and a range of methods for manipulating them, such as **genetic engineering.** Historically, the first use of biotechnology took place centuries ago with the discovery of **fermentation** of wine and beer, and later bread. Fermentation still remains at the centre of biotechnology. However, modern biotechnology came of age with genetic engineering, an advance that has given scientists the ability to alter nature, and take advantage of the processes and substances that **microbes,** and anything else with genes, produce naturally without any prompting. The new technologies, developed in 1973, allowed scientists, for the first time, to work in detail on the genetic changes of cells, genes and DNA (deoxyribonucleic acid), the chemical chains containing the genetic information of most organisms. This meant that genes could be transplanted to another plant or animal. This has led to long debates about the implications of the release of genetically engineered microbes into the environment. It also pioneered the way forward into **gene therapy.**

BIOTECHNOLOGY, see **biotech.**

BIOTIC refers to the living organisms, plants and animals in an **ecosystem.**

BITUMINOUS COAL is a good quality soft **coal,** that was misnamed originally because of an incorrect assumption that it contained bitumen. The coal contains between 75% and 92% **carbon.**

The **BLACK SEA** is the large inland sea bordered by Bulgaria, Romania, Ukraine, Russia, Georgia and Turkey. It has some of the world's largest deposits of several mineral gases, including **ammonia, methane** and ethane. Tons of highly toxic hydrogen sulphide, the debris of decomposed remains of animals and vegetation, are thought to be rising to the surface. The Black Sea is also severely polluted by the dumping of waste and by waste released from the factories on its shores. There are plans to build a power station to use the deposits of mineral gas for electricity generation. The plan also includes the development of a pumping station designed to separate water saturated with the gases and the hydrogen sulphide, which will be used to produce **fertilizers,** with clean water being returned to the sea. It is estimated that money spent on purifying the sea will amount to only 20% of the resources accruing from the power that could be generated.

The **BLOOD-BRAIN BARRIER** is the mechanism by which the body usually keeps poisonous substances that have invaded the blood stream from getting into the brain and central nervous system. There are substances that manage to breach the barrier, such as the toxic **heavy metals.**

BOD (biological oxygen demand) is an official measure of how polluting a liquid is when discharged into a river, stream, brook or lake. Sewage is treated with the main aim of reducing the BOD of the liquid, so that it can be discharged without causing serious pollution. Those authorities concerned with water pollution usually set limits on the volume and BOD of effluent before it is permitted to be discharged. The organic matter in healthy streams and lakes gives the water a BOD of a few parts per million, which the system can handle easily. Strict pollution controls, mostly on discharges of sewage, have greatly reduced the BOD levels in some of the principal rivers in Western Europe and North America in recent years. The River Thames, for example, was so polluted in the late 1950s that few fish could survive in it. In the 1960s new sewage works were built and others modernized, and the fish began to return to the river. The number of birds, including swans, teal, shelduck, dunlin and tern, has also increased since the 1960s. However, BOD cannot measure the extent to which a waterway may be polluted with harmful chemicals used in **pesticides, nitrates** from farmland or **leachates** from toxic waste dumps unless they raise the oxygen demand of the water.

The **BONN CONVENTION** is the Convention of the Conservation of Migratory Species of Wild Animals, which was drawn up to protect all migratory species considered to be endangered through all or part of their range. The Convention came into effect in 1983 and imposes strict conservation responsibilities. Hunting or trapping of 51 species threatened with **extinction,** half of them birds, is strictly prohibited. The Convention also aims to encourage countries along migration routes to draw up agreements to protect another 2,000 threatened species. Animals have always migrated, either to find food, an agreeable climate or to spawn. However, as they migrate along established routes, across land, through the air or the sea, many species are seriously threatened, either during the journey or on arrival at their destination. Apart from hunting, trapping and netting, many long-established habitats have been destroyed or damaged by pollution. **Oil spills** pose a serious threat for migrating birds, and, at the end of their journey, breeding, resting and feeding places are often threatened because marshes, wetlands and estuaries have been drained to make way for agriculture or industry, or for recreational facilities.

BOREAL FOREST, also known as **taiga,** describes the largely coniferous forest areas in the northern hemisphere between the latitudes of 50° and 70°. Large areas of boreal forest are found in Scandinavia, Siberia and North America. Severe winters and heavy snowfalls mean that the growing season is short, sometimes as little as three months. As well as the spruce, pine, larch and fir trees, a few hardwoods also grow among the conifers and make up a valuable timber resource. Boreal forests are home to the beaver, sable, mink and fox, as well as larger mammals, such as the grey wolf and its prey, caribou, reindeer and moose. Bears also have ranges that spread northwards into the boreal forests. Species that do not hibernate during the winter have adapted to make it easier to move over snow. For example, the reindeer and moose have evolved splayed-out feet to support their weight. Boreal forests also provide industrial resources, such as coal, natural gas, oil and mineral ores. The needle-like leaves of the conifers retain heat and do not radiate heat quickly or lose it to the wind. The conical shape of the trees makes it easier for them to shed snow before its weight causes structural damage. There are frequent frosts in the boreal forests and, in fact, only some two months in the year are free of frost. While some boreal forest has been over-exploited by man for timber and wood pulp, there are, because of their enormous economic value, many areas that have strict forest management schemes, with major re-planting programmes designed to replace trees lost to forestry.

BORON is an unusual element in that it occurs in a non-metallic and a metallic form. The element is never found free in nature, but is present in volcanic springs and rock formations of volcanic origin. In one form it is regarded as a mineral element, commonly found as boric acid in soil, providing an essential **trace element** for plant growth, but not, apparently, essential to animals.

BOVINE SOMATOTROPHIN, see **BST.**

BP is the abbreviated form of the expression 'before present' day. It is widely used in geochronology, the study of dating various substances and layers from the Earth, especially with regard to **carbon dating.**

BRACKISH is the term used to describe water which is **sea-water** mixed with **freshwater.** It is any water that contains too much salt for drinking, but not enough to be sea-water.

The **BRAER** was the Liberian-registered supertanker that went aground on the Shetland coast in January 1993. It was carrying 85,000 metric tons of Norwegian light crude oil, making it, potentially, one of the largest oil spills in history, and having the capability of causing more environmental damage than the **Exxon Valdez,** which went aground in Alaska in 1989. However, the light crude oil was dispersed by one of the longest storms in many years. Nevertheless, hundreds of birds died and thousands of salmon from Shetland's fish farms were contaminated. Most of the pollution dispersed into the sea, but much of it was blown onto the land, contaminating fields where sheep grazed.

The **BRANDT COMMISSION** is the popular name for the Independent Commission on International Development Issues (ICIDI). It was named after the late chancellor of the former West Germany, Willy Brandt. The 18-member Commission, which sat in 1977, was given the brief to examine economic and social inequalities and find solutions to them. Its first report, *North–South: a Programme for Survival,* published in 1980, recommended urgent action to improve relations between the developed and the developing worlds, covering aid, food, trade and energy. The Commission was disbanded after publishing a second report, *Common Crisis,* in 1983.

BRINE is any water which is strongly impregnated with salt.

The **BRITISH ANTARCTIC SURVEY (BAS)** has been carrying out scientific research in Antarctica for many years and is responsible for nearly all the UK's research in the region. It has five permanent research stations strategically placed in **Antarctica,** carrying out year-round research and monitoring. The BAS collaborates with 21 other countries active in Antarctic research, as part of the Antarctic Treaty. The Survey operates under the auspices of the Natural Environment Research Council (NERC) and its research covers the interdependent physical, chemical and biological processes of the entire Antarctic system. The BAS investigates, for example, pattern and change in the physical environment of Antarctica, the geological evolution, and the dynamics of freshwater and land ecosystems. Research into polar **ice sheets** aims to understand their potential to raise global sea levels in a warmer climate brought about by the **greenhouse effect.** The BAS is administered from its headquarters in Cambridge, UK, which is responsible for the maintenance and operation of the five research stations and for transporting more than 100 expedition field personnel and 1,100 metric tons of equipment and essential supplies to the polar continent. Two ice-strengthened research ships are used, one of them providing the deep-sea platform for marine projects.

The **BRITISH ECOLOGICAL SOCIETY** is the oldest ecological society in the world. It was formed in 1913 and since 1980 it has operated as a charitable company, limited by guarantee. The membership is open to all who are genuinely interested in **ecology,** in the UK or abroad. There are some 5,000 members, about half of them from overseas. The Society holds a number of meetings every year, including the winter meeting at which a wide range of papers are delivered, often by research students. The spring symposium and other smaller meetings are more specialized and feature lectures by world-renowned speakers. The Society produces four scientific journals and a bulletin free to its members. It also supports expeditions, small projects and ecological education. The Society is becoming increasingly involved in policy matters, and in the promotion of ecology in education and government.

BRITISH NUCLEAR FUELS (BNFL) was formed in 1971 when the British Government decided to hive off the production activities of the **United Kingdom Atomic Energy Authority** into a separate company. BNFL manufactures nuclear fuel at Springfields, conducts uranium **enrichment** at Capenhurst, and undertakes reprocessing and high-level waste storage at **Sellafield.** It owns and controls the only low-level radioactive waste dump in Britain, at Drigg, near Sellafield.

The **BROADS** are the freshwater lakes of East Anglia, UK. Broads are usually edged with reeds and are connected to a slow-flowing river near its estuary. It is thought that the Norfolk and the Suffolk Broads were created primarily by the removal of peat in medieval times and subsequent flooding by the neighbouring rivers, which often had linking channels.

BROMINE is a reddish-brown liquid element that gives off suffocating vapours.

BROWN COAL is a brown fibrous material that is at the intermediate stage between **peat** and **bituminous coal,** which is good quality soft coal. The term 'brown' is sometimes used as an alternative name for **lignite.** In practice, brown coal is nearer bituminous, and the plant debris that has been compressed over thousands of years is unrecognizable as such and has changed into an amorphous mass. Brown coal is usually extracted in open cast operations to yield a heavy soft fuel. It is usually used in power stations located near the coal head. It tends to have a high sulphur content, and so is dirty coal to burn if used in a power station without **flue gas desulphurization** equipment.

The **BRUNDTLAND REPORT** is the report of the World Commission on Environment and Development, set up by the General Assembly of the United Nations in 1983. It was named after Gro Harlem Brundtland, the Norwegian Prime Minister, who chaired the commission. The brief of the commission was to draw up a global agenda for change and, in particular, to devise a strategy for achieving **sustainable development** for the year 2000 and beyond. It urged the replacement of chemically dependent agriculture by **sustainable agriculture** and that there be a shift from export crops to food crops for local consumption. It recommended that 'soft' energy systems (**renewable energy** plus energy conservation measures) should be adopted, describing them as the best way towards a sustainable future. The report also urged industry to assess the potential impact of new technologies before their employment, so that their production, use and disposal do not put too much stress on environmental resources. The report set out seven goals for the future: to revive economic growth; change the quality of growth; meet essential needs for jobs, food, energy, water and sanitation; ensure a sustainable level of population; conserve and enhance the resource base; re-orientate technology and management risk; and to merge environment and economics in decision-making.

BST (bovine somatotrophin) is the name of a hormone found in cattle that is responsible for growth and the quality of milk yields. BST was one of the substances to be produced by **biotech.** Animal welfare groups are opposed to the introduction of BST as an additive in the cow's diet, arguing that some BST might be passed on to children through milk.

BUMBLEBEE, or humble bee, is the common name for the **Bombini** insect species. They are found in most of the world, but usually in temperate climates. They have been introduced into Australia and New Zealand to help in the pollination of several flowering plants. They are hardy insects and can stand extremes of climate. However, there is now concern that, certainly in the UK, numbers are dwindling, primarily because of the introduction of **intensive agriculture,** the use of **pesticides,** the destruction of their nesting areas and the removal of plants, particularly wildflower habitats, on which they depend. Bumblebees usually carry more pollen than honey bees and are, therefore, more effective in pollinating crops. Recent research suggests that farmers should set aside strips of land which can return to their natural state so that the bumblebee will survive. The research indicates that clover and bean yields are increased when the crops are pollinated by bumblebees. Gardeners were also encouraged to help by growing plants rich in nectar, such as rhododendron, honeysuckle and foxgloves.

BURN-UP is the amount of energy that can be extracted from **nuclear fuel** before it is considered spent.

C

C3 PLANTS are a group of plants including the soya bean, wheat and cotton. They are classified as C3 because in the process of carbon fixation, during **photosynthesis,** they produce a compound essential for growth that consists of molecules, each containing three atoms of carbon. The photosynthesizing productivity of C3 plants is expected to increase if the **carbon dioxide** concentrations in the atmosphere continue to grow.

C4 PLANTS are a group of plants which include maize and sorghum. The classification refers to a process of carbon fixation usually found in tropical plants with high growth rates. These plants have adapted to high temperatures, strong light, low water supply and low **carbon dioxide** levels. They are classified as C4 because in the process of carbon fixation, during **photosynthesis,** they produce a compound essential for growth that consists of molecules that each contain four atoms of carbon. Compared with C3 plants, C4 plants show little change in their rate of photosynthesis as a result of an increase of carbon dioxide concentrations in the atmosphere.

CADMIUM is one of the toxic **heavy metals** which has caused deaths and permanent illnesses in a series of major pollution incidents around the world. Cadmium has no useful biological purpose. However, it has wide industrial applications. It has been used for decades in metal plating to prevent corrosion, in rechargeable batteries and as a pigment in certain plastics and paints. Special care is taken in the industrial smelting of ores and subsequent handling of cadmium, because occupational exposure is known to have caused heart, chest and kidney disorders. **Environmental health** problems have come from exposure to various sources of pollution. The worst recorded incident was in Japan. A whole village was stricken and gave cadmium poisoning the name by which it has become known, 'itai-itai' disease, meaning 'it hurts, it hurts'. The prolonged exposure, which came from polluted effluent water that steadily contaminated rice crops, caused painful brittle bones, as well as kidney failure.

CAESIUM (Cs) is a soft silvery-white and highly reactive metal belonging to the alkali group of metals. It is a radiation hazard, because it can occur in two radioactive forms. Caesium-134 is produced in nuclear reactors, not directly by **fission,** but by the reaction. It emits **beta-** and **gamma-radiation** and has a **half-life** of 2.06 years. Caesium-137 is a fission product of **uranium** and occurs in the fallout from nuclear weapons. It emits beta- and gamma-rays and has a half-life of 30 years. Caesium-137 was the principal product released into the atmosphere, and hence the **food chain,** from atmospheric testing of nuclear weapons and from the **Windscale fire** and **Chernobyl** nuclear accidents. After the Chernobyl accident, which spread a radiation cloud across Europe, the European Commission proposed new and more restrictive limits on levels of caesium in food and drinking water.

CAM PLANTS include plants such as cacti and other succulents. They are distinctive in that, unlike C3 and C4 plants, they are able to carry out carbon fixation in the dark, as well as being able to **photosynthesize** in sunlight. CAM stands for crassulacean acid metabolism.

CAMELFORD is a town in Cornwall that gained prominence in 1989 because of a water pollution episode. A lorry driver made a mistake with a delivery to the local water authority. He unloaded a batch of **aluminium** sulphate directly into the public drinking water supply. The compound should have been tipped into a storage tank for later use in an effluent treatment process. The pollution caused prolonged cases of stomach illness. However, there was concern about the possible long-term effects because the accident coincided with medical research suggesting aluminium was a cause of **Alzheimer's** disease.

CAMPYLOBACTER are a widespread group of **bacteria,** some of which cause serious enteritis and systemic disease.

CAPTIVE BREEDING is a method of breeding in zoos or reserves those species of animals, plants, birds and fish that are close to **extinction.** Animals and plants in the wild are threatened with extinction as their natural habitats are destroyed, or fragmented, to clear land for agriculture and to build houses for a growing population. Hunting is also a threat to habitats, as is pollution. It is recognized that captive breeding is not the ideal way of preserving a species, but, if habitats cannot be protected, it is argued it is at least a way to save and protect certain species until they can be safely released back into the wild. The breeding programmes take a great deal of management, particularly during the transition back to the wild, and there have been some notable successes, including the Arabian **oryx,** the red wolf, the black-footed ferret, the **rhinoceros,** the Californian condor, the **golden lion tamarin** monkey and the bearded vultures of the Alps. Great care is taken while the animals are in captivity to keep them as isolated as possible in surroundings that resemble as much as possible their original habitats. Efforts are made to ensure that they do not become too trusting of humans. The bald eagle, which has been in decline since the 1940s, is also being bred in captivity, from eggs recovered in the wild, and some have been released back into the wild. In Oklahoma, at the Sutton Avian Research Centre, elaborate steps are taken to ensure that the chicks do not 'imprint' humans as parents, which would mean they would not be able to survive alone in the wild. The chicks are fed via a glove puppet with an eagle head. Programmes to prepare animals for their release include public meetings to inform and reassure the local human population, and training for the animals in how to go about hunting live prey in readiness for the wild. Once released animals are kept in reserves and a strict monitoring programme is maintained, using radio collars in some cases so that straying animals can be returned to their reserve, or taken back into captivity if they are thought to be in danger.

CAR-FREE CITIES was a study undertaken for the European Commission which was designed to answer a general question. Is it possible, and if so to what extent, to conceive of a city which will operate more efficiently than the type of cities we have at present, using alternative means of transport to the private car? The study was predicated on the belief that the boom in car ownership has brought a crisis in the relationship between cars and cities. According to the Commission, in European cities it is the car which, more than any other means of transport, consumes an increasingly precious resource—space. The car was reproached additionally as an environmental nuisance because of noise, air and visual pollution, and urban sprawl. The study concluded that the car-free city would cost between two to five times less to run. The transition from a city with cars to a city without has to take place gradually, with action at several levels. New transport services must be available close to where people live, making mobility a choice and not a necessity. Public transport had to be more efficient and more competitive, and had to include experiments with alternative means of transport, such as moving pavements and escalators.

CARBOHYDRATES are one of three major components from which all organic matter is formed. The other two are **proteins** and **lipids.** Carbohydrates are a vast family of compounds made from various arrangements of carbon, hydrogen and oxygen. They get their name (watered carbon) because the hydrogen and oxygen are in the correct proportions to form water: that is, two atoms of hydrogen to one of oxygen. The most important carbohydrates are starches, sugars, cellulose and gums.

CARBON (C) is a non-metallic element which is unique in the number of compounds it can form and the way in which they are formed. Its molecules form part of many other elements. Chemically, carbon is unique because it can form many compounds containing chains and rings of carbon atoms. Pure carbon occurs naturally as diamond. However, it is carbon's role as the universal constituent of all living things—**microbes,** plants and animals—which makes it so important. The primary deposits of carbon compounds, such as the carbonates (chalk and limestone) and the **fossil fuels** (coal, oil and natural gas), are derived from living things. Only tiny quantities occur as graphite and diamond. Carbon burns in oxygen to form **carbon dioxide.** Before the extensive burning of fossil fuels by industrialized countries and the massive burning of trees, bringing about **deforestation,** both of which are primary **carbon sinks,** the **carbon budget** was in balance; meaning the annual exchange of carbon between the atmosphere and the oceans and flora and fauna was equal. There are two stable **isotopes** of carbon and four radioactive ones. One of the latter, Carbon 14, is used in **carbon dating.**

The **CARBON BUDGET** is the balance in the exchange of **carbon** between the atmosphere and living things in the **biosphere.** Burning of **fossil fuels** and **deforestation,** which release carbon into the atmosphere, threaten to upset the balance.

The **CARBON CYCLE** explains how every living thing from the simplest microbe to human beings obtains its carbon, the element that occurs in all the compounds that make up living organisms. The carbon cycle shows how green plants are the starting-point for the myriad of **food chains** that support life. Green plants are the primary producers of food, animals the consumers, bacteria and fungi are the decomposers. The chain begins when plants absorb carbon dioxide from the air in the process of **photosynthesis,** in order to make starch, sugars, proteins and cellulose. When eaten by animals, the organic plant material is converted into molecules for building animal-type tissues. However, in making the molecules which are the building blocks of tissues, both plants and animals use energy that comes from breaking down particular carbohydrates. The process, **respiration,** converts the carbohydrates back into carbon dioxide and water, which are then excreted back into the environment. Even small disruption to a food chain, by pollution or a natural phenomenon, can have catastrophic effects on plants and animals. As well as absorbing carbon dioxide, plants also release carbon dioxide during respiration and burning. Scientists calculate that 100,000 million tons of carbon dioxide a year is removed from the atmosphere by green plants during photosynthesis, 50,000 million tons is put back into the air from respiration, and 2,000 million tons are released when trees are burned. Another 50,000 million tons of carbon dioxide is released through decomposition of vegetation of all types. Marine life depends on the carbon cycle, while the oceans absorb an estimated 104,000 millions tons of carbon dioxide a year and release 100,000 million tons a year by biological and chemical processes.

CARBON DATING, or radiocarbon dating, was a method of determining the age of organic remains such as wood, bone and shells, which was devised for archaeological research. However, it has become an important technique for studies of the environment, especially for investigating objects that contain evidence of past **climatic change.** The objects include terrestrial fossils, sediments taken from the ocean and ice cores drilled from glaciers and the polar ice fields. Because of the steady bombardment of the Earth by **cosmic radiation,** a small proportion of **nitrogen** atoms in the atmosphere are transformed by neutron bombardment into the radioactive nuclei of carbon-14, which has a **half-life** of 5,600 years. The atmosphere also contains normal carbon-12. Both forms of carbon atoms are absorbed by living trees, or other plants or animals, in the form of **carbon dioxide,** and during **photosynthesis** or **respiration** they are converted into plant tissue or bone. When the plant or animal dies, the carbon-14 becomes locked in any buried, preserved material, and the **radioactivity** of the buried carbon-14 continues to decay. The carbon-12 is stable and therefore remains in its original concentrations. An analysis of the proportions of carbon-14 to carbon-12 is used to calculate the age of an object.

CARBON DIOXIDE is a natural, essential component of the air and is the vital ingredient of the **carbon cycle.** It becomes a form of pollution when excessive amounts are discharged into the air, primarily from burning coal, oil, natural gas and by deforestation. Measurements made of gases trapped in air bubbles in the polar ice sheets and glaciers show that, prior to the intensive coal-burning that fuelled the Industrial Revolution, the concentration of carbon dioxide in the atmosphere was 280 parts per million. This had increased by over 25% by 1990.

CARBON MONOXIDE is a colourless, poisonous gas which is slightly lighter than air. The main source of carbon monoxide is from the **combustion** processes of industrial furnaces and car engines, when the carbon in the coal or petrol is burned incompletely. Consequently, some of the carbon combines with oxygen in the air to form carbon monoxide. Carbon monoxide is a source of pollution in traffic-congested urban streets. When carbon monoxide is inhaled, it combines with red blood cells by replacing the useful oxygen in the molecules that give healthy blood its bright red colour. The tissues of the body are then deprived of oxygen.

The **CARBON SINK** refers to any part of the environment, from trees to the oceans, that acts as a reservoir or store for carbon. Some carbon sinks, like those in the shells of former marine organisms or fossil fuel deposits, can be many millions of years old. Others may be seasonal, like the bloom of **phytoplankton** in the ocean.

The **CARBON TAX** is a strategy to limit **carbon dioxide** emissions and improve fuel efficiency by putting an energy tax on petrol. It was proposed to the EC Environment Ministers in October 1991 in a recommendation for action that could take effect by the year 2000. The scheme was based on the scientific uncertainty about the impact of human activity on **climate change,** and on the risks associated with phenomena which, if not controlled in time, would be irreversible. The proposition for an energy tax reflected a 'no regret' strategy, which the European Commission adopted in proposals to curb pollution in other areas in addition to **climatic change.** The EC considered phasing in a carbon tax of US $10 on a barrel of oil by the year 2000.

CARBON TETRACHLORIDE is an industrial solvent made from chlorinating methane. Moist carbon tetrachloride breaks down to the poisonous gas, phosgene, and hydrogen chloride. Safer alternatives are being developed.

A **CARCINOGEN** is a substance that causes cancer in humans and animals. The first carcinogen to be discovered was **arsenic,** followed later by **cadmium,** cobalt, chromium and **asbestos.** A wide range of substances are suspected or have been proved to be carcinogenic. They include the chemicals in **tobacco** and tobacco smoke, synthetic polychlorinated biphenyls (PCBs) used in plastics, electrical components and brake fluids in automobiles, and many of the synthetic compounds used in industry, such as resins and adhesives.

The **CARIBBEAN** is believed to be one of the **regional seas** most polluted by **pesticide** run-off and industrial discharges. Industrial pollution, from sugar refineries, mining, offshore oil drilling and industrial waste all contribute to the pollution. It also contains some of the busiest oil tanker routes in the world. The increased volume of **tourism** is another threat to the Caribbean. Fish in the **coral reefs** in the area are being depleted and coral is being taken to sell to tourists. It is subject to a convention signed under the United Nations Environment Programme's Regional Seas Programme.

CARING FOR THE EARTH, A STRATEGY FOR SUSTAINABLE LIVING, is the revised and more practical version of the **World Conservation Strategy,** published in 1980, which outlined a single, united approach to global problems. The updated version, like the previous document, was sponsored by major international conservation organizations. It was launched in 1991 in more than 60 countries.

The **CARRYING CAPACITY** is the ideal population that a habitat can support indefinitely within a given set of environmental conditions.

CARSON, RACHEL, see ***Silent Spring***.

CASH CROPS are, literally, any crops that are grown for sale in the town markets or for export. They include coffee, cocoa, sugar, vegetables, peanuts and non-foods, like **tobacco** and cotton. Huge areas of countries in the developing world have been turned over to cash crops. Those countries with no mineral or oil resources depend on cash crops for foreign money, so that they can import materials to develop roads, for construction, or to buy Western consumer goods and, indeed, food. However, critics argue that cash crops are planted on land that would otherwise be used to grow food for the local community and say this is a cause of world famine. Cash crops, such as peanuts, can ruin the land if it is not left fallow after six years of harvests. Moreover, if the best agricultural land is used for cash crops, local farmers are forced to use marginal land to grow food for local consumption, and this has a further dramatic effect on the environment.

CATALYTIC CONVERTERS are designed to clean up the exhaust fumes from petrol-driven vehicles, which are otherwise the major threat to **air quality** standards in congested urban streets and on motorways. Converters remove **carbon monoxide,** the unburned hydrocarbons and the oxides of nitrogen. These compounds are damaging to human health and the environment in a variety of ways. Scientists estimated that in London alone in 1990 1.3 million tons of pollutants were being pumped into the air each year. More than 80 per cent of this came from cars. Carbon monoxide is poisonous to humans and also contributes to the **greenhouse effect.** Hydrocarbons irritate the lungs and react with sunlight to produce **photochemical smog.** Nitrogen oxides react with water vapour in the atmosphere to produce **acid rain,** and also help form photochemical smog. Catalytic converters, often referred to simply as 'cats', can remove over 90% of the pollutants if they are a 'full three-way system'. Although catalytic converters are contained in a cylindrical tube about 30 cm long and 23 cm wide, the technology is an extension of the industrial type of **catalytic cracking,** which is used to break down crude oil into its constituent parts: petrol, paraffin, diesel and heavy engine oil. The converter is attached to the vehicle's exhaust near the engine. Exhaust gases pass through the cellular ceramic substrate, a honeycomb-like filter. While compact, the intricate honeycomb structure provides a surface area of 23,000 square metres. This is coated with a thin layer of platinum, palladium and rhodium metals, which act as catalysts that stimulate a reaction to changes in the chemical composition of the gases. Platinum and palladium convert hydrocarbons and carbon monoxide into carbon dioxide and water vapour. Rhodium changes nitrogen oxides and hydrocarbons into nitrogen and water, which are harmless.

CATALYTIC CRACKING is the process for treating thick, heavy crude oil in order to separate out the light fractions, like petrol and those which provide the raw material from which the petrochemical industry makes plastics, paints and pesticides. The basic process of cracking or separation of the various fractions by heating the crude oil, at a high pressure and a high temperature, is made more efficient by the use of a catalyst; hence the name catalytic cracking.

CETACEANS are aquatic mammals that include **whales, porpoises,** and **dolphins.** They are found in practically every marine environment. In 1990, during the Third International Conference on the Protection of the North Sea, the ministers of the countries adjoining the North Sea signed a memorandum of understanding because of concern about the effects of pollution and habitat deterioration on the small cetacean population. The Institute of Zoology, an institute of London University, which carries out fundamental and applied research in conservation biology, is conducting a research project in collaboration with the Natural History Museum and the Sea Mammal Research Unit, and other institutions. The project, funded by the UK Department of the Environment, aims to increase knowledge about diseases in marine mammals and the effects of pollution and other human influences and, in co-ordination with other North Sea countries, to establish a system for long-term monitoring of the mortality of cetaceans.

CFCs, see **Chlorofluorocarbons.**

CHAIN REACTION is the process by which energy is released by splitting atoms of **uranium-235,** by the process of **fission,** either in a controlled form in a nuclear reactor or in the uncontrolled destructive form of an atomic bomb. The process of fission can be produced in **fissile** substances.

CHEMISTERILANTS are chemical compounds that cause sterilization or prevent reproduction in insects or birds and rodents.

CHERNOBYL is a small town in the Ukraine which gave its name to the world's worst accident at a civil nuclear power plant. A nuclear reactor did what experts had claimed was impossible. On 26 April 1986 it blew up. The reactor of Chernobyl's No. 4 nuclear power station exploded when an uncontrolled reaction called a **prompt criticality** took place in a rod of nuclear fuel. The accident blew the reinforced concrete top from the Russian Graphite Moderated Channel Tube Reactor (RBMK), releasing a cloud of radioactive waste which spread across the whole of northern Europe. As a result of the far reaching health, social, political and economic consequences of the accident, experts from the then Soviet Union presented a full report on the accident to an open international meeting, known as the Chernobyl Post-accident Review. It was held from 25–29 August 1986, at the **International Atomic Energy Agency's** headquarters in Vienna. Soviet experts explained that the impossible happened during an experiment, when the safety systems were deliberately switched off so that the reactor could be operated for a short time in an unauthorized way for the purposes of the test. Thirty-one people died and more than 200 others suffered acute radiation sickness. Hundreds more were exposed to lower doses of radiation that are expected to cause long-term radiation illnesses, such as leukaemia. Over 135,000 people had to be evacuated from the immediate area and large amounts of fresh food had to be destroyed. Rural areas near Chernobyl will remain in quarantine for decades. Once the fire and radioactive emissions from the No. 4 RMBK reactor had been halted, the reactor was sealed in concrete.

The radioactivity from Chernobyl spread across Europe. In Scandinavia herds of reindeer were found to be suffering from the effects of radiation; in the northern areas of Scotland, England and Wales, hill farm pastureland, where farmers were grazing sheep, was contaminated by radioactive **fallout.** The British Government banned the movement and sale of sheep in affected areas. Estimates vary as to the number of people who are likely to die of cancer because of the explosion. A study by the National Radiological Protection Board, a government organization in Britain, known as the watchdog on nuclear safety, calculated a long-term figure of 2,000 extra deaths from cancer covering all European Community countries. Others have put the figure much higher. In 1987 six former senior officials and technicians at the plant were sent to labour camps for blatant violation of the safety regulations.

CHLORDANE is a toxic chlorinated hydrocarbon **insecticide,** now banned in the UK, France and Germany, and proscribed in the EC. Its use is also severely restricted in the USA. However, it is still in agricultural use in some parts of the developing world. Chlordane was mostly used to protect telephone and electricity poles, fences and other wood in close contact with the ground, from fungal decay micro-organisms and insects, particularly **termites.** When absorbed through the skin it affects the nervous and the respiratory systems. It can also lead to liver and kidney damage. Chlordane accumulates in the **food chain.**

CHLORINATED HYDROCARBONS are toxic chemicals that are almost entirely synthetic and do not occur in nature. Because of this the micro-organisms that bring about natural decay can not break down the compounds, and they persist in the environment. Products made from chlorinated hydrocarbons are now largely banned in industrialized countries, but are still manufactured and used in developing countries. Hydrocarbons are organic chemicals containing only **carbon** and **hydrogen** in their molecules. It is when the hydrogen is replaced with **chlorine** that chlorinated hydrocarbons are produced. The most infamous chlorinated hydrocarbon is the **insecticide, DDT,** which caused widespread environmental pollution and destroyed wildlife. Chlorinated hydrocarbons are stored in body fat and move up the **food chain.** Small amounts ingested regularly build to high levels in the tissues. A predator eating several small animals with large amounts of chlorinated hydrocarbons could well absorb a fatal dose. As well as insecticides, the contamination of water by **chlorinated solvents** is a growing problem as many of these substances are toxic and suspected **carcinogens.**

CHLORINATED SOLVENTS are the **chlorinated hydrocarbons** that are widely used in industry and domestic products. Solvents are liquids capable of dissolving materials for many purposes. Chlorinated solvents are causing a growing environmental problem because most are poisonous to some extent. Some are used in dry-cleaning processes to remove grease. Chloroform was used as an industrial degreaser and carbon tetrachloride, which is closely related to chloroform, has been used in fire extinguishers. These solvents have also been used in paint removers and thinners for typewriter correcting fluid. Inhaling chlorinated solvents can cause drowsiness or unconsciousness. Putting newly dry-cleaned clothes in a warm car or confined space can lead to a damaging build up of fumes. Chloroform has been banned as an ingredient of medicines in North America and some European countries. Chlorinated solvents linger in the environment because they are not easily broken down by micro-organisms. If they are dumped in **landfill** sites they may evaporate and cause damage to the **ozone layer,** or penetrate **groundwater.** Chlorinated solvents do not burn easily. If their vapours are exposed to a naked flame, phosphene, a poisonous gas, forms. If they are burned at too low a temperature, **dioxin** is created.

CHLORINE (Cl) is a very reactive and highly toxic green, gaseous element, belonging to the **halogen** family of substances. It is one of the most widespread elements, as it occurs naturally in sea-water, salt lakes and underground deposits, but usually occurs in a safe form as common salt (NaCl). It accounts for about 0.055% of the Earth's crust. Commercially it is used in large quantities by the chemical industry both as an element to produce chlorinated organic solvents, like polychlorinated biphenyls (PCBs), and for the manufacture of polyvinyl chloride plastics, thermoplastics and hypochlorite bleaches. Chlorine was the basis for the organochlorine pesticides, like **DDT** and other agricultural chemicals that have killed wildlife. The reactivity of chlorine has proved disastrous for the **ozone layer** and has been the cause of the creation of the **ozone hole,** which was first detected in the Southern Hemisphere over Antarctica and then over the Northern Hemisphere.

CHLOROFLUOROCARBONS (CFCs) are a family of chemicals, discovered in 1930 by the American chemist Thomas Midgley, which revolutionized industrial and domestic **refrigeration.** They were hailed as wonder chemicals. Aside from their thermodynamic properties as refrigerant gases, they were non-toxic, non-flammable, non-corrosive and stable. Development of the convenience food business relied heavily on the deep freezers cooled by CFCs. Soon after, other applications followed with CFC propellants for aerosol spray cans and the 'blowing' of foam plastics from which the lightweight, heat insulating packaging associated with fast foods is made. CFCs mainly comprise methyl chloroform and carbon tetrachloride, the solvent cleaner. A combination of the stability of the CFC compounds and the effect of the chlorine molecules they contained, once they were released in the upper atmosphere and came into contact with the Earth's protective **ozone layer,** became the source of the environmental alarm. CFCs are also potent **greenhouse gases.** However, concern about destruction of the ozone layer led to the **Montreal Protocol,** agreed in 1987, to reduce CFC consumption by 50% by the end of the century, while alternatives were devised.

CHOLERA is a disease transmitted primarily through contaminated water and food, especially raw vegetables and seafood. If drinking water supplies become contaminated, particularly in overpopulated areas with bad sanitation, infection spreads rapidly. The disease-causing organism is a bacterium called *vibrio cholerae.*

CITES, see **Convention on International Trade in Endangered Species of Wild Flora and Fauna.**

The **CLEAN AIR ACT** was first introduced in the UK in 1956 to reduce **air pollution,** following the death of 4,000 Londoners during the smog in 1952. It controlled the emissions of smoke, grit and dust into the atmosphere, but not **sulphur dioxide.** It was later reinforced by the 1968 Clean Air Act. As a result of the acts, smokeless zones were established and there was a dramatic improvement in general **air quality standards.** The USA brought in similar legislation in 1977.

CLEARING-HOUSE was launched in 1982 by the **United Nations Environment Programme (UNEP)** to help developing countries to manage their own environmental problems. The programme also helps to work out, monitor and manage priority programmes and projects, and finds funding and other resources. It also acts as a 'broker', matching up potential donors with identified projects and forging links between developing countries and donors. Clearing-house operates in projects throughout the world, including a study of the impact of **sea-level** rise in the **Maldives,** a cleaning-up programme for the heavily populated Manila Bay in the Philippines, and reducing water pollution in the Chinese canal city of Shaoxing. It also organizes regional workshops on hazardous **waste disposal** and management. Clearing-house relies on voluntary contributions and support from the international community.

CLIMATE is the average weather condition in a region of the world. Many aspects of the Earth's geography affect the climate. Equatorial, or low, latitudes are hotter than the polar latitudes because of the angle at which the rays of sunlight arrive at the Earth's surface. The difference in temperature at the equator and at the poles has an influence on the **global circulation** of huge masses of air. Cool air at the poles sinks and spreads along the surface of the Earth towards the equator. Cool air forces its way under the lower density warmer air in the lower regions, pushing the lighter air up and toward the poles, where it will cool and descend. These movements of air are, in effect, conveyor belts carrying prodigious quantities of energy. They also have a dominant effect on the climate. Equatorial regions with rising air have low-pressure conditions, and the hot humid air, cooling as it rises, sheds a lot of rain.

CLIMATE MODELS are attempts to describe mathematically how the climate system functions and how the various factors driving it interact. The models are converted into immensely complex computer programmes, which can only be analysed with super-computers. The purpose is to attempt to predict the chain reaction that would happen after changing one factor, such as the level of one of the **greenhouse gases,** and the change it would cause in **global warming.** Climate models are complex because, in addition to predicting the interaction of the greenhouse gases, they have to analyse the cumulative effects of other factors; the extent of sea and ice on the earth and the effect this has on the level of water vapour in the atmosphere; what happens when the amount, types and height of clouds change, because **clouds** differ in their effect on global warming; how different parts of the ocean absorb and release carbon dioxide according to the seasons and temperature of the water. Since the physics and chemistry of some of these processes are not fully understood, climate modelling is surrounded by uncertainties, in much the same way as economic modelling.

CLIMATIC CHANGE usually alludes to the natural short-term and long-term changes in the weather patterns. The possibility that man-made pollution might provoke climate change comes from concern about the **greenhouse effect.** Other human activities have the potential to effect the climate. A change in **albedo** (reflectivity) of the land brought about by **desertification** and **deforestation** affects the amount of solar energy absorbed at the Earth's surface. Man-made **aerosols** produced from the sulphur released from power stations can modify clouds. Changes in **ozone** levels in the stratosphere due to **chlorofluorocarbons (CFCs)** may influence climate. The danger of climate change is its unpredictability. Climate models designed to predict climate change have difficulty in forecasting averages. It is not the averages but the extremes that kill, through severe droughts and storms. The Earth has known many climates. They have left their prehistoric marks on the shape of the land by erosion of rocks, in fossils showing changes in the habitats of plant and animal species in the sediment of deep oceans and in centuries-old ice sheets, with gas bubbles and particles trapped from ancient atmospheres. Early historical records contain evidence of droughts, floods and severe winters indicating the different climates experienced by our ancestors. The first systematic meteorological observations, a relatively new source of data, have their origins in records started over 250 years ago. The definitive climatic history of the planet has yet to be written. However, speculation that the discharge of **carbon dioxide** gases from industrial activities might cause climatic change by **global warming** was made well over a century ago by American scientist J. Tyndall in 1863. In 1896 the Swedish chemist Svante Arrhenius calculated that if the carbon dioxide levels in the atmosphere were doubled, as a result of coal burning, temperatures would increase by six degrees Celsius. In the 1930s and 1950s scientists recognized the importance of **water vapour** as an even more copious greenhouse gas, which could physically interact with carbon dioxide in a warming process. Greater attention was focused on the issue in the mid-1960s, prompted by publication of the first long-term records of changing carbon dioxide concentrations. They were amassed at an observatory at Mauna Loa, on Hawaii, by Charles Keeling and are still being measured; between 1958 and 1988 Keeling measured an increase in atmospheric carbon dioxide of 11%, from 315 parts per million to 351 parts per million.

CLOUDS are one of the largest areas of uncertainty for scientists devising **climate models** with which to simulate the ways in which global warming might feed on itself, like a runaway chain reaction. For example, a rise in air temperature will accelerate evaporation from the ocean. The **water vapour** produced may form high-altitude cirrus clouds. They have a significant **greenhouse effect,** trapping heat radiated from the Earth. Other clouds, such as the stratocumuli that cover about one-third of the ocean, have a predominantly cooling effect, because they reflect a good deal of sunlight back into space. Much research time has been spent trying to decide how to describe what are the realistic cloud effects to be built into the models. Oceans also have an enormous thermal inertia. While scientists accept that the way oceans absorb and release their heat has a profound influence on weather patterns, some experts believe there might be a lag of at least 20 years between the time greenhouse gases reach a particular concentration and the time the total increase in temperature is achieved.

The **CLUB OF ROME** is an association formed in 1968 by the Italian industrialist Aurelio Peccei, which united people concerned at the state of the world, which was drawing too heavily on finite resources, and who thought alternative policies should be adopted. Its members came from business, politics and the world of social and environmental sciences. The Club commissioned research into the problems of pollution, urban planning, inflation, unemployment, and the increasing gap between the developed and developing countries. In a 1972 report, ***Limits to Growth,*** the Club of Rome predicted that if the then present trends continued, the limit to economic growth could be reached within 100 years and would result in over-development and collapse. Ecologists were on the whole in favour of the report, but it was criticized by economists and technologists. The Club of Rome was also part of the movement that recommended a New International Economic Order (NIEO), to lessen the inequality between the developed and the developing world. This concept was included in proposals adopted by the United Nations in 1974. By the 1980s the possibility of a NIEO had faded, primarily because of the economic situation in the world.

COAL tends to be regarded as one of the major sources of **carbon dioxide** pollution, which is believed to be enhancing the **greenhouse effect** and, consequently, increasing the likelihood of **global warming.** The carbon, released together with oxygen with which it combines when burned in power stations, was trapped as coal aeons ago, when trees and plants were buried in swamps. Coal is part of a family of hydrocarbon **fossil fuels.** The lowest form, both in terms of its carbon content and use as an energy source, is **peat,** which contains the highest amounts of water and other non-combustible materials. **Lignite,** or brown coal, comes next, and is in plentiful supply in opencast sites; then hard coal and, finally, the highest ranking by carbon energy content, **anthracite.** When burned, one ton of hard coal yields three times the energy of a ton of lignite; they are the two most widely used grades world-wide. The additional environmental problem comes from coal with a high sulphur content. When this coal is burned, a chemical reaction occurs with oxygen to produce **sulphur dioxide,** the principal source of **acid rain.** These emissions can be greatly reduced on existing types of coal burning stations by filters and **flue gas desulphurization** technology, such as electron beam desulphurization. However, a more direct method lies in newer coal-burning processes such as **fluidized bed combustion** and by energy conservation measures, like **combined heat and power** energy systems, that drastically reduce the amount of coal burnt for a given amount of energy.

COALBED METHANE is the methane generated in coal seams, which would benefit the environment, and safety, if it was extracted and used as a fuel. Otherwise, methane leaking into the air presents an environmental hazard, because it is a powerful **greenhouse gas.** Methane absorbs about 21 times as much heat energy as carbon dioxide.

COASTAL AREAS describe the areas where the land masses meet the seas. Some 60% of the world's population live within 100 km of the coast; coastal waters include many essential ecosystems and provide breeding areas for an estimated 90% of the world's fish catch. Coastal areas throughout the world are under enormous environmental stress, which is caused by a wide range of factors, including pollution and the destruction and deterioration of marine habitats. Coastal pollution, from land-based industry, logging or mining, agricultural run-off and human sewage, is a major problem. Untreated sewage can lead to increased growth of poisonous **algal bloom,** which often flows back with the tide to pollute the beaches and shorelines. There is also a threat to human health from the bacteria and viruses in sewage that cause stomach upsets, hepatitis, cholera and typhoid. These are caught either by swimming in contaminated waters or by eating seafood caught in them. The Group of Experts on the Scientific Aspects of Marine Pollution (GESAMP), an advisory panel of the United Nations, reported in 1990 that pathogenic organisms which are discharged in domestic sewage from estuaries, drainage canals and rivers into the sea are 'the principal problem for human health on a world-wide scale'. Accidental oil spills also threaten marine organisms and can cause serious damage in coastal areas. Another serious problem is the devastation of coastal habitats, in particular the wetlands, **mangrove forests, salt marshes** and **sea-grasses** in order to make way for industrial, urban and recreational developments. Coral reefs, home to a third of the world's fish species, are also gradually being eroded by pollution and are being over-exploited. The reefs, together with the mangrove forests, provide valuable protection against coastal erosion. Another threat is **global warming** which could lead to a rise in **sea-levels.** Many countries would then face the danger of their coastal areas disappearing under the rising sea.

COCKROACHES are among the most primitive of the living winged insects. It is thought they have been unchanged for more than 300 million years, and are among the oldest fossil insects. Cockroaches are usually found in tropical climates, but a few species, out of the total 3,500 known species, have become pests. They are common household pests in many countries, imported by ship and carried home in grocery bags. Cockroaches eat plant and animal products, including food, paper, clothing and soiled hospital waste, fouling everything they touch with their droppings and unpleasant odour, to which many people are allergic. They are a major health hazard and carry harmful bacteria, protozoan parasites and faunal pathogens, including those that cause typhoid, **leprosy** and **salmonella.** Given the right conditions, such as dark, humid places like the corners of kitchens, bakeries and hospitals, they breed prolifically all year round. Many thousands of millions of pounds have been spent throughout the world in an attempt to eradicate cockroaches, but cockroaches have survived because they breed so quickly and quickly develop resistance to most of the pesticides that have been used.

COLIFORM BACTERIA refers to the strains of bacteria, particularly in contaminated water, that can cause gastro-intestinal complaints. Analyses of the quality of drinking water supplies include strict tests to detect the presence of coliform organisms.

COMBINED HEAT AND POWER (CHP) systems are power stations which use the heat that would be wasted by conventional power stations to provide hot water and space heating in domestic and industrial premises. Conventional power stations produce electricity with about 35 per cent efficiency. The remainder is discharged into the environment as waste hot water. With CHPs the energy efficiency level can be as high as 75 per cent or more. Many American and European cities, for example New York, Copenhagen and Hamburg, have CHP schemes for **district heating,** which by improving energy use saves fuel and reduces energy costs. An early CHP system was established at Battersea power station in London in 1950. The heat was piped under the Thames to provide heat and hot water for blocks of flats on the opposite side of the river until the power station was closed down in 1984.

COMBUSTION, or burning, occurs when carbon-containing fuels, such as wood, **coal, oil** and **natural gas,** are burned and the carbon is oxidized to form carbon dioxide, which is widely regarded as the primary greenhouse gas. The hydrocarbon fuels, like coal and oil, come from ancient plants, which grew by absorbing carbon from the air before being decomposed and were buried millions of years ago. Similarly, an atom of carbon which is part of a carbon dioxide molecule in the air may well become a molecule in the cellulose of a blade of grass or other plant. When the grass is eaten, it is digested and absorbed into the cow's bloodstream as an energy-providing glucose molecule. When the glucose molecule has done its job and has been metabolized in the process of **respiration,** the carbon atom may be breathed out into the air again as carbon dioxide, in one tiny example of the **carbon cycle.**

COMMENSAL RELATIONSHIP describes that between two living organisms of different species, in which one is unaffected by the relationship, while the other benefits. See also **symbiosis** and **parasitism.**

The **COMMISSION ON HEALTH AND ENVIRONMENT** was established by Dr Hiroshi Nakajima, Director-General of the **World Health Organization (WHO),** as an independent body in 1990 to respond to the concern about the impact that environmental change can have on human health. The WHO Commission, chaired by Simone Veil, a French Member of the European Parliament, was supported by four expert panels to cover the areas of energy, food and agriculture, industry and **urbanization.** In March 1993 a draft Global Strategy for Health and Environment was published by the WHO, based on the findings and recommendations of the Commission and the outcome of the **Earth Summit.** The draft strategy outlines the WHO's objectives and proposed action programmes to achieve them. The priority areas decided upon were: a broader programme to promote environmental health; an expanded programme for the promotion of chemical safety; wider action and collaboration throughout the WHO on issues related to health and the environment; and stronger partnerships with other international and **non-governmental organizations (NGOs).** The focus was to be on three main areas of **environmental health:** rural, urban and global. The strategy for global matters will place particular emphasis on urban areas where the expanding population and centres of modern economic development result in special health and environmental problems.

The **COMMON AGRICULTURAL POLICY (CAP)** was adopted by the European Community to protect agricultural production and farm prices. Incentives and subsidies, paid during the 1970s and 1980s, led to food 'mountains'. Many small farms were encouraged to switch to **intensive agriculture** and this led to the exhaustion of the soil and widespread environmental damage throughout Europe, causing loss of habitats for wildlife. **Grasslands** and **wetlands** have also been put under threat. In the late 1980s several measures were taken to redress some of the environmental and financial problems caused by the CAP. The CAP now funds Environmentally Sensitive Areas, where farmers are paid to farm in ways that benefit the environment and its wildlife. The **set-aside policy** was a measure, introduced in 1988, which pays the farmer to take at least 20% of his land out of production and use it for more environmentally beneficial purposes.

The **COMPRESSION RATIO** is the characteristic of car engines that has led to the use of **anti-knock agents** to improve the **octane rating** of petrol.

CONSERVATION means preservation from destructive influences, decay or waste. It applies as much to the protection of valuable paintings and antiquities in museums as it does to old buildings and all forms of animal and plant wildlife. As a result, the conservation movement consists of groups which focus on separate issues. For some, the protection of nature is most important and is achieved by creating **national parks** and **nature reserves.** For others, it is the sustained production of valuable materials from the living resources of the land and sea, or the prudent use of fuels and minerals. In recent years conservation has become synonymous with actions to improve the quality of life for people by the control or elimination of environmental pollution. Conservation is seen as critical to our survival because life depends on the health of the **biosphere,** the air, water, soil and rock, on which all life on Earth exists. The modern conservation movement has its roots in the mid-19th century in the USA, where national parks such as **Yosemite** were created. However, in the years after the Second World War the world population exploded, increasing pressure on land and resources. Synthetic pesticides began to have a severe effect on the environment and became a target of the conservationists. By 1970 the problems of pollution of the air, land, rivers and sea became a global concern. However, no single country could contain the situation. It was obvious that to solve the problems there had to be an international authority with powers to co-ordinate environmental protection activities. Many recommendations were suggested for conserving wildlife and halting pollution. A lot of them were eventually incorporated in the Action Plan for the Human Environment produced by the **United Nations Conference on the Human Environment.** However, critics argue that unless countries are willing to delegate more authority to international organizations, progress towards the solution of global problems will be slow.

The **CONTAINMENT** is the reinforced steel or concrete vessel that encloses a nuclear reactor. It is designed to withstand minor explosions in the core, to keep **radionuclides** from escaping into the environment, and to be safe against terrorist attack.

The **CONTINENTAL SHELF** is the gently sloping seabed of the shallow water nearest to a continent, covering about 45 miles from the shore and deepening over the sloping sea floor to an average depth of 400 ft. It continues until it reaches the **continental slope.** The continental shelf contains most of the important fishing grounds and a range of resources, including gas and oil, sand and gravel. However, the shelf is, in general, a structural extension of the continent, and so may also be a source of minerals found in that region, such as tin, gold and platinum.

The **CONTINENTAL SLOPE** is the steep slope which descends from the edge of the continental shelf until it reaches the **abyssal depths.**

The **CONVENTION ON CLIMATE CHANGE** (see also **climatic change**) began with a meeting of government representatives, called in Geneva in September 1990 by the **World Meteorological Organization (WMO)** and the **United Nations Environment Programme (UNEP)** to prepare for negotiations on a convention, based on the work of the **Intergovernmental Panel on Climate Change (IPCC)** which had begun in November 1988. The results of the study were produced as three reports: **IPCC First Assessment Report: WG1 (Science); IPCC First Assessment Report: WG2 (Impacts); IPCC First Assessment Report: WG3 (Response strategies).**

The **CONVENTION ON INTERNATIONAL TRADE IN ENDANGERED SPECIES OF WILD FLORA AND FAUNA (CITES)** came into force in July 1973. It was drawn up to protect a wide variety of animals and plants thought to be at risk of extinction. The treaty prohibits commercial trade in 375 living wild animals which are considered to be **endangered species,** and trade in any products derived from the animals, such as hides, fur or horn. Trade in another 239 species is allowed only if a permit is granted by the importing and exporting countries. The treaty also covers endangered plants. Countries wanting to protect their native populations of specific flora and fauna can place them on a third register.

The **CONVENTION ON THE ESTABLISHMENT OF AN INTERNATIONAL FUND FOR COMPENSATION FOR OIL POLLUTION DAMAGE** was adopted at an **International Maritime Organization** conference in 1971 in order to provide compensation for victims of oil pollution damage. It came into effect in 1978. See also **Torrey Canyon.**

COOLING PONDS are the large pools at nuclear power stations in which spent **nuclear fuel** is stored for one to two years after its removal from the nuclear reactor. One third of the nuclear fuel rods of a nuclear power station are replaced each year. The used fuel is stored in water to cool down thermally and radioactively before it can be transported for **nuclear reprocessing** or permanent storage.

COPEPODS are an immense sub-class of tiny marine crustacea, a few millimetres long, found in **plankton.**

COPPER is very toxic in high concentrations, but is also an essential **trace element,** because it is a component of many proteins. A deficiency of copper in the body leads to anaemia, since iron can not be used without it.

CORAL BLEACHING is potentially a new threat to coral reefs. It has been observed in the **Caribbean** and the eastern Pacific and takes place when the **algae,** which give the coral its pink colour, abandon the reef. Without the algae the reef will eventually die. Pollution and **sediments** are known to cause coral bleaching, but some observers think that water temperature rises of one or two degrees could cause the bleaching—an early warning that reefs might become a victim of global warming.

CORAL REEFS have been built up from the skeletons of reef-building coral, a small primitive marine animal, and other marine animals and **algae** over thousands of years. They occur in clear, shallow and sunlit seas. Coral reefs are one of the most productive and diverse ecosystems and are estimated to yield about 12% of the world's fish catch. They are very vulnerable to any change in their environment, especially pollution, because it makes the water opaque. They must have light in order that **photosynthesis** by the algae can take place. Like trees, corals reflect the environmental conditions in which they grow, indicating marine pollution, sea-surface temperature and other aquatic conditions. Corals can also indicate previous sea-levels. For example, raised reefs off Bermuda, Australia, the Bahamas and Hawaii indicate that the sea level was some six metres higher during the warmest time between the two ice ages than it is today. Many reefs are being destroyed by sewage pollution and by the proliferation of the Crown of Thorns starfish, which feeds on the living coral. Some reefs have been mined for limestone to provide materials for increased building programmes on tropical islands. Moreover, as a result of tourism, some reefs have been depleted to make jewellery or decorations for aquaria. See also the **Great Barrier Reef.**

COSMIC RADIATION is the radiation that comes from outer space, much of which is blocked by the upper layers of the **atmosphere.** Some cosmic radiation is very energetic and is able to penetrate a mile or more into the Earth.

CRUSTACEANS are found in every type of marine environment. There are more than 25,000 species, including shrimps, crabs, lobsters and **barnacles.** The larvae of most crustaceans are **filter feeders.** Adults develop into a remarkable variety of diggers, borers or predators. Most graze or scavenge, but the larger, more advanced lobsters, crabs and shrimp, are more often predators.

The **CURIE** is the unit of measurement of the amount of **radioactivity** of a substance. One curie is the amount of a substance that it takes to produce 37,000 million atomic disintegrations a second. There are significant differences between the physical quantities of a substance and its radioactivity. For example, 10,000 curies of newly produced, lethal, radioactive **iodine**-130 would be scarcely noticeable, consisting of only a few milligrams of material. If the 10,000 curies were newly produced **caesium,** the substance would weigh about 150 grams and be about the size of a tangerine. However, iodine has a short **half-life,** and every 12 hours it loses half its radioactivity. Whereas an exposure of three minutes to newly produced iodine-130 would be lethal, three minutes' exposure to the same substance 40 days later would be within acceptable recommended dose limits. In contrast, ten minutes exposure to caesium, which has a half-life of 30 years, would be lethal even ten years after it was produced. There are also other factors which make some substances more biologically dangerous. One curie, and in some cases a small fraction of one curie, may cause genetic abnormalities, cancer or death, depending on the length and circumstances of exposure and whether the substance is an emitter of **alpha-, beta-** or **gamma-radiation.** The curie is being replaced by the **becquerel,** which is equal to the decay, or atomic disintegration, of just one nucleus a second.

CYANOBACTERIA are a large and varied group of bacteria, and are remarkable in that they possess chlorophyll, carry out **photosynthesis** in the presence of sunlight and air, and release oxygen. Fossil evidence suggests cyanobacteria existed over 3,000 million years ago. They are believed to have been the first oxygen-producing organisms and to have been responsible for generating the oxygen in the atmosphere on which life depends. Cyanobacteria can also carry out **nitrogen** fixation. They are the only organisms known to fix both carbon and nitrogen. The organisms are widely distributed in freshwater, the marine environment, rocks and soil.

D

DAMS are one of the most environmentally controversial of major man-made developments. Dams are built to provide **irrigation** schemes, **hydroelectric power,** and to control flooding. Dams have been built throughout history but, since the second world war, several ambitious and extremely large dams have been constructed in a relatively short period of time, primarily because at the time they offered solutions to two great needs, food and energy. However, sometimes dams create problems which are more severe than those they are intended to remedy. For instance, millions of people have been uprooted from their homes to make way for the reservoirs, and **water-borne diseases** have increased in many areas around large dams. Impoundment also stops river silt, which generally has a high nutrient value, from being carried to the sea. In the short term the building of a dam and reservoir often dramatically increases fish yields. However, there is evidence that, in later years, because submerged vegetation and soils soon rot down and the amount of nutrient is reduced, the fish stocks are depleted. There is also an adverse effect on fish life within the river basin and in the seas immediately beyond its estuary. Aquatic weeds which often invade reservoirs, and the pesticides subsequently used to destroy them, also affect fish populations. The flooding of forests, agricultural land and bush has also led to a loss of habitats for thousands of species, some of them already endangered. Where dams have been built on waters that cross territorial boundaries, there are often bitter international disputes. There is also a suspicion that very large dams could trigger earthquakes.

The **DANUBE** is the second largest river in Europe, following only the Volga in length and volume. It flows through Germany, Austria, Slovakia, Hungary, Croatia, the Federal Republic of Yugoslavia, Romania, Bulgaria and Ukraine. Much of the pollution of the **Black Sea** has its source in the Danube. A European Community (EC) clean-up project of the river basin was announced in 1993. Under the PHARE (Poland, Hungary Action for Economic Restructuring) programme, funds will be used in a long-term work plan involving the Community, those countries whose land is drained by the Danube, and other countries. In the upper reaches of the river a hydroelectric power project at Gabčikovo-Nagymarós was seen as the largest water diversion scheme in Europe. The project aroused widespread opposition and the EC has been mediating in the disagreement between Slovakia and Hungary about water use.

DDT was banned as a pesticide in the USA in 1973 and in the UK and elsewhere in Europe in 1984, after a long campaign by environmentalists that began with the publication of ***Silent Spring*** in 1962. DDT is still used in many developing countries for spraying swamps that are the breeding ground of the larvae of the malaria-carrying **anopheles** mosquitoes. The pesticide was declared an environmental hazard because of its harm to wildlife and because it was accumulating in the **food chain.** Because DDT is not readily metabolized into harmless waste products by animals, including humans, it becomes stored in body fat. Wildlife such as worms, snails and voles absorb DDT from eating plants and seeds. The DDT then accumulates in their predators, eventually reaching animals that prey on other animals, and also in animals that eat plant tissue. However, from the post-war years until the late 1960s it had been accepted as the miracle pesticide. DDT was cheap to produce. It killed body lice, which can carry typhus, and the mosquito that transmits **malaria** and yellow fever. It was seen as both a public health benefit for combating malaria and other insect-borne diseases, and in boosting crop production by killing the Colorado beetle and hosts of other plant-destroying insects. In the industrialized world it became an integral part of **intensive agriculture.** Scarcely a home was without a DDT product to deal with flies, clothes moths, pet fleas and garden insects. DDT was first synthesized in 1873 by a German graduate student, but he had no idea of its insecticidal value, so it was forgotten. It was rediscovered by the Swiss entomologist, Dr Paul Müller, when he was looking for a long-lasting insecticide against the clothes moth and found that it was extremely effective against disease-carrying flies and mosquitoes. This discovery earned him the Nobel Prize for Medicine in 1948.

DEADTIME is the time it takes waste to decompose, provided it is **biodegradable.** The description is employed when assessing the types of waste that could be disposed of in a **landfill** site. Depending on the material, calculations of deadtime hinge on the amount of water and air there is seeping through the site. Decomposition takes longer when water and air are scarce.

DEBT-FOR-NATURE SWAPS are innovative schemes designed to help alleviate the debt burden of developing countries and, at the same time, provide financial support for nature conservation in countries with environmental problems. Several banks began to accept that many of the loans they had made could never be repaid and conservation organizations turned the debt crisis to ecological good by buying the debt at a discount so that valuable natural resources under pressure could be saved. Under the debt-for-nature scheme, the debt of a developing country is bought by a conservation organization, at a discount on the secondary debt market. In return, the debtor country has an obligation to carry out an agreed conservation programme. For the developing country it means a reduction of the debt they have to service and, instead of having to pay an outside source in hard currency, it can use local currency to invest in conservation projects. The funds of the purchasing organization are, in effect, converted at a more favourable exchange rate. As a result of the scheme, which was launched in 1987, environmental programmes are in place in several countries, including Madagascar, Zambia, Costa Rica, Ecuador and Poland. For example, the US environmental group Conservation International bought US $650,000 of Bolivia's debt in return for nearly 6,000 square miles of **tropical rainforest.** The **World Wide Fund for Nature (WWF)** has been responsible for several debt-for-nature swaps with several countries, including Zambia. The local currency equivalent will, among other conservation programmes, go towards protection of **rhinoceros** and **African elephant** populations. Other programmes include conserving wetlands, forests, establishing nature reserves, and halting soil erosion and habitat destruction. There are provisions, too, for education and training programmes.

DECAY refers both to biological decomposition of organic matter to create natural fertilizer and **humus,** and to the very important feature of all radioactive materials determining how their **radioactivity** declines with time. Radioactive nuclei change spontaneously into stable non-radioactive materials in one or two stages. Each step is accompanied by the emission of radiation. The decay process is defined by the time it takes for the activity of a substance to drop by 50%. Each radioactive species has its own fixed half-life, which varies from fractions of a second to millions of years.

DECIBELS (dBs) are the units commonly used as measurements of sound power or loudness. The decibel equals one-tenth of a bel; the bel got its name from Alexander Graham Bell, inventor of the telephone. The laws to control **noise pollution** set limits on decibels. However, they are not always easy to interpret because a change of 1 decibel is equivalent to a change of 26% in the intensity of sound, and a doubling in loudness or power of a sound adds 3 decibels to its rating. Hence, a sound of 21 decibels is twice as loud as a sound of 18 decibels.

DECIDUOUS FOREST are the temperate forests comprised of trees that seasonally shed their leaves, located in the east of the USA, in Western Europe from the **Alps** to Scandinavia, and in eastern Asia. The hardwoods of these forests have been exploited since the 16th century. The trees of deciduous forests usually produce nuts and winged seeds. A relatively small number of mammals live in the forests, including bears, badgers, deer, wild pigs and squirrels, but their seed and nut-gathering activities have a considerable effect on the distribution of seeds and, therefore, the survival of deciduous forest trees. In spring and summer migrant birds return from their winter quarters in warmer climates.

The **DECLARATION OF THE HAGUE** was the first document drawn up by an intergovernmental meeting in which national leaders expressed a willingness to diminish their sovereignty in specific areas for the sake of the planet. The meeting was arranged in March 1989 by the Prime Ministers of the Netherlands, Norway and France. As the implications of the twin problems of **global warming** and **ozone** depletion emerged from the latest research, the governments of these countries were concerned about the difficulties of obtaining the unprecedented degree of international co-operation needed to head off ecological disaster. The declaration was a statement of principles, not an action plan. However, it proposed new institutions that would be needed to finance and facilitate activities such as the transfer of energy-efficient technologies between countries and provide **chlorofluorocarbon (CFC)** substitutes to developing countries in exchange for their agreement to limit carbon dioxide and CFC emissions.

DECOMMISSIONING is the process of making a nuclear power station safe at the end of its useful life, after the last batch of spent **nuclear fuel** is removed and the power station's cooling fluids are drained. Once a reactor is retired, there are three options. The first, called 'immediate dismantlement', involves prompt decontamination and disassembly of the plant. The second, 'mothballing', needs some initial clean-up, followed by several decades of quarantine, before the parts of the nuclear plant are dismantled, cut up and sent to a disposal site. The third, known as 'entombment', involves encasing the reactor in concrete. Although nuclear power was providing over 15% of the world's electricity by 1990, not a single large commercial station had been dismantled. Therefore, it was not known how much it would cost. However, work had begun on development of technologies for decontamination, dismantling and transporting of the **activation products,** the metal and concrete structures. Estimates suggested that decommissioning the average **pressurized water reactor (PWR),** the most commonly used nuclear power system, would create 18,000 cubic metres of **low-level wastes** alone—enough to bury a football pitch under four metres of radioactive debris. This was also 70 times as much as was produced each each year by all the world's operating nuclear reactors. The biggest stumbling block for all nations with nuclear plant is the lack of permanent disposal facilities, especially for **intermediate-level wastes** and **high-level wastes.** In the three decades after 2005, more than 350 civil nuclear reactors will reach the end of their lives.

DEFOLIANTS are **herbicides** used to remove leaves from plants and trees. See also **agent orange.**

DEFORESTATION has been going on for thousands of years as people cut down trees for fuel and clear space for arable land. In the past few decades the activity has been more concentrated in **tropical rainforests.** The potential impact on climate comes about through altering the **carbon cycle** and **nitrogen cycle.** Destruction of the forest releases carbon dioxide into the air. Forest resources are essential to both development and the preservation of the global environment. Mismanagement of forests, inadequate fire control, unsustainable commercial logging, overgrazing and the harmful effects of airborne pollutants are linked to degradation of soil and water sources, loss of wildlife and **biological diversity** and the aggravation of global warming. **Agenda 21** has said that urgent action is necessary to conserve and expand existing forests, and has proposed the creation of National Forestry Action Programmes for **sustainable development** of forests by governments.

DEGRADABLE PLASTICS were developed as an answer to criticisms of the pollution problem created by disposable plastic bags, cups, cartons and bottles. The assumed virtue of degradable plastics is that they are formulated either to break down in sunlight (photodegradable) or to be decomposed by bacteria **(biodegradable).** The biodegradable variety is made from a chemical polymer which contains 6%–10% starch, binding it together. Bacteria feed on the starch, breaking the plastic into small pieces. Some environmetalists regard degradable plastic made from traditional petrochemical raw materials as a sleight of hand, rather than the development of a genuinely 'greener' product. They are said to work at cross purposes to **recycling.**

DELIBERATE RELEASE describes the process of intentionally releasing something into the environment. In particular, it refers to the release of **genetically modified organisms (GMOs),** which may be **biopesticides** or genetically modified seeds of, say, the potato, tomato or wheat plant which have been designed for better pest-resistance or improved growth. The first GMO was released in 1986 and was a genetically engineered frost-proofing bacterial strain called ice-minus.

A **DELTA** is a vast, fan-shaped creation of land, or low-lying plain, formed from successive layers of sediment washed from uplands to the mouth of some rivers, such as the Nile, the Mississippi and the **Ganges.** The nutrient-rich sediment is deposited by rivers at the point where, or before which, the river flows into the sea. Deltas are formed when rivers supply and deposit sediments more quickly than they can be removed by waves or ocean currents. The importance of deltas was first discovered by prehistoric man, who was attracted to them because of their abundant animal and plant life. Connecting waterways through the deltas later provided natural routes for navigation and trade, and opened up access to the interior. Deltas are highly fertile and often highly populated areas. They would be under serious threat of flooding from any **sea-level** rise.

DENDROCHRONOLOGY is the science of dating the age of a tree by studying annual growth rings. It is also employed to interpret previous environments and climatic variations by examining certain kinds of trees. It is based on the theory that the width of the growth ring reflects the amount of rainfall and the temperature of the year in which it was formed. In the USA, a study of a bristlecone pine tree traced its history back more than 3,500 years. The technique is, however, somewhat imprecise because the tree rings may also reflect changes in other environmental factors, such as soil, forest fires, etc.

DENGUE FEVER is a viral disease transmitted to man by the bite of the mosquito, ***aedes aegypti,*** which is found in the tropics and subtropics. An estimated 30 million or more people suffer from dengue. It is uncomfortable, but rarely fatal. Patients are debilitated and need a long convalescence. The incidence of the disease is increasing, particularly in the Caribbean area.

DENITRIFICATION is the loss of **nitrogen** from soil by biological or chemical means. It is a gaseous loss, unrelated to loss by physical processes such as through **leachates.**

DESALINATION is a process for converting salt or **brackish** water into a source of safe drinking water. Parts of the world with severe water shortages are looking to desalination plants to solve their problems. Already use of desalination plants is increasing in the richer countries, such as Saudi Arabia, Kuwait, the United Arab Emirates and the USA. Desalination of water is still nearly four times more expensive than obtaining water from conventional sources. However, technology is improving and costs are likely to decrease slightly in the future. Balanced against the increased cost of traditional **freshwater** it should mean that desalination plants will gradually become more popular. There is also now more interest in building distillation plants beside electric installations so that the waste heat from power generation can be used to drive the desalination process. The largest proportion of desalination plants treat sea water, while a smaller number are devoted to treating brackish water. There are other applications: the treatment of effluent water; the treatment of river water for boilers; and the removal of pollutants, like **nitrates** and pesticides, from **groundwater.**

DESERT VEGETATION has to develop ingenious ways to protect itself from the extremes of almost permanent drought and high daytime temperatures. Some have deep and widespread roots, others develop a fibrous network in the surface soil to catch the dew, like many cacti which store the moisture in their fleshy stems. Many desert plants are able to grow very quickly if it suddenly rains and often seeds can remain dormant, sometimes for many years, waiting for rain to provide the right conditions for germination. **Lichen,** which grow on rocks, can play a major role because they can capture and fix **nitrogen** from the atmosphere. The nitrogen is passed back into the soil by snails, which absorb the lichen from the rocks. As they escape to the shade of the rocks to avoid the daytime heat they excrete the nitrogen into the soil with their faeces, which enriches the soil to encourage plant growth.

DESERTIFICATION is the process of land damage which allows the soil to spread like a desert in arid and semi-arid regions. There is a loss of vegetative cover and the soil deteriorates in texture, nutrient content and fertility. Desertification affects the lives of three-quarters of the world's population, 70% of all drylands and one quarter of the total land area of the planet. The results of it include poverty, a decline in soil fertility and the degradation of rangeland, rainfed crops and irrigated land. There are many reasons for desertification, but the majority are caused by human activities—**overgrazing, deforestation,** poor land management and over-exploitation. However, if people cause the destruction they are also its victims. They have to leave the countryside for the cities and, as refugees, face intolerable living conditions and are threatened by disease, disability, malnutrition and death. There are some remedial and preventive actions against desertification, among them **irrigation,** planting of trees and grasses, and the erection of fences to secure sand dunes. **Agenda 21** states that the priority in combating desertification should be establishing preventative measures for lands that are not yet, or are only slightly, degraded. At the same time, severely degraded land should not be neglected. For regions prone to desertification and drought, better information and monitoring systems are needed to identify the priority areas for action.

DESERTS are spread across more than a third of the world's land surface. They are regions formed because there is an extreme lack of water, either because there is almost no rainfall, or because they are isolated from the sea, or cut off by high mountains so that moisture in the atmosphere is lost before it can reach the area. The main concern now is that desert regions are spreading across wider areas of the world because of the process of **desertification,** in which land becomes desert-like because of a combination of: overcultivation of poor soils; **overgrazing** by sheep, camels, cattle and goats on impoverished land; cutting down too much **fuelwood;** or because of large irrigation schemes which lead to **salinization** of the land. Traditional deserts can be 'hot', like the Sahara, or 'cold' like the deserts north of the **Himalayas** or in Afghanistan. Some deserts support no living matter at all, in others the **desert vegetation** is sparse. Most desert animals are nocturnal or active during the cooler periods of dawn and dusk to avoid the daytime heat. They have developed a remarkable adaptation to the arid conditions. The camel and the Arabian **oryx,** for example, survive on very little water and small mammals get their water solely from the foliage they eat. At the edges of many of the world's deserts there are stretches of less extreme climate and some woody scrub vegetation survives, but they are now under increasing pressure as a result of overgrazing by local populations.

DESICCATION is a gradual increase of dryness, primarily because of climate change. It can be caused by natural changes, like a drop in rainfall, or alterations to a river's flow. It is also caused by human actions, such as **overgrazing, irrigation** schemes or **deforestation,** which all result in a drying-out of the soil and a deterioration in the amount and quality of vegetation.

DETERGENTS have been used for decades to remove dirt and grease from clothes, dishes and surfaces. Until 1918 soap was the main product used, but then synthetic chemicals were added to improve performance. It was in the 1960s when these additives caused recognizable problems, associated with the huge increase in the volume of detergents used by industry and in washing machines. They had ingredients that failed to break down in water. Effluent containing these chemicals disrupted sewage treatment processes, and the release of partially treated effluent into rivers caused the detergents to produce large quantities of foam. The foam was dispersed on the wind and spread **bacteria** that caused infection in humans and animals. In the 1960s the ingredients of detergents were altered to make them **biodegradable.** However, these contain a great amount of **phosphate** materials which can lead to **eutrophication.**

DIATOMS are single cell organisms which are abundant in marine and freshwater **plankton** and **benthos,** seabed-dwelling organisms.

DIELDRIN, see **aldrin.**

The **DIESEL ENGINE** is a combustion engine which environmentalists find more acceptable for vehicles than petrol-driven ones. Diesel engines cause less problems with air pollution and do not need additives, such as **anti-knock agents.**

DIOXIN has never been deliberately manufactured by the chemical industry. It is referred to as an **intermediate chemical,** because it is a temporary substance that forms for a short time in one of the manufacturing stages when other more complex materials, usually pesticides, are made. However, it is very poisonous and traces of it usually remain in the final bulk chemical. It can also be created as a by-product of fires of electrical insulating materials that contain polychlorinated biphenyls (PCBs). Dioxin came to prominence as a serious environmental hazard when it was released in an accident at a chemical plant at **Seveso,** in Italy. An explosion at the factory happened at the critical stage of the production process, releasing a huge **aerosol** cloud containing neat dioxin. The compound belongs to a family of 75 substances. The most notorious is perhaps tetrachlorodibenzo-paradioxin (TCDD), which was the contaminant of **Agent Orange.** Objections to **incineration** for burning waste usually involve concern about the possibility of dioxin being created and discharged if certain plastics and chemicals are burnt.

The **DIRTY DOZEN** is, in fact, a group of 18 chemicals identified by the **Pesticide Action Network (PAN)** as examples of the most dangerous pesticides which threaten human health. The Dirty Dozen includes campheclor (toxaphene), **chlordane**/heptachlor, chlordimeform, **DDT,** dibromochloropropane, the 'drins' (**aldrin,** dieldrin and endrin), **lindane,** ethyl parathion, **paraquat,** pentachlorophenol and **2,4,5-T.** All of them accumulate in the **food chain** and cause death and disease in many species of wildlife. Traces of some of these chemicals have been found in Antarctica, many miles from sites where the pesticides were used. By 1990 the Dirty Dozen had been banned or severely restricted in at least 60 countries, but there is evidence that some are still being used in developing countries.

DISASTERS are caused by natural phenomena like earthquakes, volcanic eruptions, hurricanes, typhoons and tidal waves in which there are heavy fatalities or destruction of vast areas of habitation. Disasters are also caused by humans and there has been a steady increase in the number of man-made disasters over recent years. **Environmental degradation,** particularly in countries where there is heavy **deforestation** or **desertification,** or where marginal land has been overcultivated or overgrazed, can trigger disasters like floods or drought, landslides or mudslides, or make the effects of a natural disaster worse. For example, deforestation in the Himalayas has led to increased flooding in **Bangladesh,** a country which is already a victim of floods. In the Sahel region of Africa widespread desertification has increased the prolonged drought. Other man-made disasters include oil spills, chemical plant leakages or explosions, and nuclear radiation. A growth in population means that, should an accident occur in a crowded city, many more people are killed or injured because they are forced to live closer to the locations of potentially dangerous sites, as was the case in **Bhopal** in India when a chemical plant exploded.

DISINFECTANTS are chemicals, or other agents, that kill or render inactive micro-organisms that cause disease, in animals, seeds or other plant parts. The term also refers to chemicals used to clean or sterilize inanimate objects.

DISTILLATION is the start of the oil-refining process, when crude oil is split into a number of parts or cuts. The separation is made on the basis of the boiling points of the various hydrocarbons in oil, enabling them to be divided at specific temperatures.

DISTRICT HEATING schemes use the surplus hot water from power stations by channelling it to provide space heating for buildings in the vicinity. District heating systems are also capable of burning virtually any type of fuel, including municipal refuse. There are many district heating schemes in mainland Europe, Scandinavia, the former Soviet Union, the UK and the USA. Some power stations generate electricity and hot water, and are referred to as **combined heat and power.** Others just generate heat.

DNOC (4,6-dinitro-o-cresol) was the first chemical insecticide, discovered in 1892. Later, it was found it could kill annual weeds in cereal crops without damaging the crop. DNOC did not appear to leave toxin residues in the soil, so did not enter the **food chain.** However, it was later found to be harmful to wildlife, animals and humans, if inhaled or absorbed through the skin. It was replaced by other herbicides, including **2,4-D.**

DOBSON UNITS are used for measuring the total amount of **ozone** in a vertical column of air above the Earth's surface; the method was perfected by George Dobson in 1956. He suspected seasonal changes occurred in the **ozone layer,** but he made his first measurements at Halley Bay, in **Antarctica,** almost 25 years before the **ozone hole** was observed by his colleagues. When Dobson made his first measurements the ozone level was 150 Dobson units less than calculations suggested it should have been. He initially suspected faulty equipment. However, readings made the following year confirmed the pattern. Measurements made elsewhere in the world did not repeat the pattern. However, throughout the 1960s the seasonal decrease in ozone levels at Halley Bay were monitored. Then the depletion, now known as the ozone hole, was confirmed by Joe Farman and Jonathan Shanklin of the **British Antarctic Survey,** between 1982 and 1985. A reassessment of satellite photographs revealed the development of a hole every spring since 1970. Another hole was then confirmed over the Arctic, showing that ozone depletion was not restricted to the Southern Hemisphere.

DOLPHINS are the most agile and intelligent of the **cetaceans.** They are able to communicate both with other members of their group and with humans, with high-pitched barks and squeaks. They appear to have some affinity with humans, which has led to legendary tales of dolphins saving shipwrecked sailors. Most often seen are the common and the bottle-nosed dolphins. They move in herds, following herring or sardine shoals. There is evidence that dolphins live up to 35 years in the wild. However, they are under threat because they swim and feed with tuna, and frequently get trapped in the **drift nets** used to catch the tuna. Pollution is another hazard. For example, a bottle-nose dolphin found stranded in Cardigan Bay in Wales in 1988 had levels of dieldrin, **DDT,** and polychlorinated biphenyls (PCBs) in its blubber more than ten times higher than those found in other cetaceans sampled around the UK coasts in previous years.

DRIFT NETS were formerly made with natural fibres such as hemp, cotton or flax, and were easily visible and acoustically detected by marine mammals. In the 1950s, fishing nets began to be made with cheaper, very thin, but extremely strong plastic. They are now made from a synthetic material which is invisible to marine mammals and is not **biodegradable.** One net can be nearly 40 miles long. They are primarily used by fishing fleets in Japan, Taiwan and Korea. It is estimated that 4,350 miles of net are used every year in the Pacific alone. An estimated 1,200 metric tons or more of commercial drift nets are lost or discarded at sea every year. Drift nets, or broken-off pieces, continue to be a hazard for many years by trapping and killing marine animals, such as turtles, **dolphins** and **porpoises.** In 1989 the United Nations General Assembly adopted a resolution calling for the banning of ocean drift nets. The International Convention for the Prevention of Pollution from Ships (Marpol), which came into effect in 1988, aims to cut down the number of nets lost or discarded, by banning the disposal at sea of all plastics including, but not limited to, synthetic ropes, synthetic fishing nets and plastic garbage bags.

DROUGHT can be the result of natural weather conditions or of a change in the environment produced when trees are cut down or vegetation is destroyed. Droughts can have a disastrous and long-term effect. For example, the drought in Africa's Sahel region, which started in 1968, caused some 100,000 deaths over five years and irreversibly changed the lives of people in the region. The Ethiopian drought of 1973–74 was responsible for the deaths of an estimated 200,000 people. Five continents had serious droughts in 1982. As well as the droughts in northern and southern Africa, Australia, Italy and Spain had their worst droughts for over 50 years. Crops were also destroyed in Nicaragua, when a dry spell came in the middle of what is normally the rainy season. Moreover, in India, a delay in the annual **monsoon** set back crop production. In hot and cold desert regions there is usually an almost permanent drought. Plants adapt to the harsh environment and crops can be produced only by continual **irrigation,** using **groundwater** resources.

A **DRY SPELL** is regarded as 14 consecutive days without measurable rainfall in the USA. In the UK, a similar meaning is signified by **absolute drought.**

DRY/WET DEPOSITION: dry deposition is the pollution on the ground resulting from the **acid rain** formed in the atmosphere by dry phase conversion of **sulphur dioxide** or nitrogen oxides into their respective acids. Wet deposition is the acid content of rain, snow, sleet or mist by the pollutant formed by the **wet phase conversion** of sulphur dioxide or nitrogen oxides to the respective acids.

DUNG FUEL is dried animal excrement that is used as fuel in developing countries, especially when **fuelwood** is scarce or not available. Millions of tons are burned each year, mostly for cooking. However, dung is a valuable natural **fertilizer,** especially in those countries where there are already food shortages, and the over-use of it for burning deprives the soil of vital nourishment.

A **DUST DEVIL** is a short-lived whirl of dust particles which are lifted to a considerable height above the ground. It is caused by intense solar heat on the ground surface in arid lands and can reach speeds of 20 miles per hour, but dies away without causing any serious damage.

DUST STORMS happen in arid or semi-arid regions, when a large volume of dust is whipped up by the wind and causes a lack of visibility, or sometimes zero visibility. Dust can be carried to great heights and over great distances. Storms usually occur when there are excessively high temperatures, low humidity, and a high electric force. Sometimes they appear as a wall of dust or as a whirlwind. Dust storms can cause havoc to crops in areas that are dependent on cultivating fragile ecosystems. They also pose a serious danger to aircraft.

The **DUST VEIL INDEX** was the scale devised by the climatologist Hubert H. Lamb to rank the severity of volcanic dust in terms of the mass of material thrown into the **atmosphere,** the length of time it takes to be washed out by rain and the extent of its spread round the planet in the meantime.

DUSTBOWL was the term used to describe a vast area of the Great Plains in the USA, which is primarily arable land dedicated to large-scale cereal farming. Agriculture spread into areas where rainfall was frequently low and this, combined with extensive droughts of the 1930s, severely damaged the soil. There was massive **soil erosion** and large quantities of precious topsoil were blown away, creating frequent dust storms. Millions of hectares of land were laid to waste across five states. Huge quantities of agricultural chemicals and water were applied to the land and restored much of it to a state where it was able to support crops again.

DUTCH ELM DISEASE is a serious tree disease caused by a fungus carried by two species of bark beetles. It was first recognized in the Netherlands in 1919 and spread rapidly. Ten years later it appeared in the USA and Canada. The beetles use the bark of trees killed by the disease as breeding sites. Flying from tree to tree they carry spores of the fungus with them, and these block the vessels that carry water to the leaves of the tree. The leaves wilt and eventually die. Dutch elm disease was found in England in 1931 and, over six weeks, 20% of trees in the south of the country were killed. In the 1960s a more virulent strain of the fungus developed in the northern USA and Canada, and this strain was believed to have been introduced into the UK by beetles carried on imported logs from Canada in 1964. Within two years nine million British elms had died. Almost all the elms in southern England were wiped out.

E

E-NUMBERS are an internationally accepted coding system to identify substances added to food during its manufacture and processing. Food additives have been suspected of provoking symptoms of hyperactivity, asthma, eczema and migraine. The numbers cover six categories of food additives: preservatives; colourings; emulsifiers and stabilizers; antioxidants; sweeteners; and other miscellaneous additives. E-numbers were introduced by the European Community to reassure consumers that any substance added to food has been thoroughly tested and pronounced safe. To get an E-number, the EC's seal of approval, a food additive must pass stringent safety tests and be approved by all member countries. Synthetic flavourings are not yet regulated by the EC, but the USA, Japan, Spain, Germany and Italy have a list of permitted flavourings. There are plans to draw up similar lists in the EC.

EARTH DAY is celebrated on 22 April throughout the world to draw attention to environmental issues. It grew from an American event in 1970, when environmental and conservation groups nominated the day as part of their campaigning efforts to focus awareness on the increasing pollution of the air, water and land. At that time, anxiety about the environment had come to prominence with the realization that the bald eagle, peregrine falcon and osprey were among the scores of birds and other wildlife at risk of extinction from **DDT** poisoning. Public awareness of the environment and the existence of ecology had been triggered in North America and Europe by the publication of *Silent Spring* by Rachel Carson. Her catalogue of evidence of environmental pollution was reinforced by studies showing that areas such as Lake Erie (see **Great Lakes**) were dying and industrial pollution was destroying ecosystems established thousands of years ago. The Cuyahoga river in Cleveland, Ohio, was so polluted it caught fire. Subsequently, interest has focused on the destruction of the **ozone layer** by **chlorofluorocarbons (CFCs)** and the increasing threat of global warming.

The **EARTH OBSERVATION SYSTEM (EOS)** is the network of satellites which now scan the surface of the Earth and its atmosphere from space. Interpreting the information gathered by one of the **Earth Resources Technology Satellites** involves very advanced computer analysis on the ground. There are immense advantages, from the environmental point of view, in exploring from space with a combination of **infrared radiation** sensors, radar scanners and photographic and TV cameras. These instruments show seasonal and longer-term events. They also provide an early warning of potential short-term disasters: developing hurricanes, volcanic eruptions, locust swarms, forest fires. Earth observation began with the first meteorological satellite, **Tiros 1,** which was launched in 1960. An international network of spacecraft for meteorological forecasting now forms a permanent World Weather Watch system. A second group of satellites was developed in parallel with the weather system. It began with the US family of specialist reconnaissance Landsats, and now includes the French SPOT satellites and the European Space Agency's ERS (Earth Research Satellite) series. These specialist spacecraft can read the signature of the reflected radiation from objects on the ground. They can tell the difference between various rocks, soil, plants and trees, and, from the colour of the vegetation, show whether a crop or forest is suffering from drought or disease. The European Remote Sensing satellite ERS-1 has unique instruments for surveying oceans and sea ice.

EARTH RESOURCES TECHNOLOGY SATELLITES are the family of spacecraft that monitor atmospheric, terrestrial and ocean conditions. They have become the world's primary early warning systems of global scale pollution and ecological changes, such as the impact of deforestation. The first of the satellites launched between 1972 and 1984 by the American National Aeronautics and Space Administration from 1972 to 1978 were called Landsat. They were designed to collect information about the Earth's natural resources, particularly the location of mineral deposits, and to conduct inventories of crops. From the mid-1980s the American Earth resources satellite programme was taken over by the Earth Observation Satellite Corpn (Eosat). A parallel European programme is known as the European Remote Sensing (ERS) satellite.

The **EARTH SUMMIT,** the United Nations Conference on Environment and Development (UNCED), was the biggest and most ambitious world environmental conference. It was held over 12 days in June 1992 in Rio de Janeiro, Brazil. Sixty world leaders gathered to formulate a series of agreements and treaties. The 27-point Rio Declaration was signed as a commitment to pursue **sustainable development** and the eradication of poverty. It ranges over major environmental issues, such as land, air, water, living things and management of toxic waste. It covers the principles of **polluter-must-pay,** free trade and environmental protection, and upholds the rights of women, children and indigenous people. The key to this is that countries have sovereign rights to exploit their resources, but an equal responsibility to make sure that their actions do not harm the environment for others. Two international conventions were also signed at the end of the summit: the **World Climate Convention,** to reduce **greenhouse gas** emissions in developed countries to their 1990 levels by 2000; and a **Bio-Diversity Convention,** to protect the world's animal and plant life. See also **Agenda 21.**

EARTHWATCH is a global environmental assessment programme, which was included in the Action Plan for the Human Environment adopted by the **United Nations Conference on the Human Environment** in 1972. It has four main aims: to provide a basis on which to identify specific knowledge and to take the steps to find it; to generate the new knowledge needed to guide decision-making; to gather and evaluate specific data in order to discern and predict important environmental conditions and trends; and to distribute knowledge to scientists and technologists and provide useful and up-to-date information to decision-makers at all levels. The **United Nations Environment Programme's** Earthwatch programme has four main arms: the **Global Environment Monitoring System (GEMS)**; the **Global Resource Information Database (GRID)**; the **International Register of Potentially Toxic Chemicals (IRPTC)**; and **INFOTERRA,** the global information system.

ECO-LABELLING is the European Community's (EC) initiative to encourage the promotion of environmentally friendly products. The scheme came into operation in late 1992 and was designed to identify products which are less harmful to the environment than equivalent brands. It was hoped that by buying labelled goods, consumers would be able to put pressure on manufacturers and retailers both to make and to stock 'greener' products. This includes the effect they have on the environment at all stages. For example, eco-labels will be awarded to products that do not contain **chlorofluorocarbons (CFCs)** which damage the **ozone layer,** to those products that can be, or are, recycled, and to those that are energy efficient. The labels are awarded on environmental criteria set by the EC. These cover the whole life cycle of a product, from the extraction of raw materials, through manufacture, distribution, use and disposal of the product. The first products to carry the EC eco-labels were washing machines, paper towels, writing paper, light bulbs and hairsprays.

ECOLOGICAL NICHE refers to the characteristics of an environment that provides all the essential food and protection for the continued survival of a particular species of flora or fauna. In addition to food and shelter, there is no long-term threat to existence in that place from potential predators, parasites and competitors. The concept of the ecological niche goes a long way beyond the idea of the species habitat.

ECOLOGIST refers both to a specialist who studies the relationship between plants and animals and their environments, that is ecology, and to the magazine founded more than 25 years ago to report on ecological and environmental science, and to defend the rights of those people most affected by environmentally damaging development projects. It has dealt with difficult and unfashionable issues, many of which have been borne out during its years of publication. The principal founders of the magazine were Edward Goldsmith, one of the pioneers of the modern green movement and co-author of the *Blueprint for Survival*, and Peter Bunyard, Robert Allen and Michael Allerby. In 1984, Goldsmith and Nicholas Hildyard produced *Social and Environmental Effects of Large Dams*, a book which highlighted the environmental effects that large **dams** were having, particularly in developing countries. With other environmentalists throughout the world, *The Ecologist* has also been active in campaigning against the vast complex of dams planned for the Narmada River in India.

ECOLOGY is the study of living things as they exist in their own natural surroundings, or habitats. The term was adopted by the German naturalist, Ernst Haeckel, in 1869, and comes from the Greek *oikos,* meaning home or dwelling. He spelt it 'oecology', and it covered work in biology in which the observations and experiments of the habitats of plants and animals were treated as separate studies. Ecology has come to be more broadly defined as the interrelationship between all living things and their environments. The ecologist needs to have a command of chemistry, physics and statistics, as well as biology. Hence, the **environmental sciences,** which depend on ecological research, were the first interdisciplinary sciences.

ECONOMIC BOTANY is the study of plants that could one day be useful to humanity. It was started by Sir William Hooker, the first director of the **Royal Botanic Gardens** at Kew. It was studies at Kew which led to the development of the viscose-rayon and acetate fibre industries. Vital investigations into cures for AIDS, finding plant sources for the creation of safer **insecticides,** and developments of new programmes in molecular biology are among the many research projects that are being carried out at the Royal Botanic Gardens.

ECOSYSTEMS are resources which are formed by a combination of interacting living and inanimate, but biologically active, components. An ecosystem may refer to a particular type of freshwater habitat in which it is ideal for trout populations to flourish, because of the availability of food and the quality of food. The list of ecosystems is almost endless since the term refers to the particular biological and physical attributes which characterize every large or small woodland, forest, hedgerow, pasture, heathland, wetland, beach, desert or pond, lake, river or ocean. An ecosystem cannot be separated into its living and non-living parts because the whole thing is an interdependent, dynamic system.

ECOTOXICOLOGY is the study of the harmful biological effects of specific pollutants or combinations of pollutants that are discharged into the environment. It refers especially to the effect caused by the discharge of 'cocktails' of contaminants into rivers, estuaries and coastal waters. Common pollutants that are toxic to aquatic life in **freshwater** include **ammonia,** chromium, **copper,** cyanide, nickel, **nitrite, phenol** and **zinc.** Studies in ecotoxicology have led to a new approach to the control of effluent. Ammonia, for instance, is one of the most frequent pollutants of freshwater in European countries, but its toxicity to aquatic life is very variable and depends on many factors. Local water quality conditions have a considerable effect on its toxicity. Extensive research has focused on the complex chemical and biological processes influencing the toxicity of ammonia to fish and other aquatic life in order to produce standards of practical control for their protection. The information on lethal and sub-lethal effects of pollutants is used in computer models employed to predict the impact of particular concentrations of contaminants on fish. The development of ecotoxicology methods for setting water quality standards has depended on development of new biological assessment techniques, both for laboratory screening of water samples and for routine monitoring of effluent.

EFFLUENT is the waste liquid from domestic sewage, industrial sites or from agricultural processes. Effluents are harmful when they enter the environment, especially in **freshwater,** because of their polluting chemical composition.

EIA, see **Environmental Impact Assessment.**

EL NIÑO is a climatic phenomenon in which the trade winds and ocean currents that normally flow across the tropical Pacific Ocean from the Americas to Asia abruptly reverse. El Niño has been blamed for many disasters much further afield, such as bush fires in Australia arising from **drought,** rain storms in California, storms in normally calm Pacific Islands, and also for interrupting the Indian and East African monsoons. The abnormal retention of water of a high temperature along the South American coast has led to the demise of **plankton,** squid and fish that prefer to live in cold waters. With the death of the anchovy and sardine, hake, herring and mackerel disappear or die. Sea-birds and **seals** that feed on them abandon their nests and lose their young. The decaying fish are washed on to the beaches producing large quantities of sulphurous gas. It was thought that El Niño was predictable, arriving normally every three years, and lasting for about 18 months. However, the 1990 episode reached maturity during early 1992 and began to weaken in mid-1992, only to increase unexpectedly in November 1992. Some researchers are now warning that El Niño could be affected by **global warming** and, if this should increase, it could have an influence on this destructive climatic force.

ELECTRIC VEHICLES are attractive to some environmentalists because they offer a method of road transport that eliminates air pollution from car exhausts in congested towns. Several manufacturers in Japan, the USA and Europe are striving to develop commercially attractive electric vehicles. The main technical problem is to develop batteries for electric vehicles that are cheaper, last longer, weigh less and have faster recharge times than those currently available.

ELECTROMAGNETIC RADIATION usually refers to the low-level electromagnetic radiation emitted from high-voltage power lines, microwave ovens, radar dishes and radios. All living organisms generate minute electrical currents. Indeed, in animals they are important regulators of the central nervous system. Magnetic fields develop when an electric current is passed. The brain, and all other organs, produce very low-level magnetic fields, which can be detected outside the body. Low-frequency electromagnetic radiation (microwaves) was thought to be harmless because it is non-ionizing, but in the 1980s its effects began to be more fully understood. The **World Health Organization (WHO),** in a report on the effects on health of microwaves, said that exposure led to the appearance of autonomic and central nervous disturbances, which produced damaging effects. In 1982 it was reported that cows on a dairy farm located near a high-voltage electricity power line had given birth to stillborn or defective calves, and that chickens were laying abnormal eggs.

ELECTRON BEAM DESULPHURIZATION is a high technology device for extracting **sulphur dioxide** from gases produced in a power station or factory furnace before they are released into the atmosphere.

ELEPHANTIASIS is one of the infections caused by filarian roundworms. Lymphatic **filariasis,** prevalent in the tropics and subtropics, leads to elephantiasis when the parasitic worms are transmitted to humans by the bite of various types of **mosquito.** The disease causes legs and the genital areas to swell to enormous proportions.

ENDANGERED SPECIES are those of the planet's flora and fauna which are threatened with **extinction.** Hunting and poaching to fuel the trade in ivory, horn, skins, fur and feathers have long been a threat to already endangered species. Pollution, agricultural expansion, loss of **wetlands, deforestation** and other erosion of habitats have been added to the hazards. Breaches of the **Convention on International Trade in Endangered Species (CITES)** is another menace, brought about by the illegal, direct trade in live animals by collectors. Extinction, like the dinosaurs' demise, is also a feature of natural environmental change or catastrophe. However, human activity was responsible for most of the animals and plants known to have been lost in the past two centuries. They include antelope, 17 species or subspecies of bears, four of cats, three of deer, foxes, goats, horses, wolves and zebras. The last dodo was killed in Mauritius in 1681. Just over 100 years ago the passenger pigeon, which gathered in flocks by the hundred in North America, was killed off for food. Threats to large animals like the **African elephant, rhinoceros,** gorilla, **tiger, giant panda, polar bear, otter** and **whales,** have come to public attention because of the work of conservation groups, such as the **World Wide Fund for Nature (WWF)**; and the campaigning of the environmental groups like **Greenpeace** and **Friends of the Earth.** Species have been rescued from the brink of extinction by successful, controlled breeding of managed herds, as in the case of the **oryx,** and by creating protected reserves, as in the case of the Indian tiger.

The **ENERGY BALANCE** is the way the Earth and **atmosphere** interact to absorb incoming solar energy. The Earth-atmosphere system absorbs about 68% of the solar energy which reaches the outer layer of the atmosphere. Some of the solar energy is bounced directly back into space rather than penetrating to lower layers of the atmosphere. Some is absorbed by the various gases in the air and turned into heat. Some is absorbed by the ground and some is reflected back into the atmosphere. Some sunlight stimulates **photosynthesis** and so is stored temporarily in vegetation. Solar energy is stored for longer periods in the trees and plants that were turned into **fossil fuels.** Objects respond differently. Dark rocks readily absorb solar energy, because they have a low **albedo.** Ice and snow reflect most of it back. Water is better at absorbing heat than land because of convection currents. These exchanges between the Earth and atmosphere balance out through the year to give the Earth a fairly consistent environment, season by season. Consequently, global temperatures have varied very little over centuries. While there is an overall balance between energy absorbed and energy lost, the processes by which it is maintained are fairly turbulent. The turbulence is not restricted to the atmosphere. The oceans carry immense quantities of energy in currents from one part of the globe to another. See also **global circulation.**

ENERGY CONSERVATION is a far broader issue than just the realization of the benefits of saving fossil fuel by greater energy efficiency. Energy conservation covers the relationships between energy, **ecology** and the environment. For example, the methods of extracting and using the sources of energy, from chopping trees for fuelwood to coal, oil and nuclear power generation, can adversely affect the ecology and environment of the world. Energy conservation includes the development of the renewable, non-polluting sources of power like **solar energy, wind power** and tidal energy. Energy conservation also embraces developments like the **genetic engineering** of crops, so that plants can flourish without the need for high levels of manufactured fertilizer, which is an energy-intensive process.

ENERGY CROPS are those plants grown wholly or partly for their energy content, and in effect providing a means of storing solar energy in the form of **carbohydrate.** Both cellulose and starch can then be fermented to produce **ethanol** and **methane.**

ENERGY EFFICIENCY refers to actions to save fuels by better building design, the modification of production processes, better selection of road vehicles and transport policies, the adoption of **district heating** schemes in conjunction with electrical power generation, and the use of domestic insulation and double glazing in homes.

ENRICHMENT is the process of taking natural uranium ore, which consists of uranium-238 (99.3%) and **uranium-235** (0.7%), and changing these proportions. The amount of U-235 is raised to 3.5%–4% in order to provide uranium fuel for a nuclear reactor, and to almost 100% for the fuel for an atomic bomb. For 30 years enrichment was done by the difficult and expensive gaseous diffusion method, based on a discovery by Lars Onsager, a Norwegian-American chemist working at Yale University. He received the Nobel Prize for his research. Gaseous diffusion is both difficult and expensive. The method depends first on making uranium hexafluoride gas and pumping it through thousands of special filters. Because of the tiny difference in the nucleus of the atoms, slightly more U-235 gets through each filter. As the gas stream is pumped through a lot of filters, the proportion of U-235 steadily increases. In the 1970s, scientists at the United Kingdom Atomic Energy Authority's research establishment at Harwell invented a gas centrifuge method of enrichment that was both quicker and cheaper.

ENTROPY refers to the amount of disorder in a system. It involves the factors which define the energy state of a substance, and relates to the unavailability of the energy because of the disorder in a system.

The word **ENVIRONMENT** suggests different things to different people. To the meteorologist and climatologist, for instance, it usually means the **atmosphere.** To the environmental engineer, it usually means the atmosphere in an enclosed space, such as a factory, office block or hospital, which the engineer has to maintain in a safe, healthy and comfortable state. To the **ecologist,** the environment is synonymous with the term habitat, within which plants and animals live. In the context of this publication environment has the broadest possible connotation. It refers to the entire range of external influences, natural and man-made, that can impinge on the **life-support systems** essential for health and survival.

An **ENVIRONMENTAL AUDIT** (environmental accounting) takes into account the real value of natural resources, such as clean air and water, before they become degraded. For example, when a forest is cut down, a company counts timber sales as income, which later appears as growth in industry and national accounts. No one, in strict accountancy terms, pays the environmental costs that might ensue, except those who suffer because of increased **soil erosion** and flooding, loss of fuelwood and wildlife habitats, damaged watersheds and increased global warming. Environmental audits call for assessments by governments or an individual company of the existence and effectiveness of policies to prevent damage to the environment from their processes, products and services. The United Nations and other organizations are working to produce a manual for environmentally adjusted accounting. See also **polluter-must-pay.**

ENVIRONMENTAL DEGRADATION is the process by which the **soil,** a rich but fragile ecosystem, containing millions of micro-organisms to sustain fertility, is destroyed. It can be sterilized by salts, contaminated by chemicals, under-fertilized, buried under buildings or swamps, damaged by a change in the traditional patterns of grazing and shifting cultivation which gave soil time to recover between crops, or even blown away by the wind. In arid, semi-arid and dry sub-humid areas the process of **desertification** begins. People ekeing out a living from the land overgraze cattle and cut down trees and shrubs and, when **fuelwood** runs out, they are forced to burn dung, one of the only fertilizers that is affordable. In dry areas productive land turns into desert and in more humid areas it is debased and becomes unproductive wasteland. Environmental degradation is also caused by air- or water-borne pollution to rivers, seas and forests, and contributes to the corrosion of buildings. The Sahel drought between 1968 and 1974 impelled the United Nations to convene the Conference on Desertification. It adopted a plan of action to combat desertification. Ninety-four countries pledged support at the conference, but progress has been slow. See also **soil erosion.**

ENVIRONMENTAL HEALTH refers to the causes of ill health and diseases in people which are produced or exacerbated by environmental conditions. The impact on health of weather, climate, water supply, eating habits and mode of life generally is not a modern-day discovery. In the 4th century BC they were taken to be essential considerations for 'whoever wishes to investigate medicine properly'. Nowadays a multiplicity of additional environmental factors must be added to the list. One is the overcrowding of populations in big cities and the tempo and tensions of life in them that cause stress. Others include: **air pollution** from cars, industrial emissions and nuclear radiation; pollution of land by industrial wastes and pesticides; and pollution of streams, rivers, lakes and seas by sewage, industrial effluent and oil. Some of the changes are obvious because of blatant large-scale pollution, others are being wrought by imperceptible alterations in the environment such as **low-level radiation** releases and accumulation of low levels of toxic **heavy metals** in the **food chain.** The traditional approach to medicine has always been that the primary role of the doctor is the treatment of the individual patient. Now the practice of medicine has moved into an era where an understanding of the social and environmental factors in relation to disease has become of paramount importance.

An **ENVIRONMENTAL IMPACT ASSESSMENT (EIA)** is an audit of the environmental consequences of any planned major development before it leaves the drawing board stage. It should provide both an ecological and amenity balance sheet. The idea of EIA techniques was adopted originally in the 1970s by oil companies, to cope with tougher legislation on all their industrial and shipping activities. An EIA directive from the European Commission lists 35 obvious developments for an audit, ranging from power stations and airports to scrapyards and toxic waste dumps. However, there are more than 200 EC directives that involve environmental controls. EIAs pose questions like: will a development affect the local topography? Will it sterilize agricultural land? What pollutants will be unleashed? What are the potential threats to flora and fauna and for **environmental health?**

ENVIRONMENTAL LAW covers a wide spectrum of options from binding 'hard' laws, such as international treaties and national legislation, to 'soft' laws, covering guiding principles, recommended practices and procedures, and standards. Environmental law also attempts to reconcile international considerations with concerns that focus on very specific problems such as soil degradation, marine pollution or the depletion of non-renewable resources. It also embraces **environmental impact assessment,** which in some countries is becoming 'hard' law and in others remains a 'soft' option. Until 1972 there were 58 international treaties and other agreements on the environment. By 1992, 96 such agreements had been adopted regionally or internationally.

ENVIRONMENTAL LEAGUE, see **Lega per l'Ambiente.**

ENVIRONMENTAL LIABILITY is the penalty to be paid by an organization for the damage caused by pollution and restoration necessary as a result of that damage, whether by accidental spillages from tankers, industrial waste discharges into waterways or land, or deliberate or accidental release of radioactive materials. The concept is part of the **polluter-must-pay** principle. However, the philosophy is difficult to put into practice. Ultimately, the cost of pollution is met by the consumer through the price of goods or taxes for public services. Progress in enforcing liability rests on the adoption of an increasing number of environmental laws on the statute book and negotiation of international agreements. The potential scale of oil pollution, which became apparent with the **Torrey Canyon** oil tanker spillage in 1967, led to the first major understandings on liability. These resulted in the establishment of the 1969 International Convention on Civil Liability for Oil Pollution Damage and, in 1971, to the International Fund for Compensation for Oil Pollution Damage. There are now over 200 EC directives alone governing the environment.

The **ENVIRONMENTAL PROTECTION AGENCY (EPA)** is the US Government's watchdog agency responsible for controlling the pollution of air and water, pesticides, radiation hazards and noise pollution. The agency is also involved in research to examine the effects of pollution.

ENVIRONMENTAL SCIENCE is the general label given to the vast range of research into the natural environment.

ENVIRONMENTALLY FRIENDLY, see **green consumerism.**

ERS-1, see **Earth resources technology satellite.**

ERTS, see **Earth resources technology satellite.**

ESSENTIAL ELEMENTS are the basic elements, sometimes referred to as the macronutrients, that are needed to produce the basic molecules of life. There are ten: they are **carbon, oxygen, hydrogen, nitrogen, phosphorus, sulphur,** potassium, magnesium, calcium and iron. In addition, there are other vital minerals that are needed in small amounts, referred to as the trace elements. They include **boron, manganese, zinc, copper, molybdenum** and others. Excess amounts of trace elements can also be toxic.

An **ESTUARY** is the area at the mouth of a river where it broadens into the sea, and where fresh and sea water intermingle to produce **brackish** water. The estuarine environment is very rich in wildlife, particularly aquatic, but it is very vulnerable to damage as a result of the actions of humans. Experience has shown these ecosystems are at risk if there is an artificial change in the deposition of **sediments** and salinity caused by the abstraction of water upstream, or by the construction and operation of flood-control barrages on the estuary. Chemical pollution from domestic and industrial sources can cause serious problems, as can pollution from ships or the discharges of coolant water from power stations. See also **delta.**

ETHANOL is the alcohol best known for its intoxicating qualities, and for centuries it has been made from grain and other crops. However, from the mid-1970s it has attracted the interest of environmentalists, as the fuel of the future for cars and lorries (see **biofuels**). It can be blended with motor fuel or used on its own in modified engines. The possibilities of using ethanol to replace oil-derived motor fuel have been demonstrated extensively in Brazil. A huge government-subsidized industry was set up to produce enough ethanol to fuel half the country's cars. The question of expanding the global use of ethanol was not a technical but an economic one. Crop-derived ethanol had to compete with oil, and fuel crops must be grown on the planet's limited amount of fertile land at a time when many people are starving. Scientific research in 1989 showed ethanol could be produced from wood and cellulose wastes, instead of valuable food crops like grain.

EUROENVIRON is an extension of the 19-nation European Research Co-ordination Agency (EUREKA) research programme, created to establish a European programme of research and development to counterbalance technological development in the USA and Japan. Euroenviron extends this to green technologies for halting pollution emissions, reducing waste discharges and saving energy.

The **EUROPEAN CONVENTION ON CIVIL LIABILITY FOR DAMAGE RESULTING FROM ACTIVITIES DANGEROUS TO THE ENVIRONMENT** is an agreement between the member countries of the Council of Europe which defines legal liability for environmental damage. It represents the first international legal instrument to set out the principle of civil liability for all damage caused in the environment, with the exception of damage resulting from the transport of dangerous products and nuclear accidents, which are covered by other international conventions. The convention requires that each signatory 'shall ensure that, where appropriate, taking account of the risks of the activity, operators conducting a dangerous activity on its territory, be required to participate in a financial security scheme or to have and maintain a financial guarantee up to a certain limit, of such type and terms specified by international law, to cover the liability under this Convention'. This is interpreted by the Council as a requirement that operators must 'take out compulsory insurance as soon as the environment insurance market is sufficiently developed'. Ultimately, the Convention aims to make insurance cover or payments into national compensation funds compulsory in order to pay for damage. The agreement was opened for formal signature in June 1993 and was to come into force soon after.

The Convention is open to other states, as well as members of the Council of Europe. It deals with compensation for damage to the environment, to persons and property, as well as compensation for economic loss following impairment of the environment; for example, the loss of earnings to a tourist industry following pollution of a beach. The phrase 'dangerous activity' covers actions which produce or make use of a provisional list of over 100 dangerous substances. It also concerns the use of **genetically modified organisms (GMOs)** or micro-organisms, and also involves the treatment of waste and waste dumps. Liability rests on the person who controls the dangerous activity. This may be an individual or a public authority, so that a local authority operating a refuse dump, an industrialist manufacturing **fertilizers,** a farmer using them, and a laboratory making GMOs, will be liable for any damage caused by their activities. The Convention will extend to creeping pollution, such as harmful substances that are deposited in a waste dump and then seep into the ground contaminating **groundwater** and drinking water sources.

The **EUROPEAN ENVIRONMENT AGENCY** is being set up to provide the European Community and its member states with objective, reliable and standardized information on the environment. It will assess the success of existing environmental policies and the data will be used to develop new policies for environmental protection measures. The Agency, with necessary scientific and technical support, will come into force when it is decided where it is to be located. It will gather information covering the present, and foreseeable, state of the environment. The priority areas are: **air quality** and emissions; water quality, pollutants and resources; soil quality, flora and fauna, and biotopes; land use and **natural resources;** waste management; **noise pollution;** chemicals; and protection of **coastal areas.** The Agency will also take into account the socio-economic dimension, cover transboundary and international matters, and avoid the duplication of the activities of other bodies. Among its aims are the promotion of European information into international environment monitoring programmes, for example United Nations agencies, to co-operate with other Community and international bodies and programmes, and to publish a report on the state of the environment every three years. All environmental data coming to or from the Agency may be published and available to the public, subject to the European Commission or national rules on confidentiality.

The **EUROPEAN ENVIRONMENTAL BUREAU (EEB)** is a **non-governmental organization (NGO)** formed in 1974 to co-ordinate the activities of European NGOs in order to maximize their impact on the environmental projects and policies of the European Community (EC). Based in Brussels, its main areas of interest are environmental policy, education, nature protection, pollution control, land use, planning, transport, energy and agriculture. Another area is the environmental aspect of European relations with developing countries. It operates through lobbying, general and specific publications, and also runs a series of seminars and conferences.

EUTROPHICATION is a process of pollution that occurs when a lake or stream becomes over-rich in plant nutrients. It becomes overgrown in **algae** and other aquatic plants. The plants die and decompose. In decomposing the plants rob the water of oxygen and the lake, river or stream becomes lifeless. **Nitrate** fertilizers which drain from the fields, nutrients from animal wastes and human sewage are the primary causes of eutrophication. They have a high biological oxygen demand (**BOD**). This is the figure showing how much oxygen is needed to purify water affected by a particular form of contamination.

The **EVERGLADES** are the subtropical marshy region of southern Florida, covering some 4,000 square miles, which stretches down to the **mangrove forest** swamps that border the Gulf of Mexico and Florida Bay. The Everglades contain a variety of ecosystems, providing habitats for a range of wildlife and flora and fauna. Much of the area is covered with saw grass and, on low-lying islands, trees and shrubs, including cypresses, palms and other tropical species, which form a natural habitat for deer, panthers and a variety of reptiles. The mild climate, combined with open areas of water, creates ideal conditions for wading birds, and for many kinds of snake, turtle and alligator. The Everglades also house a wide variety of insects. In the 19th century regulations were introduced to protect birds which were being hunted for their feathers. Subsequently, similar restrictions were placed on the killing of alligators for their skins. In 1947 the central and southern regions of the glades were designated as the Everglades **national park** to protect wildlife habitats. However, a great part of the Everglades has been damaged by the fertilizers that are discharged from nearby agricultural areas. They have led to the growth of cat-tails and furry grasses, which have taken over some 6,000 acres, depleting the oxygen in the water and therefore the number of wading birds. The Everglades are also under constant pressure from a large increase in population and from the growth of **tourism.**

EXTINCTION describes the complete disappearance of a species of plant or animal from the planet. It is a natural process, part of the general evolutionary pattern. There have been mass extinctions throughout history, but these were lengthy, mostly gradual processes. Species are now being lost at a much more rapid pace, entirely due to human activities, such as the growing levels of pollution and hunting, excessive fishing, and the destruction of habitats. Species are lost because of **deforestation** of the tropical forests and by the damage being done to **coral reefs** and **wetlands.** It is estimated that wildlife species are being lost at a rate 400 times faster than ever before. The **International Union for the Conservation of Nature and Natural Resources (IUCN)** estimated in 1988 that about 800 species of mammals, reptiles and amphibians were threatened with extinction. Another 1,073 birds, 596 species of fish and more than 2,000 species of invertebrates were also in danger of being lost forever. The IUCN regularly produces up-to-date lists of threatened species, and by the mid-1990s it expects to be able to produce comprehensive analytical data.

The **EXXON VALDEZ** was the oil tanker responsible for one of the worst oil pollution disasters. It devastated the ecology of hundreds of miles of pristine Alaskan coastline and the breeding grounds of marine life, fish and mammals centred on Smith Island. It became grounded off the island in 1989. The oil slick spread over some 3,000 square miles, contaminating 350 miles of beaches. It is estimated that the **oil spill** was responsible for the deaths of some 33,000 birds and 1,000 **otters.** The pollution of the shores of Prince William Sound destroyed marine organisms and vegetation and disturbed the local fishing industry. Four years later scientists discovered brain-damaged seals, birds that do not breed, fish with malformed jaws and a dwindling number of **whales** in the area. It is believed the whales inhaled toxic vapours from the oil slick and died.

F

FALL-OUT refers to both the airborne and waterborne **radioactivity** from nuclear weapons testing and emissions, in both accidental and planned circumstances, and from nuclear power plants. The most intensive testing of nuclear weapons in the atmosphere occurred during the periods 1955–58 and 1961–62. After the **Nuclear Test Ban Treaty** on atmospheric testing was agreed between the USA, the former Soviet Union and Britain in 1963, only France and China continued with atmospheric tests. The most important **radionuclides** in weapons' fall-out are caesium-90 and strontium-90, which are both fissile products of uranium and plutonium. The two most serious cases of accidental fall-out were from the **Windscale fire** in 1957 and the explosion of the **Chernobyl** nuclear reactor in 1986. The **fission products** released from Windscale were iodine-131, caesium-137, strontium-89, strontium-90 and polonium-210. A large area around Windscale became contaminated with iodine and the soil had to be removed. In the Chernobyl accident, the fall-out also included aerosol particles of plutonium.

FAO, see **Food and Agriculture Organization.**

The **FAST BREEDER REACTOR (FBR)** is a type of nuclear reactor that breeds more **fissile** fuel while it is consuming **nuclear fuel** to generate power. The first FBR was built by the United Kingdom Atomic Energy Authority at Dounreay in Scotland. The reactor is called fast because it has no moderator to slow down the neutrons that are generated once the chain reaction has begun. This allows some neutrons to escape from the fuel core to be absorbed by a blanket of uranium-238. Some of the U-238 is then converted to **plutonium,** which is extracted by the **nuclear reprocessing** of the uranium.

The **FAUNA AND FLORA PRESERVATION SOCIETY (FFPS)** is the oldest international **conservation** organization. It is a voluntary organization, with its headquarters in London, which promotes international conservation of wildlife, especially **endangered species** of plants and animals. The Society, an independent body, increases public awareness about endangered wildlife by holding meetings and exhibitions and publishing information. It also promotes the establishment and management of wildlife reserves and the implementation of laws covering the protection and conservation of wild animals and plants.

FERMENTATION is a process of breaking down plant material and food without the use of oxygen. In other words, it is an anaerobic process. The most well-known fermentation process involves adding yeast to the raw materials from which alcohol is derived, in the making of wine and brewing of beer. It is also used to cause the dough to rise in baking. However, fermentation is also the basis of the **biotech** industry. Many pharmaceutical products, like antibiotics, employ fermentation at the preliminary stage to make the active ingredient that is eventually packaged as a tablet, capsule or liquid. It is the process by which waste materials, such as cellulose and straw, and organic refuse, are deliberately decomposed to generate **biogas.**

FERRUGINOUS DISCHARGES are a type of effluent, like that in the floodwater from abandoned coal mines, which is contaminated with iron.

FERTILE nuclear materials, such as **uranium-238** and thorium, are those that can be converted into fissile material in a nuclear reactor.

FERTILIZERS are any material that is added to the **soil** in order to provide and replace vital nutrients for crop growth. That material can be natural or organic, such as compost and manure. The term is more commonly associated today with manufactured fertilizers, made from substances derived from chemical processes or mining. Most manufactured fertilizers are made of a compound of salts of nitrogen, phosphorus and potassium. The use of artificial fertilizers increased rapidly throughout the world from the 1950s when farmers realized that fertilizers meant they no longer had to leave fields fallow to regain fertility or, indeed, keep farmyard animals just for their manure.

FIFTH ENVIRONMENTAL ACTION PROGRAMME, see **Towards Sustainability.**

FILARIASIS describes several different infections caused by various species of filarian roundworms. The two most prominent are **onchocerciasis,** also known as river blindness, and lymphatic filariasis, which causes **elephantiasis.** It is estimated that the lives of some 1,000m. people are threatened by these diseases, mostly in Africa and south-east Asia. Lymphatic filariasis is carried by the larvae of worms transmitted by various types of **mosquitoes,** including ***aedes aegypti*** and **anopheles.** The parasite larvae get into the bloodstream and develop into adult worms, and this leads to elephantiasis. An estimated 90m. people suffer from lymphatic filariasis and another 900m. are at risk from it in Africa, the eastern Mediterranean, south-east Asia and Latin America. Drugs are available, but they have unpleasant side-effects. Vector-control and drugs programmes have reduced the infection rates where the disease is endemic.

FILTER FEEDERS are a variety of animal, critical to the **food chain,** which live mostly on the ocean floor or in **plankton.** They feed by creating a current running through a filter, a scraping device to move food from the filter, and a water exit. The filter absorbs particles of an acceptable size and rejects those that are too large. In sponges and sea cucumbers, the entire body acts as the filter, water inlet and water exit. Mussels and oysters are filter feeders. However, the baleen **whales** are the largest filter feeders. They have blowholes that expel the water that has been ingested along with the **krill** that they feed on.

FINITE FUELS are the deposits of **coal,** oil, gas (see **fossil fuels**) and **uranium** ores that were formed millions of years ago by geological and biogeological processes. As such they are non-renewable reserves, and are finite as it is possible to calculate their date of exhaustion according to certain rates of exploitation.

FIRE is one of the greatest tools of mankind. Once early man had learned to control it, he was able to provide warmth and light, to cook, to kill insects and to clear forests of undergrowth so that game could be more easily hunted. Later, he found that this burning produced more richly abundant **grasslands.** Agriculture brought a new urgency to clear brush and trees, and used ash as a fertilizer. This procedure, **slash-and-burn agriculture,** continues in many tropical and temperate areas throughout the world today. Fire also permitted the opening up of industrial and manufacturing processes. However, uncontrolled and spontaneous fire is destructive. Fires in forests, grassland or brushland are a constant problem, causing loss of vegetation, trees and crops, and sometimes the lives of wildlife or humans. Dried grass, leaves and light branches ignite easily and this often generates enough heat to set fire to tree stumps and the matted undergrowth of the forest floor. Evergreens, such as pine, cedar, fir and spruce, have flammable fuels that burst into flames if they are heated sufficiently by a forest fire.

FISH FARMING, see **aquaculture.**

FISSILE describes a substance which is capable of **fission.**

FISSION is the process of atom-splitting employed in a nuclear reactor and an atomic bomb. Splitting an atom is accompanied by the release of an immense burst of raw energy. Fission occurs when a neutron strikes the nucleus of an atom of a fissile substance, such as **uranium-235** or **plutonium**-239, causing it to split into more or less equal halves. In addition to the release of energy, many more neutrons are also produced. They strike other uranium or plutonium atoms, generating further neutrons and, under special circumstances, sustaining the fission process in a **chain reaction.** In a nuclear reactor the chain reaction is controlled to generate heat. In a bomb, energy is released as an uncontrolled destructive force.

FISSION PRODUCTS are the substances formed by splitting atoms of **uranium-235** or any other fissile element. Fission products are contained in the fall-out from nuclear weapons and in the various forms of **nuclear waste** from nuclear power stations and from **nuclear reprocessing** of spent fuel. Fission products cover a wide range of radioactive elements, or radionuclides, from **americium, caesium** and **iodine** to strontium and **zinc.** One year's operation of an average nuclear power station produces waste fission products of 5,000m. **curies** of radioactivity. Many of these products have an extremely short **half-life.** A significant proportion of them are rendered safe simply by storing spent nuclear fuel for one to two years in cooling ponds at the nuclear power station, during which time a large amount of **radioactivity** decays. Hence, the 180m. curies of radioactivity in each ton of spent nuclear fuel drops 260 times to 693,000 curies after a year. The remaining fission products take much longer to disintegrate. Some will still be dangerous in 10,000 years time. Accidents such as the **Windscale fire** in 1957 and the **Chernobyl** explosion were disastrous because they pumped fission products in their most radioactive state into the atmosphere.

FLATFISH is a generic term for any of the 600 species of oval-shaped, flattened fish. Flatfish include halibut, plaice, sole, turbot, dab and flounder, and they live in tropical and cold waters. Most are marine fish and are found along the **continental shelf,** but some are found in freshwater. Flatfish are important to environmental monitoring because they can give early warning signs when chemical contamination is becoming toxic. As well as studying the effects of contamination on the flatfish immune system, scientists at the National Environment Research Council have been investigating the impact on liver cells. The fish act as a continuous monitor because it is possible to link subtle changes in their liver cell function, and health, to specific groups of contaminants. These techniques are now being used by scientists in Europe and the USA. They are likely to be much more cost-effective than conventional analysis, and demonstrate whether the contaminants are actually harming fish.

FLIP is the acronym for the Floating Instrument Platform, which looks rather like an oil rig. It is part of the equipment used to make acoustical soundings in the open ocean. The platform was designed by the **Scripps Institution of Oceanography** in La Jolla, California. Once towed to a research site, the ballast tanks are flooded and the platform flips into an upright position.

FLOODS are caused either by tidal waves (or **tsunami**), when rivers overflow their natural or artificial banks on to dry land, or by excessively high rainfall over a short space of time, or when large amounts of ice formed on rivers melt in the spring. Sometimes this is a natural process, such as when a river submerges the surrounding floodplain. Egypt's Nile Basin is an example of flooding being beneficial to the environment, because agricultural productivity is maintained by the deposits of fertile sediment left on the land after flooding. Severe and sudden flooding can cause large-scale loss of life and damage to property and land. **Environmental degradation,** especially where there is large-scale **deforestation** or urbanization, also contributes to a cycle of severe drought followed by equally disastrous floods, which have been experienced in some regions in recent years. Forests control the run-off to rivers and most of the annual rainfall in the forested watershed of rivers in the tropics is trapped in the sponge-like network of roots beneath the ground. If the forest is uprooted, and therefore the 'sponge' is taken away, only a small amount of water remains in the soil and the greatest proportion goes into local rivers in a short space of time, resulting in serious flooding. Another problem is that **global warming** could lead to a rise in **sea-levels** and flood low-lying **coastal areas.** A rise of more than 10 cm in sea-levels would have a serious impact on humans, plants and animals in low-lying areas like **Bangladesh,** the **Maldives,** Florida and parts of the UK.

FLORA AND FAUNA are, respectively, the plants or plant life in a region or at a point of time, and a collective name for animals or animal life of any region or point of time.

FLORAS are full catalogues and checklists of known plant species. Many of the world's plants are disappearing at a rapid rate, particularly in the tropical and subtropical forests, where habitats are being destroyed. It is now a major priority to explore these regions and bring back for cultivation the most rare and scientifically interesting species, some of which are potentially of great value for food sources or medicines. The **Royal Botanic Gardens** at Kew have been involved in the production of floras for Africa, Australia, India, Cyprus and Iraq. New studies are concentrating on regions where the vegetation is particularly diverse, or of special concern for the conservation of genetic resources. Small samples are collected by plant-hunters from Kew and brought back to be cultivated and studied, and eventually to be used as a reservoir to support wild populations so that they do not become extinct.

FLUE GAS DESULPHURIZATION is the necessary, but costly, method of removing **sulphur dioxide** and nitrogen oxides from the gas in power stations when **fossil fuels** are being burned. The control technologies include scrubbers and electrostatic precipitators, which remove emissions from flue gases before they are discharged into the atmosphere.

FLUIDIZED BED COMBUSTION is the most efficient way of burning all fuels, but particularly coal. The furnace contains a layer or bed of fine inert particles such as sand or ash. When hot air is blown up from below the fuel and inert particules behave like a fluid. The fuel burns very efficiently and traps waste products in the bed instead of in gas discharges through the chimney.

FLUORIDES are substances containing **fluorine** that are added to water supplies as a way of stopping dental caries. This has caused great concern among people who object to the addition of fluoride to water. When fluorine is combined with calcium, it is an important factor in developing the hardness of bones and the enamel of teeth. Although fluorides are usually present naturally in the environment and water supplies, some areas have no fluoride-bearing rocks.

FLUORINE, in the form of **fluoride,** is incorporated in the structure of bone and teeth. Individuals growing up in areas where the natural water is short of fluoride minerals have poor teeth. In areas of abundant fluoride, the teeth have strong markings. The first really vocal health and environmental movement was probably the anti-fluoride lobby, which still objects to the treatment of water supplies with fluoride to protect children's teeth.

A **FLUX** is a general state of change. In the environmental context, it refers to processes by which **carbon** is exchanged between the atmosphere and oceans.

FOG is a cloud-like body of water droplets suspended in the atmosphere. It reduces visibility to less than one kilometre. It is formed when the lower layers of the atmosphere cool and **water vapour** condenses.

The **FOOD AND AGRICULTURE ORGANIZATION (FAO)** is the oldest of the permanent specialized UN agencies. It was established in 1945 to co-ordinate food, agriculture, forestry and fisheries development programmes, to improve standards of living in rural communities and to fight malnutrition and hunger at the end of World War II. It advises governments on policy and agricultural planning (giving technical assistance to individual countries on a project basis), on nutrition, improved food storage and distribution, fertilizer production, pest and disease control, **soil** fertility and **irrigation.** It also runs educational programmes, keeps statistics on world production, trade and consumption of agricultural products, and publishes yearbooks, periodicals and research bulletins. Its headquarters are in Rome and there are other offices throughout the world. It is governed by a council elected by the biennial FAO conference of member countries. In 1985 it began to concentrate on the problems of tropical forestry, with the introduction of measures to promote conservation and the increased use of **afforestation** in agricultural and industrial planning.

The **FOOD CHAIN** is a process that begins with the simplest green plants and ends in a large animal which has no predators relying on it for food. Examples of food chains could include: green plant, aphids, ladybird, insectivorous bird, hawk. Or marine **algae, copepod,** small fish, large fish, man. Or again, green plant, rabbit, fox. The lower members of the food chain have to support the higher members. Consequently, they have to reproduce at a rate greater than that necessary to maintain their species, because they provide a margin to feed those above them in the chain. Any serious fluctuations in the numbers at any point in the chain, either as a result of natural causes or contamination from pollution, can have disastrous repercussions for every member. Of course, feeding is the way animals get the energy they need to exist. In the process the energy moving along a food chain is described as passing from one trophic level to the next. Plants form the first, producer, trophic level. Then come the consumer levels of herbivores, carnivores and parasites. Ecologists have shown that the number of trophic levels is limited to five or six.

FOOD POISONING is the general term to describe illness caused by eating food which is infected with bacteria or contaminated with toxins. In recent years outbreaks of certain types of infection have increased, particularly poultry-borne **bacteria** such as **salmonella,** and also **listeria,** which is responsible for much of the contamination of salads and dairy produce. Precise figures for the incidence of food poisoning are not known, because most people have it in such a mild form that they do not report it. In some cases, however, it can be serious and even life-threatening. The bacterium salmonella and **campylobacter** are killed when they are thoroughly heated. Outbreaks of food poisoning usually occur when there is inadequate hygiene in the preparation and handling of food, or if food is stored at the wrong temperature.

FORAMINIFERA are an abundant family of simple, mostly marine, organisms which are important to scientists who are studying **climatic change.** Most have a shell which falls to the bottom of the sea when the organisms die and which becomes incorporated into rock, appearing as chalk or limestone. About a third of all sediment at the bottom of the sea is made up of discarded shells. The living foraminiferans either crawl around or attach themselves to rocks, feeding on **bacteria, diatoms** and other algae.

FORESTS AND WOODLANDS cover about a third of the land area on Earth. Forests are indispensable to humanity but they are at the heart of environmental concern. Forests and woodlands play many roles in the social and material well-being of mankind in general and particularly of rural populations. Pine forests in the north, the **tropical rainforests** on the equator and the temperate forests of the southern hemisphere regulate the climate, protect water resources and provide forest products. Half the world's population depends on them for fuel, they are home to millions of plant and animal species and they also secure water supplies by withholding rainfall. The forests are also a source of new medicines, chemicals, foods, commercial products and services. Trees recycle moisture through their leaves, absorb the sun and soak up carbon dioxide, but as more and more trees are cut down **deforestation** contributes an estimated 10% to global warming. Loss of forest land is particularly rapid in developing countries, and many species are being driven towards **extinction** with the removal of their habitat. It is calculated that two-fifths of the developing world's remaining forests will have disappeared by 2000, taking with them thousands of species. The forests are depleted by logging for timber, firewood and charcoal, or cleared for cattle. Another threat is the hundreds of millions of people who, driven by the shortage of more suitable land, cut down the forests to build houses and to farm. They then move on because the land loses its fertility. The **Earth Summit** adopted and recommended for endorsement the **Statement of Principles on the Management, Conservation and Sustainable Development of all Types of Forest.** In 1985 the UN Conference on Tropical Timber ratified the **International Tropical Timber Agreement,** to regulate the tropical timber trade. It encourages co-operation between producing and consuming countries.

FOSSIL FUELS are the energy-containing materials which were converted over many thousands of years from their original form of trees, plants and other organisms after being buried in the ground. Physical and chemical processes occurred in the Earth's crust that changed them into coal, peat, oil or natural gas. Fossil fuel use, when expressed for clarity in terms of coal equivalent energy, has risen from 1,000m. tons a year at the beginning of the century to 12,000m. tons in 1990. If fossil fuels are completely burned, or oxidized, they are converted into **carbon dioxide,** water and heat energy. Therefore, burning fossil fuels destroys the energy source. Fossil fuels are used for other purposes. Coal and oil are used for the manufacture of a wide variety of carbon-containing compounds such as plastics, synthetic fibres, medicines, food and pesticides. Although **nuclear fuels** are inorganic substances, unlike the biological origins of fossil fuels, they are also destroyed when they are used in the generation of heat in nuclear reactors or in nuclear weapons.

FRAMEWORK CONVENTION ON CLIMATE CHANGE, see **World Climate Convention.**

FRESH KILLS LANDFILL is the world's largest **landfill** waste disposal site and is located on Staten Island, off New York. It also became the subject in 1993 of a major research project, applicable to all such sites suitable for **methane** generation, into why even readily degradable items, such as food, decompose very slowly in solid waste landfills. Converting solid wastes to methane is an attractive way to degrade refuse and collect a **renewable energy** source. The researchers' report, entitled *Environmental Factors Influencing Methanogenesis from Refuse in Landfill Samples,* shows that microbial activity is inhibited wherever the landfill's moisture content is low, the pH is either very acidic or alkaline (low or high), and, surprisingly, where sulphate ion levels are high. Analysis of the site revealed that paper and textiles, which compose over 40% of the landfill's volume, contain high levels of sulphate. However, this ion is not found in fresh paper and textile wastes. Instead, the authors suspect that construction and demolition debris which contain gypsum and constitute up to 14% of the landfill may be the source of the sulphate. As sulphate is leached from the gypsum the paper and textiles may soak up the water and serve as a reservoir.

FRESHWATER is an essential element for life. Some 70% of the Earth is covered by water, but only a fraction of that is readily available as freshwater. Ninety-four per cent of the water is salty and is found in **oceans** or as **sea-water,** and of the rest, 99% is out of reach, frozen in icecaps and **glaciers,** or buried deep underground. Only about 8% of freshwater supplies is withdrawn as drinking water or for domestic or municipal use; the rest is used in agriculture and industry. Freshwater supplies are coming under intense environmental pressure from two main sources: pollution of supplies and the fact that during the last 40 years world water use has more than trebled, and is predicted to increase into the next century. And, although the annual global rainfall is more than adequate to satisfy the need for freshwater, its distribution throughout the world is uneven, leading to severe **droughts** in several regions. Pollution of freshwater stems from industrial and human wastes, agricultural fertilizers, **acid rain,** and seepage from toxic waste dumps. Excess nutrients from sewage and soil erosion cause **algal blooms** which deplete the oxygen in the water; **heavy metals** and synthetic organic compounds from industry, mining and agriculture **bioaccumulate** in aquatic organisms; and pathogens from sewage spread disease. The **Commission on Health and Environment,** set up by the World Health Organization, reported in 1992 that freshwater should be considered as a scarce, valuable and finite resource. An estimated 2,500m. people, the report said, suffer from illnesses linked to insufficient or contaminated water and lack of sanitation. The United Nations Environment Programme (UNEP) made the 1980s the **International Drinking Water Supply and Sanitation Decade** because of global concerns about water supplies. At the end of the 1980s there were still about 1,232m. people without safe water, and about 1,600m. without sanitation.

FRIENDS OF THE EARTH (FOE) is an international pressure group which operates in 28 countries. It started in the USA as an off-shoot of the conservation group the **Sierra Club.** It is a non-party group that lobbies governments and politicians on all environmental issues and works to increase public awareness. FoE is very active in campaigning against the construction of new nuclear power stations and in establishing safety levels for existing nuclear facilities. The organization campaigns against: air, sea and land pollution, ozone depletion, acid rain and carbon dioxide build-up; and the use of whale products, aerosols containing CFCs, leaded petrol, food additives and non-returnable bottles. FoE's activities include campaigns for recycling, increased use of bicycles and public transport, alternatives to tropical hardwoods, better energy conservation and home insulation and the adoption of catalytic converters in cars to reduce emissions of carbon monoxide, unburned hydrocarbons and nitrogen oxides that cause photochemical smog.

FUELWOOD accounts for the greater part of the energy consumption of many developing countries. Supplies of wood and charcoal are running dangerously low in some countries, and fuelwood is being consumed at a faster rate than it can be replanted. The **Food and Agriculture Organization (FAO)** has predicted that, by 2000, some 1,000m. people will be faced with chronic fuelwood shortages. Often there are no alternatives to fuelwood, or if there are they are too expensive. There are many reasons for the shortage of fuelwood: **deforestation,** as forests have been over-exploited; populations have increased; and the need for energy, to heat and cook, has increased. However, as the shortages reach crisis point, this sets in motion a chain of other events. Without wood water cannot be boiled, so disease spreads, people are forced to eat only one cooked meal a day, and they burn **dung fuel** and crop residues that would be better used to fertilize the soil. In rural communities village women often take a whole day to collect wood to last only a day or two; time that used to be spent growing food.

FUNGI are a group of plants without chlorophyll, leaves, actual stems and roots. They include mould, mildews, rusts, yeasts and mushrooms, and do a remarkable job at breaking down, or decomposing, almost anything from wood to leather, and bread to cloth. The by-products of this decomposition process are sometimes toxic, causing great problems to the environment, while others, at the same time, are a boon, producing some cancer-curing toxins. Others have been the basis of beneficial agents like cyclosporin, the immune suppressant 'miracle drug' that has made organ transplants possible.

FUNGICIDES are chemicals used to kill or halt the development of fungi that cause plant disease, such as: storage rot; seedling diseases; root rots; vascular wilts; leaf blights, rusts, smuts and mildews, and viral diseases. These can be controlled by the early and continued application of selected fungicides that either kill the pathogens or restrict their development. Fungal diseases are more difficult to control than insects because the fungus lives in close quarters with its host. It is difficult to find chemicals that will kill the fungus without harming the plant and, because **fungi** may also go through secondary cycles rapidly during a three-month growing season, repeated applications of fungicides may be necessary. They must also be applied to plants during stages when they are vulnerable to inoculation by pathogens, before there is any evidence of disease. There are about 225 fungicidal materials, most of which are recently discovered organic compounds. **Sulphur** is probably the oldest effective fungicide and it is still used in gardens.

FUSION is a potential clean source of energy using a machine which it is believed will harness on the ground the nuclear reactions that take place uncontrollably in the sun and in the hydrogen bomb. However, even at the earliest estimate, the commercial generation of fusion energy is at least 50 years away. The attraction of fusion is that it avoids the pollution associated with nuclear power and the burning of **fossil fuels.** Whereas nuclear **fission,** the principle on which nuclear power stations operate, splits heavy atoms like **uranium** to create lighter ones, fusion begins with hydrogen and forms heavier ones. The atoms of hydrogen, in the form of tritium, are squeezed by immense electromagnetic fields until two of them fuse to form helium, and in the process release large amounts of energy. Experimental fusion reactors are operating in Europe, the USA and Russia. The drawback is their immense research cost.

FUSION TORCH is the most efficient incinerator for neutralizing even the most hazardous and toxic compounds. It uses a burner called a plasma torch which heats the wastes to thousands of degrees Celsius to reconvert them to their natural constituent elements.

G

G7 is a group of leading industrialized countries; it held the world's first Green Summit, in Paris, in July 1989. The seven countries that participate in annual summits contain 12% of the world's population but contribute nearly half of the pollutants that cause global warming, the environmental issue singled out for the agenda of the Paris gathering. While omitting specific targets the meeting, which included the leaders from Canada, France, Italy, the USA, Britain, Germany and Japan, stated that it 'strongly advocated common efforts to limit emissions of carbon dioxide and other greenhouse gases that threaten to induce climate change'. The ideas prepared the foundation for a **World Climate Convention** adopted at the **Earth Summit,** at Rio de Janeiro, in June 1992. The convention commits countries that ratify it to prepare national programmes to contain emissions of **greenhouse gases.** The aim is to restore emissions of greenhouse gases to 1990 levels by the year 2000. Between the Paris and Rio summits a report from experts on the **Intergovernmental Panel on Climate Change (IPCC)** concluded that, overall, mean global temperature had increased by 0.3°–0.6°C. At the

G7 summit held in Munich in July 1992 the industrialized countries urged others to join them in ratifying the convention.

G15 is the group of 15 semi-industrialized countries that meets to discuss development and other concerns for the developing world. It was formed in 1990 by the member states: Algeria, Argentina, Brazil, Egypt, India, Indonesia, Jamaica, Malaysia, Mexico, Nigeria, Peru, Senegal, Venezuela, the former Yugoslavia and Zimbabwe.

GAIA is the name of the ancient Greeks' all-pervading Earth goddess. The name was suggested by the novelist William Golding to describe the theory expounded by the scientist James Lovelock and now shared by many other biologists and chemists, which depicts the whole biological, physical and chemical world as a self-adjusting living organism. *Gaia* is now the title of a book by Lovelock.

GAMMA RAYS are the most intense form of nuclear **radiation;** they can penetrate deep into the body, or even pass through it, without losing their energy.

The **GANGES** river rises in the Himalayas in Nepal and northern India and flows into the Bay of Bengal, passing through the world's largest **delta** in **Bangladesh.** It irrigates the Ganges plain, which contains several of India's principal cities, and is the world's second largest agricultural area, supporting more than 300m. people. However, it is also a major depository for sewage and industrial discharges. The Ganges is a sacred river for the Hindus, who bathe in its waters and are cremated on its banks. Their ashes are then scattered over the river. As a result of the rising cost of wood, some bodies are thrown directly into the river because their relatives cannot afford cremations. **Topsoil** has been eroded and the silt levels in the Ganges have risen. This has largely been caused by **deforestation** in the Himalayas and on the river banks, which has caused flooding in the low-lying areas of Bangladesh.

GARP is the acronym for Global Atmospheric Research Project, an international group created to map and follow meteorological phenomena. It also monitors their effects on the world weather system.

GENE BANKS are storehouses of seeds or vegetative tissue, kept in low humidity and temperature, to help maintain genetic diversity. Sometimes known as seed banks or germ plasm banks, their contents mostly originate from a wide range of primitive strains and wild crop varieties. The International Board for Plant Genetic Resources (IBPGR), which was established in 1974, promotes the collection, documentation, evaluation, conservation and eventual use of genetic resources of significant plant species. Gene banks are the subject of international controversy because they contain seeds that have mostly been acquired from the developing countries by the industrially rich countries, where they have been used in breeding programmes to develop new strains. Instead of taking decades over a traditional plant breeding programme by fertilization, it is now possible to manipulate directly the genes of plants, creating **genetically modified organisms (GMOs),** which are plants modified to give a higher resistance to disease and improved growth and yields and, therefore, increase the profit of the plant breeder and farmer.

GENE SYNTHESIS is the complete manufacture of a gene using chemical reagents from the laboratory shelf to create a segment of DNA (see **nucleic acids**). The synthesis is done in a machine and is an alternative to using biochemical scissors, **restriction enzymes,** to snip a gene from a cell and transfer it to a micro-organism, which is then grown in fermentation tanks.

GENE THERAPY is genetic manipulation that specifically relates to changing the genetic make-up of a cell by deleting or inserting genetic material. There are two approaches: germline gene therapy and somatic cell gene therapy. The former changes the 'germ cells' of the sperm and ova, and would have a permanent effect on all descendants of whoever had the therapy. It is banned by most countries. Somatic cells are all the other cells of the body. Changing these does not affect the germ cells, only the individual who has been engineered.

GENERAL CIRCULATION is the day-to-day weather changes in the middle latitudes, which can vary very rapidly over a short period of time.

GENETIC ENGINEERING is the modification of the genetic characteristic of a microbe, plant or animal by inserting a gene from one species into another in order to: make fruit and vegetables more disease-resistant, for example, or make them grow bigger; extend the colour range of flowers; breed cows that produce extra vitamins in their milk; alter yeast or other cells so they become biological factories to produce antibiotics and other pharmaceuticals in **fermentation** tanks. Studies into the potential environmental risks from **genetically modified organisms (GMOs)** were made in preparing European Community regulations. EC advisers concluded that the risk posed by GMO technology varies greatly, according to the proposed products and their applications. Many are expected to offer no risk, but an environmental benefit. For example, the use of an ecologically harmful pesticide might be reduced because of the growing of stronger, pest-resistant plants, or as a result of the biological control rather than chemical control of pests.

GENETIC PEST RESISTANCE IN PLANTS is the potential alternative to using chemical **pesticides.** Genetic engineers are working toward identifying and introducing into plants the genes that give resistance to **pests.** There are two possible ways of doing it. The first is to locate existing genes that give protection and seem to be more abundant in hardy wild varieties of flora. The object is then to transfer them to valuable crop plants which are susceptible to attack and disease. This approach seems more promising in the search for resistance to bacterial and fungal **pathogens.** The trick is to compare the genes of the resistant plant and the pathogen, and to identify in the pathogen the 'avirulence genes', which make it so infectious, and the corresponding plant genes evolved to stop them. The second approach is to introduce a completely new type of gene into the plant. For example, the bacterial organism *Bacillus thuringiensis* produces a substance that stops insects absorbing food when it gets into their gut. The gene for the *Bacillus thuringiensis* toxin, called BTK protein, has been inserted into plants. Consequently, when the insect nibbles the leaf, the insect dies.

GENETICALLY MODIFIED ORGANISMS (GMOs) are genetically engineered organisms released into the environment. They may be micro-organisms designed for use as **biopesticides,** or seeds that have been altered genetically to give a plant better disease resistance or growth.

GEOCHEMISTRY is the study of the chemistry of the Earth, particularly its crust, waters and atmosphere. The chemistry of the land affects that of the oceans, because a number of chemical particles are present in both sea and **freshwater.** The particles in **sea-water** are often the result of run-off from the land, while others come from volcanic activity.

A **GEOLOGICAL PERIOD** is the measure used as a geological timescale to divide the ages of the rock formations into epochs.

GEOTHERMAL ENERGY taps the vast reservoir of heat that lies beneath the Earth's surface and is created by the energy welling up from beneath the Earth's crust. Some of this energy can be captured directly, when it emerges at the surface in the form of **geysers** or hot springs, or is pumped from natural underground reservoirs of hot water. Other ways are being tested of extracting energy from hot rocks by drilling deep boreholes, pumping water down one and extracting it from another once it has been heated in the rock.

GEYSERS are hot springs that spout jets of steam and water intermittently. Geysers, whose name is derived from the Icelandic word for 'to rush forth', are produced when underground waters are heated by coming into contact with, or being very close to, molten volcanic rock. They are commonly associated with recent volcanic activity and often make a dramatic spectacle as they throw a column of steam and boiling water high above the surface of the ground. Some geysers discharge continuously, others may erupt violently for a few minutes and then remain inactive for hours or days. Sometimes a geyser, like Old Faithful in the **Yellowstone National Park** in the USA, has a regular and predictable pattern and erupts for a few minutes every hour. It is possible in some areas to harness and control the geyser steam to generate electricity, as with the geysers near San Francisco. However, the disadvantage is that often the power station has to be built on a site near the geyser which is prone to damage either from volcanic activity or earthquakes.

GIANT PANDAS, despite legal protection, special reserves, harsh penalties for poaching and a well-funded conservation programme, continue to decline in number. It has been estimated that there are only about 1,000 in the wild, with another 100 in captivity. They have become an **endangered species** for three reasons. As the human population expands, more and more of the pandas' habitat has been cleared and they have been forced to live in isolated areas, cut off from each other and from other sources of bamboo, their only food. When one variety of bamboo dies off, the pandas die if they cannot find another kind to eat. Secondly, although China has severe penalties for killing pandas and almost every country prohibits the sale of panda skins, poaching still goes on. Some pandas have also been killed accidentally in traps set for other animals. Thirdly, the panda does not breed very well in captivity.

GLACIER ICE now covers about 10% of the Earth's surface. However, at the height of the **Pleistocene** ice age it probably covered about 29%. Geological exploration shows the effect of the ice in shaping land forms, grinding rocks and shifting **sediments** over large distances.

GLACIERS are slow moving masses of ice which have accumulated either on mountains or in polar regions. They are found where warm, moist air or warm water meets cold air or water. They move, influenced by the force of gravity and the pressure of the ice, above the underlying slush layers and slide downhill, eventually melting at lower levels to form rivers or reaching sea-level, where they form **ice shelves** or fall into the water as icebergs.

GLASOD (Global Assessment of Soil Degradation) is a three-year global study sponsored by the **United Nations Environment Programme (UNEP)** and co-ordinated by the International Soil Reference and Information Centre (ISRIC) in the Netherlands. More than 250 soil scientists and 21 regional co-ordinators have been asked to estimate the degree, type and causes of soil degradation caused by human intervention since 1945. Parts of the survey will be combined with information on population, climatology and vegetation loss in world desertification. GLASOD is also involved with a more detailed global soils project, the World Soils and Terrain Digital Database (SOTER), which is also co-ordinated by ISRIC for the International Society of Soil Science (ISSS). This project was due to be completed within 15 to 20 years. UNEP is sponsoring the GLASOD study because it is politically important to have an assessment of good quality now, rather than an assessment of very good quality in 15 or 20 years' time.

GLASSPHALT is glass recovered from the ash of waste incinerators and other furnaces. It provides an alternative to gravel as an aggregate and has been used in road construction materials in the USA.

GLOBAL 2000 was the study commissioned by US President Jimmy Carter in 1977, in order to gather information about what life would be like in the run-up to the year 2000. A report published in 1980, *The Global Report to the President,* concluded that by the beginning of the next century the world would be more crowded, more polluted, less ecologically stable and more vulnerable to destruction than in 1980, and that for hundreds of millions of the desperately poor, the outlook for food and other necessities would be no better, and for many it would be worse. It attracted a lot of support, but President Carter was defeated in the 1980 election by President Reagan, who did not take the project further. Jimmy Carter subsequently started the Global 2000 project, an independent centre where steps are taken to act on implementing some of the report's recommendations.

GLOBAL ATMOSPHERIC RESEARCH PROJECT, see **GARP.**

A **GLOBAL ATMOSPHERE WATCH** station, the world's first inland, is to be established in the People's Republic of China, a country which experiences frequent disasters relating to weather and climate. The country is co-operating with the **World Meteorological Organization,** UNEP and the UNDP in the establishment of the station, as part of the Global Climate Observing System, which was supported at the Second World Climate Conference.

GLOBAL CIRCULATION is the key factor in maintaining the different climatic zones of the Earth: the desert, temperate agricultural areas, forest, prairie or ice-field. Day-to-day climate changes in middle latitudes are often so great that it seems unlikely that there is an overall pattern to the weather. Yet there are large-scale wind circulations over most of the Earth that are persistent and predictable, and they drive the weather machine. They form the general global circulation of the **atmosphere.** Although everyone is familiar with the life-giving sunshine and warmth from the sun, it is not so widely recognized that the Earth and its atmosphere radiate heat back into space. If they did not the planet would become increasingly hot and barren. However, the incoming solar heating of the Earth's surface is largely concentrated in the lower latitudes, while the outgoing radiation from the Earth is more uniform. This imbalance is the underlying source of the major systems of winds and ocean currents, which take excess heat from the tropical regions to the poles (see **atmospheric circulation, energy balance** and the **Hadley Cell**). The pattern of the circulating winds is influenced by the rotation of the Earth, the distribution of the oceans, continents, mountain ranges, deserts, forests, snow and ice. In addition, there are fluctuations in the sun's energy output. All these things interact to make a vastly complicated engine pushing the atmosphere around the globe, as surface winds and high-level winds, or **jet streams.** Weather and climate forecasting depends on portraying these processes mathematically, creating so-called global circulation models, or descriptions of the **global climate system,** for making computer predictions.

The **GLOBAL CLIMATE COALITION** is a group of American commercial and industrial interests, formed after President Clinton's election in the USA, intending to present a business voice on economic and social issues arising from climate change and **global warming.** Its declared goal is to ensure plans for reductions of greenhouse gas emissions in the USA are 'cost effective'. The coalition backed President Clinton's declared policy on global warming, in April 1993, calling 'not for more bureaucracy or regulation or unnecessary costs, but instead for American ingenuity and creativity to produce the best and most energy-efficient technology'. The coalition also made clear its opposition to action beyond the Framework Convention on Climate Change for reductions in emissions, which was signed at the United Nations conference in Rio de Janeiro in 1992 (see **World Climate Convention**).

GLOBAL CLIMATE OBSERVING SYSTEM, see **World Meteorological Organization.**

The **GLOBAL CLIMATE SYSTEM** is the complete web of physical, chemical and biological processes that happen within the Earth-atmosphere system in response to the energy received from the sun.

The **GLOBAL ENERGY BALANCE** is the overall balance between the heat energy received by the Earth and atmosphere and that lost into space. The process starts with the amount of energy absorbed from the sun. The Earth and the atmosphere absorb 68% of **insolation,** energy from the sun, and the rest is reflected back into space as the E-A (Earth-atmosphere) **albedo.** However, there are numerous energy transfers between different layers of the atmosphere and the Earth. See also **energy balance** and **global circulation.**

GLOBAL ENVIRONMENT FUNDS were first officially proposed by Gro Harlem Brundtland, the Norwegian Prime Minister who chaired the commission responsible for the **Brundtland Report.** The Fund would provide money for a global programme to reduce pollution and improve the environment. The United Nations Environment Programme (UNEP) has estimated that it would cost US $400m. per year throughout the 1990s for the developing countries just to be able to carry out the recommendations of the **Montreal Protocol,** which were intended to protect the **ozone layer.** Several plans for global funds to improve the environment have been suggested, but the problem has been how to finance them. Ideas have included a world-wide **carbon tax** or that industrial polluters should pay large sums for using 'free' air, soil and water (see **polluter-must-pay**). Another is a trust fund, which would be funded by levying taxes on the users of oceans, Antarctica and outer space. The United Nations Environment Programme (UNEP) and the World Bank, in a joint statement in 1990, called for the establishment of a US $1,000m. global fund.

The **GLOBAL ENVIRONMENT MONITORING SYSTEM (GEMS)** is a world-wide collective effort to monitor the global environment and assess the health of its constituents. The GEMS Programme Activity Centre (PAC) works with UN bodies, governments and national and international institutions to monitor atmosphere and climate, environmental pollution, renewable resources and environmental data. It was established in 1975 by UNEP. Data on climate gathered by GEMS has encouraged international action to save the ozone layer and to mitigate the effects of climate change. The World Glacier Monitoring Service, backed by UNEP, UNESCO and the Swiss Federal Institute of Technology, incorporates over 750 glacier stations in 21 countries. It has published the *World Glacier Inventory*, the first survey of all the world's glaciers and permanent icefields. GEMS also monitors environmental pollutants, including chemical pollutants in the air, water and even food, which pose a threat to human health.

GLOBAL ENVIRONMENTAL CHANGE refers to environmental changes that affect the total global system, and those that occur at local or regional levels but have consequences for the total Earth system. It includes any alteration of the climate system as a result of the build-up of **greenhouse gases,** depletion of the stratospheric **ozone layer, changing sea-levels,** loss of **biodiversity, desertification,** and depletion of **tropical rainforest.** It is accepted that human activities have caused, added to, or accelerated global environmental change. However, throughout history a process of continual change has been recorded. Current research, being co-ordinated by the Inter-Agency Committee on Global Environmental Change, and involving government agencies responsible for the global environment, is concentrating on the likely consequences of the present more rapid change on our weather, the natural environment, crops, and the way we live. By marrying past environmental change with the current acceleration in global environmental change, scientists are working out how best to predict, for example, the probable levels that **global warming** will reach, and how quickly these changes will come about. Ways can then be worked out of how best to respond to these changes.

The **GLOBAL ENVIRONMENTAL RADIATION MONITORING NETWORK (GERMON)** was set up after the Chernobyl disaster, in order to strengthen national and international monitoring systems and to arrange for information to be exchanged between countries should such a disaster happen again. It was established by UNEP and WHO, in co-operation with other UN bodies and national institutions. It provides information on radiation levels under normal conditions. If GERMON detects abnormally high levels it circulates reports and any other details of a nuclear accident to countries that are affected.

The **GLOBAL OCEAN OBSERVING SYSTEM (GOOS)** is a scheme which was established by the Intergovernmental Oceanographic Commission and the **World Meteorological Organization (WMO).** It is a permanent monitoring network, which feeds data about the oceans into research programmes on **climatic change.**

GLOBAL RESOURCE INFORMATION DATABASE (GRID), which comes within the framework of the **Global Environment Monitoring System (GEMS)**, uses computer technology to process and analyse information in order to produce environmental maps and print-outs for planners. Based on environmental geographic information systems (GIS), these can be used in mainframe computers to study global issues or with microcomputers to study national or local ones. GRID compiles geographically-referenced environmental data compiled by other organizations, and supplies it to other GRID users. It also helps other countries and institutions to receive GIS, with related image-analysis technology, in order to develop a global environmental information exchange network. It operates through seven regional centres, each acting as a kind of environmental switchboard to send data to users in its region. GRID has helped to set up an environmental database for Uganda to assist development planning and to estimate the number of **African elephants.** It has pinpointed suitable sites to develop aquaculture in Costa Rica and evaluated new methods to use satellite data to map and monitor the tropical forests.

GLOBAL STRATEGY FOR HEALTH AND ENVIRONMENT, see **Commission on Health and Environment.**

GLOBAL WARMING describes changes in the surface-air temperature, referred to as the global temperature, brought about by the **greenhouse effect** which is induced by emission of greenhouse gases into the air.

GLOSS is the acronym for Global Sea-Level Observing System, a global network of stations that measure the level of the sea. There are more than 300 tide gauges around the world, set up either on and around **coastal areas,** or in the middle of the **oceans.** Average **sea-level** figures are relayed monthly to the Permanent Service for Mean Sea Level at the Proudman Oceanographic Laboratory (POL) on Merseyside, in the UK. These ground measurements are combined with satellite images of the surface of the sea in order to produce a rapid global monitoring system.

The **GOLDEN LION TAMARIN** is a small monkey that lives in the **tropical rainforest** of Brazil, close to Rio de Janeiro. However, numbers have declined, along with those of many other species, as all but a tiny fraction of the forest has been cleared for agriculture and to allow for the expansion of urbanization north and south of Rio de Janeiro. There are now an estimated 500 golden lion tamarins left in the wild. The majority of these are in a reserve near Rio, but that is threatened by the expansion of cattle-farming on adjoining ranches and by squatters. A **captive breeding** programme has ensured that there are still enough tamarins to make a viable population.

GRASSLANDS cover nearly one-fifth of the Earth's land surface. They include **savannah,** the prairies of North America, and the steppes of Russia and Central Asia. Grassland ecosystems support thousands of different species, above and below the ground, and have a vital part to play in maintaining the ecological balance of the world. Ruminants supported by grasslands—beef and dairy cattle, water buffalo, goats and sheep—supply most of the world's meat, milk and leather. As the world's population and the demand for food grows, the grasslands have come under environmental pressure from **overgrazing,** and because many grasslands are being converted into agricultural croplands.

GRAVITY is governed by the Earth's size, and it is the factor that prevents the **atmosphere** from either drifting off into space or pressing with a crushing force.

The **GREAT BARRIER REEF** spreads for more than 1,250 miles off the north-east coast of Australia. It consists of thousands of individual reefs, shoals and small islands. There are at least 350 species of coral, and living among them are anemones, worms, gastropods, lobsters, crayfish, prawns, crabs and numerous fish and birds. In the 1960s, when there were proposals to drill for oil and minerals in the reef, environmentalists brought the beauty and biological signifance of the Great Barrier Reef to the attention of the Australian people. In 1975 the Great Barrier Reef Marine Park Act was passed, banning mining from the reef. The reef is now managed by a joint Federal-State government authority and is zoned to allow various groups to visit the reef for commercial fishing, shell collecting, **tourism** and recreational and scientific activities. In 1981 it became a **World Heritage Site.**

The **GREAT LAKES,** in North America, contain 18% of the world's liquid fresh water and supply drinking water to 21m. people in the USA and Canada. The lakes, Superior, Michigan, Huron, Erie and Ontario, were in danger of becoming dead lakes—especially Lake Erie—until strict pollution controls were brought in to remove sources of pollution, particularly sewage. There is still concern, however, about the level of toxic chemicals in the lakes and **acid rain** damage to the streams that feed the lakes. In 1985, the eight US states and two Canadian provinces that border the lakes signed the Great Lakes Charter. Signatories agreed to consult each other over any plans to divert water from the lakes. A year later the Toxic Substances Control Agreement was signed. This aims to reduce toxic discharges as much as possible. It was agreed that the Niagara River discharges into Lake Superior were to be cut by half by 1995. The Agreement also bans oil drilling in all the lakes.

GREEN CONSUMERISM is the public awareness of green issues and the wish to replace hazardous products, such as conventional detergents, bleaches, leaded petrol and insecticides with more acceptable or 'friendlier' alternatives. It also relates to people choosing to buy goods that have least impact on the environment in terms of waste disposal, designing goods that avoid the use of scarce resources, tropical hardwoods from the tropical rainforests, and avoiding any product that threatens wildlife or health. The 'green' consumer chooses to buy, where possible: products that have been recycled, or that can be recycled or re-used (see **recycling (domestic)**); cosmetics and toiletries that have not been tested on animals; and energy-efficient household goods, such as long-life light bulbs. Moreover, the 'green' consumer avoids any product that contains **chlorofluorocarbons (CFCs),** which harm the **ozone layer,** and selects foods that are organically grown.

GREEN PLANTS have the unique ability to transform **solar energy** into food by **photosynthesis,** thereby providing the food that sustains all other forms of life.

GREEN POLITICS describes any political activity that is concerned with conservation and promotes the protection of the environment. It is thought that the first 'green' party was the Values Party, launched in New Zealand in 1972. It did not win any seats in national elections, but in 1979 the West German ecology party (Die Grünen) won seats in local elections and went on to win seats in the Federal Parliament. The movement grew through the 1980s, particularly in the European Community (EC). In the UK, the Green Party arose out of the former Ecology Party in 1985, and in the EC election of 1989 the UK Green Party won 15% of the votes cast. Green politicians were successful in pushing serious environmental issues into mainstream politics. However, their representation at parliamentary level began to diminish in the late 1980s because other main political parties, beginning to realize the significance of environmental issues and the number of voters who were concerned about them, began to incorporate environmental protection and awareness into their own manifestos.

The **GREEN REVOLUTION** was the name given to the widespread development of high-yield strains of wheat, corn and rice during the 1960s and early 1970s. It was more formally known as the Indicative World Plan for Agricultural Development. The revolution came after the Food and Agriculture Organization held the World Food Congress in 1963. A 'Freedom from Hunger' campaign was set up with the goal of increasing food supplies and solving the world's hunger problems. The hybrid cereals were short-stemmed, so they were not blown over by the wind, and were quick to mature, making it possible to grow three crops on the same land every year. Norman Borlaug, the American agricultural scientist and plant pathologist, developed the high-yield strains that greatly increased crop yields. He laid the groundwork for the green revolution and was awarded the Nobel Prize for Peace in 1970. In spite of its success in increasing yields, the programme came in for a great deal of criticism for a variety of reasons: in order to get the high yields seeds were used with a high addition of **fertilizers** and **pesticides** and also needed a great deal of water, which often led to large **irrigation** schemes; storage of the high-yield strains was also a problem, and because of their high water content they were susceptible to moulds, such as **aflatoxins;** as a result of changing from traditional agriculture patterns, particularly in the developing world, there was a loss of genetic diversity in many areas, which was followed by pest infestations; the hybrids were also vulnerable to pests and diseases, which led to an increase in the use of pesticides, which has had an effect on those countries which depend on fish as a major source of animal protein.

The **GREEN TURTLE** is a large quadruped sea reptile which is in danger of **extinction.** It has been hunted extensively in the Caribbean Sea for meat and for its shell, which is used to make ornaments and jewellery. The female does not produce eggs every year, but when she does she always returns to her native beach to lay her eggs. Eggs are often removed from the beach because they are thought to have aphrodisiac qualities, and when eggs are allowed to hatch they are prey to shore birds. However, the increased use of the beaches for **tourism** and tourist facilities has now destroyed many of the turtle's nesting sites. This destruction of habitat has led to dwindling populations of the green turtle throughout the 20th century and threatens the turtle's survival. One project to protect the turtle's future involves taking eggs from nests, which are scooped out of the sand by the mother turtle, and moving some of them to another beach where species of green turtles have been seen in the past. It is hoped that the hatched female turtles will subsequently return to lay their eggs at the new 'native' beach.

The **GREENHOUSE EFFECT** is a gradual raising of the temperature of the air, produced by changes in the gases in the **atmosphere.** Many scientists and environmentalists believe this temperature rise threatens animal and plant life. Strictly speaking, the greenhouse effect should be called an enhanced effect because a natural one exists, which already keeps the Earth warmer than it would otherwise be. Without the blanket of gases forming the atmosphere, the temperature of the Earth would be −18°C. The anxiety derives from changes in the gaseous composition of the atmosphere, which has been stable for thousands of years. The gases act like a pane of glass in a greenhouse. Useful light-energy from the sun (short-wave solar radiation) passes through the clear atmosphere relatively unimpeded. Terrestrial heat-energy (long-wave radiation), emitted by the warm surface of the Earth, is partially absorbed by the trace gases and water droplets, which form the greenhouse gases, and some of the heat-energy escapes into space (see **radiation window**). Air pollution from industrial and intensive agricultural activities is changing the atmosphere by increasing the concentration of trace gases—**carbon dioxide, methane, chlorofluorocarbons (CFCs), ozone** and **water vapour.** Predictions about the repercussions of the greenhouse effect have produced contrasting forecasts about its possible impact on the future of all animal and plant life, and on **climatic change, desertification** and rising **sea-levels.**

GREENHOUSE GASES are so-called because of their heat-absorbing properties, which vary enormously. While most attention is focused on **carbon dioxide,** other gases are far more potent even if present in far smaller quantities: **methane** is 25 times more potent, molecule-for-molecule, than carbon dioxide, **chlorofluorocarbons (CFCs)** are 15,000 times more thermodynamically effective than carbon dioxide, and nitrous oxide is 230 times more heat-absorbent. **Water vapour** is the most abundant of all greenhouse gases, but is largely naturally present rather than created by human activity.

GREENPEACE, one of the first of the environmental pressure groups, was founded in 1971, dedicated to campaigning against abuse of the natural world by non-violent direct action protests. All its protests are backed by scientific research. Greenpeace began when a group of Americans and Canadians chartered a ship and went to Amchitka Island, Alaska, where a series of nuclear tests was planned. As a result only one test was carried out. The island has since become a bird sanctuary. Greenpeace have continued in this active style, protesting about the slaughter of **seals,** against whaling (see **whales**) and the dumping of waste at sea. Their methods, which often put campaigners in great personal danger, have made headlines around the world, bringing remote issues directly to the public's attention. Its action against the slaughter of seals contributed to an EC ban on imports of seal pelts and products while the number of seals slaughtered dropped to a sixth of its previous level. In 1987 it raised its World Park Antarctica flag in an attempt to prevent exploitation of the continent. Greenpeace has also carried out successful campaigns to prevent the dumping of radioactive waste and waste chemicals.

GROUNDNUTS, also known as peanuts, are very nutritious and can be eaten raw or roasted, used in cooking, or made into peanut butter. They also produce an oil which can be used for cooking, as well as to make animal feed. They are native to South America. The seeds develop underground to prevent **desiccation.**

GROUNDWATER is the water below the surface and is a valuable part of the **hydrological cycle.** It comes from two sources: as mineral springs rising up from deep in the earth, or from rain. **Artesian wells** are sometimes used to bring groundwater to the surface and seepage also occurs. In many parts of the world a large proportion of groundwater is brought up for domestic, industrial and agricultural use, and in arid areas it is used for **irrigation.** But there has been a sharp increase in the use of groundwater in recent years, and supplies throughout the world are being depleted and polluted at a disturbing rate. Significant drops in the level of the water table are having a serious effect on the environment. A drop in the water table in coastal areas can lead to the incursion of seawater into groundwater. Inland, it can cause subsidence. Another concern is that groundwater is becoming polluted by leaks from **landfill** sites, waste dumps, effluent in rivers, and from nuclear waste sites.

GROWTH REGULATORS are organic substances used in minute amounts to stimulate or retard growth of plants or insects. An insect-growth regulator is a chemical substance that disrupts the action of insect hormones controlling maturity from pupal stage to adulthood and other growth functions. A plant regulator is a chemical that decreases, increases or changes a plant's normal growth or reproduction.

GUANO is a deposit of droppings of fish-eating birds, such as the brown pelicans of Peru. On many islands and coasts of the world, guano deposits have built up to considerable depths. Guano is rich in **nitrogen** and **phosphorus** and in many areas it is mined and used in the manufacture of **fertilizer.** On the Pacific island of Nauru extensive mining of guano deposits has rendered 80% of the island uninhabitable. The guano of **bats** is also prized as a fertilizer.

H

The **HABER PROCESS** is a chemical manufacturing process on which the modern **agrochemical** nitrogen-based fertilizer industry was founded. The Haber process makes **ammonia** by taking **nitrogen** from the air and combining it with hydrogen under pressure. Prior to its invention by the German chemist Fritz Haber, in 1908, there was no industrially useful way of extracting atmospheric nitrogen as a raw material, even though the air around us is composed of 78% nitrogen. The best commercial sources of nitrogen at the beginning of the century were **nitrate** deposits in the desert of northern Chile. Production of ammonia by the Haber process provided a molecule containing nitrogen in a suitable form for making fertilizer. All plant life needs nitrogen, which is necessary to build up **amino acids** and **proteins.** Animal life also requires proteins, but acquires them by eating plants. Although there is plenty of nitrogen in the atmosphere, plants and animals have no means of making use of gaseous nitrogen, or fixing it to form compounds, because it is chemically inert. Plant life obtains it from nitrates in the soil. However, as nitrates are inorganic salts that are soluble in water, they are liable to be washed rapidly from the soil after rainfall in the form of **leachates.** However, there are nitrogen-fixing bacteria that are capable of converting atmospheric nitrogen into ammonia. Once ammonia is formed, it is converted into nitrates by nitrifying bacteria. Without such activity, life on earth would be impossible. However, the natural rate of conversion of atmospheric nitrogen to nitrates by bacteria in the soil was not enough to support the level of crop-growing farmers implemented under the **intensive agriculture** schemes which were started in the 1950s and then encouraged by the **Common Agricultural Policy (CAP).** Industrial fixation of nitrogen has grown at such a pace that there is now concern about whether another natural bacterial process, denitrification or reconversion of surplus nitrates back to nitrogen, can keep up. The consequence has been the over-accumulation of nitrates, which have leached into lakes and rivers causing **eutrophication.**

A **HABITAT** is the locality in which a plant or animal naturally grows or lives. It can be either the geographical area over which it extends, or the particular station in which a specimen is found. It is the term mainly used to indicate the type of locality, such as the sea, the mountains, forest, etc.

The **HABITATS DIRECTIVE** was adopted by the EC in May 1992. It establishes a legal framework for the conservation and preservation of Europe's most threatened species. It commits members of the Community to maintain wildlife and habitats with a favourable conservation status. The habitats directive requires nations to set aside about 10% of their land and sea for conservation. Besides nominating important habitats for protection, governments must also protect sites containing named species, including, at sea, **dolphins, otters** and common and grey **seals.**

The **HADLEY CELL** is a system of **atmospheric circulation** that ultimately distributes air from the tropics to the poles. It is sustained by large-scale convection currents in which hot air is replaced by cooled air. The process was first described by George Hadley, in 1735, to explain the Trade Winds. See also **global circulation.**

HALF-LIFE is the period of time it takes half the number of atoms of a radionuclide to disintegrate. Over each successive half-life, half the atoms remaining decay, so that after two half-lives one quarter of the original amount is left. The half-life of radionuclides varies enormously from minutes to millions of years.

HALOGENATED COMPOUNDS are the vast family of substances containing **halogens.** They embrace **chlorofluorocarbons (CFCs), DDT** and a wide range of other organochlorine **pesticides** and **herbicides,** as well as the PCB (polychlorinated biphenyl) stabilizers which have been added to plastics and lubricating oils for the past 65 years.

HALOGENS are a group of naturally occurring elements which include **fluorine, chlorine** and **bromine.** They cause lasting damage when released into the environment because of their high chemical reactivity with other molecules in the atmosphere or with tissues if they are ingested by animals.

HAY FEVER, or allergic rhinitis, causes irritation of the eyes, ears, nose and palate. It is an allergy triggered, usually in the spring, by airborne pollen from trees, flowers, grasses and weeds. Allergic rhinitis can also be caused by any inhaled allergen, including house dust, feathers, fungus spores and animal dandruff (see also **indoor allergens**). There is now growing medical concern about the role of **air pollution,** which is thought to be aggravating the symptoms of allergic respiratory diseases, such as hay fever.

The **HAZARDOUS WASTE REDUCTION AND COMBUSTION STRATEGY** was a major programme announced by the US **Environmental Protection Agency (EPA)** in the spring of 1993, which was designed to protect public health and the environment by reducing the amount of hazardous waste produced in the country and by strengthening federal controls governing hazardous waste incinerators and industrial furnaces. It included the first-ever hazardous waste reduction guidelines the EPA has ever published. The purpose was to change the national approach to hazardous waste management. A temporary freeze was introduced on the expansion of waste handling capacity in order to examine the country's waste strategy. The strategy's goals were: to overhaul the federal rules governing waste **incineration;** to introduce immediate use of full risk assessments, including those for indirect exposure, in the scientific evidence backing applications for permits; to specify tougher **dioxin** emission standards and more stringent control for metals with new **waste disposal** permits; and to conduct a national review of the relative roles of waste combustion and waste reduction in hazardous waste management. Under the federal hazardous waste law, the Resource Conservation and Recovery Act (RCRA), hazardous waste producers had been required since 1984 to have waste reduction programmes in place, but no specific guidance had been issued to define those programmes. According to the EPA, nearly five million tons of hazardous waste, equivalent to the amount of waste a 2,400 mile-long line of 6,000 gallon trucks would contain, are burned each year in 184 incinerators and 171 industrial furnaces, including 34 cement kilns. About half of the five million tons are burned each year at 15 large commercial incinerators and the 34 cement kilns, which also take commercial waste. The remaining incinerators and industrial furnaces are 'on-site' facilities, permitted for non-commercial use only.

HEALTH FOR ALL 2000 is the target set by the **World Health Organization (WHO).** It was the aim of WHO to raise the standard of health for the world's population by 2000. Since WHO was established in 1948 there has been a marked general improvement in health, and life expectancy has risen. However, in 1990 it was estimated that some 1,000m. people in the world were diseased, in poor health, or malnourished. In developing countries, millions suffer from a range of diseases, from **malaria, tuberculosis** and respiratory infections, to **schistosomiasis, onchocerciasis** and **leishmaniasis.** Since WHO was formed other diseases, too, have significantly increased, such as heart disease from unhealthy diets, smoking-related illnesses and AIDS.

HEAT ISLANDS describe the sprawling urban and industrial areas where air and surface temperatures are higher than in remote places because of the amount of surplus heat energy discharged from buildings and vehicles. Although they are called islands, these areas may cover hundreds of square miles in industrialized countries. These regions also include some of the major meteorological monitoring stations from which data is collected to look for signs of **global warming.** Some experts believe that heat islands distort the overall picture of global warming.

HEAVY METAL POISONING refers to the health hazard represented by a number of **heavy metals** that are similar in the effects they cause at low levels of contamination. The main ones are **antimony, arsenic, lead, mercury** and **thallium,** which all interfere with one of the normal enzyme activities of the body. The group also includes **cadmium,** which damages the joints and kidneys.

HEAVY METALS are a group of metals that have no useful biological function if absorbed into the body. However, when ingested even in small amounts from food, air or water, they may cause irreversible damage to the brain and the central nervous system, or to the kidneys and liver. They include **mercury, lead** and **cadmium.** One of the most important natural safeguards against this form of **heavy metal poisoning** is usually provided by the **blood-brain barrier.** It is a remarkable filtering system evolved by the body to protect its most vital organ, allowing the blood to carry essential nutrients to the brain while blocking harmful molecules. Unfortunately, when heavy metals are transformed into certain complex compounds, like the tetra-ethyl lead that is added to petrol to increase its octane number, the metal can cross the blood-brain barrier. Conversion of metals into the more toxic forms can occur in the environment or in the body. Metal pollution first came to prominence in the **Minimata Bay** incident in Japan in the 1950s. Since the 1930s the Chisso chemical plant had been discharging methyl mercury into the bay. Instead of dispersing to sea as expected, it slowly accumulated in the **food chain,** until the fish on which the local population largely depended for food were contaminated.

HEDGEROWS play an important part in the environment, providing protection against erosion by wind or rain and giving shelter to livestock and crops. They control the drainage of water from fields and provide a valuable habitat for wildlife. There are some insects and fungi that harm the crops, but, in the main, hedgerows house wildlife that play a valuable part in the environment, such as the **bumblebee** which pollinates flowering crops, while birds of prey and insect predators keep farm **pests** under control. However, over a period of 45 years thousands of miles of hedgerows have disappeared in order to clear the fields for more extensive farming or because there was a lack of long-term management. In Britain alone, government research showed that between 1984 and 1990 53,000 miles of hedgerows were lost. There is evidence that this trend is being halted. The Hedgerow Incentive Scheme was launched by the Countryside Commission as part of the government's 'Action for the Countryside' initiative. The scheme provides aid for the improved management of environmentally important hedges.

HEPATITIS, inflammation of the liver, is a major cause of human ill health and death. Hepatitis A is caused by a virus that is usually contracted by eating food contaminated by faeces, especially water, milk and shellfish. However, it is also an **environmental health** problem due to poor sanitation, particularly as a result of overcrowding in human settlements, which frequently leads to epidemics, especially among the young. Adults who have not previously been exposed to the virus are particularly at risk when they travel in, or move to, an area of high infection where there is poor sanitation. There is evidence that the Hepatitis A virus is transmitted not only from man to primates, but also from infected primates to animal handlers. Hepatitis B is transmitted in blood and other body fluids. Hypodermic needles, dental and surgical instruments, tattoo needles and razors have been implicated in the transmission of infection.

HER MAJESTY'S INSPECTORATE OF POLLUTION (HMIP) is the central monitoring unit in Britain which oversees the regulation of wastes and pollution at a national level. The HMIP and the Health and Safety Executive (HSE) are the two national bodies with responsibility to ensure good practice in hazardous waste management. The National Rivers Authority has a similar role in water quality. Since 1989, when the British government introduced new laws for protection of the environment, under the Environmental Protection Act, the HMIP was required to replace a range of piecemeal pollution control schemes with a system of integrated pollution control with centralized responsibility for environmental policing. From 1991 all major emissions to air, land and water were covered by the same legislation and the same regulations. It replaced the **Alkali Inspectorate** which enforced clean air laws, particularly **aluminium** smelting, the petrochemical industry and mineral grit and dust produced by industrial processes. The Alkali Acts of 1863 and 1906 said that, as far as possible, gases should be made harmless and inoffensive before they were discharged into the atmosphere.

HERBICIDES are chemicals used to kill weeds and other undesirable plants. They have played a part in the increase of agricultural productivity, allowing a greatly improved quality and quantity of food. They are also used on industrial sites, roadsides, ditches, irrigation canals, railway embankments and power lines to remove plants that might cause damage or a fire risk. Some, known as contact herbicides, destroy only the plant tissue they touch. Others, translocated herbicides, are absorbed by the leaves and carried within the plant so that it dies. Herbicides are not usually acutely toxic to humans and other mammals, although some, such as **paraquat,** dinoseb and **DNOC,** are poisonous.

HIGH-LEVEL WASTE refers to the **fission products** that contain relatively large quantities of **alpha-emitters** and have a long **half-life,** and are left behind after treating spent nuclear fuel. Because of the intensity of its radioactivity, high-level waste also generates a lot of heat. It is often left in either liquid form in water-cooled tanks or cast into glass ingots, in order to cool for 50–100 years before disposal into some form of underground repository.

The **HIMALAYAS** form the highest mountain range in the world, generally characterized by snowy peaks and **glaciers.** However, the Himalayas also contain rich arable land, extensive grasslands and forest, as well as mineral deposits. The mountains act as a climatic divide and influence air and water circulation. Because of its height the range prevents the cold continental air from the north getting to India in winter; to the south-west, the rain-bearing monsoon winds give up most of their moisture before crossing the mountains. The Himalayas are prone to erosion. A rapidly increasing population in the Himalayas, **deforestation,** and unsustainable farming on the mountains' steep slopes mean that the land cannot absorb water. Part of the major flood problems of **Bangladesh** is a direct effect of water flooding from the Himalayas.

HINKLEY POINT is the site, on the Somerset coast in Britain, of both a **Magnox** and **advanced gas-cooled reactor (AGR)** nuclear power station. In 1989 a public inquiry was held into a planning application for a third station, an American-type **pressurized water reactor (PWR),** but it was not built.

The **HUMAN ENVIRONMENT** refers to any of the conditions in which people live. However, human environment has come to mean all human settlements in villages, towns or major cities, which require environmental management to provide water, public spaces, remove public wastes, and protect water and **air quality.** At the turn of the century 13.6% of the world's population were urban dwellers. By 1960 this figure had risen to 33.6% and by 1985 it had risen to 41.6%. While urbanization has been the most significant aspect of changes in the distribution of the global population, the dry statistics can obscure the tremendous diversity in the patterns of human settlement.

HUMUS is the decomposed organic matter that gives **soil** both its dark brown colour and its nutrient richness. High quality soils rich in humus have often taken centuries to build up, only to be depleted of their goodness in a few years by **intensive agriculture.**

HYDROELECTRIC POWER describes the production of electricity by generators driven by water. The generators can be contained in barrages built across rivers or in the wall of a **dam** constructed specifically to provide hydroelectric power. In most cases, dams are constructed for the combined purpose of regulating river flows for agricultural development and for providing energy.

HYDROGEN seems an ideal source of alternative or **renewable energy** at first glance. Its energy density—the amount of energy released when it is burned—is three times as much as the equivalent weight of petrol. It has no carbon, so the only chemical by-product of combustion is water. It is renewable because it can be extracted from water. Experimental hydrogen-fuelled cars are operating. The drawback with hydrogen is storage. At normal temperatures it is a gas and must be stored under pressure in heavy cylinders. Scientists are experimenting with methods of storing hydrogen in devices which can soak it up like a sponge.

HYDROGRAPHIC AND CHEMICALS TRACER STUDY is the name of a research project that is part of the **World Ocean Circulation Experiment (WOCE).** The research is designed to show exactly how the vast quantity of energy absorbed by the oceans from the sun regulates the world's climate (see **ocean circulation**). The hydrographic part of the study involves cruise ships and satellites to monitor hundreds of different types of instruments, some radio-controlled and some sonar devices, which record the pattern of currents flowing across the oceans and their temperatures. In practice, these ocean currents are vast conveyor belts of energy. The hydrographic experts are employing the latest computer and microchip technology for studies that the pioneers, 100 years ago, did by dropping plumb lines and towing buoys behind sailing ships. However, the oceanographers are also using a new technique of trace analysis. They are measuring the chemical constituents of **sea-water** because they contain telltale markers of the recent source of the water. **Oxygen** is a standard indicator together with the nutrients silicate, **phosphate** and **nitrate.** For example, the cold water recently discharged from polar waters into the Southern Oceans are high in oxygen and nutrients, particularly silicates. In addition, the man-made **chlorofluorocarbon (CFC)** pollutant gases also provide chemical tracers. Because it is known when they must have entered the ocean from the atmosphere, scientists can calculate the flow of some ocean currents which carried them to the places where they were detected. These measurements are particularly important in calculating the time it takes, in the mixing process, for surface water to get carried to the ocean bed. Oceanographers can 'age' a particular parcel of water and work out its recent history.

HYDROLOGICAL CYCLE is the way that water transfers from its gaseous, liquid and solid states from the sea to land and back again. Water evaporates from the **oceans** into the **atmosphere** and is carried by the wind to the land where it condenses and falls as rain, snow, sleet, hail or dew. Some evaporates as it falls and returns to the oceans by winds moving in the opposite direction. Water that reaches the ground goes back to the sea when it evaporates from the land, or as **water vapour** from plants and vegetation, or as the run-off from rivers, or **groundwater** movements.

HYDROLOGY is the study of the occurrence, properties, movement and use of water and ice on or under the Earth's surface. It follows the **hydrological cycle** from precipitation, as rainfall and snow, to the discharge of water by rivers into the sea, or its return by evaporation back into the **atmosphere.**

HYDROPONICS is the method of cultivating plants by growing them without soil, using a **nutrient** solution. The plants' roots are held in gravel or sand. Crops can be grown hydroponically in arid regions, where soil is infertile or toxic. Large-scale cultivation is a very expensive operation, so its use is restricted. It is a method, though, that can be used on a small scale, at low cost, in areas of the world where there are frequent water shortages, including cities. Hydroponics is being used in research experiments by scientists from the National Aeronautics and Space Administration (NASA) in Florida to prepare for a future colonization of Mars, when space crews will have to grow their own food and recycle air and water in order to survive. Crops, such as potatoes, wheat, soya bean and strawberries, are grown in an airtight chamber. The experiment will help scientists to understand how plants would survive in a sealed environment on Mars. The Biomass Production Chamber is designed to provide food for future pioneering astronauts and also to monitor the changes in temperature, humidity, light, carbon dioxide and oxygen. Scientists are also looking at how plants cope with increased carbon dioxide, which contributes to the **greenhouse effect.** Another experiment, Biosphere-2, in the Arizona desert, has been studying miniature **tropical rainforest, savannah** and **coral reefs** in a large airtight greenhouse.

HYDROSPHERE refers to all the water on the planet, whether it is in the oceans, lakes, rivers or frozen in glaciers and the ice sheets. It includes the **water vapour** in the atmosphere. The movement of moisture between the hydrosphere, the **atmosphere,** and the **lithosphere** is the **hydrological cycle.**

I

ICE SHEETS form over the polar land masses. The two great ice sheets, Greenland and the **Antarctic,** between them hold more than 90% of the Earth's **freshwater.** The thickness of the ice sheets continually changes as ice is redistributed within the sheet. Falling snow forms ice on the central plateaux of the ice sheets. This is then balanced by loss at the coastal region in the formation of **ice shelves** and the calving of icebergs. This can take many thousands of years. Monitoring with radar altimeter data is crucially important in providing very precise and repeated measures of the topography of the ice sheets. This data will provide answers to a fundamental question which is still unanswered after decades of fieldwork: are the ice sheets shrinking or growing? If they are shrinking this could be one of the most sensitive indicators of global **climatic change.** More than two million measurements made with the Geosat altimeter over the Antarctic ice sheet show surface flow features and reflect the underlying topography of the Antarctic landmass. Data from **Earth Resources Technology Satellites,** which have a wider range, will provide a greatly improved coverage of Antarctica.

ICE SHELVES are formed over the open seas, as opposed to **ice sheets** which form over polar land masses. When ice sheets and **glaciers** reach the edge of the land, ice shelves extending hundreds of kilometres into the ocean may be formed. These are subject to dramatic changes. For example, the Wordie ice shelf, off the western side of the Antarctic peninsula, has, over 10–15 years, rapidly disintegrated because of a rise in mean air temperature. The increased amounts of melt-water, combined with underlying fracture processes, led to the shelf breaking up. Even more catastrophically, in 1986, at the edge of the much larger Larsen ice shelf which is at the most northerly edge of the Antarctic peninsula, a large promontory, weighing some 1,000,000m. metric tons, became detached as an enormous iceberg. The iceberg, now slowly melting, equals the total annual loss of ice from Antarctica during a normal summer.

INCINERATION of hazardous wastes in an incinerator provides, in theory, a better option of disposal than burial in the ground or dumping at sea. In practice, however, many substances undergo transformations into more harmful agents, such as **dioxin,** and are discharged into the air if the furnace is not controlled properly. Incinerators are usually operated by local authorities and private waste contractors.

INDOOR ALLERGENS are those allergens which play a key role in triggering some cases of asthma. Sources of indoor allergens include house dust mites, **fungi,** domestic pets and **cockroaches.** The dust mite, a microscopic organism that lives primarily in carpeting, pillows, mattresses and upholstery, is responsible for significant residential allergen exposure. The allergens are contained in fecal balls that accumulate in bedding and other surfaces. Cat dandruff, another common allergen, becomes airborne in very small particles and can linger in buildings long after the cat has gone. Environmental control is important in reducing disease severity, which may mean removing carpeting and replacing it with polished floors, reducing clutter, and lowering the relative humidity on which allergens, such as fungi, thrive. Covering mattresses and pillows with impermeable materials and washing bed linens weekly in hot water will help to reduce allergic illnesses, and is particularly useful in limiting exposure during childhood. Concern about indoor allergens as a public health threat coincides not only with the trend towards spending more time indoors, but also with the move to construct more energy-efficient offices and residences. Buildings are increasingly sealed tightly and weather-stripped, and with central heating and air-conditioning, there may be a reduction in the exchange of air from the outside. That, plus a relatively high level of interior humidity, particularly in homes with excess moisture in carpets and carpet padding, combines to encourage the growth of airborne allergens.

INDUSTRIAL COMPETITIVENESS AND PROTECTION OF THE ENVIRONMENT was a proposition in a communiqué made in November 1992 by the European Commission to the Council of Ministers of the European Community and the European Parliament. It asserted that 'what is good business can also be good for the environment'. The object of the message was to raise awareness that environmental protection had become a pre-condition for industrial expansion, and had to be incorporated in measures for stimulating industrial development. In asserting that industrial development and environmental protection were not necessarily incompatible, industries that made environmental concerns an integral part of their strategy could derive significant competitive advantages from developing lean technologies which cut energy costs or space requirements. Beyond the production process, there was greater consumer awareness of ecologically sound products like unleaded petrol and **phosphate-free detergents.** By 1990 the EC environmental protection market had expanded to over Ecu 50,000m. and was expected to grow by between 50% and 100% before the turn of the century.

INFOTERRA was formed as a result of the **United Nations Conference on the Human Environment,** held in Stockholm in 1972, and operates as an exchange of environmental information, tapping a vast reservoir of information and experience that has been accumulated over the years. Governments, industry and researchers across the world have access to the information. It is now the largest environmental information system in the world, linking national and international institutions and experts across 139 countries. It works through a network of institutions, which are designated by governments as a focal point for the scheme. Each one prepares a listing of environmental expertise within its borders and chooses the best sources for inclusion in INFOTERRA's International Directory, which has more than 6,500 institutions listed. INFOTERRA deals with 21,500 queries on the environment each year, and it has proved a very effective way of supplying environmental information to users world-wide.

INFRARED RADIATION is sometimes referred to as the long-wave heat radiation that is emitted by any hot object, including people. It is also fully 60% of the sun's radiation. Infrared is also a significant factor in the process of **global warming.** In practice, infrared refers to the band of radiation energy with wavelengths ranging from 3.0 micrometres to 100 micrometres. Like any other hot object, when the Earth's surface has been warmed by solar radiation during the day it emits long-wave infrared radiation. Most of it is absorbed by the various gases in the atmosphere. However, there is a band of infrared of 8–11 micrometres that passes through a cloudless sky and escapes into space. This is called the **radiation window.** However, this window is partially closed by **clouds** and atmospheric pollutants, which means the atmosphere absorbs the heat rather than letting it escape back into space.

INSECTICIDES are a class of **pesticides** used to destroy insects. A variety of substances have been used over the years: hot water, soapy water, whitewash, vinegar, turpentine, brine, lye and fish oil. Before the Second World War arsenicals, nicotine, **sulphur,** cryolite, and hydrogen cyanide gas were used as insecticides, but then new chemicals, the synthetic organic insecticides, were discovered, the first of which was **DDT.**

INSOLATION refers to the energy from the sun that is received and trapped by the **atmosphere** and Earth, and which comes mostly in the form of radiation ranging from **ultraviolet radiation** wavelengths, through visible light to the **infrared radiation.**

INTEGRATED PEST MANAGEMENT (IPM) describes a method that combines the judicious use of **pesticides** with **biological control** to control pests. It is a strategy being worked out by the United Nations Environment Programme (UNEP), the Food and Agricultural Organization (FAO), the United Nations Development Programme (UNDP) and others, using the pests' natural predators, pest-resistant crop varieties, crop rotation and traditional means of control. Studies are in progress in regional programmes in some 40 countries and include cotton production in North and East Africa, Latin America and the Near East, rice fields in South East and South Asia, and small farmers' crops in West Africa.

INTENSIVE AGRICULTURE is a farming technique which uses high levels of capital, labour and **agrochemicals** to achieve high outputs and yields from a unit of land.

The **INTERGOVERNMENTAL PANEL ON CLIMATE CHANGE (IPCC)** was sponsored by two UN agencies: the **World Meteorological Organization** and the **United Nations Environment Programme (UNEP).** It was set up to assess the potential threat of global warming as a result of pollution of the air by **greenhouse gases** and other human activities that might alter the climate. The IPCC first met in November 1988. Three expert groups were set up (see **IPCC First Assessment Report**): one to provide a scientific assessment of the current knowledge of **climatic change;** another to assess the potential impact of climatic change on agriculture and changes in **sea-level;** and a third to consider response strategies. All three groups reported to the meeting of the IPCC in Sweden in August 1990. The results of their work were presented to the second World Climate Conference at the end of October 1990.

INTERMEDIATE CHEMICALS, often referred to simply as intermediates, are substances that are created temporarily during the synthesis of complicated chemical compounds. If the chemical reaction goes wrong, or is interrupted, the intermediate may be released as a pollutant. Public attention was first drawn to the hazards of intermediates, and the existence of **dioxin** as an intermediate, by the **Seveso** chemical works explosion in Italy. It was not the first industrial accident involving dioxin. However, the earlier ones had been restricted to contamination of workers, and so were listed as occupational accidents and not environmental pollution accidents. Seveso was the first major environmental disaster attributed to an intermediate organic chemical. Other environmental pollution accidents have happened with other intermediates at chemical works. Manufacturing of products like pesticides has gone wrong, and the resulting **effluent** containing toxic intermediates has poured into rivers. The problem is not limited to the manufacture of new stocks of chemicals. Destruction of long-established substances, such as polychlorinated biphenyls (PCBs), by **incineration,** must be carried out meticulously, to make sure intermediates, including dioxin, are properly neutralized. However, the PCB family of substances, which have been used extensively in industry for 50 years, are very stable and highly soluble in fat once in the food chain. Since the PCBs are so stable, rather than being metabolized they accumulate in fat. This resistance to **metabolism** which allows a pollutant to accumulate, and includes many other organochlorine agents such as **DDT,** is referred to as **biological magnification.**

INTERMEDIATE-LEVEL WASTE is the radioactive scrap produced by the nuclear industry in the form of contaminated metal, discarded components and pipework from nuclear reactors and the cladding stripped from spent **nuclear fuel** elements.

The **INTERNATIONAL ATOMIC ENERGY AGENCY (IAEA)** is an autonomous inter-governmental agency (although it is administratively part of the United Nations, and reports to the UN General Assembly) set up to develop the civil and peaceful uses of atomic energy. However, it is also the repository of the **Non-Proliferation Treaty.** The agency was established in 1957 and its work includes research on the applications of nuclear energy to medicine, agriculture, water location and industry. It runs conferences and training programmes, sends out publications to promote the exchange of technical information and skills, and gives technical assistance, in particular to developing countries. It establishes and administers radiation safeguards. Its headquarters are in Vienna.

The **INTERNATIONAL BANK FOR RECONSTRUCTION AND DEVELOPMENT (IBRD)** is known as the World Bank, but it is, in fact, one arm of the World Bank group. It has over 170 member states and operates as an independent organization, but with the status of a United Nations body. Since 1948 most of its loans have been to projects that secure economic growth in developing countries, foster educational and social schemes, and resettlement programmes. It advises on development programmes in developing countries, sponsors consultative groups and meetings, and publishes annual statistics.

The **INTERNATIONAL CODE OF CONDUCT ON THE DISTRIBUTION AND USE OF PESTICIDES** aims to establish voluntary standards of conduct for governments and all those involved in the distribution and use of pesticides. The Code was published in 1986 by the Food and Agriculture Organization (FAO) and developed with the United Nations Environment Programme (UNEP), the World Health Organization (WHO) and other agencies. It is aimed especially for use in countries where national controls are weak or do not exist. UNEP backs up the Code with training programmes.

INTERNATIONAL CONVENTION FOR THE PREVENTION OF POLLUTION FROM SHIPS (MARPOL), see **International Maritime Organization (IMO).**

The **INTERNATIONAL COUNCIL OF SCIENTIFIC UNIONS (ICSU)** is a non-governmental group of scientific organizations representing 80 National Academies of Science, 20 International Unions and 26 other bodies called Scientific Associates. The object of ICSU is to achieve the free exchange of scientific information and co-operation between the world's scientific communities. To accommodate the spread of activities in which many of the member unions are involved, ICSU has established 13 scientific committees with experts from a multiplicity of scientific disciplines. One of them is the **Scientific Committee on Problems of the Environment (SCOPE),** which directs attention to developing countries. SCOPE was formed in 1969 in response to environmental concerns emerging at the time. Its first task was to prepare a report on **Global Environment Monitoring** for the **United Nations Conference on the Human Environment** in Stockholm. The mandate of SCOPE is: to assemble, review and assess the information available on human-induced environmental changes and the effects of these changes on humans; to assess and evaluate the methods of measuring environmental parameters; and to provide an intelligence service on current research and a body of informed advice.

The **INTERNATIONAL DRINKING WATER SUPPLY AND SANITATION DECADE** began in 1980. It was launched by the United Nations with the aim of achieving 'clean water and adequate sanitation for all by 1990'. Hundreds of millions of people did receive safe water and proper sanitation for the first time, in many parts of the developing world. However, it was always recognized to be an ambitious programme and by the end of the decade there were almost as many people without access to safe water and sanitation as there were when it began. **Population** increases, **droughts** and economic crises all contributed to its lack of overall success, together with the fact that the millions of dollars that the programme required every day to make it work were not forthcoming from the industrialized and developed countries. Some initiatives did emerge from the decade: national action plans were drawn up even though many had to be shelved because of lack of funds. As technologies become cheaper there is hope that some schemes could be completed at a fraction of the previous costs for items like hand pumps and borehole equipment. The UN's International Conference on Water and the Environment, held in Dublin in 1992, reported that the cost of keeping the original decade target would be US $36,000m. per year.

The **INTERNATIONAL FUND FOR PLANT GENETIC RESOURCES** was set up in 1989 under the auspices of the **Food and Agriculture Organization (FAO).** It is administered by the FAO and was launched to conserve and promote the use of plant genetic resources at a global level. The Fund was intended to assist developing countries in the identification and survey of plant resources, as well as to assist in plant breeding and seed multiplication and distribution. It provides a way for countries, inter-governmental and **non-governmental organizations (NGOs),** and private industries and individuals to maintain the genetic diversity of the world's plants.

INTERNATIONAL GEOPHYSICAL YEAR (IGY) was the international geophysical research programme in which more than 70 countries took part. IGY began in July 1957 and ended in December 1958, and during that time several important projects were carried out simultaneously. These projects included the exploration of the variability of solar radiation and **sunspots,** and gathering information on **cosmic radiation,** the nature of the upper atmosphere, and ice in **glaciers** and **ice sheets.** The programme's data was interpolated with weather fluctuations and used orbiting satellites for the first time. Ocean currents were mapped, including the subsurface currents which have profound effects on all life on Earth (see **ocean circulation**). A variety of charting devices were used during IGY to study undersea landslides and currents, volcanic eruptions and their output. Much of the work carried out during IGY is still the subject for on-going research.

The **INTERNATIONAL HYDROLOGICAL DECADE** was a research programme on water problems. It began on 1 January 1965, following a resolution adopted at the general conference of the United Nations Educational, Scientific and Cultural Organization (UNESCO). The scientific programme covered all aspects of water: its characteristics; its movement; and the use of water on and below the Earth and in the atmosphere (see **hydrology**). The information acquired over the period of the Decade was important in preparing development projects of widely different kinds. National committees were set up in 96 countries, bringing together, in many cases for the first time, representatives of different national organizations concerned with water problems.

The **INTERNATIONAL INSTITUTE FOR ENVIRONMENT AND DEVELOPMENT (IIED)** is an international, independent policy research institute. It works mostly in the developing world, undertaking research and analysis on environmental and development topics, including environmental planning, human settlements, forestry, and the use and production of energy. It is a charitable, non-profit making organization based in London and in Buenos Aires. It is funded by international organizations, governments, private and corporate foundations, and from public donations. IIED also co-operates with other environmental development agencies, including the **United Nations Environment Programme (UNEP),** the **International Union for the Conservation of Nature and Natural Resources,** and the **World Resources Institute.**

The **INTERNATIONAL MARITIME ORGANIZATION (IMO)** is one of the specialized agencies of the UN and was the first international body devoted exclusively to maritime matters. Its prime objectives are safety at sea and prevention of marine pollution. The IMO came into force in 1958 and has since promoted 30 conventions and protocols, and adopted more than 700 codes and recommendations on safety at sea. In the 1970s the IMO became the agency responsible for negotiating international agreements controlling pollution from ships. Its members have adopted the international convention on Oil Pollution Preparedness, Response and Co-operation (OPRC), designed to improve the ability of nations to cope with emergencies, such as tanker accidents. It does not yet have mandatory status, but many of its provisions were used as the basis for the IMO's response to the massive pollution of the Persian Gulf in 1991. The organization was able to provide assistance that saved many ecologically important and sensitive areas from major damage. The first important IMO agreement on controlling pollution from ships was the International Convention for the Prevention of Pollution from Ships (MARPOL), which deals not only with pollution by oil, but also chemicals, garbage and sewage. More recently the IMO has been the forum for international negotiations on dumping at sea, which has been proposed as a means of routine disposal of industrial wastes. The negotiations led to the convention on 'Prevention of Marine Pollution by Dumping of Wastes and Other Matter', better known as the **London Convention,** which halted dumping of radioactive waste at sea.

The **INTERNATIONAL MONETARY FUND (IMF)** is a specialized agency of the United Nations. It was established in 1947 to promote international monetary stability, the growth of world trade and the stability of foreign exchange. It also gives financial assistance to the 172 member countries when they have balance of payments problems and smooths multilateral payments and financial arrangements among member states. The IMF works in close collaboration with the **World Bank.**

The **INTERNATIONAL PROGRAMME ON CHEMICAL SAFETY (IPCS)** was formed in 1980 to assess the risks that specific chemicals pose to human health and the environment. It was set up by the World Health Organization (WHO), the United Nations Environment Programme (UNEP) and the International Labour Organization. It liaises with scientists to ensure that international research methods are comparable and reliable. The organizations work with the **International Register of Potentially Toxic Chemicals (IRPTC),** which provides the basis for chemical safety decisions. IPCS, which is based at WHO's headquarters in Geneva, processes the information and publishes evaluations of health and environmental risks, and advice on control and exposure levels. IPCS produces detailed environmental information for scientists and non-technical health and safety guides for administrators, managers and decision-makers. It also provides international chemical safety cards for easy reference in the workplace and poisons information booklets for medical use. IPCS also works with the IRPTC on a training programme on chemical control for developing countries. A computerized poisons information package (INTOX) is also available, in three languages, to countries without the necessary facilities for coping with poisonings.

The **INTERNATIONAL REGISTER OF POTENTIALLY TOXIC CHEMICALS (IRPTC)** operates through a network of national and international organizations, industries and external contractors. Its computerized central data files have profiles of more than 800 chemicals, as well as special files on **waste disposal** and management, on chemicals currently being tested for toxic effects and a file of national regulations covering some 9,000 substances. The files can be accessed via computers through telephone lines, and the IRPTC is working to make them available on personal computers. The IRPTC plays an important part in the drive to ensure that substances which are banned or severely restricted in developed countries are not dumped in developing countries. Since 1950 world production of organic chemicals has increased from 7,000m. metric tons to around 250,000m. metric tons. With some 70,000 chemicals in common use, the greatest threat is ignorance about their use or dangers. By 1992, a total of 149 national authorities had been nominated to act as channels for a procedure known as a prior informed consent (PIC). Initially, PIC will be applied to all chemicals banned or severely restricted in 10 or more countries. Eventually it will be applied to all chemicals banned or severely restricted in any country.

The **INTERNATIONAL TROPICAL TIMBER AGREEMENT (ITTA)** came into force in 1985 and is managed by the International Tropical Timber Organization. The agreement's main role is in marketing and promotion, but the agreement also includes the goal that producer countries should be helped to develop better techniques for reforestation and forest management. It also works to encourage more timber processing in producing countries, so adding to the value of the products, and supports research and development programmes. Environmental organizations, among them the United Nations Environment Programme, the World Wide Fund for Nature, and the **International Union for the Conservation of Nature and Natural Resources,** co-operated with ITTA to build ecological considerations into the agreement. Some countries have decided only to import wood from sustainable sources by 1995. Others have already banned the import of unsustainably harvested timber. There has been some criticism that ITTA achievements are uneven and restricted to tropical forests, but it is felt the agreement could form the model for a revised agreement to take in all types of **forests and woodland.**

The **INTERNATIONAL UNDERTAKING ON PLANT GENETIC RESOURCES** was adopted by the member states of the **Food and Agricultural Organization (FAO)** in 1984. The agreement proposed that all germ plasm, including seeds improved by **biotech,** should be freely available to all. However, there is some dispute about this because of considerable world-wide disagreement about ownership, and therefore the rights to commercial use, of the seeds contained in **gene banks.** The dispute hinges on the fact that more than 90% of the genetic material held in gene banks around the world, almost all of it in developed countries, came originally from farmers in the developing world. In 1986, Brazil, Ethiopia, India and Taiwan banned the export of germ plasm from specific crops. The USA also refused to allow genetic material held in a gene bank to be sent to Nicaragua. There was increasing pressure to allow farmers in the developing world who have developed strains after years of cultivation to share in the rewards. The Undertaking was reaffirmed in 1989 and a resolution endorsed the concept of farmers' rights. The Undertaking was drawn up to ensure that plant genetic resources are preserved and made as widely available as possible for plant breeding for the benefit of present and future objectives, and to conserve plant genetic resources of economic and/or social interest, particularly for agriculture. Another objective focuses on cultivated plants, plants or varieties which have been in cultivation in the past, primitive versions of cultivated plants, wild relatives of such plants and certain special genetic stocks. The agreement is not legally binding on the signatories. There is also no mechanism to settle disputes. In 1989, at the FAO conference, the International Fund for Plant Genetic Resources was set up.

The **INTERNATIONAL UNION FOR THE CONSERVATION OF NATURE AND NATURAL RESOURCES (IUCN),** also known by the title World Conservation Union, is an independent, global organization which promotes the world-wide conservation of wildlife, habitats and natural resources. It was founded in 1948 to alleviate the worst effects of human influence on the environment from industrial and urban development. It is also concerned about excessive exploitation of the Earth's natural resources. The IUCN's members include government agencies and conservation organizations representing more than 100 countries. It has close links with several UN agencies. The organization has six commissions, five of which cover **ecology,** education, legislation, landscape planning and **national parks.** The sixth commission is the Survival Service Commission, which collects information on all plant and animal species in danger of **extinction** and takes action to prevent this happening. This group produces a series of **Red Data Books,** giving information about the threat to rare plant and animal species, mammals, birds, amphibians, reptiles and higher plants both in individual countries and on continents.

The **INTERNATIONAL WHALING COMMISSION (IWC)** was set up in 1946 after the signing of the International Convention for the Regulation of Whaling. This was the first effort to control whaling. However, because the commission has no regulatory powers to enforce its decisions, **whales** continued to be hunted, some to near-**extinction.** By the end of the 1960s scientists and large numbers of the general public concerned about conservation of wildlife were becoming alarmed at the depletion of whale stocks. A series of campaigns to save the whale were conducted against the IWC, co-ordinated principally by **Greenpeace.**

The **INTERTIDAL MARINE ENVIRONMENT,** sometimes known as the littoral zone, describes the area between land and sea which is regularly exposed to the air by the tidal movement of the sea. Sand beaches, rocky bottoms, protected bays, **salt marshes, mangrove forests,** and **coral reefs** are all part of the intertidal marine environment. Marine organisms that inhabit the intertidal marine environment have to adapt to periods of exposure to air and to the waves created by wind, which makes it the most physically demanding of the marine habitats. In this semi-terrestrial region, plants and animals have to adjust their breathing to survive for most of the time spent out of the water. Different species survive at various levels within the intertidal region, depending on their adaptability and whether the area is sandy, muddy or rocky. Species found within these levels include limpets, molluscs and **barnacles,** winkles, snails, and burrowing molluscs, like clams, and also sea urchins and **crustaceans,** as well as predators, such as starfish, dog whelks and crabs.

INVERSION, see **temperature inversion.**

IODINE was the first essential **trace element** to be discovered as such. It is indispensable to life because it is essential for the synthesis of thyroid hormones which regulate the metabolic rate in all cells, it controls growth and development, and helps burn excess fat. It is present in sea-water and was previously extracted from seaweed. It is still generally available in kelp tablets from chemists. In addition to the natural stable form, there are 14 radioactive isotopes of iodine. Radioactive iodine-131, a **fission product** released in the fall-out from weapons testing, and in nuclear accidents such as the **Windscale fire** and **Chernobyl** explosion, is indistinguishable to our bodies from stable iodine. Because it is a very volatile element, iodine is one of the radioactive **isotopes** most likely to be released in large quantities in an accident. If it contaminates pasture land, it is absorbed rapidly into the **food chain** and the milk supply. As was discovered after the Chernobyl disaster, cattle, sheep and goats grazing over large areas of ground will concentrate radioactive fall-out in their bodies. So, an element that is normally essential to our health becomes concentrated in the thyroid gland. In children iodine contamination can cause serious developmental problems.

IONIZATION is the process of producing ions when an atom or molecule temporarily loses some of its electrons. There are several ways in which ions may be formed from atoms. Ionization occurs in certain chemical reactions and when some substances are in a dissolved state in solution. Ions are produced by **ionizing radiation** if the rays carry enough energy to break up a molecule or detach electrons from an atom.

IONIZING RADIATION refers particularly to **radiation** that causes electrons to be stripped from an atom, and so converts an atom from its neutral state into an electrically charged state. The process temporarily alters the chemistry of the atom. If a lot of ionization happens to atoms that form organic molecules in the cells of the body, the change can lead to disruption of the normal biochemistry and the death of a cell or its transformation into a cancerous one. Ionizing radiation includes X-rays, **gamma-rays,** and **alpha** and **beta radiation.**

The **IPCC FIRST ASSESSMENT REPORT: WG1 (SCIENCE)** gave, in 1990, the conclusions of an international group of experts set up by the **Intergovernmental Panel on Climate Change (IPCC)** to assess all available scientific evidence on factors affecting **climatic change.** These included analysis of **greenhouse gases** and assessment of the capabilities of **climate models** in monitoring global and regional climate change and their predictability. The group also examined past climate records and contemporary climatic anomalies, projections of future climate and **sea-levels,** and the timing of the changes. They concluded that emissions were substantially increasing the atmospheric concentration of the greenhouse gases: carbon dioxide, methane, nitrous oxide and chlorofluorocarbons (CFCs). These increases must have caused some additional warming of the Earth's surface. They calculated that by continuing with the current level of emissions, there will be a doubling of equivalent carbon dioxide concentrations by 2025, with a concurrent temperature rise of 0.3°C. Without curbs on emissions **global warming** would lead to average temperatures being 3°C higher than the rate for the early 1990s by the end of the 21st century.

The **IPCC FIRST ASSESSMENT REPORT: WG2 (IMPACTS)** gave, in 1990, the conclusions of an international group of experts set up by the **Intergovernmental Panel on Climate Change (IPCC)** to review the environmental and socio-economic impacts of **climatic change** in an integrated manner. The experts also examined the impacts on a regional and national scale of **global warming** and **sea-level** rise on agriculture, forestry, health, energy and water resources, with special regard to **floods** and **desertification.** The group observed that large-scale natural events, such as **El Niño,** caused significant impacts on agriculture and human settlement, and they emphasized that the predicted population explosion would impose severe impacts on land use and on demands for energy, freshwater, food and housing. The impacts will fall most heavily on regions already under stress. Climate change, pollution and **ultraviolet radiation** caused by ozone depletion can interact, reinforcing their damaging effects on plants and organisms.

The **IPCC FIRST ASSESSMENT REPORT: WG3 (RESPONSE STRATEGIES)** gave, in 1990, the conclusions of an international group of experts set up by the **Intergovernmental Panel on Climate Change (IPCC)** to consider the policy dimension of **climatic change.** The group emphasized that the information which existed was inadequate for policy-making and was riddled with uncertainties. A major dilemma was that actions needed to counter climate change caused by emission of **greenhouse gases** might be required well before many of the specific issues under scrutiny could be analysed properly by further research. The energy sector accounted for an estimated 46% of the global warming effect due to human activities. **Fossil fuel** burning caused 70%–90% of the total emissions of carbon dioxide by human activities. The remaining 10%–30% was caused by disruption of ecosystems by activities like **deforestation.**

IRRIGATION is one of the most effective ways of increasing agricultural productivity known, and irrigated land produces more than 30% of the world's food. Irrigation has been practised for centuries, especially in parts of Africa, Asia and South America. In arid areas, traditional methods of irrigation draw on centuries of experience of the best ways to ensure the proper distribution of water, the clearing of drainage ditches, and observing fallow periods. Some modern irrigation schemes, encouraged because they were seen as a way to ease world hunger, are built on a vast scale and many are now under criticism. Apart from the huge expense, they mostly disregard fallow seasons and are organized to grow three crops on the same ground every year. Consequently, the mismanagement of some schemes has led to waterlogged land, leaching out of soil nutrients, and **salinization,** which sterilizes the land. There is also concern that a large number of crops are grown as **cash crops** for export to earn the money to pay for the installation of irrigation schemes. The worst fear is that irrigation projects create new and permanent places for the carriers of diseases such as **malaria, filariasis,** and **schistosomiasis.** Agricultural pests, insects and rats thrive in newly irrigated areas because of the whole-year-round food supplies.

ISLANDS are unique but fragile and vulnerable ecosystems because the evolution of their flora and fauna has taken place in relative isolation. Many remote islands have some of the most unique flora in the world; some have species of plants and animals that are not found anywhere else, which have evolved in a specialized way, sheltered from the fierce competition that species face on mainland. Endemic island animals have little defence when predators, such as cats, dogs or pigs, are introduced into an island, because they have developed without any natural predators. It is thought that the dodo became extinct because during its long evolution on the island of Mauritius it lost its ability to fly, and its sense of fear, because these things had become unnecessary during its evolution. **Vanishing habitats,** lost because of growing populations, island development, agriculture, **deforestation,** mining and quarrying, the introduction of non-native species, and **tourism** are a big threat to many islands. Most of the recorded **extinctions** of species have occurred on islands: 75% of all mammals driven to extinction have been island dwellers, as have 97 out of 108 extinct bird species. The tropical rainforests of Madagascar, in the Indian Ocean, have an estimated 12,000 different plant species and some 70% of them are unique to the island, including the Madagascar rosy periwinkle, which is a valuable component of a drug used to fight leukaemia and Hodgkin's disease. Madagascar's remaining rainforest is under constant pressure to provide land for its growing population and to accommodate tourists, thus threatening the survival of these species.

ISOTOPE refers to a species of a chemical element with a different atomic mass, but the same **atomic number,** or characteristic, which determines which element it is. The chemical nature of a particular element is a property of the number of protons in the nucleus. However, the other particles which make up the nucleus of an atom, the neutrons, may vary and so change the atomic mass. Atoms of elements with differing neutron numbers are known as isotopes. Uranium-238, for example, has 92 protons and 146 neutrons, whereas **uranium-235** has 92 protons but only 143 neutrons. Some isotopes are stable: carbon-12, for instance. Others, such as carbon-14, are unstable.

ISOTOPIC ANALYSIS is a primary method of dating the origin of samples of material that provide evidence of the state of the Earth's climate hundreds or thousands of years ago. The samples may be fossils of primitive marine organisms retrieved from sediments on the sea bed or bubbles of atmospheric gas trapped for thousands of years in a glacier. The technique depends on the fact that there are two forms, or isotopes, of **oxygen.** One is stable and the other slowly decays. A measurement of the proportion of the two forms of oxygen from a specimen can be used to calculate when it became buried or trapped in an ice sheet.

J

The **JET STREAM** is a strong and enduring belt of westerly winds that travels at 12,000 metres in the **troposphere.** Their typical speed varies from 110 kilometres per hour in summer to 180 kilometres per hour in winter.

The **JOINT GLOBAL OCEAN FLUX STUDY (JGOFS)** is an international investigation that scientists believe will help them understand by the turn of the century the links between the ocean **carbon cycle** and **climatic change.** The world's oceans have been referred to as the 'sleeping giant' of global carbon dioxide control. There is an estimated 20 times more carbon dissolved in **sea-water** than occurs on land in plants, animals and soil. Release of just 2% of the carbon stored in the ocean would more than double the level of carbon dioxide in the atmosphere. Each year about 15 times more carbon dioxide is taken up and released by natural marine processes than the total produced by the burning of fossil fuel, deforestation and other human activities. The amounts of carbon dioxide entering and leaving the oceans, over an annual cycle, are estimated to be in balance. Research suggests that balance was reached not long after the last ice age, 10,000 years ago. Nevertheless, the exchanges of carbon dioxide between the air and sea are controlled by many different processes. They are influenced by regional and seasonal factors, with direct and indirect links to climatic factors. Consequently, scientists are concerned that the effects of **global warming** could have a major influence on some of the processes involved in carbon dioxide exchange, either accelerating or slowing down temperature changes.

The JGOFS study is investigating the physical conditions of the oceans and the atmosphere which govern the rate at which carbon dioxide is dissolved from or released into the air, and the biological processes whereby marine plants absorb and release carbon dioxide. Carbon dioxide is extracted from the air by **photosynthesis** of the **phytoplankton,** which flourish in the top layers of the sea, and is released by **respiration.** Uncertainties about global warming arise because scientists do not know exactly what happens to the carbon being absorbed and released from many sources. Human activities add about 5,000–6,000m. tons of carbon to the atmosphere a year, but the annual increase in atmospheric carbon is only 3,000m. tons. It was thought the balance must be absorbed by the oceans. Recent estimates of the processes for absorbing carbon by the oceans leave over 1,000m. tons unaccounted for. Improvements in understanding the mechanisms of processing carbon are urgently needed for reliable predictions of future increases in atmospheric carbon dioxide, and hence, global warming.

K

KARIN B was a ship that was refused entry to European ports with its cargo of hazardous chemical waste, containing polychlorinated biphenyls (PCBs), in 1988. The cargo had been dumped illegally in Nigeria by an Italian chemical company, and the Karin B was trying to take it for **incineration** in a chemical disposal plant. The waste cargo was eventually returned to Italy. The incident exposed the trade in toxic waste from rich countries to the poor, developing countries. It paved the way for the **Basel Convention** which came into effect in 1992.

KELP is brown **seaweed** that grows below low-level tides, usually anchored to the sea-bed. Kelp, when burned, provides ash that is used as a **fertilizer.** It is also a valuable source of **iodine,** potash and soda.

KINETIC ENERGY is the energy of a moving object due to its motion. A fast-flowing river or a waterfall possess kinetic energy. This energy can be harnessed to drive electric generators for **hydroelectric power.**

KRILL is the collective name for some 80 species of shrimp-like crustaceans, about 5 cms long, which live exclusively in the cold waters around the poles, particularly in **Antarctica.** They feed on **plankton** and form the second level of the marine **food chain.** Krill have a high level of protein and are an important food source for fish, five species of baleen **whales,** some species of seal and several species of birds, including penguins. Krill shoals weighing as much as two million metric tons are not unknown. Whales consume several million tons of krill every year, but when the moratorium on whale hunting was declared, krill became more abundant as a food source for sea birds and seals. However, they are now being caught in vast quantities by fishing fleets, Russian and Japanese, for animal feedstock and, in some cases, as a nutritious food source for humans. However, pollution of the oceans presents a threat to krill, because it could destroy the plankton on which they feed.

L

LAKE BAIKAL, in Siberia, is the oldest lake in the world. Its depth of 5,250 feet also makes it the world's deepest lake. More than 1,500 species of animals and 1,000 plants have been found in Baikal and some 1,300 of these have not been recorded anywhere else in the world. The local community, many of them Buddhists, regard the lake as holy and, until the 1950s, its water was noted for its purity. Large-scale development after the Second World War brought factories to the shores of the lake, including the Baikalski paper mill, which has poured massive amounts of industrial effluent into the lake every day for more than 30 years. Large areas of the surrounding forests have been destroyed by pollution from the emissions from the factories' chimneys. Since the fall of the Communist regime in the former Soviet Union, environmental campaigns have been set in motion to clean up the lake and some remedial programmes have been started, including a new pipeline from the paper mill taking discharges to a nearby river. Lake Baikal is favoured by scientists because of its wealth of unique flora and fauna. The Russian Academy of Sciences is negotiating new collaborative agreements with overseas organizations. For example, the US Geological Survey has co-operated with Russian researchers to collect sediment from the floor of the lake.

LAKE ERIE, see **Great Lakes.**

LAND-OCEAN INTERACTION STUDY (LOIS) is the name of a research project to collect data so that scientists will be able to predict the likely impact on the UK coastal zone of natural and man-induced environmental changes over the next 50 to 100 years. The possible changes under investigation are fluctuations in **sea-level,** tidal patterns and the wind-wave climate. The researchers are examining the possible effects of local influences such as altered patterns of agriculture, the use of waterways and water pollution; the impact of trends in **climate** attributed to **global warming;** and the results of changes in land-use for coastal defence, or for development and leisure. LOIS came about because the UK lies on a wide **continental shelf** bordering the Atlantic Ocean. It possesses an extremely long and intricate coastal zone of great physical and ecological diversity, on which lie many centres of large population, commerce and industry. Interaction between human and natural factors is causing the physical and biological environments of the coastal zone to change rapidly. In many places cliffs and dunes are being eroded. In others, the coast is being built out and estuaries are being filled up as a result of accumulation of sediment. Some of the **sediment** may be supplied by rivers and carried along the coastline, or carried shoreward from the sea. Human activities not only influence the physical character of this zone, they have an especially strong influence on its ecosystems. These have been significantly modified over the last 200 years by an increasing flow to the coastal zone of an ever more diverse range of substances connected with agriculture, urban growth, the burning of fossil fuels and industrial development. While some of these topics have been explored separately, the Natural Environment Research Council decided that a piecemeal approach was no longer a satisfactory way to attempt to understand how best to manage coastal zones. Consequently, a large number of research groups from universities and other institutes were invited to conduct the LOIS project. However, LOIS itself forms only one piece in the jigsaw of a particular area of environmental research. LOIS will obtain information on which to work from larger studies such as BOFS (Biogeochemical Ocean Flux Study) and TIGER (**Terrestrial Initiative in Global Environmental Research**). See also **coastal areas.**

LANDFILL is the oldest method of **waste disposal** for the solid matter discarded in the domestic dustbin, along with the packaging material and paper from high street shops and offices. Landfill sites are usually disused quarries and gravel pits. When they were filled, previous practice was to cover them up with soil and forget about them. Housing estates have been built, often with disastrous consequences, on old landfill dumps. Waste burial has now become a serious technology and a potential source of energy. Landfill sites can be designed to be bioreactors, which deliberately produce **methane** gas as a source of **biofuel** or alternative energy. Traditionally, waste tips remained exposed to air and aerobic microbes—those which thrive in air—in order to turn some of the waste into compost. However, open tips also encourage vermin, smell in hot weather and disfigure the landscape. In the 1960s, as a tidier and safer option, landfill operators began to seal each day's waste in a clay cell. While excluding vermin, the clay also excluded air. Decomposition relied on anaerobic microbes, which die in air. However, the process produced methane (natural gas), which was a safety hazard. The methane is now extracted by sinking a network of perforated pipes into the site. See also **Fresh Kills Landfill.**

LANDSAT, see **Earth Resources Technology Satellite.**

The **LAW OF THE SEA CONFERENCES** are a series of United Nations conferences which began in 1958. The first was convened to standardize the national limits of territorial waters. Traditionally, the limit of territorial waters was three miles, because this had been the longest range of naval artillery in the 17th century. Chile was the first nation to claim 200 miles along its shoreline and other nations soon followed with similar claims. The conference tried to negotiate claims to territorial waters and to investigate questions of national rights to coastal mineral deposits. The second conference took place in 1960 and was unable to reach agreement. In 1973 the third conference began and continued until 1982 when the Law of the Sea Treaty was agreed.

The **LAW OF THE SEA (UNCLOS III)** was the treaty agreed in 1982 at the end of the third of the United Nations **Law of the Sea Conferences** which lasted from 1973 to 1982. It was an attempt to establish controls over a wide range of maritime problems, including fisheries, navigation, **continental shelves,** the deep sea-bed, scientific research and marine pollution. Out of 150 nations who took part in the conference, only a few have ratified the Treaty. Thirty-three countries rejected it, including the USA, the UK and West Germany. Others have signed but not ratified the Treaty. However, many of the Treaty's provisions to regulate maritime activities within defined zones and boundaries have been widely accepted as customary international law.

The Treaty was criticized because the question of areas of overlap between adjoining countries was not resolved, and because the arrangements could have meant that six countries would have a claim to more than half of desirable coastal and continental shelf areas of the entire world. Under the terms of the Treaty, 40% of the seas come under the control of countries with **coastal areas.** The treaty proposes: a territorial zone, extending 12 nautical miles from land, with a universal right to total sovereignty given to coastal states; a contiguous zone of 24 miles, with restricted rights; an exclusive economic zone (EEZ), a 200-mile area, with rights over scientific research, environmental preservation and economic operations; and a continental shelf over which states are allowed to explore and exploit resources as long as they do not infringe the legal status of water and air above. Sixty per cent of the seas, the 'high seas', are outside the jurisdiction of any one nation, but there is a proposal that there should be no restrictions on fishing or passage, although countries are bound by the Treaty to co-operate in preventive measures to control pollution. It is proposed that 42% of the 'high seas' are designated the 'common heritage of mankind'. Resources of these sea-bed areas would be controlled by the United Nations and it is proposed that an International Sea-Bed Authority (ISA) would be appointed.

LEACHATE is the liquid that drains from a waste dump when water percolates through the waste and gathers up polluting materials. The water may be from rain or streams flowing over or through the surface. A poorly managed **landfill** site can produce large amounts of leachate. Fresh waste produces the strongest leachate, with a very high biochemical oxygen demand (**BOD**). If there are residues of **ammonia** or **heavy metals** in the dump, the leachate could cause severe pollution if a large quantity were to get into rivers. Older leachates have lower BODs, but still contain a large amount of dissolved ammonia, which is especially toxic to fish.

LEAD is a poisonous metal, but one that has been used widely for thousands of years—in batteries, alloys, shot, paints, petrol, in construction and in plumbing. Lead in paint, especially the older high-lead forms, has caused many deaths. Adding tetra-ethyl lead to petrol as an **anti-knock agent** to stop the engine knocking has led to distribution of lead into the atmosphere, and high concentrations appear in urban dust and air. This also increases levels of lead in food. The problem is exacerbated when lead solder is used in tins for food, but this is being phased out in many countries. **Lead poisoning** causes stomach pains, headaches, tremors and irritability. In severe cases it causes coma and death. There is controversy about the effects of low levels of lead on the development of a child's brain. It affects the nerves and the brain at very low concentrations. Research in the USA has found that the mental development of babies is affected by lead levels in the blood as low as 10 micrograms per decilitre. Lead-free petrol is now available.

LEAD-FREE PETROL is petrol free from the addition of organic lead compound, which has traditionally been introduced during blending as the most convenient way to obtain a desired **octane rating.**

LEAD POISONING can occur at any age, but it refers particularly to the problem for children younger than six years old. Accumulated exposure can affect both the growth and intellectual development of an infant. Contamination by very tiny levels appears to be enough to impair intelligence if it happens in the very young. The primary hazard is from lead-based paint, ancient lead plumbing and airborne emission from exhausts of cars using leaded petrol. However, cases of contamination have also been attributed to lead workers bringing home lead dust on their clothes.

LEGA PER L'AMBIENTE (ENVIRONMENTAL LEAGUE) is the name of the youngest of the three main environmental movements active in Italy. It is thought to be the one most focused on grass-roots action. Since 1986 there has been a dramatic increase in awareness in Italy of the threat to the environment. It was provoked by a series of ecological scandals from the radioactive fall-out from **Chernobyl,** to the pollution of the Adriatic Sea and the highly publicized wanderings in 1988 of Italian toxic waste barges, looking for somewhere to dump their cargoes (see **Karin B**). A national referendum in 1987 rejected the country's nuclear energy programme. As a country almost entirely surrounded by sea Italy has a great reliance on marine resources and maritime activities. Its geography also creates particularly difficult pollution problems. Water in the **Mediterranean Sea** is renewed only once every 80 years so pollution levels build up more quickly than in the open oceans. The Adriatic presents additional, serious problems, since it is a smaller and shallower sea with narrow access to the Mediterranean. Large rivers from Italy and the former Yugoslavia discharge a variety of industrial, agricultural and domestic wastes into the Adriatic. In 1990 more than 80% of sewage from Italy's coastal cities and towns was dumped into the sea without treatment. Organic nutrients from sewage and agricultural fertilizer have caused devastating **eutrophication** in the Adriatic, as a result of the proliferation of a vast blanket of **algae.** The algae then decay, absorbing the water's oxygen supply, turning coastal waters into oxygen-less pools that kill fish and force the closure of beaches.

LEGIONNAIRES' DISEASE is an airborne bacterial respiratory infection which occurs mostly in North America and Europe. It is transmitted primarily through hot water and air-conditioning systems, such as showers, baths and health spas. It was identified in 1976 when 182 delegates at a State American Legion convention in Philadelphia became infected. Thirty people later died. There was widespread speculation that this was a new and virulent infection, but after tests it was found that the disease was not new and that there had been earlier outbreaks which had been put down to unexplained cases of pneumonia. Other outbreaks have been recorded since the Philadelphia incident, resulting in several deaths. The disease most often affects males aged around 50 years and those who are heavy smokers or drinkers. In 1989 the World Health Organization warned that, although Legionnaires' Disease is so far only known to occur in the industrialized world, it could soon emerge in developing countries as they develop more facilities for tourism.

LEGUME INOCULANTS, one of the most straightforward applications of **biotech,** involve injecting **nitrogen**-fixing bacteria into the **soil** around the roots of legumes like peas and beans, in order to increase the soil population of the naturally occurring **rhizobia,** the nitrogen-fixing bacteria that live symbiotically with legumes (see **symbiosis**). Nitrogen is an essential macronutrient, a type of substance people need in their diet. Though 80% of the air is nitrogen, plants and animals cannot harness it to convert it into **amino acids** and protein. They rely on other sources of nitrogen, such as natural or man-made sources of **ammonia** and **nitrate** fertilizer for plants, which in turn provide amino acids and protein for animals. Only a few organisms can extract nitrogen from the air and convert it into a form that plants can assimilate. The rate at which this process of nitrogen fixation can supply nitrogen to a plant puts a limit on plant growth. In addition to inoculating soils, biologists are experimenting with ways of improving the efficiency with which plants process nitrogen.

LEISHMANIASIS is a parasitic tropical disease transmitted to humans by the bite of sandflies. In tropical and subtropical regions an estimated 350m. people from 80 countries, including Israel and the former Soviet Union, are at risk from the disease. Nearly half a million new cases are reported each year. There are two forms of the disease: one damages the internal organisms, particularly the liver and spleen; the other form affects the skin tissues and causes open sores or ulcers, which can eat away the nose, mouth and throat.

LEPROSY is a disease that is widespread in tropical countries. An estimated 11m. people in the world have leprosy and more than 1,600m. people could be at risk from it. It is transmitted by sneezing or by skin contact and is particularly prevalent in poor, overcrowded conditions. It can be controlled, but not cured, by drugs.

LICHENS are composite organisms formed by the **symbiosis** between species of **fungi** and an **algae.** They are either crusty patches or bushy growths on tree trunks, stone walls, roofs or garden paths. Because they have no actual roots, they get their sustenance from the atmosphere and rainwater. Lichens play an important role in the detection and monitoring of pollution, especially **sulphur dioxide,** as they are highly sensitive to pollution and different species disappear if pollution reaches specific levels.

LIFE-SUPPORT SYSTEMS are those environmental conditions which are essential for human life: air, water, food and shelter. The environment can also present a hazard to health. According to the World Health Organization (WHO), 'good health is a state of complete physical, mental and social well-being and not merely an absence of disease or infirmity'. It represents a balanced relationship of the body and the mind and complete adjustment to the total environment. The WHO says, conversely, that disease is maladjustment or maladaptation in an environment. It is a reaction for the worse between a person and the hazards or adverse influences in that individual's external environment. For the purposes of **environmental health,** WHO puts environmental influences into three broad categories: the physical environment, the biological environment and the human environment. The atmosphere is the primary part of the physical environment. A person suffocates in seconds without the oxygen in air. On the other hand, exposure to excess solar radiation can cause sunburn, skin cancer and genetic mutations. While water is essential for life—a person can survive at most for a week without drinking water—it also presents a range of health hazards. Many diseases are caused by water-borne organisms. Bacteria causing **cholera,** dysentery, paratyphoid and typhoid are transmitted by contaminated water. Water-borne parasites cause **malaria,** tapeworm, yellow fever and **schistosomiasis.** Viruses which cause infectious **hepatitis** and poliomyelitis are other **water-borne diseases.** In addition to pathogens, there are other impurities in water. They include pesticides, herbicides, lead, zinc, mercury, arsenic, nitrates, fluorides, selenium, molybdenum, cadmium and sodium.

LIGNITE is a low-grade brown-black **coal** which gives off the most smoke and least heat. It is used mostly for generating electricity. It is an intermediate stage of formation between **peat** and **bituminous coal.**

LIMITS TO GROWTH was a study commissioned by the **Club of Rome** and published in 1972 by a team of economists and futurologists working with Professor Dennis Meadows at the Massachusetts Institute of Technology, in the USA. The Club of Rome was formed by a group of international businessmen who were concerned by speculation about the impact on society and the environment of emerging trends in science and technology. The study employed elaborate computer calculations to process the most complex mathematical model so far devised to predict future developments from current trends. It projected a **Malthusian** vision that predicted the collapse of world order if **population** growth, industrial expansion, increasing pollution, insufficient food production and exploitation of mineral resources continued at current rates. *Limits to Growth* proposed a series of alternative developments to offset disaster. It called for 'a Copernican revolution of the mind' that rejected the belief in endless growth and the tacit acceptance of waste in a throw-away society. It proposed **zero population growth,** a levelling-off on industrial production, pollution control, **recycling** of materials, the manufacture of more durable and repairable goods, and a shift from consumer goods to a more service-oriented economy.

LINDANE is an odourless and highly volatile **insecticide.** It was popular for use as a household fumigant and was sold in pellet form to attach to light bulbs or electric wall vaporizers. Tests later proved that it was dangerous to humans and pets.

LISTERIA is a bacteria, known properly as *Listeria monocytogenes,* that causes the fever- and flu-like illness listeriosis, which is especially dangerous during pregnancy. The infection is often transmitted by contaminated dairy products, such as soft cheese and yoghurt. Normally the symptoms are so mild the person is unaware of the infection. However, in pregnancy it can lead to miscarriage, still birth, neonatal death and serious neurological damage, including mental retardation.

The **LITHOSPHERE** is the Earth's solid upper shell on which all species live. It is also the source of raw materials and energy. Increased industry and agriculture put a strain on the lithosphere and can lead to landslides, mudflows, **groundwater** pollution and other geological disasters.

The **LONDON CONVENTION** is the familiar name for the convention on Prevention of Marine Pollution by Dumping of Wastes and Other Matter. It was adopted in 1972 at a conference held under the auspices of the UK and came into force in 1975. The convention, previously the London Dumping Convention, controls and regulates on a global scale the disposal at sea of wastes and other material of any kind, including ships and platforms. It covers all the world's oceans, aiming to control the dumping of waste from ships and aircraft. There is a specific blacklist of substances most dangerous to the environment: **mercury** and mercury compounds, **cadmium** and cadmium compounds, organohalogen compounds (see **halogenated compounds**), persistent plastics, crude oil, fuel oil, heavy diesel oil, hydraulic fluids and lubricating oils (if taken on board to be dumped) and high-level radioactive waste and other radioactive materials if the **International Atomic Energy Agency** rules they are unsuitable. Materials produced for chemical or biological warfare are also on the black list. A 'grey' list of substances that can only be dumped with a special permit includes: wastes with significant amounts of arsenic, lead, copper and zinc, organosilicon compounds, cyanides, fluorides and pesticides. The convention has specific regulations on the dumping of several other materials which may present a risk to the marine environment and human health. It also controls the **incineration** of wastes on board ship and, in 1988, contracting parties agreed to take steps to minimize or substantially reduce the use of incineration of noxious liquid waste by January 1994.

LONDON DUMPING CONVENTION, see **London Convention.**

LOVE CANAL was a housing estate in Niagara City, upper New York state. It was built in the late 1950s on a **landfill** site, used during the 1940s and early 1950s to dump thousands of tons of chemical waste. In 1977 toxic gases released from buried and corroding drums of chemical waste began to bubble through the earth. The waste, mainly **dioxin,** seeped into the houses. The site was declared a Federal Disaster Area, the first time that a national emergency had been declared as a result of chemical pollution. It was more than 12 years later that the site was finally declared free from contamination.

LOW-LEVEL RADIATION can refer to the radiation from **low-level wastes,** or to a long-standing controversy over whether such a thing as a safe threshold of **radiation** exists.

Most studies of the response of people to radiation have been conducted on individuals known to have received doses of 100 rems, the equivalent of one **sievert** in the new standard units. Information about the effect of doses of radiation below 10 rems, or 0.1 of a sievert, is less reliable. Firstly, there are no radiation dose records for people who live near nuclear power stations or **nuclear reprocessing** and waste handling plants (see **Three Mile Island**). Secondly, researchers believe a cancer induced by low-level radiation may not appear for 30 or more years. The main source of information on the effects of low-level radiation are from the continuing studies of the survivors of the nuclear bombs dropped on Japan.

LOW-LEVEL WASTE is the most bulky kind of radioactive waste. It ranges from laboratory rubbish, such as bits of broken glassware, rubber gloves and pieces of plastic, to the cleaning materials used in the nuclear industry. Much of it was formerly sealed in concrete drums and dumped at sea, a practice that was stopped under the conditions of the **London Convention.**

M

MACHYNLLETH, in mid-Wales, is the site of the Centre for Alternative Technology, where a community of 30 to 40 people provide a living example of 'environmentally friendly' behaviour. The community uses **wind power,** water turbines and **solar energy** to provide energy for their washing and cleaning, and to drive the usual household devices. Wood stoves are used to heat houses and for cooking.

MAGNOX REACTORS were the nuclear reactors that powered Britain's first nuclear power programme. They used natural **uranium** as fuel encased in magnesium alloy, or Magnox, canisters. The reactors were cooled by carbon dioxide. The versions built for the electricity utilities were based on the Calder Hall reactor, built at Windscale (see **Sellafield**). Calder Hall was called the first commercial nuclear power station when it was opened in 1956. In fact, it was built to produce **plutonium** for the British atomic weapons programme. The electricity it generated was a by-product that was sold to the national electricity grid.

MALARIA is a disease associated with bad environmental conditions producing stagnant water near homes. The stagnant water may be the result of schemes intended to improve irrigation by a network of ditches or water storage impoundments. Control calls for elimination of the breeding sites of the **mosquitoes** that carry the disease-causing parasite. Hopes of eliminating the mosquito with **DDT** and other pesticides have faded. The insects appear to have developed resistance to the pesticides and this has further aggravated the problem of infestations.

The Republic of the **MALDIVES** hosted the first Small States Conference on Sea Level Rise. As a result of the conference, the Malé Declaration on Global Warming and Sea Level Rise was made, setting out the steps small island states should take to combat the effects of the rise in sea-level. The Maldive Islands face a serious threat from any rise in sea-levels. The average height of the islands is about two metres above mean sea level and a rise of 50 cm would have serious consequences for the Maldives and other low-lying islands. Some islands would be flooded while others could be submerged. As a result of stronger wave action, the infrastructure at some resorts could also be affected. As the Maldive economy relies mainly on tourism and fisheries, **climatic change** could have a damaging effect. It is feared that a 1°C rise in global temperature could affect the migration patterns of some species of fish. An action group, formed at the Malé conference, with representatives from the Caribbean, South Pacific, Mediterranean and Indian Ocean regions, recommended: public information and education campaigns; encouraging the international media to promote the views of small states on climate change and the rise in sea-level; increasing the part played by small states in global climate programmes; intensifying efforts to mobilize funds, so that small states can play a wider part in international meetings and activities; and seeking international help to upgrade the technical capabilities of national climate and sea-level monitoring institutions so that they can reach adequate standards.

MALTHUSIAN refers to the theory of the classical economist, Thomas Malthus (1766–1834), who maintained that population growth would always outrun food supply and that in the end human numbers would have to be kept down by famine, disease or war. He was the first person in Britain to be appointed a professor of political economy and he first formulated what is now called the law of diminishing returns.

MAN AND THE BIOSPHERE (MAB) was a **United Nations Educational, Scientific and Cultural Organization (UNESCO)** project started in the 1970s. It was launched to further long-term conservation of the Earth's **ecosystems,** based on sound scientific foundations. MAB results have increased scientific understanding of ecosystems and their inter-relationships. The detailed research opened up new avenues for tropical ecosystems in particular. New species were discovered, and many forest plant species were seen to have potential medicinal benefits. By the 1980s about 25% of Western medicines contained at least one ingredient supplied by the **tropical rainforests.**

MAN-MADE HABITATS are those areas developed and influenced by man over the years to provide shelter and food for the population. See **agricultural habitats** and **urban habitats.**

MANGANESE is an essential **trace element,** required both by animals and plants. In plants it is involved in the process of **photosynthesis.** The leaves of plants deficient in manganese become pale green or yellow. In animals it is a growth factor necessary to encourage bone development. Critical enzymes need manganese. However, inhalation of manganese dust as a pollutant also causes poisoning, including brain damage.

MANGROVE FORESTS or swamps form one of the most diverse ecosystems throughout the tropics. They are also one of the most endangered. Mangroves grow at the edge of the sea and provide a unique habitat for some 2,000 species of fish, invertebrates and plants. They are a valuable source of timber, pulpwood, fuel and charcoal, and of raw materials used in dyes, glues, rayon and tannin. They are also the breeding ground for a variety of fish, shrimps, crabs and molluscs. Mangrove forests play an important part in the desalination of **sea-water.** Because the roots of the trees are in the sea-water they protect shorelines from erosion and act as a bulwark against storms. However, over the years many acres of mangrove forests have been destroyed. They have either been cut down for charcoal production, fish farming development, or to clear areas for agricultural or urban expansion. They have also been affected by pesticides leached from agricultural areas and choked by sediment from coastal development and mining operations. The mangroves of the Philippines, for example, have been reduced to only 380 square kilometres from an estimated 5,000 square kilometres in the 1920s. Most of the lost area has been cut for its timber, or converted into fish farms or rice fields. In many areas throughout the tropics the destruction of the mangrove forests has started a chain of coastal degradation. However, in some countries, including the USA and Thailand, mangrove forests are being cultivated to protect coastlines from erosion, to trap the silt draining off the land and to filter out pollution.

MARPOL, see **International Maritime Organization (IMO).**

MAUNA LOA OBSERVATORY, in Hawaii, was the world's first observation station to measure long-term changes in **carbon dioxide** concentrations in the **atmosphere.** It was established by an American scientist, who began work in 1957. Subsequently, carbon dioxide measurements have been made by scores of stations across the world. The carbon dioxide content of the atmosphere has increased by about 25% since the burning of oil and coal on a large scale began during the Industrial Revolution, before which the carbon dioxide concentration was 280 ppm (parts per million). When observations began at Mauna Loa, carbon dioxide concentrations had reached 315 ppm, rising to 350 ppm by 1990. This annual increase, at a rate of 1.8 ppm per year, corresponds to an increase of 33,000m. tons in the total mass of carbon contained in the atmosphere as carbon dioxide. This has to be compared with the total annual emissions into the atmosphere, which are estimated at 67,000m. tons, and have been brought about by the burning of **fossil fuels** and **deforestation.** Thus, only 50% of the **anthropogenic** carbon dioxide remains in the atmosphere. The other human-induced emissions are a small part of the large transfers of carbon dioxide conducted between the atmosphere, the ocean and the terrestrial **biosphere** during seasonal growing cycles. Relatively small changes in the world **ocean circulation** and chemistry, the temperature of the oceans, or in the life cycle of terrestrial vegetation, could dramatically affect the levels of carbon dioxide in the atmosphere.

The **MEDITERRANEAN SEA** has been the dumping ground for millions of metric tons of waste, mostly emanating from land sources, for several decades. The Mediterranean has a coastal population of some 100m. Another 100m. tourists visit the area every year. A high proportion of raw sewage is dumped into the sea. There are 60 oil refineries on the coastline, producing substances that are very destructive to the marine environment. The direct dumping of wastes from ships is banned, but the Mediterranean is still one of the most polluted seas in the world. To compound the situation, the sea has a high rate of evaporation and a slow rate of water renewal, and, consequently pollutants accumulate but do not break down. It is believed that it takes about 80 years for the Mediterranean to renew its water. Shellfish from the sea are unsafe for human consumption, except for those from a few recognized growing areas. A Mediterranean Action Plan has been drawn up in an attempt to halt pollution and clean up the sea. The action plan, part of the United Nations Environment Programme (UNEP) Regional Seas Programme, involves 16 Mediterranean states which border the sea. These states have initiated water improvement projects.

MERCURY is widely used in industry, but it has long been known that it is highly poisonous and, hence, an occupational hazard, causing brain damage and death if as little as one gram is ingested. It is a highly toxic **heavy metal** which attacks the nervous system. Small doses cause irritability and headaches, while large doses lead to convulsions, coma and death. Like all heavy metals it accumulates through **food chains.** The increased toxicity of mercury was recognized relatively recently once it was realized that it could be transformed in the environment into an organic compound and then absorbed in food, such as fish.

The **METABOLISM** is the sum of all the physical and chemical routes by which the body processes foodstuffs and other compounds. As foodstuffs and other substances are broken, the new compounds are referred to as metabolites. Sometimes metabolites are poisonous. For example, the processes of metabolism bind strontium to bone and allow **lead, mercury** and **cadmium** to cross the **blood-brain barrier.**

METHANE (CH4) is the main constituent of natural gas. It is the simplest hydrocarbon and a valuable raw material for making organic chemical compounds. Methane is produced by decaying organic matter. So, aside from extracting it from deep natural gas wells, methane also bubbles up in marshes and bogs, is produced in the flatulence of ruminant animals like cows and in paddy-field cultivation, and is generated by termites as they eat their way through forest wastes. Methane is also a **greenhouse gas** and an **ozone layer** destroyer. As a greenhouse gas, atmospheric methane has a high capacity for absorbing heat to increase the risk of **global warming.** Molecule for molecule, methane is 20 times more effective as a greenhouse gas than carbon dioxide, which is widely regarded as the most critical greenhouse gas.

METHYL-MERCURY is an **organometallic compound,** which is a substance produced when a metal atom, like **mercury,** is combined with an **organic group** by the process of **methylation.** Organometallic compounds may be man-made or produced naturally in the environment when a metal pollutant is caused to react with an organic group. If an environmentally produced agent, like methyl-mercury, gets into the **food chain,** it can be highly toxic to people. The **Minimata Bay** pollution disaster was caused by factory effluent containing mercury, which was converted on the sea-bed into methyl-mercury and then entered the food chain. The methylation process was performed by marine micro-organisms, which in turn were eaten by other marine life. The toxin accumulated in shellfish eaten by the local population, with tragic results.

METHYL-TERT-BUTYL ETHER (MTBE) is the chemical name of an additive that can be put into petrol as an alternative to the environmentally hazardous **organometallic compound** tetra-ethyl lead. Additives are used in petrol as a cheaper means of increasing the **octane rating** of automobile fuel in preference to more expensive additional refining. In other words, the performance of low octane-rated blends of petrol can be boosted by additives like MTBE. These additives are also referred to as oxygenates, which have the purpose of ensuring better combustion of the hydrocarbon fuel. They also reduce the amount of carbon dioxide in the exhaust gases. There are still by-products of combustion, such as nitrogen oxides, that can cause **photochemical smog.** The by-products can be removed by devices like **catalytic converters.** However, converters are damaged by lead. The impetus for MTBE-type additives began in California and they were then adopted in the US Clean Air Act. Demand for MTBE was eight million tons in 1991, with demand forecast for 35m. tons by 2001.

METHYLATION is a process by which compounds are modified by the addition of a methyl group, usually by a reaction involving methyl alcohol. It is a vital process, which is exploited by the chemical and pharmaceutical industry and in research. However, methylation can happen naturally, and a relatively innocuous substance can be converted in the environment into a substance that is highly toxic if it gets into the food chain, as happened in the **Minimata Bay** tragedy. Besides being the product of important industrial processes, many essential methyl compounds are made by animals and plants.

MICROBES are micro-organisms, single cells, or clusters of cells, that can survive on their own if they have nourishment. In more common parlance they are called 'bugs'. They include **bacteria, algae, fungi** and yeasts. They are independent cells, unlike brain cells, for example, which can only survive in the brain, or leaf cells, which can only survive in the plant leaf. They can adapt to any niche in any environment around the world. They have in fact been used over the centuries, albeit sometimes unknowingly, by farmers, animal and plant breeders, vintners, brewers and bakers.

The **MICROBIOLOGICAL RESOURCES CENTRES (MIRCENs)** are a network of regional institutions which maintain regional centres for **microbiology** research and development of micro-organisms. The six centres were started by the United Nations Environment Programme (UNEP) in collaboration with the United Nations Educational, Scientific and Cultural Organization (UNESCO) in 1974. It was the first attempt to extend **biotech** to the developing world, and maximize the potential of using micro-organisms to solve regional and national environmental problems. Biotechnology offers developing countries a way to tap their enormous **biological diversity** for economic development. The centres were established in Brazil, Guatemala, Egypt, Kenya, Thailand and Senegal. Each MIRCEN is a network of collaborating laboratories and institutions, maintaining a regional **gene bank** of microbial resources, and serves as a centre of excellence for conservation, research and training. In Guatemala, MIRCEN concentrates on the use of microbes to convert coffee wastes into **biogas** and organic fertilizer; in Thailand research is carried out on turning cassava wastes into power alcohol; in Cairo they are examining the use of micro-organisms in pest and vector control and in degrading persistent pollutants. Other MIRCENs are exploring biological fertilizers, based on nitrogen-fixing bacteria.

MICROBIOLOGY is a branch of biology which deals with micro-organisms, and the major scientific discipline from which principles are applied for use in **biotech,** which produces vaccines, hormones, vitamins, organic solvents, dairy products, wine and beer, and compost. Microbiology has contributed in various ways to such advances as the identification, in the 1940s, of DNA (see **nucleic acids**) as the genetic material, the revelation of the universal genetic code in the 1950s, and the production of human insulin from genetically engineered bacteria in the 1980s.

MIGRATION, see **Bonn Convention.**

Milutin **MILANKOVITCH** and James Croll conceived an astronomical theory of **climatic change.** They analysed the pattern of the last five ice ages, the **Pleistocene** ice ages, and concluded that the climate had been influenced by the variation in the Earth's orbit of the sun. The Earth does not revolve in a circular orbit round the sun at a constant velocity, but over a period of years changes occur which affect the amount and distribution of solar radiation received by the Earth. Firstly, the tilt of the Earth's axis of rotation relative to the plane of its orbit varies between 21.80 and 24.40 degrees over a period of 40,000 years. This effects the seasonality or thermal range between summer and winter. Secondly, the ellipticity or eccentricity of the Earth's orbit varies slightly, and in the 1930s Milankovitch, a mathematician, calculated a periodicity of 96,000 years for this variation. In other words, the orbit becomes alternately more circular and more elliptical. Thirdly, the season when the Earth is nearest the sun (perihelion) changes over a period of 21,000 years. At present it is nearest in the Northern Hemisphere winter, on 7 January, and farthest in the Northern Hemisphere summer, on 7 July. Milankovitch analysed these cycles and computed the times when he argued they conspired to trigger the start of ice ages.

The **MILLIMETRE-WAVE ATMOSPHERIC SOUNDER (MAS)** is a special instrument developed as part of a long-term, international environmental monitoring programme for analysing the **atmosphere** from the US space shuttle. The MAS searches the atmosphere for global changes in levels of **ozone,** chlorine monoxide, **water vapour,** temperature and pressure. The MAS is, in effect, a receiver that 'listens' to the emissions from molecules in the atmosphere for their unique signatures. Measurements began on space flight STS-45, on 24 March 1992, at the start of a ten-year programme. The same instrument was due to go into orbit every year for 10 years, covering a solar cycle, in order to sort out natural variations, such as **sunspot** fluctuations, from man-induced ones. Since scientists have yet to devise a long-term method of permanently measuring anything from orbit whose accuracy did not soon degrade, the MAS instrument has to be flown back and forth in the Shuttle. As little as a 1% change in accuracy invalidates any readings. By retrieving the MAS after every flight and recalibrating it, scientists can be assured that readings are comparable year after year. The researchers are looking for changes of a fraction of a per cent in ozone, chlorine monoxide and water vapour levels for indications of global **climatic change.** They are also looking for temperature changes because as the Earth's surface heats up, the upper atmosphere cools down. The atmosphere should cool more than the surface heats up, and it might be easier to spot changes in the atmosphere because that change occurs at a faster rate. It was initially decided to go into orbit because ground-based instruments have to look through all the layers of the atmosphere and it is difficult to separate them out and interpret the measurements. In orbit, the MAS looks tangentially across the edge of the atmosphere and can scan much thinner layers of it. Measurements can also be made on a global scale, rather than just from one location.

MINIMATA BAY in Japan had the misfortune to bring the health hazards of **heavy metal poisoning** to public attention. The danger of **mercury** poisoning was well-known during the last century as an occupational hazard in industries handling the substance. The accident which shocked the world into recognizing its dangers if discharged into the environment happened at Minimata. Since the 1930s methyl mercury had been discharged from the Chisso chemical plant into the bay, which was the source of shellfish for local villagers. Mercury discharges were thought to be diluted and dispersed to sea. In practice, the mercury settled on the seabed and was absorbed into the **food chain,** especially by **filter-feeders** such as shellfish. Once in the food chain, mercury gradually accumulated in the bodies of local people, causing a life-threatening and painful disease. By the 1950s it was causing mental retardation, seizures, loss of co-ordination and partial or total loss of vision and hearing. At first, the mystery illness, which is now known as Minimata disease, was simply called itai-itai, resulting from the cry of pain from its victims when they tried to move their limbs. Mothers who were apparently healthy gave birth to children with severe physical and mental disabilities. The last report on the incident, made by the **Worldwatch Institute,** recorded 649 deaths and 1,385 other sufferers from Minimata disease.

MISSION TO PLANET EARTH is the long-term, international research effort to study the Earth as a global environmental system, which was launched by the US space agency, NASA. The venture involves satellites, such as NASA's **Nimbus-7,** Space Shuttle missions and airborne and ground-based studies.

MOLLUSCICIDES are used to control snail populations, both of the type which inhabit gardens, fields or greenhouses, and those which are the intermediate hosts of parasites that cause disease in humans. Some species of snail, especially freshwater snails, are intermediate hosts to organisms which cause **schistosomiasis** in humans, and liver and lung fluke in humans, dogs, cats, and domestic animals. One molluscicide, metaldehyde, has been used as a bait for slugs and snails since it was discovered in 1936. (See also **pesticides**).

MOLYBDENUM is an essential **trace element** that is a component of a number of enzymes. Excess amounts of it are injurious to the human body.

The **MONSOON** is the intense seasonal rainfall which falls largely in India, China and South-East Asia. It is caused by a large-scale change of wind direction in the tropics, largely brought about because of changes in the heating of **oceans.**

The **MONTREAL PROTOCOL** was signed in 1987 by 24 countries, who agreed to cut **chlorofluorocarbon (CFC)** production by 50% by 1999. Other countries signed the Protocol later. Global production of CFCs had been growing at a rate of 30% a year. The reductions in CFC levels will not be reflected in global temperatures for at least 10 years. Therefore, even with the cuts that were agreed and the development of alternatives to CFCs, it will be well after the year 2000 that the benefits will come. The harmful **chlorine** molecules released from the CFCs remain in the atmosphere for decades until they are gradually washed out. CFCs have two harmful effects: they destroy the protective **ozone layer,** which filters out the **ultraviolet radiation** which causes skin cancer and, in excess, damage plant life; and, because they are also **greenhouse gases,** they increase global warming.

MOSQUITOES are among the worst disease-carrying pests known to mankind. There are some 3,000 species, distributed almost world-wide. In the tropics they are particularly abundant. The males of most species feed on plant juices, but the females need blood to reproduce. They get this by biting and sucking the blood of humans and animals and, at the same time, often transmit serious diseases to both humans and livestock. The ***aedes aegypti*** breeds in rivers and coastal areas, near human settlements. It is the principal carrier responsible for transmitting **dengue fever** and yellow fever. The female lays her eggs in whatever stagnant water she finds: puddles, discarded tin cans, a hole in a tree, old tyres or coconut shells. Other species of aedes act as vectors of **filariasis** and other viruses that cause forms of encephalitis. **Anopheles** mosquitoes also inhabit both tropical and temperate regions. The **malaria** parasite, *anopheles plasmodium,* is transmitted only through the bite of a female. Other species of anopheles transmit the parasites which cause Bancroftian filariasis, which can lead to blindness.

MOUNT PINATUBO is a volcano in the Philippines which erupted in June 1991. Scientists believe it may have caused the low **ozone** values of 1992 and 1993 recorded in the Northern Hemisphere. While the exact cause is unknown, the low ozone level may be related to the continuing presence of particles produced in the upper atmosphere following the eruption.

MOUNTAINS are an important source of water, energy, minerals, forest and agricultural products, and recreation. They are storehouses of **biological diversity** and **endangered species** and an essential part of the global ecosystem. About 10% of the world's population depend on mountain resources and nearly half of these people are affected by the degradation of mountain watershed areas. Mountain ecosystems are very vulnerable. They are increasingly sensitive to **soil erosion,** landslides and rapid loss of habitat and genetic diversity. Widespread poverty and an increase in the numbers of mountain inhabitants lead to **deforestation,** cultivation of marginal lands, excessive livestock grazing, loss of **biomass** cover and other forms of **environmental degradation.** Because little is known about mountain ecosystems, **Agenda 21** has proposed the establishment of a global mountain database. This is essential for the launch of programmes that would contribute to the **sustainable development** of mountain ecosystems. The proposals also focus on promoting watershed development and alternative employment for people whose livelihoods are linked to practices that degrade mountains.

MUSSELS are bivalve marine molluscs found in most waters, but more especially in cool seas. There are some 1,000 types of freshwater mussels in streams, lakes and ponds throughout most of the world. Some species, such as the blue mussel, are an important food source in Europe and are often farmed. The mussel is also extremely important as a biological indicator of water quality. Scientists have now developed new, sensitive techniques which indicate when human activity is causing damage to marine life. Using mussels, researchers are currently working on a programme which uses as a yardstick the amount of energy available for growth. The presence of toxins in the water may also influence the animal's internal defence mechanisms. As a result of the use of specific blood cells from mussels and fish, there are now simple tests to assess the immune system efficiently. Mussel populations studied from 26 sites along the east coast of Britain show that there is considerable variation in the mussels' ability to resist infection between sites. Those from the generally less polluted Scottish coast, for example, tend to show higher resistance than those from the more polluted Thames and Teesmouth areas.

A **MUTAGEN** is an agent that causes changes to plants and animals, particularly to their genetic material and especially at the time of reproduction. Certain chemicals and forms of radiation are powerul mutagens that damage the DNA, or genetic material (see **nucleid acids**), in the centre of every cell of a living organism.

MUTATIONS come about as a result of changes in the DNA (see **nucleic acids**) which forms the genes in a nucleus of a cell. Mutations in sperm or egg cells may lead to inherited defects in children. Mutations in the non-germ (or somatic) cells which make up the body may lead to disease in an individual. The DNA of a person contains about 50,000 genes. A chemical pollutant or exposure to radiation may damage any gene.

N

NADPH, see **photosynthesis** and **respiration.**

NATIONAL PARKS are areas of outstanding natural beauty, set aside for the **conservation** of flora, fauna and scenery, and for recreation, if this does not conflict with the conservation objectives of the parks and their landscapes. Hunting, logging, mining, commercial fishing, agriculture and livestock grazing are all controlled within national parks, as is industrial activity. The first national park was established at **Yellowstone** in Wyoming in 1872. It was regarded as one of the first national efforts to protect wild nature. National parks are usually selected for their unique characteristics: outstanding beauty, unusual geological factors, exceptional wild animals or plant life. National parks tend to be found in more densely populated countries, such as the UK, and are usually areas of natural beauty which have been changed by long periods of human occupancy. More emphasis is also placed on the protection of traditional landscapes, which also include facilities for sailing, walking, climbing and fishing. National parks have had amazing success in restoring plant and animal species. The **International Union for the Conservation of Nature and Natural Resources (IUCN)** has encouraged all countries to set aside at least 10% of national land and water resources for conservation, preferably as national parks. In fact, although they vary in size and character, and have different styles of management, more than 100 countries have already established national parks.

NATURAL GAS is an attractive fuel for small- to medium-sized power stations from an environmental viewpoint, because it almost eliminates emissions of **sulphur dioxide,** reduces nitrogen oxides by about 45% and discharges about 50% less carbon dioxide than coal.

NATURAL RESOURCES used to refer to the living resources such as forests, wildlife and plants that were able to regenerate through reproduction. They were regarded by the early conservationists as separate from the inanimate, non-renewable resources such as mineral and **fossil fuel** deposits that, once used, were irreplaceable. Anxiety about the environment, and particularly concern generated by ***Limits to Growth,*** which was published in 1972, focused attention on the need to conserve the inanimate natural resources. **Recycling** is one of the most attractive methods of re-using or refining waste materials instead of producing new materials. Consequently, recycling can reduce the volume of mineral extraction. This is also applicable to products made from living resources, such as wood and paper, and to the use of sewage concentrate as a fertilizer as an alternative to having it manufactured in a chemical plant.

NATURE RESERVES are areas allocated to preserve and protect certain animals and plants, or both. They differ from **national parks,** which are largely a place for public recreation, because they are provided exclusively to protect species for their own sake. **Endangered species** are increasingly being kept in nature reserves to prevent them from extinction, particularly in India, Indonesia and some African countries. Nature reserves were used once to preserve the animals that landowners hunted, but, in the 19th century, they became places where animals were kept to prevent them from dying out. Special refuges and sanctuaries are also often designated to protect certain species or groups of wild animals or plants, especially if their numbers and distribution have been significantly reduced. They also serve as a place for more plentiful species to rest, breed or winter. Many parts of the world also have marine and aquatic reserves to protect different species of sea or freshwater plant and animal life, as, for example, on Australia's **Great Barrier Reef.**

NEKTON describes animals which swim powerfully enough to make them independent of tides and currents. Fish predominate: demersal fish are those which live on or near the sea-bed. They include round-fish such as cod, haddock, hake and dogfish. The best-known demersals are probably the flatfish which lie on the sea-bed: plaice, sole, turbot, brill, skate and ray. Most of these demersal fish lay huge numbers of eggs which float to the surface. The subsequent fry add to the spring **plankton.**

NEPTUNIUM (Np) is a radioactive, silvery element of the **actinide** series of substances. In 1940, it became the first **transuranic** element to be synthesized. It forms part of the **high-level waste** products that are produced in spent nuclear fuel. Neptunium is particularly hazardous because it is an **alpha-emitter.**

NGANA is a parasitic disease found in cattle in tropical Africa and spread by the **tsetse** fly. As a consequence of the disease millions of square miles in the sub-Saharan region cannot safely be used for rearing livestock.

NIMBUS-7 is a meteorological satellite, operated by NASA, the American space agency, which found in the second half of 1992 that global **ozone** levels were 2%–3% lower than in any previous year and 4% lower than normal. The very lowest levels were observed in December 1992 when the global average of ozone levels was approximately 280 **Dobson units.** By comparison, a normal December value is about 293 Dobson units. Previously, the lowest level, of 286 Dobson units, had been observed in December 1987. The 1992 ozone levels were especially low in the mid-latitudes of the Northern Hemisphere. The December 1992 mid-latitude ozone levels were 9% below normal. The low mid-latitude ozone values continued into 1993. The January 1993 ozone levels were 13%–14% below normal. Scientists can only speculate on the cause of the low ozone values in 1992. While the exact cause is unknown, the low ozone may be related to the continuing presence of particles produced in the upper atmosphere following the eruption of **Mount Pinatubo** in the Philippines in June 1991. Nimbus-7 has measured ozone levels since November 1978 and continues to be the primary monitor of global ozone levels. Its measurements are backed up by the NASA TOMS instrument orbiting on the Russian Meteor-3 satellite, which was launched in August 1991, and the **Shuttle Solar Backscatter Ultraviolet (SSBUV)** instruments flown on the space shuttle and the NOAA-11 satellite.

NITRATE-SENSITIVE AREAS have water sources with high levels of **nitrates,** which are thought to have an affect on health. High nitrate levels are a problem and add to the costs of providing drinking water. Research has identified adjustments that could solve the problem of nitrate **leachates** and a five-year pilot project in the UK was studying 10 farms within nitrate-sensitive areas, each within the **groundwater** catchment area of a public supply source. Within these areas, ranging in size from a few hundred to several thousand hectares, a variety of voluntary agricultural methods were being carried out. Each area has different hydro-geological, land-use and climatic conditions.

NITRATES/NITRATE RELEASE FERTILIZERS are man-made fertilizers, based on ammonium nitrate, that have been spread intensively in the past 40 years to provide the **nitrogen** needed for rapid growth of crops. Surplus nitrates are washed into rivers and threaten water supplies, causing a potential health hazard for both children and adults.

NITRIC ACID is one of the most widely used reagents within the chemical laboratory and industry. It is formed in the atmosphere by chemical reactions involving the nitrogen oxides discharged from coal- and oil-fired power stations and petrol-driven vehicles to produce a damaging and corrosive environmental pollutant.

NITRIFICATION is the process by which **ammonia** compounds, including man-made **fertilizers** and the **humus** provided by organic matter of plant and animal origin, are converted into **nitrites** and then **nitrates,** which are then absorbed as a nutrient by crops. Excess nitrate can be leached into surface waters and groundwaters, causing pollution. Excess nitrate may also be converted by microbes back into gaseous **nitrogen,** which is an important greenhouse gas, and released back into the atmosphere. The ultimate source of nitrogen in the ecosystem is the molecular nitrogen in the atmosphere. To a very limited extent, some dissolves in water. However, none is found in rock.

A **NITRITE** is a compound which is formed as an **intermediate substance** when **ammonia**-based fertilizers, whether man-made or of natural origin, are converted by microbes in the soil to nitrate. On heavily fertilized land, the rate of formation of nitrite may be too rapid and some of it gets leached into waterways as pollution.

NITROGEN is an essential **nutrient** in the food supply of plants and the diets of animals. Animals obtain it in nitrogen-containing compounds, particularly **amino acids.** Although the atmosphere is nearly 80% gaseous nitrogen, very few organisms have the ability to use it in this form. The higher plants normally obtain it from the **soil** after micro-organisms have converted the nitrogen into **ammonia** or **nitrates,** which they can then absorb. This conversion of nitrogen, known as nitrogen fixation, is essential for the formation of amino acids which, in turn, are the building blocks of **proteins.** Nitrogen starvation of a plant is indicated by yellowness of its green leaves and shoots, and retarded flowering and fruiting. Some plants have special root nodules which are better at absorbing the nitrogen substances. In modern intensive agriculture, farmers have increased the production of crops by spreading nitrate-based **agrochemicals** on the land. While productivity has risen, the excess fertilizers have become environmentally damaging. They drain into waterways and cause **eutrophication.**

The **NITROGEN CYCLE** is the complex set of processes by which crops acquire the large amount of **nitrogen** they need to make **proteins, nucleic acids** and other biochemicals of which they are composed, and how the nitrogen returns to the atmosphere. The ultimate source of nitrogen in the global ecosystem is the molecular nitrogen in the atmosphere. To a very limited extent, some dissolves in water. However, there is none in rock. Yet many arable crops will remove at least a ton of nitrogen, in the form of **nitrate,** from each five hectares of land during the growing season. Molecular nitrogen from the atmosphere first enters biological pathways by assimilation, or nitrogen fixation, by a variety of micro-organisms which live freely in **soil.** Once the nitrogen has been captured by one of the micro-organisms, there are many different routes, or processes, by which it is handled. So the nitrogen cycle can be divided broadly into processes which add nitrogen in the form of nitrate to the soil and the processes that remove it. Since the recycling of nitrogen has been going on for millions of years, there are, in fact, various places where nitrogen is temporarily stored. Some is retained in the soil as decomposing plant matter, waiting to be released and taken up by new crops as nitrate. That nitrate is produced by nitrifying bacteria living in the soil that thrive on **ammonia,** which is produced by decaying plant and animal material. Nitrogen is an important element in ammonia, and in processing it the nitrifying bacteria produce nitrate that can be absorbed by the roots of plants. In addition to the nutrient-rich **humus** of decomposed plant matter or animal urea, man-made fertilizer is added to the soil in manufactured forms of nitrates and urea. The absorption of nitrates by plants is the major process of removing nitrogen from the soil. However, some nitrogen is released from nitrates by denitrifying bacteria, and escapes in gaseous form from the soil back into the atmosphere. Some drains away in the form of **leachates** of nitrates into waterways. Leaching of nitrates into surface and groundwaters is of particular concern. EC directives set a limit of 50 milligrams per litre of nitrate in drinking water or in any water where there is a risk of **eutrophication.**

NOISE POLLUTION is perhaps the most obvious example of environmental pollution, caused when noise from traffic, aircraft, factories, loud radios, or neighbouring houses and gardens, becomes a nuisance or detrimental to health. Noise can be a constant intrusion or merely an irritant. It can also cause severe damage to health and hearing, and at a high level can rupture eardrums. The level of noise is measured in **decibels** (dBs), which is the unit used for classifying the intensity of sound and allows different sources of noise to be compared. A report in the New England Journal of Medicine suggested that for 25 years noise levels had risen by about 1 dB a year. Noise is rapidly becoming a serious polluting factor. An increase in the amount of aircraft, road and rail traffic, portable radios and tape recorders, and household gadgets all contribute to growing noise pollution. There are laws covering the control of noise for road traffic, aircraft and airports. For example, the European Community issued a directive on aircraft noise which was due to come into force in mid-1993. New regulations also cover a new noise limit for buses, coaches and heavy goods vehicles.

NON-GOVERNMENTAL ORGANIZATIONS (NGOs) are private, voluntary, non-profit organizations, acting as pressure groups. Throughout the world there are more than 5,000 international NGOs which are concerned with the environment and development, with millions of supporters. In all, it is estimated that globally there are some 12,000 such NGOs, ranging from small grass-roots agencies to influential international groups like **Greenpeace** and **Friends of the Earth.** The development of such NGOs started primarily during the post-war period when movements were formed because of growing concern for the environment. After the **United Nations Conference on the Human Environment** in Stockholm in 1972, these movements expanded into developed and also developing countries. The NGOs operate on a regional, national or international level as an independent voice, sometimes acting as the focus of opposition to government policies. They are growing, both in numbers and influence, and are increasingly involved in decision-making processes at both national and international level. Some NGOs are now beginning to work with the **World Bank** as consultants during the planning and evaluation stages of bank-financed projects in areas like rural and urban development, and health and population. The NGO/World Bank Committee, which includes senior bank managers and 26 leaders of international NGOs, was formed in 1982. There are now also strong ties between NGOs and United Nations agencies. The **United Nations Environment Programme (UNEP)** and the **United Nations Development Programme (UNDP)** are working with a number of NGOs. Some NGOs can attend all UNEP meetings and co-operate on planning and carrying out projects.

The **NON-PROLIFERATION TREATY (NPT)** is an arrangement by which countries that signed the treaty allow their nuclear power stations, and other nuclear sites, to be inspected in order to demonstrate that no materials are being diverted for weapons purposes. The treaty is supervised by the **International Atomic Energy Agency,** which is based in Vienna.

NORTH SEA, see **Oslo Treaty.**

The **NUCLEAR FUEL CYCLE** indicates the fact that the production of useful power from nuclear **fission** in a nuclear reactor is only one part of a complex system. The process begins with the mining and refining of **uranium** ore. Next comes a uranium enrichment stage. Metal rods of uranium fuel are then fabricated and placed in a reactor. When removed from a reactor where they have been subjected to months of bombardment by neutrons, the fuel rods are still predominantly uranium; in addition, they now contain potentially valuable **plutonium** and potentially hazardous long-lived radioactive wastes, called **actinides,** and scores of short- and medium-term radioactive wastes. Reactors were built originally to produce plutonium for weapons. Only a small proportion of the uranium fuel rods was converted into plutonium, which was then extracted from the spent fuel by chemical **nuclear reprocessing.** When the first nuclear power stations came into operation in the mid-1950s, reserves of uranium ore were still in short supply. Therefore, it made sense to continue with the idea of reprocessing to recover the useful uranium left in the fuel. There was also the possibility of using plutonium in a new type of reactor. However, the disadvantage of reprocessing comes with the problems of handling and disposing of a wide variety of **low-level, intermediate-level** and **high-level waste** streams. All the activities from mining of uranium ore and fuel fabrication to reprocessing and waste disposal constitute the nuclear fuel cycle.

NUCLEAR FUELS are obtained from inorganic minerals extracted by mining. Although they are at least partially consumed when used in nuclear reactors for the production of heat, they differ from **fossil fuels** in the way they release energy. Burning of fossil fuels, such as coal, oil and gas, is a chemical reaction. Nuclear fuels, such as **uranium,** are destroyed by a process of spontaneous disintegration, called **fission,** and prompted by natural radioactivity. If the process is left to occur naturally in **uranium**-bearing rock, the rate of change is imperceptibly small. In a man-made nuclear reactor the energy-releasing processes of disintegration, which in the natural state happen slowly over thousands of millions of years, are compressed into minutes. The release of energy is harnessed to generate steam which drives electricity generators.

NUCLEAR RADIATION, see **radiation.**

NUCLEAR REPROCESSING is the method by which unused **uranium** and newly created **plutonium** can be recovered from spent nuclear fuel rods after they are removed from a nuclear reactor. Nuclear power technology is separated from all other energy technologies by one particular feature. Unlike the ash of residues from coal or oil burning, the spent **nuclear fuel** from a nuclear power station contains both potentially valuable material and uniquely hazardous wastes. The first nuclear reactors and reprocessing plants were not built as sources of electrical energy. The reactors were built expressly for the 'transmutation' of **fertile** uranium-238 into weapons-grade plutonium-239, which is a fissile material, meaning it is able to undergo nuclear fission, in order to achieve the splitting of the atom. However, only a small proportion of the uranium-238 was converted into plutonium, and so plutonium had to be extracted chemically. The method became called reprocessing.

In principle, the idea is straightforward. Fuel rods are dissolved in a vat of nitric acid, creating a mixture of uranium nitrate, plutonium nitrate and comparable salts of the hundreds of other by-products in spent fuel. Separation of these is the crucial stage in the process. Although the activity is a chemical engineering task, reprocessing works are far from ordinary chemical plants. Because the materials are so radioactive they must be kept behind heavy shielding at all times and every operation has to be carried out by remote control. Plutonium is still created automatically in the uranium fuel used in the reactors of commercial power stations. One-third of the nuclear fuel rods in a nuclear reactor are replaced each year. The old rods are stored for over a year in cooling ponds at nuclear power stations, to allow the bulk of the short-lived radioactive substances to disintegrate or **decay.** The medium- and long-lived radioactive substances remain in the fuel rods which are then transported in protected heavy-duty, water-filled nuclear fuel flasks to a reprocessing or waste storage plant. In Britain they are taken to the **Sellafield** site operated by **British Nuclear Fuels,** and in France to a site at Cap la Hague operated by Cogema. Reprocessing is a controversial activity because of the wastes, especially the **high-level wastes,** generated in recovering the uranium and plutonium. There are several objections by environmentalists against creating stockpiles of plutonium. It is a potential weapons material and it is highly dangerous, **carcinogenic,** and presents unusual risks when transported.

The **NUCLEAR TEST BAN TREATY** is an agreement between the nuclear superpowers to stop the testing of nuclear weapons in the atmosphere.

NUCLEAR WASTE is any unwanted by-products of controlled nuclear reactions, particularly the waste products of nuclear power stations and nuclear reprocessing. Controlled nuclear reactions involve the **fission** of **uranium** and **plutonium,** which gives rise to a host of fission products that must be kept isolated from the environment until they have decayed to a harmless state. See also **low-level, intermediate-level** and **high-level wastes.**

NUCLEAR WINTER was the phrase used to describe the potential atmospheric and climatic consequences of the dust, gases, smoke and **radioactivity** that would be generated by a nuclear war. The discovery by researchers, in the 1970s, that dense clouds of soil particles may have played a role in mass extinctions of fauna and flora on Earth refocused attention on the possible consequences of nuclear war. Leading members of the international scientific community compiled a report, *Environmental Consequences of Nuclear War,* describing the effects of widespread nuclear explosions on climate, ecosystems and food supply. Sponsored by the **International Council of Scientific Unions,** the scientists held workshops in London, Stockholm, Leningrad (now St Petersburg), New Delhi, Delft, Paris, Hiroshima, Toronto, Melbourne and Caracas. They assessed the biological effects resulting from likely changes in sunlight, temperature, rainfall, atmospheric chemistry, **ionizing radiation, ultraviolet radiation,** plant and animal growth, and the resultant agricultural productivity.

NUCLEIC ACIDS are sometimes referred to as the molecules of life. There are two types, DNA (deoxyribonucleic acid), which carries the genetic code, and RNA (ribonucleic acid). They consist of long chains of nucleotides.

NUMERICAL FORECASTING is the forecasting of the behaviour of the weather using mathematical models.

NUTRIENTS have a particular meaning for the ecologist. They refer to those chemical elements which are involved in the construction of living tissue and which are needed by both plant and animal. The most important in terms of bulk are **carbon,** hydrogen and **oxygen,** with other essential ones including **nitrogen,** potassium, calcium, sulphur and **phosphorus.** Plants get their nutrients either from the atmosphere (carbon and oxygen) or by extracting them from the soil with their roots (nitrogen, potassium, calcium and other salts). Animals depend on their food and water intake for their nutrient supply.

NUTRITION is the process by which an organism provides itself with, or is provided with, all the materials and nutrients needed for healthy growth and survival. The raw materials needed by green plants are water, carbon dioxide and a variety of inorganic substances, but particularly the so-called **essential elements** or macronutrients. There are ten of these: **nitrogen, phosphorus, sulphur,** potassium, magnesium, calcium and iron together with **carbon,** hydrogen and **oxygen.** In addition, plants require other elements in such tiny amounts that they are referred to as **trace elements,** or micronutrients. They include **boron, manganese, zinc, copper, molybdenum** and other elements. A concentration in water of a few parts per million of the trace elements is nutritionally sufficient. The sources of all these materials vary with the environment of the plant, so survival can be threatened by disruption of its ecosystem or habitat. All these substances, except carbon dioxide, enter from the soil solution surrounding the roots of the plant. Carbon enters the plant, in combination with oxygen, as carbon dioxide molecules. Water is the source of hydrogen, which, like carbon, forms part of the structural materials of plant tissue, which are synthesized in growth. Oxygen enters the plant in two ways: as part of the carbon dioxide molecule and as part of the water molecule. It is also used in making plant tissue, but in its gaseous form it is also needed for aerobic respiration.

O

OCEAN covers 70% of the Earth's surface and is very important to the human population as a source of minerals, fish, shellfish, sand and gravel, and oil and gas reserves. However, the marine environment is under threat from rising pollution, from agricultural chemicals, litter, plastic, chemical and bacterial pollution, oil spills and untreated human sewage. Shipping and dumping at sea contribute about 10% of the oceans' pollution. This pollution, in turn, enters the **food chain** causing serious food poisoning. Over-fishing has dangerously depleted fish stocks. Fishing gear, such as large-scale **drift nets** that are sometimes 60 kilometres long and not selective in the fish that are trapped in them, have caused particular concern. There are fears that by the year 2000 annual fish catches may fall short of demand by an estimated 10m.–15m. metric tons. Marine mammals have been threatened because their habitats have been damaged by pollution and competition over food supplies. **Agenda 21** has called for improved monitoring of pollution levels and compliance with shipping protocols and agreements, and stricter international regulations to reduce the risk of accidents. It has suggested that ocean spill response centres and a global database on marine pollution should be set up and that ports should have facilities for collecting rubbish, oil and chemical residues from ships. It has also said that marine species should be restored and maintained at sustainable levels and that selective fishing gear should be used to minimize waste. See also **London Convention** and **sea-water.**

OCEAN CIRCULATION is the way energy from the sun, stored in the sea, is transported around the world. The currents explain, for example, why the UK has ice-free ports in winter, while St Petersburg, at the same latitude as the Shetland Islands, needs ice breakers. Evidence is growing that the world's ocean circulation was very different during the last ice age and has changed several times in the distant past, with dramatic effects on climate. Scientific research programmes, such as that of the **World Ocean Circulation Experiment (WOCE),** are now working to tie down the critical links between circulation and climate. The long-term aim is to forecast the **climatic change** caused by the **greenhouse effect** up to 100 years in the future. The oceans are vital as storehouses, as they absorb more than half the sun's heat reaching the Earth. This heat, which is primarily absorbed near the equator, is carried around the world and released elsewhere, creating currents which last up to 1,000 years. Oceans warm slowly. As the Earth rotates and the wind acts upon the surface, currents carry warm tropical water to the cooler parts of the world. The strength and direction of the currents are affected by landmasses, bottlenecks through narrow straits, and even the shape of the sea-bed. When the warm water reaches polar regions its heat evaporates into the atmosphere, reducing its temperature and increasing its density. When **sea-water** freezes it leaves salt behind in the unfrozen water and this cold water sinks into the ocean and begins to flow back to the tropics. Eventually it is heated and begins the cycle all over again. As the current of warm tropical water in the North Atlantic heads towards the **Arctic** it passes north-west Europe, keeping the ports ice-free and the winters warm and mild. However, the warm current misses the west side of the Atlantic, so that Canada and Greenland have cold, icy winters.

OCEAN CIRCULATION MODELLING refers to a major international research effort designed to produce methods and data for predicting what is happening in the oceans, and is comparable to methods used for weather forecasting. The ocean modelling work is the ultimate goal of the many projects that form the **World Ocean Circulation Experiment (WOCE).** The WOCE experiments have been designed to meet the needs of the World Climate Research Programme, intended to improve understanding of the pattern of the currents that move vast amounts of energy, known as the global **ocean circulation.** Planet Earth is unique in having two separate but interacting fluid envelopes: the **atmosphere** and the **ocean**. Both influence the climate through the way they interact with each other. The effect of the winds on the surface layer of the ocean is referred to as one of the atmospheric forces driving the oceans; but the strength of the winds depends on how much energy the atmosphere has obtained from the ocean. Thus, while atmospheric 'forcing' drives the oceans, the ocean in turn regulates the climate of the atmosphere, until a balance is reached. The traditional way of dividing the world into three broad regions, tropical, temperate and polar zones, is possible because of that balance. This balance was reached 10,000 years ago at the end of the last ice age. Global warming now looks set to disrupt it.

Against this background, WOCE is the largest investigation ever undertaken into understanding the physical characteristics of the oceans. However, in order to produce a computerized mathematical model for forecasting the future behaviour of the oceans, it is necessary to have an accurate knowledge of their present state. The purpose of WOCE is to understand precisely how the vast amount of the sun's energy which is absorbed into the oceans is then released to regulate the climate. Although the great ocean currents act as conveyor belts carrying energy round the globe, the process of energy transfer is done unevenly. The way warm near-surface water is transported up the Atlantic to the high northern latitudes is not repeated in the same way in the southern Pacific with a comparable flow towards **Antarctic** latitudes. The monitoring of the oceans' conveyor belts is done by measuring physical characteristics, such as the heat and density of the various layers of water at different depths. However, changes to these measurements, which could come from global warming, could in turn be linked to biological characteristics of the oceans. The marine **biota** are intimately involved in the cycles of nutrients and carbon within the oceans. Concern about man-made emissions of carbon dioxide has focused attention on the role of biology in the global **carbon cycle.** Two biological processes are of key importance. Firstly, **photosynthesis** by **phytoplankton** (marine plants), which consumes carbon dioxide, thus reducing the partial pressure of the gas in sea-water and increasing its transfer (flux) from atmosphere to ocean. Secondly, living organisms die and sink to the ocean depths, a process that can be thought of as a pump maintaining a high carbon dioxide content in deep water, and a low content in surface waters and the atmosphere. The marine ecosystem is, however, not that simple. Growth of phytoplankton is limited by nutrients other than carbon; the interaction between different elements within the **biogeochemical cycle** causes a great deal of complexity. Phytoplankton compete with **bacteria** for inorganic nutrients, and are themselves consumed by higher predators, such as fish and mammals. Biological models are aimed at predicting carbon fluxes between the atmosphere and deep ocean.

The **OCEAN DRILLING PROGRAMME (ODP)** is the name of the largest multinational research programme for ocean exploration. It was started in 1968 by the world's largest oceanographic research centre, the **Scripps Institution for Oceanography,** which is based at La Jolla, California. It became international in 1975 and provides a continuous flow of discoveries about the Earth's past climate which have relevance in predicting its future climate. A temperature history of the ocean's deep waters between 70m. and 20m. years ago, based on measurements of benthic foraminifera (see **benthos**) from deep-sea drilling cores, shows a prominent warming phase and two cooling phases in the Earth's climate. Changes in climate leave their mark in the contents of individual organic compounds and species locked in **sediments.**

The **OCTANE RATING** or octane number is a measurement that indicates the properties of a particular blend of petrol (motor fuel). The primary purpose is to show that a fuel will behave efficiently in a car engine with a high **compression ratio:** hence, the regular and premium grades on sale at filling stations. Ideally, the combustion of the fuel-air mixture in an engine proceeds smoothly from the spark plug across the cylinder chamber. Since the fuel is highly compressed before ignition, the mixture may explode in several places to produce a distinctive knocking sound and loss of power. Knock resistance is achieved by a careful blending of fuels. The perfect anti-knock fuel comes by judicious mixing of a blend of iso-octane and normal-heptane, both pure hydrocarbons. A blend of 90% iso-octane and 10% heptane would have an octane number of 90. In practice, petrol with the equivalent properties is obtained by producing more complex blends at the refinery and use of **anti-knock agents** additives like tetra-ethyl-lead and tetra-methyl-lead (see **organometallic compounds**). The additives are cheap, as well as being effective as knock depressants. Leaded petrol is also an environmental hazard and is being phased out under environmental laws.

The **OFFICE RECYCLING BLUEPRINT** encompasses methods for extending the use of office stationery, photocopying and computer paper, newsprint, aluminium cans, plastic and glass. The idea is to: make two-sided copies and use the reduction feature on the copy machine; use less paper by cutting full sheets for short memos; use the back of poor or botched photocopies for notes and telephone messages; circulate periodicals and other 'FYI' (for your information) materials instead of mass distribution of photocopies; use china coffee mugs instead of paper or styrofoam cups; if you must use disposable products, substitute paper for plastic products; use white paper and colour code by using felt tip markers as coloured paper is not recyclable; recycle white or buff-coloured office stationery, corrugated cardboard and computer paper; use manila folders instead of plastic ones; use refillable pens instead of disposable ones; substitute water-based for solvent-based correction fluid; conserve energy, reduce heat and air conditioning use and general electrical use; use a plant maintenance company that minimizes the use of insecticides or uses plant-derived products.

OIL POLLUTION PREPAREDNESS, RESPONSE AND CO-OPERATION CONVENTION (OPRC), see **International Maritime Organization.**

OIL SPILLS, caused when tankers collide at sea or break up, disgorging their load in coastal waters, are a major source of pollution, causing death to sea-birds, fish and wildlife. However, these accidental spillages, dramatic though they are, are not the major cause of pollution by oil at sea. Most marine pollution from ships is the result of routine operations by tankers at sea, such as washing out crude oil tanks. An estimated one million tons of oil also seep into the seas from freighters, tankers and offshore drilling rigs each year, as do millions more tons of crude oil products, such as gasoline solvents and waste engine oil. The International Convention for the Prevention of Pollution from Ships (MARPOL) was the first comprehensive anti-pollution convention adopted by the **International Maritime Organization (IMO),** and came into effect in 1983. It also deals with pollution from chemicals, other harmful substances, garbage and sewage. The MARPOL convention greatly reduces the amount of oil which can be discharged into the sea by ships. It bans discharges in already polluted areas, such as the **Black Sea** and the Red Sea. Residues from tank cleaning operations must be discharged in port reception areas and not the sea. The convention also includes a number of modified construction requirements, such as protective and segregated ballast tanks.

ONCHOCERCIASIS is most commonly known as river blindness. It is a tropical disease found mostly in West and Central Africa and in parts of Latin America and Arabia. Onchocerciasis is a form of **filariasis** and is caused by penetration of the skin by a parasitic worm.

OPEN-CAST MINING, sometimes called strip-mining, refers to the method of excavating large quantities of **brown coal,** and iron and copper ores, by removing topsoil to expose hydrocarbon and mineral deposits near the surface.

The **ORGANIC** method is a way of growing food and husbanding eggs and livestock naturally, without the use of any manufactured additives, **fertilizers,** or **pesticides.** Organic fertilizers, such as compost and **humus,** which are produced entirely from plant sources, are used as materials for building soil fertility on farms, allotments and gardens.

The **ORGANIC GROUP** are compounds containing **carbon** and, usually, hydrogen, which are the starting point of the branch of chemistry classed as organic chemistry. When an organic group combines with elements such as, for example, the metals **lead** or **mercury,** they form **organometallic compounds** like tetra-ethyl lead and **methyl-mercury.** Organometallic substances include some of the most hazardous environmental toxins. They can be very poisonous if they are absorbed into the body by inhaling polluted air or eating contaminated food because they tend to react with the central nervous system. Originally, organic chemicals were those chemical compounds based on carbon that were found in living organisms. Subsequently, the modern chemical and pharmaceutical industry was largely created by making complex carbon-based compounds, such as plastics, synthetic fibres and drugs, from petrochemicals.

ORGANIC RECYCLING relates to the use of crop residues and manure in low-cost methods of non-intensive agriculture, which are employed in developing countries.

The **ORGANIZATION FOR ECONOMIC CO-OPERATION AND DEVELOPMENT (OECD)** is a group of 24 industrialized nations; it has drawn up rules to protect the **recycling** industry. It estimated that in 1990 40,000m. tons of waste were exported for recycling throughout the world, a large part of it to developing countries. The OECD's rules divide hazardous wastes into three categories, red, amber and green, for trade between its member countries. On the red list are dangerous wastes, such as **asbestos,** polychlorinated biphenyls (PCBs) and chlorinated **dioxin.** Ash from iron and steel manufacture, petrol, coke and bitumen, lead acid batteries, waste oils and oil/water mixtures, **phenols, thallium** compounds and hydrogen peroxide solutions are on the amber list. The green list—regarded as non-hazardous—includes scrap metals, gold, platinum, silver, iron and steel, copper, nickel, lead, **cadmium,** plastics, ceramics, mining wastes, glass, rubber wastes, textile and tannery wastes, and certain ashes from coal-fired power stations. See also **Basel Convention.**

An **ORGANOMETALLIC COMPOUND** is a compound which consists of a metal atom combined with an organic group substance. For example, tetra-ethyl lead is an organometallic compound. Industrially, organometallic compounds have an increasingly important role as catalysts. In the environment, organometallic compounds tend to be hazardous substances. Tetra-ethyl lead is used as an additive to petrol to improve its **octane rating.** However, tetra-ethyl lead emitted from car exhausts is a particularly dangerous contributor to **air pollution.** If absorbed into the body, it gets into the bloodstream and can then cross the **blood-brain barrier** to cause brain damage.

The **ORYX** once ranged over the Arabian Peninsula, but they had been hunted to near-**extinction** by the early 1960s. Herds of the Arabian oryx have been returned to Oman after what is recognized as one of the most successful results of **captive breeding.** What were thought to be the last few surviving animals were rounded up and taken to the Phoenix Zoo in the USA. The animals bred well and were distributed to European zoos. In 1980, 18 animals were returned to Oman. The animals were penned so that they could gradually acclimatize to their new surroundings and be studied. The Sultan of Oman and the Harasis tribe welcomed the oryx back and became their protectors, wardens and observers. They were released from the pens after a few months and settled down into coherent herds. By 1991 the Omani herd was still thriving, while other herds, both from other captive breeding programmes and from the small native population which was unknown to the original rescuers, continued to thrive in parks in Qatar, Abu Dhabi, Jordan, Saudi Arabia, Bahrain and Dubai.

The **OSLO TREATY** came into force in 1974 and banned the dumping of certain wastes, such as **mercury, cadmium** and persistent plastics, in the North Sea. The Oslo Treaty has similar black lists and grey lists to the **London Convention** of substances that are banned from dumping. The treaty was signed by the governments of Belgium, Denmark, France, the former West Germany, the Netherlands, Norway, Sweden and the UK. Three North Sea Conferences were subsequently organized. At the first, in 1984, it was agreed that the phasing out of the use and discharge of polychlorinated biphenyls (PCBs) should be stepped up, that wastes, including sewage sludge, should not be dumped in the North Sea and that specified hazardous waste should come under more stringent control. In 1987, at the second conference, it was agreed to reduce by half discharges of 'substances that are toxic, persistent and liable to bioaccumulate' by 1995 (see **bioaccumulation**). Restrictions on the use of marine **incineration** and the dumping of sewage sludge were also agreed. At the third conference, in 1990, the eight signatories to the original treaty and Switzerland had agreed to stop dumping sewage sludge at sea, but the UK said it would only phase this out by 1998. The dumping of PCBs was causing some disagreement among the signatories of the Oslo Treaty. Denmark, Sweden and Norway wanted this to be prohibited by 1995, Germany by 1998, Belgium, the UK and the Netherlands by 2005, and France by 2010. There is a problem in curbing the discharge by the UK and Norway of oil-contaminated waste from offshore installations in the North Sea.

OTTERS are semi-aquatic mammals which prefer water, but can also travel across land faster than a man can run. Their fur is highly prized and because of it some species have been hunted to near-**extinction.** The fur is critical to the survival of the otter because, unlike other northern marine mammals which rely on blubber to protect them from the cold, the fur is its only insulation. Otters are particularly vulnerable to **oil spills** because when oil penetrates their fur, cold water gets to the skin and the otter freezes to death. During the 1989 oil spill from the **Exxon Valdez,** many hundreds of sea otters died. Otters are also seen as enemies of fishermen because they eat abalone, clams and mussels, a popular food of many humans. The giant sea otter population had dwindled by 1910 and few thought it would survive. However, in the mid-1930s they were sighted again. River otters were among Europe's most endangered mammals, badly affected by the pollution of rivers which has contaminated the eels and fish on which otters depend. In the UK they have been protected since the 1981 Wildlife and Countryside Act was passed. As a result of a breeding scheme implemented by English Nature and the Otter Trust in Earsham on the Norfolk-Suffolk border, otters are now returning to England and Wales. Out of 51 otters released into the wild under the scheme, there is evidence that at least 27 have bred successfully in the wild.

OUR PLANET, OUR HEALTH was the name of the report of the World Health Organization (WHO) Commission on Health and the Environment, prepared for the **Earth Summit** in Rio. The report said that the deterioration of the environment, with its huge daily toll in human lives, was one of the most serious threats facing the world. Its key message was that the environmental well-being of the planet and the health of humanity depend on each other. It argued that health rarely receives high priority in environmental policies and development plans, despite the fact that the quality of the environment and the nature of development are major determinants of health.

OVERGRAZING is intensive grazing by animals, for example cattle, sheep or goats, on an area of pasture. It has become a serious threat to the world's rangelands and **grasslands.** Several factors have led to overgrazing, which leads to the soil being degraded and becoming liable to erosion by wind and rain, and even to **desertification.** The main pressures leading to widespread overgrazing have been the need to increase the size and numbers of herds to produce more food for an increasing human **population,** and the transformation of traditional pasture land into plantations to grow **cash crops.** Throughout the dry tropics, where traditionally herds ranged over vast areas, intensive livestock-rearing schemes have taken over, mostly to provide meat for the export market. Well-digging operations have also led to heavy concentrations of animals in small areas.

OXYGEN comprises about two-thirds of the body weight of plants and animals. It is an essential constituent of carbohydrates, lipids (fatty and oily substances) and most **proteins.** It is necessary, either directly or indirectly, for **respiration** in all forms of life. Some eminent biologists have said that most, if not all, the oxygen in the atmosphere is derived from the **photosynthesis** of green plants. In its natural state it is an odourless, colourless gas, which occurs in the **atmosphere,** dissolved in the oceans and other water (the **hydrosphere**) and is chemically combined in the Earth's strata (the **lithosphere**). The composition of dry air is over 20% oxygen.

OXYGEN/PEROXIDE-BLEACHED products are an environmentally softer alternative to the synthetic bleaches, which are found in conventional washing powders or used in industrial processes for bleaching paper products, tampons and disposable nappies, and can leave traces of **dioxin** contamination.

OXYGENATES are the additives put into petrol, like MTBE (**methyl tert-butyl ether**) and tert-amyl ether, and which are preferred as **anti-knock agents** to lead additives (see **organometallic compounds**).

OZONE (O_3) is a molecule that consists of three atoms of oxygen stuck together, in contrast to the oxygen in the atmosphere which only has two atoms of oxygen. The ozone form is very reactive and is toxic to humans and living matter. However, in the **stratosphere,** ozone provides a protective layer that blocks harmful **ultraviolet radiation** from the sun. A fall in the level of stratospheric ozone allows radiation to reach the ground, causing skin cancers and killing sensitive plants. Ozone destruction occurs when man-made gases, particularly chlorinated chemicals, **chlorofluorocarbons (CFCs)** and halons, are discharged into the atmosphere. When they reach the stratosphere they decompose and release **chlorine** atoms that trigger an ozone-destroying reaction. The chlorine remains unchanged and can cause 100,000 destructive reactions before it eventually leaves the atmosphere. So 100 grams of CFCs emitted from aerosol sprays, in which they are used as propellants, can destroy over 3 tons of ozone. Ozone is also formed at street level by the action of sunlight on the nitrogen oxides and unburned hydrocarbons released from vehicle exhausts. Low-level ozone is a nose and chest irritant that can precipitate severe bouts of **hay fever** and provoke bronchitis among the elderly.

The **OZONE HOLE** is the description given to the discovery by Dr Joe Farman, in 1982, of an alarming deterioration in the **ozone layer.** This was formally confirmed and published in the scientific magazine *Nature* in 1984. Measurements of ozone had been routinely made for decades at the start of spring at the **British Antarctic Survey** base, at Halley Bay. Regular seasonal records are kept of **Dobson units.** The units are measurements made with a special ground-based instrument showing the total amount of ozone in a vertical column of the atmosphere directly overhead. Each subsequent spring the hole in the ozone layer, which forms a protective screen in the sky to filter harmful **ultraviolet radiation** from the sun, had got bigger.

The **OZONE LAYER** is the protective layer of ozone in the upper **stratosphere,** found at a level between 15 and 30 km above the Earth, which prevents harmful ultraviolet B radiation, which causes skin cancer and threatens plant life, from reaching the ground. The fragile shield is being damaged by chemicals released on Earth. The main chemicals that are depleting stratospheric ozone are **chlorofluorocarbons (CFCs),** which are used in refrigerators, **aerosols,** and as cleaners in many industries, and halons, which are used in fire extinguishers. The damage is caused when these chemicals release highly reactive forms of **chlorine** and **bromine.** Over the past 30 years ozone levels over parts of **Antarctica** have dropped by almost 40% during some months and a 'hole' in ozone concentrations is clearly visible in satellite observations. However, it was not until the early 1980s that the first evidence of this damage was discovered.

P

PALAEOCLIMATOLOGY is the study of climate conditions in the geological past. While most scientists are sure that an increase in the levels of carbon dioxide and other **greenhouse gases** in the atmosphere would result in increased **global warming,** they are still uncertain about the significance for the effects on the climate of natural events: **sunspot** activity, volcanic dust and phenomena like the **Milankovitch** cycle and the onset of ice ages. Consequently, studies of past climates are intended to help remove some of the uncertainties about the factors influencing climatic change. Although there are few simple and accurate indicators of very ancient climates, proxy data is available from which to assemble a picture of the environment. The remains of pollens, beetles and molluscs are valuable because of their profusion. Knowledge of the habitats and ecosystems in which today's flora and fauna flourish, and of how the various species are distributed, provides a yardstick for comparison. The occurrence of the fossil remains of those species similar to today's species of animals and plants is indirect evidence of what the temperature and seasonal conditions must have been like at the time those remains were formed. However, the biological data is often inconclusive and needs confirmation by lithological, geochemical and palaeomagnetic indicators. The lithological evidence is the record contained in sediments and rocks, which show the type of weathering that created those formations. Geochemical indicators include evidence of the composition of the calcium carbonate taken from marine fossil shells in ocean cores. The technique is to measure the ratio of the **isotopes** of oxygen in the fossil shells. This reflects the temperature of the water when the shell was formed during the lifetime of the animal, and the chemical composition of the **sea-water** at the time. Palaeomagnetism is a study of the Earth's geomagnetic field in the geological past.

PALAEONTOLOGY is the study of fossil plants and animals and of the rock formations where they are found. The information is used increasingly in **environmental sciences** to build up a picture of past environments, the patterns of **climatic change,** the species of plants and animals that existed, and those that may already have disappeared through **extinction.** As concern grows about the increasing loss of species because of human activities, it is essential to compare past data with the present to assess what damage is being done to the world's **biological diversity.** The effect of losing species is not known, but past studies should help to identify the danger signs when a species or population drops dangerously low, or is killed off by pollution, for example, or by predators.

The **PANEL OF EXPERTS ON ENVIRONMENTAL MANAGEMENT FOR VECTOR CONTROL (PEEM)** is the organization which promotes alternatives to **pesticides** in controlling **vector-borne diseases.** Since it was formed in 1980 it has focused specifically on the various aspects of water resources development associated with **water-borne diseases.** The United Nations Environment Programme, the Food and Agriculture Organization and the World Health Organization use the PEEM to promote environmental means of controlling the vectors of diseases such as **malaria, schistosomiasis** and **sleeping sickness,** which affect millions of people every year.

PARAQUAT is a **herbicide** and **defoliant.** Unlike most other herbicides it is highly toxic to mammals and has been responsible for poisonings, some of them fatal. Less than a teaspoonful of paraquat is likely to be lethal if swallowed, while it can also be absorbed through the skin. There is no known antidote. Concentrated paraquat is listed as a poison in the UK and is only sold to professionals. The USA has restricted its use to licensed operators and it is also severely restricted in many other parts of the world but is widely used in developing countries. When used properly as a spray in agriculture or horticulture it is regarded as fairly harmless because it sticks to the soil and cannot poison other plants. However, there is some evidence that some soils do not hold it so tightly and that it can drain into rivers where it becomes extremely toxic.

PARASITISM describes the relationship between two organisms in which one, the parasite, lives on or within the other, the host. Unlike **symbiosis,** where two organisms have relationships which are mutually beneficial, parasites depend on the host in a variety of ways, and usually inflict harm by causing severe illness and even death. Humans are infected by diseases transmitted or caused by almost every kind of parasitic organism, from viral infections, such as the common cold and AIDS, to skin diseases contracted from **fungi** and ringworm, to **malaria** and **sleeping sickness,** in which the parasites, respectively the **mosquito** and the **tsetse** fly, invade many hosts during their life cycles. Malaria, for example, is spread when an already infected mosquito bites a person to feed on their blood and passes on the disease they have picked up after biting an infected person. Several of the most serious tropical illnesses are caused by macroparasites, microscopic organisms that include parasitic viruses, **bacteria** and fungi. Larger organisms, such as worms, insects, ticks, vertebrates and plants, often give birth to offspring on a host and leave it to establish new infections. The tropical disease **filariasis** is caused by nematode roundworms which are spread from person to person by mosquitoes. The worms cause inflammation and eventually block the lymph vessels which, in severe cases, results in **elephantiasis,** an extremely disfiguring disease.

PAST ENVIRONMENTAL CHANGE is the record of past **climates** and **climatic changes** which have been preserved in rocks, ice and historical records. They give a unique insight into the environmental processes which have controlled the Earth in the past. This geological record provides a partial record of past events, their causes and the impact they had. Generally, the older the rocks are, the more general is the information they contain. Records cover the last 300,000 years and they are particularly good for the last few thousand years. Oceans and lakes provide sinks for **sediments** and fossil material, which give a continuous, detailed record of the physical, chemical and biological processes associated with environmental change. Similar records are also found in ice cores, **permafrost, peat,** cave deposits and tree-rings. This information is used by applying methods from **palaeoclimatology** to examine the impact and rate of influence on climate systems of identified forcing factors, particularly those which are not captured in instrumental or historical records. A variety of geological information on a single event, or era, makes it possible to reconstruct past climate systems. These past eras may, to some extent, mirror conditions we could expect in the future. By understanding these past processes and their impact, it is, therefore, possible to predict future processes and their impact.

A **PATHOGEN** is any disease-producing organism.

PEAT is a dense accumulation of only partially decomposed plant materials, which has formed in **wetlands** over a period of thousands of years. It collects in still water, which does not have enough oxygen to complete the total process of decay. Peat, in fact, is the first stage in the formation of **coal** and **lignite.** It is used as fuel in Ireland, Scotland, the Netherlands and Scandinavia, and traditionally rural communities have cut the peat over the years without causing any environmental harm. However, mining the peatlands destroys them faster than the rate at which they can regenerate. There is concern now that too many peatlands, a vital element in the **water cycle,** are being destroyed. Peatlands are home to leeches, worms, molluscs, arthropods, fish, amphibians and wildfowl. They are important as habitats for birds and for many endangered species. **Air pollution** is degrading the peatlands and the danger is that when peatlands are oxidized they release tons of carbon dioxide, a major contributory factor in the **greenhouse effect.**

PEDOLOGY is the scientific study of the origin, structure, composition, distribution and classification of **soil** types.

PENTACHLOROPHENOL (PCP), one of the universally toxic phenolic compounds, is a general purpose agent that is used as a **fungicide, herbicide** and **molluscicide,** particularly in Egypt where it is used to control snails that carry the larval human blood flukes that cause **schistosomiasis.** It is also used in wood preservatives and is very poisonous.

PERMAFROST is permanently frozen surface ground that has been below freezing point, 0°C (32°F), for more than two years. It is estimated to cover 20% of the world's land surface, including glaciers. It is found in the **Arctic,** sub-Arctic and **Antarctica.** It can penetrate to depths of up to 700 metres. In the summer months the upper layers of the ground may thaw, but there is concern now that if the temperatures rise as a result of **global warming,** this could have a dramatic impact on the Arctic permafrost. The permafrost is a vast storehouse of rotten organic matter that would produce vast quantities of the greenhouse gas, **methane,** if defrosted. The fear is that global warming would thaw the permafrost, and the huge surge in methane would in turn accelerate global warming even more. See also **polar tundra.**

PEST is the term used to describe an animal, insect, plant, or other organism that is considered dangerous or inconvenient to man because it either spreads disease or destroys crops. To the farmer, pests include insects and mites that damage crops, weeds that compete with field crops for **nutrients** and water, plants that choke **irrigation** channels or drainage systems, rodents that eat young plants and grain, and birds that eat seedlings or stored foodstuffs. In the home, flies, **cockroaches,** moths, slugs, snails, **termites,** beetles and bugs, rats and mice are all regarded as destructive pests. In tropical regions **mosquitoes** and the **tsetse** fly carry a threat because of the serious diseases they transmit. The development of a wide range of **pesticides** led to the eradication or control of many pests, but several brands have now been withdrawn or restricted because they were found to cause cancer, or because they were persistent or non-selective and caused widespread pollution. Many species have also developed **pesticide resistance.** Other methods of pest control are now being developed or expanded. **Integrated pest management** uses a variety of methods to control pests, including the introduction of natural predators, crop rotations and the use of pest-resistant varieties of plants; **biological control** relies on nature's own sets of checks and balances, by using natural predators or releasing sterilized males.

The term **PESTICIDE** has come to be synonymous with synthetic chemical compounds. They are produced to kill insects, vegetation, fungi and rodents that are regarded as a nuisance or health threat, or as an agricultural **pest** or weed. In practice, attempts to control pests with chemicals are centuries old. Three hundred years ago **arsenic** was used in honey as an ant bait. Various substances like Paris green (copper acetoarsenite—see **arsenic**), **mercury** formulations and other compounds, such as kerosene mixed with soap, were developed to protect crops. The progress of the synthetic chemical industry during the 1930s, particularly following the belated recognition of **DDT,** which had been discovered 60 years earlier, gave a new dimension to synthetic pesticide development. Modern pesticides are divided into categories which describe their main targets, such as **insecticides, herbicides, fungicides, rodenticides** and **molluscicides.** Less obvious are: acaricides, that kill mites; pediculicides, that kill lice; silvicides, that kill trees and bushes; and slimicides, that kill slimes. Agricultural and industrial experts who favour their use argue that pesticides are essential to prevent crops from being destroyed by insects. Environmentalists claim they cause an unacceptable level of environmental pollution and that not enough is known about their effect on humans. Their use steadily increased in the post-war years, but by the 1970s many, including **aldrin,** benzene hexachloride, **chlordane,** DDT, dieldrin, dinitrophenol, endrin and heptachlor, had been banned or restricted. See also: **attractants, chemosterilants, defoliants, disinfectants, growth regulators, pheromones** and **repellents.**

The **PESTICIDE ACTION NETWORK (PAN)** is a worldwide affiliation of groups and individuals opposed to the misuse and spread of **pesticides.** It has campaigned to have the use of the chemicals known as the **Dirty Dozen** phased out.

PESTICIDE POISONING is caused by the widespread and indiscriminate use or abuse of **pesticides.** The World Health Organization (WHO) estimates that 40,000 people, mostly in developing countries, die every year and another two million suffer some sort of injury as a result of pesticide poisoning. A number of factors have led to the high incidence of pesticide poisoning. Many pesticides are persistent and, even where they have been proscribed, people still ingest minute amounts of the chemicals. In some countries people do not always understand the instructions giving the safety warnings or the recommended dosage on pesticide containers, either because the labels are misleading, or simply because the operator cannot read. Some people do not wear protective clothing when using pesticides, others wash the application equipment in irrigation channels or rivers. Since many homes do not have running water, workers and their families often wash in streams contaminated with pesticides. Careless and profligate aerial spraying has exacerbated the problem. See also the **International Programme on Chemical Safety (IPCS).**

PESTICIDE RESISTANCE is the ability of a **pest** to develop an immunity to **pesticides.** The Food and Agricultural Organization (FAO) reported in 1990 that 1,600 species had developed resistance to the most widely used pesticides. Insects, mites, fungi and rodents have all developed some resistance to both chemical and biological methods of control. In the 1950s, rats developed resistance to the **rodenticide warfarin.** Thirty years later, there was evidence that they were becoming resistant to the second-generation rodenticides. The United Nations Environment Programme (UNEP) stated in 1989 that there were signs that chemicals were losing their edge in the fight against pests. Natural predators, as well as the pests they were intended for, were also being wiped out by pesticides, leaving the field open to new pest species.

PHENOL, or carbolic acid, is perhaps best known as an antiseptic. However, it is the basis of a variety of chemical industry products. Phenol was formerly obtained by the distillation of coal tar, which was a source of river pollution. Other manufacturing processes are now used, such as **benzene,** for example.

A **PHEROMONE** is a very powerful chemical produced by insects for communication. Pheromones have been synthesized in the laboratory to trap some species.

PHOSPHATE is the form of mineral compound in which plants obtain **phosphorus,** which is an essential nutrient. It also refers to any of a large group of chemical compounds derived either naturally as phosphate rocks, including sediments containing the calcium phosphate shells of former marine invertebrates, bird droppings or **guano,** or artificially, including fertilizers made by an industrial process from **phosphoric acid,** an acid derivative of phosphorus.

PHOSPHATE-FREE DETERGENTS are detergents in which a **phosphate** ingredient has been replaced by an environmentally safer substance, such as zeolite. The substitute will not act as a nutrient, and so will not cause **eutrophication** as a result of the accelerated growth of plants and micro-organisms if it is released into waterways.

PHOSPHORIC ACID is the most important acid that is derived from **phosphorus** and it is used to make phosphate salts for fertilizers. It is also used in the food processing and textile industries.

PHOSPHORUS is a vital mineral that is a major constituent of bones and teeth, **nucleic acids** (DNA and RNA), cell membranes, and energy transfer systems in the body involving **adenosine triphosphate (ATP).** In animals, 80%–85% of phosphorus is found in bone and teeth, in combination with calcium. The phosphorus cycle is less complicated than the **nitrogen** and **carbon cycles.** Phosphorus does not enter the atmosphere in any form other than occasional dust. Hence the phosphorus cycle involves only the soil and aquatic parts of the natural environment. The influx of phosphorus to rivers and lakes, together with the formation of sewage and run-off from fertilized land, stimulates the growth of small aquatic plants and micro-organisms, blue-green **algae,** with disastrous consequences for the ecosystem. Plants take up phosphorus as phosphate from the soil or water. Animals eliminate excess organic phosphorus in their diets by excreting phosphorus salts in their urine.

PHOTOCHEMICAL REACTIONS are the chemical reactions which need light or **ultraviolet radiation** to occur. The reactions range from those that are absolutely essential to life, such as sunlight-driven **photosynthesis,** and those that are a threat to life when ultraviolet radiation reacts with nitrogen oxides and hydrocarbons to produce **photochemical smog.**

PHOTOCHEMICAL SMOG is the **air pollution** caused by the action of sunlight on unburned hydrocarbons and nitrogen oxides, mostly from car exhausts. It occurs over large industrial areas and urban complexes, and causes eye irritations, breathing problems and damage to plant life. It was first reported in Los Angeles County in 1942, and later throughout California. Strict controls were established in the late 1960s in California, including the compulsory installation of **catalytic converters** and the use of **lead-free petrol.** The measures did ease the problem, but by 1989 there had been such an increase in the number of motor vehicles that incidences of smog were back to pre-1967 levels. In 1989 more stringent regulations were introduced, to be implemented in three phases. The regulations included a proposal that all cars would have to be powered by electricity or other 'clean' technologies by 2007. Free parking for petrol-driven vehicles would be abolished and, by 1994, a limit would be set on the numbers of cars per family. Petrol-powered lawnmowers and solid-fuel barbecues would be banned. Photochemical smog has been recorded in half the states in the USA and also in Australia, India, Belgium and the UK.

PHOTOCHEMISTRY is the branch of chemistry dealing with photochemical reactions.

PHOTOLYSIS is a chemical reaction produced by exposure to light or **ultraviolet radiation.** Research scientists specializing in this branch of photochemistry are experimenting with photolysis as a way of storing **solar energy** in a novel kind of **renewable energy** system.

PHOTOSYNTHESIS is the process by which plants manage to tap the sun's energy. It is a complex process, which essentially consists of two stages. Firstly, the green pigment chlorophyll absorbs some of the energy of sunlight, which activates it in such a way that part of the energy is captured chemically in the formation of a substance known as **adenosine triphosphate (ATP).** ATP becomes a temporary store of energy and, consequently, is known as an energy carrier.

The reaction also produces a chemical-reducing agent called NADPH. Both of these compounds are needed at a later stage. A second stage begins that is not dependent on sunlight. **Carbon dioxide** is drawn into the plant and converted into organic compounds in a reaction which uses the energy supplied by the ATP and the chemical-reducing properties of NADPH. The organic compounds produced are various types of sugars, needed for building new plant tissues, food storage and the **respiration** needed to generate the energy to keep the plant cells operating.

PHOTOVOLTAIC SOLAR POWER SYSTEMS use semiconductors, similar to the materials such as silicon used for microchips, for the direct conversion of sunlight into electricity (see **solar energy**). The crystals can be made into solar cells smaller than a postage stamp, in order to power pocket calculators, or large roof-top **solar panels** of several square metres. The electricity generated by solar panels may be used to charge a car-type battery, which may, in turn, provide power for household lighting, refrigeration, telephones, radio and television, and perhaps water heating. Solar power systems are environmentally clean sources of energy and have the potential to provide an inexhaustible supply of electricity, provided it can be stored in some form of battery. Although photovoltaic solar power systems were invented a long time ago, the prohibitive costs of the semiconductor materials restricted their use. So, initially, their main use was in space technology, producing electricity for satellites and spacecraft. In the 1980s manufacturers developed cheaper technologies for mass-producing solar cells, similar to printing on rolls of paper. Small photovoltaic solar power systems could be very important for families in rural parts of the developing world which are a long way from an electricity grid. **Dams** and **hydroelectric power** once seemed to be the most promising developments for bringing irrigation agriculture and electricity to developing countries. However, experience shows such schemes usually mean flooding fertile valleys and the displacement of large numbers of people in parts of the world where there are no alternative settlements available. So, although it was thought that the advent of hydroelectric power would make rural populations more productive, in practice rural economies could not afford to build the networks that would link them into the hydroelectric distribution grid.

PHYTOPLANKTON are drifting microscopic plants that, with an animal component **zooplankton,** make up **plankton.** They are so small that in a teaspoonful of sea-water there could be thousands of phytoplankton plants. They are at the root of the ocean **food chain** and play an important part in the regulation of the climate by capturing heat gases in the atmosphere, such as carbon dioxide, methane and water vapour. See also **ocean circulation modelling.**

PLANKTON form one of the most diverse assemblies of organisms and are found at the start of the **food chain** in the oceans and in freshwater. Having little or no power of locomotion, plankton drift with the tides and currents, but in incredible numbers. There are **phytoplankton** (plants) and **zooplankton** (animals). The most important of the organisms are **diatoms,** which are microscopic green plants with beautiful siliceous skeletons. Diatoms and other tiny green plants form the marine pastures for the herbivores of the sea, which in turn fall prey to the carnivores. There are great seasonal blooms of minute green plants in spring. By early summer, myriads of larvae hatch from their eggs. In early autumn, the diatoms flourish once again. Plankton are vulnerable to **ultraviolet radiation,** which is increasingly prevalent as a result of the destruction of the **ozone layer** and the growth of chemical pollutants. The phytoplankton, which need sunlight for photosynthesis, flourish down to depths of 80 m, and survive, although in diminishing numbers, down to 180 m, below which there is insufficient light for **photosynthesis.**

PLEISTOCENE describes the geological period which included the last four major ice ages, and which covered the interval from one and a half million to 10,000 years ago. The subsequent period, which includes the present day, is the Holocene.

PLUTONIUM is a man-made **transuranic** element created for the first time at Berkeley University, California, in February 1941. Only a few grams were made. Four years later several kilograms had been produced to fuel two atomic bombs. The first was tested over the desert at Alamogordo, in New Mexico, and produced the explosive power of 18,500 tons of TNT. The second was dropped over Nagasaki, Japan, on 9 August 1945 (the fissile material used in the bombing of Hiroshima three days before this was **uranium-235**). As with the first test a mushroom cloud of radioactive **fall-out** rose 40,000 feet into the atmosphere, and over a period of months it spread round the Earth. Plutonium is created in a nuclear reactor. It is made when fuel rods of uranium are bombarded with neutrons. A fission **chain reaction** follows, in which some of the uranium atoms are split into atoms of lower **atomic number,** accompanied by an immense release of energy. Other atoms, such as plutonium, are produced by a transmutation of uranium to a higher atomic number by the absorption of a neutron into its nucleus. Plutonium is extracted from the reactor fuel by a chemical separation process known as **nuclear reprocessing.** It is started by dissolving uranium fuel rods in nitric acid after their removal from the reactor. As the plutonium is extracted, great care is needed to keep the material in small quantities to avoid creation of a critical mass. Plutonium is protected for a second reason. It is regarded as the most poisonous substance on Earth. One millionth of a gram is capable of causing cancer of the lung or bone marrow if absorbed into the body.

POLAR BEARS have no natural enemies except humans, but there are nevertheless threats to the survival of the species. They need a large area in which to hunt and also reproduce at a very slow rate. They have also been hunted for many years for food, fur and sport. There are two other major threats to their survival: **vanishing habitat** and **oil spills.** Even minor changes in the food chain have a dramatic effect in a fragile environment like that of Alaska. When the **Exxon Valdez** spilled 11m. tons of oil into Alaska's Prince William Sound in 1989, severe damage was caused to the breeding grounds of the polar bear. There are only an estimated 25,000 wild polar bears left in the world in spite of the measures employed to protect them. The former Soviet Union banned the hunting of polar bears in 1956, while the USA, Canada, Norway and Greenland took measures to protect them some years later. However, in Canada and Alaska the native Inuit (Eskimos) are allowed to hunt and kill polar bears for food and fur. There are debates about whether this practice should be allowed to continue. See **endangered species.**

The **POLAR ICE SHEET PROGRAMME (PISP)** is an international project involving 23 countries. It is the largest comprehensive research study of the polar **ice sheets** ever carried out.

The **POLAR ORBITING SATELLITE** is one of the most important developments space technology brought to meteorology and **numerical forecasting.** The satellite, which circles the Earth over the poles, at right angles to the equator, provides scientists with observations and measurements of parts of the globe that were inaccessible for routine monitoring. The polar satellites enabled more accurate global measurements to be made of the amount of snow cover and likely changes that could be caused by human activity.

The **POLAR TUNDRA** are the treeless regions south of the **Arctic** polar ice, the northern areas of Siberia, Scandinavia, Canada and Alaska. The extreme cold of the long winters, with temperatures sometimes as low as −36°C, limits the vegetation in the tundra to mosses, **lichen,** sedges and rushes, with some grasses and flowering herbs. Cold, drying winds and **permafrost** just below the ground surface also restrict plant growth. The tundra are particularly fragile environments and the flat, almost featureless expanse of the tundra leaves animals exposed to the harsh climatic conditions; many avoid the worst tundra conditions in the winter. The only large mammal which stays in the tundra throughout the winter is the musk ox, which is protected from the cold by long, matted, windproof hair over thick fat layers. They huddle together for warmth and for protection against predators. During the long daylight hours of the brief summer plant growth is sufficient to attract insects, migrant birds and migrating herbivores, such as moose, caribou and elk, moving northwards from the **boreal forests.** When oil was discovered in the North American tundra, the number of people arriving as a result put a great strain on the region. Driving vehicles over the frozen ground resulted in erosion and damaged the vegetation, which can only be restored slowly. Pipelines disrupted the migratory routes of the caribou because the animals were afraid of the alien structures. The possibility of **global warming** presents a real threat to the tundra. In a warmer climate it is feared that the **boreal forest** could encroach into the region and deplete the present tundra.

A **POLLUTANT** is any biologically harmful material or agent that contaminates or degrades a natural environment, causing **pollution** of air, land or water.

The **POLLUTER-MUST-PAY** principle is the notion that those who cause industrial **pollution** should pay for its effects, either by finding ways to avoid pollution in the first place, or by providing compensation for its effects. The principle has not been easy to enforce, particularly where atmospheric pollutants are concerned, because there are so many sources and it is not always possible to trace the guilty parties. Pollution can also be blown away from its source at a great speed. However, there has been more success following disasters, such as **oil spills** from tankers or nuclear incidents. **Air quality standards,** usually fixed regionally, are often more effective in controlling pollution levels. There is concern that the cost of preventing, repairing, or compensating for pollution would, eventually, be paid for by the consumer.

POLLUTION is the act of discharging a substance, either accidentally or deliberately, that in any way damages or poses a threat to the environment.

POLONIUM-210 is a radioactive element and an **alpha-emitter.** It is produced by irradiating bismuth-209 in a nuclear reactor. The polonium is then used in combination with beryllium as the neutron trigger for atomic bombs. Although it was 30 years before the truth was admitted, polonium was released in the radioactive **fall-out** from the **Windscale fire** in 1957. Because the British government withheld the truth that bismuth was being irradiated in one section of the reactor, the cancer-causing risks of fall-out containing polonium from the Windscale fire were underestimated.

POLYMERS have become synonymous with man-made plastics derived from the petrochemical industry. In effect, polymers are long-chain molecules. They are made by linking together a relatively simple chemical compound, a monomer, to form a complex one, in a process known as polymerization. Essential parts of animals and plants, such as **proteins,** cellulose and **nucleic acids,** are polymers. In general, synthetic polymers, such as nylon, polyethylene, polypropylene and polyurethane, are stable compounds. Hence their attraction for use in textiles and fabrics, and engineering components and packaging materials, among other applications. Polymers are also the basis of concrete, glass, rubber and paper. Because they are lighter than metals, polymer-reinforced materials are used increasingly in boats, vehicles and aircraft. The disadvantages of polymers include the problem of **waste disposal.** Scientists are devising polymers that will decompose harmlessly in sunlight. However, the bulk of discarded plastics raises disposal problems.

POPULATION growth forecasts indicate that by the year 2000 there will be 6,000m. people living on Earth, and that this figure will increase throughout the next century, assuming that there are no major global disasters, such as wars, famines or epidemics of disease that might bring about drastic reductions in human numbers. A growing population puts a great strain on environmental resources, food, energy, transport and land. As more and more people compete for these resources there is a danger that they will move into previously uninhabited areas, that more trees will be cut down to provide fuel, causing **deforestation,** and that more **fossil fuels** will be used up. At present, most people still live in rural areas, but the pattern of human settlements is changing. It is estimated that by 2025 about 60% of the population will live in overcrowded urban areas, as farmers migrate to the city to seek better jobs and an easier life. Services, communications, transport and water supplies may break down under this strain. About one-third of city dwellers in developing countries live in slums and shanty towns, where clean water is scarce, fuel is expensive and refuse is rarely collected. Disease becomes endemic and violence common. The United Nations Environment Programme (UNEP), in co-operation with the UN Centre for Human Settlements (Habitat), has developed environmental guidelines for urban development which are being applied in cities in order to make urban management more responsive to the environment. **Agenda 21** has made many proposals for improving the social, economic and environmental quality of human settlements and the living and working environments of all people, particularly the urban and rural poor, through planning and management. It has advocated that developing countries should promote **reafforestation, biomass** energy and make more use of **solar energy,** and of **hydroelectric** and **wind power** as energy sources. It has also stated that all countries should adopt transport plans which encourage high-occupancy public transport, safe cycle paths and footways, and development patterns which reduce the demand on transport.

PORPOISES are small, playful and gregarious **cetaceans.** Porpoises are smaller than **dolphins** and are recognizable by the loud noise they make when they emerge from the water. They eat fish and **crustaceans,** but prefer squid.

POVERTY is linked in a vicious circle with **environmental degradation** and **population** growth. The majority of people who are unable to meet their own basic needs for nutrition, shelter and health (see **life-support systems**) live in environmentally vulnerable areas, where even minor changes in climate or land use can have a dramatic effect on the quality of the local environment and its ability to support the local population. Rapid population growth adds to the problem.

PRESSURIZED WATER REACTORS (PWRs) are the most commonly built version of the light water reactor (LWR). The PWR was a direct offshoot of the development of nuclear propulsion by the US Navy for submarines. The PWR uses ordinary water both as a moderator and coolant, and it uses enriched uranium in which the more **fissile** isotope of uranium, **uranium-235,** has been increased to form 3% of the uranium fuel. The reactor is operated under pressure to prevent the water from boiling. The reactors have a relatively small, compact core and a high power density compared with the British **Magnox** and **advanced gas-cooled reactors (AGRs).** However, the design is a source of controversy among environmentalists, who believe that an accident resulting in a sudden loss of coolant could degenerate rapidly into a meltdown of the nuclear fuel. Their fears were reinforced by the **Three Mile Island** accident.

PREVENTION OF MARINE POLLUTION BY DUMPING OF WASTES AND OTHER MATTER, see **London Convention.**

PROMPT CRITICALITY is the type of explosion in a nuclear reactor that caused the **Chernobyl** disaster, but which the nuclear industry had believed impossible. The nuclear industry had always claimed that the physical conditions of all types of nuclear reactor would make a nuclear explosion impossible.

PROTEINS are one of the three major compounds of which all organisms are formed. Thus, any change in the environment or pollutant that affects the synthesis of proteins is a threat to the survival of animals or plants.

PROTOZOA are single-cell animals, and the smallest and most profuse of all micro-organisms. They often feed on other **microbes,** primarily **bacteria** and some small **algae.** They are best known as insect and animal parasites, and some can cause **malaria** and other debilitating diseases. This, scientists think, could be turned to advantage by **genetic engineering** to develop non-chemical pesticides. The first protozoan pesticide, which was approved by the **Environmental Protection Agency** in the USA, kills grasshoppers, which can destroy millions of dollars' worth of crops every year.

PURE AIR only exists in laboratory conditions. In the natural environment the atmosphere is laden with dust particles which come from volcanic ash, fine sand from **dust storms** and sea salt, plus the particles and gases from chimneys and cars. A laboratory sample of pure, dry air comprises 78.08% **nitrogen,** 20.94% **oxygen,** 0.03% **carbon dioxide,** an absolutely vital 0.0006% of **ozone,** and then the unimportant 'also-rans' of 0.093% argon, 0.0018% neon, 0.00005% hydrogen and even smaller traces of krypton and xenon.

PYRITE is the most widespread sulphide mineral, combining a metal and **sulphur.** When combined with iron, forming iron pyrites, it superficially resembles gold in appearance and is known as 'fool's gold'. If wetland regions are drained artificially and the land aerated, pyrites locked in the soil can oxidize to create **sulphuric acid** that pollutes the land.

Q

The QUATERNARY is the geological period, still in progress, that has lasted less than two million years. In geological terms it spans very little time, but it includes **climatic change** of an unparalleled abruptness and magnitude in the Earth's history. The hallmark of the quaternary is a series of episodes of dramatic cooling and spread of **ice sheets,** interspersed with a retreat of the ice sheets, or interglacial periods. The period covering the past 10,000 years is referred to as the Holocene epoch, or the postglacial time. Some experts suggest this may be a misnomer because there is no evidence to show that we are not simply in an interglacial period. The question has obvious implications in studying climatic change and separating natural and man-made influences on the climate. Consequently, great effort is devoted to the study of quaternary climates. One of the most common techniques is that of pollen analysis, which involves analysing the proportion of different types of pollen in non-glacial sediments, to show whether the plant species was characteristic of a warm or cold climate. Faunal remains, ranging from large mammals to beetles, have also been used frequently in establishing patterns of climate change. Since 1980, **foraminifera,** fossils of primitive marine organisms from the **sediment** of the sea-bed, have been used widely to divine ocean temperatures. They have been used with **isotopic analysis** of oxygen from sediments and from glacier ice from cores taken from the continental ice sheets of Greenland and Antarctica. Isotopic analysis from ice sheets can date back more than 100,000 years. **Carbon dating** of organic material is effective for analysing material between 30,000–50,000 years old. For timescales of a few thousand years, or less, environmental scientists study **lichens,** tree rings and varves. The study of deep-sea sediment shows there have been 17 glacials (ice ages) and 17 interglacials in Europe in the last 1,700,000 years. The climatic pattern shown by these analyses closely follows the fluctuations in the **insolation** curve, the solar energy received per unit of area of the Earth.

R

RABIES is one of the oldest diseases known to man. In the 16th century it was described in the Americas as a bat-transmitted disease which was fatal to man. Dogs, cats, rats, squirrels, foxes, jackals, cattle and **bats** are reported as the main source of rabies in several countries. It is an acutely infectious disease of the central nervous system to which all warm-blooded animals, including humans, are susceptible. It is usually transmitted by bites and licks from an infected animal. It has been known to be an **airborne disease** in caves inhabited by rabid bats.

RADIATION refers, in general, to the emission of any rays from either natural or man-made origins, such as radio waves, the sun's rays, medical X-rays and the **fall-out** and **nuclear wastes** produced by nuclear weapons and nuclear energy production. Radiation is usually divided between non-ionizing radiation, such as thermal radiation (heat) and light, and nuclear radiation. Non-ionizing radiation includes **ultraviolet radiation** from the sun which, although it can damage cells and tissues, does not involve the **ionization** events of nuclear radiation. Nuclear radiations only became a cause for concern during the 1950s and 1960s. Routine monitoring of the air found that radioactive materials from atmospheric tests of atomic and hydrogen bombs by the nuclear powers spread radioactive fall-out across the whole Northern Hemisphere. Radiation exposure is taken to mean exposure to **alpha** and **beta radiation, gamma-rays,** X-rays and neutrons emitted by nuclear power plants or the wastes they release.

The **RADIATION WINDOW** is the condition that exists when the energy emitted from the Earth is in the form of radiation that passes through a cloudless sky and escapes into space. The radiation window is partially closed by **clouds** and atmospheric pollutants, which absorb its energy. As a rule, a clear night sky means a colder night, because energy escapes into space instead of warming the **atmosphere.** Since the gases in the atmosphere control the loss of energy from Earth to space, changes in the concentration of atmospheric gases can alter the **global energy balance.** The surface of the Earth absorbs short wavelength solar radiation and gets warm. Like any object that gets hot, the Earth then radiates some energy, but as radiation of a longer wavelength which, in the lower atmosphere, is readily absorbed by water vapour and water droplets, forming clouds and carbon dioxide gas. These in turn heat up. In stopping the energy escaping into space, the clouds and gases are said to close the radiation window. The result is what is called the **greenhouse effect.**

RADIOACTIVITY is a phenomenon that is associated with the nucleus of an element that happens to be in an unstable form. Radioactivity is emitted when an atom tries to convert from the radioactive to the stable form or when, in special circumstances, it undergoes **fission.** The nuclei of most naturally occurring elements are stable, with the exception of **uranium-235, radium, radon** and thorium. Most artificially produced nuclei are radioactive.

A **RADIONUCLIDE** is any element that is radioactive.

RADIUM is a naturally occurring radioactive element obtained from pitchblende, uranium ore. It was discovered by Pierre and Marie Curie in 1898 and named radium by them. It was used as the luminous additive in paint for the dials of clocks and watches, until it was discovered workers using it contracted radiation-induced cancers.

RADON is a naturally occurring radioactive gas which presents a definable cancer risk in some regions, especially those where there is a lot of granite. The origin of radon is **uranium**-238 that is dispersed throughout soil and rock. Uranium-238 slowly **decays** at the start of the creation of a long series of radionuclides which result in the production of lead in a process measured in a **half-life** of 4,500m. years. One of the decay products is radon, which can accumulate in poorly ventilated buildings.

The **RAMSAR CONVENTION** is the Convention on Wetlands of International Importance, Especially as Waterfowl Habitat. It is named after the town in Iran (now known as Sakht-Sar) where it was signed in 1971. The Convention, which came into force in 1975, was one of the first international agreements to deal with environmental problems. It is also unique because it protects a specific type of ecosystem, the **wetlands,** on a world-wide basis. Wetlands play an important role in purifying water and as a source of food and fuel. They are important breeding grounds and nursery areas for a great proportion of the world's fish catch. They are under threat from pollution, both from industry and agricultural fertilizers and pesticides, and by many human activities, including the construction of dams and canals, the removal of **peat** and forestry. By 1990, 58 countries had signed the Convention, each agreeing to specify at least one wetland which would either be protected or replaced with one of equal worth if the original site was destroyed. More than 500 wetland sites throughout the world were designated. The work of the Ramsar Convention is shared by the **World Wide Fund for Nature,** the **International Union for the Conservation of Nature and Natural Resources,** and the International Wildfowl Research Bureau. There have been moves to set up a permanent, funded secretariat.

REAFFORESTATION is the planting of trees in forest areas which have been cleared. See also **forests and woodlands.**

RECYCLING (DOMESTIC) of glass, paper and cans has three benefits for the environment: it saves valuable raw materials; it saves energy which would otherwise have been consumed making original products; and it cuts down on litter and waste refuse. Collection points for waste glass, paper and cans are becoming familiar fixtures in the UK, Europe and the USA. Glass-making is very energy-intensive and uses a great deal of sand, limestone and soda ash, but, once made, glass can be recycled virtually forever. In Switzerland, for example, more than 40% of glass is recycled and this accounts for 60% of glass production. It is estimated that every year we each consume paper or cardboard to the equivalent of two trees. It takes about two tons of wood to produce one ton of paper and the manufacture of paper products requires a great deal of energy, water and chemicals. Over 30% of paper products are made from recycled paper, including stationery, paper towels, toilet paper, egg boxes and newspapers. Making paper from recycled fibre uses 40% less energy than producing it from virgin pulp. The recycled paper industry has also found ways to improve the quality of its products. The smelting of **aluminium** cans saves 95% of the energy used to make aluminium from raw materials. The **Worldwatch Insti-**

tute reported in 1987 that if recovery rates of aluminium were doubled worldwide, 'over a million tons of air pollutants, including toxic fluoride, would be eliminated'.

RECYCLING (INDUSTRIAL) is a long-established practice, in that scrap metal, glass and paper have been mixed with virgin raw materials for decades. One estimate suggests steel produced from scrap reduces air pollution by 85%, cuts water pollution by 75% and eliminates mining wastes altogether. Similarly, paper made from recycled pulp reduces air pollutants by 75% and water effluent by 35%, and minimizes pressure on the forests. Only 5% as much energy is needed to produce **aluminium** from scrap, compared to production from bauxite ore. Of course, the classic example of recycling is the car. In 1990 Europeans sent 15m. old cars to the breaker's yard, with the figure set to pass 20m. in 2001. While scrap metal merchants thrived long before the environmental appreciation of the need for recycling, it was only with the planning of cars for the 1990s that designers of vehicles began thinking about the most economic way of dismantling as well as assembling them. A car has about 20,000 components. A traditional breaker's yard would reclaim between 50%–75% of the weight of materials in a car, depending on how much of the vehicle could be stripped. An average pre-1985 car comprises 40% sheet metal, 13% cast iron, 13% machine-forged steel parts, 10% electrical and mechanical equipment, 9% plastic, 4% rubber, 3.5% glass, 1.5% fabrics, 1% oil and grease, 0.5% copper and zinc, and miscellaneous ceramics and non-ferrous metal components, etc. Since 1960, the proportion of plastic has risen from 2%, and by the year 2000 it could exceed 20%.

RED DATA BOOKS are the series of publications produced by the **International Union for the Conservation of Nature and Natural Resources (IUCN).** They provide an inventory on the threat to rare plants and animal species. Information includes status, geographical distribution, population size, habitat and breeding rate. The books also contain the **conservation** measures, if any, that have been taken to protect the species. There are five categories of rarity status: **endangered species;** vulnerable organisms, which are those unlikely to adapt to major environmental effects; rare organisms, which are those at risk because there are few of them in the world, such as plants which only grow on mountain peaks or on islands; out of danger species, which were formerly in the above categories, but have had the threat removed because of conservation actions; and indeterminate species, which are the plants and animals probably at risk, although not enough is known about them to assess their status.

RED TIDE is **sea-water** which is covered or discoloured by the sudden growth of **algal bloom** or by a great increase in single-celled organisms, dinoflagellates. Red tides are often fatal to many forms of marine life and, in some cases, can result in human deaths because the dinoflagellates are eaten by clams and mussels which concentrate the paralysing toxins which they produce.

The **REDWOOD TREE** is the tallest living coniferous evergreen tree, and is known to reach a height of over 300 feet. The *sequoia sempervirens* (the redwood tree is also very long-lived) are found from south-west Oregon down to central California, in the USA. Their reddish brown bark is resistant to threats such as insects, **fungi** and **fire,** and can be a foot or more thick on an old tree. When a tree is cut, it sprouts and natural reproduction occurs through seed production. Redwood timber is used in carpentry and for furniture, fence posts and panelling. Knots cut from the trunk are used for bowls, trays and veneers.

REFRIGERATION was the discovery that did much to make food safer, enabling it to be kept longer, and also kept fresh during international export journeys. However, refrigerators also use **chlorofluorocarbons (CFCs),** which contribute to the depletion of the **ozone layer.** Refrigeration is responsible for the largest and fastest-growing use of CFCs in the developing world. The industrial countries, and some developing countries, have taken exceptional steps to control and, eventually, ban the production of CFCs and other ozone-depleting materials by the year 2000. However, many developing nations have not signed the **Montreal Protocol** because they are afraid that the cost of changing over to alternative, ozone-friendly technology will be too high. In response, the Protocol was amended in 1990 to create a fund, paid for by the largest CFC users, to compensate for extra costs. It is recognized that, unless the developing countries comply with the Protocol, it cannot be successful. It is expected that, given time, the CFC substitutes will become cheaper and easier to adapt to local conditions. Meanwhile, some developing countries are already being innovative in the search for alternatives to CFCs.

REGIONAL SEAS are most affected by pollution because they are either enclosed or partly enclosed and have a slow rate of water renewal. Because of this, they do not have the self-cleansing and restorative properties of oceans. Many of these seas support dense human populations and cater for an enormous number of tourists. The **Mediterranean, Caribbean** and **Baltic Seas** are especially vulnerable. The Regional Seas Programme, co-ordinated by UNEP, was started in the mid-1970s to bring together coastal nations in order to prevent further decline in ten of the the world's **coastal areas,** inshore waters and open **oceans.** The programme drew up a series of individual action plans which set a strategy for co-operation on research, monitoring, pollution control and protection, and the rehabilitation and development of coastal and marine resources. Each action plan contained a legally binding convention covering general commitments and detailed protocols dealing with specific issues, such as **oil spills,** dumping, emergency co-operation and protected areas. The Regional Seas Programme is funded by the United Nations Environment Programme (UNEP) and trust funds set up by the governments involved. Although the programme has had some success in taking steps to reduce pollution—from industrial and agricultural discharges, direct dumping from ships, oil pollution and agricultural run-off—the scale of the problem is enormous.

RENEWABLE ENERGY, or alternative energy, refers to those sources of power which produce energy without depleting the Earth's non-renewable resources such as the **finite fuels** of coal, oil, natural gas and uranium. The supplies of renewable resources come from sunlight, wind, flowing water, plants and forests, and hot rocks beneath the ground. Alternative energies include **solar energy, wind,** tidal, **wave** and **hydroelectric power, biomass** and **geothermal energy.**

REPELLENTS are substances used to repel ticks, gnats, **mosquitoes** and flies, birds, dogs, deer, moles, rabbits and rodents. They act because they are either foul-smelling or unpleasant to touch or taste. They are, therefore, less environmentally invasive than the various '-cides' which kill the irritant or **pest.**

RESPIRATION has two meanings. It is a vital process by which plants and animals obtain their biological energy, by breaking down organic molecules created in earlier chemical reactions. For example, in a primary process like **photosynthesis,** a reaction driven by sunlight involving **carbon** and **oxygen** creates complex molecules, such as **carbohydrates** and glucose, which then provide a store of energy. When this energy is needed for other purposes, both plants and animals can undo the results of photosynthesis to unlock the stored energy by respiration. During this transformation of a complex molecule, the molecule returns to a simple form like carbon dioxide when the energy is released. The essence of respiration is that the energy stored in compounds like starch and sugar is available on demand when needed by the organism's fundamental power units, known as **adenosine triphosphate (ATP)** and NADPH. They need the energy for fuelling the movement of muscles or for making some other molecule. The other meaning of respiration is a physical-chemical process of transporting oxygen to, and carbon dioxide from, tissues, and is part of breathing.

RESTRICTION ENZYMES are the special agents which, in 1971, two American microbiologists, Daniel Nathans and Hamilton Othanel Smith, discovered could be used as biochemical scissors to snip genes from DNA (see **nucleic acids**). This discovery became the foundation of genetic engineering and the modern biotechnology industry.

The **RHINE** is the largest, and reputedly the most polluted, Western European river. It has its source in Switzerland and flows through France, Germany, the Netherlands and into the North Sea. The river passes through major urban and industrial areas and for many years it has suffered from severe pollution by the large volumes of industrial waste discharged into it from chemical plants on its banks (see also **Sandoz**). Concern about the increased salinity of the river goes back to the 1950s. In 1963, France, the former West Germany, Luxembourg, the Netherlands and Switzerland formed the International Commission for the Protection of the Rhine against Pollution (ICPRP), which was given the responsibility of monitoring pollution in the river. In 1976, in an effort to curb increasing pollution, members of the ICPRP signed the Convention for the Protection of the Rhine against Chemical Pollution. This was followed in 1986 by the Rhine Action Programme, which aimed to reduce sediment pollution, restore the environment of the river so that indigenous aquatic life would return and produce safe drinking water from the Rhine. The action plan agreed that the discharge of 30 priority pollutants would be reduced to half the pre-1985 levels by 1995. In 1991 the German chemical industry federation agreed to reduce the flow of toxic chemicals into the Rhine. At the same time, domestic pollution controls have been strengthened. These international initiatives have already produced some improvements in the water quality of the Rhine: the concentration of **heavy metals** has fallen and the biological treatment of organic waste has reduced depletion of oxygen and the number of fish killed.

RHINOCEROS populations in the past 25 years have decreased by as much as 85%. The number of black rhinoceroses, for example, has dwindled from 65,000 to less than 5,000. It is now a critically **endangered species** and a losing battle is being fought against poachers who kill for the rhinoceros's horn. Several African countries brought in strict measures to prevent the killing of rhinoceros, but still the rhinoceros is hunted down, mostly because these measures have increased the price of the scarce rhinoceros horn. Trade in all rhinoceros products is banned in most countries. Education programmes to reduce the demand for rhinoceros horn have been implemented, as have **captive breeding** programmes. Rhinoceroses have been moved to protected reserves and ranches in Africa and 10 black rhinoceroses have been taken to the USA. A count of black rhinoceroses in Zimbabwe in 1992 revealed that there were only 250 left in the wild; another 160 were in special game ranches. Zimbabwe is also trying to discourage poaching by de-horning black rhinoceroses. The horns grow back at a rate of about three inches a year.

RHIZOBIA are a genus of bacteria that live in the roots of leguminous plants, such as the pea family, in a symbiotic relationship (see **symbiosis**).

RISK MANAGEMENT is the process by which early efforts and assessments are taken to prevent environmental risks or accidents. It is a major part of **Towards Sustainability,** the European Community's Fifth Environmental Action Programme, and emphasizes four key areas of concern. One is industrial accidents and hazards, and another covers chemicals. The number of chemicals has greatly increased, so that now there are some 100,000 used in manufacturing industries. Most of these can be hazardous if they are incorrectly applied or released in large quantities, and their effects are either unknown or long-lasting. The European Commission has been assessing the potential risks of existing chemicals and has targeted 50 specific chemicals in comprehensive programmes to reduce risks. Another area of concern has been product-labelling, so that dangerous substances can be identified quickly in an emergency. The fourth area of concern is **biotech,** because it is not always known what effect certain **genetically modified organisms** could have on the environment in the long term. The Fifth Environmental Action Programme planned to set up a standardized set of environment risk assessment requirements and safety measures for biotechnology products.

RODENTICIDES are **pesticides** used to destroy several small mammals, especially rodents, which damage either houses, stored products or cultivated crops. Rodents breed prolifically and are constantly competing with people for food. In the developing countries, where grain is stored in the open, rats are known to eat as much as 20% of grain crops. There are other ways to destroy rodents, such as shooting or trapping, but poisoning is the most widely used, because it is both economical and effective. Rodenticides differ in their chemical make-up and also in the way they affect animals. The most successful group are the coumarins, or anticoagulants, a classic example of which is **warfarin.** They require repeated ingestion over several days, making the rodent weaker by the day. Coumarins inhibit the material in the blood that causes blood to clot, finally resulting in internal bleeding.

The **ROYAL BOTANIC GARDENS,** at Kew in Surrey, have been at the forefront of botanical research and plant conservation for some 200 years. They house the largest and most diverse collection of living plants, and the most comprehensive research collection of preserved plant material in the world. Kew's botanists and horticulturalists advise on environmental management, preserve **endangered species** as seeds or living plants, and carry out research programmes to identify alternative crops as sources of medicines, food, and fuel. Kew Gardens' first director, Sir William Hooker, started the programme of **'economic botany',** the study of plants that were 'either eminently curious or in any way serviceable to mankind'. He established Kew as a centre of experimental scientific research. Regular expeditions are mounted to remote regions of the world by botanists from Kew, working in close co-operation with scientists from the host country, to collect seeds, seedlings or cuttings of plants that are endangered, or thought to be near **extinction,** and some that have not been discovered before. They are grown in conditions as much like their native habitat as possible. The plant collection is used for research, and cultivated plants are exchanged with other botanic gardens, scientific institutes and universities for a whole range of studies. Many are then reintroduced into the wild. Today, as natural plant habitats throughout the world are being destroyed and plants are becoming extinct, a priority of the research at Kew is to assess the value of what is being lost, so that a way can be found to reduce and reverse the rate of destruction of the world's plant species. The **World Seed Bank,** which houses the seeds of over 2,500 species of flowering plants—one of the world's largest collections—is at the Royal Botanic Gardens' annexe at Wakehurst Place in Sussex.

S

SALINIZATION occurs when salt concentrations in the **soil** build up and leave it unable to support plant life. All soils contain a small amount of salt, but if there is regular rainfall it is flushed out of the soil into the **groundwater** or flow through streams and rivers to the sea and so do not accumulate in the soil. However, when rainfall is low and evaporation rates high, as they are in arid areas, the soil tends to have a naturally high salt content. Salinization occurs when the delicate salt balance is upset. The salt builds up in the root zone of crops, or sometimes forms a salt crust on the surface. A major cause of salinization in arid zones is **irrigation** schemes which are continued year after year without the land being left fallow. If the land is not well drained, the **water table** rises and brings the salt to the surface. The UN conference on **desertification** reported that one-tenth of the world's irrigated land is waterlogged. Salinization and waterlogging is badly affecting many countries, including Pakistan, Egypt, Iraq, India, Argentina, Peru and the USA.

SALMONELLA is the general name for a family of micro-organisms, one of the largest groups of bacteria, that includes those most frequently implicated in **food poisoning** and gastro-enteritis. Unhygienic handling and inadequate cooking of poultry and meat, improper storage of cold meats and, more recently, contamination of battery-reared hen eggs, are the most common sources of salmonella infections.

SALT MARSHES are areas of **brackish,** shallow water usually found in **coastal areas** and in **deltas.** There are also inland marshes in arid areas where the water has a high salt level because of evaporation. They were once considered to be waste lands, but recently the importance of the salt marsh ecosystem has been recognized. They are environmentally delicate areas, extremely vulnerable to pollution by industrial or agricultural chemicals, or to **thermal pollution,** which often results when river water has been used as the coolant in power stations and industrial plants. The water temperature of the salt marsh is usually higher than that of the adjacent sea, and the water is usually shallower. These factors contribute to increased **photosynthesis** and, therefore, a flourishing plant growth that supports a large number of animals. Soil brought to the marshes by rivers and the droppings of animals that feed there also increase the fertility of the marsh. Many animals that live in the continental shelf region, especially fish and **crustaceans,** spend part of their lives in salt marshes, either as larvae or as feeding adults.

The **SANDOZ** chemical factory, near Basel in Switzerland, caught fire in the early hours of 1 November 1986. More than 30 metric tons of **herbicides** and **insecticides** from exploding drums were washed into the River **Rhine.** The fire started with an electrical fault in a warehouse which contained 1,246 metric tons of chemicals, principally insecticides. A slick spread 30 miles down the river, destroying hundreds of thousands of fish and threatening the drinking water supply for towns on the river. Two hundred miles of the upper Rhine lost most or all of its marine life and the West German government ordered that all wells along the Rhine should be closed. People in towns near Bonn had to have water ferried in by tanker. Fish, birds, insects and river plants were killed between Basel and Karlsruhe in West Germany.

SATELLITE TRACKING TECHNOLOGY is being used to track sperm **whales** in the Gulf of Mexico. The venture started in the summer of 1993. Twelve sperm whales are being followed with the aid of space satellites by a team of marine biologists and marine mammal specialists from Oregon State University. Concern about the sperm whale's status as an **endangered species** has persuaded the US Department of the Interior's Mineral Management Service to fund the research. The department is also responsible for leasing offshore gas and oil exploration areas. With oil and gas activities in the Gulf of Mexico moving into deeper water, it has been accepted that there is a need to know more about sperm whale activities in these deep water areas in order to prevent any conflicts. Sperm whales are one of the deepest- and longest-diving of all species of whales. They can reach depths of more than two miles, remaining underwater for more than an hour, and have even become entangled in transatlantic telephone cables. The presence of sperm whales in the Gulf of Mexico was confirmed in 1990 during aerial surveys. For the satellite tracking of whales, a radio transmitter is attached to the mammal. Two types of radio tags were being used. One, about the size of a flashlight, records the depth and duration of each dive, and water temperature. The other tag, a miniaturized version one-fifth the size of the first, will allow the researchers to track the whale's location for a much longer period of time. Sperm whales are the largest of the toothed whales. Adult males reach as much as 60 feet in length and may weigh 42 tons. Adult females grow to 38 feet in length and weigh upwards of 20 tons. Sperm whales feed almost exclusively on squid, including giant squid more than 50 feet long, and reach ages of 65 years and older. Until 1980, when hunting was banned, sperm whales accounted for much of the annual world whale harvest.

The **SATELLITES AND SURFACE METEOROLOGY PROJECT** is an enormous investigation that forms part of an even wider research venture, the **World Ocean Circulation Experiment (WOCE).** The overall purpose of WOCE is to understand exactly how the **oceans** regulate the world's climate and to accumulate data in order to produce computer models for predicting the effects of global warming on the oceans, and vice versa. The oceans contain vast currents that are, in effect, huge conveyor belts, carrying energy absorbed from the sun from the equator to the polar regions (see **ocean circulation**). However, the oceans also release large quantities of energy into the atmosphere, and these releases are primarily responsible for the world's weather patterns (see **atmospheric circulation**). The purpose of the Satellites and Surface Meteorology Project is to understand more fully the interaction between the sea surface and the atmosphere. Remote sensing of the sea surface from space has become an established technique for regular, large-scale observations. Measurements from satellites can include sea surface temperatures, wind speeds, wave height and slopes in sea level, which are all related to ocean currents. The main spacecraft used for satellite measurements are the European Space Agency's Earth Research Satellite (ERS) series, the joint US/French TOPEX/POSEIDON programme and the US National Aeronautic and Space Administration's **Earth Observation System (EOS)** programme. The most important measurements are the transfers (or fluxes) of heat, water and momentum between the air and sea, which are needed for driving ocean models and for verifying climate models of the 'coupled' ocean-atmosphere system. Measurements of these fluxes include the use of a special instrument called the scatterometer on the ERS-1.

A **SAVANNAH** is typically an area of tropical **grassland,** with a combination of abundant high grass and low tree growth in Africa. Savannahs have a seasonal climate with wet and dry seasons, and they extend across a wider area than tropical forest. They are the grazing ground for wild animals and, more recently, for domesticated species, such as cattle and goats.

SCHISTOSOMIASIS, or bilharzia, is a debilitating parasitic **water-borne disease.** It is transmitted by a species of water snail, which acts as host to the first larval stage of flukes, schistosoma. It is estimated that more than 200m. people in 74 tropical countries suffer from the disease and at least 600m. more are at risk of contracting it. Approximately 200,000 people die from schistosomiasis every year. Most at risk are those who come into contact every day with water, through swimming, fishing or washing clothes. When the flukes leave the snail, they pass through human skin and produce large quantities of eggs, which pass into the intestine or bladder. These cause anaemia and inflammation. Cirrhosis of the liver and enlargement of the spleen can follow. A large number of the eggs also pass out of the body in urine or faeces and, consequently, the cycle continues. **Molluscicides** in the water can control the snail hosts, while a vaccine is in the development stage.

The **SCIENTIFIC COMMITTEE ON PROBLEMS OF THE ENVIRONMENT (SCOPE)** is a committee established in 1969 by the **International Council of Scientific Unions (ICSU).** Its brief is: to evaluate the information available on man-made environmental changes and the effects of these changes on the population; to assess the methods of measuring environmental parameters; and to provide an international information network on current research and a freely available body of specialist advice.

The **SCRIPPS INSTITUTION OF OCEANOGRAPHY** is a centre for research in marine sciences, especially marine geology, **sediments,** air-sea interactions, **ecology** and **palaeontology.** The Institution, founded as the Marine Biological Station in 1903, launched one of the largest environmental projects with the international **Ocean Drilling Programme.** The Institution was also responsible for the establishment of the **Mauna Loa Observatory** in Hawaii, where accurate monitoring of carbon dioxide levels in the atmosphere began in 1957, providing the most accurate, long-term, modern record on which **global warming** calculations are based. The Institution is part of the University of California at San Diego, California. The Institution uses research vessels as well as satellite receiving equipment.

SEA-GRASSES grow close to the shores in the shallow waters of temperate or tropical seas. They play a valuable role in trapping **sediment,** slowing down erosion and helping to keep waters clear. Many species of fish and shellfish are dependent on sea-grass beds at some stage in their lives, either as nurseries or as a life-long habitat. Sea-grasses are very vulnerable to pollution, especially from sedimentation, when shallow coastal waters are affected by the erosion of agricultural land. They are also under threat because of the depletion of **mangrove forests,** and due to dredging, the development of **tourism** and oil production, and the release of industrial wastes.

SEA ICE plays an important role in substantially modifying interactions between the ocean and the atmosphere (see **atmospheric circulation**). Ice disrupts the normal exchange of energy and momentum between the two. It can have a very significant effect on the climate because it effectively insulates the sea from the cold atmosphere. The melting and freezing of sea ice influences the structure of the ocean, which can affect **ocean circulation** and the ability of oceans to remove carbon dioxide from the atmosphere. Data from a range of remote sensing systems can now determine the expanse, age and, to some degree, the thickness of sea ice. Altimeter data and microwave imagers are used to map the extent of ice cover.

SEA-LEVEL rises are a possible consequence of **global warming.** As the amount of free water in the ocean increases, and as the water becomes warmer, global warming will increase. In addition, according to theory, the heating at the poles may reduce the amount of water trapped in glaciers and ice caps. By the year 3000, the seas could rise between one and two metres. Such an event would clearly threaten low-lying areas, particularly in Asia, where millions of people live and farm on river deltas and flood plains. However, there are also developed **coastal areas** of Europe and the USA which would need massive coastal sea defences to protect them.

SEA-WATER accounts for more than 98% of the mass of the **hydrosphere** and covers just over 70% of the globe. Because of the composition and stability of the **oceans,** and the way they are controlled, they are of great importance to the climate, and great attention has been given to studying the effects of pollution. Long-term experiments include the **Joint Global Ocean Flux Study** and the **Ocean Drilling Progamme.** Man's activities are believed to be accelerating the change in the composition of sea-water. Although the effects are mainly obvious in estuaries and coastal regions, there are signs of the global consequences of pollution. These consequences have begun to be found by the ocean studies that are analysing tens of thousands of samples of sea-water, from which marine chemists have been compiling a global map of the chemical composition of sea-water. As there are no specific preserved records to provide direct evidence about the composition of sea-water in earlier geological periods, a picture of how sea-water evolved to its present-day composition has been assembled by indirect evidence. The primary sources of information have come from the analysis of preserved ocean **sediments,** which contain fossils and can be investigated using **isotopic analysis.** The sediments can usually be accurately dated. Interpreting the relationships between the characteristics of sea-water and features contained in the sedimentary record requires very complex analysis.

SEALS are marine mammals found in most of the world's marine environments, but more often in cool or cold coastal waters. Their bodies contain a great deal of fat which insulates them and acts as a pressure regulator as they dive rapidly to depths of over 250 metres and stay submerged for up to 30 minutes at a time. One group of seals, the eared seals, includes the sea lions and fur seals which are often seen in circuses and aquaria. Elephant seals, named because they have large noses, were hunted for their fur to near **extinction** in the last century, but a rescue operation carried out by the USA and Mexico has resulted in a large population of elephant seals being established on the islands off California. In 1982, 25,000 pups were born, compared with only six in 1911.

SEAWEED is the red, brown or green **algae** growing in or by the sea. Many forms are edible and are used as food, particularly in China and Japan, and some of its extracts are used in medical products. It is a source of many substances, such as **iodine** and the gelling agents in food processing. Combined with straw it can provide an alternative to burning **peat,** which is scarce. Experts have stated that seaweed has a far greater potential for health and nutritional purposes. The European Community launched a four-year project in April 1993 in which researchers across Europe were carrying out a variety of tests on the possible future uses of seaweed. These range from an extract to reduce blood cholesterol levels to the development of a variety of foodstuffs.

SECOND WORLD CLIMATE CONFERENCE, see **World Meteorological Organization.**

SEDIMENTS are the particles of soil and rock fragments that are washed from mountains and uplands to be deposited in rivers, pushed along by **glaciers,** carried out to sea and blown by wind. Sediments were first seriously studied by scientists on the *HMS Challenger* expedition between 1872 and 1876, the first completely scientific circumnavigation of the world, which collected specimens of plants and animal life. Sediments are divided into two groups: those that are usually land-based, consisting of rock, gravel, sand and mud; and those that drop from all levels of water and finally land on the bottom of the sea, such as the shells or skeletal remains of **plankton.**

SELLAFIELD is the name of the spent nuclear fuel reprocessing plant operated by British Nuclear Fuels, in Cumbria. The plant was known for 30 years as Windscale, the site in 1957 of the first known nuclear power accident. A nuclear pile producing **plutonium** for Britain's nuclear weapons programme caught fire, discharging a cloud of radioactive **iodine,** strontium and **caesium.** Surrounding grassland was contaminated and milk from dairy herds had to be thrown away. Spent nuclear fuel from Britain's first generation of **Magnox** nuclear power stations is stored at Sellafield. When it is 'cool' enough it is reprocessed to extract the plutonium and remaining **uranium.** The process also creates three batches of waste classed as **low-level waste, intermediate-level waste** and **high-level waste.** In 1976 the first public inquiry, called the **Windscale Inquiry,** into the use of nuclear energy was forced by environmental organizations, led by **Friends of the Earth.** The Windscale Inquiry was concerned with a plan to build a new type of plant for handling the waste from spent nuclear fuel, called THORP (thermal oxide reprocessing plant). Controversy has centred on claims by opponents of Sellafield, who allege that discharges from the plant have caused leukaemia among children in the region and have been responsible for radiation-induced cancers among the workforce.

The **SET-ASIDE POLICY** was introduced in 1988 by the European Community both to reduce cereal output and as an alternative to subsidized storage and the build-up of grain 'mountains' in Europe. It is also designed to improve the environment, which had been damaged during years of **intensive agriculture.** Under the policy, farmers are paid to take at least 20% of their land out of production and use it in one of four ways to improve the environment. It can be set aside, for permanent fallow, allowing gamebirds to flourish and attracting seed-eating birds, such as finches, to eat any thistle, teasel or cereal heads left after mowing. Small mammals, mice, voles and shrews, also thrive, leading to increased numbers of birds of prey and predatory animals. Rotational fallow allows some of the rare wild flowers of arable ground to restore depleted seed banks in the soil. In some areas it may also encourage ground-nesting birds to return. The set-aside land can also be used for woodland, which has long-term value to wildlife and the landscape. The fourth option is for the farmer to use the land for non-agricultural purposes.

SEVESO, in northern Italy, was the place that alerted the world to the dangers of **dioxin.** A vat used in the manufacture of the bacteriocide hexachlorophene exploded at a chemical plant in Seveso in 1976, spreading a huge cloud of lethal dioxin over the countryside. Seven square miles of surrounding countryside were contaminated and more than 900 people had to be evacuated. In the year following the accident, the rate of birth defects increased by more than 40%. The soil in the most heavily contaminated area was so polluted that the top eight inches had to be removed and buried in a massive plastic-lined pit. The factory was dismantled and some two tons of dioxin-contaminated waste was removed for disposal. Many years later soil samples from the area were still 'dioxin-positive'.

SEWAGE TREATMENT comes in two stages—primary and secondary treatment. The primary stage involves a process of screening solids from sewage, leaving a sludge and relatively clear water for further treatment or for disposal into rivers, the sea or on to the land. In the secondary stage the sludge is stirred constantly in vast tanks to get more oxygen into the mixture, allowing bacteria to break down the organic matter and leave a harmless residue that falls as a sediment to the bottom of the tank. After processing, the clear water on top of the tank is discharged into rivers and the sediment is used as **landfill** or discharged at sea.

The **SHUTTLE SOLAR BACKSCATTER ULTRAVIOLET (SSBUV)** instrument was an instrument used as part of an American space shuttle mission, in the spring of 1993, which confirmed the decline in **ozone levels** in the Northern Hemisphere which was first found by the **Nimbus-7** meteorological satellite. Only in the equatorial region were ozone values well within the range of the previous year's data. The SSBUV has flown annually on the space shuttle since 1989. The shuttle-borne instrument complements measurements from the one on the NOAA-11 SBUV/2 satellite, which has also monitored ozone since January 1989.

The **SIERRA CLUB** is one of the world's most influential conservation groups. The organization was created in 1892, in the USA, by the naturalist John Muir, and can claim to be the founder of the modern environmental movement. Muir was a leading advocate of wilderness preservation. The ideas of conservation of the environment and appreciation of the nature of ecosystems on which the club is based have been adopted by environmental activists like **Friends of the Earth, Greenpeace** and the parties which launched the **green politics** movement in Europe. These ideas are enshrined in Muir's book *My First Summer in the Sierra,* a diary of camping and exploration in 1869, which is regarded as one of the classics of American geographical writing. The club gradually extended its activities to encompass its present concern for preservation of the scenic and amenity resources and the wildlife of both the sierras and of other comparable areas in North America.

The **SIEVERT** is the modern unit used for expressing the dose of radiation to which a person has been exposed. One sievert is equivalent to 100 rem (roentgen equivalent man), an older measure. The average dose to which a person is exposed through **background radiation** is only 2 millisieverts (thousandths of a sievert).

SILENT SPRING was the book that gave the first warnings about the harmful effects of synthetic chemical **pesticides** on wildlife, particularly on species of birds. Written by Rachel Carson, a trained biologist and ecologist, and first published in the USA in 1962, *Silent Spring* cited more than 600 research papers showing the destruction of ecosystems as a result of human activity. In particular, Carson indicated 23 substances belonging to the families of the organochlorine and organophosphate pesticide compounds, the use of which had become commonplace in intensive agriculture. All but nine of these substances have since been banned in most of the industrialized countries. For many years, pesticides were very successful in killing or controlling insects and mites that damage crops, and in saving human lives by killing off pests such as the larvae of the malaria-carrying mosquito. Carson pointed to the evidence of accumulation of pesticides in the **food chain.** As an example, she described how, in the 1950s, robins were dying in the grounds of Michigan State University. The culprit was **DDT,** which had been sprayed on American elm trees to protect them from **dutch elm disease,** imported by a fungus-carrying beetle. Earthworms ate the leaves when they fell to the ground, and were then eaten by birds and other predators. The clue came from research in which earthworms were fed to crayfish, which subsequently died. The poison was shown to be DDT. Subsequent scientific tests showed that 11 contaminated worms was a lethal dose for a robin. Carson documented other deaths: of kestrels, hawks and larks in Norfolk, Virginia; of migrating salmon in a Canadian river, killed by forestry spraying; and of thousands of fish on the Colorado river.

SILT is the fine mineral material formed from the erosion of rock fragments and deposited by rivers and lakes. Its particles are the intermediate form between sand and clay. The particles can range in size from 0.01–0.05 mm in diameter.

A **SITE OF SPECIAL SCIENTIFIC INTEREST (SSSI)** is a classified area in the UK, designated by the Nature Conservancy Council (NCC) as having flora, fauna, geology or natural phenomena that merit conservation. Since 1949 more than 3,500 SSSI sites have been established. They are graded according to the scientific value of the site. A proposed change of land use which is likely to affect an SSSI has to pass stringent tests. If agreement is not reached between land developers, the planning authority and the NCC, then a public inquiry has to be held to decide the future use of the site. Controversies over SSSIs have tended to arise as a result of the extension of agricultural land under the **Common Agricultural Policy (CAP).**

SLASH-AND-BURN AGRICULTURE is a traditional farming system, also known as shifting cultivation or swidden agriculture, that has been used by generations of farmers in tropical forests and the **savannah** of north and east Africa. It is known to be an ecologically sound form of cultivation, and because the soil is poor in **tropical rainforests** it is a sustainable method of farming. It is still practised today, primarily in the developing countries. Small areas of bush or forest are cleared and the smaller trees burned. This unlocks the nutrients in the vegetation and gives the soil fertilizer that is easily taken up by plants. A few years later the soil is degraded and the farmer moves on to do the same at another site. The original ground is left fallow for anything up to 20 years so that the forest can regenerate. With the growth in **population** and in the subsequent need for more farming land to produce food, the method is increasingly being used today to clear large areas of tropical forests for cattle ranching, and in most cases the ground is not left fallow for long enough and, with modern mechanized farming systems, not enough tree stumps or suitable habitats for plant life are left to start the regeneration process.

SLEEPING SICKNESS is one of the **trypanosomiasis** group of diseases and one of the six major tropical diseases. It is caused by a parasitic organism and transmitted by the bite of the **tsetse** fly.

SMOG, see **photochemical smog.**

SOIL means very specific but different things to different people. Farmers and gardeners regard it as the top 25–30 cm of the Earth's surface. Good soil is dark brown, a sign that it is nutrient-rich in **humus** resulting from decades, or even centuries, of decomposing organic matter. To geologists and soil scientists (pedologists) it consists of layers of variable depth, any of which may be affected physically or chemically by the atmosphere. The various layers are referred to as soil horizons. Early soil scientists regarded soil simply as fragmented rock mixed with organic compounds from decomposed plants. With the development of **ecology** in the 1930s, awareness grew of the influence of the living components in the soil like micro-organisms, insects and worms as well as plants, and of the large variations in soil formations resulting from climatic differences and human activities. The pioneers of modern soil science identified seven major factors in soil formation: climate; organisms; parent material; relief or physical shape; **hydrology;** human factors; and time. Soil temperatures are fairly constant below one metre, approximating to the mean annual air temperature. However, in higher levels the temperatures are more variable and the rates of many processes, like the decomposition of organic matter, are closely linked to maximum air temperature. Consequently, soil fertility could be altered significantly by global warming. The effect of human activity on soils has been significant since at least the Neolithic period. Changes vary from the beneficial, increasing crop yields, to the destructive, leading to pollution and land degradation. Artificial drainage of **wetlands** can improve aeration and let crops grow on previously unproductive swamps. Conversely, unwholesome common minerals may be locked in the waterlogged soil, such as **pyrite.** Pyrite reacts with the oxygen from the air to produce **sulphuric acid,** increasing acidity of the soil. Similarly, **deforestation** for **slash-and-burn agriculture,** followed by soil cultivation and growth of arable crops, can result in the gradual loss of organic matter and nutrient reserves accumulated in soil during the forested period. Cultivation may be followed by weakening of the soil structure and susceptibility to erosion by run-off and wind, and a steady decline in soil fertility because of leaching.

SOIL EROSION refers to any of a multiplicity of factors that are slowly undermining a large proportion of the world's arable and growing land. Every year the world's **population** increases by over 80m., but there is over 20,000m. tons less **topsoil** on which to produce food. **Soil** is a rich and fertile, but fragile, ecosystem. A centimetre of soil can take centuries to develop but be lost in a season: blown away by wind, scoured from deforested slopes by rain or **environmental degradation,** sterilized by salts (see **salinization**), poisoned by chemicals, exhausted of its nutrients or buried under swamps and buildings. The problem afflicts industrialized countries as well as the developing countries, but for different reasons. The human disaster provoked when soil erosion strikes was first described in the 1930s by John Steinbeck, in his novel *The Grapes of Wrath.* It was a tragedy of families trapped by circumstances when drought turned the Great Plains of the USA into a **dustbowl.**

SOLAR DOME describes the shape of hemispherical glasshouses used for research into the **climatic change** caused by global warming. The air inside a solar dome is controlled in order to simulate what would happen in the natural environment if levels of carbon dioxide increased. Artificially mixed atmospheres containing varying levels of carbon dioxide are created and pumped into the solar domes. The air inside each glasshouse is kept under a small positive pressure, meaning the pressure in the glasshouse is always slightly higher than outside. Consequently, air can only flow from the dome into the atmosphere and not in reverse.

The **SOLAR ENERGY** falling just on the Earth's surface each year is equal to about four times the total estimated **fossil fuel** reserves of the planet. Solar energy is harnessed in many ways, including as forms of **renewable energy** which are derivatives of solar power. In addition, there are **biofuels,** which are dependent on the sun for growing wood for burning, and crops like sugar cane that can be converted into **ethanol** and alcohol fuels for cars and lorries. However, passive solar energy systems form the method that humankind has used for 2,000 years to harness the sun's rays. Passive solar energy refers to the way houses and buildings are designed and built to take the most advantage of the sun. Features such as a patio, verandah and conservatory, which are built as sun traps, are taken for granted. Most of the world's architectural styles have been shaped by the prevailing pattern of sunshine. Traditional designs of buildings in hot regions position the openings for windows and doors with great care to allow a breeze to cool the inhabitants. Highly reflective colours help to reduce the solar heating of the buildings during the day and the dissipation of the stored heat of the building at night to the low temperatures of the cloudless sky. In the higher latitudes, large areas of glass on the south-facing side of buildings capture the sun. Active solar energy systems go beyond architecture and judicious choice of building materials, by employing pumps, valves, motors and electronic control mechanisms for applications like water heating, space heating and providing heat for industrial processes. **Photovoltaic solar power systems** convert sunlight directly to electricity and are used for remote navigation lights and beacons, automatic weather stations and remote telecommunications links.

SOLAR PANELS are one of various devices for collecting **solar energy,** either by direct heating of water or direct conversion of sunlight to electricity.

SOLVENTS are capable of dissolving other liquids, solids and gases and form the bulk of chemical compounds, such as paint, plastic and **pesticides.** By their very nature, solvents tend to be damaging to the environment.

The **SPECIAL PROGRAMME FOR RESEARCH AND TRAINING IN TROPICAL DISEASES** is a specialized programme combining the work of 5,000 researchers throughout the world. It is co-ordinated by the World Health Organization (WHO), the United Nations Environment Programme (UNEP) and the World Bank. The researchers work to develop vaccines and drugs, and to find ways of preventing the six major tropical diseases: **malaria, schistosomiasis, filariasis, trypanosomiasis, leishmaniasis** and **leprosy.**

SSBUV, see **Shuttle Solar Backscatter Ultraviolet.**

The **STATEMENT OF PRINCIPLES ON THE MANAGEMENT, CONSERVATION AND SUSTAINABLE DEVELOPMENT OF ALL TYPES OF FOREST** was a document agreed at the **Earth Summit** and regarded as the first global consensus on the management, conservation and **sustainable development** of forests. The statement calls for a recognition of the vital role of all types of forests in maintaining the ecological processes and balance at local, national, regional and global levels, in protecting fragile ecosystems, watersheds and freshwater resources, and as sources of genetic material for **biotech** products, as well as **photosynthesis.** The statement says that all types of forests embody complex and unique ecological processes which are the basis for their present and potential capacity to provide resources to satisfy human needs as well as environmental values. Their sound management and conservation is of concern to the governments of the countries to which they belong and they are of value to local communities and to the environment as a whole. The Principles state that the costs of achieving benefits associated with forest conservation needs and increased international co-operation should be shared by the international community. They also call for appropriate measures to protect forests against harmful effects of pollution, including air-borne pollution, fires, pests and diseases. All countries, especially the developed countries, should take positive action towards reforestation, **afforestation,** and forest conservation.

The **STEPPELANDS** are the steppes of Eurasia and the prairies of North America, which occupy the heartland of the two great land masses of the Northern Hemisphere and are the granaries of these two parts of the world. Their climatic characteristics are light rainfall in the spring and early summer, which are the growing seasons, and dry, sunny conditions in the rest of the summer, which are ideal for ripening and harvesting. Since they are inland and far away from the moderating influences of the oceans, the steppelands' climate is one of large diurnal, or 24-hour, variation and some large annual ranges of temperature and precipitation. During the warm summers, monthly mean temperatures vary between

17°C (63°F) and 20°C (68°F). Winter is long, with freezing temperatures well below 0°C (32°F) for months. However, the vulnerability of the agriculture of the steppes has been demonstrated more than once this century. Climatic fluctuations have caused a decrease in the moisture-bearing westerly winds that nourish crops.

STOMATA are the tiny pores or holes in the leaves of plants that let in the **carbon dioxide** used in **photosynthesis,** and release oxygen and water vapour. Comparisons of preserved specimens with today's plants suggest that leaves now have fewer stomata than those of the same but earlier species. The investigations began with plants of 200 years ago, but have been extended to cover intervals over the past 160,000 years. Long-term environmental studies are examining whether these changes might be related to changes in the level of carbon dioxide in the atmosphere. The investigations include experiments in growing trees under differing carbon dioxide levels.

The **STRATOSPHERE** is the layer of the **atmosphere** which is sandwiched between the **troposphere** and mesosphere. Of the energy that reaches the Earth from the sun, only 3% is absorbed in the stratosphere, but that includes the vitally important process of absorption of **ultraviolet radiation** by the stratospheric **ozone layer.** The stratosphere is cloudless and dust free, and almost unaffected by the turbulent conditions of the underlying level of the atmosphere.

STUBBLE BURNING is the practice of burning the straw on fields after harvesting. It began in the late 1960s, as arable farming intensified and artificial fertilizer began to be preferred to the ploughing in of straw to renew the structure and humus in soil. A demand for stubble burning to be phased out grew in the late 1970s, by which time the practice had become widespread and presented at the very least an environmental nuisance, but frequently also a threat to **hedgerow** ecosystems and occasionally a serious safety hazard. However, it took more than 15 years for campaigners against stubble burning to get action from the British government on the issue.

SUCCESSION is the term employed by ecologists to describe how a new ecosystem develops when an area of land has been disrupted. When, for example, land is exposed by **deforestation,** erosion, the drying up of a lake or the retreat of a glacier. Succession refers to the process by which the new vegetation and animals become established in this new habitat.

SULPHUR is an important basic raw material in the chemical industry for the manufacture of **sulphuric acid** and many other substances. However, it is also known as the oldest effective **insecticide,** going back 3,000 years, when brimstone was burned as a fumigant, and, going back 250 years, when its use was first recorded as a **fungicide** against mildew. Sulphur and sulphur candles were used regularly in the early part of the 20th century for all kinds of purposes, from bedbug fumigation to the purification of a house just removed from medical quarantine of smallpox and other infectious diseases. Sulphur dusts and sprays are still effective measures against mites and insects.

SULPHUR DIOXIDE usually refers to the emissions of the gas given off during the burning of **fossil fuels** in power stations and other boilers. Sulphur dioxide is created because sulphur is an impurity in most coal and oils. When the fuel is burned the hot sulphur reacts with oxygen in the atmosphere to form sulphur dioxide. The gas is colourless, but it has a pungent odour and is an irritant to humans, especially those with chest complaints. The deaths resulting from the London fogs, particularly during a four-day episode in December 1952, from bronchitis and related diseases, were due largely to the combined effect of high concentrations of smoke and sulphur dioxide. Over the four days, the mortality rate more than doubled, to almost 4,000. The fog was exceptionally thick because of climatic conditions that produced a **temperature inversion,** which traps emissions of gases in the lower layers of the atmosphere. As a result, powers were included in a **Clean Air Act,** produced in 1956, allowing local authorities to create zones in which only smokeless fuels could be burned. Subsequent research has shown that people react to concentrations of sulphur dioxide at levels of five parts per million (ppm), with sensitive individuals responding at one ppm. The sulphur content of **coal** varies enormously. **Brown coal** and **lignite** have a high ash content and sulphur contents that may range from 0.5% to over 5%. However, the energy yields are only half that of hard coal, which means more is burned in the regions where brown coal is most accessible, thus increasing emissions of sulphur dioxide. Most brown coal is surface-mined, so there is the added environmental deterioration of **open-cast mining.**

The **SULPHUR PROTOCOL** was the arrangement proposed by the **United Nations Economic Commission for Europe** (**ECE**) in order to reduce emissions of **sulphur dioxide** from power stations and boilers in factories and ships. For the first time an agreement, which was due to be signed by the end of 1993, would set different clean-up targets for different countries, based on assessments of the 'critical loads' of **acid rain** that ecosystems of forests and lakes can endure before they suffer damage. European government scientists have concluded that it would be too expensive to bring acid rain to less than the calculated critical loads for each area, so they have proposed targets that would lower the gap between existing deposition and the critical load by at least 50%. If agreed, the protocol will replace the 30% Club, an agreement which most European countries signed over 10 years ago. ECE figures show that the UK, for example, would be required to reduce its acid emissions by 80% from 1990 levels.

SULPHURIC ACID is one of the most basic chemistry laboratory reagents. It is found in car batteries and is a raw material of a myriad of industrial processes. When created in the atmosphere by chemical reactions on the **sulphur dioxide** discharged from coal- and oil-fired power stations, sulphuric acid becomes a very damaging pollutant to people, soil, plants and building fabrics.

SUNSPOTS are the dark areas that appear periodically on the surface of the sun. Each one is between 2,000–3,000 km (1,250 to 1,850 miles) in diameter, and they are darkest at the centre. Sunspots appear in the mid-latitudes of the disc and migrate toward the equator. They seldom occur individually but appear in groups of two or three, and tend to be short-lived, forming and disappearing again over two or three weeks. However, the frequency and size of the development of sunspots fluctuates, reaching a peak roughly every 11 years. There are also minor peaks within the 11-year cycle, and a 22-year cycle (consisting of two of the 11-year cycles). Many scientists suspect that sunspot activity has a greater influence on **climatic change** than variations attributed to the **greenhouse effect.** A survey of 1,500 experts in climatology carried out in 1990 for the scientific magazine *Nature* found that 71% believed that temperature rises experienced over the last 100 years are within a natural range of fluctuation. Consequently, they must ascribe the change to a natural cause of **global**

warming. The most obvious natural cause is solar activity. The Earth depends on the sun for its existence as a planet hospitable to life, and **solar energy** is the major factor determining the climate. Hence, conditions on the sun and conditions on Earth are inextricably linked.

Although the sun's rays may appear unchanging, its radiation does vary. Whether sunspots are an accurate indication of the change, and whether the fluctuations have a greater effect on climate than **greenhouse gases,** is a matter of controversy. Sunspots have been observed for centuries and they do vary dramatically over decades. At the beginning of the 19th century only 45 sunspots a year were recorded, whereas in the middle of the 1950s over 190 sunspots a year were observed. Temperatures on Earth in the mid-1950s were notably higher than at the start of the 19th century. Between 1645 and 1715 very few sunspots were seen by astronomers. This period of diminished solar activity became known as the Maunder minimum, after Walter Maunder, who was the first to notice their reduction and a reduction in solar radiation. Until recently, the Maunder minimum was not taken too seriously until it was realized that it coincided with a particularly cold period, known as the Little Ice Age, which lasted until the early 19th century. It has been estimated that the Little Ice Age experienced a drop in temperature large enough to correlate with a 0.5% fall in solar radiation, an enormous drop in terms of the consequences for the Earth's climate.

SUSTAINABLE AGRICULTURE is the subject of extensive research, with the objective of devising systems of agriculture that provide what is needed in ways that are non-polluting, meet the needs of farmers to diversify and are compatible with **conservation.** The issue of sustainable agriculture tends to be regarded as a problem for developing countries. Poor rural communities suffer from overgrazing of land, **soil erosion** and other factors that reduce agricultural productivity. However, it is also a critical problem for industrialized countries, which have put wildlife and public health at risk as a result of **intensive agricultural** practices, which are based on widespread use of **pesticides** and destruction of wildlife habitats for the sake of extending arable land.

SUSTAINABLE DEVELOPMENT was the central issue debated at the **Earth Summit.** It is development that meets the needs of the present without compromising the ability of future generations to meet their own needs. It does not mean a return to a pre-industrial or pre-technological era, but calls for continued economic growth and for business and industry to play a key role in achieving sustainable livelihoods for all people, thus alleviating **poverty** and improving living standards while maintaining the integrity of the global environment. Business and industry have traditionally regarded the environment as an almost limitless source of energy and raw materials, with the environmental costs of business shifted to society at large or to future generations. Industrial and business activities are the engines that drive the world economy but they are also a major cause of **environmental degradation.** Oil refineries, chemical factories, nuclear power and weapons plants, metal mining, plastics factories and other manufacturing companies release hundreds of millions of tons of toxic waste into the land, air and water every year. This contributes to **global warming, air pollution** and water pollution, **acid rain,** depletion of the **ozone layer,** and the build-up of toxic and hazardous waste. Logging companies cutting down trees in tropical rainforests and northern temperate areas cause **deforestation,** which leads to global warming and loss of species. Throw-away products are still produced, swamping the world's **waste disposal** facilities. Typical industrial processes and business operations make inefficient use of energy and rely too much on **fossil fuels.** There is a growing awareness of the links between commerce and the environment, and there is general agreement that there will have to be major changes in business if the current level of commercial activity is to be sustainable in the long-term. See also **Agenda 21.**

SYMBIOSIS describes the relationship between members of two different species in which both benefit. Some species depend on a symbiotic relationship to survive in a hostile environment. These two-way relationships occur in an assorted range of pairings. Legumes, the group of plants that includes peas, beans and acacias, are able to grow in soil that is deficient in **nitrates,** because their roots have nodules inhabited by nitrogen-fixing bacteria. In return, the bacteria receive **nutrients** from the plant tissue. Many symbiotic partnerships are based on the need for pollination of plants. Insects, and sometimes birds, are rewarded by a feast of nectar that the plant produces to lure the pollinator. Many types of aphids are protected by their association with ants and, in return, the ants obtain food by stroking the abdomen of the aphid, which secretes a drop of a sugary essence for the ant to eat. **Fungi** and **algae** asexually reproduce through a symbiotic association to form **lichens.** In another example, large marine animals, especially fish, regularly have their teeth cleaned by smaller fish who feed on the morsels of food or parasites in the mouth. In the intestines of many animals, including humans, are symbiotic micro-organisms that are necessary to aid digestion. **Termites** and cows can only digest their cellulose-rich diet because of these micro-organisms, which can only survive in the oxygen-free environment of the gut.

T

TAIGA, which is also known as the **boreal forest,** is a Siberian term that refers to the severity of the climatic conditions of the northern forests, where the coldest temperatures in January average −47°F (−53°F), and the warmest summer month, July, averages 16°C (61°F): a difference of 62°C (114°F) between the two extremes. The taiga stretches in a band across Eurasia and North America. Its northern boundary is marked by the so-called 10°C (50°F) July isotherm and its southern boundary occurs where the regions begin which have more than four months of the year above this temperature. Since trees and plants do not begin growth until the air temperature gets above 6°C (43°F), the growing season of the taiga is short and lasts between one and three months. Even the hardy coniferous trees have had to adapt to an environment in which one growing season is not enough to produce fruit and seeds. Their life-cycle has been modified so the conifers can survive without bearing fruit. Instead, the seeds are carried on the outside of cones, which take about one and a half seasons to ripen. So, by early summer of alternate years the cones are ready to cast seeds from which the next generation of trees will grow. The taiga provides a habitat for a wide variety of wildlife, including the bear, wolf, fox, lynx, deer, and elk or moose. Valuable fur-bearing animals such as the sable and marten are now found mainly in the remoter parts of the taiga because of the decline caused by hunting. The taiga almost rivals the deserts and the polar regions in lack of human habitation, making up 10% of the world's land area and containing less than 1% of its inhabitants.

TALL STACKS are the very tall chimneys built on large modern coal-fired power stations in the 1960s to carry **sulphur dioxide,** nitrogen oxides and any other gases that might contribute to **acid rain,** into a higher part of the atmosphere. The theory was that polluting gas clouds would, firstly, become diluted as they dispersed in the atmosphere, and secondly, be gradually neutralized by weathering processes. In practice, the gas clouds travelled great distances to create acid rain and became a source of **transboundary pollution.**

TAXONOMY is the science of classification of organisms, sometimes referred to as 'systematic biology'. It involves, as well as the cataloguing and naming of plants and animals, the clarification of evolutionary relationships among surviving organisms and those which lived at different times in the past. Taxonomy was generally lowly regarded among the biological sciences and has been in decline for a number of reasons, including the belief that organisms have, on the whole, already been classified. It is now experiencing a revival as a science central to addressing the problems of **biological diversity,** the protection of the world's natural variety of flora and fauna and their habitats, in a time when animals and plants are threatened with **extinction** because of human activities. The rough estimate of the numbers of species in the world is between three and 30 million, and only some 1.8 million have so far been identified. Documentation is essential because knowledge of many plant and animal groups is still very poor. It is necessary to have precise details of the evolution of species and their extinctions in the past in order to develop a picture of the future. New techniques, including the advances in 'genetic finger-printing', combined with information technology, make the task of documentation of species much easier. Plants and animals contain genetic resources that can be captured or modified by man for use in agriculture, industry and medicine (see **genetic engineering**), because the familiar crops and livestock that we grow for food, clothing and shelter were all originally derived from the habitats around us.

TECHNAZEN is a chemical sprayed on potatoes to prevent rotting and sprouting during storage. It caused river pollution when supermarkets started to request bags of washed potatoes, which caused the chemical to be leached into rivers.

TEMPERATURE CONSTRAINTS refer to the fact that where temperatures are continuously below −10°C (14°F) or above 45°C (113°F), life is absent.

TEMPERATURE INVERSION is a phenomenon of the weather in which there is a temperature increase, the higher the altitude in the atmosphere, instead of the usual decrease. The event can have serious consequences for pollution. Temperature inversions can cause layers of air to remain trapped near the ground, rather than circulating upwards to take polluted air away. Consequently, pollution from car exhausts and chimneys accumulates and, especially in the summer, this produces **photochemical smog.**

TERMITES are found in many parts of the world, but primarily in **tropical rainforests,** where they live in highly organized colonies to a strict caste system. The nests are built to provide central heating and air-conditioning. They are soft-bodied insects and play an important role because they help to convert plant cellulose into substances that can be recycled into the ecosystem to support new growth. There are some 2,000 species of termites, and some of them are regarded as **pests** because they eat all kinds of vegetation, especially the wood used to build houses. Many species have been unintentionally transported from their native habitats to other parts of the world in shipping crates, boat timbers, lumber and furniture. Dry wood termites live in small colonies in wood and can survive in seasoned wood and furniture for long journeys. These feed on the wood and are difficult to control, except by fumigation. Other methods of control are the use of chemically treated wood in buildings and paint to seal cracks in the surface of the wood. Subterranean termites depend on contact with moisture in the soil and get into man-made structures through the foundations. One way to control them is to fill a trench with insecticide around the affected building and cover it with soil. Another solution is to design buildings with foundations that have reinforced concrete foundations at least six inches above the ground, with no cracks and no contact with any outside wood. It is also thought that the world's population of termites may contribute to the production of **greenhouse gases** because as they digest wood they release carbon dioxide and methane into the atmosphere. It is estimated that several millions of pounds of damage are caused by termites around the world every year. Enormous amounts of money, research and technology are devoted to finding ways to destroy them.

The **TERRESTRIAL INITIATIVE IN GLOBAL ENVIRONMENTAL RESEARCH (TIGER)** is a major research project, organized by the British Natural Environment Research Council (NERC), designed to obtain a better understanding of the complexity of the links between the terrestrial environment and **climatic change.** Changes to the climate will modify the availability of water, the growth of vegetation and ecosystems, all of which alter the terrestrial environment. Alterations of the terrestrial environment will, in turn, change the surface **energy balance,** the **flux** and storage of carbon and the generation of trace greenhouse gases, all of which will modify the climate. The objective of TIGER is to establish the precise role of any process occurring in the **biosphere** that influences this never-ending cycle of events. See **Tiger 1, Tiger 2, Tiger 3** and **Tiger 4**.

TETANUS, often known as lockjaw, is caused by a micro-organism in the soil, *Clostridium tetani*, which, if it enters the body through an open cut or wound, causes severe muscle spasms. It is a major health problem in developing countries. It is believed that more than 400,000 people die every year from tetanus, although the figure could be higher. The disease is common in warm climates and in rural areas that are highly cultivated with a large population of humans and animals. Agricultural workers are particularly at risk because injuries can often be contaminated.

THALLIUM is an extremely toxic and cumulative **heavy metal poison** with a lethal dose of 0.2–1.0 grams. Thallium is absorbed through the gut or through contact with the skin.

THERMAL EXPANSION OF THE OCEANS refers to the fact that at constant mass the volume of the oceans, and hence **sea-level,** will vary with changes in the density of **sea-water.**

Density is inversely related to temperature. In other words, as the oceans warm and density decreases, the oceans expand.

THERMAL POLLUTION is the rise in temperature of water in rivers, lakes or estuaries brought about by the discharge of warm waste cooling waters from power stations and factories. This causes **algae** growth, which robs the water of oxygen and hence threatens fish and aquatic life.

THERMODYNAMICS is the study of the relationship between heat and all the other forms into which energy is transformed in natural and man-made processes. It is only concerned with the changes in energy, not the mechanisms. The first law of thermodynamics states that in a closed system the total amount of energy is constant. To the dismay of inventors determined to create perpetual motion, the first law, in effect, says it is impossible to create a continuously operating machine that does not obtain energy from an external source. The second law deals with the direction of any chemical or physical process involving energy that takes place. In nature, heat is never found to proceed up a temperature gradient of its own accord. In mechanical systems, heat cannot flow of its own accord from a cold to a hotter body. Thermodynamics lies at the heart of the efficiency of all energy conversion processes, whether it is the conversion of **nutrients** in the diet into **proteins** or muscular movement, or the generation of electricity from **fossil fuels** or of power to drive a vehicle.

THIS COMMON INHERITANCE, known more popularly as the 'Green Bill', is the British government's White Paper setting out legislation to control discharges to land, water and air from industry. In addition to tightening controls on the dumping of trade and toxic wastes, it covers laws to prevent littering and canine faeces in public places. *This Common Inheritance* also explains the British government's attitude to the role of devices such as economic incentives and taxes to control pollution. These include financial incentives designed to make individuals and organizations more environmentally-friendly and aware of benefits from, for example, energy conservation in the home, office and factory, and in transport. The White Paper discusses issues ranging from the global warming effect of greenhouse gas emissions to the problems of waste recycling and litter, radioactive wastes, the health hazards of congestion in towns, the hazards of pesticides and **genetically modified organisms,** and the need to protect the countryside and coastline.

THORP, see **Sellafield** and **Windscale Inquiry.**

THREE MILE ISLAND is the power station located on the Shenandoah River, near Harrisburg, Pennsylvania, USA, which has come closest to a nuclear core meltdown, the most feared accident that environmental groups had predicted could happen with **pressurized water reactors** in civil nuclear power stations. The accident happened on 28 March 1979: the first acknowledged large-scale nuclear disaster (unacknowledged, until many years after their occurrence, were the **Windscale fire** and the accident in the **Ural Mountains** region of Russia, both in 1957). Until **Chernobyl,** Three Mile Island was seen as a critical example of what could go wrong with nuclear power. There were no direct casualties in the incident but a report commissioned by the Nuclear Regulatory Commission concluded that the reactor had been one hour from meltdown and that a catastrophe had been averted only by sheer good fortune. After the accident, young children and pregnant women within five miles of the power station were advised to leave the area. Forty thousand people left the area within a few hours and another 150,000 were eventually evacuated. The accident is believed to be responsible for significant health problems. A rise in the death rate among old people living in adjoining states was blamed on the accident. Between 1979 and 1982 at least 50,000 more elderly people died in the affected states than expected. The death rate in Pennsylvania was more than three-and-a-half times higher than the national average. It is thought that **low-level radiation** released during and after the accident could have damaged the immune system of elderly people and speeded their deaths from apparently 'natural' causes.

TIDAL WAVES, see **Tsunami.**

TIGER 1 (see also **Terrestrial Initiative in Global Environmental Research**) is concerned with the movements of carbon between the soil, plants and atmosphere that contribute to a higher level of carbon dioxide and a warmer environment (see **carbon cycle**). Man-made changes to the natural cycling of carbon lie behind the increased **greenhouse effect,** largely due to the burning of fossil fuels and deforestation, which together release over 6,000m. tons of carbon into the atmosphere each year. This is a modest amount compared with soil, which provides a **carbon sink** for 1,500,000m. tons of carbon, and with vegetation which provides a carbon sink for 550,000m. tons. Nevertheless, fossil fuels and deforestation have been responsible for an increase of nearly 30% in atmospheric carbon dioxide since pre-industrial times. Scientists estimate that the Earth's surface and the atmosphere annually exchange 100,000m. tons of carbon. **Photosynthesis** draws carbon from the atmosphere, while decomposition and **respiration** return carbon to the atmosphere. The goal of TIGER 1 is to improve the estimates of the rates of exchange of this carbon cycling through the world's different **biomes,** such as temperate zone soils and tropical forests. Ultimately, mathematicians should be able to create computer models of the terrestrial climate cycle that will form a critical part of the bigger models for predicting global climate change.

TIGER 2 (see also **Terrestrial Initiative in Global Environmental Research**) is a research project designed to investigate trace **greenhouse gases** released from soil into the atmosphere. The project is especially concerned with **methane** and nitrous oxide, which are estimated to account for nearly one-fifth of **global warming.** Both have terrestrial sources and have shown marked rises in their atmospheric concentration. Both gases are produced by micro-organisms in the soil: in methane's case, production comes particularly from **wetlands** and rice paddies, and in the case of nitrous oxide it comes from agricultural soils. Once in the air, the methane molecule has the same greenhouse effect as 30 molecules of carbon dioxide. The increased terrestrial source of nitrous oxides is believed to be primarily **ammonia**-based fertilizers, both in their man-made chemical form and as the natural waste of farm animals. It was hoped to be able to feed the data about the release of methane and nitrogen oxides into other larger studies on trace gases and their climatic consequences.

TIGER 3 (see also **Terrestrial Initiative in Global Environmental Research**) is a research project that uses ground-based, airborne and satellite instruments to investigate the **flux** of water, energy and carbon between different terrestrial environments and the atmosphere. The instruments explore the differences between these transfers of water and energy in temperate soils, various kinds of vegetation, tropical forests, tropical drylands and boreal forests. TIGER 3 is concerned with the water-dominant processes of the terrestrial environment: whether water, as a liquid, limits plant growth, or, in the form of water vapour, it is an important greenhouse gas, or, when solid in ice sheets, snow and glaciers, it plays a key role in the Earth's **radiation** balance, which is the difference between incoming and outgoing radiation. Of the approximately 340 watts of solar radiation received on average by a square metre of the Earth, about half is used in warming the surface. This warming gives rise to an outward emission from the ground of long-wave, **infrared radiation.** Some of this radiation penetrates the atmosphere and escapes into space, but a significant proportion is trapped by greenhouse gases and warms the lower layers of the atmosphere. Climate and atmospheric changes affect the radiation balance, soil moisture conditions and vegetation. Successful climate predictions, on a global or regional scale, would depend on understanding the interplay between the energy balance, water availability and vegetation, and the realization that carbon uptake by **photosynthesis** and evaporation by plants are intrinsically linked functions.

TIGER 4 (see also **Terrestrial Initiative in Global Environmental Research**) is a research project into ecosystems, studying the effects of increasing carbon dioxide and climate change on plants and animals. The change in climate and the carbon dioxide composition of the atmosphere will alter the geographical range and the mixture of species in ecosystems, particularly the sensitivity of individual animal and plant species to changes in elevated carbon dioxide levels and higher temperatures. The responses of plant and animal populations are expected to result in changes in rates of birth, death and dispersal. Although **climatic change** may restrict one plant's potential range, the environmental change may open the way for replacement species to move into a **biome.** A feature of the predicted environment, which has no counterpart with the recent past, is an atmosphere enriched by carbon dioxide, which acts like a fertilizer to many plants. The critical, unknown factors resulting from this are how this may affect the established competition and relationships between plants, alter the food quality for herbivores, and influence the decomposition rates of plant material and the levels of organic matter in soil.

TIGERS once roamed all over Asia. Today three subspecies, the Caspian, the Javan and the Balinese are extinct. The Chinese tiger has practically vanished from the wild and the Sumatran, the Indo-Chinese and the Siberian are on the brink of **extinction.** Hunting of tigers was halted when Asian countries realized they were disappearing. The USSR banned hunting in the 1950s and the People's Republic of China followed suit in the early 1960s. The Bengal tiger of India is among the **endangered species.** In 1973 the **World Wide Fund for Nature** launched Project Tiger, an international campaign designed to raise money to set up reserves for tigers throughout Asia. The project envisaged moving villages out of certain areas to give the tigers more room to range and also allowing the tiger's prey to move back into these areas. However, on 21 January 1993, the 20th anniversary of the project, India warned that its tiger population could be extinct by the end of the century because mining, poaching, grazing and logging, and an increase in population and livestock, had put more pressure on forest areas.

TIROS-1 was the name of the world's first weather observation satellite, launched on 1 April 1960. It marked the beginning of a new era in meteorology and in understanding the Earth's environment. The tiny 100 kg spacecraft was equipped with only a modified black and white television camera. It was designed to find out whether cloud cover could be measured from space. Tiros was the first of a family of weather probes that were followed by more advanced meteorological satellites, which today provide a continuous stream of television pictures of weather systems. Tiros was also the forerunner of other **Earth observation satellites (EOS),** such as Landsat and ERS-1, designed to track other changes in the terrestrial and ocean environments.

TOBACCO is known to damage health and pollute the atmosphere when it is smoked. There is now evidence that non-smokers may be affected by 'passive smoking', by inhaling the smoke from others' cigarettes. However, tobacco is also a crop that degrades the soil when it is being grown, while vast areas of forests are cut down for the fuel used in the curing process, thus encouraging **desertification.** The plant denudes the soil of nitrogen, phosphorus and potassium. However, for many developing countries it is a major export **cash crop.**

The **TOPPING CYCLE** is a process for burning **coal** which belongs to a category referred to as clean coal technology. It is a method of burning coal in power stations to extract more energy from the fuel and to discharge less waste. The process is a combination of two techniques, **fluidized bed combustion** and coal gasification. Fluidized bed combustion works by first crushing a mixture of fuel and limestone into very small particles. A stream of air, powerful enough to suspend the mixture, is pumped from the bottom of the furnace through the fuel, causing the mixture to behave as if it was in a liquid state. The limestone reacts with the coal to remove sulphur and other substances in the ash.

TOPSOIL refers to the fertile surface layers of the land essential for healthy pasture, arable crops and trees to grow (see also **soil**). Experts estimate that between 1970 and 1990 the world's farmers lost 480,000m. tons of topsoil, roughly the amount covering the USA's agricultural areas, as a result of **soil erosion, salinization** of soils and the effects of **deforestation.** In the same 20 years, the Earth's forests shrank by 500m. acres, while deserts expanded by 300m. acres.

The **TORONTO CONFERENCE ON THE CHANGING ATMOSPHERE** was held in 1988, under the auspices of the **United Nations Environment Programme (UNEP),** and called for a global cut in **carbon** emissions by 2005. The meeting concentrated on **energy efficiency.** Estimates examined at the meeting suggested energy efficiency could make a 3,000m. ton, or 30%, difference in the amount of carbon released into the air each year. One of the fastest ways of reducing carbon emissions would be to improve the fuel economy of cars. The average car adds its own weight in carbon to the atmosphere each year. Policies discussed at Toronto which were designed to reduce carbon emissions included a gradual increase in tax on motor vehicle fuel and the introduction of a **carbon tax** weighted to the carbon content of all **fossil fuels.** Coal, for example, would be taxed more heavily than natural gas because of its greater carbon content.

The **TORREY CANYON** was a tanker that sank in the English Channel in 1967 off the Cornish coast, leaking 117,000 tons of Kuwaiti oil. Oil slicks spread from Cornwall to Guernsey in the Channel Islands, killing sea-birds and destroying fish stocks. It was the first major accident that alerted the world to the dangers to the marine environment of massive **oil spills** from tankers. Two years after the Torrey Canyon disaster the **International Maritime Organization (IMO)** produced two conventions to address the problem. The first,

Intervention on the High Seas in Cases of Oil Pollution Casualties, established the right of coastal states to intervene in incidents on the high seas which are likely to result in oil pollution. It came into effect in 1975. The second convention, the International Convention on Civil Liability for Oil Pollution Damage, deals with the civil liability of the owner of a ship, or cargo, for damage as a result of an oil pollution incident. This also came into force in 1975 and is intended to ensure that adequate compensation will be readily available to victims of pollution. The obligation for paying compensation is with the shipowner. Because some governments felt that the established liability limits were too low and that the compensation could prove to be inadequate, the IMO called another conference in 1971. The **Convention on the Establishment of an International Fund for Compensation for Oil Pollution Damage (IOPC)** was adopted and came into force in 1978. The Fund was drawn up to provide compensation for victims of oil pollution damage which exceeds the compensation available under the Civil Liability Convention. The Fund, operated by an International Oil Pollution Compensation Fund Organization based in London, spreads the burden of compensation evenly between the shipowner and those with cargo interests.

The **TOTAL OZONE MAPPING SPECTROMETER (TOMS)** is an American-designed instrument on the Russian satellite Meteor-3, launched in 1991. It confirmed especially low **ozone** levels in 1992 and 1993 in the mid-latitudes of the Northern Hemisphere, which were first measured by the US space agency's **Nimbus-7** satellite.

TOURISM is one of the world's growth industries and is predicted to expand even more in the 21st century. In 1990 there were more than 500m. international tourists, spending their holidays in the countryside and cities throughout the world, on mountains, rivers, lakes and, particularly, in **coastal areas.** The tourist and travel industry is one of the largest employers in the world and, economically, many countries rely on the foreign currency that tourism brings in. For some countries tourism is practically the only industry that enables them to survive. However, the sheer number of tourists is beginning to extract a toll on the environment, in the developed and developing worlds alike, as more and more land is cleared for tourist accommodation and amenities, and for access roads and airports. The number of tourists in India, for example, is expected to double by 2000. On the **Mediterranean** shorelines the deluge of visitors in the summer doubles the resident population and, therefore, the amount of human waste that flows into the sea, which is already dangerously polluted. In mountain ranges such as the **Alps,** trees have been cut down to make way for ski-runs, car parks and hotels, and wildlife habitats have been damaged by tourists. The remote **Himalayas** and the Andes are now showing signs of environmental damage because of the large numbers of tourists, and the debris they leave behind. On the slopes of Mount Everest, for example, tons of spent oxygen cylinders, tents, equipment and even bodies have been left behind by scores of expeditions. In **Antarctica** the number of cruise ships taking tourists to visit scientific stations and penguin rookeries has increased and since the 1970s there have been commercial airline sightseeing tours. There are plans to increase the scope of tourism in the Antarctic, but ways have to be found to ensure that the flora and fauna are not disturbed and that there are adequate provisions for the disposal of waste and sewage, so as not to affect sensitive local ecosystems. There are signs that some travel companies are beginning to develop 'softer', more environmentally friendly forms of tourism, where the visitor properly appreciates the wildlife and the environment.

TOWARDS SUSTAINABILITY is the title of the European Community's fifth environmental action programme to improve and protect the quality of the environment. It was adopted by the European Commission in 1992 and was scheduled to run from 1993 to 2000. Unlike the previous environmental action programmes, which responded to environmental problems after they had happened, the fifth action programme was the first to establish a framework to tackle environmental degradation before it became a problem. It is based on the recommendations of the 1987 **Brundtland Report.** The main impetus behind the programme is that environmental damage would not be halted or prevented if producers, consumers, governments and people did not give the environment greater priority. The main goal of Towards Sustainability was to raise public awareness about the problems facing the environment and to change people's attitudes and behaviour, so that they would become more environmentally friendly. It was aimed at three main groups: central and local government, which have to enforce legislation and also make planning decisions, and inform and educate the public; private and public enterprises, large consumers of materials and waste, who were encouraged and obliged to take their responsibilities towards the environment more seriously; and the general public, who, as voters, producers of pollution and consumers of goods and services, could exert an influence on the future quality of the environment.

TOXIC SHOCK SYNDROME is an acute life-threatening allergic reaction and collapse of a person exposed to traces of a highly potent poison like **dioxin.** Tampons contaminated with dioxin residues from the bleach used in manufacture have provoked toxic shock syndrome and deaths.

TRACE ELEMENTS are natural chemical elements that are necessary for health, but which make up a very small part of the body weight of a person. A deficiency can cause disease, but an excess caused by pollution of the **food chain** can also cause death. Trace elements can be divided into three groups. Firstly, a group which forms less than 1% of body weight and includes sodium, potassium, magnesium, sulphur, chlorine, calcium and iron. The second group is equally essential, but present in even smaller amounts, comprising less than 0.05% of body weight. These are **boron, fluorine,** silicon, **manganese, copper** and **iodine.** Precise differences between the trace elements occurring in vegetables and water supplies reflect the characteristics of local rocks and soils. However, vegetables, water supplies and the air can be contaminated with trace elements which have no biological usefulness whatsoever and are very dangerous in tiny concentrations. These include **mercury, lead, cadmium** and selenium.

TRAFFIC is the acronym for the Trade Records Analysis of Flora and Fauna in Commerce. It is the global network set up by the **International Union for the Conservation of Nature and Natural Resources (IUCN)** and the **World Wide Fund for Nature** to monitor trade in wild animals, plants and wildlife products. It helps to carry out the **Convention on International Trade in Endangered Species (CITES).** TRAFFIC has offices in 15 major wildlife trade areas throughout the world and carries out it own investigations into illegal smuggling activities, keeping CITES and customs officials informed of all illegal activities.

TRANSBOUNDARY POLLUTION describes polluted air and water, or any other contaminated waste, that is generated in one country and transmitted to others. The most pervasive pollutant is **sulphur dioxide,** a by-product of the burning of **fossil fuels** in power stations. This problem has been exacerbated by the so-called **tall stacks** policy of some countries, especially Britain, which chose to build high chimneys on its power stations in preference to installing elaborate and expensive equipment for cleaning flue gases. The idea was to discharge gases into a higher part of the upper atmosphere where they would be gradually diluted. In practice, sulphur dioxide clouds tend to be carried long distances before they are washed back to earth as **acid rain.** Similarly, countries on the upper reaches of the **Rhine** and the **Danube,** among other rivers, dump pollutants into the rivers that affect countries downstream.

TRANSPIRATION is the process by which a plant loses water by evaporation from its tissues. It is the effect of a wet surface losing water vapour into an unsaturated atmosphere. However, different plants lose water by transpiration at different rates.

TRANSPORT usually refers in environmental discussions to the impact of road vehicles on **air quality,** land use and the 'road versus rail' debate. In principle, it also embraces air, sea, inland waterways and the bicycle. The issues raised by air travel focus on development of airports and noise pollution from aircraft. However, the effect of road vehicles tends to dominate transport concerns. At the start of the 1990s there were about 400m. cars on the roads of the world, with forecasts of steadily rising numbers. The problems they raise include emission of nitrogen oxides and hydrocarbons, which are the cause of low-level ozone and **photochemical smog.** They also produce congestion in towns. Other transport-related issues include the disappearance of green belt land due to more road building. A difference in attitudes toward specific projects, the NIMBY ('not-in-my-backyard') argument is perhaps most obvious with transport schemes. The disappearance of green belt land due to more road building, and the building of the Channel Tunnel and its associated rail links and infrastructures, provoked greater opposition in the heavily populated south-east corner of England than its corresponding development in France.

TRANSURANIC ELEMENTS are synthetic elements which belong to the **actinide** family of substances. They did not exist before nuclear **fission** was discovered. They form the **high-level wastes** in spent nuclear fuel. They are called transuranics because they have **atomic numbers** greater than 92, which means they lie beyond uranium, which until their creation was the heaviest element in the periodic table.

TREGRO, short for tree growth, is the name of a computer model that simulates the effect of environmental damage on tree growth. It predicts the impact of **acid rain, drought** and **ozone** pollution on trees, from their roots to the top of the canopy. TREGRO was developed at the Institute for Plant Research, Cornell University, USA, to predict as far as 100 years ahead the changes in the pattern of growth for any combination of environmental conditions. As a result, the technique has been used in long-term planning of conservation and tree-planting programmes in the USA by the US government's **Environmental Protection Agency,** the National Forest Service and the National Park Service; it has also been used to do research into environmental sciences at universities and national laboratories. TREGRO simulates the mechanisms by which plants regulate their carbon, water and nutrient cycles to mitigate damage from stress. Computer simulation has produced new insights into the role of environmental stresses on tree growth. For example, although a tree in poor health, because of inadequate nutrients, would seem especially susceptible to ozone damage, TREGRO predicted the opposite because such a tree would take in less ozone than normal. In a confirmation of the accuracy of the computer model, scientists examined in the living world the effects of severe defoliation of red spruce trees, an event occasionally produced by bitter winter cold, and compared them with computer simulations. TREGRO predicted a 41% reduction in growth, by volume, in the first three years after a 50% defoliation. The prediction accorded well with the field observations of **forests and woodland.** TREGRO has been used to predict the possible reductions in growth caused by pollution within five major forest types around the USA, from the sugar maples and red oaks common in eastern forests to the north-western ponderosa pine and white fir forests. In addition to stresses from pollution and drought, researchers have been looking at the possible effects of global warming on the giant sequoia forests (see **redwood tree**) in the west, with the intention of learning, for instance, whether lowland species would migrate to higher altitudes and displace sequoia if the average temperature was to rise. The model was being used to examine questions where experiments were difficult or impossible to perform. One investigation showed that the susceptibility of red spruce to ozone damage depended more than anything else on the rate of root growth—an area where experimental data was in short supply.

TRIBUTYL-TIN is an **organometallic compound** used as the active ingredient in anti-fouling paint for ships and boats.

TROPICAL RAINFORESTS are the most valuable and the richest ecosystems on Earth. They play a critical part in the Earth's life support systems and house 50%, and possibly as much as 90%, of all the species on Earth. They are a key storehouse of foods, oils and minerals, and a source of ingredients that make up a range of medical treatments. They also represent home and livelihood for many people. However, more than half of the rainforests have disappeared, chopped down for valuable tropical hardwoods, or cleared to provide areas for cattle grazing or human habitation. It is feared that if the widespread deforestation continues at current levels, the rainforests will have disappeared by 2050, and with them many thousands of species, some not yet even discovered. The forests play an important part in climate patterns, and **deforestation** is thought to be responsible for 18% of global warming. Furthermore, as they disappear there is also an **albedo** effect—a damaging increase in the sunlight reflected—which affects wind and rainfall patterns. The demand for tropical hardwoods such as mahogany, teak and ebony by the developed countries has resulted in the massive decline of the forests. Other areas have been burned to make way for the growing of cassava, imported by the European Community to feed cattle, pigs and poultry, for the grazing of cattle or for the cultivation of cash crops. Efforts to save the rainforests are at the top of the agenda for most international conservation organizations and environmental agencies. Several **reafforestation** initiatives have been established. Forests were also a central issue debated at the **Earth Summit.** See also **International Tropical Timber Agreement.**

The **TROPOSPHERE** is the lowest part of the atmosphere, rising from ground level to about 10–15 km in altitude. Structurally, the atmosphere is divided into a number of layers that are related to their temperature profile and the effects of the absorption of energy from the sun. The troposphere is dominated by convection as a result of the heating of the Earth's surface by solar radiation. At the 10–15 km level the troposphere gives way to the **stratosphere,** which gets its name because there is no strong mixing current. However, it is heated by absorbing the ultraviolet rays in sunlight. Absorption of ultraviolet radiation also brings chemical changes, the most important being the production of the **ozone layer,** or the **ozonosphere,** found between 15–30 km above the Earth's surface.

The name **TRYPANOSOMIASIS** covers a group of diseases caused by a parasitic, microscopic organism that is transmitted by the **tsetse** fly. **Sleeping sickness** and the cattle disease **ngana** are among the group. African sleeping sickness is severe and often fatal. It is estimated that 50m. people are at risk from the trypanosomiasis diseases.

The **TSETSE** is a blood-sucking fly which feeds on humans and animals, transmitting **sleeping sickness** to humans and **ngana** to animals. They are generally found close to rivers or on the edge of areas of **savannah** and are extremely difficult to control because as larvae reach maturity in the female she deposits them directly into the soil to pupate. Consequently, any control methods can only be effective on adults. Aerial **pesticide** sprays are not enough on their own to control the adult tsetse flies because they spend the day sheltering under branches or leaves. In order to control them, a double programme of aerial spraying and intensive applications of pesticides to the ground has to be carried out. The threat of the tsetse fly means that vast areas of Africa cannot produce livestock, and it is not possible to put arable land into full use. Small-scale eradication schemes have been carried out and led to an influx of people, **overgrazing,** and land degradation. Conservation groups and ecologists fear that if other proposed large-scale programmes go ahead the same detrimental result will be achieved, only on a much larger scale. In the battle to control the tsetse fly many dangerous chemicals have been used, including **DDT,** which have badly affected wildlife. If a new concerted effort to control the tsetse was to be tried the conservationists believe that safe, odour-baited traps, which have been successful in field trials, should be used, and that there should also be stringent land-use plans.

A **TSUNAMI,** sometimes known as a tidal wave, is a fast-moving, large sea wave caused by submarine earthquakes, landslides or volcanoes. They can travel at up to 90 km per hour in deep water and, as they approach coastal areas, can reach a height of 30 metres. These waves can cause serious floods in low-lying areas, resulting in loss of life and extensive damage to property.

TUBERCULOSIS (TB) is an **airborne disease,** the extent of which was declared a global emergency by the **World Health Organization (WHO)** in April 1993. The WHO estimated that if immediate action was not taken to control the disease, 30m. people would die in the next decade. Tuberculosis, which has made a resurgence after many years when it was thought to have been under control, is spread when an infected person coughs. Residues of droplets from the cough can remain in the air for a prolonged period and so can continue to infect other people in the environment, even though the infected person has left. Overcrowded living conditions play a part in the spread of the disease. Once a person is infected, he or she risks developing active TB, and that risk continues for life. It is triggered when the body's immune system is damaged, hence the link between TB and HIV, the human immunodeficiency virus that leads to AIDS. The dual infection of TB and HIV is a major cause for concern.

2,4-D is the chemical description for a synthetic **herbicide** that has the common name Nitrofen. It is a relative of the **agrochemicals** which were based on **phenols.** Although rather old now, the phenol-derived pesticides were the first mainstay of modern intensive farming. Nitrofen was a general purpose agent intended for use in protecting a wide range of vegetable crops, sugar beets and rice. It behaves like a hormone, rather like those of natural plant regulators. It causes overgrowth of annual weeds, contorting the stems and leaves, and produces minute roots within a short time of application. This results in a collapse of the plant's natural growth regulating system. The herbicide is cheap and easy to use, it is **biodegradable** and almost entirely non-toxic to humans and animals.

U

ULTRAVIOLET RADIATION is very short-wave **electromagnetic radiation** emissions coming from the sun. In terms of the spectrum of **radiation,** it comes between X-rays and visible light. Most ultraviolet radiation is absorbed by the ozone molecules which provide the protective **ozone layer** in the upper atmosphere, but some reaches the Earth's surface in sunlight. The highest levels are recorded in mountainous areas. In small amounts ultraviolet radiation is useful in playing an important part as a photochemical agent in some life processes. For instance, it acts on the skin of some animals, including people, to produce vitamin D. However, an excess of ultraviolet radiation is lethal for plant and animal life. If humans are over-exposed to ultraviolet radiation by sunbathing in hot climates, the production of highly malignant skin cancers or melanomas can be triggered; these look like very dark moles on the skin. Previously, animal and plant life has been protected from excessive ultraviolet radiation by a complex cycle of chemical reactions of the normal gases in the upper atmosphere. The process regenerates ozone molecules to absorb the incoming ultraviolet rays continuously. This delicate balance is being disrupted by pollution caused by **chlorofluorocarbon (CFC)** aerosols and other gases that are also destroying ozone molecules. Hence, the process of ozone destruction is going faster than the rate at which natural chemical reactions can recreate new ozone molecules.

UNBLEACHED PAPER has become an important part of the effort to remove an environmental health hazard and pollutant. Where once only whiter-than-white, **chlorine**-bleached products were the norm in supermarkets and shops, they are being replaced, for example, by brown coffee filters and cream-coloured napkins, milk cartons and tampons. The idea that brilliant whiteness might be associated with pollution rather than purity first took root most strongly in Scandinavia. The eastern coast of Sweden, in particular, suffered from a concentration of pulp mills. Their effluent, which was pumped into the **Baltic Sea** and Gulf of Bothnia, contained the by-products of chlorine-based bleaching compounds, including a compound known as TCDD. It is a member of the **dioxin** family of compounds, which include the most toxic substances ever made. Because these chlorine-based substances are very stable, they accumulate in the environment. They become concentrated in marine wildlife and then move up the **food chain** to fish-eating mammals, including humans, which ingest them at thousands of times their concentration in the water.

UNCED, see **Earth Summit.**

UNDP, see **United Nations Development Programme.**

UNEP, see **United Nations Environment Programme.**

UNESCO, see **United Nations Educational, Scientific and Cultural Organization.**

The **UNITED KINGDOM ATOMIC ENERGY AUTHORITY (UKAEA)/AEA TECHNOLOGY** came into being in 1954 as the umbrella organization for the development of nuclear weapons and civil nuclear power in Britain. The first research establishment was the Atomic Experimental Research Establishment at Harwell, Berkshire. The UKAEA had responsibility for all the subsequent nuclear sites: the **plutonium** production plant at Windscale (now **Sellafield**); weapons at Aldermaston; tritium production at Chapelcross in Scotland; the weapons-grade **uranium** at Capenhurst, Cheshire; uranium fuel manufacturing at Springfields, Lancashire; **fast breeder reactor** research at Dounreay, Caithness; the Atomic Energy Establishment at Winfrith Heath, Dorset; the Fusion Research Centre at Culham, Oxfordshire; and the Radiochemical Centre at Amersham, Buckinghamshire. The sites were divided between the UKAEA, the Ministry of Defence and British Nuclear Fuels in 1970 and 1971. The UKAEA adopted the name AEA Technology in 1989 for trading reasons, in order to sell its non-nuclear services.

UNITED NATIONS CONFERENCE ON ENVIRONMENT AND DEVELOPMENT, see **Earth Summit.**

The **UNITED NATIONS CONFERENCE ON THE HUMAN ENVIRONMENT** was held in Stockholm in 1972. Its aim was to focus the attention of governments and people on the importance and urgency of environmental issues and to identify the places where international co-operation and agreement were needed. The recommendations that came out of the conference are regarded as a turning point, representing the most advanced understanding of environmental problems achieved to that date. Technological advances in the first half of the 20th century had raised fundamental questions about whether the Earth could continue to support its rapidly growing **population** and whether the technology itself was appropriate. The questions were first debated by scientists, but soon the general public joined in the concern. By the late 1960s environmentalism began to concentrate on the interrelation of the environment and human beings. Environmental issues appeared on the political agenda. In 1968 the Swedish government proposed a conference to debate the problems of the human environment. In the four years preceding the conference a great deal of information was gathered about global environmental problems and the links between environment and development. As a result of the conference 26 principles for guiding national and international action were drawn up, together with the 109 specific recommendations of the Action Plan for the Human Environment, which provided a framework for environmental preservation and enhancement. After the conference, the **United Nations Environment Programme (UNEP)** was set up to catalyse, stimulate and co-ordinate a programme of action designed to protect and improve the human environment.

The **UNITED NATIONS DEVELOPMENT PROGRAMME (UNDP)** was established as a United Nations organization in 1965, following the merger of the Expanded Programme of Technical Assistance and the UN Special Fund. UNDP's objective is to build more productive societies and economies in low-income nations by helping them develop natural resources and human capabilities. It assists in making experts available to study growth potentials, as well as to help provide facilities and applied scientific research. It administers aid through five-year Country Programmes. These fund projects are intended to attract development capital and train skilled workers, and to expand commerce, industry and communications.

The **UNITED NATIONS ECONOMIC COMMISSION FOR EUROPE (ECE)** was set up in 1947 and encouraged regular meetings between the countries of Eastern and Western Europe. It has been mainly concerned in recent years in reaching decisions and establishing conventions on environmental problems, and in facilitating **sustainable development.**

The **UNITED NATIONS EDUCATIONAL, SCIENTIFIC AND CULTURAL ORGANIZATION (UNESCO)** was formed in 1945 to ease international tension in the post-war years. UNESCO is responsible for the International Hydrological Programme, and declared the ten years from 1965 the **International Hydrological Decade.** It fostered the interchange of scientific and cultural ideas and set out to improve education. At the end of the 1960s UNESCO launched the **Man and the Biosphere** programme, an international project which broke new ground. It concentrated on long-term scientific studies to monitor small, cumulative changes in the environment. UNESCO's World Heritage Programme was launched in 1978 to protect historic sites and natural landmarks of outstanding universal significance (see **World Heritage Site**).

The **UNITED NATIONS ENVIRONMENT PROGRAMME (UNEP)** evolved from the **United Nations Conference on the Human Environment,** which was held in Stockholm in 1972. UNEP is a programme that includes all the activities within the United Nations that relate to the environment. It was set up as a catalyst scheme designed to co-ordinate and stimulate action within the UN, not to execute or finance it. It encourages and supports an integrated approach to the planning and management of development, taking account of environmental consequences in order to achieve maximum social, economic and environmental benefits. It helps all countries, especially developing countries, to deal with environmental problems and also helps to encourage additional financial resources in order to provide the necessary technical assistance, education, training, flow of information and exchange of ideas. UNEP works closely with other UN agencies and has forged links with scientists and decision-makers, industrialists and environmental activists. It actively promotes environmentally sound development which maintains and improves economic progress without damaging the environment and natural resource bases upon which future development depends.

UNLEADED PETROL, see **lead-free petrol.**

The **URAL MOUNTAINS,** in Russia, were the site of a serious explosion in a nuclear weapons production plant, located between Yekaterinburg (formerly Sverdlovsk) and Chelyabinsk. The accident was almost as serious as that at **Chernobyl.** Although the accident happened in 1957, it was kept secret until 1976. A storage tank, containing 80 metric tons of waste from the **plutonium** plant, had exploded. Twenty million **curies** of radiation were released into the air, most of it falling on the plant itself, contaminating roads and laboratories. Two million curies were blown across 8,000 square miles. Nearly 11,000 people were evacuated and 23 villages were bulldozed into the ground. Some 9,000 metric tons of crops were destroyed and thousands of animals were slaughtered.

URANIUM is an element with 92 protons in its nucleus. It is one of the heaviest elements found naturally on Earth. The uranium extracted from mineral ores by miners is a mixture of two forms: one form comprising over 99.3% of the ore is the **isotope** uranium-238, the other 0.7% is the isotope **uranium-235.** Although a rare commodity, uranium-235 provided the key to producing nuclear weapons and atomic power stations. Uranium-235 is the only natural substance able to undergo nuclear **fission,** and so is referred to as a fissionable material. On the other hand, uranium-238 has properties which make it a **fertile** material.

URANIUM ENRICHMENT, see **enrichment.**

URANIUM-235 is the part of **uranium** ore which makes the element so dramatically different from any other mineral. Fuel for nuclear reactors uses uranium in which the proportion of U-235 has been increased to about 3.5%–4%. Atomic bombs employ nearly 100% U-235. The two forms, or **isotopes,** of uranium can not be separated from each other by a chemical process. The task of increasing the proportion of U-235 is done by **enrichment.**

URBAN BIOMES describe the towns and cities which replace natural ecosystems with man-made ecosystems organized for human survival. Urban sprawl has an obvious impact on the visual environment and drastically alters the wildlife of the area, simply because animals and plants can no longer compete in the way the land is used. Urban development also modifies the local climate. The concentration of heat for space heating of buildings makes towns and cities **heat islands.** The lowest winter temperatures in urban areas are at least 2°–3°C higher than in the surrounding countryside. Consequently, although towns and cities receive an average of 5%–10% more rainfall than their surroundings, there is continuous evaporation and water loss. Traffic exhausts and fumes from commercial and domestic boilers emit **sulphur dioxide,** nitrogen oxides, unburned hydrocarbons and other pollutants into the atmosphere. Urban air has on average ten times the concentration of dust particles of the neighbouring rural atmosphere. There is 25 times as much **carbon monoxide,** ten times more **carbon dioxide** and five times as much sulphur dioxide. Cities have up to 10% more cloud cover and lose up to 20% of sunshine. The clean air of rural areas is signalled by the appearance of **lichens** and moss on the roofs of houses.

URBAN HABITATS are man-made habitats that replace natural ecosystems during the process of urbanization. As people crowd into towns and cities, the wildlife of the area is radically altered, mainly because most plants and animals have no place in the new environment. Some animals adapt to the new conditions, such as feral pigeons, rats and **cockroaches.** Gardens, parks, and green areas at the edges of towns and cemeteries are often colonized by animals moving in from surrounding countryside, such as foxes. City-dwelling foxes, for example, are now commonplace in the UK, and have shown an amazing adaptability to urban living. They mostly hide by day in the corners of churchyards, on derelict land, in cellars or under garden sheds. At night they are frequently spotted in the most built-up areas. Fleas, lice and bedbugs also invade overcrowded cities, as do **ants,** flies, moth caterpillars, mites, and silverfish. Buildings provide living space for some animals: ledges provide roosting or nesting sites for birds; ventilator shafts and roof spaces are used by **bats** and swifts; rats live in the drains and sewers, and mice in the wall spaces. Large scavenging animals, particularly those whose natural habitat has been destroyed or disturbed, including racoons, bears and elk, come into cities to find food.

URBANIZATION describes the current trend of a growing part of the world's **population** to migrate from rural areas to the towns and cities. It has been estimated that by 2010 the world's urban population will be more than 4,000m., an increase of 70% since 1990. Urbanization in the industrialized world was spread over several decades, allowing economic, social and political systems and public services to evolve gradually to cope with the problem of an expanding population. However, the rapid growth of urban populations is a cause for concern in many countries, particularly the developing countries, where it has been predicted that 60% of city dwellers will be squatters by the end of the century. Overcrowding, **poverty,** malnutrition, poor hygiene and sanitation, lack of safe water, inadequate shelter and housing, and illiteracy all contribute to a mounting and serious public health problem. The rate of urban growth, combined with economic decline, overwhelms the capacity of governments to provide an urban infrastructure and essential public services, especially in areas where the poor live. The Panel on Urbanization reporting to the **Commission on Health and Environment,** set up by the World Health Organization to investigate the impact of the environment on health, has said that in cities in the industrialized world, improvements in the standards of living, general sanitation and average incomes for most urban dwellers have been associated with a drastic fall in death and illnesses from communicable diseases over the past 50 years. In the developing world, however, cities are plagued by communicable diseases, malnutrition, parasitic diseases, illnesses associated with unsafe water supplies and poor sanitation. Filthy water around houses is ideal breeding ground for **vector-borne diseases** such as **malaria** and many other severe diseases. **Air pollution** is a major problem linked to a high incidence of bronchitis, asthma and other respiratory diseases. Overcrowding and lack of basic hygiene mean that **airborne diseases** such as influenza, meningitis and **tuberculosis** are easily transmitted. Animal and insect carriers of disease are adapting to new habitats and introducing new infections which spread rapidly under crowded and insanitary conditions.

V

VANISHING HABITATS are the greatest threat to the wide variety of natural wildlife species, both animals and plants, on the planet. Habitats have been, and continue to be, destroyed in most parts of the world by human activities, taking away **biological diversity** and the natural variety of the world's flora and fauna as well as their habitats. Extensive hunting and a degree of exploitation was the main cause of the depletion of species until the 1950s. The destruction of natural habitats has been the cause for the overwhelming reduction in numbers of thousands of forms of wildlife. Many animals and plants have faced **extinction,** or have become extinct. As the **tropical rainforest** is cleared for human habitation, logging, roads or mining, more species are threatened, and some disappear forever. Most of these are animals and plants which are essential to ensure an ecological balance. Pollution destroys the habitats of wildlife in rivers, at sea and on land. Draining land for agricultural purposes or converting it to cropland or pasture has meant that natural habitats are lost or irreversibly damaged. Large **dams** and **irrigation** schemes, in both the developed and undeveloped world, have frequently caused floods, destroying or washing away established habitats. At the **Earth Summit,** the United Nations Conference on Environment and Development held in 1992, the **Biological Diversity Convention** was agreed and signed by 154 states. Its provisions are intended to ensure effective national action to curb the vanishing species, habitats and ecosystems.

VECTOR-BORNE DISEASES are those carried by other organisms, and are persistent in some regions of the world because of adverse environmental factors. Such diseases include **malaria, schistosomiasis, filariasis** and **onchocerciasis.**

The **VIENNA CONVENTION FOR THE PROTECTION OF THE OZONE LAYER** was the first drawn up to deal with the destruction of the **ozone layer.** The international agreement came after a conference in 1985, and was adopted by 21 states and the European Community. It called for a halt in the production of **chlorofluorocarbon (CFC)** gases and procedures to reduce all emissions of CFC gases. The producers of CFCs, mainly found in the developing world, would not sign the Convention until they received guarantees over compensation. The Convention led to the **Montreal Protocol** in 1987, under which signatories agreed to freeze the use of CFCs at 1986 levels with a 20% cut by 1994 and the eventual goal of a 50% reduction by 2000. This treaty was amended in 1990 after research showed that the depletion of ozone was more serious than had been previously thought. CFCs (used in **refrigeration** and **aerosols,** and as cleaners in industry), halons (used in fire extinguishers) and **carbon tetrachloride** (which is used as a solvent and is also a chemical that depletes the ozone layer) would be phased out by 2000. A timetable has been set for phasing out other harmful substances. A multilateral fund has been set up by the United Nations Environment Programme (UNEP), the United Nations Development Programme (UNDP) and the World Bank to help developing countries with the costs of complying with the revised protocol and provide for the necessary technology transfer.

VITRIFICATION is a way of handling radioactive **high-level wastes,** which are predominantly liquid wastes produced from **nuclear reprocessing** of fuel. They have accumulated since reprocessing was invented to extract **plutonium** from nuclear fuel for weapons programmes. The high-level wastes have high concentrations of radioactive substances, which remain active for a long time, especially **actinides,** which are exceptionally hazardous. High-level wastes increased rapidly with reprocessing of spent nuclear fuel from the **Magnox** commercial nuclear power stations. The waste is currently stored in stainless steel, water-cooled tanks housed in underground bunkers. However, in the 1990s work began on a vitrification process for converting high-level liquid wastes into glass blocks, using a technique developed by the French nuclear reprocessing company, Cogema. Nuclear developers hope to create deep underground repositories for the glass blocks by the turn of the century.

A **VOL2 PHOTOMETER** is an instrument used in air pollution monitoring that measures the concentration of **photochemical smog** in the atmosphere.

W

WARFARIN belongs to a family of synthetic chemicals called coumarins, which are the most successful **rodenticides** (poisons for the control of rats and other rodents and small mammals). Warfarin is the most widely used of these. They are anticoagulants and encourage internal bleeding. In addition to its use as a rodenticide, warfarin is used to treat humans, by dissolving blood clots that could cause heart attacks and thromboses.

WARREN SPRING is the British government's research laboratory which monitors levels of **air pollution** in the UK and devises anti-pollution strategies, such as methods of cleaning **oil spills** and methods of **recycling.**

WASH-OUT removes particles from the atmosphere when it rains. Wash-out is beneficial in clearing pollen and other substances to which people are allergic from the air. The disadvantage is that acidic molecules that have formed in the air from the discharge of gases from power stations, factories and cars, are also washed out, causing **acid rain.**

WASTE DISPOSAL experts divide the world's sources of rubbish into domestic, industrial and agricultural wastes. The contents of the domestic dustbin are referred to as municipal solid waste. Every day the average Briton is responsible for one kilogram of waste in the form of paper, plastic cartons and other packaging, tins, polystyrene bottles, disposable nappies, kitchen waste and other objects. Municipal solid waste is handled by one of three processes: **landfill, incineration** or **recycling.** Each year, industry in Britain produces 78m. tons of waste. Some of it, like paper and cardboard, ends up in landfill sites. However, these are small volumes compared with the millions of tons of waste water and sewage effluent produced every day. Britain finds more space for landfill—about 88% of the municipal solid rubbish is disposed of in this way—than Sweden, Germany, Japan and the USA, which have a preference for incineration. Attitudes are changing with increasing environmental awareness. There is a rising interest in recycling. Waste is also a source of alternative energy and **biofuel.** New incinerators incorporate boilers that raise steam to generate electricity or supply hot water to warm buildings, instead of simply releasing waste heat into the atmosphere. Established landfill sites are also a source of **methane** gas, produced by decomposition of organic material. Methane is collected by sinking perforated pipes into old landfill sites. Scientists are developing ways of deliberately turning landfill sites into bioreactors, so they produce the maximum amount of useful methane.

The **WATER CYCLE** is the constant recycling of water over the face of the Earth. Water evaporates into the air from the soil, **transpiration** by plants, lakes and most of all the oceans, which contribute about 84% of the total moisture from evaporation. The **atmosphere** only accounts for one part in 10,000 of all the global water. Yet water is the only substance to occur naturally in the atmosphere as a solid, liquid and gas; the energy it absorbs or releases in changing from one state to another influences both local weather patterns and acts as a driving force to the **general circulation** of the atmosphere. Water that evaporates to the air is eventually returned as rain, mist and snow in a cycle that completely replaces the atmospheric moisture on average every ten days. In temperate or tropical coastal zones, the complete **hydrological cycle** can be as short as one hour, whereas in icy polar wastes it can take thousands of years. However, only 75% of the atmospheric evaporation falls on the ocean, in other words much less than evaporates from it. The balance falls on the land and eventually gets back to the sea via the rivers and, over much longer time-scales, when glaciers and edges of ice sheets disintegrate. In all, the oceans contain 97% of the world's water and the rest is locked up or stored in glaciers, ice sheets, rivers and lakes, underground **aquifers** and as **groundwater.**

The **WATER TABLE** is the upper surface of **groundwater.** The depth varies with the seasons, climatic factors and the nature of the bedrock. Major engineering works, such as **dams** and canals, can have serious effects on the level of the water table. Water percolates into the surrounding rock, the water table in the area rises and low-lying land may be flooded, damaging housing and agriculture. However, in dry areas the rise in the water table can be very advantageous because farmers can use the water to irrigate land that was formerly arid.

WATER VAPOUR is the most abundant **greenhouse gas** in the atmosphere. It gets little attention in debates about **global warming** and **climatic change** because there is little to be done about it. Water vapour plays other roles, equally crucial for maintaining the health of the planet, in addition to supplying fresh water. Water vapour traps heat and keeps the planet warm. A dilemma occurs if other greenhouse gases in the atmosphere, like carbon dioxide, are increased by man's activities and cause a rise in temperature. Then more water vapour may evaporate, perhaps increasing the temperature still further in a process called 'positive feedback'.

WATER-BORNE DISEASES are those transmitted by **bacteria,** insects and other organisms that live or breed in water. These diseases are caused by contaminated water or food, by lack of hygienic washing or sanitation facilities, or by insects which breed or live near water. An estimated 25,000 people die as a result of these diseases every day. Another 160m. suffer from **malaria,** 200m. from **schistosomiasis** and 250m. from **elephantiasis.** Water-borne diseases also include **cholera, leprosy, onchocerciasis** (river blindness), yellow fever, scabies, polio and diarrhoea. The **World Resources Institute** has estimated that some five million children a year under the age of five die from diarrhoea. Other diseases associated with water-borne pathogens include gastroenteritis, dysentery, typhoid, tapeworm, yellow fever, infectious **hepatitis** and polio. Impurities in water can also have serious adverse effects on health. These include pesticides, lead, zinc, nitrates, fluorides and cadmium. The **World Health Organization** estimates that 80% of all illnesses in the developing world are caused by lack of safe water and adequate sanitation.

WAVE POWER uses devices in the open **ocean,** but anchored to the sea-bed, which can exploit the undulating motion of the waves to drive an electrical generator. Several devices, the best known of which is perhaps the 'nodding duck' invented at Edinburgh University, have been tried successfully, but all have failed to attract the necessary support for development as a major source of alternative energy.

WET (AQUEOUS) PHASE CONVERSION is the chemical interaction of gases, particularly **sulphur dioxide** and nitrogen oxides, which derive from coal-burning power stations and road vehicles, in damp atmospheric conditions, to produce **acid rain.** The formation of the acid is a slow, steady process and can happen far away from the source of pollution, even hundreds of miles away.

WETLANDS exist in both inland and coastal areas, covering 6% of the Earth's land surface. Throughout both the industrialized and the developing world wetlands are threatened by drainage schemes, land reclamation and the reservoirs of large dams. Wetlands consist of saltwater or freshwater marshes, swamps, bogs, wet meadows, fens and potholes. Wetland describes any area of low-lying land where the **water table** is at or near the surface for most of the time, producing habitats appropriate for extensive shallow, open water and waterlogged areas; they are typically found in **estuaries** and along lowland rivers. They do occur in upland areas, where the natural drainage is blocked by boulder clay deposited in the ice ages. In their natural state, wetlands perform several functions. They provide food, breeding grounds, water and sanctuary for many forms of fish, birds and other animal and plant life. Wetlands are among the most productive ecosystems, producing timber, **peat** moss and crops such as rice and a variety of berries. **Mangrove forests** are an important type of wetland, providing roofing materials, fibres for cloth, fodder for animals, tannin for leather, and sugar. Wetlands also act as a buffer to regulate flood-waters, which spread out before reaching higher ground as a flood or storm tide. In Africa, nomadic herdsmen rely on seasonal flood-plains for grazing cattle, sheep and goats. In Europe, the coastal wetlands of the Waddenzee (on the North Sea coast of the Netherlands) support 60% of the North Sea's brown shrimp, and are the spawning grounds for 50% of the sole, 80% of the plaice and most of the herring. Wetlands play a vital role in cleansing water by filtering out pollutants. Wetland plants are able to absorb excess **nutrients** without succumbing to **eutrophication.** Some of the best known wetlands include the **Everglades** in Florida and the Mississippi river basin. Since 1971 wetlands have been protected by the Convention on Wetlands of International Importance, Especially as Waterfowl Habitats (see **Ramsar Convention**).

WHALES have been hunted for their meat, bone, blubber and other substances which are used in the manufacture of textiles, candles, perfumes and cosmetics. Many species have been hunted to near-extinction. In 1946 whalers began to regulate their catches, but by 1970 eight whale species were listed as endangered. In 1982, the **International Whaling Commission** declared a moratorium on commercial whaling which took effect in 1986 and has led to a dramatic drop in hunting. However, it has few regulatory powers and has been unable to prevent some whaling nations, notably Japan, Norway and Iceland, from variously defying, or resigning from, the IWC or announcing plans to resume whaling; Japan has continued to catch whales in large numbers. In 1963 the UK stopped commercial whaling and the USA followed in 1972. One of the most **endangered species** of whale is the humpback, known for its 'songs' and acrobatics. While it is not now hunted, other factors are making it difficult for it to survive: pollution of its feeding grounds and a scarcity of food; **drift nets,** which are almost invisible and trap the whales, so they are unable to surface to breathe; and tourists who go out on whale-watching excursions to the feeding and breeding grounds. There is a fear that too much noise interferes with the communication between the whales. Some sanctuaries have now been created for the whales where no boats are allowed.

WIND POWER is mechanical or electrical power generated by a windmill, a machine with wind-driven sails or blades, which turns a rotor either providing mechanical power, as in traditional flour grinding mills, or electrical energy, as in modern wind turbogenerators. It is a leading potential source of **renewable energy.** One example of windmill design is the Savonius Rotor, a device named after its Finnish inventor which provides one of the easiest and most reliable windmills. The Savonius rotor is a cylindrical-shaped object turning in the wind; a primitive type can be made from an oil drum. Another design is the Darrieus generator, a vertical axis windmill that consists of thin aerodynamic strips.

The **WINDSCALE FIRE** was an accident that occurred in 1957 in a British nuclear reactor, known as the Number 1 pile, at Windscale, in Cumbria. The site was then operated by the **United Kingdom Atomic Energy Authority (UKAEA).** Subsequently, the UKAEA was split up. A new group, British Nuclear Fuels, was formed in 1971 to operate the Windscale site but under the changed name of **Sellafield.** The Number 1 pile consisted of a stack of graphite bricks with channels, into which **uranium** fuel rods were slotted. The pile was used to convert some of the uranium into **plutonium.** In the nuclear process a large amount of energy, called Wigner energy, built up in the graphite blocks. Periodically it was released in a surge of heat that was blown up a tall chimney. On 8 October 1957 the process went wrong. Graphite blocks overheated, caught fire and caused 11 tons of uranium to burn. At these super temperatures, attempts to blanket the fire with carbon dioxide only fed the flames. The reactor core became a raging furnace. By the time the fire was extinguished 20,000 **curies** of radioactive **iodine**-131 had escaped, along with quantities of strontium-90 and **polonium-210.** Sales of milk from farms surrounding Windscale for 500 square kilometres were banned and one million gallons was poured away. The most controversial issue surrounds the number of cancers which were probably caused by the accident. The fall-out produced a rise in exposure to low levels of radiation over a very wide area. However, no records were made to keep track of people who were near Windscale at the time of the fire.

The **WINDSCALE INQUIRY,** held in 1977, was the first public inquiry in Britain into a planning proposal for a nuclear power development. The plan was for a $2,500m. nuclear fuel **reprocessing plant** called THORP (thermal oxide reprocessing plant), on the coast of Cumbria. The name of the site was changed later to **Sellafield.** The decision to hold the first public inquiry into a nuclear power scheme can be attributed to a campaign conducted by the *Daily Mirror* newspaper in 1974. It described the Windscale proposal as a 'plan to make Britain the world's nuclear dustbin'. The story triggered a national uproar. It was seized upon by **Friends of the Earth,** which was the principal opponent to further development of nuclear power. THORP was designed by **British Nuclear Fuels (BNFL)** to reprocess spent nuclear fuel rods after they had been removed from **pressurized water reactors** and **advanced gas-cooled reactors** in nuclear power stations in Britain, Japan, Germany and elsewhere. THORP would chemically extract the **plutonium.** The environmentalists lost the inquiry. Permission to build was given in 1978. The vast plant was finished at the end of 1992, but plans to start operating early in 1993 were stopped by the British government's Nuclear Installation Inspectorate. It refused to issue an operating licence because it regarded the level of radioactive emissions BNFL intended to discharge with its wastes as too high.

The **WORLD BANK** is made up of three bodies: the **International Bank for Reconstruction and Development (IBRD)** raises money on international financial markets and lends at commercial rates of interest; the International Development Agency provides loans at concessional rates of interest to the poorest countries from member subscriptions, donations and grants from the IBRD; and the International Finance Corporation lends money to the private sector with government guarantees. It operates as a multinational development agency with over 170 member nations, and was formed to give loans and technical assistance to the poorer countries of the world. It was established in 1945 after the Bretton Woods conference of 1944, at which Canada, the United Kingdom and the USA established a system of international financial rules. The World Bank, with its headquarters in Washington, is the largest of the multilateral development banks, investing more than US $16,000m. a year in developing countries. It is controlled by member countries who vote according to the size of their donations. Because it is a commercial operation any loans it makes must be repaid, so that funding is mostly given to those projects that are likely to be profitable. The World Bank works in collaboration with its sister organization, the **International Monetary Fund (IMF),** which lends bridging finance to countries with short-term balance-of-payments difficulties. The World Bank has been criticized by environmental organizations because many of the projects it has funded have had adverse effects on the environment and on local people, including many large **dams** and **irrigation** schemes, road construction through forested areas, and livestock-rearing programmes. In 1983 the pressure on the World Bank began to mount, and was spearheaded by **Non-governmental Organizations (NGOs).** In 1986 there was a co-ordinated campaign against the Polonoreste road-building and colonization project in Brazil. This resulted in a vote against a loan to Brazil, the first time that environmental reasons were cited for such a delay. In 1987 the World Bank published a report, *Environment: Growth and Development*, setting out new environmental guidelines for future activities. Future projects, it said, will be obliged to have three objectives: economic growth, poverty alleviation and environmental protection. A new enlarged Environment Department was formed in 1990.

The **WORLD CLIMATE CONVENTION,** the United Nations Framework Convention on Climate Change, was adopted at the **Earth Summit** in 1992 after fifteen months of preparation. Its aim is to protect the **atmosphere** from a build-up of gases that trap heat from the sun, causing the **greenhouse effect.** The document does not provide specific timetables and targets for limiting emissions of **greenhouse gases** for industrialized countries. It sets out, as an objective, the stabilization of greenhouse gas concentrations in the atmosphere at a level that would prevent dangerous interference with the climate system. Such a level, it says, should be reached within a time-frame sufficient to allow ecosystems to adapt naturally to **climatic change.** As a first step, developed countries specified in the text recognized the importance of returning to their 1990 levels of greenhouse gas emissions by the end of the 1990s. The Convention provides for the establishment of machinery, in the form of a conference of the parties, to take decisions regarding subsequent steps, as needed. It also provides for a review, before 2000, of the progress of the commitments, as well as for amendments.

WORLD COMMISSION ON ENVIRONMENT AND DEVELOPMENT, see **Brundtland Report.**

The **WORLD CONSERVATION STRATEGY (WCS)** was the policy document published in 1980 which outlined a single united approach to global problems. It was sponsored by key international conservation organizations, the **International Union for the Conservation of Nature and Natural Resources (IUCN),** the **World Wide Fund for Nature (WWF),** the UNESCO and the FAO. The document warned that the basic **life-support systems** of the Earth must be preserved intact if life is to continue. It stressed that genetic diversity is the major key to the future health of the planet and that mineral and fossil fuels were finite, and if they were exploited further by industrial development and led to pollution, it would lead to a serious degeneration of the quality of the **biosphere.** A revised version of the strategy, **'Caring for the Earth: A Strategy for Sustainable Living',** was produced in 1991.

The **WORLD CONSERVATION UNION** is the simpler name adopted by the **International Union for the Conservation of Nature and Natural Resources.** The acronym IUCN remains.

WORLD CULTURAL AND NATURAL HERITAGE SITES, see **World Heritage Site.**

The **WORLD HEALTH ORGANIZATION (WHO)** is a specialized agency of the United Nations, formed in 1948 to foster international co-operation for improved health conditions. It took over epidemic control, quarantine measures, and drug standardization from the Health Organization of the League of Nations and the International Office of Public Health. The WHO aims to promote the highest possible level of health for all peoples, defined as a state of complete physical, mental and social well-being, not just the absence of disease or infirmity. It is funded by member governments and an allocation from the technical assistance programme of the UN. Its administrative headquarters are in Geneva. The central secretariat keeps member countries informed of the latest developments in the use of vaccines, cancer research, nutritional discoveries, control of drug addiction and the health hazards of nuclear **radiation.** The WHO sponsors the control of epidemic and endemic disease by promoting mass campaigns involving nation-wide vaccination programmes, and gives instruction in the use of antibiotics and **insecticides.** The WHO was largely responsible for the eradication of smallpox in 1980, while other campaigns have been successful in combating **tuberculosis, malaria,** venereal disease, cholera, trachoma and yellow fever. On request, the WHO also gives technical advice to governments in the preparation of long-term national health plans and sends out international teams of experts to conduct field surveys and demonstration projects. The WHO works with the **United Nations Environment Programme (UNEP)** and the **Food and Agricultural Organization (FAO)** to promote environmental means of controlling **vector-borne diseases** such as **malaria, schistosomiasis** and **sleeping sickness,** through the joint **Panel of Experts on Environmental Management for Vector Control (PEEM).** The three organizations also work together on the assessment, monitoring and control of mycotoxins, which are released by fungi into poorly stored food.

WORLD HERITAGE SITE refers to any natural area or historic site designated by the **United Nations Educational, Scientific and Cultural Organization (UNESCO)** as being one of outstanding and universal importance. Countries belonging to the 1972 Convention Covering the Protection of the World Cultural and Natural Heritage (usually referred to as the World Heritage UNESCO Convention) nominate the sites. To be accepted, a natural site must have either: major characteristics of the Earth's evolutionary history; a habitat for endangered or rare species; a unique example of natural phenomena; or a place where there is a continuing geological process. The World Heritage Programme was launched in 1978, and by mid-1992 the World Heritage List comprised a total of 358 sites, which included the **Great Barrier Reef** (Australia), the Galapagos Islands (Ecuador) and the Serengeti National Park (Tanzania).

The **WORLD METEOROLOGICAL ORGANIZATION (WMO)** was established in 1951 as a specialized agency of the United Nations, and took over the responsibilities of the International Meteorological Organization, which had been founded in 1873. The WMO continued to intensify international co-ordination of measurements of atmospheric composition, including **ozone, carbon dioxide,** and other **greenhouse gases.** Much of our knowledge of climate comes from the global scientific and technical programmes co-ordinated by the WMO. The First World Climate Conference was organized by the WMO in 1979, in collaboration with other UN organizations and the **International Council of Scientific Unions.** A World Climate Programme was established, recognizing that there was an immediate need for nations to use existing knowledge of climate and climate variations in the planning of social development. In 1988, the **Intergovernmental Panel on Climate Change (IPCC)** was established jointly by the WMO and the **United Nations Environment Programme (UNEP).** The IPCC was to study all aspects of possible **climatic changes,** including the socio-economic implications. The Second World Climate Conference was held in late 1990. The conference supported the establishment of a Global Climate Observing System to integrate the existing

World Weather Watch, the **Global Ocean Observing System** and the Global Atmosphere Watch. The conference acknowledged and endorsed the activities carried out within the World Climate Programme and IPCC, and called for urgent action, including negotiations on a framework for a **Convention on Climate Change.** That convention was agreed at the **Earth Summit.**

The **WORLD OCEAN CIRCULATION EXPERIMENT (WOCE)** is a critical component of the **World Meteorological Organization's** World Climate Research Programme. WOCE is the largest ever international study of the physics of the oceans and their role in regulating the climate of the planet. The ocean has an enormous capacity to store heat. The top three metres of the ocean have a greater heat capacity than the entire atmosphere. The oceans determine the timescale and regional patterns of climate change. Ocean currents transport huge amounts of energy, absorbed from the sun, from the tropical latitudes toward the poles (see **ocean circulation**). The surface layer of the ocean acts as an energy buffer. Some of its heat is transferred back into the atmosphere and some of it is transferred into the ocean interior. Huge swathes of water at great depths then move like a gigantic conveyor belt, forming vast currents of thermal energy that move toward the poles, while immense 'rivers' of cold water flow toward the equator from the Arctic and Antarctic. Precise measurements of these great ocean currents are difficult to make; hence, one aspect of WOCE is a **Hydrographic and Chemicals Tracer Study,** which analyses samples of **sea-water** from thousands of different areas and depths. However, vast amounts of energy are also released into the atmosphere, and this heat influences the winds, rainfall patterns and regional temperatures. Consequently, a critical part of WOCE is a **Satellites and Surface Meteorology Project,** measuring interactions between the sea and atmosphere. Through WOCE, scientists are collecting vital data in order to develop realistic computerized **ocean circulation modelling** experiments, in an **Ocean Modelling Experiment,** demonstrating how the ocean links with the atmosphere to influence climate. By the end of the 1990s scientists are hoping to produce forecasts on the changes in energy levels in the oceans and what they mean for the climate pattern of the world. WOCE leads on to a longer-term plan for systematic monitoring of the ocean on a global basis, the **Global Ocean Observing System.**

WORLD PARKS ENDOWMENT is one of the innovative schemes designed to help the spread of **conservation** programmes. It was launched in 1989 to support conservation in 250 environmentally rich areas of the world. It raises funds and channels them into protected area projects in tropical countries. The endowment is administered by the **International Union for the Conservation of Nature and Natural Resources (IUCN).** It has already bought an area of tropical rainforest in Guatemala and endowed a watershed protection area in Belize.

The **WORLD RESOURCES INSTITUTE (WRI)** was established in 1982 in Washington, DC, as a policy research centre. The WRI's research agenda embraces agriculture, **tropical rainforests, climatic change,** health care and environmental and resource information systems. It is funded by private donations, industry, the United Nations and government agencies. The WRI's function is to help international organizations, governments and the private sector to foster economic growth without affecting natural resources and the environment.

The **WORLD SEED BANK** is housed in the UK at Wakehurst Place in Sussex, the annexe of the **Royal Botanic Gardens** at Kew, Surrey. It contains the seeds of more than 2,500 species of flowering plants and is one of the largest collections in the world. Seeds, collected from most parts of the world, are dried by exposure to low relative humidity and then stored in airtight rooms and deep-frozen. This improves and extends the lifespan of the seeds; in the case of barley it could be as long as 20,000 years. During expeditions, as many types of plant seeds as possible are collected in an area, because it is impossible to know which ones will eventually prove useful in the future, and also to make sure that there is a broad genetic base. Sufficient seeds are left to regenerate in future years. Collectors also gather seeds from plants that local people use, particularly those that have medicinal qualities. In a recent programme, collectors concentrated on plants in arid and semi-arid countries, like Mexico, Australia and Morocco. When a rare spell of rain produces flowers and seed, the collectors set out at a moment's notice to remote deserts. Remote-sensing satellites are to be used in the future to pinpoint places where rain has fallen.

The **WORLD WIDE FUND FOR NATURE (WWF)** is one of the largest international nature **conservation** organizations. It was formed in 1961 and until 1988 was known as the World Wildlife Fund. The WWF raises funds, mostly from the general public, for conservation projects. Its aim is to conserve nature by preserving genetic species and ecosystem diversity. The WWF's international campaigns have brought about a growing awareness of environmental issues and of the threats to the natural environment. The WWF works closely with the **International Union for the Conservation of Nature and Natural Resources (IUCN),** and has played a part in implementing international laws and agreements on conservation, including the **Convention on International Trade in Endangered Species (CITES).** In collaboration with IUCN and the United Nations Environment Programme (UNEP), the WWF launched the **World Conservation Strategy (WCS)** in 1980.

The **WORLDWATCH INSTITUTE** is an independent, non-profit-making research organization based in Washington, DC. It was set up to analyse and focus attention on global problems. It is funded by private foundations and United Nations organizations. It distributes regular papers to a world-wide audience of decision-makers, scholars and the general public. The topics range from energy, the environment, technology and space, to employment and economics, among many others. It also produces a bi-monthly magazine, *World Watch,* which aims to raise public awareness of environmental threats to the planet. Using the Institute's world-wide information-gathering system, the magazine is designed to provide a global framework for the thousands of organizations in the world that work on energy, environmental, food, population and peace issues.

X

XEROPHYTE is the general name for plants and trees adapted to a desert-dry habitat, and which can survive in alkaline, acid, salt or dry soils.

Y

YELLOWSTONE NATIONAL PARK was the first **national park** to be designated in the USA. It was established in 1872 and is the largest national park in the USA. It covers 898,353 hectares in the states of Wyoming, Montana and Idaho. Among its attractions are 30,000 **geysers,** many of which gush to a height of 100 feet or more. The most famous geyser is Old Faithful which erupts regularly every 33 to 93 minutes. The park has more than 500 miles of roads and more than 1,000 miles of trails. Yellowstone has a number of lakes and rivers, including the Yellowstone River which has several spectacular waterfalls. Hundreds of species of birds live in the park, including many waterfowl. Animals found in the park include buffalo, elk, deer, moose, coyotes and bears.

YOSEMITE NATIONAL PARK is the area in California that was first set aside as a state park in 1864. In 1890, combined with surrounding land, it was made a **national park.** It covers 308,106 hectares in the Sierra Nevada range. Giant sequoia groves include trees that are thousands of years old (see **redwood trees**). The park's animal life includes various squirrels, chipmunks, mule deer and black bears.

Z

ZERO POPULATION GROWTH refers to a balance between birth and mortality rates that would produce a stabilized **population.** The world population increases by over 80m. a year, and if this acceleration remained unchecked the world's population would reach 10,000m. by 2030.

ZINC is an **essential element** that is as important to life as the principal **vitamins.** Zinc deficiency in plants stops leaves expanding, giving a rosette-like appearance. In humans, deficiency of zinc in the diet causes disorders like anaemia, short stature and poor healing of wounds.

ZOOGEOGRAPHIC REGIONS are those areas of the world, usually continents, where unique, endemic animals have evolved.

ZOOGEOGRAPHY is the study of the geographical distribution of animals within **zoogeographic regions.** It analyses the present distribution, how the animals came into existence, and how they have evolved. Each of the world's continents has its own unique animals that are found nowhere else, such as kangaroos, koalas and the duck-billed platypus in Australia, and the zebras and wildebeest in Africa. Most of the zoogeographic regions are single continents that are completely or nearly isolated by a large expanse of ocean or sea.

ZOOPLANKTON are the animal components that, together with the plant components, **phytoplankton,** form **plankton.** They consist mainly of small crustaceans and fish larvae, and are intermixed with the phytoplankton, on which they feed. They are a prime source of food for herring, sprat and pilchard. One group of zooplankton occur in vast numbers in the Antarctic ocean, forming **krill,** which is the main source of food for whales, penguins, seals and squid.

DIRECTORY
of
ENVIRONMENTAL ORGANIZATIONS

INTERNATIONAL ENVIRONMENTAL ORGANIZATIONS

Advisory Committee on Pollution of the Sea (ACOPS): 57 Duke St, Grosvenor Sq, London W1M 5DH, UK; tel (71) 499-0704; telex 261681; fax (71) 493-3092; Chair: Lord Clinton-Davis; study of all forms and sources of marine pollution.

Asian Development Bank: 6 ADB Ave, 1501 Mandaluyong, Metro Manila, Philippines; tel (2) 7113851; telex 63587; fax (2) 7417961; Chair, Bd and Pres: Kimimasa Tarumizu; Vice-Pres: In Yong Chung, William R Thomson, Günther Schulz; raising funds from private and public sources for Asian development; assistance to member states in the co-ordination of economic policies; technical assistance for development projects.

Asian Forum of Environmental Journalists (AFEJ): c/o Press Development Institute, 647 Mooban, Happyland Townhomes, Klongjan, Bangkapi, Bangkok 10240, Thailand; tel (2) 378-0981; fax (2) 378-0981.

Asociación Latinoamericana de Instituciones Financieras de Desarrollo (ALIDE—Latin American Association of Development Financing Institutions): Apdo 3988, Lima 100, Peru; Paseo de la Republica 3211, Lima 27, Peru; tel (14) 422400; telex 21037; fax (14) 428105; Pres: Oscar Espinosa Villarreal; Sec-Gen: Carlos Garatea Yori; information exchange; research; representative for Latin American development banks.

Association of Marine Laboratories of the Caribbean (AMLC): c/o Department of Marine Sciences, University of Puerto Rico, Mayagüez, PR 00680, Puerto Rico; tel (809) 899-2048; telex 3452024; fax (809) 899-5500; Exec Dir: Paul M Yoshioka; Sec and Treas: Beverly B Buchanan; meetings; newsletter.

Atlantic Salmon Federation: POB 429, St Andrews, NB E0G 2X0, Canada; tel (506) 529-4581; fax (506) 529-5985; Pres: David Clark; Chair: Donal C O'Brien, Jr; international non-profit organization dedicated to the preservation and wise management of the Atlantic salmon and its habitat through research, management, education and international co-operation.

Caribbean Community and Common Market (CARICOM): Bank of Guyana Bldg, POB 10827, Georgetown, Guyana; tel (2) 69280; telex 2263; fax (2) 56194; Sec-Gen: Edwin Carrington; Dep Sec-Gen: Frank O Abdulah; 'Caribbean Co-operation in Health' undertakes an environmental protection programme, including the control of disease-bearing pests; Port of Spain Accord on the Management and Conservation of the Environment drawn up in 1990.

Caribbean Conservation Association (CCA): Savannah Lodge, The Garrison, St Michael, Barbados; tel 429-5373; fax 426-8483; Exec Dir: Calvin A Howell; information; environmental projects in the Caribbean.

Caribbean Tourism Organization (CTO): Sir Frank Walcott Bldg, 2nd Floor, Culloden Farm Complex, St Michael, Barbados; Mer Vue, Marine Gardens, Christ Church, Barbados; tel 427-5242; telex 2488; fax 429-3065; international development agency; objectives include ensuring harmonious interaction between tourism and the social and natural environment; programme of technical conferences, workshops and seminars at national and regional levels covers environmental and socio-cultural issues; publications include results of environmental studies conducted or commissioned by CTO.

Centro Agronómico Tropical de Investigación y Enseñanza (CATIE—Tropical Agricultural Research and Training Centre): Programa de Manejo Integrado de Recursos Naturales, Apdo 7170, Turrialba, Costa Rica; tel 56-6431; telex 8005; fax 56-1533; Dir: Dr José Flores Rodas; research and training in tropical cultivars and animal production, forestry and agroforestry, and integrated natural resources management.

Cercle International pour la Promotion de la Création (CIPCRE—International Society for the Promotion of Creation): BP 1256, Bafoussam, Cameroon; tel 44-62-67; fax 44-51-91; Dir-Gen: Jean-Blaise Kenmogne; environment protection; sustainable development.

Coalition Clean Baltic (CCB): c/o SSSN POB 4625, Asögatan 115, 116 91 Stockholm, Sweden; tel (8) 42-20-15; fax (8) 42-21-21; Exec Sec: Gunnar Norén; network of environmental NGOs from all countries bordering the Baltic Sea.

Comité Maritime International (International Maritime Committee): 203 Mechelsesteenweg, B6, 2018 Antwerp, Belgium; tel (3) 218-48-87; telex 31653; fax (3) 218-67-21; Pres: Allan Philip; Sec-Gen: Dr Norbert Trotz; Administrator and Treas: Henri Voet-Genicot; unification of all aspects of maritime law; establishment of national associations of maritime law.

Comité Permanent Inter-Etats de Lutte contre la Sécheresse dans le Sahel (CILSS—Permanent Inter-State Committee for Drought Control in the Sahel): 03 BP 7049, Ouagadougou 03, Burkina Faso; tel 30-67-58; telex 5263; fax 30-72-47; Exec Sec: Ali Djalbord Diard; concerned with water resources, agriculture, plant protection, livestock production, energy, demography, food supplies and human health.

Commission du Bassin du Lac Tchad (Lake Chad Basin Commission): BP 727, N'Djamena, Chad; tel 51-41-45; telex 5251; fax 51-41-37; Exec Sec: Abubakar B Jauro; programmes of the Commission emphasize prevention of desertification, saving Lake Chad, sound environmental management of Lake Chad Basin, improvement of communication links between member countries, sub-regional co-ordination of development projects and activities relating to crops, animals and forestry.

Commission Internationale pour l'Exploration Scientifique de la Mer Méditerranée (CIESM—International Commission for Scientific Exploration of the Mediterranean Sea): 16 blvd de Suisse, MC 98000, Monaco; tel 93-30-38-79; fax 92-16-11-95; Pres: HSH Prince Rainier III of Monaco; Sec: Prof François Doumenge; intergovernmental and multilateral co-operation in marine sciences and coastal resource management.

Commission Internationale pour la Protection des Alpes (CIPRA—International Commission for the Protection of the Alps): Heiligkreuz 52, 9490 Vaduz, Liechtenstein; tel (75) 81166; fax (75) 82819; Pres: Dr Mario F Broggi; concerned with problems of nature and countryside protection and with the upkeep and management of protected areas and land.

Commission of the European Communities, Directorate-General XI (Environment, Civil Protection and Nuclear Safety): 34 ave Belliard, 1049 Brussels, Belgium; tel. (2) 299-11-11; telex 21877; fax (2) 299-01-22; Dir-Gen: Laurens Jan Brinkhorst; environmental action programme 'Towards Sustainability' (1993–2000) aims to address the root causes of environmental degradation by raising environmental awareness and changing the behaviour of authorities, enterprises and the general public; programme of research and technological development on the environment; research into marine science and technology; monitoring of the implementation of environmental legislation; establishment of a European Environment Agency.

The Commonwealth: Commonwealth Secretariat, Marlborough House, Pall Mall, London SW1Y 5HX, UK; tel (71) 839-3411; telex 27678; fax (71) 930-0827; Sec-Gen: Chief E Chukwuemeka (Emeka) Anyaoku; Food Production and Rural Development Division; Women and Development Programme.

CAB International: Wallingford, Oxon OX10 8DE, UK; tel (491) 832111; telex 847964; fax (491) 833508; Dir-Gen: Dr D Laing; intergovernmental organization dedicated to improving human welfare, particularly in developing countries; promotes research into agriculture, forestry, health and natural resource management.

Commonwealth Forestry Association: c/o Oxford Forestry Institute, South Parks Rd, Oxford OX1 3RB, UK; tel (865) 275072; fax (865) 275074; Admin: E G Rutherford; production, collection and circulation of information relating to temperate, sub-tropical and tropical forestry and the commercial utilization of forest products; various publications.

Commonwealth Human Ecology Council (CHEC): 57/58 Stanhope Gardens, London SW7 5RF, UK; tel (71) 373-6761; telex 8951182; fax (81) 318-1439; promotes understanding of the complex relationships of man, society and nature and develops responsible policies and programmes in the Commonwealth and internationally.

Commonwealth Secretariat Group on Environment: Marlborough House, Pall Mall, London SW1Y 5HX, UK; tel (71) 839-3411; fax (71) 930-0827; Co-ordinator: Chandrashekhar Krishnan; helps developing Commonwealth countries to formulate and implement effective environmental policies, and prepares for effective participation in international environmental discussions and negotiations.

Standing Committee on Commonwealth Forestry: Forestry Commission, 231 Corstorphine Rd, Edinburgh, EH12 7AT, UK; tel (31) 334-0303; telex 727879; fax (31) 334-0442; Sec: P Baylis; forum for discussion on any forestry matters of common interest to member governments which may be brought to the Committee's notice by any member country or organization.

Conseil International de la Chasse et de la Conservation du Gibier (International Council for Hunting and the Preservation of Game): 30 rue de Miromesnil, 75008 Paris, France; tel (1) 47-42-13-60; telex 285602; fax (1) 47-42-13-48; Admin: Jean François Mahe; promotion of sustainable management of wildlife.

Conservation International: 1015 18th St, NW, Suite 1000, Washington, DC 20036, USA; tel (202) 429-5660; fax (202) 887-5188; Pres: Dr Russell A Mittermeier; conservation of ecosystems and biodiversity.

Coordination Européene des Amis de la Terre (Friends of the Earth Europe): 29 rue Blanche, 1050 Brussels, Belgium; tel (2) 537-72-28; fax (2) 537-55-96; co-ordination of national member groups in Europe.

EarthAction Network, International Office: 39 rue René Christiaens, 1160 Brussels, Belgium; tel (2) 675-58-01; Int Co-ordinators: Lois Barber, Nicholas Dunlop; world-wide public pressure on global decision-making relating to environmental, development and peace issues.

Earthwatch Europe: Belsyre Court, 57 Woodstock Rd, Oxford OX2 6HU, UK; tel (865) 311600; fax (865) 311383; Exec Dir: Brian Wilson Walker; human and capital resources for field science in support of sustainable development.

Economic Community of West African States (ECOWAS): Abuja, Nigeria; Exec Sec: Dr Abass Bundu; promotion of trade, co-operation and self-reliance in West Africa; programme for energy, to increase energy efficiency and find alternative sources; dissemination of information on renewable energy.

Ecosystems Conservation Group: c/o UNEP, POB 30552, Nairobi, Kenya; formed by FAO, IBRD, IUCN, UNEP, UNESCO and WWF to co-ordinate their environmental activities.

Europa Nostra/International Castles Institute (IBI): Lange Voorhout 35, 2514 EC The Hague, Netherlands; tel (70) 3560333; fax (70) 3617865; Pres: HRH the Prince Consort of Denmark; Exec Pres: Daniel Cardon de Lichtbuer; protection and enhancement of Europe's architecture and nature; encouragement of high standards in architecture and town and country planning; improvement of Europe's environment.

European Association for Aquatic Mammals: c/o Chartley House, Swannell's Wood, Studham, Dunstable LU6 2QB, UK; tel (582) 872487; fax (582) 872315; Pres Elect: Dr Bernhard Neurohr; Sec and Treas: V J A Manton; improvement of captive facilities and encouragement of related student activities.

European Association of Environment and Resource Economists: c/o Fondazione Mattei, Via Santa Sofia 27, 20122 Milan, Italy; tel (2) 52036934; fax (2) 52036946; Pres: Henk Folmer; Vice-Pres: Ignazio Musu; annual conference; workshop in environmental economics; newsletter; journal.

European Bureau for Conservation and Development (EBCD): 9 rue de la Science, 1040 Brussels, Belgium; tel (2) 230-30-70; fax (2) 230-82-72; Dir: Despina Symons; promotion of the World Conservation Strategy and, particularly, the wise use of renewable natural resources.

European Cetacean Society (ECS): c/o Sea Mammal Research Unit, Natural Environment Research Council, High Cross, Madingley Rd, Cambridge CB3 0ET, UK; tel (223) 311354; fax (223) 328927; Pres: Giuseppe Notarbartolo di Sciara; Sec: Geneviève Desportes; promotes and co-ordinates scientific study and conservation of cetaceans; gathers and disseminates information.

European Ecological Federation (EEF): Department of Ecology, University of Lund, Lund, Sweden; tel (46) 10-70-00; fax (46) 11-95-52; Pres: Prof R J Berry; Sec-Gen: Dr Pehr H Enckell; Vice-Pres: Dr P Jouventin; promotion of co-operation throughout Europe in the science of ecology.

European Environmental Bureau (EEB): 22-26 rue de la Victoire, 1060 Brussels, Belgium; tel (2) 511-02-06; fax (2) 511-77-21; federation of 120 major environmental NGOs from the 12 member states of the EC; representative of environmental interests with EC institutions.

European Federation for the Protection of Waters: c/o OMEGAM, Amstelveenseweg 88, 1075 XJ Amsterdam, Netherlands.

European Union for Coastal Conservation (EUCC): Stationsweg 12-14, POB 11059, 2301 EB Leiden, Netherlands; tel (71) 122900; fax (71) 124069; Sec-Gen: A H P M Salman; promotion of nature conservation and the wise use of coastal zones of Europe.

European Youth Forest Action: POB 566, 6130 AN Sittard, Netherlands; tel (46) 513045; fax (46) 516460; Principal Officers: Pascus Calta, Michel Verschoor; campaigns and action for a better environment.

EUROSITE: 9 rue de la Collégiale, 59800 Lille, France; tel 55-90-44; fax 06-29-62; Programme Officer: Peter L Nowicki; association of public and private bodies which manage the

European natural heritage; co-operation for the conservation of European natural sites.

Fauna and Flora Preservation Society: 1 Kensington Gore, London, SW7 2AR, UK; tel (71) 823-8899; fax (71) 823-9690; Admin: Amanda Hillier; protection of endangered species and habitats; protection of mountain gorillas.

Fédération des Associations de Chasseurs de la CE (FACE—Federation of Field Sports Associations of the EC): 82 rue F Pelletier, 1040 Brussels, Belgium; tel (2) 732-69-00; fax (2) 732-70-72; Pres: J van den Driest; Sec-Gen: Dr Y Lecocq; representation of member national hunting associations at EC and international levels; promotion and defence of field sports in accordance with wise use of natural resources.

Federation of Nature and National Parks of Europe (FNNPE): 94481 Grafenau, Kröllstr 5, Germany; tel (49) 85522839; fax (49) 85523242; Principal Officers: Prof Aitken Clark, Dr Hans Bibelriether, Eva Pongratz; encouragement of practical co-operation and exchange of information between national and nature parks and other organizations and individuals responsible for nature conservation; improvement of the management of existing parks; dissemination of information and technical knowledge on the management of natural and cultural heritage.

Fondation Internationale pour la Sauvegarde du Gibier (FISG—International Foundation for the Conservation of Game): 15 rue de Téhéran, 75008 Paris, France; tel (1) 45-63-51-33; telex 640430; fax (1) 45-63-32-94; Pres: HIH Prince Abdorreza of Iran; Dir: B des Clers; conservation of wildlife and nature around the world; environmental policy, management and planning; monitoring of migratory species.

Food and Agriculture Organization of the United Nations (FAO): Via delle Terme di Caracalla, 00100 Rome, Italy; tel (6) 57971; telex 610181; fax (6) 57973152; Dir-Gen: Jacques Diouf (Senegal); various units within certain divisions and departments of the FAO deal with the environmental issues relating to their field of activity; member of Ecosystems Conservation Group.

Environment and Sustainable Development Co-ordinating Centre (AGRE): Head: L S Botero; formerly Environment and Energy Programmes Co-ordinating Centre; Secretariat of IDWG on Environment and Sustainable Development; reports directly to Special Adviser/Assistant Director-General for Environment and Sustainable Development.

Indo-Pacific Fishery Commission (IPFC): FAO Regional Office for Asia and the Pacific, Maliwan Mansion, 39 Phra Atit Rd, Bangkok 10200, Thailand; tel (2) 281-7844; telex 82815; fax (2) 280-0445; Sec: Dr Veravat Hongskul; area of interest includes socio-economic aspects of coastal fisheries management.

Interdepartmental Working Group (IDWG) on Environment and Sustainable Development: founded in 1969 as the IDWG on Natural Resources and the Human Environment, then became the IDWG on Environment and Energy in 1982, before being renamed in 1990; reports to the Steering Committee for Environment and Sustainable Development; proposals for multidisciplinary activities; co-ordination of proposals endorsed by the Steering Committee.

Foundation for European Environmental Policy: 51 rue de la Concorde, 1050 Brussels, Belgium; tel (2) 514-01-24; telex 21504; fax (2) 512-32-65; Rep: Dr Gerrit Vonkeman; partner institutes in France, Germany, the Netherlands and the UK.

Friends of the Earth International (FoEI): POB 19199, 1000 GD Amsterdam, Netherlands; Prins Hendrikkade 48, Amsterdam, Netherlands; tel (20) 6221369; telex 918023; fax (20) 6392181; Chair: John Hontelez; International Co-ordinator: Bert van Pinxteren; Information Officer: Eka Morgan; worldwide federation of national environmental organizations which promotes the conservation, restoration and rational use of the environment and natural resources through public education and campaigning; dedicated to repairing damage to the environment caused by human activity or negligence and to preventing further deterioration; preservation of the world's ecological, cultural and ethnic diversity; promotion of public participation and democratic decision-making, primarily by those most directly affected, in the protection of the environment and the management of natural resources; ensuring social, economic and political justice and equal access to resources and opportunities at a local and international level; promotion of environmentally sustainable development on a local and global scale.

Global Legislators Organization for a Balanced Environment (GLOBE) International: 50 rue du Taciturne, 1040 Brussels, Belgium; tel (2) 230-65-89; fax (2) 230-95-30; Pres: Hemmo Muntingh; Vice-Pres: Carlos Pimenta; exchange of information between GLOBE members, in order to compare and improve the environmental legislation of their respective countries and regions; proposals for similar new legislation; support for individuals and groups working for global conservation.

Green Cross International: 160 route de Florissant, 1231 Geneva, Switzerland; tel (22) 7891662; fax (22) 7891695; Exec Dir: Roland Wiederkehr; Dir: Andreas Eggenberg; established in 1993; contribution to the non-governmental implementation of Agenda 21 (programme of activities to promote sustainable development adopted at the UNCED); responses to industrial disasters; development of ecological law.

The Hague Office: Lange Vijverberg 16, AC The Hague, Netherlands; tel (70) 3633636; fax (70) 3624973; Dir: Bart Jan Krouwel.

Greenpeace International: Keizersgracht 176, 1016 DW Amsterdam, Netherlands; tel (20) 5236555; telex 18775; fax (20) 5236500; Exec Dir: Paul Gilding; Chair, Int Bd: Uta Bellion; independent, non-political organization dedicated to the protection of the environment through peaceful means; publicizing environmental damage; suggestions for alternative practices where possible; campaigns are divided into five categories: atmosphere and energy, nuclear power and weapons, ocean ecology, toxic substances, and tropical rainforests; offices in 24 countries.

Greenway—Central and East European Network of Environmental NGOs: POB 163, 814 99 Bratislava, Slovakia; tel (7) 827-213; fax (7) 827-213; Principal Officer: Dr Elena Vartíková; environmental education in the form of seminars, workshops and training courses on energy and waste; information exchange for environmental NGOs in Central and Eastern Europe; co-ordination of their activities; information; newsletter.

Helsinki Commission (HELCOM), Baltic Marine Environment Protection Commission: Katajanokanlaituri 6B, 00160 Helsinki, Finland; tel (90) 1397031; telex 125105; fax (90) 13970339; Chair: Fleming Otzen; Exec Sec: Ulf Ehlin; governing body of the Convention on the Protection of the Marine Environment of the Baltic Sea Area; responsibilities include prevention of airborne and land-based pollution; member states: Denmark, Estonia, Finland, Germany, Lithuania, Poland, Russia and Sweden; Latvia has observer status.

Humedales para las Americas (Wetlands for the Americas): Monroe 2142, 1428 Buenos Aires, Argentina; tel (1) 781-9171; fax (1) 781-6115; Dir: Pablo Canevari; Tech Co-ordinator: Daniel Blanco; conservation and sustainable use of wetland ecosystems in the Americas.

Institute for European Environmental Policy (IEEP): 158 Buckingham Palace Rd, London SW1W 9TR, UK; tel (71) 824-8787; fax (71) 824-8145; Dir: Nigel Haigh; Sr Fellow: David Wilkinson; independent body for the analysis of environmental policies in Europe, aiming primarily to promote development and implementation of EC environmental policy; partner institutes in Belgium, France, Germany and the Netherlands.

Instituut voor Europees Milieubelied (Institute for European Environmental Policy): Jansbuitensingel 14, 6811 AB Arnhem, Netherlands; tel (85) 422929; fax (85) 453210; Dir: Dr Graham Bennett; Research Assoc: B Verhoeve; independent body; analysis and advancement of environment policy in Europe; research on environmental protection, focusing on development, implementation and evaluation of the EC's environmental policy; partner institutes in Belgium, France, Germany and the UK.

Inter-American Institute for Co-operation on Agriculture (IICA): Apdo 55, 2200 Coronado, Costa Rica; tel 29-0222; telex 2144; fax 29-4741; Dir-Gen: Dr Martin Piñeiro; 33 offices throughout South America, Central America and the Caribbean; five programmes: Agricultural Policy Analysis and Planning, Technology Generation and Transfer, Organization and Management for Rural Development, Trade and Agroindustry, and Agricultural Health.

Inter-American Parliamentary Group on Population and Development (IAPG): 902 Broadway, 10th Floor, New York, NY 10010, USA; tel (212) 995-8860; telex 620661; fax (212) 995-8853; Chair: Blanca Esponda; Exec Co-ordinator: Hernán Sanhueza; Sr Adviser: Francisco Di Blasi; network for the exchange of information on population and development legislation; sponsorship of parliamentary conferences, seminars, regional dialogues and publications; support for research projects; technical assistance to members; public information about population and development issues.

Inter-American Tropical Tuna Commission: Scripps Inst of Oceanography, California Univ, 8604 La Jolla Shores Drive, La Jolla, CA 92093-1508, USA; tel (619) 546-7100; fax (619) 546-7133; Dir: James Joseph; Chief Scientists: Richard B Deriso, Martin A Hall; open to adherence by governments whose nationals fish for tropical tuna in the eastern Pacific Ocean; study of tuna and related species; assessment of abundance; recommendations for conservation measures when necessary, while maintaining production; protection of porpoise stocks; aerial surveys; dolphin tagging and research; members: Costa Rica, France, Japan, Nicaragua, Panama, the USA, Vanuatu and Venezuela.

Intergovernmental Authority on Drought and Development/Autorité Intergouvernementale sur la Sécheresse et le Développement (IGADD): rue de Georges Clemenceau, BP 2653, Djibouti, Djibouti; tel 354050; telex 5978; fax 356994; Exec Sec: Dr David Stephen Muduuli; concerned with the effects of drought and other related natural disasters; support for the development programmes of member states (Djibouti, Ethiopia, Kenya, Somalia, Sudan and Uganda); the main programmes are concerned with control of desertification, environment protection, food security and water resources.

Internationaal Centrum voor Bodem-Referentie en -Informatie (ISRIC—International Soil Reference and Information Centre): Divendaal 9, POB 353, 6700 AJ Wageningen, Netherlands; tel (8370) 71711; telex 45888; fax (8370) 24460; Dir: Dr L R Oldeman; Deputy Dir: Dr J H V van Baren; collection and dissemination of information about soils and their sustainable use around the world; promotion of applied research and improved research methods; advice; encouragement of geographically referenced soil and terrain digital databases.

International Association of Botanic Gardens: Botanic Gardens of Adelaide, North Terrace, Adelaide, SA 5000, Australia; tel (8) 228-2320; fax (8) 223-1809; Sec-Gen: Dr B Morley; Pres: Prof P Ashton; promotion of collaboration between botanic gardens and arboreta around the world.

International Association for Ecology (INTECOL): Savannah River Ecology Lab, Drawer E, Aiken, SC 29802, USA; tel (803) 725-2472; fax (803) 725-3309; Pres: Prof W Haber; Vice-Pres: Prof J Lee; four-yearly International Congress of Ecology; newsletter; annual journal; involved with efforts to create resources for ecologists in developing countries.

International Association for Environmental Hydrology (IAEH): POB 1088, Alexandria, VA 22313, USA; tel (703) 683-9768; fax (703) 683-6137; Pres: Dr R W Peebles; exchange of information on the prevention and clean-up of water pollution internationally and the development of simple, environmentally sound solutions to water pollution, particularly in developing countries; solutions to the effects of climatic change on freshwater resources; publications.

International Association of Environmental Testing Laboratories: 1911 N Fort Myer Drive, Suite 400, Arlington, VA 22209, USA; tel (703) 524-2427; fax (703) 524-1453; Principal Officers: Robert Walker, G Wade Miller; representing the interests of the environmental analytical services industry.

International Association of Fish and Wildlife Agencies: 444 North Capitol St, NW, Suite 544, Washington, DC 20001, USA; tel (202) 624-7890; fax (202) 624-7891; Exec Vice-Pres: R Max Peterson.

International Association of Hydrogeologists (IAH): POB 9, Kenilworth CV8 1JG, UK; tel (926) 56561; telex 366748; fax (926) 56561; Pres: E Romijn; Sec-Gen: A Skinner; publications; congresses; programme for developing nations; participation in the International Hydrological Programme.

International Association for Impact Assessment (IAIA): POB 70, Belhaven, NC 27810, USA; tel (919) 964-2338; fax (919) 964-2340; Pres: Prof F Larry Leistritz; Exec Dir: Dr Maurice E Voland; promotion of local and global applications of impact assessment and international capacity to anticipate, plan and manage the consequences of development.

International Association for Indoor Air Quality (Indoor Air International): Postfach 2, 4467 Rothenfluh, Switzerland; tel (767) 318474 (Pres, UK); fax (767) 313929 (Pres, UK); Pres: G B Leslie (UK); Pres-Elect: R Perry (UK); concerned with health, comfort and efficiency matters; conferences; publication of books and periodicals.

International Association for the Protection of the Environment in Africa (ENVIRO-PROTECT): POB 13623, Yaoundé, Cameroon; tel 23-54-35; fax 22-18-73; Pres: François Tiani Keou; Gen Sec: Augustin Youmbi.

International Association for the Study of Traditional Environments (IASTE): c/o Center for Environmental Design Research (CEDR), California Univ, Berkeley, 390 Wurster Hall, CA 94720, USA; tel (510) 642-2896; telex 3667114; fax (510) 643-5571; Exec Dir: Nezar al Sayyad; Chair, Bd: Jean-Paul Bourdier; publications; organization of biennial conferences (in Paris, France 1992 and in Tunis, Tunisia 1994).

International Association on Water Quality (IAWQ): 1 Queen Anne's Gate, London SW1H 9BT, UK; tel (71) 222-3848; telex 918518; fax (71) 233-1197; Exec Dir: Tony Milburn; Pres: Prof Petr Grau; formerly International Association on Water Pollution Research and Control (IAWPRC); membership association; conferences; publications; 28 specialist groups.

International Association of Wildland Fire (IAWF): POB 328, Fairfield, WA 99012-0328, USA; tel (509) 283-2397; fax (509) 283-2264; Exec Dir: Dr Jason M Greenlee; Dir: Dr Andrea L Koonce; Chief Scientific Ed: Dr Michael G Weber; publications on forestry and forest fire issues.

International Atomic Energy Agency (IAEA): 1400 Vienna, POB 100, Wagramerstr 5, Austria; tel (1) 2360-1275; telex 112645; fax (1) 234564; Dir-Gen: Hans Blix; technical co-operation and training; food and agriculture programmes; life sciences programmes; physical sciences programmes, including the IAEA Marine Environmental Laboratory in Monaco, which studies radionuclides and other ocean pollutants; advice on nuclear power; radioactive waste management advisory programme (WAMAP); nuclear safety programme; international nuclear information system (INIS); international safeguards; International Centre for Theoretical Physics at Trieste, Italy; advice on the nuclear fuel cycle; publication of reports, directories and bulletins.

International Bank for Reconstruction and Development (IBRD—World Bank): 1818 H St, NW, Washington, DC 20433, USA; tel (202) 477-1234; telex 248423; fax (202) 477-6391; Pres and Chair of Exec Dirs: Lewis T Preston; Man Dirs: Attila Karaosmanoglu, Sven Sandstrom, Ernest Stern; primary objectives are the achievement of sustainable economic growth and the reduction of poverty in developing countries and the protection of the environment; jointly established the Global Environment Facility (GEF) with UNDP and UNEP as a programme designed to provide grants for investment projects and technical assistance; its aim is to assist developing countries in implementing projects that benefit the global (not just the local) environment; administration of the Global Trust Fund; responsible for the implementation of investment projects.

International Development Association (IDA): affiliated to the IBRD; assistance is aimed at poorer developing countries on more flexible terms than those offered by the IBRD; priority given to the reduction of poverty, support for 'sound macroeconomic and sectoral policies' and the protection of the environment.

International Board for Plant Genetic Resources (IBPGR): Via delle Sette Chiese 142, 00145 Rome, Italy; tel (6) 518921; fax (6) 5750309; Dir: Geoffrey C Hawtin; Dep Programme Dir: Masa Iwanaga; advancement of the use and conservation of plant genetic resources.

International Centre for Alpine Environments (ICALPE): Technolac, BP 230, 73374 Le Bourget du Lac, France; tel 79-25-20-04; fax 79-25-28-48; Principal Officers: Michel Dubost, Pier Carlo Zingari; co-ordination and networking of research on European mountain environments and development.

International Centre for Integrated Mountain Development (ICIMOD): Jawalakhel, Lalitpur, POB 3226, Kathmandu, Nepal; tel (1) 525313; telex 2439; fax (1) 524509; Dir-Gen: Dr E F Tacke; research and development into mountain farming systems, population and employment, infrastructure and technology, and environmental management; information exchange; natural resource assessment and monitoring; development planning and implementation; institutional and professional development.

International Centre for Living Aquatic Resources Management (ICLARM): POB 1501, Makati, Metro Manila 1299, Philippines; tel (2) 8189283; telex 64794; fax (2) 8163183; Dir, Aquaculture Programme: Dr Roger S V Pullin; research into fisheries and aquaculture in tropical developing countries; ecological monitoring; marine resource management and conservation; captive breeding of species; resource assessment.

International Centre for Research in Agroforestry (ICRAF): UN Ave, POB 30677, Gigiri, Nairobi, Kenya; tel (2) 521450; telex 22048; fax (2) 521001; Dir-Gen: Dr Pedro A Sánchez; Dep Dir-Gen: R B Scott; mitigation of tropical deforestation, land depletion and rural poverty through improved agroforestry systems.

Cameroon Office: IRA/ICRAF Project, BP 2067, Yaoundé, Cameroon; tel 23-75-60; telex 1140; fax 23-74-40; research and training in aspects of agroforestry.

International Coalition for Development Action (ICDA): 115 rue Stévin, 1040 Brussels, Belgium; tel (2) 230-04-30; fax (2) 230-03-48; support for NGO activities in the fields of international trade and sustainable development.

International Commission on Irrigation and Drainage: 48 Nyaya Marg, Chanakyapuri, New Delhi 110021, India; tel (11) 3015679; telex 65920; fax (11) 3015962; Sec-Gen: Dr M A Chitali; Sec: S P Kaushish; dissemination of knowledge through publications.

International Commission for the Protection of the Rhine against Pollution: 5400 Koblenz, Hohenzollernstr 18, POB 309, Germany; tel (261) 12495; telex 8624899; fax (261) 36572; Pres: Dr M Caspari; Sec: D Hogervorst; preparation and commission of research to establish the nature of the pollution of the Rhine; proposal of measures of protection and ecological rehabilitation of the Rhine to the signatory governments; publication of annual report; 23 delegates from France, Germany, Luxembourg, the Netherlands, Switzerland and the EC.

International Conferences on Environmental Issues and Management of Waste in Energy and Minerals Production: c/o IJSM, POB 68002, Ranchlands RPO 7750, Ranch View Drive, NW, Calgary, AB T3G 3N8, Canada; tel (403) 239-3849; fax (403) 241-9460; Principal Officer: Dr Raj K Singhal; organization of conferences in Perth, Australia in 1994, Cagliari, Italy in 1996, Ankara, Turkey in 1998, and Calgary, Canada in 2000.

International Conferences on Headwater Control: c/o Inst of Applied Ecology, Prague Agricultural Univ, Kostelec n Č 1, 281 63 Prague, Czech Republic; Principal Officer: Dr J Krecek; organization of conferences in the Czech Republic in 1992 and New Delhi, India in 1995, for professionals in forestry, hydrology and environmental management.

International Council for Bird Preservation: 32 Cambridge Rd, Girton CB3 0PJ, UK; tel (223) 277318; telex 818794; fax (223) 277200; Pres: Donal C O'Brien, Jr; Dir: Dr Christopher Imboden; determining the status of bird species throughout the world; compilation of data on all endangered species; identification of conservation problems and priorities; initiation and co-ordination of conservation projects and international conventions; various publications; representatives in 47 countries; national sections in 63 countries.

International Council of Environmental Law: 53113 Bonn, Adenauerallee 214, Germany; tel (228) 2692240; fax (228) 2692250; Exec Gov: Dr Wolfgang E Burhenne.

International Council on Metals and the Environment: 300 Albert St, Suite 1550, Ottawa, ON K1R 7X7, Canada; tel (613) 235-4263; fax (613) 235-2865; Chair: Keith Hendrick; Sec-Gen: Gary Nash; founded in 1991 by mining companies to promote responsible environmental practices in the industry; members drawn from 20 companies in 10 countries.

International Council on Monuments and Sites (ICOMOS): Hôtel St Aignan, 75 rue de Temple, 75004 Paris, France; tel (1) 42-77-35-76; fax (1) 42-77-57-42; Pres: Dr Roland Silva; Sec-Gen: Jean Louis Luxen; promotion of the application of theory, methodology and scientific techniques to the conservation of architectural heritage.

International Council of Scientific Unions (ICSU): 51 blvd de Montmorency, 75016 Paris, France; tel (1) 45-25-03-29; telex 630553; fax (1) 42-88-94-31; Pres: Prof M G K Menon; Sec-Gen: Prof J W M La Rivière; co-ordination of theoretical and applied sciences and promotion of national scientific research through the intermediary of affiliated national organizations; General Assembly of representatives of national and scientific members meets every three years to formulate policy; established committees include: Scientific Committee on the Application of Science to Agriculture, Fisheries and Aquaculture, Scientific Committee on Antarctic Research, Scientific Committee on Oceanic Research, ISCU-UATI Co-ordinating Committee on Water Research, Committee on Science and Technology in Developing Countries, Scientific Committee on Problems of the Environment, Committee on Biotechnology and Scientific Committee on the International Geosphere/Biosphere Programme; national members drawn from academies or research councils in 80 countries.

Scientific Committee on Problems of the Environment (SCOPE): tel (1) 45-25-04-98; telex 64554; fax (1) 42-88-14-66; Pres: Prof John W B Stewart; Sec-Gen: Prof Philippe Bourdeau; advancement of the knowledge of the influence of humans on their environment; research into the effects of environmental change.

International Federation of Organic Agriculture Movements (IFOAM): 66636 Tholey-Theley, Ökozentrum Imsbach, Germany; tel (6853) 5190; fax (6853) 30110; Pres: Thomas B Harding, Jr; Gen Sec: Bernhard Geier; promotion of organic agriculture as an ecologically and socially sound and sustainable method of food production.

International Foundation for Development Alternatives (IFDA): 5 rue du Marché, CP 9, 1260 Nyon, Switzerland; tel (22) 3618281; fax (22) 3610525; Council Pres: Rajni Kothari; Pres, Steering Cttee: Marc Nerfin.

International Fund for Agricultural Development (IFAD): Via del Serafico 107, 00142 Rome, Italy; tel (6) 54591; telex 620330; fax (6) 5043463; Pres and Chair, Exec Bd: Idriss Jazairy; facilitation of agricultural and rural development; emphasis given to environmental conservation in order to alleviate poverty that results from the deterioration of natural resources; special programme for Sub-Saharan Africa (SPA).

International Geosphere/Biosphere Programme: A Study of Global Change (IGBP): c/o Royal Swedish Acad of Sciences, POB 50005, Lilla Frescativägen 4, 104 05 Stockholm, Sweden; tel (8) 16-64-48; telex 17509; fax (8) 16-64-05; Exec Dir: Prof Thomas Rosswall; Chair, Scientific Cttee: Prof Peter Liss; programme to guide research on global change and ensure that research results are acted upon; description and understanding of the Earth's interactive physical, chemical and biological processes, the unique environment that it provides for life, the changes that are occurring and the manner in which they are influenced by human activities.

International Institute for Applied Systems Analysis: 2361 Laxenburg, Schlossplatz 1, Austria; tel (223) 671-52-10; telex 079137; fax (223) 671-13-13; Dir: Dr Peter E de Janosi; Chair: George S Golitsyn; Project Dir: David Victor; research into the effectiveness of international environmental commitments.

International Institute for Infrastructural Hydraulic and Environmental Engineering (IHE): POB 3015, 2601 DA Delft, Netherlands; Westvest 7, 2611 AX Delft, Netherlands; tel (15) 151715; telex 38099; fax (15) 122921; Principal Officer: Prof W A Segeren; education, training and consultancy on infrastructure, hydraulics and the environment.

International Institute for Land Reclamation and Improvement: Lawickse Allee 11, POB 45, 6700 AA Wageningen, Netherlands; tel (8370) 90144; telex 45888; fax (8370) 17187; Dir: M J M P Pinkers; Research: Dr M G Bos; research; publications; training; information; consultancy; active in Egypt, India, Kenya, Pakistan and Indonesia.

International Institute for Sustainable Development (IISD)/Institut International du Développement Durable (IIDD): 161 Portage Ave East, 6th Floor, Winnipeg, MB R3B 0Y4, Canada; tel (204) 958-7700; fax (204) 958-7710.

International Irrigation Management Institute (IIMI): POB 2075, 127 Sunil Mawatha, Pelawatte via Colombo, Sri Lanka; tel (1) 567404; telex 22318; fax (1) 566854; Dir-Gen: Dr Roberto L Lenton; Dir of Research: Dr Jacob Kijne; transnational and country-specific research in collaboration with national agencies; networking with other organizations; exchange and dissemination of information; training based on research results; formulation of policies and management strategies for the irrigated agriculture sector.

International Juridical Organization for Environment and Development: Via Barberini 3, 00187 Rome, Italy; tel (6) 4742117; fax (6) 4745779; Pres: Mario Guttieres; Exec Dir: Mary Ellen Sikabonyi; study and discussion of legal aspects of significant transnational problems such as environmental protection, sustainable development, East–West and North–South co-operation, and the evolution and harmonization of international environmental law and national legislation.

International Labour Organisation (ILO), International Occupational Safety and Health Information Centre (CIS): 4 route des Morillons, 1211 Geneva 22, Switzerland; tel (22) 7996740; telex 415647; fax (22) 7986253; Head: Dr Jukka Takala; scanning of relevant international publications with the help of national centres in more than 70 countries; the CISDOC database produces publications and provides computerized searches in English and French; direct on-line access and CD-ROM.

International Lake Environment Committee Foundation (ILEC): 3-13-25 Kusatsu, Kusatsu City, Shiga 525, Japan; tel (775) 67-2155; fax (775) 67-2156; Dir-Gen: Kei Yamazaki; Sec-Gen: Kanji Yamano; promotion of environmentally sound management of lake and reservoir basins; co-operation with the UNEP International Environmental Technology Centre.

International League for Protection of Cetaceans (ILPC): Podere Il Falco, 06062 Città della Pieve, Italy; tel (578) 298186; fax (578) 299186; Exec Dir: Dr Sidney J Holt; Exec Sec: Leslie A Busby.

International Oil Pollution Compensation Fund: Inco House, 4 Albert Embankment, London SE1 7SR, UK; tel (71) 582-2606; telex 23588; fax (71) 735-0326; Dir: M Jacobsson.

International Organization for Biological Control of Noxious Animals and Plants (IOBC): Inst de Phytomédecine, ETH-Zentrum, Clausiusstr 21, 8092 Zürich, Switzerland; c/o Sec-Gen, Swiss Fed Research Station for Agronomy (FAP), 8046 Zürich, Switzerland; tel (1) 3777111 (Sec-Gen); fax (1) 3777201 (Sec-Gen); Pres: E Delfoss; Sec-Gen: F Bigler; scientific meetings and working groups; publications; European insect identification service; training in biological control; a commission of the International Union of Biological Sciences, it has liaison status with the FAO.

International Organization for Standardization (ISO): CP 56, 1211 Geneva 20, Switzerland; 1 rue de Varembé, 1204 Geneva, Switzerland; tel (22) 7490111; telex 412205; fax (22) 7333430; Pres: John A Hinds; Sec-Gen: Dr Lawrence D Eicher; responsible for standardization in all fields except electrical and electronic engineering; ISO international environmental standards cover air, water and soil quality, road vehicle exhaust emissions, noise levels and vibrations; work has begun in the fields of environmental management and wastes.

International Ozone Association (IOA), Pan-American Group (PAG): 31 Strawberry Hill Ave, Stamford, CT 06902-2608, USA; tel (203) 348-3542; fax (203) 967-4845; Pres: L Joseph Bollyky; Chair, PAG: Bruce Long; Exec Dir, PAG: Margit Istok; non-profit professional association for individuals and companies involved with ozone technology; promotion of the beneficial use of ozone; publications; congresses and seminars.

International Primate Protection League: PO Drawer 766, Summerville, SC 29484, USA; tel (803) 871-2280; fax (803) 871-7988; Co-Chair: Dr Shirley McGreal; conservation and protection of primates, both in captivity and in the wild, through public education and fund-raising.

International Rivers Network: 1847 Berkeley Way, Berkeley, CA 94703, USA; tel (510) 848-1155; fax (510) 848-1008; Exec Dir: Chris Harris; Research Dir: Leonard Sklar; Information Officer: Petra J Yee; network of technical experts, environmentalists and academics specialized in water and river issues, dedicated to saving the world's rivers and watersheds; proposals for alternatives to destructive large-scale hydroelectric development.

International Social Sciences and Environment Network/Réseau International Sciences Sociales et Environnement (RISE): Institut de Politique Internationale et Européenne, 31 rue de Chazelles, 75017 Paris, France; 101 rue des Tuilières, 87100 Limoges, France; tel (1) 46-22-40-23; Founder and Co-ordinator: Dominique Allan Michaud; meetings of organizations and of researchers with different specialities, in order to popularize and disseminate research results.

International Society of Complex Environmental Studies (ISCES): c/o Public Health Dept, Tampere Univ, Haiharankatu 17B 11, 33710 Tampere, Finland; POB 607, 33101 Tampere, Finland; tel (31) 633207; fax (31) 633207; Chair: Dr Olavi Manninen; conference every two years.

International Society for Ecological Modelling: Langkaer Vaenge 9, 3500 Vaerlose, Denmark; tel 42-48-06-00; Principal Officer: Sven Erik Jorgensen; publication of various scientific and environmental journals.

International Society of Naturalists (INSONA): Oza Bldg, Salatwada, Baroda 390 001, India; tel (265) 558759; telex 175272; fax (265) 66984; Gen Sec: Dr G M Oza; research; environmental conservation; education; workshops and conferences.

International Society of Soil Science: 1180 Vienna, Soil Science Inst, Agric Univ, Gregor Mendel-Str 33, Austria; tel (1) 310-60-26; fax (1) 310-60-27; Pres: Dr Andrés Aguilar-Santelises; Sec-Gen: Prof Dr W E H Blum; Deputy Sec-Gen: J H V van Baren; furthering of all aspects of soil science.

International Solid Wastes and Public Cleansing Association (ISWA): Bremerholm 1, 1069 Copenhagen K, Denmark; tel 33-91-44-91; fax 33-91-91-88.

International Tropical Timber Organization (ITTO): Int Orgs Center, 5th Floor, Pacifico-Yokohama, 1-1-1 Minato-Mirai, Nishi-ku, Yokohama 220, Japan; tel (45) 223-1110; telex 3822480; fax (45) 223-1111; Exec Dir: Freezailah Bin Che Yeom; consultation and co-operation between countries producing and consuming tropical timber in order to strike a balance between utilization and conservation; campaigns to ensure that all trade in tropical timber is from sustainably managed resources by the year 2000; promotion of research and development; reafforestation.

International Union for the Conservation of Nature and Natural Resources (IUCN): see World Conservation Union.

International Union of Directors of Zoological Gardens—World Zoo Organization (IUDZG): POB 3036, Station B, Calgary, AB T2M 4R8, Canada; tel (403) 232-9300; fax (403) 237-7582; Pres: Peter Karsten; Vice-Pres: Bent Jorgensen.

International Union of Forestry Research Organizations (IUFRO): 1131 Vienna, Seckendorff-Gudent-Weg 8, Austria; tel (1) 877-01-51; telex 312646; fax (1) 977-93-55; Pres: M N Salleh; Sec: H Schmutzenhofer; non-profit, non-governmental association for organizations and individuals involved in forestry research; international co-operation in the whole field of scientific research related to forestry.

International Union of Societies of Foresters (IUSF): c/o Inst of Foresters of Australia, POB 1697, Brisbane, Qld 4001, Australia; tel (7) 234-0011; telex 43988; fax (7) 234-0271; Pres: A Leslie; Vice-Pres: Dr Gary Bacon; Exec Dir: Keith Jennings; assistance for formation of national forestry societies; participation in formulation of forestry policy.

International Water and Sanitation Centre (IRC): Prinses Margrietplantsoen 20, POB 93190, 2509 AD The Hague, Netherlands; tel (70) 3314133; telex 33296; fax (70) 3814034; Dir: J M G van Damme; collaboration with the WHO; documentation services; publications; research; advice; training programmes on water supply, sanitation and related environmental issues in developing countries.

International Water Supply Association: 1 Queen Anne's Gate, London SW1H 9BT, UK; tel (71) 957-4567; fax (71) 222-7243; Sec-Gen: L R Bays; Pres: Dr H Tessendorff; exchange of information on water supply and related issues by means of world congresses, specialized and regional conferences and workshops, and scientific and technical publications.

International Water Tribunal Foundation (IWT): Damrak 83, 1012 LN Amsterdam, Netherlands; tel (20) 6240610; fax (20) 6228384; Exec Dir: Arthur van Norden; Legal Officer: Liesbeth Bijvoet; Environmental Officer: Esther de Lange; support for alternative methods of settling environmental disputes; support for sustainable use of natural resources; organization of international tribunals, public hearings and debates; independent research.

International Waterfowl and Wetlands Research Bureau: Slimbridge GL2 7BX, UK; tel (453) 890624; telex 437145; fax (453) 890697; Dir: Dr Michael Moser; stimulation and co-ordination of research on and conservation of waterfowl and their wetland habitats, particularly through the Ramsar Convention; co-ordination of research, wetland and waterfowl management, and training by institutes and individuals (professional and amateur) throughout the world; alerting governments and organizations when wetlands are threatened; publication of various reports.

International Wheat Council, Food Aid Committee (FAC): 1 Canada Sq, Canary Wharf, London E14 5AE, UK; tel (71) 513-1122; telex 8813241; fax (71) 712-0071; Exec Dir: D G Amstutz; administration of the Food Aid Convention of the International Wheat Agreement (1986); annual food aid in grain to developing countries; reviewing food aid policies; forum for donors' discussions.

International Youth Federation for Environmental Studies and Conservation (IYF): 36/626, Thiruvananthapuram 695 008, India; tel (471) 450877; fax (471) 446859; Chair: David Santa; Sec-Gen: N Suresh; federation of self-governing youth organizations concerned with the study, understanding, awareness and conservation of the environment; regional offices in Benin, Costa Rica, Denmark, the Netherlands and the Philippines.

Denmark Office: Astridsvej 7-5-1, 8220 Braband, Denmark; tel 86-25-01-90; fax 86-25-01-90; Treas: Mohan M Mathews; encouragement of the increase in knowledge, understanding and awareness of environmental problems among young people world-wide.

Internationale Gewässerschutzkommission für den Bodensee (IGKB—International Commission for the Protection of Lake Constance): c/o Fed Ministry of Agric and Forestry, 1012 Vienna, Stubenring 1, Austria; tel (1) 71-1-00; fax (1) 51-5-10; no permanent secretariat as the two-year presidency of the Commission rotates between Austria, Germany and Switzerland, the nations bordering Lake Constance; presiding country 1992–94: Austria; research and investigations into the state of the lake; reports; recommendations for measures to be implemented in the rivers feeding the lake.

IUCN: see World Conservation Union.

Nature 2000 International: International Youth Union for Conservation of Nature: Nature 2000 House, 33 Bervoetstraat, 2000 Antwerp, Belgium; tel (3) 231-26-04; telex 34710; fax (3) 233-64-99; Pres: Julius Smeyers; Sec-Gen: Dirk Ostes;

research; nature conservation; environmental education; information.

Nordic Council of Ministers: Store Strandstraede 18, 1255 Copenhagen K, Denmark; tel 33-11-47-11; telex 15544; fax 33-93-89-55; Sec-Gen: Pät Stenbäck; administration of the Nordic Convention on the protection of the Environment and the Convention on the Marine Environment of the Baltic; member states harmonize regulations for protection of the environment and assess measures affecting neighbouring countries; development of a programme of long-term goals for Nordic co-operation; members: Denmark, Finland, Iceland, Norway, Sweden, Greenland, the Faeroe Islands and the Åland Islands.

North Atlantic Treaty Organisation (NATO), Division of Scientific and Environmental Affairs: 1110 Brussels, Belgium; tel (2) 728-41-11; telex 23867; fax (2) 728-41-17; Asst Sec-Gen: Jacques Ducuing; advice to the Secretary-General on scientific matters; scientific exchange programmes between member countries; research fellowships; advanced study institutes; support for less advanced member countries.

Committee on the Challenges of Modern Society (CCMS): tel (2) 728-48-50; fax (2) 728-42-32; meetings twice annually in plenary session; dedicated to improving the exchange of views and experience among member countries with the aim of creating a better environment for their societies; consideration of specific problems of the human environment with the objective of stimulating action by member governments; commissioning pilot studies; co-sponsoring of international conferences and a fellowships programme.

Oil Companies International Marine Forum (OCIMF): 15th Floor, 96 Victoria St, London SW1E 5JW, UK; tel (71) 828-7696; telex 24942; fax (71) 245-2921; Dir: E J M Ball; promotion of pollution prevention and safe transportation by sea of crude petroleum and its products.

Organisation for Economic Co-operation and Development (OECD): 2 rue André-Pascal, 75775 Paris Cedex 16, France; tel (1) 45-24-82-00; telex 620160; fax (1) 45-24-85-00; Sec-Gen: Jean-Claude Paye; Dep Secs-Gen: R A Cornell, Pierre Vinde, Makoto Taniguchi; OECD's Environment Policy Committee is responsible for the economic and policy aspects of environmental policy; the Chemicals Programme promotes co-operation and mutual assistance in the control of the 80,000 chemicals on the commercial market; the OECD also deals with air and water management, noise abatement and transfrontier pollution; review of member states' environmental policies; a Group on Urban Affairs deals with ecological aspects of the built-up environment.

International Energy Agency (IEA): telex 630190; fax (1) 45-24-79-21; Exec Dir: Helga Steeg; improvement of the world energy supply and demand structure; efficient use of energy; development of alternative energy sources; integration of environmental and energy policies; promotion of co-operation between petroleum-producing and petroleum-consuming countries.

Organization of African Unity (OAU): POB 3243, Addis Ababa, Ethiopia; tel (1) 517700; telex 21046; fax (1) 512622; Chair: Abdou Diouf; Sec-Gen: Salim Ahmed Salim; formulation and execution of policy; advisory services for member states.

Scientific, Technical and Research Commission (STRC): NP PLC Bldg, 26-28 Marina, PMB 2359, Lagos, Nigeria; tel (1) 2633289; telex 22199; fax (1) 2636093; the area of activity of the Agriculture, Forestry, Fisheries and Oceanography section includes food and sustainable agricultural development, and soil science and research; the Natural Resources, Science and Technology section is concerned with science and technology for research and development, industrial and energy research, environmental protection and training.

Pan African Institute for Development (PAID)/Institut Panafricain pour le Développement (IPD): BP 4056, Douala, Cameroon; tel 42-10-61; telex 6048; fax 42-43-35; Acting Sec-Gen: Dr Laban N Kirya; Deputy Sec-Gen: Anthony L Hagan.

Pan American Health Organization, Division of Health and Environment (HPE): 525 23rd St, NW, Washington, DC 20037, USA; tel (202) 861-3311; fax (202) 861-8462; Dir: Horst Otterstetter; support for national efforts to prevent and control environmental conditions adverse to health.

Program of Basic Sanitation (HES): Co-ordinator: Francisco Zepeda; responsible for urban and rural water supply, disposal of wastewater and excreta, solid waste management and domestic sanitation.

Program of Environmental Quality (HEQ): Co-ordinator: Carlos Cúneo; responsible for evaluation, monitoring and control of environmental risks to health, safety of chemical substances, management of natural resources and health, urban development and occupational environments.

Centro Panamericano de Ecología Humana y Salud (ECO—Pan American Center for Human Ecology and Health): Apdo 37-473, 06696 México, DF, Mexico; Apdo 249, 50000 Toluca, Méx, Mexico; tel (5) 16-4344; fax (5) 16-4404; Dir: Dr Jacobo Finkelman; regional centre supporting execution of the Program of Basic Sanitation and the Program of Environmental Quality; evaluation of environmental health risks.

Centro Panamericano de Ingeniería Sanitaria y Ciencias del Ambiente (CEPIS—Pan American Center for Sanitary Engineering and Environmental Sciences): Calle Los Pinos 259, Urbanización Camacho, Lima 12, Peru; Casilla Postal, Lima 100, Peru; tel (14) 371077; fax (14) 378289; Dir: Alberto Flórez; regional centre supporting execution of the Program of Basic Sanitation and the Program of Environmental Quality; control of environmental health risks.

Population Council: 1 Dag Hammarskjöld Plaza, New York, NY 10017, USA; tel (212) 339-0500; telex 234722; fax (212) 755-6052; Pres: Margaret Catley-Carlson; non-profit organization; application of science and technology to the solution of population problems in developing countries; programmes in 50 countries world-wide; research; advice; technical assistance; information; publications; conferences; workshops.

Ramsar Convention on Wetlands of International Importance, Especially as Waterfowl Habitat: 28 rue Mauverney, 1196 Gland, Switzerland; tel (22) 9990170; telex 419624; fax (22) 9990169; Sec-Gen: Daniel Navid; Asst Sec-Gen: Michael Smart; fostering international co-operation; technical assistance for the conservation and wise use of wetlands.

Regional Organization for the Protection of the Marine Environment (ROPME): POB 26388, 13124 Safat, Kuwait; tel 5312140; fax 5324172; Exec Sec: Dr Abdul Rahman al-Awadi; Tech and Admin Co-ordinator: Dr Badria al-Awadi; implementation of marine pollution monitoring and research programmes in the sea area surrounded by the states of the region: Bahrain, Iran, Iraq, Kuwait, Oman, Qatar, Saudi Arabia and the United Arab Emirates.

Society for Conservation Biology: Florida Museum of Natural History, University of Florida, Gainesville, FL 32611, USA; tel (904) 392-8372; Chief Financial Officer: Dr S R Humphrey.

South Asian Association for Regional Co-operation: POB 4222, Kathmandu, Nepal; Sec-Gen: I H Zaki; regional organization for Bangladesh, Bhutan, India, the Maldives, Nepal, Pakistan and Sri Lanka; promotion of economic and social development.

South Pacific Commission (SPC): BP D5, Nouméa Cedex, New Caledonia; tel 262000; telex 3139; fax 263818; Sec-Gen: Ati George Sokomanu; Dir of Programmes: Paloma Komiti; largest single activity is assistance to member countries in fisheries development; concerned with the environmental factors which affect national fisheries departments as well as the development, management and rational exploitation of the renewable marine resources of the region.

South Pacific Forum Fisheries Agency (FFA): POB 629, Honiara, Solomon Islands; tel 21124; telex 66336; fax 23995; Dir: Sir Peter Kenilorea; promotion of co-operation in fisheries among coastal states of the region; collection and dissemination of information and advice on the living marine resources of the region, including their management, development and exploitation.

South Pacific Regional Environment Programme (SPREP): POB 240, Apia, Western Samoa; tel 21929; fax 20231; Dir: Dr Vili Fuavao; Deputy Dir: Don Stewart; formerly part of South Pacific Commission (SPC), independent since 1991; co-ordination of national and regional efforts to protect the environment, encouraging NGOs and research organizations to assist in these efforts; regional forum for information exchange and specialists; members: American Samoa, Australia, the Cook Islands, Fiji, France, French Polynesia, Guam, Kiribati, the Marshall Islands, Micronesia, Nauru, New Caledonia, New Zealand, Niue, the Northern Mariana Islands, Palau, Papua New Guinea, the Solomon Islands, Tokelau, Tonga, Tuvalu, the UK (representing the Pitcairn Islands), the USA, Vanuatu, the Wallis and Futuna Islands and Western Samoa.

Southern African Development Community (SADC), Environment and Land Management Sector Co-ordination Unit: Ministry of Agriculture, Co-operatives and Marketing, POB 24, Maseru 100, Lesotho; tel 322158; telex 4414; fax 310190; Dir: B Leleka.

Southern African Development Community (SADC), Fisheries, Wildlife and Forestry: Ministry of Forestry and Natural Resources, Private Bag 350, Lilongwe 3, Malawi; tel 731322; telex 4465.

Survival International: 310 Edgware Rd, London W2 1BY, UK; tel (71) 723-5535; fax (71) 723-4059; Dir-Gen: Stephen Corry; Pres: Robin Hanbury-Tenison; Chair: Richard Bourne; world-wide movement to support tribal peoples, their right to determine their own future and efforts to protect their lands, environment and way of life; members in 73 countries.

Tourism Council of the South Pacific (TCSP): 35 Loftus St, POB 13119, Suva, Fiji; tel 304177; telex 2306; fax 301995; Dir: Levani Tuinabua; Head, Planning and Devt Dept: Donald Kudy; regional, intergovernmental organization; fostering environmental protection and cultural preservation.

TRAFFIC International (Trade Records Analysis of Fauna and Flora in Commerce): 219c Huntingdon Rd, Cambridge CB3 0DL, UK; tel (223) 277314; telex 817036; fax (223) 277237; Dir: Jorgen Thomsen; monitoring trade in plant and animal wildlife by compiling data from numerous sources; reports and bulletins; co-ordination of the activities of 17 offices world-wide; a programme of IUCN and WWF.

United Nations Centre for Human Settlements (UNCHS—Habitat): POB 30030, Nairobi, Kenya; tel (2) 520600; telex 22996; fax (2) 226473; Exec Dir: Dr Arcot Ramanchandran; assistance to governments in activities related to human settlements; promotion of research; provision of technical co-operation; dissemination of information; sub-programmes include: Settlement Policies and Strategies, Settlements Planning, Shelter and Community Services, Development of the Indigenous Construction Sector, Low-Cost Infrastructure for Human Settlements, Land, Mobilization of Finance, and Human Settlements Institutions and Management.

United Nations Children's Fund (UNICEF): 3 UN Plaza, New York, NY 10017, USA; tel (212) 326-7000; telex 7607848; fax (212) 888-7465; Exec Dir: James P Grant; co-ordination with governments, local communities and other aid organizations on programmes in health, nutrition, education, water and sanitation, the environment, women in development and other fields of importance to children; publication of reports.

United Nations Conference on Trade and Development (UNCTAD): Palais des Nations, 1211 Geneva 10, Switzerland; tel (22) 7891676; telex 289696; fax (22) 7336542; Sec-Gen: Kenneth Dadzie; Dep Secs-Gen: Yves Berthelot, Carlos Fortin; responsible for the Generalized System of Preferences (GSP) whereby a certain proportion of manufactured goods that are exported by developing countries recieve preferential tariff treatment by developed countries; provision of assistance to developing countries in the area of debt-management and seeking debt-relief from creditors.

United Nations Development Programme (UNDP): One United Nations Plaza, New York, NY 10017, USA; tel (212) 906-5000; fax (212) 826-2057; Admin: William H Draper; Assoc Admin: Luís María Gómez; established in 1965 by the UN General Assembly to help the developing countries increase the wealth-producing capabilities of their human and natural resources; joint management of the Global Environment Facility (GEF), which began operations in 1991, over a three-year period, aiming to support projects for reducing emissions of 'greenhouse' gases, to preserve biological diversity, to protect international waters and reduce ozone layer depletion; in 1991 UNDP established a panel of experts, the Environmental and Natural Resources Group, to monitor the environmental implications of UNDP's own work; UNDP assisted developing country governments in preparing for the UN Conference on Environment and Development (UNCED) held in Rio de Janeiro in June 1992.

United Nations Development Fund for Women (UNIFEM): Dir: Sharon Capeling-Alakija; provision of direct financial and technical support to enable low-income women in developing countries to increase earnings, gain access to labour-saving technologies and otherwise improve the quality of their lives; funding of activities that include women in decision-making related to mainstream development projects.

United Nations Revolving Fund for Natural Resources Exploration (UNRFNRE): Dir: Shigeaki Tomita; established in 1974 to provide risk capital to finance exploration for natural resources (particularly minerals) in developing countries and to attract investment; the revolving charter of the Fund lies in the undertaking of contributing governments to make replenishment contributions to the Fund when the projects that it finances lead to commercial production.

United Nations Sudano-Sahelian Office (UNSO): UN Sahelian Regional Office, BP 366, ave Dimdolobsom, Ouagadougou, Burkina Faso; tel 367-81; telex 5262; Dir: Peter Branner; established in 1973 to assist 22 countries across the Sudano-Sahelian belt of Africa in combating drought and desertification; involved in the sustainable management of natural resources; special emphasis is given to strengthening the environmental planning and management capacities of national institutions.

United Nations Economic Commission for Africa (UN/ECA): Africa Hall, POB 3001, Addis Ababa, Ethiopia; tel (1) 517200; telex 21029; fax (1) 514416; Exec Sec: Layashi Yaker; the Social Development, Environment and Human Settlements Division of the ECA undertakes study, production of technical publications and the organization of meetings; advisory services are provided to member states on the formulation of policies relating to overall social development, integrated rural development, and the development of mineral resources and their exploitation; environmental activities in Africa are co-ordinated with UNEP and other agencies; member states are assisted in the assessment and use of water resources and the development of lakes and river-basins common to more than one country.

United Nations Economic Commission for Europe (UN/ECE): Palais des Nations, 1211 Geneva 10, Switzerland; tel (22) 9171234; telex 412962; fax (22) 9170036; Exec Sec: G Hinteregger; Pres: Andrei Ozadovski; development of regional environmental policies; elaboration and implementation of international environmental legislation; promotion of co-operation on environmentally sound technology and products and management of hazardous chemicals; development of environmental management capabilities in countries in transition to a market economy; concerned with energy efficiency, sustainable industrial development and the environmental aspects of transport and agriculture; subsidiary bodies include: the Committee on Agricultural Problems, the Timber Committee, the Working Party on Coal, the Working Party on Electric Power, the Working Party on Gas, the Committee on Human Settlements, Senior Advisers to ECE Governments on Environmental and Water Problems, and Senior Advisers to ECE Governments on Science and Technology.

United Nations Economic Commission for Latin America and the Caribbean (UN/ECLAC): Edif Naciones Unidas, Avda Dag Hammerskjöld, Casilla 179D, Santiago, Chile; tel (2) 208-5051; telex 340295; fax (2) 208-0252; Exec Sec: Gert Rosenthal; founded in 1948 to co-ordinate policies for the promotion of economic development in the Latin America region; collaboration with regional governments in the investigation and analysis of regional and national economic problems; guidance in the formulation of development plans; the Environment Programme is jointly funded and organized with UNEP's Environment and Development Unit.

United Nations Economic and Social Commission for Asia and the Pacific (UN/ESCAP): United Nations Bldg, Rajadamnern Ave, Bangkok 10200, Thailand; tel (2) 282-9161; telex 82392; fax (2) 282-9602; Exec Sec: Rafeeduddin Ahmed; Dep Exec Sec: Seiko Takahashi; founded in 1947 to encourage the economic and social development of Asia and the Far East; agriculture and rural development; development planning; economic co-operation; support for industrial development, human settlements and the environment; ESCAP assists the countries of the region to benefit from their mineral resources by exploration and assessment of land and marine areas, conducting studies of mineral commodities, promoting interdisciplinary research, sponsoring technical co-operation and stressing the environmental impact of the exploitation of these resources; regional network of information and advisory services on solar, wind and biomass energy; aid for countries in the efficient use of fossil energy supplies; sponsorship of the investigation of new and renewable energy resources in the region; development of water resources through national water plans.

ESCAP/WMO Typhoon Committee: POB 7285, Domestic Airport Post Office Lock Box, Pasay City, Metro Manila, Philippines; tel (2) 9228055; telex 66682; fax (2) 9218313; Co-ordinator of Secretariat: Dr Roman L Kintanar; intergovernmental body sponsored by ESCAP and WMO for mitigation of typhoon damage; establishment of efficient typhoon and flood warning systems through improved meteorological and telecommunication facilities.

WMO/ESCAP Panel on Tropical Cyclones: Technical Support Unit, Abhawa Bhaban, Agargon, Dhaka 1207, Bangladesh; founded in 1973 to mitigate damage caused by tropical cyclones in the Bay of Bengal and the Arabian Sea.

United Nations Economic and Social Commission for Western Asia (UN/ESCWA): POB 927115, Amman 11192, Jordan; tel 694351; telex 21691; fax 694981; Exec Sec: Sabah Bakjaji; established in 1974 by a resolution of the UN Economic and Social Council (ECOSOC) to provide facilities of a wider scope for those countries previously served by the UN Economic and Social Office in Beirut (UNESOB); study of economic and technological problems of the region; collection and dissemination of information and provision of advisory services; co-operation with the FAO in regional planning, food security and management of agricultural resources; work is also undertaken with UNFPA in population programmes, with ILO in statistical surveys on labour, with UNCTAD in development planning, and with UNEP in integrating environmental considerations (particularly control of desertification) into development programmes; members: Bahrain, Egypt, Iraq, Jordan, Kuwait, Lebanon, Oman, the Palestine Liberation Organization (PLO), Qatar, Saudi Arabia, Syria, the United Arab Emirates and Yemen.

United Nations Educational, Scientific and Cultural Organization (UNESCO): 7 place de Fontenoy, 75352 Paris, France; tel (1) 45-68-10-00; telex 204461; fax (1) 45-67-16-90; Dir-Gen: Federico Mayor Zaragoza; work includes research in the field of environmental sciences and natural resources; examples include the Man and Biosphere Programme which involved local people in solving practical problems of environmental resource management; International Geological Correlation Programme; International Hydrological Programme; Intergovernmental Oceanographic Commission; organization of training and research programmes in environmental sciences; environmental education programmes; the World Heritage Programme aims to protect historic sites and natural landmarks by the provision of financial aid for restoration, technical assistance, training and management planning; regional offices world-wide.

Office for the Pacific States: POB 5766, Matautu Uta Post Office, Apia, Western Samoa; tel 24276; telex 209; fax 22253; Chief of Mission: Fred Griffiths; Science Adviser: Trevor Sankey; programmes in the fields of education, science, culture and communication, including promotion of science and environmental education, application of microbial biotechnology, natural hazard reduction, water sciences, ecological and marine sciences and development of media for the dissemination of environmental information.

United Nations Environment Programme (UNEP): POB 30552, Nairobi, Kenya; tel (2) 230800; telex 22068; fax (2) 226890; Exec Dir: Elizabeth Dowdeswell; established in 1972 by the UN General Assembly following recommendations of the 1972 UN Conference on the Human Environment, to encourage international co-operation in matters relating to the human environment; maintenance of a constant watch on the changing state of the environment; analysis of environmental trends; assessment of problems using a wide range of data and techniques; promotion of projects leading to environmentally sound development; encouragement of the inclusion of environmental issues in education; the joint UNEP/UNESCO International Programme in Environmental Education includes the training of teachers, publications and technical assistance to governments for the incorporation of environmental education at all levels of general education; UNEP's environmental assessment programme, Earthwatch, aims to study the interaction between people and the environment, provide early warning of potential environmental hazards and determine the state of natural resources; the Global Environmental Monitoring System (GEMS) collects data on renewable resources, the atmosphere, environmental pollution, long-range transport of pollutants, integrated monitoring of pollutants and ecosystems, and oceans; a global resource information data-base (GRID) converts the data collected into information usable by decision-makers; in 1987 the Montréal Protocol to the Vienna Convention for the Protection of the Ozone Layer (1985) provided for a 50% reduction in the production of chlorofluorocarbons (CFCs) by the year 2000; in 1990 an amendment was passed which required manufacture of ozone-depleting substances to be completely phased out by 2000; in 1989 the Basel Convention on the Control of Transboundary Movements of Hazardous Waste and their Disposal was adopted; UNEP was instrumental in the drafting of the two Conventions which were signed by over 150 countries at the UN Conference on Environment and Development (UNCED or the 'Earth Summit') in 1992; the Climate Change Convention deals with the emissions of 'greenhouse gases' and the resultant warming of the Earth, while the Convention on Biological Diversity aims to preserve the variety of plant and animal species on the planet; Agenda 21 was also adopted at the conference, a programme of activities to promote sustainable development; UNEP is also involved in collaboration with the IUCN in the protection of endangered species and habitats, and provides a secretariat for the Ecosystem Conservation Group (consisting of UNEP, the FAO, UNESCO and the IUCN).

Africa Office: UNEP HQ (see above).

Asia and the Pacific Office: UN Bldg, 10th Floor, Rajadamnern Ave, Bangkok 10200, Thailand; tel (2) 282-9161; telex 82392; fax (2) 280-3829.

Europe Office: Palais des Nations, 8-14 ave de la Paix, 1211 Geneva 10, Switzerland; tel (22) 7988400; telex 415465; fax (22) 7983945.

Latin America and the Caribbean Office: Pres Mazaryk 29, 5°, Col Chaputlepec Morales, 11570 México, DF, Mexico; tel (5) 203-4579; telex 1771055; fax (5) 203-4465.

North America Office: UNDC Two Bldg, Rm 0803, 2 United Nations Plaza, New York, NY 10017, USA; tel (212) 963-8138; telex 422311; fax (212) 963-7341.

West Asia Office: 1083 Rd 425, Jufair 342, Manama, Bahrain; tel 276072; telex 7457; fax 276072.

Committee of International Development Institutions on the Environment (CIDIE): Nairobi, Kenya; tel (2) 230800; telex 22068; fax (2) 226886; Co-ordinator: Elizabeth Dowdeswell; promotion of full consideration for environmental concerns in the formulation of all policies and programmes within member institutions.

Convention on the Conservation of Migratory Species of Wild Animals (CMS): 5300 Bonn 2, Mallwitzstr 1-3, Germany; tel (228) 9543501; telex 885556; fax (228) 9543500; Co-ordinator: Arnulf Müller-Helmbrecht.

Convention on International Trade in Endangered Species of Wild Fauna and Flora (CITES): 15 chemin des Anémones, CP 456, 1219 Châtelaine/Geneva, Switzerland; tel (22) 9799139; telex 415391; fax (22) 7973417; Sec-Gen: I Topkov; Deputy Sec-Gen: J Berney; also referred to as the Washington Convention; assistance in the implementation of the convention, including technical assistance and training activities; co-operation with international environmental and non-environmental organizations; publications.

Harmonization of Environmental Measurement (HEM): 85758 Oberschleißheim, c/o GSF, Neuherberg, Postfach 1129, Germany; tel (89) 31875488; fax (89) 31873325; Dir: Dr Hartmut Keune; Sr Scientist: Dr A B Murray; harmonization of monitoring programmes and classification systems in selected environmental sectors; compilation of indexes related to environmental measurement; development of the HEMIS information system.

Industry and Environment Programme Activity Centre (IE/PAC): Tour Mirabeau, 39-43 quai André Citroen, 75539 Paris Cedex 15, France; tel (1) 40-58-88-50; telex 204997; fax (1) 40-58-88-74.

INFOTERRA: Nairobi, Kenya; tel (2) 621234; telex 22068; fax (2) 226949; Dir (acting): Linda Spencer; Chief, Information Service: Feodor Kozak; Chief for Computer and Telecommunications: Roger Johnson; UNEP scientific and technical information exchange system; network of 158 national focal points, 11 regional service centres and a Programme Activity Centre (PAC); INFOTERRA database compiled from over 6,000 sources of environmental expertise; publication of various bulletins and directories.

Interim Secretariat to the Basel Convention (ISBC): CP 59, 1292 Chambéry-Geneva, Switzerland; tel (22) 7582510; telex 415465; fax (22) 7581189; Co-ordinator: I Rummel-Bulska; the Basel Convention on the Control of Transboundary Movements of Hazardous Wastes and their Disposal was adopted in 1989, and governs the transport and disposal of hazardous wastes, with the aim of preventing the dumping of wastes from industrialized countries in countries that have no processing facilities; 31 members.

International Register of Potentially Toxic Chemicals Programme (IRPTC/PAC): Palais des Nations, 1211 Geneva 10, Switzerland; tel (22) 7988400; telex 28877; fax (22) 7332673; Dir: Dr J W Huismans; development of chemical data profiles; operation of an information service; implementation of the London Guidelines for the exchange of information on chemicals in international trade; provision of training and workshops on the control of chemical hazards; publication of a bulletin.

Regional Co-ordination Unit for the West and Central African Action Plan (WACAF-RCU): Nairobi, Kenya; tel (2) 520600; telex 25164; fax (2) 230127; Sr Programme Officer: Richard Congar; technical co-ordination of the Action Plan for protection of the marine environment and coastal areas, under the supervision of the Oceans and Coastal Areas Programme Activity Centre (OCA/PAC); secretariat for the Action Plan and for the Abidjan Convention.

United Nations Industrial Development Organization (UNIDO): 1400 Vienna, POB 300, Austria; tel (1) 21-13-10; telex 135612; fax (1) 23-21-56; Dir-Gen: Domingo L Siazón, Jr; concerned with the macro- and micro-economic aspects of industrial growth; technical assistance to developing countries; energy and environmental management in Brazil; publication of reports and newsletters.

United Nations Population Fund (UNFPA): 220 East 42nd St, New York, NY 10017, USA; tel (212) 297-5000; telex 422031; fax (212) 370-0201; Exec Dir: Dr Nafis Sadik; provision of financial and technical assistance to developing countries; family planning; information, education and communication; basic data collection; utilization of population data and research for policy formulation and development planning; women, population and development.

Wildlife Preservation Trust International: 3400 West Girard Ave, Philadelphia, PA 19104-1196, USA; tel (215) 222-3636; fax (215) 222-2191; Exec Dir: Dr Jerry A Eberhart; affiliated with the Jersey Wildlife Preservation Fund; preservation of endangered animal species by captive propagation, education and field work.

World Conservation Monitoring Centre: 219 Huntingdon Rd, Cambridge CB3 0DL, UK; tel (223) 277314; fax (223) 277136; Dir: Dr Robin Pellew; Dir of Programmes: Dr Mark Collins; scientific research and data analysis in support of active conservation, identification of conservation priorities and sustainable management of biological diversity; conservation information; technical services.

World Conservation Union (IUCN): 28 rue Mauverney, 1196 Gland, Switzerland; tel (22) 9990001; telex 419624; fax (22) 9990002; Pres: Sir Shridath Ramphal; Dir-Gen: Dr Martin W Holdgate; independent union of governments, government agencies and NGOs; founded in 1948 as the International Union for Conservation of Nature and Natural Resources, to promote the conservation of nature, and especially of biodiversity, as an essential foundation for the future; ensuring wise use of the Earth's natural resources in an equitable and sustainable way; development of programmes to protect and sustain the most important and threatened species and ecosystems; maintenance of a conservation library and documentation centre and units for monitoring traffic in wildlife; members include 62 governments, 99 government agencies, 526 national and 49 international NGOs and 37 affiliates; member of Ecosystems Conservation Group.

Commission on National Parks and Protected Areas: fax (22) 9990015; Head, Protected Areas Programme: David Sheppard; global network of professionals specialized in protected area management; technical advice.

Societas Europaea Herpetologica (European Herpetological Society): 64283 Darmstadt, Zoology Dept, Hessen Museum, Friedensplatz 1, Germany; tel (6151) 165781; fax (6151) 28942; Pres: Prof Dr B Lanza; Gen-Sec: Dr U Joger; Chair, Conservation Cttee: Keith Corbett; IUCN specialist group for European amphibians and reptiles; development of action plans for conservation of European amphibians and reptiles on behalf of the EC and the Council of Europe.

World Energy Council (WEC): 34 St James's St, London SW1A 1HD, UK; tel (71) 930-3966; telex 264707; fax (71) 925-0452; Pres: Dr M Gómez de Pablos; Chair, Exec Ass: Dr G Ott; Sec-Gen: I D Lindsay; promotion of the sustainable supply and use of energy; 100 countries are members; congresses; publication of technical and reference materials; energy study programme.

World Food Council (WFC): Via delle Terme di Caracalla, 00100 Rome, Italy; tel (6) 57971; telex 610181; fax (6) 5745091; Pres: Dr Issa Kalantari; Exec Dir: Gerald I Trant; increasing food production in developing countries; resolution of problems which affect the hungry poor; promotion of agricultural development strategies.

World Food Programme (WFP): Via Cristoforo Colombo 426, 00145 Rome, Italy; tel (6) 57971; telex 626675; fax (6) 5127400; Exec Dir: Catherine A Bertini; provision of food to people in low-income, food-deficit countries; assistance in the implementation of economic and social development projects; publication of journals.

World Health Organization (WHO), Division of Environmental Health: ave Appia, 1211 Geneva 27, Switzerland; tel (22) 7913582; telex 415416; fax (22) 7884218; Dir: Dr W Kreisel; global co-ordination of environmental health matters; information on environmental health matters; support for regional offices; concerned with water supply and quality, sanitation, air quality, rural and urban development, and nuclear safety.

World Information Service on Energy: POB 5627, 1007 AP Amsterdam, Netherlands; tel (20) 6392681.

World Meteorological Organization (WMO): CP 2800, 41 ave Giuseppe Motta, 1211 Geneva 2, Switzerland; tel (22) 7308111; telex 414199; fax (22) 7342326; Sec-Gen: Prof G O P Obasi; Dep Sec-Gen: David Axford; the World Weather Watch Programme includes the Global Observing System, the Instruments and Methods of Observation Programme (IMOP), the Global Data Processing System, the Global Telecommunication System, Data Management, System Support Activity, the Executive Council Working Group on Antarctic Meteorology, the Executive Council Panel of Experts / CBS Working Group on Satellites and the Tropical Cyclone Programme; the World Climate Programme includes the World Climate Data and Monitoring Programme (WCDMP), the World Climate Applications and Services Programme (WCASP), the World Climate Research Programme (WCRP) and the World Climate Impact Assessment amd Response Strategies Programme (WCIRP); the Atmospheric Research and Environment Programme includes Global Atmosphere Watch (GAW), Weather Prediction Research Programmes, the Tropical Meteorology Research Programme and the Physics and Chemistry of Clouds and Weather Modification Research Programme; Applications of Meteorology Programme; the Hydrology and Water Resources Programme includes the Operational Hydrology Programme and the Programme on Water-Related Issues; Education and Training Programme; the Technical Co-operation Programme with UNDP includes the Voluntary Co-operation Programme (VCP); co-operation with other bodies; publication of reports and bulletins.

World Pheasant Association: POB 5, Lower Basildon, Reading RG8 9PF, UK; tel (734) 845140; fax (734) 843369; Chair: Edward Dickenson; Dir-Gen: Keith Howman; conservation of galliformes world-wide.

World Society for the Protection of Animals (WSPA): Park Place, 10 Lawn Lane, Vauxhall, London SW8 1UD, UK; tel (71) 793-0540; fax (71) 793-0208; Dir-Gen: Gordon Walwyn; incorporating the World Federation for the Protection of Animals and the International Society for the Protection of Animals; promotion of animal welfare and conservation; dissemination of literature to encourage humane management and slaughter of food animals; control of domestic and wild animal communities.

World Tourism Organization (WTO): Calle Capitan Haya 42, 28020 Madrid, Spain; tel (91) 5710628; telex 42188; fax (91) 5713733; Sec-Gen: Antonio Enríquez Savignac; development assistance; education; training; statistics; market research; documentation; health and safety aspects of tourism; the Environment Committee deals with environment and planning issues.

World Wide Fund for Nature (WWF) International: ave du Mont Blanc, 1196 Gland, Switzerland; tel (22) 3649111; fax (22) 3643239; Dir-Gen: Claude Martin; formerly the World Wildlife Fund; conservation of nature and ecological processes by preserving biodiversity; ensuring the sustainable use of renewable natural resources in the short and long term; promotion of actions to reduce pollution and wasteful exploitation and consumption of resources; raising awareness of threats to the natural environment; affiliate organizations in Australia, Austria, Belgium, Canada, Denmark, Finland, France, Germany, Hong Kong, India, Italy, Japan, Malaysia, the Netherlands, New Zealand, Norway, Pakistan, South Africa, Spain, Sweden, Switzerland, the United Kingdom and the USA (the WWF is still known as the World Wildlife Fund in Canada and the USA); associate organizations in Argentina, Ecuador, Nigeria, Thailand and Venezuela; member of the Ecosystems Conservation Group.

Youth and Environment Europe (YEE): Oudegracht 42, 3511 AR Utrecht, Netherlands; tel (30) 311537; fax (30) 343986; nature study; training courses and seminars on traffic, landscape management, alternative energy, forestry and other issues; European working groups on environmental education, the North Sea, the Baltic Sea, motorways and other issues; annual meeting in August.

AFGHANISTAN

Governmental Organizations

Ministry of Agriculture and Land Reform, Department of Environmental Protection: Jamal Mina, Kabul; tel (93) 41151.

Ministry of Water Resources, Development and Irrigation: Darulaman Wat, Kabul; tel (93) 40743; Minister: Sayd Mohammad Ayub.

ALBANIA

Governmental Organizations

Ministry of Health and Environmental Protection: Tirana; telex 4205; Minister: Tritan Shehu.

Committee of Environmental Preservation and Protection (CEPP): Tirana; tel (42) 27907; fax (42) 27907; Chair: Dr Lirim Selfo; Vice-Chair: Gani Deliu; the highest environmental body in Albania, responsible for all relevant issues.

Non-Governmental Organizations

Albanian Academy of Sciences, Hydrometeorology Institute: Academy of Science of the Republic of Albania, Rruga Kongresi i Përmetit 26, Tirana; tel (42) 22439; fax (42) 23518; compilation of data on water and climate using statistics from the national hydrometeorological network; local studies and chemical analyses of surface and marine water; produces technical reports in the field of hydrology and climatology; six sections cover meteorology, climatology, agrometeorology, hydrometry, oceanography and hydrology.

Instituti i Kërkimeve Bujqësore (Institute of Agricultural Research): Lushnjë; tel (24) 982194; Dir: Dr Milto Hyso; Vice-Dir: Dr Hiqmet Demiri; breeding of new varieties of wheat, cotton, sunflower and dry bean; study and conservation of their germ-plasm.

Partia e Blerte Shqiptare (Albanian Green Party): c/o Kuvendi Popullor (People's Assembly), Tirana; Chair: Nasi Bozhegu; campaigns on environmental issues.

Protection and Preservation of the Natural Environment in Albania (PPNEA): c/o Department of Zoology, Faculty of Natural Sciences, Tirana University, Tirana; tel (42) 27669; Head: Lekë Gjikmuri; Deputy Head: Algi Çullaj; Sec: Dhurata Frashëri; active in the protection of Albania's forests and the botanical gardens of Tirana.

ALGERIA

Governmental Organizations

Agence Nationale pour la Protection de l'Environnement (ANPE—National Agency for Environmental Protection): BP 154, el-Amnaser.

Ministry of Water Resources, Forestry and Fishing: Grand Séminaire, Kouba, Algiers; tel (2) 58-95-00; telex 54125.

Non-Governmental Organizations

Institut des Sciences de la Mer et de l'Aménagement du Littoral (ISMAL—Institute of Marine Sciences and Coastal Management): Villa 4, Plage Ouest, Sidi-Predj, BP 54, Staoueli, Tipaza; tel (2) 39-19-13; telex 67523; fax (2) 39-35-38; Dir: Dr Rachid Semroud; oceanographic research; fish biology; aquaculture; marine pollution and chemistry; geomorphology, microbiology; algology and plankton research.

ANDORRA

Governmental Organizations

Ministry of Agriculture and Natural Heritage: Andorra la Vella; Minister: Marc Moles.

Non-Governmental Organizations

Associació per a la Defensa de la Natura (ADN—Association for the Defence of Nature): Apartat de Correus Espanyols, 96, Andorra La Vella; tel 43248; Pres: Alain Moreno; Sec: Amelia Blanch; Treas: Esther Bartumeu; ornithology; botany; campaigning for protected areas; publication of field guides; popularizing scientific knowledge about nature.

ANGOLA

Governmental Organizations

Direccão Nacional da Conservação de Natureza (National Nature Conservation Department): BP 74, Luanda.

Non-Governmental Organizations

National Ecological Party of Angola (PNEA): Luanda; Leader: Sukawa Dizizeko Ricardo; member of National Opposition Council (CNO).

ANTIGUA AND BARBUDA

Governmental Organizations

Ministry of Agriculture, Fisheries, Lands and Housing: Long St, St John's; tel 462-1007; Minister: Hilroy Humphreys.

ARGENTINA

Governmental Organizations

Ministry of Health and Social Action, Secretariat of Housing and Environmental Management: Avda Santa Fé 1548, 10°, 1060 Buenos Aires; tel (1) 44-9018; Minister: Avelino José Porto.

Administración de Parques Nacionales, Secretaría de Agricultura y Ganadería (National Parks Administration, Agriculture and Livestock Secretariat): Avda Santa Fé 690, 1059 Buenos Aires; tel (1) 311-6633; telex 21535; Pres: Jorge Aumedes; management of national parks, national reserves and natural monuments; scientific research; monitoring planning and use of parks and recreational areas.

Dirección de Intereses Maritimos y Pesca Continental (Department of Marine Affairs and Continental Fishing): 9 de Julio 280, 9103 Rawson; tel (965) 82604; fax (965) 82604; Principal Officers: Gabriel Punta, José Saravia; responsible for management of natural resources, including sea-birds.

Dirección de Producción Forestal (DPF—Department of Forestry Production): Avda Paseo Colón 982, Anexo Jardín, 1063 Buenos Aires; tel (1) 362-4803; fax (1) 362-4803.

Dirección de Recursos Naturales Renovables (Office of Renewable Natural Resources): Boulogne sur Mer esq Avda El Libertador, Parque General, San Martín, 5500 Mendoza; tel (61) 25-7208; fax (61) 38-0523; Dir: Antonio Daniel Dalmasso; management of fauna, native flora, forests and protected natural areas.

Dirección Nacional de Fauna Silvestre (National Wildlife Department): Paseo Colón 922, 2°, 1063 Buenos Aires; tel (1) 362-1207.

Secretaría de Recursos Naturales y Ambiente Humano (Secretariat of Natural Resources and Human Environment): Balcarce 50, 1064 Buenos Aires; Minister: María Julia Alsogaray.

Non-Governmental Organizations

Amigos de la Tierra Argentina (Friends of the Earth—FoE Argentina): Zapata 343, 1426 Buenos Aires, Argentina; Casilla 3560, Correo Central, 1000 Buenos Aires; tel (1) 773-5947; fax (1) 331-4323; Principal Officers: Osvaldo B Corvalán, María Belén Villa, Roberto Reisinger, Maximo Winkler, María del C Cogurno, Ana Ganopol, Anahi Magdaleno; environmental education; research into flora and fauna; protection of natural areas, environmental management; processing of household waste.

Asociación Ornitológica del Plata (River Plate Ornithological Association): Avda 25 de Mayo 749, 2-6, Casilla 3368, 1002 Buenos Aires; tel (1) 312-8958; fax (1) 312-8958; Principal Officers: Diego Gallegos, Pablo Tubaro; publication of journal and magazine; courses for ornithologists; Argentinian Naturalists School; camps; outings; library.

Centro Interdisciplinario de Estudios sobre el Desarrollo Latinoamericano (CIEDLA—Interdisciplinary Research Centre on Latin American Development): L N Alem 690, 20°, 1001 Buenos Aires; tel (1) 313-3522; fax (1) 311-2902; Principal Officers: Dr Hermann Schneider, Dr Thomas Klöckner; research projects; seminars and conferences on Latin America; library of approx 5,000 volumes, containing documentation on Latin America and Europe.

Centro Latinoamericano de Estudios Ambientales (CLEA—Latin American Environmental Research Centre): Callao No 1103, 10°, Departamento 'I', 1023 Buenos Aires; tel (1) 42-0968; fax (1) 42-0968; Chair: Pablo Quiroga; Sec: Dr Daniel Silva; environmental impact assessment, education, training and legislation; regional planning.

EcoBios: Alberdi 745, Olivos, 1636 Buenos Aires; tel (1) 790-9713; Principal Officers: Dr Patricia Alejandra Gandini, Dr Esteban Frere; conservation of sea-birds in Patagonia.

ECOSOLAR: Posadas 1996, 3260 Concepción del Uruguay; tel (442) 28221; fax (442) 26565; Principal Officers: Dr Reynaldo Jorge Tubán Felix, Alicia Magistris, Sonia Cot Sanchez; appropriate technology; sustainable bio-dynamic and organic agriculture.

Foro Ecologista (Ecologist Forum): Casilla 1296, Correo Central, 5000 Córdoba; tel (51) 51-2631; telex 51894; fax (51) 89-4353; Dir: Walter Atilio Fontana; journalistic and educational activities in fields related to the environment and development; educational material; radio programmes on environmental themes.

Fundación Ambiente y Recursos Naturales (FARN—Foundation for Environment and Natural Resources): Monroe 2142, 1428 Buenos Aires; tel (1) 781-9171; telex 22088; fax (1) 781-6115; Exec Dir: Pedro Tarak; research, promotion, consultation and training in the areas of environmental policy, legislation, economics and administration.

Fundación Bariloche (Bariloche Foundation): Casilla 138, Avda 12 de Octubre 1915, 8400 San Carlos de Bariloche; tel (944) 22050; fax (944) 22050; Pres: Carlos E Suárez; Vice-Pres: Dr Edgardo Catterberg; energy economics; environment; quality of life; epistemology; environment and energy.

Buenos Aires Office: Piedras 482, 1070 Buenos Aires; tel (1) 331-1649; fax (1) 334-4717.

Fundación Cruzada Patagonica (Patagonian Crusade Foundation): Felix San Martín 678, 8371 Junin de los Andes; tel (944) 91262; fax (944) 91262; Pres: Germán A Pollitzer; education, demography, agriculture and public transport serving Campeñiuen Araucanian communities.

Fundación de Acción Ecológica (Ecological Action Foundation): Casilla 519, Correo Central, 5000 Córdoba; Pres: Walter Atilio Fontana; educational campaigns under the titles Los Niños y el Ambiente (Children and the Environment) and Los Jóvenes y el Ambiente (Youth and the Environment), and training courses; environmental bills; management of specific environmental problems.

Fundación para el Desarrollo en Justicia y Paz (FUNDAPAZ—Foundation for Development in Justice and Peace): San Martín 617, 2°, Departamento 'F', 1004 Buenos Aires; tel (1) 394-2610; fax (1) 311-4677; Principal Officers: Gastón Bordelois, Esteban de Nevares, Juan Luis Díaz, Tomás Nelson, Silvia S de Pereda, Jorge Williams; involved in the development of poor, rural communities in northern Argentina by means of social organization, education, communication, and technical and financial assistance for agriculture, with particular emphasis—taking into consideration natural resources and biodiversity—on finding self-sustaining alternatives for areas dependent on a single crop.

Fundación Vida Silvestre Argentina (FVSA—Argentine Wildlife Foundation): Defensa 245/51, 6°, Dto K, 1065 Buenos Aires; tel (1) 331-4864; fax (1) 331-3631; Pres: Dr Teodosio Brea; Vice-Pres: Federico Zorraquin; national associate organization of World Wide Fund for Nature (WWF) International; promotion of conservation and wise use of natural resources; environmental education.

Greenpeace Argentina/Cono Sur (Greenpeace Argentina/Chile/Uruguay): Bartolome Mitre 226, 4°, 1036 Buenos Aires; tel (1) 331-1032; fax (1) 331-1033; Exec Dir: (vacant); Asst to Exec Dir: Stella Maris Higa; concerned with pesticides, nuclear energy and trade in toxic materials.

Grupo de Educadores Ambientalistas (G de EA—Environmentalist Teachers Group): Laprida 1373, 1642 San Isidro; tel (1) 743-9467; fax (1) 812-1350; Principal Officers: Vivian Lavalle de Vignaroli, Monica Ramirez de Spinardi; training in environmental education for teachers.

Grupo Ecologista Universitario Misionero (GEUM—Misiones University Ecologist Group): Félix de Azara 174, 7° F, 3300 Posadas, Misiones; tel (752) 27498; fax (752) 25414; Chair: Hugo A Chaves; Sec: Alberto Goñi; Treas: Tamara Piñero; establishment of parks or reserves in threatened environments or those representative of ecosystems in the province of Misiones; study and preservation of endangered animal and plant species; lectures and conferences; publication of articles in newspapers, with the aim of making the public aware of the need for environmental preservation.

Grupo Esquel Argentina (Esquel Group Argentina): Córdoba 652, 4°, Depto F, 1054 Buenos Aires.

Instituto Argentino de Investigaciones de las Zonas Aridas (IADIZA—Argentinian Institute for Research into the Arid Zones): Casilla 507, CRICYT, Bajada del Cerro s/n, 5500 Mendoza; tel (61) 24-1995; telex 55438; fax (61) 38-0370; Acting Dir: Juan Carlos Guevara; Research Dir: Virgilio G Roig; attached to National Research Council of Argentina; involved in research and education in the areas of ecology and taxonomy of desert flora and fauna, management of Argentina's desert regions, global change and biodiversity of arid lands and assessment and control of desertification.

Instituto de Estudios e Investigaciones sobre el Medio Ambiente (IEIMA—Institute for Environmental Study and Research): Alsina 1816, 1090 Buenos Aires; tel (1) 46-2951; fax (1) 476-1850; Pres: Elva Roulet; Dir: Dr Osvaldo F Canziani; research and documentation in all areas of environmental science, technology and development.

INTEC: Guemes 3450, 3000 Santa Fé; tel (42) 55-9175; fax (42) 55-0944; Principal Officer: Dr Argelia Lenardon; concerned with chemistry and the environment.

International Institute for Environment and Development (IIED) América Latina: Avda Corrientes 2835, Cuerpo A, 6°, 1193 Buenos Aires; tel (1) 961-3050; fax (1) 961-1854.

Partners of the Americas, Northern Argentina: Maipú 672, 1006 Buenos Aires; tel (1) 322-3855; telex 18630; fax (1) 322-2106; Pres: Blanca Arazi; Vice-Pres: Marta Carbó; private voluntary organization operating programmes in the areas of culture, natural resources management, agriculture and rural development, training and health; paired with Southern California, USA.

Red Informatica Ecologista Argentina (Ecologist Computer Network of Argentina): Mascagni 2185 e Dante y Albeniz, 1686 Hurlingham, BA; Principal Officer: Amanda Bertolutti; information; publications (technical and scientific research, pamphlets for children and teachers, and directories); ecological research into flora, fauna and natural areas; collaboration on TV programmes; co-operation with government agencies.

Sociedad Argentina para el Derecho y la Administración del Ambiente y de los Recursos Naturales (SADARN—Argentinian Society for Environmental Law and Administration of Natural Resources): Uruguay 467, 10°, Of B, 1015 Buenos Aires; tel (1) 40-5867; fax (1) 812-1575; Pres: Dr Eduardo A Pigretti; Vice-Pres: Dr Pascual V Politi; Sec: Dr Graciela Berra Estrada; participation in environmental projects and conferences; working out of proposals for legislation; creation of Comité Argentino pro Tribunal Internacional del Ambiente (Argentinian Committee for an International Environmental Tribunal).

Sociedad Central de Arquitectos (Central Society of Architects): Montevideo 938, 1019 Buenos Aires; tel (1) 812-3644; fax (1) 42-6629; Pres: Julio Keselman; professional assocation.

Taller Ecologista (Environmental Workshop): Casilla 441, 2000 Rosario, Santa Fé; tel (41) 56-4201; fax (41) 56-0179; Principal Officers: Elba Stancich, Juan Carlos Villalonga, Daniel Raposo; energy; urban refuse; medicinal plants; education; maintenance of documentation centre.

TRAFFIC Sudamérica—Oficina Argentina (TRAFFIC South America—Argentina Office): Ayacucho 1477, 9° B, 1111 Buenos Aires; tel (1) 811-4348; fax (1) 811-4348; Dir: Tomás Waller; Consultants: Juan X Gruss, Patricio A Micucci; field surveys; statistics on trade in wildlife; concerned with sustainable management of wildlife and commercially valuable species.

ARMENIA

Governmental Organizations

Ministry of Nature and Environmental Protection: 375002 Yerevan, Moskovyan St 35; tel (8852) 53-07-41; fax (8852) 53-49-02; Minister: Vahan Melkonyan; Deputy Ministers: Suren Avetisian, Karen Shahinian; nature conservation strategy; co-ordination of ecological research; environmental monitoring; preventive measures against pollution; regulation of the use and reproduction of natural resources; international co-operation.

Non-Governmental Organizations

Greens' Union of Armenia: 375093 Yerevan, Furmanov St 47-13; tel (8852) 52-54-41; telex 243240; fax (873) 111-07-60; Pres: Hakob Sanasarian; Foreign Relations Co-ordinator: Anahit Allakhverdian; environmental education; establishment and development of ecologically clean technologies; monitoring and preservation of food, water, soil and air quality; rational use of resources; recycling; waste treatment; alternatives to nuclear power; protection of fauna and flora.

AUSTRALIA

Governmental Organizations

Department of the Arts, Sport, the Environment and Territories: POB 787, Canberra, ACT 2601; tel (6) 274-1111; telex 62960; fax (6) 274-1123; Minister: Ros Kelly; advice on and implementation of policies and programmes for environmental conservation and ecologically sustainable development.

Alligator Rivers Region Research Institute: Locked Bag 2, Jabiru, NT 0886; tel (89) 799711; fax (89) 792499; Dir: Dr Arthur Johnston; research into the effects of uranium mining on the environment of the Alligator Rivers Region.

Australian Capital Territory Parks and Conservation Service: POB 1119, Canberra, ACT 2601; Tuggeranong, ACT 2901; tel (6) 293-5295; fax (6) 293-5315; nature conservation and land management in the ACT; establishment and maintenance of nature reserves and a national park.

Australian Heritage Commission: POB 1567, Canberra, ACT 2601; tel (6) 271-2111; fax (6) 273-2395; identification of places of natural and cultural value; advice on protection.

Australian International Development Assistance Bureau (AIDAB): POB 887, Canberra, ACT 2601; tel (6) 276-4000; telex 62631; fax (6) 276-4880; Dir: David Luck; international development assistance.

Australian National Parks and Wildlife Service: POB 636, Canberra, ACT 2601; tel (6) 246-6211; telex 62971; fax (6) 250-0228; Dir: Peter Bridgewater; advice to government on national nature conservation policies; management of parks and reserves under national conservation legislation, including Aboriginal parks and World Heritage Sites; administration of international agreements.

Bureau of Rural Resources: POB E11, Queen Victoria Terrace, Parkes, ACT 2602; tel (6) 272-3933; fax (6) 272-4747; Dir: Dr George Wilson; collection of scientific and technical information on key rural resource issues in primary production; scientific advice for management of government programmes and policy.

Great Barrier Reef Marine Park Authority (GBRMPA): POB 1379, 1-37 Flinders St, Townsville, Qld 4810; tel (77) 81-8811; telex 47332; fax (77) 72-6093; Chair and CEO: Prof Graeme Kelleher; Exec Officer: Dr Wendy Craik; provision for the protection, wise use, understanding and enjoyment of the Great Barrier Reef through care and development of the Marine Park, while recognizing that any use of the Reef or associated areas should not threaten the Reef's essential ecological characteristics and processes; Great Barrier Reef Ministerial Council co-ordinates policy between the Commonwealth and Queensland governments and comprises two ministers from each government; the Great Barrier Reef Consultative Committee represents public and private sector interests and advises the GBRMPA and ministers; publication of twice-yearly newsletter.

New South Wales Forestry Commission: Locked Bag 23, Pennant Hills, NSW 2120; tel (2) 980-4559; fax (2) 980-7042; Environmental Assessment Man: Peter S Smith; forest management; research; environmental management; services to forestry industry.

New South Wales National Parks and Wildlife Service: 43 Bridge St, POB 1967, Hurstville, NSW 2220; tel (2) 585-6444; telex 26034; fax (2) 585-6555; Dir: W J Gillooly; conservation and management of the natural and cultural heritage, including national parks, historic sites, Aboriginal heritage and reserves.

Northern Territory Conservation Commission: POB 496, Palmerston, NT 0831; tel (89) 895511; telex 85336; fax (89) 323849; Dir: Matti Urvet; environmental conservation research, strategies and education; impact assessment; ecological monitoring; management and planning of protected areas; soil conservation.

Office of the Supervising Scientist (OSS): POB E348, Queen Victoria Terrace, Parkes, ACT 2600; tel (6) 274-1222; fax (6) 273-5019; Supervising Scientist (acting): Dr Glen Riley; research and monitoring organization established to protect the environment of the Alligator Rivers Region of the Northern Territory from the effects of uranium mining.

Queensland Department of Environment and Heritage: POB 155, Albert St, Brisbane, Qld 4002; tel (7) 227-7111; fax (7) 227-6534; Dir-Gen: Dr Craig Emerson; Deputy Dir-Gen: Dr R F I Smith; promotion of ecologically sustainable development; protection and wise use of the natural and cultural heritage and a cleaner and safer environment.

Queensland Energy Information Centre (EIC): POB 194, Brisbane, Qld 4001; Minerals and Energy Centre, Ground Floor, 61 Mary St, Brisbane, Qld 4000; tel (7) 234-9807; fax (7) 234-9969; Sr Energy Information Officer: Des Ede; Energy Officer: Geoff Cook; an initiative of the Queensland Department of Minerals and Energy; impartial information and advice on the efficient use of all forms of energy in the domestic and small business sectors; solar passive house design; solar energy; water heating; insulation; energy rating of appliances.

Queensland Forest Service, Department of Primary Industries: Forestry House, 160 Mary St, POB 944, Brisbane, Qld 4001; tel (7) 234-0145; fax (7) 234-0326; Principal Officer: T Ryan; forestry; afforestation; research.

Resource Assessment Commission (RAC): Locked Bag 1, Queen Victoria Terrace, Canberra, ACT 2600; tel (6) 271-5888; fax (6) 273-4196; Chair: D G Stewart; Head of Office: G Linnane; advice to the Prime Minister; inquiries into the environmental, cultural, social, industrial and economic implications of the use of Australia's natural resources.

South Australia Department of Environment and Planning: POB 667, Adelaide, SA 5000; tel (8) 207-2000; Acting Dir-Gen: Bruce Leaver; environmental management and planning; landscape protection; conservation strategies; protected areas; resource management; land use planning; environmental policy; regulation control; sustainable development; national parks; integration of conservation and development; pollution control.

Tasmania Department of Environment and Land Management: POB 44A, Hobart, Tas 7001; tel (02) 33-3230; fax (02) 23-5905; Sec: John Ramsay; land information; property services; surveying; valuation; planning; environmental management; land titles office; parks and wildlife office.

Victoria Department of Conservation and Environment: 8/240 Victoria Parade, E Melbourne, Vic 3002; tel (3) 412-4363; fax (3) 412-4540; Dir-Gen: Alan Thompson; management of public land and natural resources in Victoria, including national parks, forests, fisheries, fire protection, protection of land from degradation and protection of flora and fauna.

Victoria Renewable Energy Authority: 176 Wellington Parade, Level 6, East Melbourne, Vic 3002; tel (3) 651-7775; fax (3) 651-7367; f 1990; energy conservation; use of renewable energy in Victoria.

Western Australia Department of Conservation and Land Management: Hayman Rd, POB 104, Como, WA 6152; tel (9) 386-8811; telex 94858; fax (9) 386-1578; Exec Dir: Dr Syd Shea; conservation and management of wildlife; planning and management of national parks, nature reserves, marine parks, marine nature reserves, conservation parks, state forests and timber reserves.

Western Australia Environmental Protection Authority: Western Sq, 141 St George's Terrace, Perth, WA 6000; tel (9) 222-7000; fax (9) 322-1598; Principal Officer: Barry Carbon.

Non-Governmental Organizations

ANUTECH Pty Ltd: Australian National University, POB 4, Canberra, ACT 2601; tel (6) 249-0617; telex 62760; fax (6) 249-5875; Man Dir: John Morphett; Man, Forestry and Environment Div: Dr Ken Shepherd; commercial company wholly owned by Australian National University; intended to facilitate consultancies, market discoveries and contract research; consultancy services in environmental impact assessment, forestry and rural development; professional courses for overseas students; co-operation with School of Resource and Environmental Management.

Association for Research, Exploration and Environmental Aid (AREA): POB A529, Sydney S, NSW 2000; tel (2) 262-6983; Chair: Peter Wolfe; provision of environmental and development assistance to developing countries; monitoring; resource assessment; forestry; use of wildlife resources; use of marine resources; international trade; pollution control; energy; impact assessment; agriculture; human ecology; endangered animal species; design and planning of protected areas; watershed management.

Australian and New Zealand Solar Energy Society: POB 124, Caulfield East, Victoria 3145; tel (3) 571-7557; telex 154087; fax (3) 563-5173; Chair: Dr Steven V Szokolay; promotion of the science, technology and application of solar energy.

Australian Committee for IUCN: c/o Dept of Environment and Heritage, POB 155, N Quay, Qld 4002; tel (7) 227-7111; fax (7) 227-7676; Pres: Dr Craig Emerson.

Australian Conservation Foundation: 340 Gore St, Fitzroy, Vic 3065; tel (3) 416-1166; fax (3) 416-0767; Exec Dir: Patricia J Caswell; campaigns for an ecologically sustainable society by protecting biodiversity, promoting global change and managing natural resources; protection of forests and endangered species; campaigns on waste minimization and reduction of ozone layer depletion, greenhouse gas production, pollution and land degradation.

Australian Council of National Trusts: POB 1002, Civic Square, Canberra, ACT 2608; tel (6) 247-6766; fax (6) 249-1395; Chair: K Newman; Exec Officer: A Graham; management of lands and buildings of aesthetic, historic, scientific, social or other value; lobbying of governments and other national and international bodies, to conserve such sites and promote public appreciation, knowledge and enjoyment of them.

Australian Geographic Society: POB 321, Terrey Hills, NSW 2084; tel (2) 450-2344; telex 176203; fax (2) 450-2990; Chief Exec: Ike Bain; Publisher and Editor-in-Chief, Australian Geographic Journal: Dick Smith; Editor: Howard Whelan; encouragement and financial support for research into Australia's flora, fauna and the environment; fund-raising through sales of magazines and Australian Geographic Shops.

Australian Littoral Society (ALS): POB 49, Moorooka, Qld 4105; tel (7) 848-5235; fax (7) 892-5814; Exec Officer: Diane Tarte; study and conservation of aquatic life.

Australian Museum: 6-8 College St, POB A285, Sydney South, NSW 2000; tel (2) 339-8111; fax (2) 630-4350; Dir: Dr Des Griffin; Deputy Dir: Dr H C Cogger; natural history and human studies; promotion of understanding of Australia's natural environment and cultural heritage through public programmes and exhibitions and scientific research; taxonomic and environmental research based on the museum's collections and field studies; commercial work for environmental surveys.

Australian National Parks Council (ANPC): POB 2227, Canberra, ACT 2601; tel (6) 257-1063; Pres: Neville Esau; protection, promotion and extension of the national park systems; NGO forum.

Australian National University, Energy Research Centre: Research School of the Physical Sciences and Engineering, Institute of Advanced Studies, POB 4, Canberra, ACT 2601; tel (6) 249-5135; fax (6) 249-1884.

Australian National University, Water Research Foundation of Australia: Centre for Resource and Environmental Studies, Canberra, ACT 0200; tel (6) 249-0651; fax (6) 249-0757; Chair: J G Beale; research into Australia's water and land resources.

Biodiversity Coalition: c/o Post Office, Cygnet, Tas 7112; tel (02) 951745; fax (02) 951964; Co-ordinator: Alistair Graham; newsletter and networking information.

Canberra and South-East Region Environment Centre: POB 1875, Canberra, ACT 2601; tel (6) 248-0885; Pres, Man Cttee: David Pfanner; Librarian: Kerrie Tucker; environmental library.

Conservation Council of South Australia, Inc: 120 Wakefield St, Adelaide, SA 5000; tel (8) 223-5155; fax (8) 232-4782; Admin Officer: Michelle Grady; Environmental Consultant: Marcus Beresford; organization for over 60 environment groups in South Australia; environmental advocacy; environmental advice, information and education services; liaison with government, industry, trade unions and community groups on environmental matters.

Conservation Council of Victoria: 247 Flinders Lane, Melbourne, Victoria 3000; tel (3) 654-4833; fax (3) 650-5684; Pres: Peter Harley; Co-ordinator: Linda Parlane; campaigns for transport alternatives to the car and against logging in native forests; research into environmentally-friendly sewerage systems.

Conservation Council of Western Australia: 79 Stirling St, Perth, WA 6000; tel (9) 220-0652; fax (9) 220-0653; Pres: Prof Philip Jennings; environmental conservation; information; education.

Ecological Society of Australia, Inc: POB 1564, Canberra, ACT 2601; tel (6) 249-5092; fax (6) 249-5095; Pres: Dr Ian Noble; Sec: Dr Tony Norton; Treas: Dr Geoff Hope; scientific study of organisms in relation to their environment; application of ecological principles; reservation and sound management of natural areas; advice to governmental and other agencies where appropriate.

Environment Centre of the Northern Territory (NT), Inc: Home Centre, 24 Cavenagh St, Darwin, NT 0800; POB 2120, Darwin, NT 0801; tel (89) 81-1984; fax (89) 41-0387; Co-ordinator: Sue Jackson; promotion of environmental awareness; environmental information; vehicle for Northern Territory conservationists nationally and internationally; research; lobbying governments and industry.

Environment Centre of Western Australia: 587 Wellington St, Perth, WA 6000; tel (9) 321-5942; fax (9) 322-3045; Principal Officers: Luci Ravi, Carina Calzoni; information resource centre for environmental issues; contact point for a range of environmental initiatives.

Fraser Island Defenders Organization: 36 Kemp St, POB 71, Gladesville, NSW 2111; tel (2) 817-4660; Project Officer: J Sinclair; promotion of wise use of natural resources; conservation of flora and fauna.

Gippsland Waters Coalition (GWC): POB 463, Moe, Vic 3825; tel (51) 27-6508; fax (51) 27-7560; Sec: Neil Grigg; organization of community groups; protection of aquatic environments; promotion of ecologically sustainable allocation of resources.

Greening Australia: POB 9868, Canberra, ACT 2601; tel (6) 285-8590; fax (6) 285-8585; Nat Man: Jan Davis; national community organization; conservation of trees and vegetation.

Greenpeace Australia (GPA): POB 51, Suite 14, 37 Nicholson St, Balmain, NSW 2041; tel (2) 555-7044; fax (2) 555-7154; Exec Dir: Paul Gilding; campaigns on marine environments, sea pollution prevention, coastal ecosystems, mineral exploitation, whaling, marine ecology, Antarctic ecosystems, dumping at sea, Pacific ecosystems, wildlife, use of marine resources and atmosphere and climate.

Institute of Foresters of Australia: POB 675, Nedlands, WA 6009; Pres: Dr Frank McKinnell; forestry; afforestation; agroforestry; forestry research.

Jacobs Well Environmental Education Centre: MS 1372, Beenleigh, Qld 4207; tel (75) 46-2317; fax (75) 46-2317; Primary Principal: Glenn Leiper; Secondary Principal: Jim Rebgetz; environmental education activities for students of all ages and levels of ability in the education system; school day trips and camps; programmes focus on local environmental issues and are conducted in coastal wetlands and remnant forests.

Macquarie University, School of Law, Environmental Law Centre (ELC): Sydney, NSW 2109; tel (2) 805-7065; telex 122377; fax (2) 805-7686; Dir: Zada Lipman; undergraduate and postgraduate teaching, research; consultancies in environmental law, local government, pollution and heritage.

Men of the Trees Inc (Queensland): POB 283, Clayfield, Qld 4011; tel (7) 262-1096; tree-planting; raising awareness of the importance of trees.

Movement against Uranium Mining, New South Wales: POB K133, Haymarket, NSW 2000; tel (2) 212-4538; fax (2) 281-5216; Principal Officers: Steve Broadbent, Murray Matson, Claire Gerson.

Murray Darling Association, Inc: POB 359, Albury, NSW 2640; tel (60) 21-3655; fax (60) 21-2025; Pres: Max Moor; Gen Man: Leon Broster; association of local governments, groups, corporations and incividuals with an interest in maintaining the environmental interests of the Murray and Darling rivers while also encouraging sustainable development of the river valleys' resources.

National Parks Association of New South Wales, Inc: POB A96, Sydney S, NSW 2000; tel (2) 264-7994; fax (2) 264-7160; Exec Officer: Rod Bennison; landscape protection; environmental research; resource management; environmental education; lobbying government; land use planning; management and planning of protected areas; camps and excursions.

National Trust of Australia: Observatory Hill, Watson Rd, Rocks, NSW 2000; tel (2) 258-0123; fax (2) 258-1264; Exec Dir: Wendy McCarthy; Pres: Mr Justice B S J O'Keefe; conservation of the natural and built environment.

Nature Conservation Council of New South Wales: 39 George St, Rocks, NSW 2000; tel (2) 247-2228; fax (2) 247-5945; Chair: Dr Judy Messer; association of environment organizations; campaigns on forests, protected areas, land use, pollution, waste management, environmental research, urban planning, national parks, wildlife, energy, urban bushland and environmental law.

North Queensland Conservation Council: POB 364, Townsville, Qld 4810; tel (77) 71-6226; fax (77) 21-1713; Co-ordinators: Jasper Taylor, Shelagh Curmi; conservation of the natural and human environments; bookshop; responsible for an area between the towns of Mackay and Innisfail and the border with the Northern Territory to the west.

Organic Growers' Association of Western Australia: POB 213, Wembley, Perth, WA; collection and dissemination of information on organic growing; demonstrations of practical alternatives to chemicals in agriculture; restoration of living soil.

Pelican Lagoon Research and Wildlife Centre: Penneshaw, Kangaroo Island, SA 5222; tel (8) 483-3174; fax (8) 483-1204; Dir: Mike McKelvey; Resident Biologist: Dr Peggy Rismiller; long-term environmental and biological field research into native Australian flora and fauna.

Project Jonah Australia: POB 234, Baulkham Hills, NSW 2153; tel (2) 899-2111; telex 26064; fax (2) 899-2151; Sec: Patricia Lawson; whale conservation; environmental education; lobbying government.

Queensland Conservation Council: POB 12046, Elizabeth St, Brisbane, Qld 4002; tel (7) 221-0188; fax (7) 229-7992; Co-ordinator: Nicky Hungerford; co-ordination of the activities of interest groups; liaison with government and media on behalf of interest groups; environmental information centre.

Rainforest Conservation Society, Inc (RCS): 19 Colorado Ave, Bardon, Qld 4065; tel (7) 368-1318; fax (7) 368-3938; Pres: Dr Aila Keto; Dir: Dr Keith Scott; rainforest conservation; ecologically sustainable management of native forests; energy issues; identification and management of world heritage sites.

Rainforest Information Centre: POB 368, Lismore, NSW; 7 Wotherspoon St, Lismore, NSW 2480; tel (66) 21-8505; fax (66) 22-2339; Principal Officers: John Seed, John Revington, Brenda Crosby; concerned with global rainforest protection and reforestation; information on rainforests; Goodwood Project; tropical timber imports campaign.

Royal Society for the Prevention of Cruelty to Animals, New South Wales: 201 Rookwood Rd, POB 34, Yagoona, Sydney, NSW 2199; tel (2) 709-5433; fax (2) 796-2258; Acting Chief Exec Officer: Raeleen Redford.

Royal Zoological Society of South Australia: Adelaide Zoological Gardens, Frome Rd, Adelaide, SA 5000; tel (8) 267-3255; fax (8) 239-0637; Principal Officer: David J Langdon; promotion of awareness and understanding of animal life; preservation and propagation of rare and endangered species.

Soil and Water Conservation Association of Australia (SAWCAA): POB W21, West Pennant Hills, NSW 2125; tel (2) 484-3703; fax (2) 980-6728; Pres: R A Farley; Exec Officer: R W Roberts; promotion of sustainable land use, in order to protect soil and water resources for present and future generations; education and information; research; state branches; federal council; publication of quarterly journal.

Tasmanian Conservation Trust, Inc: 102 Bathurst St, Hobart, Tas 7004; tel (02) 34-3552; fax (02) 31-2491; Dir: Michael Lynch; conservation of the natural and cultural heritage of Tasmania.

Total Environment Centre (TEC): 18 Argyle St, Sydney, NSW 2000; tel (2) 247-4714; fax (2) 247-7118; Dir: Milo Kanangra Dunphy; non-profit community organization; environmental action, education, research and fund-raising.

Western Australia University, Department of Zoology: Nedlands, Perth, WA 6009; tel (9) 380-2237; Head of Dept: Assoc Prof W J Bailey; Chair of Zoology: Prof S D Bradshaw; undergraduate and postgraduate teaching; research.

Wild Life Preservation Society of Australia: POB 3248, Sydney, NSW 2001; tel (2) 043-4708; Pres: Vincent Serventy; sustainable development; conservation of biodiversity; international co-operation.

Wilderness Society: POB K29, Haymarket, Sydney, NSW 2000; tel (2) 267-7929; fax (2) 264-2673; Dir: Karenne Jurd; national, non-profit, community-based organization dedicated to the protection and conservation of wilderness areas in Australia, involved in local level activism, public education and lobbying of politicians and industry; current campaigns focus on forests, arid lands, Cape York (Queensland) and federal and state legislation affecting wilderness areas.

Tasmania Office: 130 Davey St, Hobart, Tas 7000; tel (02) 34-9799; fax (02) 23-5112.

Wildlife Preservation Society of Queensland, Inc: Level 4, 160 Edward St, Brisbane, Qld 4000; tel (7) 221-0194; fax (7) 221-0701; Pres: Dr Glen Ingram; Dir: Adrian Jeffreys; community-based volunteer organization committed to creating an ecologically sustainable future for people and wildlife through advocacy and action; environmental education; consultancy; campaigns for environmental legislation; publications.

Wildlife Survival: POB R1519, Royal Exchange, NSW 2000; tel (2) 247-6300; fax (2) 247-8778; Principal Officer: Pam Eiser; campaigns on kangaroo slaughter, endangered species habitat protection, Antarctic conservation and preservation and management of wildlife.

World Wide Fund for Nature (WWF) Australia: Level 10, 8-12 Bridge St, POB 528, Sydney, NSW 2001; tel (2) 247-6300; telex 24328; fax (2) 247-8778; Pres: Prof Alistair Gilmour; Vice-Pres: K McInnes; national affiliate organization of World Wide Fund for Nature (WWF) International; support for the World Conservation Strategy; conservation projects for endangered species and habitats; environmental education projects.

Melbourne Office: 20 Joseph St, Blackburn N, Melbourne, Vic 3130; tel (3) 899-0802; fax (3) 898-9878; Man: Lesley Ryall.

AUSTRALIAN EXTERNAL TERRITORIES

Norfolk Island

Australian Nature Conservation Agency, Norfolk Island Conservancy: POB 310, Norfolk Island 2899; tel 22695; fax 23397.

AUSTRIA

Governmental Organizations

Ministry of Environment, Youth and Family: 1031 Vienna, Radetzkystr 2; tel (1) 71-1-58; telex 3221371; fax (1) 711-58-42-21; Minister: Dr Ruth Feldgrill-Zankel.

Umweltbundesamt (Federal Environmental Agency): 1090 Vienna, Spittelauer Lände 5; tel (1) 31-30-40; fax (1) 31-30-44-00; Dir: Dr Wolgang Struwe; f 1985; specialist institution of the Austrian Minister for the Environment; responsible for environmental monitoring and elaboration of federal environmental policies.

Non-Governmental Organizations

Die Grüne Alternative (Grüne—Green Alternative): 1070 Vienna, Stiftgasse 6; tel (1) 52-125-0; fax (1) 52-125-40; Chairs: Franz Floss, Franz Renkin; campaigns for environmental protection, peace and social justice.

Global 2000: 1120 Vienna, Flurschützstr 13; tel (1) 81-25-73-00; fax (1) 81-25-72-8; Exec Dir: Alexander Egit; Media Spokesperson: Linda Führing; campaigning organization concerned with ecological reconstruction of Central and Eastern Europe, rainforests, nuclear power and ozone.

Nationalpark-Institut Donau-Auen (NPI—Danube-Auen National Park Institute): 1080 Vienna, Wickenburg 14/9; tel (1) 40-26-498; fax (1) 40-27-800; Principal Officer: Prof Dr Bernd Lötsch; promotion of a Danube National Park in accordance with IUCN regulations; research; fund-raising; public relations; audiovisual productions; visitor centres.

Österreicher Naturschutzbund (Austrian League for Nature Conservation): 5020 Salzburg, Bundesgeschäftsstelle, Haus der Natur, Arenbergstr 10; tel (662) 64-29-09; fax (662) 64-37-34; Pres: Prof Dr E Stüber; environmental conservation.

Österreichische Gesellschaft für Natur- und Umweltschutz (Austrian Society for the Protection of Nature and the Environment): 1010 Vienna, Hegelgasse 21; tel (1) 51-32-962; Principal Officer: Walter Scharf; lobbying government, aid agencies and industry; teacher training; publications; protected areas; education; conferences.

Salzburger Nationalparkfonds (NPF—Salzburg National Park Fund): 5741 Neukirchen, Markt 306; tel (6565) 6620; fax (6565) 6558-0; Principal Officer: Martin Neureiter; promotion of the National Park; management of protected areas; landscape protection; scientific research.

Umweltverband WWF Österreich (World Wide Fund for Nature (WWF) Austria): 1162 Vienna, Ottakringerstr 114-116/9, Postfach 1; tel (1) 409-16-41; telex 047114900; fax (1) 409-16-41-29; Principal Officers: Winfried Walter, Karl Wagner; national affiliate organization of World Wide Fund for Nature (WWF) International; nature conservation in Austria and abroad.

Vereinte Grüne Österreichs (United Green Party of Austria): 4020 Linz, Göthestr 9; tel (732) 66-83-91; fax (732) 50668; Chair: Josef Buchner; Gen Secs: Wolfgang Pelikan, Günter Ofner; ecologist party.

AZERBAIJAN

Governmental Organizations

State Committee for Ecology and Use of Nature: Baku; Chair of the State Committee for Ecology and Use of Nature: A E Mansurov.

Non-Governmental Organizations

Green Party of Azerbaijan: Baku.

THE BAHAMAS

Governmental Organizations

Ministry of Health, Department of Environmental Health: PO Bldg, E Hill St, POB N-3729, Nassau; tel 322-7425; telex 20516; fax 322-7788; Minister of Health and Environment: Ivy Leona Dumont.

Ministry of Agriculture, Trade and Industries: E Bay St, POB N-3028, Nassau; tel 323-1777; telex 38220264; fax 328-8212; Minister of Agriculture and Fisheries: Tennyson Roscoe Gabriel Wells.

Non-Governmental Organizations

Bahamas National Trust: POB N-4105, Nassau; tel 393-1317; fax 393-4978; Pres: Pericles Maillis; Deputy Pres: John F Bethell; non-profit organization dedicated to resource conservation, historic preservation and environmental education, particularly of schoolchildren; administration of national parks and protected areas.

Bahamas Red Cross Society: POB N-8331, John F Kennedy Drive, Nassau, NP; tel 323-7370; telex 20657; fax 323-7404; Pres: Ishmael Lightbourne; Dir-Gen: Marina Glinton; involved in emergency relief, training, social welfare and youth development.

BAHRAIN

Governmental Organizations

Ministry of Health: POB 12, Sheikh Sulman Rd, Manama; tel 250834; telex 8511; Minister: Jawad Salim al-Arrayedh.

Non-Governmental Organizations

Murshidat Al Bahrain (Girl Guides of Bahrain): POB 43, Manama; tel 259249; telex 9094; fax 271468; Chair: Fawzia Amin; Int Commr: Faika Ameen; education on family life, the environment and community service.

BANGLADESH

Governmental Organizations

Ministry of Environment and Forests: Bana Bhaban, Gulshan Rd, Mohaklahi, Dhaka 1212; tel (2) 605923; Minister of the Environment, Forests, Fisheries and Livestock: Abdullah al-Noman.

Ministry of Local Government, Rural Development and Co-operatives: Bangladesh Secr, Bhaban 7, 1st 9-Storey Bldg, 6th Floor, Dhaka; tel (2) 318682; Minister: Abdus Salam Talukder; includes Department of Environment Pollution Control.

Chief Conservator of Forests: Bana Bhaban, Gulshan Rd, Mohakhali, Dhaka 1212; tel (2) 605845.

Non-Governmental Organizations

Bangladesh Centre for Advanced Studies: 620 Road 10A, Dhanmondi, Dhaka; tel (2) 815829; telex 657634; fax (2) 811366; Exec Dir: Dr Saleemul Huq; research on the environment.

Bangladesh National Herbarium: Bangladesh Agricultural Research Council, New Airport Rd, Farm Gate, Dhaka 1215; tel (2) 311273; Dir: Dr M Matiur Rahman; Principal Scientific Officer: Manzur-ul-Kadir Mia.

Bangladesh Poush: 5-7 Sir Syed Rd, Block A, Mohammadpur, Dhaka 1207; tel (2) 812430; telex 642986; fax (2) 863060; Pres: Haroun er-Rashid; afforestation; health and sanitation; environmental education; disaster management; environmental conservation; preservation of biodiversity; fund-raising; research; publishing.

Comilla Proshika (Centre for Development): House 12, Rd 12, Dhanmondi R/A, Dhaka 1209; tel (2) 311462; telex 642940; Exec Dir: Rahat Uddin Ahmed; rural development programmes; sustainable development.

Institute for Environment and Development Studies (IEDS—FoE Bangladesh): 29 Purana Paltan, 3rd Floor, POB 3691, Dhaka 1000; tel (2) 231464; telex 632170; fax (2) 863797; Principal Officers: Chowdhury M Farouque, Amir Hussain Chowdhry, Mofizur Rahman; part of Friends of the Earth International Network; active through research, campaigns, demonstrations, training and publications; concerned with environmental issues affecting land, water and air; issues include the Himalayan environment, banning of trade in frog legs, organic farming, tropical rainforests, mangrove swamps, global warming and climate change, waterlogged land and air and water pollution.

IUCN Bangladesh: 76 Satmasjid Rd, Dhanmondi, Dhaka 1209; tel (2) 815061; fax (2) 813466; Nat Rep: A M Choudhury; Sr Investigator: Monoj K Roy; preparation with government of National Conservation Strategy; collaboration with other IUCN branches, NGOs and government in identification and development of projects for conservation and sustainable management of natural resources and biodiversity; implementation of projects; securing donor support; symposia and workshops to raise public awareness of environment and development issues.

Multidisciplinary Action Research Centre: House 12, Rd 12, Dhanmondi R/A, Dhaka 1209; tel (2) 311462; telex 642940; Man Dir: Monowar Hossain; sustainable development.

Wildlife Society of Bangladesh: c/o Zoology Dept, Dhaka Univ, Dhaka 2; tel (2) 506330; Pres: Prof K Z Husain; organization of wildlife biologists and other students of wildlife in Bangladesh; environmental education.

BARBADOS

Governmental Organizations

Ministry of the Environment, Housing and Lands: Sir Frank Walcott Bldg, Culloden Rd, St Michael; tel 431-7681; fax 437-8859; Minister: Harcourt Lewis; Perm Sec: Fozlo Brewster; environmental policy, research, education, awareness-raising and legislation.

Non-Governmental Organizations

Barbados Environmental Association: 210 Courts Plaza, St George St, POB 132, Bridgetown; tel 427-0619; fax 427-01619; Pres: Gordon Bispham; Vice-Pres: Leo Brewster; Sec: Stephen Boyce; awareness-raising on environmental issues and the need for conservation and management of natural resources; research on environmental issues.

BELARUS

Governmental Organizations

State Committee for Nature Protection: Minsk, vul Killektornaya 10.

State Committee for the Problems of the Consequences of the Disaster at the Chernobyl Nuclear Power Plant: 220010 Minsk, Government House; tel (0172) 29-64-60; telex 252457; fax (0172) 29-34-39; Chair: Ivan A Kenik; Deputy Chairs: Igor V Rolevich, Nikolay M Markovskyi; reduction of the impact of radiation on the country; resettlement of people away from particularly dangerous areas; provision of uncontaminated food and medical treatment; decontamination measures.

Non-Governmental Organizations

Belaja Rus Youth Ecological Movement: 220600 Minsk, K Marx St 40; tel (0172) 39-91-29; fax (0172) 34-53-50; Co-ordinator: Sergei Dorozhko; dissemination of environmental information; co-ordination of the activities of environmental organizations in Belarus; ecological camps and trips; seminars; conferences; cultural exchange.

Belarusian Green Movement: 220030 Minsk; Principal Officer: Sergei Vladimirovich Dorozhko.

Belarusian Scientific Research Institute of Land Reclamation and Meadow Management: 220040 Minsk, vul M Bogdanovich 153; tel (0172) 32-47-14; fax (0172) 32-64-96; Dir: Karlovsky Vladislav Filippovich; Deputy Dirs: Chernik Petr Konstantinovich, Meerovsky Anatoly Semenovich; reclamation of waterlogged soils and development and maintenance of reclaimed land.

Chernobyl Socio-Ecological Union: 220000 Minsk, Myasnikob Str 39; tel (0172) 20-39-04; fax (0172) 71-58-19; Pres: Vasily Yakavenka; Vice-Pres: Yuri Voronzhtsev; social welfare of victims of the Chernobyl disaster and cleaning up of the environment.

Polytechnic Academy, Ecology Department: 220027 Minsk, Skaryna Ave 65; tel (0172) 39-91-29; telex 252193; fax (0172) 32-83-16; Head: Dr Sergei Dorozhko; environmental education for engineers; ecological training programmes; scientific research; student exchange.

BELGIUM

Governmental Organizations

Ministry of Social Integration, Public Health and the Environment: Bâtiment Vesalius, ave Pacheco 19, 1010 Brussels; tel (2) 230-01-70; Minister: Laurette Onkelinckx.

Ministry of National Education, Administration of Community Affairs and State Scientific Establishments: 8 rue de la Science, 1040 Brussels; tel (2) 64-82-11; Dir: M J Decrock-Beyrus.

Ministry for the Walloon Region, Department of Regional Development and Housing, DOCATEL Service: 1 rue des Brigades d'Irlande, 5100 Jambes; tel (81) 33-24-72; fax (81) 33-21-10; Principal Officer: A-F Goffaux; documentation on regional development, town planning, architecture, cartography, building techniques and materials, housing, heritage and economic development.

Non-Governmental Organizations

Anders Gaan Leven (Agalev—Ecologist Party): 78 Tweekerkenstraat, 1040 Brussels; tel (2) 230-66-66; fax (2) 230-47-86; Political Sec: Johan Malcorps; Flemish ecologist political party with members in the Flemish and national parliaments.

Belgian National Union for Conservation of Nature: 84 Italiëlei, 2000 Antwerp; tel (3) 231-26-04; fax (3) 233-64-99; Sec-Gen: Julius-Anton Smeyers; federation of the main nature conservation NGOs in Belgium; nature conservation; environmental education.

Bond Beter Leefmilieu (BBL—League for a Better Environment): 26 Overwinningsstraat, 1060 Brussels; tel (2) 539-22-17; fax (2) 539-09-21; Pres: Luc Hens; Gen Sec: Lieve Grauls; federation of over 90 environmental organizations in Flemish-speaking Belgium; protection of the environment from pollution and insensitive development by encouraging co-operation between the federation and its member organizations, representing member organizations where necessary, acting as legal adviser to the environmental movement, political lobbying, influencing public opinion through the media, and collaborating with other NGOs in Europe and world-wide.

Centre Permanent d'Education à la Conservation de la Nature (Permanent Centre for Conservation Education): Parc de Mariemont, 29 rue du Parc, 6518 La Hestre; tel (64) 21-69-55; Dir: Jean-Pierre Cokelberghs; conservation education and strategies.

Ecolo (Ecologist Party): 28 rue Basse-Marcelle, 5000 Namur; tel (81) 22-78-71; fax (81) 28-06-03; Fed Secs: Daniel Burnotte, Gérard Lambert, Vincent Decroix, Anne-Marie Lateur, Roland Libois; French-speaking ecologist party.

Fonds Mondial pour la Nature (World Wide Fund for Nature WWF Belgium): 608 chaussée de Waterloo, 1060 Brussels; tel (2) 347-30-30; fax (2) 344-05-11; Pres: Christiane Linet Verlinden; Dir: Yves Boulpaep; national affiliate organization of World Wide Fund for Nature (WWF) International; conservation of fauna, flora, land, waters, soil and natural resources; fund-raising.

Inter-Environnement Wallonie (IEW): 26 rue de la Victoire, 1060 Brussels; tel (2) 539-09-78; fax (2) 539-09-21; Pres: Thérèse Snoy; Gen Sec: Jean-Luc Roland; independent, non-profit federation of over 100 environmental organizations in French-speaking Belgium; member of European Environmental Bureau.

Koninklijke Maatschappij voor Dierkunde van Antwerpen (Royal Zoological Society of Antwerp): 26 Koningin Astridplein, 2018 Antwerp; tel (3) 231-16-40; telex 32147; fax (3) 231-00-18; Dir: F J Daman; environmental education; research.

Ligue Royale Belge pour la Protection des Oiseaux (Royal Belgian League for the Protection of Birds): 43-45 rue de Veeweyde, 1070 Brussels; tel (2) 521-28-50; fax (2) 527-09-89; Chair: Roger Arnhem; Dir: Jan Rodts; Deputy Dir: Hugues Fanal; active in the protection of wild birds throughout Europe by opposing hunting, trapping and trade, and conservation of their natural habitats.

Ruralité-Environnement-Développement (RED—Rurality-Environment-Development): 2 rue des Potiers, 6717 Attert; tel (63) 22-37-02; fax (63) 21-98-70; Pres: Georges Christophe; Dir: Patrice Collignon; international conferences on rural development in Europe; research; publications; telematic service; rural architecture competition; organization of Centre Européen d'Intérêt Rural (European Rural Interests Centre), Strasbourg, France.

Stichting Leefmilieu (Environment Foundation): 11 Kipdorp, 2000 Antwerp; tel (3) 231-64-48; fax (3) 232-63-98; information; education; forum on environmental problems and solutions.

Université Libre de Bruxelles (ULB—Brussels Free University): 50 ave F D Roosevelt, 1050 Brussels; tel (2) 650-23-17; telex 23069; fax (2) 650-36-30; Principal Officer: Prof Jean-Paul Harroy; environmental research and education; ecology.

Université de Liège, Faculté des Sciences, Institut E van Beneden (Liège University, Sciences Faculty, E van Beneden Institute): 22 quai van Beneden, 4020 Liège; tel (41) 66-50-81; telex 41397; fax (41) 41-50-10; Principal Officer: Prof J C Ruwet; wildlife preservation; management of natural resources.

BELIZE

Governmental Organizations

Ministry of Tourism and the Environment: E Block, Belmopan; tel (8) 23393; fax (8) 23815; Minister: Glenn Godfrey; environmental impact assessments; environmental monitoring; public awareness-raising; data collection and distribution.

Non-Governmental Organizations

Belize Audubon Society: 49 Southern Foreshore, POB 1001, Belize City; tel (2) 77369; fax (2) 78562; Pres: Therese Rath; Vice-Pres: David Craig, Lombardo Riverol; management of reserves; environmental education; community information programme.

Belize Center for Environmental Studies (BCES): POB 666, 55 Eve St, Belize City; tel (2) 45545; fax (2) 32347; Man Dir: Lou Nicolait; Administrator: Frances Griffith; non-profit organization; integration of natural resource conservation and environmentally sound planning with sustainable economic development in Belize, through education, research, publication of findings, technical assistance and training, policy evaluation, environmental impact assessments, dissemination of information, networking and use of a resource centre.

Programme for Belize: POB 749, Belize City; tel (2) 74616; fax (2) 74635; Man Dir: Joy Grant; Technical Adviser: Roger Wilson; project of World Wide Land Conservation Trust; management of large areas of tropical forest; conservation and limited exploitation; co-operation with local people to save threatened habitats.

BENIN

Governmental Organizations

Ministry of the Environment, Housing and Town Planning: BP 01-3621, Cotonou; tel 32-21-00; Minister: Eustache Sarre.

Department of Land Management and Urban Development: BP 239, Cotonou; telex 5200.

Department of Water Resources, Forests and Hunting: BP 393, Cotonou; tel 33-06-62.

Non-Governmental Organizations

Friends of the Earth: BP 03-1162, Cotonou; tel 30-21-05; fax 30-22-05; Principal Officer: Venance Dassi; environment; development; community health.

International Youth Federation for Environmental Studies and Conservation (IYF)—Regional Office for West Africa: Club UNESCO Espoir, BP 04-0539, Cotonou; tel 30-11-72; Regional Co-ordinator: Antonin Moussouvikpo; world secretariat in India and regional offices in Costa Rica, the Netherlands and the Philippines.

BHUTAN

Governmental Organizations

Ministry of Agriculture and Forests: POB 130, Thimphu; tel 2504 (Dept of Agriculture), 2395 (Dept of Forestry); HM Rep in the Ministry of Agriculture: HRH Ashi (Princess) Sonam Chhoden Wangchuck.

Planning Commission: Thimphu; tel 2304; telex 204.

Non-Governmental Organizations

World Wide Fund for Nature (WWF) Bhutan/Nepal Office: POB 210, Thimphu; tel 23528; fax 23518; Bhutan and Nepal Country Rep: Mingma Norbu Sherpa.

BOLIVIA

Governmental Organizations

Secretaría Nacional del Medio Ambiente (SENMA—National Environmental Secretariat): Plaza España esq Mendez Arcos, Edif Zamorano, La Paz; tel (2) 36-9383; fax (2) 36-9304; Minister: Gerardo Aguirre Ulloa; environmental regulation, monitoring and management.

Ministry of Planning and Co-ordination, Science and Technology Department: Avda Arce 2147, La Paz; tel (2) 37-2060; telex 5321; Minister: Samuel Doria Medina.

National Office of Wildlife, National Parks, Hunting and Fishing, Forest Development Centre: Avda Camacho 1471, 6°, Casilla 1862, La Paz; tel (2) 37-1268.

Non-Governmental Organizations

Apoyo para el Campesino Indígena del Oriente Boliviano (APCOB—Support for the Native Farmer of Eastern Bolivia): Avda 26 de Febrero 1058, Casilla 4213, Villa Primero de Mayo s/n, Santa Cruz de la Sierra; tel (3) 33-6033; fax (3) 33-0708; Principal Officers: Jürgen Riester, Graciela Zolezzi; consolidation and protection of inter-municipal territory and sustained and integral management of natural resources; efficient development and management of forestry industry; support to native organizations and promotion of women's issues through development programmes.

Asociación Conservacionista de Torotoro (Torotoro Conservationist Association): Casilla 1749, Calle Mercado No 1328, Edif Mariscal Ballivián, 17°, Of 1706, La Paz; tel (2) 32-1619; Pres: Dr Rodolfo Becerra de la Roca; conservation, protection and management of speleological, archaeological and palaeontological sites, wild flora and fauna of the Andean valleys and natural resources of the Torotoro region.

Asociación Cristiana Femenina Potosí (ACF—Potosí Women's Christian Association): Calle Padilla 45, Casilla 473, Potosí; tel (62) 26921; fax (62) 22045; Pres: Dr Hilda R de Salazar; promotion and development of women's issues, environmental education, health and the environment and leadership.

Asociación Ecológica del Oriente (ASEO—Eastern Ecological Association): Casilla 4831, Ballivián 157, Santa Cruz de la Sierra; tel (3) 33-9252; fax (3) 33-9252; Pres: Ester Ballerstaedt; environmental education, monitoring and awareness-raising.

Bolivian Committee for IUCN: c/o Museo Nacional de Historia Natural, Casilla 8706, Calle 26, Cota Cota, La Paz; tel (2) 36-0583; fax (2) 36-5903; Deputy Dir: Alejandra Sánchez de Lozada.

Catholic Relief Services, Bolivia: Casilla 2561, Jacinto Benavente 2190, 2°, La Paz; tel (2) 32-3335; fax (2) 39-2228; Dir: Barbara Myers; Project Co-ordinator: Gonzalo Solares; financial and technical support for sustainable agriculture, environmental protection, preventive health care and microenterprise development.

Centro de Comunicación y Desarrollo Andino (CENDA—Centre for Communication and Andean Development): Casilla 3226, Adela Zamudio 1736, Cochabamba; tel (42) 43412; fax (42) 44131; Dir: Pablo Regalsky; education, research, communication and agroforestry programmes for Quechua communities.

Centro de Datos para la Conservación (CDC Bolivia—Conservation Data Centre): Calle 26 y Avda Muñoz Reyes s/n, Cota Cota, Casilla 11250, La Paz; tel (2) 79-7399; fax (2) 79-7399; Exec Dir: Maria R de Marconi; provision of scientific and technical bases for conservation; compilation, analysis and processing of information; technical assistance.

Centro de Investigación Agrícola Tropical (Centre for Tropical Agricultural Research): Casilla 247, Avda Ejercito 131, Santa Cruz de la Sierra; tel (3) 34-2996; fax (3) 34-2996; Exec Dir: César Samur; Dir of Research: José Abela; research into agroforestry and technology transfer, farm and community forestry and silviculture of native tree species.

Centro de Investigaciones y Estudio de la Capacidad de Uso Mayor de la Tierra (CUMAT—Research Centre for Optimum Land Use): Casilla 10387, Avda Busch 1897, Miraflores, La Paz; tel (2) 32-3419; fax (2) 37-5496; Exec Dir: Juan Carlos Quiroga M; concerned with environmental impact assessment, land use, deforestation, agricultural alternatives, image processing, watershed management and agricultural management projects; research; technical consultancy.

Centro de Promoción y Fomento al Cooperativismo (CEPROCOOP—Centre for Promotion of Co-operatives): Casilla 747, Calle Potosí 1814 y Murguía, Oruro; tel (52) 55256; Director: Fernando Sandalio Viñola; courses and seminars on co-operative training and education; advice to rural, urban and mining co-operatives; distribution and publication of information on co-operatives, environmental education, farming and Bolivian Altiplano affairs; Campito Emaus public library.

Centro de Servicios Múltiples de Tecnologías Apropiadas (SEMTA—Appropriate Technology Centre): Casilla 20410, Alfredo Ascarrunz 2675, La Paz; tel (2) 36-0042; telex 2220; fax (2) 39-1458; Exec Dir: Oscar Aguilar C; non-profit, independent institution working for sustainable development, particularly in rural areas of Bolivia, based on agroecology and self-sufficiency.

Fundación Amigos de la Naturaleza (FAN—Friends of Nature Foundation): POB 2241, Avda Irala 421, 41327 Santa Cruz; tel (33) 33806; fax (33) 41327; Exec Dir: Hermes Justiniano; private non-profit organization which aims to protect the biodiversity of Bolivia.

Liga de Defensa del Medio Ambiente (LIDEMA—League for the Defence of the Environment): Casilla 11237, Avda 20 de Octubre 1763, La Paz; tel (2) 35-3352; fax (2) 39-2321; Exec Dir: Dr Luis A Rodrigo Gainza; private non-profit organization which co-ordinates the activities of 16 Bolivian conservation organizations; development of national projects in education, basic and applied research, promotion and public relations.

Museo Nacional de Historia Natural (MNHN—National Museum of Natural History): Casilla 8706, La Paz; tel (2) 79-5364; telex 2526; fax (2) 39-1016; Deputy Dir: Alexandra Sanchez de Lozada; environmental research; conservation; education; publications.

POLEN: Tiquipaya-Apote, Casilla 1836, Cochabamba; tel (42) 88820; fax (42) 88381; consultancy in agroecology and rural development.

Protección del Medio Ambiente Tarija (PROMETA—Protection of the Tarija Environment): O'Connor 559, Casilla 59, Tarija; tel (66) 33873; fax (66) 45865; Principal Officers: Gabriel Baracatt S, Alfonso Blanco L; biodiversity conservation; research; environmental education, policy and lobbying.

BOSNIA AND HERZEGOVINA

Governmental Organizations

Ministry of Town Planning, Construction and Environmental Protection: 71000 Sarajevo, Vojvode Putnika 3; tel (71) 213777; fax (71) 653592; Minister: Mustafa Dizdarević.

BOTSWANA

Governmental Organizations

Ministry of Agriculture: Private Bag 003, Gaborone; tel 350581; telex 2543; fax 356027; Minister: Kebathlamang Pitseyosi Morake.

Ministry of Local Government, Lands and Housing: Private Bag 006, Gaborone; tel 352091; Minister: C J Butale.

Ministry of Mineral Resources and Water Affairs: Private Bag 0018, Gaborone; tel 352454; telex 2503; Minister: Archie M Mogwe.

Department of Wildlife and National Parks: POB 131, Gaborone; tel 51461; telex 2414.

Non-Governmental Organizations

Chobe Wildlife Trust: POB 43, Kasane; tel 650340; telex 2765; fax 650280; Chair: Lt Gen S K I Khama; Vice-Chair: J M Gibson; conservation in Northern Botswana; environmental education; assistance to Department of Wildlife; campaign against poaching; support for protection of the rhinoceros.

IUCN Botswana: Private Bag 00300, Plot 712, Ext 2, Gaborone; tel 371584; telex 2412; fax 371584; Projects Officer: Peter Hancock.

Kalahari Conservation Society: POB 859, Ground Floor, Botsalano House, The Mall, Gaborone; tel 374557; telex 2410; fax 374557; CEO: P Venson; Conservation Officer: I Pilane; Public Relations Officer: Esther van de Put; conservation of the environment; sustainable use of natural resources, particularly where this benefits local communities.

Francistown Branch: POB 40, Francistown; tel 214211.

Maun Branch: Private Bag 28, Maun; tel 660599.

Palapye Branch: Private Bag 7, Palapye; tel 420684.

Serowe Branch: POB 60, Serowe; tel 430420.

Tsabong Branch: POB 266, Tsabong.

BRAZIL

Governmental Organizations

Ministry of the Environment: SAIN, Av L4 N, Ed Sede Terrep, 70.800 Brasília, DF; tel (61) 226-8221; telex 2120; fax (61) 322-1058; Minister: Fernando Coutinho Jorge.

Ministry of Agriculture, Supplies and Land Reform, Brazilian Institute of Forest Development (IBDF), National Parks and Reserves Department: Palácio do Desenvolvimento, 12°, Setor de Autarquias N, Brasília, DF; tel (61) 225-3241; telex 611711.

Instituto Brasileiro de Meio Ambiente e dos Recursos Naturais Renováveis (IBAMA—Brazilian Institute of Environment and Renewable Natural Resources): SAIN, Avda L4 N, Lote 04, 70.800 Brasília, DF; tel (61) 226-8492; telex 611711; fax (61) 224-5206; Pres: Maria Tereza Jorge-Padua; management and conservation of renewable natural resources, particularly flora and fauna; administration of national parks and nature reserves.

Secretariat for the Environment, São Paulo: Rua Tabapua 81, 04.533 São Paulo, SP; tel (11) 883-0766; telex 1132621; fax (11) 280-5468; Environment Sec: Dr Alaor Caffé Alves; preservation of parks and protected areas; environmental education, planning, research and information; Botanic Institute; Forest Institute; Geological Institute.

Companhia Energética de São Paulo, Departamento de Meio Ambiente e Recursos Naturais (CESP—Energy Corporation of São Paulo, Environment and Natural Resources Department): Avda Paulista 2064, 8°, 01.310 São Paulo, SP; tel (11) 256-5064; telex 31004; fax (11) 255-3649; Man: Luiz Fernando Galli; environmental protection; environmental education; wildlife management.

Non-Governmental Organizations

Amparo ao Menor Carente (AMENCAR—Support for Impoverished Minors): CP 433, 93.001-970 São Leopoldo, RS; Rua André Ebling 234-B, Santo André, 93.042-320 São Leopoldo, RS; tel (51) 592-8111; fax (51) 592-8623; Pres: Elcias Alves de Mello; Exec Sec: Sigmar Reichel.

Associação Brasileira de Caça e Conservação (ABC—Brazilian Hunting and Conservation Association): Rua Mourato Coelho 1372, 05.417-002 São Paulo, SP; tel (11) 813-8238; fax (11) 813-8238; Pres: Maria Clara do A Cambrai; Exec Dir: Ricardo Freire; conservation of renewable natural resources; land restoration; resource management; environmental research; workshops and conferences.

Associação Brasileira de Ecologia (Brazilian Association of Ecology): Avda Atlantica 734, Apto 1201, Rio de Janeiro,

RJ; tel (21) 275-0111; Pres: A Monteiro da Silva; campaigns on ecology, quality of life, food and agriculture, traditional knowledge, wild flora and fauna and health; scientific research; botanical gardens.

Associação Catarinense de Preservação da Natureza (ACAPRENA—Santa Catarina Nature Preservation Association): Rua Antonio da Veiga 140, CP 1507, 89.010-971 Blumenau, SC; tel (473) 26-8288; telex 473302; fax (473) 22-8818; nature preservation association; environmental education and monitoring.

Associação de Defesa do Meio Ambiente—São Paulo (ADEMASP—Association for the Defence of the Environment—São Paulo): CP 832, 04.531 São Paulo, SP; tel (11) 280-7244; telex 33654; fax (11) 280-5468; Pres: Dr Paulo Nogueira-Neto; campaigns on wild fauna and flora, endangered animal and plant species and ecology; environmental and scientific research.

Associação dos Pescadores com Isca Artificial (APIA—Brazilian Game Fishing Association): Rua Berta 72, 04.120 São Paulo, SP; tel (11) 570-8088; Pres: Amedeo Giovanni Cataldi; campaigns on use of wildlife resources; environmental research; conservation education; workshops and conferences; international co-operation.

Associação de Preservação do Meio Ambiente do Alto Vale do Itajaí (APREMAVI—Alto Vale do Itajaí Environmental Preservation Association): CP 628, 89.160-000 Rio do Sul, SC; tel (478) 22-2497; fax (478) 22-2497; Principal Officers: Miriam Prochnow, Wigold Schäffer.

Brazilian Committee for IUCN: c/o Rua Conselheiro Carrao 640, Bela Vista, 01.328 São Paulo, SP; tel (11) 289-5031; telex 03833654; fax (11) 280-5468; Regional Councillor: Dr José Pedro de Oliveira Costa.

Centro de Cultura Luiz Freire (Luiz Freire Cultural Centre): Rua 27 de Janeiro 181, 53.020-020 Carmo, Olinda, PE; tel (81) 429-3444; fax (81) 429-4881; communication; human rights; environmental education; politics.

Centro de Estudos Alternativos e Vida Natural (CEAVN—Centre for Alternative Studies and Natural Living): Rua Jornalista Augusto Vaz Filho 1012, Farol, 57.057-150 Maceió, AL; tel (82) 241-2516; Principal Officers: Mariano Paredes y Correia, Rosália Ribeiro Paredes; alternative technology; ecology; cultural issues.

Centro de Pesquisas para Conservação das Aves Silvestres (CEMAVE—Research Centre for Conservation of Wild Birds): Parque Nacional de Brasília, Via Epia, SMU, 70.630 Brasília, DF; CP 04/034, 70.312-970 Brasília, DF; tel (61) 233-3251; fax (61) 233-5543; Head: João Luiz Xavier do Nascimento; Acting Head: Inês de Lima S do Nascimento.

Conselho Indigenista Missionário (CIMI Norte I—Missionary Council for Indigenous Peoples, North): CP 984, 69.011-970 Manaus, AM; Rua Tapajós 54, Centro, 69.025-140 Manaus, AM; tel (92) 233-5020; fax (92) 232-7347; Co-ordinator: Miguel Feeney; support for indigenous peoples of Amazonas and Roraima; land issues and self-sustainment.

Engenharia de Proteção Ambiental Ltda (EPA—Environmental Protection Engineering Ltd): Rua Dr Paulo Vieira 153, 01.257-000 São Paulo, SP; tel (11) 263-0555; fax (11) 62-0089; Principal Officers: Camal Abdon Salomão Rameh, Rita de Cássia Egydio Rameh; environment and pollution control consultancy.

Fundação Brasileira para a Conservação da Natureza (FBCN—Brazilian Foundation for Conservation of Nature): Rua Miranda Valverde 103, Botafogo, 22.281-000 Rio de Janeiro, RJ; tel (21) 266-5008; telex 2137984; fax (21) 537-1343; Pres: Jairo Costa; Exec Officer: Anthenor Navarro; research into Brazilian biodiversity; protection of wildlife habitats; environmental education programmes.

Fundação Gaia (Gaia Foundation): Rua Jacinto Gomes 39, 90.040-270 Porto Alegre, RS; tel (51) 330-3567; fax (51) 330-3567; Pres: José A Lutzenberger; promotion of ecological consciousness, sustainable development, regenerative agriculture and democracy; defence of the cultural identity of minorities; protection of ecosystems, flora and fauna.

Fundação Grupo Esquel (Esquel Group Foundation): SAS, Quadra 6, Bloco K, Edif Belvedere, 8°, Sala 801-A, 70.070 Brasília, DF; tel (61) 322-2062; fax (61) 322-1063; Pres: César Aguiar; Co-ordinator: Silvio Rocha Sant'Ana; organization working for sustainable development in Brazil.

Fundação Museu do Homem Americano (FUMDHAM—Museum of American Man Foundation): Abdias Neves 551, 64.770 São Raimundo Nonato, PI; tel (86) 582-1612; fax (86) 582-1656; Scientific Dir: Dr Anne-Marie Pessis; private scientific research institute which is responsible for the protection of the Serra da Capivara National Park.

Fundação Pró-Natureza (FUNATURA—Foundation for Nature): SCLN 107, Edif Gemini Center II, Bloco B, Salas 201/13, 70.743 Brasília, DF; tel (61) 274-5449; fax (61) 274-5324; Pres: Dr Paulo de Tarso Zuquim Antas; promotion of the role of the private sector in the protection of Brazilian biodiversity; environmental education.

Fundação SOS Mata Atlântica (Brazilian Foundation for the Conservation of the Atlantic Rainforest): Rua Manoel dâ Nóbrega 456, Paraíso, 04.001-001 São Paulo, SP; tel (11) 887-1195; fax (11) 885-1680; Pres: Roberto L Klabin; Gen Manager: Luiz Son.

Fundação Vitória Amazônica: Parque Adrianópolis, Quadra C, Casa 3, 69.060-370 Manaus, AM; tel (92) 642-1336; fax (92) 236-3257; Exec Dir: Carlos Miller; conservation project in Rio Negro basin; consolidation of Jaú National Park; environmental education of students in Manaus; socio-economic projects.

Gritos da Terra Entidade Ecológica (Gritos da Terra Ecological Group): Conj Alfredo G de Mendonça, Bloco 16, Apto 302, Jacarecica, 57.030-020 Maceió, AL; Pres: José Raimundo dos Santos; general and human ecology; environmental education; lectures.

Grupo de Trabalho Amazônico (GTA—Amazon Working Group): SRTVS Quadra 701, Conj L, Bloco 1, 38, Salas 709-711, 70.340-906 Brasília, DF; tel (61) 224-6735; fax (61) 322-3055; Principal Officer: Juan Carlos Carrasco Rueda; co-ordination of NGOs developing local-level, socio-environmental projects of relevance to the Amazon.

IBAMA/CNIA—Centro Nacional de Informação Ambiental (National Environmental Information Centre): SAIN Avda L4 N, Bloco C sub solo, 70.800-200 Brasília, DF; tel (61) 316-1205; telex 614304; fax (61) 224-5206; diffusion of information on the environment and agriculture.

Instituto Brasileiro de Análises Sociais e Econômicas (IBASE—Brazilian Institute for Social and Economic Analysis): Rua Vicente de Souza 29, Botafago, 22.251-070 Rio de Janeiro, RJ; tel (21) 286-6161; telex 36466; fax (21) 286-0541; Principal Officers: Herbert de Souza, Candido Grzybowski, Carlos Afonso.

Instituto para o Homem, Agricultura e Ecologia (IPHAE—Institute for Mankind, Agriculture and Ecology): CP 585, Rua Senador Alvaro Maia 1034, Olaria, 78.900-970 Porto Velho, RO; tel (69) 223-3945; fax (69) 221-8931; Exec Dir: Willem Pieter Groeneveld; Tech Dir: Thomas E Cochrane; Administrator: Lilian M Moser.

Instituto de Pesquisas Antropológicas do Rio de Janeiro (IPARJ—Rio de Janeiro Institute of Anthropological Research): Rua do Riachuelo 42, Salas 704-705, Centro, Lapa, Rio de Janeiro, RJ; tel (21) 242-6721; Pres: Mercio Gomes;

Research Dir: Darcy Ribeiro; consultancy for native Brazilian issues, the Amazon and Brazilian culture.

Instituto de Pesquisas Ecológicas (IPE—Institute of Ecological Research): Avda dos Operários 587, 13.416-460 Piracicaba, SP; tel (194) 38-7259; fax (194) 38-7259; Pres: Suzana Machado Padua; Vice-Pres: Claudio V Padua.

Instituto PNUMA Brasil (Brazil UNEP Institute): Avda Nilo Peçanha 50/1313, 20.044-900 Rio de Janeiro, RJ; tel (21) 220-1820; fax (21) 262-4233; Principal Officers: Haroldo Mattos de Lemos, Carlos Raja Gabaglia Penna; promotion of aims and programmes of UNEP; distribution of translations of UNEP literature; conferences on relevant issues.

Instituto Sul Mineiro de Estudos e Conservação da Natureza (South Minas Gerais Research and Nature Conservation Institute): Fazenda Lagoa, 37127-000 Monte Belo, MG; tel (35) 571-1500; fax (35) 571-1500; geographical and biological research; natural resources conservation; environmental education; international relations (forest conservation and sustainable development).

International Wildlife Coalition (Brazil): POB 5087, 88.040-970 Florianópolis, SC; tel (482) 82-1150; fax (482) 34-1580; Pres: José Truda Palazzo; Vice-Pres: Dr Maria do Carmo Both; research, education and consultancy in wildlife and nature protection in Brazil and abroad; projects on whales, environmental restoration and establishment and enforcement of wildlife sanctuaries.

Magüta: Centro de Documentação e Pesquisa do Alto Solimões (Magüta Documentation and Research Centre, Alto Solimões): Avda Castelo Branco 396, 69.630-000 Benjamin Constant, AM; tel (92) 415-5624; fax (92) 415-5624; Pres: Pedro Inácio Pinheiro; First Sec: Nino Fernandes; research and documentation centre for health, education, sustainable development and soil quality.

MARI, Grupo de Educação Indigena (MARI Indigenous Education Group): CP 8105, Departamento de Antropologia, Universidade de São Paulo, Cidade Universitária, 05.508-900 São Paulo, SP; tel (11) 211-2552; Gen Co-ordinator: Dr Aracy Lopes da Silva; Administrative Co-ordinator: Luis Donisete Benzi Grupioni; technical advisers to indigenous communities' education projects.

Movimento de Apôio à Resistência dos Waimiri Atroari (MAREWA—Support Movement for Waimiri Atroari Resistance): BR-174, Km 107, 69.735-000 Presidente Figueiredo, AM; tel (92) 324-1117; fax (92) 324-1117; Principal Officers: Doroti Alice Müller, Egídio Schwade; ethnohistorical archive; support for rights of indigenous peoples.

Movimento de Educação de Base (CNBB—Basic Education Movement): SCS, Quadra 3, Bloco A 79, Edif João Paulo II, 2°, 70.300-500 Brasília, DF; tel (61) 225-2999; fax (61) 225-2943.

Movimento dos Trabalhadores Rurais Sem Terra (MST—Landless Rural Workers Movement): rua Ministro Godoy 1484, Perdizes, 05.015-001 São Paulo, SP; tel (11) 864-8977; telex 82153; fax (11) 871-4612; rural workers' organization.

Movimento pela Vida (MOVIDA—Movement for Life): CP 230, Rua Sargento Jaime 370, Prado, 57.020-970 Maceió, AL; tel (82) 231-7459; fax (82) 223-8791; Pres: Anivaldo de Miranda Pinto; Vice-Pres: Arthur Frassy; opposition to chemical pollution; campaigns in defence of rainforests in Alagoas; environmental education.

Research Centre on Human Population and Wetlands in Brazil (NUPAUB): Rua do Anfiteatro 181, Colmeias, Favo 06, Cidade Universitária, Butantã, 05.508-90 São Paulo, SP; tel (11) 211-0011; fax (11) 813-5819; Principal Officers: Dr Antonio Carlos S Diegues, André Cotti Moreira, Lucila P Vianna.

Sociedade Brasileira de Paisagismo (Brazilian Landscape Association): CP 19-178, 04.599-970 São Paulo, SP; tel (11) 280-4279; fax (11) 280-4462; Principal Officers: Rodolfo Geiser, Paulo del Pichia; research, training courses and conferences in the fields of landscape and agronomy.

Sociedade de Defesa do Pantanal (SODEPAN—Society for the Defence of the Pantanal): Parque de Exposições Laucidio Coelho, Avda Américo Carlos da Costa 320, 79.080-170 Campo Grande, MS; tel (67) 742-1891; fax (67) 742-3388; Pres: Beatriz Diacopulos Rondon; Exec Dir: Joaquim Rocha Azevedo; non-profit organization; conservation of the environment of the Paraguay river basin and tributaries; protection of flora and fauna.

União Protetora do Ambiente Natural (UPAN—Union for Protection of the Natural Environment): CP 189, Lindolfo Collor 560, 93.001-970 São Leopoldo, RS; tel (51) 592-7933; fax (51) 592-6617; Exec Sec: Carlos Cardoso Aveline; Principal Officer: Aline dos Santos.

Universidade Estadual Paulista, Instituto de Biociencias, Departamento de Ecologia (São Paulo State University, Biosciences Institute, Ecology Department): Avda 24A 1515, CP 199, 13.506-900 Rio Claro, SP; tel (195) 34-0244; fax (195) 34-4433; education and research in pollution, biodiversity, ecosystems, restoration ecology, remote sensing and applied ecology.

World Wide Fund for Nature (WWF) Brazil: SHIS EQ QL 6/8, Conj E, 2°, 71.620-430 Brasília, DF; tel (61) 248-2899; fax (61) 248-7176; Principal Officer: Eduardo Martins.

BRUNEI

Governmental Organizations

Ministry of Development: Old Airport, Berakas, Bandar Seri Begawan; tel (2) 241911; telex 2722; Minister of Development: Pengiran Dato' Dr Haji Ismail; includes Directorate of Town and Country Planning.

Ministry of Culture, Youth and Sport, Museums Department: Kota Batu, Bandar Seri Begawan 2018; tel (2) 244545; fax (2) 242727; Dir: Haji Matussin Omar; Librarian: Fatimah Haji Aji; museum of history, culture and natural history.

BULGARIA

Governmental Organizations

Ministry of the Environment: 1000 Sofia, ul William Gladstone 67; tel (2) 87-61-51; telex 22145; fax (2) 52-16-34; Minister: Valentin Bossevski; environmental protection; nature conservation; environmental education.

Ministry of Agriculture: 1040 Sofia, Blvd Chr Botev 55; tel (2) 51-09-57; Minister: Dr Georgi Tanev; Deputy Minister: Dr Rumen Christov; geological mapping; biological agriculture.

Non-Governmental Organizations

Association of Bulgarian Ecologists (ABECOL): 1000 Sofia, Slaveykov Pl 7; 1231 Sofia, POB 52; tel (2) 37-23-50; fax (2) 80-36-33; Pres: Prof Dr Simeon Nedialkov; Vice-Pres: Prof Dr Nikola Nestorov; Scientific Sec: Assoc Prof Marin Mehandjiev.

Bulgarian Academy of Sciences, Forest Research Institute: 1756 Sofia, St Kliment Ohridski Blvd 132; tel (2) 62-29-61; fax (2) 62-29-65; Dir: Prof Petko Borissov Petkov; Deputy Dir: Prof Groud Dickov Popov; forestry; forest inventory; forest soil science and microbiology; ecology and environmental conservation; forest conservation; forest genetics and tree-breeding; tree physiology; wildlife management; forest harvesting.

Bulgarian Academy of Sciences, Institute of Botany: Sofia 1113, Acad Georgi Bonchev St, Bl 23; tel (2) 71-82-59; Dir: Prof Dr Emanuil Palamarev; Scientific Sec: Prof Dr Simeon Vanev.

Bulgarian Academy of Sciences, National Institute of Meteorology and Hydrology: 1784 Sofia, Tsarigradsko Chaussée 66; tel (2) 72-22-72; telex 22490; fax (2) 88-44-94; Dir: Dr Vassil Andreev; area of activity includes monitoring of air, soil and water pollution and radioactivity.

Bulgarian Society for the Conservation of the Rhodope Mountains: 1113 Sofia, Gagarin St 2; tel (2) 70-51-78; fax (2) 70-54-98; Chair: Assoc Prof Jordan Kirilov Danchev; conservation of the natural and cultural heritage of the Rhodope Mountains.

Field Office: 4850 Chepelare, Stojo Shishcov St 9; tel (3051) 26-28; fax (3051) 26-28.

Conservative Ecological Party: c/o 1000 Sofia, Rakovski Blvd 134; tel (2) 88-25-01; Chair: Khristo Bisserov; member of the Soyuz na Demokratichnite Sili (SDS—Union of Democratic Forces).

Ecoglasnost National Movement: 1000 Sofia, Kniaz Dondukov Blvd 39, 4th Floor, Rm 45; tel (2) 80-23-23; fax (2) 88-15-30; Pres: Edwin Sugarev; Sec: Lutchezarr Toshev; campaigns against nuclear power and pollution; promotion of human rights and democracy; protection of nature in the Black Sea and the River Danube; sustainable development; renewable energy; recycling; waste management.

National Ecological Club: 1000 Sofia, Acad George Bontchev Str, Bl 11, Rm 327, 4th Floor, POB 1653; tel (2) 713-25-23; fax (2) 70-50-24; Principal Officers: Prof V Sgurev, Prof G Bliznakov, Dr S Rakovski, Prof V Lichev, Dr D Panayotov, Dr E Russeva, Prof S Batov, R Kostov, T Georgieva; environmental conservation; co-operation with government agencies and NGOs; ecological education.

Smolyan Club: 4700 Smolyan, Rajkovo Dist, ZhK Izvor, etr B, Flat 4; tel (301) 228-80; Principal Officer: George Kazaliev.

United Front for Ecological Saving of Kardjali: 6600 Kardjali, 30 Republikanska Str; tel (361) 300-07; preservation of the life and health of the people and environment of Kardjali.

Wilderness Fund: 1113 Sofia, Gagarin St 2; tel (2) 71-91-38; Public Relations Officer: Lyubomira Mileva; non-profit organization dealing with conservation of nature areas and the self-recovery of nature.

BURKINA FASO

Governmental Organizations

Ministry of Environment and Tourism: BP 7044, Ouagadougou; tel 33-41-65; telex 5555; fax 30-77-64; Minister: Anatole G Tiendrébéogo.

Ministry of Agriculture and Animal Resources: BP 7005, Ouagadougou; tel 33-33-26; telex 5555; Minister: Jean-Paul Sawadogo; includes Plant Protection Service.

Non-Governmental Organizations

IUCN Burkina Faso: 01 BP 31 33, Ouagadougou 01; tel 30-12-25; telex 5371; fax 30-13-51; Rep: Remy Fontannaz.

BURUNDI

Governmental Organizations

Institut National pour l'Environnement et la Conservation de la Nature (INECN—National Institute for the Environment and Nature Conservation): BP 56, Gitega; tel (40) 2071; telex 3000; fax (40) 2075; responsibility includes management of national parks and nature reserves.

Ministry of Agriculture and Animal Husbandry: BP 631, Bujumbura; tel (2) 22087; Minister: Jumaine Hussein; includes Department of Water Resources and Forests.

CAMBODIA

Governmental Organizations

Ministry of Agriculture: Phnom-Penh; Minister: Nguon Nhel.

CAMEROON

Governmental Organizations

Ministry of Environment and Forests: Yaoundé; Minister: Dr Dji Ngoar Bava.

Ministry of Agriculture, Department of Water Resources and Forests: BP 194, Yaoundé; tel 22-10-46; telex 8325; Minister of Agriculture: Stephen Ndiyam.

Ministry of Tourism, Wildlife and Protected Areas Department: POB 266, Yaoundé; tel 22-44-11; telex 8318; Minister of Tourism: Pierre Souman; nature conservation; wildlife management.

Ecole pour la Formation de Spécialistes de Faune (School for the Training of Wildlife Specialists): BP 271, Garoua; tel 27-11-25; fax 27-20-49; training in wildlife management for French-speaking African states; affiliated to Ministry of Tourism.

Permanent Secretariat of the National Commission on Man and the Biosphere: BP 4742, Yaoundé; tel 22-05-62.

Wildlife and National Parks Department, General Delegation for Tourism: Yaoundé; tel 23-11-37; telex 8318.

Non-Governmental Organizations

IUCN Cameroon: Waza-Logone, c/o CEDC, BP 410, Maroua; tel 29-21-00; fax 29-33-56; Project Leader: Michael Allen.

World Wide Fund for Nature (WWF) Cameroon: BP 2417, Douala; tel 43-21-71; fax 43-21-71; Rep: Dr Steve J Gartlan.

CANADA

Governmental Organizations

Environment Canada: 10 Wellington St, 28th Floor, Hull, PQ K1A 0H3; tel (819) 997-1441; telex 0533608; fax (819) 953-3457; Minister of the Environment: P H Vincent; Deputy Minister: N Mulder.

Canadian Association on Water Quality (CAWQ): Technology Development Branch, 425 St Joseph Blvd, 4th Floor, Hull, PQ K1A 0H3; tel (819) 953-9365; fax (819) 953-9029; Pres: Dr J D Norman; Vice-Pres: M Lywood, Dr D V Ellis; Exec Dir: Dr H R Eisenhauer; Sec: Dr Y Comeau; Treas: S G Nutt; formerly Canadian Association on Water Pollution Research and Control; research on the scientific, technological, legal and administrative aspects of water quality and water pollution control; exchange of information; practical application of such research for public benefit.

Canadian Parks Service: Jules Leger Bldg, Ottawa, ON K1A 0H4; tel (819) 994-1871; telex 911308; fax (819) 953-4908; Asst Deputy Minister: Aimée Lefebvre-Anglin; protection of natural and cultural heritage, including 34 national parks and over 100 historic sites (among these 8 World Heritage Sites and 2 Biosphere Reserves); regional offices in Halifax, Quebec, Cornwall, Winnipeg and Calgary.

Canadian Wildlife Service (CWS): 17th Floor, Vincent Massey Sq, Ottawa, ON K1A 0H3; tel (819) 997-1095; telex 0533608; fax (819) 953-6283; Dir-Gen: D B Brackett; protection and management of migratory birds, nationally significant habitats and endangered species; biological research; studies on the socio-economic importance of wildlife.

Atlantic Region: 21 W Main St, POB 1590, Sackville, NB E0A 3C0; tel (506) 536-3025.

Ontario Region: 49 Camelot Drive, Nepean, ON K1A 0H3; tel (613) 952-2418.

Pacific and Yukon Region: 5421 Robertson Rd, POB 340, Delta, BC V4K 3Y3; tel (604) 946-8546.

Québec Region: 1141 route de l'Eglise, CP 10100, Sainte-Foy, PQ G1V 4H5; tel (418) 648-7225.

Western and Northern Region: Twin Atria Bldg 2, Suite 210, 4999 98th Ave, Edmonton, AB T6B 2X3; tel (403) 468-8903.

Conservation and Protection Service: 351 St Joseph Blvd, Hull, PQ K1A 0H3; Asst Deputy Minister: H A Clarke.

Ecosystem Sciences and Evaluation Directorate (ESED): Ottawa, ON K1A 0H3; Dir-Gen: D L Egar; promotion of sustainable management of water resources, in keeping with Federal Water Policy and national water objectives; research; data collection; protection of aquatic and marine resources; 2 research institutes for multidisciplinary water science, focusing on health and productivity of aquatic ecosystems; co-operation with provinces, industry and the public on river basin management and cleaning up existing pollution; administration of federal-provincial agreements; implementation of the Canada Water Act, the Boundary Water Treaty Act and the International River Improvements Act; environmental information; environmental assessment; prevention of and response to environmental emergencies.

National Hydrology Research Institute: 11 Innovation Blvd, Saskatoon, SK S7H 3H5; Dir: Dr T M Dick; research into climate and hydrology, land/atmospheric processes, the northern environment, nutrient impacts and ecosystem rehabilitation and groundwater and contaminants; interdisciplinary research with Saskatchewan University and other national and international groups; science liaison division disseminates scientific information to a national and international audience.

National Laboratory for Environmental Testing: Canada Centre for Inland Waters, 867 Lakeshore Rd, POB 5050, Burlington, ON L7R 4A6; Chief: R C J Sampson; analytical, scientific and advisory services; quality management programme; technology advancement and transfer, focusing on automation in sampling and analysis; liaison with academic and industrial laboratories.

National Water Research Institute: Canada Centre for Inland Waters, 867 Lakeshore Rd, POB 5050, Burlington, ON L7R 4A6; Exec Dir: Dr R J Daley; research and development in aquatic sciences; Lakes Research Branch; Rivers Research Branch; Research and Applications Branch; International Programs Group.

Department of Fisheries and Oceans: 200 Kent St, Ottawa, ON K1A 0E6; tel (613) 990-0279; fax (613) 954-0807; Minister for Fisheries and Oceans and Minister for the Atlantic Canada Opportunities Agency: John Crosbie; conservation, development and sustained economic use of fisheries resources.

Department of Natural Resources: 580 Booth St, Ottawa, ON K1A 0E4; tel (613) 995-3065; telex 0533117; fax (613) 996-9094; Minister: Barbara Jane Sparrow.

Alberta Environment: Oxbridge Pl, 9820 106th St, Edmonton, AB T5K 2J6; tel (403) 427-6267; fax (403) 422-3571; Minister of Environmental Protection and Deputy Govt House Leader: Brian Evans.

Atmospheric Environment Service (AES): 4905 Dufferin St, Downsview, ON M3H 5T4; tel (416) 739-4770; fax (416) 739-4232; Asst Deputy Minister: Dr D K Dawson; weather forecasting; meteorological research; data collection and archiving; policy development; climate and air quality services.

Information Resource Centre (Library): 4905 Dufferin St, Downsview, ON M3H 5T4; tel (416) 739-5702; fax (416) 739-4212; Head: Maria A Latyszewskyj; Reference Asst: Roberta McCarthy; intended primarily for AES research staff, but open for public reference use; collection available to users world-wide by inter-library loan; serial collection and 20,000 books covering meteorology, climate research, air quality, severe weather, Canadian meteorological data and world observational weather data; depository of AES and WMO publications and reports.

British Columbia Ministry of Environment, Lands and Parks: Parliament Bldgs, Victoria, BC V8V 1X5; tel (604) 387-5002; fax (604) 387-5669; Minister of Environment, Lands and Parks: John Cashore.

Canadian Council of Ministers of the Environment (CCME): 326 Broadway, Suite 400, Winnipeg, MB R3C 0S5; tel (204) 948-2090; fax (204) 948-2125; Principal Officers: Les Cooke, Norm Brandson, Nick Mulder, Joe Handley, Julian Walker, Dr Eva L J Rosinger; Canadian intergovernmental forum meeting twice a year for discussion and joint action on environmental issues of national and international concern; 13 environment ministers from federal, provincial and territorial governments; two steering committees advise the Council and co-ordinate projects assigned to intergovernmental task groups; projects cover issues including air quality, packaging, waste management and integration of environment and economy.

Canadian Museum of Nature: POB 3443, Station D, Ottawa, ON K1P 6P4; tel (613) 996-9281; fax (613) 955-3040; Dir: Dr Alan R Emery; natural science museum with departments of zoology, botany and mineralogy; education.

Department of Economic Development and Tourism/Provincial Parks of Prince Edward Island: POB 2000, Charlottetown, PE C1A 7N8; tel (902) 368-4250; fax (902) 368-5737; Deputy Minister: Stirling Breedon.

Manitoba Natural Resources: Legislative Bldg, Rm 314, Winnipeg, MB R3C 0V8; tel (204) 945-3785; fax (204) 945-3586; Minister of Natural Resources: Harry John Enns.

New Brunswick Department of the Environment: POB 6000, Fredericton, NB E3B 5H1; tel (506) 453-3700; fax (506) 453-3843; Minister of Environment: Jane Barry; Deputy Minister: Donald Dennison.

Newfoundland Department of Environment and Lands: Confederation Bldg, POB 8700, St John's, NF A1B 4J6; tel (709) 729-2574; fax (709) 729-1930; Minister of Environment and Lands and Minister responsible for Status of Women: Patricia Anne Cowan.

Northwest Territories Ministry of Renewable Resources: POB 1320, Yellowknife, NT X1A 2L9; fax (403) 873-0110; Minister of Renewable Resources and Minister of Municipal and Community Affairs: Titus Allooloo.

Nova Scotia Department of the Environment: POB 2107, Halifax, NS B3J 3B7; tel (902) 424-5300; fax (902) 424-0503; Minister of Environment: Robert S Harrison; Deputy Minister: Wayne Grady.

Nova Scotia Environmental Control Council: POB 2107, Halifax, NS B3J 3B7; tel (902) 424-6387; fax (902) 424-0503; CEO: Shirley L Nicholson; advisory body reporting to the Nova Scotia Minister of the Environment; public consultation review process.

Ontario Ministry of Natural Resources: Whitney Block, Rm 6521, Queen's Park, 99 Wellesley St West, Toronto, ON M7A 1W3; tel (416) 314-1990; fax (416) 314-1944; Asst Deputy Minister: David Balsillie; provision for resource development and outdoor recreation facilities; administration and protection of public lands and waters.

Ontario Ministry of the Environment: 135 St Clair Ave West, Toronto, ON M4V 1P5; tel (416) 323-4321; fax (416) 323-4264; Minister of Environment and Minister responsible for Greater Toronto Area: Ruth Grier.

Prince Edward Island Department of Agriculture, Fisheries and Forestry: Jones Bldg, 5th Floor, 11 Kent St, Charlottetown, PE C1A 7N8; tel (902) 368-4880; fax (902) 368-4857; Minister of Agriculture, Fisheries and Forestry: Walter Bradley.

Prince Edward Island Department of Environmental Resources: Jones Bldg, 4th Floor, 11 Kent St, Charlottetown, PE C1A 7N8; tel (902) 368-5000; fax (902) 368-5830; Minister of Environmental Resources: Barry Hicken.

Québec Ministry of the Environment: 3900 rue Marly, Sainte-Foy, PQ G1X 4E4; tel (418) 643-6071; fax (418) 643-2268; Minister for Environment and Govt House Leader: Pierre Paradis; responsible for protection of wildlife, terrestrial and freshwater environments, protected areas, human health, endangered animal and plant species, land use and development planning, pollution control, quality of life and landscape protection; environmental research; education and training.

Saskatchewan Department of Environment and Public Safety: 3085 Albert St, Regina, SK S4S 0B1; tel (306) 787-6113; fax (306) 787-0197; Minister of Environment and Public Safety: Berny Wiens.

Yukon Territory Ministry of Renewable Resources: POB 2703, Whitehorse, YT Y1A 2C6; tel (403) 667-5811; fax (403) 667-3518; Deputy Govt Leader and Minister of Renewable Resources: Bill Brewster.

Non-Governmental Organizations

African Wildlife Husbandry Development Association: POB 528, Station R, Toronto, ON M4G 4E1; tel (416) 961-8001; fax (416) 922-0712; Projects Dir: Clark Lungren; promotion of rational use of wildlife and habitat resources to improve living standards and self-sufficiency of local people.

Alberta Environmental Network Society: 10511 Saskatchewan Drive, Edmonton, AB T6E 4S1; tel (403) 433-9302; Pres: Judy Huntley; Treas: Howard Samoil; support for networking, communication and resource-sharing between non-governmental environmental organizations.

Alberta University, Canadian Circumpolar Institute (CCI): Old St Stephen's College, 8820 112th St, 3rd Floor, North Wing, Edmonton, AB T6G 2E2; tel (403) 492-4512; fax (403) 492-1152; Dir: Dr Clifford G Hickey; Asst Dir (Programs): E L Maloney.

Alberta University, Environmental Research and Studies Centre: CW-401L Biological Sciences Bldg, Edmonton, AB T6G 2E9; tel (403) 492-6659; fax (403) 492-8160; Dir: Dr Ellie E Prepas; promotion of new environmental research; funding for researchers; facilitation of interdisciplinary collaboration on research and teaching of environmental issues, using current expertise and resources; raising the profile of environmental research and courses at the university and facilitating the exchange of environmental knowledge between the university and governments, the private sector and the general public.

Aquatic Ecosystem Health and Management Society: POB 85388, Brant Plaza Postal Outlet, Burlington, ON L7R 4K5; tel (416) 336-4867; Pres: Dr M Munawar; Sec: Dr P Ross; Treas: Dr D Malley; concerned with the health and recovery of marine, freshwater and estuarine ecosystems; scientific initiatives to protect the global aquatic environment from toxic contamination and nutrient enrichment; environmental information through public awareness programmes and education.

Banff Centre for Management, Resource and Environment Management Program: POB 1020, Banff, AB T0L 0C0; tel (403) 762-6137; telex 03826657; fax (403) 762-6422; Sr Program Man: Felicity N Edwards; resource management; environmental management and planning; environmental education.

BC Environmental Network: 1672 E 10th Ave, Vancouver, BC V5N 1X5; tel (604) 879-2279; fax (604) 879-2272; grouping of the majority of local environmental organizations in the province of British Columbia; controlled by 12-member elected executive board.

Calgary University, Arctic Institute of North America: Library Tower, 2500 University Drive, NW, Calgary, AB T2N 1N4; tel (403) 220-7515; fax (403) 282-4609; Exec Dir: Michael Robinson; Assoc Dir: Gerald Thompson; community-based economic planning and development; joint ventures with northern communities using participatory action research to preserve traditional knowledge, train communities and encourage self-reliance.

Calgary University, Faculty of Environmental Design: 2500 University Drive NW, Calgary, AB T2N 1N4; tel (403) 220-6601; fax (403) 284-4399; Dean: Dr Robert Page; Assoc Dean: Prof Philip Elder.

Calgary Zoological Society: Calgary Zoo, Botanical Garden and Prehistoric Park, POB 3036, Station B, Calgary, AB T2M 4R8; tel (403) 232-9300; fax (403) 237-7582; Dir: Peter Karsten; education; recreation; conservation; scientific study.

Canadian Arctic Resources Committee: 1 Nicholas St, Suite 401, Ottawa, ON K1N 7B7; tel (613) 236-7379; fax (613) 232-4665; Chair: Nigel Bankes; Research Dir: Glen Okrainetz; environmental campaign organization.

Canadian Association for Humane Trapping: 142 Davenport Rd, Suite 200, Toronto, ON M5R 1J2; tel (416) 324-8726; fax (416) 324-8912; Exec Dir: Jane Vinet; liaison with trappers, the fur industry, governments and animal welfare organizations.

Canadian Coalition for Nuclear Responsibility, Inc/Regroupement pour la Surveillance du Nucléaire: CP 236, Snowdon Station, Montréal, PQ H3X 3T4; tel (514) 987-8023; fax (514) 489-5118; Pres: Gordon Edwards; Sec and Treas: Marc Chénier; research and educational activities relating to all aspects of nuclear technology and the alternatives, including uranium mining, reactor safety, energy policy, weapons proliferation, radioactive waste management, medical and environmental effects of radiation and standard-setting.

Canadian Committee for IUCN: c/o Canadian Museum of Nature, POB 3443, Station D, Ottawa, ON K1P 6P4; tel (613) 996-9606; fax (613) 952-9693; Sec: Richard Pratt.

Canadian Environmental Law Association: 517 College St, Suite 401, Toronto, ON M6G 4A2; tel (416) 960-2284; fax (416) 960-9392; Exec Dir (acting): Michelle Swenarchuk; Pres: Graham Rempe; concerned with environmental law reform and litigation.

Canadian Environmental Network: POB 1289, Station B, Ottawa, ON K1P 5R3; tel (613) 563-2078; Exec Dir: Eva Schacherl.

Environment and Development Support Program (EDSP): POB 1289, Station B, 251 Laurier Avenue West, Suite 1004, Ottawa, ON K1P 5R3; tel (613) 563-2078; fax (613) 563-7236; initiated and funded by Canadian International Development Agency (CIDA).

Canadian Forestry Association—A Federation of Provincial Forestry Associations: 185 Somerset St West, Suite 203, Ottawa, ON K2P 0J2; tel (613) 232-1815; fax (613) 232-4210; Pres: John R Hreno; Exec Dir: Glen Blouin; wise use and sustainable development of forests and their resources; National Forest Week, wildfire prevention campaign featuring the character 'Smokey Bear', National Forest Strategy, forest education, National Forest Congress and first Canadian Urban Forests Conference.

Canadian Global Change Program: Royal Society of Canada, POB 9734, Ottawa, ON K1G 5J4; tel (613) 991-5639; fax (613) 991-6996; Chair, Bd of Dirs: Dr Hugh Morris; Dir: Dr Brian Bornhold; co-ordination of research planning activities; conferences and seminars; policy analysis related to global change issues.

Canadian Nature Federation (CNF): 520-521 Nicholas St, Ottawa, ON K1N 7B7; tel (613) 562-3447; fax (613) 562-3371; Exec Dir: Julie Gelfand; Marketing Dir: Lesley MacGregor; formerly Canadian Audubon Society; campaigns on parks, wildlife, endangered species and forest management; publication of magazine.

Canadian Parks and Wilderness Society (CPAWS): 160 Bloor St East, Suite 1150, Toronto, ON M4W 1B9; tel (416) 972-0868; fax (416) 928-4514; Trustee: Ted Mosquin; preservation of wild ecosystems in parks, wilderness and natural areas; preservation of biodiversity, habitats and species.

Canadian Solar Industries Association (CanSIA): 2415 Holly Lane, Suite 250, Ottawa, ON K1V 7P2; tel (613) 736-9077; fax (613) 736-8938; Dir of Admin: Suzan Tuke; dissemination of technical and commercial information relating to solar energy; representative of the interests of the solar industry to Canadian municipal, provincial and federal governments in order to ensure favourable legislative and economic conditions.

Canadian Wildlife Federation: 2740 Queensview Drive, Ottawa, ON K2B 1A2; tel (613) 721-2286; fax (613) 721-2902; Exec Vice-Pres: Colin Maxwell; Gen Man: Richard Leitch; conservation of fauna and wildlife and wise use of Canada's renewable resources; liaison with individuals, provincial governments, municipalities and businesses; funding of research on species at risk and wildlife conservation projects; promotion of environmental legislation.

Centre Canadien d'Etude et de Coopération Internationale (CECI—Canadian Centre for International Research and Co-operation): 180 Ste-Cathérine Est, Montréal, PQ H2X 1K9; tel (514) 875-9911; telex 05562484; fax (514) 875-6469; Head: Claude Perras; non-profit international development aid corporation which assigns voluntary workers to developing countries; campaigns on afforestation, wildlife, land restoration, soil conservation, resource protection and forest conservation.

Citizens for a Safe Environment: 765 Queen St East, Toronto, ON M4M 1H3; tel (416) 462-3860; fax (416) 392-0025; Pres: Paula Greaves; Sec: Bill Brown; Exec Dir: Aine Suttle; promotion of safe alternatives to mixed waste landfill and incineration, and to incineration of sewage sludge.

Clean Nova Scotia Foundation: 1675 Bedford Row, POB 2528, Station M, Halifax, NS B3J 3N5; tel (902) 420-3474; fax (902) 424-5334; Chair, Bd of Dirs: Alan Parish; Exec Dir: Martin Janowitz.

Conservation Council of New Brunswick: 180 St John St, Fredericton, NB E3B 4A9; tel (506) 458-8747; fax (506) 458-1047; Pres: Janice Harvey; Exec Dir: Merredith Brewer; Vice-Pres: Inka Milewski, Louise Boldon; Policy Dir: David Coon; promotion of ecological approaches to forestry, agriculture, fisheries, energy and economic development.

Conservation Council of Ontario: 489 College St, Suite 506, Toronto, ON M6G 1A5; tel (416) 969-9637; fax (416) 960-8053; Pres: Duncan MacDonald; Exec Dir: Chris Winter; association of major environmental organizations.

Dalhousie University, School for Resources and Environmental Studies (SRES): 1312 Robie St, Halifax, NS B3H 3E2; tel (902) 494-3632; telex 01921863; fax (902) 494-3728; Dir: Dr Gordon E Beanlands; Int Programme Co-ordinator: Dr Robert Clarke; interdisciplinary research and education in natural resource management and environmental problems.

Environment and Resource Management Project, Philippines: tel (902) 494-3632; telex 01921863; fax (902) 494-2859; Canadian Project Dir: Dr Shirley Conover; Philippines Project Dir: Dr Nicomedes Briones.

Ducks Unlimited Canada: 1190 Waverley St, Winnipeg, MB R3T 2E2; tel (204) 477-1760; fax (204) 452-7560; Pres: Claude H Wilson; Exec Vice-Pres: D Stewart Morrison; management of waterfowl breeding habitats in Canada; waterfowl and wetland research; public education.

EcoCity Society: 9523 Jasper Ave, Edmonton, AB T5H 3V2; POB 205, Edmonton, AB T5J 2J1; tel (403) 429-3659; fax (403) 497-7617; Principal Officer: Angela Bischoff; urban environmental research, advocacy and education.

Ecology Action Centre: 3115 Veith St, 3rd Floor, Halifax, NS B3K 3G9; tel (902) 454-7828; fax (902) 454-4766; Dir: Howard Epstein; Chair Bd: Anita Regan; environmental education and advocacy on issues relevant to Nova Scotia; reference library.

Environmental Law Centre (Alberta) Society: 10350 124 St, Suite 201, Edmonton, Alberta T5N 3V9; tel (403) 482-4891; fax (403) 488-6779; Exec Dir: Donna G Tingley; public information and assistance; legal research; monitoring and reform.

Federation of Alberta Naturalists: POB 1472, Edmonton, AB T5J 2N5; tel (403) 453-8639; fax (403) 453-8553; Pres: Barry Jenkins; Exec Dir: Glen Semenchuk.

Federation of Ontario Naturalists (FON): Conservation Centre, Moatfield Park, 355 Lesmill Rd, Don Mills, ON M3B 2W8; tel (416) 444-8419; Environment Dir: Marion Taylor; campaigns on freshwater ecology, national parks, wildlife, waterfowl, birds, environmental education, migratory species, wetland ecosystems, wild flora, endangered animal species, poaching and management of protected areas.

Fédération québécoise de la faune (Québec Wildlife Federation): 319 rue St-Zotique Est, Montréal, PQ H2S 1L5; tel (514) 271-2487; fax (514) 271-9262; Pres: André Pelletier; First Vice-Pres: Bernard Arsenault.

Forest Management and Policy: School of Forestry, Lakehead University, 955 Oliver Rd, Thunder Bay, ON P7B 5E1; tel (807) 343-8508; telex 073-4594; fax (807) 343-8116; Assoc Prof and Chairholder: Dr Peter N Duinker.

Friends of the Earth Canada: 251 Laurier Ave West, Suite 701, Ottawa, ON K1P 5J6; tel (613) 230-3352; fax (613) 232-4354; Exec Dir: Susan Tanner; Policy Dir: Jeremy Byatt.

Friends of the Oldman River: POB 1288, Rocky Mountain House, AB T0M 1T0; tel (403) 845-4667; fax (403) 845-5377; Pres: Cliff Wallis; Vice-Pres: Martha Kostuch; opposition to the Oldman Dam.

Gaia Group: 2258 Rae St, Regina, SK S4T 2E9; tel (306) 352-4804; Pres: Jim Elliott; promotion of energy efficiency, waste minimization, and protection and restoration of wildlife habitats.

Green Party of Canada/Canadian Greens: 831 Commercial Drive, Vancouver, BC V5L 3W6; tel (604) 254-8165; fax (604) 254-8166; Exec Sec: Steve Kirby; Leader: Chris Lea; environmentalist political organization.

Le Parti Vert du Canada: CP 262, Montréal, PQ H1S 2Z2; 545 rue Lemay, Montréal, PQ H1T 2L8; tel (514) 256-6870; fax (514) 843-3757; Principal Officer: Rolf Bramann.

Green World Center: POB 29, Philipsburg, PQ J0J 1N0; Principal Officer: Walter Miale; affiliated to Earth Island Institute, USA.

Inuit Tapiriiksat Kanatami (Inuit Brotherhood of Canada): 170 Laurier Ave West, Suite 510, Ottawa, ON K1P 5V5; tel (613) 238-8181; fax (613) 234-1991; Pres: Rosemarie Kuptana; Sec and Treas: Martha Flaherty; advocacy, research and consultation on issues affecting Canadian Inuit, including human rights, justice, autonomy, social policy, housing, environmental health and contaminants, environmental assessment and planning, traditional skills and wildlife management.

Manitoba Eco-Network, Inc: POB 3125, Winnipeg, MB R3C 4E6; tel (204) 956-1468; fax (204) 949-9052; Chair: Jack Dubois; Sec: Harry Messman.

Manitoba Naturalists Society: 128 James Ave, Rm 302, Winnipeg, MB R3B 0N8; tel (204) 943-9029; fax (204) 949-9052; Pres: George Holland; Exec Dir: Herta Gudauskas.

Metropolitan Toronto Zoological Society: POB 280, West Hill, ON M1E 4R5; tel (416) 392-5900; fax (416) 392-5934; Exec Dir: William A Rapley; environmental education; research into zoology, conservation and wildlife.

National Energy Conservation Association: POB 3214, Winnipeg, MB R3C 4E7; tel (204) 783-1273; fax (204) 774-6702; CEO, Sec and Treas: Laverne Dalgleish.

National Round Table on the Environment and the Economy: 1 Nicholas St, Suite 1500, Ottawa, ON K1N 7B7; tel (613) 992-7189; fax (613) 992-7385; Chair: Dr George Connell; Exec Dir: Ronald L Doering; independent body reporting to Prime Minister of Canada.

New Brunswick Environmental Network: RR 4, Sussex, NB E0E 1P0; tel (506) 433-6101; fax (506) 433-6101; Principal Officers: Marc Spence, Elizabeth Dacombe, Adrien Theriault.

Newfoundland and Labrador Environmental Network: POB 944, Corner Brook, NF A2H 6J2; tel (709) 634-2520; fax (709) 634-2520; Contact: Martin von Mirbach.

Ocean Voice International/L'Echo de l'Océan: 2883 Otterson Drive, Ottawa, ON K1V 7B2; tel (613) 990-2207; fax (613) 521-4205; Pres: Don E McAllister; Vice-Pres: Gary Spiller; conservation of diversity of marine life and ecosystems; promotion of sustainable marine resource management through research and education; projects in Atlantic, Indian and Pacific Oceans.

Ontario Environment Network (OEN): 69 Wyndham St North, Suite 200, Guelph, ON; tel (519) 837-2565; fax (519) 837-8113; Co-ordinator: Tom Klein Beernink.

Ontario Federation of Anglers and Hunters (OFAH): POB 2800, Peterborough, ON K9Y 8L5; tel (705) 748-6324; fax (705) 748-9577; Exec Vice-Pres: Richard G Morgan; co-ordination of conservation activities of angling and hunting associations; support for scientific research.

Paper and Paperboard Packaging Environmental Council (PPEC): 701 Evans Ave, Suite 400, Etobicoke, ON M9C 1A3; tel (416) 626-0350; fax (416) 626-7054; Exec Dir: John Mullinder.

Pitch-In Canada: 1676 Martin Drive, Suite 200, White Rock, BC V4A 6E7; tel (604) 538-0577; fax (604) 538-3497; Chair: Bette Ballhorn; Pres: Allard van Veen; non-profit organization encouraging volunteers from the general public to initiate recycling and composting programmes and to become involved in the clean-up of streams, countryside and urban areas; member of Clean World International.

Planetary Association for Clean Energy: 100 Bronson Ave, Suite 1000, Ottawa, ON K1R 6G8; tel (613) 236-6265; fax (615) 235-5876; Principal Officer: Dr Andrew Michrowski.

Prince Edward Island Environmental Network: 126 Richmond St, Charlottetown, PE C1A 1H9; tel (902) 566-4170; fax (902) 566-4037; Co-Chairs: Dr Irené Novaczek, Jeremy Stiles.

Recycling Council of Ontario: 489 College St, Suite 504, Toronto, ON M6G 1A5; tel (416) 960-1025; fax (416) 960-8053; Exec Dir: John Hanson; Chair: Glenda Gies; information on waste reduction; annual conference and awards; educational projects.

Saskatchewan Eco-Network: 219 22nd St East, Suite 103, Saskatoon, SK S7K 0G4; tel (306) 665-1915; fax (306) 665-2128; Chair: Larry Morris; Co-ordinator: Lynn Brown.

Saskatchewan Environmental Society: POB 1372, Saskatoon, SK S7K 3N9; tel (306) 665-1915; fax (306) 665-2128; Pres: Margaret Asmuss; Program Co-ordinator: Ann Coxworth; information, education and research on environmental issues; lectures for community groups; various environmental publications.

Services d'Information Vent du Nord, Inc (North Wind Information Services, Inc): CP 38, Succursale place du Parc, Montréal, PQ H2W 2M9; tel (514) 842-0552; fax (514) 287-7355; Pres and Editor-in-Chief: Carol Karamessines; publication of series of short, academic texts on issues related to hydroelectric development in north-eastern Canada.

Sierra Club Canada: 1 Nicholas St, Suite 420, Ottawa, ON K1N 7B7; tel (613) 233-1906; conservation organization based in the USA.

Society Promoting Environmental Conservation (SPEC): 2150 Maple St, Vancouver, BC V6J 3T3; tel (604) 732-7736; fax (604) 736-7115; Principal Officer: Dermot Foley;

information and practical demonstrations on energy saving strategies, organic gardening, composting, waste reduction and recycling; network facility for environmental organizations; proposal of alternatives to use of hazardous pesticides and herbicides; campaigns for protection of wetlands.

Toronto University Peace and Conflict Studies Program: University College, A102, 5 King's College Circle, Toronto, ON M5S 1A1; tel (416) 978-8148; fax (416) 978-8416; Co-ordinator: Prof Thomas Homer-Dixon; research projects on environmental change and acute conflict, and environmental scarcity, state capacity and civil violence.

Union Québécoise pour la Conservation de la Nature (UQCN—Quebec Union for Nature Conservation): 160 76e rue Est, 2e Etage, Charlesbourg, PQ G1H 7H6; tel (418) 628-9600; fax (418) 626-3050; Pres (acting): André Desrochers; Gen Man: Christian Simard; union of societies concerned with natural sciences and the environment; environmental education; promotion of rational use of resources.

Water Environment Association of Ontario: 63 Hollyberry Trail, North York, ON M2H 2N9; tel (416) 502-1440; fax (416) 502-1786; Pres: J Greenshields; Vice-Pres: B Evans; formerly Pollution Control Association of Ontario; member of Water Environment Federation.

Western Canada Pollution Control Association: POB 6168, Station A, Calgary, AB T2H 2L4; tel (403) 259-4041; fax (403) 258-1631; Manager: M M Janice Taylor; member association of Water Environment Federation, USA.

Western Canada Wilderness Committee: 20 Water St, Vancouver, BC V6B 1A4; tel (604) 683-8220; fax (604) 683-8229; Principal Officers: Adriane Carr, Paul George; public education; promotion of wilderness preservation.

Wood Bison Re-Establishment Foundation: 253-375 York Ave, Winnipeg, MB R3C 3J3; tel (204) 943-9581; Chair: George Clavelle; re-establishment of the endangered wood bison herd in protected areas.

World Wildlife Fund (WWF) Canada: 90 Eglinton Ave, E, Suite 504, Toronto, ON M4P 2Z7; tel (416) 489-8800; fax (416) 489-3611; Chair: James W McCutcheon; Chief Exec and Pres: Monte Hummel; national affiliate organization of World Wide Fund for Nature (WWF) International; conservation of wild animals, plants and habitats; implementation of the World Conservation Strategy.

CAPE VERDE

Governmental Organizations

Ministry of Fisheries, Agriculture and Rural Activity, National Institute for Research and Agrarian Development (INIDA): CP 50, Praia, São Tiago; tel 61-15-70; fax 61-17-70; Minister: Dr António Gualberto do Rosário; Pres: Horácio Silva Soares; Dir Office for Studies and Project Development: Carlos Pinheiro Silva; research in the fields of agriculture, forestry, livestock and environmental sciences.

National Institute of Agrarian Research: Office of the Pres, CP 84, Praia, São Tiago; tel 61-15-70; telex 6072.

THE CENTRAL AFRICAN REPUBLIC

Governmental Organizations

Ministry of Water, Forests, Wildlife, Fisheries and Tourism, Directorate-General of Water Resources, Forests, Hunting and Fishing: BP 830, Bangui; tel 61-24-31; Minister: Raymond Mbitikon.

Ministry of Public Works, Directorate-General of Land Management, Environment Service: BP 129, Bangui; tel 61-28-00; telex 5267; Minister: Dieudonné Nana.

CHAD

Governmental Organizations

Ministry of Tourism and the Environment: N'Djamena; Minister: Pierre Tokinou.

Ministry of Mines, Energy and Water Resources: N'Djamena; tel 51-44-76; Minister: Hassan Djanbe.

Department of Forests and Protection of the Environment: BP 447, N'Djamena; tel 51-50-32; reafforestation; prevention of desertification; protection of plants and wildlife.

CHILE

Governmental Organizations

Ministry of Public Health, Department of Environmental Programmes: Enrique McIver 541, 1°, Santiago; tel (2) 39-4001; telex 240136; fax (2) 698-6622; Minister of Public Health: Julio Montt Momberg.

Ministry of National Properties, Ecological Consultancy: Avda B O'Higgins 280, Santiago; tel (2) 222-4669; fax (2) 222-0404; Minister of National Properties: Luis Alvarado.

Corporación Nacional Forestal (CONAF—National Forestry Association): Avda Bulnes 285, Of 501, Santiago; tel (2) 696-6749; telex 240001; fax (2) 671-5881; Exec Dir: Juan Moya Cerpa; Head, Natural Heritage Dept: Pedro Araya Rosas; conservation, management and optimum use of national forest resources.

Servicio Agrícola y Ganadero (Agriculture and Livestock Office): Avda Gen Bulnes 140, Casilla 4088, Santiago; tel (2) 71-2323; telex 240745.

Non-Governmental Organizations

Asociación para la Difusión de los Programas de las Naciones Unidas sobre Medio Ambiente (ADNUMA—Association for the Dissemination of United Nations Programmes on the Environment): Casilla 102, 2 Sur 1014, Of 7, Talca; tel (71) 22-7046; fax (71) 22-7046.

Centro de Estudios Agrarios (CEA—Centre for Agrarian Studies): Casilla 164, Valdivia; tel (63) 21-5846; fax (63) 21-3149; Dir: Patricia Möller; environmental education; promotion of sustainable development; environmental impact studies; wildlife management.

Centro de Investigación y Planificación del Medio Ambiente (CIPMA—Centre for Research on Environment and Planning): Holanda 1515, Providencia, Santiago; tel (2) 274-9600; fax (2) 223-2028; Pres: Guillermo Geisse G; Research Co-ordinator: Michael Nelson.

Red Regional de Información sobre Medio Ambiente en Chile (REDMA—Regional Network of Environmental Information in Chile): Dir: María Teresa Lladser.

Centro Latinoamericano de Demografía (CELADE—Latin American Demographic Centre): Edif Naciones Unidas, Avda Dag Hammarskjold, Casilla 91, Santiago; tel (2) 208-5051; fax (2) 208-0252; Dir: Reynaldo F Bajraj; Co-ordinator, Population and Development Section: Miguel Villa.

Colegio de Ingenieros Forestales (College of Forestry Experts): San Isidro 22, Of 503, Santiago; tel (2) 639-3289; fax (2) 638-5280; Pres: Nobil Kuncar A; First Vice-Pres: Jorge Gómez A; trade association.

Comisión Permanente del Pacifico Sur (CPPS—Permanent South Pacific Commission): Teatino 20, Of 93, Santiago; Casilla 16638, Agencia 6400-9, Santiago 9; tel (2) 672-6652; fax (2) 695-1100; Sec-Gen: Hugo Llanos; Deputy Sec-Gen for Scientific Affairs: Romulo Jordan.

Comité Nacional Pro-Defensa de la Fauna y Flora (CODEFF—National Committee for the Defence of Flora and Fauna): Casilla 3675, Antonia Lope de Bello 24, Bellavista, Santiago 1; tel (2) 737-7290; fax (2) 737-7290; Mem, Bd: Hernán Verscheure; environmental education, research and legal action; improvement of environmental legislation and enforcement.

Grupo de Investigaciones Agrarias (GIA—Agricultural Research Group): Universidad Academia de Humanismo Cristiano, Ricardo Matte Perez 0342, 6122 Santiago; tel (2) 225-5636; fax (2) 223-5249; Dir: Cecilia Leiva; research on rural development and training.

Juventudes para el Desarrollo y la Producción (JUNDEP—Young People for Development and Production): Fanor Velasco 27, Santiago; tel (2) 698-5728; fax (2) 698-0520; Pres: Jaime Valdes Castro; independent social development organization; programmes based on organization, education and technical development, aimed at small rural and urban communities.

Los Verdes (The Greens): Casilla 16879, Santiago; Pres: Andrés R Koryzma; environmentalist party.

Partners of the Americas—Chile, Environment and Emergencies Committee: Casilla 9286, Moneda 1467, Santiago; tel (2) 698-0400; fax (2) 698-0400; Pres: Augusta Crino; organization of community projects.

Red Ecológica Anti-Smog (Anti-Smog Ecological Network): Casilla 4012, Santiago 1; fax (2) 697-3937; Principal Officer: Juan Carlos Araya; public information and education on air pollution and its effects on humans.

Red Nacional de Acción Ecológica (RENACE—National Ecological Action Network): Seminario 774, Santiago; tel (2) 223-4483; fax (2) 223-4522; Pres: Luis Fernandez; Perm Secs: Victor Santibañez, Ines Valvekens; exchange of information and expertise; ecological development.

Silber Editores Ltda: Girardi 1325, Providencia, Santiago; tel (2) 204-3108; fax (2) 204-3108; Publr: Cecilia Aguayo Ipinza; publishers of directories.

Sociedad de Vida Silvestre de Chile (Chile Wildlife Society): Avda Alemania 0422, Casilla 1705, Temuco; tel (45) 21-0773; telex 267018; fax (45) 23-4126; Pres: Darcy Ríos Leal; environmental education; publication of various environmental bulletins; biannual International Congress on Natural Resources Management.

THE PEOPLE'S REPUBLIC OF CHINA

Governmental Organizations

National Environmental Protection Agency (NEPA): 115 Xizhimennei Nanx, Beijing; tel (1) 896673; telex 222359.

Non-Governmental Organizations

Beidaihe Bird Conservation Society (BBCS): Apt 1-1-302, Beijing Commission for Science and Technology, Balizhuang, Haidian District, Beijing 100037; tel (1) 8417773; Vice-Pres: Prof Weishu Hsu, Dr Martin Williams.

China Wildlife Conservation Association: Hepingli, Beijing 100714; tel (1) 4216343; telex 22237; fax (1) 4214717; Gen Dir: Qing Tianhua; resource protection; wildlife management; co-ordination of environmental trainers and administrators.

Chinese Academy of Sciences (CAS), Commission for Integrated Survey of Natural Resources (CISNAR): POB 767, Beijing 100101; tel (1) 4913580; telex 4844; fax (1) 4914230; Deputy Dir: Prof Li Wenhua; co-ordination of survey teams for rational use of natural resources; multidisciplinary research into natural resources.

Chinese Academy of Sciences (CAS), Institute of Botany: 141 Xizhimenwai St, Beijing 100044; tel (1) 8312840; fax (1) 8319534; Deputy Dir: Prof Wang Xianpu; multidisciplinary research organization; botanical garden; conservation of rare and endangered plant species; study of plant genetic resources.

Chinese Academy of Sciences (CAS), Institute of Hydrobiology, Department of River Dolphin Research: Luojiashan, Wuhan, Hubei 430072; tel (27) 723481; fax (27) 725132; Dir: Liu Renjun; Deputy Dir: Zhang Xianfeng.

Chinese Academy of Sciences (CAS), Research Centre for Eco-Environmental Sciences: POB 2871, Beijing 100085; tel (1) 2555019; fax (1) 2555381; formerly Institute of Environmental Chemistry.

Chinese Society for Environmental Sciences: 115 Xizhimennei Nanxiaojie, Beijing 100035; telex 222359; fax (1) 6015641; Pres: Qu Geping; promotion of environmental chemistry; exchange of scientific information; advice.

Sichuan Environmental Monitoring Centre (SEMC): 18 Section 4, Renmin Nanlu, Chengdu 610041, Sichuan Province; tel (28) 551286; telex 4148; Principal Officer: Wan Xianjin; technical and data centre.

World Wide Fund for Nature (WWF) China Programme: Co-ordinator: Stuart Parkins; see chapter on Hong Kong.

CHINA (TAIWAN)

Governmental Organizations

Environmental Protection Administration: Taipei; Dir-Gen: Chang Lung-Sheng.

Non-Governmental Organizations

Anti-Nuclear Coalition for Taiwan: POB 843, Tunghai University, Taichung 40704; tel (4) 359-5622; fax (4) 359-5622; Pres: Jun-Yi Lin; Vice-Pres: Pi-Yao Lin.

COLOMBIA

Governmental Organizations

Ministry of Agriculture, National Institute of Renewable Natural Resources and the Environment: Carrera 10, 20-30, Apdo 13458, Santa Fe de Bogotá, DC; tel (1) 284-1700; telex 44428; fax (1) 283-5834; Dir-Gen: Dr Manuel Rodriguez-Becerra; Asst Dir for Environment: Maria Teresa Szauer; research into conservation and use of renewable natural resources; environmental protection; management of parks and natural areas.

Comisión de Protección del Medio Ambiente (Environmental Protection Commission): Ministerio Público, Proviradúria Gen de la Nación, Santa Fe de Bogotá, DC.

Corporación Autónoma Regional del Cauca (CVC—Autonomous Regional Corporation of the Cauca): Apdo 2366, Cali; tel (23) 39-8949; telex 55708; fax (23) 30-4080; Deputy Dir: Juan Gabriel Casas Lozada; land reclamation; water resource management; sustainable resource use; conservation aid; land use.

Non-Governmental Organizations

Centro de Asistencia Legal Ambiental (CELA—Centre for Environmental Legal Assistance): Calle 2a Oeste, 2-46, Apdo 2741, Cali; tel (23) 82-1723; fax (23) 82-1723; Dir: José M Borrero; Sec: Martha C Velásquez; non-profit organization providing advice and assistance to NGOs, communities and local-level groups on environmental legal issues.

Centro Internacional de Agricultura Tropical (CIAT—International Centre for Tropical Agriculture): Apdo 6713, Cali; tel (23) 67-5050; telex 05769; fax (23) 64-7243; Dir-Gen: Gustavo A Nores; Deputy Dir-Gen for Resource Management Research: Filemón Torres; Deputy Dir-Gen for Germ-Plasm Development Research: William Scowcroft.

Comité Colombiano para la Información Ambiental (Colombian Committee for Environmental Information): Apdo 16914, Santa Fe de Bogotá, DC; tel (1) 283-1240; Sec: Alberto Donadio; Chair: Prof J M Idrobo.

Comité Colombiano de la UICN (Colombian Committee for IUCN): Carrera 12, 70-96, 3°, Santa Fe de Bogotá, DC; tel (1) 310-0026; fax (1) 210-4515; Co-ordinator: Dr Elsa Matilda Escobar.

Corporación Colombiana para la Amazonia 'Araracuara' (COA—Araracuara Colombian Association for the Amazon): Calle 20, 5-44, Apdo 034174, Santa Fé de Bogotá, DC; tel (1) 283-6755; fax (1) 286-2418; Man: Dr Dario Fajardo Montaña; Tech Dir: Dr Braulio Gutierrez; scientific research on biological, social and ecological aspects of the Amazon region of Colombia.

Corporación para la Investigación Ecológica, la Biotecnología y el Desarrollo Ambiental (ECOBIAM—Association for Ecological Research, Biotechnology and Environmental Development): Calle 57C, 45A-14, Of 201, Apdo 20991, Santa Fe de Bogotá, DC; tel (1) 222-3217; Dir: Edmundo Rodríguez; Advisor: Dario A Rodríguez.

Corporación Nacional de Investigación y Fomento Forestal (CONIF—National Forest Research and Promotion Association): Carrera 50, 27-70, Módulo 1, Of 604, Apdo 091676 ó 095153, Santa Fe de Bogotá, DC; tel (1) 221-3473; fax (1) 221-8624; Pres: Alberto Leguizamo Barbosa.

Fondo para la Protección del Medio Ambiente 'José Celestino Mutis' en Colombia (FEN Colombia—José Celestino Mutis Foundation for Protection of the Environment in Colombia): Calle 62, 4-68, 20°, Apdo 052936, Santa Fé de Bogotá, DC; tel (1) 248-1333; fax (1) 249-0437; Dir: Angel Guarnizo; Research Co-ordinator: Eduardo Guerrero; promotion of research into protection of natural resources and the environment.

Fundación Andina Bosques y Desarrollo (Andean Forests and Development Foundation): Calle 14, No 7-33, Of 403, Apdo 2672, Santa Fe de Bogotá, DC; tel (1) 285-8564; fax (1) 281-4912; Exec Dir: Edgar Fernando Cortes Saenz; advice and consultation on forest development projects; reports of experiments in environmental and forest resource management.

Fundación para la Comunicación Popular (FUNCOP-Cauca—Foundation for Popular Communication): Calle 5, No 8-77, Apdo 2096, Popayán, Cauca; tel (28) 24-4806; fax (28) 24-0409; Exec Dir: Laura Simmonds M; organization concerned with democracy, rural development, training of rural communities, health and ecology.

Fundación para la Defensa del Interés Público (FUNDEPUBLICO—Foundation for Defence of the Public Interest): Calle 71, 5-83, Santa Fe de Bogotá, DC; tel (1) 210-4685; fax (1) 210-4685; Pres: Dr German Sarmiento Palacio; promotion of environmental legislation.

Fundación Herencia Verde (FHV—Green Heritage Foundation): Calle 4a Oeste, 3 A-32, Apdo 32802, Cali; tel (23) 80-8484; fax (23) 81-3257; Principal Officer: Emilio Constantino; conservation of natural resources; environmental education; planning and management of nature reserves; sustainable development; conservation of biodiversity.

Fundación para un Mejor Ambiente (FMA—Foundation for a Better Environment): Calle 8a, 3-14, 1°, Apdo 1565, Cali; tel (23) 82-3271; telex 55724; fax (23) 82-4627; Exec Dir: Dr Humberto Swann Barona; non-profit organization for conservation of the environment.

Fundación Natura (FN—Nature Foundation): Carrera 12, 70-96, 3°, Santa Fe de Bogotá, DC; tel (1) 249-7590; Exec Dir: Dr Juan Pablo Ruiz; management of protected area projects; co-operation with the government and private sector; conservation of biological resources.

Fundación Puerto Rastrojo (FPR—Puerto Rastrojo Foundation): Apdo 241438, Santa Fe de Bogotá, DC; tel (1) 284-9010; Dir: Thomas Walschburger; conservation and sustainable management of resources; research.

Fundación pro-Sierra Nevada de Santa Marta (Foundation for the Sierra Nevada of Santa Marta): Apdo 5000, Santa Fe de Bogotá, DC; tel (1) 217-3487; fax (1) 218-3256; Exec Dir: Juan Mayr; conservation of the ecological and cultural heritage of the Sierra Nevada; research; sustainable development; co-operation with the government and the Indian and Colono communities.

Innovaciones y Redes para el Desarrollo, Oficina Regional América Latina (IRED-AL—Development Innovations and Networks, Latin America Regional Office): Calle 54, No 10-81, Of 501, Apdo Aéreo 38013, Santa Fe de Bogotá, DC; fax (1) 255-9581.

Popayán Office: Carrera 9, No 5-81, Apdo 2007, Popayán; tel (28) 24-2416; fax (28) 24-0409; Regional Dir: John Jairo Cardenas M; issues covered by programmes include sustainable development and disaster management.

Instituto de Estudios Ambientales (IDEAS—Environmental Research Institute): Universidad Nacional de Colombia, Carrera 50, 27-70, Edif Camilo Torres, Bl B-7, Apdo 14490, Santa Fe de Bogotá, DC; tel (1) 268-0569; fax (1) 222-5263; Principal Officers: Margarita Pacheco, Augusto Angel Maya.

Instituto de Investigaciones Ecológicas y Energéticas (ICECOL—Institute of Ecological and Energy Research): Apdo 050326, Barranquilla; tel (58) 31-8318; fax (58) 58-0083; Principal Officer: Hernando Jímenez Xiques.

Partners of the Americas, South-Western Colombia: Carrera 39, 46-54, Cali; tel (23) 52-4438; fax (23) 52-4439; Pres: Hernando Concha; Dir: Oscar Mejia; paired with South Carolina, USA.

Sociedad Colombiana de Ecología (Colombian Society of Ecology): Carrera 50, 27-70, Edif Camilo Torres, Bl C, Módulo 4, 5°, Apdo 8674, Santa Fe de Bogotá, DC; tel (1) 221-9701; fax (1) 221-6950; Pres: Dr J M Idrobo; promotion of conservation and rational use of natural resources; preservation of wildlife habitats and biodiversity.

Universidad Pontificia Bolivariana, Centro de Investigaciones para el Desarrollo Integral (Bolivian Pontifical University, Research Centre for Integrated Development): Apdo 1178, Medellín; fax (4) 411-2372; Head: Elena Agudelo M; research; consultancy.

THE COMOROS

Governmental Organizations

Ministry of Transport, Tourism and the Environment: Moroni; tel 2098; telex 244; Minister: Saïd el Had Mohamed.

Ministry of Production, Industry, Fishing, Crafts and Rural Development, Forest Development Department: BP 41, Moroni; tel 2292; telex 240; Minister: Ahmed Saïd Issilame.

THE CONGO

Governmental Organizations

Ministry of Tourism and the Environment, Department of the Environment: BP 958, Brazzaville; tel 83-30-46; telex 5282; fax 83-71-50; Gen Dir: Marius Issanga-Ngamissimi; nature protection, environment and development.

Conseil Supérieur de l'Environnement (Higher Council for the Environment): BP 2499, Brazzaville; tel 83-39-99; telex 5325.

Non-Governmental Organizations

Institut Français de Recherche Scientifique pour le Développement en Coopération (ORSTOM—French Scientific Research Institute for Co-operation in Development): Centre ORSTOM-DGRST, BP 181, Brazzaville; tel 83-26-80; fax 83-29-77; Dir: Claude Reichenfeld; fields of research include bioclimatology, hydrology, pedology, agricultural and medical entomology, nutrition, major endemic diseases, microbiology, phytopathology and social sciences.

COSTA RICA

Governmental Organizations

Ministry of the Environment: Avda 8-10, Calle 25, Apdo 10.104, 1000 San José; tel 57-1417; telex 2363; fax 57-0697; Minister: Hernán Bravo Trejos.

Ministry of Agriculture, Directorate-General of Fish Resources, Forestry and Wildlife: Apdo 10.094, 1000 San José; tel 32-4496; telex 3558; Minister: Juan Rafael Lizano Sáenz.

Oficina de Planificación Nacional y Politica Económica, Coordinación de Asuntos Ambientales (Office of National Planning and Political Economy, Co-ordination of Environmental Affairs): Apdo 10.127, San José; tel 23-7833; telex 2962.

Non-Governmental Organizations

Asociación ANAI (ANAI Association): Apdo 170, 2070 Sabanilla; tel 24-8815; fax 53-7524; Co-Dirs: William McLarney, James R Lynch, Robert B Mack; promotion of sustainable development amongst rural communities and local organizations in the Talamanca region.

Asociación Centro Ecológico La Pacifica (La Pacifica Ecological Centre): Apdo 1124, 1000 San José; tel 34-2274; fax 34-6837; Exec Dir: Lorena San Roman Johanning; conservation; research; education; encouragement of the development of ecological tourism.

Asociación para la Conservación de los Recursos Naturales de Costa Rica (ACORENA-CR—Costa Rica Association for Conservation of Natural Resources): 150m oeste Bomberos de Guápiles, Apdo 8-4330, San José 1000; tel 35-8552; telex 1018; fax 87-0545; Vice-Pres: Franklin Carmiol Umaña; private conservation group; environmental education in rural areas; protection of tropical forests.

Asociación Cultural SEJEKTO de Costa Rica (SEJEKTO Cultural Association of Costa Rica): Calle 7 y Avda 2, Apdo 906-2150, Moravia, San José; tel 36-1427; fax 40-2275; Pres: José Duálok Rojas; nature conservation; scientific research.

Asociación Demográfica Costarricense (Costa Rican Demographic Association): Apdo 10.203, 1000 San José; tel 31-4425; fax 31-4430; Exec Dir: Victor H Morgan Alvarado; private non-profit organization; information, research, education and training in natural resources and the environment.

Asociación Ecologista Costarricense (AECO—Costa Rican Environmental Association): Paseo Colón, de la Mercedes Benz 125m al norte y 75m al oeste, calle sin salida, penultima casa, Apdo 177, Curridabat, San José; tel 33-3013; fax 23-3925; Pres: Jorge Polimeni Salinas; Dir: Oscar Fallas Baldi.

Asociación Talamanqueña de Ecoturismo y Conservación (ATEC—Talamanca Association for Ecotourism and Conservation): Puerto Viejo de Talamanca, Limón; tel 58-3844; fax 53-7524; Principal Officers: Mauricio Salazar S, Mario Alfredo González Arrieta; non-profit organization dedicated to the promotion of ecologically sound tourism, environmental education, conservation and local enterprise.

Asociación de Voluntarios en Investigación y Desarrollo Ambiental (Association of Volunteers for Research and Environmental Development): Apdo 7.350, 1000 San José; Calle Central, Avdas 9 y 11, 928, San José; tel 21-1411; fax 21-1411; Pres: Frank Garita; Exec Dir: Carlos Barquero.

Biomass Users Network: 200 oeste y 50 sur de la Capilla del Colegio Saint Francis, Moravia Los Colegios, San José; tel 40-8997; fax 40-8998; Principal Officer: José M Blanco Rodríguez; organization of projects relating to agriculture and alternative energy with groups in rural areas; exchange of experience between communities in the region and throughout the developing world.

Alajuela Office: Apdo 960, 4050 Alajuela.

Caribbean Conservation Corporation: Apdo 246, 2050 San Pedro; tel 24-9215; fax 34-1061; Chair: Landon Clay; Exec Dir: David Carr; conservation of marine and related coastal resources; long-term study of marine turtles; organization with Wildlife Conservation International of Costa Rican component of project to preserve biological diversity in Central America.

Centro Agronómico Tropical de Investigación y Enseñanza (CATIE—Tropical Agricultural Research and Training Centre): Apdo 7170, Turrialba; tel 56-6431; telex 800; fax 56-1533; Dir-Gen: Dr Rubén Guevara Moncada; Dir, Integrated Natural Resources Management Programme: Dr Carlos J Rivas P; Dir, Sustainable Agriculture Programme: Dr Joseph Saunders.

Centro Científico Tropical (Tropical Science Centre): B La Granja, Apdo 8.3870, 1000 San José; tel 53-3267; fax 53-4963; Gen Man: Raúl Solórzano; Pres: Joseph A Tosi; research on tropical ecology; management of natural areas; environmental impact assessment; environmental economics; land-use planning.

Centro de Derecho Ambiental y de los Recursos Naturales (CEDARENA—Environmental and Natural Resources Law Centre): 75m suroeste del antiguo Higuerón, Apdo 134-2050, San Pedro; tel 53-7239; fax 25-5111; Pres: Rodrigo Barahona; Treas: Robert Wells; non-profit, independent association of lawyers, law students and volunteers concerned with environmental law, land tenure and the rights of indigenous peoples in Costa Rica; established permanent centre for research, information exchange and education.

Centro de Estudio Ambiental (Environmental Education Centre): San Francisco de Guadalupe, 295 Avda Calle 21-23, Apdo 347-2120, San Jose; tel 22-6608; fax 22-5052; Principal Officer: Pablo Calderón.

Centro Feminista de Información y Acción (CEFEMINA—Feminist Centre for Information and Action): Apdo 5.355, 1000 San José; tel 57-1756; fax 57-1758; in addition to issues specific to women, CEFEMINA works to improve education, housing, health care and political representation for low-income families and is involved in environmental issues.

Corporación de Investigaciones para el Desarrollo Socio Ambiental (CIDESA—Socio-Environmental Development Research Corporation): Apdo 414, 3000 Heredia; tel 33-1072; fax 23-1609; Vice-Pres: Florangel Villegas Verdu; environmental management and planning; scientific research; resource management; conservation education.

Fondo Mundial para la Naturaleza, Coordinación Regional para Centroamerica (WWF—World Wide Fund for Nature, Central America Regional Co-ordination): c/o Centro Agronómico Tropical de Investigación y Enseñanza (CATIE—Centre for Research and Training in Tropical Agriculture), Apdo 7170, Turrialba; tel 56-1383; telex 8005; fax 56-1421; Regional Co-ordinator: Miguel Cifuentes; Programme Officer: Oscar Brenes.

Fundación Ambio (Ambio Foundation): Apdo 1604, 1000 San José; tel 24-8782; fax 24-9169; Principal Officers: Roxana Salazar, Mario Carazo; resolution of environmental legal problems through analysis and conciliation.

Fundación Neotrópica (Neotropical Foundation): Apdo 236, 1002 San José; tel 33-6771; fax 33-0617; Acting Dir-Gen: Vera Varela; private foundation dedicated to conservation and sustainable development of biological resources; environmental education.

Fundación de Parques Nacionales (National Parks Foundation): c/o Universidad Latinoamericana de Ciencia y Tecnología, Apdo 10.235, 1000 San José; tel 55-1150; fax 22-4542; Pres: Dr Pedro León; protection and development of the National Park System and equivalent reserves; environmental education.

International Youth Federation for Environmental Studies and Conservation (IYF)—Regional Office for Latin America: Apdo 3.153, 1000 San José; tel 36-4274; fax 57-2273; Regional Co-ordinator: Mario León; world secretariat in India; regional offices in Benin, the Netherlands and the Philippines.

IUCN Costa Rica: Oficina Regional de América Central (ORCA—Central America Regional Office), Apdo 1161-2150, Moravia, San José; tel 36-2733; fax 35-6568; Regional Rep: Virgilio Cozzi.

Junta Nacional Forestal Campesina (JUNAFORCA—National Farmers' Forestry Assembly): Avda 1, Calles 1 y 3, Edif Cristal, 3°, Of 309, San José; Barrio México, Costado Norte de la Iglesia de Barrio México, Apdo 315-1005, San José; tel 22-5809; fax 22-5809; national and local meetings and workshops on themes related to forestry and sustainable development.

Monteverde Conservation League: Apdo 10.165, 1000 San José; tel 64-5003; fax 64-5104; Pres: Marvin Hidalgo Montero; Exec Dir: Omar Coto Loría; conservation of rainforests; reafforestation; environmental education.

CÔTE D'IVOIRE

Governmental Organizations

Ministry of the Environment, Construction and Town Planning, Posts and Telecommunications: ave Jean Paul II, BP V6, Abidjan 01; tel 29-13-67; telex 22108; Minister: Akèle Ezan.

Ministry of Agriculture and Animal Husbandry, Directorate-General of Water and Forests: BP V82, Abidjan; telex 23612; Minister: Lambert Kouassi Konan.

Ministry of Industry and Trade, Department of Environment, Normalization and Technology: BP V65, Abidjan; tel 33-13-71.

Non-Governmental Organizations

Parti pour la Protection de l'Environnement (PPE—Environmental Protection Party): Abidjan; Sec-Gen: Dioba Coulibaly.

World Wide Fund for Nature (WWF) Côte d'Ivoire: 01 BP 932, Abidjan 01; tel 37-04-65; fax 24-13-60; Rep: Dr Francis Lauginie.

CROATIA

Governmental Organizations

Ministry of Environmental Protection and Construction: 41000 Zagreb, Ave Vukovar 78; tel (41) 633444; fax (41) 612929; Minister: Zdenko Karakaš.

Institute for Nature Protection: 41000 Zagreb, Ilica 44/II; tel (41) 432022; fax (41) 431515; Dir: Miho Miljanić.

Non-Governmental Organizations

Hrvatsko Biolosko Drustvo (Croatian Biological Society): 41000 Zagreb, c/o Medical School, Zagreb University, Salata 3; tel (41) 271188; fax (41) 424001; Pres: Prof D Serman; environmental and conservation research; ecology; environmental impact assessment; workshops and conferences; environmental policy; conservation strategies; study of marine mammals.

Hrvatsko Udruženje Za Zaštitu Prirodnog i Kulturno-Povijesnog Nasljeda (Croation Society for the Conservation of the Natural and Cultural Heritage/Europe Conservation Croatia): 41000 Zagreb, Trg Kralja Tomislava 19; tel (41) 274809; fax (41) 427676; Pres: Dr Goran Sušić.

Institut za Medicinska Istrazivanja i Medicinu Rada (Institute for Medical Research and Occupational Health): 41001 Zagreb, Zagreb University, Ksaverska Cesta 2, POB 291; tel (41) 434537; fax (41) 274572; Dir: Dr L Štilinovič (acting); Head of Scientific Council: Dr E Reiner.

Javno Poduzeće Uprava Nacionalnog Parka Krka (Krka National Park Management): 59000 Šibenik, Kačića 9; tel (59) 27720; fax (59) 36836; Gen Man: Gorana Deljac-Stipaničev; Man, Protection Dept: Vinko Matošin; protection of natural and other resources of the park; education; promotion of natural and cultural heritage.

Javno Poduzeće Uprava Nacionalnog Parka Mljet (Mljet National Park Management): 50226 Govedjari, Pristanište 2; tel (50) 744054; fax (50) 744043; Dir: Ing Karlo Renko.

Javno Poduzeće Uprava Nacionalnog Parka Paklenica (Paklenica National Park Management): 57244 Starigrad Paklenica; tel (57) 79155; fax (57) 79002; Dir: Zoran Šikić; Head of Promotion: Josipa Marasović; protection, maintenence and promotion of Paklenica national park.

Savez Zelenih Hrvatske (Green Alliance of Croatia): 41000 Zagreb, Dordićeva 7; tel (41) 430119; fax (41) 430119; Chair: Dr Inge Perko Šeparović; waste minimization; sustainable energy policy; environmental education.

Zelena Akcija Split (Green Action Split): 58000 Split, Zrtava Fašizma 8; tel (58) 44421; Pres: Zoran Pokrovac.

Zelena Akcija Zagreb (Green Action Zagreb): 41000 Zagreb, Radnicka cesta 22, pp 876.

CUBA

Governmental Organizations

Comisión Nacional de Protección Medio Ambiente y Aprovechamiento de Recursos Naturales (National Commission for Environmental Protection and Development of Natural Resources): 17 5008 E/50 Y52, Miramar, Havana; tel (7) 22-3594; telex 0511290.

Ministry of Agriculture: Avda Independencia, entre Conill y Santa Ana, Havana; tel (7) 70-1434; telex 511966; Minister: Carlos Pérez León.

Ministry of the Fishing Industry: Barlovento, Santa Fé, Havana; tel (7) 22-7474; telex 511444; Minister: Capt Jorge A Fernández-Cuervo Vinent.

Centro de Información y Documentación Agropecuario (CIDA—Agricultural Information and Documentation Centre): Apdo 4149, Havana 4; tel (7) 81-8808; telex 1219; fax (7) 33-5080; Dir: Dr David Williams Cantero; information on farming and forestry.

Comité Estatal de Colaboración Económica (State Committee for Economic Co-operation): Calle 1a 201, Vedado, Havana; tel (7) 3-6661; telex 511297; Minister, State Committee for Economic Co-operation: Ernesto Meléndez Bach.

Non-Governmental Organizations

Instituto Cubano de Hidrografía (Cuban Institute of Hydrography): Calle 4, 304, Apdo 606, Marianao 13, Playa, Havana 11300; tel (7) 23-8316; telex 511421; fax (7) 33-2869; Principal Officer: Capt Pedro Perera Ruiz.

Laboratorio de Dinámica Costera de Holguín (Holguín Coastal Dynamics Laboratory): Apdo 246, Holguín 80100; tel (24) 46-2774; Dir: Cosme Casals Corella; multidisciplinary research relating to conservation and restoration of beaches.

CYPRUS

Governmental Organizations

Ministry of Agriculture and Natural Resources, Environmental Service: Loukis Akritas Ave, Nicosia; tel (2) 302883; telex 4660; fax (2) 445156; Dir: Nicos Georgiades; Environmental Officers: Antonis Antoniou, Costas Papastavrou, Costas Hadjipanayiotou; co-ordination of service for government environmental protection programmes; secretariat of Environmental Council and Environmental Committee of Cyprus; administrative section of Environmental Committee; head of technical committee on environmental impact assessment; advice; implementation of environmental policy.

Non-Governmental Organizations

Cyprus Conservation Foundation: POB 257, Limassol; tel (5) 369475; fax (5) 352657; Exec Dirs: Adrian Akers-Douglas, Costas Kleanthous, Dolores Savvides, Artemis Yiordamli; environmental education and lobbying; acquisition, management and protection of sites of natural or cultural importance; Environment Studies Centre.

Cyprus Ornithological Society/Birdlife International, Cyprus: 4 Kanaris St, Strovolos 154; tel (2) 420703; fax (2) 493689; Pres: Dr Kyriacos Pierides; Sec: P Neophytou.

Friends of the Akámas: POB 5204, Nicosia; tel (2) 314079; fax (2) 313053; Principal Officers: Yiannos Ioannoy, Niki Kypragora; campaigns to have the Akámas Peninsula in north-western Cyprus declared a national park; collaboration with Greenpeace on nature conservation, particularly the protection of turtles.

Laona Project: POB 257, Limassol; tel (5) 369475; telex 5516; fax (5) 352657; Dir: Artemis Yiordamli; Project Architect: Elena Kalliri; promotion of sustainable development in rural areas of north-western Cyprus, near the proposed Akámas national park.

'TURKISH REPUBLIC OF NORTHERN CYPRUS'

Ministry of the Interior, Rural Affairs and Environment: Lefkoşa (Nicosia), Mersin 10, Turkey; tel (520) 83344; Minister of the Interior, Rural Affairs and Environment: Günay Caymaz.

THE CZECH REPUBLIC

Governmental Organizations

Ministry of the Environment: Vršovická 65, 100 10 Prague 10; tel (2) 7121111; fax (2) 731357; Minister: František Benda.

Cesky Ústav Ochrany Prírody (CUOP—Czech Institute for Nature Conservation): Slezska 9, 120 29 Prague 2; tel (2) 2151111; fax (2) 254555; Dir: Dr Jaroslav Hromas; management and legislation for protected areas, protected buildings and monuments and archaeological sites; scientific research; nature conservation; environmental education; remote sensing.

Správa Krkonosského Národního Parku (Krkonose National Park Administration): 543 11 Vrchlabí-Zamek; tel (2) 7702110; Dir: Jan Stursa; conservation of the fauna, flora, soils and landscape of the National Park; monitoring of traffic and construction.

Správa Národního Parku A Chráněné Krajinné Oblasti Šumava (Šumava National Park and Landscape Protected Area Administration): Zamek, 385 01 Vimperk; tel (339) 21334; fax (339) 22947; Dir: Jiří Kec; nature conservation; administration on behalf of the state; forest management; environmental education, research and biomonitoring.

Ustredie Státnej Ochrany Prírody (Central Office for State Nature Conservation): Leninova 11, 031 01 Liptovsky Mikulás; tel (2) 24177; Dir: Anton Lucinkiewicz; conservation strategies; environmental management and planning; database; protection of endangered animal and plant species; management and planning of protected areas; environmental statistics; resource, impact and damage assessments; promotion of environmental law; resource management.

Non-Governmental Organizations

Centre for Environmental Scholarship: Univerzita Karlova, Petrská 3, 110 00 Prague 1; tel (2) 2315334; fax (2) 2315324; Dir: Dr Bedřich Moldan; Deputy Dir: Dr Ivan Rynda.

Česky Svaz Ochráncu Přírody (CSOP—Czech Union for Nature Conservation): u Michelského lesa 366, 140 00 Prague 4; tel (2) 2195; fax (2) 496619; Pres: Assoc Prof Dr Bedřich Moldan; First Vice-Pres: Dr Petr Slovak; CEO: Zuzana Maresova; campaigns on protected areas, national parks, nature reserves, endangered animal and plant species, wildlife habitats, resource protection, forest management, pollution, waste disposal, protected monuments, and enforcement and promotion of environmental law; environmental education and training.

Český Ústav Ochrany Přírody (Czech Institute for Nature Conservation): Slezská 9, 120 00 Prague 2; tel (2) 2151111; fax (2) 254555; Dir: Dr Jaroslav Hromas; Deputy Dir: Dr Bohumil Kučera; Vice-Pres IUCN: Dr Jan Čeřovský.

Czech Academy of Sciences, Institute of Landscape Ecology: Na Sádkách 7, 370 05 České Budějovice; tel (38) 40183; fax (38) 45719; Principal Officers: Prof Dr V Mejstřik, J Těšitel, Dr P Cudlin.

Czech Co-ordinating Committee for IUCN: c/o State Institute for Protection of Monuments and Nature Conservation, Valdstejn 1, 118 01 Prague 1; Principal Officer: Dr Jan Čeřovský.

Dřípatka Centre for Ecological Education: Rumpálova 402, 383 01 Prachatice; tel (338) 22806; Principal Officers: Helena Klimešová, Karel Tříska; programmes for children; workshops and courses on nature conservation for students and teachers.

Ecopoint Foundation Environment Liaison Centre: Slezská 9, 120 29 Prague 2; tel (2) 24172227; fax (2) 254555; Pres, Bd of Dirs: Dr Jan Čeřovský; Sr Exec Officer: Milan Biba; liaison between environmental NGOs, government and international organizations; seminars and training courses; active in Czech Republic and Slovakia.

Environmental Partnership for Central Europe: Brehova 3, 110 00 Prague 1; tel (2) 2324317; fax (2) 2324317; Country Dir: Ivan Indracek; see German Marshall Fund, USA.

Environment and Quality of Life in Central Europe: Problems of Transition—International Geographic Union Regional Conference (IGU RC) 1994: Conference Secretariat, Albertov 6, 128 43 Prague 2; tel (2) 24912060; fax (2) 24915817; Sec-Gen: Dr Tomáš Kučera; Chair of Organizing Cttee: Dr Jan Kára.

Green Party: nám Republiky 7, 111 49 Prague 1; Chair: Jan Jecmínek.

Hnuti Brontosaurus (Brontosaurus Movement): Bubenská 6, 170 00 Prague 7; tel (2) 802910; fax (2) 802906; Principal Officer: Helena Izbická; voluntary youth environmental movement in Bohemia and Moravia; environmental education and campaigning; conservation work.

DENMARK

Governmental Organizations

Ministry of the Environment: Højbro Plads 4, 1200 Copenhagen K; tel 33-92-76-00; telex 42230; fax 33-32-22-27; Minister: Svend Auken; Permanent Sec of State: Leo Bjørnskov; environmental and nature protection.

Geological Survey of Denmark.

National Agency of Environmental Protection: Strandgade 29, Copenhagen K.

National Environmental Research Institute (NERI): Frederiksborgvej 399, POB 358, 4000 Roskilde; tel 46-30-12-00; fax 46-30-11-14; Dir-Gen: Henrik Sandbech; Deputy Dir-Gen: John Tychsen.

NERI, Department of Freshwater Ecology and Department of Terrestrial Ecology: Vejlsøvej 25, POB 314, 8600 Silkeborg; tel 89-20-14-00; fax 89-20-14-14.

NERI, Department of Wildlife Ecology: Grenåvej 12, Kalø, 8410 Rønde; tel 89-20-14-00; fax 89-20-15-14.

National Forest and Nature Agency: Slotsmarken 13, 2970 Hoersholm; tel 45-76-53-76; fax 45-76-54-77; Head: Veit Koester.

Department of Meteorology and Wind Energy: Risø National Laboratory, POB 49, 4000 Roskilde; tel 46-77-50-12; telex 43116; fax 46-75-56-19; Head of Meteorology Section: Dr S E Larsen; activities include air pollution meteorology and wind resource assessment.

Naturbeskyttelsesraadet (Nature Preservation Council): Slotsmarken 13, 2970 Hoersholm; tel 45-76-53-76.

Non-Governmental Organizations

Biologforbundet (Biologists Association): Østergade 9, 8450 Hammel; tel 86-96-36-35; Editor: Jens Ole Ravn-Nielsen; national association for teachers, biologists and others interested in biology; campaigns for more teaching on ecological and environmental issues in Danish schools.

Danmarks Naturfredningsforening (Danish Society for the Conservation of Nature): Nørregade 2, 1165 Copenhagen K; tel 33-32-20-21; fax 33-32-22-02; Dir: David Rehling; protection of wild flora and fauna; resource protection; protection of endangered animal and plant species; conservation of freshwater environments, forests, terrestrial environments and protected areas.

Dansk Ornitologisk Forening (DOF—Danish Ornithological Society): Vesterbrogade 140, 1620 Copenhagen V; tel 31-31-44-04; fax 31-31-24-35; Conservation Officer: Knud Flensted; awareness-raising; bird and nature conservation.

Dansk Skovforening (Danish Forestry Society): Amalievej 20, 1875 Frederiksberg C; tel 31-24-42-66; fax 31-24-02-42; Dir: Jens Thomsen; objectives include conservation of the natural resources of Danish forests.

Dansk Vandteknisk Forening (Danish Water Supply Association): Vilh Becks Vej 60, 8260 Viby J; tel 86-11-23-33; fax 86-11-79-39; Dir: Anders Baekgaard; promotion of the common interests of Danish water suppliers; co-ordination and promotion of the interests of the members in international water-supply co-operation; support for research and development projects.

De Grønne (Green Environmentalists' Party): Nat Secr, Vester Skerning, Fyn; Int Secr, Slugten 10, 3300 Frederiksvaerk; tel 42-34-89-19.

Foreningen til Dyrenes Beskyttelse i Danmark (Danish Animal Welfare Society): Alhambravej 15, 1826 Frederiksberg C; tel 31-22-32-22; Exec Dir: Peter Mollerup; campaigns on animal rights, prevention of cruelty to animals, control of breeding and slaughter, enforcement of environmental law; environmental education.

Friluftsraadet (Open Air Council): Olof Palmes Gade 10, 2100 Copenhagen 0; tel 31-42-32-32; fax 31-42-32-32; Man Dir: Jan Eriksen; conservation strategies; development planning.

Greenpeace Danmark: Linnesgade 25, 1361 Copenhagen K; tel 33-93-53-44; fax 33-93-53-99.

NOAH Friends of the Earth Denmark: Nørrebrogade 39, 1, 2200 Copenhagen N; tel 35-36-12-12; fax 35-36-12-17; Nat Officers: Ingrid Hind, Gert Børsen; Int Officer: Mads Teisen; issues include traffic pollution, water resources, ozone depletion, global warming, food additives, sustainable agriculture and the UNCED process.

OOA Energibevægelsen (Movement on Energy and Environment): Ryesgade 19, 2200 Copenhagen N; tel 31-35-55-07; fax 31-35-45-65; information on energy and the environment and opposition to nuclear power.

Selskabet for Vand og Miljø (Society of Water and Environment): Laboratory for Technical Hygiene, Technical University of Denmark, Bldg 115, 2800 Lyngby; tel 45-93-39-08; fax 45-93-28-50; Chair: Mogens Henze.

Verdensnaturfonden (World Wide Fund for Nature WWF Denmark): Ryesgade 3F, 2200 Copenhagen N; tel 35-36-36-35; fax 31-39-20-62; Sec-Gen: Lene Witte; Head, Conservation Dept: Kim Carstensen; Head of Campaigns: Mik Steenberger; national affiliate organization of World Wide Fund for Nature (WWF) International; campaigns on tropical forests, wetland ecosystems, biodiversity, wildlife conservation, use of wildlife resources and protected areas; environmental education, training, management and planning.

Zoologisk Have (Zoological Gardens of Copenhagen): Sdr Fasanvej 79, 2000 Copenhagen F; tel 36-30-25-55; fax 36-44-24-55; Man Dir: Bent Jorgensen; non-commercial scientific and educational institution; conservation education; participation in international projects to save endangered species.

DANISH EXTERNAL TERRITORIES

The Faeroe Islands

Fiskivinnuumsitingin (Fisheries Administration): POB 87, 110 Tórshavn; tel 13068; telex 81235; fax 14942; Minister of Fisheries: John Petersen.

Ministry of Transport and Energy: Føroya Landsstýri, POB 64, 110 Tórshavn; tel 11080; telex 81310; fax 14942; Minister: Svend Aage Ellefsen.

Greenland

Ministry for Health and the Environment: c/o Statsministeriet, Grønlandsafdelingen (Prime Minister's Office, Greenland Section), Hausergade 3, 1128 Copenhagen K, Denmark; tel 33-92-33-00; telex 27125; fax 33-93-68-15; Minister: Ove Rosing Olsen.

Greenland Environmental Research Institute (GERI): Tagensvej 135, 4th Floor, 2200 Copenhagen N; tel 35-82-14-15; fax 35-82-14-20; Dir: Hanne Petersen; assesses environmental impact of exploitation of mineral resources and hydroelectric power in Greenland and participates in approval procedure of Mineral Resources Administration for Greenland.

DJIBOUTI

Governmental Organizations

Ministry of Agriculture and Nomadic Development: BP 453, Djibouti; tel 351297; telex 5871; Minister: Umar Chirdon Abbas.

DOMINICA

Governmental Organizations

Ministry of Agriculture: Government Headquarters, Kennedy Ave, Roseau; tel 4482401; fax 4487999; Minister: Maynard Joseph.

Forestry, Wildlife and National Parks Division: Botanical Gardens, Roseau; environmental advice to government; management and implementation of forestry, wildlife, national parks and environmental education programmes.

Non-Governmental Organizations

Cottage Forest Industries Project: Hertford, Gimmit, 10 Upper Lane, POB 436, Roseau; tel 4492131; fax 4487486; Principal Officers: Randau A LaRonde, Tighe Geoghegan, Yves Renard; timber extraction, rainforest management and reafforestation.

THE DOMINICAN REPUBLIC

Governmental Organizations

Comisión Ambiental (Environmental Commission): Calle Dr Delgado 164, Santo Domingo, DN; tel 689-3808; telex 3460461.

Departamento Planificación Ambiental (Environmental Planning Department): Santo Domingo, DN; tel 687-2508; telex 3460140.

Secretariat of State for Agriculture, Under-Secretariat of State for Natural Resources: Centro de los Héroes de Constanza, Santo Domingo, DN; tel 532-5935; telex 3460393; fax 535-6940; Sec of State: Nicolás Concepción García.

Non-Governmental Organizations

Fundación para el Mejoramiento Humano PROGRESSIO (PROGRESSIO Foundation for Human Improvement): Avda Máximo Gómez, Esq 27 de Febrero, Edif APAP, 3°, Santo Domingo, DN; tel 685-9562; telex 3460669; fax 686-9882; Pres: Enrique Armenteros Rius; Gen Man: Fernando Domínguez Aquino; reafforestation; management of protected areas; environmental education; conservation of endangered species.

Universidad Autónoma de Santo Domingo, Centro de Investigaciones de Biología Marina (CIBIMA—Marine Biology Research Centre): Facultad de Ciencias, Apdo 721-2, Santo Domingo, DN; tel 685-6682; fax 533-1106; Dir: Lic Venecia Alvarez; marine sciences and aquaculture.

ECUADOR

Governmental Organizations

Ministry of Agriculture and Livestock: Avda Eloy Alfaro y Amazones, Quito; tel (2) 548-708; telex 2291; Minister: Mariano González.

Ministry of Education and Culture, Programme for Education in Environment: Mejía 322, Quito; tel (2) 216-224; telex 1267; Minister: Dr Eduardo Peña Triviño.

Ministry of Energy and Mines, General Directorate for Environment (DIGEMA): Santa Prisca 223 y Manuel Larrea, Quito; tel (2) 570-141; telex 22271; Minister: Francisco Acosta Coloma.

Ministry of Foreign Affairs, Department of International Co-operation and Organizations, Environment Unit: Avda 10 de Agosto y Carrión, Quito; tel (2) 501-261; fax (2) 504-933; Minister: Dr Diego Parades; conservation of ecosystems; sustainable use of natural resources; exchange of information with government and non-government institutions in Ecuador and abroad.

Ministry of Industry, Trade, Integration and Fisheries: Roca 582 y Juan León Mera, Quito; tel (2) 527-988; telex 2166; Minister: Mauricio Pinto Mancheno.

Ministry of Public Health: Juan Larrea 444, Quito; tel (2) 529-163; telex 22677; fax (2) 569-786; Minister: Dr Leonardo Viteri Molinari.

Ministry of Public Works and Communications, Environment Unit: Avda 6 de Diciembre 1184, Quito; tel (2) 561-180; telex 2663; Minister: Pedro López Torres.

Dirección Nacional Forestal (National Institute of Forestry and Natural Areas): Avda Amazonas y Eloy Alfaro, Quito; tel (2) 548-924; telex 22291.

Instituto Ecuatoriano de Obras Sanitarias (IEOS—Ecuadorean Institute of Sanitation): Toledo 684 y Lérida, Quito; tel (2) 522-738; telex 22489.

Secretaría de Planificación, Consejo Nacional de Desarrollo, Departamento de Recursos Naturales (CONADE—Secretariat of Planning, National Development Council, Directorate of Natural Resources): Edif Consejo Provincial, Calle Manuel Larrea y Arenas, Quito; tel (2) 550-033; telex 308022114.

Non-Governmental Organizations

Catholic Relief Services (CRS) Ecuador and Colombia: Avda América 1805 y La Gasca, Casilla 226A, Quito; tel (2) 500-808; fax (2) 501-429; Programme Dir: Dieter Wittkowski; Agriculture Programme Manager: Hugo Ortiz; development of agriculture, health and microenterprise programmes.

Centro Andino de Acción Popular (CAAP—Andean Popular Action Centre): Martín de Utreras 733 y Selva Alegre, Casilla 17-15-173-B, Quito; tel (2) 522-763; fax (2) 568-452; Principal Officers: Francisco Rhon Dávila, Jaime Borja; promotion of alternative policies for national and rural development programmes; research into low input agriculture and natural resource management; provides support for local development projects and environmental policy implementation.

Centro de Educación y Promoción Popular (CEPP—Centre for Education and Popular Development): Sucursal 12 de octubre, Avda Atahualpa 427 y Hernando de la Cruz, Casilla 870, Quito; tel (2) 456-133; fax (2) 501-177; education, research and promotion of local development.

Centro de Investigaciones Sociales Alternativas (CISA—Alternative Social Research Centre): Vargas 760 y Matovelle, 3°, Casilla 17-15-167C, Quito; tel (2) 218-460; fax (2) 509-202; sustainable development programmes for Machalilla national park; national water collection system and other alternative technology; proposals for ecotourism in the Ecuadorean Amazon; self-management projects for indigenous communities.

Comité Ecológico ESPOL (ESPOL Ecological Committee): Escuela Superior Politécnica del Litoral (ESPOL), Malecón 100 y Loja, Casilla 09-04-15098, Guayaquil; tel (4) 563-042; telex 43509; fax (4) 563-042; Principal Officer: José A Delgado-Mendoza; research into petroleum in Amazonia, water quality, tropical forests and the urban environment; organization of Red de Acción e Información para la Defensa del Bosque Tropical (INFOTROP—Action and Information Network for the Defence of Tropical Forests).

Consejo Ecuatoriano para la Conservación e Investigación de las Aves (CECIA—Ecuadorean Council for Bird Research and Conservation): Avda Los Shyris 2030 y la Tierra, Apdo 17-17-906, Quito; tel (2) 244-734; fax (2) 244-734.

Consejo Latinoamericano de Iglesias (CLAI): Avda Patria 640 y Amazonas, Edif Banco International, 11°, Apdo 85-22, Quito; tel (2) 561-539; fax (2) 504-377; activities cover health, environment, indigenous communities, refugees, peace plans, liturgy, evangelism, women, children and youth issues; operation of an environmental programme.

Corporación de Defensa de la Vida (CORDAVI—Life Defence Association): Juan Montalvo 200 y Avda 6 de Diciembre, Edif Ponce Larrea, Of 001, Casilla 17-12-309, Quito; tel (2) 232-029; fax (2) 232-029; research and dissemination of information on environmental law; legal consultants to organizations, indigenous groups and rural communities.

Corporación de Estudios Regionales-Guayaquil (CER-G—Guayaquil Regional Studies Association): Victor Manuel Rendon 460 y Baquerizo Moreno, 1°, Casilla 9421, Guayaquil; tel (4) 307-388; telex 3697; fax (4) 563-477; Pres: Gaitan Villavicencio L; Exec Sec: Roberto Vernimmen B; social and economic projects; development of environmental and ecological research and training programmes.

Corporación Ornitológica del Ecuador (CECIA—Ornithological Corporation of Ecuador): Avda Los Shyris 20-30 y La Tierra, Casilla 17-17-906, Quito; tel (2) 468-876; telex 2849; fax (2) 468-876; Exec Dir: Juan Carlos Matheus; bird and bird habitat conservation; monitoring bird populations; research and information.

EcoCiencia, Fundación Ecuatoriana de Estudios Ecológicos (ECOC—EcoCiencia, Ecuadorean Foundation for Ecological Studies): Avda 12 de Octubre 959 y Roca, Edif Mariana de Jesús, Of 701, Apdo 17-12-257, Quito; tel (2) 526-802; fax (2) 502-409; Exec Dir: Danilo Silva Chiriboga; Pres: Patricio Mena; non-profit organization dedicated to conservation of wildlife and habitats; environmental education.

Estudios de Estructura y Administración del Estado (ESTADE—Studies of State Structure and Administration): Rusia y Shyris, Edif Shyris I, 4°, Of 401, Casilla 17-17-8, Quito; tel (2) 242-080; fax (2) 467-830; Exec Dir: Efraín Pérez; Assoc Dir: Rosa Elena de la Torre; provision of assistance to governments in implementation of public policies and conducts studies of legislation on the environment and national resources.

Fundación Antisana: Mariana de Jesús s/n y Carvajal, Casilla 17-03-1486, Quito; tel (2) 433-849; fax (2) 433-851; Pres: David Parra; Exec Dir: María-Helena Jervis; Tech Dir: Tarsicio Granizo; co-operation with communities on sustainable development projects in the protection zone of the Antisana Ecological Reserve; support for the Ecuadorean government in the management and protection of the Reserve.

Fundación Ayuda en Acción Ecuador (Actionaid Ecuador): Polonia 339 y Vancouver, Casilla 17-03-769, Quito; tel (2) 548-501; fax (2) 222-691; Exec Dir and Legal Rep: Fernando Páez Franco; aims include securing lasting improvements in the quality of life of Ecuador's poorest communities; activities include environmental preservation and promotion of sustainable development.

Fundación de Capacitación e Investigaciones para el Desarrollo Socio-Ambiental Ecuador (Fundación CIDESA—Training and Research Foundation for Socio-Environmental Development, Ecuador): República 2561 y Veracruz, Apdo 608, Sucursal 12 de Octubre, Quito; tel (2) 467-684; fax (2) 502-399; Exec Dir: Rodrigo Calero; development of marginal rural and urban communities through sustainable use of natural resources.

Fundación Charles Darwin para las Islas Galápagos (FChD—Charles Darwin Foundation for the Galapagos Islands): Avda Colón y 6 de Diciembre, 6°, Casilla 17-01-3891, Quito; tel (2) 244-803; fax (2) 443-935; Pres: Craig MacFarland; Sec-Gen: Alfredo Carrasco; international scientific body dedicated to conservation and research in the Galapagos Islands; management of the Charles Darwin Research Station.

Estación Científica Charles Darwin (ECChD—Charles Darwin Research Station): Puerto Ayora, Isla Santa Cruz, Galápagos; Dir: Chantal Blanton; operational arm of FChD; conservation of native Galápagos flora and fauna; scientific advisor to the Ecuadorean authorities on conservation matters in Galápagos.

Fundación Ecológica Arcoiris (Arcoiris Ecological Foundation): Casilla 11-01-860, Loja; tel (7) 572-926; fax (7) 572-926; Exec Dir: Fausto López Rodríguez; Vice-Pres: Rodrigo Tapia Castro; environmental education, scientific research and public action in defence of the forests of southern Ecuador.

Fundación Ecológica Podocarpus, Fundación para la Investigación y Conservación del Sur del Ecuador (Podocarpus Ecological Foundation for Research and Conservation in Southern Ecuador): Calle Sucre 08-43 y 10 de Agosto, Casilla 11-01-436, Loja; tel (4) 571-734; Dir: Gustavo Costa Jaramillo; activities focus on Podocarpus National Park in southern Ecuador.

Fundación Ecuatoriana para el Estudio de Mamíferos Marinos (FEMM—Ecuadorean Foundation for the Study of Marine Mammals): Vélez 911 y 6 de Marzo, Edif El Forum, 5°, Of 5-16, Casilla 09-01-6637, Guayaquil; tel (4) 524-608; Pres: Fernando Félix; Vice-Pres: Ben Haase; scientific research, education and conservation.

Fundación Maquipucuna (Maquipucuna Foundation): Baquerizo 238 y Tamayo, Casilla 17-12-167, Quito; tel (2) 507-200; fax (2) 507-201; Pres: Rodrigo Ontaneda; Exec Dir: Rebeca Justicia; implementation of conservation and management projects; research; sustainable development projects; environmental education; technical assistance.

Fundación Natura (Nature Foundation): Avda América 5653 y Voz Andes, Casilla 17-01-253, Quito; tel (2) 447-341; telex 030821211; fax (2) 434-449; Pres: Rodolfo Rendon; Chief Exec and Nat Exec Dir: Dr Roberto Troya Villacorta; national associate organization of World Wide Fund for Nature (WWF) International; promotion of environmental awareness and education; co-ordination of public and private organizations; campaigns on environmental policy, legislation and pesticides.

Fundación de Protección de los Recursos Naturales (FUNDEPRENA—Foundation for Protection of Natural Resources): CC Bauhauz, POB 01-09-884, Avda Fco de Orelana (frente a Dicentro), Planta Baja, Of 16, Guayaquil; tel (2) 294-114; fax (2) 292-448; protection of natural resources; reforestation; rural development in the Santa Elena peninsula, Guayas.

Instituto de Estrategías Agropecuarias (IDEA—Agricultural Policy Institute): Bossano 617 y Coronel C Guerrero, Casilla 17-12-853, Sucursal 12 de Octubre, Quito; tel (2) 245-344; fax (2) 449-473; Exec Dir: Neptalí Bonifaz; private non-profit foundation; forum for dialogue between public and private sectors on management of agriculture and natural resources.

Instituto de Estudios Ecuatorianos (IEE—Institute of Ecuadorean Studies): Carrión 555-B y Juan León Mera, Casilla 17-12-147, Quito; tel (2) 232-123; Principal Officers: Jorge

Trujillo, Hernán Ibarra; rural and urban research, assessment of development projects, environmental impact assessment and research, cultural analysis and studies of ethnic groups.

International Council for Bird Preservation (ICBP) Ecuador: Avda Los Shyris 2030 y la Tierra, Apdo 17-17-906, Quito; tel (2) 244-734; fax (2) 244-734.

IUCN Ecuador—Oficina Regional para América del Sur (South America Regional Office): Avda Atahualpa 955 y República, Edif Digicom, 4°, Casilla 17-17-626, Quito; tel (2) 430-974; fax (2) 430-974; Regional Rep: Luis Manuel Castello.

IUCN National Committee—Ecuador: Casilla 257, Sucursal 12 de Octubre, Quito; tel (2) 502-409; Sec: Danilo Silva.

MAP Internacional, Oficina Regional para América Latina (Regional Office for Latin America): Los Shyris 3517 y Tomás de Berlanga, Casilla 17-08-8184, Quito; tel (2) 458-263; fax (2) 435-500; Regional Dir: Gregory Rake; Christian, non-profit organization operating health and development programmes.

EGYPT

Governmental Organizations

Egyptian Environmental Affairs Agency (EEAA): Cabinet of Ministers, 11A Hassan Sabry St, Zamalek, Cairo; tel (2) 21328; telex 93794.

Ministry of Electricity and Energy: Cairo (Nasr City); tel (2) 829565; Minister: Muhammad Mahir Abazah.

Ministry of Irrigation: Sharia Qasr el-Eini, Cairo; tel (2) 3552120; Minister of Works and Water Resources: Isam Radi Abd al-Hamid Radi.

Ministry of Land Reclamation: Land Reclamation Bldg, Dokki, Giza; tel (2) 703011; Minister: Dr Yousuf Amin Wali.

Ministry of Petroleum and Mineral Resources: el-Mokhayem el-Dayem St, Cairo (Nasr City); tel (2) 2622237; telex 92197; fax (2) 2636060; Minister: Dr Hamdy al-Banbi.

Academy of Scientific Research and Technology (ASRT): Dept of Scientific Socs and Int Unions, 101 Kasr el-Aini St, Cairo; tel (2) 3541044; telex 93069; fax (2) 3562820; Vice-Pres: Prof Dr Hamdy A A Moursy; scientific research for industry and agriculture.

Egyptian Committee for IUCN: tel (2) 727022; Dir: Sayed A Kansouh.

Non-Governmental Organizations

Association for the Protection of the Environment (APE): POB 32, Qalá, Cairo; Chair Bd of Dirs: Yousriyia Loza; Vice-Chair: Dr George Stino; non-profit environment and development organization dependent on local-level participation; programmes cover issues such as solid waste management, recycling, employment, income generation, functional literacy, health, skill training and crisis management.

Green Party: c/o Majlis ash-Sha'ab (People's Assembly), Cairo; Chair: Hassan Rageb.

Tree Lovers' Association: POB 592, El-Maadi, 11728 Cairo; tel (2) 3530621; Principal Officer: Prof Dr Ibrahim Kelany; research, policy planning, information dissemination and education on trees and the environment; organization of meetings on environmental problems and ways in which NGOs can contribute to their solution.

EL SALVADOR

Governmental Organizations

Ministry of Agriculture and Livestock: OSPA 31 Avda Sur 627, San Salvador; tel 23-2598; telex 20228; Minister: Antonio Cabrales; incorporates Centre of Renewable Natural Resources.

Ministry of Public Health and Social Security: Calle Arce 827, San Salvador; tel 21-5008; Minister: Dr Lisandro Vásquez Sosa.

Consejo Nacional de Planificación y Coordinación Económica (National Council of Planning and Economic Co-ordination): Casa Presidencial, San Salvador.

Servicio de Parques Nacionales y Vida Silvestre (National Parks and Wildlife Service): Calle y Cantón El Matasano, Soyapango; tel 77-0622; fax 77-0490; Head: Carlos Roberto Hasbún; Head, Wildlife Dept: Daniel Roberto Burgos; Head, Natural Areas Dept: Andrés Sánchez Martínez.

Non-Governmental Organizations

Asociación Amigos del Arbol (Friends of the Tree): Calle Los Granados 9, Col Las Mercedes, San Salvador; tel 23-1841; Pres: Francisco Rivas Méndez; forestry; ecology; conservation of flora and fauna; resource protection; promotion of environmental law; campaigns on air pollution, protected areas and land use; education and training.

Asociación Salvadoreña de Conservación del Medio Ambiente (ASACMA—Salvadorean Association for Environmental Conservation): Calle Maquilishaut 208, Urb Buenos Aires 3, San Salvador; tel 26-5514; Dir: Néstor Herrera; wildlife conservation; management of natural areas; environmental education.

Centro Salvadoreño de Tecnología Apropiada (CESTA—Salvadorean Centre for Appropriate Technology): 33 Calle Poniente 316, Apdo 3065, San Salvador; tel 25-6746; fax 26-6903; Dir: Dr Ricardo A Navarro; promotion of sustainable development through appropriate technology, environmental education and skills-training workshops.

Desarrollo Juvenil Comunitario (DJC—Community Youth Development/Save the Children Federation, El Salvador): 35 Avda Sur 626, Col Flor Blanca, Apdo 05 (95), San Salvador; tel 71-4900; fax 71-0750; Dir: James D Criste; Deputy Dirs: Salvador Diaz, Rolando Aguilar; provides developing communities with basic training in three areas: education; health and nutrition, including environmental sanitation and provision of drinking water; and natural resources, including soil and water conservation, alternative crops, agroforestry and dam construction.

Fundación Ecológica Salvadoreña: 79 Avda N 509, Col Escalón, San Salvador; tel 23-8947; fax 23-3620; Exec Pres: Juan Marco Alvarez; conservation of El Imposible National Park; fund-raising; land acquisition.

Fundación Montecristo (Montecristo Foundation): Col Las Mercedes, Calle Los Eucaliptos 23, San Salvador; tel 26-9219; fax 23-5267; Pres: Ana Celia Domínquez; environmental conservation and education; promotion of rational use of natural resources.

IUCN El Salvador: c/o Fundación Friedrich Ebert, Calle Nueva 2, Casa 9, Col Escalón, San Salvador; tel 23-5385; fax 24-2478; Activities Co-ordinator: Manuel Benitez.

EQUATORIAL GUINEA

Governmental Organizations

Ministry of Agriculture and Forestry: Bata; Minister: Angel Alogo Nchama; incorporates Technical Forestry Department.

Ministry of Mines and Hydrocarbons: Malabo; Minister: Juan Olo Mba Nseng.

ERITREA

Governmental Organizations

Ministry of Agriculture: Asmara; Minister: Tesfay Girma Tzion.

Ministry of Energy, Mines and Water Resources: Asmara; Minister: Tesfay Gebre Selassie.

ESTONIA

Governmental Organizations

Ministry of the Environment: Toompuiestee 24, Tallinn 0100; tel (2) 452-507; telex 173238; fax (2) 453-310; Minister: Andres Tarand; Chancellor: Rein Ratas.

Matsalu State Nature Reserve: 3190 Lihula; tel (47) 78-413; fax (47) 78-413; Dir: Taivo Kastepõld; Head of Scientific Staff: Alex Lotman; preservation of wetland biodiversity, semi-natural communities and bird habitats; scientific research.

Vilsandi State Nature Reserve: Kihelkonna 3334, Saaremaa; tel (45) 76-604; bird protection; preservation of marine ecosystems and biological monitoring.

Non-Governmental Organizations

Balti Rahvusparkide Assotsiatsioon (Association of Baltic National Parks—ABNP): Secretariat, Lääne-Virumaa, 2128 Viitna; tel (32) 45-759; fax (32) 45-759; Sec: Teet Koitjärv.

Baltic Information Centre for Protected Areas: provided jointly with Union of Protected Areas of Estonia; information on laws of the Baltic Republics, Moldova, Russia and Ukraine relating to the environment and nature protection, and on the protected areas and national parks of the Baltic Republics, Armenia, Belarus, Georgia, Moldova, Russia and Ukraine.

Baltic Commission for the Study of Bird Migration: Vanemuise 21, Tartu 2400; tel (34) 30-331; fax (34) 33-472; Chair: Vilju Lilleleht; research into the effects of pollution in the Baltic Sea on migratory birds' habitats.

Biosphere Reserve of the West Estonian Archipelago, Hiiumaa Centre: Vabrikuväljak 1, Kärdla 3200; tel (46) 96-260; fax (46) 96-269; Research Dir: Toomas Kokovkin; Man Dir: Ruuben Post; research on insular biosphere reserve development, nature conservation, ecological education and environmental monitoring.

Saaremaa Centre: Kalevi 10, Kuressaare 3300; tel (45) 57-768; Dir: Mart Herman; Head of Nature Education and Information Dept: Tiina Talvi.

Commission for Nature Conservation: Vanemuise 21, 2400 Tartu; fax (34) 33-472.

Ecofiller Scientific-Creative Association: Teaduse 18-4, Saku 3400, Harju; tel (2) 721-564; fax (2) 771-385; Chair: Edvin

Nugis; Sec: Jelena Homyakova; development of environmentally friendly agricultural, forest and energy technologies; promotion of sustainable soil use; scientific research.

Eesti Kaitsealade Liit (EKAL—Union of Protected Areas of Estonia): Lääne-Virumaa, Viitna 2128; tel (32) 49-253; fax (32) 45-759; Chair: Arne Kaasik; co-ordination of activities of protected areas; represents interests of protected areas; exchange of information between members; training courses; information on nature conservation.

Eesti Looduseuurijate Selts (Estonian Naturalists' Society): 2400 Tartu, POB 43, Kompanii 3; tel (34) 32-000; Pres: Kalevi Kull; Vice-Pres: Mati Martin, Raivo Mänd; natural sciences publications; co-ordination of activities of professional and amateur naturalists; promotion of nature conservation.

Eesti Roheline Liikumine (Estonian Green Movement): POB 300, Tartu 2400; tel (34) 32-986; telex 173243; fax (34) 35-440; Chair: Tonu Oja; campaigns on environmental issues.

Eestimaa Looduse Fond (ELF—Estonian Fund for Nature): Struwe 2, POB 43, Tartu 2400; tel (34) 30-198; fax (34) 32-433; Pres: Fred Jüssi; Chair, Exec Cttee and Sec-Gen: Rein Kuresoo; independent organization for protection of biodiversity and promotion of sustainable use of natural resources in Estonia.

Endla Nature Reserve: Jógevamaa, 2340 Tooma; tel 46-429; Dir: Ärni Kuusik; Scientific Dir: Kai Kimmel; preservation of Endla marshes and their ecosystems; includes Tooma hydrometeorological station.

Estonian Academy of Sciences, Institute of Ecology: Kevade Str 2, Tallinn 0001; tel (2) 451-634; fax (2) 453-748; Dir: Dr Prof Jaan-Mati Punning.

Estonian Academy of Sciences, Institute of Zoology and Botany: Vanemuise Str 21, Tartu 2400; tel (34) 31-331; fax (34) 33-472; Dir: Andres Koppel; field of activity includes plant and animal ecology.

Estonian Research Institute of Forestry and Nature Conservation: Rôômu tee 2, Tartu 2400; tel (34) 36-381; fax (34) 36-375; Dir: Dr Kalle Karoles; Research Dir: Dr Tônu Terasmaa.

Haanja Nature Park: PK 37, Vôrumaa Post, Vôru 2710; tel (241) 21-209; landscape protection; co-ordination of tourism; scientific research on nature and nature protection; conservation education.

InnoCent (Vitsu Society for Nature Education): Laheääre tee 7, Virtsu 3195; fax (47) 78-113; Chair: Peeter Vissak; Scientific Manager: Üllas Ehrlich; ecological education, renewable energy, energy efficiency, environmental economy, ecotourism and scientific research.

Institute of Chemical Physics and Biophysics: Rävala puiestee 10, Tallinn 0100; tel (2) 441-304; fax (2) 440-640; Dir: Endel Lippmaa; Deputy Dir: Tiit Saluvere; Scientific Sec: Jüri Siigur; area of interest includes environmental research, biotechnology and bioenergetics.

Kõrvemaa Riiklik Maastikukaitseala (Kõrvemaa Landscape Reserve): Pärnu 75, Paide 2820; tel 43-139; Head: Iver Asarov; preservation of natural landscape of forests, bogs and drumlins.

Lahemaa Rahvuspark (Lahemaa National Park): Lääne-Virumaa, Viitna 2128; tel (32) 49-253; fax (32) 45-759; Dir: Arne Kaasik; protection of nature, ecosystems and biodiversity in northern Estonia, conservation education and ecotourism.

Otepää Maastikukaitseala (Otepää Protected Landscape Area): Kolga tee 28, Otepää 2513; tel (42) 55-876; Dir: Madis Kivilo; Forestry Advisor: Ene Vahter; Environmental Advisor: Tarmo Evestus; Nature Conservationist: Meelis Linnamägi; acts as nature advisory service.

Rebala Kaitseala (Rebala Historical-Cultural Protected Area): Harjumaa, Jóelähtme sjsk 3003; tel (2) 723-897; Dir: Mehis Pever; facilitates research, mainly historical, but also in geology, soil science and other fields; responsible for tourism in the protected area.

Taasen Renewable Energy (Re-En) Centre: Akadeemia 23A, Tallinn 0026; tel (2) 537-295; fax (2) 527-555; Dir: Tõnu Lausmaa; research into renewable energy sources, such as wind and solar energy, and promotion of their use.

Tallinn Zoo: Paldiski Rd 145, Tallinn 20035; tel (2) 559-944; fax (2) 393-049; Dir: Mati Kaal; conservation; education.

Tartu Ecocentre: POB 318 (Veski 35), Tartu 2400; tel (34) 62-532; fax (34) 62-084; Chair: Tõnu Oja; environmental education, and environmental and ecological consulting.

Tartu Student Nature Protection Club: Veski 35, Tartu 2400; tel (34) 62532; fax (34) 62084; environmental education, practical action, research and publications on nature reserves and forest protection.

Viidumäe Riiklik Looduskaitseala (Viidumäe Nature Reserve): Lümanda 3335, Saaremaa; tel 76-321; Dir: Mari Reitalu; Vice-Dir (Research): Tõnu Talvi; nature conservation, research and long-term monitoring of rare plant species.

Virtsu-Laelatu-Puhtu (VLP) Nature Reserve: Laheääre tee 7, PK 138, Virtsu 3195; fax (47) 78-113; Dir: Peeter Vissak; protection of natural and semi-natural ecosystems and of water-fowl nesting and feeding sites; scientific research on the ecology of species-rich plant communities and bird population ecology.

ETHIOPIA

Governmental Organizations

Ministry of Development of Natural Resources and Environmental Protection: POB 1034, Addis Ababa; tel (1) 514725; fax (1) 513042; Minister: Dr Mesfin Abebe; Deputy Minister: Ato Gedion Asfaw; formulation of policies and strategy; drafting of laws concerning the development, protection and utilization of Ethiopia's water, forest, soil and wildlife resources; sewerage; land use; meteorology; research.

Ethiopian Valley Development Studies Authority (EVDSA): POB 1086, Addis Ababa; tel (1) 611322; telex 21278; Gen Man: Ato Workalemahu Tesfa; Asst Head Environment Desk: Ato Kemal Zekaria; preparation of River Basin Development Master Plan, preliminary financial studies of medium- and large-scale irrigation projects and related environmental studies; operation of national surface-water hydrological network and policy formulation on trans-boundary river issues.

Ministry of Agriculture: POB 1223, Addis Ababa; tel (1) 448040; fax (1) 513042; Minister: Elias Negasa.

Ethiopian Wildlife Conservation Organization (EWCO): POB 386, Addis Ababa; tel (1) 445970; telex 21460; Man: Tadesse Michael.

Forestry and Wildlife Conservation Development Authority: POB 386, Addis Ababa.

National Conservation Strategy (NCS) Secretariat: Ministry of Planning and Economic Development, POB 1037, Addis Ababa; tel (1) 128973; fax (1) 553844; Principal Officers: Dr Tewolde Berhan, J P Sutcliffe; formulation of national and sub-national strategies for sustainable development and management of natural resources.

Office of the National Committee for Central Planning, Department of Construction and Physical Planning: POB 1037, Addis Ababa; tel (1) 128952.

Non-Governmental Organizations

Ethiopian Wildlife and Natural History Society: POB 60074, Addis Ababa; tel (1) 164045; Hon Chair: Prof Shibru Tedla; Hon Vice-Chair: John Atkins; study and promotion of the interests of the fauna, flora and natural environment of Ethiopia; dissemination of the knowledge gained and support for legislation to protect these resources.

IUCN Ethiopia: NCS Secr, ONCCP, POB 1037, Addis Ababa; tel (1) 128973; fax (1) 513133; Sr Technical Adviser: Peter Sutcliffe.

FIJI

Governmental Organizations

Ministry of Lands and Mineral Resources: Govt Bldgs, 3 Maafu St, Suva; Minister: Ratu Ovini Bokini.

Ministry of Housing and Urban Development: Govt Bldgs, 3 Maafu St, POB 2350, Suva; tel 211792; telex 2167; Minister: Col Jonetani Kaukimoce.

Non-Governmental Organizations

National Trust for Fiji: 3 Maafu St, POB 2089, Govt Bldgs, Suva; tel 301807; fax 305092; Chair: Ratu Josefa Iloilo; management of national parks, protected areas, tropical forests, wild and endangered fauna and flora, historic buildings and archaeological sites.

South Pacific University, South Pacific Action Committee for Human Ecology and the Environment (SPACHEE): c/o Inst of Natural Resources, POB 1168, Suva; tel 313900; fax 301305; Vice-Chair: William Aarlbersberg; co-ordination of organizations dedicated to environmental management, research and education; implementation of the South Pacific Action Plan.

FINLAND

Governmental Organizations

Ministry of the Environment: POB 399, Ratakatu 3, 00121 Helsinki; tel (0) 19911; telex 123717; fax (0) 1991499; Minister of the Environment: Sirpa Pietikäinen; Sec-Gen: Lauri Tarasti; environmental protection; nature conservation; wildlife protection; waste management; pollution control; land use planning; co-ordination of environmental policy; international co-operation.

Finnish IUCN Working Group: Chair: Esko Jaakkola.

Ministry of Foreign Affairs, Section for Environmental Protection: POB 276, 00170 Helsinki; tel (0) 1601; telex 124636; Minister of Foreign Affairs: Paavo Väyrynen.

Forest and Park Service—Nature Protection: Vernissakatu 4, POB 94, 01301 Vantaa; tel (0) 857841; fax (0) 85784234; Dir-Gen: Pentti Takala; Dir of Nature Protection: Matti Helminen; management of state-owned land and waters, including majority of national parks and other protected areas; recreational services; promotion of sustainable use of forest resources.

National Board of Waters and the Environment: POB 250, 00101 Helsinki; tel (0) 40281; telex 126086; fax (0) 4028345; Dir-Gen: Kaj Bärlund; management and protection of water and other natural resources; reparation of damage caused by water; research on water and other aspects of the environment.

Non-Governmental Organizations

Finnish Forest Research Institute (METLA): Unioninkatu 40A, 00170 Helsinki; tel (0) 857051; fax (0) 625308; Dir-Gen: Eljas Pohtila; comprised of Depts of Forest Ecology, Forest Production and Forest Resources.

Ilmamsuojeluyhdistys ry (Finnish Air Pollution Prevention Society—FAPPS): POB 335, 00131 Helsinki; tel (0) 631442; fax (0) 631442; Chair: Kari Larjava; Vice-Pres: Martti Muurinen; Sec: Kristina Saarinen; public education against air pollution.

Lintutieteellisten Yhdistysten Liitto (Birdlife Finland): POB 17, 18101 Heinola; tel (18) 152579; fax (18) 147682; Dir: Tapani Veistola; Sec: Osmo Ruottinen; bird protection.

Luonto-Liitto (The Nature League): Tallberginkatu, ID 113, 00180 Helsinki; tel (0) 6947899; Gen Sec: Anu Rinkenen; Sec: Jaana Hinkkanen; youth organization for environmental protection.

Natur och Miljo (Finnish Society for Nature and Environment): POB 240, 00151 Helsinki 15; tel (0) 644731; fax (0) 605850; Sec Gen: Ulrica Cronstrom; Swedish-speaking; campaigns on air and water pollution, environmental education, island and wetland ecosystems, protected areas and national parks, energy and sustainable development.

Suomen Luonnonsuojeluliitto (Finnish Association for Nature Conservation): POB 169, 00150 Helsinki; tel (0) 642881; fax (0) 6221815; Int Co-ordinator: Tuula Varis; conservation of flora, fauna and natural areas; planning for use of natural resources; ecological planning for cities and the countryside.

Vihreä Liitto (Green League): Eerikinkatu 24, 00100 Helsinki; tel (0) 6933877; fax (0) 6933799; Chair: Pekka Sauri; Sec: Dave Pemberton.

World Wide Fund for Nature (WWF) Finland: Uudenmaankatu 40, 00120 Helsinki 12; tel (0) 644511; fax (0) 602239; Sec-Gen: Meri Saarnilahti; Head, Conservation Dept: Lassi Karivalo; national affiliate organization of World Wide Fund for Nature (WWF) International; organization of conservation projects with government and municipal authorities, universities, research bodies and other conservation organizations.

FINNISH EXTERNAL TERRITORY

Åland Islands

Gröna på Åland (Green Party): 13 Mariegatan, 22100 Mariehamn; tel (28) 11528; Chair and Leader: Christina Hedman-Jaakkola.

FRANCE

Governmental Organizations

Ministry of the Environment: 45 ave Georges Mendel, 75016 Paris; Minister: Ségolène Royal.

Ministry of Foreign Affairs, Economic Department: 37 quai d'Orsay, 75007 Paris; tel (1) 47-53-53-53; fax (1) 45-51-60-12; Deputy Dir for Environment: F R Cazala.

Institut Français de Recherche Scientifique pour le Développement en Coopération (ORSTOM—French Scientific Research Institute for Co-operation in Development): 213 rue La Fayette, 75480 Paris Cedex 10; tel (1) 48-03-77-77; telex 214627; fax (1) 48-03-08-29; Dir of Research: Gérard Morel; international research organization concerned with developing countries; scientific research into terrestrial and marine environments, agriculture, climate, urbanization, health and development.

Office National de la Chasse (National Office of Hunting): 85 bis ave de Wagram, 75017 Paris; tel (1) 44-15-17-17; fax (1) 47-63-79-13; Dir: Christian Mary; research into wildlife and endangered species.

Non-Governmental Organizations

Aide et Action (Aid and Action): 67 blvd Soult, 75012 Paris; tel (1) 40-19-04-14; fax (1) 40-19-06-62; Chief Exec: Jean Claude Buchet; Chair Bd of Dirs: Jean Ratredat; improvement of primary education, community organization and quality of life in developing countries.

Amis de la Terre (Friends of the Earth—FoE) France: 38 rue Meslay, 75003 Paris; tel (1) 48-87-33-44; fax (1) 48-87-28-23; Pres: Philippe Lequenne.

Association Française du Fonds Mondial pour la Nature (World Wide Fund for Nature WWF France): 151 blvd de la Reine, 78000 Versailles; tel (1) 39-24-24-24; fax (1) 39-53-04-46; Pres: Philippe Poiret; Vice-Pres: François Tiger; national affiliate organization of World Wide Fund for Nature (WWF) International; protection of fauna, flora, natural sites, water, soil and resources; liaison with public and private organizations.

Association pour la Prévention de la Pollution Atmosphérique: 58 rue du Rocher, 75008 Paris; tel (1) 43-93-62-07; fax (1) 42-93-41-99; Pres: Pierre Gaussens; reduction and prevention of all forms of atmospheric pollution.

Association pour la Protection des Animaux Sauvages et du Patrimoine Naturel (ASPAS—Association for the Protection of Wild Animals and the Natural Heritage): BP 34, 26270 Loriol; tel 75-62-64-86; fax 75-62-66-00; Pres: Alain Clement; Asst: Philippe Boré.

Centre Ecologique Européen (European Ecological Centre): Chôteau de Vaugran, St Paul Lacoste, 30480 Cendras; tel 66-55-67-57; Principal Officers: Jacques Vecker, Klaus von Bresinski; environmental protection, energy and biology.

Centre International d'Art et Culture (International Art and Culture Centre): 13 ave Aristide Briand, 68200 Mulhouse; tel 89-42-26-93; Man: Gaby Rochette; publications with IUCN, UNESCO and environmental organizations.

Centre International de la Construction en Terre (CRATerre-EAG—International Centre for Earth Construction): Centre Simone Signoret, BP 53, 38090 Villefontaine; tel 74-96-60-56; telex 308658; fax 74-96-04-63; Pres: Patrice Doat; Sec-Gen: Hubert Guillaud; research, education, consultancy and information on earthen architecture in developed and developing countries.

Centre International de Droit Comparé de l'Environnement (International Centre of Comparative Environmental Law): c/o Centre de Recherches Interdisciplinaires en Droit de l'Environnement, de l'Aménagement et de l'Urbanisme (CRIDEAU), URA-CNRS No 967, Université de Limoges, 143 rue François-Perrin, 87000 Limoges; 4 Place du presidial, 87031 Limoges Cedex; tel 55-50-54-21; fax 55-33-67-18; Dir: Michel Prieur.

Centre Naturopa: Conseil de l'Europe, 67075 Strasbourg Cedex; tel 88-41-20-00; telex 870943; fax 88-41-27-15; information and documentation on the state of the environment.

Club Alpin Français (CAF—French Alpine Club): 24 rue Laumière, 75019 Paris; tel (1) 42-02-68-64; fax (1) 42-03-55-60; Pres: Louis Volle; research into highland and Alpine ecosystems, management of protected areas and landscape protection; field surveys; workshops and conferences; camps and excursions.

Fédération des Parcs Naturels Régionaux de France: 4 rue de Stockholm, 75008 Paris; tel (1) 42-94-90-84; fax (1) 45-22-70-78; Pres: Jean-Paul Fuchs; Dir: Jean-Luc Sadorge.

Fonds Français pour la Nature et l'Environnement (French Fund for Nature and the Environment): 2 rue Washington, 75008 Paris; tel (1) 42-25-60-19; fax (1) 43-59-00-51; Pres: Jacques Maillet; development of commercial patronage of the environment (Nature & Entreprises) by creating partnerships between businesses and environmental management organizations.

France Nature Environnement (FFSPN—France Nature Environment): 57 rue Cuvier, 75005 Paris; tel (1) 70-73-19-5; telex 260921; Principal Officer: A Gille; conservation of wildlife, habitats and coastal, marine and wetland ecosystems; campaigns on water pollution, use of wildlife resources, resource protection, animal rights, endangered animal species, prevention of cruelty, hunting and conservation economics, strategies and education; research; international co-operation; financial assistance to NGOs.

Friends of Animals: 16 rue Beaugrenelle, 75015 Paris; Principal Officer: Judith Clark.

Génération Ecologie: 57 blvd de la Villette, 75010 Paris; Leader: Brice Lalonde; ecologist political party.

Groupe Energies Renouvelables et Environnement (GERES—Study Group for Renewable Energy and Environment): 73 ave Corot, 13013 Marseille; tel 91-70-92-93; fax 91-06-19-46; Gen Man: Alain Guinebault; evaluation and initiation of renewable energy and environmental projects in developing countries; technical and financial assistance; co-ordination of similar programmes in Mediterranean nations.

Groupe d'Etudes et de Services pour l'Economie des Ressources (GEYSER—Research and Services Group for Resource Economy): rue Grande, 04870 St Michel-l'Observatoire; tel 92-76-62-44; fax 92-76-65-50; genetic resources, the environment, and ecological and sustainable agriculture.

Groupement de Recherche et de Développement sur le Patrimoine Génétique Végétal et Animal de la Région Provence Alpes Côte d'Azur (PAGE PACA—Research and Development Group for the Plant and Animal Genetic Heritage of the Provence-Alpes-Côte d'Azur Region): Secretariat, La Thomassine, 04100 Manosque; tel 92-72-40-24; fax 92-87-31-12.

Institut national de l'environnement industriel et des risques (INERIS—National Institute of the Industrial Environment and of Hazards): Parc Technologique ALATA, BP 2, 60550 Verneuil-en-Halatte; tel 44-55-66-77; telex 140094; fax 44-55-66-99; Dir-Gen: Michel Laverie; Deputy Dir-Gen: Georges Labroye; measurement and analysis of the effects of industry on the natural environment, risk assessment and development of safety systems.

Institut pour une Politique Européenne de l'Environnement (Institute for European Environmental Policy): 55 rue de Varenne, 75341 Paris Cedex 7; tel (1) 42-22-12-34; telex 201220; fax (1) 45-44-18-31; Dir: Thierry Lavoux; partner institutes in Belgium, Germany, the Netherlands and the United Kingdom.

Institut des Sciences et des Techniques de l'Equipement et de l'Environnement pour le Développement: La Grande Arche, 92055 Paris-La Défense Cedex 04; tel (1) 40-81-24-06; fax (1) 40-81-23-31; Pres: Jean Baudoin; Gen Delegate: Hervé Laine; organization of national and international exhibitions, dissemination of information and application of French experience relating to the development of transport and urban environments in developing countries.

IUCN France: 2 rue de Bel Air, 37400 Amboise; tel 47-57-30-23; fax 47-23-12-82; Attaché for French-Speaking Countries: Gérard Sournia; co-ordination of activities of IUCN in francophone countries and of IUCN members in France.

Comité Français de l'UICN (French Committee for IUCN): c/o Ministry of the Environment, Nature Protection Section, 14 blvd Gén Leclerc, 92524 Neuilly-sur-Seine Cedex; tel (1) 47-58-12-12; fax (1) 47-45-23-60; Dir: François Letourneux.

Journalistes-Ecrivains pour la Nature et l'Ecologie (JNE—Journalist-Writers for Nature and Ecology): 51 rue Laugier, 75017 Paris; tel (1) 60-08-27-18; fax (1) 60-08-27-18; Pres: Nicole Lauroy; Sec: Jean-Paul Thorez; members include journalists and writers from all mass media; organizes conferences, symposia and awareness campaigns.

Laboratoire de Microbiologie Marine: Centre National de la Recherche Scientifique UPR 223, Campus de Luminy, CP 907, 13288 Marseille Cedex 9; tel 91-26-90-52; fax 91-26-90-49; Principal Officers: Micheline Bianchi, Armand Bianchi; microbial activity and diversity in marine environments.

Ligue pour la Protection des Oiseaux (LPO—League for the Protection of Birds): La Corderie Royale, BP 263, 17305 Rochefort Cedex; tel 46-82-12-34; fax 46-83-95-86; Chair: Allain Bougrain Dubourg; Dir: Michel Metais; member of BirdLife International.

Mouvement National de Lutte pour l'Environnement (MNLE—National Movement of the Struggle for the Environment): 106 ave Jean Lolive, BP 79, 93503 Pantin Cedex; tel (1) 48-46-04-14; fax (1) 48-46-44-53; Sec-Gen: Guy Léger; acts on a local, national and international level to ensure that the progress of science and technology reconciles man and the environment.

Muséum National d'Histoire Naturelle (National Museum of Natural History): 57 rue Cuvier, 75005 Paris; tel (1) 40-79-30-00; fax (1) 40-79-34-84; Dir: Jacques Fabries; research; conservation of natural history collections.

Laboratoire d'Ecologie Générale: 4 ave du Petit Château, 91800 Brunoy; tel 60-46-48-51; fax 60-46-57-19; Dir: Prof Patrick Blandin.

Conservation des Espèces Animales: Ménagerie du Jardin des Plantes, 57 rue Cuvier, 75005 Paris; tel (1) 40-79-37-94; fax (1) 40-79-37-93; Dir: Prof J-J Fetter.

Laboratoire de l'Evolution des Systèmes Naturels et Modifiés: 36 rue Geoffroy St-Hilaire, 75005 Paris; tel (1) 40-79-32-59; fax (1) 40-79-32-71; Dir: Prof Jean-Claude Lefeuvre.

Secrétariat Faune et Flore: Ménagerie du Jardin des Plantes, 57 rue Cuvier, 75005 Paris; tel (1) 40-79-32-74; fax (1) 43-36-13-39; Dir: Hervé Maurin.

Société Nationale de Protection de la Nature (SNPN—National Nature Protection Society): 57 rue Cuvier, BP 405, 75221 Paris Cedex 05; tel (1) 47-07-31-95; fax (1) 47-07-07-16; Pres: François Ramade; protection of species and natural resources; research into freshwater and terrestrial environments, forestry, wild and endangered flora and fauna, use of wildlife resources and protected areas.

Office pour l'Information Eco-Entomologique (Office for Eco-Entomological Information): BP 9, 78041 Guyancourt Cedex; tel (1) 30-44-13-43; Sec-Gen: Robert Guilbot; information; awareness-raising; advice on eco-entomology.

Panos Paris: 53 rue Turbigo, 75003 Paris; tel (1) 42-71-20-21; fax (1) 42-71-21-55; Exec Dir: Charles Condamines; independent information organization promoting socially, environmentally and economically sustainable development world-wide; research into key global issues; production of radio programme *Chronique Environnement*; offices in the United Kingdom and the USA.

Programme Solidarité Eau (Water Network): GRET, 213 rue Lafayette, 75010 Paris; tel (1) 40-35-13-14; fax (1) 40-35-08-39; Pres: Stéphane Hessel; Exec Sec: Pierre-Marie Grondin; promotion of co-operation in the areas of water, health, agriculture and the environment.

SILVA, Arbres, Forêts et Sociétés (SILVA, Trees, Forests and People): 21 rue Paul Bert, 94130 Nogent-sur-Marne; tel (1) 48-75-59-44; fax (1) 48-76-31-93; Principal Officers: Jean Clement, Lucie de Framond, François Besse, Viviane Appora; information on forests and trees around the world through publications and training courses, in order to combine the efforts of those who wish to defend forests.

Société Française pour le Droit de l'Environnement (SFDE—French Society for Environmental Law): Univ Robert Schuman, place d'Athènes, 67084 Strasbourg Cedex; tel 88-41-42-57; fax 88-61-30-37; Pres: Gilles Martin; non-profit association of French lawyers, dedicated to the development of environmental law as an instrument for nature protection.

Structuration Spatiale et Evolution des Diversités Biologiques (Spatial Structure and Evolution of Biological Diversity): Laboratoire de Zoogéographie, Université Montpellier III, BP 5043, 34032 Montpellier Cedex 1; tel 67-14-23-15; telex 490067; fax 67-14-20-52; Principal Officer: Prof Charles P Blanc; bio-geographic genetics, ecology, palaeoecology and biodiversity.

Survival International (France): 45 rue du Faubourg du Temple, 75010 Paris: tel (1) 42-41-47-62; fax (1) 42-45-34-51; Principal Officers: Jean Patrick Razon, Dominique Verjut.

Les Verts (Green Party): 50 rue Benoît Malon, 94250 Gentilly; tel (1) 49-08-91-31; fax (1) 49-08-97-44; Nat Sec: Guy Cambot; ecologist party.

FRENCH OVERSEAS DEPARTMENTS

Guadeloupe

Coopération Franco-Caraïbe: 36 Marina Pieds dans l'Eau, 97110 Pointe-à-Pitre; tel 81-0040; fax 93-0854; Pres: Max Vincent; Treas: Catherine Ciriac; regional development organization working from Guadeloupe.

Martinique

Comité de Résistance à la Destruction de l'Environnement Martiniquais (CORDEM—Committee of Resistance against Nature Destruction in Martinique): rue du Prof Garcin, voie 5, Fort-de-France, 97200 Martinique; tel 641824; Pres: Georges Negouai; campaigns against waste and environmental destruction; research; intervention against unlawful planning and building.

Réunion

Ecologie Réunion (Réunion Ecology): 14 ave Roland Garros Gillot, 97438 Sainte-Marie; tel 53-51-83; Pres: Guy Clergue; Vice-Pres: Rose May Duchenoy; environmental conservation; promotion of ecological ideas; social research; direct action.

Société Réunionnaise pour l'Etude et la Protection de l'Environnement (Réunion Society for Study and Protection of the Environment): 695 les Calebassiers, B T2, 97490 Ste Clotilde; tel 28-10-89; Sec-Gen: Max Vigreux; scientific research; environmental impact assessment; campaigns on wild and endangered flora and fauna and development planning; environmental education.

FRENCH OVERSEAS TERRITORIES

French Polynesia

Délégation à l'Environnement (Environment Delegation): rue des Poilus Tahitiens, BP 4562, Papeete; tel 432409; fax 419252; Head of Pollution Prevention: Laurent Borde; Head of Nature Protection: Michel Guérin; monitoring and management of the natural environment; co-ordination of all environmental activity; training and information relating to environmental awareness.

New Caledonia

Association pour la Sauvegarde de la Nature Néo-Calédonienne (Association for the Protection of the Nature of New Caledonia): 37 rue Georges Clemenceau, BP 1772, Nouméa; tel 283275; fax 283275; Pres: Jean-Louis d'Auzon; Vice-Pres: Marie-Josée Fiori, Tony Willmott; guided walks; publications; conferences; exhibitions; education in schools, villages and tribes; study of marine turtles, local birds and wetlands; creation of natural reserves; campaign against bush fires; reforestation; tree days; cleaning islands and beaches; competitions and prizes for environmentalists.

GABON

Governmental Organizations

Ministry of Water, Forestry Resources, Fishing and the Environment: BP 2275, Libreville; Minister: Eugène Capito.

Ministry of Agriculture, Livestock and Rural Development, Department of Water Resources and Forests: BP 551, Libreville; tel 76-29-43; telex 5587; Minister: Eugène Mayaza Kakou.

Non-Governmental Organizations

World Wide Fund for Nature (WWF) Gabon: c/o Inter Service, BP 5555, Libreville; tel 72-47-43; telex 5802; fax 76-68-08.

THE GAMBIA

Governmental Organizations

Ministry of National Resources and the Environment: 5 Marina Parade, Banjul; tel 27431; telex 2204; Minister of National Resources and the Environment: Sarjo Touray.

GEORGIA

Governmental Organizations

Ministry of Environmental Protection: 380071 Tbilisi, Mindeli St 9; tel (8832) 38-58-39; Minister of Environmental Protection: Shota Adamia.

GERMANY

Governmental Organizations

Federal Ministry for the Environment, Nature Conservation and Nuclear Safety: 5300 Bonn 2, Kennedyallee 5; tel (228) 305-0; telex 885790; fax (228) 3053225; Minister of the Environment, Nature Conservation and Nuclear Safety: Prof Dr Klaus Töpfer; environmental conservation; inter-ministerial co-ordination; research into nature conservation and ecology; technical advice on nature conservation and landscape preservation; liaison with regional governments.

Bùndesforschungsanstalt für Naturschutz und Landschaftsökologie (Federal Research Centre for Nature Conservation and Landscape Ecology): 5300 Bonn 2, Konstantinstr 110; tel (228) 84910; fax (228) 8491200; Dir: Prof Dr Walter Mrass; scientific research into wild flora and fauna, biology, environmental impact assessment, landscape protection and management, conservation strategies, habitat protection and protected areas; international co-operation; environmental statistics; funding for conservation projects.

Landesanstalt für Umweltschutz Baden-Württemberg (Baden-Württemberg Regional Institute for Environmental Protection): 76157 Karlsruhe, Postfach 21 07 52; tel (721) 9831424; fax (721) 9831414; Head of Dept: Dr Winfried Krahl; Head of Publications and Ed: Dr Günter Schmid; nature conservation.

Non-Governmental Organizations

Arbeitskreis Verkehr und Umwelt eV (UMKEHR—Working Group for Transportation and the Environment): 13357 Berlin-Wedding, Exerzierstr 20; tel (30) 4927473; fax (30) 4927972; Principal Officer: Karl-Heinz Ludewig; promotion of pedestrianism, cycling and public transport.

Big Mountain Aktionsgruppe eV: Frohschammerstr 14, 80807 München; tel (89) 35651836; fax (89) 35651837; Exec Dir: Monica Seiller; supports native Americans in the USA and Canada on issues related to human rights and environmental protection.

Brehm Fonds für Internationalen Vogelschutz (Brehm Fund for the International Conservation of Birds): 5300 Bonn 1, Adenauerallee 214, Postfach 120369; tel (228) 213453; Chair: Wolf W Brehm; scientific conservation of birds, animals and habitats through international co-operation.

Bund Naturschutz in Bayern (Bavarian Federation for Nature Conservation): 8000 München 80, Kirchenstr 88; tel (89) 459918; fax (89) 485866; Leader: Helmut Steininger; conservation strategies for wildlife; sustainable development; environmental education.

Bund für Umwelt- und Naturschütz Deutschland (German Association for Environment and Nature Conservation): 5300 Bonn 3, Im Rheingarten 7; tel (228) 400970; fax (228) 4009740; Principal Officer: Onno Poppinga; campaigns on wild and endangered flora and fauna, forests, lakes, land use, atmospheric pollution, renewable energy, ecology, sustainable development, environmental education, hazardous substances and waste disposal; international co-operation; camps and excursions.

Reutlingen Office: 72764 Reutlingen, Lederstr 86; tel (7121) 320993; Regional Sec: Thilo Schäfer.

Deutscher Heimatbund (German Heritage Federation): 53113 Bonn, Adenauerallee 68; tel (228) 224091; fax (228) 215503; Pres: Dr Hans Tiedeken; Man: Dieter Hornung; environmental conservation; preservation of historical monuments.

Deutscher Jagdschutzverband (German Hunting Association): 5300 Bonn 1, Johannes-Henry Str 26; tel (228) 214172; fax (228) 214185; Dir: Martin Wiese; 16 regional branches; control of hunting, trapping and poaching; conservation of nature and species; protection of habitats and wild flora and fauna.

Deutscher Naturschutzring (DNR—German Association for the Conservation of Nature): 5300 Bonn 3, Kalkuhlstr 24; tel (228) 441505; telex 8861170; fax (228) 444290; Dir: Helmut Röscheisen; organization of nature conservation groups.

Die Grünen (The Greens—West): 5303 Bornheim-Roisdorf, Im Ehrental 2-4, Bundesgeschäftsstelle; tel (2222) 7008-0; fax (2222) 7008-99; Spokesperson: Ludger Volmer; Gen Sec:

Heide Rühle; comprised of Grüne Aktion Zukunft (Future Green Action), Grüne Liste Umweltschutz (Environmental Protection Green List), Aktionsgemeinschaft Unabhängiger Deutschen (Action Community of Independent Germans) and other groups; left-wing; campaigns on ecological issues, dissolution of NATO, breaking down of large economic concerns into smaller units, a 35-hour working week and unlimited rights to strike; linked with the Green Party of the former German Democratic Republic.

Environmental Protection Encouragement Agency (EPEA): 20147 Hamburg, Feldstr 36; tel (40) 4392081; fax (40) 4392085; Head: Dr Michael Braungart; proposals for waste minimization and handling; consultation and education on environmental protection.

Tübingen Office: 72070 Tübingen, Neue Str 4; tel (7071) 26143; fax (7071) 26144; Principal Officer: Dorothea Häußer.

Fachgruppe Wasserchemie in der Gesellschaft Deutscher Chemiker (Water Chemistry Specialist Group of the German Chemical Society): Universität Karlsruhe, 76128 Karlsruhe, Richard-Willstätter-Allee 5; tel (721) 6082597; fax (721) 606989; Chair of Bd: Prof Dr F H Frimmel; co-ordination of water chemistry research activities, standardization of water analytics, study of relevant environmental problems.

Fussgängerschutzverein eV (FUSS—Association to Protect Pedestrians): 13357 Berlin-Wedding, Exerzierstr 20; tel (30) 4927473; fax (30) 4927972; Principal Officer: Bernd Herzog-Schlagk; promotion of pedestrianism.

German Co-ordination Committee for IUCN: c/o Bund für Naturschutz, 5300 Bonn 2, Konstantinstr 110; tel (228) 8491-0; Dir: Prof Dr Walter Mrass.

Gesellschaft für Ökologie (Society for Ecology): 100 Berlin 41, Institut für Ökologie, Technische Universität Berlin, Rothenburgstrasse 12; tel (314) 71350; fax (314) 71355; Pres: Prof Dr Reinhard Bornkamm; Vice-Pres: Prof Dr Herbert Sukopp; promotion of ecological science.

Hamburger Umwelt-Institut, Zentrum für Soziale und Ökologische Technik eV (Hamburg Environment Institute, Centre for Social and Ecological Technology): 20147 Hamburg, Feldstr 36; tel (40) 4392091; fax (40) 4392085; Founder and Principal Officer: Dr Michael Braungart.

Institut für Europäische Umweltpolitik (Institute for European Environmental Policy): 55300 Bonn, Roonstr 7; tel (228) 914970; fax (228) 9149722; Dir: Dr Jan Bongaerts; partner institutes in Belgium, France, the Netherlands and the United Kingdom.

Institut für Ökologisches Recycling (IfÖR-Institut—Institute for Ecological Recycling): 10785 Berlin, Kurfürstenstr 14; tel (30) 2628021; fax (30) 2650366; Principal Officers: Simone Martzloff, Jürgen Halbekath; research, information and publications on ecological waste prevention and management.

IUCN Germany: Environmental Law Centre, Adenauerallee 214, 5300 Bonn 1; tel (228) 2692231; fax (228) 2692250; Head: Françoise Burhenne-Guilmin.

Naturschutzbund Deutschland eV (NABU—Nature Preservation Society): 53225 Bonn, Herbert-Rabius-Str 26; tel (228) 975610; fax (228) 9756190; Pres: Jochen Flasbarth; Dir: Uwe Hüser; protection of indigenous and migratory birds and their habitats.

Niedersächsischer Heimatbund eV (Federation for Lower Saxony): 3000 Hannover 1, Goseriede 15; tel (511) 131565; fax (511) 17475; Principal Officer: Dr Roswitha Sommer; preservation and promotion of the culture, history and environment of Lower Saxony.

Schutzgemeinschaft Deutscher Wald—Bundesverband (German Association for Protection of Forests and Woodlands): 5300 Bonn 1, Meckenheimer Allee 79; tel (228) 658462; Principal Officer: Dr H Köpp; forestry; land use; landscape protection; environmental education; teacher training; awareness-raising.

Schutzgemeinschaft Deutsches Wild (German Wildlife Conservation Association): 5300 Bonn 1, Postfach 120371; tel (228) 2692217; fax (228) 2692252; Chair, Bd: H J Rohr; conservation of wild animals and their habitats.

Society for Ecotoxicology and Environmental Safety (SECOTOX): 8042 Neuherberg, Institut für Ökologische Chemie, GSF München, Ingolstädter Landstr 1; tel (89) 31872690; fax (89) 31873371; Pres: Prof Dr J H Koeman; Sec: Prof Dr A Kettrup; promulgation of knowledge through the international exchange of views and experience and promotion of research.

Stattauto Car-Sharing GmbH—European Car Sharing (ECS): 1000 Berlin 36, Manteuffelstr 40; tel (30) 6113527; fax (30) 6113727; Business Dir (ECS): Carsten Petersen.

Umweltstiftung WWF Deutschland (World Wide Fund for Nature WWF Germany): 60591 Frankfurt am Main, Hedderichstr 110, Postfach 701127; tel (69) 6050030; telex 041505990217; fax (69) 617221; Chair, Bd: Carl-Albrecht von Treuenfels; Dir, Conservation Bd: Dr Hartmut Jungius; national affiliate organization of World Wide Fund for Nature (WWF) International; protection of wild and endangered flora, fauna and habitats, marine environments and wetland ecosystems; conservation strategies and education; aid to developing countries; environmental policy.

Wattenmeerstelle Schleswig-Holstein (Schleswig-Holstein Mud-flat Station): 25813 Husum, Norderstr 3; tel (4841) 62073; fax (4841) 4736; Principal Officers: Hans-Joachim Augst, Hans-Ulrich Rösher; programmes for the conservation of the Waddenzee and Baltic Sea and of the East Atlantic habitats of coastal birds; monitoring of migratory birds in the Waddenzee.

Verband Deutscher Sportfischer (German Fishermen's Association): 6050 Offenbach a M, Siemenstr 11-13; tel (69) 855006; fax (69) 873770; Leader: U Schuller; campaigns on freshwater environments, lakes, rivers, wildlife, animal rights and landscape protection; international co-operation.

Vereinigung Deutscher Gewässerschutz (VDG—German Water Protection Association): 5300 Bonn 2, Matthias-Grünewald-Str 1-3; tel (228) 375007; Dir: Prof Dr Dieter Flinspach; campaigns on freshwater environments, lakes, rivers, watersheds and water pollution and management.

Weltwirtschaft, Ökologie und Entwicklung eV (WEED—World Economy, Ecology and Development): 53111 Bonn, Berliner Platz 1; tel (228) 696479; fax (228) 696470; Pres: Barbara Unmüßig; Exec Dir: Peter Wahl; information and consultancy organization dedicated to the restructuring of relations between developed and developing countries.

Wildbiologische Gesellschaft München eV (Munich Wildlife Society): 82488 Ettal, Linderhof 2; tel (8822) 6363; fax (8822) 6161; Pres: Prof Dr Wolfgang Schröder; Exec Sec: Dr Doris Hofer.

Wuppertal-Institut für Klima, Umwelt, Energie GmbH (Wuppertal Institute for Climate, Environment and Energy): 42103 Wuppertal, Döppersberg 19, Postfach 100480; tel (202) 24920; fax (202) 2492108; Pres: Prof Dr Ernst Ulrich Freiherr von Weizsäcker; Principal Officer: Lorenz Kneser.

Zoologische Gesellschaft Frankfurt von 1858, Hilfe für die Bedrohte Tierwelt (Frankfurt Zoological Society of 1858, Help for Threatened Wildlife): 6000 Frankfurt a M 1, Alfred-Brehm-Platz 16; tel (69) 21233727; telex 4170246; fax (69) 21237855; Pres: Dr Richard Faust; wildlife conservation strategies and education.

GHANA

Governmental Organizations

Ministry of the Environment, Science and Technology: POB M47, Accra; tel (21) 665421; Minister: Dr Christine Amoako-Numah.

Ministry of Lands and Forestry: POB M212, Accra; tel (21) 665421; Minister of Lands and Forestry: Dr Stephen Ayidiah.

Ministry of Local Government and Rural Development, Environmental Protection Council: POB M326, Accra; tel (21) 664697; telex 2609; fax (21) 662690; environmental management, co-ordination and education; research.

Department of Game and Wildlife: POB M239, Accra; tel (21) 664654; telex 2851; fax (21) 666476.

Forestry Commission: 4 3rd Ave Ridge, POB M434, Accra; tel (21) 221315; Principal Officers: Prof Kwabena Tufuor, Dr E Dwumfour; advice on national forest policy and monitoring of the impact of its implementation; database on forests and wildlife resources for agencies, NGOs, universities and others.

Non-Governmental Organizations

Ghana Association for Conservation of Nature (GACON): c/o Institute of Renewable Natural Resources, University of Science and Technology (UST), Kumasi; tel (51) 5351 (UST); telex 2555; Pres: Dr W K Oduro; Sec: P K Epke.

Ghana University, Regional Institute for Population Studies (RIPS): POB 96, Legon; tel (21) 774070; telex 2164; Dir: Dr P O Ohadike; Admin Sec: A Kwaa; training, research, publishing and advisory services.

IUCN Ghana: Forest Resource Management Project, Game and Wildlife Dept, Ministry of Lands and Natural Resources, POB M239, Ministry PO, Accra; tel (21) 666129; fax (21) 666476; Project Leader: Dr John Grainger.

GREECE

Governmental Organizations

Ministry of the Environment, Physical Planning and Public Works: Odos Amaliados 17, 115 23 Athens; tel (1) 6431461; telex 216734; fax (1) 6425300; Minister of Environment, Physical Planning and Public Works: Achilles Karamanlis.

Non-Governmental Organizations

Ecologist Alternative: Athens; green party.

Elliniki Etairia (Hellenic Society for the Protection of the Environment and Cultural Heritage): Tripodon St 28, 105 58 Athens; tel (1) 3225245; fax (1) 3225240; Pres: Dr Michael J Scoullos; awareness-raising; environmental protection; lobbying government on environmental legislation.

Greenpeace Greece: Kallidromiou 44, 114 73 Athens; tel (1) 3640774; fax (1) 3604008; Exec Dir: Elias Efthymiopoulos.

Hellenic Organization for the Protection of the Environment (HOPE): Asomaton 9, 105 33 Athens; tel (1) 4125808; telex 241321; fax (1) 4179664; Chair: Prof B N Metaxas; Sec: Dr M Pazarzis; environmental research; organization of conventions, seminars, lectures and meetings; advice to environmental authorities.

Hellenic Society for the Protection of Nature: Nikis St 24, 105 57 Athens; tel (1) 3224944; fax (1) 3225285; Sec: George Sfikas; conservation of wild flora and fauna, forests, protected areas, coasts and sea, wetland ecosystems and national parks.

Sea Turtle Protection Society of Greece: Solomou 35, 106 82 Athens; tel (1) 3644146; fax (1) 3644146; Principal Officer: Dimitris Margaritoulis.

World Wide Fund for Nature (WWF) Greece: Asklepiou St 14, 106 80 Athens; tel (1) 362-3342; fax (1) 362-3342; Rep: Dr Georgia Valaoras.

GRENADA

Governmental Organizations

Ministry of Health, Housing and the Environment: Carenage, St George's; tel (440) 3485; fax (440) 4127; Minister of Health, Housing and the Environment: Phinsley St Louis; Perm Sec: Shirley Mathlin; co-ordination of public and private sector environmental management agencies.

Non-Governmental Organizations

Friends of the Earth (FoE) Grenada: Victoria Post Office, Church Hill Rd, Victoria, St Mark's; tel (444) 8838; fax (444) 2802; Pres: Joseph Antoine; Gen Sec: Dereck Allard.

GUATEMALA

Governmental Organizations

Ministry of Foreign Affairs, Ministerial Commission for Conservation and Improvement of the Human Environment: Palacio Nacional, Guatemala City; telex 5321; Minister of Foreign Affairs: Gonzalo Menéndez Park.

Presidency of the Republic, National Committee on Environment: 7a Avda 4-35, Zona 1, Guatemala City; Co-ordinator: Antonio Ferrate.

Instituto Guatemalteco de Turismo (INGUAT—Guatemala Tourist Commission): 7a Avda 1-17, Zona 4, Centro Civico, Guatemala City 01004; tel (2) 31-1333; telex 5532; fax (2) 31-8893; Dir: E A Reimers Galvez; management and planning of natural areas.

Instituto Nacional Forestal (National Forestry Institute): Edif Galerías España, 6°, 7 Avda 11-63, Zone 9, Guatemala City; tel (2) 31-9141.

Non-Governmental Organizations

Asociación Amigos del Bosque (Friends of the Forest Association): 9a Calle 2-23, Zona 1, 01001 Guatemala City; tel (2) 83-486; Pres: Ann Bennaton; protection of natural areas, water resources and flora and fauna through rational use of resources; afforestation; education.

Asociación Audubon de Guatemala: 10a Avda 2-44, Zona 14, 01014 Guatemala City; tel (2) 33-7491; fax (2) 33-7490; Principal Officers: Nini de Berger, Roberto Dorión, Dr Roberto Labee; conservation of tropical forest through private protected areas.

Asociación de Entidades de Desarrollo y Servicio No Gubernamentales de Guatemala (ASINDES-ONG—Association of Non-Governmental Development and Service Organizations of Guatemala): 4a Avda 'A' 7-70, Zona 10, 01010 Guatemala City; Apdo 504-A, 01909 Guatemala City; tel (2) 32-1828; fax (2) 31-9931; Exec Dir: Ing Abel Girón; Dir of Tech Assistance and Training: Licda Ligia Orantes de Aguirre.

Asociación Rescate y Conservación de Vida Silvestre (ARCAS—Wildlife Rescue and Conservation Association): 11 Calle 6-66, Zona 2, Guatemala City; tel (2) 53-5329; fax (2) 53-5329; Pres: Alfonso Muralles; Exec Dir: Myriam Monterroso; rescue and rehabilitation of wildlife; environmental education; community development; ecotourism; conservation of protected areas; management of scientific research stations.

Centro de Estudios Conservacionistas (CECON—Centre for Conservation Research): Avda de la Reforma 0-63, Zona 10, 01010 Guatemala City; tel (2) 31-0904; fax (2) 34-7664; Dir: Billy T Alquijay Cruz; multidisciplinary centre for management of renewable natural resources; conservation; sustainable development.

Centro Mesoamericano de Estudios sobre Tecnología Apropiada (CEMAT—Central American Centre of Studies on Appropriate Technology): Apdo 1160, 01901 Guatemala City; 1a Avda 32-21, Zone 12, 01012 Guatemala City; tel (2) 76-2018; fax (2) 71-8380; Gen Dir: Dr Roberto Cáceres; Admin Dir: Kajsa de Asturias; support for economic development, technical co-operation and social progress in Guatemala and Central America; information exchange; promotion of scientific research; development of projects with local groups; sustainable development; publications.

Comisión Centroamericana de Ambiente y Desarrollo (Central American Commission for Environment and Development): 7a Avda 13–01, Zona 9, Edif La Cupula, 2°, Guatemala City; tel (2) 32-0684; fax (2) 34-3876; Exec Sec: Jorge Cabrera Hidalgo; promotion and co-ordination of regional system of co-operation relating to environment and development.

Defensores de la Naturaleza (Defenders of Nature): 7a Avda y 13 Calle, Edif La Cúpula Z9, Guatemala City 01009; tel (2) 32-5064; Pres: Héctor Centeno Bolaños; non-profit organization dedicated to protection of natural resources; preservation of biosphere reserves; education; lobbying government.

Fundación del Centavo, Fundación Guatemalteca para el Desarrollo (FUNDACEN—Guatemalan Foundation for Development): 8a Calle 5–09, Zona 9, Guatemala City; tel (2) 32-2443; fax (2) 31-0754; Principal Officers: Antonio Aycinena Arrivillaga, Ing Miguel Hernan Quan Berducido; agricultural finance, credit facilities to enable farmers to buy land, technical assistance and training in health and education, and environmental conservation.

Fundación para la Conservación del Medio Ambiente y de los Recursos Naturales Mario Dary Rivera (Mario Dary Rivera Foundation for Conservation of the Environment and Natural Resources): Avda de la Reforma 0-63, Zona 10, 01010 Guatemala City; tel (2) 31-0904; fax (2) 34-7664; Treas: Juan Mario Dary Fuentes; promotion of nature conservation and development of natural resources; support for related organizations.

Greenpeace América Latina, Oficina Centro América (Greenpeace Latin America, Central America Office): 10 Calle 3-15, Zona 1, 01001 Guatemala City; tel (2) 53-4172; fax (2) 53-4206; Exec Dir: Beatriz Barraza.

Instituto de Nutrición de Centro América y Panamá (INCAP—Institute of Nutrition of Central America and Panama): Carretera Roosevelt, Zona 11, Apdo 1188, 01901

Guatemala City; tel (2) 72-3762; telex 5696; fax (2) 73-6529; Dir: Dr Hernán L Delgado.

Unión Mundial para la Naturaleza—UICN (World Conservation Union—IUCN): 31 Avda 1-12, Zona 7, Utatlán I, Guatemala City; tel (2) 94-6849; fax (2) 94-6849; National Rep: Juan Carlos Godoy; Wetlands Co-ordinator: Néstor Windevoxhel.

Wildlife Preservation Trust International: Parque Zoológico 'La Aurora', Zona 13, Guatemala City; tel (2) 72-0507; fax (2) 71-5286; Regional Co-ordinator for Central America: Lorena Calvo; organizes training courses on environmental education and conservation, and field research on endangered species.

GUINEA

Governmental Organizations

Ministry of Natural Resources, Energy and the Environment: BP 295, Conakry; tel 44-50-01; telex 22350; Minister of Natural Resources, Energy and the Environment: Dakoun Toumany Sakho.

Ministry of Agriculture and Animal Resources: BP 576, Conakry; tel 44-19-66; Minister: Ibrahima Sory Sow; agriculture and animal resources; water resources, forestry and hunting; meteorology; rural engineering and hydraulics.

Secrétariat d'Etat à la Recherche Scientifique, Centre de Recherche Océanographique de Rogbane (Secretariat of State for Scientific Research, Rogbane Oceanographic Research Centre): BP 561, Conakry.

Non-Governmental Organizations

Association des Amis de la Nature et de l'Environnement: BP 206 bis, Conakry.

Parti Guinéen des Ecologistes (Guinean Ecologists' Party): Conakry; Leader: Oumar Sylla.

GUINEA-BISSAU

Governmental Organizations

Ministry of Natural Resources, Department of Forestry and Hunting: CP 399, Bissau; Minister of Natural Resources: João Cardoso.

Non-Governmental Organizations

IUCN Guinea-Bissau: BP 23, 1031 Bissau Cedex; tel 201230; fax 201168; Project Leader: Pierre Campredon.

GUYANA

Governmental Organizations

Ministry of Agriculture: POB 1001, Regent and Vlissingen Rds, Georgetown; tel (2) 69154; Minister: Reepu Daman Persaud.

HAITI

Governmental Organizations

Ministry of Agriculture, Natural Resources and Rural Development: BP 1441, Damien, Port-au-Prince; tel 22-3457; Minister: Jacques Becker.

Non-Governmental Organizations

Pan-American Development Foundation, Productive Land Use Systems Project (PLUS): Delmas 31, 27 Port-au-Prince; tel 46-3938; fax 46-4616; Dir: Arlin Hunsberger; PLUS Project Dir: Bertrand Laurent; research into methods of increasing farmers' income and productivity through environmentally sound interventions; soil and water conservation; training of farmers; tree-planting.

Service de Prochains Haitiens (SPH—Haitian Neighbours Service): c/o Lynx Air, POB 407139, Fort Lauderdale, FL 33340, USA; ave Frecyneau et rue Christophe 54, BP 9, Saint-Marc; Exec Sec and Founder: Edouard Jacob Toussaint; non-profit, Christian community development organization; activities include reafforestation programme and improvement of farming practices in order to increase yields.

HONDURAS

Governmental Organizations

Ministry of Natural Resources: Blvd Miraflores, Avda La Fao, Tegucigalpa, DC; tel 32-8688; fax 32-5375; Minister of Natural Resources: Mario David Nufio Gamero.

Secretariat of Natural Resources, State Fisheries Office, International Co-operation Department: Apdo 309, Tegucigalpa; tel 32-8600; telex 8071; fax 32-5375; Gen Dir: Dr Gabino Zuniga; Int Officer: Bessy Aspia; management and conservation of marine, coastal and freshwater resources.

Non-Governmental Organizations

Agua para el Pueblo (Water for the People): Apdo 1149, Tegucigalpa; tel 31-4359; fax 39-0651; Dir: Jacobo Nuñez; Assoc Dir: Anthony Brand; construction of infrastructure for water and sanitation, training of local water boards and associations, and protection and recuperation of catchment basins.

Asociación Hondureña de Ecología para la Conservación de la Naturaleza (AHE—Honduran Ecology Association for Conservation of Nature): Apdo T50, Toncontín, Centro Comercial CA B27, Tegucigalpa, DC; tel 32-9018; fax 31-0976; Exec Dir: Roberto Vallejo Larios; non-profit organization dedicated to research into aquatic and land ecoystems and protection of endangered plant and animal species.

Centro de Desarrollo Comunitario (CEDEC—Community Development Centre): Edif Everest No 301, 4 Avda 4-5 Calles No 428, Comayaguela; Apdo 1495, Tegucigalpa; tel 37-4759; fax 38-7676.

Centro de Investigación y Promoción Agrícola-Ambiental (CIPA-A—Agricultural-Environmental Research and Promotion Centre): Apdo 156, El Progreso, Yoro; tel 66-1110; Gen Co-ordinator: Juan Antonio Mejía Guerra.

Christian Children's Fund of Honduras: Col Matamoros, 5a Avda, Calle Almeria, Paseo La Campaña, 3002, Apdo 3263, Tegucigalpa; tel 32-0265; fax 32-9314; Nat Dir: Norma S de Sierra; Human Resources Manager: Fredy Acosta; works for the survival, development and protection of children through the provision of health care, education, a safe environment and water.

Comité para la Defensa y Desarrollo de la Flora y Fauna del Golfo de Fonseca (CODDEFFAGOLF—Committee for the Defence and Development of the Flora and Fauna of the Fonseca Gulf): Apdo 3663, Tegucigalpa; tel 37-2655; fax 37-2655; Exec Dir: Jorge Varela Márquez; Head of Projects: Bonifacio Sánchez; preservation of wetlands in Southern Honduras; organic farming projects; ecotourism programmes.

Cooperativa Americana de Remesas al Exterior Internacional en Honduras (CARE International, Honduras): Apdo 729, Tegucigalpa; tel 32-8601; fax 32-0913; Dir: Marge Tsitouris; Asst Dir: Peter Heffron; community development activities include construction of rural aqueducts and protection of forests and water sources.

Instituto Hondureño de Bienestar Rural (INHBIER—Honduran Institute of Rural Welfare): Barrio la Concordia, entre 10 y 11 Avda, frente a la Gasolinera Dippsa, Apdo 2339, Tegucigalpa; tel 22-5441; development of low-income farmers and their environment through mutual aid programmes.

HUNGARY

Governmental Organizations

Ministry of Environmental Protection and Urban Development: 1011 Budapest, Fő u 44-50, POB 351; tel (1) 115-4840; telex 224879; fax (1) 136-2198; Minister of Environmental Protection and Urban Devt: János Gyurko; conservation of freshwater and terrestrial environments, forests, wild flora and fauna, national parks and protected areas, landscape, habitats and resources; formulation of environmental policy; conservation strategies; monitoring and impact assessment.

Non-Governmental Organizations

Danube Commission: 1068 Budapest, Benczúr u 25; tel (1) 122-8083; fax (1) 142-3312; Dir of Secr: Dr H Strasser; Asst Dirs: P Moralyiski, G Balasoiu, M Szathmáry; responsibilities include standardization of sanitary monitoring, co-ordination of hydrometeorological services on the River Danube and publication of a hydrological newsletter and long-and short-term hydrological forecasts.

Environmental Partnership for Central Europe: 1023 Budapest, Frankel Leo ut 102, IV, 40; tel (1) 136-3370; fax (1) 136-3370; Country Dir: Zsuzsa Foltanyi; see German Marshall Fund, USA.

ISTER, East European Environmental Research Unit: 1025 Budapest, Kapy u 6; tel (1) 176-2288; fax (1) 176-2288; Pres: János Vargha; Exec: György Droppa; protection of the environment through research, advisory work and publications.

Magyar Madartani es Termeszetvedelmi Egyesulet (MMF—Hungarian Ornithological and Nature Conservation Society): 1121 Budapest, Kolto u 21; tel (1) 156-2133; telex 226115; fax (1) 175-8327; Gen Sec: Gyorgy Kallay; environmental education; nature conservation; conservation of birds, including migratory species; publications.

Magyar Természetvédok Szovetsége (Hungarian National Society of Conservationists): 1121 Budapest, Kolto u 21; tel (1) 175-0684; fax (1) 175-0684; Pres: Dr Erzsébet Schmuck; organization of local conservation and environmental groups and regional and national societies.

Magyar Tudományos Akadémia Ökológiai és Botanikai Kutatóintézete (Ecological and Botanical Research Institute of the Hungarian Academy of Sciences): 2163 Vácrátót, Alkotmány u 2-4; tel (27) 36-0122; telex 282201; fax (27) 36-0110; Dir: Dr Edit Kovács-Láng.

World Wide Fund for Nature (WWF) Hungary: 1124 Budapest, Németvölgyi ú 78B; tel (1) 175-4790; fax (1) 175-4790; Principal Officer: László Haraszthy.

ICELAND

Governmental Organizations

Ministry of the Environment: Vonarstraeti 4, 150 Reykjavík; tel (1) 609600; fax (1) 624566; Minister: Eidur Guđnason.

Ministry of Education and Culture: Söllvhólsgötu 4, 150 Reykjavík; tel (1) 609500; telex 623068; fax (1) 623068; Minister: Ólafur G Einarsson.

Ministry of Health and Social Security: Laugavegi 116, 150 Reykjavík; tel (1) 25000; Minister: Sighvatur Björgvinsson.

Ministry of Social Affairs: Hafnarhúsinu viđ Tryggvagötu, 150 Reykjavík; tel (1) 25000; telex 3000; Minister: Jóhanna Sigurđardóttir.

Directorate of Shipping: Hringbraut 121, 107 Reykjavík; tel (1) 25844; fax (1) 29835; Dir: B E Gudmundsson; Head, Environmental Section: D Egilson; inspection of ships; maritime pollution control; emergency response.

Nature Conservation Council: Laugavegi 13, 101 Reykjavík; tel (1) 25000.

Non-Governmental Organizations

Kirkens Nödhjælp i Island (Icelandic Church Aid): Tjarnargata 10, 150 Reykjavík; tel (1) 624400; fax (1) 624495; Dir: Jonas Thorisson; Public Relations Officer: Johannes Tomasson; long-term relief and development programmes.

Skógraektarfélag Íslands (Forest Association of Iceland): Ránargata 18, 101 Reykjavík; tel (1) 18150; fax (1) 627131; Chair: Hulda Valtýsdóttir; Dir: Brynjólfur Jónsson.

Wildlife Management Institute: Hlemmur 3, POB 5032, 125 Reykjavík; tel (1) 626990; fax (1) 627790; Dir: Dr Pall Hersteinsson; wildlife management, wildlife management research and animal damage control.

INDIA

Governmental Organizations

Ministry of the Environment and Forests: Paryavaran Bhavan, CGO Complex Phase II, Lodi Rd, New Delhi 110 003; tel (11) 4360721; telex 66185; fax (11) 4360678; Minister of State for Environment: Kamal Nath; Sec: R Rajamani; environmental and forestry programmes; conservation and survey of flora, fauna, forests and wildlife; prevention and control of pollution; afforestation; regeneration of degraded areas; environmental impact assessment; research; education and training; international co-operation; awareness-raising.

Indian Board for Wildlife: tel (11) 389288; telex 315265; Sec: Shri Samar Singh; conservation of wildlife; forestry; environmental management and policy; conservation strategies; setting environmental standards; protected areas; land use.

Wildlife Institute of India (WII): New Forest PO, Dehra Dun 248 006, Uttar Pradesh; tel (9) 28760; telex 337; Dir: H S Panwar; training in wildlife biology and management; conservation research.

Non-Governmental Organizations

Asian Elephant Specialist Group of IUCN/SSC: 45 Suhasini Ganguly Sarani, Calcutta 700 025; tel (33) 481144; Chair: Lyn De Alwis; Vice-Chair: R Sukumar; conservation of the Asian elephant and its habitat.

Bombay Natural History Society: Hornbill House, Shaheed Bhagat Singh Rd, Bombay 400 023; tel (22) 244085; fax (22) 2830300; Dir: Dr Jay S Samany; natural history; wildlife conservation; conservation education; environmental impact assessment.

Centre for Science and Environment: F-6, Kailash Colony, New Delhi 110 048; tel (11) 6433394; fax (11) 6475879; Dir: Anil Agiarwal; environmental research.

Chemtech Foundation: Taj Mahal Bldg, 3rd Floor, 210 Dr D N Rd, Fort, Bombay 400 001; tel (22) 2042044; telex 1182775; fax (22) 2850170; Pres: Jasu Shah; Sec-Gen: Indra Mohan; organization of trade fairs, conferences and workshops focusing on environmental issues facing industry; provision of marketing base for pollution-control companies.

Data Centre for Natural Resources: 143 Infantry Rd, Kamala Mansion, Bangalore 560 001; tel (812) 573206; Principal Officer: C Sathyendra; collection and dissemination of information relating to environmental conservation; workshops and seminars; conservation of biodiversity in Western Ghāts and eastern Himalaya; affiliated to WWF India.

Development Alternatives: Environment Systems Branch, B-32 Tara Crescent, Qutab Institutional Area, New Mehrauli Rd, New Delhi 110 016; tel (11) 665370; telex 3173216; fax (11) 6866031; Pres: A Khosla; non-profit organization dedicated to sustainable development in India.

Environment Society of Chandigarh: Karuna Sadan, Sector 11-B, Chandigarh 160 011; tel (172) 22807; Sec: S K Sharma; monitoring of birds and wildlife, including migratory species; conservation education; environmental management and planning; resource management; research into human settlements.

Himalayan Research Group: POB 1, Tallital, Nainital 263 002, Uttar Pradesh; Pres: Dr S C Joshi; publications on environment and development for colleges and primary schools in central Himalaya.

Indian Academy for Environmental Law, Conservation and Research: 6 Akbar Rd, New Delhi 110 011; tel (11) 373258; nature conservation; research into pollution and destruction of the environment.

India Committee for IUCN: c/o Dept of Environment, Forests and Wildlife, B-11 Block, Paryavaran Bhavan, CGO Complex Phase II, Lodi Rd, New Delhi 110 003; tel (11) 384461; telex 315265; Sec: Shri T N Seshan.

IUCN Commission on Education and Communication: c/o 1 12th Cross, Rajmahal Vilas Ext, Bangalore 560 080; tel (812) 345595; telex 0818452334; fax (812) 341674; Chair: M A Partha Sarathy; Deputy Chair: Frits Hesselink.

Indian Ecological Society: c/o Bungalow 51, Sector 4, Chandigarh 160 001; tel (11) 540814; telex 0386473; Pres: Prof Avtar Singh Atwal; ecological research and education; nature conservation; pollution control; waste recycling.

Indian National Trust for Art and Cultural Heritage (INTACH): 71 Lodhi Estate, New Delhi 110 003; tel (11) 4631818; Sec: B K Thapar; awareness-raising; preservation of the cultural and natural heritage of India, including places of archaeological, historical, artistic and scientific value.

Jawaharlal Nehru University, School of Environmental Sciences: New Delhi 110 067; tel (11) 652438; telex 3173167; fax (11) 6865886.

Kalyani University, Environment and Ecology Communication Center: Fisheries Laboratory, Department of Zoology, Kalyani 741 235, West Bengal; tel (3162) 98220; Dir: Prof S K Konar; research on all aspects of ecology.

National Women's Welfare Centre: Vanitha Bhavan, Ariyancode, Ottasekharamangalam PO, Thiruvananthapuram 695 125, Kerala; tel (471) 255209; Principal Officers: S Lilly Genet, T Kamalam, L Indiradevi; training, consultancy, publishing, research and documentation.

North-Eastern Hill University (NEHU), Department of Zoology: St Edmund's College, Shillong, Meghalaya 793 003; fax (364) 22976; Principal Officer: Dr Abhik Gupta; protection of local forests and streams from pollution and over-exploitation.

Ranthambhore Foundation: 19 Kautilya Marg, Chanakyapuri, New Delhi 110 021; tel (11) 3016261; fax (11) 3019457; Exec Dir: Valmik Thapar; Field Dir: Dr G S Rathore; Projects Co-ordinator: Sunny Phillip; work with communities around Ranthambhore National Park and Project Tiger Reserve in the fields of primary health care, family planning, income generation from traditional skills, dairy development, animal husbandry, education, afforestation, agroforestry, land-use, alternative energy and wildlife research.

M S Swaminathan Research Foundation—Centre for Research on Sustainable Agricultural and Rural Development: Third Cross Rd, Institutional Area, Taramani, Madras 600 113; tel (44) 2351229; telex 415119; fax (44) 2351319; Principal Officers: Dr M S Swaminathan, Mina Swaminathan.

Tata Energy Research Institute (TERI): 102 Jor Bagh, New Delhi 110 003; tel (11) 4632639; telex 3161593; fax (11) 4621770; Dir: Dr R K Pachauri; autonomous, non-profit organization engaged in research on Energy Policy, Energy Engineering and Technology, Industrial Energy, Biosciences, and Information and Computer Sciences; training programmes on energy and environment for senior officials in the government, private and public sectors.

UNESCO Regional Office for Science and Technology for South and Central Asia: UNESCO House, 8 Poorvi Marg,

Vasant Vihar, New Delhi 110 057; tel (11) 677310; telex 3165896; fax (11) 6873351; Dir: Dr John V Kingston.

Wild Society for Conservation of Forest and Wildlife: 21A Asmita, Swanand Soc, Sahakar Nagar 2, 411 009 Pune; tel (212) 430877; Founder Pres: Shekhar Devidas Nanajkar; Sec: Rajiv Shardkumar Pandit; education through slide shows, films, lectures, posters, handbills, the media and exhibitions; campaign to make Koyana Valley a national park; Peshwa Park zoo; plantation and plant nursery managed by handicapped people; monthly magazine; wildlife training; organization of treks and expeditions; campaigns on waste and noise and air pollution.

Wildlife, Environment and Tourism Institute: 3/16 Murai Tola, Ayodhya, Faizabad 224 123, Uttar Pradesh; tel (5276) 2214; Founder, Pres and Dir: Dr M R Yadav; education, training, conferences, management, research and conservation.

Turtle Research and Conservation Centre: c/o Department of Zoology, K S Saket Postgraduate College, Avadh University, Ayodhya, Faizabad 224 123, Uttar Pradesh.

World Wide Fund for Nature (WWF) India: POB 3058, 172B Lodhi Rd, New Delhi 110 003; tel (11) 616532; fax (11) 4626837; Pres: Dr Monkombu S Swaminathan; Vice-Pres: Sohrabj P Godrej; national affiliate organization of World Wide Fund for Nature (WWF) International; fund-raising; conservation; research; education; support for conservation agencies; international co-operation.

Y C College of Science, Department of Pollution Studies: Karad 415 110; tel (2164) 2356; Principal Officer: Dr R K Trivedy; undergraduate and postgraduate studies, and consultancy.

INDONESIA

Governmental Organizations

Ministry for Population and the Environment: Jalan Medan Merdeka Barat 15, Jakarta Pusat; tel (21) 371295; telex 46143; Minister of State: Prof Dr Emil Salim.

Ministry of Forestry, Directorate-General of Forest Protection and Nature Conservation: Jalan Jenderal Gatot Subroto, Jakarta 10270; tel (21) 581820; telex 45996; fax (21) 5700226; Minister of Forestry: Dr Hasrul Harahap; wise use of natural resources; wetland conservation; wildlife conservation; conservation of marine environments; impact assessments.

Research Institute for Water Resources Development: Jl Ir H Juanda 193, Bandung; tel (22) 2504053; telex 28283; fax (22) 2500163; Principal Officer: Dr Badruddin Machbub; research and development; preparation of standards for water resources and environmental quality; irrigation; sediment control; river engineering.

Non-Governmental Organizations

Bogor Agricultural University, Faculty of Fisheries, Laboratory of Oceanography: Jalan Rasamala, Campus IPB Darmaga, Bogor 16680; tel (251) 21171; fax (251) 311868; Principal Officers: Dr Mulia Purba, Dr John Pariwono; courses and research on marine sciences and fisheries management.

UNESCO Regional Office for Science and Technology for South-East Asia (ROSTSEA): UN Bldg, 2nd Floor, Jalan M H Thamrin 14, Jakarta; tel (21) 321308; telex 61464; fax (21) 3150380; Dir: Dr J Hillig; Environment Co-ordinator: Dr J R E Harger; training courses, seminars and fellowships in all aspects of environmental affairs.

Wahana Lingkungan Hidup Indonesia (WAHLI—Indonesian Environmental Forum): Jalan Penjernihan I, Kompleks Keuangan 15, Pejompongan, Jakarta 10210; tel (21) 586820; telex 45141; fax (21) 586181; Exec Dir: Agus Purnomo; network of environmental NGOs; liaison with government.

World Wide Fund for Nature (WWF) Indonesia: Jalan Pela 3, Gandaria Utara, POB 7928 JKSKM, Jakarta Selatan 12079; tel (21) 7203095; telex 47243; fax (21) 7395907; Rep: Dr Russell Betts.

IRAN

Governmental Organizations

Department of the Environment: Ostad Nejotollahi Ave 187, POB 5181-15875, Teheran; tel (21) 898233; telex 215064; fax (21) 898246; Dir: Dr Hadi Manafi Rasi; scientific research on all environmental issues, including air, water, soil and noise pollution and wildlife conservation; issue of regulations and setting of environmental standards; promotion of environmental education; monitoring, control and law enforcement by rangers; establishment and expansion of natural history museums and environmental exhibitions.

IRAQ

Governmental Organizations

Council for Environment Protection: Aqaba bin Nafi Sq, POB 2419, Baghdad; tel 719-7438.

Ministry of Agriculture and Irrigation: Khulafa St, Khullani Sq, Baghdad; tel 887-3251; telex 212222; Minister: Abd al-Wahab Mahmoud Abdullah.

Ministry of Health, Labour and Social Affairs, Directorate of Environmental Protection Services: Khulafa St, Khullani Sq, Baghdad; tel 887-1881; telex 212621; Minister: Umeed Madhat Mubarak.

Solar Energy Research Council: Scientific Research Centre, Jadiriyah, POB 13026, Baghdad; telex 213976.

IRELAND

Governmental Organizations

Department of the Environment: Custom House, Dublin 1; tel (1) 6793377; telex 31014; fax (1) 8742710; Minister for the Environment: Michael Smith.

Department of the Marine: Leeson Lane, Dublin 2; tel (1) 6785444; telex 91798; fax (1) 6618214; Minister for the Marine: David Andrews; Minister for State: Gerry O'Sullivan; responsible for regulation and development of the marine sector.

Office of Public Works, National Parks and Wildlife Service: 51 St Stephen's Green, Dublin 2; tel (1) 613111; fax (1) 610747; establishment and management of national parks and nature reserves; implementation and enforcement of legislation for the protection of wildlife and habitats.

Non-Governmental Organizations

Bord Iascaigh Mhara (Irish Sea Fisheries Board): POB 12, Crofton Rd, Dun Laoghaire, Co Dublin; tel (1) 2841544; fax (1) 2841123; Chief Exec: A O Gannon; Sec: E Boland; development of sea fishing industry, including aquaculture.

Comhaontas Glas (The Green Party): 5A Upper Fownes St, Dublin 2; tel (1) 6771436; fax (1) 6797168; Co-ordinator: Bronwen Maher; co-operation with members in local government and the Dáil (House of Representatives) to ensure that environmental issues are high on the political agenda.

Dublin Naturalists' Field Club: c/o 35 Nutley Park, Donnybrook, Dublin 4; tel (1) 2697469; Pres: Patrick W Jackson; Sec: David Nash; study of all aspects of the natural history of the Greater Dublin area.

Earthwatch—Friends of the Earth (FoE) Ireland: Harbour View, Bantry, Co Cork; tel (27) 50968; telex 265871; fax (27) 50545; Co-ordinator: Jeremy Wates; campaigning membership organization working for environmental protection.

Environment Policy Committee: Irish Business and Employers' Confederation, Confederation House, Kildare St, Dublin 2; tel (1) 6779801; fax (1) 6777823; Dir-Gen: John Dunne.

Greenpeace Ireland: 44 Upper Mount St, Dublin 2; tel (1) 6619836; fax (1) 6605258; Campaigns Dir: John Bowler; Development Dir: Margaret Ward.

Irish Deer Society: c/o P Cronin, Mangerton Rd, Mucross, Killarney, Co Kerry; Pres: Matt Hyland; Chair: Dr H McKeague; conservation of free-living deer and their habitat; dissemination of information and promotion of humane culling of deer.

Irish Landscape Institute: 8 Merrion Sq, Dublin 2; tel (1) 6761703; fax (1) 6610948; represents interests of professional landscape architects and landscape managers.

Irish Peatland Conservation Council: Capel Chambers, 119 Capel St, Dublin 1; tel (1) 8722397; Chair: Dr Peter Foss; Information Officer: Hilary O'Brien.

Organic Trust Ltd: Islands, Urlingford, Kilkenny; tel (56) 31411; Principal Officers: Clare Mooney, Alan Tyrrell; organic farming standards, inspection and certification.

Royal Zoological Society of Ireland: Phoenix Park, Dublin 8; tel (1) 6771425; fax (1) 6771660; Dir: P Wilson; Pres:

Prof David McConnell; maintenance, exhibition, breeding, conservation and study for educational purposes of all species of living animals; promotion of study and knowledge of zoology and cultivation of interest in the conservation of animals.

An Taisce—The National Trust for Ireland: Tailors' Hall, Back Lane, Dublin 8; tel (1) 541786; fax (1) 533255; Pres: David Rowe; Chair: Patricia Oliver; independent, voluntary organization concerned with conservation of the natural and man-made environment; campaigns; lobbying; education; participation in planning.

Tree Council of Ireland—Comhairle Crann na h Éireann: 33 Botanic Rd, Glasnevin, Dublin 9; tel (1) 306946; fax (1) 306948; Pres: Fergal Mulloy; Vice-Pres and Public Relations Officer: Margaret Gormley; promotes planting, propagation, conservation and maintenance of trees in urban and rural areas; dissemination of knowledge on trees and tree-care; advisory functions; organization of annual National Tree Week.

University College, Environmental Institute: Richview, Clonskeagh, Dublin 14; tel (1) 2697988; fax (1) 2837009; Dir: Prof Frank J Convery; Assoc Dir: Prof Vincent Dodd; research and development, and teaching.

Young Environmentalist Federation: c/o Ballinakella Lodge, Whitegate, Co Clare; tel (61) 927030; fax (61) 927030; Principal Officers: Hugh W L Weir, Francis O'Dwyer; federation of affiliated groups.

ISRAEL

Governmental Organizations

Ministry of the Environment: POB 6234, Kiryat Ben-Gurion, Jerusalem 91061; tel (2) 701606; telex 25623; fax (2) 513945; Minister: Yossi Sarid; Dir-Gen: Dr Israel Peleg; environmental policy; strategies, standards and priorities for environmental protection; air, marine and water quality control; monitoring of solid waste, hazardous substances, radiation and noise; pest control; agro-ecology; land use planning; legislation; international co-operation; education and information.

Ministry of the Interior, Environmental Protection Service: POB 6158, 2 Rehov Kaplan, Kiryat Ben-Gurion, Jerusalem 91061; tel (2) 701411; fax (2) 639368; Minister: Arye Deri.

Nature Reserves Authority: 78 Yirmeyahu St, Jerusalem 94467; tel (2) 387471; fax (2) 374887; Dir-Gen: Dan Perry; Chair: Uri Baidatz; creation, restoration and preservation of nature reserves; protection of wildlife and other natural resources through scientific monitoring, education and law enforcement.

Non-Governmental Organizations

Ben-Gurion University of the Negev, Mitrani Centre for Desert Ecology (MCDE): Jacob Blaustein Institute for Desert Research, Sede Boqer Campus, Beersheba 84990; tel (57) 56507; fax (57) 555058; Head: Uriel N Safriel.

EcoNet Israel, the Ecological Network: POB 581, Karkur 37105; tel (6) 377072; fax (6) 377072; Dir: Shirley Benjamin.

Guardians of the Environment: POB 1488, Rechovot 76110; tel (8) 491062; fax (3) 5232901; Chair: Dany Montag; Sec: Nechama Chen; monitoring of the environmental status of industrial plants and companies; issues own seal of approval.

Hai-Bar Society for the Establishment of Biblical National Wildlife Reserves in Israel: c/o Nature Reserves Authority, 78 Yirmeyahu St, Jerusalem 94467; tel (2) 387471; fax (2) 383405; Chair: Uri Baidats; establishment of national wildlife reserves for animals referred to in the Bible; wildlife breeding and release into natural habitats.

Histadrut Consumers and Environment Protection Authority: Histadrut Bldg, 93 Arlosoroff St, Tel-Aviv 62098; tel (3) 6963582; fax (3) 5441285; Project Manager: Moshe Friedman; branch of General Federation of Labour in Israel.

Israel Union for Environmental Defence: 317 Hayarkon St, Tel-Aviv 63504; tel (3) 5468099; fax (3) 449941; Exec Dir: Dr Alon Tal; Chair Bd of Dirs: Eilon Schwartz; legal defence of the environment in Israel; facilitation of co-operation with neighbouring countries to protect the coral reefs of the Red Sea; Citizen Laboratory provides scientific data on environmental hazards.

Jerusalem Hebrew University, Mayrock Centre for Russian, Eurasian and East European Research: Mount Scopus, Jerusalem 91905; tel (2) 883180; telex 26458; fax (2) 322545; Dir: Dr Stefani Hoffman; ecological research.

Jewish National Fund: POB 283, Jerusalem 91002; tel (2) 291333; fax (2) 291311; World Chair: Moshe Rivlin; Dir-Gen: Yitzhak Eliasheyv; land reclamation; infrastructure; afforestation; ecological conservation; development of forest recreation; drainage; water resources.

Life and Environment: POB 20040, Tel-Aviv 61200; tel (3) 6919241; fax (3) 5223785; Chair: Josef Tamir; Hon Sec: Miriam Zibner; co-ordinatation of activities of member environmental organizations in Israel; public representative of National Planning and Building Board; lobbying, campaigns and environmental education.

Society for the Protection of Nature in Israel (SPNI): 4 Hashfela St, Tel-Aviv 66183; tel (3) 375063; fax (3) 377695; Exec Dir: Yossi Leshem; Chair: Yoav Sagi; non-profit organization; conservation of landscape and ancient relics; protection of plant and animal life; environmental protection; research; education.

Tel-Aviv University, George S Wise Faculty of Life Sciences, Institute for Nature Conservation Research: Ramat Aviv 69978; tel (3) 6409813; fax (3) 6407304; Dir: Prof Avital Gasith; study of the environment in Israel with an emphasis on endangered species and ecosystems.

ITALY

Governmental Organizations

Ministry of the Environment: Piazza Venezia 11, 00187 Rome; tel (6) 703713204; fax (6) 703713203; Minister: Valdo Spini.

Ministry of Agriculture, Department for Mountain Economy and Forests: Via XX Settembre, 00187 Rome; tel (6) 46651; telex 610148; fax (6) 4742314; Minister: Gianni Fontana.

Ministry of Health, Public Hygiene Services Department: Via dell'Industria 20 EUR, 00144 Rome; tel (6) 5994; fax (6) 5934774; Minister of Health: Raffaele Costa.

Ministry of the Merchant Marine, Central Inspectorate of Sea Defence: Via dell'Arte 18 EUR, 00144 Rome; tel (6) 59081; Minister of Transport and the Merchant Marine: Dr Giancarlo Tesini.

Centro di Documentazione Internazionale Parchi della Provincia di Firenze (CEDIP—International Park Documentation Centre, Province of Florence): Villa Demidoff, Pratolino, 50036 Florence; tel (55) 409051; fax (55) 476116; Co-ordinator: Prof Giovanni Valdrè; information on protected area issues; participation in development of protected areas system in Italy.

Comitato Nazionale per la Ricerca e per lo Sviluppo dell'Energia Nucleare e delle Energie Alternative (ENEA—National Committee for Research and Development of Nuclear and Alternative Energy): Viale Regina Margherita 125, 00198 Rome; tel (6) 85282041; telex 610183; fax (6) 85285875; Pres: Prof Nicola Cabibbo; Dir-Gen: Dr Fabio Pistella; environmental protection; energy saving; use of renewable energy sources.

Consiglio Nazionale Delle Ricerche (National Research Council): Via Salaria 221, 00199 Rome; tel (6) 49931; telex 610076; fax (6) 4957241; environmental research.

Coordinamento Parchi e Reserve Regionali Italiane (CPRRI—Italian Regional Parks and Reserves Co-ordination): Parco dell'Etna, Via della Regione 25, 90035 Nicolosi; Pres: Bino L I Calsi; co-ordination of administration of Italian regional and national parks and protected areas.

Ente Autonomo del Parco Nazionale d'Abruzzo (Abruzzo National Park Office): Viale delle Medaglie d'Oro 141, 00136 Rome; tel (6) 3496993; telex 624550; fax (6) 3497594; Dir: Prof Franco Tassi; protection of the environment of the Abruzzo National Park; management of habitats; education in wise use of natural resources; control of tourism.

Non-Governmental Organizations

Advisory Committee on Protection of the Sea (ACOPS), Mediterranean Office: Castello 5856, Campo S Maria Formosa, Venice; tel (41) 5331351.

Centro Interdisciplinare Studi Ambientali (CISA—Centre for Interdisciplinary Environmental Studies): Via Masaccio 67, 50132 Florence; tel (55) 2345754; fax (55) 244300; International Co-ordinator: Paolo G Meozzi; Tech Dir: Andrea Bertelli; environmental impact analysis, hazardous waste management, agriculture and forestry projects with minimum impact on natural resources.

DocTer Institute of Environmental Studies: Foro Bonaparte 12, Milan; non-profit institute; research and documentation on planning and the environment; office in the United Kingdom.

Federazione Nazionale per le Liste Verdi (Green Party): Rome; tel (6) 4957383; promotion of environmentalist and anti-nuclear policies; branch of the European Green movement.

Fondo Mondiale per la Natura—World Wide Fund for Nature (WWF) Italy: Via Salaria 290, 00199 Rome; tel (6) 8417634; fax (6) 8554410; Pres: Grazia Francescato; Vice-Pres: Carlo Galli; Sec-Gen: Valerio Neri; conservation; education; management of wildlife areas (64 natural reserves); proposals for environmental policy (energy; transport; consumer affairs; land management); 300 local offices; national affiliate organization of World Wide Fund for Nature (WWF) International.

Milan Office: Via Donatello 5B, 20131 Milan; tel (2) 29404260; fax (2) 29404903.

Istituto Nazionale per la Fauna Selvatica (INFS—National Institute for Wildlife: Via Cà Fornacetta 9, Ozzano Emilia, 40064 Bologna; tel (51) 798746; fax (51) 796628; Admin Officer: Rina Signani; national research and advisory agency for issues concerning wildlife conservation.

Lega per l'Abolizione della Caccia (LAC—League for the Abolition of Hunting): Via Carlo Alberto 39, 00185 Rome; Pres: Prof Carlo Consiglio; campaigns on wildlife, abolition of hunting, animal rights, trapping, poaching and endangered species.

Lega Italiana Protezzione Uccelli (LIPU—Italian Bird Protection League): Vicolo S Tiburzio 5, 43100 Parma; tel (521) 233414; fax (521) 287116; Head, Int Dept: Barbara Lombatti; protection of birds and their habitats, and other fauna; education.

Società Italiana di Ecologia (SITE—Italian Ecological Society): c/o Istituto di Ecologia, Viale delle Scienze, 43100 Parma; tel (521) 905659; fax (521) 905665; Pres: Prof Oscar Ravera; Gen Sec: Prof Angelo Tursi; promotion of theoretical and applied ecological research; dissemination of ecological knowledge; development of exchange between researchers and facilitation of national and international co-operation through conferences, symposia and publications.

Survival International Italian Office: CP 1194, 20101 Milan; tel (2) 8900671; fax (2) 8900674; Principal Officer: Francesca Cassella.

Università di Catania, Istituto Dipartimentale di Biologia Animale e di Genetica (Catania University, Departmental Institute of Animal Biology and Genetics): Via Androne 81, 95124 Catania, Sicily; tel (95) 327221; fax (95) 4463865; ecology; biochemistry; biology; conservation biology; zoology.

Università di Padova, Istituto di Botanica e Fisiologia Vegetale, Rete Fenologica Italiana (University of Padova, Institute of Botany and Vegetable Physiology, Italian Phenological Network): Via Orto Botanico 15, 35123 Padova; tel (49) 26922; botany; biology; agricultural research; horticulture; research into herbicides, soil pollution, pesticides and insecticides.

JAMAICA

Governmental Organizations

Ministry of Tourism and the Environment: Petrojam Bldg, 36 Trafalgar Rd, Kingston 10; tel 926-9170; Minister: John Junor.

Ministry of Agriculture, Natural Resources Conservation Department: Hope Gardens, Kingston 6; tel 927-1731; Minister: Seymour Mullings.

Non-Governmental Organizations

Gosse Bird Club: 2 Starlight Ave, Kingston 6; tel 927-8444; fax 927-8444; Pres: John Fletcher; Sec: Marcia Mundle; study of birds; lobbying for conservation of birds and their habitats.

Inter-American Institute for Co-operation on Agriculture (IICA) Jamaica: POB 349, 11 Fairway Avenue, Kingston 6.

Jamaica Conservation and Development Trust (JCDT): POB 1225, 46 Duke St, Kingston 8; tel 922-0668; fax 922-0665; Exec Dir: Dr David Smith; Programmes Dir: Susan Anderson; planning and management of protected areas, including Blue and John Crow Mountains National Park and Montego Bay Marine Park; environmental education; liaison with government and the private sector.

Negril Chamber of Commerce (NCC): POB 55, Negril, Westmoreland; tel 957-4067; fax 957-4591; Chair, Conservation Cttee: Sylvie Grizzle; preservation of the marine environment, coral reefs and coastal and marine protected areas; prevention of sea pollution; lobbying government.

Portland Environment Protection Association (PEPA): Railway Bldg, 32 West St, POB 199, Port Antonio, Portland; tel 993-3407; Pres and Founder: Marguerite Gauron; Sec and Treas: Boyd Lewis; environmental education; active conservation; environmental monitoring; promotion of national and international environmental media events.

Society of Caribbean Ornithology: 2 Starlight Ave, Kingston 6; tel 927-8444; fax 927-8444; Pres: Catherine Levy; Vice-Pres: Dr J M Wunderle, Jr; promotion of scientific study and conservation of Caribbean birds and their habitats.

United Nations Association of Jamaica (UNAJ): 33 Anthurium Drive, Kingston 6; tel 962-9771; fax 962-9333; Pres: Glenn Bowen; Gen Sec: Iris Henry; activities include advocacy of environmental management.

West Indies University, Electron Microscopy Unit: Mona Campus, Kingston 7; tel 977-1076; fax 927-1640; Principal Officers: Prof V B Meyer-Rochow, W A Reid; understanding and protecting cave environments and their inhabitants.

JAPAN

Governmental Organizations

Environment Agency of Japan: 1-2-2 Kasumigaseki, Chiyoda-ku, Tokyo; tel (3) 3581-9111; telex 33855; fax (3) 3504-1634; Minister of State and Dir-Gen: Taikan Hayashi; environmental management; establishment of environmental policy; protection of wildlife; sustainable development; research; statistics; impact assessment; protection of marine, freshwater and terrestrial environments; pollution control.

Non-Governmental Organizations

Ecological Society of Japan: c/o Biological Inst, Science Faculty, Tohoku Univ, Aoba-ku, Sendai 980; tel (22) 225-0636; fax (22) 225-0636; Pres: Yasuyuki Oshima; Sec-Gen: Moritaka Nishihira; research; publication of journals; conferences and meetings.

Elsa Nature Conservancy: Tsukauba-Gakuen, POB 2, Ibaraki 305; tel (298) 51-1637; fax (298) 51-1637; Pres: Eiji Fujiwara; conservation of fauna and flora; education.

Environmental Technology Research Institute: Matsumo Bldg 2-1-20, Kita-ku, Osaka 530; tel (6) 357-7611; fax (6) 357-7612; Pres: Shigehisa Iwai; monthly magazine; book publication; research into environmental conservation; international conferences.

Global Environmental Forum: 1-9-7 Iikura Bldg, Azabudai, Minato-ku, Tokyo 106; tel (3) 5561-9735; fax (3) 5561-9737; Exec Dir: Takashi Hirano; Sec-Gen: Shosaku Kamei; scientific research on domestic and international environmental issues; environmental policy research; dissemination of environmental information and research results by means of symposia, conferences and publications.

Tsukuba Office: 16-2, Onagawa, Tsukuba City, Ibaragi Prefecture; tel (298) 58-1366; fax (298) 58-1346.

Greenpeace Japan: Yoyogikaikan 4F, 1-35-1 Yoyog, Shibuya-ku, Tokyo 151; tel (3) 5351-5400; fax (3) 5351-5417; Exec Dir: Naoki Ohara; Admin Asst: Sanae Shida.

Hokkaido Shizen Hogo Kyokai (Nature Conservation Society of Hokkaido): Kamori Bldg, Central Ward, North 3, West 11, Sapporo 060; tel (11) 251-5465; fax (11) 251-5465; Principal Officer: Dr Kogure Tokuo; research; publications.

Institute for Himalayan Conservation: 3-5-7-111, Yoyogi, Shibuya-ku, Tokyo 151; tel (3) 5350-8458; fax (3) 5350-8459; Chair: Jiro Kawakita; environmental conservation and development in the mountainous regions of Nepal.

Japan Air Pollution Victims Association: Suny Sity Shinjyukugyoenn 10F, 2-1-3 Shinjyuku, Shinjyuku-ku, Tokyo; tel (3) 3352-9475; fax (3) 3352-9476; Chair: Kimio Moriwaki; eradication of environmental pollution and support for victims.

Japan Committee for IUCN: c/o Nature Conservation Soc of Japan, Toranomon Denki Bldg, 2-8-1 Toranomon, Minato-ku, Tokyo 105; tel (3) 502-0488; Chair: Prof Dr Makoto Numata.

Japan Environment Association: Toranoman-1 Bldg, 5-8 Toranoman-1, Minato-ku, Tokyo 105; tel (3) 508-2651; Pres: Dr Kiyoo Wadachi; environmental protection; education.

Japan Environmental Exchange: Tomitaya Bldg 4F, Teramachi-dori, Oike-agaru, Nakagyo-ku, Kyoto 604; Principal Officers: Susumo Sato, Maura Hurley, Hikaru Furutaka; project of the Earth Island Institute, USA; exchange of volunteers between USA and Japan to work on environmental protection and restoration projects.

Japan Network for Earth Environment and Prevention of Pollution (JNEP): 201 Miyanaga Bldg, 1-5-1 Motoakasaka, Minato-ku, Tokyo 107; tel (3) 3470-0335; fax (3) 3470-0335; Principal Officer: Sin Honnma.

Japan Society of Air Pollution: Shinjuku 1-29-8, Shinjuku-ku, Tokyo 160; tel (3) 3341-5632; fax (3) 3341-8224; Principal Officer: Dr Naoomi Yamaki; annual conference; bi-monthly journal; annual prize; international co-operation.

Japanese Association of Zoological Gardens and Aquariums (JAZGA): Parkside Bldg, Ueno 2-13-2, Taito-ku, Tokyo 110; tel (3) 837-0211; Chair: Dr Shiro Nakagawa; environmental education; resource protection.

Kaichu Koyen Centre (Marine Parks Centre of Japan): Toranomon Denki Bldg, 2-8-1 Toranomon, Minato-ku, Tokyo; tel (3) 3591-5518; fax (3) 3502-1377; Chair: Tasuku Uno; conservation of marine environments, cetaceans, aquatic plants, coral reefs and protected areas; international co-operation; research.

Kokuritsu Kohen Kyokai (National Parks Association of Japan): c/o Toranomon Denki Bldg, 2-8-1 Toranomon, Minato-ku, Tokyo 105; tel (3) 3502-0488; fax (3) 3502-1377; Pres: Osamu Ikenouye; conservation of nature, areas of natural beauty and recreation areas, including national parks; environmental protection; education.

Nature Conservation Society of Japan (NACS-J): Toranomon Denki Bldg, 2-8-1 Toranomon, Minato-ku, Tokyo 105; tel (3) 503-4896; fax (3) 592-0496; Chair: Takao Fukiwara; conservation of primeval beech forest, coral reefs, rivers and wildlife.

Nettairin Kodo Network (JATAN—Japan Tropical Forest Action Network): 6-5 (1F) Uguisudani-cho, Shibuya-ku, Tokyo 150; tel (3) 3770-6308; fax (3) 3770-0727; Co-ordinator: Yoichi Kuroda; campaigns against Japanese corporate and government involvement in destruction of tropical, temperate and boreal forests around the world; concerned with Japan's contributions to development institutions.

Nihon Yaseiseibutsu Kenkyu Senta (JWRC—Japan Wildlife Research Centre): Yushima 2-29-3, Bunkyo-ku, Tokyo 113; tel (3) 3813-8806; fax (3) 3813-8898; Pres: Prof Dr Taisitiroo Satoo; Exec Dir: Kazuhiro Yamase; research on wildlife management, conservation and ecology; compilation and management of a database on the natural environment of Japan.

Okinawa Daigaku Chiiki Kenkyusyo—Okidai Chiikiken (Okinawa University, Research Institute on Regional Studies): Kokuba Naha 902 747, Okinawa; tel (988) 33-1578; fax (988) 32-3216; Principal Officer: Prof Jun Ui; study of the environment of Okinawa and of Asian and Pacific countries.

Sahel no Kai (Sahel Association): 3-21-4 Nishihara, Shibuya-ku, Tokyo 151; tel (3) 3460-3912; telex 34799; Principal Officer: Tadayoshi Miyao; prevention of desertification and its effects on the people of the Sahel region.

Science Council of Japan, National Committee on Nature Conservation: 7-22-34 Roppingi, Minato-ku, Tokyo 106; tel (3) 403-6921; Chair: Makoto Honma; ecology; biochemistry; biology; zoology; botany; genetic engineering; research.

Tokyo Metropolitan University, Law Faculty, Japan Centre for Human Environmental Problems: Room 304, 1-1 Minami-Ohsawa, Hachioji City, Tokyo 100; tel (3) 717-0111; Pres: Dr Ichiro Kato; research; international exchange of information; development of environmental law.

Toyama-ken Shizen Hogo Kyoukai (Toyama Prefectural Association for Nature Conservation): Shinrin Suisan Kaikan, 4-19, Funabashi-kita-machi, Toyama-shi, Toyama-ken 930; tel (764) 41-6166; fax (764) 41-5737; Pres: K Wakabayashi; Vice-Pres: M Fujihira, M Nagai; research on nature in the region; education through publications and lectures; co-ordination with other organizations.

Wild Bird Society of Japan: Aoyama Flower Bldg, 1-1-4 Shibuya, Shibuya-ku, Tokyo 150; tel (3) 3406-7141; fax (3) 3406-7144; Dir: Noritaka Ichida; protection of birds and wildlife; research; education and training; sustainable development.

World Wide Fund for Nature (WWF) Japan: Nihonseimei Akabanebashi Bldg, 3-1-14 Shiba, Minato-ku, Tokyo 105; tel (3) 3769-1711; fax (3) 3769-1795; Chair: Nobuya Hagura; Vice-Chairs: Hisako Hatakeyama, Keizo Saji; national affiliate organization of World Wide Fund for Nature (WWF) International; campaigns on wildlife, pollution, environmental policy and management, sustainable development, energy and protected areas; education and training.

JORDAN

Governmental Organizations

Ministry of Municipal, Rural and Environmental Affairs, Department of Environment: POB 1799, Amman; tel (6) 641393; fax (6) 672135; Minister: Abdel-Razzaq Tabeishat; environmental management; water and air pollution control; prevention of soil erosion; disposal of solid wastes; enforcement of environmental law.

Non-Governmental Organizations

Jordanian Society for Control of Environmental Pollution: POB 922821, Amman; tel (6) 699844; fax (6) 695857; Principal Officer: Ahmed Obeidat; public environmental awareness campaigns and environmental publications.

Royal Scientific Society (RSS): POB 925819, Amman; tel (6) 844701; telex 21276; fax (6) 844806; comprised of Environmental Research Centre, Renewable Energy Research Centre and Industrial Chemistry Centre, which has an Environmental Division.

Environmental Research Centre (ERC): Dir: Dr Murad Bino; Officer: Dr Yaseen Khayyat; studies and applied research; provision of specialized technical services, advice and training in the field of the environment for the public and private sectors in Jordan and neighbouring Arab countries; comprised of Water and Soil Division, Air and Hazardous Chemicals Division and Ecology Division.

Royal Society for the Conservation of Nature (RSCN): POB 6354, Amman; tel (6) 811689; telex 21456; fax (6) 628258; Pres: HE Anis Mouasher; protection of wild flora and fauna and their habitats; co-operation with government and international and regional agencies.

KAZAKHSTAN

Governmental Organizations

Ministry of Ecology and Bioresources: 480091 Almaty, Panfilov Str 106; Minister: S A Medvedev.

Ministry of the Forestry Industry: 480091 Almaty, pr ul Zheltoksan 112; tel (3272) 62-63-20; telex 251232; fax (3272) 62-20-26; Minister: Aleksandr K Amanbayev; concerned with forestry and hunting.

Non-Governmental Organizations

Fund in Support of Ecological Education: 480005 Almaty, S Kovalevskaya 63, Apt 13, POB 898; Almaty, Kazakh State University, Biological Department, Timiryazeva 46; tel (3272) 41-29-91; telex 251232; fax (3272) 63-12-07; Co-ordinator: Zharas A Takenov; Principal Officers: Gulmira Djamanova, Raushan Kryldakova.

Green Movement Socio-Ecological Centre (GMC): 484006 Dzhambul, ul Lunacharsky 42-2; tel (32622) 52-483; fax (32622) 75-500; Chair: Alexander Zagribelny; Exec Sec: Valeriy Kuklin; opposition to manufacture of red phosphorus fertilizer; publications and documentary films; participated in UNCED, Brazil in 1992.

Zelonoye Spaseniye (Green Salvation Ecological Society): Almaty, ul Shagabutdinova 133, kv 66; tel (3272) 67-89-06; Principal Officers: Sergei Kuratov, Artem Salin, Serge Solijnik, Alexandr Anisenkov; ecological database; media campaigns; ecological education; direct action.

KENYA

Governmental Organizations

Ministry of Environment and Natural Resources: Kencom House, POB 30126, Nairobi; tel (2) 29261; Minister: John Sambu.

Ministry of Land Reclamation, Regional and Water Development: Maji House, Ngong Rd, POB 49720, Nairobi; tel (2) 723103.

Ministry of Tourism and Wildlife, Wildlife Conservation and Management Department: Utalii House, 5th Floor, Uhuru Highway, POB 54666, Nairobi; tel (2) 331030; telex 25016; Minister of Tourism: Noah Katana Ngala; conservation of water, soil, forests and wildlife.

National Environment and Human Settlements Secretariat: POB 67839, Nairobi; tel (2) 332383.

Non-Governmental Organizations

African Centre for Technology Studies (ACTS): POB 45917, Nairobi; tel (2) 744047; fax (2) 743995; policy research centre.

African Social and Environmental Studies Programme (ASESP): POB 44777, Nairobi; tel (2) 740817; fax (2) 747960; Chair: Prof W Senteza Kajubi; Exec Dir: Dr Peter Muyanda-Mutebi; research, training, publishing and networking in social studies and environmental education.

African Wildlife Foundation: POB 48177, Nairobi; tel (2) 710367; telex 22152; fax (2) 710372; Principal Officers: Mark Stanley Price, Deborah Snelson.

British Association for Better Land Husbandry: POB 39042, Parklands Post Office, Nairobi; tel (2) 582413; fax (2) 582413; Pres: Prof Norman Hudson; Gen Sec: R Jim Cheatle; networks, projects and workshops for farmers and land managers.

Climate Network Africa: POB 21136, Nairobi; tel (2) 214898; fax (2) 214898; Co-ordinator: Grace Akumu; promotion of information exchange on issues related to climate change between NGOs, scientists and policy-makers in Africa.

East African Wildlife Society: POB 20110, Nairobi; tel (2) 748170; fax (2) 746868; Exec Officer: N K Rotich; non-profit organization dedicated to conservation of wildlife and habitats.

Eastern Africa Environmental Network (EAEN): POB 20110, Nairobi; tel (2) 748170; fax (2) 746868; Dir: Nathaniel Arap Chumo; Chair of Bd: Belinda Rego; environmental education.

Econews Africa: POB 21136, Nairobi; tel (2) 214898; fax (2) 214898; Co-ordinator: Wagaki Mwangi; environmental and development organization concerned with the analysis of global issues from a local perspective.

Environment Liaison Centre International: POB 72461, Ndemi Rd, off Ngong Rd, Nairobi; tel (2) 562015; telex 23240; fax (2) 562175; Librarian and Information Man: Dianah Macharia.

Global Coalition for Environment and Development: POB 72461, Ndemi Rd, off Ngong Rd, Nairobi; tel (2) 562015; telex 23240; fax (2) 562175; Librarian and Information Man: Dianah Macharia.

Hydrological Research Institute (HRI): POB 26581, Nairobi; Chair: Albert Kenyani Inima; research, training and consultancy organization.

IUCN Kenya: Eastern Africa Regional Office, POB 68200, Nairobi; tel (2) 502650; telex 25190; fax (2) 608026; Regional Rep: Dr Rob Malpas.

Kenya Energy and Environment Organizations (KENGO): Mwanzi Rd, Westlands, POB 48197, Nairobi; tel (2) 749747; telex 25222; fax (2) 749382; Exec Dir: Achoka Awori; Assoc Dir: Dominic Walubengo; development and promotion of renewable energy technology, environmental conservation and management of natural resources; policy advocacy.

Kenya Wildlife Service: POB 40241, Nairobi; tel (2) 501081; telex 22804; fax (2) 501752; Dir: Dr Richard E Leakey; conservation and protection of Kenya's wildlife.

Wildlife Clubs of Kenya: POB 20184, Nairobi; tel (2) 891903; fax (2) 891906; environmental education for young people through publications, teachers' workshops, student seminars and awareness-raising campaigns.

World Wide Fund for Nature (WWF) Eastern Africa Regional Office: POB 62440, Nairobi; tel (2) 332963; telex 098725495; fax (2) 332878; Regional Rep: Dr Jean-Pierre d'Huart.

KIRIBATI

Governmental Organizations

Ministry of the Environment and Natural Resource Development: POB 64, Bairiki, Tarawa; tel 21099; telex 77039; Minister: Nakibae Teuatabo.

Ministry of Home Affairs and Rural Development: POB 75, Bairiki, Tarawa; tel 21092; fax 21133; Minister: Binata Tataeka.

DEMOCRATIC PEOPLE'S REPUBLIC OF KOREA

Governmental Organizations

Co-ordinating Committee of Environmental Science: Changsan St, Sosung Dist, Pyongyang; tel (2) 51956.

Non-Governmental Organizations

Democratic People's Republic of Korea Academy of Sciences, Institute of Environmental Protection: Central District, Pyongyang.

Natural Conservation Union: 220-93-7-24 Dongsong St, Cen Dist, Pyongyang; Sec Gen: Han Sok-Yuk; conservation of wild flora and fauna; afforestation and forestry; conservation of protected areas, public lands and national parks; scientific research; management and planning of protected areas; environmental management and planning; flood control; training for management of protected areas; education in resource management.

REPUBLIC OF KOREA

Governmental Organizations

Ministry of the Environment: 7-16 Sincheon-dong, Songpa-ku, Seoul 138-240; tel (2) 421-0217; telex 25783; fax (2) 421-0280; Minister: Hwang San-Sung; formulation of environmental policies; enforcement of legislation.

Ministry of Home Affairs, Division of Nature Preservation: c/o 280-17 Bulkwang-dong, Eunpyung-ku, Seoul 122-040; tel (2) 383-0694; fax (2) 383-0695; Dir: Lee Sangyong; environmental protection; co-operation with government, organizations and individuals; education.

Forest Management Bureau: Chungyang San 1, Dongdaemun-ku, Seoul 131.

National Parks Authority: Gun-Sol-Hwi-Kwan 71-2, Nanhyun-dong, Kangnam-gu, Seoul 135-010; Dir of Public Relations: Hyong-Woo Lee; administration of 15 national parks, monuments, historic sites and recreation areas; conservation of natural resources.

Non-Governmental Organizations

Han Guk Ja Yon Bo Jon Hyop Hoe (Korean Association for Conservation of Nature): 207 Ch'ongnyangri-dong, Dongdaemun-gu, Seoul 130-012; tel (2) 962-2894; fax (2) 963-3160; Pres: Dr Won Pyong-Oh; preservation of land and natural resources; research; resource assessment; statistics; ecological monitoring.

Korean Central Council for Nature Preservation (KCCNP): 44-2 Chuja-dong, Chung-gu, Seoul 100-240; tel (2) 267-8451; fax (2) 277-5726; Pres: Dr Chung Yung Ho; planning and deployment of national conservation strategies; promotion of scientific research; environmental education; publishing; provision of consulting services; environmental monitoring; participation in international conferences.

National Parks Association of Korea: Rm 706, Sahak Bldg 19, Naeja-dong, Chongno-ku, Seoul 110-053; tel (2) 736-9656; fax (2) 720-5421; Pres: Lee Hyun Jae; Vice-Pres: Lee Hwa Young, Oh Whee Young; development of public parks and tourist sites.

KUWAIT

Governmental Organizations

Environment Protection Council: POB 24395, 13104 Safat; tel 456836; telex 46408.

Ministry of Finance and Planning, General Department of Technical Co-operation (Environmental Affairs): POB 9, 13001 Safat, al-Morkab St, Ministries Complex, Kuwait City; tel 2468200; telex 22527; fax 2404025; Minister: Nasser Abdullah ar-Rodhan.

Non-Governmental Organizations

Jamiyat Himayat al-Bea' al-Kuwaitiah (KEPS—Kuwait Environment Protection Society): POB 1896, 13019 Safat; tel 4848256; fax 4837856; Sec-Gen: Dr Badria Abdullah al-Awadi; campaigns against pollution; awareness-raising.

KYRGYZSTAN

Governmental Organizations

State Committee for the Protection of Nature: Bishkek; Chair: I S Muratalin.

LAOS

Governmental Organizations

Ministry of Agriculture and Forestry: Vientiane; Minister: Gen Sisavat Keobounphan.

Ministry of Foreign Affairs: rue That Luang, Vientiane; Deputy Prime Minister and Minister of Foreign Affairs: Gen Phoune Sipraseuth; co-operation with international environmental organizations.

Non-Governmental Organizations

IUCN Laos: c/o Forest Conservation Project, POB 807, Vientiane; tel 4600; telex 4491; Sr Conservation Adviser: Klaus Berkmuller.

LATVIA

Governmental Organizations

Ministry of the Environment and Regional Development: Straumes St 2, Jurmala, LV 2015; tel (2) 762-282; telex 161175; fax (2) 228-159; Dir: Dr Guntis Pukitis; environmental monitoring.

Ministry of Agriculture: Republikas lauk 2, Rīga, LV 1981; tel (2) 325-107; telex 161142; fax (2) 320-593; Minister: Dainis Ģegers.

Ministry of Forestry: Smilšu iela 1, Rīga, LV 1932; tel (2) 228-873; fax (2) 211-176; Minister: Kazimirs Šlakota.

Non-Governmental Organizations

Latvijas Zaļa Partija (Latvian Green Party): Kaļķu iela 8-1A, Rīga, LV 1050; tel (2) 228-892; Chair: Oļegs Batarevskis.

World Wide Fund for Nature (WWF) Latvia: Kristapa str 30, Rīga, LV 1046; tel (2) 348-962; fax (2) 348-962; Principal Officer: Ugis Rotbergs.

Ziemelvidzemes Regionalais Dabas Aizsardzibas Komplekss (ZRDAK—Northern Vidzeme Regional Nature Protection Complex): Ostas St, Salacgriva 4033, Limbazu Dist; tel (40) 41944; fax (40) 41933; Dir: Andris Urtans; Chief Ecologist: Maris Klavins; management of protected areas and natural resources; environmental education; monitoring; research; data analysis.

LEBANON

Governmental Organizations

Ministry of State for the Environment: Beirut; Minister: Samir Mokbel.

Council for Development and Reconstruction: Office of the President, Ba'abda, Beirut.

Non-Governmental Organizations

Centre de Recherches Marines (Marine Research Centre): POB 123, Beirut.

Mahmyiat Bentaël (Bentaël National Park): Pres: Dr Simon Sakr; protection of fauna and flora; reintroduction of endangered species.

LESOTHO

Governmental Organizations

Ministry of Agriculture, Co-operatives and Marketing, Department of Conservation and Forestry, Environmental Division: POB 24, Maseru 100; tel 323600; telex 4330; Minister: Ntsukunyane Mphanya.

Ministry of Natural Resources: POB 772, Maseru 100; tel 322491; telex 4253; Minister: Monyane Moleleki.

LIBERIA

Governmental Organizations

Ministry of Lands, Mines and Energy: Dukor Palace Hotel, Monrovia; Minister: Jonathan Mason.

Forestry Development Authority (FDA): POB 10-3010, 1000 Monrovia 10; tel 224459; Man Dir: Benson S Gwyan; Deputy Man Dir for Operations: Boakai Kawah; establishment of permanent forest reserves; prevention of destruction of forest and waste of associated natural resources; ensuring sustainable use of forest products.

Non-Governmental Organizations

Society for the Conservation of Nature of Liberia (SCNL): POB 2628, Monrovia; tel 261061; fax 262849; Sec-Gen: Wynfred Russell; promotion of nature conservation and environmental protection.

LIBYA

Governmental Organizations

National Committee for the Protection of Environment: Tripoli; tel (21) 30838; telex 10122.

Secretariat for Agricultural Reform, Land Reclamation and Animal Resources: Sirte; Sec: Isa Abd al-Kafi as-Sid.

Technical Centre for Environment Protection: POB 83618, Tripoli; tel (21) 48452; telex 20138; fax (21) 38098; Dir: Dr Abdulghader Ali; planning, strategy development, monitoring, legislation, training, awareness-raising, international co-operation and research and development in the field of the environment.

Non-Governmental Organizations

Al Fateh University, Environment Engineering Division: POB 13626, Tripoli; tel (21) 36010; telex 20629; Co-ordinator: Dr Bashir Faris; Technical Co-ordinator: Dr A Abufayed; undergraduate and postgraduate education, research and development; consulting and technical assistance.

LIECHTENSTEIN

Governmental Organizations

Ministry of Agriculture and Forestry: Regierungsgebäude, 9490 Vaduz; tel (75) 2366111.

Ministry of Public Health: Regierungsgebäude, 9490 Vaduz; tel (75) 2366111.

Landesforstamt (National Forestry Office): 9490 Vaduz; tel (75) 66400; fax (75) 66411; Head of National Forestry: Felix Näscher.

Non-Governmental Organizations

Freie Liste (FL—Free List): Postfach 177, 9494 Schaan; tel (75) 2320218; Principal Officers: Evelyne Bermann, Nicolaus Ruther; progressive ecological party.

Liechtensteinische Gesellschaft für Umweltschutz (LGU—Liechtenstein Society for Environmental Protection): Heiligkreuz 52, 9490 Vaduz; tel (75) 2325262; fax (75) 2321177; Principal Officer: Wilfried Marxer-Schädler.

LITHUANIA

Governmental Organizations

Ministry of Agriculture: Gedimino 19, Vilnius 2025; tel (2) 625-438; fax (2) 224-440; Minister of Agric: Rimantas Karazija.

Ministry of Forests: Gedimino 56, Vilnius 2685; tel (2) 626-864; fax (2) 612-178; Minister: Gintautas Kovalčikas.

Non-Governmental Organizations

Geographical Society of Lithuania: Zirmunu 128-9, Vilnius 2012; tel (2) 771-482; fax (2) 56-526; Pres: Prof Dr Stasys Vaitekunas; area of interest covers the ecology of the Baltic Sea and of former Soviet military bases, and rural and urban population and environment.

Institute of Botany: Želiųjų ežerų 47, Vilnius 2021; tel (2) 736-462; fax (2) 359-950; Dir: Dr Romas Pakalnis; Chair of Bd: Dr Juozas Staniulis.

Institute of Ecology: Akademijos 2, Vilnius 2600; tel (2) 359-275; fax (2) 359-257; Dir: Prof Dr Juozas Virbickas; research into functioning of aquatic and terrestrial ecosystems and fauna of Lithuania.

Lithuanian Green Party: Pylimo 38-1, Vilnius 2001; tel (2) 224-215; Chair: Irena Ignatavičienė; seeks a demilitarized, neutral and ecologically sound Lithuania.

Vilnius University, Department of Zoology: Čiurlionio 21-27, Vilnius 2009; tel (2) 661-229; Rector: Prof Dr R Pavilionis; Head of Dept: Dr R Rakauskas; courses includes hydrobiology and animal ecology; fields of research include hydrobiology and nature protection.

LUXEMBOURG

Governmental Organizations

Ministry of the Environment: 18 montée de la Pétrusse, 2918 Luxembourg; tel 47-86-82-4; telex 2536; fax 40-04-10; Minister of Environment, Territorial Admin, Communications, Energy and Information Technology: Alex Bodry.

Direction des Eaux et Forêts (Department of Water Resources and Forests): Admin des Eaux et Forêts, 67 rue Michel Welter, 2730 Luxembourg; tel 40-22-01; forest management; protection of nature, resources and habitats; restoration of degraded lands; management of protected areas; study of flora and fauna; public information; law enforcement.

Service de la Conservation de la Nature (Nature Conservation Service): Principal Officers: Jean-Marie Sinner, Frank Erasmy.

Non-Governmental Organizations

Greenpeace Luxembourg: 24 rue Dicks, 4081 Esch/Alzette; POB 229, 4003 Esch/Alzette; tel 54-62-52; fax 54-54-05; Exec Dir: Roger Spautz.

Déi Gréng Alternativ (GAP—Green Alternative): BP 454, 2014 Luxembourg; tel 46-37-40; fax 46-37-43; Sec: Abbes Jacoby; ecological political party; advocates 'grass-roots' democracy, environmental protection, social concern and increased aid to developing countries.

Gréng Lëscht Ekologesch Initiativ (Green List Ecological Initiative): BP 1567, 1015 Luxembourg; tel 22-77-67; fax 46-37-41; ecology party.

Ligue Luxembourgeoise pour la Protection de la Nature et de l'Environnement (NATURA—Luxembourg League for the Protection of Nature and the Environment): 6 blvd F D Roosevelt, 2450 Luxembourg; tel 47-23-12; fax 47-41-27; Pres: René Schmitt; protection of nature and the environment; wise use of natural resources; information; education.

MACEDONIA

Governmental Organizations

Central Institute for Safety at Work and Environmental Protection: 91000 Skopje, 13 November Quay, Bldg I/1; tel (91) 115398; fax (91) 229706; Dir: Dimce Isjanovski.

Ministry of Agriculture, Forestry and Water Resources Management: 91000 Skopje, Vasil Dorgov bb; tel (91) 113045; fax (91) 221997; Minister: Prof Dr Eftim Ančev; Dir: Voislav Popovoski; protection of water resources from chemical and toxic pollution; protection of forests and agricultural land from the harmful effects of pesticides and pollution; protection of fauna and flora.

Ministry of Development: 91000 Skopje, Bote Botevski St 9; tel (91) 236318; fax (91) 223027; Minister: Dr Sofija Todorova; analysis of the conditions, problems and effects of economic development; formulation of proposals for development policy and annual macroeconomic policy.

Ministry of Foreign Relations: 91000 Skopje, Dame Grueva bb; tel (91) 115832; fax (91) 115790; Acting Minister of Foreign Relations and Deputy Prime Minister: Stevo Crvenkovski; Special Adviser: Saško Todorovski; co-ordination of environmental activities of the multilateral organizations.

Ministry of Urban Planning, Construction, Transport and Environmental Protection: 91000 Skopje, Dame Gruev St 14; tel (91) 117288; telex 51119; fax (91) 117163; Minister: Antoni Pešev; Assistant Ministers: Jorgo Šundovski, Nikola Srbinoski; monitoring of air quality, gas emissions and water quality; environmental protection against pollution; urban planning; information about air and water pollution and chemical pollution; co-operation with European ecological network.

Institute of Occupational Medicine: 91000 Skopje, II Makedonska Brigada Br 43; tel (91) 264805; fax (91) 264805; Principal Officers: Prof Vladimir Cvetanov, Prof Elisaveta Stikova, Prof Jovanka Bislimovska; monitoring of the environment; monitoring of the working environment and health screening; assessment of the effects of the environment on human health.

Institute for the Protection of Natural Rarities: 91000 Skopje, ul Rudžer Bošković bb; tel (91) 251133; Acting Dir: Vasil Anastasovski; research on natural rarities; education.

Macedonian Public Health Institute: 91000 Skopje, 50-ta Divizija 6; tel (91) 223354; fax (91) 223354; Dir: Dr Danail Simovski; Head, Ecology Dept: Dr Dragan Gjorgjev; management of air quality, drinking and surface water and food; monitoring of the environment; educational and scientific activities in the field of environmental medicine.

Non-Governmental Organizations

Badar Ecological Association: 91000 Skopje, Ognjan Prica 26-2-6; tel (91) 235697; fax (91) 113362; Principal Officer: Dragutin Nakovski; environmental protection in the Badar region; afforestation; conservation of the River Pcinja; water conservation; forest fire alarm systems.

Čair Environmental Section: 91000 Skopje, ul Boro Menkov 2; tel (91) 274074; Principal Officer: Ante Jončevski; environmental conservation of the Skopje-Čair region.

Ecological Movement of Macedonia (DEM): 91000 Skopje, Bul Ilinden bb, POB 558; tel (91) 220518; fax (91) 117612; Principal Officers: Dr Josif Tanevski, Prof Pandora Nikuševa, Prof Dr Mihail Tokarev; co-operation with Macedonian environmental NGOs; organization of environmental events and workshops; environmental education.

Enhalon Ecological Association: 96330 Struga, Vojdan Cernodrinski 24; tel (96) 71308; Principal Officer: Mate Gogoski; nature conservation.

Flora Environmental Association: Kriva Palanka, ul M Tito bb; tel (901) 75131; Principal Officer: Dr Gorgi Kerovski; environmental conservation.

Forum of Young Ecologists of Macedonia (FOYEC): 91000 Skopje, bul Ilinden, Block 12; tel (91) 234373; fax (91) 116545; Pres: Marjan Madžovski; Vice-Pres: Katerina Cvetkovska; environmental education for young people; monitoring the environment; environmental conservation; organization of international eco-camps in co-operation with young people from the Balkans, Europe and elsewhere; researching ecological criteria for economic reform; co-operation with related organizations.

Grašnica Ecological Association: 96000 Ohrid, ul Mahmud Rifat 45; tel (96) 25872; Principal Officer: Živko Martinovski; environmental protection, especially of Lake Ohrid.

Horticultural Society of Macedonia: 91000 Skopje, Dame Gruev 14; tel (91) 229040; Principal Officer: Biljana Aleksievska; education in horticulture, urban ecosystems and environmental protection.

I Love Nature, I Love Life Committee: 91000 Skopje, Bul Ilinden bb, POB 558; tel (91) 220518; fax (91) 117612; Principal Officer: Milunka Pejovska; environmental education.

Institute of Engineering and Scientific Societies and Unions of Macedonia: 91000 Skopje, 14A Dame Gruev St, IT House, POB 139; tel (91) 229040; Pres: Prof G M Dimirovski; Sec-Gen: T Dimitrov; promotion of engineering and technology; conferences and seminars in the fields of engineering, technical sciences, architecture, agriculture and environment and ecology; safeguarding the status and professional interests of engineers.

Kliment Ohridski National University Library: 91000 Skopje, Blvd Goce Delčev 6; tel (91) 115177; fax (91) 230874; Dir: Vera Kalajlieva; editing of bibliographies; information about international databases; international exchange and loan of books and printed materials.

Korali Underwater Activities Club: 91000 Skopje, Radisanska 114; tel (91) 277481; Principal Officer: Nikola Isajlovski; protection of underwater life from pollution.

Macedonian Ecologists' Society: 91000 Skopje, Gazi Baba bb, POB 162; tel (91) 261330; Principal Officers: Dr Ljupčo Grupče, Ljupčo Melovski; organization of workshops; publications; education.

Macedonian Hydrometeorological Institute: 91000 Skopje, Skuppi bb; tel (91) 363299; telex 51392; fax (91) 226839; Principal Officers: Vitomir Dimitrovski, Ilija Panovski, Slavko Kirovski, Zoran Karamanolevski; environmental conservation (air, water, chemical pollution statistics).

Macedonian Independent Association for Environmental Protection: 91000 Skopje, Ilindenska bb; tel (91) 239470; Principal Officers: Jelena Dimitrijević, Stefanka Hadžipecova; environmental conservation.

Macedonian Union of Clean Air Associations: 91000 Skopje, ul Dame Gruev 14; tel (91) 231307; telex 51119; fax (91) 117163; Principal Officers: Nikola Srbinovski, Slavčo Nočev; monitoring of air quality, gas emissions and meteorological effects of air pollution.

Mariovo Ecological Association: 97508 Vitolište, E D Mariovo; tel (98) 25126; Pres: Dr Panče Ilioski; Sec: Risto Nikoloski; preservation of the Mariovo region; development of agriculture, cattle-breeding, fruit-growing and bee-keeping; development of tourism.

Mikrobiotika: 91000 Skopje, Durmitorska 23b; tel (91) 513779; Principal Officers: Feri Hristov, Tatjana Nedelkovska; seminars on healthy eating and living and microbiotic food; production of food.

Odek Ecological Association: 91430 Kavadarci, Šiška bb; tel (93) 76160; Principal Officers: Prof Tome Lisičanec, Duško Petrov; co-ordination of ecological activities in Kavadarci; organization of environmental events, workshops and education for young people.

Ozon Ecological Association: 92400 Strumica, ul Vasil Surčev br 11, POB 5; tel (902) 25053; fax (902) 23020; Principal Officer: George Ristov; monitoring the environment.

Prilep Ecological Association: Prilep, Pionerska 1; tel (97) 21104; Principal Officer: Dr Ilija Bašeski; environmental conservation of the Prilep region.

Priroda Ecological Association: 91440 Negotino, Jane Sandanski bb, POB 5; tel (93) 64655; conservation of water resources and the environment.

Resen Eco-Initiative: 96310 Resen, Boris Kidrich 95; tel (96) 42803; Co-ordinator: Aron Prtsulovski; ecological education in schools; exhibitions; broadcasting; events.

Skopje Centre for the Application of Radioisotopes in Science and Industry: 91000 Skopje, Bulevar Oktomvriska, Revolucija bb, POB 274; tel (91) 252525; fax (91) 256435; Man: Nebojša Jovanovski; research into environmental protection, electrochemistry, corrosion and quality control.

Survival Ecological Association: 91000 Skopje, Bul Ilinden bb; tel (91) 220518; fax (91) 117612; Principal Officers: Sveto Stefanovski, Vasilka Cvetkovska, Pero Stojčevski; monitoring of air and water pollution and industrial and urban pollution in the Skopje region; forestation; environmental education; organization of workshops, protests and events.

Vila Zora Ecological Association: 91400 Titov Veles, Pajak Planina 13; tel (93) 24830; Principal Officer: Pandora Nikuševa; environmental protection of the Beles region.

Zavod Za Vodostopanstvo Na Republika Makedonija (Water Development Institute of Macedonia): 91000 Skopje, Železnička St 62; tel (91) 228028; fax (91) 239401; Principal Officer: Ratko Tasev; projects and research in the fields of water development, ecology, applied hydraulics and erosion.

Zdravko Cvetkovski Building Education Centre: 91000 Skopje, Partizanski Odredi; tel (91) 362964; Principal Officers: Vasilka Cvetkovska, Evka Topuzovska; tree-planting; exhibitions; collaboration with schools; maintenance of schools and public monuments.

MADAGASCAR

Governmental Organizations

Ministry of Agricultural Production: BP 500, Anosy, 101 Antananarivo; tel (2) 24710; telex 22508; fax (2) 26561; Minister of State for Agriculture and Rural Development: Emmanuel Rakotovahiny.

Office National de l'Environnement (ONE): Direction Générale, Immeuble Super Aslaujo, Lot IG 161 A, Ambalavao-Isotry, 101 Antananarivo; tel (2) 25999; fax (2) 30693; Gen Dir: Joseph Andriamampianina; co-ordination of various organizations responsible for implementation of the Environmental Action Programme (PAE); development of national environmental policy.

Ministry of Foreign Affairs: BP 836, 101 Antananarivo; tel (2) 21198; telex 22236; fax (2) 34484; Minister of Foreign Affairs: Césaire Rabenoro.

Ministry of Livestock, Water and Forests: BP 484, Anosy, 101 Antananarivo; tel (2) 24026; telex 22520; fax (2) 27434; Minister of Livestock: Tsialetra; Minister of Water and Forests: Etsifosaina.

Ministry of Scientific and Technological Research for Development, Department for Research into Environmental Sciences: 27 rue Fernand Kasanga, BP 6224, Andoharano-Tsimbazaza, 101 Antananarivo; tel (2) 33288; telex 22539; fax (2) 24075; Minister of Scientific and Technological Research for Devt: Pierre Andriananteanaina.

Direction des Eaux et Forêts (Department of Water Resources and Forests): BP 243, Nanisana, 101 Antananarivo; tel (2) 40610; fax (2) 40230; Principal Officers: Estelle Raharinaivosoa, Georges Rakotonarivo, André Rakotoarivelo, Radioela Rahamalivony; technical support (statistics, documentation and research); maintenance of forest ecosystems and campaign against bush fires; soil conservation; exploitation of forest resources; forest development.

Non-Governmental Organizations

Biodiversity and Development (BIODEV): Lot VX 18, Andrefandrova, 101 Antananarivo; tel (2) 28651; telex 22485; fax (2) 28651; Dir: Olivier Nicolas Paul Behra; Programme Officer: Christine Lippai; crocodile farming, management and conservation expertise, formulation of conservation and development programmes and studies of wildlife utilization and its impact.

National Association for Environmental Actions (NAEA): BP 5092, Antananarivo; tel (2) 33289; fax (2) 22720; Gen Man: Koto Rabemananjara; Admin Officer: Pierrot Andriamparany; funding of small-scale projects to reduce soil erosion and alleviate rural poverty; sponsorship of training for local NGOs and farmers' associations.

World Wide Fund for Nature (WWF) Madagascar: près Lot II M 85 ter, Antsakaviro, BP 738, 101 Antananarivo; tel (2) 34638; telex 098622459; fax (2) 34888; Head of Programme: Dr Sheila O'Connor; Admin: Lalao Andrianifahanana.

MALAWI

Governmental Organizations

Ministry of Forestry and Natural Resources: Private Bag 350, Capital City, Lilongwe 3; telex 44465; fax 731452; Minister of Forestry and Natural Resources: Dr E Salabo.

Department of National Parks and Wildlife: POB 30131, Lilongwe 3; tel 723566; telex 4465; fax 723089; wildlife management; research; education.

Ministry of Agriculture: POB 30134, Capital City, Lilongwe 3; tel 733300; telex 44648; Pres and Minister of Agriculture: Dr Hastings Kamuzu Banda.

Ministry of Health: POB 30377, Capital City, Lilongwe 3; telex 44558; Minister: Dr Hetherwick M Ntaba.

National Research Council: Private Bag 301, Lilongwe 3.

Non-Governmental Organizations

Wildlife Society of Malawi: POB 1429, Blantyre; tel 643502; telex 44539; fax 651664; Nat Chair: Dr John Wilson; Exec Dir: Carl Bruessow; forestry; environmental education; research; fund-raising.

MALAYSIA

Governmental Organizations

Ministry of Science, Technology and the Environment: Wisma Sime Darby, 13th Floor, Jalan Raja Laut, 50662 Kuala Lumpur; tel (3) 2938955; telex 28154; fax (3) 2931480; Minister: Law Hieng Ding.

Department of Wildlife and National Parks: Km 10, Jalan Cheras, 56100 Kuala Lumpur; tel (3) 9052872; telex 8428154; fax (3) 9052873; Dir: Mohd Khan bin Momin Khan; wildlife protection; conservation of biodiversity.

Sabah Parks Board of Trustees: Sabah Parks, POB 10626, 88806 Kota Kinabalu, Sabah; tel (88) 211524; Dir: Lamri Ali Alisaputra; management of protected areas, national parks and wildlife.

Sabah Wildlife Department: Menara Sabah Bank Bhd, 7th Floor, Wisma Tun Fuad Stephens, 88300 Kota Kinabalu; tel (88) 211645; fax (88) 222476; Dir: Patrick Andau Mahedi; forestry; management of wildlife, habitats and protected areas; control of hunting, trapping and poaching.

Non-Governmental Organizations

Forest Research Institute of Malaysia (FRIM): Karung Berkunci 201, Kepong, 52109 Kuala Lumpur; tel (3) 6342633; telex 27007; fax (3) 6367753; Dir-Gen: Dr Salleh Mohd Nor; Deputy Dir-Gen: Dr Abdul Razak Mohd Ali.

Malaysian Committee for IUCN: c/o Dept of Wildlife and Nat Parks, Malaysia Barat-Komplek Pejabat, Pejabat Kerajaan, Blok 20, Jalan Duta, Kuala Lumpur; Chair: Mohd Khan bin Momin Khan.

Persatuan Pencinta Alam (Malaysian Nature Society): POB 10750, 50724 Kuala Lumpur; tel (3) 7912185; fax (3) 7917722; Pres: Dato' Dr Salleh Mohd Nor; Vice-Pres: Prof P N Avadhani, Dr Leong Yueh Kwong; Exec Officer: Lee Su Win; conservation of the environment and natural resources through education, publications and awareness-raising activities.

TRAFFIC Southeast Asia: Locked Bag No 911, Jalan Sultan PO, 46990 Petaling Jaya, Selangor; tel (3) 7913159; fax (3) 7175405; Dir: Stephen V Nash; regional office of Trade Records Analysis of Fauna and Flora in Commerce (TRAFFIC) International; monitoring of wildlife trade and utilization.

University of Malaya, Institute of Advanced Studies: 59100 Kuala Lumpur; tel (3) 7570322; telex 39845; fax (3) 7568940; Dean: Prof Mohd Ali Hashim; postgraduate teaching, interdisciplinary and multidisciplinary research, and consultancy.

World Health Organization (WHO) Western Pacific Regional Environmental Health Centre (EHC): POB 12550, 50782 Kuala Lumpur; tel (3) 9480311; fax (3) 9482349; Dir: Dr Paul Guo; fmrly WHO Western Pacific Regional Centre for the Promotion of Environmental Planning and Applied Studies (PEPAS); policy development, education and training of personnel, provision of technical information, environmental monitoring, information processing and promotion of appropriate technology.

World Wide Fund for Nature (WWF) Malaysia: Wisma IJM Annexe, 3rd Floor, Jalan Yong Shook Lin, Locked Bag 911, 46990 Petaling Jaya; tel (3) 7579192; fax (3) 7565594; Pres: Y B Tan Sri Mohamed Khir Johari; Chair: Prof Mohamad Nordin Hj Hasan; national affiliate organization of World Wide Fund for Nature (WWF) International; fund-raising for conservation of nature and natural resources; education; field projects.

MALDIVES

Governmental Organizations

Ministry of Planning and the Environment: Ghaazee Bldg, Ameer Ahmed Magu, Malé 20-05; tel 322965; telex 66110; fax 327351; Minister: Ismail Shafeeu.

Ministry of Health and Welfare: Ghaazee Bldg, Ameer Ahmed Magu, Malé 20-05; tel 323216; telex 66181; Minister: (vacant).

Ministry of Home Affairs and Sports: Huravee Bldg, Ameer Ahmed Magu, Malé 20-05; tel 323821; telex 77039; fax 324739; Minister: Umar Zahir.

MALI

Governmental Organizations

Ministry of Rural Development and the Environment: BP 1676, Bamako; tel 22-60-24; Minister of Environment: Mohamed Ag Erlaf.

Ministry of Foreign Affairs, Malians Abroad and African Integration: Koulouba, Bamako; tel 22-54-89; telex 2560; Minister: Mohamed Aloussine Touré.

Direction Nationale des Eaux et Forêts (National Department of Water Resources and Forests): BP 275, Bamako; tel 22-20-22.

Non-Governmental Organizations

Association ELWAN pour la Protection des Éléphants du Gourma (ELWAN Association for the Protection of the Elephants of Gourma): BP 23, Sevare Mopti; tel 42-01-10; fax 42-01-90; Head of Programmes: Dr Noumou Diakite; protection of the elephants and ecosystems of Gourma.

Institut du Sahel (Sahel Institute): BP 1530, Bamako; tel 22-21-48; fax 22-23-37; Dir-Gen: M S Sompo-Ceesay; Scientific and Technical Co-ordinator: Gadussou Tradlé; research on demography, agriculture and the environment.

IUCN Mali: BP 1567, Bamako; tel 22-75-72; fax 22-75-72; Rep: Loriana Riccarelli Dembele.

MALTA

Governmental Organizations

Ministry for the Environment: Floriana; tel 222378; telex 1861; fax 231293; Minister: Michael Falzon.

Non-Governmental Organizations

Alternattiva Demokratika: 12 Triq Antonio Agius, Floriana VLT 14; tel 240334; fax 224745; Chair: Dr Wenzu Mintoff; Environmental Spokesperson: M Borg; political party concerned with the environment and human rights; campaigns against hunting, coastal development, environmental hazards in the workplace and water shortage; involved in revelation of corruption in granting of permits for land development.

Malta Ornithological Society (MOS): POB 498, Valletta CMR 01; tel 230684; fax 225665; Pres: Joe A Doublet; Dir: Paul Portelli; protection and study of birds and their natural habitat.

Society for the Study and Conservation of Nature (SSCN): POB 459, Valletta CMR 01; tel 248558; fax 487332; Pres: Vincent Attard; Gen Sec: Annick Bonello.

University of Malta, Foundation for International Studies, Euro-Mediterranean Centre on Marine Contamination Hazards: St Paul St, Valletta; tel 240746; telex 1673; fax 230551; Dir: A Micallef; Project Officer: C Galdies; dissemination of information, training, organization of workshops and symposia, and scientifiic research.

MARSHALL ISLANDS

Governmental Organizations

Ministry of Health and Environment: Majuro; Minister: Evelyn Konou.

MAURITANIA

Governmental Organizations

Ministry of Rural Development and the Environment: BP 366, Nouakchott; tel 520-20; Minister: Mouloud Ould Sidi Abdallah.

MAURITIUS

Governmental Organizations

Ministry of the Environment and Quality of Life: Ken Lee Tower, 2nd Floor, Barracks St, Port Louis; tel 2126080; fax 2126671; Minister: Bashir Khodabux.

Ministry of Agriculture, Fisheries and Natural Resources: Government Centre, Port Louis; tel 4644016; fax 4648749; Minister: Murlidas Dulloo; preservation and propagation of endangered and endemic animal and plant species; management of Black River Gorges national park and native forests.

Ministry of Energy, Water Resources and Postal Services, Waste Water Authority: Aisha Bldg, Louis Pasteur St, Port Louis; tel 2426444; fax 2086497; Minister: Mahyendrah Utchanah.

Ministry of Health, Environmental Health Unit: Atchia Bldg, cnr Suffren and Mgr Gonin Sts, Port Louis; tel 2122110; fax 2080376; Deputy Prime Minister and Minister of Health: Dr Paramhansa Nababsing.

Ministry of Tourism: Emmanuel Anquetil Bldg, Sir Seewoosagur Ramgoolam St, Port Louis; tel 2012286; fax 2086776; Minister: Noe Ah-Qwet Lee Cheeong Lem.

Non-Governmental Organizations

Action for Development (AFORD): POB 1042, Port Louis; tel 2616145; telex 4484; fax 2082634; Principal Officers: Chooramun Gunput, Ramsingh Hurrynac; raising of rural standard of living, protection of the environment, raising public awareness of sustainable development and organization of workshops and conferences on environment and development.

Council for Development, Environmental Studies and Conservation (MAUDESCO): POB 1124, Port Louis; tel 2121886; fax 2128877; Exec Chair: R Awotar; Sec: B Seebaluck; national, non-profit organization.

Fleurir Maurice Committee: c/o MGIO, Emmanuel Anquetil Bldg, Port Louis; Chair: C K Bhackory.

Friends of the Environment (FOEM): 25 St Georges St, Port Louis; tel 2126783; fax 2088414; Exec Sec: Dr Monique Koenig; environmental conservation; research; monitoring of pollution; promotion of environmental law; statistics; workshops and conferences; awareness-raising.

Indian Ocean Commission/Commission de l'Océan Indien: Q4 ave Sir Guy Forget, BP 7, Quatre Bornes; tel 4259564; telex 5273; fax 4251209; Gen Sec: Henri Rasolondraibe; regional co-operation in the fields of environment, fisheries, tourism and crafts; development of trade and scientific co-operation between islands.

Mauritius Council for Environmental Studies, Conservation and Youth Agenda (MAUCESCOYA): POB 304, Port Louis; tel 2125593; Gen Sec: Dr Luchmedes Ramsaha; prevention of pollution of sea, soil and air; protection of wildlife; education in resource management; promotion of environmental law.

Society for the Protection and Conservation of the Environment (SPACE): 3 ave Osman, Port Louis; tel 4643800; Pres: N Burn.

MEXICO

Governmental Organizations

Secretariat of State for Social Development (SEDESOL—Secretaría de Desarrollo Social): Avda Constituyentes 947, Col Belem de las Flores, 01110 México, DF; tel (5) 277-4394; fax (5) 271-6614; Secretary of State: Luis Donaldo Colosio Murrieta; Technical Sec: José Luis Grijalva Otero; implementation of policies and programmes on social development, human settlements, regional and urban development, housing and ecology.

National Institute of Ecology: Rio Elba 20, 16°, Col Cuauhtemoc, 06500 México, DF; tel (5) 553-9968; fax (5) 286-6872; Pres: Sergio Reyes Lujan; Sec: Hector Rodriguez Morales; programmes for sustainable use of land and natural resources; formulation of technical standards to prevent and control pollution and reduce the environmental impact and risks of socio-economic activity.

Office of the Federal Attorney for Environmental Protection: observance of standards and regulations on environmental matters; public information service.

Secretariat of Urban Development and Ecology: Río Elba 20, 16°, Col Cuauhtémoc, 06500 México, DF; tel (5) 553-9647; telex 11771198; fax (5) 277-6574; Sec: Patricio Chirinos Calero.

Consejo Nacional de la Fauna (CNF—National Wildlife Council): Fray Payo de Rivera 320, Lomas de Chapultec Virreyes, 11000 México, DF; tel (5) 540-7390; fax (5) 520-1881; Pres: Victor Hugo O'Farrill Avila; Exec Pres: Marco Antonio Pastrana de la Portilla; conservation, development and exploitation of woodland wildlife and habitats as renewable natural resources.

Instituto Nacional de Investigaciones Forestales y Agropecuarias, Campo Experimental de Los Cañones (National Institute of Forest and Agricultural Research, Los Cañones Test Station): Km 174, Carretera Guadalajara-Saltillo, Apdo 69, 99600 Jalpa, Zac; tel (495) 52206; fax (495) 52206; Principal Officer: Ramon Gutierrez Luna; research into hydrology, erosion, grazing.

Instituto Nacional de la Pesca, Programa Nacional de Investigación de Tortugas Marinas (National Fisheries Institute, National Sea Turtle Programme): Apdo 591, 28200 Manzanillo, Col; tel (333) 23750; fax (333) 23751; Principal Officers: René Marquez, Aristoteles Villanueva O, Miguel A Carrasco A; sea turtles research and conservation in beach breeding grounds; stock assessment; life cycle studies; research into behaviour and migration; international collaboration for conservation programmes.

Non-Governmental Organizations

Amigos de Sian Ka'an AC (Friends of Sian Ka'an): Apdo 770, 77500 Cancún, Q Roo; Exec Dir: Juan E Bezaury Creel; non-profit organization dedicated to preservation of Sian Ka'an Biosphere Reserve, wildlife management, rural development, management of coral reefs and environmental education.

Asociación Mexicana Pro-Conservación de la Naturaleza AC (PRONATURA—Mexican Association for the Conservation of Nature): Avda Nuevo León 144, Col Hipodromo-Condesa, 06100 México, DF; tel (5) 286-9642; fax (5) 286-9480; Exec Dir: Hans Herrmann; Pres: Mauricio Ruiz Galindo; conservation of biodiversity, environmental education and sustainable use of natural resources.

Asociación Mexicana para la Rehabilitación y Conservación de los Psitácidos, AC (AMERYCOP—Mexican Association for the Rehabilitation and Conservation of Parrots): Apdo 4-174, Itzimná, 97101 Mérida, Yuc; fax (99) 27-9571; Dir: Dr John P Ehrenberg; Pres: Marlen Ahrenberg; struggle against the destruction of tropical ecosystems in Mexico through environmental education and promotion of sustainable development; conservation of parrots chosen as case model for preservation of an endangered tropical wildlife resource.

Brigada de Acción Ecológica 'Huaxyacac' AC (Huaxyacac Ecological Action Squad): Avda Juárez 302, 68000 Oaxaca, Oax; tel (951) 64884; Principal Officers: Dr Sergio Gómez Nuñez, Juan Montes Lara; concerns include atmospheric pollution, promotion of ecological alternatives and reafforestation.

Canacintra—Gerencia de Ecología (Canacintra Ecological Management): Cámara Nacional de la Industria de Transformación, Avda San Antonio 256, Col Ampliación Nápoles, 03849 México, DF; tel (5) 611-3269; fax (5) 598-3882; Chair Environmental and Ecological Control Cttee: Juan Alvarez Barroso; representation and defence of the interests of the industrial sector; promotion of sustainable industrial development compatible with environmental protection.

Casa Ecológica Teotihuacán (Teotihuacán Ecology House): Illinois 17, Departamento 1, Col Napoles, 03810 México, DF (Head Office); Calle Reforma 15, San Francisco Mazapa, 55800 San Juan de Teotihuacán, Méx; tel (5) 687-2354; education, research and promotion in the fields of health, nutrition and building.

Centro de Agroecología San Francisco de Asis AC (Saint Francis of Assisi Agroecology Centre): 1a Avda Sur 3, 30900 Motozintla, Chis; tel (964) 10029; attempts to overcome poverty in the region by promoting self-sufficiency through ecologically-sound agricultural methods.

Centro de Educación Ambiental y Acción Ecológica, AC (CEDUAM—Centre for Environmental Education and Ecological Action): Cerrada del Ferrocarril 3, Apdo 30, 90200 Calpulalpan, Tlax; Co-ordinator: Rogelio Cova Juárez; Officers: Fernando García Flores, Gabriel Sánchez Ledezma.

Centro Tropical de Capacitación Agropecuaria y Forestal (CETROCAF—Tropical Centre for Training in Agriculture and Forestry): Km 68.5 Carretera Villahermosa-Escárcega, Apdo 45, 86700 Macuspana, Tab; tel (936) 20358; telex 76401; fax (936) 21519; Principal Officers: Dr Arturo Romero Villanueva, Dr Angel Tárano Bande; agricultural training and applied research for the development of smallholders in five regions of Mexico.

Comisión Interinstitucional para el Desarrollo e Investigación sobre Contaminación Ambiental por Plaguicidas (CIDICAP—Interinstitutional Commission for Research and Development on Pesticide Pollution): Apdo 18, 98500 Calera de V R, Zac; tel (492) 50198; fax (492) 50363; Principal Officers: Dr Lilia Albert, Dr Fernando Rincón Valdez, Dr Manuel I Díaz Cervantes; training and information dissemination.

Comité Miembros Mexicanos de la UICN (Committee of Mexican Members of the World Conservation Union—IUCN): Campos Eliseos 400, 40°, Lomas de Chapultepec, Delegación Miguel Hidalgo, 11000 México, DF; tel (5) 876-0222; fax (5) 876-0217; Principal Officers: Lic Rodolfo Ogarrio Ramirez España, Ramón Pérez Gil Salcido, Ing Carlos Gottfried.

Consejo Ecológico de Participación Ciudadana del Estado de Puebla, AC (Ecological Council of the State of Puebla for Civic Participation): Avda Reforma 1906, 72000 Centro Puebla, Pue; tel (22) 46-1980; fax (22) 32-0051; Pres: Ing Herberto Rodríguez Concha; Dir-Gen: Luis Armin Stein Stein; monitoring of reafforestation and the environment; organization of conferences and ecological exhibitions.

Ducks Unlimited de México AC (DUMAC): Avda Vasconcelos 209 Ote, Apdo 776, Residencial San Agustin, Garza García, 64000 Monterrey, NL; tel (83) 78-6757; fax (83) 78-6439; Principal Officer: Dr Eric W Gustafson C; conservation of migratory aquatic birds.

Ecosolar: Jardín 5, Col Tlacopac, 01040 México, DF; tel (5) 682-0507; fax (5) 682-0507; Principal Officer: Aurelio Ahumada R; sustainable development projects; community-based projects; research and development of environmental technology; organic farming.

Familia y Sociedad AC (Family and Society): Avda San Antonio 36-9, Col Noche Buena, 03720 Del Benito Juárez, México, DF; tel (5) 523-3475; fax (5) 682-2556; Principal Officers: Ruben Aguilar Valenzuela, Laura Becerra Pozos, Luis Pineda; improvement of urban environment; promotion of social projects; urban education.

Federación Conservacionista Mexicana (FECOMEX—Mexican Conservationist Federation): Apdo 6-1092, 44420 Guadalajara, Jal; tel (36) 30-1679; fax (36) 15-7797; Pres: René Solinis N; independent organization of 28 conservation groups; integration of conservation and development; impact assessment; resource management; co-ordination of research; international co-operation.

Fundación Miguel Alemán AC (Miguel Alemán Foundation): Ruben Darío 187, 11570 México, DF; tel (5) 531-1936; fax (5) 250-1043; Pres: Miguel Alemán Velazco; Dir: Fernando Castro y Castro; the Environmental Programme supports NGO conservation and sustainable development projects in rural areas by means of equipment or publications.

Fundación Chiapaneca para el Manejo de Areas Tropicales, AC (FUNDAMAT—Chiapas Foundation for the Management of Tropical Areas): Apdo 970, Tuxtla Gutiérrez, Chis; tel (961) 33362; telex 74445; fax (961) 10750; Pres: Ramón Pérez-Gil Salcido; conservation projects in Chiapas; management of protected areas, tropical forests and ecosystems and wildlife; conservation strategies; monitoring; statistics; education; protection of biodiversity, mangroves, mammals, amphibians, reptiles and fish; lobbying aid agencies and government; campaign against hunting.

Fundación Mexicana para la Educación Ambiental AC (Mexican Foundation for Environmental Education): Apdo 44, Carretera Circunvalación s/n, 54600 Tepotzotlán, Méx; tel (5) 876-0222; telex 1771390; fax (5) 876-0217; Pres: Manuel Arango A; Dir-Gen: Rodolfo Ogarrio; non-profit institution; nature conservation; improvement of environmental quality; based in Xochitla Gardens, the Environmental Training Centre and the Environmental Information Centre.

Gestión de Ecosistemas AC (Ecosystems Management): Apdo 19-182, Crater 21, Jardines del Pedregal, 03910 México, DF; tel (5) 593-3988; fax (5) 593-3988; Dir-Gen: Dr Silvia del Arno; management, conservation and restoration of ecosystems; research into natural resources, their experimental development and sustainable use; promotion of ecologically-sound production methods which benefit local communities; information bank on the use and management of resources; newsletter.

Grupo Ecológico Aguascalientes AC (Aguascalientes Ecological Group): Cerro de Laurel 505, 20120 Aguascalientes, Ags; tel (491) 14-0189; Principal Officer: Carlos González García; environmental education; environmental engineering.

Grupo Ecologista de Tlalpan: Carretera Libre a Cuernavaca Km 25-7000, San Miguel Xicalco, Deleg Tlalpan, 14490 México, DF; tel (5) 846-0150; fax (5) 254-0314; Pres: Sergio Novelo-Galindo; tree-planting campaigns; creation of gardens in derelict urban sites; environmental education in elementary and secondary schools in the Tlalpan area; waste management training (recycling, reduction, re-use and compost-making); environmental meetings; Guardias de Cuicuilco nature camps for teenagers.

Grupo de Tecnología Alternativa SC (Alternative Technology Group): Alamo 8-16, Col Los Alamos, San Mateo, Naucalpan de Juárez, 53230 México, DF; tel (5) 393-0939; fax (5) 393-0939; Principal Officers: Josefina Mena Abraham, Pedro Rubio Alvirde; design of recycling systems for liquid and solid waste, based on indigenous cultures; design, building and research for ecological housing schemes; manufacture of prefabricated units; research into social, technical and economic aspects of recycling systems.

Habitat International Coalition: Cordobanes 24, Col San José Insurgentes, 03900 México DF; tel (5) 651-6807; telex 1776454; fax (5) 651-2371; Pres: Han van Putten; Exec Sec: Enrique Ortiz; promotion of public awareness of human settlements' problems.

Instituto de Historia Natural (Natural History Institute): Calzada Cerro Hueco s/n, Zoológico Regional, Apdo 60, Tuxtla Gutierrez, Chis; tel (961) 23754; fax (961) 29943; Dir: Miguel Alvarez del Toro; environmental education; exhibition of flora and fauna; protection of natural areas; research; publications.

Instituto Manantlán de Ecología y Conservación de la Biodiversidad (Manantlán Institute for Ecology and Biodiversity Conservation): Valentín Velasco 253, 48740 El Grullo, Jal; tel (338) 72748; telex 681744; fax (338) 72749; ecology; conservation; resource management; rural development.

Instituto Mexicano de Recursos Naturales Renovables (IMERNAR—Mexican Institute of Renewable Natural Resources): Avda Dr Vertiz 724, Col Narvarte, 03020 México, DF; tel (5) 519-4505; telex 1771390; fax (5) 538-4520; Principal Officer: Enrique Beltrán G; ecological research; protection, promotion and use of renewable natural resources; library; advises related institutions.

Instituto de la Naturaleza y la Sociedad de Oaxaca (Oaxaca Institute of Nature and Society): M Bravo 210 Altos, Col Centro, 68000 Oaxaca, Oax; tel (951) 46490; fax (951) 46490; Principal Officers: Alejandro de Avila Blomberg, Juan José Consejo Dueñas, Javier de la Maza Elvira, Alvaro Glez Rios, Miguel Medina Zenon, Alberto Sanchez Lopez, Alfredo Saynes Vazquez; nature protection; information; ecological planning; alternative technology workshop; tropical forests programme; ecological programmes; environmental education; urban ecology; waste management; reforestation.

Instituto Mexicano de Tecnologías Apropiadas AC (IMETA—Mexican Institute for Appropriate Technology): Farallones 60-B, Col Acueducto de Guadeloupe, Apdo 63-254, 07270 México, DF; tel (5) 391-1888; fax (5) 391-1888; Principal Officer: Dr Edmundo Arias Torres; consulting and technological research, mainly in energy and waste management.

IUCN Ecotourism Consultancy Programme: Camino al Ajusco 551, Tepepan, Xochimilco, 16020 México, DF; tel (5) 676-8734; fax (5) 676-5285; Dir: Hector Ceballos-Lascurain; consultancy services to governments, development agencies, the tourism industry and NGOs on the planning, development and promotion of ecotourism projects.

Monarca: Avda Constituyentes 345-806, Col Daniel Garza, 11830 México, DF; tel (5) 515-9910; telex 1772829; fax (5) 271-3665; Pres: Carlos F Gottfried Joy; protection of the Monarch butterfly and its winter habitat; research; awareness-raising.

Movimiento Ecologista Mexicano de Puebla AC: Chapolco 1, desp 101, Col La Paz, Puebla, Pue; tel (22) 30-1357; fax (22) 48-9441; Principal Officer: José Ernesto Yitani Damian; education; reforestation; waste management training.

Museo de la Isla de Cozumel (Cozumel Island Museum): Avda Rafael Melgar y Calle 2 Norte, 77000 Cozumel, Q Roo; tel (987) 21475; fax (987) 21545; environmental education; sea turtle conservation programme; beach-cleaning campaigns; reforestation; ecological theatre; information on environmental problem-solving.

Network Espacios: Tlaloc 40-3, Col Tlaxpana, 11370 México, DF; tel (5) 566-4265; fax (5) 592-1989; Principal Officers: Cristina Lavalle, Rogelio Cova, Luis López Llera, Ignacio Peón; formulation of proposals for development.

Partido Ecologista Mexicano (PEM—Mexican Ecologist Party): Guadeloupe I Ramírez 78, Tepepan, Xochimilco, 16020 México, DF; ecologist party.

PGT Consultores SC: Apdo 1000, 29000 Tuxtla Gutierrez, Chis; tel (961) 51536; fax (961) 51097; Principal Officers: Ramón Pérez-Gil Salcido, Ana María Muñiz Salcedo, Fernando Jaranillo Monroy; advice on conservation, ecology, protected natural areas and wildlife.

Pro-Ecología de Colima AC (Colima Ecology Group): Apdo 29, F Medina 111 Altos, 28000 Colima, Col; tel (331) 29810; fax (331) 21160; Pres: Francisco J Cárdenas Munguia; Sec: Laura de la Mora Marti; environmental research; education through lectures, festivals and tourism; implementation of government conservation schemes.

Programa de Acción Forestal Tropical AC (Tropical Forest Action Programme): Schiller 417 PB, Polanco, 11560 México, DF; tel (5) 254-3514; fax (5) 250-2341; Principal Officers: Dr Arturo Gómez Pompa, Dr Silvia del Arno; promotion of sustainable rural development and conservation of natural resources; research; training.

Promoción Ecologica Campesina AC (PROE—Rural Ecology Movement): Apdo 86, Avda Malintzi 195 Pte, San Pedro Muñoztla, 90800 Santa Ana Chiautempan, Tlaxcala, Tlax; tel (246) 43125; fax (246) 43125; Principal Officers: Jesús Quirós Quirós, M del Carmen Olivera de H; promotion of ecological self-sufficiency; training in ecologically-sound technologies; advice at regional and national levels; bioagricultural research; ecological meetings for children.

Provida de Ecologistas en el Estado de Campeche para la Conservación de los Recursos Naturales AC (Campeche State Ecologists for Conservation of Natural Resources): Manzana K, Lote 9, Fracc Lavalle Urbina, Campeche, Cam; tel (981) 50837; environmental conservation; advice; awareness-raising; environmental education; research; information; support for ethnic groups; lobbying; bulletin, reviews and publications; documentation centre.

Rescate Infantil Ecológico AC (Children's Ecological Rescue): Apdo 166A, 48311 Puerto Vallarta, Jal; tel (322) 48506; Principal Officers: Flavio Pimienta, Adriana Sandoval; environmental education for children.

Servicio Desarrollo y Paz AC (SEDEPAC—Development and Peace Service): Calle Huatusco 39, Col Roma Sur, 06760 México, DF; tel (5) 574-0892; fax (5) 584-3895; Exec Sec: Rafael Reygadas Robles Gil; rural development programme; women's campaign; education and training of young people; support centre for urban social movements; support centre for refugees from Central America.

Servicios de Apoyo Local al Desarrollo de Base en México AC (SALDEBAS—Local Support Services for Basic Development in Mexico): Cantera 58-1, Barrio del Niño Jesús, Coyoacán, 04000 México, DF; tel (5) 554-4153; fax (5) 659-5439; Principal Officer: Sergio Martínez Vázquez; monitoring, evaluation and research; training courses.

Sociedad Botánica de México AC (Mexican Botanical Society): Apdo 70-385, Ciudad Universitaria, 04510 México, DF; tel (5) 622-4835; fax (5) 622-4828; Pres: Dr Javier Alvarez Sánchez; Vice-Pres: Cristina Mapes; promotion of co-operation between botanists; promotion of botanical research.

Sociedad Mexicana de Médicos Veterinarios Zootecnistas Especialistas en Ecozootecnía (SMMVZEE—Mexican Society of Veterinary Specialists in Ecozootechnology): Cultivos 212, Col Valle del Sur, 09810 México, DF; tel (5) 670-4181; fax (5) 530-1024; Pres: Marco Aurelio Ocampo Ruiz; Sec: Guadalupe Araceli Cid Ibarra; advice and research on economic, productive, financial and ecological aspects of wildlife management and environmental protection; education; publications.

Unión de Grupos Ambientalistas (IAP—Union of Environmentalist Groups): Callejón San Miguel 46, Barrio San Lucas Coyoacán, 04030 México, DF; tel (5) 532-2717; fax (5) 532-5639; Pres: Regina Barba Pérez; Vice-Pres: Luis M Robles Gil; concerns include environmental education, sustainable agriculture, ecotourism, ecotechnology and pollution.

Universidad Nacional Autónoma de México, Centro de Ciencias de la Atmosfera (UNAM, CCA—National Autonomous University of Mexico, Centre for Atmospheric Sciences): Circuito Investigación Científica, Ciudad Universitaria, 04510 México, DF; tel (5) 622-4059; fax (5) 616-0789; Dir: Dr Julian Adem Chahin; Academic Sec: Dr Fernando García García; fields of research include theory of climate and climatology, atmospheric chemistry, air pollution, urban climatology, aerobiology and agrometeorology.

Universidad Nacional Autónoma de México, Grupo Etnobotánico Latinoamericano (UNAM, GELA—National Autonomous University of Mexico, Latin American Ethnobotanical Group): Jardín Botánico, Apdo 70-614, Coyoacán, 04510 México, DF; tel (5) 622-9045; fax (5) 616-2326; Principal Officers: Javier Caballero, Monserrat Gispert, Abigail Aguilar, Victor M Toledo, Olga Lucía Sanabria, Sonia Lagos-Witte; promotion of scientific exchange between ethnobotanical researchers in Latin America; newsletter.

Universidad Nacional Autonóma de México, Programa Universitario de Medio Ambiente (UNAM, PUMA—National Autonomous University of Mexico, University Environment Programme): Coordinación de la Investigación Científica, Ciudad Universitaria, 04510 México, DF; tel (5) 550-8815; fax (5) 550-8834; Co-ordinator: Dr Octavio Rivero Serrano; environmental engineering; ecological monitoring of the university campus; bioecology; environmental chemistry; environmental health; environmental information; environmental events; research; training.

World Wide Fund for Nature (WWF) Mexico: 38 Goldsmith 4, Polanco, 11560 México, DF; tel (5) 280-3366; fax (5) 280-2727; Rep: Martin Goebel.

MICRONESIA

Governmental Organizations

Department of Resources and Development: POB PS-53, Palikir, Pohnpei, Eastern Caroline Islands, FM 96941; tel 320-2649; telex 7296807; fax 320-2785; Sec: Bernard Helgenberger.

MOLDOVA

Governmental Organizations

Ministry of Agriculture and the Food Industry: Chişinău, Ştefan cel Mare; tel (3732) 23-35-36; Minister: Vitalie Gorincioi.

Non-Governmental Organizations

Moldovan Academy of Sciences: Chişinău, bul Ştefan cel Mare 1; tel (3732) 26-14-78; fax (3732) 26-20-91; research in natural sciences; collaboration with other scientific centres; recommendations to government on social, economic and scientific problems.

Moldovan Society of Animal Protection: Chişinău, str Serghei Lazo 17, ap 6; tel (3732) 24-75-99; Pres: Prof P I Nesterov; Assts: Dr Z Z Neculiseanu, Dr N M Trofimenco; education; co-operation with government agencies and NGOs with interests in animal welfare and the environment.

MONACO

Governmental Organizations

Centre Scientifique de Monaco, Observatoire Océanologique Européen (Monaco Scientific Centre, European Oceanological Observatory): 16 blvd de Suisse, MC 98030; tel 93-25-89-54; telex 469796; fax 93-30-24-74; Delegate: Prof François Doumenge; oceanography; marine ecology; environmental research; conservation strategies; management of protected areas, wildlife, endangered animal species and aquariums; international co-operation; captive breeding.

Département des Travaux Publics (Department of Public Works): BP 522, MC 98015 Monte Carlo; tel 93-30-19-21; telex 469942.

Service des Relations Extérieures, Affaires Techniques (External Relations Office, Technical Affairs Division): Villa Girasole, 16 blvd de Suisse, MC 98030; tel 93-30-33-71; telex 469796; fax 93-30-24-74; Perm Delegate to Int Orgs: César Charles Solamito; co-operation with international environmental organizations.

Non-Governmental Organizations

IUCN Commission on Ecology: c/o Musée Océanographique, ave St Martin, MC 98000; tel 93-15-36-00; telex 042469037; fax 93-50-52-97; Chair: Prof François Doumenge; Deputy Chair: Prof Hans Lundberg.

MONGOLIA

Governmental Organizations

Ministry of Nature and Environmental Protection: October St 5, Ulan Bator 11; tel 329619; telex 79343; fax 321401; Minister of the Environment: Dzambyn Batjargal.

Department of Environmental Science, Monitoring and International Co-operation: Khudaldaany Gudamj 5, Ulan Bator 11; tel 329619; telex 79343; fax 321401; Dir-Gen: Dr Tsokhiogyn Adyasuren; research on nature conservation and rehabilitation and sustainable use of natural resources; monitoring of the environment; hydrometeorology; prevention of natural disasters; seminars and workshops on the environment.

State Committee for Environmental Control: Khudaldaany Gudamj 5, Ulan Bator 11; tel 320943; telex 79343; fax 321401; formulation of policy on environmental protection and management of natural resources; environmental legislation; institution of environmental standards; economic evaluation of natural resources; environmental science projects.

Ministry of Foreign Relations: Ulan Bator; tel 321870; telex 245; Minister: Tserenpiliyn Gombosüren; co-operation with international environmental organizations.

Institute of Water Policy: Baruun Selbe St 13, 210638 Ulan Bator; tel 321862; Dir: M Gantimur; Scientific Sec: B Myagmarjav; water research programmes; formulation of proposals on water policy.

Non-Governmental Organizations

Mongolian Greens Party: POB 1089, Ulan Bator; tel 323871; telex 79236; Chair: L Nyam; political wing of the Alliance of Greens.

MOROCCO

Governmental Organizations

Ministry of Agriculture and Agrarian Reform, Nature Protection Division, Department of Water Resources, Forests and Soil Conservation: Quartier Administratif, Rabat; tel (7) 60993; telex 31038; Minister: Abd al-Aziz Meziane.

Forestry Research and Experimentation Division: Charia Omar Ibn el Khattab, BP 763, Rabat-Agdal; tel (7) 672759; telex 36992; fax (7) 671151.

Ministry of Foreign Affairs and Co-operation: ave Franklin Roosevelt, Rabat; tel (7) 62841; telex 31007; Minister of State: Abd al-Latif Filali; co-operation with international environmental organizations.

Ministry of Ocean Fisheries and the Merchant Navy, Training and International Relations Department: 63 blvd Moulay Youssef, Rabat; tel (7) 63366; telex 32679; Minister of Ocean Fisheries and Merchant Navy: Ben-Salem Smili.

Non-Governmental Organizations

Société Marocaine pour le Droit de l'Environnement (Moroccan Society for Environmental Law): c/o Faculté de Droit, route d'el-Jadida, Km 8, BP 8110, Oasis, Casablanca; tel (2) 231100; Pres: Prof Mohamed Ali Mekouar; scientific association dedicated to the development of environmental law; international co-operation; workshops and conferences; training; research.

MOZAMBIQUE

Governmental Organizations

Ministry of Agriculture, Forestry and Wildlife Department: Praça dos Heróis Moçambicanos, CP 1406, Maputo; tel 460010; telex 6195; fax 460145; Minister: Alexandre José Zandamela; forestry; afforestation; research; land use; agricultural policy.

National Environment Commission: Avda Acordos de Lusaka 2115, CP 2020, Maputo; tel 465843; telex 6962; fax 465849; Pres: John Kachamila; Vice-Pres: João Mário Salomão; co-ordination and promotion of environmental activities; management, preservation and use of natural resources; environmental education.

State Secretariat for Physical Planning: Avda Acordos de Lusaka 2115, Maputo; tel 732168; telex 6500.

MYANMAR

Governmental Organizations

Ministry of Agriculture and Forests: Ministers' Office, Yangon; tel (1) 86597; Minister for Agriculture: Maj-Gen Myint Aung; Minister for Forestry: Lt-Gen Chit Swe.

Ministry of Health: Ministers' Office, Yangon; tel (1) 85896; Minister for Health and for Education: Rear-Adm Than Nyunt.

Agricultural Mechanization Department, Rural Water Supply Division: 459 Prome Rd, Yangon; tel (1) 30584.

Irrigation Department: 9/21 Strand Rd, Yangon; tel (1) 80028; telex 21217.

NAMIBIA

Governmental Organizations

Ministry of Wildlife, Conservation and Tourism: Private Bag 13306, Windhoek 9000; tel (61) 2849111; telex 509083109; fax (61) 229936; Minister: Nico Bessinger; Dir of Wildlife Conservation and Research: P Swart; agricultural policy; land use; conservation strategies; sustainable development; environmental quality standards; management of wildlife and protected areas.

Directorate of Environmental Affairs: tel (61) 63131; fax (61) 63195; Principal Officer: Dr C J Brown; environmental protection, planning and co-ordination at national, regional and ministerial levels.

Etosha Ecological Institute: Okaukuejo PO, Via Outjo; Principal Officers: Dr M Lindeque, Dr P M Lindeque, J L Scheepers, W du Plessis, Dr C Brain; ecological research for wildlife management in Etosha National Park.

Non-Governmental Organizations

Desert Research Foundation of Namibia: POB 1592, Swakopmund; tel (61) 229855; fax (61) 230172; Exec Dir: Dr M K Seely; ecological research in arid zones; research training; environmental education.

Integrated Rural Development and Nature Conservation (IRDNC): POB 339, Swakopmund; tel (641) 62661; fax (641) 4664; Principal Officers: Garth Owen-Smith, Margaret Jacobsohn; community-based natural resource management programmes; integration of conservation and development.

Namibia Nature Foundation: POB 245, Windhoek 9000; tel (61) 35111; telex 50908700; fax (61) 34021; Dir: Douglas F Reissner; fund-raising; nature conservation; education.

Rössing Foundation: POB 20746, Windhoek; tel (61) 221721; fax (61) 221273; Exec Dir: David Godfrey; Deputy Dir: Jen le Roux; social, agricultural and community development; environmental education.

NAURU

Governmental Organizations

Office of the President, Chief Secretariat: Yaren; telex 33100; Pres: Bernard Dowiyogo; Minister Assisting the Pres: Vinson Detenamo.

NEPAL

Governmental Organizations

Ministry of Forests and the Environment: Babar Mahal, Kathmandu; Minister of State: Bir Mani Dhakal; implementation of forestry policy; enforcement of forestry law; administration of forestry resources.

Department of Forests: Kathmandu; 75 district forest offices; forest administration, use and planning; community forestry.

Department of National Parks and Wildlife Conservation: Baneswor, POB 860, Babar Mahal, Kathmandu; tel (1) 215850; Chief Ecologist: Dr Tirtha Maskey; management of national parks and wildlife reserves.

Ministry of Water Resources: Kathmandu; telex 2312; Minister of State: Laxman Prasad Ghimire.

Department of Soil Conservation and Watershed Management: Kathmandu; support for soil conservation by farmers and other land users through integrated rural development projects.

Secretariat of the National Commission for Conservation of Natural Resources, National Planning Commission: POB 1284, Singha Durbar, Kathmandu; tel (1) 14846.

Non-Governmental Organizations

Environment Camps for Conservation Awareness (ECCA): POB 3923, Kathmandu; tel (1) 526391; telex 2464; fax (1) 521506; Project Co-ordinator: Anil Chitrakar; Programme Dir: Prachet Shrestha; counsellor training camp; camps for children organized with a network of over 500 counsellors in Nepal and Bhutan; hikes; liaison with local, national and international bodies.

IUCN Nepal: POB 3923, Kathmandu; tel (1) 521506; telex 2566; fax (1) 226820; Sr Adviser: John McEachern.

King Mahendra Trust for Nature Conservation (KMTNC): Babar Mahal, POB 3712, Kathmandu; tel (1) 526571; telex 2567; fax (1) 526570; Exec Officer: Chandra P Gurung; conservation of wildlife, resources, national parks and nature reserves; research; afforestation; soil conservation; water resource management; workshops and conferences.

Mount Everest Environment Conservation Foundation (EECF): Kathmandu Municipality 32, Maitidevi, POB 3508, Kathmandu; tel (1) 417757; fax (1) 414324; Founder and Chair: Laghu Dhan Rai; disposal of waste left on Mount Everest and throughout the Himalayas by mountaineers, tourists and trekkers.

Nepal Batabaran Patrakar Samuha (NEBAPASA—Nepal Forum of Environmental Journalists): Thapathali, POB 5143, Kathmandu; tel (1) 227691; fax (1) 226820; Pres: Binaya Guru-acharya; training for journalists in investigative reporting; liaison with politicians, decision-makers and opinion leaders; radio programme; wall bulletin; publications.

Nepal Heritage Society: POB 1041, Kathmandu; tel (1) 211438; Exec Mem: Ukesh Raj Bhuju; nature conservation; awareness-raising; education and training; lobbying government; monitoring pollution.

THE NETHERLANDS

Governmental Organizations

Ministry of Housing, Physical Planning and the Environment (VROM): POB 30945, 2500 GX The Hague; Van Alkemadelaan 85, 2597 AC The Hague; tel (70) 3393939; telex 34429; Minister: Hans Alders.

Directorate-General for Environmental Protection (DGM): Rijnstraat 8, 2515 XP The Hague.

Ministry of Agriculture and Fisheries, Directorate for Nature Protection, Environment and Fauna Management: POB 20401, 2500 EK The Hague; tel (70) 3793911; telex 32040; fax (70) 3793600; Minister: Piet Bukman; administration of the Nature Conservation Act, designating protected nature monuments and amimal and plant species; administration of the Birds Act, the Landscape Protection Act, the Endangered Exotic Animal Species Act and the Game Act; protection of natural areas, nature reserves, forests and national parks.

Afval Overleg Orgaan (AOO—Waste Management Council): POB 19015, 3501 DA Utrecht; tel (30) 342800; fax (30) 342260; Chair: H G Ouwerkerk; Dir: R Rense; organization of national waste management.

Agricultural Research Department (DLO-NL): Bornsesteeg 53, 6708 PD Wageningen; tel (8370) 74000; fax (8370) 24060; Dir: M Heuver; Deputy Dir and Dir of Research: Prof Dr L van Vloten-Doting; research into sustainable, healthy and competitive agriculture, forestry and fisheries and attainment of optimal development in rural areas; approximately 150 research programmes in 17 research institutes; co-operation with other research bodies around the world.

Commission for Environmental Impact Assessment: Arthur van Schendelstraat 800, POB 2345, 3500 GH Utrecht; tel (30) 347666; fax (30) 331295; Chair: P van Duursen; Sec-Gen: J J Scholten; independent body advising competent authorities on the required content of environmental statements; assessment of the scientific quality and completeness of environmental statements.

Natuurbeschermingsraad (Nature Conservation Council): Maliebaan 12, 3581 CN Utrecht; tel (30) 331441; Dir: T Klumpers; advisers to the Ministry of Agriculture and Fisheries; conservation of wild fauna and flora; conservation strategies; promotion of environmental law; resource management; land use and restoration; environmental management, planning and policy; education and training.

Non-Governmental Organizations

Allantie voor Duurzame Ontwikkeling (Platform for Sustainable Development): POB 19245, 3501 DE Utrecht; tel (30) 367997; fax (30) 367998; Man Dir: Kees Waagmeester; Int Affairs: Tineke van der Schoor; External Relations: Gerrit Vledder; forum for discussion between Dutch environmental organizations, development organizations, industry, trade unions, scientists, farmers, women, young people, the peace movement, consumers and the church; implementation of sustainable development; evaluation of Dutch progress in implementing UNCED Agenda 21; stimulation of public and political debate and action.

Bureau Milieu en Ruimtelijke Ordening (Office of Environmental and Physical Planning): Prinses Beatrixlaan 5, 2508 AB The Hague; tel (70) 3497475; fax (70) 3850707; Dir: J J de Graeff; seeks to influence government environmental policy on behalf of industry association (VNO) and employers' association (NCW).

Centrale Organisatie voor Radio-actief Afval (Central Organization for Radioactive Waste): POB 20, 1755 ZO Petten.

Centrum voor Energiebesparing (Centre for Energy Conservation and Environmental Technology): Oude Delft 180, 2611 HH Delft; tel (15) 150150; fax (15) 150151; Man Dir: Jan Paul van Soest; research and advice for governments and industry on environmental policy and energy conservation.

Centrum voor Landbouw en Milieu (Centre for Agriculture and Environment): POB 10015, 3505 AA Utrecht; tel (30) 441301; fax (30) 441318; Dir-Gen: G K Horst; Dir of Research and Devt: W J v d Weyden; independent platform for farmers and environmentalists; promotion of sustainable agriculture which meets the needs of consumers and conservation; promotion of co-operation through research and policy proposals; information.

Communicatis en Adviesbureau over Energie en Milieu (CEA): POB 21421, Leeuwenstraat 9-11, 3001 AK Rotterdam; tel (10) 4046080; fax (10) 4048022; Man Dir: R W Boerée; Deputy Dir: A F L Slob; consultancy and research on energy conservation, transport demand management, environmental management (products, waste and emissions), consumer behaviour and communication planning.

DLO Winand Staring Centre for Integrated Land, Soil and Water Research (SC-DLO): Marijkeweg 11, POB 125, 6700 AC Wageningen; tel (8370) 74200; telex 75230; fax (8370) 24812; Dir: G A Oosterbaan; Information Officer: A Reijmerink; focus on the problems of physical planning, water and soil management and environmental protection in rural areas; research into the efficient, sustainable and socially acceptable use of rural resources nationally and internationally.

Energie Onderzoekscentrum Nederland (ECN—Netherlands Energy Research Foundation): POB 1, 1755 ZG Petten; tel (2246) 4949; telex 57211; fax (2246) 4480; Gen Man: Prof Dr H H van den Kroonenberg; Deputy Man Dir: Dr R L J Scholte; energy research; development of technologies for safe, clean, reliable, efficient, viable and environmentally friendly energy; work on renewable energy (solar and wind sources), nuclear energy and fossil fuels; policy studies.

Erasmus Universiteit Rotterdam, Studiecentrum voor Milieukunde (Erasmus University, Rotterdam, Centre for Environmental Studies): POB 1738, 3000 DR Rotterdam; tel (10) 4082050; fax (10) 2120834; environmental research and studies.

Fietsersbond enfb (Cyclists Federation): POB 2150, 3440 DD Woerden; tel (3480) 23119; fax (3480) 17058; Co-ordinator: Jim A Schouten; promotion of use of bicycles; development of bicycle policy; information for cyclists.

It Fryske Gea (Het Friese Landschap—Frisian Landscape): POB 3, 9244 ZN Beetsterszwaag; tel (5126) 1448; fax (5126) 2973; Principal Officers: Henk Kroes, Dr Ultsje G Hosper; nature conservation.

Gelderse Milieufederatie (Gelderland Environmental Federation): Jansbuitensingel 14, 6811 AB Arnhem; tel (85) 515069; fax (85) 453210; Principal Officer: G Veeman; nature conservation; landscape and environmental protection.

Greenpeace Netherlands: Keizersgracht 176, 1016 DW Amsterdam; tel (20) 5236555; telex 18775; fax (20) 5236500; environmental pressure group; direct action; education; publications; information; awareness-raising.

De Groenen (Green Party): POB 3244, 1001 AA Amsterdam; tel (30) 341545; fax (30) 341545; Pres: K S Kuipers; Sec: P A Buitenhuis; ecologist political party.

Initiatiefgroep Wijs op Weg (Transport Initiative Group): POB 2000, 3800 CA Amersfoort; tel (33) 724224; fax (33) 751496; Co-ordinator: Dr G Knibbeler; survey of transport issues; campaigns for local and national government to introduce a more sustainable transport system.

Instituut voor Natuurbeschermingseducatie (IVN—Institute of Nature Conservation Education): POB 20123, 1000 HC Amsterdam; tel (20) 6228115; fax (20) 6266091; Co-ordinator: G J Jutten; association of volunteers and professionals working to promote environmental awareness.

International Youth Federation for Environmental Studies and Conservation (IYF), Regional Office for Europe: Nieuwe Markt 36, 7411 PC Deventer; tel (5700) 1784; Regional Co-ordinator: David Santa; world secretariat in India and regional offices in Benin, Costa Rica and the Philippines.

Katholieke Universiteit Brabant, Interuniversitair Milieu-instituut (Brabant Catholic University, Inter-University Environment Institute): Celebeslaan 4, 5641 AG Eindhoven.

Koninklijk Zoologisch Genootschap (Royal Zoological Society): Plantage Kerklaan 38-40, 1018 CZ Amsterdam; tel (20) 5233404; fax (20) 5233419; Dir: Dr M T Frankenhuis.

Koninklijke Nederlandse Natuurhistorische Vereniging (KNNV—Royal Dutch Natural History Society): Oude Gracht 237, 3511 NK Utrecht; tel (30) 314797; study of nature and conservation of wildlife; ecology; conservation strategies.

Koninklijke Nederlandse Toeristenbond (ANWB—Royal Dutch Touring Club): POB 93200, 2509 BA The Hague; tel (70) 3146435; telex 32032; fax (70) 3146969; Principal Officer: R Sinke; nature conservation; education.

Landbouw Universiteit Wageningen, Centrum voor Milieustudies (Wageningen Agriculture University, Centre for Environment and Climate Studies): POB 9101, 6700 HB Wageningen; tel (8370) 84812; telex 45015; fax (8370) 84839; Dir: Prof L Hordijk; Environment Co-ordinator: Dr H de Jager; Climate Co-ordinator: R S de Groot; interdisciplinary research groups on regional environmental studies, global change, acidification, environmental impacts of agriculture, integrated control of the production chain; undergraduate courses on environmental protection and management.

Landelijk Platform tegen Kernenergie (National Anti-Nuclear Platform): Damrak 26, 1012 LJ Amsterdam; POB 19199, 1000 GD Amsterdam; tel (20) 6221366; fax (20) 6275287; Principal Officers: Wim Kersten, Peer de Rijk; co-ordination of national activities against nuclear power.

Landelijke Vereniging tot Behoud van de Waddenzee (Dutch Society for the Preservation of the Waddensea): Voorstraat 16-18, POB 90, 8860 AB Harlingen; tel (5178) 15541; fax (5178) 17977; Pres: W H Kuiper; Sec: B Oost; legal action; raising public awareness; environmental education; publications.

Milieufederatie Noord-Holland (North Holland Environment Federation): Nicolaasstraat 2B, 1506 BB Zaandam; tel (75) 351598; fax (75) 704636; Dir: Dr H P Gallacher; nature conservation; environmental policy.

Milieukompas: POB 19199, 1000 GD Amsterdam; tel (20) 6221366; fax (20) 6275287; Principal Officer: A P A M

Bruyns; support for environmental action groups; public information and education on environmental issues.

Milieukontakt Oost-Europa (Foundation for Environmental Contacts with Eastern Europe): POB 18185, 1001 ZB Amsterdam; tel (20) 6392716; fax (20) 6391379; liaison with environmental groups in Central and Eastern Europe.

MilieuTelefoon (Environment Telephone): Damrak 26, 1012 LJ Amsterdam; tel (20) 6262620; fax (20) 6275287; Principal Officer: Simone de Jong; information by telephone and mail; environmental publications for consumers in the Netherlands and Belgium.

Nederlandse Vereniging van Dierentuinen (Dutch Society of Zoological Gardens): Plantage Kerklaan 38-40, 1018 CZ Amsterdam; tel (20) 5233400; fax (20) 5233419; Principal Officer: Dr D Dekker; awareness-raising; national and international agreements on zoos and wildlife protection.

Netherlands Committee for IUCN: c/o Plantage Middenlaan 2B, 1018 DD Amsterdam; tel (20) 6261732; fax (20) 6279349; Principal Officer: Dr Wim Bergmans.

Open Universiteit, Leerstofgebied Natuurwetenschappen (Open University of the Netherlands, Natural Sciences Department): Valkenburgerweg 167, POB 2960, 6401 DL Heerlen; tel (45) 762359; telex 56559; fax (45) 711486; provision of open and distance learning; M SC programmes in Environmental Science and Environmental Management.

Prince Bernhard Foundation: POB 80084, 3508 TB Utrecht; tel (30) 536835; telex 76288; fax (30) 518366; Principal Officer: Dr W Dijkman; support for education and academic research in the international aspects of nature conservation.

Raad voor het Milieu- en Natuuronderzoek (Advisory Council for Research on Nature and Environment): POB 5306, 2280 HH Rijswijk; tel (70) 3364300; fax (70) 3364310; publication of reports on necessary programmes of research into nature and the environment; co-ordination of research; monitoring gaps and overlapping in research.

Rijksuniversiteit Groningen, Faculteit der Rechtsgeleerdheid, Vakgroep Milieurecht (Groningen University, Law Faculty, Environmental Law Group): Turftorenstraat 13, 9712 BM Groningen; tel (50) 639111; telex 53410; Dean: Prof Dr A Nooteboom.

Rijksuniversiteit Groningen, Interfacultaire Vakgroep Energie- en Milieukunde (Groningen University, Centre for Energy and Environmental Studies): Nijenborgh 4, 9747 AG Groningen; tel (50) 634609; fax (50) 637168; Principal Officers: Prof Dr A J M Schoot Uiterkamp, Dr W Biesiot; research into environmental issues including sustainability, environmental quality, and depletion of energy and other resources.

Rijksuniversiteit Leiden, Centrum voor Milieukunde (CML—Leiden University, Centre for Environmental Science): POB 9518, Garenmarkt 1A, 2300 RA Leiden; tel (71) 277486; telex 39427; fax (71) 277496; Dir: Prof Dr H A Udo de Haes; interdisciplinary research and education in landscape ecology, ecotoxicology, life cycle assessment, substance flow analysis, policy instruments, environment and development in developing countries, and wetlands; contract research for government departments, industry, the Commission of the European Communities and other organizations.

Environment and Development Programme: Principal Officers: Prof Dr Wouter T de Groot, Dr Gerard A Persoon, Carel A Dryven, Hans H de Iongh; research and education in social and ecological dynamics of deforestation (especially in the Philippines and Indonesia) and local resource management (especially in the Sahel region).

Rijksuniversiteit Utrecht, Interfacultaire Vakgroep Milieukunde (Utrecht University, Department of Environmental Studies): Heidelberglaan 2, POB 80.115, 3508 TC Utrecht; tel (30) 532359; fax (30) 540604; Principal Officer: Dr R Klaassen; research; graduate activities.

Seas at Risk: Vossiusstraat 20-III, 1071 AD Amsterdam; tel (20) 6754336; fax (20) 6753806; Chair: Ward Dossche; Co-ordinator: John J Maggs; lobbying internationally for protection of European seas, especially the North Sea.

Stichting het Drentse Landschap (Drentse Landscape Foundation): Kloosterstraat 5, 9401 KD Assen; tel (5920) 13552; fax (5920) 18089; Dir: A Ernst; management of nature reserves and buildings; seeks to influence government and province policy on local nature conservation.

Stichting het Flevolandschap (Flevoland Landscape Foundation): Zuiderwagenplein 2, 8224 AD Lelystad; POB 600, 8200 AP Lelystad; tel (3200) 98989; fax (3200) 34300; Principal Officer: J van Olst; protection, monitoring, preservation and development of nature in Flevoland; information for visitors to nature reserves.

Stichting Greenpeace Council (Greenpeace Council Foundation): Keizersgracht 176, 1016 DW Amsterdam; tel (20) 5236555; telex 18775; fax (20) 5236500.

Stichting IUCN-Ledencontact (Netherlands National Committee for IUCN): Plantage Middenlaan 2B, 1018 DD Amsterdam; tel (20) 6261732; fax (20) 6279349; Principal Officers: Wim Bergmans, Simone Bilderbeek, Christine Enthoven, Wouter Veening; research; lobbying the public, business, politicians and government departments responsible for the conservation of biodiversity and sustainable use of nature and natural resources; focus on Dutch activities in developing countries.

Stichting Koninklijke Rotterdamse Diergaarde (Royal Rotterdam Zoological and Botanical Gardens): POB 532, 3000 AM Rotterdam; tel (10) 4654333; fax (10) 4677811; Dir: Dr A H Dorresteyn; environmental legislation and improvement; training; research; promotion of study of nature and conservation of wildlife.

Stichting Kritisch Faunabeheer (Wildlife Management Foundation): Amsteldijk Noord 135, 1183 TJ Amstelveen; tel (20) 6410798; fax (20) 6473700; promotion of scientific and ethical wildlife management and minimal intervention.

Stichting Milieu Educatie (SME—Environmental Education Foundation): POB 13030, Australiëlaan 14, 3507 LA Utrecht; tel (30) 802444; fax (30) 801345; Man Dir: Frits Hasselink; Head of Operational Services: Kees Kil; non-profit consultancy on environmental communication and education; promotion of a sustainable society; information; support for dialogue between government and other interested parties; advice; courses; educational materials.

Stichting Milieufederatie (Environment Federation Foundation): Bunnikseweg 39, 3732 HV De Bilt.

Stichting Milieuraad Drenthe (Drenthe Environment Council Foundation): Hertenkamp 6, 9401 HL Assen; tel (5920) 11150; fax (5920) 30201; promotion of the care and optimal management of nature and landscapes; public education in environmentally responsible behaviour.

Stichting Mondiaal Alternatief (Foundation for Ecodevelopment): POB 151, 2130 AD Hoofddorp; tel (2503) 32305; fax (2503) 41359; Sec: Dr R Gerrits; Co-ordinator: Dr J Weverling; promotes the idea of a need for a worldwide social and ecological conservation strategy; campaigns on sustainable living and other conservation issues; publication of ecological material.

Stichting Natuur en Milieu (Society for Nature and Environment): Donkerstraat 17, 3511 KB Utrecht; tel (30) 331328; fax (30) 331311; Dir-Gen: P Nijhoff; Dir: A J M van den Biggelaar; organization of national and provincial environment groups; lobbying government.

Stichting Nederlandse Milieugroep Alpen (Alps Environment Group Foundation): Keucheniushof 15, 5631 NG Eindhoven; tel (40) 817484; fax (15) 616032; Chair: H Folmer; Vice-Chair and Sec: P van der Zee; information, education and advice on tourism and the environment in the Alps and other mountain regions.

Stichting het Noordbrabants Landschap (Northern Brabant Landscape Foundation): Kasteellaan 4, 5076 RE Haaren; POB 80, 5076 ZH Haaren; tel (4117) 2775; fax (4117) 3365; Dir: A B Gijtenbeek; Asst Dir: J Baan; acquisition, conservation and management of natural sites in Northern Brabant province; quarterly information bulletin; twinning projects with nature conservation organizations in other European countries.

Stichting Noorderbreedte (Noorderbreedte Foundation): Ossenmarkt 9, 9712 NZ Groningen; tel (50) 141956; fax (50) 142374; Principal Officers: Jan Abrahamse, Johan Visser; publication of a magazine concerned with the landscape, nature and environment of the three northern provinces of the Netherlands.

Stichting het Noordhollands Landschap (Northern Holland Landscape Foundation): POB 257, 1900 AG Castricum; tel (2518) 59750; fax (2518) 52422; protection of nature reserves in Northern Holland province.

Stichting Reinwater (Clean Water Foundation): Vossuistraat 20-II, 1071 AD Amsterdam; tel (20) 6719322; fax (20) 6753806; Co-ordinator: José Houweling; Water Quality and Research: Wim Verhoog; Research: Marc de Rooy; Legal Affairs: Joost Rutteman; Education: Bowine Wijffels; water quality research; collection of water samples at points of discharge for industrial waste water; legal action over unlawful pollution of surface waters, particularly the Rhine, Meuse and Scheldt rivers; environmental education in water issues.

Stichting Reservaten Przewalski Paard Nederland (Foundation for Reserves for the Przewalski Horse): Da Costastraat 27, 3027 JC Rotterdam; tel (10) 4370447; Treas: J M Grosjean; preservation of the Przewalski horse and its reintroduction into the wild.

Stichting Veldwerk Nederland (Netherlands Field Work Foundation): Bosweg 24A, 7314 AP Apeldoorn; tel (55) 214835; fax (55) 786057; Man Dir: J W H van Ginkel; nature work weeks and field study courses for students from elementary, secondary and vocational schools; teacher training; study materials.

Stichting het Wereld Natuur Fonds WNF Nederland (World Wide Fund for Nature WWF Netherlands): POB 7, 3700 AA Zeist; tel (3404) 37333; telex 76122; fax (3404) 12064; Chair: Dr Ed H T M Nijpels; Chief Exec and Exec Dir: Dr Siegfried Woldhek; national affiliate organization of World Wide Fund for Nature (WWF) International; conservation of marine, freshwater and terrestrial environments and wildlife; forestry; protected areas; land use; climatology; education and training; international co-operation.

TNO Study Centre for Environmental Research: POB 6013, 2600 JA Delft; tel (15) 696884; telex 38071; fax (15) 613186; Principal Officers: Prof Dr F B de Walle, G van Ee, Dr J van Ham, Dr H T U van Rijn, Dr A C Leget; multidisciplinary environmental studies for governments, institutes and private companies on waste management, soil protection, policy evaluation and environmental impact assessment, risk management, sustainable development and air pollution.

Unie van Provinciale Landschappen (Union of Provincial Landscape Organizations): POB 80, 5076 ZH Haaren; tel (4117) 3225; fax (4117) 3365; Pres: Dr M C A Klinkenbergh; Dir: P C von Meijenfeldt; union of 12 regional nature protection organizations which own and manage 65,000 ha of land; lobbying government and national and international organizations.

Universiteit van Amsterdam, Wetenschappenlijk Instituut voor Milieumanagement, Stichting Envicon (Amsterdam University, Economic Institute for Environmental Management, Envicon Foundation): Roetersstraat 11, 1018 WB Amsterdam; tel (20) 5254077; fax (20) 5255281; Principal Officer: J L P Piet; education and research on environmental management and environmental accounting.

Van Tienhoven Stichting tot Internationale Natuurbescherming (Netherlands Foundation for International Nature Protection): c/o Inst of Taxonomic Zoology, Mauritskade 61, 1092 AD Amsterdam; tel (20) 5255437; fax (20) 5257238; Chair: Dr A C van Bruggen; Sec: Dr P J H van Bree; promotion of international nature conservation projects, particularly those carried out by Dutch citizens.

Vereniging tot Behoud van Natuurmonumenten in Nederland (Dutch Society for the Preservation of Nature Monuments): Schaep en Burgh, Noordereinde 60, 1243 JJ 's-Graveland; tel (35) 62004; fax (35) 63174; conservation of buildings, monuments, archaeological sites, landscape, national parks and protected areas; land use.

Vereniging tegen Milieubederf in en om het Nieuwe Waterweggebied (New Canal Environment Association): POB 252, 3130 AG Vlaardingen; tel (10) 4350696; Chair: A van Leeuwen; exhibitions, education and projects on environmental issues for secondary schools.

Vereniging Milieudefensie (FoE Netherlands): Damrak 26, POB 19199, 1000 GD Amsterdam; tel (20) 6221366; fax (20) 6275287; Chair: Bram van der Lek; waste; energy; the greenhouse effect; sustainable development; transport; agriculture; rainforests.

Vereniging voor Natuur- en Milieueducatie (IVN—Association for Natural and Environmental Education): Plantage Middenlaan 2C, POB 20123, 1000 HC Amsterdam; tel (20) 6228115; fax (20) 6266091; Dir: Gerard Jutten; Deputy Dir: Hugo Bunte; environmental education in primary schools; conservation of the environment by volunteers.

Vogelbescherming Nederland (Netherlands Society for the Protection of Birds): Driebergseweg 16C, 3708 JB Zeist; tel (3404) 25406; fax (3404) 18844; conservation of wild birds, including marine birds, migratory species and waterfowl; freshwater ecology; conservation of wetland ecosystems and protected areas.

Youth Federation for Nature and Environmental Studies: Oudegracht 42, 3511 AR Utrecht; tel (30) 360925; fax (30) 343986; organization of environmental activities for members, including study camps; youth education.

Zuid-Hollandsie Milieufederatie (South Holland Environment Federation): G W Burgerplein 5, 3021 AS Rotterdam; tel (10) 4765355; fax (10) 4775562; Dir: N v Brussel; Water: H Muilerman; Waste: B Bakker; Industry: W Berends; Agriculture: P Schaap; Planning: H Wiersema; Press: A Steekelenburg; promotion of sustainable development as provincial policy in South Holland (Rotterdam, The Hague, Leiden and Dordrecht); advice and support for approximately 100 voluntary organizations in South Holland; lobbying bulb growers, market gardeners, chemical and petrochemical industry, builders and others.

NETHERLANDS DEPENDENCIES

Aruba

Ministry of Public Works and Public Health: L G Smith Blvd, Oranjestad; tel (8) 24900; telex 5060; fax (8) 26826; Deputy Prime Minister and Minister of Public Works and Public Health: Martinus Flanegin.

The Netherlands Antilles

Ministry of Health and the Environment: Heelsumstraat, Willemstad, Curaçao; tel (9) 61-4555; fax (9) 61-2388; Minister: C H Statia.

Saba Marine Park and Saba Conservation Foundation: Fort Bay, POB 18, Bottom, Saba; tel (4) 63295; fax (4) 63435; Chair: Tom van't Hof; Sec: Kathy Hassell; conservation of resources through sustainable use; education about marine and terrestrial ecosystems.

Stichting Caraïbisch Marien Biologisch Instituut (CARMABI—Caribbean Marine Biology Institute Foundation): POB 2090, Piscaderabaai z/n, Curaçao; tel (9) 62-4242; fax (9) 62-7680; Dir: Dr W L Bakhuis; Deputy Dir: Dr A O Debrot; marine biological research; conservation research; advice to local government.

Stichting Nationale Parken Nederlandse Antillen (STINAPA—National Parks Foundation of the Netherlands Antilles): POB 368, Bonaire; tel (7) 8444; fax (7) 8444; Chair: R R Hensen; Sec: V G Croes; protection and conservation of the natural environment of Bonaire.

NEW ZEALAND

Governmental Organizations

Ministry for the Environment: 84 Boulcott St, POB 10362, Wellington; tel (4) 473-4090; fax (4) 471-0195; Minister: Rob Storey; Sec: Dr Roger Blakely; advice to government on environmental administration; environmental research and information.

Department of Conservation: 59 Boulcott St, POB 10420, Wellington; tel (4) 471-0726; telex 3588; fax (4) 471-1082; Minister: Denis Marshall; Dir-Gen: Bill Mansfield; conservation of 12 national parks, three maritime parks, two World Heritage Sites, 21 conservation parks, reserves, forest parks and wildlife; production of educational material.

New Zealand Conservation Authority: Chair: James K Guthrie; Sec: Gwenda Harris; development of national conservation policies.

Ministry of Agriculture and Fisheries: POB 2526, Wellington; tel (4) 472-0367; telex 31532; fax (4) 472-9071; Minister: John Falloon.

MAFTech Organic Production Information Library (OPIL): computer database.

New Zealand Association of Crown Research Institutes (ACRI): POB 1578, Wellington; tel (4) 472-9979; fax (4) 472-4025; Pres: Dr Alan Frampton; Exec Dir: Mike Collins; Int Science Co-ordinator: Wayne Robertson; formerly Department of Scientific and Industrial Research (DSIR); association of government-owned research institutes registered as companies.

Forest Research Institute Ltd: POB 3020, Rotorua; tel (7) 347-5899; fax (7) 347-9380; Chief Exec: Dr Frank Wood; research into production forestry and environmentally friendly wood products.

Institute for Crop and Food Research Ltd: Private Bag 4704, Christchurch; tel (3) 325-2511; fax (3) 325-2074; CEO: Dr M W Dunbier; multidisciplinary research for production and processing of arable crops, vegetables, seed crops, flowers, ornamentals, essential oils, medicinal crops and seafood; plant breeding; soil science; aquaculture; post-harvest physiology; molecular biology; agronomy; entomology; research into fishing methods.

Institute of Environmental Science and Research (ESR): Level 6, Clear Centre, 15-17 Murphy St, Thorndon, POB 12-444, Wellington; tel (4) 499-0540; fax (4) 499-0541; CEO: Mark Templeton; Chief Financial Officer and Co-Sec: Neil Wanden; Nat Science Man: Craig Stevenson; research, analysis and consultancy in public health, environmental health and forensic sciences for the public and private sectors.

Institute of Geological and Nuclear Sciences Ltd: 1 Gracefield Rd, POB 30-368, Lower Hutt, Wellington; tel (4) 570-1444; fax (4) 569-0600; CEO: Dr David Ross; Business Man: Dr J Varcoe; research on the core and crust of the planet, providing a scientific basis for the management of mineral and energy resources; assessment and mitigation of geological hazards.

Landcare Research New Zealand Ltd: POB 40, Lincoln; tel (3) 325-2125; fax (3) 325-2127; Chief Exec: Dr Andrew Pearce; research into management of national land environments through sustainable land use and prediction of the impacts of major resource utilization proposals.

National Institute of Water and Atmospheric Research: POB 14 901, Kilbirnie, Wellington; tel (4) 386-1189; fax (4) 386-3442; Chief Exec: Dr Malcolm Grant; research on rivers and lakes, the oceans and atmosphere and their interactions and effect on national climate; provision of weather forecasting services.

National Radiation Laboratory: POB 25099, Christchurch; tel (3) 366-5059; fax (3) 366-1156; Dir: Dr A C McEwan; Head Environmental Section: Dr K M Matthews; administration of New Zealand radiation protection legislation; monitoring of environmental radioactivity in New Zealand and the South Pacific region.

Non-Governmental Organizations

Bio-Dynamic Farming and Gardening Association: POB 306, Napier; tel (6) 352428.

Environment and Conservation Organisations of New Zealand (ECO): POB 11057, Wellington; tel (4) 384-6971; fax (4) 384-6971; Co-Chair: Catherine Wallace; Co-Vice-Chair:

Clive Monds; lobbying of government and officials; information; international networking.

Green Party of Aotearoa—New Zealand: POB 11-652, Wellington; tel (4) 801-6262; Convenors: Carolynne Stone, Harry Parke; formerly Values Party; ecologist socialist party.

Greenpeace New Zealand: Private Bag, Wellesley St, Auckland; tel (9) 377-6128; fax (9) 303-2676; Exec Dir: Cindy Kiro; environmental campaigns on climate change and energy, forestry, fisheries and ocean ecology, ozone, Antarctica and nuclear issues.

IUCN Commission on National Parks and Protected Areas: c/o 1/268 Main Rd, Tawa; tel (4) 232-5581; fax (4) 232-9129; Chair: P H C Lucas; Deputy Chair: Adrian Phillips.

Lincoln University, Centre for Resource Management: POB 56, Canterbury; tel (3) 25-2811; fax (3) 25-2944; Principal Officer: Prof Kevin F O'Connor.

New Zealand Biological Producers and Consumers Council: POB 36-170, Northcote, Auckland 9; tel (9) 443-8435; fax (9) 443-8436; Principal Officers: Perry Spiller, Lyn Watson; certification of biological food producers.

New Zealand Committee for IUCN: c/o Dept of Zoology, Otago Univ, POB 56, Dunedin; tel (24) 735022; telex 3588; Principal Officer: Prof Carolyn Burns.

New Zealand Institute of Forestry: POB 19840, Christchurch; tel (3) 384-2432; fax (3) 384-2430; Pres: Peter F Olsen; Sec: Robert J van Rossen; independent advocate for forestry.

New Zealand Water and Wastes Association: POB 27356, Auckland; tel (9) 620-4651; fax (9) 620-4651; Pres: Jeff Booth; technical meetings; seminars; conferences; publications.

Pacific Institute of Resource Management: POB 10 123, The Terrace, Wellington; tel (4) 473-8312; fax (4) 472-6374; Pres: George Porter; Exec Chair: Derek Wilson; education and awareness-raising on global and national social, environmental and economic policy issues; lobbying; publishing; seminars; international networking.

Royal Forest and Bird Protection Society of New Zealand Inc: POB 631, Wellington; tel (4) 385-7374; fax (4) 385-7373; Conservation Dir: Kevin Smith; Nat Sec: Rosemarie van Dam; conservation of indigenous forest; protection of endangered and native birds and marine life; campaign against dangerous fishing methods; protection of plants.

Soil and Health Association of New Zealand Inc: POB 36-170, Northcote, Auckland 9; tel (9) 443-8435; fax (9) 443-8436; Principal Officers: Perry Spiller, Lyn Watson; publication of bimonthly journal; promotion of organic cultivation.

Waitomo Caves Museum Society: POB 12, Waitomo (via Te Kuiti); tel (813) 87640; Principal Officer: Peter Dimond; informing the public about the caves and their inhabitants.

World Wide Fund for Nature (WWF) New Zealand: Botanic Garden, POB 6237, Wellington; tel (4) 499-2930; fax (4) 499-2954; Exec Dir: Ian Higgins; Conservation Scientist: Sue Miller; Communications Officer: Tonya St George; Educ Officer: Jenny Jones; Marketing Dir: Dave Boardman; national affiliate organization of World Wide Fund for Nature (WWF) International; campaigns on threatened species and habitats, sustainable resource use, ecological restoration and environmental education.

NEW ZEALAND'S DEPENDENT AND ASSOCIATED TERRITORIES

Ross Dependency

Ross Dependency Research Committee: c/o Antarctic Policy Div, Ministry of Foreign Affairs and Trade, Private Bag 18901, Wellington; tel (4) 472-8877; fax (4) 472-8039; Chair: Sir Robin Irvine; Exec Officer: Louise Sparrer; advice to government on the scientific component of the annual Antarctic Programme; co-ordination of the Antarctic Programme.

Cook Islands

Ministry of Post and Telecommunications and Conservation and for the House of Ariki and Koutu Nui: Avarua, Rarotonga; tel 29300; telex 62006; fax 20856; Minister: Teanua Kamana.

Ministry of Agriculture, Marine Resources, Trade, Labour and Transport and Civil Aviation: Avarua, Rarotonga; tel 29300; telex 62006; fax 20856; Minister: Vaine Tairea.

Niue

Ministry for Agriculture, Forestry and Fisheries, Public Works, Posts and Telecommunications and Business Relations: Alofi; tel 4224; telex 67014; fax 4232; Minister: Terry D Coe.

NICARAGUA

Governmental Organizations

Instituto de Recursos Naturales (IRENA—Institute for Natural Resources): Apdo 5123, Managua, JR; tel (2) 263-1273; telex 1328; fax (2) 263-1274; Dir: Dr Jaime Incer; nature conservation; implementation of national planning policy; management of renewable natural resources.

Departamento de Regulación y Control de Recursos Naturales y del Ambiente (Department for Regulation and Control of Natural Resources and the Environment): Km 12½ Carretera Norte, Apdo 5123, Managua, JR; tel (2) 31110; telex 1328.

Ministry of Foreign Affairs: Detrás de Los Ranchos, Managua, JR; tel (2) 96563; fax (2) 66-3971; Minister: Ernesto Leal Sánchez; co-operation with international environmental organizations.

Instituto Nicaraguense de Estudios Territoriales (INETER—Nicaraguan Institute of Land Studies): Apdo 4596, Managua, JR; tel (2) 26540.

Non-Governmental Organizations

Asociación de Biólogos y Ecólogos de Nicaragua (ABEN—Nicaraguan Association of Biologists and Ecologists): Apdo 2431, Managua, JR; tel (2) 74563; fax (2) 66-3303; Pres: Juan José Montiel; research into environmental policy; advice to government.

CARE International Nicaragua: Apdo 3084, Managua; tel (2) 78-0018; fax (2) 67-0386; Country Dir: Swaleh Karanja; Deputy Dir: Michel Belisle; Sector Co-ordinator: Allan J Hruska; development and emergency work; sustainable agriculture; pesticide reduction; agroforestry; drinking water supply; community organization.

Centro Alexander von Humboldt para la Promoción del Desarrollo y la Gestión del Ambiente (Alexander von Humboldt Centre for the Promotion of Development and Management of the Environment): Bello Horizonte R-IV-5, De la Iglesia Pio X 1c al Sur, 20 Varas al Este, Managua; tel (2) 40788; fax (2) 40788; Principal Officer: Amado Isabel Ordóñez Mejía; land management; local development; monitoring environmental politics; information on land management; urban planning.

Fundación del Río para la Conservación y el Desarrollo del Sureste de Nicaragua (River Foundation for the Conservation and Development of South-East Nicaragua): San Carlos, Río San Juan, Detras del Hospital, Managua; tel (2) 11-0154; fax (2) 78-4863; Principal Officer: Antonio Ruiz Melendez; environmental education; sustainable eco-agriculture projects; cultural and environmental development.

IUCN Nicaragua: Apdo 1989, Managua 5; tel (2) 71256; fax (2) 71256; Rep: Maria Virginia Sandino.

Movimiento Ambientalista Nicaragüense (MAN—Nicaraguan Environmentalist Movement): Apdo A-99, Paseo Tiscapa, Casa del Puente, Costado sur Hosp Militar, Managua, JR; tel (2) 25611; Dir: Silvia Ayon Ruiz; environmental education and training; education in resource management; land restoration; sustainable development; monitoring pollution; promotion of environmental law.

NIGER

Governmental Organizations

Ministry of Water Resources and the Environment: BP 241, Niamey; tel 73-30-44; Minister: Soumana Billo.

Direction de l'Environnement (Department of the Environment): BP 578, Niamey; tel 73-33-29; telex 5509; management of the environment, wildlife and fisheries; scientific information on the environment (forests, wildlife, fish, pollution, pests and other issues).

Office de l'Energie Solaire (ONERSOL—Solar Energy Office): BP 621, Niamey; tel 72-39-23; Dir: Abdoussalam Ba; Head of Production: Yahaya Saley; research and development; use of renewable sources of energy; training.

Non-Governmental Organizations

Centre Régional de Formation et d'Application en Agrométéorologie et Hydrologie Opérationnelle (AGRHYMET—Regional Centre for Training and Application in Agrometeorology and Operational Hydrology): BP 11011, Niamey; tel 73-31-16; telex 5448; fax 73-24-35; training in agrometeorology, hydrology and plant protection; information about agriculture in the Sahel region; application of remote sensing to natural resources.

IUCN Niger: BP 10933, Niamey; tel 73-33-38; telex 5505; fax 73-22-15; Dir: Anada Tiega; Social Sciences Advisor: Thomas G Price; Programme Officer: Peter J Kristensen; wildlife management; environmental education; conservation and management of natural resources; technical assistance; workshops.

Organisation Nigérienne de Volontaires pour la Préservation de l'Environnement (ONVPE—Niger Organization of Volunteers for the Protection of the Environment): BP 2842, Niamey; tel 73-45-73; Pres: Albert Ferral; preservation and rational use of water, soil, fauna, flora, energy and ecosystems; training; information.

Université Abdou Moumouni Dioffo, Faculté d'Agronomie (Abdou Moumouni Dioffo University, Agronomy Faculty): BP 10960, Niamey; tel 73-22-90; telex 5258; fax 73-38-62; Dean: Dr A Gouro; training and research in agronomy; specialization in Sahel agriculture and environmental protection.

Centre Régional d'Enseignement Spécialisé en Agriculture (CRESA—Regional Centre for Specialized Training in Agriculture): selection of students for agricultural training courses; research programmes; joint collaborative projects with interested parties.

NIGERIA

Governmental Organizations

Federal Ministry of Agriculture, Water Resources and Rural Development: PMB 12613, 6 Ijeh Villages, Obalende, Lagos; tel (1) 684178; Sec of Agric, Water Resources and Rural Devt: Alhaji Garuba Abdulkadir; incorporates Department of Forestry.

Federal Ministry of Works and Housing: 24 Keffi St, Obalende, PMB 12698, Lagos; tel (1) 682625; Sec of Works and Housing: Barnabas Germade; incorporates Environmental Planning and Protection Division.

National Centre for Energy Research and Development: c/o Nigeria Univ, Nsukka, Enugu State; tel (42) 771911; telex 51496; Dir: Prof O C Iloeje; research and development of technologies for the utilization of all forms of energy resources, particularly solar and other renewable energy resources.

Nigeria National Park Service: PMB 0258, Garki-Abuja; tel (9) 5230429; fax (9) 5230409; Dir and Chief Exec: Lawan B Marguba; conservation of biodiversity; protection of historic and archaeological sites; promotion of tourism.

North East Arid Zone Development Programme (NEAZDP): PMB 18, Garin Alkali, via Gashua, Yobe State; tel (76) 232898; telex 82160; fax (76) 232989; Principal Officers: Jens H Christensen, Ali A Abatcha; integrated rural development.

Non-Governmental Organizations

Chad Basin National Park: PMB 1026, Maiduguri, Borno State; tel (76) 234120; conservation of indigenous flora and fauna; promotion of tourism; anti-poaching patrols; education campaigns; development programmes to reduce human dependence on available natural resources.

IUCN Nigeria: Hadejia-Nguru Wetlands Conservation Project, POB 32, Kano; Project Advisor: Henry Thompson.

Nigerian Conservation Foundation: Plot 5, Moseley Rd, Ikoyi, POB 74638, Victoria Island, Lagos; tel (1) 683563; fax (1) 684717; Pres: Chief S L Edu; Chair, Council: Amb Aduke Alakija; national associate organization of World Wide Fund for Nature (WWF) International; campaigns against pollution and wasteful use of renewable and non-renewable resources; conservation of nature and natural resources; improvement of the quality of human life; promotion of conservation and sustainable development through public awareness-raising and environmental education.

NORWAY

Governmental Organizations

Ministry of the Environment: Myntgt 2, POB 8013 Dep, 0030 Oslo; tel 22-34-90-90; telex 21480; fax 22-34-95-60; Minister: Thorbjørn Berntsen.

Avdeling for Naturvern og Kulturminner (Department for Nature Conservation and Cultural Heritage): POB 8013 Dep, 0030 Oslo; tel 22-34-58-53; telex 21480; fax 22-34-27-56; conservation of biodiversity and landscape; wise use of natural resources.

SFT (State Pollution Control Authority): POB 8100 Dep, Strømsvn 96, 0032 Oslo; tel 22-57-34-00; telex 76684; fax 22-67-67-06; Dir: Harald Rensvik.

Directorate for Nature Management: Tungasletta 2, 7005 Trondheim; tel 73-58-05-00; fax 73-91-54-33; Principal Officer: Peter Johan Schei.

Non-Governmental Organizations

Fylkeslistene for Miljø og Solidaritet (Local Candidates for the Environment and Solidarity): Oslo; Chair: Kåre Andre Nilsen; Sec: Gunnar Wahl; formerly Norges Kommunistiske Parti (Communist Party of Norway); name changed and merged with Rød ValgalliANSE (Red Electoral Alliance) in 1989; aims to supersede capitalism through a peaceful transition to a parliamentary socialist republic; favours environmental protection.

Institutt for Naturanalyse (INA—Institute for Environmental Analysis): 3800 Boe i Telemark; tel 395-04-99; fax 395-19-87; Researcher: Dr Faye Benedict; private foundation providing public and private agencies with environmental and social information for use in planning and rational use of natural resources.

Miljøpartiet de Grønne (Green Environmental Party): Grensen 8, 0159 Oslo; tel 22-42-97-58.

Norges Naturvernforbund (Norwegian Society for Conservation of Nature): Nedregate 5B, POB 2113, Grünerløkka, 0505 Oslo 5; tel 22-71-55-20; fax 22-71-56-40; Sec: Guro Tarjem; member of FoE International; lobbying government; information; newsletters and bimonthly journal.

Norsk Ornitologisk Forening (Norwegian Ornithological Society): Innherredsvn 67A, 7000 Trondheim.

Norsk Polarinstitutt (Norwegian Polar Research Institute): POB 5072, Middelthuns Gate 29, Majorstua, 0301 Oslo; tel 22-95-95-00; fax 22-95-95-01; Dir: Nils A Øritsland; Deputy Dir: Knut Svendsen; mapping; biological, geological and geophysical research; management of Norwegian areas at the North and South Poles.

Norway Agricultural University, Department of Biology and Nature Conservation: POB 14, 1432 Ås NLH; tel 994-81-22; fax 994-13-10; Principal Officer: Prof Sigmund Hagvar; education and research in nature conservation, wildlife and management of inland fisheries.

Norway Agricultural University, Norwegian Centre for International Agricultural Development (NORAGRIC): POB 2, 1432 Ås NLH; tel 994-99-50; telex 77384; fax 994-07-60; Dir: Dr Stein W Bie; scientific research for Norwegian aid policy in food production and rural development; advice to government, NGOs and international organizations.

Prosjekt Alternativ Framtid (Alternative Future Project): Sognsvn 70, 0855 Oslo; tel 22-18-11-70; fax 22-18-20-77;

Dir: William M Lafferty; Head of Admin: Mari H Ruge; Information Officer: Runar Malkenes; research, documentation and information on the environment, peace and development.

Universitetet i Bergen, Senter for Utviklingsstudier (SFU—Bergen University, Centre for Development Studies): Stromgt 54, 5007 Bergen; tel 531-72-50; telex 8441023; fax 532-26-86; Dir: Gunnar M Sorbo; research and training in development and environmental issues.

WWF i Norge (World Wide Fund for Nature WWF Norway): Kristian Augustsgt 7A, POB 6784, St Olavs Plass, 0130 Oslo; tel 22-20-37-77; telex 05678425; fax 22-20-06-66; Chair, Gen Ass: Jørgen Randers; Chair, Bd: Morten Werring; Sec-Gen: Vera Selnes; field projects for the conservation of biodiversity; national affiliate organization of World Wide Fund for Nature (WWF) International.

NORWEGIAN EXTERNAL TERRITORIES

Svalbard

Norsk Polarinstitutts Forskningsstasjon (Norwegian Polar Institute Research Station): Ny-Ålesund; tel (80) 27-115; fax (80) 27-002; permanent research base.

OMAN

Governmental Organizations

Ministry of Regional Municipalities and Environment: POB 323, Muscat; tel 696444; telex 5404; fax 602320; Minister: Sheikh Amir bin Shuwain al-Hosni.

Council for Conservation of Environment and Water Resources: POB 5310, Ruwi; tel 704344; telex 3590.

PAKISTAN

Governmental Organizations

Ministry of Food, Agriculture and Co-operatives, National Council for Conservation of Wildlife in Pakistan: Bungalow 485, St 84, Sector G-6/4, Islamabad; tel (51) 829756; telex 5844; fax (51) 216909; Minister: Lt-Gen Abdul Majid Malik; Wildlife Conservator: Syed Naseer A Gillani; formulation of policy; co-ordination of provincial activities; liaison with NGOs; wildlife management.

Ministry of Housing and Works, Environment and Urban Affairs Division: Block B, Pakistan Secr, Islamabad; tel (51) 825941; Minister of Housing and Works: Javed Hashmi.

Pakistan Space and Upper Atmosphere Research Commission (SUPARCO): Secr Bldg, off Univ Rd, POB 8402, Karachi 75270; tel (21) 474261; telex 25720; fax (21) 466902; Chair: Sikandar Zaman; atmospheric research; environment, pollution and ozone monitoring; satellite meteorology; remote sensing.

Sindh Wildlife Management Board: Stratchen Rd, POB 3722, Karachi 1; tel (21) 5683176; Wildlife Conservator: Abrar Husain Mirza; wildlife conservation; research; education; publications.

Non-Governmental Organizations

Belour Advisory and Social Development Organization (BASDO): 543 Shahrah-e-Quaid-e-Azam, Jutial, Gilgit 15100; tel (57) 22543; Gen Sec: Ghulam Rasool; environmental education and training; lobbying government; advice; workshops and conferences.

IUCN Commission on Environmental Law: c/o Paaf Bldg, 7D Kashmir Egerton Rd, Lahore; tel (42) 360800; telex 08244805; fax (42) 303825; Chair: Dr Parvez Hassan; Deputy Chair: Dr Nicholas Robinson.

IUCN Pakistan: 1 Bath Island Rd, Karachi 75530; tel (21) 573079; telex 24154; fax (21) 530976; Country Rep: Aban Marker Kabraji; formulation of policy; training; field programmes; raising public awareness; environmental education.

Islamabad Office: 22 Bazar Rd, G-6/4, Islamabad; tel (51) 812725.

Pakistan Institute for Environment-Development Action Research (PIEDAR): 2nd Floor, Yasin Plaza 74W, Blue Area, Islamabad; tel (51) 211015; telex 54199; fax (51) 216507; Pres: Syed Ayub Qutub; Vice-Pres: Dr Akmal Hussain; promotion of community organizations for sustainable development; sustainable development policy advice to governments and business; research.

Lahore Office: 40A Ahmad Block, New Garden Town, Lahore.

Scientific and Cultural Society of Pakistan: B-7 St 25, Model Colony, Karachi 75100; tel (21) 409336; fax (21) 409336; Pres: Muhammad Zaheer Khan; organization of seminars, conferences, educational training courses and workshops in environmental science, wildlife and organic agriculture.

Sub-Office: 31 Writers Chamber, Dunolly Rd, Karachi.

Sub-Office: Zoology Dept (Toxicology), Karachi Univ, Karachi; tel (21) 479001.

Shehri: Citizens for a Better Environment (CBE): 206-G, Block 2, PECHS, Karachi 75400; tel (21) 441769; fax (21) 438226; Co-ordinator: Arifa Kapadia; Asst Co-ordinator: Arif Usman; environmental protection and improvement; sustainable development; legal action against polluters on behalf of the public; lobbying; architectural conservation; publications; information.

Shirkat Gah (Women's Resource Centre): 1 Bath Island Rd, Karachi 75530; tel (21) 578067; telex 24154; fax (21) 530976; Co-ordinator: Meher M Noshirwani; environmental publications and information; education and training; workshops and conferences.

Society for Conservation and Protection of Environment (SCOPE): 511 al-Amin Tower (NIPA), Gulshan-e-Iqbal, Univ Rd, Karachi 75300; tel (21) 4965042; fax (21) 4965042; Pres: Tanveer Arif; water purification; environmental action; rural development; control of industrial pollution; consultancy.

Sungi Development Foundation: 19 St 59, Sector F-8/4, Islamabad; tel (51) 256323; fax (51) 818020; Chair: Omar Asghar Khan; prevention of environmental degradation through awareness-raising and co-operation with NGOs; community-based programmes.

Teachers' Resource Centre (TRC): 67B Garden Rd, Karachi; tel (21) 717967; telex 28018; Dir: Mehrunnisa Ahmad Ali; teacher training in ecology and the environment; workshops and conferences.

World Wide Fund for Nature (WWF) Pakistan: Ali Inst of Educ, Ferozepur Rd, POB 5180, Lahore 54600; tel (42) 856177; telex 44866; fax (42) 852810; Pres: Babar Ali; Dir: Dawood Ghaznavi; national affiliate organization of World Wide Fund for Nature (WWF) International; conservation; environmental education; wildlife management.

Islamabad Office: c/o Milkpak Ltd, 1st Floor, Yaseen Plaza, Blue Area, Islamabad; tel (51) 817748; fax (51) 821899.

Karachi Office: c/o Crescent Group of Industries, 9th Floor, Sidco Ave Centre, 264-RA Lines, Karachi; tel (21) 5689604; telex 28880; fax (21) 5680476; Dir and CEO: Dawood Ghaznavi.

Peshawar Office: Dept of Environmental Studies, Peshawar Univ, Peshawar; tel (521) 841593; fax (521) 841594.

Quetta Office: Game Bird Breeding Farm, Sabzal Rd, POB 432, Quetta; tel (81) 42803.

PANAMA

Governmental Organizations

Comisión Nacional de Medio Ambiente (CONAMA—National Environment Commission): Edif 1069, Curundu, Area Revertida, Apdo 10120, Zone 4, Panamá; tel 32-6055; telex 3683.

Instituto Nacional de Recursos Naturales Renovables (INRENARE—National Institute for Renewable Natural Resources): Apdo 2016, Paraíso, Ancón; tel 32-4518; telex 3697; fax 32-4975; Dir-Gen: Luis Fedel Narvaez R; planning and implementation of policy for use, conservation and development of renewable natural resources.

Non-Governmental Organizations

Asociación Chiricana para la Protección Ambiental (ACPPA—Chiriquí Association for Environmental Protection): Entrega Gen, David, Chiriquí; tel 74-1041; Pres: Naval Andrade C; conservation of natural resources and the environment; sea turtles project; wildlife breeding project; tapir project; environmental education.

Asociación Conservacionista Caribaró (ACCA—Caribaro Conservation Association): Apdo 17, Bocas del Toro, Isla Panamá; tel 78-9488; fax 78-9488; Exec Dir: José G Thomas; Deputy Dir: Luis L Mou; support for development of protected areas in Bocas del Toro and the Panamanian Caribbean; promotion of environmental education, applied research and exchange of related expertise in the conservation of natural resources; eco-development; preservation of the cultural heritage of the Panamanian Caribbean.

Asociación para la Investigación y Propagación de Especies Panameñas (AIPEP—Association for the Investigation and Propagation of Panamanian Species): Apdo 2320, Balboa, Ancón; tel 32-6125; Pres: Brooke Alfaro; wildlife protection; conservation of endangered species; support for creation of natural wildlife parks and organization of zoological parks.

Asociación Kunas Unidos por Napguana (Kunas United Association for Mother Earth): Via España, Edif Domino, 2°, Of 31, Apdo 536, Panamá 1; tel 69-6525; fax 69-3514; ecology; development; integration of indigenous people.

Asociación Nacional para la Conservación de la Naturaleza (ANCON—National Association for Nature Conservation): Apdo 1387, Panamá 1; tel 64-8100; fax 64-1836; Exec Dir: Juan Carlos Navarro Q; Dir of Conservation and Science: Raúl Fletcher; Dir of Operations: Oscar Vallarino; conservation of national parks with the INRENARE government institution responsible for the management of natural resources; reforestation; establishment of buffer zones next to national parks and protected areas; environmental training in schools and higher education establishments.

Centro de Estudios y Acción Social Panameña (CEASPA—Panamanian Centre for Research and Social Action): Apdo 6-133, El Dorado; tel 26-6602; fax 26-5320; Dir: Raul Leis; Research Co-ordinator: Charlotte Elton; promotion of development alternatives; public education; research into sustainable development, rights of indigenous people, women's rights and peasant organizations.

Círculo de Estudios Científicos Aplicados (CECA—Applied Scientific Studies Circle): Calle Enrique Geenzier, Edif Fino-Fino 2, Panamá; tel 96-4853; telex 9776; fax 96-1559; Gen Dir: Milton Villarreal; Projects Chair: René Chang; environmental education, research and planning in wild areas; liaison with local groups; campaign against pesticides; conservation in mangrove swamps, rainforests and marine ecosystems (estuaries and islands).

Comité Ambiental de Alanje (CAA—Alanje Environment Committee): Barqueta Beach, Alanje, Chiriquí; tel 74-0078; Head: Teofilo Diaz Muñoz; collection and hatching of turtle eggs; education.

Fundación Dobbo Yala (Dobbo Yala Foundation): Apdo 83-0308, Zone 3, Calle 42 y Avda México, Edif Guadalupe, Of 27, Panamá; tel 25-6895; fax 32-5978; Exec Pres: Eligio Alvarado P; Exec Sec: José Mendoza Acosta; sustainable development for indigenous people; environmental conservation.

Fundación para el Desarrollo de los Jóvenes (FUNDEJOVEN—Foundation for Youth Development): Apdo 87-1740, Zone 7, Villa Cáceres, Calle 77B, Casa 26 Vereda, Panamá; tel 61-6896; fax 29-1051; Pres: A Lourdes Alvarado; youth development; environmental education; small business; recycling; community development.

Fundación de Parques Nacionales y Medio Ambiente (PANAMA—National Parks and Environment Foundation): Apdo 6-6623, El Dorado; tel 27-5370; Exec Dir: Francisco Delgado; participation in decision-making, planning and management of natural resources.

Grupo de Ciencia y Tecnología para el Desarrollo (GRUCITED—Science and Technology for Development Group): Apdo 934, David, Chiriquí; tel 75-1652; fax 75-1652; Principal Officer: Manuel Humberto Ruiloba; organization of farmers; training and technical assistance; finance for small farmers; support for sustainable agricultural production.

IUCN Panama: Apdo 9-053, Zone 7, Panamá; tel 23-4931; fax 23-1458; Rep: Ramon Alvarado.

Proyecto de Estudio para el Manejo de Areas Silvestres de Kuna Yala (PEMASKY—Study Project for Management of the Wildlands of Kuna Yala): c/o Asociación de Empleados Kunas (AEK), Apdo 2012, Paraíso, Ancón; tel 27-6022; Dir: Guillermo Archibold; human ecology; resource management; development planning; environmental management and planning; support for community conservation projects.

PAPUA NEW GUINEA

Governmental Organizations

Department of Environment and Conservation: POB 6601, Boroko; tel 271788; fax 271900; Minister: Perry Zeipi; Sec: Iamo Ila; environmental planning and monitoring; regulation of pesticides, hazardous substances, ozone-depleting substances and contaminants; conservation of water resources; establishment and management of protected areas; conservation of crocodiles and butterflies; design of management strategies for endangered species; marine and coastal resource management.

Papua New Guinea Forestry College: POB 92, Bulolo, Morobe Province; tel 445236; fax 445311; Principal Officer: T Nahuet; 3-year training programme for foresters and park rangers.

Non-Governmental Organizations

Foundation of the Peoples of the South Pacific/Papua New Guinea, Inc (FSP/PNG): POB 1119, Boroko, National Capital District; tel 258470; fax 252670; Exec Dir: Yati A Bun; Chair, Bd of Dirs: Sir Paulias Matane; ecoforestry; Awareness Community Theatre programme (ACT); Literacy Education Awareness Development programme (LEAD); Farm Areas Resources Management programme (FARM).

PARAGUAY

Governmental Organizations

Ministry of Agriculture and Livestock, Department of Agriculture, Forestry and Conservation of Natural Resources: Ptde Franco 472, Asunción; tel (21) 49385; telex 324; Minister: Raúl Torres Segovia.

Ministry of Public Health and Social Welfare, National Environmental Health Service: Avda Pettirossi y Brasil, Asunción; Minister: María Cynthia Prieto Conti.

Non-Governmental Organizations

Centro de Estudios y Formación para el Ecodesarrollo (ALTER VIDA—Centre for Ecodevelopment Studies and Training): Casilla 2334, Avda Artigas 960, Asunción; tel (21) 21-3246; Admin: Jorge Abbate; environmental research; education in resource management; monitoring the environment; impact assessment.

Centro de Estudios Rurales Interdisciplinarios (CERI—Centre for Interdisciplinary Rural Studies): Mcal Estigarribia 1050 e Brasil y Estados Unidos, Asunción; tel (21) 21-1779; fax (21) 44-4842; Dir: Ramón Fogel; Sec: José Angel Benítez; training in development with Asunción National University; environmental research and action; member of the Red Internacional de Medio Ambiente (International Environment Network).

Centro Interdisciplinario de Derecho Social y Economía Política (CIDSEP/UC—Interdisciplinary Centre of Social Law and Political Economy): Independencia Nacional y Comuneros, Asunción; tel (21) 49-5752; fax (21) 49-5752; Dir: Dr Carlos A Gonzalez; Gen Co-ordinator: Dr Gustavo Becker M; advice to government; education in politics; research; publications.

Fundación Moisés Bertoni (Moisés Bertoni Foundation): Rodriguez de Francia 770, Calle Tacuary, Asunción; tel (21) 44-0238; fax (21) 44-0239; Exec Dir: Raul Gauto; conservation of nature and biodiversity; management of protected areas; sustainable use of resources; environmental education; training in environmental management.

PRONATURA Sociedad Paraguaya para la Protección de la Naturaleza (PRONATURA Society of Paraguay for Nature Protection): Casilla 24 97, Alférez Silva 857, Asunción; Pres: Ricardo Careaga Boggino; Vice-Pres: Dr Albina Velázquez; environmental awareness-raising; environmental projects with public and private institutions; environmental research.

Universidad Católica Nuestra Señora de la Asunción, Centro para la Preservación del Patrimonio Ambiental (CPPA—Our Lady of the Assumption Catholic University, Centre for Preservation of Environmental Heritage): Tomas Romero Pereira y Antequera, Encarnación; tel (71) 3549; fax (71) 3165; Dir: Carmelo Netto; regional heritage database; technical assistance; research; conservation of cultural heritage and monuments.

Universidad Nacional de Asunción, Instituto de Desarrollo Ambiental (Asunción National University, Environmental Development Institute): Campus Universitario, Ciudad de San Lorenzo, Asunción; tel (21) 50-8919; research into environmental development.

PERU

Governmental Organizations

Instituto Nacional de Recursos Naturales (INRENA—National Institute for Natural Resources): Calle 17 355, Urb El Palomar, San Isidro, Lima; tel (14) 410425; fax (14) 414606; Head: Dr Wilder Valenzuela Andrade; conservation of natural resources; research into viable renewable resources; co-ordination with the public and private sector; policy proposals; evaluation of the environmental impact of agricultural projects.

Oficina Nacional Evaluación Recursos Naturales (ONERN—National Office for Evaluation of Natural Resources): POB 4992, Calle 17 355, Urb El Palomar, San Isidro, Lima; tel (14) 410425.

Ministry of Agriculture and Food, Department of Forests and Wildlife: Avda Salaverry s/n, Edif Ministerio de Trabajo, Lima; tel (14) 324040; telex 25835; Minister: Absalón Vásquez Villanueva.

Non-Governmental Organizations

Asociación Arariwa para la Promoción Técnico Cultural Andina (ARARIWA—Arariwa Association for Andean Technical and Cultural Progress): Apdo 872, Avda Los Incas 1606, Wanchaq, Cuzco; tel (84) 236887; fax (84) 236889; Pres: Francisco Javier Lou Burriel; Exec Dir: Rosario Valdeavellano Roca Rey; agriculture (seed-sowing and Andean agriculture and natural resources); fisheries (research and sustainable use of stocks); women's development; family and child health; training.

Asociación para la Conservación de la Naturaleza Amazónica (Association for the Conservation of Amazonian Wildlife): Avda Mariscal Cáceres 666, Iquitos; tel (94) 237408; fax (94) 232951; Pres: Filomeno Encarnación; Environmental Educ Co-ordinator: Armando Coral; Research and Management Co-ordinator: César Ascorra; environmental education and community participation; biodiversity research and management; information; awareness-raising and mobilization.

Asociación de Conservación para la Selva Sur (Southern Forest Conservation Association): Centro Comercial, Los Risueñores 123, Portal de Panes, Of 305, Piza de Armas, Cuzco; tel (84) 226392; fax (84) 226392; Principal Officers: David Ricalde, José Palomino, Daniel Blanco, Patricia Villafuerte; conservation of protected areas in South-Eastern Peru; environmental education; development projects.

Asociación de Ecología y Conservación (ECCO—Ecology and Conservation Association): Casilla 18-0359, Lima 18; tel (14) 472369; telex 21506; fax (14) 300615; Pres: Tony Luscombe; conservation strategies; environmental and conservation research; legal assistance; planning of land use and use of wildlife resources; environmental education; promotion of environmental law.

Asociación Evangélica Luterana de Ayuda para el Desarrollo Comunal (DIACONIA—Lutheran Evangelical Aid Association for Communal Development): Gen Orbegoso 650, Breña, Lima 5; tel (14) 235245; fax (14) 318010; development of rural infrastructure; agricultural improvement; livestock management; forestation; family support; training in food production and nutrition.

Asociación Interetnica de Desarrollo de la Selva Peruana (AIDESEP—Interethnic Association for the Development of Peruvian Amazonia): Avda San Eugenio 981, Urb Santa Catalina, La Victoria, Lima 13; tel (14) 726621; fax (14) 724605; Pres: Miqueas Mishari Mofat; Vice-Pres: Francisco Shajian Sakejat; promotion of the legal, social and economic interests of the indigenous people of Peruvian Amazonia.

Asociación Nacional de Centros de Investigación, Promoción Social y Desarrollo (ANC—National Association of Centres for Research, Social Progress and Development): Avda Pablo Bermúdez 234, Lima 11; tel (14) 333477; fax (14) 333470; Pres: Federico Arnillas Lafert; exchange of information and co-ordination between centres for national development.

Asociación Peruana para la Conservación de la Naturaleza (APECO—Peruvian Association for Conservation of Nature): Parque José de Acosta 187, Lima 17; tel (14) 625410; fax (14) 633048; Principal Officers: Silvia Sanchez, Victor Merino; wildlife conservation; environmental education; field training.

Biosfera Asociación Cultural (BAC—Biosphere Cultural Association): Avda del Ejercito 735, Of 201, Miraflores, Lima 18; tel (14) 425855; fax (14) 425855; Principal Officers: Ana María Pérez Villarreal, Jorge Eduardo Suárez Rivas; production of audio-visual material on environmental science and technology for the cinema and TV.

Centro para el Desarrollo del Indígena Amazónica (Centre for the Development of the Amazonian Indian): Psje Bonifacio 166, Urb Los Rosales de Santa Rosa, La Perla, Callao, Lima; tel (14) 650708; fax (14) 650708; Principal Officer: Lelis Rivera Chávez; land rights programmes; training for native communities in the Southern Peruvian forest.

Centro para el Desarrollo de los Pueblos Ayllu (CEDEP-AYLLU—Centre for the Development of the Ayllu Peoples): Calle Resbalosa 520, Barrio San Cristobal, Cuzco; tel (84) 235436; fax (84) 235436; rural development with Andean peasant communities in Southern Peru; agricultural production (crop rotation, seeds, warehousing); natural resource management (irrigation, soil conservation, reforestation, production of native and exotic plants); social development.

Sub-Office: Calle Vigil 246, Pisac, Cuzco; tel (84) 203007.

Centro Eori de Investigación y Promoción Regional (Eori Regional Research and Promotion Centre): Jr Loreto 101, Puerto Maldonado, Madre de Dios; tel (84) 571380; fax (84) 571183; advice to organizations of indigenous people in the Madre de Dios region; project organization and management; legal defence of the land rights of indigenous people; advice to farmers' organizations; agroforestry and conservation.

Centro de Estudios y Prevención de Desastres (Centre for the Study and Prevention of Disasters): Avda San Felipe 370, Jesús María, Lima.

Centro de Estudios Sociales Solidaridad (CESS—Solidarity Social Studies Centre): Apdo 212, Leoncio Prado 443, Chiclayo; tel (74) 235013; fax (74) 243034; Pres: German Torre Villafane; management training; marketing; environmental health; technical research; research into human systems, small-scale farming, computer technology and small-scale businesses and women's development; global research; publications; documentation centre.

Centro de Investigación y Capacitación Campesina (CICCA—Rural Research and Training Centre): Avda Mariño 101, Abancay, Apurímac; tel (84) 321503; fax (84) 321503.

Lima Office: Avda Mariscal Miller 2214, Calle Lince, Apdo 11-0175, Lima 11; tel (14) 700836.

Centro de Investigación, Educación y Desarrollo (CIED—Research, Education and Development Centre): Avda El Buen Retiro 231, Urb Monterrico Chico, Lima 33; tel (14) 378327; fax (14) 378327; Pres: Juan Sánchez Barba; Dir: Eduardo Burga B; rural development; local and regional development; sustainable development; agroecology.

Centro Panamericana de Ingeniería Sanitaria y Ciencias (CEPIS—Pan American Centre for Sanitary Engineering and Environmental Sciences): Apdo 4337, Lima 100; tel (14) 371077; fax (14) 378289; Dir: Alberto Flórez Muñoz; environmental conservation and improvement; improvement of drinking water; re-use of waste water; information; technical training.

Centro Peruano de Estudios Internacionales (CEPEI—Peruvian Centre for International Studies): San Ignacio de Loyola 554, Lima 18; tel (14) 457225; fax (14) 451094; Exec Chair: Dr Eduardo Ferrero Costa; research; conferences; co-operation and co-ordination with universities.

Centro Privado para el Desarrollo del Campesinado y del Poblador Urbano Marginal (CEDECUM—Private Centre for the Development of the Rural and Urban Poor): Apdo 542, Jr Libertad 137, Puno; tel (54) 353461; fax (54) 353461; Principal Officer: Dr Arturo Vásquez Salazar; rural organization and management; environmental management; production methods; infrastructure; credit; marketing.

Comercio Alternativo de Productos No Tradicionales y Desarrollo en Latinoamérica, Perú (CANDELA Perú—Alternative Trade in Non-Traditional Products and Development in Latin America, Peru): Apdo 14-0233, Lima 14; Comandante Gustavo Jimenez 153, Lima 17; tel (14) 614556; fax (14) 614556; Dir: Gaston Vizcarra; Gen Co-ordinator: Guadalupe Lanao; processing and marketing of Brazil nuts in Peruvian Amazonia; processing and marketing of agricultural products; research and development of new sustainable products in Amazonia; marketing of Peruvian crafts.

Comité Peruano—UICN (Peruvian Committee for IUCN): Plaza Arróspide 9, San Isidro, Lima 27; tel (14) 400549; fax (14) 424365; Principal Officer: Dr Carlos Chirinos.

Consorcio Mochica Chimú (Mochica Chimú Association): Atahualpa 490, Trujillo, La Libertad; tel (44) 252692; fax (44) 252692; Pres: Prof Demetrio Ramos Rau; Vice-Pres: Prof César Campos; integrated rural development; urban development (health, environmental health, small-scale business).

Coordinadora de las Organizaciones Indígenas de la Cuenca Amazónica (COICA—Co-ordination of Indian Organizations of the Amazon Basin): Jirón Larco Herrera 1057, Lima 17; tel (14) 619228; fax (14) 619228; Co-ordinator: Evaristo Nugkuag; defence of the rights of the native peoples of the Amazon.

Escuela de Pintura Amazónica Usko-Ayar (Usko-Ayar Amazonian School of Painting): Jr L M Sánchez Cerro 465-467, Pucallpa; tel (64) 573088; fax (64) 573088; Principal Officers: Pablo César Amaringo Shuña, Luis Eduardo Luna Porres; encouragement of respect for the Amazon environment through artistic education for young people; artistic documentation of Amazon flora, fauna and ways of living; information about Amazon nature, art and culture.

Fundación Peruana para la Conservación de la Naturaleza (FPCN—Peruvian Foundation for the Conservation of Nature): Apdo 18-1393, Lima 18; tel (14) 426616; telex 25129; fax (14) 427853; Exec Dir: Gustavo Suárez de Freitas C; conservation of soils, water resources, fauna, flora and other renewable natural resources.

Grupo Ecológico Natura (GEN—Natura Ecological Group): Apdo 3051, Lima 100.

Grupo de Investigación y Extensión de Tecnología Popular (TALPUY—Group for Research and Extension of Popular Technology): Apdo 222, Avda Centenario 589, San Carlos.

Grupo de Tecnología Intermedia para el Desarrollo (ITDG—Intermediate Technology Development Group): Casilla 18-0620, Avda Jorge Chávez 275, Lima 18; tel (14) 467324; fax (14) 466621; Dir: Andrew Maskrey; Project Dir: Alfonso Carrasco; international technical co-operation; research and development of alternative technologies for developing countries.

Instituto de Desarrollo y Medio Ambiente (IDMA—Development and Environment Institute): Apdo 11-0384, Lima 11; Calle Durero 301, Lima 41; tel (14) 362688; fax (14) 753041; Exec Dir: Juan Vaccari; Deputy Exec Dir: Abel Sanchez; sustainable rural development; agroecology; environmental education; conservation of biodiversity and protected natural areas; rural community development.

Instituto de Desarrollo Urbano Ciudad (IDU CIUDAD—City Institute for Urban Development): Bronsino 119, Of 301, San Broja, Lima.

Instituto Nor Peruano de Desarrollo Económico Social (INDES—North Peruvian Institute for Economic and Social Development): Atahualpa 490, Trujillo, La Libertad; tel (44) 252692; fax (44) 252692; Exec Dir: Prof Demetrio Ramos Rau; environmental health; health and nutrition; agroecology.

Pequeñas Empresas, Tecnología y Comunidad (PEMTEC—Small Businesses, Technology and Community): Avda José Pardo 741, 4°, Miraflores; tel (14) 451741; fax (14) 469945; Exec Pres: Fernando Villarán de la Puente; research into pollution.

Pro Defensa de la Naturaleza (Pro Nature Defence): Avda Nicolás de Piérola 742, Of 703, Edif Int, Lima; tel (14) 234507; telex 20338; Exec Pres: Augusto Dammert León;

conservation of wild and endangered fauna and flora and coastal and marine protected areas; education.

Proterra: Avda Esteban Campodónico 208, Urb Santa Catalina, La Victoria, Lima 13; tel (14) 723800; fax (14) 703371; Pres: Dr Antonio Andaluz Westreicher; Technical Dir: Dr Walter Valdez Muñoz; Exec Dir: Dr Carlos Andaluz Westreicher; research, information, training and lobbying in the fields of environmental law and sustainable development in Peru and Latin America, particularly in the Andes and Amazonia; publications; co-operation with politicians, environmental workers and other interested groups.

Red de Agricultura Ecológica del Perú (RAE—Ecological Agriculture Network of Peru): Avda Arenales 645, Lima 1; tel (14) 247773; fax (14) 331073; Pres: Fernando Alvarado de la Fuente; information; training; technical advice.

Sociedad Pachamama (SP—Pachamama Society): Camino Real 479, 8°, San Isidro, Lima 27; tel (14) 416031; fax (14) 411990; Technical Dir: Eric Cardich; organization of business people and environmental professionals; awareness-raising; sustainable development; co-operation with developed countries.

Sociedad Peruana de Derecho Ambiental (SPDA—Peruvian Society for Environmental Law): Plazuela Arróspide 9, San Isidro, Lima 27; tel (14) 224033; telex 20147; fax (14) 424365; Pres: Jorge Caillaux; Exec Dir: Enrique Ferrando; Gen Co-ordinator: Manuel Pulgar-Vidal; advice; legal support; environmental defence; information; training and support for specialists in environmental law.

Sociedad Zoológica del Perú (Peruvian Zoological Society): Casilla 17-0132, Lima 17; Avda la Marina Cuadra 24, Parque de las Leyendas, Lima 32; tel (14) 642282; research; conservation of national wildlife; organization of zoos.

Technoserve Perú Inc: Apdo 11-0257, Sinchi Roca 2633, Lince, Lima 11; tel (14) 408864; fax (14) 425195; Country Dir: Alberto Pérez; improvement of the economic and social welfare of the poor in developing countries by enterprise development; agricultural programmes; rural development; environmental advice; sustainable development.

Universidad Nacional Agraria La Molina, Centro de Datos para la Conservación (CDC-Perú—La Molina National Agriculture University, Conservation Data Centre): Apdo 456, Lima 100; tel (14) 371143; fax (14) 371143; Dir: Manuel A Ríos; Technical Co-ordinator: Pedro G Vásquez; Admin: Jorge M Chavez; Zoologist: Antonio Tovar; computerized database on Peruvian biodiversity; collaboration with conservation groups on decisions about natural resources.

Vecinos Perú (VP—Neighbours Peru): Paseo de los Eucaliptos 328, Urb Camacho, La Molina, Lima 12; tel (14) 364173; fax (14) 364173; Exec Dir: Andres Solari Andrade; support for community agricultural development programmes in towns and in the countryside.

THE PHILIPPINES

Governmental Organizations

Department of the Environment and Natural Resources: Visayas Ave, Diliman, Quezon City, Metro Manila; tel (2) 976626; Sec: Angel Alcata.

Forest Management Bureau: tel (2) 984788; fax (2) 9219060; Dir: Lope D Reyes; Asst Dir: Doroteo U Antonio; planning and formulation of policy; technical and economic research into forestry; monitoring of forestry programmes and forest users.

Protected Areas and Wildlife Bureau: Quezon Ave, Diliman, Quezon City 1100; tel (2) 978511; telex 7572000; fax (2) 9240109; Dir: Samuel R Penafiel; conservation of protected areas, national parks, wildlife and biodiversity.

Bureau of Forest Development: Visayas Ave, Diliman, Quezon City, Metro Manila; tel (2) 964826.

Philippine Council for Agriculture, Forestry and Natural Resources Research and Development (PCARRD): Los Banos, Laguna 4030; tel (94) 50014; Exec Dir: Cledualdo B Perez; Deputy Dir of Research: Beatriz del Rosario; planning, monitoring and co-ordination of research into agriculture and natural resources.

Non-Governmental Organizations

Benguet State University, Highland Agriculture and Resources Research and Development Consortium (HARRDEC): La Trinidad, Benguet 2601; tel (74) 4322176; fax (74) 6322176; Exec Dir: Dr Percival B Alipit; Chair: Dr Faustino K Maslan; monitoring and co-ordination of research and development projects in the highlands of the Philippines.

Ecological Society of the Philippines: c/o 53 Tamarind Rd, Forbes Park, Makati, Metro Manila; tel (2) 8108957; telex 23636; fax (2) 790655; Pres: Antonio M Claparols; conservation of marine and coastal environments; resource management; prevention of sea pollution and oil spills; protection of cetaceans, aquatic plants and coral reefs.

Haribon Foundation for the Conservation of Natural Resources (HARIBON): Suite 901, Richbelt Tower, 17 Annapolis St, Greenhills, 1502 San Juan, Metro Manila; tel (2) 8108957; telex 23636; fax (2) 790655; Science Programme Officer: Danilo S Balete; conservation of natural resources; management of protected areas; community-based resource management training; legal assistance; research; environmental information and education campaigns.

International Youth Federation for Environmental Studies and Conservation (IYF), Regional Office for Asia and the Pacific: 55 Olive St, SSS Village, Marikina, 1800 Metro Manila; tel (2) 7227180; fax (2) 7226357; Regional Co-ordinator: Daniel M Ocampo; world secretariat in India and regional offices in Benin, Costa Rica and the Netherlands.

World Ecologists Foundation: Gold Bldg, 15 Annapolis St, Greenhills, Metro Manila; tel (2) 7224016; fax (2) 7213356; Pres: Charley Barretto; education and training; conservation of energy; protection of resources; lobbying government.

World Wide Fund for Nature (WWF) Philippines: Richbelt Tower Condominium, 17 Annapolis St, Suite 501, Green Hills, San Juan, Metro Manila; tel (2) 7227759; fax (2) 7226455; Rep: Dick Edwards.

POLAND

Governmental Organizations

Ministry of Environmental Protection, Natural Resources and Forestry: 00-922 Warsaw, ul Wawelska 52-54; tel (22) 253355; telex 812816; fax (22) 253355; Minister: Zygmunt Hortmanowicz.

Non-Governmental Organizations

Biuro Obsługi Ruchu Ekologicznego (BORE—Service Office for the Environmental Movement—SOEM): 00-420 Warsaw, ul Szara 14 m 34; tel (22) 296433; fax (22) 296433; Principal Officer: Jolanta Pawlak; organization of public meetings with the Ministry of Environmental Protection; publication of monthly information bulletin; organization of courses; support for environmental campaigns.

Ecological Library Foundation: 61-715 Poznań, ul Kościuszki 79; tel (61) 521325; fax (61) 528276; Pres: Jacek Purat; Dir: Jarek Fiszer; ecological library; publications; book distribution; protection of endangered species in Poland; film presentations.

Ecology and Health Foundation: 00-215 Warsaw, al Ujazdowskie 13; tel (2) 6942153; fax (2) 6942153; Principal Officers: Dr Marek Siemiński, Stefan Bogusławski, Janusz Olton; information on environmental threats to health; education; information through bulletins, brochures and books; advice and research for local communities.

Environmental Partnership for Central Europe: 50-020 Wrocław, Dom Technika NOT, ul Pilsudskiego 74, Room 129; tel (71) 31611; fax (71) 31611; Rep: Krystyna Wolniakowski; Asst: Elzbieta Harok; see German Marshall Fund, USA.

European Association of Environment and Resource Economists, Polish Division: 31-510 Kraków, ul Rakowicka 27, Room 155; tel (12) 210099; telex 0325414; fax (12) 212182; Principal Officer: Dr Kazimierz Gurko; refer to section on Italy.

International Union of Anthropological and Ethnological Sciences, Commission of Human Ecology (CHE/IUAES): c/o Dept of Human Ecology, Acad of Sciences, 00-330 Warsaw, Nowy Swiat 72; tel (22) 268312; fax (22) 513100; Chair: Dr N Wolanski; Sec: Dr F Dickinson; support for research in human ecology; information exchange; co-ordination of international research programme on the biosocial status of human populations as an indicator of environmental changes.

IUCN Poland: c/o 00-922 Warsaw, ul Wawelska; tel (22) 253355; telex 812816; fax (22) 253355.

Liga Ochrony Przyrody (League for the Conservation of Nature in Poland): 00-922 Warsaw, ul Reja 3/5; tel (22) 253593; fax (22) 253593; Pres: Dr Janusz Janecki; Gen Sec: Zygmunt Trzebiatowski; nature conservation; ecological education, especially in schools; protest group; publications; co-operation with other organizations in Poland and abroad.

Panstwowa Rada Ochrony Przyrody (National Council for Protection of Nature in Poland): 00-922 Warsaw, ul Wawelska 52-54; tel (22) 251114; telex 812816; Sec: Jan Lapaczewski; conservation of forests, highlands, birds, endangered flora and fauna, landscape and national parks; design, planning and management of protected areas.

Polish Academy of Sciences, Committee of Nature Protection: 31-512 Kraków, ul Lubicz 46; Chair: Prof Dr Kazimierz Zarzycki.

Polish Academy of Sciences, Department of Human Ecology (DHE/PAN): 00-330 Warsaw, Nowy Swiat 72; tel (22) 268312; fax (22) 513100; Chair: Prof Dr N Wolanski; Deputies: Dr A Siniarska, Dr C Dziewiecki; field research into environmental factors in human development and structure of human populations; studies in conceptual human ecology; biomedical research.

Polish Academy of Sciences, Institute of Ecology: 05-092 Łomianki, Dziekanów Leśny; tel (22) 513157; telex 817378; fax (22) 513100; Dir: Prof Dr Kazimierz Dobrowolski.

Department of Vertebrate Ecology: tel (22) 513046; Head: Prof Dr Jan Pinowski; research on the effects of pollution, parasites and diseases; research into populations and distribution of birds, bats and rodents in relation to forest structure; research into movement and dispersal of house mice.

Gdańsk Ornithological Station: Górki Wschodnie, 80-860 Gdańsk 40, Nadwiślańska str 108; tel (58) 380759; fax (58) 380759.

Lublin Research Station: 20-080 Lublin, Sławińskiego str 18; tel (81) 26182.

Mikołajki Hydrobiological Station: 11-730 Mikołajki, Leśna str 14; tel (55) 16051; telex 522547.

Świdwie Ornithological Station: 72-003 Dobra Szczecińska.

Polish Academy of Sciences, Institute of Nature Conservation: 31-512 Kraków, ul Lubicz 46; tel (12) 210348; telex 322630; fax (12) 210348; Dir: Prof Dr Zygmunt Denisiuk; Deputy Dir: Prof Dr Zbigniew Witkowski; scientific research into natural biological processes; preservation of biodiversity and rare and threatened plant and animal species; research into the degradation of natural resources, changing ecosystems, water resources, landscape and the Earth's surface.

Polish Academy of Sciences, Mammals Research Institute: 17-230 Białowieza; tel (835) 12278; telex 853491; fax (835) 12289; Dir: Prof Dr Zdzisław Pucek; Deputy Dir: Dr Małgorzata Krasińska; research into population biology, genetics and ecology of mammals; ecology of carnivore communities; mammal conservation; mammal phylogeny; journal.

Polski Klub Ekologiczny (PKE—Polish Ecological Club): 31-031 Kraków, ul Garbarska 9; tel (12) 222111; Mem, Bd: Prof Stanislaw Juchnowicz; promotion of sustainable development; human ecology; environmental education; landscape protection; energy conservation; waste recycling; biological pest control; organic farming; conservation strategies; campaigns on urban planning and human health.

PORTUGAL

Governmental Organizations

Ministry of the Environment and Natural Resources: Av D Carlos I 126-6°, 1200 Lisbon; tel (1) 362751; Minister: Teresa Patricio Gouveia.

Instituto da Conservação da Natureza (Nature Conservation Institute): Rua da Lapa 73, 1200 Lisbon; tel (1) 675962; nature conservation; protected areas.

Serviço Nacional de Parques, Reservas e de Conservacão da Natureza (National Parks, Reserves and Nature Conservation Service): Rua da Lapa 73, 1200 Lisbon; tel (1) 3950464; telex 44089; fax (1) 601048; Pres: Joaquim Marques Ferreira; conservation strategies; landscape protection; conservation of wildlife and habitats; environmental education; impact assessments; management and planning of protected areas; international co-operation; landscape management; scientific research.

Non-Governmental Organizations

Amigos da Terra/Associação Portuguesa de Ecologistas (FoE/Portuguese Ecologists' Association): Travessa Marquês de Sampaio 42 r/c esq, 1200 Lisbon; tel (1) 3479599; fax (1) 3473586; Principal Officers: Mario Alves, Nicole Mueller; projects on tourism, biotechnology, packaging, environmental education, tropical rainforests and oceans and the coastal environment.

Grupo de Estudos de Ordenamento do Território e Ambiente (GEOTA—Research Group for Land and Environment Management): Rua Ricardo Espírito Santo 1 r/c esq, 1200 Lisbon; tel (1) 3956120; fax (1) 3956120; Principal Officers: João Joanaz de Melo, Conceição Martins, Luís Rosmaninho, Marlene Marques; formulation and assessment of environmental policy (drafting and analysis of regulations and legislation on environmental impact assessment, land management, water resource management, forests, nature conservation and energy planning); lobbying; environmental education; environmental conservation (co-ordination of Coast Watch Portugal).

Liga para a Protecção da Natureza (League for the Protection of Nature): Estrada do Calhariz de Benfica 187, 1500 Lisbon; tel (1) 7780097; telex 15095; fax (1) 7783208; Vice-Pres: Luis Miguel Vieira; conservation of marine, freshwater and terrestrial environments, endangered wildlife and plants, protected areas, national parks and forests; environmental policy; conservation strategies.

Núcleo Português de Estudo e Protecção da Vida Selvagem (NPEPVS—Portuguese Centre for the Study and Protection of Wildlife): Bairro Fundo de Fomento de Habitação, Bloco D r/c, 5300 Bragança; tel (2) 24632; fax (2) 7326514; Pres: Armando José Pereira; study of flora and fauna; awareness-raising; advice on conservation issues.

Oikos Cooperação e Desenvolvimento (Oikos Co-operation and Development): Avda Visconde de Valmor 35, 3°, 1000 Lisbon; tel (1) 7964719; fax (1) 7939791; Principal Officers: Agostinho Jardim Gonçalves, Jeremias Carvalho, David Valente; development projects in developing countries; development education; international co-operation.

Quercus—Associação Nacional de Conservação da Natureza (Quercus National Association for Nature Conservation): Nat Secr, Apdo 230, 7802 Beja Codex; tel (84) 321326; fax (84) 321326; Chair: Prof Dr Viriato Soromenho Marques; Nat Sec: Dr José Paulo Martins; study of Portuguese flora, fauna and ecosystems; nature conservation; environmental education; promotion of sustainable development in political and economic decision-making.

Os Verdes (The Greens): Lisbon; Leader: Maria Santos; ecological party.

QATAR

Governmental Organizations

Permanent Committee for the Protection of Environment: POB 7634, Doha.

ROMANIA

Governmental Organizations

Ministry of Waters, Forestry and Environmental Protection: 70005 Bucharest, Blvd Libertatii 12, Tronson 5; tel (1) 6316146; telex 0650011457; fax (1) 124227; Minister: Aurel Constantin Ilie; establishment of guidelines for environmental policy; monitoring activities affecting waters, forests and the environment.

Institutul Delta Dunarii (Danube Delta Institute): 8800 Tulcea, Str Babadag 165; tel (915) 24546; fax (915) 24547; Dir: Romulus Stiuca; Science Dir: Mircea Staras; monitoring of flora and fauna; sustainable resource use (fisheries, tourism, hunting, grazing); ecological rehabilitation; monitoring of water and soil; agriculture, fisheries and forestry.

National Council for Environment: 70081 Bucharest, Negustori Str 3, Dist 2.

Romanian Marine Research Institute: 8700 Constanţa 3, Blvd Mamaia 300; tel (91) 643288; telex 14418; fax (91) 831274; Dir: Dr Simion Nicolaev; Deputy Dir: Dr Alexandru Bologa; physical and chemical oceanography; ecology; pollution research; ecological rehabilitation; research into natural resources; biochemistry; technological research.

Non-Governmental Organizations

Partidul Ecologist Român (Romanian Ecological Party): Bucharest, Blvd George Casbuc 1, Block P58; tel (1) 6312350; Chair: Otto Weber; merged with Mişcarea Ecologistă din Romania (Romanian Ecological Movement) 1993; support for environmental protection and pursuit of democratic, pacifist and humanist values.

Tineretul Ecologist din Romania (Ecologist Youth of Romania): 71243 Bucharest, Av Kiseleff 2, Sector 1; tel (1) 6593865; fax (1) 3124263; Pres: Calin Georgescu; Exec Dir: Ovidiu Oancea; development of environmental NGOs; nature conservation; public information.

THE RUSSIAN FEDERATION

Governmental Organizations

Ministry of Environmental Protection and Natural Resources: 123812 Moscow, Bolshaya Gruzinskaya ul 4-6; tel (95) 254-78-63; telex 411692; Minister: Viktor I Danilov-Danilyan.

Research Institute for Building Physics (NIISF): 127238 Moscow, proezd Locomotivni 21; tel (95) 482-40-76; fax (95) 482-40-60; Dir: George L Osipov; Deputy Dir: Vladimir A Mogutov; heat engineering for buildings; energy conservation; climatology; research into thermal transmission, acoustics, noise control and natural and artificial lighting; publications.

State Committee for Hydrometeorology and Environmental Control: Moscow, ul Kuybysheva 4.

State Committee for Social Protection of Citizens and the Rehabilitation of Territories Affected by Chernobyl and Other Radiation Accidents: 103132 Moscow, Staraya pl 8/2, pod 3; tel (95) 206-48-81; Chair: Vasily Y Voznyak.

Non-Governmental Organizations

All Russian Society for Conservation of Nature: 103012 Moscow, proezd Kuibychev 3; tel (95) 924-77-65; Chair: I F Bazishpol; environmental conservation and information.

Association of Ecological Centres: 107078 Moscow, Novo-Basmannaya d ul 10.

Environmental Education Association: 249020 Obninsk 1, Kaluga Region, POB 152; tel (8439) 3-86-48; telex 64412633; fax (95) 255-22-25; Co-ordinator: Vadim Kalinin; environmental education in the former USSR; support for teachers; development of an information network; improvement of teaching skills; projects and workshops.

Greenpeace Russia: 103006 Moscow, ul Dolgorukovskaya 21; tel (95) 251-90-73; fax (95) 251-90-88; Exec Dir: Alexander Knorre; Int Campaign Co-ordinator: Dima Litvinov; campaign against nuclear energy; nuclear-free seas campaign; fisheries campaign; forestry campaign; toxic substances campaign.

Moscow Ecological Federation: 121019 Moscow, POB 211; tel (95) 206-88-94; Co-Chairs: Lyubov Rubinchik, Vadim Damie; Co-ordinator: Valery Sevostianov; concerned with Moscow's ecological problems; dissemination of information; assistance with urban development plans.

Partiya Zelyenykh Prikamya (Prikamya Green Party): 614006 Perm, ul Kuybysheva 14, KDTs Komn 318; Chair: Marcel Akhmetov; Deputy Chair: Ivan Ezhikov; environmental protection; involving the public in ecological, peace-making and humanitarian activities.

Rostovskiy Oblastnoy Ekologicheskiy Tsentr (Rostov Regional Ecological Centre): 344007 Rostov, ul Stanislavskogo 114, kv 1; tel (863) 62-14-47; Pres: Valery Privalenko; ecology; environmental safety.

Russian Academy of Sciences, Institute of Atmospheric Physics: 109017 Moscow, per Pyzhevskii 3; tel (95) 231-55-65; fax (95) 233-16-52; Dir: Prof G S Golitsyn; Deputy Dirs: Dr I P Malkov, Dr Y A Volkov; monitoring, modelling and measurement of urban air pollution; modelling of the growth of air pollution in the environment; modelling and forecasting global change in climate and ecosystems.

Russian Academy of Sciences, Institute of Evolutionary Animal Morphology and Ecology (IEAME): Moscow V-71, Leninsky pr 33; tel (95) 232-20-88; telex 411682; fax (95) 129-13-54; Dir: V E Sokolov; animal morphology; impact assessment; ecology; biology; conservation and use of wildlife resources; remote sensing.

Socio-Ecological Union (SEU): 121019 Moscow, POB 211; 125319 Moscow, ul Krasnoarmeyskaya 25, kv 85; tel (95) 928-76-08; fax (95) 206-82-73; Chair, Council: Sviatoslav Zabelin; Dir, Information Centre: Lyubov Rubinchik; co-ordination of 180 environmental committees, clubs and societies in 11 of the former republics of the USSR; international co-operation; campaigns on the environment, human rights, biodiversity protection, energy efficiency, nuclear energy and radioactive pollution and environmental education.

Centre for Co-ordination and Information (CCI): Moscow; facilitation of communication and collaboration between members of the Socio-Ecological Union (SEU); international co-operation.

Centre for Independent Ecological Programmes (CIEP): Moscow; anti-pollution programmes; environmental health programmes; international co-operation.

Perm Department: 614081 Perm, POB 5786; tel (3422) 33-48-58; Chair: Yuly Schipakin; Vice-Chair: Larisa Sholina; formulation of sustainable development policy; information; debates; lobbying the Perm City Soviet; scientific research.

Tambov Green Party: 392032 g Tambov, bul Entuziastov 32, kv 47; tel (752) 35-01-33; Principal Officer: Ludmila Spiridonova.

Zelyoni Mir (Green World Environmental Association): 603047 Nizhny Novgorod, ul Krasniye Zori 15, kv 409; tel (831) 224-39-41; fax (831) 244-02-85; Gen Co-ordinator: Valentina Malakhova; Task Co-ordinator: Fyodor Krylov; nature conservation; environmental education; conferences.

RWANDA

Governmental Organizations

Ministry of Environment and Tourism: POB 2378, Kigali; tel 77415; fax 72237; Minister: Ernest Rukangira; formulation of environmental policy; environmental legislation; research; planning; awareness-raising.

Ministry of Agriculture and Livestock, Department of Rural Engineering and Soil Conservation: BP 621, Kigali; tel 75324; Minister: Frédéric Nzamurambaho.

Ministry of Health, Department of Public Hygiene and Environment: BP 84, Kigali; tel 76681; telex 502; Minister: Dr Casimir Bizimungu.

Ministry of Public Works and Energy, Department of Habitat and Human Settlements: BP 24, Kigali; tel 86649; telex 502; Minister: Félicien Gatabazi.

Office Rwandais du Tourisme et des Parcs Nationaux (ORTPN—Rwandan Office of Tourism and National Parks): BP 905, Kigali; tel 76514; telex 22542; fax 76512; Dir: Juvenal Uwilingiyimana; tourism; conservation of nature and national parks.

Non-Governmental Organizations

Association Rwandaise des Ecologistes (ARECO Rwanda Nziza—Rwandan Ecologists' Association): BP 1754, Kigali; tel 77932; fax 72237; Pres: Ernest Rukangira; Sec: Dangilla Mukakamari; environmental education and awareness-raising; nature conservation; support for consumers.

Fondation pour l'Environnement et le Développement des Mille Collines (FEDMICO—Mille Collines Environment and Development Foundation): BP 1754, Kigali; tel 77932; fax 72237; Pres: Ernest Rukangira; environmental impact assessment; management of natural resources; rural development.

Parti Ecologiste (Peco—Ecologist Party): Kigali; member of the Alliance pour le Renforcement de la Démocratie (Alliance for the Reinforcement of Democracy) with the ruling Mouvement Républicain National pour la Démocratie et le Développement (MRNDD—National Republican Movement for Democracy and Development).

Wildlife Conservation International, Projet Conservation de la Forêt de Nyungwe (Nyungwe Forest Conservation Project): BP 363, Cyangugu; tel 37193; fax 37193; assistance for the government in the conservation and management of the Nyungwe Forest Reserve (ecotourism, applied ecological research, conservation education).

SAINT CHRISTOPHER AND NEVIS

Governmental Organizations

Ministry of Natural Resources and the Environment: Basseterre; Minister: Simeon Daniel.

Non-Governmental Organizations

Nevis Environmental Education Committee: POB 563, Charlestown, Nevis; tel 469-5786; fax 469-0274; Vice-Chair: Valarie Sargeant; Sec and Treas: Lornette Hanley.

Nevis Historical and Conservation Society: POB 563, Charlestown, Nevis; tel 469-5786; fax 469-0274; Pres: Vincent Hubbard; Vice-Pres: Clara Walters; environmental conservation; preservation of historic monuments; museum of Nevis history; Horatio Nelson museum; library; archives; environmental education.

SAINT LUCIA

Governmental Organizations

Ministry of Agriculture, Lands, Fisheries and Forestry: Waterfront, Castries; tel 452-2526; telex 6243; fax 453-6314; Minister: Ira d'Auvergne.

Ministry of Health, Information, Broadcasting and Local Government: Chaussee Rd, Castries; tel 452-2827; telex 6394; Minister: Stephenson King.

Ministry of Planning, Personnel, Establishment and Training: POB 709, Castries; tel 452-3688; telex 6243; fax 452-2506; Minister: John G M Compton; Perm Sec: Ausbert d'Auvergne; formulation of economic policy; monitoring development; formulation of environmental policy.

Non-Governmental Organizations

Saint Lucia Red Cross Society: POB 271, Castries; tel 452-5582; fax 453-7811; Pres: Kenneth Monplaisir; Sec: Doris Boland; Dir-Gen: Terencia Gaillard; training in first aid, disaster-preparedness, health education, youth services and water safety.

SAINT VINCENT AND THE GRENADINES

Governmental Organizations

Ministry of Health and Environment: Kingstown; tel 456-1111; Minister: Burton Williams; Minister of State: Alpian Allen.

Ministry of Finance, Central Planning Unit: Kingstown; tel 456-1111; Prime Minister and Minister of Finance: James F Mitchell.

Non-Governmental Organizations

Saint Vincent and the Grenadines National Trust: POB 1538; tel 456-1060; fax 456-2142; Chair: Dr I A E Kirby; Vice-Chair: Morrison Baisden; Sec: Norma Keizer; environmental conservation; preservation of the cultural heritage; research; documentation; public awareness-raising.

SAN MARINO

Governmental Organizations

Secretariat of State for the Environment: San Marino; Minister of State: Piernatalino Mularoni.

SÃO TOMÉ AND PRÍNCIPE

Governmental Organizations

Ministry of Agriculture and Fisheries: São Tomé; tel 22714; Minister of Agriculture and Rural Development: Xavier Mendes; Minister of Commerce, Industry, Fisheries and Tourism: Arzemiro dos Prazeres.

SAUDI ARABIA

Governmental Organizations

Ministry of Defence and Civil Aviation, Meteorology and Environmental Protection Administration (MEPA): POB 1358, Jeddah 21431; tel (2) 651-2312; telex 401236; fax (2) 651-1424; Pres: Dr Abdulbar A al-Gain; formulation of criteria, standards, guidelines and policies for environmental protection and prevention of pollution; development of the National Oil Spill and Hazardous Substance Pollution Contingency Plan.

National Commission for Wildlife Conservation and Development (NCWCD): POB 61681, Riyadh 11575; tel (1) 441-8700; telex 405930; fax (1) 441-0797; Sec-Gen: Dr Abdulaziz Abu-Zinada; preservation of wildlife; creation and management of protected areas.

Assir Principality, Tourism Development Department: Abha; tel (7) 224-7437; telex 901013; Gen Dir: Ibrahim A N al-Sayed; development of national parks, forests and gardens.

Non-Governmental Organizations

King Abdulaziz University, Environmental Science Department: POB 9034, Jeddah 21413; tel (2) 695-2368; telex 601141; fax (2) 695-2364; Principal Officers: Dr M H Hashim, Dr A S Radady, Dr M M Mohanna, Dr G M Hachim, Dr A S Arafa, Dr M H Goknil; teaching and research in water pollution and water quality management, air pollution and air quality management, solid waste management, noise control, environmental toxicology, food hygiene and consumer protection and occupational health.

King Saud University, Centre for Desert Studies: POB 2454, Riyadh 11451; tel (1) 467-5571; fax (1) 467-5574; Principal Officer: Dr Khaled al-Hamoudi; desert research; remote sensing and geographic information systems (GIS); afforestation; desertification control; range management.

Red Sea and Gulf of Aden Environment Programme (PERSGA): POB 1358, Jeddah 21431; tel (2) 651-9868; telex 601236; fax (2) 651-9868; Exec Dir: Dr Nizar Tawfiq; environmental conservation and natural resources management in the marine and coastal areas of the Red Sea and the Gulf of Aden.

SENEGAL

Governmental Organizations

Ministry of Tourism and Air Transport and of the Environment and Nature Conservation: 23 rue Calmette, BP 4049, Dakar; tel 21-11-26; Minister of Environment and Nature Conservation: Abdoulaye Bathily.

Direction des Eaux, Forêts et Chasses, et de la Conservation des Sols (Department of Waters, Forests and Hunting, and Soil Conservation): BP 1831, Dakar; tel 32-06-28; fax 32-04-26; Nat Dir: Bocar Oumar Sall; Deputy Nat Dir: Ibrahima Gueye; reforestation; forest management; wildlife conservation; integrated land management; dune management; soil and water conservation.

Non-Governmental Organizations

Association pour la Promotion de l'Agriculture Biologique (ASPAB—Association for the Promotion of Biological Agriculture): BP 412, Thiès; tel 51-20-56; Principal Officer: Cheikh Tidiane Drame.

Centre de Recherche et de Documentation sur le Sahel (Centre for Research and Documentation on the Sahel): BP 369, Saint-Louis; tel 61-12-00; Principal Officer: Souleymane G Diouf.

Conseil des ONG d'Appui au Développement (CONGAD—NGO Council for Development Support): BP 4109, km 1, ave Cheikh Anta Diop, Dakar; tel 22-46-94; fax 24-24-31; Sec-Gen: Malamine Savane; co-ordination of activities of member NGOs; information for members; improvement of relations between NGOs and government.

ENDA—Environmental Development Action: POB 3370, Dakar; tel 21-60-27; telex 51456; fax 22-26-95; Exec Sec: Jacques Bugnicourt; Co-ordinators: Youba Sokona, Raphaël Ndiaye; co-ordination with other environmental groups; research into alternative development possibilities; training programmes; assistance with the implementation of development programmes in various countries.

Environnement et Développement du Tiers-Monde (ENDA Tiers-Monde—Environmental Development Action in the Third World): 4-5 rue Kleber, BP 3370, Dakar; tel 22-42-29; telex 51456; fax 22-26-95; Principal Officers: Jacques Bugnicourt, Cheikh Hamidou Kane; action and information on the environment, development, urbanization, energy, nature conservation, human rights, refugees, children, appropriate technology and health.

Forum du Tiers-Monde (Third World Forum): BP 3501, Dakar; tel 21-11-44; telex 1300; fax 22-12-01; Dir: Samir Amin; Dir of Programmes: Bernard Founou; research network on development issues.

Gambia River Basin Development Organisation: BP 2353, 13 passage Le Blanc, Dakar; tel 22-31-59; telex 51487; fax 22-59-26; Exec Sec: Nassirou Diallo; integrated development of the Gambia, Kayanga-Geba and Koliba-Corubal river basins.

Groupe de Recherches et de Réalisations pour l'Eco-Développement (GRED—Eco-Development Research and Achievement Group): Bakel; tel 83-51-91; Principal Officer: Jacques Dubois.

IUCN Senegal: Western Africa Regional Office, BP 3215, Dakar; tel 32-05-45; telex 3103; fax 32-92-46; Regional Rep: Patrick Bouland.

Parti Africain des Ecologistes—Sénégal (PAES—African Ecologists' Party, Senegal): Ecole Normale Germaine Legoff, Dakar; Sec-Gen: Aboubacry Dia.

Rodale International, Centre des Ressources pour une Agriculture Régénératrice (RI—Rodale International Resource Centre for Regenerating Agriculture): BP A237, Quartier Carrières, Thiès; tel 51-10-28; Principal Officer: Amadou Makhtar Diop.

Satis Global Union for Technologies for Environment and Sustainable Development: BP 2664, Dakar; tel 21-75-95; telex 090651456; fax 22-26-95; Principal Officer: Djimingue Nanasta; work in the fields of appropriate technology, the environment and sustainable development; information.

SEYCHELLES

Governmental Organizations

Ministry of the Environment, Economic Planning and External Relations: POB 656, Mont Fleuri; tel 24688; telex 2260; fax 24845; Minister: Danielle de St Jorre; conservation of nature; management of national parks and nature reserves; control of refuse disposal; impact assessment; education; reafforestation; forestry; monitoring of protected trees; fire control; landscaping; maintenance of parks and gardens.

SIERRA LEONE

Governmental Organizations

Department of Lands, Housing and the Environment: Youyi Bldg, Brookfields, Freetown; Sec of State: Lt-Col S F Y Koroma.

Non-Governmental Organizations

Concerned Environmentalists Action Group (CEAG): 22A Hindowa St, Kenema; Founder and Chair: Cyril E Wilson; Sec-Gen: Irene Carew; campaign against destruction of tropical rainforest; community education on environmental health and pollution of air, land and water.

Conservation Society of Sierra Leone: 4 Sanders St, POB 1292, Freetown; tel 229716; telex 3638; fax 224439; promotion of wise use of natural resources; protection of wildlife and habitats; protection and management of national nature reserves; co-ordination of school nature clubs; prevention and control of bush fires; annual Sierra Leone Wildlife Week.

Future in Our Hands Movement (FIOH): PMB 950, 33 Robert St, Freetown; Dir: Olatunde Johnson; Chair: Abiodun Savage; agricultural projects (including Food for All by the Year 2000—FFA2000); campaigns on bush fires, tree-planting and pollution; primary health care for women and children; education and training.

SINGAPORE

Governmental Organizations

Ministry of the Environment: 40 Scotts Rd, Environment Bldg, Singapore 0922; tel 7327733; telex 34365; fax 7319456; Minister: Mah Bow Tan.

National Parks Board: Singapore Botanic Gardens, Cluny Rd, Singapore 1025; tel 4709914; fax 4754295; Exec Dir: Dr Tan Wee Kiat; Deputy Exec Dir: Dr Leong Chee Chiew; management of Singapore Botanic Gardens, Fort Canning Park and nature reserves for recreation, conservation, research and education purposes; scientific research on nature conservation.

Non-Governmental Organizations

Nature Society (Singapore): 7 Asimont Lane, 01-14, Singapore 1130; tel 2551708; fax 2532179; Pres: Prof Wee Yeow Chin; Sec: Evelyn Lim-Eng Lee Choo; Treas: Allan Owyang; conservation of natural areas; environmental education; surveys of flora and fauna; environmental impact assessment.

Singapore Zoological Gardens: Mandai Lake Rd, Singapore 2572; tel 2693411; telex 56430; fax 3672974; Exec Dir: Bernard Harrison; conservation of endangered animal species; captive breeding.

SLOVAKIA

Governmental Organizations

Ministry for the Environment: Hlboká 2, 812 35 Bratislava; tel (7) 492-002; fax (7) 311-368; Minister: Jozef Zlocha.

Ustredie Stätnej Ochrany Prirody (Central Office for State Nature Conservation): Leninova 11, 031 01 Liptovsky Mikulås; tel (2) 201-589; Dir: Anton Lucinkiewicz.

Národny Park Nízke Tatry (NAPANT—Low Tatras National Park): Horná 67, 974 00 Banská Bystrica; tel (88) 543-72; fax (88) 247-28; Dir: Dr Luboš Čillag.

Non-Governmental Organizations

Comenius University, Centre for Environmental Impact Assessment: Faculty of Natural Sciences, Mlynská Dolina B-2, 842 15 Bratislava; tel (7) 720-462; fax (7) 729-064; Principal Officers: Dr Mária Kozová, Dr Katarína Pavličková; co-operation with Ministry of Environment in preparation of a strategy for implementation of environmental impact assessment regulations; training courses and manuals; co-operation with relevant organizations in other countries.

Environmental Partnership for Central Europe: Skvteckeho 30, 974 00 Banská Bystrica; tel (88) 294-259; fax (88) 294-220; Country Dir: Juraj Mesik; see German Marshall Fund, USA.

Greenpeace Slovakia: POB 58, 815 99 Bratislava.

Nadácia Zelená Alternatíva (Green Alternative Foundation): Vajanského 3, 921 01 Piešt'any; tel (838) 239-27; Pres: Dr Peter Sabo; Vice-Pres: Vlasta Körnerová; promotes ecologically and spiritually sound alternatives to current ways of living; involved in environmental education and nature protection; publication of books; library.

Nadácia Zelená Nádej (Green Perspective Foundation): Levočská 5, 080 01 Prešov; tel (91) 345-00; fax (91) 314-45; Principal Officers: Juraj Lukáč, Radoslav Potočný, Dagmar Balážová; campaigns for environmental conservation and responsible use of biosystems; catalyst for individual and collective action; campaigns to save the Čergov forest in north-east Slovakia; information centre; publications.

Pavol J Šafárik University, Botanical Garden: Mánesova 23, 043 52 Košice; tel (95) 315-55; fax (95) 373-53; Principal Officer: Dr Sergej Mochnacký; collection of information about endangered Central European plant species; study of the biology of autochthonous species; cultivation of tropical, subtropical, protected and medicinal plants.

Slovak Academy of Sciences, Institute of Botany: Dúbravská cesta 14, 842 23 Bratislava; tel (7) 373-507; fax (7) 371-948; Dir: Dr Otília Gašparíková; Deputy Dir: Dr Ivan Jarolímek; Sec: Dr Milada Čiamporová; experimental research in the fields of plant physiology, embryology and pathology.

Slovak Academy of Sciences, Institute of Evolutionary and Applied Landscape Ecology: Andreja Khet'a 13, 969 00 Banská Štiavnica; tel (859) 218-06; fax (859) 238-16; Principal Officers: Dr Peter Múdry, Dr Yozef Šteffek; solutions to theoretical and practical problems associated with the ecological carrying capacity of the landscape and territorial systems of ecological stability.

Slovak Academy of Sciences, Institute of Forest Ecology: Štúrova 2, 960 53 Zvolen; tel (855) 223-12; fax (855) 274-85; Principal Officers: Dr Olga Kontrišová, Vincent Štrba.

Slovak Academy of Sciences, Institute of Landscape Ecology: Štefánikova 3, POB 254, 814 99 Bratislava; tel (7) 334-311; fax (7) 332-560; Dir: Prof Dr Milan Ružička; Deputy Dir: Dr Tatiana Hrnčiarová; basic research in landscape ecology, including optimal organization and use of land, ecological problems associated with urban landscapes, effects of crop protection and nutrition, evolutionary changes and ecological modelling.

Slovak Technical University, Department of Environmental Science: Faculty of Chemical Technology, Radlinského 9, 812 37 Bratislava; tel (7) 495-243; fax (7) 493-198; Principal Officer: Prof Juraj Tölgyessy; environmental chemistry; water and air protection technology; radioecology.

Slovenská Ekologická Spoločnost' (SEKOS—Slovak Ecological Society): POB 254, Štefánikova 3, 814 99 Bratislava; tel (7) 332-560; fax (7) 332-560; Pres: Prof Dr Milan Ružička; Vice-Pres: Dr Pavol Eliáš; Sec: Július Oszlányi; promotion of the science of ecology as a means of solving environmental problems; ecological education and programmes for young ecologists; symposia and workshops for the co-ordination of research programmes; establishment of specialist groups and an information service; member of the European Ecological Federation.

Slovensky Hydrometeorologicky Ústav (Slovak Hydrometeorological Institute): Jeséniova 17, 833 15 Bratislava; tel (7) 377-336; telex 93265; fax (7) 375-670; Dir: Štefan Škulec; Deputy Dir for Environment: Dr Ivan Zuzula; meteorology; hydrology; climatology; environmental science; air protection; water supply and quality; environmental monitoring.

Slovensky Zvaz Ochrancov Prirody a Krajiny (SZOPK—Slovak Union of Nature and Landscape Protectors): Gorkého 6, 811 01 Bratislava; tel (7) 364-665; fax (7) 335-422; Principal Officers: Dr Miroslav Fulín, Dr Jozef Gregor; environmental education; protection of natural resources and biodiversity; protection of historical monuments and the landscape; 353 local groups; 14,000 individual members.

Spoločnost' pre Trvalo Udržatel'ný Život v SR (Society for Sustainable Living in Slovakia): Staroturský chodník 1, 811 01 Bratislava; tel (7) 363-462; fax (7) 313-968; Chair: Dr Mikuláš Huba; Sec: Andrea Chorváthová; promotion of research projects and sustainable technologies and practices; lobbying; public meetings; translation and distribution of relevant materials; increasing public awareness; involved in creation of a Chamber of Environmentalists.

Strom Života (Tree of Life): Pražská 11, 813 36 Bratislava; tel (7) 498-473; fax (7) 498-703; Principal Officers: Oto Makýš, Peter Hrabina; environmental education; landscape conservation; sustainable tourism; work with young people.

Združenie Národných Parkov a Chránených Území Slovenska (Association of National Parks and Protected Areas of Slovakia): 059 60 Tatranská Lomnica; tel (969) 967-248; fax (969) 967-958; Principal Officer: Dr Ivan Voloscuk; secretariat of Carpathian Association of National Parks, Nature Parks and Reserves (ACANAP), which comprises relevant administrations in Hungary, Poland, Romania and Ukraine.

SLOVENIA

Governmental Organizations

Ministry of Environment and Regional Planning: 61000 Ljubljana, Župančičeva 6; tel (61) 154208; fax (61) 224548; Minister: Miha Jazbinšek.

Non-Governmental Organizations

Zelena Slovenije (ZS—Greens of Slovenia): 61000 Ljubljana, Komenskega 11; tel (61) 311629; fax (61) 313065; Pres: Dr Dušan Plut; Parl Group Leader: Dr Božidar Volč; one of the first opposition groups to the Communists, advocating environmental protection and demilitarization of an independent Slovenia.

SOLOMON ISLANDS

Governmental Organizations

Ministry of Natural Resources: POB G24, Honiara; tel 22944; telex 66306; Minister: Victor Ngele.

Environment and Conservation Division: tel 21521; telex 66306; fax 21245; Chief Environment and Conservation Officer: Moses Biliki; Principal Conservation Officer: Henry Isa; Sr Environment Officer: Bernard Telei; environmental impact assessment; wildlife management; development of protected areas; environmental education and awareness-raising; formulation and enforcement of relevant legislation; development of a resource data base (GIS).

Ministry of Agriculture and Lands: POB G13, Honiara; tel 21621 (Agriculture), 21511 (Lands); telex 66417; Minister: George Luialamo.

SOMALIA

Governmental Organizations

National Environmental Co-ordination Committee: Department of Co-ordination, Ministry of Planning and Co-ordination, Mogadishu; tel 80384; telex 715.

Agricultural Development Committee: c/o Ministry of Agriculture, POB 930, Mogadishu; tel 80598.

SOUTH AFRICA

Governmental Organizations

Ministry of National Education and of Environment Affairs: Fed Life Forum Bldg, 4th Floor, 315 Pretorius St, Pretoria 0002; POB 3572, Pretoria 0001; tel (12) 3204210; Minister: Jacob (Japie) van Wyk.

Department of Environment Affairs: 315 Pretorius St, Private Bag X447, Pretoria 0001; tel (12) 3103911; telex 320142; fax (12) 3222682; Dir-Gen: Dr C M Cameron; aims to ensure effective protection and sustainable utilization of the environment; field of responsibilty includes sea fisheries and the Antarctic.

Cape Nature Conservation and Museums: Utilitas Bldg, 1 Dorp St, Private Bag X9086, Cape Town 8000; tel (71) 4834227; telex 522368; fax (71) 230939; Chief Dir: Dr J H Deethling; nature conservation; maintaining ecological systems and processes; conserving genetic diversity; ensuring sustainable utilization.

National Parks Board of South Africa: POB 787, Pretoria 0001; tel (12) 3430907; telex 21324; fax (12) 3439958; Exec Dir: Dr G A Robinson; conservation and management of national parks; research; ecotourism; environmental education.

Orange Free State Provincial Administration, Nature and Environmental Conservation Directorate: POB 517, Bloemfontein 9300; tel (51) 4054260; fax (51) 4054873; Dir: Dr L D Stoltz; creation and management of nature reserves; conservation of biodiversity; advice on maintaining acceptable environmental standards; environmental education; research; formulation and enforcement of relevant legislation.

Non-Governmental Organizations

Botanical Society of South Africa: Kirstenbosch, Claremont 7735; tel (21) 7972090; fax (21) 7972376; Pres: Prof O A M Lewis; Exec Officer: Daphne Barends; wise use, cultivation and study of the indigenous flora of southern Africa; co-operation with the National Botanical Institute.

Earthlife Africa: POB 11383, Johannesburg 2000; tel (11) 8375365; fax (11) 8393764; voluntary, non-profit organization; education; campaigns against environmental destruction; 14 branches throughout South Africa.

Endangered Wildlife Trust: Private Bag X11, Parkview 2122; tel (11) 4861102; fax (11) 4861506; Chair, Bd of Trustees: David Mitchell; Dir: Dr John Ledger; conservation of biological diversity in central and southern Africa; research; awareness-raising; practical conservation campaigns; publications.

Foundation for Research Development: POB 2600, Pretoria 0001; tel (12) 8412911; telex 21312; fax (12) 8412418; Man, Environmental Programmes: Dr R D Walmsley; stimulation and co-ordination of research needed to develop ecological principles and practices necessary for the conservation of indigenous species.

Kangwane Parks Corporation: POB 1990, Nelspruit 1200; tel (1311) 53931; fax (1311) 23153; Dir: Dr J L Anderson; preservation of biodiversity through wildlife management in nature reserves.

Kruger National Park: Private Bag X402, Skukuza 1350; Exec Dir: Dr S Joubert.

Natal Parks Board: POB 662, 3200 Pietermaritzburg, Natal; tel (331) 471961; telex 643481; fax (331) 471037; CEO: Dr G R Hughes.

Natal University, Department of Zoology and Entomology: POB 375, Pietermaritzburg 3200; tel (331) 955104; telex 643719; fax (331) 955105; Principal Officers: Prof M Perrin, Prof G Maclean; teaching and research in the areas of ecophysiology, limnology, population ecology, landscape ecology, behavioural ecology, ornithology, parasitology and desert ecology.

Natal University, Institute of Natural Resources: 67 St Patricks Rd, POB 375, Pietermaritzburg 3200; tel (331) 68317; fax (331) 68891; Principal Officers: Prof Charles M Breen, Prof John Erskine, Prof Dan Archer; active in the fields of integrated rural development and environmental policy and management.

National Association for Clean Air: POB 5777, Johannesburg 2000; tel (11) 6462210; technical and educational organization for air pollution management; information; workshops and conferences; publication of materials for young people.

National Botanical Institute: Private Bag X7, Claremont 7735; tel (21) 7621166; telex 21807; fax (21) 7623229; Chief Dir: Prof Brian John Huntley; concerned with the study and conservation of South Africa's vegetation and flora.

National Zoological Gardens of South Africa: POB 754, Pretoria 0001; tel (12) 283285; fax (12) 3234540; Dir: Welhelmus Labuschagne; indigenous and exotic animals.

Nature and Environmental Conservation Cape Town: Private Bag X9086, Cape Town 8000; tel (21) 4834094; telex 522368; fax (21) 230939; Chief Dir: Dr J H Neethling.

Okavango Wildlife Society (OWLS): POB 52362, Saxonwold 2132; tel (11) 8803833; fax (11) 7935400; Chair: Mike Hardisty; conservation of the Okavango Delta in northern Botswana, including the Kalahari Desert, the Delta and the Chobe National Park.

Pretoria University, Centre for Wildlife Management: Pretoria 0002; tel (12) 4202569; telex 322723; fax (12) 433432; Principal Officers: Prof W van Hoven, Prof J du P Bothma; concerned with wildlife management research, wildlife nutrition and ecotourism.

Rhino and Elephant Foundation: POB 381, Bedfordview 2008; tel (11) 8820606; fax (11) 8820666; Chair: Clive Walker; Man: Petra Mengel; funding for conservation and research projects; practical support to wildlife management agencies.

South African Association for Marine Biological Research: 2 West St, Durban 4001; POB 10712, Marine Parade, Durban 4056; tel (31) 373536; fax (31) 372132; Exec Dir: Prof Antonio J de Freitas; comprised of Durban Sea World marine park, the Oceanographic Research Institute and an education department; stimulating public awareness of the marine environment and ensuring wise, sustainable exploitation of marine resources.

Oceanographic Research Institute (ORI): Deputy Dir for Research: Rudy van der Elst; research focuses on marine conservation.

South African Committee for IUCN: Wildlife Society of South Africa, 3 Delta Park Rd, Victory Park, POB 44189, Lindern 2104; tel (11) 7824716; fax (11) 7827216; Exec Dir: A A Ferrar.

Southern African Wildlife Management Association: POB 324, Bellville 7535; tel (2231) 70180; fax (2231) 71606; Pres: N Fairall; Vice-Pres: A F Boshoff; Sec: N G Palmer; symposia and workshops; publication of periodicals.

Transvaal Nature Conservation Division: Private Bag X 209, Pretoria; tel (12) 2012469; telex 321170; fax (12) 3253869; Dir-Gen: Dr P F S Mulder; promotion of the conservation of natural resources and ecosystems.

Umgeni Valley Project: POB 394, Howick 3290; tel (332) 303931; fax (332) 304576; Dir: Malcolm Powell; Principal Education Officer: Tim Wright; Reserve Manager: Bruce Ducasse; environmental education courses for children and adults; environmental education resources.

Wilderness Trust of Southern Africa: POB 577, Bedfordview 2008; tel (11) 4537645; fax (11) 4537649; Principal Officer: Clive Hatton Walker; environmental education trails for adults and children throughout Southern Africa.

Wildlife Society of Southern Africa: POB 44189, Linden 2104; Northwards, 21 Rockridge Rd, Parktown, Johannesburg; tel (11) 4821670; fax (11) 4822436; Pres: T I Steenkamp; Exec Dir: A A Ferrar; environmental education; conservation of natural areas and wildlife; promotion of sustainable development and public participation in caring for the environment; environmental publishing; lobbying to influence public policies; founder member of IUCN.

Border Branch: East London; tel (431) 439409; Contact: Colin Budworth.

Eastern Cape Province Branch: Port Elizabeth; tel (41) 559606; Contact: Valerie Hunt.

Natal Branch: 100 Brand Rd, Durban 4001; tel (31) 213126; fax (31) 219525; Chair: S Craven; Chair of Publications: B H Harper.

Northern Cape Province Branch: Kimberley; tel (531) 22171; Contact: Graham Main.

Orange Free State Branch: Bloemfontein; tel (51) 366909; Contact: Mathilda Roos.

Transvaal Branch: Johannesburg; tel (11) 4822430; Contact: Di Jones.

Western Cape Province Branch: Cape Town; tel (21) 7011397; Contact: Di Wilson.

World of Birds Wildlife Sanctuary: Valley Rd, 7800 Hout Bay, Cape Town; tel (21) 7902730; fax (21) 7904839; Dir: Walter Mangold; 450 species of birds and small animals.

World Wide Fund for Nature (WWF) South Africa—Southern African Nature Federation (SANF): Voorgelegen, 116 Dorp St, Stellenbosch 7600; POB 456, Stellenbosch 7599; tel (2231) 72801; telex 555421; fax (2231) 79517; Pres: Dr Anton E Rupert; Chair: Gavin W H Relly; Chief Exec: Dr John Hanks; national affiliate organization of WWF International.

SPAIN

Governmental Organizations

Ministry of Public Works and Transport: Paseo de la Castellana 67, 28071 Madrid; tel (1) 5334900; telex 46388; Minister: José Borrell Fontelles.

Secretaría de Estado para las Políticas del Agua y el Medio Ambiente, Dirección General de Política Ambiental (Department of State for Water Policies and the Environment, State Environmental Policy Office): tel (1) 5531600; fax (1) 5330711.

Ministry of Agriculture, Fisheries and Food: Paseo Infanta Isabel 1, 28014 Madrid; tel (1) 3475000; telex 47062; Minister: Vicente Albero Sillas.

Instituto Nacional para la Conservación de la Naturaleza (ICONA—National Institute for the Conservation of Nature): Gran Via de San Francisco 4, 28005 Madrid; tel (1) 3476000; fax (1) 3476258; Dir: Santiago Marraco Solana; liaison agency between the Spanish government and the IUCN; co-ordination of nature conservation activities.

Agencia de Medio Ambiente, Consejera de Cultura y Medio Ambiente (Environment Agency, Government of Andalusia): Avda de Eritana 2, 41013 Sevilla; tel (54) 550550; telex 73071; fax (54) 623800; Pres: Fernando Martínez Salcedo.

Comisión de Defensa de la Naturaleza, Medio Ambiente y Repoblación Forestal de Valencia (Commission for Nature Protection, Environment and Reforestation of Valencia): Plaza de Manises 2, Valencia; Pres: Francisco Ruíz Leon.

Consejería de Medio Ambiente y Urbanismo, Principado de Asturias (Department of Environment and Land Planning, Principality of Asturias): Sol 8, 33071 Oviedo, Asturias; tel (85) 242700; telex 84112; fax (85) 243432; Principal Officer: María Luisa Carcedo Roces.

Consejería de Política Territorial de la Comunidad Autónoma de Canarias (Land Use Board of the Autonomous Community of the Canaries): Rambla General Franco 57 (Villa Clara), 38006 Santa Cruz de Tenerife, Islas Canarias; tel (22) 288362; fax (22) 286334; Environment Dir: Victor Pérez Borrego.

Departament de Medi Ambient Generalitat de Catalunya (Ministry of Environment Government of Catalonia): Passeig de Gràcia 94, 08008 Barcelona; tel (3) 4872234; fax (3) 4873180; Int Relations Co-ordinator: Kenty Richardson; concerned with environmental planning and management.

Servei de Parcs Naturals, Diputació de Barcelona (Natural Parks Service, Government of Barcelona): Edifici del Rellotge, 3°, calle Urgell 37, 080036 Barcelona; tel (3) 4022222; fax (3) 4022493; Dir: Joan I Caetelló i Vidal; management of national parks in the province of Barcelona; fire prevention and conservation education programmes.

Non-Governmental Organizations

Amigos de la Tierra España (Friends of the Earth FoE Spain): Juan Pradillo 26, 1°, 28039 Madrid; tel (1) 3112186; fax (1) 3114874; Sec-Gen: Humberto da Cruz.

Asociación Amigos de Doñana (Friends of Doñana Association): Apdo 2182, 41080 Sevilla; tel (54) 332445; telex 72230; fax (54) 630859; Pres: Javier Castroviejo Bolibar.

Asociación para la Defensa de la Naturaleza (ADENA—Association for Defense of Nature): WWF-España, Santa Engracia 6, 28010 Madrid; tel (1) 3082309; fax (1) 3083293; Sec-Gen: Cristina García-Orcoyen.

Asociación para la Defensa de la Naturaleza y los Recursos de Extremadura (ADENEX—Association for the Defence of the Nature and Resources of Extremadura): Cuba 10, 06800 Mérida; tel (24) 371202; fax (24) 373118; Pres: Dr Santiago Hernández Fernández; preparation of reports and scientific studies, scientific and information publications; environmental education.

Asociación Española para el Ecodesarrollo y la Defensa del Medio Ambiente (AEMA—Spanish Association for Ecodevelopment and Environmental Defence): Memendez Pelayo 50, 3 F, 41003 Sevilla; tel (54) 426457; fax (54) 426907; Pres: Prof Dr Fernando Díaz del Olmo; investigation, promotion, diffusion and implementation of ecodevelopment planning; conservation of the natural and cultural heritage.

Asociación Malagueña para la Protección de la Vida Silvestre (SILVEMA—Malaga Association for the Protection of Wildlife): Mariblanca 21 bajo, Apdo 4046, 29080 Malaga; tel (52) 229595; fax (52) 601691; Pres: Saturnino Moreno Borrell; non-profit organization concerned with the protection of the natural and urban environment and the conservation of Malaga and Andalusia's natural resources.

Asociación para la Supervivencia de la Naturaleza y el Medio Ambiente de Andalucía (ANDALUS—Association for the Survival of the Nature and Environment of Andalusia): Marqués de Paradas 18, 1°, 7, Apdo 143, 41080 Sevilla; tel (54) 214251; fax (54) 214251; Pres: Carlos Segovia Espiau; Gen Sec: Honorio Inés; conservation of coasts and wetlands; protection of endangered species; conservation and management of Mediterranean forests; planning and land use; landscape ecology; habitat fragmentation.

Catalonian League for the Defence of the Natural Heritage: Calle Aragón 281, 2°, 2 A, 08009 Barcelona; tel (3) 2151484; fax (3) 4871560; Principal Officers: Enric Carrera, Lluís Motjè; protection and restoration of natural areas in Catalonia; environmental education; legal advice; increasing public environmental awareness.

Centro Europeo de Formación Ambiental y Turística (European Centre for Professional Training in Environment and Tourism): Viriato 20, 3° D, 28010 Madrid; tel (1) 5933186; fax (1) 5930980; Principal Officer: Arturo Crosby; concerned with environmental and rural tourism; training; information; computerized communication service; technical advice.

Comité Español para la UICN (Spanish Committee for the IUCN): Instituto Nacional para la Conservación de la Naturaleza, Gran Vía de San Francisco 35, 28079 Madrid; tel (1) 2668914; telex 48566; Sec: Dr Antonio Franco.

Coordinadora de Organizaciones de Defensa Ambiental (CODA—Co-ordinator of Environmental Defence Organizations): Plaza Santo Domingo 7, 7° B, 28013 Madrid; tel (1) 5596025; fax (1) 5597897; Chair: Juan Manuel Gallego Luque;

Gen Sec: Theo Oberhuber; federation of 160 Spanish ecological groups; campaigns; publications.

Estación Biológica Doñana (Doñana Biological Station): Pabellón del Perú, Avda de María Luísa s/n, 41013 Sevilla; tel (54) 232340; fax (54) 621125; Dir: Dr Miguel Delibes; research in the fields of nature conservation, animal ecology, ethology, game biology and evolutionary ecology.

Federación de Grupos para la Defensa de la Naturaleza (Federation of Groups for the Defence of Nature): Plaza de Santo Domingo 7, 7°, 28013 Madrid; tel (1) 5596025; fax (1) 5597897; Pres: Guillermo Díaz; federation of non-profit organizations; concerned with conservation, nature protection and sustainable development.

Fundación Bios para la Conservación de la Naturaleza (Bios Foundation for Nature Conservation): Avda Manuel Siurot 3, Edif San Leandro 6-2, 41013 Sevilla; Apdo 1250, 41080 Sevilla; tel (54) 615679; fax (54) 214251; Pres: Carlos Segovia Espiau; Gen Sec: Enrique E Alés; research and management of Mediterranean woodland and wetland ecosystems; involved in forest conservation at Puerto Moral biological reserve and bird research and conservation at Anastasio Senra ornithological station; member of IUCN.

Fundación José María Blanc (José María Blanc Foundation): Fortuny 27, 1° D, 28010 Madrid; tel (1) 3197979; telex 45854; fax (1) 3198984; Pres: José María Blanc Díaz; study, conservation and promotion of wild fauna and game and their natural habitat.

Genetic Resources Action International (GRAIN): Jonqueres 16, 6° D, 08003 Barcelona; tel (3) 3105909; fax (3) 3105952; Principal Officers: H Hobbelink, Renée Veliré; concerns include sustainable use and conservation of genetic diversity, policies to support and improve local farming systems and the impact of biotechnology on farmers in developing countries; activities include European and global networking, monitoring and research, lobbying, campaigns and development of alternative strategies.

Greenpeace Spain: Rodriguez San Pedro 58, 4°, 28015 Madrid.

Grup Balear d'Ornitologia i Defensa de la Naturalesa (GOB—Balearic Group for Ornithology and Defence of Nature): Verí 1, 3R, 07001 Palma de Mallorca, Balears; tel (71) 721105; fax (71) 711375; Pres: Anthony Bonner; a non-profit and politically independent association.

Jardín Botánico de Córdoba (Cordoba Botanical Garden): Avda de Linneo s/n, 14004 Córdoba; tel (57) 200077; fax (57) 295333; Dir: Prof D J Esteban Hernández Bermejo; promotion of cultural, educational and scientific activities; protection and conservation of the plant genetic resources of Spain.

Lliga per a la Defensa del Patrimoni Natural (DEPANA—League for Natural Heritage Defence): Aragó, 281 2°, 2 A, 08080 Barcelona; tel (3) 2151484; fax (3) 2576219; Pres: Josep del Hoyo Calduch; preservation of natural resources, endangered habitats and wildlife in Catalonia through pressure campaigns and education.

Real Sociedad Española de Historia Natural (Royal Spanish Natural History Society): Facultad de Biología y Geología, Universidad Complutense, Ciudad Universitaria, 28040 Madrid; tel (1) 3945000; conferences; scientific meetings; exchange of scientific publications.

Sociedad Biosfera (Biosphere Society): Pedro del Toro 2, Apdo 10043, 41080 Sevilla; tel (54) 217655; fax (54) 217655; Pres: Jesús Vozmediano y Gómez-Feu; organization dedicated to nature and the international environment; protection and conservation of wetlands and the Doñana reserve; campaigns in defence of African forests; environmental education; denunciation of environmental infractions.

Survival International, Spanish Office: Calle Princípe 12, 3°, Oficina 2, 28012 Madrid; tel (1) 5217283; fax (1) 5231420; Principal Officer: Rosario Santamaría Martín.

Los Verdes (The Greens): Apdo 565, 38400 Puerto de la Cruz, Tenerife; green political party; opposes nuclear energy and development of tourism in the Canary Islands; in favour of progressive abolition of the armed forces; formed by the amalgamation of 19 ecological groups in January 1993.

World Wide Fund for Nature (WWF) Spain: ADENA, Santa Engracia 6, 28010 Madrid; tel (1) 3082309; fax (1) 3083293; Pres: Manuel de Prado y Colón de Carvajal; Sec-Gen: Cristina Garcia-Orcoyen Tormo; national affiliate organization of WWF International.

SRI LANKA

Governmental Organizations

Central Environmental Authority: Maligawatte New Town, Colombo 10; tel (1) 549455.

Ministry of Fisheries and Aquatic Resources: New Secretariat, Maligawatte, Colombo 10; tel (1) 546183; telex 21419; Minister: Joseph Michael Perera.

Ministry of Lands, Irrigation and Mahaweli Development: 500 T B Jayah Mawatha, Colombo 10; tel (1) 36271; Minister: Gamini Athukorale.

Department of Wildlife Conservation: 82 Rajamalwatte Rd, Battaramulla; tel (1) 33012; telex 22146.

Forest Department: Rajamalwatta Rd, Battatamulla; tel (1) 566616; fax (1) 566633; Conservator of Forests: K P Sri Bharathie; management of forest resources in Sri Lanka.

Ministry of Policy Planning and Implementation: 7th and 8th Floors, North Tower, Central Bank Bldg, Colombo 1; tel (1) 547358; Minister: Dr Wimal Wickramasinghe.

Department of Coast Conservation: New Secretariat, Maligawatte, Colombo 10; tel (1) 549755.

Department of Wildlife Conservation: Transworks House, Lower Chatham St, Colombo 01; tel (1) 33012.

National Aquatic Resources Agency: Crow Island, Mattakkuliya, Colombo 15; tel (1) 522007.

Natural Resources, Energy and Science Authority: 47-5 Maitland Place, Colombo 07.

Non-Governmental Organizations

Ceylon Bird Club: POB 11, Colombo 1; tel (1) 3288944; telex 21204; fax (1) 44848; Chair and Sec: T W Hoffmann; concerned with the study and protection of Sri Lankan birds and their habitats; information; publications.

Environmental Foundation Ltd: 3 Campbell Terrace, Colombo 10; tel (1) 697226; telex 22894; fax (1) 446518; Chair: Ravi Algama; Exec Dir: Lalanath de Silva; Sr Environmental Scientist: Hemantha Withanage; legal aid and free information on environmental issues for changes in environmental law.

IUCN Sri Lanka: 82 Rajamalwatta Rd, Battaramulla, Colombo; tel (1) 568048; fax (1) 566656; monitoring of the Forest Resources Development Project; co-ordination of implementing agencies and other relevant organizations; dispensing project funds; reviewing new proposals for donor assistance.

March for Conservation: c/o University of Colombo, Thurstan Rd, Colombo 3; tel (1) 5003399; telex 22756; Gen Sec: Nirmalie Pallewatta; promotion of environmental education and conservation of Sri Lanka's natural resources through sustainable development.

National Herbarium: POB 15, Royal Botanic Gardens, Peradeniya; tel (1) 88053; Principal Officers: A H M Jayasuriya, D S A Wijesundara; taxonomic research (including revision of the flora of Ceylon); surveys of biodiversity and of underexploited plants; conservation education.

Sarvodaya Shramadana Movementka: 98 Rawatawatte Rd, Moratuwa; tel (1) 507159; telex 22803; fax (1) 507084; Pres: Dr A T Ariyaratne; Gen Sec: D A Perera; Buddhist movement seeking peace and justice through community education and personal development; village development projects; environmental conservation with particular emphasis on forests; institutional care for orphans and people with disabilities.

Sri Lanka Environmental Journalists Forum (SLEJF): POB 30, 5 Fourth Lane/Nawala, Sri Jayawardenapura; tel (1) 566717; telex 22144; fax (1) 698315; Chair: Dharman Wickremeratne; professional body of print and electronic media journalists engaged in environmental communication and awareness-building; education; training workshops; conferences; Environmental Information Centre; publications; elected to UNEP Global 500 Roll of Honour in 1992.

Wildlife and Nature Protection Society of Sri Lanka: Chaitiya Rd, Marine Drive, Fort, Colombo 1; tel (1) 501842; fax (1) 580721; Pres: Dr Ranjen Fernando; concerned with environmental education and its relevance to everyday life.

SUDAN

Governmental Organizations

Ministry of Agriculture and Natural Resources: Khartoum; tel 72300; Minister: Dr Ahmad Ali Genief.

Ministry of Irrigation and Water Resources: Khartoum; tel 77533; Minister: Dr Yaqub Abu Shura Musa.

Relief and Rehabilitation Commission: POB 1975, Khartoum; tel 74925; telex 22022.

Non-Governmental Organizations

Energy Research Institute: POB 4032, Khartoum; Principal Officer: Shomo Shaa el Din; research on biomass, solar and wind energy, and water power.

Environmentalists Society: c/o University of Khartoum, POB 321, Khartoum; tel 80993; Principal Officers: Dr Yagoub Abdalla Mohamed, Izat Mirghani Taha; voluntary organization dedicated to development of rural areas, increasing environmental awareness among the mass media and women's development through income-generating activities.

Sudanese Environmental Conservation Society: POB 16, El Morda Omdurman; tel 76925; telex 22738; Exec Dir: Dr Mutasim Bashir Nimir; active in public awareness, environmental education and environmental rehabilitation.

SURINAME

Governmental Organizations

Ministry of Agriculture, Livestock, Fisheries and Forestry: Cultuurtuinlaan 10, POB 1807, Paramaribo; tel 74177; telex 118; Minister: Soeratno Setroredjo.

Ministry of Natural Resources and Energy: Dr J C de Mirandastraat 13-15, Paramaribo; tel 74666; Minister: Harold Pollack.

Hoofd Dienst s' Lands Bosbeheer: c/o Head of Nature Conservation Department, POB 436, Cornelis Jongbawstraat 10, Paramaribo; tel 75845.

Non-Governmental Organizations

Stichting Natuurbehoud Suriname (STINASU—Foundation for Nature Preservation in Suriname): POB 436, Cornelis Jongbawstraat 14, Paramaribo; tel 476597; fax 410256; Dir: Muriel M Held; nature education, promoting public awareness of nature issues; nature tourism.

SWAZILAND

Governmental Organizations

Ministry of Agriculture and Co-operatives: POB 162, Mbabane; tel 42731; telex 2343; fax 44700; Minister: Themba M Masuku.

Ministry of Natural Resources, Land Utilization and Energy: POB 57, Mbabane; tel 46244; telex 2301; fax 42436; Minister: Ernest Richard Senzenjani Shabalala.

SWEDEN

Governmental Organizations

Ministry of the Environment and Natural Resources: Tegelbacken 2, 103 33 Stockholm; tel (8) 763-10-00; fax (8) 24-16-29; Minister: Olof Johansson; Minister for Physical Planning: Görel Thurdin.

Domän AB: 791 81 Falun; tel (23) 84-000; telex 74130; fax (23) 84-186; Chief, Land Use Dept: Elis Hedén; state-owned forestry company, responsible for the care and administration of the main part of the state-owned forests in Sweden, as well as several parks and nature reserves.

Fiskeriverket (National Board of Fisheries): Lilla Bommen 6, POB 423, 401 26 Göteborg; tel (31) 63-03-00; fax (31) 15-65-77; fisheries conservation; commercial fishing and aquaculture; recreational fishing; marine, coastal and freshwater research; assistance to developing countries.

Statens Naturvardsverk (Swedish Environmental Protection Agency): Smidesvägen 5, 171 85 Solna; tel (8) 799-10-00; telex 11131; fax (8) 29-23-82; Dir-Gen: Rolf Annerberg; Deputy Dir-Gen: Mats Olsson; co-ordination and promotion of environmental policies; active in the fields of conservation, air and sea pollution prevention, environmental research, national park management, transport, industrial regulation, public information, education and training.

Statistics Sweden: Karlavägen 100, 115 81 Stockholm; tel (8) 783-40-00; telex 15261; fax (8) 783-47-63; Dir-Gen: Jan Carling; Head, Environmental Statistics Unit: Inger Öhman; responsible for production of official statistics according to government resolutions; in the field of the environment, current statistics are produced for emissions and discharges to air and water, solid waste, air quality, pollution levels in marine waters, agricultural pesticides, industrial investment in environmental protection, local authority environmental expenditure and the implementation of environmental protection laws.

Non-Governmental Organizations

Fältbiologerna (Swedish Youth Association for Nature Studies and Environmental Protection): POB 6022, 191 06 Sollentuna; tel (8) 754-46-40; fax (8) 754-53-63; Principal Officer: Magnus Johansson; activities include bird-watching, summer camps, botanical excursions, political lobbying, public information and co-operation with similar organizations.

Friends of the Earth (FoE) Sweden: Fjallgatan 23A, 116 45 Stockholm.

Greenpeace Sweden: Industrigatan 2A 11tr, 112 46 Stockholm.

Göteberg Office: Herkulesgatan 3A, 402 73 Götenberg.

Miljöförbundet (Environmental Federation of Sweden): POB 7048, 402 31 Göteborg; tel (31) 12-18-08; fax (31) 12-18-17; Principal Officer: Olis Oljelund; local, national and global environmental activities; conservation of resources and international co-operation.

Miljöpartiet de Gröna (MpG—Green Party): Kristinebergs Slott, POB 25014, 100 23 Stockholm; tel (8) 695-02-02; fax (8) 656-00-11; Spokespersons: Marianne Samuelsson, Birger Schlaug; green political party.

Sportfiskarna (Swedish Anglers' National Association): POB 2, 163 21 Spånga; tel (8) 795-33-50; has a representative on the board of the Swedish NGO Secretariat on Acid Rain.

Stiftelsen Skansen (Skansen Foundation): Zoological Department, Hazeliusbaken 11, 115 21 Stockholm; tel (8) 463-05-00; fax (8) 463-05-00; Dir: Per-Olaf Palm.

Stockholm Environment Institute (SEI), International Institute for Environmental Technology and Management: POB 2142, 103 14 Stockholm; Järntorget 84, Stockholm; tel (8) 723-02-60; telex 19580; fax (8) 723-03-48; Dir: Prof M J Chadwick; Vice-Dir: Dr L Kristoferson; international policy research institute specializing in global environment and development issues, including energy futures, climate change, acidic depositions, clean technology, urban environments, water resources, sustainability, technology transfer, capacity building, environmental economics and agricultural biotechnology; offices in Estonia, the United Kingdom and the USA.

Svenska Naturskyddsföreningen (Swedish Society for the Conservation of Nature): POB 4625, Asogatan 115, 116 91 Stockholm; tel (8) 702-65-00; fax (8) 702-08-55; Dir: Gunmar Landborn; consultant to government agencies on environmental matters.

Sveriges Ornitologiska Förening (Swedish Ornithological Society): Skeppargatan 19, POB 14219, 104 40 Stockholm; tel (8) 662-64-34; fax (8) 662-69-88; Principal Officer: Gustaf Aulen; promotion of protection of birds and nature, and of public interest in birds; research.

Swedish NGO Secretariat on Acid Rain: POB 245, 401 24 Göteborg; tel (31) 15-39-55; fax (31) 15-09-33; Principal Officers: Christer Ågren, Reinhold Pape; dedicated to increasing awareness of air pollution problems and to reducing emissions of atmospheric pollutants to environmentally acceptable levels; information centre; co-ordination of international action; lobbying; quarterly magazine; board contains representatives from Miljöförbundet, Fältbiologerna, Svenska Naturskyddsförening, Sportfiskarna and World Wide Fund for Nature (WWF) Sweden.

World Wide Fund for Nature (WWF) Sweden: Ulriksdals Slott, 170 71 Solna; tel (8) 85-01-20; fax (8) 85-13-29; Pres: HM King Carl XVI Gustaf; Chair, Bd: Ingemar Öhrn; Sec-Gen: Jens Wahlstedt; national affiliate organization of WWF International.

SWITZERLAND

Governmental Organizations

Federal Office for the Protection of the Environment: c/o Federal Department of Home Affairs, Bundeshaus, Inselgasse, 3003 Berne; tel (31) 619111; telex 33500; fax (31) 618032; Head: Ruth Dreifuss.

Federal Department of Transport, Communications and Energy: Bundeshaus-Nord, 3003 Berne; tel (31) 614111; telex 911282; fax (31) 229576; Head: Adolf Ogi.

Bundesamt für Energiewirtschaft/Office Fédéral de l'Energie/Ufficio Federale dell'Energia (Federal Energy Office): 3003 Berne; tel (31) 615611; fax (31) 264307; Dir: Dr E Kiener; Vice-Dir: H-L Schmid; legislation, research, technology and information relating to energy.

Non-Governmental Organizations

Académie Internationale de l'Environnement (International Academy of the Environment): 4 chemin des Conches, Conches 1231, Geneva; tel (22) 7891311; fax (22) 7892538; Dir: Prof Bernard Giovannini; education and training in environment and development; research; active in policy dialogues and capacity building.

Alliance Internationale de Tourisme (AIT—International Tourism Alliance): 2 quai Gustave-Ador, 1207 Geneva; tel (22) 7352727; telex 413103; fax (22) 7352326; Dir-Gen: Peter Doggwiler; co-operation with UN agencies and the EC on issues related to motor transport and tourism, including environmental protection.

Alp Action—International Corporate Funding for the Alpine Environment: 1 rue Muzy, 1207 Geneva; tel (22) 7359295; fax (22) 7368060; Pres: Prince Sadruddin Aga Khan; Dir: Marc Sursock; Exec Dir: Javed Ahmad; encouraging investment by corporate partners in Alpine conservation projects; promotion of traditional mountain agriculture; protection of mountain lakes, rivers, forests and species; preservation of the Alpine cultural heritage.

Association pour l'Appel de Genève (APAG—Geneva Appeal Association): CP 113, Malagnou, 1211 Geneva 17; fax (21) 7350942; Principal Officers: Joel Jakubec, Ivo Rens; promotion of exchange of information on fast-breeding of plutonium, nuclear power generation and 'soft' energy alternatives; publication of journal.

Berne University, Group for Development and Environment (GDE): Institute of Geography, Hallerstr 12, 3012 Berne; tel (31) 6318822; fax (31) 6318511; Dirs: Dr H Hurni, Dr U Wiesmann; applied research on sustainable resource use in developing countries; concerned with soil conservation, applied ecology, ecological and socio-economic monitoring, rural household strategies and geographic information systems; education; project support.

Centre for our Common Future: Palais Wilson, 52 rue des Pâquis, 1201 Geneva; tel (22) 7327117; telex 412436; fax (22) 7385046; Exec Dir: W H Lindner; charitable foundation; promotion of sustainable development; information on the UNCED process; advice and materials; conferences; periodicals.

Comité d'Action pour la Défense des Animaux en Péril (Action Committee for the Protection of Endangered Species): 59 chemin des Falquets, Cologny 1223, Geneva; tel (22) 522391; fax (22) 496458; Principal Officer: Jeanne Marchig; prevention of the extinction of animals imperilled by unrestrained commercial exploitation, particularly fur-bearing animals; campaigns against inhuman methods for trapping birds and animals.

Greenpeace Switzerland: Muellerstrasse 37, Postfach 276, 8026 Zürich.

Grüne Partei der Schweiz/Parti Ecologiste Suisse (GPS/PES—Green Party of Switzerland): Marienstr 11, 3005 Berne; tel (31) 3521441; fax (31) 3521408; Chair: Verena Diener; Gen Sec: Bernhard Pulver; Leader, Parliamentary Group: Hanspeter Thür.

INTERCOOPERATION—Organisation pour le Développement et la Coopération (Swiss Organization for Development and Co-operation): Maulbeerstrasse 10, BP 6724, 3001 Berne; tel (31) 260861; fax (31) 263605; Chief, Forestry Section: Jean Laurent.

International Centre for Environmental Education and Ethics: c/o Assisi Nature Council, 3 ave de Jaman, 1005 Lausanne; Principal Officer: Maria Luisa Cohen; affiliated to Earth Island Institute, USA.

Schweizer Heimatschutz/Ligue Suisse du Patrimoine National (Swiss League for the Preservation of the National Heritage): Secretariat, Postfach, Merkurstr 45, 8032 Zürich; tel (1) 2522660; fax (1) 2522870; Pres: Ronald Grisard; Sec-Gen: Hans Gattiker; environmental protection, with particular emphasis on the built environment; participation in development decisions; influencing public opinion; using legal means to oppose construction projects; building restoration.

Schweizer Vogelschutz/Association Suisse pour la Protection des Oiseaux (Swiss Association for the Protection of Birds): Postfach, 8036 Zürich; tel (1) 4637271; fax (1) 4614778; Pres: Fritz Hirt; protection of birds and their habitats.

Schweizerische Vogelwarte Sempach/Station Ornithologique Suisse (Swiss Institute of Ornithology): 6204 Sempach; tel (41) 990022; fax (41) 994007; Principal Officers: Dr Bruno Bruderer, Dr Niklaus Zbinden; general and applied ornithology, including bird censuses, bird ringing, radar ornithology, avifaunistics and habitat evaluation; co-operation with amateur ornithologists.

Schweizerischer Bund für Naturschutz/Ligue Suisse pour la Protection de la Nature (Swiss League for Nature Protection): Postfach, 4020 Basel; tel (61) 13127442; fax (61) 13127447; Pres: Jacques Morier-Genoud; Sec: Otto Sieber; members' organization relying on voluntary activity; management of more than 500 nature reserves, including the Swiss National Park; Field Study Centres near the Aletschwald and at Champ-Pittet; production of books, periodicals, teaching materials for schools and other documents.

World Wide Fund for Nature (WWF) Switzerland: Förrlibuckstr 66, 8037 Zürich; tel (1) 2722044; fax (1) 2722844; Pres: Dr Alain Thierstein; Dirs: Dr Ulrich Halder, Dr Peter Uhr; national affiliate organization of WWF International.

Geneva Office: 14 chemin de Poussy, Vernier, 1214 Geneva; tel (22) 7827151; fax (22) 3412784.

SYRIA

Governmental Organizations

Office of Environmental Affairs: Office of the Prime Minister, Damascus.

Ministry of Agriculture and Agrarian Reform: 29 rue Ayar, Damascus; tel (11) 113613; Minister: Assad Mustafa.

Directorate of Agricultural Extension: Dar al-Mouhandissin, front of Sham Palace, 4th Floor, Damascus; tel (11) 232190; Dir: Rajab Ali; promotion of agricultural, social and economic issues.

Ministry of Public Works and Water Resources: rue Saadallah Jabri, Damascus; Minister: Abd ar-Rahman Madani.

Non-Governmental Organizations

Arab Centre for the Study of Arid Zones and Dry Lands (ACSAD): POB 2440, Damascus; tel (11) 755713; telex 412697; fax (11) 755712; Dir-Gen: Dr Mohamed el Khash; study of problems relating to the management, conservation and development of agricultural resources, including water, soil, plant and animal resources; surveying and assessment of resources and the causes of degradation and desertification.

TAJIKISTAN

Governmental Organizations

Ministry of Environmental Protection: Dushanbe; Minister: Saydullo Khayrulloyev.

TANZANIA

Governmental Organizations

Ministry of Tourism, Natural Resources and Environment: POB 9372, Dar es Salaam; tel (51) 21241; Minister: Juma Hamad Umar.

Wildlife Division: Tancot House, POB 1994, Dar es Salaam; tel (51) 27271; Dir: C A Mlay.

Ministry of Energy, Minerals and Water: POB 9153, Dar es Salaam; tel (51) 31433; Minister: Jakaya Kiwete.

National Environment Management Council: POB 63154, Dar es Salaam; tel (51) 27815; telex 41959; fax (51) 34603; advisers to the government on all matters relating to the environment; formulation of environmental management policy; co-ordination of the activities of other environmental organizations in Tanzania.

Non-Governmental Organizations

College of African Wildlife Management: POB 3031, Moshi; tel (55) 50063; telex 43082; provision of professional and technical training.

Malihai Clubs of Tanzania: POB 1541, Arusha; tel (57) 6026; telex 42130; Principal Officer: Peter A Ottaru; co-ordination of Malihai clubs established through conservation education in secondary schools and colleges.

Tanzania National Parks: POB 3134, Arusha; tel (57) 3471; telex 42130; fax (57) 8216; Dir-Gen: Lota Melamari; Dir of Park Management and Conservation: Benjamin Kanza; conservation of flora and fauna.

World Wide Fund for Nature (WWF) Tanzania: POB 63117, Dar es Salaam; tel (51) 22664; telex 81117; fax (51) 46232; Principal Officer: John I Boshe.

THAILAND

Governmental Organizations

Ministry of Science, Technology and Environment: Thanon Phra Ram Hok, Bangkok 10400; tel (2) 246-0064; fax (2) 246-8106; Minister: Phisan Moonlasartsathorn.

Ministry of Agriculture and Co-operatives: Thanon Ratchad-amnoen Nok, Bangkok 19299; tel (2) 281-5955; Minister: Niphon Phromphan.

National Environment Board: Soi Pracha-Sumpun 4, Rama VI Rd, Bangkok 10400; tel (2) 278-5465.

Office of Environmental Policy and Planning: 60/1 Soi Phibun Wattana 7, Rama VI Rd, Bangkok 10400; tel (2) 279-7180; fax (2) 271-3226; Sec-Gen: Sunthad Somchevita; Dir of International Environmental Affairs Division: Orapin Wongchumpit; formulation of policy and plans, including evaluation measures, to develop and maintain the quality of the national environment in accordance with other national policy.

Pollution Control Department: 539/2 Gypsum Metropolitan Bldg, 16th and 17th Floors, Sri-Ayudhya Rd, Rajdhavee, Bangkok 10400; tel (2) 245-0929; fax (2) 245-9953; formulation of conservation policies and environmental recommendations; monitoring of pollution and development of management systems; comprises Water Quality Management Division, Air Quality and Noise Management Division, Hazardous Substances and Waste Management Division, Legal and Petition Division, and Pollution Management Co-ordination Division.

Royal Forest Department: 61 Phaholyothin Rd, Chatuchak, Bangkok 10900; tel (2) 579-4842; fax (2) 579-8611; Dir-Gen: Tiwa Sapakit; responsible for the establishment, administration and management of wildlife sanctuaries, habitat and population.

Natural Resources Conservation Bureau: tel (2) 579-1565; fax (2) 579-8611; implementation of conservation and resource management measures in the protected forest areas of Thailand, including 77 national parks, 50 forest parks, 35 wildlife sanctuaries and 48 non-hunting areas.

Non-Governmental Organizations

Asian Institute of Technology: POB 2754, Bangkok 10501; Km 42 Paholyothin Highway, Klong Luang, Pathumthani 12120; tel (2) 516-0110; telex 84276; fax (2) 516-2126; Pres: Prof Alastair M North; Vice-Pres for Development: Dr Suvit Yodmani; Vice-Pres for Academic Affairs: Prof Ricardo P Pama; postgraduate education, research and continuing education.

Asian Society for Environmental Protection (ASEP): c/o Carl Duisberg Gesellschaft (CDG) South East Asia Program Office (SEAPO), Asian Institute of Technology; tel (2) 524-5363; telex 84276; fax (2) 524-5390; Pres: Henry Chiu; Sec-Gen: Günter Tharun; international, non-profit association of professionals and institutions committed to the sound management of the Asian environment and its natural resources, and to sustainable development; promotion of public awareness of environmental priorities and the exchange and dissemination of information and experience in the field of environmental management and technology; collaboration with similar organizations.

Environmental Systems Information Center (ENSIC): tel (2) 524-5863; fax (2) 524-5870; Dir: H A Vespry; Tech Adviser: Dr C Visvanathan; collection, collation and dissemination of information on water supply, the environment and sanitation; provision of reference and referral services, and consultancy and training in information technology.

Regional Energy Resources Information Center (RERIC): tel (2) 524-5866; fax (2) 524-5870; Dir: H A Vespry; Assoc Dir: Prof R H B Exell; Manager and Information Specialist: On-Anong Suraniranat; collection, collation and dissemination of information on energy conservation and planning and renewable energy; provision of reference and referral services, and consultancy and training in information technology and services.

IUCN Thailand: Thailand Office, c/o Asian Institute of Technology, POB 2754, Bangkok 10501; tel (2) 524-5393; telex 84276; fax (2) 524-5392; Regional Co-ordinator, Wetlands Programme: M Zakir Hussain.

King Mongkut's Institute of Technology, Thonburi—Division of Environmental Technology: School of Energy and Materials, Rutburana, Bangkok 10140; tel (2) 427-0039; fax (2) 427-9062; Principal Officer: Asst Prof Dr Sirintornthep Towprayoon; teaching on energy-related environmental problems and remedial technology including environmental assessment; supervision of postgraduate study and research in the fields of air pollution, waste and hazardous waste treatment.

Wildlife Fund Thailand: 251/88-90 Phaholyothin Rd, Bangkhen, Bangkok 10220; tel (2) 521-3435; fax (2) 552-6083; Pres: Dr Nadda Sriyabhaya; Sec-Gen: Pisit na Patalung; national associate organization of WWF International.

TOGO

Governmental Organizations

Ministry of the Environment and Tourism: ave de Sarakawa, BP 3114, Lomé; tel 21-53-52; telex 5007; Minister of Rural Development, Environment and Tourism: Kossi Nicolas Monindji.

TONGA

Governmental Organizations

Office of the Minister of Lands, Survey and Natural Resources: POB 5, Nuku'alofa; tel 22655; telex 66269; Minister: Dr Ta'afu Tupou.

Office of the Prime Minister, Minister of Agriculture, Forestry and Fisheries, and Marine Affairs: Nuku'alofa; tel 21300; telex 66269; fax 23888; Prime Minister: Baron Vaea of Houma.

TRINIDAD AND TOBAGO

Governmental Organizations

Ministry of Health and Environment: 35-37 Sackville St, Port of Spain; tel 625-4611; fax 638-8620; Minister of Health: John Eckstein.

Ministry of Agriculture, Land and Marine Resources: St Clair Circle, Port of Spain; tel 622-2652; fax 622-4246; Minister: Keith Rowley.

Non-Governmental Organizations

Crusoe Reef Society: POB 890, Port of Spain; tel 674-8104; fax 674-8966; Pres: Steven P Hale; Sec: Stephen Abiraj; monitoring and protection of the marine and coastal environment of Tobago.

Institute of Marine Affairs: POB 3160, Carenage Post Office, Carenage; tel 634-4291; fax 634-4433; Chair, Bd of Govs: Prof R Saunders; Dir: L F Ballah; basic and applied research in marine and environmental disciplines, including fisheries and aquaculture, physical oceanography, marine pollution, coastal dynamics, marine policy and law, and environmental impact assessment; public education and information.

Zoological Society of Trinidad and Tobago Inc: Emperor Valley Zoo, Royal Botanical Gardens, Port of Spain; tel 622-3530; fax 622-7808; Pres: Frank McKenzie Cook; Curator: Hans Boos; management of Emperor Valley zoological garden and promotion of zoology and animal physiology.

TUNISIA

Governmental Organizations

Ministry of the Environment and Land Planning: 15 rue Montplaisir, BP 52, Le Belvédère, 1002 Tunis; tel (1) 660-088; Minister: Mehdi Melika.

Agence Nationale de Protection de l'Environnement (ANPE—National Environmental Protection Agency): 12 ave Kheireddine Pacha, BP 52, Tunis 1002; tel (1) 799-201; telex 17190; fax (1) 189-844; Pres and Dir-Gen: Abderrahmen Gannoun; protection of the environment and management of natural resources; prevention and control of pollution; promotion of study and research into the conservation of the environment.

Ministry of Agriculture: 30 rue Alain Savery, Tunis; tel (1) 660-088; telex 13378; Minister: Muhammad ben Rejeb; Secretary of State for Hydraulic Resources: Ameur Horchani.

Ministry of Education, Higher Studies and Scientific Research: place de la Kasbah, Tunis; tel (1) 660-088; telex 13004; Minister: Muhammad Charfi; Secretary of State for Scientific Research: Mongi Safra.

Ministry of Equipment and Urban Development: Cité Jardin, Tunis; tel (1) 680-088; telex 13565; Minister: Ali Chaouch.

Ministry of Planning and Regional Development: rue de Kairouan, Tunis; tel (1) 660-088; telex 13448; Minister: Mustapha Nabli.

Non-Governmental Organizations

Association Tunisienne pour la Protection de la Nature et de l'Environnement (ATPNE—Tunisian Association for the Protection of Nature and the Environment): National Headquarters, 12 rue Tantaoui El Jaouahiri, El Omrane, 1005

Tunis; tel (1) 288-141; fax (1) 797-295; Pres: Dr Abrougui Mohamed Ali; campaigns for the conservation of flora and fauna; opposition to all forms of pollution; environmental awareness-raising; participation in legislative decision-making; international representation of Tunisian environmentalists.

Ain Draham Regional Office: Ain Draham; tel (8) 47169; Contact: Bouazizi Abderrahman.

Béja Regional Office: Béja; tel (8) 51872; Contact: Dr Hallaj Noureddine.

El Jem Regional Office: El Jem; tel (3) 90224; Contact: Mohsen Haddad.

El Mourouj Regional Office: El Mourouj; tel (1) 255-239; Contact: Ahmed Barkallah.

Hammamet Regional Office: Hammamet; tel (2) 82584; Contact: Dr Sahli Salem.

Kébili Regional Office: Kébili; tel (5) 90012; Contact: Jelassi Mohamed Lahbib.

Kélibia Regional Office: Kélibia; tel (2) 95985; Contact: Dr Lassaad Dridi.

Korba Regional Office: Korba; tel (2) 239-893; Contact: Belgacem Haddad.

La Marsa Regional Office: La Marsa; tel (1) 742-400; Contact: Arfaoui Khédija.

Medjez El Bab Regional Office: Medjez El Bab; tel (8) 60174; Contact: Mnaouar Kouki.

Nabeul Regional Office: Nabeul; tel (2) 86110; Contact: Bouina Kamel.

Radès Regional Office: Radès; tel (1) 296-910; Contact: Ali Gharbi.

Sousse Regional Office: Sousse; tel (3) 31177; Contact: Amor Ernez.

Tajerouine Regional Office: Tajerouine; tel (8) 87108; Contact: Mouelhi Faouzi.

Tébourba Regional Office: Tébourba; Contact: Dr Trabelsi Amor.

Zaghouan Regional Office: Zaghouan; tel (2) 75008; Contact: Amri Mohamed.

Zarzis Regional Office: Zarzis; tel (5) 81082; Contact: Dr Ridha Jarmoud.

Greenpeace Tunisia: 51 ave Abdelaziz Dhalbi, El Manar 2, 2092 Tunis.

TURKEY

Governmental Organizations

General Directorate for the Environment: Office of the Prime Minister, Karanfil Sokak No 8, Kizilay, Ankara; tel (4) 1184531; fax (4) 1177971.

Ministry of Agriculture and Rural Affairs: Tarim Orman ve Köyişleri Bakanliği, Bakanliklar, Ankara; tel (4) 4191677; fax (4) 4177168; Minister: Necmettin Cevheri.

Ministry of Energy and Natural Resources: Enerji ve Tabii Kaynaklar Bakanliği, Ankara; tel (4) 2231109; fax (4) 2229405; Minister: Ersin Faralyali.

Ministry of Forestry: Orman Bakanliği, APK Kurul Bşk 194, Bakanliklar, Ankara; tel (4) 4178499; fax (4) 4179160; Minister: Vefa Tanir.

Non-Governmental Organizations

Arkadaş: Z Gökalp Cad 171, 21950 Ergani; tel (8371) 5032; fax (8371) 5525; Sec-Gen: Yaşar Öztürk; Chair: Ümit Öztürk; independent non-profit organization dedicated to environment and development issues such as wildlife, afforestation, developing countries, human rights, peace and disarmament; involved in campaigns, lobbying and public awareness-raising; seminars and conferences, training courses and field projects.

Doğal Hayati Koruma Derneği (DHKD—Turkish Society for the Protection of Nature): PK 18, 80810 Bebek, Istanbul; tel (1) 2790139; fax (1) 2795544; Gen Dir: Nergis Yazgan; educational activities, research, projects and field work related to the conservation of nature in Turkey.

Türkiye Tabiatini Koruma Dernegi (Turkish Association for the Conservation of Nature and Natural Resources): Menekse Sokak 29/4, Kizilay, Ankara; tel (4) 4251944; fax (4) 4179552; Pres: Hasan Asmaz; concerned with general environmental problems as well as the conservation of nature and natural resources.

TURKMENISTAN

Governmental Organizations

State Committee for the Preservation of Nature: Ashkhabad; Chair: T Chorekliyev.

Ministry of Forestry: Ashkhabad; Minister: T Chorekliyev.

Ministry of Irrigation and Water Economy: Ashkhabad; Minister: A Ilamanov.

Non-Governmental Organizations

Turkmenian Society for Conservation of Nature: 1st of May St 62, Ashkhabad; Principal Officer: Prof A K Rustamov.

TUVALU

Governmental Organizations

Ministry for Natural Resources and Home Affairs: Vaiaku, Funafuti; Minister: Tomu Sione.

UGANDA

Governmental Organizations

Ministry of Water, Energy, Minerals and Environment Protection: POB 7270, Kampala; tel (41) 234995; telex 61098; Minister: Henry Kajura.

Ministry of Tourism, Wildlife and Antiquities: Parliament Ave, POB 4241, Kampala; tel (41) 232971; telex 62218; Minister: James W Wapakhabulo.

Uganda Game Department: POB 4, Entebbe; tel (42) 20073; fax (42) 20073; Chief Game Warden: J Moses Okua; conservation and sustainable economic development of wildlife resources; conservation education; captive species management; enforcement of international treaties dealing with wildlife trade.

Uganda National Parks: Plot 31, Kanjokya St, POB 3530, Kampala; tel (41) 256534; fax (41) 258351; Dir: Dr Eric L Edroma; management and preservation of Uganda's seven national parks.

Non-Governmental Organizations

IUCN Uganda: IUCN Uganda Liaison Office, POB 10950, Kampala; tel (41) 233738; fax (41) 242298; Liaison Officer: Dorothy Etoori.

Mbarara University of Science and Technology—Institute of Tropical Forest Conservation: POB 7487, Kampala; fax (41) 245597; Principal Officer: Prof Dr Jonathan Baranga; compilation of inventories of biodiversity; monitoring of mountain gorilla population; environmental impact assessment; study of the relationships between humans and nature parks; management planning; WWF Impenetrable Forest Conservation Project.

Uganda Institute of Ecology: POB 3530, Kampala; Headquarters, Mweya, Queen Elizabeth National Park; fmrly Nuffield Unit of Tropical Animal Ecology (NUTAE); research arm of Uganda National Parks; conducts research which helps Uganda National Parks make informed management decisions about the conservation of wildlife and the environment; collaboration with local and national research institutions and with natural resources departments of Makerere University; postgraduate training in natural resource management and conservation.

Wildlife Clubs of Uganda (WCU): Uganda National Parks Bldg, Plot 31, Kanjokya St, POB 4596, Kampala; tel (41) 256534; Chair: Dr Eric Edroma; Gen Sec: Dr Violet Kajubiri Froelich; conservation of wildlife and the environment; education to raise public awareness of the need to protect and conserve natural resources.

UKRAINE

Governmental Organizations

Ministry of Environmental Protection: 252001 Kiev, vul Kreshchatik 5; tel (044) 228-06-44; fax (044) 229-83-83; Minister: Dr Yuriy I Kostenko; First Deputy Minister: Dr Yuriy G Ruban; implementation of environmental legislation; establishment of environmental standards; financing of scientific and technical programmes; operation of state monitoring system; dissemination of information on the state of the environment; co-operation with institutions from other countries.

Ministry for Protection of the Population against the Consequences of Chernobyl: 254655 Kiev, Lvivska pl 8; tel (044) 212-50-49; fax (044) 212-50-03; Minister: Georgy Olexandrovych Gotovchits; First Deputy Minister: Borys S Prister.

Ministry of Forestry: 252001 Kiev, vul Kreshchatik 5; tel (044) 226-32-53; fax (044) 228-77-94; Minister: Valeriy I Samoplavskiy.

Non-Governmental Organizations

Advisory Committee on Protection of the Sea (ACOPS), CIS Office: c/o Institute of Biology of Southern Seas (IBSS), Ukrainian Academy of Sciences, 335000 Sevastopol, pr Nakhimova 2; tel (0692) 52-48-27; telex 187124; fax (0871) 140-23-43.

Ekostroy Industrial-Ecological Firm: 310168 Kharkov, ul Geroev Truda, d 12, kv 461; tel (0572) 67-30-81; fax (0572) 47-50-95; Principal Officer: A V Ryazanova; Kharkov division of the Socio-Ecological Union; involved in the promotion of industrial waste water treatment and recycling, and investigation of the viability of recycling used tyres as high quality hermetic materials.

Greenpeace Ukraine: 254025 Kiev, vul V Zhytomyrskaya 27, Apt 8; tel (044) 212-29-43; fax (044) 212-29-43; Exec Dir: Olga Savran; Deputy Exec Dir: Andrey Pleskonos.

Greens' Party of Ukraine: 252024 Kiev, Luteranska str 24; tel (044) 293-69-09; fax (044) 293-52-36; Pres: Vitaliy Kononov; Speaker of Political Council: Serghiy Kurykin; Head of Secr: Yuriy Samiylenko; f 1990; independent political organization; founded by members of Zeleny Svit (Green World, Ukrainian Environmental Association); party branches in 18 administrative regions of Ukraine; 3,600 mems.

National Commission on Radiological Protection: 252014 Kiev, vul Bastionna 6-A; tel (044) 296-29-72; fax (044) 267-10-50; Principal Officers: D Grodzinsky, N Omelyanets; development of standards and supervision of programmes for radiological protection; establishment of criteria for the determination of zones of unacceptably high radiation and of tolerable levels of human internal and external exposure to radiation.

Zeleny Svit (Green World—Ukrainian Environmental Association): 254070 Kiev, Kontraktova pl 4; tel (044) 417-02-83; fax (044) 417-43-83; Exec Dir: Serghiy Fedorynchyk; Sec: Liubov Malashenko; monitoring of acid rain and the consequences of the Chernobyl disaster; involved in discussions on the future of the energy industry; planting of trees for river protection; promotion of recycling; protection of nature reserves.

Union for Rescue from Chernobyl: 252207 Kiev, vul Glushkov 32, Apt 115; tel (044) 416-39-50; fax (044) 265-44-63; Pres: Evgen Korbetskiy; Principal Officer: Misha Magal; elaboration of proposals for improvement of the legal and social position of inhabitants of contaminated areas, urgent programme of public radio-ecological education; establishment of an independent public dosimetry control service; co-operation with organizations from other countries in the provision of humanitarian aid.

UNITED ARAB EMIRATES

Governmental Organizations

Ministry of Agriculture and Fisheries: POB 213, Abu Dhabi; tel (2) 662781; Minister: Said Muhammad ar-Raqabani.

Ministry of Electricity and Water: POB 629, Abu Dhabi; tel (2) 335099; telex 46453; fax (2) 213738; Minister: Humaid Nasser al-Owais.

Ministry of Petroleum and Mineral Resources: POB 59, Abu Dhabi; tel (2) 651810; telex 22544; fax (2) 663414; Minister: Yousuf bin Omeir bin Yousuf.

UNITED KINGDOM

Governmental Organizations

Department of the Environment: 2 Marsham St, London SW1P 3EB; tel (71) 276-3000; telex 22221; fax (71) 276-0818; Secretary of State for the Environment: John Selwyn Gummer; Minister for the Environment and Countryside: Timothy Yeo; Minister for Housing and Planning: Sir George Young; Minister for Local Government and Inner Cities: David Curry.

Department of the Environment for Northern Ireland: Stormont, Belfast BT4 3SS; tel (232) 763210; Minister of State responsible for the Department of the Environment: Robert Atkins.

Ministry of Agriculture, Fisheries and Food: Whitehall Place, London SW1A 2HH; tel (71) 270-3000; telex 889351; fax (71) 270-8151; Minister: Gillian Shephard; Minister of State: Michael Jack.

Department of Agriculture for Northern Ireland: Dundonald House, Belfast BT4 3SF; tel (232) 650111; Parliamentary Under-Secretary of State responsible for the Department of Agriculture: The Earl of Arran.

AEA Technology: Bldg 7.12, Harwell, Didcot, Oxon OX11 0RA; tel (235) 433484; telex 83135; fax (235) 432854; government-owned public corporation; formerly Atomic Energy Authority; concerned with environmental safety and technology transfer; consultancy, training and scientific and technical services; publications; national environmental technology centre.

English Nature—Nature Conservancy Council for England: Northminster House, Peterborough, Cambs PE1 1UA; tel (733) 340345; fax (733) 68834; Chair: The Earl of Cranbrook; Chief Exec: Dr D L Langslow; statutory adviser to the government on nature conservation in England.

Environment Select Committee: Committee Office, House of Commons, London SW1A 0AA; tel (71) 219-5462; fax (71) 219-2731; Chair: Robert Jones, MP; Clerk: Steve Priestley; monitoring of policy, expenditure and administration of the Department of the Environment.

Forestry Commission: Department of Forestry, 231 Corstorphine Road, Edinburgh EH12 7AT; tel (31) 334-0303; fax (31) 334-3047; Chair: Sir Raymond Johnstone; Dir-Gen: T R Cutler; advises on and implements government forestry policy, which encourages multi-purpose use of Britain's woodland.

Laboratory of the Government Chemists (LGC): Queen's Rd, Teddington, Middx TW11 0LY; tel (81) 943-7000; telex 9312122476; fax (81) 943-2767; Govt Chemist: Dr Richard Worswick; consultancy and analysis for the private and public sectors; project management from initial surveys to interpretation and recommendations; active in the fields of radiochemistry and microbiology.

National Radiological Protection Board: Chilton, Didcot, Oxon OX11 0RQ; tel (235) 831600; telex 837124; fax (235) 833891; Dir: Prof R H Clarke; Sec: M C O'Riordan; Chair: Sir Richard Southwood; independent body; information, advice, technical services, monitoring and field and laboratory research on radiological hazards.

Scottish Natural Heritage: 12 Hope Terrace, Edinburgh EH9 2AS; tel (31) 447-4784; fax (31) 446-2277; Chair: Magnus Magnusson; Chief Exec: Roger Crofts; government agency responsible to Secretary of State for Scotland; works to conserve and enhance the wildlife, habitats and landscapes that constitute Scotland's natural heritage and to promote awareness of it.

Scottish Office Environment Department: Rm 6/115, New St Andrew's House, Edinburgh EH1 3TG; tel (31) 244-4042; fax (31) 244-4822; Sec: H H Mills.

Non-Governmental Organizations

Advisory Committee on Protection of the Sea (ACOPS): 11 Dartmouth St, London SW1H 9BN; tel (71) 799-3033; fax (71) 799-2933; Pres: Rt Hon Lord Callaghan of Cardiff; Exec Dir: Dr Viktor Sebek; offices in Italy and Ukraine.

Alarm UK—The Alliance Against Roadbuilding: 13 Stockwell Rd, London SW9 9AU.; tel (71) 737-6641; fax (71) 737-4065; Principal Officers: John Stewart, Jonathan Bray; provision of information and assistance to groups opposing road-building schemes.

Ark Environmental Foundation: 8 Bourdon St, London W1X 9HX; tel (71) 409-2638; fax (71) 409-2663; Dir: Roisin Orosz; multi-media environmental education for consumers; marketing of own, non-profit brand of household products.

Association for the Conservation of Energy: 9 Sherlock Mews, London W1M 3RH; tel (71) 935-1495; fax (71) 935-8346; Dir: Andrew Warren; Head of Policy Development: Linda Taylor; lobby group; carries out policy research into energy conservation; membership limited to 24 United Kingdom-based companies with substantial interests in energy conservation equipment and services.

Association for the Protection of Rural Scotland (APRS): Gladstone's Land, 3rd Floor, 483 Lawnmarket, Edinburgh EH1 2NT; tel (31) 225-7012; fax (31) 225-6592; Chair: P Playfair-Hannay of Kingsmuir; Dir (designate): Elizabeth Garland.

Bat Conservation Trust: London Ecology Centre, 45 Shelton St, London WC2H 9HJ; tel (71) 240-0933; fax (71) 240-4525.

Bedfordshire and Cambridgeshire Wildlife Trust: Enterprise House, Maris Lane, Trumpington, Cambridge CB2 2LE; tel (223) 846363; fax (223) 846085; Dir: Mark Rose; Conservation Dir: Richard Woolnough.

Bio-Dynamic Agricultural Association: Goethean House, Woodman Lane, Clent, Stourbridge, West Midlands DY9 9PX; tel (562) 884933; fax (562) 886219; Chair: Alan Brockman; Sec: Ann Parsons; promotion of the bio-dynamic approach to farming, gardening and forestry, based on the principles of Rudolf Steiner.

Born Free Foundation: Coldharbour, Dorking, Surrey RH5 6HA; tel (306) 712091; fax (306) 713350; Dir: Bill Travers; Founders: Bill Travers, Virginia McKenna; co-ordinates and develops effective campaigns to prevent animal suffering, protect endangered species and their habitats and keep wildlife in the wild.

Into the Blue: programme for the rescue, rehabilitation and release of captive dolphins.

Zoo Check: monitors conditions of wildlife in zoos, safari parks and circuses, with the aim of preventing captive animal abuse.

ELEFRIENDS Elephant Protection Group: 162 Boundaries Rd, London SW12; tel (81) 682-1818; fax (81) 767-5459; aims to combat poachers and end trade in ivory.

British Antarctic Survey: High Cross, Madingley Road, Cambridge CB3 0ET; tel (223) 61188; telex 817725; fax (223) 62616; Dir: Dr D J Drewry; Deputy Dir: Dr R B Heywood; government-funded research body responsible for study of the natural environment of Antarctica, S Georgia and the S Sandwich Islands; also involved in meteorological research; played an important role in the discovery of ozone layer depletion.

British Association for Shooting and Conservation (BASC): Marford Mill, Rossett, Wrexham, Clwyd; tel (244) 570881; fax (244) 571678; Dir: John Swift; membership association promoting and defending quarry shooting; research, conservation, education, training, firearms and public affairs departments.

British Ecological Society: 26 Blades Court, Deodar Rd, Putney, London SW15 2NU; tel (81) 871-9797; fax (81) 871-9779; Pres-Elect: I Newton; Exec Sec: R J Hammond; promotes the science of ecology through publications and scientific meetings; encourages research and ecological education in schools.

British Effluent and Water Association (BEWA): 5 Castle St, High Wycombe, Bucks HP13 6RZ; tel (494) 444544; trade association of plant and equipment manufacturers, contractors and designers in the water and effluent treatment industry.

British Herpetological Society: c/o Zoological Society of London, Regent's Park, London NW1 4RY; tel (81) 452-9578; Pres: Prof J L Cloudsley-Thompson; Chair: Dr T Beebee; Sec and Treas: M Green; conservation and captive breeding of amphibians and reptiles.

British Library Environmental Information Service: 25 Southampton Bldgs, London WC2A 1AW; tel (71) 327-7955; telex 266959; fax (71) 323-7954; Head of Section: John Sweeney; Information Officer: Nigel Lees; enquiry and research service in all areas of environmental information; publications; training courses on environmental information.

British Naturalists' Association: 48 Russell Way, Higham Ferrers, Rushden, Northants NN10 8EJ; tel (933) 314672; fax (933) 314672; Gen Sec: J F Pearton; natural history society; promotion and organization of Blake Shield competition for young people; publications.

British Organic Farmers'/Organic Growers' Association (BOF/OGA): 86 Colston St, Bristol BS1 5BB; tel (272) 299800; fax (272) 252504; Dir: Patrick Holden; Chair: Helen Browning.

British Society of Dowsers: Sycamore Barn, Hastingleigh, Ashford, Kent TN25 5HW; tel (233) 750253; Sec: Michael Rust; lectures, short courses, conferences and production of a quarterly journal on divination.

British Steel Save-a-Can: POB 18, Ebbw Vale, Gwent NP3 6YL; tel (495) 350818; fax (495) 350988; works with local authorities to establish recycling schemes for all metal cans; provides assistance, advice and support material.

British Trust for Conservation Volunteers (BTCV): 36 St Mary's St, Wallingford, Oxon OX10 0EU; tel (491) 839766; fax (491) 839646; Chief Exec: Tom Flood; Pres: Lord Norrie.

British Waste Paper Association: Alexander House Business Centre, Station Rd, Aldershot, Hants GU11 1BQ; tel (252) 344454; fax (252) 23417; Nat Sec: G L Jones; Pres: I C Berridge; trade association representing waste paper recovery and processing industry; acts as a centre for national and international exchange of ideas and knowledge.

Butterfly Conservation: POB 222, Dedham, Colchester, Essex CO7 6EY; tel (206) 322342; fax (206) 322739; Pres: Gordon Beningfield; Chair: Dr Harold Hughes; raises awareness, funds research and establishes reserves.

BWCC: 57 King Henry's Walk, London N1 4NH; tel (71) 254-2929; fax (71) 923-4702; Exec Dir: Kate Black; formerly Beauty Without Cruelty International; publicizes the suffering which animals undergo for the sake of fashion; promotes non-cruel alternatives.

Campaign against the Trade in Endangered Species (CATES): 23 Clifton Rd, Henlow, Beds SG16 6BL; tel (462) 814672; Principal Officers: Robert Edward Piller, Carole Mary Piller; also campaigns on issues such as rainforests, habitats, pesticides and architecture.

Campaign for Nuclear Disarmament (CND): 162 Holloway Rd, London N7 8DQ; tel (71) 700-2393; fax (71) 700-2357; Chair: Marjorie Thompson; Gen Sec: Gary Lefley; campaigns against the use of nuclear power to process weapons-grade plutonium and the development, production, transport and deployment of all weapons of mass destruction, on environmental, political, economic and moral grounds.

Campaign for the Protection of Rural Wales/Ymgyrch Diogelu Cymru Wledig: Tŷ Gwyn, 31 High St, Welshpool, Powys SY21 7JP; tel (938) 552525; fax (938) 552741; Dir: Dr Neil Caldwell; works to protect the coast and countryside of Wales and encourages sustainable development.

Cat Survival Trust: The Centre, Codicote Rd, Welwyn, Herts AL6 9TU; tel (438) 716873; fax (483) 717535; Hon Chair: P Smith; Hon Dir: Dr T Moore; Hon Sec: J E Moore; charity promoting the preservation of all species of wild cat, ideally in their natural habitats; operates breeding programmes and assists research into conservation of cats.

The Earth: Hon Sec: A Johnson; general environmental branch; promotes collaboration between environmental societies world-wide and maintains a database of such societies; seeks to further the right of humans to live in an unpolluted environment.

Mission Rainforest Foundation: Hon Sec: R Cheeseman; administers rainforests and sites of scientific interest for The Earth, principally in South America; collaborates with Cat Survival Trust on Mission Rainforest Project.

Centre for Alternative Technology: Machynlleth, Powys SY20 9AZ; tel (654) 702400; fax (654) 702782; Dir: Roger Kelly; education and display centre demonstrating renewable energy, organic agriculture and environmental building; provides residential courses, information and consultancy services.

Centre for Environmental Initiatives (CEI): 24 Rosebery Rd, Cheam, Sutton, Surrey SM1 2BW; tel (81) 642-3030; Dir: Vera Elliott; formerly Centre for Environmental Information; charitable trust; acts as adviser to local government, produces reports and feasibility studies and encourages public environmental action.

Centre for Environmental Interpretation: Manchester Metropolitan University, St Augustines, Lower Chatham St, Manchester M15 6BY; tel (61) 247-1067; fax (61) 236-7383; Dirs: Graham C Barrow, Gillian Binks; promotion and development of environmental interpretation in the fields of recreation, tourism and heritage; provides advice, information, consultancy and specialist training courses in all aspects of environmental interpretation.

Scottish Office: University of Edinburgh, 20 Chambers St, Edinburgh EH1 1JZ; tel (31) 650-8017; fax (31) 650-6517; Contact: James Carter.

Chemical and Oil Recycling Association: 9 Parc Luned, Kinmel Bay, Rhyl, Clwyd LL18 5JG; tel (745) 332427; fax (745) 343300; Sec: J Looker.

Christian Ecology Link: 20 Carlton Rd, Harrogate, N Yorks HG2 8DD; tel (423) 871616; Nat Co-ordinators: George E Dent, Rev Gary A Lawson, Dr Judith M Allinson; seeks to involve Christians and churches in concern for the environment; conferences, publications and membership network.

Civic Trust: 17 Carlton House Terrace, London SW1Y 5AW; tel (71) 930-0914; fax (71) 321-0180; Patron: HRH The Prince of Wales; Dir: Martin Bradshaw; charitable federation of nearly 1,000 local groups; aims to improve living and working environments through promotion of high standards of planning and conservation; administers Local Project Grant Fund on behalf of Dept of the Environment, grants annual awards for good development, operates Regeneration Unit and Campaign for Liveable Places; regional offices in Bristol, Liverpool and Manchester.

Cleveland Wildlife Trust: Bellamy House, Brighouse Business Village, Riverside Park, Middlesbrough, Cleveland TS2 1RT; tel (642) 253716; fax (642) 251072; Dir: Joe Cole; Sr Conservation Officer: Jeremy Garside; Sr Education Officer: Kathryn Hardy; voluntary conservation organization; member of Royal Society for Nature Conservation Wildlife Trusts Partnership.

Combined Heat and Power Association: 3rd Floor, 35/37 Grosvenor Gardens, London SW1W 0BS; tel (71) 828-4077; fax (71) 828-0310; Dir: David Green; promotion of combined heat and power and community heating.

Confederation of British Industry (CBI), Environment Management Unit: Centre Point, 103 New Oxford St, London WC1A 1DU; tel (71) 379-7400; fax (71) 240-0988; provides information to help companies improve their environmental performance; CBI Environment Business Forum promotes voluntary action on environment.

Conservation Foundation: Lowther Lodge, 1 Kensington Gore, London SW7 2AR; tel (71) 823-8842; fax (71) 823-8791; Principal Officers: Prof David Bellamy, David Shreeve; creation and management of a variety of environmental initiatives in the United Kingdom and internationally, often with commercial support.

Consumers against Nuclear Energy (CANE): POB 697, London NW1 8YQ; campaigns for phasing out of nuclear power, encourages energy conservation and development of renewable energy sources.

Council for Education in World Citizenship: Seymour Mews House, Seymour Mews, London W1H 9PE; tel (71) 935-1752; fax (71) 935-5548; Dir: Patricia Rogers; Office Manager: Rolant Moelwyn-Hughes; raises young people's awareness of international issues through publications, events and information; membership open to individuals and institutions.

Council for Environmental Education (CEE): University of Reading, London Rd, Reading, Berks RG1 5AQ; tel (734) 756061; fax (734) 756264; Dir: Dr Ewan McLeish; Head of Information: Christine Midgley; co-ordinating and policy development body for England, Wales and Northern Ireland; represents and promotes members' interests; information, advice, research and publications.

Council for National Parks: 246 Lavender Hill, London SW11 1LJ; tel (71) 924-4077; fax (71) 924-5761; Pres: Chris Bonington; Dir: Amanda Nobbs; independent charity; campaigns to protect the 11 national parks of England and Wales.

Council for Posterity: 20 Heber Rd, London NW2 6AA; tel (81) 208-2853; fax (81) 452-6434; Gen Sec: Nicholas Albery; voluntary organization to protect and represent interests of future generations; launched Natural Step UK, national plan for long-term sustainability.

Council for the Protection of Rural England (CPRE): 25 Buckingham Palace Rd, London SW1W 0PP; tel (71) 976-6433; fax (71) 976-6373; Pres: Jonathan Dimbleby; Dir: Fiona Reynolds.

Countryside Commission: John Dower House, Crescent Place, Cheltenham, Glos GL50 3LX; tel (242) 521381; fax (242) 584270; Chair: Sir John Johnson; Dir-Gen: Michael Dower; adviser to government on countryside issues; works to conserve English countryside, designates national parks, areas of outstanding natural beauty and 'heritage coasts'.

CSV Environment: 17 Midland Rd, St Philips, Bristol BS2 0JT; tel (272) 411114; fax (272) 412990; Principal Officer: Steve Smith; encourages public participation in the preservation of the environment.

DocTer International UK Ltd—International Institute for Environmental Studies: Hyde Park House, 5 Manfred Rd, London SW15 2RS; tel (81) 877-1080; fax (81) 874-2150; Principal Officers: A Cutrera, L Mottola, Colinette Craig; compilation of environmental documentation; office in Italy.

Earthscan Publications Ltd: 120 Pentonville Rd, London N1 9JN; tel (71) 278-0433; fax (71) 278-1142; Publishing Dir: J Sinclair Wilson; Sales Dir: K Griffin; general, academic and reference books on environment and development issues.

Earthwatch Europe: Belsyre Court, 57 Woodstock Rd, Oxford OX2 6HU; tel (865) 311600; fax (865) 311383; Chair: Sir Crispin Tickell; Exec Dir: Brian Walker; Deputy Dir: Andrew W Mitchell; Chair, Science Advisers Group: Sir Richard Southwood; European branch of organization based in the USA.

Ecology Building Society: 18 Station Rd, Cross Hills, Keighley, W Yorks BD20 7EH; tel (535) 635933; fax (535) 636166; Chief Exec: Bob Lowman; Chair: Gus Smith; provides means of finance for the purchase of ecologically sound properties only; opposes the wasteful use of land and resources and aims to regenerate rural areas and inner cities.

Ecoropa: Henbant Fach, Llanbedr, Crickhowell, Powys NP8 1TA; tel (873) 810758; fax (873) 810758; Dirs: Edward Goldsmith, Fern Morgan-Grenville; member of European Group for Ecological Action; aims to inform world leaders about environmental problems and disseminate information to the public, enabling them to pressurize governments to change their policies.

Energy Action Scotland: 21 W Nile St, Glasgow G1 2PJ; tel (41) 226-3064; fax (41) 221-2788; Dir: Ann Marno (acting); Energy Audit Trainer: Richard McCarthy; Development Officer: Andrew Fairbairn; promotes energy efficiency and conservation; provides training in energy awareness and home energy auditing.

Environment Council: 21 Elizabeth St, London SW1W 9RP; tel (71) 824-8411; fax (71) 730-9941; Chair: Dr Malcolm Aickin; CEO: Steve Robinson; independent charitable organization; aims to find solutions to environmental problems, to help the public understand environmental issues and to act effectively to protect the environment; operates Business and Environment Programme and Information Programme.

Environment Foundation: Ibex House, 15 Minories, London EC3N 1NJ; tel (71) 481-3232; telex 8813411; fax (71) 480-6137; Chair of Trustees: M D Martin; Exec Dir: T C M O'Donovan; charitable organization.

Environmental Forum: 12A Ennis Rd, Finsbury Park, London N4 3HD; tel (71) 263-8505; Co-ordinating Dir: Criton Tomazos; aims to increase public awareness of environmental issues; promotes related artistic, cultural and social events and organizes discussions.

Environmental Investigation Agency: 2 Pear Tree Court, London EC1R 0DS; tel (71) 490-7040; fax (71) 490-0436; Chair: Allan Thornton; Exec Dir: Dave Currey; investigation of environmental abuses world-wide and campaigning for environmental protection legislation.

Fauna and Flora Preservation Society: 1 Kensington Gore, London SW7 2AR; tel (71) 823-8899; fax (71) 823-9690; Chair: David Jones; Vice-Chair: Dr David Chivers; charitable membership organization working to protect endangered plant and animal species world-wide.

Field Studies Council (FSC): Central Services, Preston Montford, Montford Bridge, Shrewsbury, Salop SY4 1HW; tel (743) 850674; fax (743) 850178; Principal Officers: Chris Bayliss, Tony Thomas; environmental courses, publications, consultancy and research.

Field Studies Council Research Centre: Fort Popton, Angle, Pembroke, Dyfed SA71 5AD; tel (646) 641404; fax (646) 641425; Principal Officers: Dr C I Morgan, D Levell, G Hobbs; environmental research and monitoring, with particular emphasis on the effects of offshore petroleum exploitation; provides specialist environmental training services.

Fish Conservation Centre: Easter Cringate, Stirling FK7 9QX; tel (786) 451312; Principal Officer: Dr Peter S Maitland; research on fish conservation and ecology, and consultancy work in freshwater ecology and impact assessment.

Forth Naturalist and Historian: University of Stirling, Stirling FK9 4LA; tel (259) 215091; telex 777557; fax (786) 63000; Sec: L Corbett; charitable body formed by collaboration between the University of Stirling and Central Regional Council; aims to promote the environment and heritage of Central Scotland through publications and annual symposia.

Friends of the Earth (FoE) England, Wales and Northern Ireland: 26-28 Underwood St, London N1 7JQ; tel (71) 490-1555; fax (71) 490-0881; Exec Dir: Charles Secrett; Campaigns Dir: Andrew Lees; national office for 250 local groups.

Membership and Publications Section: 56-58 Alma St, Luton, Beds LU1 2Y2; tel (582) 482297; fax (582) 401106.

Friends of the Earth Birmingham Ltd: 54-57 Allison St, Digbeth, Birmingham B5 5TH; tel (21) 632-6909; Campaigns Co-ordinator: Chris Crean; Information Officer: Dave Clare.

Friends of the Earth Cymru: 33 The Balcony, Castle Arcade, Cardiff, S Glamorgan CF1 2BY; tel (222) 229577; fax (222) 228775; Co-ordinator: Margaret Minhinnick; Information Officer: Robert Minhinnick.

Friends of the Earth Northern Ireland: Bradbury Bldgs, 56 Bradbury Pl, Belfast BT7 1RU; tel (232) 311555; fax (232) 313322; Local Group Devt Officer: Teresa O'Hara.

Friends of the Earth Scotland: 1 Bonnington Mill Business Centre, 70-72 Newhaven Rd, Edinburgh EH6 5QG; tel (31) 554-9977; fax (31) 554-8656; Dir: Kevin Dunion; Campaign Organizer: Johnny Woods; Information Officer: Louie Fooks.

Game Conservancy Trust: Burgate Manor, Fordingbridge, Hants SP6 1EF; tel (425) 652381; fax (425) 655848; Dir-Gen: G R Potts; Dir for Research: S C Tapper; researches methods of conservation of game species and associated wildlife; encourages less intensive rural land use and offers management advice to landowners and farmers.

Global Commons Institute: 60 Lordship Park, London N16 5OA; tel (81) 800-6075; fax (81) 800-6075; Principal Officers: Aubrey Meyer, Anandi Sharan; socio-economic analysis of global climate change; advice to UN intergovernmental negotiations.

Grass Roots Conservation Corps: c/o 24 Deramore Park, Belfast BT9 5JU; tel (232) 666108; Principal Officers: Dr A Hardcastle, D Brown; practical conservation activities such as tree planting and dry stone wall building.

Green Alliance: 49 Wellington St, London WC2E 7BN; tel (71) 836-0341; fax (71) 240-9205; Dir: Julie Hill; Chair: Nigel Haigh; aims to increase the prominence of the environment on the agendas of the major policy-making institutions in the United Kingdom.

Green Party: 10 Station Parade, Balham High Rd, London SW12 9AZ; tel (81) 673-0045; fax (81) 675-4434; Chair: John Norris; formerly the Ecology Party; contests local, national and European Parliament elections, campaigning on environmental, peace and economic issues; aims to increase public awareness of green politics and the need for a sustainable economy.

GreenNet: 23 Bevenden St, London N1 6BH; tel (71) 608-3040; fax (71) 253-0801; global computer communications network for environment, peace, human rights and development groups; member of Association for Progressive Communications (APC).

Greenpeace UK: Canonbury Villas, London N1 2PN; tel (71) 354-5100; fax (71) 696-0012; Exec Dir: Peter Melchett; national office of Greenpeace International; limited liability company.

Greenpeace Environmental Trust: charitable trust.

Greenpeace Communications Ltd: 5 Baker's Row, London EC1R 3DB; tel (71) 833-0600; media contact office; video library.

Habitat Scotland: Hazelmount, Heron Place, Portree, Isle of Skye IV51 9EU; tel (478) 612898; fax (478) 613254; Dir: Graeme Robertson; independent research charity established to investigate conservation issues arising from varied use of land and sea in the Scottish highlands and islands.

Henry Doubleday Research Association (HDRA): Ryton Organic Gardens, Ryton-on-Dunsmore, Coventry CV8 3LG; tel (203) 303517; fax (203) 639229; Chief Exec: Alan Gear; Exec Dir: Jackie Gear; educational charity; researches, demonstrates and promotes environmentally friendly growing techniques.

Hertfordshire University, Division of Environmental Sciences: College Lane, Hatfield, Herts AL10 9AB; tel (707) 284516; fax (707) 284514; Principal Officer: Prof Keith S Bardon; undergraduate and postgraduate interdisciplinary courses, postgraduate research, contract research and consultancy, and short courses.

Industry Nature Conservation Association (INCA): 1 Belasis Court, Belasis Hall Technical Park, Billingham, Cleveland TS23 4AZ; tel (642) 370319; Contact: D H Robinson; association of industrial companies, local authorities and nature conservation organizations; encourages industry to consider the environment and undertake habitat improvement schemes; offers advice on practical conservation schemes.

Institute of Biology: 20-22 Queensbury Place, London SW7 2DZ; tel (71) 581-8333; fax (71) 823-9409; Pres (1992–94): Prof C R W Spedding; professional body; activities include formulation and submission to government and other bodies of policy documents; publications and symposia.

Institute of Chartered Foresters (ICF): 7A St Colme St, Edinburgh EH3 6AA; tel (31) 225-2705; fax (31) 220-6128; Pres: D W G Taylor; Sec and Treas: M W Dick; safeguards the public interest in forestry matters, advises the government on matters of policy and maintains the standards regulating entry into the profession.

Institute of Ecology and Environmental Management: 36 Kingfisher Court, Hambridge Rd, Newbury, Berks RG14 5SJ; tel (635) 37715; fax (635) 550230; Exec Dir: Paul Goriup; Development Officer: Sue Everett; professional membership organization; aims to raise the profile of the profession of ecology and environmental management, maintain and enhance professional standards and promote an ethic of care for the environment among members and their clients and employers.

Institute of Energy: 18 Devonshire St, London W1N 2AU; tel (71) 580-7124; fax (71) 580-4424; Sec: J E H Leach; professional body for specialists actively engaged in production, conversion and use of all forms of energy; encourages prudent use of resources and protection of the environment.

Institute for Social Inventions: 20 Heber Rd, London NW2 6AA; tel (81) 208-2853; fax (81) 452-6434; Chair: Nicholas Albery; international body aiming to find original solutions to social problems.

Natural Death Centre: Co-ordinator: Josefine Speyer; Dirs: Nicholas Albery, Christianne Heal; concerns include promotion of environmentally friendly burial in nature reserves or on the deceased's own land, using cardboard or recycled coffins.

Institute of Wastes Management (IWM): 9 Saxon Court, St Peter's Gardens, Northampton NN1 1SX; tel (604) 20426; fax (604) 21339; Sec-Gen: R A Bispham; professional association.

Institution of Water and Environmental Management (IWEM): 15 John St, London WC1N 2EB; tel (71) 831-3110; fax (71) 405-4967; Exec Dir and Sec: Howard R Evans; Pres: Ken Clarke; multidisciplinary professional and examining body for engineers, scientists and others engaged in water and environmental management; conferences and seminars; publications on all aspects of water and wastewater treatment.

Intermediate Technology (it): Myson House, Railway Terrace, Rugby, Warwicks CV21 3HT; tel (788) 560631; telex 317466; fax (788) 540270; Chief Exec: Frank Almond; aims to enable people with little or no income to develop productive technologies and methods which contribute to the long-term development of their communities.

International Centre for Conservation Education (ICCE): Greenfield House, Guiting Power, Cheltenham, Glos GL54 5TZ; tel (451) 850777; telex 94081169; fax (451) 850705; Dir: Mark Boulton; Deputy Dir: Phil Steele; environmental education training, consultancy, design and production of resource materials.

International Fur Trade Federation: 2 The Quintet, Churchfield Rd, Walton-on-Thames, Surrey KT12 2TZ; tel (932) 232866; telex 917513; fax (932) 232656; Exec Officer: Juliette Bailey; promotion and protection of trade in furs; collaboration with conservation organizations to ensure future supply of fur-bearing animals; investigation of introduction of humane trapping methods.

International Glaciological Society: Lensfield Rd, Cambridge CB2 1ER; tel (223) 355974; fax (223) 336543; Pres: B Wold; Sec-Gen: C S L Ommanney; encourages interest in and research on global scientific and technical problems related to snow and ice.

International Institute for Environment and Development (IIED): 3 Endsleigh Street, London WC1H 0DD; tel (71) 388-2117; fax (71) 388-2826; Chair: Sir Crispin Tickell; Exec Dir: Richard Sandbrook; aims to ensure public and political acceptance of the integral relationship between environment and development and to translate it into practical actions; acts to implement findings of the 1992 Earth Summit (UNCED) in Brazil.

Drylands Programme: Dir: Camilla Toulmin; Sr Research Associate: Charles Lane; aims to achieve more effective and equitable management of natural resources in arid and semi-arid regions of Africa; collaborative research programmes, training workshops, networking, advocacy and publications.

International Planned Parenthood Federation (IPPF): Regent's College, Inner Circle, Regent's Park, London NW1 4NS; tel (71) 486-0741; telex 919573; fax (71) 487-7950; Sec-Gen: Dr Halfdan Mahler; Pres: Dr Fred Sai; provision of family planning services and education, and limited involvement in environmental projects through voluntary national family planning associations.

International Tree Foundation: Sandy Lane, Crawley Down, W Sussex RH10 4HS; tel (342) 712536; fax (342) 718282; Chair: J Caunce; Deputy Chair: Lt-Col E L K A Carr; formerly Men of the Trees; plants trees world-wide to increase land fertility; particularly active in desert areas and tropical rainforests.

Irish Sea Forum: Oceanography Laboratories, University of Liverpool, POB 147, Liverpool L69 3BX; tel (51) 794-4089; fax (51) 794-4099; Dir and Chair: Dr D F Shaw; Administrator: H Davies; formerly Irish Sea Study Group; organizes seminars on issues related to the environmental health of the Irish Sea and the sustainable use of its resources.

IUCN United Kingdom: c/o World Conservation Monitoring Centre, 219 Huntingdon Rd, Cambridge CB3 0DL; tel (223) 277314; fax (223) 277136.

IUCN Plant Conservation Office: Botanic Gardens Conservation Secretariat, Descanso House, 199 Kew Rd, Richmond, Surrey TW9 3BW; tel (81) 948-4363; fax (81) 940-4547.

Joint Committee for the Conservation of British Invertebrates (JCCBI): c/o Royal Entomological Society, 41 Queen's Gate, London SW7 5HR; tel (71) 938-8905; fax (71) 938-8937; Principal Officers: Dr P E S Whalley, S J Brooks.

King's College London, Foundation for International Environmental Law and Development (FIELD): Manresa Rd, London SW3 6LX; tel (71) 352-8123; fax (71) 351-6435; Principal Officers: James Cameron, Philippe Sands; formerly Centre for International Environmental Law (CIEL); research, publishing, teaching and legal assistance.

Landlife: The Old Police Station, Lark Lane, Liverpool L17 8UU; tel (51) 728-7011; fax (51) 728-8413; Dir: Grant Luscombe; formerly Rural Preservation Association; works to bring city dwellers, particularly children, into contact with wildlife.

Landscape Institute: 6-7 Barnard Mews, London SW11 1QU; tel (71) 738-9166; Pres: Prof M J Ellison; Registrar: P R Broadbent; professional body for landscape architects, managers and scientists.

Living Water: 15 Buccleuch Place, Edinburgh EH8 9LN; tel (31) 331-2406; telex 727442; fax (31) 331-2406; develops ecological water and effluent treatment systems; Living Water Charitable Trust (LWCT) supports projects to protect and improve the environment and create learning opportunities, particularly in developing countries.

London Ecology Centre: 45 Shelton St, London WC2H 9HJ; tel (71) 379-4324; Information Officer: Judy Chaitram; promotes awareness of environmental issues in London.

London Energy Group: c/o Lockhart Ball Associates, 934 Garratt Lane, London SW1 0ND; Chair: Hugh Lockhart Ball; Sec: Laurance L Gill; promotes the efficient use of energy in the built environment.

London Green Belt Council: 13 Oakleigh Park Ave, Chislehurst, Kent BR7 5PB; tel (81) 467-5346; ensures preservation of areas of open land surrounding Greater London; consultants to Department of the Environment; involved in planning applications, inquiries and local plan consultations.

London Greenpeace: 5 Caledonian Rd, London N1 9DX; tel (71) 837-7557; independent collective organization; not affiliated to Greenpeace International; public education; meetings; campaigns against environmentally unsound multinational corporations; networking with other NGOs.

London School of Economics and Political Science, Centre for Environmental Law and Policy: Houghton St, London WC2A 2AE; tel (71) 955-7247; fax (71) 955-7366; Dir: Dr G Plant; research and consultancy; current programmes cover the sustainable city, environmental protection and the law of war, the marine environment and environmental institutions.

London School of Hygiene and Tropical Medicine, Centre for Population Studies: 99 Gower St, London WC1E 6AZ; tel (71) 388-3071; fax (71) 388-3076; Head: John Simons; Staff: Prof John Cleland, Basia Zaba; research and advanced teaching on measurement of population change, effective family planning provision and interactions between population and environment.

London Wildlife Trust: 80 York Way, London N1 9AG; tel (71) 278-6612; fax (71) 837-8060; Dir: Graham Turnbull; Head of Conservation: Ralph Gaines; protection and improvement of London's green spaces for the benefit of people and wildlife, through management of 52 wildlife sites; also organizes over 400 free events annually.

Lothian and Edinburgh Environmental Partnership: Bonnington Mill, 72 Newhaven Rd, Edinburgh EH6 5QJ; tel (31) 555-4010; fax (31) 555-2768; Chair: Ian Perry; Dir: Mike Thornton; supports initiatives to supply or promote green products and services and environmental technology; particularly active in the areas of recycling, energy conservation and transport.

Mammal Conservation Trust: Kentshill Farm House, Landermere, Thorpe-le-Soken, Essex CO16 0NN; tel (255) 861976; Principal Officers: Timothy Dansie, Dr Oliver Dansie, Dr Josephine Pemberton; mainly concerned with wild and park fallow deer.

Mammal Society: Zoology Department, University of Bristol, Bristol BS8 1UG; tel (272) 272300; fax (272) 257374; Chair: Dr P A Morris; Pres: W N Bonner; Hon Sec: Dr J Gurnell; promotes study and conservation of mammals; publications and meetings.

Manchester Metropolitan University, Atmospheric Research and Information Centre: Department of Environmental and Geographical Studies, Chester St, Manchester M1 5GD; tel (61) 247-1592; fax (61) 247-6332; Principal Officers: Dr Longhurst, Sue Hare, Paula Owen; specializes in air pollution, acid rain, urban air quality and global climate changes; also acts as an environmental consultancy and conducts research into the chemistry of acid deposition.

Marine Biological Association of the United Kingdom: The Laboratory, Citadel Hill, Plymouth, Devon PL1 2PB; tel (752) 222772; fax (752) 226865; Pres: Sir Crispin Tickell; Dir and Sec: Dr Michael Whitfield; research, publications, scientific meetings and an educational programme through Plymouth Aquarium.

Marine Conservation Society (MCS): 9 Gloucester Rd, Ross-on-Wye, Hereford HR9 5BU; tel (989) 66017; fax (989) 67815; Pres: HRH The Prince of Wales; Vice-Pres: Prof David Bellamy; protection of the marine environment through promotion of research, information dissemination and lobbying.

Marine Forum for Environmental Issues (MF): c/o Department of Zoology, Natural History Museum, Cromwell Rd, London SW7 5BD; tel (71) 938-9114; fax (71) 938-9158; Administrator: S-A Brodie Cooper; 'umbrella' organization aiming to promote effective communication between users of the marine environment and encourage constructive debate on topical issues; ensures that members' views are publicized and conveyed to the government.

Media Natura: 21 Tower St, London WC2H 9NS; tel (71) 240-4936; fax (71) 240-2291; Dir: Michael Keating; charitable media company providing communication services to environmental groups, charities and NGOs.

National Advisory Unit: 35 Fountain St, Piccadilly, Manchester M2 2AF; tel (61) 236-5581; fax (61) 237-5981; involved in accessible community transport research, evaluation of accessibility investments and the impact of law on service development.

National Association for Urban Studies (NAUS): Lewis Cohen Urban Studies Centre, University of Brighton, 68 Grand Parade, Brighton BN2 2JY; tel (273) 673416; fax (273) 673416; Chair: Graham Ranger; Deputy Chair: Pat Mortimer; promotes urban environmental education at all levels, to encourage public participation in working towards sustainability in cities.

National Association of Waste Disposal Contractors (NAWDC): Mountbarrow House, 6-20 Elizabeth St, London SW1W 9RB; tel (71) 824-8882; fax (71) 824-8753; Exec Dir: B E Dennis; Policy Dir: S Webb; Operations Dir: S Taylor; industrial representative body; training and campaigns in the United Kingdom and Europe for higher standards and better enforcement of waste disposal legislation.

National Council for the Conservation of Plants and Gardens (NCCPG): Wisley Garden, Woking, Surrey GU23 6QB; tel (483) 2114665; fax (483) 211750; Gen Sec: R A W Lowe; Horticultural Adviser: G A Pattison; conservation of British garden plants using the skills of amateurs and professionals.

National Council for Voluntary Organisations (NCVO): Regent's Wharf, 8 All Saints St, London N1 9RL; tel (71) 713-6161; fax (71) 713-6300; Dir: Judy Weleminsky; Environment Support Team provides information and assistance for environmental voluntary action by working with networks, producing publications and operating a referrals service.

National Federation of City Farms (NFCF): AMF House, 93 Whitby Rd, Brislington, Bristol BS4 3QF; tel (272) 719109; fax (272) 721838; Dir: Ian Egginton-Metters; supports and advises city farms and community gardens; provides technical, managerial, financial and legal assistance and training programmes, and prepares feasibility studies.

National Rivers Authority (NRA): Eastbury House, 30-34 Albert Embankment, London SE1 7TL; tel (71) 820-0101; fax (71) 820-1603; Chief Exec: Ed Gallagher; Chief Scientist and Dir of Water Quality: Dr Jan Pentreath; independent body; statutory responsibilities in England and Wales include water resource management, pollution control, conservation and the coastal marine environment.

Sub-Office: Rivers House, Waterside Drive, Aztec West, Almondsbury, Bristol BS12 4UD; tel (454) 624400; fax (454) 624409.

National Society for Clean Air and Environmental Protection: 136 North St, Brighton BN1 1RG; tel (273) 326313; fax (273) 735802; Sec-Gen: Dr Tom Crossett.

National Trust (National Trust for Places of Historic Interest or Natural Beauty): 36 Queen Anne's Gate, London SW1H 9AS; tel (71) 222-9251; fax (71) 222-5097; Dir-Gen: Angus Stirling; Chair: Lord Chorley; charitable membership organization; acquires and preserves historic buildings and areas of natural beauty in England, Wales and Northern Ireland.

Northern Ireland Regional Office: Rowallane House, Saintfield, Ballynahinch, Co Down BT24 7LH; tel (238) 510721; fax (238) 511242.

National Trust for Scotland: 5 Charlotte Sq, Edinburgh EH2 4DU; tel (31) 226-5922; fax (31) 243-9501; Dir: Douglas Dow; Pres: The Marquess of Bute; Chair: R C Tyrrell; promotes the preservation of places of historic and architectural interest or natural beauty.

Natural Environment Research Council (NERC): Polaris House, North Star Ave, Swindon SN2 1EU; tel (793) 411500; telex 444293; fax (793) 411501; Chief Exec (acting): Dr Eileen Buttle; Chair: Prof J Knill; Dir: Prof C Arme; research and consultancy in environmental aspects of the terrestrial, marine, freshwater, atmospheric and biological sciences.

Institute of Oceanographic Sciences—Deacon Laboratory (IOSDL): Brook Rd, Wormley, Godalming, Surrey GU8 5UB; tel (428) 684141; telex 858833; fax (428) 683066; Dir: Dr C P Summerhayes; Head of Information: P Simpson; formerly National Institute of Oceanography (NIO); national centre for deep sea oceanography; physical, chemical, geological, geophysical and biological research, consultancy and information services; houses national oceanographic library.

Neighbourhood Energy Action: St Andrew's House, 90-92 Pilgrim St, Newcastle upon Tyne NE1 6SG; tel (91) 261-5677; fax (91) 261-6496; Dir: Andrea Cook; promotes efficient use of energy to relieve poverty and protect health; disseminates information to government and the private sector.

Network for Alternative Technology and Technology Assessment (NATTA): c/o Energy and Environment Research Unit, Faculty of Technology, The Open University, Walton Hall, Milton Keynes MK7 6AA; tel (908) 654638; fax (908) 653744; Principal Officers: Dr David Elliott, Tam Dougan; promotes renewable energy technologies and offers information, advice and a careers service.

New Consumer: 52 Elswick Rd, Newcastle upon Tyne NE4 6JH; tel (91) 272-1148; fax (91) 272-1615; Dir: Richard Adams; public interest research into the social, ethical and environmental policies of major business corporations; publications and research on corporate social and environmental responsibility.

New Economics Foundation: 89/94 Wentworth St, London E1 7SA; tel (71) 377-5696; disseminates information to members, supports research into viable alternatives to unrestrained growth and promotes the formulation of alternative economic indicators and the conservation of energy.

Northern Ireland Environment Link: Armagh House, Ormeau Ave, Belfast BT2 8HB; tel (232) 234393; fax (232) 234023; Dir: Dr Susan Christie; Chair: Dr Maurna Crozier; forum for voluntary conservation organizations, serving as a link between the voluntary sector and government and providing environmental information.

Odonata Specialist Group of the IUCN Species Survival Commission: c/o The Farm House, Swavesey, Cambs CB4 5RA; tel (954) 30233; Chair: Dr Norman W Moore; promotes conservation of dragonflies world-wide and advises the IUCN on conservation research priorities.

Open Spaces Society: 25A Bell St, Henley-on-Thames, Oxon RG9 2AB; tel (491) 573535; Principal Officers: Kate Ashbrook, Karin Groeneveld; protects common land, public paths, etc, and advises local authorities on the management and preservation of open spaces acquired by gift or purchase.

Open University, Environmental Education and Training Programme: Walton Hall, Milton Keynes MK7 6AA; tel (908) 652029; fax (908) 655159; Liaison Officers: Mark Yoxon, Christine Blackmore; co-ordination of environmental education and training across the Open University and provision of new materials.

Oxfam UK/I: Oxfam House, 274 Banbury Rd, Summertown, Oxford OX2 7DZ; tel (865) 312116; telex 83610; fax (865) 312600; Dir: David Bryer; Overseas Dir: Stuart Wallis; involved in sustainable development and emergency relief projects in approximately 70 countries.

Panos London: 9 White Lion St, London N1 9PD; tel (71) 278-1111; telex 9419293; fax (71) 278-0345; Exec Dir: Simon Burne; independent information organization promoting socially, environmentally and economically sustainable global development; research into key global issues; numerous publications; photographic library, Panos Pictures; offices in France and the USA.

Peak Park Joint Planning Board: Aldern House, Baslow Rd, Bakewell, Derbyshire DE45 1AE; tel (629) 814321; fax (629) 812659; National Park Officer: C J Harrison; Dir of Admin and Sec: K M Francis; protection of the natural beauty of the Peak national park.

Pedestrians' Association: 126 Aldersgate St, London EC1A 4JQ; tel (71) 490-0750; Pres: Terence Bendixson; Chair: Faith Lawson; Sec: Felicity Rea; encourages provision of safe and convenient footways, promotes road safety and provides advice and information for pedestrians.

Permaculture Association (UK): POB 1, Buckfastleigh, Devon TQ11 0LH; tel (3643) 333; fax (3643) 333; Co-ordinator: George Sobol; educational charity concerned with Permaculture design system for sustainable land use and regional and community development; provides training courses; member of international network.

Plantlife: Natural History Museum, Cromwell Rd, London SW7 5BD; tel (71) 938-9111; fax (71) 938-9112; Dir: Dr Jane Smart; Chair (acting) and Vice-Pres: Sir William Wilkinson; Pres: Prof David Bellamy; conservation of wild plants and their habitats.

Policy Studies Institute: 100 Park Village East, London NW1 3SR; tel (71) 387-2171; fax (71) 388-0914; Dir: Pamela Meadows; Sec: Eileen Reid; Head of External Relations: Nicholas Evans; independent research organization aiming to inform public policy; undertakes studies of economic, industrial and social policy and of matters relating to the environment.

Population Concern: 231 Tottenham Court Rd, London W1P 9AE; tel (71) 631-1546; telex 919573; fax (71) 436-2143; Dir: Rajamani Rowley; Information Officer: Fiona Barr; seeks to raise awareness of the related issues of population, development and the environment through UK Advocacy Programme; operates family planning programmes in 18 developing countries.

Railway Development Society (RDS): 15 Clapham Rd, Lowestoft, Suffolk NR32 1RQ; tel (502) 581721; Chair: Steve Wilkinson; Gen Sec: Trevor Garrod; campaigns for the retention, expansion and greater use of rail transport for passengers and freight.

Rainbow Resource Centre: 180-182 Mansfield Rd, Nottingham NG1 3HW; tel (602) 585666; Principal Officers: Patrick Smith, Tracy Lean; provision of resources for local groups campaigning for human and animal rights, environmental issues, peace and global co-operation.

Ramblers' Association: 1-5 Wandsworth Rd, London SW8 2XX; tel (71) 582-6878; fax (71) 587-3799; Dir: Alan Mattingly; promotes walking for pleasure, protects rights of way, campaigns for access to open country and defends the beauty of the countryside.

Reforesting Scotland: Duartbeg, Scourie, Sutherland IV27 4TJ; tel (971) 502406; fax (971) 502442; Dir: Donald McPhillimy; Development Officer: Andy Wightman; raises awareness of the deforestation of Scotland and its implications; promotes ecological restoration and development through reafforestation and the concepts of sustainable forestry and integrated land use; collaborates with similar organizations world-wide.

Rendel Science and Environment: Great Guildford House, 30 Great Guildford St, London SE1 0ES; tel (71) 962-9884; fax (71) 962-9885; Managing Dir: John Simson; Dir: Mike Schwar; provides services including environmental management, assessments and audits, waste management, environmental radiation monitoring, and consultancy; concerned with ecology, air and water quality, noise and vibration.

Replanting Vietnam: POB 6, Diss, Norfolk IP22 2LB; tel (81) 341-9392; fax (81) 341-4879; Campaign Co-ordinator: J Edward Milner; fund-raising for forest protection and for reafforestation in Vietnam.

Royal Botanic Gardens: Kew, Richmond, Surrey TW9 3AB; tel (81) 940-1171; fax (81) 332-5197; Dir: Prof G T Prance; Deputy Dir for Science: Dr Charles Stirton; Deputy Dir for Operations: John Lavin; scientific research institute and tourist attraction.

Royal Entomological Society: 41 Queen's Gate, London SW7 5HR; tel (71) 584-8361; fax (71) 581-8505; Pres: Dr Hon M Rothschild; Registrar: G G Bentley; improvement and diffusion of entomological science.

Royal Geographical Society: 1 Kensington Gore, London SW7 2AR; tel (71) 589-5466; fax (71) 584-4447; Pres: Earl Jellicoe; Dir: Dr J Hemming; promotion and diffusion of geographical knowledge.

Royal Scottish Forestry Society (RSFS): Camsie House, Charlestown, Fife KY11 3EE; tel (383) 373014; fax (383) 872863; Dir: Michael Osborne; membership organization; involved in forest education and training, field excursions, formulation of forest policy, research, forest industry development and dissemination of information.

Royal Society: 6 Carlton House Terrace, London SW1Y 5AG; tel (71) 839-5561; telex 917876; fax (71) 930-2170; Exec Sec: Dr P T Warren; national academy of science; promotes and advances all fields of physical, biological, medical and agricultural science, mathematics and engineering, and their applications.

Royal Society for Nature Conservation (RSNC) Wildlife Trusts Partnership: The Green, Witham Park, Waterside South, Lincoln LN5 7JR; tel (522) 544400; fax (522) 511616; Chief Exec: Tim Cordy; Dir of Conservation: Tim Sands; voluntary organization concerned with all aspects of wildlife protection; partnership of 47 wildlife trusts, 47 urban wildlife groups and a junior wing; protects almost 2,000 sites nationally.

Royal Society for the Protection of Birds (RSPB): The Lodge, Sandy, Beds SG19 2DL; tel (767) 680551; telex 82469; fax (767) 692365; Chief Exec: Barbara Young; conservation of wild birds and their environment; creation of nature reserves.

Young Ornithologists' Club (YOC): Nat Organizer: Peter Holden; junior section of RSPB; leisure and educational activities and publications.

Rural Forum Scotland: Highland House, 46 St Catherine's Rd, Perth PH1 5RY; tel (738) 34565; fax (738) 38699; Chair: Deirdre Hutton; Pres: Barbara Kelly; Dir: Dermot Grimson; policy development, research, conferences and advice on all rural issues, including economic development and the environment.

Safety Course Unit: POB 3, Brighton BN2 6RW; tel (273) 696472; fax (273) 670009; Dir: Dr Alan Sherratt; Course Organizer: Geoff Stephens; courses and seminars on health and safety at work and in laboratories, fire safety, manual handling of hazardous substances, biological hazards and environmental protection, with special emphasis on the needs of education establishments, hospitals and research institutes.

Salmon and Trout Association: Fishmongers' Hall, London Bridge, London EC4R 9EL; tel (71) 283-5838; fax (71) 929-1389; Dir: C W Poupard; campaigns to safeguard the game fishing environment and represents the views and concerns of members.

Scott Polar Research Institute: University of Cambridge, Lensfield Rd, Cambridge CB2 1ER; tel (223) 336540; telex 81240; fax (223) 336549; Dir: Dr J Heap; study of the Arctic, Antarctica and glaciology, with particular emphasis on environmental change; offers one-year postgraduate course in polar studies.

Scottish Association for Marine Science (SAMS): Dunstaffnage Marine Research Laboratory, POB 3, Oban, Argyll PA34 4AD; tel (631) 62244; fax (631) 65518; Dir: Prof J B L Matthews; Pres: Prof A D McIntyre; formerly Scottish Marine Biological Association; promotes research and education in marine science.

Scottish Campaign to Resist the Atomic Menace (SCRAM): 11 Forth St, Edinburgh EH1 3LE; tel (31) 557-4283; fax (31) 557-4284; Principal Officers: Mike Townsley, Graham Stein; campaigns against nuclear power and for renewable energy and energy efficiency throughout the United Kingdom.

Scottish Council for National Parks: 15 Park Terrace, Stirling FK8 2JT; tel (786) 465714; fax (786) 473843; Chair: The Countess of Glasgow; Vice-Chair and Hon Sec: Brian K Parnell; organizes meetings and produces publications with the aim of securing legislation for the establishment of national parks in Scotland.

Scottish Field Studies Association: Kindrogan Field Centre, Enochdhu, Blairgowrie, Perthshire PH10 7PG; tel (250) 881286; fax (250) 881433; Pres: Dr Alan Pyke; Dir: Dr Alastair Lavery; provides courses on aspects of natural history for educational establishments and groups of adults, with the aim of instilling understanding of the Scottish countryside.

Scottish River Purification Boards' Association (SRPBA): 1 South St, Perth PH2 8NJ; tel (738) 27989; fax (738) 30997; Sec and Treas: R L Cowan; co-ordination of the interests of the ten Scottish river purification authorities.

Scottish Solar Energy Group: Mechanical Engineering Dept, Napier University, 10 Colinton Rd, Edinburgh EH10 5DT; tel (31) 455-2660; fax (31) 447-8046; Chair: Dr Colin Porteous; Head of Membership: Kerr MacGregor; promotion of research, development and use of solar energy in Scotland through meetings, visits and conferences.

Scottish Wildlife and Countryside Link (SWCL): POB 64, Perth PH2 0TF; tel (738) 30804; fax (738) 43290; Sec: Jennifer Anderson; Co-ordinator (acting): Pat Wilson; liaison body for voluntary organizations concerned with the conservation of wildlife and the countryside of Scotland; member of IUCN.

Council for Scottish Archaeology (CSA): c/o National Museums of Scotland, York Bldgs, Queen St, Edinburgh EH2 1JD; tel (31) 225-7534; Pres: Dr W S Hanson; Dir: Jill Harden; encourages study and conservation of Scotland's archaeological heritage and the promotion of archaeology among environmental issues.

John Muir Trust: Gardenhurst, Newbigging, Dundee DD5 3RH; tel (82) 623315; fax (82) 623315; Dir: Dr Terry Isles; Sec: Bill Wallace; Chair: Nicholas Luard; protects wild land and raises awareness of its value, while considering the needs of those who live there; conducts practical conservation work.

Scottish Environmental Education Council (SEEC): Department of Environmental Science, University of Stirling, Stirling FK9 4LA; tel (786) 467867; fax (786) 467864; Dir: Kate Sankey; Information Officer: Sheila Winstone; Development Officer: Pete Clark; membership organization for environmental educators; promotes national policy and good practice, organizes Regional Environmental Education Forums (REEFs) and conferences, produces publications and provides an enquiry service.

Socialist Environment and Resources Association (SERA): 11 Goodwin St, London N4 3HQ; tel (71) 263-7424; Vice-President: Chris Smith, MP; Co-ordinator: Tony Gray; Labour Party environmental pressure group; aims to ensure that an environmental perspective is given to all aspects of policy and decision-making.

Society of Chemical Industry (SCI): 15 Belgrave Sq, London SW1X 8PS; tel (71) 235-3681; fax (71) 823-1698; Pres: Jean-Marc Bruel; Chair of Council: Alan Hayes; learned society aiming to advance the application of chemistry and related sciences for public benefit; produces publications and holds meetings on diverse subjects, including environmental issues.

Society for Environmental Exploration (Frontier): 77 Leonard St, London EC2A 4QS; tel (71) 613-2422; fax (71) 613-2992; Dir: Eibleis Fanning; non-profit company; aims to provide a channel through which individuals can make a significant contribution to urgent research and conservation projects; projects are developed in collaboration with institutions in co-operating countries and operate under the name

Frontier; teams of self-financing volunteers assembled in the United Kingdom implement specific programmes of field research and practical conservation work in Tanzania, Uganda and Vietnam.

Society for the Protection of Animals Abroad (SPANA): 15 Buckingham Gate, London SW1E 6LB; tel (71) 828-0997; fax (71) 630-5776; Pres: Lord Newall; Chair: Prof G H Arthur; protection of working animals and other animals in North Africa and the Middle East; education programme promotes the concepts of animal welfare and care for the environment.

Soil Association: Organic Food and Farming Centre, 86 Colston St, Bristol BS1 5BB; tel (272) 290661; fax (272) 252504; educates the public, offers advice, produces publications, lobbies the authorities and conducts research on organic agriculture and food; sets independent standards for organic farming based on international guidelines; subsidiary company inspects and licenses growers, food processors and manufacturers.

Solar Trade Association Ltd: Brackenhurst, Greenham Common S, Newbury, Berks RG15 8HH; tel (635) 46561; fax (635) 46561; Chair: F E Partlett; Dir: Dr R E H Swayne; acts as a national enquiry centre on solar energy and a forum for organizations involved with the solar industry; regulates the activities and standards of member organizations.

Sunseed Trust: POB 2000, Cambridge CB3 0JF; tel (480) 411784; fax (480) 411784; Man Trustees: Tim Eiloart, Nance Pepin, Arnold Pease; research and development and education on ways of containing or reclaiming deserts and on technologies such as tree-planting and solar ovens which help peoples living on desert margins.

SUSTRANS: 35 King St, Bristol BS1 4DZ; tel (272) 268893; fax (272) 294173; Chief Engineer: J Grimshaw; Information Officer: P Insall; design and construction of routes for cyclists, pedestrians and wheelchair users, advice and consultancy on sustainable transport issues for local authorities, businesses and other organizations.

Tourism Concern: Proebel College, Roehampton Lane, London SW15 5PU; tel (81) 878-9053; fax (81) 392-3300; Co-ordinator: Tricia Barnett; membership organization promoting awareness of the impact of tourism on host communities and environments and advocating tourism that is just, sustainable and participatory.

Town and Country Planning Association: 17 Carlton House Terrace, London SW1Y 5AS; tel (71) 930-8904; fax (71) 930-3280; Pres: Lord Rippon of Hexham; Dir: David Hall; policy and research work on planning and environmental matters; reports, conferences and lobbying.

Transport 2000: Walkden House, 10 Melton St, London NW1 2EJ; tel (71) 388-8386; fax (71) 388-2481; campaigns for environmentally sensitive transport policies.

Tree Council: 35 Belgrave Square, London SW1X 8QN; tel (71) 235-8854; fax (71) 235-2023; Dir: Robert Osborne; independent registered charity; aims to improve the urban and rural environment in the United Kingdom by promoting the planting and conservation of trees, to disseminate information about tree management and to act as a forum for similar organizations.

Trees for Life: Findhorn Foundation, The Park, Findhorn Bay, Forres IV36 0TZ; tel (309) 691292; fax (309) 691155; Dir: Alan Watson; Field Officer: Adam Powell; conducts practical conservation work to regenerate the Caledonian Forest in Scotland; operates a volunteer programme and produces publications.

Turning Point: The Old Bakehouse, Cholsey, Oxon OX10 9NU; tel (491) 652346; Principal Officers: James Robertson, Alison Pritchard; campaigns through seminars and publications for the creation of a globally sustainable political economy in the 21st century.

UK2000 Scotland: c/o British Gas plc (Scotland), 60 Maxwell Rd, Glasgow G41 1QX; tel (41) 304-3420; fax (41) 304-3392; Chair: Sam Toy; Exec Dir: David Westwood; supports network of environmental projects, by grant aid and other means, providing opportunities for training, employment and voluntary activity.

UK Centre for Economic and Environmental Development (UK CEED): 3E King's Parade, Cambridge CB2 1SJ; tel (223) 67799; fax (223) 67794; Exec Dir: David Cope; independent, charitable research institute specializing in economic analysis of environmental issues.

UK Dependent Territories Conservation Forum: The Old Plough, 2 Caxton Rd, Great Gransden, Sandy, Beds SG19 3BE; tel (767) 677558; fax (767) 677964; Chair: Sara Oldfield; Co-ordinator: Sara Cross; promotion of the conservation of biodiversity in British Dependent Territories through liaison with local conservation agencies, NGOs in the United Kingdom, experts and government departments; provides information and co-ordinates project activities.

UK Foundation for the Peoples of the South Pacific: 32 Howe Park, Edinburgh EH10 7HF; tel (31) 445-5010; fax (31) 445-5255; Dir: Dorothy McIntosh; promotes small-scale, community-based, sustainable forestry projects and sustainable use of marine resources; opposes commercial logging operations, pesticide abuse and destruction of coral reefs.

UK Global Environmental Research (GER) Office: Polaris House, North Star Ave, Swindon SN2 1EU; tel (793) 411734; fax (793) 411691; Principal Officer: D A Brown; founded by five national research councils; provides information on national and international science and policy developments, maintains a database of current national global environmental research activities and acts as secretariat for national Inter-Agency Committee on Global Environmental Change (IACGEC).

University College of Wales, International Centre for Protected Landscapes: Unit 8, Science Park, Aberystwyth, Dyfed SY23 3AH; tel (970) 622617; fax (970) 622619; Dirs: Prof John Aitchison, Michael Beresford; resource centre, research, consultancy, training and conferences on protected landscape planning and management; aims to promote concept of protected landscapes and sustainable development around the world.

University of Dundee, Centre for Social and Environmental Accounting Research (CSEAR): c/o Dept of Accountancy and Business Finance, Dundee DD1 4HN; tel (382) 307789; fax (382) 24419; Dir: Prof R H Gray; networking institution which gathers and disseminates information; enables academics and practitioners to exchange ideas and experience.

University of East Anglia, School of Environmental Sciences: Norwich NR4 7TJ; tel (603) 592994; fax (603) 507719; Dean: Prof F J Vine; Research Administrator: Dr J P Darch; teaching and research in environmental sciences, including atmospheric chemistry, hydrology, hydrogeology, earth sciences, ecology, soil science, environmental economics, risk assessment, health studies, natural hazards, oceanography and geophysics.

Climatic Research Unit: tel (603) 56161; fax (603) 507784; Dir: Prof T D Davies; research includes tropospheric ozone climatology, global temperature variations, global precipitation changes, nuclear waste disposal, wind power generation, acid precipitation and CFC concentration projections.

University of Edinburgh, Centre for Human Ecology: 15 Buccleuch Place, Edinburgh EH8 9LN; tel (31) 650-3470; fax (31) 650-6520; Dir: Dr Ulrich Loening; Development Dir: Alastair McIntosh; research and MSc course on Human Ecology, international PhD in Ecological Economics, short courses on sustainability; seminars and conferences.

University of Kent at Canterbury, Durrell Institute of Conservation and Ecology: Faculty of Natural Sciences, Canterbury CT2 7NX; tel (227) 475480; telex 965449; fax (227) 475481; Founder and Research Dir: Dr Ian R Swingland; Exec Dir: Dr Michael Walkey; teaching, research and consultancy.

University of Surrey, Robens Institute of Industrial and Environmental Health and Safety: Guildford GU2 5XH; tel (483) 509203; telex 859331; fax (483) 503517; Principal Officers: Prof D A Stubbs, Prof J W Bridges; aims to advance human and environmental health and well-being through the provision of independent scientific research, investigation, advice and training world-wide; activities cover environmental health, human factors and ergonomics, and chemical safety.

University of York, Institute of Advanced Architectural Studies, Centre for the Conservation of Historic Parks and Gardens: The King's Manor, York YO1 2EP; tel (904) 433966; fax (904) 433949; Principal Officer: Peter Goodchild; educational courses, research, publications, information exchange and consultancy.

Volunteer Centre UK: 29 Lower King's Rd, Berkhamsted, Herts HP4 2AB; tel (442) 873311; fax (442) 870852; Dir: Andrea Kelmanson; Environmental Development Officer: Jonathan Pinkney-Baird; provides information and advice to volunteer organizers and campaigns for policy changes to make volunteering more accessible.

Wadebridge Ecological Centre (WEC): Worthyvale Manor Farm, Camelford, Cornwall PL32 9TT; tel (840) 212711; fax (840) 212808; Principal Officers: Edward Goldsmith, Nicholas Hildyard, Peter Bunyard; publication of books on environmental issues and conferences on subjects such as Gaia.

Wales Green Party/Plaid Werdd Cymru: Trewylan, Brynymor Rd, Aberystwyth, Dyfed SY23 4LJ; tel (970) 625143; fax (970) 625143; political campaigning on social and environmental issues in local, general and European Parliament elections; research and promotion of relevant educational material.

Warmer Campaign: 83 Mount Ephraim, Tunbridge Wells, Kent TN4 8BS; tel (892) 524626; fax (892) 527287; Dir: Kit Strange; collects and disseminates information on the minimization, re-use and recycling of waste and the generation of energy from waste.

Waste Watch: 68 Grafton Way, London W1P 5LE; tel (71) 383-3320; fax (71) 383-3364; Dir: Jo Gordon; Public Relations and Press Officer: Jonathan Wooding; Information Officer: Penny Pitty; national agency working with local and national government, industry and other groups to promote waste reduction and recycling.

Welsh Development Agency (WDA): Pearl House, Greyfriars, Cardiff CF1 3XX; tel (222) 222666; fax (222) 390752; Chief Exec: P Head; stimulation of economic development and environmental improvement.

West Wales Energy Group: Trefelin, Cilgwyn, Newport, Dyfed SA42 0QN; tel (239) 820470; fax (239) 820470; Chair: Glyn Jones-Davies; Vice-Chair: Brian John; Sec and Treas: Martyn Payne; charitable organization concerned with energy conservation, promotion of renewable energy, the environmental impact of various energy technologies, planning issues and fuel poverty; free advice service and publications.

Whale and Dolphin Conservation Society: Alexander House, James St W, Bath BA1 2BT; tel (225) 334511; fax (225) 480097; Principal Officer: Sean R Whyte.

Wildfowl and Wetlands Trust: Slimbridge, Glos GL2 7BT; tel (453) 890333; fax (453) 890827; Dir-Gen: Dr Myrfyn Owen; charitable organization; maintains eight public centres; concerned with wetland habitats, wildfowl biology and conservation; conducts research on wetland ecology and provides public education.

Women's Environmental Network: Aberdeen Studios, 22 Highbury Grove, London N5 2EA; tel (71) 354-8823; fax (71) 354-0464; Dir: Bernadette Vallely; Information Co-ordinator: Lin Collins; information and campaigns to educate and empower women who care about the environment.

Woodland Trust: Autumn Park, Dysart Rd, Grantham, Lincs NG31 6LL; tel (476) 74297; fax (476) 590808; Chief Exec: John James; Financial Dir: Julian Purvis; charity concerned with the creation and conservation of native and broad-leaved woodland areas as places of recreation and wildlife habitats; manages over 580 woods of varying size.

Working Weekends on Organic Farms (WWOOF): 19 Bradford Rd, Lewes, E Sussex BN7 1RB; tel (273) 476286; Principal Officers: Don Pynches, Maureen Pynches; membership organization for voluntary workers in the organic agriculture movement.

World Association of Soil and Water Conservation Europe: Oxford Brookes University, School of Social Sciences, Gipsy Lane, Headington, Oxford OX3 0BP; tel (865) 483750; fax (865) 483937; Vice-Pres for Europe: Prof M Haigh; promotion of soil and water conservation as better land husbandry, through symposia, workshops, publications and practical demonstration projects; international headquarters in the USA.

World Wide Fund for Nature (WWF) Antarctic and Cetacean Programme: 2 The Old Rectory, Dumbleton, Evesham WR11 6TG; tel (386) 882055; fax (386) 882055; Antarctic and Cetacean Officer: Cassandra Phillips.

World Wide Fund for Nature (WWF) United Kingdom: Panda House, Weyside Park, Godalming, Surrey GU7 1XR; tel (483) 426444; telex 859602; fax (483) 426409; Pres: HRH Princess Alexandra: Chair: J M K Laing; Dir: George Medley; national affiliate organization of WWF International.

World Wide Fund for Nature (WWF) Scotland: 1 Crieff Rd, Aberfeldy, Perthshire; tel (887) 820449; fax (887) 829453; Head of Operations: Simon Pepper.

World Wide Land Conservation Trust: The Old Mission Hall, Sibton Green, Saxmundham, Suffolk IP17 2JY; tel (728) 668501; fax (728) 668680; Chief Exec: John A Burton; registered charity; involved in conservation initiatives, including land purchase and management of threatened areas.

Programme for Belize (PFB): POB 99, Saxmundham, Suffolk IP17 2LB; UK Rep: John A Burton; principally involved in fundraising and public awareness campaigns for the head office in Belize.

Rainforest Action Costa Rica: POB 99, Saxmundham, Suffolk IP17 2LB; project of the World Land Conservation Trust.

WRC plc: Henley Rd, Medmenham, Marlow, Bucks; POB 16, Marlow, Bucks SL7 2HD; tel (491) 571531; fax (491) 579094; Managing Dir: Dr J Moss; formerly Water Research Centre; independent research and development and consultancy organization specializing in water, waste water and total environmental management.

Wye College (University of London), Global Environmental Change (GEC) Programme: Wye, Ashford, Kent TN25 5AH; tel (223) 812401; fax (223) 813187; Research Co-ordinator: Prof Michael Redclift; Asst Co-ordinator: Alister Scott; research programme of the Economic and Social Research Council (ESRC); social science element of UK Global Environmental Research (GER) programme; aims to understand the societal and economic forces behind global environmental change and to determine suitable policy objectives to address environmental problems.

Wye College (University of London), Sustainable Agriculture Group: Wye, Ashford, Kent TN25 5AH; tel (233) 812401; telex 94017832; fax (233) 813320; Principal Officers: Prof David Leaver, Dr Howard Lee; research and teaching at undergraduate and postgraduate level in aspects of sustainable crop and animal production; organization of conferences.

Yorkshire Naturalists' Union: c/o University of Bradford, Bradford BD7 1DP; tel (274) 384212; fax (274) 384231; Hon Sec: D Bramley; dedicated to the study of natural history in England, particularly Yorkshire; issues publications.

Young People's Trust for the Environment and Nature Conservation (YPTENC): 95 Woodbridge Rd, Guildford, Surrey GU1 4PY; tel (483) 39600; fax (483) 301992; Founder and Dir: Cyril Littlewood; Chair of Trustees: Ian S MacPhail; independent charity dedicated to educating young people about conservation of the world's natural resources and the environment; lectures, publications and projects.

Zoological Society of Glasgow and West of Scotland: Glasgow Zoo, Calderpark, Uddingston, Glasgow G71 7RZ; tel (41) 771-1185; fax (41) 771-2615; Dir and Sec: Richard O'Grady; Pres: William R S MacKenzie; development and maintenance of zoological gardens, promotion of conservation through education, and participation in national and European captive breeding programmes.

Zoological Society of London: Regent's Park, London NW1 4RY; tel (71) 722-3333; telex 265247; fax (71) 483-4436.

BRITISH DEPENDENT TERRITORIES

Anguilla

Office of the Chief Minister: The Secretariat, The Valley; tel 2518; telex 9313; fax 3389; Chief Minister and Minister of Home Affairs, Tourism, Agriculture and Fisheries: Emile R Gumbs.

Bermuda

Ministry of the Environment: Government Administration Bldg, 30 Parliament St, Hamilton HM 12; tel 295-5151; fax 292-2349; Deputy Premier and Minister of the Environment: Ann Cartwright DeCouto.

Bermuda Biological Station for Research, Inc: 17 Biological Lane, Ferry Reach GE 01; tel 297-1880; fax 297-8143; Dir: Dr A H Knap; marine science research, oceanography, study of the local environment, educational programmes, and summer courses for graduates and undergraduates.

British Virgin Islands

Ministry of Natural Resources and Labour: Road Town, Tortola; tel 43701; fax 44435; Deputy Chief Minister and Minister for Natural Resources and Labour: Ralph O'Neal.

Cayman Islands

Executive Council: c/o Office of the Governor, Government Administration Bldg, Grand Cayman; tel 949-7900; telex 4260; fax 949-7544; Elected Member responsible for Tourism, the Environment and Planning: Thomas C Jefferson.

Gibraltar

Ministry for the Environment: Hygeia House, 37-41 Town Range; tel 70620; fax 74119; Minister for Tourism and the Environment: Hon J E Pilcher; Head of Ministry and Chief Environmental Health Officer: Alexis J G Almeda; Deputy Chief Environmental Health Officer: Joseph L Delgado.

Gibraltar Botanic Gardens: The Alameda, Red Sands Rd, POB 843; tel 72639; fax 74022; Dir: Dr John E Cortés; Curator: Brian M Lamb; Horticulturalist: Andrew Abrines; conservation of plants and environmental education.

Gibraltar Ornithological and Natural History Society: Gibraltar Natural History Field Centre, Jews' Gate, Upper Rock Nature Reserve, POB 843; tel 72639; fax 74022; Gen Sec: Dr John E Cortés; conservation and research on botany, mammalogy, entomology, malacology, herpetology and marine biology.

Hong Kong

Environmental Protection Department: 24th-28th Floors, Southorn Centre, 130 Hennessy Rd, Wan Chai, Hong Kong; tel 8351001; fax 8382155; Dir of Environmental Protection: Dr S B Reed; enforcement of environmental protection legislation; environmental monitoring; strategic planning; policy development.

Agriculture and Fisheries Department: Canton Rd Government Offices, 393 Canton Rd, 12th Floor, Kowloon; tel 7332211; fax 3113731; Dir: Dr Lawrence Lee Hay-yue; activities include monitoring and enforcement of legislation pertaining to the health and welfare of animals and plants, conservation of endangered animal and plant species and control of international trade in them, protection and management of country parks and nature reserves, control of pesticides and reduction of rural pollution.

University of Hong Kong, Centre of Urban Planning and Environmental Management: Pokfulam Rd; tel 8592720; telex 71919; fax 5590468; Dir: Prof Peter Hills; provides professional training and doctoral programmes; offers consultancy and training services to government, international agencies and the private sector; produces publications.

University of Hong Kong, Faculty of Science: Pokfulam Rd; tel 8592819; telex 71919; fax 8584620; Dean: Dr I J Hodgkiss; offers B SC course in Environmental Science, M SC course in Environmental Management and research degrees in Environmental Studies; research and consultancy in marine biology, environmental physiology, ecotoxicology, microbial ecology, and freshwater and terrestrial ecology.

World Wide Fund for Nature (WWF) Hong Kong: POB 12721, Tramway Path, Central, Hong Kong; tel 5261011; telex 66711; fax 8452734; Chair: Sir David Akers-Jones; Pres: The Hon Sir Kenneth Ping-fan Fung; Exec Dir: David Melville; national affiliate organization of WWF International.

Montserrat

Ministry of Agriculture, Trade and Environment: The Grove, Plymouth; tel 491-2545; Minister: Charles Kirnon.

Montserrat National Trust: Parliament St, POB 86, Plymouth; tel 491-3086; fax 491-3599; activities include conservation of natural beauty, preservation of historic sites and buildings, acquisition of property for the benefit of the island, compilation of records, public education on the island's heritage and fund-raising.

St Helena

Council Committee on Agriculture and Natural Resources: c/o Office of the Governor, Plantation House; telex 202; Chair: P Peters.

St Helena Heritage Society: Broadway House, Jamestown; tel 4525; Chair: John Bailey; Vice-Chair: Cathy Hopkins; activities include liaison with government departments on heritage matters.

Turks and Caicos Islands

Ministry of Natural Resources: c/o Office of the Chief Minister, Government Compound, Grand Turk; tel 946-2801; fax 946-2777; Minister: Arabella Smith.

Department of Environment, Heritage and Parks: tel 946-2855; fax 946-1145; Dir: Judith L Garland; Chief Parks Officer: Eric Lightbourne; management of 33 marine and terrestrial protected areas, including national parks, nature reserves, sanctuaries and historical sites; protection of the marine and terrestrial environment; acts in close co-operation with National Trust.

National Trust of the Turks and Caicos Islands: POB 66, Grand Turk; tel 946-2220; fax 946-2448; Dir: Ken Wiley; Chair: Oswald Williams; preservation of sites of historic or cultural significance; conservation of lands, natural features and submarine areas of beauty or environmental importance; protection of native flora and fauna.

UNITED STATES OF AMERICA

Governmental Organizations

Environmental Protection Agency: 401 M St, SW, Washington, DC 20460; tel (202) 382-4700; telex 892757; fax (202) 382-7886; Dir: Carol M Browner.

Environmental Research Laboratory: 1 Sabine Island Drive, Gulf Breeze, FL 32561-5299; tel (904) 934-9200; fax (904) 934-9201; Dir: Dr Robert E Menzer; Deputy Dir: Dr Raymond G Wilhour; development, evaluation and application of strategies for the management of hazardous and biotechnological materials in near-coastal environments, particularly coastal wetlands and estuaries; quantification of toxicological, pathological and microbial processes to predict, monitor and reverse the impact of pollutants and other stressors; investigations primarily involve chemical compounds and biological products regulated by the EPA.

Environmental Research Laboratory—Corvallis (ERL-C): 200 SW 35th St, Corvallis, OR 97333; tel (503) 754-4600; fax (503) 754-4799; Dir: Dr T A Murphy; Deputy Dir: Dr R Lackey; EPA national centre for research on plant and wildlife ecology and inland ecosystems.

National Air and Radiation Environmental Laboratory: 1504 Ave A, Montgomery, AL 36115-2601; tel (205) 270-3400; fax (205) 270-3454; Dir: Samuel T Windham; environmental radiation monitoring, radiochemical analysis, radon studies, electromagnetic field measurement, mixed waste analysis, emergency response to nuclear accidents, and site characterization and assessment.

Department of Agriculture: 14th St and Independent Ave, SW, Washington, DC 20250-0001; tel (202) 447-2791; fax (202) 447-5340; Sec: Mike Espy.

Agricultural Research Service—Southwest Watershed Research Center: 2000 E Allen Rd, Tucson, AZ 85719; tel (602) 670-6381; fax (602) 670-5550; Research Leader: Leonard J Lane; interdisciplinary research in hydrology, erosion and sedimentation, water quality, global change and its impact on land use, and decision support systems; development of technology for the utilization and conservation of US agricultural and natural resources; promotion of sustainable resource management.

Forest Service: POB 96090, Washington, DC 20250; Asst Deputy Chief Int Forestry: David A Harcharik.

Department of Commerce, National Oceanic and Atmospheric Administration (NOAA): 14th St and Constitution Ave, NW, Washington, DC 20230; tel (202) 482-3436; fax (202) 482-6203; Sec of Commerce: Ronald H Brown; Admin: Dr D James Baker; Asst Admin: Douglas K Hall; weather forecasting, oceanic and atmospheric research, management of marine fisheries, protection of endangered species, sea and sky charting.

Department of Energy, Office of Scientific and Technical Information (OSTI): POB 62, Oak Ridge, TN 37831; tel (615) 576-1188; fax (615) 576-2865; Dir: Elizabeth Buffum; Asst Dirs: David Havrill, Dora Moneyhun; responsibilities include maintenence of databases on energy-related environmental, safety and health issues, and management of the Department of Energy's files on world-wide energy research and development.

Department of the Interior, Fish and Wildlife Service: 1849 C St, NW, Washington, DC 20240; tel (202) 208-4717; Dir: John F Turner; conservation of biodiversity for the continuing enjoyment and benefit of mankind; concerned with issues such as endangered species, acquisition and management of refuge land, fish hatcheries, mitigation of the adverse impacts of federal activities, enforcement of conservation law, conservation and management of migratory birds and anadromous fish, and research related to fish and wildlife and their habitats.

Department of the Interior, National Park Service: 18th and C Sts, NW, POB 37127, Washington, DC 20240-0001; tel (202) 343-7063; fax (202) 343-7059.

Department of State, Bureau of Oceans and International Environmental and Scientific Affairs (OES): Rm 7831, 2201 C St, NW, Washington, DC 20520-7818; tel (202) 647-4000; fax (202) 632-5854; Sec of State: Warren M Christopher; OES is the central contact point with other US government departments and agencies for liaison on environmental issues.

Agency for International Development (AID): 320 21st St, NW, Washington, DC 20523; tel (703) 875-4106; Dir: Dr Twig Johnson.

Alabama Department of Environmental Management: Congressman W L Dickinson Drive, Montgomery, AL 36130; tel (205) 271-7700; fax (205) 271-7950; Dir: Leigh Pegues.

Alaska Department of Environmental Conservation: 410 Willoughby Ave, Suite 105, Juneau, AK 99801-1795; tel (907) 465-5050; fax (907) 465-5070; Commr: John Sandor; Deputy Commr: Mead Treadwell; state agency responsible for writing environmental policy and regulations enforced by regional offices in Anchorage, Fairbanks and Juneau; maintains a library.

Alaska Department of Natural Resources: 400 Willoughby Ave, 5th Floor, Juneau, AK 99801; tel (907) 465-2400; fax (907) 586-2754; Commr: Glen A Olds; incorporates Division of Agriculture.

Arizona Commission on the Environment: 1645 W Jefferson, Suite 416, Phoenix, AZ 85007; tel (602) 542-2102; fax (602) 542-2104; Exec Dir: Charlene Ledet; Chair: T J Harrison; publications, workshops and conferences on environmental topics.

Arizona Department of Environmental Quality: 3033 N Central Ave, Phoenix, AZ 85012; tel (602) 207-2300; fax (602) 207-2218; Dir: Edward Z Fox; Deputy Dir: William D Wiley; environmental monitoring and regulation, covering air and water quality and waste management.

Arizona Division of Natural Resources: Department of Land, 1616 W Adams St, Rm 330, Phoenix, AZ 85014; tel (602) 542-4621; fax (602) 542-2590; Dir: Robert E Yount.

Arkansas Department of Pollution Control and Ecology: 8001 National Dr, POB 8913, Little Rock, AR 72219-8913; tel (501) 562-7444; fax (501) 562-4632; Dir: Randall Mathis.

California Energy Resources, Conservation and Development Commission: 1516 9th St, Sacramento, CA 95814; tel (916) 654-4420; fax (916) 654-4420; Chair: Charles R Imbrect.

California Environmental Protection Agency: 555 Capitol Mall, Suites 235 and 525, Sacramento, CA 95814; tel (916) 445-3846; fax (916) 445-6401; Sec: James M Strock.

Colorado Department of Natural Resources: Centennial Bldg, 1313 Sherman St, Rm 718, Denver, CO 80203; tel (303) 866-3311; fax (303) 866-2115; Exec Dir: Kenneth L Salazar; Deputy Dir: Ron Cattany.

Division of Wildlife: 6060 Broadway, Denver, CO 80216; management of state wildlife resources.

Colorado Environment Office: Department of Health, 4300 Cherry Creek Dr South, Bldg A, 5th Floor, Denver, CO 80222-1530; tel (303) 692-2000; fax (303) 782-0095; Exec Dir: Patricia A Nolan.

Connecticut Department of Environmental Protection: State Office Bldg, 165 Capitol Ave, Hartford, CT 06106; tel (203) 566-2110; fax (203) 566-7932.

Council on Environmental Quality: 722 Jackson Place, NW, Washington, DC 20006; tel (202) 395-5780; fax (202) 395-3744.

Delaware Department of Natural Resources and Environmental Control: 89 Kings Highway, POB 1401, Dover, DE 19903-1401; tel (302) 739-4403; fax (302) 739-6242; Sec: Christopher A G Tulou.

Florida Department of Environmental Regulation: Twin Towers, 2600 Blair Stone Rd, Rm 618, Tallahassee, FL 32399-2400; tel (904) 488-4805; fax (904) 487-4938; Sec: Virginia Wetherall.

Florida Department of Natural Resources: Marjory Stoneman Douglas Bldg, 3900 Commonwealth Blvd, Tallahassee, FL 32399-3000; tel (904) 488-0024; fax (904) 487-1469; Exec Dir (acting): Don Duden.

Florida Game and Fresh Water Fish Commission: 620 S Meridian St, Tallahassee, FL 32399-1600; tel (904) 488-4676; fax (904) 488-6988; Commrs: Ouinton L Hedgepeth, Joe M Hilliard; wildlife law enforcement, management and research.

Georgia Environmental Protection Division: Department of Natural Resources, East Tower, 205 Butler St, Suite 1252, Atlanta, GA 30334; tel (404) 656-3500; fax (404) 656-0770; Commr: Joe D Tanner.

Hawaii Department of Land and Natural Resources: Kalanimoku Bldg, 1151 Punchbowl St, Honolulu, HI 96813; tel (808) 587-0400; fax (808) 587-0360; Chair: Keith W Ahue.

Health and Environmental Sciences Division: Office of Technology Assessment (OTA), Congress of the United States, Washington, DC 20510-8025; tel (202) 224-3695; Asst Dir: Clyde Behney; provision of early indications to congressional committees of the probable beneficial and adverse impacts of the applications of technology; development of other co-ordinate information to assist the Congress.

Food and Renewable Resources Program: tel (202) 228-6510; Programme Man: Walter Parham.

Oceans and Environment Program: tel (202) 228-6840; fax (202) 228-6098; Programme Man: Robert W Niblock; responsible for all ocean-related questions, including ocean resources and maritime policy, and for large-scale environmental issues, such as climate change and water pollution.

Idaho Department of Lands: 1215 W State St, Boise, ID 83720-7000; tel (208) 334-0200; fax (208) 334-2339; Dir: Stanley F Hamilton; Asst Dir for Forestry and Fire: Winston Wiggins; Asst Dir for Lands, Range and Minerals: Jay Biladeau; management of Idaho's endowment trust lands for maximum long-term benefit through forest, fire and minerals management, rangeland and surface leasing, and soil conservation.

Idaho Division of Environmental Quality: Department of Health and Welfare, Towers Bldg, 450 W State St, Boise, ID 83720; tel (208) 334-5500; fax (208) 334-6558; Dir: Jerry L Harris.

Illinois Department of Energy and Natural Resources: 325 W Adams St, 3rd Floor, Springfield, IL 62704; tel (217) 522-2666; fax (217) 785-2618; Dir: John S Moore.

Illinois Environmental Protection Agency (IEPA): 2200 Churchill Rd, POB 19276, Springfield, IL 62706-9276; tel (217) 782-3397; fax (217) 782-9039; Dir: Mary A Gade; Deputy Dir: Bernard P Killian; Assoc Dir: David J Engel; regulatory agency responsible for implementation of Illinois Environmental Protection Act and of federal environmental laws at state level.

Indiana Department of Environmental Management: 105 S Meridian St, Rm 519A, Indianapolis, IN 46225; tel (317) 232-8162; fax (317) 232-5539; Commr: Kathy Prosser.

Indiana Department of Natural Resources: Government Center South, 402 W Washington St, Rm West 256, Indianapolis, IN 46204-2212; tel (317) 232-4020; fax (317) 232-8036; Dir: Patrick R Ralston.

Iowa Environmental Protection Division: Department of Natural Resources, Wallace State Office Bldg, Des Moines, IA 50319; tel (515) 281-5385; fax (515) 281-6794; Dir: Larry J Wilson.

Kansas Department of Health and Environment: Landon State Office Bldg, 900 SW Jackson St, Topeka, KS 66612-1290; tel (913) 296-1500; fax (913) 296-6247; Sec: Robert C Harder.

Kentucky Natural Resources and Environmental Protection Cabinet: Capitol Plaza Tower, 5th Floor, Frankfort, KY 40601; tel (502) 564-3350; fax (502) 564-8131; Sec: Phillip J Shepherd.

Louisiana Department of Agriculture and Forestry: POB 631, Baton Rouge, LA 70821-0631; tel (504) 922-1234; fax (504) 922-1253; Commr: Bob Odom.

Louisiana Department of Natural Resources: POB 44396, Baton Rouge, LA 70804; tel (504) 342-4500; fax (504) 342-2707; Sec: John F Ales.

Maine Department of Environmental Protection: State House, Station 17, Augusta, ME 04333; tel (207) 287-2812; fax (207) 626-7826; Commr: Dean C Marriott.

Maine Department of Inland Fisheries and Wildlife: 284 State St, Station 41, Augusta, ME 04333-0041; tel (207) 287-5202; fax (207) 287-6395; Commr (acting): Norman E Trask; management and protection of all species of wildlife and of aquatic resources in the state of Maine.

Maryland Department of Natural Resources: Tawes State Office Bldg, 580 Taylor Ave, Annapolis, MD 21401; tel (410) 974-3987; fax (410) 974-2383; Sec: Torrey C Brown; Deputy Sec: John R Griffin.

Maryland Department of the Environment—Air and Radiation Management Administration: 2500 Broening Highway, Baltimore, MD 21224; tel (410) 631-3255; fax (410) 631-3202; Sec: Robert Perciasepe; Dir: Merrylin Zaw-Mon; monitoring of air quality; development of plans and regulations and establishment of emission standards to prevent air pollution; implementation of state and federal laws; issue and review of permits for air pollution sources; regulation of all sources of ionizing radiation; responsible for asbestos.

Massachusetts Division of Fisheries and Wildlife: Field Headquarters, 75 North St, Westborough, MA 01581; tel (508) 792-7270; fax (508) 792-7275.

Massachusetts Executive Office of Environmental Affairs: 100 Cambridge St, Rm 2000, Boston, MA 02202; tel (617) 727-9800; fax (617) 727-2754; Sec: Gertrude Coxe.

Michigan Department of Natural Resources: POB 30028, Lansing, MI 48909; tel (517) 373-2329; fax (517) 373-1547; Dir: Roland Harmes.

Minnesota Department of Natural Resources: 500 Lafayette Rd, POB 7, St Paul, MN 55155-4001; tel (612) 296-6157; fax (612) 297-3500; Commr: Rodney W Sando.

Minnesota County Biological Survey (MCBS): tel (612) 296-3344; fax (612) 297-4961; Co-ordinator: Carmen K Converse; systematic identification of the locations of Minnesota's rare natural ecosystems, their component natural communities and rare species.

Natural Heritage Program: tel (612) 296-8217; fax (612) 297-4961; Principal Officer: Bonita Eliason; identification and protection of Minnesota's rare native plants and endangered natural communities.

Mississippi Department of Environmental Quality: POB 20305, Jackson, MS 39289-1305; tel (601) 961-5000; fax (601) 354-6965; Exec Dir: James I Palmer.

Missouri Department of Natural Resources: POB 176, 205 Jefferson St, Jefferson City, MO 65102; tel (314) 334-6946; fax (314) 751-7627; Dir: David A Shorr; Deputy Dir of Operations: Sherry L Boldt; comprises Division of Environmental Quality, Division of State Parks, Division of Energy and Division of Geology and Land Survey.

Division of Environmental Quality: Dir: John A Young; programmes concerned with air and water pollution control, soil and water conservation, public drinking water, hazardous waste, solid waste management and land reclamation.

Montana Department of Health and Environmental Sciences: Cogswell Bldg, Helena, MT 59620; tel (406) 444-2544; fax (406) 444-1804; Dir: Robert J Robinson.

Montana Department of Natural Resources and Conservation: 1520 South Ewing East 6th Ave, Helena, MT 59620-2301; tel (406) 444-6699; fax (406) 444-6721; Dir: Mark Simonich; incorporates Division of Energy.

Nebraska Department of Environmental Quality: 1200 North St, Suite 400, POB 98922, Lincoln, NE 68509-8922; tel (402) 471-2133; fax (402) 471-2909; Dir: Randy Wood.

Nevada Department of Conservation and Natural Resources: 123 West Nye Lane, Carson City, NV 89710; tel (702) 687-4360; fax (702) 687-6972; Dir: Peter G Morros.

New Hampshire Department of Environmental Services: 6 Hazen Dr, Concord, NH 03301; tel (603) 271-3503; fax (603) 271-1953; Commr: Robert W Varney.

New Jersey Department of Environmental Protection and Energy: 401 E State St, Trenton, NJ 08625-0402; tel (609) 292-2885; fax (609) 777-1781; Commr: Scott A Weiner.

New Mexico Department of Energy, Minerals and Natural Resources: 2040 S Pacheo St, Santa Fe, NM 87505; tel (505) 827-5950; fax (505) 438-3855; Sec: Anita Lockwood.

New Mexico Department of the Environment: Harold Runnels Bldg, POB 26110, Santa Fe, NM 87503; tel (505) 827-5950; fax (505) 827-2836; Sec: Judith Espinosa.

New York Department of Environmental Conservation: 50 Wolf Rd, Albany, NY 12233; tel (518) 457-3446; fax (518) 457-1088; Commr: Thomas C Jorling; incorporates Office of Natural Resources.

North Carolina Department of Environment, Health and Natural Resources: Archdale Bldg, 512 North Salisbury St, POB 72687, Raleigh, NC 27611; tel (919) 733-4984; fax (919) 733-0513; Sec: Jonathan B Howes.

North Dakota Division of Environmental Health: Department of Health and Consolidated Laboratories, 1200 Missouri Ave, Bismarck, ND 58502-5520; tel (701) 221-5150; fax (701) 221-5200; Principal Officer: Francis J Schwindt.

Ohio Department of Natural Resources: Bldg D, Fountain Sq, Columbus, OH 43224-1387; tel (614) 265-6879; fax (614) 261-9601; Dir: Francis Seiberling Buchholzer.

Oklahoma Division of Oil and Gas Conservation: Corporation Commission, Jim Thorpe Bldg, Oklahoma City, OK 73105; tel (405) 521-3771; fax (405) 521-6045; Principal Officer: Mike Battles.

Oregon Department of Environmental Quality: 811 6th St, SW, Portland, OR 97204; tel (503) 229-5696; fax (503) 229-6124; Dir: Fred J Hansen.

Pennsylvania Department of Environmental Resources: Fulton Bldg, POB 2063, Harrisburg, PA 17105-2063; tel (717) 787-2814; fax (717) 783-8926; Sec: Arthur A Davis.

Rhode Island Department of Environmental Management: 9 Hayes St, Providence, RI 02908; tel (401) 277-2231; fax (401) 277-6548; Dir: Louise C Durfee; incorporates Division of Agriculture.

Office of Environmental Co-ordination—Pollution Prevention Section: tel (401) 277-3434; fax (401) 277-2591; Principal Officers: Richard Enander, Janet Keller; provision of technical assistance to Rhode Island industry on prevention of pollution.

South Carolina Department of Health and Environmental Control: J Marion Sims Bldg and R J Aycock Bldg, 2600 Bull St, Columbia, SC 29201; tel (803) 734-4880; fax (803) 734-4777; Commr: Thomas E Brown; incorporates Division of Energy.

South Carolina Land Resources Conservation Committee: 2221 Devine St, Suite 222, Columbia, SC 29205; tel (803) 734-9100; fax (803) 734-9200; Exec Dir: John W Parris.

South Dakota Department of Environment and Natural Resources: 523 E Capitol Ave, Pierre, SD 57501-3181; tel (605) 773-5559; fax (605) 773-6035; Sec: Robert E Roberts; responsible for management of water resources, evaluation of mineral resources, regulation of mineral, oil and gas production, and environmental quality.

Tennessee Department of the Environment and Conservation: Life and Casualty Tower, 21st Floor, 401 Church St, Nashville, TN 37243-0435; tel (615) 532-0109; fax (615) 532-0120; Commr: J W Luna.

Texas Bureau of Environmental Health: 1100 W 49th St, Austin, TX 78756; tel (512) 834-6640; fax (512) 834-6707; Head of Div: Charles R Maddox.

Utah Department of Environmental Quality: 168 N 1950 West St, Salt Lake City, UT 84116; tel (801) 536-4402; fax (801) 536-4401; Exec Dir: Dianne R Neilson.

Utah Department of Natural Resources: 1636 W North Temple St, Rm 316, Salt Lake City, UT 84116-3193; tel (801) 538-7200; fax (801) 538-7315; Exec Dir: Ted Stewart.

Office of Energy and Resource Planning: 3 Triad Center, Suite 450, 355 W North Temple St, Salt Lake City, UT 84180-1204; tel (801) 538-5428; fax (801) 521-0657; Dir: Dr Richard Anderson; research, analysis and demonstration of energy technologies and energy efficiency.

Vermont Agency of Natural Resources: State Office Complex, 103 S Wissell, Waterbury, VT 05676; tel (802) 244-7347; fax (802) 244-1102; Sec: Charles Clark.

Virginia Council on the Environment: 202 N 9th St, Rm 900, Richmond, VA 23219; tel (804) 786-4500; fax (804) 225-3933; Admin: Keith J Buttleman; part of the Secretariat of Natural Resources.

Washington Department of Ecology: POB 47600, Olympia, WA 98504-8711; tel (206) 459-6168; fax (206) 459-6007; Dir: Mary Riveland.

Washington Department of Natural Resources: 111 Washington St, SE, POB 44000, Olympia, WA 98504-4000; tel (206) 902-1000; fax (206) 902-1775; Commr of Public Lands: Jennifer Belcher.

West Virginia Department of Commerce, Labor and Environmental Resources: State Capitol Bldg, Rm 151, 1900 Kanawha East Blvd, Charleston, WV 25305-0310; tel (304) 348-0350; fax (304) 558-4983; Sec: John M Ranson.

West Virginia Tourism and Parks Division: 2101 Washington St East, Charleston, WV 25305; tel (304) 558-2286; fax (304) 558-0108; Commr: John H Brown, Jr.

Wisconsin Department of Natural Resources: 101 S Webster St, POB 7921, Madison, WI 53707; tel (608) 266-2121; fax (608) 267-3579; Sec: George C Meyer.

Wyoming Department of Environmental Quality: Herschler Bldg, 122 W 25th St, Cheyenne, WY 82002; tel (307) 777-7938; fax (307) 777-7682; Dir: Dennis Hemmer.

Non-Governmental Organizations

Abalone Alliance Safe Energy Clearinghouse: 2940 16th St, Suite 310, San Francisco, CA 94103; tel (415) 861-0592; fax (415) 558-8135; Principal Officers: Don L Eichelberger, Roger Herried; source of information on nuclear power, alternative energy and conservation.

Abundant Life Seed Foundation: 1029 Lawrence St, Port Townsend, WA 98368; tel (206) 385-5660; fax (206) 385-7455; Principal Officers: Robert Greenway, Lynn Moser; non-profit organization dedicated to preservation of biodiversity through propagation and distribution of open-pollinated 'heirloom' seeds; operates World Seed Fund which donates seed worldwide.

Acid Rain Foundation: 1410 Varsity Drive, Raleigh, NC 27606-2010; tel (919) 828-9443; fax (919) 515-3593; Principal Officer: Dr Harriett Stubbs; public awareness-raising; provision of educational resources; support for relevant research.

Adirondack Council: 2 Church St, POB D-2, Elizabethtown, NY 12932-0640; tel (518) 873-2240; fax (518) 873-6675; Chair: Peter R Borelli; Vice-Chairs: John L Ernst, Thomas D Thacher, II; environmental advocacy organization working to protect Adirondack Park, NY; informs public, policymakers and media of the park's conservation needs; liaison with state government; research and policy analysis.

Adopt-A-Stream Foundation (AASF): POB 5558, Everett, WA 98206; tel (206) 388-3487; Exec Dir: Tom Murdoch; Chair of Bd: Deberra Stecher; provision of environmental education materials and training for those interested in caring for their local streams; promotion of habitat protection and stream restoration.

African Wildlife Foundation: 1717 Massachusetts Ave, NW, Washington, DC 20036; tel (202) 265-8393; fax (202) 265-2361; Dir of Admin: Elizabeth McCorkle; member of Earth Share.

African Wildlife News Service: POB 546, Olympia, WA 98507-0546; tel (206) 459-8862; fax (206) 459-8862; Pres and Treas: Stephen R Mishkin; Sec: Dr Paula J Schweich.

Air and Waste Management Association: POB 2861, Pittsburgh, PA 15230; tel (412) 232-3444; fax (412) 232-3450; Pres: Douglas Fox; Exec Vice-Pres: Martin Rivers; non-profit organization promoting practical environmental management, exchange of technological information and development of cleaner technology through conferences, courses and publications.

Akron Zoological Park: 500 Edgewood Ave, Akron, OH 44307; Pres and CEO: L Patricia Simmons; professional zoo specializing in local public education about the environment and the dignity of animals.

Alaska Environmental Lobby, Inc: POB 22151, Juneau, AK 99801; tel (907) 463-3366; fax (907) 463-3320; Pres: Claire Holland; Vice-Pres: George Matz; lobbying, educating Alaskans about the legislative process and citizen activism and informing environmental groups in Alaska about legislative issues.

American Association of Zoological Parks and Aquariums (AAZPA): AAZPA Conservation Center, 7970-D Old Georgetown Rd, Bethseda, MD 20814-2493; tel (301) 907-7777; fax (301) 907-2980; Dir Conservation and Science: Michael Hutchins; non-profit organization dedicated to the advancement of zoological parks and aquariums for conservation, education, scientific studies and recreation.

American Committee for International Conservation (ACIC): c/o Chicago Zoological Society, Brookfield, IL 60513; tel (708) 485-0263; fax (708) 485-3532; Chair: Dr George Rabb; association of non-governmental organizations concerned with international conservation; promotion of research and conservation activities.

American Ecological Research Institute (AERIE): POB 380, Fort Collins, CO 80522; tel (303) 224-5307; Dir: Jay W Tischendorf; Sec: Joyce Wiland; private natural resource research and consultation company; concerned with predators, endangered species and cryptozoology, and the use of non-invasive, minimum impact investigative techniques.

American Fisheries Society: 5410 Grosvenor Lane, Suite 100, Bethesda, MD 20814; tel (301) 897-8616; fax (301) 897-8096; President elect: Raymond C Hubley, Jr; First Vice-Pres: Lee C Redmond; promotion of conservation and wise use of fisheries and the development and teaching of all branches of fisheries science; collection and dissemination of information about fisheries science and practice.

American Forests: 1516 P St, NW, Washington, DC 20005; POB 2000, Washington, DC 20013; tel (202) 667-3300; fax (202) 667-7751; Pres: Donald C Willeke; Exec Vice-Pres: R Neil Sampson; environmental education on trees and forests; assessment and dissemination of scientific information; encouraging wise public policy; magazine publishing; conducts Global ReLeaf citizen education and action campaign; member of Earth Share.

American Geographical Society: 156 5th Ave, Suite 600, New York, NY 10010-7002; tel (212) 242-0214; fax (212) 989-1583; Pres: John E Gould; Chair: Richard H Nolte; Exec Dir: Mary Lynne Bird; publications, lectures and geographical education reform.

American Ground Water Trust: 6375 Riverside Drive, Dublin, OH 43017; tel (614) 761-2215; fax (614) 761-3446; Exec Dir: Andrew W Stone; public education and information about ground water issues.

American Nature Study Society: 5881 Cold Brook Rd, Homer, NY 13077; tel (607) 749-3655; fax (607) 753-3837; Pres: Joy Finlay; Sec: Betty McKnight; promotion of nature and environmental education through publication of periodicals; organizes workshops and field excursions and offers an award for excellence in children's nature literature.

American Ornithologists' Union: Division of Birds, NHB E607, MRC-116, Smithsonian Institution, Washington, DC 20560; tel (202) 357-2051; fax (202) 357-1932; Pres-elect: Dr Richard C Banks; Sec: Dr Mary V McDonald; publication of scientific information relating to birds.

American Society for Environmental History: Center for Technology Studies, New Jersey Institute of Technology, Newark, NJ 07102; tel (201) 596-3291; fax (201) 565-0586; Founder: John Opie; Pres: William Cronon.

American Society for the Protection of Nature in Israel: 25 W 45th St, Suite 1409, New York, NY 10036; tel (212) 398-6750; fax (212) 398-1665; Co-Chairs: Edward Geffner, Russell Rothman; Treas: Irving Cantor; educates North Americans about the environment in Israel and raises money for its protection.

American Solar Energy Society: 2400 Central Ave, Suite G-1, Boulder, CO 80301; tel (303) 443-3130; fax (303) 443-3212; Exec Dir: Larry Sherwood; organization of technical conferences and education programmes and publication of a magazine.

American Wildlands (AWL): 3609 S Wadsworth Blvd, Suite 123, Lakewood, CO 80235; tel (303) 988-2291; fax (303) 988-6573; Pres: Sally A Ranney; Chair: Jerry Mallett; Assoc Exec Dir: Clif Merritt; non-profit organization; promotion of protection and ecologically sustainable use of the wildland resources of the USA, including fresh water, wildlife, fisheries and forests.

Ancient Forest International: POB 1850, Redway, CA 95560; tel (707) 923-3015; fax (707) 923-3015; Dir: Rick Klein; promotion of world-wide conservation of native forest ecosystems, particularly old-growth forests; education, research and publications.

Animal Welfare Institute: POB 3650, Washington, DC 20007; tel (202) 337-2332; fax (202) 338-9478; Pres: Christine Stevens; Sec: Freeborn Jewett, Jr; campaigns against animal testing by laboratories, cruelty to wildlife and farm animals and trade in exotic birds born in the wild; preservation of rare species and encouragement of humane science teaching.

Antarctic and Southern Ocean Coalition: POB 76920, Washington, DC 20013; tel (202) 544-0236; fax (202) 544-8483; Principal Officer: James N Barnes; coalition of NGOs working for the full and permanent protection of Antarctica.

Antarctica Project: POB 76920, Washington, DC 20013; tel (202) 544-0236; fax (202) 544-8483; Dir: Beth Marks; secretariat of Antarctic and Southern Ocean Coalition.

Appalachian Mountain Club: 5 Joy St, Boston, MA 02108; tel (617) 523-0636; fax (617) 523-0722; Pres: Preston H Saunders; Sec: John G Casagrande; non-profit organization campaigning for conservation and responsible recreation; promotion of land and river protection, educational and urban youth programmes, air quality research, trail building and maintenance.

Archbold Tropical Research Center: 130 Lehotsky Hall, POB 341019, Clemson University, Clemson, SC 29634-1019; tel (803) 656-0457; fax (803) 656-0231; Dir: Dr Thomas E Lacher, Jr; Field Station Man: Phillip C Alexander; Project Man: Dr Ross Wagenseil; research covers tropical ecology, conservation and agriculture, agroforestry and secondary forest management, and nature-based tourism.

Arctic to Amazonia Alliance: POB 73, Strafford, VT 05072; tel (802) 765-4337; Pres and Dir: Erik van Lennep; Vice-Pres and Co-Founder: Kim Rheinlander; promotion of understanding between indigenous and non-indigenous peoples, support for human rights and social justice for indigenous peoples and promotion of education and viable alternatives to industrialization.

Association of Environmental and Resource Economists (AERE): 1616 P St, NW, Rm 507, Washington, DC 20036; tel (202) 328-5077; Sec: Dr Paul R Portney; Pres: Dr Clifford S Russell; journal and annual summer workshops.

Association of New Jersey Environmental Commissions: POB 157, 300 Mendham Rd, Route 24, Mendham, NJ 07945; tel (201) 539-7547; fax (201) 539-7713; Pres of Trustees: Candace Ashmun; Exec Dir: Sally Dudley; environmental education and training for local authority officials.

Association of State Wetland Managers, Inc: POB 2463, Berne, NY 12023-9746; tel (518) 872-1804; fax (518) 872-2171; Principal Officer: Jon Kusler; non-profit, federal membership organization; application of scientific methods to the protection and management of wetlands resources; exchange network for conservation projects; meetings and publications.

Atlantic Center for the Environment: 39 S Main St, Ipswich, MA 01938; tel (508) 356-0038; fax (508) 356-7322; Dir Int Programs: Jessica L Brown; encouragement of environmental awareness and sound resource management in rural areas of northern New England and eastern Canada through education, training, research, policy development and technical assistance.

Bicycle and Pedestrian Program: Transportation Division, Seattle Engineering Department, 708 Municipal Bldg, 600 4th Ave, Seattle, WA 98104-1879; tel (206) 684-7583; fax (206) 684-8581; Co-ordinator: Peter A Lagerwey; promotion of safe cycling and pedestrianism and creation of routes.

Bicycle Federation of America (BFA): 1506 21st St, NW, Washington, DC 20036; tel (202) 463-6622; Exec Dir: Bill Wilkinson; promotion of safe cycling and walking through advocacy, contract work for government agencies, publications and conferences.

Biodiversity Resource Center: California Academy of Sciences, Golden Gate Park, San Francisco, CA 94118-4599; tel (415) 750-7361; fax (415) 750-7106; Co-ordinator: Anne Marie Malley; provision of direct access to biodiversity information resources, including an environmental reference library and database.

Birdlife International, US Office: POB 57242, Washington, DC 20037-7242; tel (202) 778-9563; fax (202) 293-9342; Pres: Howard P Brokaw; Sec: Kimberley Young; conservation of birds and their habitat, education, research and advocacy.

Caribbean Conservation Corporation: POB 2866, Gainesville, FL 32602; tel (904) 373-6441; fax (904) 375-2449; Exec Dir: David Carr; Pres Bd of Dirs: Landon Clay; non-profit organization dedicated to the conservation of sea turtles and other coastal and marine wildlife through protection of natural areas, research, education and training.

Carnegie Mellon University, Department of Engineering and Public Policy: 5000 Forbes Ave, Pittsburgh, PA 15213; tel (412) 268-2670; fax (412) 268-3757; Head of Dept: M Granger Morgan; education and research on problems of technology and public policy, such as energy and environmental systems, risk analysis, management and communication.

Carnivore Preservation Trust (CPT): Route 6, Hank's Chapel Rd, POB 326, Pittsboro, NC 27312; tel (919) 542-6627; fax (919) 542-4454; Exec Dir: Dr Michael A Bleyman; volunteer organization providing for genetically sound survival of mammals in tropical forests.

Center for Environmental Information, Inc: 46 Prince St, Rochester, NY 14607-1016; tel (716) 271-3550; fax (716) 271-0606; Pres: Jean Thompson; clearing-house for environmental information.

Center for Environmental Study: 143 Bostwick NE, Grand Rapids, MI 49503; tel (616) 771-3935; fax (616) 771-4005; Pres: Dr Kay T Dodge; promotion and translation of environmental research; environmental education and teacher training; resource centre for environmental information; Tree Amigos intercultural environmental education and action programme; publications and conferences.

Center for Global Sustainability: 1817 White Ave, Knoxville, TN 37916; Principal Officer: Steve Smith; affiliated to Earth Island Institute.

Center for Interdisciplinary Environmental Studies (CIES): POB 26, Babson Park Branch, Boston, MA 02157-0026; International Co-ordinator: Paolo G Meozzi; research centre based in Italy; environmental impact analysis, hazardous waste management, agriculture and forestry projects with minimum impact on natural resources.

Center for Marine Conservation: 1725 De Sales St, NW, Suite 500, Washington, DC 20036; tel (202) 429-5609; fax (202) 872-0619; Pres: Roger E McManus; conservation of marine ecosystems and resources; protection of endangered and threatened species; member of Earth Share.

Center for Plant Conservation: c/o Missouri Botanical Garden, POB 299, St Louis, MO 63166; tel (314) 577-5100; Exec Dir: Donald A Falk; conservation of rare and endangered species of plants in the USA through cultivation and research.

Center for the Study of Tropical Birds, Inc: 218 Conway Drive, San Antonio, TX 78209-1716; tel (210) 828-5306; fax (210) 828-5911; Principal Officers: Jack Clinton-Eithiear, Michael Gartside; conservation of tropical birds through co-operative research and education programmes.

Center for Whale Research: 1359 Smuggler's Cove, Friday Harbor, WA 98250; tel (206) 378-5835; fax (206) 378-5954; Principal Officers: Kenneth Balcomb, Prentice Bloedel; photo-identification of whales and dolphins, and acoustic identification and localization of whales.

Central Coast Conservation Center: 790 Main St, Suite E, Half Moon Bay, CA 94019; Principal Officer: Kathy Van Velsor; affiliated to Earth Island Institute.

Charles A Lindbergh Fund, Inc: 708 S 3rd St, Suite 110, Minneapolis, MN 55415; tel (612) 338-1703; fax (612) 338-6826; Pres: Dr Sylvia A Earle; Vice-Pres: Reeve Lindbergh; furtherance of Charles Lindbergh's vision of a balance between technological advances and environmental preservation through a programme of grants, the Lindbergh Award and other educational and incentive projects.

Chesapeake Bay Foundation (CBF): Headquarters, 162 Prince George St, Annapolis, MD 21401; tel (410) 268-8816; fax (410) 268-6687; Pres: William C Baker; Sr Vice-Pres: Robert Hoyt; non-profit organization working for the conservation and restoration of Chesapeake Bay and its tributaries; environmental education, land management, environmental advocacy, lobbying and publications.

Maryland Office: 14 Market Place, Annapolis, MD 21401; tel (301) 268-8833; Exec Dir: Jane T Nishida.

Pennsylvania Office: 214 State St, Harrisburg, PA 17101; tel (717) 234-5550; Exec Dir and Sr Scientist: Jolene E Chinchilli.

Virginia Office: 1001 E Main St, Suite 815, Richmond, VA 23219; tel (804) 780-1392; Exec Dir: Joseph H Maroon.

Virginia Sub-Office: 100 W Plume Center, Suite 701, Norfolk, VA 23510; tel (804) 622-1964.

Citizens Energy Council: POB U, Hewitt, NJ 07421; tel (201) 728-7835; fax (201) 728-7664; Principal Officers: Karin Westdyk, Sara Shannon; information resource for 'sane' energy technologies; networking with similar organizations; periodicals.

Citizens for Animals, Resources and Environment (CARE): POB 18772, Milwaukee, WI 53218; tel (414) 466-1250; Founder and Dir: Debi Zweifel; awareness-raising network; meetings, public lectures, publications and liaison with other groups.

Clean Sites, Inc: 1199 N Fairfax St, Suite 400, Alexandria, VA 22314; tel (703) 683-8522; fax (703) 548-8773; Pres and Treas: Edwin H Clark; Sec: Mary Clark; non-profit company; information exchange for the solution of hazardous waste problems; technical management services, cost allocations, waste removal recommendations and reports.

Chicago Office: Monadnock Bldg, 53 W Jackson Blvd, Suite 1632, Chicago, IL 60604; tel (312) 554-0900; fax (312) 554-0193.

Clean Water Fund (CWF): 1320 18th St, NW, 3rd Floor, Washington, DC 20036; tel (202) 457-0336; fax (202) 457-0287; Exec Vice-Pres and Treas: David R Zwick; Sec and Financial Manager: Wendy Taylor; involved in training, research and education intended to ensure safe drinking water and clean seas; promotion of safe solid waste management and conservation of natural resources; member of Earth Share.

Coalition for Environmentally Responsible Economies (CERES): 711 Atlantic Ave, Boston, MA 02111; tel (617) 451-0927; fax (617) 482-2028; Co-Chairs: Joan Bavaria, Denis Hayes; promotion of responsible environmental activities on the part of businesses; signatory companies are required to follow a ten-point code of conduct (the CERES Principles, formerly the Valdez Principles) and to complete an annual environmental report, which is made public.

Committee on the Environment: League of Women Voters of New Castle, POB 364, Chappaqua, NY 10514; tel (914) 241-7242; Chair: Roberta Wiernik; consumer education on solid waste, and avoidance, reduction and proper disposal of household hazardous waste; development of 'least toxic policy' for schools and public facilities.

Community Conservation Consultants—Howlers Forever, Inc: RD 1, POB 96, Gays Mills, WI 54631; tel (608) 735-4717; fax (608) 735-4717; Dir: Dr Robert H Horwich; Asst

Dir: Jonathan Lyon; advice on protection of natural resources, sustainable resource use and development; involved in sustainable agriculture, conservation of forest products, ecotourism and public education; mainly active in Wisconsin (USA) and Belize.

Community Environmental Council (CEC): 930 Miramonte Drive, Santa Barbara, CA 93109; tel (805) 963-0583; fax (805) 962-9080; Principal Officer: Jon Clark; non-profit research and education organization dedicated to design and implementation of innovative environmental management systems.

CONCERN, Inc: 1794 Columbia Rd, NW, Washington, DC 20009; tel (202) 328-8160; fax (202) 387-3378; Chair: Burks Lapham; Exec Dir: Susan Boyd; non-profit educational organization promoting environmental literacy and action by supplying information about environmental quality and public health.

Connecticut Council on Environmental Quality: 165 Capitol Ave, Rm 239, Hartford, CT 06106; tel (203) 566-3510; fax (203) 566-6024; Chair: John A Millington; Exec Dir: Karl J Wagener; investigation of citizens' complaints; reviewing of environmental assessments of state construction activities and preparation of annual reports to the Governor.

Conservation Agency: 6 Swinburne St, Jamestown, RI 02835; tel (401) 423-2652; fax (401) 423-0199; Pres: Dr James Lazell; Vice-Pres: Dr Numi C Goodyear, Richard P Taylor; research, collection of information and development of management plans with the aim of preserving rare, endangered and little-known species.

Conservation Council for Hawaii (CCH): POB 2923, Honolulu, HI 96802; tel (808) 531-2963; fax (808) 531-3050; Principal Officers: Steven L Montgomery, Richard Scudder; citizens' organization promoting environmental health and education, conservation and management of Hawaii's natural resources; Hawaii affiliate of National Wildlife Federation.

Conservation Federation of Missouri: 728 W Main St, Jefferson City, MO 65101-1534; tel (314) 634-2322; fax (314) 634-8205; Pres: Jack Garrett; Exec Dir: Ed Stegner; association of citizens' organizations whose aims are to educate and enable individuals and organizations to conserve natural resources; affiliated to National Wildlife Federation.

Conservation Fund: 1800 N Kent St, Suite 1120, Arlington, VA 22209; tel (703) 525-6300; fax (703) 525-4610; Pres: Patrick F Noonan; Treas: Hadlai A Hull; promotion of land and water conservation in the USA; sustainable development projects; conservation of endangered species habitats; promotion of aquaculture research; land advisory service; member of Earth Share.

Conservation International: 1015 18th St, NW, Suite 1000, Washington, DC 20036; tel (202) 429-5660; fax (202) 887-5788; Chair: Peter Seligmann; Pres: Russell Mittermeier; conservation of tropical rainforest, sustainable development, ecotourism, research and marketing of rainforest products; member of Earth Share.

Conservation Treaty Support Fund (CTSF): 3705 Cardiff Rd, Chevy Chase, MD 20815; tel (301) 654-3150; fax (301) 652-6390; Pres: George A Furness, Jr; Int Vice-Pres: John C Goldsmith; Sec: Linda E Mannheim; non-profit organization; support for major intergovernmental conservation treaties such as the Convention on International Trade in Endangered Species (CITES) and the Ramsar Convention on Wetlands of International Importance; raises awareness of such treaties and seeks funding from a variety of sources; member of American Committee for International Conservation, Alliance for Environmental Education and National Ramsar Committee.

Coolidge Centre for Environmental Leadership: 1675 Massachusetts Ave, Suite 4, Cambridge, MA 02138-1836; tel (617) 864-5085; Exec Dir: Dr Robert Singer; promotion of environmental leadership in all professions world-wide, especially in developing countries.

Cousteau Society, Inc: 870 Greenbrier Circle, Suite 402, Chesapeake, VA 23320-2641; tel (804) 523-9335; telex 6974570; fax (804) 523-2747; Pres: Jacques-Yves Cousteau; non-profit organization dedicated to the protection and improvement of the quality of life by educating the public; films, publications, expeditions and lectures.

Crocodile Specialist Group (CSG) of the Species Survival Commission of the World Conservation Union—IUCN: Florida Museum of Natural History, Gainesville, FL 32611-2035; tel (904) 392-1721; fax (904) 392-9367; Chair: Prof Harry Messel; Exec Officer: Dr James Perran Ross; conservation of crocodilians; promotion of sustainable resource use; publications.

Defenders of Wildlife: 1244 19th St, NW, Washington, DC 20036; tel (202) 659-9510; fax (202) 833-3349; Dir Int Wildlife Trade Programme: James K Wyerman; protection of biological diversity; prevention of species endangerment; member of Earth Share.

Desert Fishes Council: POB 337, Bishop, CA 93515; tel (619) 872-8751; fax (619) 872-8751; Chair: Dr John N Rinne; Exec Sec: Edwin P Pister; preservation of the biodiversity and the biological and physical integrity of desert aquatic ecosystems throughout North America.

Desert Protective Council, Inc: POB 2312, Valley Center, CA 92082; tel (619) 749-3485; education of the public and the authorities through publications and programmes focusing on the four great desert systems of the USA; protection of desert resources by networking with other organizations, and litigation and legislation when necessary.

Ducks Unlimited, Inc: 1 Waterfowl Way, Memphis, TN 38120; tel (901) 758-3825; fax (901) 758-3850; Pres: Don Rollins; Exec Vice-Pres: Matthew Connolly; aims to fulfil the needs of North American waterfowl throughout their annual life cycle by protecting, restoring and managing important wetlands and uplands.

Duke University, Center for Tropical Conservation: 3705-C Erwin Rd, POB 90381, Durham, NC 27708-0381; tel (919) 490-9081; fax (919) 419-1433; Dir: Dr John W Terborgh; Exec Admin: Dr Julie A Johnson; Chair Exec Cttee: Dr William A Ascher; research into biodiversity, natural forest management, forestry institutions in the tropics and environmental policy in the context of political and economic change; operation of International Professional Training Program (IPTP) for those involved in environmental policy-making in all sectors; various publications.

Earth Ecology Foundation: 612 N 2nd St, Fresno, CA 93702; tel (209) 442-3034; Founder and Dir: Erik Wunstell; non-profit organization; provision of educational materials to students and institutions world-wide; promotion of the scientific and humane use of the earth's resources and corporate environmental investment; development of disaster management programmes pertaining to climatic change.

Earth Island Institute (EII): 300 Broadway, Suite 28, San Francisco, CA 94133-3312; tel (415) 788-3666; telex 829302; fax (415) 788-7324; Chair: David R Brower; Exec Dirs: John A Knox, David Phillips; non-profit organization dedicated to developing and providing organizational support for innovative projects for the conservation and restoration of the global environment; Conferences on the Fate of the Earth.

AT Work—Appropriate Technology Working Group: Principal Officers: Don Macleay, Arne Jacobson, Jay Singer; provision of materials and skilled volunteers for appropriate technology projects in developing countries.

Baikal Watch: Principal Officer: Gary Cook; international activities for the permanent protection of the biological uniqueness of Lake Baikal, Russian Federation.

Brower Fund: Principal Officers: David Brower, Jimmy Langman; provision of short-term support for creative approaches to environmental conservation and restoration and related efforts.

Green Alternative Information for Action (GAIA): Principal Officers: Roger Picklum, Terrie Schultz; production of communication materials relating to the 'green' movement.

Information for the Public Trust: improvement of public access to database information on environmental issues such as water rights.

International Green Circle: Principal Officer: Erika Rosenthal; ecological restoration programme; matches volunteers with ongoing projects around the world.

International Marine Mammal Project: Exec Dir: David Phillips; leads international effort to stop the slaughter of dolphins by the tuna-fishing industry and continues work to end commercial whaling and the use of drift nets.

Sea Turtle Restoration Project: Principal Officers: Todd Steiner, Kathy Nielsen, Christine Niven, Georgita Ruiz; investigation of threats to endangered sea turtles; co-ordination of public opposition to forces that contribute to their destruction and collaboration with similar projects in Central America.

Urban Habitat Program: Principal Officers: Carl Anthony, Kate Crockett; development of multicultural urban environmental leadership and organization of projects to restore urban neighbourhoods through creative use of public spaces and techniques such as 'urban barn-raising'.

Berkeley–Borneo Big Home Project: 1137 Hearst Ave, Suite D, Berkeley, CA 94702; tel (510) 649-8460; Principal Officer: Joe Lamb; advancement of rainforest preservation and the rights of indigenous peoples.

Climate Protection Institute: 5833 Balmoral Drive, Oakland, CA 94619; tel (510) 531-0100; Principal Officer: Richard Golden; development of public education materials and high school curricular materials on global climate change.

Earth Island Energy Program: 205 Fairlawn Drive, Berkeley, CA 94708; tel (510) 549-0557; Principal Officer: Robert J Mowris; promotion of maximum technical potential for energy efficiency, to minimize the impact that energy systems have on the global environment.

Endangered Species Project: E-205 Fort Mason Center, San Francisco, CA 94123; tel (415) 921-3140; Principal Officer: Sam LaBudde; provision of new species protection campaigns with resources to develop documentary evidence for public education.

Environmental Litigation Fund: POB 10836, Eugene, OR 97440; tel (503) 683-1378; Principal Officer: Peter Sorenson; information and conferences on ways in which citizens can ensure enforcement of environmental laws.

Green Education and Development Fund: POB 30208, Kansas City, MO 64112; tel (816) 731-9366; national educational group associated with the emergence of a green political movement in the USA.

Japan Environmental Exchange (JEE): POB 12751, Berkeley, CA 94701; sponsorship of volunteer exchanges between USA and Japan, for work on environmental protection and restoration projects; office in Kyoto, Japan.

Ben Linder Memorial Fund for Appropriate Technology: 2155 Northwest Aspen, Portland, OR 97210; tel (503) 227-5803; Principal Officers: David Linder, Elizabeth Linder; continuation of the rural hydroelectric development work that Ben Linder was pursuing when he was assassinated in Nicaragua, in 1987.

Sacred Land Film Project: POB C-151, La Honda, CA 94020; tel (415) 747-0685; Principal Officer: Christopher McLeod; production of films about the environmental effects of energy and resource development in the western USA and about the relationships to the land of indigenous peoples world-wide.

Stewards of the Earth: 598 N Fairview Ave, Goleta, CA 93117; tel (805) 967-7369; Principal Officer: Michael Ableman; production of materials on traditional and organic approaches to agriculture around the world.

Watchfire Productions: POB 917, Woodacre, CA 94973; Principal Officers: Pete Sears, Jeannette Sears; production of environmental music videos to educate and motivate the public on issues such as rainforest destruction and peace in Central America.

Yosemite Guardian: 7186 Manzanita Lane, Yosemite W, CA 95386; Principal Officer: Garrett DeBell; protection of Yosemite national park's ecosystems from local, regional and global threats.

Earth Share: 3400 International Drive, NW, Suite 2K, Washington, DC 20008; tel (202) 537-7100; fax (202) 537-7101; Chair: Erik Meyers; Vice-Chair: Paul Hayes; known as Environmental Federation of America for legal purposes; fund-raising federation of 39 non-profit environmental and conservation organizations operating workplace payroll deduction campaigns throughout the USA.

EarthSave Foundation: 706 Frederick St, Santa Cruz, CA 95062-2205; tel (408) 423-4069; fax (408) 458-0255; Exec Dir: Patricia Carney; Chair of Bd: Shams Kairys; non-profit organization committed to reducing the disease and environmental degradation caused by diets based on meat and dairy products; educational programmes and materials.

Earthtrust: 25 Kaneohe Bay Drive, Suite 205, Kailua, HI 96734-1711; tel (808) 254-2866; fax (808) 254-6409; Pres: Don White; Dir of Public Relations: K T Madison; non-profit conservation organization dedicated to the preservation of wildlife and the natural environment; organizer of international campaigns against whaling and damage to dolphin stocks through the use of driftnets; Asian Wildlife Initiative aims to save endangered rhinoceroses and tigers; Hawaii Wildlife Conservation Campaign protects monk seals and green sea turtles; research and environmental education; field offices in China (Taiwan), New Zealand and Switzerland.

Earthwatch: 680 Mount Auburn St, POB 430N, Watertown, MA 02272; tel (617) 926-8200; telex 6006452; fax (617) 926-8532; Pres: Brian Rosborough; funding of field research projects in the physical, biological and social sciences by recruiting non-specialist volunteers to share costs and labour; more than 150 expeditions world-wide each year; affiliated to Center for Field Research (CFFR), Watertown, MA; offices in Australia, Russian Federation, United Kingdom, and USA (Los Angeles); 73,000 mems, 28,000 EarthCorps volunteers.

Eco-Home Network: 4344 Russell Ave, Los Angeles, CA 90027; tel (213) 662-5207; fax (213) 388-5338; Pres: Bob Walter; Founder and Dir: Julia Russell; demonstrations of ecological urban living; publications.

Eco-Cities Council: 11171 Fairbanks Way, Culver City, CA 90249; tel (310) 915-9569; Co-ordinator: Howard Cohen; promotion of sustainable urban planning.

Ecological Society of America (ESA): Public Affairs Office, 2010 Massachusetts Ave, NW, Suite 420, Washington, DC 20036; tel (202) 833-8773; fax (202) 833-8775; Dir: Dr Marjorie M Holland; seeks to unify the science of ecology, stimulate research and promote the application of ecological data and principles to the solution of environmental problems; annual conference; publications; 7,200 mems.

Ecotourism Society: Office of the Executive Director, POB 755, N Bennington, VT 05257; Membership Resource Center, 801 Devon Place, Alexandria, VA 22314; tel (802) 447-2121; fax (802) 447-2122; Exec Dir: Megan Epler Wood; Pres: Dr Gerardo Budowski; aims to make ecotourism a viable tool for conservation and sustainable development; promotes the implementation of the principles of ecotourism; networks with other organizations world-wide; promotes training and provides information.

Elephant Research Foundation: 106 E Hickory Grove Rd, Bloomfield Hills, MI 48304; tel (313) 540-3947; fax (313) 540-3948; Principal Officers: Jeheskel Shoshani, William R Thompson, Sandra Lee L Shoshani; collection and dissemination of information on all species of elephant, for research, education and conservation; incorporates Elephant Interest Group.

Elmwood Institute: 2522 San Pablo Ave, Berkeley, CA 94702; POB 5765, Berkeley, CA 94705; tel (510) 845-4595; fax (510) 845-1439; Pres: Fritjot Capra; Sec: Dana Ullman; ecological 'think-tank' and educational institution promoting ecological literacy.

Endangered Species Coalition: 666 Pennsylvania Ave, SE, Washington, DC 20003; tel (202) 547-9009; fax (202) 547-9022; Campaign Dir: Karyn Strickler; conferences, training workshops, publications and presentations.

Environmental Action: 6930 Carroll Ave, 6th Floor, Suite 600, Takoma Park, MD 20912; tel (301) 891-1100; fax (301) 891-2218; Exec Dir: Margaret Morgan-Hubbard; lobbying, local-level action, education and publications on pollution prevention, solid waste and energy issues; member of Earth Share.

Environmental Action Coalition (EAC): 625 Broadway, 2nd Floor, New York, NY 10012; tel (212) 677-1601; Exec Dir: Stephen Richardson; non-profit organization specializing in urban forestry, education in recycling and solid waste management.

Environmental and Societal Impacts Group (ESIG): National Center for Atmospheric Research, POB 3000, Boulder, CO 80307-3000; tel (303) 497-8117; telex 989764; fax (303) 497-8125; Dir: Dr Michael H Glantz; assessment of the impact of climate change and variability on the environment and on human activity around the world; research focuses on climate and food, water and energy; publications and visitors' programme.

Environmental Assessment Association (EAA): 8383 E Evans Rd, Scottsdale, AZ 85260-3614; tel (602) 483-8100; fax (602) 998-8022; Man Dir: Troy E Johnson; Assoc Dir: Mary J Wassermann; dissemination of environmental information on international real estate; awards recognized professional credentials; approx 5,000 members.

Environmental Conservation Library (ECOL): 300 Nicollet Mall, Minneapolis, MN 55401; tel (612) 372-6570; Librarian: William Johnston; information resource on ecological and environmental matters for citizens of Minnesota and, by correspondence, for interested parties world-wide.

Environmental Defense Fund (EDF): 1875 Connecticut Ave, NW, Washington, DC 20009; tel (202) 387-3500; Int Counsel: Scott Hajost; originally established by a group of scientists who proved that DDT was harming the environment; now a national organization which seeks to co-ordinate scientific, economic and legal expertise to improve policies affecting the environment.

New York Office: 257 Park Ave S, New York, NY 10010; tel (212) 505-2100; fax (212) 505-2375; Chair of Bd: George Montgomery; Exec Dir: Fred Krupp; Dir of Development: Paula Hayes; non-profit organization; unites science, economics and law to find innovative, economically viable solutions to current environmental problems; member of Earth Share.

Texas Office: 1800 Guadalupe, Suite A, Austin, TX 78701; tel (512) 478-5161; fax (512) 478-8140; Dir: Jim Marston.

Environmental Health Center (EHC): National Safety Council, 1019 19th St, NW, Suite 401, Washington, DC 20036-5105; tel (202) 293-2270; fax (202) 293-0032; Exec Dir: Morris A Ward; fosters improved public understanding of significant environmental health risks and challenges facing modern society; provides environmental journalism and public education programmes, and public information on radon and lead poisoning; also concerned with emergency management and planning.

Environmental Information, Ltd (EI): 4801 W 81st St, Suite 119, Minneapolis, MN 55437; tel (612) 831-2473; fax (612) 831-6550; Pres: Cary L Perket; aims to inform and guide the development of environmental policies, services and technologies; conducts research and produces publications on industrial and hazardous waste management.

Environmental Law Institute: 1616 P St, NW, Suite 200, Washington, DC 20036; tel (202) 328-5150; fax (202) 328-5002; Pres: J William Futrell; Gen Counsel: Erik Meyers; acts to ensure implementation of environmental legislation; protects water and air quality, wildlife and wetlands, and reduces public health hazards; promotes research and provides training for communities, governments and businesses; member of Earth Share.

ERM Group: 855 Springdale Drive, Exton, PA 19341; tel (800) 544-3117; fax (215) 524-7335; Chair: Paul H Woodruff; Pres and CEO: Kent E Patterson; environmental resources management; activities include site remediation, solid and hazardous waste management, ground water and waste water treatment, health, safety and toxicology, risk assessment, ecological services and air quality management.

F T Stone Laboratory and Center for Lake Erie Area Research: Ohio State University, 1314 Kinnear Rd, Columbus, OH 43212-1194; tel (614) 292-8949; fax (614) 292-4364; Dir: Dr Jeffrey M Reutter; freshwater biological field station; college-level summer courses, workshops in autumn and spring, and research facilities throughout the year.

Federation of Western Outdoor Clubs: 512 Boylton Ave E, Suite 106, Seattle, WA 98102; tel (206) 322-3041; Pres: Winchell Hayward; Sec: Mary Mason; promotes proper use, enjoyment and protection of scenery, wilderness and resources for outdoor recreation in the USA.

Florida Atlantic University/Florida International University, Joint Center for Environmental and Urban Problems: 220 SE 2nd Ave, Fort Lauderdale, FL 33301; tel (305) 355-5255; fax (305) 760-5666; Dir: Dr John M DeGrove; applied research in the areas of government structure and operations, fiscal and economic analysis, protection of natural resources and the environment, and comprehensive planning to balance population growth and environmental protection; public service activities and publications.

Florida Conservation Foundation, Inc: 1251-B Miller Ave, Winter Park, FL 32789; tel (407) 644-5377; Pres: Dr John O Blackburn; Dir: Marcia Ramsdell; environmental information, education and publications.

Forest Ecosystem Rescue Network (FERN): POB 672, Dahlonega, GA 30533-0672; Co-ordinator (USA): Dan Hemenway; Co-ordinator (Canada): Mike Janssens; reafforestation of relatively unpolluted areas using threatened tree species; organization of conferences, workshops and lectures on forest decline and deforestation.

Forest History Society: 701 Vickers Ave, Durham, NC 27701; tel (919) 682-9319; fax (919) 682-2349; Dir: Harold K Steen; objectives are to document and promote understanding of the history of human interaction with the environment.

Forest Trust: POB 519, Santa Fe, NM 87504-0519; tel (505) 983-8992; fax (505) 986-0798; Dir: Henry H Carey; Exec Asst to Dir: Katherine Brewer; opposed to traditional forest management policies; offers resource protection strategies and land stewardship services.

Foresta Institute for Ocean and Mountain Studies: POB 41567, Tucson, AZ 41367; Pres: Dr Richard G Miller; sponsorship of educational projects, meetings and training in resource conservation, biodiversity and life quality.

Freshwater Foundation: Spring Hill Center, 725 County Rd 6, Wayzata, MN 55391; tel (612) 449-0092; fax (612) 449-0592; Pres and CEO: Lance Yohe; non-profit organization; encourages the protection of freshwater resources through educational programmes, resource management, demonstration projects and research; established Gray Freshwater Biological Institute of the University of Minnesota.

Friends of the Earth (FoE) USA: 218 D Street, SE, Washington, DC 20003; tel (202) 328-0098; Dir: Clifton Curtis; protection and preservation of the marine environment for the people and wildlife that depend on it; research into ensuring the wise use and protection of ocean, coastal and estuarine resources; member of Earth Share.

Friends of the Everglades: 101 Westward Dr, No 2, Miami Springs, FL 33166; tel (305) 888-1230; fax (305) 888-1230; Pres: Nancy Carroll Brown; Exec Dir: Joe Podgor; environmental education and protection, restoration of Everglades ecosystem, and water management, supply protection and planning.

Frostburg State University: Frostburg, MD 21532-1099; tel (301) 689-4166; fax (301) 689-4737; Pres: Dr Catherine Gira; Biology Dept Chair: Dr David Morton; offers courses leading to a B SC degree in Wildlife–Fisheries and M SC degrees in Applied Ecology, Conservation Biology, Fisheries Management and Wildlife Management.

Fund for Animals Inc: National Headquarters, 200 W 57th St, New York, NY 10019; tel (212) 246-2096; fax (212) 246-2633; Pres: Cleveland Amory; Sec and Treas: Marian Probst; Nat Dir: Wayne Pacelle; education, litigation, lobbying and direct action for animal protection; operates sanctuaries for rescued animals in California, South Carolina and Texas.

Maryland Office: 850 Sligo Ave, Suite 300, Silver Spring, MD 20910; tel (301) 585-2591; fax (301) 585-2595.

Game Conservation International (Game CoIn): 8626 Tesoro Drive, Suite 812, POB 17444, San Antonio, TX 78217; tel (210) 824-7509; fax (210) 829-1355; Pres: Mark Barrett; Exec Dir: Lawrence Means; funds anti-poaching programmes, habitat protection, and translocation and propagation of endangered species; promotes youth education; aims to instil understanding of and respect for wildlife and sound conservation practices.

German Marshall Fund of the United States: 11 Dupont Circle, NW, Suite 750, Washington, DC 20036; tel (202) 745-3950; fax (202) 265-1662; Pres: Frank E Loy; Dir of Programmes: Peter R Weitz; Programme Officer: Marianne Lais Ginsburg; independent institution founded with a donation from the people of Germany; promotes understanding of differences arising between Europe and the USA and stimulates exchange of practical experience on common problems; supports programmes in Central and Eastern Europe in the fields of political development, economics and the environment.

Environmental Partnership for Central Europe: 11 Dupont Circle, NW, Suite 210, Washington, DC 20036; tel (202) 588-9018; fax (202) 462-8097; US Project Dir: Irmgard Hunt; provides small grants, technical and organizational assistance to NGOs and local authorities in the Czech Republic, Hungary, Poland and Slovakia, for environmental projects and training, and maintains offices in those countries.

Global Tomorrow Coalition (GTC): 1325 G St, NW, Suite 1010, Washington, DC 20005-3104; tel (202) 628-4016; fax (202) 628-4018; Pres: Donald R Lesh; Vice-Pres: Diane G Lowrie; national, non-profit alliance of organizations and individuals working for a more sustainable, equitable future; concerned with environment, population and use of resources.

Grand Isle Heritage Zoo: 2103 W Stolley Park Rd, Grand Island, NE 68801-6941.

Grassroots Institute: 209 East D St, Belleville, IL 62220; Principal Officer: Don Pierce; affiliated to Earth Island Institute.

Greater Yellowstone Coalition (GYC): 13 S Willson, POB 1874, Bozeman, MT 59771; tel (406) 586-1593; fax (406) 586-0851; environmental advocacy promoting preservation and protection of the Greater Yellowstone ecosystem through sustainable forestry, agriculture and economic development; public education, publications, workshops and conferences.

Green Seal: POB 18237, Washington, DC 20036-8237.

Greenpeace Latin America, Co-ordination Office: 139 Townsend St, San Francisco, CA 94107.

Greenpeace USA: 1436 U St, NW, Washington, DC 20009; tel (202) 319-2444; fax (202) 462-4507; Information Co-ordinator: Sanjay Mishra.

Gulf and Caribbean Fisheries Institute (GCFI): Administrative Offices, 38 Wentworth St, Charleston, SC 29401; tel (803) 577-5697; fax (803) 577-5697; independent, non-profit corporation; promotes exchange of information on the use and management of marine resources in the region; annual meeting devoted to technical presentations and workshops on relevant issues.

Hawai'i Nature Center: 2131 Makiki Heights Dr, Honolulu, HI 96822; tel (808) 955-0100; fax (808) 955-0116; Pres: Warren G Haight; Vice-Pres: Oswald K Stender; fosters awareness, appreciation and understanding of the environment in Hawaii and wise stewardship of the islands; facilitates field studies.

Herpetologists' League: Department of Biology, University of Richmond, Richmond, VA 23173; tel (804) 289-8234; fax (804) 289-8482; Pres: Edmund D Brodie, Jr; Sec: Dr Joseph C Mitchell; promotion of herpetology through professional publications and annual meetings.

Hoosier Environmental Council: 1002 E Washington St, Suite 300, Indianapolis, IN 46202; tel (317) 685-8800; fax (317) 686-4794; Pres: Jack Gay; Vice-Pres: Bob Klawtier.

Hudson Valley Grass Roots Energy and Environmental Network (GREEN): POB 208, Red Hook, NY 12571; tel (914) 758-4484; Co-Chairs: Ruth Oja, Beverly Allyn; Sec: Marion Lockwood; non-profit organization concerned with safe energy and environmental preservation; newsletter; forums on regional and national issues; works to inform the public and increase public participation in energy and environmental policy-making.

Human Environment Center: 1930 18th St, NW, Suite 24, Washington, DC 20009-1796.

INFORM, Inc: 381 Park Ave South, New York, NY 10016-8806; tel (212) 689-4040; fax (212) 447-0689; Pres: Joanna D Underwood; research and education organization; examines business practices affecting the environment and suggests practical action to alleviate or prevent environmental damage; books and reports; member of Earth Share.

Institute for Development Anthropology (IDA): 99 Collier St, Suite 302, POB 2207, Binghamton, NY 13902-2207; tel (607) 772-6244; telex 932433; fax (607) 773-8993; Dirs: David Brokensha, Michael Horowitz, Thayer Scudder; independent, non-profit organization engaged in research, training, and project design and evaluation; issues covered include resettlement, rural household production systems, rural–urban linkages, rural co-operatives, river basin development and community forestry.

Institute for Earth Education (IEE): Cedar Grove, Greenville, WV 24945; tel (304) 832-6404; fax (304) 832-6077; Chair: Steve Van Matre; Int Programme Co-ordinator: Bruce Johnson; Int Training Co-ordinator: Mike Mayer; develops educational programmes to promote understanding and appreciation of the earth and its life; workshops, conferences and publications.

Institute of Ecosystem Studies: Research Program, POB AB, Millbrook, NY 12545; Education Program, POB R, Millbrook, NY 12545; tel (914) 677-5343 (Research), 677-5358 (Educ); fax (914) 677-5976 (Research), 677-6455 (Educ); Dir: Dr Gene E Likens; involved in research and education in areas of disturbance and recovery in northern temperate ecosystems, global ecology and environmental issues; promotes scientific literacy and provides continuing education.

Institute for Local Self-Reliance: 2425 18th St, NW, Washington, DC 20009-2096; tel (202) 232-4108; fax (202) 332-0463; Pres: Neil Seldman; Vice-Pres: David Morris; promotion of environmentally sound economic development through research, technical assistance and public education.

International Academy at Santa Barbara: 800 Garden St, Suite D, Santa Barbara, CA 93101-1552; tel (805) 965-5010; fax (805) 965-6071; Pres: John I Baum; Vice-Pres: Joanne St John; non-profit educational organization; aims to increase public awareness in the fields of environmental studies, energy, international relations and information resources; organizes seminars, lectures and conferences and issues publications; disseminates material relating to research and practical applications in its areas of interest, by written and electronic means; includes Environmental Studies Institute.

International Bicycle Fund: 4887 Columbia Drive South, Seattle, WA 98108-1919; tel (206) 628-9314; fax (206) 628-9314; Dir: David Mozer; provides general technical assistance to urban development that considers pedestrians, cyclists and the environment, and to sustainable rural transportation and economic development.

International Center for Environmental Strategy: California Institute of Public Affairs, 517 19th St, POB 189040, Sacramento, CA 95818; tel (916) 442-2472; fax (916) 442-2478; Dir: Thaddeus C Trzyha; acts as secretariat to IUCN Commission on Environmental Strategy; research and publishing.

International Crane Foundation: E-11376 Shady Lane Rd, Baraboo, WI 53913; tel (608) 356-9462; telex 297778; fax (608) 356-9465; Dir: Dr George Archibald; Deputy Dir: James Harris; conservation of cranes and their natural habitats through captive propagation, education and field ecology.

International Erosion Control Association: POB 4904, Steamboat Springs, CO 80477; tel (303) 879-3010; fax (303) 879-8563; Exec Dir: Ben Northcutt; Pres: Eric J Scherer; Tech Vice-Pres: Marc S Theisen; provides education and exchange of information on the causes and prevention of erosion world-wide.

International Green Party: POB 3413, Fullerton, CA 92631; tel (714) 675-4741; Chair: Randall Toler; ecological, anti-nuclear and pro-peace; 500,000 mems.

International Snow Leopard Trust: 4649 Sunnyside Ave North, Seattle, WA 98103; tel (206) 632-2421; fax (206) 632-3967; Pres: Helen Freeman; organization which provides an opportunity for the exchange of information on mountain species, especially the snow leopard; promotion of conservation and research in high altitude reserves; education of public on endangered species.

International Society for Endangered Cats (ISEC): 4638 Winterset Drive, Columbus, OH 43220; tel (614) 451-4460; fax (614) 451-4611; Exec Dir: Dr John Becker; Pres: Donell R Grubbs; conservation education, collection and dissemination of information on wild cats, support for conservation projects in the wild and in captivity, and establishment of a breeding centre.

International Society for the Preservation of the Tropical Rainforest: 3931 Camino de la Cumbre, Sherman Oaks, CA 91423; tel (818) 788-2002; fax (818) 990-3333; Pres and Dir: Arnold Newman; Dir: Roxanne Kremer; promotion of technologies for sustainable agriculture and timber harvesting in tropical areas and preservation of the Amazon pink river dolphin in Peru and Brazil; publications.

International Wilderness Leadership (WILD) Foundation: 211 W Magnolia, Fort Collins, CO 80521; tel (303) 498-0303; fax (303) 498-0403; Pres: Vance Martin; Vice-Pres: Ian Player, Cherri Bruggs; conducts projects in eight Southern African countries; organizes World Wilderness Congresses; protects wilderness and wildlife world-wide; promotes wise use of wildland resources and provides environmental education and training.

International Wildlife Coalition: 634 N Falmouth Highway, POB 388, N Falmouth, MA 02556; tel (508) 563-2843; fax (508) 563-2843; Pres: Dan Morast; Vice-Pres: Donna Hart, Steven Best; wildlife advocacy.

Iowa Natural Heritage Foundation: Insurance Exchange Bldg, 505 5th Ave, Suite 444, Des Moines, IA 50309; tel (515) 288-1846; fax (515) 288-0137; Exec Dir: Gerald F Schnepf; Dir of Programmes: Mark C Ackelson; Dir of Operations: John Tone; land protection, wetland restoration, reafforestation, resource planning and environmental education.

ISAR: 1601 Connecticut Ave, NW, Suite 301, Washington, DC 20009; tel (202) 387-3034; fax (202) 667-3291; Exec Dir: Eliza K Kloze; formerly the Institute for Soviet-American Relations; impartial, non-profit organization dedicated to supporting the democratic process and efforts to improve living conditions in the former USSR; focuses on practical projects in fields such as environmental protection, alternative energy, and sustainable economics and agriculture; promotes co-operation, operates environmental partnership programmes and produces publications.

International Clearinghouse on the Environment (ICE): established to promote better communication between environmentalists in the USA and the former USSR and to facilitate joint environmental activities; operates in close co-operation with the Socio-Ecological Union (SEU) in Moscow.

IUCN Commission on Environmental Strategy and Planning (CESP): Secretariat, POB 198040, Sacramento, CA 95818; tel (916) 442-2472; fax (916) 442-2478; Chair: Thaddeus C Trzyna; Deputy Chair: Liberty Mhlanga; aims to improve the formulation and implementation of policies for environmental protection and sustainable development, and to promote a world ethic of living sustainably; maintains networks of specialists, convenes meetings and disseminates information.

Ethics Working Group: Chair: J Ronald Engel.

Working Group on Landscape Conservation: Chair: Brynmor Green; development of approaches to conservation that use landscapes as units of analysis, planning and management.

Working Group on Strategies for Sustainability: Chair: Robert Prescott-Allen.

Working Group on Tools for Sustainability: Chair: Scott Slocombe.

IUCN Species Survival Commission: c/o Chicago Zoological Society, Chicago Zoological Park, Brookfield, IL 60513; tel (708) 387-0269; fax (708) 387-0269; Chair: Dr George B Babb; Deputy Chair: Dr Arturo Gómez Pompa.

Captive Breeding Specialist Group: 12101 Johnny Cake Ridge Rd, Apple Valley, MN 55124; tel (612) 431-9325; fax (612) 432-2757; Chair: Dr Ulysses S Seal; conservation assessment and management of broad groups of animals and regions; assessment of population and habitat viability for endangered plant and animal species.

IUCN USA: National Wildlife Federation, 1400 16th St, NW, Washington, DC 20036-2266; tel (202) 797-5454; fax (202) 797-5461; Exec Dir: Byron Swift.

Keep America Beautiful, Inc: 9 W Broad St, Stamford, CT 06902; tel (203) 323-8987; fax (203) 325-9199; Pres: Roger W Powers; Chair: Robert C Williams; non-profit public education organization dedicated to improving community waste-handling practices.

Keystone Center: POB 8606, Keystone, CO 80435-7998; tel (303) 468-5822; telex 403098; fax (303) 262-0152; Principal Officers: Robert W Craig, John R Ehrmann, Dr Michael T Lesnick; non-profit scientific, educational and public policy organization; promotes co-operation between the private sector, environmental citizens' organizations, academics and government; aims to address urgent problems, encourage understanding of the natural world and research and develop consensus on controversial public policy issues.

Land Institute: 2440 E Water Well Rd, Salina, KS 67401; tel (913) 823-5376; fax (913) 823-8728; promotes sustainable agriculture and offers postgraduate positions on research programmes.

League of Conservation Voters: 1707 L St, NW, Suite 550, Washington, DC 20036; tel (202) 785-8683; fax (202) 835-0491; Chair: Frank Loy; Exec Dir: Jim Maddy; supports environmental candidates for the US Congress and publicizes Congress members' environmental voting records.

Legacy International: 128 N Fayette St, Alexandria, VA 22314-2902; tel (703) 549-3630; telex 510600; fax (703) 549-0262; Exec Dir: Dr Ira Kaufman; encourages cross-cultural co-operation in order to address controversial issues; promotes environmentally sound development, peaceful resolution of ethnic, social and religious conflicts, comprehensive education and leadership training.

Legal Environmental Assistance Foundation, Inc (LEAF): 1115 N Gadsden St, Tallahassee, FL 32303-6327; tel (904) 681-2591; fax (904) 224-1275; Pres: B Suzi Ruhl; Vice-Pres for Programmes: Cynthia Valencic; provides free legal and technical assistance to citizens and local-level organizations and empowers the public to protect their health and environment from pollution by toxic chemicals; programme areas are justice and empowerment, pollution prevention and reduction, and energy advocacy.

Maine Organic Farmers and Gardeners Association (MOFGA): 283 Water St, POB 2176, Augusta, ME 04338-2176; tel (207) 622-3118; Exec Dir: Nancy Ross; provides technical information and advice, consumer education, public policy, organic certification and farm apprenticeships.

Marine Mammal Center: Marin Headlands, Golden Gate National Recreation Area, Sausalito, CA 94965; tel (415) 289-7325; Chair Bd of Dirs: Jerry Gibbons; Exec Dir: Peigin Barrett; studies, rescues and rehabilitates ill or injured marine mammals stranded along the Californian coastline; science and public education programmes.

Marine Mammal Commission: 1825 Connecticut Ave, NW, Rm 512, Washington, DC 20009; tel (202) 606-5504; fax (202) 606-5510; Exec Dir: John R Twiss, Jr; Scientific Programme Dir: Dr Robert J Hofman; furthers the purposes and policies of the US Marine Mammal Protection Act 1972.

Michigan Environmental Council (MEC): 115 W Allegan, Suite 10B, Lansing, MI 48933-1712; tel (517) 487-9539; fax (517) 487-9541; Exec Dir: Carol K Misseldine; Dir of Communications: Sharon L Riggle; coalition of environmental and conservation organizations seeking to protect the natural and human environment of Michigan and ensure its sustainability; acts through networking, advocacy, education and research.

Millennium Institute: 1611 N Kent St, Suite 204, Arlington, VA 22209-2135; tel (703) 841-0048; fax (703) 841-0050; Exec Dir: Dr Gerald O Barney; Chair Bd of Trustees: Peter H Aykroyd; formerly Institute for 21st Century Studies; non-profit organization promoting long-term, integrated global thinking, with the aim of securing a sustainable future for the planet; programmes support a world-wide network of individuals and organizations.

National Association of Conservation Districts (NACD): Headquarters, 509 Capitol Ct, NE, Washington, DC 20002; tel (202) 547-6223; fax (202) 547-6450; Pres: Gerald Digerness; Sec and Treas: Larry Beeler; non-profit organization; represents 3,000 local soil and water conservation districts (units of local government) in the USA and its territories; provides information and services.

Service Center: 408 E Main St, POB 855, League City, TX 77574-0855; tel (713) 332-3402.

North Central Regional Office: 1052 Main St, Stevens Point, WI 54481; tel (715) 341-1022.

North-eastern Regional Office: 804 S State St, POB 260, Dover, DE 19903-0260; tel (302) 734-7337.

Pacific Regional Office: 1615 NE Eastgate Blvd, Suite B, Pullman, WA 99163; tel (509) 334-1823.

Southern Regional Office: 11318 N May, Suite D, Oklahoma City, OK 73120; tel (405) 755-2288.

Western Regional Office: 9150 W Jewell, Suite 102, Lakewood, CO 80232-6469; tel (303) 988-1810.

Conservation Technology Information Center (CTIC): Purdue University Research Park, 1220 Potter Drive, Rm 170, West Lafayette, IN 47906-1334; tel (317) 494-9555; technology transfer resource centre; encourages the adoption of environmentally responsible conservation systems to improve agricultural viability; gathers and disseminates research results, publications, audio-visual materials and information on products and conservation techniques; maintains databases and publishes a newsletter.

National Association of Environmental Professionals (NAEP): 5165 MacArthur Blvd, NW, Washington, DC 20016; tel (202) 966-1500; fax (202) 966-1977; Pres: Kevin M Burger; Exec Dir: Susan Eisenberg; dedicated to the promotion of ethical practice, technical competence and professional standards.

National Audubon Society: 700 Broadway, New York, NY 10003; tel (212) 979-3000; fax (212) 353-0398; Pres and CEO: Peter A A Berle; Media Dir: Susan DeVico; preservation and protection of wildlife and its habitat through education, research, sanctuary management, litigation, lobbying and information dissemination; member of Earth Share.

Audubon Society of New Hampshire: 3 Silk Farm Rd, Concord, NH 03301; tel (603) 224-9909; fax (603) 226-0902; Pres: D Dickinson Henry; protection of wildlife and the natural environment through education, environmental advocacy, land protection and wildlife monitoring and research.

Florida Audubon Society, Inc: 460 Highway 436, Suite 200, Casselberry, FL 32707-4938; tel (407) 260-8300; fax (407) 260-9652; Pres: Dr Bernard Yokel; Vice-Pres: Charles Lee, Sr; practical conservation of wildlife and its habitats, with particular attention to birds of prey and turtles; established turtle preservation area in Guyana; environmental education, lobbying and research.

National Center for Appropriate Technology (NCAT): POB 3838, Butte, MT 59702; tel (406) 494-4572; fax (406) 494-2905; Pres: Ron Krocse; develops technical solutions that use local resources and labour to address society's problems; operates sustainable energy, resource-efficient housing and sustainable agriculture programmes; provides information and technical assistance.

National Coalition against the Misuse of Pesticides: 701 E St, SE, Washington, DC 20003; tel (202) 543-5450; fax (202) 543-4791; Exec Dir: Jay Feldman; member of Earth Share.

National Environmental Training Association: 2930 E Camelback Rd, Suite 185, Phoenix, AZ 85016; tel (602) 956-6099; fax (602) 956-6399; Exec Dir: Charles L Richardson; professional educational association; training support, materials and workshops for all environmental safety and health disciplines.

National Fish and Wildlife Foundation (NFWF): 1120 Connecticut Ave, NW, Suite 900, Washington, DC 20036; tel (202) 857-0166; fax (202) 857-0162; Exec Dir: Amos S Eno; Dir of Conservation Programmes: Whitney Tilt; non-profit organization dedicated to conservation and management of natural resources through habitat protection, environmental education, public policy development and leadership training; awards grants for conservation projects.

National Geographic Society: National Geographic Magazine, 1145 17th St, NW, Washington, DC 20036; tel (202) 828-5688; fax (202) 775-6141; Sr Writer and Environmental Specialist: Noel R Grove; founded for the increase and diffusion of geographic knowledge; concerned with public education through lectures, seminars and publications.

National Outdoor Leadership School (NOLS): 288 Main St, Lander, WY 82520-0579; tel (307) 332-6973; fax (307) 332-6973; Exec Dir: Jim Ratz; Education and Special Programmes Dir: Del Smith; wilderness education expeditions and environmental studies courses.

National Park Service, Pacific Area Office: 300 Ala Moana Blvd, Rm 6305, POB 50165, Honolulu, HI 96850; tel (808) 541-2693; fax (808) 541-3696; Dir (Pacific Area): G Bryan Harry; administration of nine units and one affiliated area of the national park system, in Hawaii, American Samoa, Guam and Saipan (Northern Mariana Islands).

National Parks and Conservation Association: 1015 31st Street, NW, Washington, DC 20007; tel (202) 223-6722; Pres: Paul C Pritchard; preservation, promotion and improvement of the national parks system; member of Earth Share.

National Speleological Society, Inc (NSS): 2813 Cave Ave, Huntsville, AL 35810-4431; tel (205) 852-1300; fax (205) 851-9241; Pres: Jeanne Gurnee; Exec Vice-Pres: Tom Rea; activities include education and research on the causes and prevention of cave damage, and co-operation with other groups dedicated to the conservation of natural areas.

National Tree Society, Inc: POB 10808, Bakersfield, CA 93389; tel (805) 589-6912; fax (805) 589-7135; Pres: Gregory W Davis; Sec and Treas: John S Bo; involved in tree-planting on public property; education of schoolchildren and the public on the importance of planting and caring for trees; promotion of recycling.

National Wildlife Federation: 1400 16th St, NW, Washington, DC 20036-2266; tel (202) 797-6000; fax (202) 797-6646; Pres: Jay D Hair; Exec Vice-Pres: William W Howard; aims to educate and assist individuals and organizations to conserve wildlife and other natural resources and protect the earth's environment, to ensure a peaceful, equitable and sustainable future; member of Earth Share.

Natural Resources Council of Maine: 271 State St, Augusta, ME 04330; tel (207) 622-3101; fax (207) 622-4343; Exec Dir: Everett B Carson; aims to conserve and restore Maine's environment and ensure wise use of its resources.

Natural Resources Defense Council (NRDC): 40 W 20th St, New York, NY 10011; tel (212) 727-2700; fax (212) 727-1773; Exec Dir: John H Adams; national, non-profit organization dedicated to protecting the world's natural resources and ensuring a safe and healthy environment; legal action; research; campaigns; 170,000 members; member of Earth Share.

Nature Conservancy: International Headquarters, 1815 N Lynn St, Arlington, VA 22209; tel (703) 841-5300; fax (703) 841-1283; Pres and CEO: John C Sawhill; Chair Bd of Governors: Joseph H Williams; private, non-profit members' organization; preservation of plants, animals and natural communities that represent the diversity of life on Earth by protecting the lands and waters they need to survive; owns and manages more than 1,300 nature reserves in the USA and operates programmes in Central and South America, the Caribbean and the Pacific; member of Earth Share.

Georgia Field Office: 1401 Peachtree St, NE, Suite 236, Atlanta, GA 30309; tel (404) 873-6946; fax (404) 873-6984; Vice-Pres and State Dir: Octavia C McCuean; Chair of Bd: Thomas H Benson.

Great Basin Field Office: POB 11486, Salt Lake City, UT 84147-0486; tel (801) 531-0999; fax (801) 531-1003; Dir: Dave Livermore; Dir of Conservation Programmes: Chris Montague.

Illinois Field Office: 79 W Monroe St, Suite 900, Chicago, IL 60603; tel (312) 346-8166; fax (312) 346-5606; Dir of Science and Stewardship: Stephen Packard.

Louisiana Field Office: POB 4125, Baton Rouge, LA 70821; tel (504) 338-1040; fax (504) 338-0103; Exec Dir: Nancy Jo Craig.

Nature Conservancy of Alabama: Pepper Place, 2821c 2nd Ave South, Birmingham, AL 35233; tel (205) 251-1155; fax (205) 251-4444; Exec Dir: Kathy Stiles Cooley.

South Carolina Chapter: POB 5475, Columbia, SC 29250; tel (803) 254-9049; fax (803) 252-7134; Exec Dir: Patrick H Morgan.

South Fork–Shelter Island Chapter: POB 2694, Sag Harbor, NY 11963; tel (516) 725-2936; fax (516) 725-3720; Chair: James D Dougherty; Sec: Bogart F Thompson.

New Forests Project (NFP): International Center for Development Policy, 731 8th St, SE, Washington, DC 20003; tel (202) 547-3800; fax (202) 546-4784; Dir: Stuart Conway; Project Co-ordinator: Jessica Goldberger; aims to prevent deforestation in developing countries; co-operation with NGOs to initiate training programmes on the importance of forests and natural resource protection; World Seed Program provides farmers with seeds for fast-growing, nitrogen-fixing tree species; newsletter.

New York Botanical Garden: Bronx, NY 10458-5126; tel (718) 817-8700; fax (718) 220-6504; Pres: Gregory Long; Vice-Pres for Botanical Science: Dr Brian Boom; Dir Inst of Systematic Botany: Dr Enrique Forero; botanical research, exploration, education and horticulture.

New York Flora Association (NYFA): c/o Biological Survey, New York State Museum, 3132 CEC, Albany, NY 12230; tel (518) 486-2027; fax (518) 473-8496; Principal Officers: Richard S Mitchell, Robert Zaremba; promotion of field botany in New York state; newsletter (3 a year); field trips; small grants for field botany projects.

North American Lake Management Society: One Progress Blvd, POB 27, Alachua, FL 32615-9536; tel (904) 462-2554; fax (904) 462-2568; promotion of understanding, protection and restoration of lakes, ponds, reservoirs and their watersheds; membership open to all interested individuals and organizations.

North Carolina State University, Department of Zoology: POB 7617, Raleigh, NC 27695-7617; tel (919) 515-2741; fax (919) 515-5327; Head of Dept: Harold Heatwone; population and community ecology, environmental physiology, fisheries, wildlife and aquaculture.

North Dakota Forest Service: First and Brander Sts, Bottineau, ND 58318; tel (701) 228-5422; fax (701) 228-5448; State Forester: Larry A Kotchman; administration of forestry programmes through a network of five forestry districts; operates a nursery specializing in the supply of conifer seedlings; technical assistance relating to management of private and state forest lands, urban and community forests, tree-planting and wildland fire protection.

Northcoast Environmental Center: 879 9th St, Arcata, CA 95521; Principal Officer: Connie Stewart; affiliated to Earth Island Institute.

Northeast Organic Farming Association: 153 Bowers Hill Rd, Oxford, CT 06478; tel (203) 888-9280; Pres of Interstate Council: Bill Duesing; Vice-Pres: Steve Gitman; support for the growth of organic farming and gardening in the north-eastern USA; publications; annual conference.

Northeast Sustainable Energy Association (NESEA): 23 Ames St, Greenfield, MA 01301; tel (413) 774-6051; fax (413) 774-6053; Exec Dir: Michael Grabscheid; Assoc Dir: Nancy Hazard; promotion of responsible energy use for a clean environment; annual Solar and Electric Vehicle Symposium; annual car race; conference for builders, architects, utilities and energy management consultants.

Northwest Conservation Act Coalition: 217 Pine St, Suite 1020, Seattle, WA 98101; tel (206) 621-0094; fax (206) 621-0097; Pres: Don Andre; Exec Dir: K C Golden; advocates the provision of energy services to the Pacific north-west USA and to the province of British Columbia, Canada, at the lowest cost to society (including the environmental cost), through use of renewable resources.

Northwest Renewable Resources Center: 1411 4th Ave, Suite 1510, Seattle, WA 98101-2216; tel (206) 623-7361; fax (206) 467-1640; Chair Bd of Dirs: James Waldo; Exec Dir: Amy C Solomon; private, non-profit organization founded by leaders of industry, native American tribes and environmental groups; dedicated to co-operative solving of natural resource and environmental policy disputes; conflict assessment; mediation; training in dispute resolution and cross-cultural communication.

Office for Interdisciplinary Earth Studies (OIES): University Corporation for Atmospheric Research (UCAR), POB 3000, Boulder, CO 80307-3000; tel (303) 497-2692; fax (303) 497-2699; Dir: Tom M L Wigley; publications and reports; meetings.

Ohio Alliance for the Environment: 445 King Ave, Columbus, OH 43201; tel (614) 421-7819; fax (614) 421-7819; Exec Dir: Irene Probasco; forums for adult environmental education; publications.

Ohio Environmental Protection Agency: 1800 Watermark Drive, POB 1049, Columbus, OH 43266-0149; tel (614) 644-3020; fax (614) 644-2329; Dir: Donald Schregardus; Asst Dir: Gerry Ioannides; responsible environmental regulation supported by scientific knowledge, effective management and comprehensive environmental education.

Ohio State University, Ohio Sea Grant College Program: 1314 Kinnear Rd, Columbus, OH 43212-1194; tel (614) 292-8949; fax (614) 292-4364; Dir: Dr Jeffrey M Reutter; funding for university-based research, extension and education projects relating to the Great Lakes and coastal issues.

Orion Society: 136 E 64th St, New York, NY 10021; tel (212) 758-6475; fax (212) 758-6784; Pres: M G H Gilliam; environmental education organization dedicated to public nature awareness.

Overseas Development Council: 1875 Connecticut Ave, NW, Suite 1012, Washington, DC 20009; tel (202) 234-8701; fax (202) 745-0067; Pres: John W Sewell; Vice-Pres: Catherine Gwin; research into the ways in which development and environmental concerns affect US domestic and international policy; research into development assistance, international and global problems and the implications of development for US economic security.

Pacific Environment and Resources Center (PERC): 1055 Fort Cronkhite, Sausalito, CA 94965; tel (415) 332-8200; fax (415) 332-2722; Pres: Armin Rosencranz; non-profit organization; policy research, public education and advocacy for global environmental protection; promotion of sustainable development world-wide and particularly in the Pacific Rim and neighbouring countries; Siberian Forests Protection Project; Tumen River Watch.

Pacific Whale Foundation: 101 N Kihei Rd, Suite 21, Kealia Beach Plaza, Kihei, Maui, HI 96753; tel (808) 879-8860; fax (808) 879-2615; Pres: Greg Kaufman; Dir of Research and Education: Paul Forestell; non-profit organization committed to research, conservation and educating the public from a scientific perspective about marine mammals and their ocean environment.

Panos Institute: 1717 Massachusetts Ave, NW, Suite 301, Washington, DC 20036; tel (202) 483-0044; fax (202) 483-3059; Exec Dir (Acting): Diane McGlyn; independent information organization promoting socially, environmentally and economically sustainable development world-wide; conducts research on key global issues and produces publications; offices in France and the United Kingdom.

Pennsylvania Environmental Council (PEC): 1211 Chestnut St, Suite 900, Philadelphia, PA 19107; tel (215) 563-0250; fax (215) 563-0528; Pres: Joanne R Denworth; Chair of Bd: Joseph M Manko; non-profit membership organization dedicated to protecting and improving environmental quality in Pennsylvania through policy-making, advocacy, education and dispute resolution; focus on land-use planning, growth management, preservation of open spaces, air and water quality and transportation issues.

Pennsylvania State University, Environmental Resources Research Institute: Land and Water Research Bldg, University Park, PA 16802; tel (814) 863-0291; fax (814) 865-3378; Dir: Archie J McDonnell; Asst to Dir: Raymond W Eyerly; research, education and technology transfer in the areas of acid mine drainage and mine land reclamation, environmental impact assessment, hydrological and environmental systems modelling, risk perception and public education, residuals control technology, health and toxicology.

Pinchot Institute for Conservation: 6118 Hibbling Ave, Springfield, VA 22150; tel (703) 912-9535; fax (703) 912-6648; Exec Dir: James W Giltmier; works for the protection of natural resources throughout the world; research into changes in environmental policies; publications.

Planet Drum Foundation: POB 31251, San Francisco, CA 94131; tel (415) 285-6556; fax (415) 285-6563; Dir: Peter Berg; Office Manager: Judy Goldhaft; ecological education group which stresses awareness of 'bioregions', areas with coherent and interconnected plant and animal communities, often defined by a watershed; publications, lectures and workshops.

Plutonium Free Future: 2018 Shattuck Ave, POB 140, Berkeley, CA 94704; tel (510) 540-7645; fax (510) 540-6159; Dir: Mayumi Oda; citizens' group based in the USA and Japan; promotion of wind and solar energy production world-wide, the phasing out of nuclear power and the abolition of nuclear weapons; works for the establishment of sustainable environmental policies in the USA and Japan.

Population Reference Bureau (PRB): 1875 Connecticut Ave, NW, Suite 520, Washington, DC 20009; tel (202) 483-1100; telex 10456; fax (202) 328-3937; Pres: Barbara Torrey; Vice-Pres: Cary Davis; private, non-profit organization which seeks to increase awareness of population trends and their implications; information on population; publications.

Primarily Primates: POB 15306, San Antonio, TX 78212-8506; tel (512) 755-4616; Dir: Wallace W Swett; offering an alternative to the destruction of primates who are abandoned by their owners.

Program on Environment: East-West Center, 1777 East-West Rd, Honolulu, HI 96848; tel (808) 944-7272; telex 989171; fax (808) 944-7298; Dir: A Terry Rambo; formerly the Environment and Policy Institute; aims to reconcile development with the environment in the Asia-Pacific region, maintaining environmental quality while improving the quality of human life; draws on the natural and social sciences to develop and disseminate new tools for policy-making, management, education and research.

Environmental Risk and Development: Co-ordinator: Kirk R Smith; aims to help countries to integrate environmental planning, regulation and control with economic development strategies; current work addresses three major categories of risk: indoor and outdoor air pollution, hazardous wastes and toxic chemicals, and global warming from greenhouse gases.

Institutions and Environmental Management: Co-ordinator: James E Nickum; programme comprises two projects: Water Management and Policy, and Urban Environmental Management.

Renewable Resources Management: Co-ordinator: Lawrence S Hamilton; programme comprises five projects: Sustainable Rural Development, Social Forestry, Mountain Development Strategies, Biodiversity and Protected Areas, and Spatial Information Systems for Resource Management.

Quebec-Labrador Foundation, Atlantic Center for the Environment: 55 S Main St, Ipswich, MA 01938; tel (508) 356-0038; fax (508) 356-7322; Chair: Robert A Bryan; Pres: Lawrence B Morris; involved in community development, environmental education, domestic and international NGO capacity-building, international training fellowships, technology transfer and research.

Rainbow Serpent: 2018 Shattuck Ave, POB 140, Berkeley, CA 94704; tel (510) 540-5917; fax (510) 540-6159; Founders (USA): Claire Greensfelder, Mayumi Oda; international network based in Japan and the USA; aims to support and bring together women around the world who are working for safe and renewable energy, use of renewable resources and abolition of nuclear power and weapons.

Rainforest Action Network (RAN): 450 Sansome St, Suite 700, San Francisco, CA 94111; tel (415) 398-4404; Dir: Randall Hayes; Dir of Operations: Tamra Peters; non-profit organization; works internationally with other environmental and human rights organizations to protect rainforests; involved in negotiation, public campaigns, research, conferences and seminars, and consumer education; publications.

Renewable Natural Resources Foundation (RNRF): 5430 Grosvenor Lane, Bethesda, MD 20814; tel (301) 493-9101; fax (301) 493-6148; Chair Bd of Dirs: Clare W Hendee; Exec Dir: Robert D Day; promotion of research and public education on renewable natural resources and of co-ordination and co-operation between professional, scientific and educational organizations; development of Renewable Natural Resources Center; conferences, publications and annual awards.

Resources for the Future: 1616 P St, NW, Washington, DC 20036; tel (202) 328-5000; fax (202) 939-3460; Pres: Robert W Fri; Vice-Pres: Paul R Portney; research and public education in the fields of natural resources and environmental quality.

Riverbanks Zoological Park: POB 1060, Columbia, SC 29202; tel (803) 779-8717; Dir: Palmer E Krantz, III; concerned with the captive reproduction of endangered species with the aim of releasing them into their natural environment at some time in the future.

Rocky Mountain Institute (RMI): 1739 Snowmass Creek Rd, Snowmass, CO 81654-9199; tel (303) 927-3851; fax (303) 927-4178; Exec Dir: L Hunter Lovins; Treas and Dir of Research: Amory B Lovins; non-profit research and educational foundation; aims to foster the efficient and sustainable use of resources as a means to ensure global security; active in the areas of agriculture, economic renewal, energy, security, transportation and water.

Rutgers University, Institute of Marine and Coastal Sciences: POB 231, New Brunswick, NJ 08903-0231; tel (908) 932-6555; fax (908) 932-8578; Principal Officer: J Frederick Grassle; research on processes governing change and stability in the ocean, particularly in the mid-Atlantic Bight region; co-manages undersea research centre.

Safe Energy Communication Council (SECC): 1717 Massachusetts Ave, NW, Suite 805, Washington, DC 20036; tel (202) 483-8491; fax (202) 234-9194; Exec Dir: Scott Denman; Research Dir: Martin Gelfand; co-ordinates and represents national energy groups; counters public relations efforts of the nuclear power industry; conducts public media workshops; disseminates information on alternative energy; develops media strategies on energy issues; member of Earth Share.

St Louis Zoological Park: 1 Goverment Dr, Forest Park, St Louis, MO 63139-1136; tel (314) 781-0900; fax (314) 647-7969; Dir: Charles H Hoessle; a park with more than 2,800 animals living on 83 acres in a variety of habitats; aims to focus public attention on animal diversity, ecology and conservation.

Savannah River Ecology Laboratory: University of Georgia, Drawer E, Aiken, SC 29802; tel (803) 725-2472; fax (803) 725-3309; Dir: Dr Michael H Smith; Assoc Dir: Dr Nat Frazer; research unit comprising three divisions: Wildlife Ecology and Toxicology, Wetlands Ecology and Biogeochemical Ecology.

Scenic America: 21 Dupont Circle, NW, Washington, DC 20036; tel (202) 833-4300; fax (202) 833-4304; Pres: Sally G Oldham; non-profit membership organization; dedicated to preserving and enhancing the scenic character of the communities and countryside of the USA through education, research and advocacy; member of Earth Share.

Sea Shepherd Conservation Society: 1314 2nd St, Santa Monica, CA 90401; tel (310) 394-3198; fax (310) 394-0360; Founder and Pres: Capt Paul Watson; investigates and documents violations of international regulations and treaties protecting marine wildlife and assists in their enforcement.

Sierra Club: 730 Polk St, San Francisco, CA 94109; tel (415) 776-2211; fax (415) 776-0350; Exec Dir: Carl Pope; Chair: J Michael McCloskey; non-profit membership organization; promotes conservation of the natural environment by influencing public policy decisions; practices and promotes responsible use of the world's ecosystems and resources; involved in campaigns, local-level activism, public education, outings and production of numerous publications; comprises 61 chapters; offices in Canada; 600,000 members.

Sierra Club Foundation: tel (415) 291-1800; Dir: Stephen M Stevick; established to fund non-legislative activities, including research, public education and publishing; member of Earth Share.

Sierra Club Political Committee: tel (415) 776-2211; separately funded political action committee; promotes environmental candidates for public office and educates the public about the Sierra Club's environmental priorities.

Sierra Club Legal Defense Fund: 180 Montgomery St, Suite 1400, San Francisco, CA 94104-4209; tel (415) 627-6700; fax (415) 627-6740; Exec Dir: Vawter Parker; provides legal services to the Sierra Club and other environmental organizations; international work includes advocacy before UN Human Rights tribunals in favour of a human right to a safe and healthy environment; member of Earth Share.

Washington, DC Office: 408 C St, NE, Washington, DC 20002; tel (202) 547-1141.

Sierra Club California: 923 12th St, Suite 200, Sacramento, CA 95814; tel (916) 557-1100.

Alaska Office: 241 E 5th Ave, Suite 205, Anchorage, AK 99501; tel (907) 276-4048.

Appalachian Office: 69 Franklin St, 2nd Floor, Annapolis, MD 21401; tel (410) 268-7411; responsible for the states of Delaware, Maryland, North Carolina, South Carolina, Tennessee, Virginia and West Virginia and the District of Columbia.

Midwest Office: 214 N Henry St, Suite 203, Madison, WI 53703; tel (608) 257-4994; fax (608) 257-3513; Regional Staff Dir: Carl Zichella; responsible for the states of Illinois, Indiana, Iowa, Kentucky, Michigan, Minnesota, Missouri, Ohio and Wisconsin.

North-East Office: 85 Washington St, Saratoga Springs, NY 12866; tel (518) 587-9166; responsible for the states of Connecticut, Maine, Massachusetts, New Hampshire, New Jersey, New York, Pennsylvania, Rhode Island and Vermont.

North-West Office: 1516 Melrose Ave, Seattle, WA 98122; tel (206) 621-1696; responsible for the states of Idaho, Oregon and Washington.

Spokane Branch Office: Route 2, POB 303A, Pullman, WA 99163; tel (509) 332-5173.

Northern California and Nevada Office: 4171 Piedmont Ave, Suite 204, Oakland, CA 94611; tel (510) 654-7847; also responsible for Hawaii.

Northern Plains Office: 23 N Scott, Rm 25, Sheridan, WY 82801; tel (307) 672-0425; responsible for the states of Montana, Nebraska, North Dakota, South Dakota and Wyoming.

South-East Office: 1330 21st Way S, Suite 100, Birmingham, AL 35205; tel (205) 933-9111; responsible for the states of Alabama, Florida, Georgia, Louisiana and Mississippi.

Florida Branch Office: 1201 N Federal Highway, Rm 250H, North Palm Beach, FL 33408; tel (407) 775-3846.

Georgia Branch Office: 1447 Peachtree St, NE, Suite 305, Atlanta, GA 30309; tel (404) 888-9778.

South-West Office: 516 E Portland St, Phoenix, AZ 85004; tel (602) 254-9330; responsible for the states of Arizona, Colorado, New Mexico and Utah.

Colorado Branch Office: 2017 10th St, Suite B, Boulder, CO 80302; tel (303) 449-5595.

Utah Branch Office: 177 E 900 S, Suite 102, Salt Lake City, UT 84111; tel (801) 355-0509.

Southern California and Nevada Office: 3345 Wilshire Blvd, Suite 508, Los Angeles, CA 90010; tel (213) 387-6528.

Southern Plains Office: 7502 Greenville Ave, Suite 670, Lock Box No 47, Dallas, TX 75231; tel (214) 369-8181; responsible for the states of Arkansas, Kansas, Oklahoma and Texas.

Sociedad Peruana de Derecho Ambiental (SPDA—Peruvian Society for Environmental Law), US Office: 636 Amberwood Lane, St Louis, MS 63021; tel (314) 391-0792; Vice-Pres: Silvana Graf.

Soil and Water Conservation Society (SWCS): 7515 NE Ankeny Rd, Ankeny, IA 50021; tel (515) 289-2331; fax (515) 289-1227; Exec Vice-Pres: Douglas Kleine; membership organization; promotes the protection, enhancement and wise use of soil, water and related natural resources; provides public education and information for researchers and analysts.

Solid Waste Association of North America (SWANA): POB 7219, Silver Spring, MD 20910; tel (301) 585-2898; fax (301) 589-7068; Exec Dir: H Lanier Hickman, Jr; promotes economically and environmentally sound municipal solid waste management in North America.

Southern Women Against Toxics (SWAT): PO Drawer 1526, 101 West Monroe St, Livingston, AL 35470; tel (205) 392-7257; fax (205) 392-7257; Founder and Dir: Linda Wallace Campbell; supports women's opposition to dumping and incineration of toxic waste; organizes lectures and media events.

Southwest Research and Information Center: POB 4524, 105 Stanford Drive, SE, Albuquerque, NM 87106; tel (505) 262-1862; fax (505) 262-1864; Dirs: Anne Albrink, David Benevides; provides technical information on environmental issues to members of the public, policy-makers, the media and national organizations, and encourages public participation in decision-making.

State and Territorial Air Pollution Program Administrators (STAPPA) and Association of Local Air Pollution Control Officials (ALAPCO): Joint Headquarters, 444 N Capitol St, NW, Suite 307, Washington, DC 20001-1512; tel (202) 624-7864; fax (202) 624-7863; Exec Dir: S William Becker; aim to improve the effectiveness of state and local air pollution control programmes and to co-ordinate the activities of officials in this field; liaise with the US Environmental Protection Agency, Congress, industrial associations and environmental organizations.

State University of New York at Buffalo, Conserve UB Energy Conservation Program: 120 John Beane Center, Amherst, NY 14260-7300; tel (716) 645-3636; fax (716) 645-5923; Energy Conservation Officer: Walter Simpson; energy conservation, recycling and monitoring of the university's impact on the environment.

State University of New York at Buffalo, Research Program in Environment and Society: Park Hall, NY 14260; tel (716) 645-2595; fax (716) 645-2166; Dir: Lester W Milbrath; research and publications on making society sustainable.

Holis: The Society for a Sustainable Future: Chair: Lester W Milbrath; Vice-Chair: Douglas McKenzie-Mohr; aims to assist the transition to sustainability and produce effective action programmes; conferences and publications.

Student Environmental Action Coalition (SEAC): POB 1168, Chapel Hill, NC 27514; tel (919) 967-4600; fax (919) 967-4648; national coalition of 2,000 high school and college environmental clubs; works for social justice and environmental protection.

Tennessee Environmental Council: 1700 Hayes St, Suite 101, Nashville, TN 37203; tel (615) 321-5075; fax (615) 321-5082; Pres: Dr Cliff Russell; Sec: Jo Ann Bennett; Exec Dir: John E Sherman; advocacy and education on the environment and public health; current projects concern prevention of toxic contamination and protection of groundwater.

Texas Technical University, International Center for Arid and Semiarid Land Studies (ICASALS): POB 41036, Lubbock, TX 79409-1036; tel (806) 742-2218; fax (806) 742-1954; Dir: Idris R Taylor; Deputy Dir: James E Jonish.

TRAFFIC USA: c/o World Wildlife Fund, 1250 24th St, NW, Washington, DC 20037; tel (202) 293-4800; fax (202) 775-8287; Dir: Ginette Hemley; Trade Records Analysis of Fauna and Flora in Commerce; monitoring of international wildlife trade; based in the United Kingdom.

Transnational Network of Appropriate Technology (TRANET): POB 567, Pond St, Rangeley, ME 04970; tel (207) 864-2252; Exec Dir: William N Ellis; Pres: Peter Thibeault; network of organizations and individuals dedicated to the conception of new social institutions, technologies and ways of life based on global sustainability.

Tree-mendous Maryland: Maryland Department of Natural Resources, Taylor Ave, Annapolis, MD 21401; tel (410) 974-3776; Exec Dir: Walter Orlinsky; Volunteer Co-ordinator: Terry Galloway; fund-raising and organization of volunteer groups to plant almost 1m. trees and seedlings each year; educational programmes on tree planting and maintenance.

Trees Foundation: POB 2202, Redway, CA 95560; tel (707) 923-4377; Administrator: Ami Goldberg; provides administrative and financial support for the protection and restoration of temperate forests.

Trees for Life: 1103 Jefferson, Wichita, KS 67203-3559; tel (316) 263-7294; fax (316) 263-5293; Pres: Balbir S Mathur; Sec: David Kimble; teaches inhabitants of villages in developing countries to plant fruit trees, as a renewable source of nutrition and to protect the environment; programmes in Guatemala, India and Nepal; offices in Austria and Canada; encourages ecological awareness among young children in the USA and promotes tree planting.

Trout Unlimited: 800 Follin Lane, SE, Suite 250, Vienna, VA 22180-4959; tel (703) 281-1100; fax (703) 281-1825; Pres and CEO: Charles F Gauvin; Chair Bd of Trustees: Robert Teufel; stream restoration, protection of salmonid habitats and environmental education.

Trumpeter Swan Society: 3800 County Rd, Suite 24, Maple Plain, MN 55359; tel (612) 476-4663; fax (612) 476-1514; Pres: Laurence N Gillette; Vice-Pres: Ruth E Shea; aims to maintain wild trumpeter swan populations and restore the bird to its original range; promotes research on the ecology and management of the trumpeter swan; collects and exchanges knowledge.

Trust for Public Land (TPL): 116 New Montgomery St, 4th Floor, San Francisco, CA 94105; tel (415) 495-4014; fax (415) 495-4103; Pres: Martin J Rosen; Exec Vice-Pres: Ralph W Benson; protects land resources for future generations; helps communities, public agencies and non-profit organizations acquire and protect open spaces; member of Earth Share.

Trustees for Alaska: 725 Christensen Drive, Suite 4, Anchorage, AK 99501; tel (907) 276-4244; fax (907) 276-7110; Exec Dir (acting): Michael M Wenig; Pres Bd of Trustees: Bob Childers; non-profit environmental law firm; represents conservation groups and native Alaskans in cases involving environmental and natural resource issues, including public lands, logging, mining, wildlife management, oil leasing, water and air pollution, and hazardous waste.

University of California, Berkeley, Center for Environmental Design Research (CEDR): 390 Wurster Hall, CA 94720; tel (510) 642-2896; telex 3667114; fax (510) 643-5571; Dir: Edward Arens; unit facilitating research in environmental planning and design; research programmes include building science, environmental assessment, urban design, and international architecture and urban development; publications.

University of Florida, Department of Wildlife and Range Sciences: POB 110430, Gainesville, FL 32611-0430; tel (904) 392-4851; fax (904) 392-6984; Chair: Dr Patricia A Werner; unit within School of Forest Resources and Conservation; research and education on understanding, managing and conserving biological resources.

University of Florida, Program for Studies in Tropical Conservation: Department of Wildlife and Range Sciences, 118 Newins-Ziegler Hall, Gainesville, FL 32611-0430; tel (904) 392-8372; fax (904) 392-6984; Dir: Dr Susan K Jacobson; Asst Dir: Dr Lyn C Branch; graduate education in tropical conservation and research in environmental education and natural resource management.

University of Maryland, Center for Global Change: Executive Bldg, Suite 401, 7100 Baltimore Ave, College Park, MD 20740; tel (301) 403-4165; fax (301) 403-4292; Principal Officer: Alan Miller; support for resolution of global environmental problems through policy research, education and public information; concerned with related issues including energy consumption, economic growth, population growth, resource allocation and development.

University of Michigan, School of Natural Resources and Environment: Dana Bldg, Ann Arbor, MI 48109-1115; tel (313) 763-8155; fax (313) 936-2195; Dean: Garry D Brewer; professional education and research on conservation biology, ecosystem management and restoration, corporate environmentalism, global change, environmental equity and urban environments.

University of Mississippi, Freshwater Biology Program: Department of Biology, MS 38677; tel (601) 232-7203; fax (601) 232-5144; Chair: James A Kushlar; environmental research and teaching in freshwater systems and graduate programmes in environmental microbiology, aquatic ecology and ecotoxicology.

University of Washington, Institute for Environmental Studies (IES): Mail Stop FM-12, Seattle, WA 98195; tel (206) 543-1812; fax (206) 543-2025; Dir: Prof James R Karr; Assoc Dir: Prof P Dee Boersma; dedicated to increasing understanding of natural systems and how they are affected by human actions; aims to prevent and reverse environmental degradation by integrating the natural, social and engineering sciences with policy and management disciplines; interdisciplinary research, teaching and public service programmes.

University of Wisconsin-Madison, Primate Center Library: Wisconsin Regional Primate Research Center, 1220 Capitol Court, Madison, WI 53715-1299; tel (608) 263-3512; fax (608) 263-4031; Dir: Larry Jacobsen; Head of Special Collections: Ray Hamel; publication of an international directory of primatology and audiovisual resources; electronic discussion and information networks.

Urban Habitat Program: 300 Broadway, Suite 28, San Francisco, CA 94133; tel (415) 788-3666; fax (415) 788-7324; Dir: Carl Anthony; promotion of multicultural environmental leadership; publication of newsletter dealing with race, poverty and the environment; development of ecological literacy project; public advocacy on issues of energy, transportation and social justice.

Utah State University, Department of Biology, Intermountain Herbarium: Logan, UT 84322-5305; tel (801) 750-1586; fax (801) 750-1575; Principal Officer: Dr Mary E Barkworth; major botanical resource for intermountain region; publishes manual of North American grasses.

Washington Environmental Council (WEC): 5200 University Way, NE, Suite 201, Seattle, WA 98105; tel (206) 527-1599; fax (206) 527-1693; Pres: Darlene Madenwald; Exec Dir: Tom Robinson; coalition of individuals and over 100 citizens' groups.

Water Environment Federation: 601 Wythe St, Alexandria, VA 22314; tel (703) 684-2400; fax (703) 684-2492; Exec Dir: Dr Q Brown; publication of periodicals, books and training materials; conferences.

Western Pennsylvania Conservancy: 316 4th Ave, Pittsburgh, PA 15222; tel (412) 288-2777; fax (412) 281-1792; Chair: Joshua C Whetzel; Pres: John C Oliver; conservation of water, land and wildlife.

Wetlands Preserve: 161 Hudson St, New York, NY 10013; Principal Officers: James Hansen, Kathy Kane; affiliated to Earth Island Institute.

Wild Wings and Underhill Foundations: 3731 W Street, NW, Washington, DC 20007; tel (202) 797-5454; Dir Tropical Conservation Program: Byron Swift; involved in environmental research, education and censuses; implementation of the Tropical Conservation Program for the protection and management of natural habitats.

Wilderness Society: 900 17th St, NW, 10th Floor, Washington, DC 20006; tel (202) 833-2300; Pres: George T Frampton, Jr; preservation of wildlife and wilderness; protection of forests, parks, rivers and coastlines; member of Earth Share.

Wildlife Center of Virginia: POB 98, Weyers Cave, VA 24486; tel (703) 234-9453; fax (703) 234-0804; Pres: Edward E Clark, Jr; Dir of Veterinary Services: Dr Stuart Porter; promotes positive behaviour towards wildlife and the environment; non-profit veterinary teaching and research hospital; rehabilitates injured wildlife with the aim of returning it to its natural habitat; environmental education.

Wildlife Conservation International: New York Zoological Society, 185th St and Southern Blvd, Bronx, NY 10460; tel (212) 220-5155; fax (212) 220-7114; Dir: Dr John Robinson; development of solutions to conservation problems through long-term field research and training; special interest in the conservation of tropical ecosystems and endangered species.

Wildlife Habitat Enhancement Council (WHEC): 1010 Wayne Ave, Suite 920, Silver Spring, MD 20910; tel (301) 588-8994; fax (301) 588-4629; Pres: Joyce M Kelly; international, non-profit organization which helps companies ensure that their land caters for the needs of wildlife by providing technical assistance for responsible corporate wildlife management practices, in addition to habitat certification, employee participation schemes and public education.

Wildlife Legislative Fund of America (WLFA) and Wildlife Conservation Fund of America (WCFA): 801 Kingsmill Parkway, Columbus, OH 43229-1137; tel (614) 888-4868; fax (614) 888-0326; Exec Dir: Rick Story; Chair: Daniel M Galbreath; non-profit organizations dedicated to the protection of American citizens' right to hunt, trap and fish, and to the promotion of scientific wildlife management practices; WLFA is responsible for lobbying and public relations and WCFA deals with education and research.

Wildlife Management Institute: 1101 14th St, NW, Suite 801, Washington, DC 20005; tel (202) 371-1808; fax (202) 408-5059; Pres: Rollin D Sparrowe; Vice-Pres: Lonnie L Williamson; Sec: Richard E McCabe; adviser to universities, professional organizations and federal, state and provincial governments on sound, scientific management of natural resources, particularly wildlife; publications and annual North American Wildlife and Natural Resources Conference.

Wildlife Society, Inc: 5410 Grosvenor Lane, Bethesda, MD 20814; tel (301) 897-9770; fax (301) 530-2471; Pres-elect: Dr Erik K Fritzell; Exec Dir: Dr Harry E Hodgdon; international scientific and educational association of professionals in the field of wildlife resources; aims to conserve biodiversity, sustain productivity and ensure wise use of wildlife resources; develops university curricula; publications, meetings and symposia.

World Bird Sanctuary: POB 270270, St Louis, MO 63127; tel (314) 938-6193; fax (314) 938-9464; Exec Dir: Walter C Crawford, Jr; activities include educational programmes, fund-raising, lectures, propagation, rehabilitation and release programmes.

World Environment Center: 419 Park Ave South, Suite 1800, New York, NY 10016; tel (212) 683-4700; telex 261290; fax (212) 683-5053; Chair of Bd: Dr Betsy Ancker-Johnson; Pres and CEO: Antony Marcil; promotes partnerships between industry, government and NGOs; forum for the exchange of information and expertise between industry, government, academia, NGOs and the public; encourages corporate environmental responsibility and leadership.

World Forestry Center: 4033 SW Canyon Rd, Portland, OR 97221; tel (503) 228-1367; fax (503) 228-3624; Pres: John L Blackwell; non-profit educational organization; promotes understanding and appreciation of forests and forest resources world-wide.

World Future Society (WFS): 7910 Woodmont Ave, Suite 450, Bethesda, MD 20814; tel (301) 656-8274; fax (301) 951-0394; Pres: Edward Cornish; association for the study of alternative futures; source of impartial information on a variety of topics, including environmental issues; conferences and publications.

World Resources Institute (WRI): 1709 New York Ave, NW, 7th Floor, Washington, DC 20006; tel (202) 638-6300; telex 64414; fax (202) 638-0036; Pres: Jonathan Lash; Sr Vice-Pres: J Alan Brewster; independent, non-profit organization; conducts international policy studies on global resources and environmental conditions; primarily concerned with issues such as forests, biodiversity, climate, energy, pollution, economics, population, technology, natural resource management strategies and assessments, and sustainable agriculture; publications, conferences and technical assistance; member of Earth Share.

Center for International Development and Environment: provides technical experts at the request of developing countries requiring assistance in the assessment or better management of natural resources; active in Africa, the Near East, South and Central America and the Caribbean.

World Wildlife Fund (WWF) USA: 1250 24th Street, NW, Suite 500, Washington, DC 20037-1175; tel (202) 293-4800; telex 64505; fax (202) 775-8287; Chair: Russell E Train; Co-Chairs: Lawrence S Huntington, Melvin B Lane; Pres (CEO): Kathryn S Fuller; national affiliate organization of WWF International; member of Earth Share.

Worldwatch Institute: 1776 Massachusetts Ave, NW, Washington, DC 20036-1904; tel (202) 452-1999; fax (202) 296-7365; Pres: Lester R Brown; Vice-Pres: Christopher Flavin, Sandra Postel; non-profit research institute concerned with global environmental issues; aims to increase public awareness of environmental threats in order to promote an effective public policy response; various publications.

WorldWIDE Network: 1331 H St, NW, Suite 903, Washington, DC 20005; tel (202) 347-1514; fax (202) 347-1524; Man Dir: Waafas Ofosu-Amaah; Co-Chairs Bd of Trustees: Kathryn Cameron Porter, Helen Freeman; non-profit membership organization; international network for women involved in development and environmental work.

Xerces Society: 10 SW Ash St, Portland, OR 97204; tel (503) 222-2788; fax (503) 222-2763; Pres: Thomas Eisner; Exec Dir: Melody Mackey Allen; Scientific Adviser: E O Wilson; international, non-profit organization dedicated to the conservation of invertebrates and the preservation of critical biosystems around the world; conducts research; provides educational materials and publications; seeks to influence public policy.

Yale University, School of Forestry and Environmental Studies, Tropical Resources Institute: 301 Prospect St, New Haven, CT 06511; tel (203) 432-3660; fax (203) 432-3817; Dir: Dr Florencia Montagnini; research, education, and information management and dissemination; concerned with sustainable utilization of resources, and restoration and conservation of tropical ecosystems.

Zoo Atlanta: 800 Cherokee Ave, SE, Atlanta, GA 30315; tel (404) 624-5600; telex 430189; fax (404) 627-7514; Dir: Dr Terry L Maple; Vice-Pres and Head of Animal Management: Dr C Dietrich Schaaf; involved in conservation research and environmental education.

UNITED STATES COMMONWEALTH TERRITORIES

Puerto Rico

Department of Natural Resources: POB 5887, Puerta de Tierra, San Juan, PR 00906; Ave Muñoz Rivera, Pda 3, San Juan, PR; tel 724-8774; fax 723-4255; Sec: Pedro A Gelabert; responsible for implementation of public policy on conservation, development and use of natural resources.

Comité Despertar Cidreño (Cidra Awakening Committee): POB 123, Cidra, PR 00739; tel 739-6091; fax 739-5492; Coordinator: Eliezer Colón Rivera; Sec: Rosita Guzmán; campaigns and education on water contamination in Puerto Rico; seminars and workshops on social ecology; support for environmental campaigns in eastern-central Puerto Rico.

UNITED STATES EXTERNAL TERRITORIES

Guam

Guam Environmental Protection Agency: IT&E Harmon Plaza, Complex Unit D-107, 130 Rojas St, Harmon, GU 96911; tel 646-8863; fax 646-9402; Admin: Fred M Castro; Deputy Admin: Joanne M Brown; maintenance of environmental quality by preventing degradation of land, water and air by pollutants, including all physical, chemical and biological agents.

Palau

Palau Environmental Quality Protection Board: POB 100, Koror, PW 96940; tel 488-1639; fax 488-2963; Chair: Marhence Madrangchar; Exec Officer: Gilbert U Demei; promulgates and enforces environmental regulations.

United States Virgin Islands

Department of Planning and Natural Resources, Division of Fish and Wildlife: 6291 Estate Nazareth, St Thomas, VI 00802; tel 775-6762; fax 775-3972; Commissioner: Roy E Adams; Dir: Ann Seiler; Chief of Environmental Education: Ralf Boulon, Jr; Chief of Wildlife Programmes: David Nellis; Chief of Fisheries Programmes: Stephen Meyers; research and surveying of fish and wildlife resources, protection and enhancement of marine and coastal habitats, research on endangered species and environmental education of residents and visitors.

Island Resources Foundation (IRF), Caribbean Headquarters: Red Hook Centre, POB 33, St Thomas, VI 00802; tel 775-6225; fax 779-6225; Pres: Dr Edward L Towle; implementation of development planning and natural resource management programmes in the Caribbean.

Virgin Islands Conservation Society: POB 3839, Christiansted, St Croix, VI 00822.

St Croix Environmental Association: 2111 Company St, POB 3839, Christiansted, St Croix, VI 00820; tel 773-1989; fax 773-7545; Principal Officers: Marcia Taylor, William Tobias; public environmental education, teacher training workshops, protection of habitats and ecosystems, promotion of recycling and responsible litter disposal, tree planting programme and water quality monitoring.

URUGUAY

Governmental Organizations

Ministry of Housing and the Environment: Zabala 1427, Montevideo; tel (2) 950211; fax (2) 962914; Minister: Manuel Romay.

Ministry of Livestock, Agriculture and Fishing: Avda Constituyentes 1476, Montevideo; tel (2) 404155; telex 937; fax (2) 499623; Minister: Pedro Saravia.

Non-Governmental Organizations

Asociación Uruguaya de Derecho Ambiental (AUDA—Uruguayan Environmental Law Association): Echeverriarza 3396, 11300 Montevideo; tel (2) 621024; fax (2) 621024; Pres and Founder: Dr Mateo J Magarinos de Mello; Dir: Dr Miguel Langón Cuñarro; research, elaboration and promotion of environmental policy and law; environmental education.

Centro Interdisciplinario de Estudios sobre el Desarrollo, Uruguay (CIEDUR—Interdisciplinary Centre for Development Studies, Uruguay): Joaquín Requena 1375, 11200 Montevideo; tel (2) 484674; fax (2) 480908; independent, non-profit organization conducting research in the social sciences; fields of interest include the environment, economic integration, employment and income, women's issues, decentralization and public participation in decision-making.

Centro de Investigación y Promoción Franciscano y Ecológico del Uruguay (CIPFE—Franciscan Ecological Research and Promotion Centre): Canelones 1164, R P Casilla 13125, Montevideo; tel (2) 907648; fax (2) 985959; Dir: P Jorge Peixoto; examines relationships between humans and nature; concerned with ecology, conservation education, deforestation, coastal and marine protected areas and endangered animal species; research in theology and history.

Centro de Investigación en Vida Silvestre (Wildlife Research Centre): Tristan Narvaja 1674, Casilla Postal 11100, Montevideo; tel (2) 803373; fax (2) 487388; Principal Officer: Francisco Rilla; wildlife research, conservation and management; involved in environmental impact assessment, consultancy and information dissemination.

Centro Latinoamericano de Ecología Social (CLAES—Latin American Centre for Social Ecology): Amezaga 1891, Casilla 13000, 11700 Montevideo; tel (2) 793132; fax (2) 793132; Co-ordinator: Eduardo Gudynas; promotion of social ecology in South and Central America and the Caribbean; co-ordination of Latin American social ecology network.

Grupo Eto-Ecologista Olimareño (GEEO—River Olimar Ethical-Environmental Group): Casa de la Cultura, Pablo Zufriateguí 412, 33000 Treinta y Tres; tel (452) 2108; fax (452) 52162; Sec: Prof Luis Alberto Miller; secondary-level environmental education, conferences, lectures and seminars; conservation of Quebrada de los Cuervos nature reserve.

Instituto de Comunicación y Desarrollo (ICD—Institute for Communication and Development)): Avda 18 de Julio 1431, Of 601, 11200 Montevideo; tel (2) 911646; fax (2) 924423; Principal Officers: Anabel Cruz, Fernando Barreiro; non-profit organization involved in information dissemination, documentation, research and technical consultancy; compilation of directory of environmental organizations in Uruguay.

NGOnet: Casilla 1539, 11000 Montevideo; tel (2) 496192; fax (2) 419222; Principal Officers: Roberto Bissio, Ricardo Carrere; environment and development information services for NGOs.

Partido del Sol (Sun Party): 18 de Julio 1235, Montevideo; tel (2) 901616; fax (2) 906739; Leader: Homero Mieres; political organization dedicated to ecology and pacifism.

Sociedad Amigos de la Preservación Ambiental (Friends of Environmental Protection Society): Fundación Naumann, Brito del Pino 1491, 11600 Montevideo; tel (2) 783445; fax (2) 792108; environmental education and legislation, protection of natural resources and prevention of pollution.

Sociedad de Conservación del Medio Ambiente (SCMA—Environmental Conservation Society): Cerro Largo 1895, 11200 Montevideo; tel (2) 499458; fax (2) 954131; Exec Dir: Carlos M Ferreira Ruella; conservation of natural resources, research and dissemination on environmental issues, and measures to improve the quality of life.

TRAFFIC Sudamérica—Oficina Regional (TRAFFIC South America—Regional Office): Carlos Roxlo 1496/301, 11200 Montevideo; tel (2) 493384; telex 23702; fax (2) 493384; Dir: Juan S Villalba-Macias; regional office of network based in the United Kingdom.

Universidad de la República, Facultad de Ciencias (University of the Republic, Faculty of Sciences): Calle Tristán Narvaja 1674, 11200 Montevideo; tel (2) 487419; fax (2) 487388; three departments involved in interdisciplinary study of wetlands and continental waters, and of the structure and function of the decomposition system; teaching.

Departamento de Botánica (Botany Department): Principal Officer: Dr Lina Bettucci.

Departamento de Ecología (Ecology Department): Principal Officers: Dr Horacio Vera, Manuela Sarasola.

Departamento de Limnología (Limnology Department): Principal Officer: Wilson Pintos.

UZBEKISTAN

Governmental Organizations

Ministry of Land Improvement and Water Resources: Tashkent, ul A Kadiry 5A; tel (3712) 41-13-53; Minister: Rim A Giniyatullin.

State Committee for Nature Conservation: Tashkent, ul A Kadiry 5A; tel (3712) 41-08-81; fax (3712) 41-39-90; Chair: Ashkat Sh Khabibullaev; First Deputy Chair: Vladimir G Kowjukhov; nature protection; promotion of the rational use of natural resources.

Non-Governmental Organizations

Ecolog Association: 700105 Tashkent, Gaidar pr 11A-10; tel (3712) 91-39-35; Co-ordinator: Oleg I Tsaruk; concerned with biodiversity and nature reserves, destruction of the Aral Sea, public health and ecological education.

International Museum of Peace and Solidarity: 703000 Samarkand, POB 76; tel (3662) 33-17-53; telex 116544; Dir: Anatoly Ionesov; exhibitions are concerned with the global challenges, including environmental protection, that the world faces in the late 20th century.

VANUATU

Governmental Organizations

Ministry of Natural Resources: POB 151, Port Vila; tel 23105; telex 1040; Minister: Paul Barthelemy Telukluk.

Department of Agriculture and Horticulture: Private Mail Bag 040, Port Vila; tel 22432; fax 25265; Minister: (vacant); activities include training, research and development.

Non-Governmental Organizations

Vanuatu Natural Science Society: POB 944, Port Vila; Pres: Nancy Stead; Sec and Treas: Barry Cook; public education through meetings, excursions and publications, on all issues related to the environment in Vanuatu.

VATICAN CITY

Governmental Organizations

Pontifical Council 'Cor Unum': 00120 Vatican City; tel (6) 69887331; telex 2030; fax (6) 69887301; Pres: Cardinal Roger Etchegaray; Sec: Mgr Iván Marín-López; responsible for international humanitarian and development aid matters.

VENEZUELA

Governmental Organizations

Ministry of the Environment and Renewable Natural Resources: Centro Simón Bolívar, Torre Sur, 18°, Of 1805, Caracas 1010; Apdo 6623, Caracas 1010-A; tel (2) 483-4683; telex 24305; fax (2) 483-2445; Minister: Adalberto Gabaldón; Deputy Minister: Norma Hernández de Gómez Mantellini.

Non-Governmental Organizations

Asociación Civíl Sociedad Conservacionista del Zulia (SOCOZULIA—Zulia Conservation Society): Calle 65, 78A-83, Urbanización Amalia, Maracaibo 4006, Zulia; tel (61) 53-7381; Pres: Marcos Portillo B; Sec for Institutional Relations: Marcos E Bracho C; research; education; information; direct action.

Asociación Educativa para la Conservación de la Naturaleza (EcoNatura—Educational Association for Nature Conservation): Apdo 63109, Caracas 1067-A; tel (2) 951-3586; fax (2) 951-3586; Gen Man: Isabel Novo; Asst Gen Man: Deborah Bigio; grants for conservation research; courses, workshops, lectures and other educational programmes on the conservation of natural resources.

Asociación para la Difusión de los Programas de las Naciones Unidas para el Medio Ambiente (ADNUMA de Venezuela—Association for the Dissemination of United Nations Environmental Programmes): Calle 65, 78A-83, Urbanización Amalia, Maracaibo 4006, Zulia; tel (61) 53-7381; Pres: Marcos Portillo; Co-ordinator: René A Bastardo; Gen

Sec: Alberto Quintero; support for the work of the UN, particularly UNEP; research; education; publications; co-operation in community work; consultancy.

Barinas Office: Apdo 181, Barinas 5201; tel (73) 21842.

Centro Conservacionista Ecológico Monagas (CECOEMO—Monagas Conservation and Ecology Centre): Avda Orinoco 157, Maturín 6201, Monagas; tel (91) 41-3904.

Centro de Recuperación de Especies en Extinción (CREE—Centre for Recuperation of Endangered Species): Calle Rivas Dávila 23, Montalbán, Ejido, Mérida; tel (74) 92-0261; fax (2) 576-1579; Gen Dir: Alexander Lobo; preservation and restoration of endangered animal and plant species; conservation of natural habitats; education; research; dissemination of information.

Fundación para la Defensa de la Naturaleza (FUDENA—Foundation for Defence of Nature): Avda Principal de Los Cortijos de Lourdes con 2A Transversal, Centro Empresarial Senderos, 6°, Ofs 611 y 612, Apdo 70376, Caracas 1071-A; tel (2) 238-2139; fax (2) 239-6547; Pres: Héctor Valencia; Exec Dir: Glenda Medina Cuervo; management of protected areas and endangered species; environmental education; dissemination of public information; national associate organization of WWF International.

Fundación La Salle de Ciencias Naturales (FLASA—La Salle Foundation for Natural Sciences): Apdo 1930, Avda Boyoca con Mariperez, Edif Fundación La Salle, Cota Mil, Caracas 1010-A; tel (2) 782-8711; fax (2) 793-7493; research; education; offices in Cojedes, Isla de Margarita and Trujillo, and in Guyana.

Fundación Polar (Polar Foundation): Calle Hans Neumann con 2a Avda de Los Cortijos de Lourdes, Edif Fundación Polar, 1°, Caracas; Apdo 70934, Los Ruices, Caracas 1071-A; tel (2) 202-3111; telex 21628; fax (2) 202-7522; Head, Documentary Information Unit: Lorena Betancourt; collaboration with other public and private development institutions; projects cover agriculture, the environment, science, culture, food economy, education, history and health.

Fundación Proyecto Paria (Paria Project Foundation): Calle Ribero 50, Rio Caribe, Sucre; tel (94) 61883; fax (94) 61883; Exec Dir: Edgardo Garcia Larralde; promotion of balanced economic and social development on the Paria Peninsula and responsible interaction between mass tourism and local development; ecotourism.

Caracas Office: Apdo 51269, Edif Galerías Bolívar, Of 22-A, Blvd Sabana Grande, Caracas 1050-A; tel (2) 762-7234; fax (2) 762-5030.

Fundación Venezolana para la Conservación de la Diversidad Biológica (BIOMA—Venezuelan Foundation for the Conservation of Biological Diversity): Edif Cámara de Comercio de Caracas, 4°, Avda Este 2, Los Caobos, Apdo 1968, Caracas 1010-A; tel (2) 571-3120; fax (2) 571-1412; Exec Dir: Dr Aldemaro Romero; identification, protection and management of significant natural areas; collation of ecological information in order to identify Venezuela's environmental problems and propose solutions; dissemination of information through courses, lectures, national and international conferences and publications.

Grupo de Estudios sobre Mujer y Ambiente (GEMA—Women and Environment Study Group): Avda Andrés Eloy Blanco (Este 2), Torre Los Caobos, 17°, Apto 171-B, Los Caobos, Caracas.

MUHITEPA Modulo de Educación Ambiental Agropecuario (MUHITEPA Agro-Environmental Education Module): Urbanización Manuel Palacio Fajardo, Bloque 1, Edif E02, Planta B0003, Apdo 181, Barinas 5201; tel (73) 21842; Principal Officers: Leidys Portillo, Dora de Martínez; support for working children and young people; defence of children's rights; technical training; psychological guidance; sports and recreation; environmental and agricultural education.

Phelps Ornithological Collection: Apdo 2009, Caracas 1010-A; tel (2) 71-9238; fax (2) 762-5921; Pres: Kathleen Phelps; Curator: Miguel Lentino R; systematic, taxonomical and biological study of birds.

PROVITA: Edif Catuche, Nivel 1, Of 106, Parque Central, Apdo 47552, Caracas 1041-A; tel (2) 576-2828; fax (2) 576-1579; Gen Dir: Franklin Rojas-Suárez; Tech Dir: Jafet Nassar; formerly Pro-Vita Animalium; conservation of wildlife and the natural environment through research, environmental education and public information.

Sociedad Conservacionista Aragua (SCA—Aragua Conservation Society): Apdo 5115, El Limón 2105-A, Maracay, Aragua; tel (43) 83-1734; Pres: Diogenes A Hermoso; Exec Sec: Carmen Valladares; environmental education; conservation of protected areas; participation in public policy-making.

Sociedad Conservacionista Audubon de Venezuela (SCAV—Audubon Conservation Society of Venezuela): Apdo 80450, Caracas 1080-A; Edif Matisco, 1°, Of 5, Calle Veracruz, Las Mercedes, Caracas; tel (2) 92-3268; fax (2) 91-0716; Pres: José Luis Mateo; Vice-Pres: Mary Lou Goodwin; conservation of wildlife and habitats; research; activism; environmental education.

Universidad del Zulia, Cátedra Libre Ambientalista (Zulia University, Chair in Environment): Apdo 4011-A-526, Maracaibo 4011, Zulia; tel (61) 51-2209; telex 64287; fax (61) 51-7252; Co-ordinator: Marcos Portillo B; part of the Faculty of Engineering.

Venezuela Committee for IUCN: Apdo 69, Caracas 1010-A; tel (2) 239-4111; telex 21535; Principal Officer: Cecilia de Blohm.

VIETNAM

Governmental Organizations

Ministry of Science, Technology and Environment: 39 Tran Hung Dao, Hanoi; tel 252731; telex 412287; Minister: Dang Hu.

Ministry of Forestry: 123 Lo Duc, Hanoi; tel 253236; Minister: Nguyen Quang Ha.

Ministry of Water Conservancy: 164 Tran Quang Khai, Hanoi; tel 268141; Minister: Nguyen Trong Nhan.

Ministry of Water Resources: 21 Lò Sú St, Hanoi; tel 4254785; telex 412281; fax 4256929; Minister: Nguyen Canh Dinh; Vice-Ministers: Dr Tran Nhon, Dr Phan Sy Ky, Dr Vu Trong Hong; plans for conservation, use and development of water resources; ensures irrigation and drainage for agriculture; meeting industrial and domestic water requirements; flood protection.

Department of Management for Science and Technology on Natural Resources and Environment: State Committee for Science and Technology, 39 Tran Dao St, Hanoi; tel 52731.

Non-Governmental Organizations

Hanoi University, Department of Environmental Science: 90 Nguyen Trai St, Dong Da, Hanoi; tel 4244734; fax 4243061; Head: Prof Dr Mai Dinh Yen; Sec: Le Huy Hoang; B SC and M SC courses in environmental sciences; research on biodiversity, environmental planning and monitoring, nature conservation and environmental impact assessment.

IUCN Vietnam: c/o UNDP Vietnam, 1 Trinh Hoai Duc, Hanoi; tel 53506; fax 59267; Project Co-ordinator: Victoria Heymell.

WESTERN SAMOA

Governmental Organizations

Agriculture Department: POB 206, Apia; tel 22561; Minister: Jack Netzler.

YEMEN

Governmental Organizations

National Council for Environment Protection in Yemen: 202 Zubeiry St, San‘a.

Ministry of Agriculture: San‘a; Minister: Sadiq Amin Abu Ras.

YUGOSLAVIA

Governmental Organizations

Co-ordinating Committee for the Environment: 11070 Belgrade, SIV, bulevar Lenjina 2; tel (11) 330349; telex 11448.

Federal Ministry of Agriculture: 11070 Belgrade, bul AVNOJ-a 104; tel (11) 602555; telex 11062; fax (11) 195244; Minister: Lovro Koviljko; Minister of the Environment: Slobodanka Djordan.

Hydrometeorological Service of the Republic of Serbia: 11000 Belgrade, Kneza Višeslava 66, POB 100; tel (11) 545240; telex 11748; fax (11) 545378; Dir: Nikola Dutina; Deputy Dir: Vladimir Dimitrijević; network of stations involved in meteorological and hydrological forecasting, agricultural meteorology, environmental control and radar monitoring.

Montenegrin Ministry of Agriculture, Forestry and Water Resources Management: Podgorica, Omladinskih Brig 2; tel (81) 31287; fax (81) 52935; Minister: Branko Abramović.

Montenegrin Ministry of Environmental Protection: Podgorica, Nemanjina Obala bb; tel (81) 23164; fax (81) 23297; Minister: Vukić Pulević; Asst to Minister: Nada Mugoša; Contact: Milica Kadić; formulation of strategy in accordance with the parliamentary declaration on the ecological state of Montenegro of September 1991; direction of economic activity according to the principles of sustainable development; legislation for environmental protection; conservation of natural resources; co-operation with organizations working for the protection of the human environment world-wide.

Serbian Ministry of Agriculture, Forestry and Water Resources Management: 11000 Belgrade, Nemamjina 26; tel (11) 642276; fax (11) 659146; Minister: Dr Jan Kisgeci.

Serbian Ministry of Environmental Protection: 11000 Belgrade, Nemanjina 22-26; tel (11) 657143; fax (11) 642242; Minister: Dr Pavle Todorović.

Non-Governmental Organizations

Yugoslav Green Party: Belgrade, Mutapova 12; tel (11) 4447030; Pres: Dragan Jovanović; environmental party open to all citizens regardless of national, religious or racial affiliation.

ZAIRE

Governmental Organizations

Department of the Environment and Nature Conservation: 15 ave de la Clinique, BP 1248, Kinshasa; tel (12) 31252; Minister of Environment and Tourism: Ramazani Malungu.

Institut Zaïrois pour la Conservation de la Nature (IZCN—Zairean Institute for Nature Conservation): 13 ave de la Clinique, BP 868, Kinshasa; tel (12) 33250; telex 21112; fax (12) 275647; Sec: Mayoke Nkongo-Nkita; official government agency responsible for the management, monitoring and administration of national parks, nature reserves, hunting areas and trapping stations.

Non-Governmental Organizations

Société Africaine d'Etudes sur l'Environnement et la Conservation de la Nature: BP 5698, Kinshasa-Gombe 10; Principal Officer: Longo Lond Efengu Isa; agroforestry; reforestation; soil protection.

ZAMBIA

Governmental Organizations

Ministry of Environment and Natural Resources: POB 34011, Lusaka; tel (1) 229624; fax (1) 252952; Minister: Dr Chuulu Kalima; conservation and management of natural resources; protection of the environment; control of pollution.

National Conservation Committee (NCC): body of environmentalists which prepared the National Conservation Strategy for Zambia.

Ministry of Energy and Water Development: Mulungushi House, Independence House, Nationalist Rd, POB 30055, Lusaka; tel (1) 214988; Minister: Edith Nawakwi.

Non-Governmental Organizations

Environment and Population Centre: POB 35614, Lusaka; tel (1) 289298; fax (1) 289298; Exec Dir: Wanga G Mumba; non-profit organization; promotion of awareness of the adverse effects of uncontrolled environmental degradation and population growth on the quality of life and on national development; information; educational workshops; family planning; public access library.

IUCN Zambia: POB 50694, Lusaka; tel (1) 232077; fax (1) 232077; Rep: Eric Feron.

Wildlife Conservation Society of Zambia: POB 30255, Lusaka; tel (1) 254226; Pres: Prof A A Siwela; Exec Dir: M Sichilongo; national environmental education and information campaign comprising publications, clubs for schoolchildren, a radio programme, newspaper articles and tourist facilities.

World Wide Fund for Nature (WWF) Zambia: POB 50551, Ridgeway, Lusaka; tel (1) 253749; fax (1) 253749; National Rep: Richard C V Jeffery; Programme Officer: Monica Ngoma Chundama; Public Relations Officer: Ann Chabala Kapapula.

Zambia University, Department of Chemistry, Environmental Group: POB 32379, Lusaka; tel (1) 213221; telex 44370; fax (1) 253952; Principal Officer: Prof Joseph Cernak; analysis of heavy metals such as copper, zinc, lead and cadmium found in the environment, particularly in mining areas; control of pesticides in the environment, waste dumping areas and lead emissions from vehicles.

Zambian Ornithological Society: POB 33944, Lusaka; monthly newletter; field meetings; sponsoring educational visits to national parks.

ZIMBABWE

Governmental Organizations

Ministry of Environment and Tourism: Karigamombe Centre, 53 Samora Machel Ave, Private Bag 7753, Causeway, Harare; tel (4) 794571; Minister: Herbert Murerwa.

Natural Resources Board: active in all areas of environmental projection and resource management.

Ministry of Lands, Agriculture and Water Development: Makombe Complex, Block 3, Harare St, Private Bag 7712, Causeway, Harare; tel (4) 707861; telex 22141; Minister: Dr Kumbirai Kangai.

Agricultural and Rural Development Authority (ARDA): POB 8439, Causeway, Harare; tel (4) 700095; telex 22272; fax (4) 705847; Principal Officers: Liberty Mhlanga, Martin Matanda, Joseph Made; rural planning; crop and livestock production; wildlife management; horticulture; large-scale irrigation.

Department of Natural Resources: Makombe Complex, Block 1, Harare St and Herbert Chitepo Ave, POB 8070, Causeway, Harare; tel (4) 705671; fax (4) 793123; promotion of sustainable use of the natural resources of Zimbabwe.

Non-Governmental Organizations

IUCN Southern Africa: Regional Office, POB 745, Harare; tel (4) 728266; telex 22670; fax (4) 720738; Regional Rep: India Musokotwane.

Ornithological Association of Zimbabwe: POB 8382, Causeway, Harare; tel (4) 882206; Pres: J Williams; encouraging interest, conservation and research in relation to the avifauna of Zimbabwe; quarterly magazine.

Wildlife Society of Zimbabwe: POB 3497, Harare; tel (4) 731596; Exec Dir: S M Brooks.

World Wide Fund for Nature (WWF) Zimbabwe: POB 8437, Causeway, Harare; tel (4) 730599; fax (4) 730599; Sr Conservation Adviser: Dr David H M Cumming.

Zambezi Society: 39 Binton Rd, Greystone Park, Harare; tel (4) 731596; Chair: Richard G Pitman; promotion of biological diversity and the ecological, aesthetic and other natural resources of the Zambezi region.

ZERO—Zimbabwe Environmental Research Organization: POB 5338, Harare; 44 Edmonds Ave, Belvedere, Harare; tel (4) 791333; fax (4) 732858; Exec Chair: Dr Yemi Katerere; Exec Sec: Sam Moyo; regional network of environmental experts; co-ordination of indigenous research on energy, environment and sustainable development issues in the Southern African Development Community (SADC).

Zimbabwe National Conservation Trust: POB 8575, Causeway, Harare; tel (4) 46105; fax (4) 46105; Exec Dir: John A Pile; local fund-raising events.

Zimbabwe Trust (Zimtrust): POB 4027, 4 Lanark Rd, Belgravia, Harare; tel (4) 730543; fax (4) 795150; Gen Sec: R H T Monro; registered welfare organization dedicated to relief of poverty and improvement of life; Communal Areas Management Programme for Indigenous Resources (CAMPFIRE) provides training support to rural communities in natural resource management; Oak Foundation provides small grants to community projects.

Zimbabwe University, Department of Biological Sciences: POB MP 167, Mount Pleasant, Harare; tel (4) 303211; fax (4) 732828; M SC course in tropical resource ecology; projects on the importance of trees, tropical soil fertility, pollution of man-made lakes and the contribution of veldt fires to global change.

BIBLIOGRAPHY
of
ENVIRONMENTAL PERIODICALS

BIBLIOGRAPHY OF ENVIRONMENTAL PERIODICALS

A3 Umwelt: A3 Zeitschriftenverlagges mbH, 2353 Gunstromsdorf, Thymianweg 1, Austria; tel (2236) 534-25; fax (2236) 534-29; Editor: Franz Summer; Advertising Manager: Friedrich Shabl; 8 a year; general environmental publication; circ 17,000.

Acid News: Swedish NGO Secretariat on Acid Rain, POB 245, 401 24 Göteborg, Sweden; tel (31) 15-39-55; fax (31) 15-09-33; Editor: Christer Aagren; 4–5 a year; newsletter of the Swedish and Norwegian NGO Secretariats on Acid Rain; covers air pollution and acidification of the environment; in English; circ approx 5,000.

Acid Precipitation: US Department of Energy, Office of Scientific and Technical Information (OSTI), POB 62, Oak Ridge, TN 37831, USA; (Subscriptions) NTIS, Springfield, VA 22161, USA; tel (615) 576-1188 (OSTI); fax (615) 576-2865; Editor: Jessie W Rushing; monthly; publishes most recent entries to relevant OSTI databases; contains abstracts of various types of sources concerning fossil energy and acid rain.

Acid Precipitation Digest: Acid Rain Information Clearinghouse, Center for Environmental Information, 46 Prince St, Rochester, NY 14607, USA; Editor: Robert W Pratt; f 1983; monthly; contains news summaries, book reviews, details of research and events; circ 1,000.

Acid Rain Update: The Acid Rain Foundation, Inc, 1410 Varsity Drive, Raleigh, NC 27606-2010, USA; tel (919) 828-9443; fax (919) 515-3593; Editor: Charlotte Shover; f 1981; quarterly; news on air pollution and its effects.

Acqua Aria: Editrice Arti Poligrafiche Europee, Via Casella 16, 20156 Milan, Italy; tel (2) 330221; fax (2) 39214341; f 1948; 10 a year; magazine dealing with ecology, pollution and the environment; circ 5,000.

Acta Theriologica: c/o Mammal Research Institute, Polish Academy of Sciences, 17-230 Białowieża, Poland; Editor-in-Chief: Zdzisław Pucek; Assoc Editor: Henryk Okarma; Asst Editor: Leszek Rychlik; f 1958; quarterly; international journal of mammalogy and related subjects including mammalian ecology; contains reports of research in over 40 countries, short articles and book reviews; many papers deal with the biology and ecology of mammals in the pristine lowland forest of Białowieża National Park; in English.

Action Alerts: Rainforest Action Network, 450 Sansome St, Suite 700, San Francisco, CA 94111, USA; tel (415) 398-4404; telex 76475; fax (415) 398-2732; monthly; members' bulletin dealing with environmental issues requiring urgent public action.

AES (Ambiente e Sicurezza): ERIS SpA, Via E Tellini 14, 20155 Milan, Italy; tel (2) 33103305; telex 323314; fax (2) 33104245; Editor: Stefano Meinardi; f 1979; monthly; magazine concerned with pollution; contains book reviews and patents; in Italian, with summaries in English; circ 7,500.

African Journal of Ecology: Blackwell Scientific Publications Ltd, Osney Mead, Oxford, Oxon OX2 0EL, UK; tel (865) 240201; telex 833355; fax (865) 721205; Editor: F I B Kayanja; f 1962; quarterly; journal of the East African Wildlife Society; former title East African Wildlife Journal; includes indexes and books reviews; circ 550.

African Wildlife: Rm 210, Monpak Centre, 22 Sir Lowry Rd, POB 58, Cape Town 8000, South Africa; tel (21) 461-7113; fax (21) 45-2607; Chair Editorial Cttee: Jan Giliomee; Editor: Sandie Anderson, Asst Editor: Pat Coles; f 1946; every 2 months; conservation magazine published by the Wildlife Society of Southern Africa; includes indexes and book reviews; in Afrikaans and English; circ 30,000.

Agazen: Ethiopian Wildlife and Natural History Society, POB 60074, Addis Ababa, Ethiopia; tel (1) 164045; educational magazine for students.

Agrifack: POB 2062, 10312 Stockholm, Sweden; Ynglingagatan 14, 11347 Stockholm, Sweden; tel (8) 613-49-10; fax (8) 20-20-81; Principal Officers: Helmer Svensson, Ingemar Holmen; f 1942; monthly; trade union issues and politics in the areas of agriculture, horticulture, forestry, food and nutrition, and the environment; circ 5,000.

Agroforestry Abstracts: CAB International, Wallingford, Oxon OX10 8DE, UK; tel (491) 32111; telex 847964; fax (491) 33508; co-published by the International Centre for Research in Agroforestry, Kenya.

Agroforestry Today: International Centre for Research in Agroforestry (ICRAF), ICRAF House, United Nations Ave, Gigiri, POB 30677, Nairobi, Kenya; tel (2) 521450; telex 22048; fax (2) 521001; quarterly magazine in English and French.

Air Currents: Bay Area Air Quality Management District, 939 Ellis St, San Francisco, CA 94109, USA; tel (415) 771-6000; fax (415) 928-8560; Editor: Teresa Galvin Lee; f 1959; monthly; aims to increase public awareness of issues related to air quality, new regulations and other local environmental questions; circ 3,000.

Air Pollution Consultant: McCoy and Associates, Inc, 13701 W Jewell Ave, Suite 202, Lakewood, CO 80228-4173, USA; tel (303) 987-0333; fax (303) 987-7917; Editor: Eric Weber; f 1991; every 2 months; information on issues related to air pollution.

Air Pollution Control: Eagle Publications, Inc, 42400 Nine Mile Rd, Suite B, Novi, MI 48375, USA; tel (800) 783-3491; fax (313) 347-3492; f 1991; every 2 months; deals with new air filtration products and techniques; circ 18,000.

Air Pollution Control: Bureau of National Affairs, Inc, 1231 25th St, NW, Washington, DC 20037, USA; 9435 Key West Ave, Rockville, MD 20850, USA (Subscriptions); tel (202) 452-4200; (800) 372-1033 (Subscriptions); telex 285656; fax (202) 822-8092; Editor: Eileen Z Joseph; f 1980; fortnightly; provides information and advice on the control of air pollution; intended for those responsible for compliance with relevant EPA and state regulations.

Air Pollution Control—Annual Report: Department of Mines, Mining Commissioner, Private Bag 0049, Gaborone, Botswana; tel 352641; telex 2557; fax 352141; provides information on air quality in certain areas of Botswana and deals with the environmental effects of pollution.

Air Pollution Management: McIlvaine Co, 2970 Maria Ave, Northbrook, IL 60062, USA; tel (708) 272-0010; fax (708) 272-9673; Editor: Robert McIlvaine; quarterly, with monthly supplements; former title *Gold Dust;* international journal providing commercial information, management advice and assistance for the air pollution control industry.

Air Pollution Titles: Pennsylvania State University, Environmental Resources Research Institute, Center for Air Environment Studies, 226 Fenske Laboratory, University Park, PA 16802, USA; tel (814) 863-0291; fax (814) 865-3378; every 2 months; listing of current air environment literature by subject, selected from government reports, abstracts, proceedings and journals.

Air and Water Pollution Control: Bureau of National Affairs, Inc, 1231 25th St, NW, Washington, DC 20037, USA; 9435 Key West Ave, Rockville, MD 20850, USA (Subscriptions); tel (202) 452-4200; (800) 372-1033 (Subscriptions); telex 285656; fax (202) 822-8092; Editor: Eileen Z Joseph; f 1986; fortnightly; contains news of recent pollution legislation and regulations, and of trends in government and industry.

Air–Water Pollution Report: Business Publishers Inc, 951 Pershing Dr, Silver Spring, MD 20910-4464, USA; tel (301) 587-6300; fax (301) 585-9075; Editor: Phil Zahodiakin; f 1963; weekly; reports on environmental protection, covering legislation, regulation, policy, relevant research and commercial information; includes statistics and book reviews.

Airmail: Indoor Air International, Postfach 2, 4467 Rothenfluh, Switzerland; tel (767) 318474 (UK office); fax (767) 313929; Publications Sec: D F Weetman; newsletter of Int Asscn for Indoor Air Quality.

Al-Mawarid Al-Tabi'iyyah—Natural Resources: Ministry of Petroleum and Minerals, POB 551, Muscat, Oman; tel 603333; telex 5280; Editor: Khalid az-Zubaidi; monthly; co-published by the Ministry of Agriculture and Fisheries; in Arabic and English.

Alberta Naturalist: Federation of Alberta Naturalists, POB 1472, Edmonton, AB T5J 2N5, Canada; fax (403) 453-8553; Editor: Anna Robertson; f 1970; quarterly; cumulative index available; circ 500.

Alert!: Immediate Edit and Design, Washington Environmental Council, 5200 University Way NE, Suite 201, Seattle, WA 98105, USA; tel (206) 343-9249; fax (206) 527-1693; Editor: Sally Van Niel; f 1975; monthly; publication of Washington Environmental Council; circ 2,000.

Alerta Ecologia e Ciencias do Ambiente: Universidade de São Paulo, Escola Superior de Agricultura 'Luiz de Queiroz', CP 9, 13.400 Piracicaba, SP, Brazil; tel (194) 33-0011; telex 191141; fax (194) 34-1611; f 1991; monthly; indexes the contents of 32 other journals covering ecology and the environmental sciences.

Alternatives: University of Waterloo, Faculty of Environmental Studies, Waterloo, ON N2L 3G1, Canada; tel (519) 885-1211; fax (519) 746-0292; Editor: Dr Robert Gibson; f 1971; quarterly; perspectives on society, technology and environment; circ 3,500.

Ambiente: c/o Publicitas Ltd, 517–523 Fulham Rd, London SW6 1HD, UK (Rep in UK); tel (71) 385-7723; fax (71) 381-8884; monthly; general environmental publication; in Italian; circ 20,000.

Ambiente Risorse Salute (Environment, Resources and Health): Centro Studi 'l'Uomo e l'Ambiente', Via delle Palme 13, 35137 Padova, Italy; tel (49) 36435; Editor: Tina Merlin; f 1982; monthly; deals with pollution, use of resources and the development of clean technologies, with particular emphasis on public health and law; in Italian, with summaries in English; circ 6,000.

Ambio: Allen Press, Inc, POB 1897, Lawrence, KS 66044, USA; tel (913) 843-1234; Editor: Elizabeth Kessler; f 1972; 8 a year; international journal of the human environment; published for the Royal Swedish Academy of Sciences; covers environmental management and technology, and the natural sciences; includes bibliographies and book reviews; circ 3,700.

Aménagement et Nature: Association pour les Espaces Naturels, 21 rue du Conseiller Collignon, 75116 Paris, France; fax (1) 45-20-45-36; Editor: Roland Bechmann; f 1966; quarterly; publication dealing with planning and the natural world; includes bibliographies, indexes and book reviews; circ 2,900.

American Environmental Laboratory: International Scientific Communications, 30 Control Drive, Shelton, CT 06484-6111, USA; tel (203) 926-9300; fax (203) 926-9310; Editor: Brian Howard; every 2 months; circ 37,050.

The American Naturalist: The University of Chicago Press, Journals Division, POB 37005, 5720 S Woodlawn Ave, Chicago, IL 60637, USA; tel (312) 702-7600; fax (312) 753-0811; Editor: Mark D Rausher; f 1867; monthly; sponsored by American Soc of Naturalists; journal of ecology, evolution and population biology; topics addressed include ecological succession, plant breeding systems, animal mating systems and adaptive behaviour.

Amicus Journal: Natural Resources Defense Council Inc, 40 W 20th St, New York, NY 10011, USA; tel (212) 727-2700; fax (212) 727-1773; Editor: Francesca Lyman; f 1979; quarterly; intended for members of the Natural Resources Defense Council and other interested parties; includes indexes and book reviews; circ 120,000.

ANJEC Report: Association of New Jersey Environmental Commissions (ANJEC), 300 Mendham Rd, Route 24, POB 157, Mendham, NJ 07945, USA; tel (201) 539-7547; fax (201) 539-7713; reports on environmental education and training for municipal officials.

Annual Environment Guide & Directory: Protec—Il Progresso e L'Ambiente, Publi & Consult SpA, Via Tagliamento 29, 00198 Rome, Italy; tel (6) 8543267; telex 622368; fax (6) 8440697; Dir: Paolo F Bancale.

Annual List of Publications: Department of the Environment and Department of Transport, Library Services, Publication Sales Unit, Bldg 1, Victoria Rd, Middx HA4 0NZ, UK; Editor: S Perrott; f 1971; bibliography; circ 2,000.

Annual Report of Alberta Department of the Environment: 9820 106th St, Edmonton, AB T5K 2J6, Canada; tel (403) 422-2813; describes the Department's programmes and services and includes annual financial accounts.

Annual Report of the Bureau of Resource Assessment and Land Use Planning: University of Dar es Salaam, POB 35097, Dar es Salaam, Tanzania; Editor: Adolfo Mascarenhas; f 1968; circ 250.

Annual Report of Chesapeake Bay Foundation: 162 Prince George St, Annapolis, MD 21401, USA; tel (410) 268-8816; fax (410) 268-6687; Managing Editor: Flannery Davis; contains interviews, profiles, accounts of ongoing programmes, lists of contributors to the Foundation and financial statements.

Annual Report of the DLO Winand Staring Centre: DLO Winand Staring Centre for Integrated Land, Soil and Water Research, Marijkeweg 11, 6709 PE Wageningen, Netherlands; POB 125, 6700 AC Wageningen, Netherlands; tel (8370) 74200; telex 75230; fax (8370) 24812; f 1990; description of research conducted at the Centre; in Dutch.

Annual Report of Environment Canada: Les Terrasses de la Chaudière, 10 Wellington St, Ottawa, ON K1A 0H3, Canada; tel (819) 997-2800; fax (819) 953-2225; annual; contains summaries of various sectors' activities; indexed; in English and French.

Annual Report of the Kerala Board for Prevention and Control of Water Pollution: Thiruvananthrapuram 685 001, Kerala, India; contents include statistical tables; in English.

Annual Report of Manitoba Environmental Council: Environmental Council, Bldg 2, 139 Tuxedo Ave, Winnipeg, MB R3N 0H6, Canada; tel (204) 945-7031; f 1973; circ 900.

Annual Report of the Ministry of the Environment: 40 Scotts Rd, Environment Bldg, Singapore 0922, Singapore; fax 7319866; f 1972; in English.

Annual Report of the National Parks Board: POB 787, Pretoria 0001, South Africa; tel (12) 343-9770; telex 21324; fax (12) 343-0153; includes statistics; circ controlled.

Annual Report of the Natural Resources Department: Government Printer, POB 136, Lusaka, Zambia; f 1964, relaunched 1976; former title *Annual Report of the Office of the Conservateur of Natural Resources*.

Annual Report of the Northern Territory Conservation Commission: Conservation Commission, POB 1046, Alice Springs, NT 0871, Australia; fax (89) 52-5390; f 1978; circ 300.

Annual Report of the Western Australia Environmental Protection Authority: Westralia Square, 141 St George's Terrace, Perth, WA 6000, Australia; tel (9) 222-7000; fax (9) 322-1598; f 1972; circ 1,000.

Annual Review of Energy and the Environment: Annual Reviews Inc, 4139 El Camino Way, POB 10139, Palo Alto, CA 94303-0897, USA; tel (415) 493-4400; fax (415) 855-9815; Editor: Robert Socolow; Assoc Editors: Dennis Anderson, John Harte; f 1976; contains approx 20 updated chapters dealing with issues in the field of energy and the environment.

Annual Summary: National Air Pollution Surveillance: Environment Canada, EP Publications, Ottawa ON K1A 0H3, Canada; fax (819) 953-5921; f 1970; produced by Environment Canada Conservation and Protection Service; monitors and assesses air quality in urban regions of Canada; in English and French; circ 600.

Applied Geography: Butterworth-Heinemann Ltd, Linacre House, Jordan Hill, Oxford OX2 8DP, UK; tel (865) 310366; fax (865) 310898; Group Editor: Barbara Clarke; Editor: Dr James Hansom; f 1981.

APT Communiqué: Association for Preservation Technology (APT), POB 2487, Station D, Ottawa, ON K1P 5W6, Canada (Publr); POB 2165, Albuquerque, NM 87103, USA (Subscriptions); tel (505) 265-3838 (Publr); Editor: Barbara L Daniels; f 1972; every 2 months; indexed; includes book reviews; mainly in English; circ 2,000.

Aquatic Conservation: Marine and Freshwater Ecosystems: John Wiley & Sons Ltd, Baffins Lane, Chichester, W Sussex PO19 1UD, UK; tel (243) 770351; fax (243) 775878; Editors: Dr P J Boon, Dr R M Mitchell; Publ Editor: H Bailey; f 1991; quarterly; international scientific journal; circ 500.

Aquatic Environment Monitoring Report: Directorate of Fisheries Research, Ministry of Agriculture, Fisheries and Food, Lowestoft, Suffolk NR33 0HT, UK; tel (502) 562244; f 1979; 2 a year.

Aquatic Sciences and Fisheries Abstracts (ASFA) Series, Part 1—Biological Sciences and Living Resources: Cambridge Scientific Abstracts, 7200 Wisconsin Ave, Bethesda, MD 20814-4823, USA; tel (301) 961-6750; fax (301) 961-6720; Publr: Ted Caris; Managing Editor: Angela Hitti.

Aquatic Sciences and Fisheries Abstracts (ASFA) Series, Part 2—Ocean Technology, Policy and Non-Living Resources: Cambridge Scientific Abstracts, 7200 Wisconsin Ave, Bethesda, MD 20814-4823, USA; tel (301) 961-6750; fax (301) 961-6720; Publr: Ted Caris; Managing Editor: Angela Hitti.

Aquatic Sciences and Fisheries Abstracts (ASFA) Series, Part 3—Aquatic Pollution and Environmental Quality: Cambridge Scientific Abstracts, 7200 Wisconsin Ave, Bethesda, MD 20814-4823, USA; tel (301) 961-6750; fax (301) 961-6720; Publr: Ted Caris; Man Editor: Angela Hitti; abstracts journal devoted to contamination problems of seas, lakes, rivers and estuaries.

Aquatic Toxicology: Elsevier Science Publishers BV, POB 181, 1000 AD Amsterdam, Netherlands; tel (20) 5803911; fax (20) 5803249; Editors: D C Malins, A Jensen; f 1981; 12 a year in 3 vols; intended for environmental and biochemical toxicologists, ecotoxicologists, marine biologists, researchers involved in environmental monitoring and conservationists; includes statistics and indexes.

Arab Water World: Chatila Publishing House, POB 13-5121, Chouran, Beirut, Lebanon; tel (1) 352413; telex 23008; fax (1) 352419; Man: Mona Chatila; Publr and Editor-in-Chief: Fathi Chatila; Asst Editor: Bassam Nakad; f 1977; trade journal of the water and waste water industries in the Middle East and Africa; intended for public and private sector firms; published in English and Arabic; circ approx 9,000.

Arbor Age: Gold Trade Publications, Valjean Ave, POB 8420, Van Nuys, CA 91409, USA; tel (818) 781-8300; Editor: Anne Goldstein; Advertising Manager: Denne Goldstein; f 1981; monthly; publication concerned with tree care; intended for professionals responsible for trees in urban environments; circ 15,000.

Arboricultural Journal: The International Journal of Urban Forestry: A B Academic Publishers, POB 42, Bicester, Oxon OX6 7NW, UK; tel (869) 320949; Editor: Dr T H R Hall; quarterly; published for Arboricultural Asscn; covers ecology and other branches of science relevant to arboriculture; includes original research papers, news and reviews; in English with summaries in French and Spanish.

Archives of Complex Environmental Studies (ACES): ACES Publishing Ltd, POB 114, 33101 Tampere, Finland; International Society of Complex Environmental Studies (ISCES), Haiharankatu 17 B 11, 33710 Tampere, Finland; tel (31) 697307 (ISCES); fax (31) 633207 (ISCES).

Archives of Environmental Contamination and Toxicology: Springer-Verlag New York Inc, 175 Fifth Ave, New York, NY 10010, USA; tel (212) 460-1627; fax (212) 473-6272; every 2 months; describes important developments and discoveries in the areas of air, soil and water pollution; circ 1,100.

Archiwum Ochrony Srodowiska: Ossolineum, Publishing House of the Polish Academy of Sciences, Wrocław, Rynek 9, Poland; telex 712771; Editor: S Godzik; quarterly; environmental protection archives; deals with natural resources and their conservation; in Polish, with summaries in English and Russian.

Arctic Circle: Canadian Arctic Resources Committee (CARC), 1 Nicholas St, Suite 412, Ottawa, ON K1P 9Z9, Canada; tel (613) 236-7379; fax (613) 232-4665; every 2 months; analysis of issues affecting the Canadian North; circ 20,000.

Arctic Journal: Arctic Institute of North America, University of Calgary, 2500 University Drive, NW, Calgary, AB T2N 1N4, Canada; tel (403) 220-7515; fax (403) 282-4609; Editor: Karen McCullough.

Arhiv za Higijenu Rada i Toksikologiju (Archives of Industrial Hygiene and Toxicology): Institute for Medical Research and Occupational Health, University of Zagreb, 41001 Zagreb, Ksaverska Cesta 2, POB 291, Croatia; tel (41) 434537; fax (41) 274572; Editor: R Plestina; f 1950; quarterly; includes abstracts, bibliographies, statistics, indexes and book reviews; in Croatian and English; circ 2,000.

Arkadaş-Bagimsiz Gazette: Arkadaş, Z. Gökalp Cad C 171, 21950 Ergani, Turkey; tel (8371) 5032; fax (8371) 5525; Editor-in-Chief: Ümit Öztürk; f 1987; weekly newsletter of Arkadaş organization; in Turkish.

Arkadaş-Independent Newspaper: Arkadaş, Z Gökalp Cad C 171, 21950 Ergani, Turkey; tel (8371) 5032; fax (8371) 5525; Editor-in-Chief: Ümit Öztürk; f 1987; monthly newsletter of Arkadaş organization; English summary of *Arkadaş-Bagimsiz Gazette*.

Ársrit Skógræktarfélag Islands: Ránargata 18, 101 Reykjavík, Iceland; tel (1) 18150; fax (1) 627131; forestry year book; published by Forest Asscn of Iceland.

ASEP Newsletter: Asian Society for Environmental Protection (ASEP), CDG-South East Asia Program Office, Asian Institute of Technology, POB 2754, Bangkok 10501, Thailand; tel (2) 524-5363; telex 84276; fax (2) 516-2126; Asst Editor: Nuanchan Sangthongsttit; Editor (acting): Jay R Wisecarver; quarterly; contains articles on environmental protection, management and technology in the form of research and project summaries, country status reports, essays, news reports and notices of current and forthcoming events; items contributed principally by members of the society.

Asian Environment: Asian Environment Journals, POB 90 MCC, Makati, Philippines; telex 63199; Editor: E A R Ouano; f 1978; quarterly; journal of environmental science and technology; published for the Pollution Control Association of the Philippines and the Philippine and Indonesian Societies of Sanitary Engineers; circ 2,000.

Asian-Pacific Environment: Asia-Pacific People's Environment Network (APPEN), c/o Sahabat Alam Malaysia (Friends of the Earth Malaysia), 43 Jalan Salween, 10050 Pulan Pinang, Malaysia; tel (4) 376930; Editor: Vasentha Sampasivam; f 1982; quarterly; members' newsletter; circ 5,000.

Asian Water and Sewage: Techni-press Asia Ltd, POB 20494, Hennessy Rd, Hong Kong; fax 5278399; Editor: A W Neill; f 1984; quarterly; news and technical articles relating to water supply and treatment industries in Asia; circ 3,300.

Attenderingsbulletin: Land, Bodem, Water: DLO Winand Staring Centre for Integrated Land, Soil and Water Research, Marijkeweg 11, 6709 PE Wageningen, Netherlands; POB 125, 6700 AC Wageningen, Netherlands; tel (8370) 74733; telex 75230; fax (8370) 24812; every 2 months; provides information on current land, soil and water research; contains abstracts.

Audubon Activist: National Audubon Society, 700 Broadway, 5th Floor, New York, NY 10003, USA; tel (212) 979-3000; fax (212) 353-0190; Vice-Pres, Conservation Information: Victoria Shaw; newspaper reporting on national environmental action and activism on a local level.

Aura: Wydawnictwo Czasopism i Ksiazek Technicznych SIGMA—NOT, 00-950 Warsaw, ul Biala 4, POB 1004, Poland; f 1973; monthly; describes problems related to conservation of the human environment; circ 14,000.

Australasian Solar Index and Buyers' Guide: Australian Syndicators Pty Ltd, POB 45, South Yarra, Vic 3141, Australia; f 1979; annual publication.

Australian Geographic Journal: Australian Geographic Pty Ltd, 321 Mona Vale Rd, POB 321, Terrey Hills, NSW 2084, Australia; tel (2) 450-2344; telex 176203; fax (2) 450-2990; Publr and Editor-in-Chief: Dick Smith; Chief Exec: Ike Bain; Editor: Howard Whelan; f 1986; quarterly; promotes respect for Australia's environment.

Australian Journal of Ecology: Blackwell Scientific Publications (Australia) Pty Ltd, 54 University St, Carlton, Vic 3053, Australia; tel (3) 347-0300; telex 10716421; fax (3) 347-5001; Editors: M Bull, D O'Dowd, P Fairweather; f 1976; quarterly; published for the Ecological Society of Australia; contains details of basic and applied ecological research conducted in Australasia, in addition to articles, reviews and news items relating to marine, freshwater and terrestrial ecosystems; circ 1,000.

Australian Waste Disposal Catalogue: Editorial and Publishing Consultants Pty Ltd, 29 First Ave, Klemzig, SA 5087, Australia; tel (8) 261-5837; fax (8) 261-2697; Editor: Frank H Schmidt; f 1979; annual; describes equipment, products and services for solid and liquid waste management; circ 2,500.

Australian Wildlife Newsletter: Wildlife Preservation Society of Australia, POB 3428, Sydney, NSW 2001, Australia; Editor: G Humphries; f 1966; quarterly; members' periodical; circ 1,000.

AWT (Abwassertechnik): Bauverlag GmbH, 6200 Wiesbaden 1, Postfach 1460, Germany; tel (611) 7910; telex 4186792; fax (611) 791285; Publr: Michael Schirmer; Editor: Hans Wendt; f 1950; every 2 months; professional journal for the wastewater treatment industry; covers sewage treatment, recycling and water pollution; circ 5,000.

Bankcheck Quarterly: International Rivers Network, 1847 Berkeley Way, Berkeley, CA 94703, USA; tel (510) 848-1155; fax (510) 848-1008; Editor: Juliette Maiot.

BBR—Wasser und Rohrbau: Rudolf Muller Verlag GmbH, 5000 Cologne, Stolberger Str 84, Postfach 410949, Germany; tel (221) 54970; fax (221) 5497326; monthly; scientific and technical journal concerned with water supply, pipeline construction, structural engineering and environmental technology; circ 3,200.

BBU Info-Dienst: Bundesverband Bürgerinitiativen Umweltschutz, 5300 Bonn 1, Prinz-Albert-Str 43, Germany; tel (228) 214032; fax (228) 214033; f 1987; every 2 months; newsletter of Federal Association for Citizens' Environmental Protection Initiatives.

Beach Conservation: Queensland Beach Protection Authority, POB 155, North Quay, Brisbane, Qld 4002, Australia; fax (7) 221-0768; Editor: A T Butler; f 1970; quarterly; circ 2,500.

Behavioral Ecology and Sociobiology: Springer-Verlag, 1000 Berlin 33, Heidelberger Platz 3, Germany; tel (30) 82071; Editor: K Eduard Linsenmair; f 1976; 12 a year in two vols.

Beifang Huanjing (Northern Environment): Heilongjiang Sheng Huanjing Baohuju, Keji Chu, 72, Dazhi Jie, Nanganggu, Harbin, Heilongjiang 150001, People's Republic of China; tel (451) 249272; Editor: Liu Hongnian; quarterly; in Chinese.

Belgian Environmental Research Index: Centre National de Documentation Scientifique et Technique, 4 blvd de l'Empéreur, 1000 Brussels, Belgium; tel (2) 519-56-50; telex 21157; fax (2) 519-56-79; Editor: L Dooms; f 1969; annual; indexes articles on environmental matters published in Belgium or by Belgian writers in foreign periodicals; also lists patents, books and conferences.

Bereginya: c/o Socio-Ecological Union (SEU), 125319 Moscow, ul Krasnoarmeyskaya 25, kv 85, Russian Federation; independent ecological newspaper published in Nizhny Novgorod; receives financial aid from the US Agency for International Development (AID) and is assisted by volunteers from ISAR (formerly the Institute for Soviet-American Relations), USA.

Bezpecnost a Hygiena Prace (Occupational Safety and Health): Vydavatelstvi a nakladatelstvi Prace, Vaclavske nam 17, 112 58 Prague 1, Czech Republic; tel (2) 2356008; fax (2) 2356043; Editor-in-Chief: Jiři Černohorský; f 1951; monthly; publication dedicated to working environments and conditions; covers developments in safety engineering, legislation, health risks and protection; in Czech and Slovak; circ 3,000.

Bilten Dokumentacije—Zastita Covekove Okoline i Iskoriscenje Otpadaka (Documentation Bulletin—Environmental Protection and Waste Utilization): Jugoslovenski Centar za Tehnicku i Naucnu Dokumentaciju, Sl Penezica-Krcuna 29–31, POB 724, 11000 Belgrade, Yugoslavia; Editor: Ljiljana Kojić-Bogdanović; f 1971; every 2 months; published by Yugoslav Centre for Technical and Scientific Documentation (YCTSD).

Biocycle, The Journal of Waste Recycling: JG Press Inc, 419 State Ave, Emmaus, PA 18049; POB 351, Emmaus, PA 18099, USA; tel (215) 967-4135; Editor: Jerome Goldstein; f 1960; monthly; contains reports and technical papers on recycling and composting of urban waste; circ 13,000.

Biodiversity and Conservation: Chapman & Hall, 2–6 Boundary Row, London SE1 8HN, UK; tel (71) 865-0066; fax (71) 522-9623; f 1991; quarterly; concerned with the description, study, conservation and management of biological diversity.

Biological Agriculture and Horticulture: An International Journal for Sustainable Production Systems: A B Academic Publishers, POB 42, Bicester, Oxon OX6 7NW, UK; tel (869) 320949; Editor: Dr R D Hodges; Deputy Editors: Dr A M Scofield, Dr J M Lopez-Real; f 1981; quarterly; original scientific and economic research reports.

Biological Conservation: Elsevier Science Publishers Ltd, Crown House, Linton Rd, Barking, Essex IG11 8JU, UK; tel (81) 594-7272; telex 896950; fax (81) 594-5942; Editor: Dr Eric Duffey; f 1969; 12 a year in 4 vols; contains original papers on wildlife conservation and the wise management of biological and other natural resources.

Biology and Fertility of Soils: Springer-Verlag, 1000 Berlin 3, Heidelberger Platz 3, Germany; tel (30) 82071; telex 185411; fax (30) 8207300; f 1985; 8 a year; deals with pure and applied soil biology and productivity studies, including techniques for assessing ecological stresses on soils; circ 500.

Biomass Bulletin: Multi-Science Publishing Co Ltd, 107 High St, Brentwood, Essex CM14 4RX, UK; tel (277) 224632; fax (277) 224632; every 2 months; includes summaries of published research papers, book reviews and news.

Biorecovery: The International Journal of Biotechnology Applied to Materials Recovery and Handling: A B Academic Publishers, POB 42, Bicester, Oxon OX6 7NW, UK; tel (869) 320949; Editor: Dr Robert Edyvean; Chief Advisory Editor: Prof V Moses; quarterly; research reports on the use of living organisms in the extraction, recovery and processing of minerals, chemicals and other materials; includes pollution control and waste treatment.

BioScience: American Institute of Biological Sciences (AIBS), 730 11th St, NW, Washington, DC 20001-4521, USA; tel (202) 628-1500; fax (202) 628-1509; Editor: Dr Julie Ann Miller; Features Editor: Anna Maria Gillis; Asst Editor: Erica Van Dommelen; f 1951; 11 a year; published by American Asscn of Biological Sciences; members' journal for professionals and advanced students covering all aspects of biology; contains essays, news, features, reviews of books and equipment; special issues focus on single topics.

Biosphère: Canadian Wildlife Federation—Fédération Canadienne de la Faune, 2740 Queensview Dr, Ottawa, ON K2B 3A2, Canada; tel (613) 721-2286; fax (613) 721-2902; Editor: Gabriel Racle; f 1985; every 2 months; magazine concerned with the environment, wildlife and the conservation of natural resources; circ 38,000.

Black Lechwe: Wildlife Conservation Society of Zambia, POB 30255, Lusaka, Zambia; tel (1) 254226; telex 44810; f 1981; quarterly; conservation magazine published primarily for members of the Society; circ 2,000.

BNA California—Environment Reporter: The Bureau of National Affairs Inc, 770 L St, Suite 910, Sacramento, CA 95814, USA; tel (916) 552-6500; fax (916) 552-6503; Managing Editor: Barrett McBride; fortnightly; provides comprehensive coverage of environmental legislation, regulation and litigation in California.

Bogong: Canberra and South-East Region Environment Centre, POB 1875, Canberra, ACT 2601, Australia; tel (6) 248-0885; Editorial Working Group: Ian Fraser, Roger Green, Michael Hissink, David Pfanner, Ann Rawson, Jane Rawson, Kerrie Tucker; f 1980; quarterly; articles emphasize regional environmental matters, but also cover national and international issues; circ 500.

Boletim da Comissão Nacional do Ambiente: Ministerio da Qualidade de Vida, Praça Duque de Saldanha 31, 1096 Lisbon Codex, Portugal; f 1974; monthly; published by National Environmental Commission; circ 5,000.

Boletín de Vida Silvestre: Sociedad de Vida Silvestre de Chile, Avda Alemania 0422, Casilla 1705, Temuco, Chile; tel (45) 21-0773; telex 267018; fax (45) 23-4126; annual scientific and technical magazine.

Books on the Environment and Related Topics: Bibliographic Press Ltd, 52 Little Paddocks, Ferring, Worthing, W Sussex BN12 5NH, UK; tel (903) 504019; f 1991; quarterly environmental bibliography.

Boomblad: De Dorschkamp Institut voor Bosbouw en Groenbeheer, POB 23, 6700 AA Wageningen, Netherlands; tel (8370) 95111; fax (8370) 24988; every 2 months; environmental publication concerned with forestry; circ 12,000.

Borealis: Canadian Parks and Wildlife Society (CPAWS), POB 1359, Edmonton, AB T5J 2N2, Canada (Publr); Canadian Parks and Wildlife Society (CPAWS), Suite 1335, 160 Bloor St E, Toronto, ON M4W 1B9, Canada (Subscriptions); tel (403) 439-8922; fax (403) 433-9646; Editor: David G Dodge; f 1965; quarterly; former title *Park News;* circ 10,100.

Bulletin de Documentation de l'Environnement: Documentation Française, 29–31 quai Voltaire, 75340 Paris Cedex 7, France (Publr); Documentation Française, 124 rue Henri Barbusse, 93308 Aubervillers, France (Subscriptions); tel (1) 40-15-70-00; 5 a year; published for Secretary of State for the Environment and Quality of Life.

Bulletin of Environmental Contamination and Toxicology: Springer-Verlag New York Inc, 175 Fifth Ave, New York, NY 10010, USA; tel (212) 460-1627; fax (212) 473-6272; monthly; deals with issues in the fields of environmental pollution, toxicology and biochemistry; circ 1,700.

Bulletin of the Australian Littoral Society: POB 49, Moorooka, Qld 4105, Australia; tel (7) 848-5235; fax (7) 892-5814; Editors: E Hegerl, D Tarte; f 1978; every 2 months; deals with marine and coastal conservation issues in Australia, with particular attention to management of the Great Barrier Reef Marine Park and Australian Mangroves; circ 600.

Business and the Environment: Cutter Information Corpn, 37 Broadway, Arlington, MA 02174-5539, USA; tel (617) 641-5125; telex 1009891; fax (617) 648-1950; Editor: Kathleen M Victory; f 1989.

Business Strategy and the Environment: European Research Press Ltd, Tayson House, 34–38 Chapel St, Little Germany, Bradford BD1 5DN, UK; tel (274) 729315; fax (274) 306981; Principal Officer: Andrew Gouldson; environmental business journal.

Buzzworm: The Environmental Journal: 2305 Canyon Blvd, Suite 206, Boulder, CO 80302, USA; tel (303) 442-1969; fax (303) 442-4875; Editor: Joseph Daniel; f 1988; independent magazine covering environmental news, culture and travel.

Bygd och Natur: Sveriges Hembygdsførbund, POB 30193, 104 25 Stockholm, Sweden; tel (8) 618-03-09; fax (8) 618-92-63; Editor: Mats Rolen; f 1920; 5 a year; published by National Association for the Preservation of Swedish Culture and Nature; circ 12,000.

Cadastro Nacional de Instituiçaos que Atuam na Area de Meio Ambiente: Instituto Brasileiro do Meio Ambiente e dos Recursos Naturais Renováveis, SAIN Av L4 Norte, 70.800 Brasília, DF, Brazil; national directory of environmental institutions; published by the Brazilian Institute for the Environment and Renewable Natural Resources.

Cahiers de la Ligue Urbaine et Rurale: 8 rue Meissonier, 75017 Paris, France; tel (1) 42-67-06-06; f 1963; quarterly; highlights major environmental problems and promotes the preservation of the architectural and natural heritage and of places of interest in France; circ 3,500.

Cahiers de la Réunion et de l'Océan Indien: 24 blvd des Cacotiers, 97434 Saint Gilles les Bains, Réunion; monthly; environmental publication for Réunion and the Indian Ocean.

Canadian Environmental Control Newsletter: CCH Canadian Ltd, 6 Garamond Court, Don Mills, ON M3C 1Z5, Canada; tel (416) 441-2992; fax (416) 444-9011; 24 a year; deals with pollution and the environmental measures intended to control it; indexed.

Canadian Environmental Law Report: Carswell Publications, Corporate Plaza, 2075 Kennedy Rd, Scarborough, ON M1T 3V4, Canada; tel (416) 609-8000; fax (416) 298-5094; Editor: Marcia A Valiante; f 1972, resumed 1987; every 2 months; former title *Canadian Environmental Law News;* published for Canadian Environmental Law Research Foundation; includes indexes and book reviews.

Canadian Environmental Mediation Newsletter: Conflict Management Resources, c/o Osgoode Hall Law School, York University, 4700 Keele St, North York, ON M3J 1P3, Canada; Editors: Doreen C Henley, D Paul Emond; f 1986; 3 a year; circ 600.

Canadian Environmental Network Bulletin/Bulletin du Réseau canadien de l'environnement: POB 1289, Station B, Ottawa, ON K1P 5R3, Canada (Subscriptions); 4 Knox St, Peterborough, ON K9H 2A8, Canada (Editorial); tel (613) 563-2078 (Subscriptions), (705) 745-3521 (Editorial); fax (613) 563-7236 (Subscriptions), (705) 745-3521 (Editorial); Editor: Clifford Maynes; quarterly; contains articles on environmental issues in Canada, news items relating to the Canadian provinces and news of forthcoming conferences and events; published in English and French editions.

Canadian Environmental Protection: Baum Publications Ltd, 1625 Ingleton Ave, Burnaby, BC V5C 4L8, Canada; tel (604) 291-9900; fax (604) 291-1906; Director: Engelbert Baum; Editor: Len Webster; f 1989; 9 a year; describes new environmental products and services available to industry; circ 27,200.

Caribbean Conservation News: Caribbean Conservation Association, Savannah Lodge, The Garrison, St Michael, Barbados; fax 429-8483; Editor: Calvin Howell; f 1975; quarterly; members' bulletin; circ 800.

CBF News: 162 Prince George St, Annapolis, MD 21401, USA; tel (410) 268-8816; fax (410) 268-6687; Editor: Flannery Davis; 2 a year; newsletter of the Chesapeake Bay Foundation (CBF).

CEP News: Caribbean Environment Programme, UNEP Regional Co-ordinating Unit, 14–20 Port Royal St, Kingston, Jamaica; tel 922-9267; telex 3672; fax 922-9292; f 1987; quarterly environmental newsletter.

CETESB Jornal: Companhia de Tecnologia de Sanemento Ambiental, Av Professor Frederico Hermann Jr 345, CEP 05459, Alto da Pinheiros-São Paulo, SP, Brazil; f 1976; monthly; journal of Company for Environmental Clean-up Technology.

Chain Reaction: POB 45, O'Connor, ACT 2601, Australia; tel and fax (6) 248-0289; Co-Editors: Clare Henderson, Larry O'Loughlin; f 1975; quarterly; national magazine of Friends of the Earth Australia.

La Chaîne: La Chaîne Bleue Mondiale (World Blue Chain for the Protection of Animals), ave de Visé 39, 1170 Brussels, Belgium; tel (2) 673-52-30; fax (2) 672-09-47; publication dedicated to the prevention of cruelty to domestic and farm animals.

Chemical Speciation and Bioavailability: Science and Technology Letters, POB 81, Northwood, Middx HA6 3AA, UK; tel (923) 823586; fax (923) 825066; f 1989; quarterly; interdisciplinary papers on the chemical, biological, ecological and physical impact of the presence of certain chemical substances in the environment.

Chemie-UmweltTechnik: Dr Alfred Hüthig Verlag GmbH, 69018 Heidelberg, Im Weiher 10, Postfach 102869, Germany; tel (6221) 489245; telex 461727; fax (6221) 489479; Publishing Dir: Friedrich Limperg; Editor-in-Chief: Dr Sieghard Neufeldt; f 1972; annual supplement to *Chemie-Technik* intended for process engineers, chemists, chemical engineers and plant managers in the chemical, pharmaceutical and food processing industries.

China Environment News: 15 A Xiaoxinglongjie, Chongwen District, Beijing 100062, People's Republic of China; tel (1) 752478; telex 222359; fax (1) 7013772; Pres: Zhang Lijun; Co-ordinator: Zhang Xiaoqing; f 1984; publicizes China's basic national policies, guidelines, laws and regulations governing environmental protection; also deals with environmental protection world-wide.

Chinese Journal of Applied Ecology: 72 Wenhua Rd, Shenyang, Liaoning Province, People's Republic of China; tel (24) 343401; telex 80095; fax (24) 391320; Editors: Shungjiang Sun, Fenglan Song, Fenqin Li; f 1990; includes original research papers, reviews and short articles on many aspects of applied ecology.

Chinese Journal of Ecology: Institute of Applied Ecology, Academia Sinica, 72 Wenhua Lu, POB 417, Shenyang 110015, Liaoning Province, People's Republic of China; tel (24) 383401; telex 80095; fax (24) 391320; Editorial Cttee: Dr Zhao Shidong, Liu Guoguang; f 1982.

Chongololo: Wildlife Conservation Society of Zambia, POB 30255, Lusaka, Zambia; tel (1) 254226; telex 44810; every 2 months; conservation magazine intended for the general public; circ 20,000.

Chongqing Huanjing Kexue (Chongqing Environmental Sciences): Chongqing Huanjing Kexue Xuehui, 212 Renmin Lu, Chongqing, Sichuan 630015, People's Republic of China; tel (811) 54425; fax (811) 350021; Editor-in-Chief: Luo Xingtao; f 1979; every 2 months; publication of Chongqing Society of Environmental Sciences; co-published by Chongqing Bureau of Environmental Protection; in Chinese.

Chrońmy Przyrodę Ojczystą: Polska Akademia Nauk, Instytut Ochrony Przyrody, 31-505 Kraków, Ul Ariańska 1, Poland; tel (12) 215144; telex 322630; fax (12) 210348; Editor-in-Chief: Zygmunt Denisiuk; Sec: Jadwiga Gawłowska; f 1945; every 2 months; publication of the Polish Acad of Sciences Inst of Nature Conservation and the State Council of Nature Protection; contains results of original research on protected and endangered plant and animal species and natural areas; deals with contemporary issues in environmental conservation, trends in relevant research in Poland and other countries, and the activities of the State and of citizens within that field; includes bibliographies, indexes and book reviews; in Polish, with summaries in English; circ 6,250.

Clean Air: Clean Air Society of Australia and New Zealand (CASANZ), POB 191, Eastwood, NSW 2122, Australia; Editor: J N O'Heare; f 1967; quarterly; includes book reviews, bibliographies and indexes; circ 1,200.

Clean Air: National Society for Clean Air and Environmental Protection, 136 North St, Brighton BN1 1RG, UK; tel (273) 326313; fax (273) 735802; f 1929; quarterly; articles and news items on air pollution and noise; includes details of new products and technology, and book reviews; circ 2,250.

Clean Water Report: CJE Associates, 237 Gretna Green Court, Alexandria, VA 22304, USA; tel (703) 823-0662; fax (703) 823-5923; Editor, Publr: Elaine R Eiserer; fortnightly (25 a year); newsletter.

Coastal Management: Taylor & Francis, 1101 Vermont Ave, NW, Suite 200, Washington, DC 20005, USA; tel (202)

289-2174; fax (202) 289-3665; Editor: Marc J Hershman; Advertising Manager: Louis M Fancher; f 1973; quarterly; discusses issues relating to the use of coastal resources and environments; describes new coastal management techniques; includes features on organizations involved in the development and conservation of coastal resources around the world; intended for public officials and researchers; circ 650.

Coastal Zone Management: Nautilus Press Inc, 1054 National Press Bldg, Washington, DC 20045, USA; tel (202) 347-6643; Editor: John R Botzum, Sr.

Code Permanent Environnement et Nuisances: Editions Législatives, 80 ave de la Marne, 92546 Montrouge Cedex, France; tel (1) 40-92-68-68; fax (1) 46-56-00-15; Man Editor: Michel Vaillant; Deputy Editor: Danièle Bonnafé; f 1973; quarterly, with monthly update bulletins; French and EC regulations affecting pollution, nature protection and all aspects of the environment.

The Cogeneration Journal: Multi-Science Publishing Co Ltd, 107 High St, Brentwood, Essex CM14 4RX, UK; tel (277) 224632; fax (277) 224632; Editor: W Payne; quarterly; includes detailed analysis of relevant environmental regulations.

Combat Nature: Edinat SARL, BP 3046, 24003 Périgueux Cedex, France; tel 53-08-29-01; Editor: Alain de Swarte; f 1966; quarterly; former titles *Revue de l'Association Ecologique et de Défense de l'Environnement* and *Maisons et Paysages—Nature et Environnement;* includes book reviews and indexes; circ 25,000.

Commonwealth Forestry Review: Commonwealth Forestry Association, Oxford Forestry Institute, South Parks Rd, Oxford OX1 3RB, UK; tel (865) 275072; fax (865) 275074; quarterly; reports of international forestry conferences, debates on government and private forest management policy, original academic and practical papers on all aspects of forestry, reports of research in progress and book reviews.

Connect: International Environmental Education Programme (IEEP), UNESCO/UNEP, 7 place de Fontenoy, 75700 Paris, France; Editor: Joseph Barry; f 1976; quarterly; UN environmental education newsletter; available in English, French, Spanish, Arabic and Russian.

Conscious Consumer: New Consumer Institute Inc, POB 51, Wauconda, IL 60084, USA; tel (708) 526-0522; Principal Officers: John Wasik, Kathleen Cealon-Wasik; newsletter concerned with environmentally sound products.

Conserva: Department of Environment Affairs, Private Bag X447, Pretoria 0001, South Africa; Editor: Louise van Niekerk; f 1974; every 2 months; former title *Environment RSA;* in Afrikaans and English; circ 30,000.

Conservation Biology: Blackwell Scientific Publications Inc, 3 Cambridge Center, Suite 208, Cambridge, MA 02142, USA (Publr); Society for Conservation Biology, c/o Prof Stanley Temple, Dept of Wildlife Ecology, University of Wisconsin, Madison, WI 53706, USA; tel (617) 225-0430 (Publr); fax (617) 494-1437; Editor: D Ehrenfeld; quarterly scientific journal.

Conservation Education: Young People's Trust for Environment and Nature Conservation, 95 Woodbridge Rd, Guildford, Surrey GU1 4PY, UK; Editor: Cyril Littlewood; f 1981; 3 a year; newsletter intended for teachers and youth leaders; circ 40,000.

Conservation Impact: Conservation Technology Information Center (CTIC), National Association of Conservation Districts, 1220 Potter Drive, Room 170, Purdue University Research Park, West Lafayette, IN 47906-1334, USA; 11 a year; members' newsletter.

Conservation Matters: Cleveland Wildlife Trust, Bellamy House, 2A Brighouse Business Village, Riverside Park, Middlesbrough, Cleveland TS2 1RT, UK; tel (642) 253716; fax (642) 251072; informs members of Cleveland Wildlife Trust of local and national wildlife issues.

Conservation News: Australian Conservation Foundation, 340 Gore St, Fitzroy, Vic 3065, Australia; tel (3) 416-1455; telex 34852; fax (3) 416-0767; Editor: David Eyre; f 1967; monthly; members' bulletin covering conservation issues in Australia; circ 22,000.

Conservation News: United Kingdom Institute for Conservation, 37 Upper Addison Gardens, London W14 8AJ, UK; tel (71) 603-5643; fax (71) 603-5643; Editor: Victoria Todd; f 1976; 3 a year; circ 1,300.

Conservation North Queensland: North Queensland Conservation Council Inc, 183 Flinders St, Townsville, Qld 4810, Australia; tel (77) 71-6226; fax (77) 21-1713; f 1975; monthly; circ 250.

Conserve: Western Pennsylvania Conservancy, 316 4th Ave, Pittsburgh, PA 15222-2075, USA; tel (412) 288-2777; fax (412) 281-1792; quarterly; members' magazine reporting on the Conservancy's efforts to preserve the water, land and wildlife of Western Pennsylvania.

The Conserver: British Trust for Conservation Volunteers, 36 St Mary's St, Wallingford, Oxon OX10 0EU, UK; tel (491) 839766; fax (491) 39646; Editor: Audrey Boyle; f 1972; quarterly; conservation news and current issues; intended for members of the Trust and others interested in practical conservation; circ 25,000.

Control of Pollution Encyclopedia: Butterworth & Co Ltd, 88 Kingsway, London WC2B 6AB, UK; tel (71) 405-6900; telex 95678; fax (71) 405-1332; in 2 vols; 3 a year, with supplementary bulletin every 2 months.

Corsonat: Corbett Society of Naturalists, 342 Shivaji Rd, Meerut 250 001, India; Editor: Y M Rai; f 1979; quarterly; deals with conservation and Indian flora and fauna; in English; circ 300.

Countryside Campaigner: Council for the Protection of Rural England (CPRE), Warwick House, 25 Buckingham Palace Rd, London SW1W 0PP, UK; tel (71) 976-6433; fax (71) 976-6373; Editor: Nicola S Frank; f 1962; 3 a year; members' bulletin covering rural issues and activities; circ 24,000.

Courants: PYC Edition, 5 ave de Verdun, BP 105, 94208 Ivry sur Seine, France; tel (1) 49-60-86-36; telex 263424; fax (1) 46-72-41-85; Publr: Pierre Benichou; Editor: Dominique Bomstein; f 1990; every 2 months; covers water supply, sewage treatment, waste management, landfill and environmental protection; circ 5,100.

Courrier de l'Antigaspillage/Bollettino sul Risparmio d'Energia/Energie-Spar-Nachrichten: Office Fédéral de l'Energie, 3003 Berne, Switzerland; tel (31) 615664; fax (31) 264307; f 1977; quarterly; government publication dealing with energy research, domestic and industrial energy efficiency and energy-efficient transport; includes statistical tables; in French, Italian and German; circ 15,000.

Current Bibliography on Science and Technology: Environmental Pollution—Kagaku Gijutsu Bunken Sokuho, Kankyo Kogai-hen: Nihon Kagaku Gijutsu Joho Senta, 5-2 Nagatacho, 2-chome, Chiyoda-ku, Tokyo 100, Japan; tel (3) 3581-6411; telex 2223604; fax (3) 3581-6446; f 1975; monthly; published by Japanese Science and Technology Information Centre; contains abstracts; circ 600.

Custos: National Parks Board—Nasionale Parkeraad, POB 787, Pretoria 0001, South Africa; tel (12) 3439770; telex 321931; fax (12) 3439958; Editor: Susan van der Merwe; f

1971; monthly; general conservation publication; in Afrikaans and English; circ 47,000.

Daily Environmental Report: Bureau of National Affairs, Inc, 1231 25th St, NW, Washington, DC 20037, USA (Publr); 9435 Key West Ave, Rockville, MD 20850, USA (Subscriptions); tel (202) 452-4200 (Publr), (800) 372-1033 (Subscriptions); telex 285656; fax (202) 822-8092; Editor: Mike Cavanagh; f 1992; national and international environmental news.

Danmarks Miljoundersogelser: Miljoministeriet, Afdeling for Miljokemi, Mørkhøj Bygade 26, 2860 Søborg, Denmark; tel 33-92-33-88; f 1971; annual; published by Department for Environmental Chemistry of the Danish Ministry of the Environment.

Daziran Tansuo (Exploration of Nature): Sichuan Kexue Jishu Chubanshe, 3 Yandao Jie, Chengdu, Sichuan 610016, People's Republic of China; tel (28) 664688; Editor: Yang Chao; quarterly; in Chinese.

Découvertes/Discoveries: Centro Nacional de Cultura, Rua António Maria Cardoso 68, 1200 Lisbon, Portugal; tel (1) 3466722; telex 18251; fax (1) 3428250; Editor: Helena Vaz da Silva; annual abridged edition of *Descobertas*, in English and French; dedicated to the promotion of Portugal's cultural and natural heritage.

Deep Ecologist: c/o POB 519, Claxton, Vic 3168, Australia; Editors: Jane Carter, Alex Shtargot; f 1982; quarterly; circ 500.

Dermatosen in Beruf und Umwelt—Occupation and Environment: Editio Cantor Verlag GmbH, 88322 Aulendorf, Postfach 1255, Germany; tel (7525) 9400; fax (7525) 940180; f 1952; every 2 months; circ 6,850.

Descobertas: Centro Nacional de Cultura, Rua António Maria Cardoso 68, 1200 Lisbon, Portugal; tel (1) 3466722; telex 18251; fax (1) 3428250; Editor: Helena Vaz de Silva; quarterly; dedicated to the promotion of Portugal's cultural and natural heritage.

Difesa Ambientale: Editoriale PEG SpA, Via Fratelli Bressan 2, 20126 Milan, Italy; tel (2) 2579841; telex 323088; fax (2) 2552779; Editor: Giancarlo Meani; Advertising Manager: Alvaro Magnani; f 1956; every 2 months; concerned with environmental protection, with particular emphasis on pollution control; intended for industrial companies, local authorities and research institutes; circ 5,100.

Directory of Hazardous Waste Services: Southam Business Communications, Inc, 1450 Don Mills Rd, Don Mills, ON M3B 2X7, Canada; tel (416) 445-6641; fax (416) 442-2200; Editor: Mary Mancini; f 1985; annual; circ 1,200.

The District Leader: National Association of Conservation Districts (NACD), 509 Capitol Court, NE, Washington, DC 20002, USA; tel (202) 547-6223; fax (202) 547-6450; monthly; newsletter on issues of concern for leaders of US Soil and Water Conservation Districts.

Diyadama: Sri Lanka Environmental Journalists Forum (SLEJF), POB 30, 5 Fourth Lane, Nawala, Sri Jayawardenapura, Sri Lanka; tel (1) 566717; telex 22144; fax (1) 698315; monthly; deals with communication and awareness-raising in the areas of environment and development; in Sinhala.

DNR-Kurier: Deutscher Naturschutzring (DNR), 5300 Bonn 3, Kalkuhlstr 24, Germany; tel (228) 441505; telex 8861170; fax (228) 444290; f 1980; fortnightly; bulletin of DNR (German Nature Conservation Group); circ 5,000.

Doch Shnati—Misrad Le-Ichut Ha-Svivah: Ministry of the Environment, POB 6234, Jerusalem 91061, Israel; tel (2) 701606; telex 25629; fax (2) 513945; Editor: Tamar Ben-Yeshaiya; f 1973; annual; in Hebrew; circ 2,500.

Documentos Taller Multidisciplinario del Medio Ambiente: Universidad Católica de Valparaíso, Vicerectoría Académica, Dirección General de Investigación, Casilla 4059, Valparaíso, Chile; f 1978; annual; abstracts from multidisciplinary environmental research papers; circ 1,000.

Dokumentation Naturschutz und Landschaft: W Kohlhammer GmbH, 7000 Stuttgart 80, Hessbrühlstr 69, Postfach 800430, Germany; tel (711) 78631; Editor: Rainer Flüeck; f 1961; quarterly; published by Bundesforschungsanstalt für Naturschutz und Landschaftsökologie (National Research Institute for Nature Conservation and Landscape Ecology); includes abstracts, indexes and book reviews; circ 800.

Du Sol à la Table: Défense de la Culture Biologique, 8 place de la République, 49320 Brissac Quince, France; tel 41-54-21-17; fax 41-54-21-75; Editor: Antoine LeMaire; f 1989; quarterly; organic agriculture publication; circ 12,415.

Ducks Unlimited: Ducks Unlimited Inc, 1 Waterfowl Way at Gilmer Rd, Long Grove, IL 60047-9153, USA; tel (708) 438-4300; f 1937; every 2 months; magazine dealing with conservation; circ 645,000.

DUMAC: Ducks Unlimited de México, AC, Apdo 776, 64000 Monterrey, NL, Mexico; Editor: Sergio Robles Puga; f 1978; every 2 months; members' magazine; circ 11,000.

E 36: Pollution de l'Eau, de l'Air et du Sol, Déchets, Bruit: PASCAL Explore, CNRS, Institut de l'Information Scientifique et Technique, BP 54, 54514 Vandoeuvre-Les-Nancy, France; tel 83-50-46-00; f 1984; 10 a year; abstracts and indexes relating to air, water, soil and noise pollution.

E-Sheet Weekly Update: Environment News Service, 3505 W 15th Ave, Vancouver, BC V6R 2Z3, Canada; tel (604) 732-4000.

Earth First! Journal: POB 1415, Eugene, OR 97440, USA; Publrs: Jim Flynn, Craig Beneviloe.

Earth Island Journal: Earth Island Institute, 300 Broadway, Suite 28, San Francisco, CA 94133-3312, USA; tel (415) 788-3666; telex 2829302; fax (415) 788-7324; Managing Editor: Justin Lowe; Editor: Gar Smith; Assoc Editor: John Sterling; quarterly.

Earthkeeper: 11 Oriole Crescent, POB 1649, Guelph, ON N1H 6R7, Canada; tel (519) 763-9357; fax (519) 836-0324; Editors: Scott Black, Patricia Brown; f 1990; every 2 months; national environmental magazine.

East African Agricultural and Forestry Journal: c/o Narc-Muguga, POB 30148, Nairobi, Kenya; tel (154) 328801; telex 25287; fax (154) 440771; Editor: J O Mugah; f 1935; findings of original research in agriculture, animal husbandry, forestry and related fields.

L'Eau, L'Industrie, Les Nuisances: Editions Johanet, 7 ave Franklin D Roosevelt, 75008 Paris, France; tel (1) 43-59-08-91; telex 649712; fax (1) 42-25-59-47; Manager: Vincent Johanet; f 1975; monthly (11 a year); journal covering water treatment, industry and water pollution; includes technical articles by experts, news items, reviews of new products, a bibliography and a calendar of relevant dates; circ 6,500.

EC Energy Monthly: Financial Times Business Enterprises, FT Newsletters, 126 Jermyn St, London SW1Y 4UJ, UK; tel (71) 411-4414; telex 296926; fax (71) 411-4415; Editor: Lucy Plaskett; f 1989; EC energy legislation and its impact on member states.

Eco-Management and Auditing: European Research Press Ltd, Tayson House, 34–38 Chapel St, Little Germany, Bradford BD1 5DN, UK; tel (274) 729315; fax (274) 306981; Principal Officer: Andrew Gouldson; environmental business journal.

Ecoalert: Conservation Council of New Brunswick Inc, 180 St John's St, Fredericton, NB E3B 4A9, Canada; tel (506) 458-8747; Editor: Lori Sinclair; every 2 months; former title *Conservation;* deals with environmental issues, particularly those affecting the province of New Brunswick; circ 550.

Ecodecision: Environment and Policy Society, 276 St Jacques St W, Suite 924, Montréal, PQ H2Y 1N3, Canada; tel (514) 282-3043; fax (514) 284-3045; Editor: Elizabeth Campbell; quarterly; circ 10,000.

Ecoforum: Environment Liaison Centre International, POB 72461, Nairobi, Kenya; tel (2) 562015; telex 23240; fax (2) 562175; Editor: William Allen; f 1974; every 2 months; serves as a forum for the exchange of information and ideas on environmental issues; in English, French, Spanish and Arabic; circ 5,000.

Ecography: Munksgaard International Publishers Ltd, 35 Nørre Søgade, POB 2148, 1016 Copenhagen K, Denmark; tel 33-12-70-30; telex 19431; fax 33-12-93-87; Editor: Rolf Vik; f 1978; quarterly; former title *Holarctic Ecology;* deals with ecology in the Holarctic region; includes statistics, book reviews and indexes; published for Scandinavian Society Oikos; in English; circ 600.

ECOL News: ECOL—Environmental Conservation Library, Minneapolis Public Library and Information Center, 300 Nicollet Mall, Minneapolis, MN 55401, USA; tel (612) 372-6570; 2 a year; newsletter on ecological and environmental matters for citizens of Minnesota.

EcoLog Canadian Pollution Legislation: Southam Business Communications Inc, 1450 Don Mills Rd, Don Mills, ON M3B 2X7, Canada; tel (416) 445-6641; fax (416) 442-2200; Editor: Mary Mancini; f 1972; monthly.

EcoLog Week: Southam Business Communications Inc, 1450 Don Mills Rd, Don Mills, ON M3B 2X7, Canada; tel (416) 445-6641; fax (416) 442-2200; Publr: Mary Mancini; Marketing Dir: Ian Rhind; Editor: Deborah Orchard; f 1972; weekly; reports on events, legislation, policy, research, technological developments, private and public sector programmes and initiatives in the environmental field; focuses on waste management and control of industrial pollution; intended for readers in business, industry and government.

Ecológica: ECOE Ediciones, Calle 26, No 5-52, Torre B, Apdo Aéreo 58308, Santa Fe de Bogotá, DC, Colombia; tel (1) 282-8974; fax (1) 235-2750; f 1989; quarterly; deals with global environmental issues, with particular attention to South and Central America; includes features, interviews and ecological news.

Ecological Abstracts: Elsevier/Geo Abstracts, Regency House, 34 Duke St, Norwich, Norfolk NR3 3AP, UK; tel (603) 626327; telex 975247; fax (603) 667934; Editor: P J Jarvis; f 1974; monthly; contains summaries of international scientific papers and conferences on all aspects of aquatic and terrestrial ecology; circ 600.

Ecological Economics: Elsevier Science Publishers BV, POB 181, 1000 AD Amsterdam, Netherlands; tel (20) 5803911; fax (20) 5803249; Editor-in-Chief: Robert Costanza; f 1989; every 2 months, in 2 vols; scientific journal published for the International Society for Ecological Economics; includes ecological impact studies and promotes investigation of the connections between ecological and economic systems.

Ecological Engineering: Elsevier Science Publishers BV, POB 181, 1000 AD Amsterdam, Netherlands; tel (20) 5803911; fax (20) 5803249; Editor-in-Chief: William J Mitsch; f 1992; quarterly; scientific journal containing papers on ecotechnology, including synthetic ecology, bioengineering, pollution control and sustainable agriculture.

Ecological Modelling: Elsevier Science Publishers BV, POB 181, 1000 AD Amsterdam, Netherlands; tel (20) 5803911; fax (20) 5803249; Editor-in-Chief: Sven E Jørgensen; f 1975; 20 a year in 5 vols; international scientific journal dealing with ecological modelling and engineering, and systems ecology; includes book reviews, bibliographies and indexes; in English.

Ecological Research: Blackwell Scientific Publications Pty Ltd, 54 University St, Carlton, Vic 3053, Australia; tel (3) 347-0300; telex 10716421; fax (3) 347-5001; Editor-in-Chief: Prof M Yasuno; Publr: M Robertson; 3 a year; members' journal published for the Ecological Society of Japan; original research papers in all fields of ecology; in English; circ 2,800.

Ecologicheski Vestnik (Ecological Herald): c/o Socio-Ecological Union (SEU), 125319 Moscow, ul Krasnoarmeyskaya 25, kv 85, Russian Federation; independent ecological newspaper published in Krasnoyarsk; receives financial aid from the US Agency for International Development (AID) and is assisted by volunteers from ISAR (formerly the Institute for Soviet-American Relations), USA.

The Ecologist: Agriculture House, Bath Rd, Sturminster Newton, Dorset DT10 1DU, UK; tel (258) 73476; fax (258) 73748; Editors: Edward Goldsmith, Nicholas Hildyard; f 1970; every 2 months; journal containing articles on social and environmental matters; includes statistics, indexes and book reviews; circ 5,000.

Ecology Abstracts: Cambridge Scientific Abstracts, 7200 Wisconsin Ave, Bethesda, MD 20814-4823, USA; tel (301) 961-6750; fax (301) 961-6720; Publr: Ted Caris; Man Editor: Angela Hitti; Editorial Advisors: F diCastri, F T Last; interdisciplinary journal of ecological research; focuses on how living organisms of all kinds interact with their environment and with other organisms.

Ecology and Environmental Management in Practice: Institute of Ecology and Environmental Management, 36 Kingfisher Court, Hambridge Rd, Newbury, Berks RG14 5SJ, UK; tel (635) 37715; fax (635) 550230; former title *Bulletin of the Institute of Ecology and Environmental Management.*

Ecology and Farming: International Federation of Organic Agriculture Movements (IFOAM), 6695 Tholey-Theley, Ökozentrum Imsbach, Germany; tel (6853) 5190; fax (6853) 30110; quarterly; provides information on the activities of IFOAM world-wide, relevant research and new agricultural publications; in English, with abstracts and articles in French and Spanish.

Ecology and Health Information Bulletin: Ecology and Health Foundation, 00-567 Warsaw, Al Ujazdowskie 13, Poland; tel (2) 6942153; fax (2) 6942153; monthly; information on environmental threats to health in Poland.

Ecology Law Quarterly: University of California Press, 2120 Berkeley Way, Berkeley, CA 94720, USA; tel (510) 642-4191; fax (510) 643-7127; f 1973; journal of environmental law and policy; intended for legal professionals, academics, scientists, engineers, planners and government officials; circ 1,300.

Ecology of Food and Nutrition: Gordon and Breach Science Publishers, POB 786, Cooper Station, New York, NY 10276, USA; tel (212) 206-8900; fax (212) 645-2459; Editors-in-Chief: Harriet V Kuhnlein, Peter L Pellett, Christine S Wilson; f 1972; international journal of the nutrition sciences.

Ecology USA: CJE Associates, 237 Gretna Green Court, Alexandria, VA 22304, USA; tel (703) 823-0662; fax (703) 823-5923; Editor, Publr: Elaine R Eiserer; fortnightly (25 a year); newsletter.

Econews: The Green Party, 10 Station Parade, Balham High Rd, London SW12 9AZ, UK; tel (81) 673-0045; fax (81) 675-4434; Editor: Dave Bradney; f 1978; every 2 months; party members' bulletin; circ 30,000.

Ecos: Commonwealth Scientific and Industrial Research Organization (CSIRO), POB 89, East Melbourne, Vic 3002,

Australia; tel (3) 418-7217; fax (3) 419-0459; Editor: R Lehane; f 1974; quarterly; reports on CSIRO research relating to the environment; circ 8,000.

Ecos: Soviet Ecology and Peace Association, 119021 Moscow, Zubovsky bul 4, Russian Federation; tel (095) 201-85-10; telex 411327; fax (095) 230-21-70; Editor: V B Rudenko; f 1990; monthly; covers nature conservation issues and the global ecological situation; includes articles by non-Russian authors; available in Russian, English and German editions.

EcoSource: EcoSource, Inc, POB 1270, Guelph, ON N1H 6N6, Canada; tel (519) 763-8888; fax (519) 763-6202; Editor: Tony Leighton; f 1990; 10 a year; overviews and summaries of articles in newspapers and periodicals describing environmental and ecological events and developments; intended for business and government.

Ecotime: Pubblindustria, Via Olindo Guerrini 5, 20133 Milan, Italy; tel (2) 2663605; fax (2) 70602499; f 1989; weekly; environmental news magazine; circ 25,000.

Ecotoxicology: Chapman & Hall, 2–6 Boundary Row, London SE1 8HN, UK; tel (71) 865-0066; fax (71) 522-9623; f 1992; quarterly journal.

Ecumene: A Journal of Environment–Culture–Meaning: Edward Arnold, Hodder & Stoughton Publishers, Mill Rd, Dunton Green, Sevenoaks, Kent TN13 2YA, UK; tel (732) 450111; fax (732) 461321; Journals Manager: Mary Altree; Editors: Denis Cosgrove, James S Duncan; first issue 1994.

Effluent and Water Treatment Journal: Thunderbird Enterprises Ltd, Omega Lodge, Troutstream Way, Rickmansworth, Herts WD3 4JN, UK; Editor: S J Tucker; f 1961; monthly; includes indexes and book reviews; circ 6,000.

EI Digest: Environmental Information Ltd, 4801 W 81st St, Suite 119, Minneapolis, MN 55437, USA; tel (612) 831-2473; fax (612) 831-6550; monthly; national journal on industrial and hazardous waste management; includes analysis of new or proposed regulations and new technologies, and documentation of specific waste services.

EI Environmental Services Directory: Environmental Information Ltd, 4801 W 81st St, Suite 119, Minneapolis, MN 55437, USA; tel (612) 831-2473; fax (612) 831-6550; annual; national directory of waste management firms, by state.

EIS: Digests of Environmental Impact Statements: Cambridge Scientific Abstracts, 7200 Wisconsin Ave, Bethesda, MD 20814-4823, USA; tel (301) 961-6750; fax (301) 961-6720; Publr: Ted Caris; Managing Editor: Angela Hitti.

EIS Cumulative: Cambridge Scientific Abstracts, 7200 Wisconsin Ave, 6th Floor, Bethesda, MD 20814, USA; tel (301) 961-6750; fax (301) 961-6720; Publr: Ted Caris; Managing Editor: Angela Hitti.

Eisei Kagaku (Japanese Journal of Toxicology and Environmental Health): Nihon Yakugakkai, 12-15, Shibuya 2-chome, Shibuya-ku, Tokyo 150, Japan; Editor: Motoyasu Osawa; f 1955; every 2 months; published by Pharmaceutical Society of Japan; includes statistics and indexes; available in Japanese and English editions, with summaries in English; circ 1,700.

Ekologia Polska (Polish Journal of Ecology): Polska Akademia Nauk, Instytut Ekologii, 05-092 Łomianki, Dziekanów Leśny k Warszawy, Poland; tel (22) 513046; telex 817378; fax (22) 513100; Editor-in-Chief: Prof Dr Kazimierz A Dobrowolski; Deputy Editor: Prof Dr Anna Hillbricht-Ilkowska; f 1952; quarterly; contains original papers on general ecology, limnology, biocoenology, population biology, soil ecology, ecological modelling, phytocoenology, landscape ecology and the ecological basis for environment protection.

Ekopanoráma: SZOP, Gorkého 6, 811 04 Bratislava, Slovakia (Publr); Križova 9, 811 04 Bratislava, Slovakia (Editorial); monthly; publication dealing with ecology and the environment.

Elainmaailma: Sanoma Corporation/Sanomaprint Magazines, POB 107, 00381 Helsinki, Finland; tel (0) 1201; telex 125848; fax (0) 1205599; Sales Manager: Kirsi-Marja Skogster; Editor-in-Chief: Jyrki Leskinen; f 1979; monthly; dedicated to aquatic and terrestrial wildlife; circ 49,500.

Electromagnetics and VDU News: POB 25, Liphook, Hampshire GU30 7SE, UK; tel (428) 751430; fax (428) 658802; Editor and Publr: Simon Best; f 1989; quarterly; newsletter on health hazards of electromagnetic fields from power lines, visual display units (VDUs) and microwave systems and the use of such fields in orthodox and alternative medicine.

EMJ—Environmental Manager's Journal: Strata Publishing Ltd, 65 Tweedy Rd, Bromley, Kent BR1 3NH, UK; tel (81) 466-0040; fax (81) 464-5637; Dir: Simon Jarvis; f 1991; news magazine for the mining and landfill industries.

Endangered Species Update: University of Michigan, School of Natural Resources, 430 E University, Dana Bldg, Ann Arbor, MI 48109-1115, USA; tel (313) 763-3243; fax (313) 936-2195; Editor: Rebekkah Lynn Gooch; f 1983; monthly; current news and information on endangered species protection; includes articles on research, management and policy issues, book reviews and technical notes; incorporates US Fish and Wildlife Service *Endangered Species Technical Bulletin*.

Endangered Wildlife: Endangered Wildlife Trust, Private Bag X11, Parkview 2122, South Africa; tel (11) 486-1102; fax (11) 486-1506; Advertising Manager: Lynn Hesketh; Editor: John Ledger; f 1983; quarterly; former title *Quagga;* journal devoted to endangered wildlife in southern Africa; contains special reports, news items, reports from southern African nations and book reviews; circ 6,000.

ENDS Report: Environmental Data Services Ltd (ENDS), Finsbury Business Centre, 40 Bowling Green Lane, London EC1R 0NE, UK; tel (71) 278-4745; fax (71) 837-7612; Editor: Marek Mayer; Subscriptions: Vivienne Plax; Advertising Manager: Maria Sztayer; f 1978; monthly; environmental information for business; intended for environmental professionals in industry, consultancy and central and local government.

Energetica: Editura Tehnică, Piața Presei Libere No 1, Bucharest ICP 71341, Romania; Bd Energeticienilor 8, Sector 3, Bucharest 79619, Romania; tel (1) 3220692; telex 10783; fax (1) 3219315; Editor-in-Chief: Dr Eugeniu Pavel; f 1953; monthly, in two alternate series; Series A discusses sources of energy and means of generation and Series B concentrates on specific problems; practical review for researchers, designers, investors, teachers and technical experts in the power sector; in Romanian, with summaries in English, French, German and Russian; circ 2,000.

Energi & Planlægning (Energy and Planning): Foreningen Sun Media, Ryesgade 19B, 2200 Copenhagen N, Denmark; tel 31-35-22-11; fax 31-35-65-45; covers energy and environmental issues.

L'Energia Elettrica: AEI, Viale Monza 259, 20126 Milan, Italy; tel (2) 2550641; fax (2) 55014919 (Advertising); Publr: Prof Giorgio Catenacci; f 1923; 11 a year; deals with the generation, supply and distribution of electricity, with particular reference to hydroelectric, nuclear, wind and solar power; circ 4,500.

Energie Alternative HTE: Editoriale PEG SpA, Via Fratelli Bressan 2, 20126 Milan, Italy; tel (2) 2579841; telex 323088; fax (2) 2552779; Chair: Solly Cohen; Managing Dir: Giancarlo Meani; Editor: Mara Portoso; f 1979; every 2 months; technical review dealing with alternative energy sources and rational use of conventional sources of energy; circ 3,000.

Energie & Milieutechnologie (E & M): Stam Tijdschriften bv, Postbus 235, 2280 AE Rijswijk, Netherlands; tel (70) 398-81-00; telex 33702; fax (70) 398-82-76; Editor-in-Chief: Edwin Kisman; International Sales Manager: Herman Voois; f 1952; monthly; concerned with energy and environmental technology; circ 5,700.

Energie Plus: Association Technique pour l'Efficacité Energétique, 47 ave Laplace, 94117 Arcueil, France; tel (1) 43-07-08-00; fax (1) 49-85-06-27; Editor: Michel Hoez; monthly; technical journal dedicated to efficient energy use and environmental protection; circ 3,300.

Energie Spektrum: Resch Verlag KG, 8032 Gräfelfing, Geigerstr 13, Postfach 1260, Germany; tel (89) 858070; fax (89) 8580762; f 1949; monthly; professional energy management publication; deals with environmental protection, solar energy, emission measurement, marketing and the ethics of the energy industry; includes statistics, indexes, book reviews, notification of events and reviews of new products; circ 9,000.

Energieanwendung: Deutscher Verlag für Grundstoffindustrie GmbH, 7031 Leipzig, Karl-Heine-Str 27B, Germany; tel (341) 4081011; fax (341) 4012571; f 1952; monthly; includes articles by experts on energy use and ways of saving energy, particularly in non-industrial contexts; intended for scientists, engineers, teachers and students; in German, with summaries in English and Russian; circ 2,000.

Energy Daily: King Communications Group Inc, 627 National Press Bldg, Washington, DC 20045, USA; tel (202) 638-4260; fax (202) 662-9744; Exec Editor: Dennis Wamstead; provides daily news of the international energy industry, covering conventional energy sources, energy technology, financial and legislative developments, alternative sources of energy and environmental issues.

Energy Data Book and Diary: Ambient Press Ltd, POB 25, Lutterworth, Leics LE17 4FF, UK; tel (455) 202281; Dir: David Bennett; Head of Admin: Margaret Wrathall; Coordinating Editor: Dr Alan Sherratt; annual.

Energy, Economics & Climate Change: Cutter Information Corpn, 37 Broadway, Arlington, MA 02174, USA; tel (617) 641-5118; telex 1009891; fax (617) 648-1950; Editor: Nick Sundt; f 1991.

Energy Economist: Financial Times Business Information, 126 Jermyn Street, London SW1Y 4UJ, UK; tel (71) 411-4414; telex 296926; fax (71) 411-4415; Editor: Chris Cragg; f 1983; news and analysis of energy issues world-wide.

Energy and Environment: Multi-Science Publishing Co Ltd, 107 High St, Brentwood, Essex CM14 4RX, UK; tel (277) 224632; fax (277) 224632; quarterly; international, interdisciplinary journal covering direct and indirect environmental impact of energy production and use.

Energy Exploration and Exploitation: Multi-Science Publishing Co Ltd, 107 High St, Brentwood, Essex CM14 4RX, UK; tel (277) 224632; fax (277) 224632; Editors: D Abbott, G Jenkins; every 2 months; technology, economics and policies of exploitation and development of all forms of energy.

Energy Policy: Butterworth Heinemann Ltd, Linacre House, Jordan Hill, Oxford OX2 8DP, UK; tel (865) 310366; fax (865) 310898; Group Editor: Robin Frampton; Editor: Nicky France; f 1973; journal containing articles on many energy issues including energy efficiency potential in the domestic sector, environmental impacts of fossil fuel use and energy demand management in developing countries; intended for managers, planners, politicians and researchers.

Energy R and D—Summary and Sources: Ferndown Publications, 302 Bramhall Lane South, Bramhall, Stockport SK7 3DL, UK; tel (61) 439-4926; Editor: Paul Smith; f 1979; reports on research and development into alternative sources of energy.

Energy Sources: Taylor & Francis, 1101 Vermont Ave, NW, Suite 200, Washington, DC 20005, USA; tel (202) 289-2174; fax (202) 289-3665; Editor: G Ali Mansoori; Advertising Manager: Louis M Fancher; f 1979; quarterly; international scientific and technical journal; seeks solutions to increasing global demand for energy and decreasing reserves of natural energy resources; circ 500.

Energy Systems and Policy: Taylor & Francis, 1101 Vermont Ave, NW, Suite 200, Washington, DC 20005, USA; tel (202) 289-2174; fax (202) 289-3665; Editor: S William Gouse, Jr; Advertising Manager: Louis M Fancher; f 1977; quarterly; addresses energy system problems, including resource assessment and energy consumption; intended for energy resource managers, policy planners and other interested parties; circ 450.

Energy World: H Howland Associates, The Martins, East St, Harrietsham, Kent ME17 1HH, UK; tel (622) 850100; fax (622) 850100; Publr: Harris Howland; Editor: Johanna Fender; 10 a year; published for the Institute of Energy.

Energy World Yearbook: H Howland Associates, The Martins, East St, Harrietsham, Kent ME17 1HH, UK; tel (622) 850100; fax (622) 850100; Publr: Harris Howland; Editor: Johanna Fender.

ENFO Newsletter: Environmental Sanitation Information Centre, Asian Institute of Technology, POB 2754, Bangkok 10501, Thailand; tel (2) 516-0100; telex 84276; fax (2) 516-2126; quarterly; in English.

Enlace: Sociedad de Vida Silvestre de Chile, Avda Alemania 0422, Casilla 1705, Temuco, Chile; tel (45) 21-0773; telex 267018; fax (45) 23-4126; quarterly newsletter.

Entomophaga: International Organization for Biological Control of Noxious Animals and Plants (IOBC), Swiss Federal Research Station for Agronomy, 8046 Zürich, Switzerland; tel (1) 3777111; fax (1) 3777201; quarterly scientific journal; newsletters and special bulletins also available.

Entsorga-Magazin/Entsorgungswirtschaft: Deutscher Fachverlag GmbH, 60326 Frankfurt am Main, Mainzer Landstr 251; 60264 Frankfurt am Main, Postfach 100606, Germany; tel (69) 759501; fax (69) 75951540; Editor: Bernd Wassmann; f 1982; 10 a year; covers waste treatment, waste water treatment, air purification and soil decontamination; circ 15,400.

EntsorgungsPraxis: Bertelsmann Fachzeitschriften GmbH, 33311 Gütersloh, Carl-Bertelsmann-Str 270, Postfach 120, Germany; tel (5241) 802241; telex 933868; fax (5241) 73055; Editor-in-Chief: Dipl-Ing Ulrich Knorra; Editor; Rainer Sürig; f 1983; monthly; trade journal dealing with developments in environmental technology affecting the refuse industry, air purification and waste water treatment; circ 15,944 (1991/92).

Entsorgungs-Technik: ecomed Verlagsgesellschaft mbH & Co KG, 86899 Landsberg, Rudolf-Diesel-Str 3; 86887 Landsberg, Postfach 1752, Germany; tel (8191) 125-563; telex 527114; fax (8191) 125-513; Editor-in-Chief: Susanne Wieland; f 1989; every 2 months; issues covered include recycling, waste management, refuse disposal, sewage, transport and clean air; circ 10,000.

Enviro: Swedish Environmental Protection Agency, 171 85 Solna, Sweden; tel (8) 799-10-87; telex 11131; fax (8) 98-45-13; Editor-in-Chief: Peter Hanneberg; f 1984; 2 a year; former title *Acid Magazine;* international environmental magazine, with particular emphasis on transboundary pollution; in English; circ 20,000.

Environ: Key Caribbean, 1 El Socorro Extension Rd, Kirpalani's Complex, San Juan, Trinidad and Tobago; (Subscriptions) POB 21, Port of Spain, Trinidad and Tobago; Editor: Roy Boyke; f 1977; annual; deals with Caribbean development trends; circ 5,000.

Environment: Environment Centre of Western Australia, Inc, 794 Hay St, Perth, WA 6000, Australia; fax (9) 322-3045; f 1975; quarterly; concerned with the social and physical environments of Western Australia; circ 1,500.

Environment: Heldref Publications Inc, 4000 Albemerle St, NW, Washington, DC 20016-1851, USA; tel (202) 362-6445; fax (202) 537-0287; f 1958; 10 a year; covers environmental and ecological issues; intended for a general readership; circ 13,000.

Environment Abstracts Annual: Bowker A & I Publishing, Reed Reference Publishing, 121 Chanlon Rd, New Providence, NJ 07974, USA; tel (908) 665-2889; fax (908) 771-7725; Exec Editor: Judith Salk; Senior Editor: Anila Rao Banerjee; f 1971; cumulative annual edition and index of *Environment Abstracts Journal* (qv).

Environment Abstracts Journal: Bowker A & I Publishing, Reed Reference Publishing, 121 Chanlon Rd, New Providence, NJ 07974, USA; tel (908) 665-2889; fax (908) 771-7725; Exec Editor: Judith Salk; Senior Editor: Anila Rao Banerjee; f 1971; monthly; journal containing approx 1,200 abstracts of articles from periodicals, conference proceedings and special reports from universities, NGOs and government agencies in the USA and abroad.

Environment Business Magazine: Information for Industry Ltd, 521 Old York Rd, London SW18 1TG, UK; tel (81) 877-9130; fax (81) 877-9938; Publr: James Benn; Editor: Ian Grant; f 1990; 2 every month; reports on pollution control, environmental protection and sustainable management of resources; intended for senior industrial managers.

Environment Business Newsletter: Information for Industry Ltd, 521 Old York Rd, London SW18 1TG, UK; tel (81) 877-9130; fax (81) 877-9938; Publr: James Benn; Editor: Ian Grant; f 1990; fortnightly.

Environment and Ecology: MKK Publications, 91A Ananda Palit Rd, Calcutta 700 014, West Bengal; Environment and Ecology Communication Centre (EECC), Dept of Zoology, Kalyani University, Kalyani 741 235, West Bengal, India; tel (3162) 98220 (EECC); Editor: Prof S K Konar; Co-Editor: Suparna Mullick; f 1983; quarterly; publishes results of research in all branches of environmental science and ecology, including agriculture, forestry, pest control and air, water and soil pollution control; circ 10,000.

Environment Features: Council of Europe, Centre Naturopa, BP 431 R6, 67006 Strasbourg Cedex, France; tel 88-41-20-00; telex 870943; fax 88-41-27-84; f 1974; every 2 months; in English and French.

Environment Management Abstracts: Referativnyi Zhurnal, Vsesoyuznyi Institut Nauchno-Tekhnicheskoi Informatsii (VINITI), A-219 Moscow, Baltiiskaya ul 14, Russian Federation; f 1987; monthly; in English.

Environment Newsletter: South Pacific Regional Environment Programme (SPREP), POB 240, Apia, Western Samoa; tel 21929; fax 20231; f 1971; quarterly; in English or French.

Environment and Planning A: Pion Ltd, 207 Brondesbury Park, London NW2 5JN, UK; tel (81) 459-0066; fax (81) 451-6454; Editors: A G Wilson, N J Thrift, R J Johnstone, W A V Clark; Reviews Editor: T Barnes; monthly; concerned with the planning of cities and regions; issues covered include inner-city regeneration and damage to the environment; contains papers, reviews of research policy and book reviews.

Environment Protection Engineering: Politechnika Wrocławska, 50-370 Wrocław, Wybrzeze Wyspianskiego 27, Poland; tel (71) 202304; telex 712559; fax (71) 223664; Editors: Tomasz Winnicki, Lucjan Pawlowski; f 1975; quarterly; contains papers on water purification, wastewater treatment, and disposal and sterilization of solid waste; in English, with summaries in Polish and Russian; circ 600.

Environment Report: Trends Publishing Inc, 1079 National Press Bldg, Washington, DC 20045, USA; tel (202) 393-0031; fax (202) 393-1732; Pres and Publr: Arthur Kranish; environmental newsletter.

Environment Risk: Euromoney Publications Plc, Nestor House, Playhouse Yard, London EC4V 5EX, UK; Quadrant Subscription Services, Oakfield House, Perrymount Rd, Haywards Heath, W Sussex RH16 2DH, UK; tel (71) 779-8660 (Publr), (444) 440421 (Subscriptions); fax (71) 779-8667; Editor: Josephine Carr; f 1992; 10 a year; addresses current environmental topics; intended for multinational companies.

Environment South Australia: Conservation Council of South Australia, Conservation Centre, 120 Wakefield St, Adelaide, SA 5000, Australia; tel (8) 223-5155; fax (8) 232-4782; Editor and Administrative Officer: Michelle Grady; f 1991; covers environmental events, news and opinions in South Australia.

Environment and Urbanisation: Human Settlements Programme, International Institute for Environment and Development (IIED), 3 Endsleigh St, London WC1H 0DD, UK; tel (71) 388-2117; telex 261681; fax (71) 388-2826; f 1989; 2 a year; dedicated to urban environmental problems in developing countries; intended for specialist and general readers; each issue focuses on an issue selected by subscribers; includes papers by experts, reviews, news, profiles of relevant NGOs and a bibliography; co-published by Instituto Internacional de Medio Ambiente y Desarrollo (IIED–América Latina).

Environment Victoria: Conservation Council of Victoria, Environment Centre, 247 Flinders Lane, Melbourne, Vic 3000, Australia; tel (3) 654-4833; fax (3) 650-5684; Editor: Harry Barber; f 1973; monthly; deals with the natural world and urban environmental issues, with particular attention to alleviation of 'greenhouse' problems; circ 2,000.

Environment Views: Department of the Environment, Communications Branch, 9820 106th St, Edmonton, AB T5K 2J6, Canada; tel (403) 422-2813; Editor: Maryhelen Vicars; f 1977; quarterly; former title *Environment News;* aims to increase public awareness of environmental matters; circ 18,500.

Environment Watch: Latin America: Cutter Information Corpn, 37 Broadway, Arlington, MA 02174, USA; tel (617) 641-5125; telex 1009891; fax (617) 648-1950; Editor: Jeff Stoub; f 1991.

Environment Watch: Western Europe: Cutter Information Corpn, 37 Broadway, Arlington, MA 02174, USA; tel (617) 641-5125; telex 1009891; fax (617) 648-1950; Editors: Stephanie Gehlen, Tony Carritt; f 1992.

Environment Week: King Communications Group Inc, 627 National Press Building, Washington, DC 20045, USA; tel (202) 638-4260; fax (202) 662-9744; Exec Editor: Dennis Wamsted; weekly; provides information on environmental policy, regulation, technology and business on a state, national and global level; covers acid precipitation, hazardous wastes, air and water quality, global warming, the greenhouse effect, and research and development aimed at reducing pollution.

Environmental Approvals in Canada: Butterworths Canada Ltd, 75 Clegg Rd, Markham, ON L6G 1A1, Canada; tel (800) 668-6481; fax (416) 479-2826; Editor: Michael I Jeffery; f 1989; 2 a year; covers the entire environmental approvals procedure, including public consultation and hearings; circ 500.

Environmental Auditor: Springer-Verlag New York Inc, 175 Fifth Ave, New York, NY 10010, USA; tel (212) 460-1627; fax (212) 473-6272; quarterly; circ 1,200.

Environmental Awareness: International Society of Naturalists (INSONA), c/o Maharaja Fatehsingh Zoo Trust, Indumati Mahal, Jawaharlal Nehru Marg, Baroda 390 001; (Subscriptions) General Secretary, Oza Bldg, Salatwada, Baroda 390 001, India; Editor: Dr G M Oza; f 1977; quarterly; members' magazine on issues relating to conservation of animals and the global environment; in English; circ 1,000.

Environmental Business Journal: Environmental Business International Inc, POB 371769, San Diego, CA 92137-1769, USA; tel (619) 295-7685; fax (619) 295-5743; Vice-Pres Publications: Lynette Thwaites.

Environmental Claims Journal: Executive Enterprises Publications Co, Inc, 22 W 21st St, 10th Floor, New York, NY 10010-6990, USA; tel (212) 645-7880; fax (212) 645-1160; Man Editor: Deborah Wenger; Editor-in-Chief: Lynne Miller; Editor: Nancy Hale; quarterly; provides information on the prevention and litigation of environmental claims.

Environmental Conservation: Elsevier Sequoia SA, ave de la Gare 50, BP 564, 1001 Lausanne 1, Switzerland; tel (21) 207381; telex 450620; fax (21) 235444; Editor: Nicholas Polunin; f 1974; quarterly; international journal dedicated to exposing and opposing environmental degradation resulting from human over-population and unsound technology; advocates urgent action to protect and improve the global environment; published for the Foundation for Environmental Conservation; in English.

Environmental Education and Information: University of Salford—Environmental Resources Unit, Peel Bldg, Salford M5 4WT, UK; tel (61) 745-5221; fax (61) 745-5999; Editor: G Ashworth; f 1981; quarterly; aims to raise awareness and promote protection of the global environment; circ 300.

Environmental Eye: Businesstek Publishing Inc, POB 250, Carleton Place, ON K7C 3P4, Canada; tel (613) 253-2833; fax (613) 253-2834; Editor: Shaun McLaughlin; f 1989; 8 a year; news and opinions on environmentally responsible business practices and relevant trends.

Environmental Geochemistry and Health: Science and Technology Letters, POB 81, Northwood, Middx HA6 3DN, UK; tel (9274) 23586; fax (9274) 25066; Editor: Brian E Davies; f 1979; quarterly; incorporates Minerals and the Environment; includes bibliographies and abstracts.

Environmental Geology and Water Sciences: Springer-Verlag New York Inc, 175 Fifth Ave, New York, NY 10010, USA; tel (212) 460-1627; fax (212) 473-6272; every 2 months; covers geological processes and hazards affecting humans, and the management of geological resources, including land, minerals, water and air; area of interest includes fuels, geological pollutants and environmental studies; circ 1,150.

Environmental Health Criteria: World Health Organization (WHO), Distribution and Sales Dept, ave Appia, 1211 Geneva 27, Switzerland; tel (22) 7912111; telex 27821; fax (22) 7910746; f 1976; approx 15 a year; discusses the dangers which certain chemicals and pollutants pose to human health and the environment; in English; circ 5,000.

Environmental Health News: Environmental Health News Ltd, Southbank House, Black Prince Rd, London SE1 7SJ, UK; tel (71) 582-3661; fax (71) 582-9405; Advertising Man: Paul Prior; Editor: Bill Randall; f 1986; weekly; journal published for environmental health profession.

Environmental Health Year Book: Environmental Health Publications, The Lansbury Estate, Knaphill, Woking, Surrey GU21 2EP, UK; tel (483) 797625; fax (483) 797839; Editor-in-Chief: Vic Quayle; Asst Editor: Pam Faithfull.

Environmental Hydrology Report: International Association for Environmental Hydrology (IAEH), POB 1088, Alexandria, VA 22313, USA; tel (703) 683-9768; fax (703) 683-6137; monthly; newsletter; intended for practising environmental hydrologists; contains research results and details of technical advances.

Environmental Impact Assessment Review: Elsevier Science Publishing Co, 655 Ave of the Americas, New York, NY 10010, USA; tel (212) 633-3806; fax (212) 633-3880; f 1985; quarterly; describes developments in environmental impact assessment, environmental decision-making and the settlement of environmental disputes.

Environmental Law Journal: Japan Centre for Human Environmental Problems, 2-17 Jinbo-cho, Kanda, Chiyoda-ku, Tokyo, Japan; tel (3) 3264-1311; f 1974; annual; in Japanese.

Environmental Management: Springer-Verlag New York Inc, 175 Fifth Ave, New York, NY 10010, USA; tel (212) 460-1627; fax (212) 473-6272; every 2 months; intended for environmental scientists, engineers, legal professionals and sociologists; serves as a forum for the assessment of techniques, ideas and research results that may be applied to specific environmental management problems; circ 1,200.

Environmental Management and Health: MCB University Press Ltd, 62 Toller Lane, Bradford, W Yorks BD8 9BY, UK; tel (274) 499821; fax (274) 547143; Editor: J Rose; f 1990; quarterly; concerned with environmental factors and their effects on human health.

Environmental Management Review: Government Institutes Inc, 4 Research Pl, Suite 200, Rockville, MD 20850-3226, USA; tel (301) 921-2300; fax (301) 921-0373; quarterly; selected seminar papers from Government Institutes' environmental compliance courses, covering environmental risk assessment, new regulatory requirements and the Environmental Protection Agency's enforcement priorities.

Environmental Manager: Executive Enterprises Publications Co, Inc, 22 W 21st St, New York, NY 10010-6990, USA; tel (212) 645-7880; fax (212) 645-1160; Man Editor: Jean Stephenson; Editor: Sarah Magee; f 1990; monthly; newsletter containing articles, case studies and news items on environmental management issues in business and industry; includes updates on environmental regulations and means of complying with them.

Environmental Monitoring and Assessment: Kluwer Academic Publishers, Postbus 17, 3300 AA Dordrecht, Netherlands; tel (78) 334911; telex 29245; fax (78) 334254; Editor: G Bruce Wiersma; f 1981; monthly.

Environmental News Briefing: Sri Lanka Environmental Journalists Forum (SLEJF), POB 30, 5 Fourth Lane, Nawala, Sri Jayawardenapura, Sri Lanka; tel (1) 566717; telex 22144; fax (1) 698315; monthly; deals with communication and awareness-raising in the areas of environment and development; in English.

Environmental News Digest: Sahabat Alam Malaysia (Friends of the Earth), 43 Jalan Salween, 10500 Pulan Pinang, Malaysia; tel (4) 376930; Editor: David Heath; f 1983; every 2 months; contains statistics, diagrams and book reviews; in English; circ 500.

Environmental Opportunities: POB 747, Mendocino, CA 95460, USA; tel and fax (707) 937-1529; Publr and Editor: Sanford Berry; f 1982; monthly; national newsletter listing primarily entry- and mid-level environmental employment opportunities in the private, non-profit sector.

Environmental Outlook: Institute for Environmental Studies (IES), University of Washington, 211 Engineering Annex, FM-12, Seattle, WA 98195, USA; tel (206) 543-1812; fax (206) 543-2025; Editor: Polly Dyer; Asst Editor: Ellen Briggs; monthly (10 a year); describes environmental activities of the IES, at the Univ of Washington and in Washington state; includes a bibliography and an environmental calendar.

Environmental Periodicals Bibliography: International Academy at Santa Barbara, Environmental Studies Institute, 800 Garden St, Suite D, Santa Barbara, CA 93101-1552, USA; tel (805) 965-5010; fax (805) 965-6071; f 1972; every 2 months, with cumulative annual index; guide to recent articles on environmental issues; covers human ecology and planning, air pollution, energy, land resources, water resources, and nutrition and health.

Environmental Planning: Israeli Association for Environmental Planning, 5 Gazit St, Tel-Aviv 69417, Israel; Editor: Dan Soen; f 1966; 3 a year; in English and Hebrew; circ 500.

Environmental Policy and Law: IOS Press, Van Diemenstraat 94, 1013 CN Amsterdam, Netherlands; tel (20) 6382189; fax (20) 6203419; Editor-in-Chief: Wolfgang C Burhenne; Editor: Marlene Jahnke; f 1975; every 2 months; published for International Council of Environmental Law, Germany; information on policy, legal and administrative matters relating to the natural environment and to sustainable development; in English and French.

Environmental Policy Review: Marjorie Mayrock Center for Russian, Eurasian and East European Research, Hebrew University of Jerusalem, Mount Scopus, Jerusalem 91905, Israel; tel (2) 883180; telex 26458; fax (2) 322545; Dir: Dr Stefani Hoffman; Editor: Dr Zeev Wolfson; f 1987; 2 a year; environmental problems and policy in the CIS and Eastern Europe, with particular attention to the consequences of the Chernobyl disaster; in English; circ 200.

Environmental Pollution: Elsevier Science Publishers Ltd, Crown House, Linton Rd, Barking, Essex IG11 8JU, UK; tel (81) 594-7272; telex 896950; fax (81) 594-5942; Editors: J P Demster, J W Manning; f 1970; 12 a year in 4 vols; international journal covering biological, chemical and physical aspects of pollution and its control; contains research papers and reviews on the distribution and environmental effects of pollution and new methods for studying and analysing it.

The Environmental Professional: National Association of Environmental Professionals (NAEP), 5165 MacArthur Blvd, NW, Washington, DC 20016, USA (Publr); c/o John Lemons, Department of Life Sciences, University of New England, Biddeford, ME 04005, USA (Editorial); tel (202) 966-1500 (Publr); fax (202) 966-1977 (Publr); Editor: John Lemons; quarterly; forum for discussion and analysis of environmental issues.

Environmental Protection: Stevens Publishing Corporation, 225 N New Rd, Waco, TX 76710, USA; POB 2573, Waco, TX 76702-2573, USA; tel (817) 776-9000; fax (817) 776-9018; Publr: Dottie LaFerney; Editor: Carol Mouché; f 1990; monthly; articles on regulations governing air, water, wastewater and the treatment, storage and disposal of hazardous materials; commentaries cover international issues, training, legal and management topics; includes annual *Buyer's Guide*; circ 91,000.

Environmental Radioactivity Annual Report: National Radiation Laboratory, New Zealand Department of Health, POB 25-099, Christchurch, New Zealand; tel (3) 3665059; fax (3) 3661156; Editor: Dr K M Matthews; f 1961; circ 300.

Environmental Research in Japan: Environment Agency, 1-2-2 Kasumigaseki, Chiyoda-ku, Tokyo 100, Japan; fax (3) 3580-3542; f 1975; annual; in English; circ 320.

Environmental Resource Economics: Kluwer Academic Publishers, Postbus 17, 3300 AA Dordrecht, Netherlands; tel (78) 334911; telex 29245; fax (78) 334254; f 1991; quarterly; deals with the application of economic theory and methods to environmental problems; in English.

Environmental Sanitation Abstracts—Low Cost Options: Environmental Sanitation Information Centre, Asian Institute of Technology, POB 2754, Bangkok 10501, Thailand; telex 84276; fax (2) 516-2126; f 1979; 3 a year; printed version of Environmental Sanitation Information Centre database; contains book reviews and abstracts; in English; circ 350.

Environmental Sanitation Reviews: Environmental Sanitation Information Centre, Asian Institute of Technology, POB 2754, Bangkok 10501, Thailand; telex 84276; fax (2) 516-0110; 2 a year; abstracts on many aspects of sanitation, with particular attention to urgent issues in developing countries; in English.

Environmental Science & Engineering: Davcom Communications Inc, 10 Petch Crescent, Aurora, ON L4G 5N7, Canada; tel (416) 727-4666; fax (416) 841-7271; Pres: Steve Davey; Publr and Editor: Tom Davey; f 1987; every 2 months; deals with industrial and local authority environmental control systems in Canada, and the treatment and supply of drinking water; circ 18,000.

Environmental Science and Technology: American Chemical Society, Drawer 1734, Atlanta, GA 30301-1734, USA; c/o Malcolm Thiele Technomedia Ltd, Wood Cottage, Shurlock Row, Reading, Berks RG10 0QE, UK (Rep); tel (800) 227-5558 (USA), (734) 343302 (UK Rep); fax (734) 343848; f 1967; monthly; concentrates on pollution control and environmental management; circ 12,400.

Environmental Software: Elsevier Science Publishers Ltd, Crown House, Linton Rd, Barking, Essex IG11 8JU, UK; tel (81) 594-7272; telex 896950; fax (81) 594-5942; Editor: Paolo Zannetti; f 1986; quarterly; assists the application of computer software to numerical and analytical methodology in the environmental sciences; includes software reviews.

Environmental Technology: Selper Ltd, 79 Rusthall Ave, Chiswick, London W4 1BN, UK; tel (81) 995-4160; Editors: J N Lester, R M Harrison; f 1980; monthly; former title *Environmental Technology Letters*; includes indexes and book reviews.

Environmental Yearbook: Tara Publishing Co Ltd, Poolbeg House, 1–2 Poolbeg St, Dublin 2, Ireland; tel (1) 719244; fax (1) 719263; Managing Editor: Fergus Farrell; Advertising Dir: Tony Cantwell; yearbook of Irish environmental health officers' association; circ 2,000.

Environmentalist: Science and Technology Letters, POB 81, Northwood, Middx HA6 3DN, UK; tel (9274) 23586; fax (9274) 25066; Editor: John F Potter; f 1980; quarterly; international journal dedicated to environmental awareness; published by the Institution of Environmental Sciences.

Environments—a journal of interdisciplinary studies: Faculty of Environmental Studies, University of Waterloo, Waterloo, ON N2L 3G1, Canada; tel (519) 885-1211; fax (519) 746-2031; Man Editor: Catherine Elstone; Editor-in-Chief: Harry S Coblentz; f 1982; 3 a year; intended for non-specialist readers; covers geography, architecture, the environment, development and urban and regional planning; includes articles and reviews; in English and French; circ 700.

Environmetrics: John Wiley & Sons Ltd, Journals, Baffins Lane, Chichester, W Sussex PO19 1UD, UK; tel (243) 779777; telex 86290; fax (243) 775878; Editors: A H El-Shaarawi, I B MacNeill; f 1990; quarterly; deals with the development of statistical methodology and its application to the environmental sciences.

Environnement: Centre d'Information sur l'Environnement, rue de la Victoire 26, boîte 15, 1060 Brussels, Belgium; tel (2) 539-09-78; fax (2) 539-09-21; every 2 months; practical and educational information; each issue focuses on a specific topic.

Environnement Africain—African Environment: ENDA Publications, Environmental Development Action in the Third World (ENDA), 54 rue Carnot, BP 3370, Dakar, Senegal; tel 22-98-90; telex 51456; fax 22-26-95; Editor: Laetitia van Drunen; f 1974; quarterly; environmental studies and regional planning bulletin; contains original papers on environment and development problems in Africa and developing countries in general, documents, book and magazine reviews, articles on technologies for ecological development and a calendar of events; published in French and English, with a small number of issues in Arabic; supplemented by occasional papers (10 a year) dealing with single issues; circ 2,200.

Environnement-Infos: Inter-Environnement Wallonie, rue de la Victoire 26, 1060 Brussels, Belgium; tel (2) 539-09-78; fax (2) 539-09-21; every 2 months; dedicated to environmental news; contains an environmental diary, reports of environmental coverage in the daily press and official Belgian and European legislative documents; occasional thematic supplements.

Environnement et Technique: Groupement d'Activités de Presse (GRAP), 21 rue d'Hauteville, 75010 Paris, France; tel (1) 42-46-32-32; fax (1) 45-23-41-44; Editor: Ginette Blery; f 1980; monthly; formerly *Information Déchets*; economic and technical strategies for resolving environmental problems; circ 8,000.

EPA Journal: Division of Editorial Services, US Environmental Protection Agency (EPA), 401 M St, SW, Washington, DC 20460, USA; tel (202) 260-6643; fax (202) 260-0231; Sr Editor: Karen Flagstad; Assoc Editor: Teresa Opheim; f 1975; quarterly; each edition deals in depth with a single national or global environmental issue; issues covered include groundwater, urban air quality, waste management, habitats, planetary warming, toxic substances, the ozone layer, desertification and biome preservation; articles do not necessarily reflect EPA policies.

Era Solar: SAPT Publicaciones Técnicas, SL, Costa Rica 13, 4°, Apdo 19295, 28016 Madrid, Spain; tel (1) 3505885; fax (1) 3459313; technical journal dealing with renewable energy, the environment and energy conservation.

Espace 90: Confédération Nationale pour l'Aménagement Rural, 129 blvd Saint-Germain, 75006 Paris, France; Editor: Jacques Fournol; f 1969; monthly; published by French National Confederation for Rural Planning.

Estudos Leopoldenses: Universidade do Vale do Rio dos Sinos—Unisinos, Av Unisinos 950, 93.010 São Leopoldo, RS, Brazil; tel (512) 92-6333; telex 5244076; fax (512) 92-1035; Editor: Rev Evaldo Heckler; f 1966; quarterly; contains abstracts, book reviews and a bibliography; in Portuguese, with English summaries; circ 1,400.

Età Verde (Green Age): Associazione l'Età Verde, CP 443, 00100 Rome, Italy; tel (6) 461849; f 1974; every 2 months.

Etudes de Pollution Atmosphérique à Paris et dans les Départements Périphériques: Préfecture de Police, Laboratoire Central, 39 bis rue de Dantzig, 75015 Paris, France; f 1964; annual; studies on air pollution in Paris and the surrounding departments.

Europe Environnement/Europe Environment: Europe Information Service SA, rue de Genève 6, 1140 Brussels, Belgium; tel (2) 242-60-20; fax (2) 242-94-10; Principal Officer: Anne Van Goethem; Editor: E Van Puyvelde; f 1973; fortnightly; in English and French.

European Bulletin: Nature and National Parks: Federation of Nature and National Parks of Europe (FNNPE), 94481 Grafenau, Kröllstr 5, Germany; tel (8552) 2839; fax (8552) 3242; information and technical knowledge relating to the management of the natural and cultural heritage in the national parks of Europe.

European Directory of Energy Efficient Building: James & James (Science Publishers) Ltd, 5 Castle Rd, London NW1 8PR, UK; tel (71) 284-3833; fax (71) 284-3737; Principal Officer: Edward Milford; annual environmental publication.

European Directory of Renewable Energy Suppliers and Services: James & James (Science Publishers) Ltd, 5 Castle Rd, London NW1 8PR, UK; tel (71) 284-3833; fax (71) 284-3737; Editor: Bruce Cross; annual; deals with all forms of renewable energy; circ 10,000.

European Environment: European Research Press Ltd, Tayson House, 34–38 Chapel St, Little Germany, Bradford BD1 5DN, UK; tel (274) 729315; fax (274) 306981; Principal Officer: Andrew Gouldson; f 1991; every 2 months; dedicated to European environmental policy and practice; intended for universities, local authorities, financial institutions and environmental pressure groups; circ 500.

European Environmental Law Review: Graham and Trotman Ltd, Sterling House, 66 Wilton Rd, London SW1V 1DE, UK; tel (71) 821-1123; telex 298878; fax (71) 630-5229; Editor: Elizabeth Bramwell; f 1992; monthly (11 a year); contains analytical articles and book reviews; reports on national jurisdictions, European Commission legislation, the European Parliament, the European Court of Justice and international organizations.

European Environmental Yearbook: DocTer International UK Ltd, Hyde Park House, 5 Manfred Rd, London SW15 2RS, UK; DocTer Institute of Environmental Studies, Foro Bonaparte 12, Milan, Italy; tel (81) 877-1080; fax (81) 874-2150; summary of information on nature conservation and town and country planning in EC member states; includes analysis of major environmental issues, a bibliography and a section dealing with EC legislation.

European Journal of Pharmacology—Environmental Toxicology and Pharmacology Section: Elsevier Science Publishers BV, POB 181, 1000 AD Amsterdam, Netherlands; tel (20) 5803911; fax (20) 5803249; Editors: J H Koeman, F P Nijkamp; f 1992; every 2 months; contains details of research into the toxic effects on man and other vertebrate animals of drugs and environmental contaminants.

European Water Pollution Control: Elsevier Science Publishers BV, POB 181, 1000 AD Amsterdam, Netherlands; tel (20) 580-39-11; fax (20) 580-32-49; Editor: J de Jongh; f 1991; every 2 months; published for the European Water Pollution Control Association; contains technical papers, reviews, information on new products and technologies, news of the Association and items intended for researchers and professionals in the field of water pollution control.

Evaluation Report of the United Nations Environment Programme: UNEP, POB 30552, Nairobi, Kenya; tel (2) 230800; telex 22068; fax (2) 226886; annual publication.

Everwild: 1089 W Broadway, No 205, Vancouver, BC V6H 1E5, Canada; tel (604) 734-3534; fax (604) 734-2760; Editor: Beverly McCormack; every 2 months; conservation magazine; circ 30,000.

Evolutionary Ecology: Chapman & Hall Ltd, 2-6 Boundary Row, London SE1 8HN, UK; tel (71) 865-0066; telex 290164; fax (71) 522-9623; Editor: M L Rosenzweig; f 1987; every 2 months.

Fauna & Flora: Transvaal Directorate for Nature and Environmental Conservation, Private Bag X209, Pretoria, South Africa; fax (12) 212530; Editor: G P Visagie; f 1950; annual; deals with nature conservation and environmental issues relevant to the Transvaal; published in Afrikaans and English editions; circ 14,000.

Federal Facilities Environmental Journal: Executive Enterprises Publications Co, Inc, 22 W 21st St, 10th Floor, New York, NY 10010-6990, USA; tel (212) 645-7880; fax (212) 645-1160; Man Editor: Deborah Wenger; Editor: John Willig; f 1990; quarterly; analytical and practical information on environmental issues at US federal facilities.

Filtration and Separation: Elsevier Advanced Technology, Mayfield House, 256 Banbury Rd, Oxford OX2 7DH, UK; tel (865) 512242; fax (865) 310981; Editor: Steve Barrett; technical magazine.

Focus: Forestry Publications, College of Forestry, Wildlife and Range Sciences, University of Idaho, Moscow, ID 83843,

USA; tel (208) 885-6673; fax (208) 885-6226; Man Editor and Dir of Information Services: George Savage; Editor: Denise Ortiz; f 1974; annual; contains feature articles describing research conducted at the College and an appendix of the year's publications and of fiscal information.

FoE Link: Friends of the Earth International (FoEI), POB 19199, 1000 GD Amsterdam, Netherlands; tel (20) 6221369; telex 918023; fax (20) 6392181; every 2 months; circ 750.

Food, Cosmetics and Drug Packaging: Elsevier Advanced Technology, Mayfield House, 256 Banbury Rd, Oxford OX2 7DH, UK; tel (865) 512242; fax (865) 310981; Publr: Chris Lloyd; f 1978; monthly; aims to enable managers to make informed, responsible decisions in relation to packaging; gives particular attention to recent legislation and other developments in the fields of health, safety and consideration of the environment.

Forest and Bird: Royal New Zealand Forest and Bird Society, 172 Taranaki St, POB 631, Wellington, New Zealand; tel (4) 835-7374; fax (4) 835-7373; Editor: Lan Close; f 1923; quarterly; includes indexes and book reviews; circ 28,000.

Forest and Conservation History: Duke University Press, College Station, POB 6697, Durham, NC 27708, USA (Publr); Forest History Society, 701 Vickers Ave, Durham, NC, USA; tel (919) 684-2173 (Publr), (919) 682-9319 (Forest History Soc); telex 802829; fax (919) 684-8644 (Publr), (919) 682-2349 (Forest History Soc); Editor: David O Percy; Managing Editor: Kevin C Foy; f 1957; quarterly; documents the history of human interaction with the environment, particularly forests, and of the conservation of natural resources; includes bibliographies, book reviews and news of current research by the Forest History Society; circ 1,900.

Forest Science: Society of American Foresters, 5400 Grosvenor Lane, Bethesda, MD 20814, USA; tel (301) 897-8720; fax (301) 897-3690; f 1955; quarterly; technical journal covering recent forestry research around the world.

Forestry: Oxford University Press, Journals Marketing Department, Walton St, Oxford OX2 6DP, UK; tel (865) 56767; fax (865) 267773; Editors: J A Petty, I R Brown; quarterly; journal of Inst of Chartered Foresters, Edinburgh; contains original international research papers and book reviews; particular emphasis on European forestry.

Forêt Conservation: Revue Forêt Conservation Inc, 145 rue St Jean, 4ème étage, Québec, PQ G1 1N4, Canada; tel (418) 529-2542; fax (418) 529-3021; Advertising Manager: Christine Côte; Editor: Serge Beaucher; f 1939; 10 a year; covers forestry, the environment and the natural sciences; in French; circ 8,000.

Forum: American Institute of Biological Sciences (AIBS), 730 11th St, NW, Washington, DC 20001-4521, USA; tel (202) 628-1500; fax (202) 628-1509; every 2 months; newsletter of AIBS public affairs programme, covering US legislation on research budgets, the environment and education.

Forum Städte-Hygiene: Patzer Verlag GmbH und Co KG, 1000 Berlin 33, Königsallee 65, Germany; Editors: K H Knoll, Bernhard Patzer; f 1949; every 2 months; publication dealing with hygiene in urban environments; circ 1,000.

Franc-Vert: Union Québécoise pour la Conservation de la Nature, 160–76 rue Est, 2nd Floor, Charlesbourg, PQ G1H 7H6, Canada; tel (418) 628-9600; fax (418) 626-3050; Editor: Jean-Pierre Drapeau; f 1984; every 2 months; former title *Franc-Nord;* conservation publication; circ 8,000.

Fresenius Environmental Bulletin: Birkhäuser Verlag, Postfach 133, 4010 Basel, Switzerland; tel (61) 737740; telex 963475; fax (61) 737950; Editor: F Korte; f 1991; monthly; published for Schweizerischer Bund für Naturschutz (Swiss League for Nature Protection); in English.

From Duck Country: Ducks Unlimited Canada, 1190 Waverley St, Winnipeg, MB R3T 2E2, Canada; tel (204) 477-1760; fax (204) 452-7566; Editor: T G Neraasen; f 1939; every 2 months; members' magazine; circ 48,000.

Fujian Huanjing (Fujian Environment): Fujiansheng Huanjing Kexue Xuehui, Hualin Lu, Fuzhou, Fujian 350003, People's Republic of China; tel (591) 570146; Editor: Gan Jinggao; every 2 months; published by Fujian Association of Environmental Science; deals with environmental management and education, and the prevention of industrial pollution; in Chinese.

Fuoristrada: G Cassini Editore Srl, Via Cavour 3, Biella, Provincia di Vercelli, Italy; tel (15) 352766; fax (15) 352766; 2 a year.

Future Survey: World Future Society, 7910 Woodmont Ave, Suite 450, Bethesda, MD 20814, USA (Publr); 5413 Webster Rd, LaFayette, NY 13084, USA (Editorial); tel (301) 656-8274 (Publr); fax (301) 951-0394; Editor and Founder: Michael Marien; f 1979; monthly; contains articles, book reviews and reports on trends, forecasts and proposals relating to alternative futures; approx 30% of items are related to the environment; supplemented by *Future Survey Annual*, containing categorized abstracts and indexes.

Gaia—Ecología y Equidad para un Mundo Sostenible: Coordinadora de Organizaciones de Defensa Ambiental (CODA), Plaza Santo Domingo 7, 7° B, 28013 Madrid, Spain; tel (1) 5596025; fax (1) 5597897; Deputy Dir: Juan Gallego; Dir: José Santamarta; independent magazine dealing with the environment and sustainable development; contains analysis, ideological debate and national and international news; produced in collaboration with *The Ecologist*.

Garbage: The Practical Journal for the Environment: POB 56519, Boulder, CO 80322-6519, USA (Editorial); Dovetale Publishers, 2 Mial St, Gloucester, MA 01930, USA (Publr); tel (800) 274-9909 (Editorial); Editor: Patricia Poore; f 1989; every 2 months; environmental journal focusing on the minimization, ecological disposal and recycling of household wastes; includes statistics, reviews of new products, news items and longer articles.

GEA: Maggioli Editore, Via Crimea 1, CP 290, 47037 Rimini, Italy; tel (541) 626777; fax (541) 622020; Editor: Ario Rupeni; f 1986; every 2 months; deals with issues related to local government and environmental economy.

GEC-O: c/o Environment Section, Wye College (University of London), Wye, Kent TN25 5AH, UK; tel (233) 812401; fax (233) 813320; Editor: Alister Scott; f 1992; quarterly; newsletter of the Economic and Social Research Council (ESRC) Global Environmental Change Programme; includes news of the most recent developments in the Programme, including information about new phases of research, events and items by researchers.

Geluid en Omgeving: CED Samsom, Kouterveld 14, 1831 Diegem, Belgium; tel (2) 723-11-11; telex 64130; fax (2) 649-84-80; quarterly; published for Nederlands Stichting Geluidshinder; describes advances in the control of noise pollution; in Flemish.

Geo: Gruner und Jahr AG und Co, 20444 Hamburg, Am Baumwall 11, Germany; tel (40) 3703-0; fax (40) 3703-5648; Editor-in-Chief: Dr Werner Funk; f 1976; monthly; environment, science and general interest magazine; circ 520,000.

Geo—Australasia's Geographical Magazine: Geo Productions Pty Ltd, 2A Blakesley St, Chatswood, NSW 2067; POB 1390, Chatswood, NSW 2057, Australia; tel (2) 41117; fax (2) 413-2689; Publr: Grant Young; Editor: Michael Hohensee; f 1978; quarterly; subjects covered include wildlife, flora and fauna, history, anthropology, travel and traditional crafts.

Gesunde Pflanzen: Verlag Paul Parey (Berlin), 1000 Berlin 42, Seelbuschring 9–17, Germany; Paul Parey Scientific Publishers, 150 E 27th St, Suite 1A, New York, NY 10016, USA; tel (30) 707840 (Germany); fax (30) 70784199 (Germany); Editor: Hansgeorg Pag; f 1947; monthly; deals with protection of plants, consumers and the environment; in German, with summaries in English; circ 3,500.

The Gildea Review: Community Environmental Council (CEC), 930 Miramonte Drive, Santa Barbara, CA 93109, USA; tel (805) 963-0583; fax (805) 962-9080; 2 a year; newsletter reporting on events and programmes at the CEC's Gildea Resource Center.

Glasnik Drustvo Konzervatora Srbija: Drustvo Konzervatora Srbija, Bozidara Adzije 11, 11000 Belgrade, Yugoslavia; tel (11) 451642; f 1978; annual; concerned with conservation issues; includes abstracts and bibliographies; in Serbo-Croat; circ 600.

Global Environmental Change: Butterworth-Heinemann Ltd, Linacre House, Jordan Hill, Oxford OX2 8DP, UK; tel (865) 310366; fax (865) 310898; Group Editor: Barbara Clarke; Editor: Prof Martin Parry; f 1990; quarterly; international journal; examines environmental issues from the point of view of human ecology and public policy.

Global Environmental Change Report: Cutter Information Corpn, 37 Broadway, Arlington, MA 02174, USA; tel (617) 641-5118; telex 1009891; fax (617) 648-1950; Editor: Bradford J Hurley; f 1989.

Globus—Begleithefte: Bund für Umwelt und Naturschutz Deutschland, 7000 Stuttgart 1, Landesverband Baden-Württemberg eV, Rotebühlstr 86-1, Germany; tel (711) 619700; fax (711) 6197070; Editor: Erich Kimmich; f 1982; monthly; published by German League for Environment and Nature Protection; supplement dealing with conservation issues; circ 15,000.

Good Earth: POB 2605, W Melbourne, Vic 3001, Australia; quarterly; general environmental magazine; circ 700.

Grasduinen: De Geïllustreerde Pers bv, Ceylonpoort 5-25, 2037 AA Haarlem, Netherlands; POB 6003, 2001 HA Haarlem, Netherlands; tel (23) 304205; fax (23) 352554; Principal Officer: Frans Buissink; f 1979; monthly; natural history magazine; covers national parks, wildlife, landscapes and nature photography; circ 70,600.

Green Brigades: 30-060 Kraków, Wydział Chemii UJ, Ingardena 3/100, Poland; tel (12) 336377; telex 322297; fax (12) 340515; Editors: Piotr Rymarowicz, Katy Wimble, Andrzej Zwawa; f 1989; quarterly; English language edition of *Zielone Brygady* (qv); intended to inform international community about activities of environmental movement in Poland.

Green Drum: Greenspur Enterprise, 18 Cofton Lake Rd, Rednal, Birmingham B45 8PL, UK; tel (21) 445-2576; Hon Editor: Ron Andrews; f 1974; issues include conservation, animal welfare, green politics, economy and human rights.

Green Engineering: Mechanical Engineering Publications Ltd, Northgate Ave, Bury St Edmunds, Suffolk IP32 6BW, UK; Institution of Mechanical Engineers (I Mech E), 1 Birdcage Walk, London SW1H 9JJ, UK; tel (284) 762277, (71) 222-7899 (I Mech E); telex 917944 (I Mech E); fax (284) 704006, (71) 222-8762 (I Mech E); Editor: Elizabeth Ellis; monthly; topical bulletin aiming to increase awareness of environmental issues among professional engineers.

Green Line: 34 Cowley Rd, Oxford OX4 1HZ, UK; Editor: Tim Andrews; f 1982; monthly; circ 1,800.

Green Magazine: United Leisure Magazines Ltd, POB 3205, 4 Selsdon Way, City Harbour, London E14 9ZR, UK (Publr); Fairwind Ltd, Mouse Lane, Steyning, W Sussex BN44 3DG, UK (Editorial); tel (71) 712-0550 (Publr), (903) 879323 (Editorial); fax (71) 712-0549 (Publr), (903) 879052 (Editorial); Publr: Robin Bines; Editor: Alistair Townley; f 1989, relaunched 1993; monthly; information, views, updates, products and reviews for members of the public with an interest in the environment.

Green Politics: Edinburgh University Press, 22 George Sq, Edinburgh EH8 9LF, UK; tel (31) 650-4218; fax (31) 662-0053; Publr: Vivian C Bone.

Green Supermarket Shopping Guide: New Consumer Institute Inc, POB 51, Wauconda, IL 60084, USA; tel (708) 526-0522; Principal Officers: John Wasik, Kathleen Cealon-Wasik; environmental consumers' guide.

The Green Work—Green World Gazette: Conservers at Work, The Environment Council, 21 Elizabeth St, London SW1W 9RP, UK; tel (71) 824-8411; fax (71) 730-9941; f 1993; every 2 months; magazine informing members of Conservers at Work of the environmental effects of their daily working practices.

Greenbits: Cleveland Wildlife Trust, Bellamy House, 2A Brighouse Business Village, Riverside Park, Middlesbrough, Cleveland TS2 1RT, UK; tel (642) 253716; fax (642) 251072; informs members of Cleveland Wildlife Trust of local and national wildlife issues.

Greenpeace: Stiftung Greenpeace Schweiz, Müllerstr 37, Postfach 4927, 8022 Zürich, Switzerland; tel (1) 2413441; quarterly; members' magazine of Switzerland Greenpeace Foundation; circ 23,000.

Greenpeace Magazin: Greenpeace Umweltschutzverlag GmbH, 2000 Hamburg 11, Vorsetzen 53, Germany; tel (40) 311860; fax (40) 31186212; f 1982; formerly *Greenpeace Nachrichten;* quarterly; circ 750,000.

Greenscene: The Vegetarian Society UK Ltd, Parkdale, Dunham Rd, Altrincham, Cheshire WA14 4QG, UK; tel (61) 928-0793; fax (61) 926-9182; Principal Officers: Sean McElherron, Carol Timperley; magazine for members of Vegetarian Society aged under 18; campaigns on issues including the environment, animal welfare, developing nations and health.

GréngeSpoun: GréngeSpoun Société Coopérative, BP 684, 2016 Luxembourg, Luxembourg; tel 221155; fax 221153; Editors: Richard Graf, Renée Wagener; f 1988; weekly newspaper presenting environmental and global ecological news and suggesting social and ecological alternatives; circ 1,500.

Gront Miljø (Green Environment): Landsforeningen Danske Anlægsgartnermestre, Linde Allé 16, 2720 Vanløse, Denmark; tel 31-74-94-00; fax 31-74-15-00; Editor: Søren Holgersen; f 1983; 7 a year; published by the Danish Landscape Contractors Association; concerned with the planning, creation and management of urban green spaces and landscapes; circ 6,400.

Grünstift (Berlin): Stiftung Naturschutz Berlin, 1000 Berlin 30, Potsdamerstr 68, Germany; tel (30) 2626001; fax (30) 2615277; Editor: Klaus-Dieter Heise; f 1983; monthly; encourages conservation of nature and environmental protection in Berlin and Brandenburg; circ 22,000.

Grünstift (Rattingen): Bund-NW eV, 4030 Rattingen 1, Graf-Adolph-Str 9, Germany; f 1989; quarterly; environmental periodical for the Düsseldorf area; circ 3,000.

GSF Mensch und Umwelt: GSF—Forschungszentrum für Umwelt und Gesundheit, 8042 Neuherberg, Ingolstädter Landstr 1, Germany; tel (89) 31872711; fax (89) 31873324; Editors: Cordula Klemm, Heinz-Jörg Haury; f 1984; annual; provides public information on environmental and health research issues; published by Research Centre for Environment and Health; circ 20,000.

H F Habitat Flag: Calegari Edizioni Srl, Via Sismondi 62, 20133 Milan, Italy; tel (2) 7389414; fax (2) 7381266; f 1984; monthly; environmental news and information; circ 20,000.

H_2O: Scelta GmbH, 6064 Rum Bei Innsbruck, Anton-Falch-Str 3, Austria; tel (5222) 62564; quarterly; environmental publication intended for the general public; circ 14,500.

Habiosphera: Ministry of the Interior, Dept of Environmental Quality, POB 6158, Jerusalem 91061, Israel; tel (2) 701411; fax (2) 639368; Editor: Yosef Hirschberg; monthly.

Habitat Australia: Australian Conservation Foundation, 340 Gore St, Fitzroy, Vic 3065, Australia; tel (3) 416-1166; fax (3) 416-0767; Editor: Merrilyn Julian; f 1973; every 2 months; covers environmental, conservation, energy and peace issues; circ 12,000.

habitat: the independent digest of environment and conservation news: The Environment Council, 21 Elizabeth St, London SW1W 9RP, UK; tel (71) 824-8411; fax (71) 730-9941; Editor: Susan Joy; fmrly published by The Council for Nature; monthly; contains local, national and international news, reports on environmental affairs concerning government and industry, and book reviews.

Haiyang Huanjing Kexue: Guojia Haiyang Ju, Haiyang Huanjing Baohu Yanjiusuo, POB 303, Dalian, Liaoning 116023, People's Republic of China; tel (411) 471429; Editor: Qian Wanying; quarterly; in Chinese.

Hakusan (Hakusan Nature Conservation Centre News): Ishikawa-ken Hakusan Shizen Hogo Senta, Kinameri, Yoshinodanimura, Ishikawa-gun, Ishikawa-ken 920-23, Japan; tel (7619) 55321; fax (7619) 55323; f 1973; quarterly; in Japanese.

Haramata: International Institute for Environment and Development (IIED), 3 Endsleigh St, London WC1H 0DD, UK; tel (71) 388-2117; telex 317210; fax (71) 388-2826; f 1988; quarterly; bulletin of IIED Drylands Programme; in English and French; each issue accompanied by two Issue Papers dealing with specific subjects in greater depth, also in English and French; circ approx 2,000.

Harrowsmith: Telemedia Publishing Inc, 7 Queen Victoria Rd, Camden E, ON K0K 1J0, Canada; tel (613) 378-6661; every 2 months; areas covered include ecology, energy efficiency, organic gardening, country life and alternatives to conventional ways of life; circ 156,000.

Hazardous Cargo Bulletin: Intapress Publishing Ltd, 38 Tavistock St, London WC2E 7PB, UK; tel (71) 240-0837; fax (71) 836-9321; Publr: Barry Perry; Editor: Michael Corkhill; monthly; international magazine on storage, transport and handling of oils, gases and chemicals world-wide.

Hazardous Cargo Bulletin Tank Guide: Intapress Publishing Ltd, 38 Tavistock St, London WC2E 7PB, UK; tel (71) 240-0837; fax (71) 836-9321; Publr: Barry Perry; supplement to *Hazardous Cargo Bulletin.*

Hazardous Materials Management: 401 Richmond St W, Suite 139, Toronto, ON M5V 1X3, Canada; tel (416) 348-9922; fax (416) 348-9744; Publr: Todd Latham; Sales Dir: Arnie Gess; Pres and Editor-in-Chief: Guy Crittenden; f 1989; every 2 months; explains new environmental regulations, waste treatment and pollution prevention technologies; intended for readers in industry and local government; circ 18,000.

Hazardous Waste Management Handbook: Southam Business Communications Inc, 1450 Don Mills Rd, Don Mills, ON M3B 2X7, Canada; tel (416) 445-6641; fax (416) 442-2200; Editors: Mary Mancini, Deborah Orchard; f 1982; annual.

Hazardous Waste News: Business Publishers, Inc, 951 Pershing Drive, Silver Spring, MD 20910-4464, USA; tel (301) 587-6300; fax (301) 587-1081; Exec Editor: Mary E Crowley; f 1979; weekly; covers testing, classification, transportation, storage, treatment, recycling and disposal of industrial hazardous waste; anticipates new regulations and changes in the status of waste substances.

Hazardous Waste Report: Aspen Publishers, c/o Swets Publishing Service, POB 825, 2160 SZ Lisse, Netherlands; tel (2521) 35111; telex 41325; fax (2521) 15888; Editor: Patricia E Papa; fortnightly; newsletter containing information on hazardous waste in the USA, policy analysis, news of legal developments and reports on the activities of the national and state governments.

Haznews: Profitastral Ltd, Park House, 140 Battersea Park Rd, London SW11 4NB, UK; tel (71) 498-2511; fax (71) 498-2343; Editor and Publr: David Coleman; f 1988; monthly; international business magazine dealing with hazardous waste management and cleaning of contaminated sites.

Health & Environment Digest: The Freshwater Foundation, Spring Hill Center, 725 County Rd 6, Wayzata, MN 55391, USA; tel (612) 449-0092; fax (612) 449-0592; Editor: Tim Burkhardt; f 1987; monthly; covers the effects upon health of environmental contaminants, not only in water; includes annual index; each issue focuses on a single environmental health problem.

Heimatschutz/Sauvegarde: Schweizer Heimatschutz/Ligue suisse du patrimoine national, Geschäftsstelle, Merkurstr 45, Postfach, 8032 Zürich, Switzerland; tel (1) 2522660; fax (1) 2522870; Editor: Marco Badilatti; f 1906; quarterly; published by Swiss League for the Preservation of the National Heritage; in German and French; circ 23,000.

Hokkaido No Shizen (Nature in Hokkaido): Hokkaido Shizen Hogo Kyokai, 064 Kamori Bldg, Nishi 1, Kita 3, Chuoh-ku, Sapporo, Japan; annual; publication of Nature Conservation Society of Hokkaido.

The Holocene: Edward Arnold, Mill Rd, Dunton Green, Sevenoaks, Kent TN13 2YA, UK; tel (732) 450111; fax (732) 461321; Journals Manager: Mary Altree; Editor: John Matthews; f 1990.

Honeyguide: Ornithological Association of Zimbabwe, POB 8382, Causeway, Harare, Zimbabwe; tel (4) 882206; Editor: M P S Irwin; quarterly; members' magazine; contains research papers, recent observations and items of general interest on birds in Zimbabwe; distributed in Africa and overseas.

Hoppo Kagaku Chosa Hokoku (Human Culture and Environmental Studies in Northern Hokkaido): Tsukuba Daigaku, 1-1, Tennodai 1-chome, Tsukuba-shi, Ibaraki-ken 305, Japan; f 1980; annual; published by University of Tsukuba; in Japanese.

La Houille Blanche: Revue Générale de l'Electricité SA, 48 rue de la Procession, 75724 Paris Cedex 15, France; f 1946; 8 a year; publication dedicated to hydroelectric power; in French and English.

Hozon Kagaku (Science for Conservation): Bunka-cho Tokyo Kokuritsu Bunkazai Kenkyujo Hozon Kagakubu, Ueno Koen, Taito-ku, Tokyo 110, Japan; f 1964; annual; published by Agency for Cultural Affairs—Tokyo National Research Institute of Cultural Property; in Japanese, with summaries in Japanese and English.

Huanjing Daobao (Environment Herald): Jiangsu Sheng Huanbao Ju, 70 Beijing Xilu, Bldg No 81, Nanjing 210013, Jiangsu Province, People's Republic of China; (Editorial) 44 Mei Yuan Xin Cun, Nanjing 210018, Jiangsu Province, People's Republic of China; tel (25) 410214; fax (25) 631720; Editor-in-Chief: Prof Yuan Kechang; Deputy Editor: Mei Changhan; f 1984; every 2 months; published by Jiangsu Environmental Protection Bureau; disseminates information on the theory and technology of practical environmental protection; in Chinese.

Huanjing Huaxue (Environmental Chemistry): Research Centre for Eco-Environmental Sciences (RCEES), Chinese Academy, POB 2871, Beijing 100083, People's Republic of China; tel (1) 2555129; telex 5499; fax (1) 2555381; Editor-in-Chief:

Jun-xiong Huang; f 1982; every 2 months; contains original papers and reviews on the theory and practice of all aspects of environmental chemistry, including atmospheric, aquatic and soil chemistry.

Huanjing Kexue (Chinese Journal of Environmental Science): Research Centre for Eco-Environmental Sciences (RCEES), Chinese Academy, POB 2871, Beijing 100085, People's Republic of China; tel (1) 2545511; telex 5499; fax (1) 2555381; Editor-in-Chief: Li Shitao; f 1976; every 2 months; interdisciplinary journal reflecting significant advances in studies of environmental protection and ecology; published by RCEES in association with Beijing Municipal Institute of Environmental Protection and Tsinghua University Institute of Environmental Engineering; in Chinese.

Huanjing Kexue Xuebao—Acta Scientiae Circumstantiae: Science Press, Marketing and Sales Dept, 16 Donghuangchenggen Beijie, Beijing 100707, People's Republic of China; tel (1) 4010642; telex 210247; fax (1) 4012180; f 1981; quarterly; published for the Ecology and Environment Research Centre of the Chinese Academy of Sciences; contains results of original research on the theory and practice of environmental science, including details of new technology and methods; in Chinese, with summaries in English; circ 11,000.

Huanjing Wuran Yu Fangzhi (Environmental Pollution and Control): Zhejiang Huanjing Baohu Kexue Yanjiusuo, 43 Tianmushan Lu, Hangzhou, Zhejiang 310007, People's Republic of China; tel (571) 882847; Editorial Bd: Liu Jian, Lin Leigu, Weng Shizhen, Chen Jingen; f 1979; every 2 months; published by Zhejiang Institute of Environmental Protection; contains papers on environmental protection law, environmental economics and management, and the technology of pollution control and environmental monitoring; in Chinese.

Huanjing yu Jiankang Zazhi (Journal of Environment and Health): Tianjin Shi Weisheng Fangbin Zhongxin, 76 Tianshanshe Dajie, Hedong-qu, Tianjin 300011, People's Republic of China; tel (22) 429908; Editor: Dong Shanheng; f 1984; every 2 months; published by Tianjin Municipal Centre of Health and Disease Prevention; contains results of scientific research into pollution of natural and man-made environments and its effects on the human body.

Hudson Valley Green Times: Hudson Valley Grass Roots Energy and Environmental Network, POB 208, Red Hook, NY 12571, USA; tel (914) 758-4484; quarterly; focuses on safe energy and environmental preservation; intended to inform the public and increase public participation in environmental and energy-related policy-making.

Hydroplus: Hydrocom, 13 rue Saint-Florentin, 75008 Paris, France; tel (1) 47-03-33-13; fax (1) 40-15-05-79; Editor: Jean-Jacques Lemoigne; f 1989; monthly; international water industry magazine; in French and English; circ 10,000.

IAIA Newsletter: International Association for Impact Assessment (IAIA), POB 70, Belhaven, NC 27810, USA; tel (919) 964-2338; fax (919) 964-2340; newsletter concerned primarily with environmental impact assessment.

IAWPC Technical Annual: Indian Association for Water Pollution Control (IAWPC), Nehru Marg, Nagpur 440 020, India; in English.

ICCET Annual Report: Imperial College Centre for Environmental Technology (ICCET), University of London, 48 Prince's Gardens, London SW7 2PE, UK; fax (71) 584-7596; Editor: H Boutwood; f 1978; circ 300.

ICDA Update: International Coalition for Development Action (ICDA), rue Stévin 115, 1040 Brussels, Belgium; tel (2) 230-04-30; fax (2) 230-03-48; covers international trade, sustainable development, and trade and the environment.

ICEL References: International Council of Environmental Law (ICEL), 5300 Bonn 1, Adenauerallee 214, Germany; fax (228) 2692250; Editor: E Howard; f 1970; quarterly; published in several languages; circ 300.

ICF News: Institute of Chartered Foresters (ICF), 7A St Colme St, Edinburgh EH3 6AA, UK; tel (31) 225-2705; fax (31) 220-6128; quarterly bulletin.

IFOAM Internal Letter: International Federation of Organic Agriculture Movements (IFOAM), 6695 Tholey-Theley, Ökozentrum Imsbach, Germany; tel (6853) 5190; fax (6853) 30110; quarterly; members' newsletter; reports on IFOAM activities world-wide.

Ilmansuojelu-uutiset (Finnish Air Pollution Prevention News): Finnish Air Pollution Prevention Society (FAPPS), POB 335, 00131 Helsinki, Finland; tel (0) 631442; fax (0) 631442; Editor-in-Chief: Kari Hämekoski; every 2 months; information bulletin; in Finnish, with English abstracts.

In-Pak: Jante-Forlaget APS, POB 15, Kongstrupvej 3, 4390 Vipperod, Denmark; tel 53-48-28-00; fax 53-48-22-05; Publr: Kjeld Krogh; Editor: Gitte Søndergaard; f 1980; monthly (10 a year); deals with all aspects of packaging, including recycling, disposal and environmental considerations; intended for the packaging industry in Denmark and its external territories, and Iceland; circ 8,500.

Indian Forester: Forest Research Institute, PO New Forest, Dehra Dun 248 006, India; tel (135) 27021; telex 348; fax (135) 23258; Publr and Advertising Manager: B S Beniwal; Hon Editor: Pankaj Khullar; f 1875; monthly; forestry science journal; intended for foresters, botanists, environmentalists, ecologists, industries relying on forestry products and research institutes; in English; circ 2,500.

Indian Journal of Ecology: Indian Ecological Society, Punjab Agricultural University, Ludhiana 141004, Punjab, India; Editor: A K Srivastava; f 1974; 2 a year; in English; circ 700.

Indian Journal of Environmental Protection: Kalpana Corporation, POB 5, Varanasi 221 010, India; Editor: Surendra Kumar; f 1981; monthly; deals with environmental pollution and its prevention; includes book reviews and indexes; circ 1,000.

Indian Journal of Power and River Valley Development: 6/2 Madan St, Calcutta 700 072, India; tel (33) 271711; Exec Editor: P K Chanda; Managing Editor: P K Menon; Advertising Manager: B N Bhattacharja; f 1950; monthly; journal dedicated to all aspects of thermal and hydroelectric power generation, irrigation and river valley projects; in English; circ 3,000.

Indian Literature in Environmental Engineering: National Environmental Engineering Research Institute, Documentation and Library Services, Nehru Marg, Nagpur 440 020, India; f 1971; annual; bibliographic review; indexed according to author and place of origin; in English.

Indoor Environment: Indoor Air International, Postfach 2, 4467 Rothenfluh, Switzerland; S Karger AG, Allschwilerstr 10, Postfach, 4009 Basel, Switzerland; tel (767) 318474 (UK office); fax (767) 313929 (UK office); Publications Sec: D F Weetman; journal of Int Asscn for Indoor Air Quality.

Industry and Environment: UNEP Industry and Environment Office, 39–43 quai André Citroen, 73739 Paris Cedex 15, France; tel (1) 40-58-88-50; telex 204997; fax (1) 40-58-88-74; Editor: Jacqueline Aloisi de Larderel; f 1978; quarterly; includes abstract and book reviews; in English; circ 5,000.

Info-Nature: Société Réunionnaise pour l'Etude et la Protection de l'Environnement, BP 1012, 97481 Saint Denis, Réunion; f 1974; nature and conservation publication of the Réunion Society for Study and Protection of the Environment.

Inform Reports: INFORM, Inc, 381 Park Ave S, New York, NY 10016, USA; tel (212) 689-4040; fax (212) 447-0689; quarterly; members' newsletter.

Information on Environmental Pollution in Foreign Countries: Tokyo Metropolitan Research Institute for Environmental Protection, 7-5, Shinsuna 1-chome, Koto-ku, Tokyo 136, Japan; f 1968; annual.

Informazione Innovativa (Inf-Inn): Centro Studi 'l'Uomo e l'Ambiente', Via delle Palme 13, 35137 Padova, Italy; tel (49) 36435; Editor: Tina Merlin; f 1985; fortnightly; examines the relationship between technological development and environmental quality, with particular attention to clean technology, energy sources, and industrial health and safety; in Italian, with summaries in English; circ 1,000.

Infoterra Bulletin: Infoterra Programme Activity Centre, United Nations Environment Programme, POB 30552, Nairobi, Kenya; tel (2) 230800; telex 22068; fax (2) 226890; every 2 months; newsletter of UNEP's international environmental information system; available in English, French, Russian or Spanish.

Inquinamento: ETAS Periodici SpA, Via Mecenate 89, 20138 Milan, Italy; tel (2) 580841; fax (2) 58012592; Dir and Editor: Paolo Berbenni; f 1959; monthly; technical journal dealing with water, air and soil quality, waste treatment, noise reduction, recycling and clean energy; circ 6,600.

Inside FERC: McGraw-Hill Inc, 1221 Ave of the Americas, New York, NY 10020, USA; 1200 G St, NW, Suite 1100, Washington, DC 20005, USA; tel (202) 383-2141; fax (202) 383-2125; Man Editor: Chris Newkumet; Editor-in-Chief: Larry Foster; f 1979; weekly; covers US Federal Energy Regulatory Commission (FERC) and the natural gas industry.

Inter-American Population News: Inter-American Parliamentary Group on Population and Development (IAPG), 902 Broadway, 10th Floor, New York, NY 10010, USA; tel (212) 995-8860; telex 620661; fax (212) 995-8853; monthly bulletin; in English, Portuguese and Spanish.

Interact: IUCN—The World Conservation Union, rue Mauverney 28, 1196 Gland, Switzerland; tel (22) 9990001; telex 419624; fax (22) 9990002; forum for exchanging views within IUCN.

International Desalination and Water Reuse Quarterly: Green Global Publications Inc, 15 Old Mill Rd, Westport, CT 06880, USA; tel (203) 227-6500; fax (203) 227-6500; Publrs: Paul Green, Irv Lineal; Editor: Floyd H Meller; quarterly; includes technical reports, case studies and reviews of new products and equipment; circ 8,200.

International Directory of Solid Waste Management: James & James (Science Publishers) Ltd, 5 Castle Rd, London NW1 8PR, UK; tel (71) 284-3833; fax (71) 284-3737; Principal Officer: Edward Milford; annual; published for the International Solid Wastes and Public Cleansing Association (ISWA), Denmark.

International Environment and Safety: Labmate Ltd, Newgate, Sandpit Lane, St Albans, Herts AL4 0BS, UK; Editor: M H Pattison; f 1978; quarterly; circ 10,500.

International Environmental Outlook: Basil Blackwell Ltd, 108 Cowley Rd, Oxford OX4 1JF, UK; tel (865) 791100; telex 837022; fax (865) 791347; 2 a year; published for the London Business School Centre for Economic Forecasting.

International Institute for Environment and Development (IIED) Annual Report: 3 Endsleigh St, London WC1H 0DD, UK; tel (71) 388-2117; telex 261681; fax (71) 388-2826; Editor: Robin Sharp; contains report from Exec Dir, programme reports and resource accounts; in English, with summaries in French and Spanish.

International Journal of Ambient Energy: Ambient Press Ltd, POB 25, Lutterworth, Leics LE17 4FF, UK; tel (455) 202281; Dir: David Bennett; Head of Admin: Margaret Wrathall; Co-ordinating Editor: A F C Sherratt; Editor: J C McVeigh; f 1980; quarterly; information on application and theory of renewable energy sources, their importance in energy policy and their impact on the environment; contains research papers, case studies and book reviews.

International Journal of Biometeorology: International Society of Biometeorology, Secrétariat, c/o Prof N N Barthakur, Département des ressources renouvelables, Université McGill, Campus MacDonald, 21 111 Lakeshore Rd, Ste Anne de Bellevue, PQ H9X 1C0, Canada; tel (514) 398-7938; fax (514) 398-7983; Editor-in-Chief: Prof H Leith; quarterly; scientific journal.

International Journal of Ecology and Environmental Sciences: International Scientific Publications, 50-B Pocket C, Sidhartha Extension, New Delhi 110 014, India; Editor: P S Ramakrishnan; f 1974; 3 a year; published for the Scientific and Environmental Educational Society; in English; circ 600.

International Journal of Environment and Pollution: Inderscience Enterprises Ltd, World Trade Centre Bldg, 110 ave Louis Casai, CP 309, 1215 Geneva-Aéroport, Switzerland; telex 289950; fax (22) 7910885; f 1991; quarterly; co-published by UNESCO; published for the European Centre for Pollution Research (UK); covers sustainable development in environmental policy and pollution control; in English.

International Journal of Environmental Health Research: Chapman & Hall Ltd, 2–6 Boundary Row, London SE1 8HN, UK; tel (71) 865-0066; telex 290164; fax (71) 522-9623; Man Dir: David Inglis; f 1991; quarterly; deals with the effects of the natural and built environments on human health, with particular attention to communicable diseases.

International Journal of Environmental Studies: Gordon and Breach Science Publishers, POB 786, Cooper Station, New York, NY 10276, USA; POB 90, Reading, Berks RG1 8JL, UK; tel (212) 206-8900 (USA); fax (212) 645-2459 (USA); Pres: Jon Gillett; Chair: Martin Gordon.

International Journal of Plant Sciences: The University of Chicago Press, Journals Division, POB 37005, Chicago, IL 60637, USA; tel (312) 702-7600; fax (312) 753-0811; Editors: Edward D Garber, Manfred Ruddat; f 1875; fmrly *Botanical Gazette*; every 2 months; presents results of original research in all areas of plant biology, including ecology.

The International Tree Crops Journal: the journal of agroforestry: AB Academic Publishers, POB 42, Bicester, Oxon OX6 7NW, UK; tel (869) 320949; Editor: Alan Grainger; Consultant Editor: Lawrence D Hills; f 1980; quarterly; journal for the communication of research results and the exchange of practical experience in the fields of multi-purpose tree crops, agroforestry, environmental management, tree crop processing, economics and forestry for local community development.

International Wildlife: Canadian Wildlife Federation, 2740 Queensview Drive, Ottawa, ON K2B 1A2, Canada; tel (613) 721-2286; fax (613) 721-2902; Editor: Gabriel Rack; f 1965; every 2 months; members' magazine; circ 100,000.

International Zoo Yearbook: Zoological Society of London, Regent's Park, London NW1 4RY, UK; tel (71) 722-3333; telex 265247; fax (71) 483-4436; Editor: P J S Olney; f 1960; international forum for exchange of information between zoos.

Intervenor: Canadian Environmental Law Association (CELA), 517 College St, Suite 401, Toronto, ON M6G 4A2, Canada; tel (416) 960-2284; fax (416) 960-9392; Editor: Jill Cameron; f 1976; every 2 months; former title *CELA Newslet-*

ter; reports on environmental issues and legal cases; news of conferences and other events; circ 1,000.

Irish Naturalists' Journal: Irish Naturalists' Journal Ltd, c/o Ulster Museum, Botanic Gardens, Belfast BT9 5AB, UK; tel (232) 245133; fax (232) 665510; Treasurer: Catherine Hackney; Sec: Dr Valerie Hall; Editor: Dr Robin Govier; f 1925; papers, articles and notes on the natural history of the Repub of Ireland and Northern Ireland; covers botany, zoology and geology.

Irrigation Journal: Gold Trade Publishing Inc, POB 8420, Van Nuys, CA 91409, USA; tel (818) 781-8300; Editor: James Gregory; Exec Editor: Anne Goldstein; f 1951; every 2 months; covers the design, construction and management of agricultural irrigation systems; circ 12,000.

Irrigazione: c/o Arthur James Chambers, British & Overseas Trade Press, 33 The Avenue, Beckenham, Kent BR3 2EE, UK; tel (81) 650-3518; quarterly; international journal of irrigation; available in Italian and four other languages; circ 12,000.

Ishikawa-ken Hakusan Shizen Hogo Senta Kenkyu Hokoku (Annual Report of Hakusan Nature Conservation Centre): Ishikawa-ken Hakusan Shizen Hogo Senta, Kinameri, Yoshinodanimura, Ishikawa-gun, Ishikawa-ken 920-23, Japan; tel (7619) 55321; fax (7619) 55323; f 1974; in Japanese, with summaries in English and Japanese.

Israel Environment Bulletin: Ministry of the Environment, POB 6234, Jerusalem 91061, Israel; tel (2) 701606; telex 25629; fax (2) 513945; Editor: Shoshana Gabbay; f 1974; quarterly; covers environmental policy, legislation, initiatives and events in Israel; in English; circ 2,000.

Issues: Underwood Publications, POB 489, Bayswater, Vic 3153, Australia; tel (3) 764-5460; Publr and Editor: Gary Underwood; Co-Publr: Shelley Underwood; f 1987; quarterly; each edition concentrates on a particular Australian or global environmental issue and features articles by experts in that field; intended for secondary school pupils and teachers, and the general public; circ 1,000.

IUCN Bulletin: IUCN—The World Conservation Union, rue Mauverney 28, 1196 Gland, Switzerland; tel (22) 9990001; telex 419624; fax (22) 9990002; Editor: Nikki Meith; f 1952; quarterly news journal; includes indexes and book reviews; published in English and French editions; circ 6,000.

IVL—Nytt: Institutet før Vatten- och Luftvaardsforskning (IVL), POB 21060, 100 31 Stockholm, Sweden; tel (8) 729-15-00; telex 155815; fax (8) 31-85-16; Editor: Erik Arnberg; f 1966; quarterly; contains articles on research conducted at the Swedish Environmental Research Institute.

IVL—Referat: Institutet før Vatten- och Luftvaardsforskning (IVL), POB 21060, 100 31 Stockholm, Sweden; tel (8) 729-15-00; telex 155815; fax (8) 31-85-16; f 1985; quarterly; summaries of research reports from the Swedish Environmental Research Institute; in English and Swedish.

IWL—Umweltbrief: Institut für gewerbliche Wasserwirtschaft und Luftreinhaltung eV (IWL), 5000 Cologne 51, Marienburger Str 22, Germany; tel (221) 376950; fax (221) 3769517; monthly; environmental newsletter published by the Institute for Industrial Water Management and Air Purification; provides business people with background information and assistance in making decisions; circ 5,000.

IWSA Yearbook: International Water Supply Association (IWSA), 1 Queen Anne's Gate, London SW1H 9BT, UK; tel (71) 957-4567; fax (71) 222-7243; scientific and technical publication intended to maximize exchange of information on water supply and related activities.

Jacht en Natuurbeheer: Koninklijke Sint-Hubertusclub van Belgie, Jan Jacobsplein 1, 1000 Brussels, Belgium; monthly; deals with hunting and nature conservation.

Jaegaren: Jaegarnas Centralorganisation, 01100 Oestersundom, Helsinki, Finland; fax (8) 777617; Editor: Paula Haenninen; f 1951; every 2 months; published by the Central Hunters' Organization; Swedish edition of *Metsastaja* (qv).

Jahrbuch des Vereins zum Schutz der Bergwelt: Verein zum Schutz der Bergwelt eV, 8000 Munich 22, Praterinsel 5, Germany; tel (89) 2350900; fax (89) 4706168; Editor: Hans Smettan; f 1900; annual; published by the Society for the Conservation of Mountain Areas; concerned with the protection of alpine plants and wildlife.

Jahresbericht des Umweltbundesamts: Umweltbundesamt, 1000 Berlin 33, Bismarckplatz 1, Germany; tel (30) 89030; telex 183756; fax (30) 89032285; Editor: Hans-Jürgen Nantke; f 1965; annual report of Federal Environment Office; circ 15,000.

Jiaotong Huanbao—Shuiyun Ban (Environmental Protection in Transportation—Water Transport Edition): Jiaotong Bu, Tianjin Shuiyun Gongcheng Keyan-suo, Xin Gang, Tanggu-qu, Tianjin 300456, People's Republic of China; tel (22) 975508; Editor: Wang Shuitian; every 2 months; in Chinese.

Journal of the Air and Waste Management Association: POB 2861, Pittsburgh, PA 15230, USA; tel (412) 232-3444; fax (412) 232-3450; Editor: Harold M Englund; f 1951; monthly; fosters the exchange of current technological information and encourages the development of new technology for a cleaner environment; circ 10,000.

Journal of Animal Ecology: Blackwell Scientific Publishing Ltd, Osney Mead, Oxford OX2 0EL, UK; tel (865) 240201; telex 83355; fax (865) 721205; Journals Dir: Keith Bowker; published for the British Ecological Society.

Journal of Aquatic Ecosystem Health: Kluwer Academic Publishers, Postbus 17, 3300 AA Dordrecht, Netherlands (Publr); Great Lakes Laboratory for Fisheries and Aquatic Science, Fisheries and Oceans Canada, Canada Centre for Inland Waters, POB 5050, 867 Lakeshore Rd, Burlington, ON L7R 4A6, Canada (Editorial); tel (78) 52-44-00 (Publr); fax (78) 18-32-73 (Publr); Tech Editor: H F Nicholson; Editorial Asst: K Gaasenbeek; Editor-in-Chief: M Munawar; f 1991; quarterly; published for Aquatic Ecosystem Health and Management Soc; contains original papers on the health of biotic components of the aquatic ecosystem; aims to deal with marine, freshwater and estuarine ecosystem health from a holistic and sustainable perspective, to promote understanding of the functioning and recovery of stressed ecosystems under the impact of contaminants, and to encourage international, interdisciplinary communication.

Journal of Clean Technology and Environmental Sciences: Princeton Scientific Publishing Co, Inc, POB 2155, Princeton, NJ 08543, USA; tel (609) 683-4750; fax (609) 683-0838; Editors: M A Mehlman, S P Maltezon; f 1991; quarterly; concerned with prevention of pollution, new techniques for identifying pollutants, toxicology and the assessment of hazardous chemicals.

Journal of Cleaner Production: Butterworth-Heinemann Ltd, Linacre House, Jordan Hill, Oxford OX2 8DP, UK; tel (865) 310366; fax (865) 310898; Gen Editor: Prof Donald Huisingh; f 1993; quarterly; international journal devoted to technologies, concepts and policies for pollution prevention, waste minimization and cleaner products.

Journal of Environment and Development: University of California (San Diego), La Jolla, CA 92093; tel (619) 534-3135; Editor: Daniel L Nielson; 2 a year.

Journal of Environmental Biology: JEB Foundation, 711 Civil Lines (South), Muzaffarnagar 251 001, India; tel (1312)

25306; Editor: Kiran Dalela; f 1980; quarterly; international, multidisciplinary journal dealing with environmental pollution and toxicology; circ 1,000.

Journal of Environmental Hydrology: International Association for Environmental Hydrology (IAEH), POB 1088, Alexandria, VA 22313, USA; tel (703) 683-9768; fax (703) 683-6137; Editor: Roger W Peebles; f 1992; quarterly; reports on the environmental aspects of hydrology and water resources, with emphasis on the prevention and cleaning up of pollution.

Journal of Environmental Law: Oxford University Press, Oxford Journals, Pinkhill House, Southfield Rd, Eynsham, Oxford OX8 1JJ, UK; tel (865) 882283; telex 837330; fax (865) 882890; Editor: Richard Macrory; f 1989; 2 a year; deals with legislation concerning pollution control, waste management, protection of habitats, biotechnology and public natural resources; circ 850.

Journal of Environmental Management: Academic Press Ltd, 24–28 Oval Rd, London NW1 7DX, UK; tel (71) 267-4466; telex 25775; fax (71) 482-2293; Editor: J N R Jeffers; f 1973; 12 a year in 2 vols; contains articles on all aspects of the management and use of the natural and built environments.

Journal of Environmental Permitting: Executive Enterprises Publications Co, Inc, 22 W 21st St, 10th Floor, New York, NY 10010-6990, USA; tel (212) 645-7880; fax (212) 645-1160; Man Editor: Isabelle Cohen; Editor: Joseph Guida; f 1991; quarterly; covers air, operating, water and RCRA permits; intended for corporate environmental engineers, design engineers, environmental managers and lawyers.

Journal of Environmental Radioactivity: Elsevier Science Publishers Ltd, Crown House, Linton Rd, Barking, Essex IG11 8JU, UK; tel (81) 594-7272; telex 896950; fax (81) 594-5942; Editor: M S Baxter; f 1984; 9 a year in 3 vols; published for the University of Glasgow Department of Chemistry; international journal containing original research papers on all aspects of radioactivity in the natural environment.

Journal of Environmental Regulation: Executive Enterprises Publications Co, Inc, 22 W 21st St, 10th Floor, New York, NY 10010-6990, USA; tel (212) 645-7880; fax (212) 645-1160; Man Editor: Deborah Wenger; Editor: Carol DiPaolo; f 1991; quarterly; provides detailed accounts of major environmental legislation, explanations of existing legislation and advice on planning for and implementing new regulations.

Journal of Environmental Science and Health, Part A—Environmental Science and Engineering: Marcel Dekker, Inc, 270 Madison Ave, New York, NY 10016, USA; tel (212) 696-9000; telex 421419; fax (212) 685-4540; Editor: James W Robinson; Assoc Editors: James R Brock, Joe E Ledbetter; 10 a year; international forum for information pertinent to environmental problems affecting air, water or soil; covers engineering innovations, effects of pollutants on health, control systems, laws and projections.

Journal of Environmental Science and Health, Part B—Pesticides, Food Contaminants, and Agricultural Wastes: Marcel Dekker, Inc, 270 Madison Ave, New York, NY 10016, USA; tel (212) 696-9000; telex 421419; fax (212) 685-4540; Editor: Shahamat U Khan; every 2 months; original research reports on advances relating to residues of pesticides, natural food contaminants and additives, other contaminants, and their metabolites in the ecosphere; covers the persistence and biodegradation of chemicals, their toxicological consequences and methods of detoxification.

Journal of Environmental Science and Health, Part C—Environmental Carcinogenesis & Ecotoxicology Reviews: Marcel Dekker, Inc, 270 Madison Ave, New York, NY 10016, USA; tel (212) 696-9000; telex 421419; fax (212) 685-4540; Editor: Yin-Tak Woo; 2 a year; multidisciplinary journal of reviews of various aspects of environmental carcinogenesis; subjects covered include synergism and antagonism, theoretical models, inhibition of carcinogenesis and interaction of physical, chemical and biological factors.

Journal of Environmental Sciences (China): Science Press, Marketing and Sales Dept, 16 Donghuangchenggen Beijie, Beijing 100707, People's Republic of China; POB 2871, Beijing 100085, People's Republic of China (Editorial); tel (1) 4012180; fax (1) 4019810; Editors-in-Chief: Liu Tungsheng, Zhuang Yahui, Pang Shuwei; Editors: Tie Xiaoshan, Zhuang Yonghui; f 1989; quarterly; specialized academic journal; English language edition of *Huanjing Kexue*.

Journal of Forestry: Society of American Foresters, 5400 Grosvenor Lane, Bethesda, MD 20814-2198, USA; tel (301) 897-8720; fax (301) 897-3690; f 1902; monthly; members' journal; concerned with forestry, forest products and issues relating to the management of natural resources.

Journal of Himalayan Studies and Regional Development: Garhwal University, Institute of Himalayan Studies and Regional Development, POB 12, Srinagar, Garhwal 246 174, India; Editor: Dr Shankar Kala; f 1977; annual; former title *Himalaya*; in English; circ 1,000.

Journal of Human Ecology: Kamla-Raj Enterprises, 2273 Gali Bari Paharwali, Chawri Bazar, Delhi 110 006, India (Publr); Indian Society for Human Ecology (ISHE), Department of Anthropology, University of Delhi, Delhi 110 007, India (Editorial); tel (11) 7257329; Managing Editor: Dr M K Bhasin; f 1990; 3 a year; contains articles, technical reports and short items on environmental issues in the fields of human ecology, biology, genetics and culture; aims to improve the relationship between humans and their environment by communicating knowledge and experience; in English; circ 150.

Journal of the Indian Association for Environmental Management: Indian Association for Environmental Management, NEERI, P O, Nagpur 440 020, India; Editor: A G Bhole; 3 a year; members' journal; in English.

Journal of Industrial Pollution Control: Enviro Media, Shaniwar Peth, Patan Colony, POB 90, Karad 415 110, India; tel (2164) 44369; Man Editor: Sukhada Trivedy.

Journal of the Japan Air Cleaning Association: Intercontinental Marketing Corporation, POB 5056, Tokyo 100-31, Japan; fax (3) 667-9646; 8 a year.

Journal of Land Use and Environmental Law: Florida State University, College of Law, Tallahassee, FL 32306, USA; tel (904) 644-4240; fax (904) 644-7781; Editor-in-Chief: Rebecca Cunningham; f 1983; law review journal for the discussion of topical environmental and land use issues.

Journal of Natural Resources (Zhiran Zhiyuan Xuebao): Science Press, Marketing and Sales Dept, 16 Donghuangchenggen Beijie, Beijing 100707, People's Republic of China; tel (1) 4012180; fax (1) 4019810; quarterly; in Chinese with English abstract.

Journal of Soil and Water Conservation: Soil and Water Conservation Society of America, 7515 NE Ankeny Rd, Ankeny, IA 50021-9764, USA; tel (515) 289-2331; fax (515) 289-1227; Editor: Max Schnepf; Managing Editor: James L Sanders; f 1946; every 2 months; members' journal; intended for landowners, professional conservationists and other interested parties; circ 11,450.

Journal of Soil and Water Conservation in India: Soil Conservation Society of India, DVC Campus, Hazaribagh, Bihar, India; Editor: K S V Raman; f 1952; annual; in English; circ 600.

Journal of Solar Energy Research: Scientific Research Centre, Solar Energy Research Council, Jadiriyah, POB 13026, Baghdad, Iraq; telex 213976; Editor: Nidhal I Al-Hamdani; f 1983; 2 a year; in Arabic and English; circ 500.

Journal of Sustainable Agriculture: Biology Department, Fairfield University, Fairfield, CT 06430-7524, USA (Editorial); Haworth Press Inc, 10 Alice St, Binghamton, NY 13904-1580, USA (Publr); tel (203) 254-4000 (Editorial); fax (203) 254-4126; Sr Editor: Raymond P Poincelot; f 1989; international journal containing peer-reviewed papers on sustainable agriculture.

Journal of Tropical Ecology: Cambridge University Press, The Edinburgh Bldg, Shaftesbury Rd, Cambridge CB2 2RU, UK; tel (223) 312393; telex 817256; fax (223) 315052; Chief Exec: Anthony K Wilson; Man Dir: Jeremy Mynott: Editor: Adrian G Marshall; quarterly; contains articles on the ecology of tropical regions, original research papers and reviews.

The Journal of World Forest Resource Management: the journal of forest policy: AB Academic Publishers, POB 42, Bicester, Oxon OX6 7NW, UK; tel (869) 320949; Editor: Alan Grainger; f 1984; 2 a year; focuses on ecology, resource assessment and all aspects of forests as a global resource; intended for those concerned with the future of the world's forest resources and with the impact of forest exploitation on the global environment; contains research papers, news items, book reviews and details of conferences.

Jumin Katsudo: Shinseikatsu Undo Kyokai, 1-3 Hibiya Koen, Chiyoda-ku, Tokyo, Japan; f 1972; quarterly.

Kanagawa-kenritsu Shizen Hogo Senta Hokoku: Kanagawa-kenritsu Shizen Hogo Senta, 657 Nanasawa, Atsugi-shi, Kanagawa-ken 243-01, Japan; f 1984; annual; reports on research conducted at Kanagawa Prefectural Nature Conservation Centre; in Japanese.

Kankyo Gijutsu (Environmental Conservation Engineering): Kankyo Gijutsu Kenkyu Kyokai, Matsumo Bldg, 2-1-20 Tenma, Kita-ku, Osaka 530, Japan; tel (6) 357-7611; fax (6) 357-7612; Editor: S Iwai; f 1972; monthly; published by Research Institute of Environmental Technology; deals with basic technical issues relating to global environmental problems; in Japanese, with summaries in English; circ 8,000.

Kobus: Wildlife Conservation Society of Zambia, POB 30255, Lusaka, Zambia; tel (1) 254226; telex 44810; f 1987; monthly; conservation magazine intended for the general public.

Koedoe: National Parks Board/Nasionale Parkeraad, POB 787, Pretoria 0001, South Africa; tel (12) 3439770; telex 321931; fax (12) 3439958; Editors: G de Graaff, J C Rautenbach; f 1958; 2 a year; journal of research relevant to the national parks of South Africa; contains original papers relating to the preservation of biodiversity and the conservation of natural resources; includes abstracts, bibliographies and cumulative indexes; circ 1,200.

Kogai Kenkyu (Research on Environmental Disruption): Iwanami Shoten, 2-5-5 Hitotsubashi, Chiyoda-ku, Tokyo 101, Japan; quarterly; in Japanese.

Kol Atar: National Park Authority, POB 7028, Tel-Aviv 61070, Israel; tel (3) 252281; Editor: S Raviv; f 1973; 2 a year; bulletin of National Park Authority; circ 5,000.

Komba: Wildlife Clubs of Kenya Association, POB 20184 and 40658, Nairobi, Kenya; Editor: Ibrahim M Ali; f 1969; 3 a year; in English; circ 6,000.

Korean Nature: Korean Association for Conservation of Nature, Pyongyang, Democratic People's Republic of Korea; f 1967; quarterly; general conservation periodical.

Kornyezetvedelmi Szakirodalmi Tajekoztato (Environmental Control Abstracts): Orszagos Muszaki Informacios Kozpont es Konyvtar (OMIKK), 1428 Budapest, Museum u 17, POB 12, Hungary (Publr); Kultura, 1389 Budapest, POB 149, Hungary (Subscriptions); Editor: Eszter Molnar; f 1983; every 2 months; published by National Technical Information Centre and Library; circ 380.

Korrespondenz Abwasser: Gesellschaft zur Förderung der Abwassertechnik, 5205 St Augustin 1, Markt 71, Postfach 1160, Germany; tel (2241) 2320; fax (2241) 23251; Editor: Rainer M Kieslinger; f 1954; monthly; published by Society for Promotion of Wastewater Technology; covers water, sewage and solid waste; in German, with summaries in German, English and French; circ 12,000.

Lago: Instituto para la Conservación del Lago de Valencia, Apdo 761, Valencia, Venezuela; Editor: Octavio Jelambi; f 1967; every 2 months; published by the Institute for the Conservation of Lake Valencia; includes diagrams and book reviews; circ 1,500.

Land Link: Department of Conservation, Forests and Lands, 5-250 Victoria Parade, E Melbourne, Vic 3002, Australia; tel (3) 412-4705; f 1986; quarterly; deals with land protection issues; circ 25,000.

Land and Water: Land and Water, Inc, 900 Central Ave, Suite 21, POB 1197, Fort Dodge, IA 50501, USA; tel (515) 576-3191; fax (515) 576-2606; Pres: Kenneth M Rasch; Editor: Teresa C Doyle; f 1974; 8 a year; magazine of natural resource management and restoration, covering erosion control and water management; circ 20,000.

Landscape Architecture Europe: Landscape Design Trust, 13A West St, Reigate, Surrey RH2 9BL, UK; tel (737) 221116; fax (737) 224206; Editor: Ken Fieldhouse; annual.

Landscape Ecology: SPB Academic Publishing bv, POB 97747, 2509 GC The Hague, Netherlands; Editor: Frank B Golley; f 1987; quarterly; contains research papers on the structure and functioning of landscapes and their interaction with humans, and relevant papers on land-use, nature management and environmental conservation; in English.

Landscape and Urban Planning: Elsevier Science Publishers BV, POB 181, 1000 AD Amsterdam, Netherlands; tel (20) 5803911; fax (20) 5803249; Editor-in-Chief: Jon E Rodiek; f 1974; 12 a year in 3 vols; international scientific journal covering landscape design, reclamation and conservation, and urban planning and ecology; includes bibliographies, book reviews and indexes.

Landschaftspflege und Naturschutz in Thüringen: Wartburg Verlag GmbH, 6900 Jena, Sophienstr 18, Germany; f 1964; quarterly; published for Institut für Landschaftsforschung und Naturschutz Halle (Halle Institute for Landscape Research and Nature Conservation); dedicated to state landscape management and nature conservation in Thuringia.

Landscope: Western Australia Department of Conservation and Land Management, POB 104, Como, WA 6152, Australia; tel (9) 367-0333; fax (9) 389-8296; f 1970; quarterly; former title *Forest Focus* (published by Forest Department); general publication dealing with conservation, forests and national parks; circ 20,000.

Leben und Umwelt: Biologie-Verlag, 6200 Wiesbaden, Postfach 1449, Germany; Editor: Herbert Bruns; f 1964; quarterly; independent journal dealing with biology, the environment and preservation of biodiversity; includes indexes and book reviews; principally in German.

Les (Forest): Redakcia Les, Pekná cesta 19, 831 05 Bratislava, Slovakia; tel (7) 284060; fax (7) 288683; Editor-in-Chief: Ján Fillo; f 1947; deals with problems relating to forestry and the natural environment; published by Ministry of Agriculture and Forestry Management.

Lettre de l'Environnement: 11 rue du Marché St Honoré, 75001 Paris, France; tel (1) 42-96-67-22; fax (1) 40-20-07-75; Editor-in-Chief: Cécile Clicquot de Hentque; 18 a year; newsletter specializing in environmental technology and industrial practices.

Links: International Society of Biometeorology, Secrétariat, c/o Prof N N Barthakur, Département des ressources renouvelables, Université McGill, Campus MacDonald, 21 111 Lakeshore Rd, Ste Anne de Bellevue, PQ H9X 1C0, Canada; tel (514) 398-7938; fax (514) 398-7983; Editor: J Bouma.

Linnut: Lintutieteellisten Yhdistysten Liitto (BirdLife Finland), PL 17, 18101 Heinola, Finland; tel (18) 152579; fax (18) 143682; f 1965; every 2 months; former title *Lintumies*; devoted to the conservation and study of birds in Finland.

London Energy News: Official Newsletter of the London Energy Group: Ambient Press Ltd, POB 25, Lutterworth, Leics LE17 4FF, UK (Publr); 10 Grovelands Court, Chase Rd, London N14 6HY, UK (Editorial); tel (455) 202281 (Publr), (81) 886-1382 (Editorial); Co-ordinating Editor A F C Sherratt; Editor: Colin Gomez; quarterly; promotes efficient use of energy in the built environment.

London Environmental Bulletin: Rendel Science and Environment, Great Guildford House, 30 Great Guildford St, London SE1 0ES, UK; tel (71) 962-9884; fax (71) 962-9885; Man Dir: John Simson; Dir: Mike Schwar.

Loris: Wildlife and Nature Protection Society of Sri Lanka, Chaitiya Rd, Fort, Colombo 1, Sri Lanka; tel (1) 25248; telex 22933; fax (1) 580721; Editor: Sirancee Gunawardena; f 1936; 2 a year; journal of wildlife in Sri Lanka; in English; circ 2,000.

Lucht en Omgeving: CED Samsom, Kouterveld 14, 1831 Diegem, Belgium; tel (2) 723-11-11; telex 64130; fax (2) 649-84-80; quarterly; covers recent developments in air pollution control; in Flemish.

Lufthygienischer Monatsbericht: Landesanstalt für Umwelt, 6200 Wiesbaden, Unter den Eichen 7, Germany; tel (611) 5810; telex 4186278; fax (611) 581221; f 1973; monthly; published by Hesse Regional Institution for the Environment; air quality reports; circ 400.

Luftverunreinigung: Deutscher Kommunal Verlag GmbH, 4000 Düsseldorf 30, Roseggerstr 5A, Germany; tel (211) 624417; fax (211) 622998; Editor: H J Schumacher; f 1973; annual; deals with air pollution.

Luonnonsuojekija: Suomen Luonnonsuojeluliitto, PL 169, 00151 Helsinki, Finland; tel (0) 642881; fax (0) 6221815; publication concerned with the conservation of nature.

Mūsų gamta (Our Nature): Rudens 33B, Vilnius 2600, Lithuania; tel (2) 696-964; Editor-in-Chief: Vytautas Klovas; f 1964; monthly; popular science, nature preservation and tourism; circ 6,000.

Madoqua: Ministry of Wildlife, Conservation and Tourism, Private Bag 13306, Windhoek, Namibia; telex 9083180; fax (61) 63131; Editor: C J Brown; f 1969; 2 a year; journal containing original papers on applied research relating to arid zone biology in the Namib Desert and the conservation of nature in Namibia and neighbouring countries; circ 750.

Magazine Recycling Benelux: POB 9000, 6800 DA Arnhem, Netherlands; tel (85) 209897; fax (85) 230283; Editor-in-Chief: M Beck; f 1966; every 2 months; publication intended for recycling and reclamation professionals; marketing information and technical reports on recycling equipment; circ 2,660.

Malaysian Forester: c/o Forestry Department Headquarters, Jalan Mahameru, 50660 Kuala Lumpur, Malaysia; quarterly; publication dedicated to technical aspects of tropical forestry and forestry products; in English.

Manitoba Naturalists Society Bulletin: 302-128 James Ave, Winnipeg, MB R3B 0N8, USA; tel (204) 943-9029; fax (204) 949-9052; Advertising Manager: René Guyot; Editor: Kevin Hill; 11 a year; members' bulletin; deals with issues related to natural history and environmental protection; includes information on outdoor pursuits, natural history lectures and workshops; circ 1,200.

Marine Conservation: Marine Conservation Society, 9 Gloucester Rd, Ross-on-Wye, Herefordshire HR9 5BU, UK; tel (989) 66017; fax (989) 67815; Editor: Bill Foster; Editorial Asst: Liz Lewis; quarterly; Marine Conservation Society members' magazine.

Marine Environmental Research: Elsevier Science Publishers Ltd, Crown House, Linton Road, Barking, Essex IG11 8JU, UK; tel (81) 594-7272; telex 896950; fax (81) 594-5942; Editors: G Roesijadi, R B Spies; f 1978; 8 a year in 2 vols; contains results of research on biological, chemical and physical interactions in marine waters; concerned with toxicology; includes the publication *Oil and Chemical Pollution*.

Marine Fish Management: Nautilus Press Inc, 1054 National Press Bldg, Washington, DC 20045, USA; tel (202) 347-6643; Editor: John R Botzum, Sr.

Marine Mammal News: Nautilus Press Inc, 1054 National Press Bldg, Washington, DC 20045, USA; tel (202) 347-6643; Editor: John R Botzum, Sr.

Marine Pollution Research Titles: Plymouth Marine Laboratory, Citadel Hill, Plymouth PL1 2PB, UK; tel (752) 222772; fax (752) 226865; Editor: D S Moulder; f 1974; monthly; bibliography and book reviews; circ 200.

Markt in Grun: Rudolf Muller Verlag GmbH, 5000 Cologne, Stolberger Str 84, Postfach 410949, Germany; tel (221) 54970; fax (221) 5497326; 8 a year; marketing magazine for green products and services; circ 10,300.

Massachusetts Wildlife: Division of Fisheries and Wildlife, Field Headquarters, 75 North St, Westborough, MA 01581, USA; tel (508) 792-7270; fax (508) 792-7275; Exec Editor: Ellie Horwitz; Editor: Peter Mirick; information on wildlife and wildlife issues in the state of Massachusetts.

Materials Reclamation Handbook: EMAP Response Publishing Ltd, Wentworth House, Wentworth St, Peterborough PE1 1DS, UK; tel (733) 63100; fax (733) 66437; Publr: D M Sanders; lists companies in the materials reclamation and recycling industries; contains details of legislation, testing procedures and identification tables.

Materials Reclamation Weekly: EMAP Maclaren, Maclaren House, 19 Scarbrook Rd, Croydon, Surrey CR9 1QH, UK; tel (81) 688-7788; telex 946665; fax (81) 760-0473; Advertising Manager: Joe Opitz; Editor: Ian Martin; business publication intended for the recycling industry and associated bodies such as local authorities; deals with the reclamation of all types of materials, including metals, waste paper, glass, plastics, textiles, chemicals and oils.

Medio Ambiente: Universidad Austral de Chile, Instituto de Ecología y Evolución, Facultad de Ciencias, Casilla 567, Valdivia, Chile; telex 71035; fax (63) 21-5012; Editor: Carlos A Moreno; f 1975; 2 a year; general environmental publication; in Spanish and English; circ 500.

Medio Ambiente en España: Ministerio de Obras Publicas y Transportes, Secretaría General Técnica, Centro de Publicaciones, Paseo de la Castellana 67, 28071 Madrid, Spain; Caja Postal de Ahorros de Madrid, Of 2088, Sucursal 9002, Madrid, Spain (Subscriptions); tel (1) 2531600; telex 22325; annual; review of the environment in Spain published by Ministry of Public Works and Transport.

Meguro-ku No Kogai (Environmental Pollution in Meguro Ward): Meguro Ward Office, Environmental Pollution Dept, 4-5, 2-chome, Chuo-Chiyou, Meguro-ku, Tokyo 152, Japan; fax (3) 3716-7169; f 1971; annual; in Japanese.

Meio Ambiente: Thesaurus Editora, SIG, Quadra 8, Lote 2356, 70.610 Brasília, DF, Brazil; tel (61) 225-3011; fax (61)

321-0401; f 1978; quarterly; general environmental publication.

Merigal: Merigal Dingo Education Centre, Australian Native Dog Training Society of NSW Ltd (ANDT), Arina Rd, Bargo, NSW 2574, Australia; tel (46) 84-1156; Editor: Berenice Walters; f 1977; every 2 months; dedicated to the conservation of the Australian dingo; circ 300.

The Messenger: Citizens Energy Council, POB U, Hewitt, NJ 07421, USA; tel (201) 728-7835; fax (201) 728-7664; every 2 months; environmental and health journal.

Messergebnisse des Zentralen Immissionsmessnetzes—Monatsbericht: Landesamt für Umweltschutz und Gewerbeaufsicht, 6500 Mainz, Rheinallee 97, Abteilung 3, Germany; tel (6131) 608306; Editor: Horst Borchert; f 1978; monthly; published by Rheinland-Palatinate Regional Office for Environmental Protection and Industrial Health and Safety; compilation of statistics relating to the environmental effects of industry; circ 500.

Metsä ja Puu: Metsaviestinta Oy, Salomonkatu 17A, 00100 Helsinki 10, Finland; tel (0) 6940300; fax (0) 6933466; Editor: Juhani Karvonen; Deputy Editor: Kristiina Vuopala; f 1877; monthly (10 a year); published by the Finnish Forestry Association; intended for forestry organizations and professionals; circ 3,500.

Metsälehti: Metsakeskus, Tapio, Maistraatinportti 4A, 00240 Helsinki, Finland; tel (0) 1562333; fax (0) 1562335; Editor: Paavo Seppänen; Advertising Manager: Timo Tuomola; f 1933; fortnightly; forestry magazine; circ 63,000.

Metsäliiton Viesti: Osuuskunta Metsäliito, Revontulentie 6, 02100 Espoo, Finland; tel (0) 46941; fax (0) 4550025; concerned with forest economy.

Metsämies: Suomen Metsänhoitajaliitto, Toolomtillomlatu 8, 00250 Helsinki, Finland; tel (0) 4771014; fax (0) 414892; covers forest economy and the forestry industry.

Metsänhoitaja: Suomen Metsänhoitajaliitto, Rautatieläisenkatu 6, 00520 Helsinki, Finland; tel (0) 1502357; fax (0) 1486960; publication dealing with forest economy and the forestry industry.

Metsastaja: Metsastajain Keskusjarjesto, 01100 Oestersundom, Helsinki, Finland; fax (8) 777617; Editor: Paula Haenninen; f 1951; every 2 months; published by the Central Hunters' Organization; in Finnish; available in Swedish edition, *Jaegaren* (qv); circ 280,000 (including *Jaegaren*).

Microbial Ecology: Springer-Verlag New York Inc, 175 Fifth Ave, New York, NY 10010, USA; tel (212) 460-1627; fax (212) 473-6272; every 2 months; original papers on all aspects of ecology involving micro-organisms; circ 1,200.

Middle East and World Water Directory: Chatila Publishing House, POB 13-5121, Chouran, Beirut, Lebanon; tel (1) 352413; telex 23008; annual directory covering all aspects of water supply.

Mie-ken Kogai Senta Nenpo (Annual Report of Mie-ken Environmental Science Institute): Kogai Senta, 8-12 Shinjo 4-chome, Yokkaichi, Mie-ken, Japan; f 1973; in Japanese.

Milieu: Environment Canada, 3 Buade, CP 6060, Québec, PQ G1R 4V7, Canada; tel (418) 648-7204; fax (418) 649-6140; f 1972; 2 a year; general environmental publication; includes bibliographies and book reviews; in English and French; circ 10,000.

Milieu en Bedrijf: CED Samsom, Kouterveld 14, 1831 Diegem, Belgium; tel (2) 723-11-11; telex 64130; fax (2) 649-84-80; fortnightly; deals with recent legislation, initiatives and research relating to environmental protection; in Flemish.

Milieudefensie: Vereniging Milieudefensie, Postbus 19199, 1000 GD Amsterdam, Netherlands; tel (20) 27-52-87; fax (20) 22-13-66; f 1972; monthly; environmental protection magazine; circ 15,000.

Miljø Aktuelt: Hovfaret 17, POB 235, Skøyen, Oslo 2, Norway; tel (2) 52-10-40; fax (2) 50-66-48; magazine covering contemporary environmental issues; intended for decision-makers and consumers of environmental products and services.

Miljø Danmark: Miljøstyrelsen, Strandgade 29, 1401 Copenhagen K, Denmark; (Subscriptions) Danske Boghandleres Kommossionsanstalt, Siljangade 4-8, 2300 Copenhagen S, Denmark; Editor: Niels-Erik Sørensen; f 1974; 8-9 a year; published by National Agency of Environmental Protection; circ 12,000.

Miljø og Teknologi (Environment and Technology): Forlaget John Vabø A/S, Hartmannsvej 47–49, 2920 Charlottenlund, Denmark; tel 39-40-80-00; fax 39-40-82-80; Man Dir: John Vabø; Sales Man: Claes Kronholm; Man Editor, Dir: Dorte Kardel; every 2 months; Scandinavian readership.

Misset's Milieu Magazine: Audet Tijdschriften BV, POB 9000, 6800 DA Arnhem, Netherlands; tel (85) 209911; Man: D D F Wiedemeijer; Editor: D van Lent; 10 a year; environmental business magazine covering environmental policy, management and technology; circ 15,000.

Mondo Sommerso: Media Sea Communication srl, Via G A Amadeo 41, 20133 Milan, Italy; tel (2) 70100020; fax (2) 70101709; Editors: Dr Fabrizio De Checchi, Augusto Guidobaldi; f 1959; monthly; contains biological reports on marine and freshwater environments and their wildlife; also covers issues related to diving; in Italian and English; circ 39,000.

Monitor: Angel Publishing Ltd, Kingsland House, 361 City Rd, London EC1V 1LR, UK; tel (71) 417-7400; fax (71) 417-7500; Editor: Bryan Rimmer; f 1991; environmental news bulletin for the packaging industry.

Monthly Summary: National Air Pollution Surveillance: Environment Canada, EP Publications, Ottawa ON K1A 0H3, Canada; fax (819) 953-5921; produced by Environment Canada Conservation and Protection Service; measurements of the concentration of pollutants in the air, by location, number of hourly samples and monthly averages; in English and French.

Monti e Boschi: Edagricole SpA, Via Emilia Levante 31/2, 40139 Bologna, Italy; tel (51) 492211; telex 510336; fax (51) 493660; Editor: Prof Umberto Bagnaresi; Advertising Manager: Giuliano Avoni; f 1949; every 2 months; forestry publication; circ 18,450.

Moonbi: Fraser Island Defender's Organization Ltd, POB 301, West End, Qld 4101, Australia; tel (7) 870-2820; Editor: J Sinclair; f 1971; quarterly; covers issues relating to the conservation of Fraser Island, Cooloola and the Great Sandy Region off the east coast of Queensland; circ 500.

Mountain Research & Development: University of California Press, 2120 Berkeley Way, Berkeley, CA 94720, USA; tel (510) 642-4191; fax (510) 643-7127; Editor: Jack D Ives; Business Manager: Sandra Whisler; quarterly; interdisciplinary journal; encourages sustainable development of mountain resources and consideration for mountain environments and peoples; circ 960.

Müllmagazin: Redaktion Müllmagazin, 10785 Berlin, Kurfürstenstr 14, Germany; tel (30) 2616854; fax (30) 2650366; Editors: Anne Wispler, Bernhard Reiser; f 1988; quarterly; published by Institut für Ökologisches Recycling eV (Institute for Ecological Recycling); specialist journal dealing with waste prevention, ecological waste management and recycling; circ 2,500.

Municipal and Industrial Water and Pollution Control: Southam Business Communications Inc, 1450 Don Mills Rd, Don Mills, ON M3B 2X7, Canada; tel (416) 445-6641; fax (416) 442-2077; Editor: Tom Davey; f 1893; every 2 months.

Nabat: Socio-Ecological Union 'Chernobyl', 220047 Minsk, Myasnikov St 39, Belarus; tel (0172) 20-39-04; fax (0172) 23-90-14; Editor-in-Chief: Vasily Yakovenko; f 1990; ecological survival and the socio-economic prosperity of peoples and nations.

Naika: Vanuatu Natural Science Society, POB 944, Port Vila, Vanuatu; journal on all issues concerning the environment in Vanuatu.

Natal Wildlife: Wildlife Society of Southern Africa, Natal Branch—Natuurleweverenigіng van Suidelike Afrika, 100 Brand Rd, Durban 4001, South Africa; tel (31) 213126; fax (31) 219525; Editor: Hazel O'Flaherty; f 1960; monthly; circ 8,000.

National Geographic: National Geographic Society, 1145 17th St, NW, Washington, DC 20036, USA; tel (202) 857-7000; fax (202) 775-6141; f 1888; monthly; popular periodical dedicated to geography, natural history, the environment and cultural anthropology; circ 10,182,900.

National Parks Journal: National Parks Association of NSW, POB A96, Sydney South, NSW 2000, Australia; tel (2) 264-7994; fax (2) 264-7160; Editor: M Rothery; f 1959; every 2 months; circ 7,000.

National Parks Magazine: National Parks and Conservation Association, 1015 31st St, NW, Washington, DC 20007-4406, USA; tel (202) 944-8530; fax (202) 944-8535; f 1919; every 2 months; concerned with national parks and other areas of natural beauty; circ 36,900.

National Trust Quarterly: National Trust of Australia (NSW), Observatory Hill, Watson Rd, The Rocks, Sydney, NSW 2000, Australia; tel (2) 258-0123; fax (2) 258-1264; Editor: M Stapleton; includes bibliographies and book reviews; circ 33,000.

National Wildlife: National Wildlife Federation, 1400 16th St, NW, Washington, DC 20036-2266, USA; tel (703) 790-4000; f 1962; every 2 months; intended for a general readership interested in nature and wildlife; circ 900,000.

Natur (Conservation): Ringier Verlag GmbH, 8000 Munich 83, Gustav-Heinemann-Ring 212, Postfach 830752, Germany; tel (89) 638180; fax (89) 63818100; Editor: Dirk Maxeiner; Advertising Manager: Marlene Mauracher; f 1981; monthly; presents information on the environment and nature preservation; distributed throughout Europe; circ 266,350.

Natur og Miljø: Danmarks Naturfredningsforening, Norregade 2, 1165 Copenhagen K, Denmark; f 1974; quarterly; published by Danish Society for the Conservation of Nature; circ 300,000.

Natur og Miljø: Norges Naturvernforbund, POB 2113, Grunerhokka, 0505 Oslo 5, Norway; Editor: Gunnar Bolstad; f 1989; every 2 months; former title *Miljømagasine;* circ 45,000.

Natur und Landschaft: Verlag W Kohlhammer GmbH, 7000 Stuttgart 80, Hessbruhlstr 69, Postfach 800430, Germany; tel (711) 78631; fax (711) 7863393; Editors: Dr W Mrass, Marlies Petzoldt; f 1925; monthly; published for the Bundesanstalt für Naturschutz und Landschaftsökologie (National Institute for Nature Conservation and Landscape Ecology); concerned with nature conservation, landscape management, botany and environmental protection; includes indexes and book reviews; circ 5,700.

Natura magazine: G & J España SA, Marques de Villamagna 4, 28028 Madrid, Spain; tel (1) 4358100; telex 43419; fax (1) 5767881; Dir: Alberto Hijerta; Editor-in-Chief: Alfredo Merino; f 1983; monthly; scientific articles on environmental issues and nature; circ 67,300.

Natura Oggi: De Agostini Rizzoli Periodici srl, Via G Gozzi 5, 20129 Milan, Italy; tel (2) 700231; monthly; deals with contemporary issues relevant to the natural world; circ 87,150.

Natura e Società: Federazione Nazionale Pro Natura, Borgo Felino 54, Parma, Italy; fax (521) 481871; Editor: Angelo De Marchi; f 1982; quarterly; ecological bulletin; contains articles on nature protection, environmental education, events and legislation, book reviews and bibliographies; circ 5,000.

Natural History: American Museum of Natural History, Central Park, W 79th St, New York, NY 10024, USA; tel (212) 873-1498; f 1900; monthly; magazine dedicated to nature and environmental issues; intended for a general readership; circ 510,000.

Natural Resources Forum: Butterworth-Heinemann Ltd, Linacre House, Jordan Hill, Oxford OX2 8DP, UK; tel (865) 310366; fax (865) 310898; Editor-in-Chief: Dr Ray Knowles; f 1977; international, multidisciplinary journal; focuses on the management of energy, mineral and water resources in developing countries; examines the economic, financial, legal and environmental aspects of natural resources development, to identify factors which influence and promote sustainable development.

Naturalist: Wildlife Society of Southern Africa, Eastern Province Branch, 2B Lawrence St, Port Elizabeth 6001, South Africa; tel (41) 564337; fax (41) 559606; Editor: L Greville-Eyres; f 1955; 3 a year; contents include indexes, diagrams and book reviews; in English; circ 2,000.

Naturalist: Yorkshire Naturalists' Union, c/o University of Bradford, Bradford BD7 1DP, UK; tel (274) 384212; fax (274) 384231; Hon Editor: Prof M R D Seaward; natural history of England, particularly Yorkshire.

Nature: Macmillan Magazines Ltd, 4 Little Essex St, London WC2R 3LF, UK; tel (71) 836-6633; fax (71) 836-9934; Editorial Dir: Roger Woodham; Editor: John Maddox; f 1869; weekly; international science journal.

Nature Canada: Canadian Nature Federation, 520-1 Nicholas St, Ottawa, ON K1N 7B7, Canada; tel (613) 562-3447; fax (613) 562-3371; Editor: Barbara Stevenson; f 1972; quarterly; former title *Canadian Audubon;* magazine focusing on Canadian natural history and conservation issues; includes statistics, bibliographies and book reviews; circ 18,500.

Nature Conservation News: Nature Conservation Council of NSW, 39 George St, Sydney, NSW 2000, Australia; fax (2) 247-5945; f 1972; 3 per month; provides up-to-date information on environmental and conservation issues in NSW and notification of the Council's activities; circ 450.

Nature et Mieux-Vivre: Sodeco, BP 3, 32350 Gavarret, France; Editor: Jean P Rousseau; f 1977; monthly; publication dedicated to the natural world and better ways of life.

Naturen: Universitetsforlaget, POB 2959, Tøyen, 0608 Oslo 6, Norway; tel (2) 27-60-60; Editor: Prof Per Min Jørgensen; f 1877; every 2 months; comprehensive natural sciences journal; intended for the general public.

Natürlich: AT Zeitschriftenverlag, Bahnhofstr 39-43, 5001 Aarau, Switzerland; tel (64) 266161; fax (64) 266213; Publr: Herbert Schatzmann; Editor-in-Chief: Walter Hess; Deputy Editor: Markus E Schmid; f 1984; monthly (11 a year); information on nature and health; covers ecology, nature preservation, education, building with environmentally friendly natural materials and enjoyment of the countryside.

Naturopa: Centre Naturopa, Council of Europe, BP 431 R6, 67006 Strasbourg Cedex, France; tel 88-41-20-00; telex 870943; fax 88-41-27-84; Editor: Hayo H Hoekstra; f 1968; 3 a year; published in English, French, German, Italian, Portuguese and Spanish editions; circ 25,000.

Die Naturwissenschaften (Natural Science): Springer-Verlag GmbH & Co KG, 1000 Berlin 33, Heidelberger Platz 3, Germany; tel (30) 8207730; fax (30) 8207300; monthly; nature magazine intended for the general public; circ 3,000.

Natuur en Milieu: Stichting Natuur en Milieu (SNM), Donkerstraat 17, 3511 KB Utrecht, Netherlands; tel (30) 33-13-28; fax (30) 33-13-11; f 1977; monthly; published by Netherlands Society for Nature and Environment; includes indexes, diagrams and book reviews; circ 10,500.

Natuurbeleving: Koninklijke Nederlandse Historische Vereniging, Afdeling Breda, Hornstraat 6, 4834 JG Breda, Netherlands; Editor: A v d Sanden; every 2 months; members' nature journal published by Royal Netherlands Historical Association; circ 450.

NERC News: Natural Environment Research Council (NERC), Polaris House, North Star Ave, Swindon SN2 1EU, UK; tel (793) 411500; telex 444293; fax (793) 411610; Editor: Leslie Jones; f 1970; quarterly; former title *NERC News Journal;* contains research papers, news of recent advances in all fields of environmental science, notification of forthcoming conferences, and book reviews; circ 6,000.

New Age News: Hadim Holdings, 1st Floor, 59 Wentworth Ave, Sydney, NSW 2000, Australia; tel (2) 281-4588; monthly; deals with nature, the environment and natural health issues; intended for the general public; circ 20,000.

New Catalyst: Catalyst Education Society, POB 189, Gabriola, BC V0R 1X0, Canada; tel (604) 247-9737; fax (604) 247-7471; Editors: Christopher Plant, Judith Plant; f 1985; quarterly; deals with alternatives to contemporary culture, politics, the environment and major issues affecting the province of British Columbia; circ 25,000.

New Cyclist: Open Road, 4 New St, York YO1 2RA, UK; tel (904) 654654; fax (904) 671707; cycling magazine with emphasis on the environment.

New Environment: Bedford House Publishing Corpn, 33 Draper St, 2nd Floor, Toronto, ON M5V 2M3, Canada; tel (416) 340-9809; fax (416) 340-9813; Editor: Nathan Rudyk; quarterly; circ 31,500.

New Forests News: New Forests Project, 731 8th St, SE, Washington, DC 20003, USA; tel (202) 547-3800; fax (202) 546-4784; quarterly newsletter dealing with reafforestation in developing countries.

New World Journal: New World Journal Corpn, 330 W 56th St, Suite 3G, New York, NY 10019-4241, USA; tel (212) 265-7970; fax (212) 265-8052; Editor: Rick Bard; f 1992; quarterly; magazine covering science, business, the arts and popular culture as they relate to the emerging new world and its total environment; circ 10,000.

New Zealand Environment: Environmental Publications Trust, 34 Norana Ave, Remuera, Auckland 1105, New Zealand; tel (9) 524-2949; Editor: Robert Mann; f 1971; quarterly; contains book reviews, statistics and a cumulative index; circ 1,000.

New Zealand Forest Industries: Profile Publishing, Level 1, 19 Hargreaves St, Ponsonby, POB 5544, Auckland, New Zealand; tel (9) 358-5455; fax (9) 358-5462; Publr: Reg Birchfield; Advertising Man: Andrew Moses; Editor: Vicki Jayne; f 1969.

New Zealand Journal of Forestry: POB 19840, Christchurch, New Zealand; tel (3) 384-2432; fax (3) 384-243; Editor: Christopher Perley; journal of New Zealand Institute of Forestry (NZIF), Inc.

NEWS (incorporating **Conservation**): The Environment Council, 21 Elizabeth St, London SW1W 9RP, UK; tel (71) 824-8411; fax (71) 730-9941; Editor: Hally Ingram; 10 a year; members' newsletter; incorporates newsletter of the Conservation Trust for Environmental Education; circ approx 4,000.

Newsletter of the Canadian Society of Environmental Biologists: POB 962, Station F, Toronto, ON M4Y 2N9, Canada; Editor: P Stewart; f 1958; quarterly; examines the theory and application of environmental management from an ecological point of view; circ 500.

Newsletter of the Centre Naturopa: Council of Europe, BP 431R6, 67006 Strasbourg Cedex, France; tel 88-41-20-00; telex 870943; fax 88-41-27-84; f 1969; 14 a year; includes bibliographies and book reviews; available in 10 European languages; circ 25,000.

Newsletter of the Environmental Resources Research Institute: Pennsylvania State University, Land and Water Research Bldg, University Park, PA 16802, USA; tel (814) 863-0291; fax (814) 865-3378; quarterly; reports on the Institute's research and on environmental issues; intended for researchers in the USA and world-wide, legislators, civil servants, libraries and the general public; circ 2,500.

Newsletter of the International Union of Societies of Foresters: c/o Institute of Foresters of Australia, POB 1697, Brisbane, Qld 4001, Australia; tel (7) 234-0011; telex 43988; fax (7) 234-0271; Editor: Bill Gordon; 2–4 a year.

Newsletter of Share the Earth: POB 18270, Boulder, CO 80308, USA; tel (303) 444-6430; fax (303) 786-9095; Editor and Publr: Jim Morris.

Newsletter of the Wildlife Conservation and Management Department: Ministry of Tourism and Wildlife, Nairobi Education Centre, POB 40241, Nairobi, Kenya; tel (2) 501081; telex 22804; fax (2) 501752; f 1979; quarterly; in English.

Nippon Seitai Gakkaishi (Japanese Journal of Ecology): Ecological Society of Japan, c/o Biological Institute, Faculty of Science, Tohoku University, Aoba-ku, Sendai 980, Japan; tel (22) 225-0636; fax (22) 225-0636; 3 a year; in Japanese with English summary; circ 2,500.

Noise and Vibration Bulletin: Multi-Science Publishing Co Ltd, 107 High St, Brentwood, Essex CM14 4RX, UK; tel (277) 224632; fax (277) 224632; Editor: B Hughes; f 1970; monthly; concerned with noise pollution, noise control, vibration in man-made structures, and instrumentation.

Noise and Vibration Worldwide: Elsevier Science Publishers Ltd, Crown House, Linton Rd, Barking, Essex IG11 8JU, UK; tel (81) 594-7272; telex 896950; fax (81) 594-5942; Editor: Steve Barrett; f 1969; 11 a year; concerned with noise control and vibration reduction engineering; includes indexes and book reviews; circ 5,850.

Nongye Huanjing Baohu (Agro-Environmental Protection): Zhongguo Nongye Shengtai Huanjing Baohu Xiehui, 31 Fukang Lu, Nankai Qu, Tianjin 300191, People's Republic of China; tel (22) 361247; Editors: Mai Yongbin, Fu Kewen; f 1982; every 2 months; journal containing reports of progress, new technology and new methods of scientific research into agro-environmental pollution and its control; published by Chinese Society of Agro-ecological Environment Protection; in Chinese.

Nonrenewable Resources: Oxford University Press—USA, 2001 Evans Rd, Cary, NC 27513, USA; tel (919) 677-0977; telex 6859654; fax (919) 677-1303; Advertising Manager: Joe Fodor; quarterly; journal of the International Association for Mathematical Geology; articles cover natural resource economics, energy and mineral management and exploration geology.

Noract: Nordinord Research Institute, Collège Universitaire de Hearst, POB 580, Hearst, ON P0L 1N0, Canada; tel (705) 362-4841; Editor: Kathryn Fournier; f 1979; every 2 months; publication covering environmental issues; includes statistical tables and book reviews; in English and French; circ 3,000.

NordRevy: Universitetsforlaget, POB 2959, Tøyen, 0608 Oslo, Norway; tel (2) 27-60-60; Editor: Lars Hedegaard Jensen; every 2 months; journal dealing with regional development, business and the environment in Norway, Denmark and Sweden; in Norwegian, Danish and Swedish, with summaries in English.

Norsk Skogbruk: Hollergt 16, Oslo 1, Norway; 20 a year; forestry publication; circ 4,500.

Northern Decisions: Canadian Arctic Resources Committee, 1 Nicholas St, Suite 412, Ottawa, ON K1N 7B7, Canada; tel (613) 236-7379; fax (613) 232-4665; Editor: Ann Ray; f 1983; 22 a year; circ 200.

Northern Perspectives: Canadian Arctic Resources Committee (CARC), 1 Nicholas St, Suite 412, Ottawa, ON K1N 7B7, Canada; tel (613) 236-7379; fax (613) 232-4665; Editor: Alan Saunders; f 1973; quarterly policy journal; includes Members' Update, supplement covering CARC activities and issues; circ 14,000.

The Northwest Environmental Journal: Institute for Environmental Studies (IES), University of Washington, Engineering Annex, FM-12, Seattle, WA 98195, USA; tel (206) 543-1812; fax (206) 543-2025; f 1984; 2 a year; presents research, policy issues and other environmental issues relating to north-western North America; contains articles, book reviews and interviews.

Notiziario dell'ENEA—Energia e Innovazione: ENEA, Viale Regina Margherita 125, 00198 Rome, Italy; tel (6) 85282041; telex 610183; fax (6) 85285875; ENEA bulletin with information on new technology, energy and the environment.

Nouvelles de PRONAT (PRONAT News): ENDA Publications, Environmental Development Action in the Third World (ENDA), 54 rue Carnot, BP 3370, Dakar, Senegal; tel 22-98-90; telex 51456; fax 22-26-95; quarterly; collaborative newsletter; covers research and training in natural crop protection and agroecology.

NSCA Pollution Handbook: National Society for Clean Air and Environmental Protection (NSCA), 136 North St, Brighton BN1 1RG, UK; tel (273) 326313; fax (273) 735802; Editor: Loveday Murley; f 1978; annual; former titles *Clean Air Year Book, NSCA Yearbook* and *NSCA Reference Book;* guide to all aspects of air pollution control, with particular attention to British legislation.

Nuclear Waste Bulletin/Bulletin sur les Déchets Nucléaires: OECD Nuclear Energy Agency, 38 blvd Suchet, 75016 Paris, France; tel (1) 45-24-96-67; f 1979; 2 a year; circ 1,000.

Nuorten Luonto: Luonto-Liitto, Tallberginkatu 1 D 113, 00180 Helsinki, Finland; tel (0) 6947899; environmental protection magazine for adolescents; published by the Nature League.

La Nuova Ecologia: Editrice Periodici Culturali SpA, Via Savoia 37, 00198 Rome, Italy (Publr); Via Nervesa 21, 20139 Milan, Italy (Advertising); tel (6) 8553177 (Publr); fax (6) 8416256 (Publr); Editor: Paolo Gentilono; f 1988; monthly; describes developments in ecology and their effects on the environment; circ 55,000.

Nyala: Wildlife Society of Malawi, POB 1429, Blantyre, Malawi; tel 643428; fax 643765; Editor: Cornell Dudley; f 1975; 2 a year; in English; circ 650.

Oasis: Industrie Grafiche Editoriali Musumeci SpA, Località Amérique 99, 11020 Quart, Italy; tel (165) 765853; telex 211359; fax (165) 765106; Publr: Sergio Musumeci; Dir: Pietro Giglio; Editor: Riccardo Oldani; Scientific Co-ordinator: Marco Ferrari; f 1985; monthly; nature, ecology, travel and photography.

Oasis: Green Movement Socio-Ecological Centre (GMC), 484006 Dzhambul, ul Lunacharsky 42-2, Kazakhstan; tel (32622) 52-483; fax (32622) 75-500; Editor: Alexander Zagribelny; monthly; independent ecological publication for NGOs.

Occupational Hazards: Penton Publishing Inc, 1100 Superior Ave, Cleveland, OH 44114, USA; tel (216) 696-7000; fax (216) 696-4135; Editor: Stephen G Minter; Publr: David Gursky; f 1938; monthly; publication concerned with health, safety and environmental management; circ 60,000.

Ocean Science News: Nautilus Press Inc, 1054 National Press Bldg, Washington, DC 20045, USA; tel (202) 347-6643; Editor: John R Botzum, Sr.

Ocean Yearbook: The University of Chicago Press, Journals Division, 5720 S Woodlawn Ave, POB 37005, Chicago, IL 60637, USA; tel (312) 702-7600; fax (312) 753-0811; Editors: Elisabeth Mann Borgese, Norton Ginsberg, Joseph Morgan; f 1978; devoted to assessment of resources, ecology and strategic importance of the world's oceans; includes reports from relevant organizations, selected documents and proceedings, and tabular data on marine activities and resources; sponsored by Int Ocean Inst, Malta.

Ochrana Ovzdusi (Air Conservation): Statni Zemedelske Nakladatelstvi, Vaclavske nam 47, 113 11 Prague 1, Czech Republic (Publr); Artia, Ve Smeckach 30, 111 27 Prague, Czech Republic (Subscriptions); tel (2) 265951 (Publr); Editor: Josef Smrtak; monthly; in Czech, with summaries in English.

Ochrona Przyrody (Protection of Nature): Polska Akademia Nauk, Instytut Ochrony Przyrody, 31-505 Kraków, ul Ariańska 1, Poland; tel (12) 215144; telex 322630; fax (12) 210348; Editor-in-Chief: Zofia Alexandrowicz; Assoc Editors: Halina Piękoś-Mirkowa, Zbigniew Witkowski; f 1920; annual; publication of the Inst of Nature Conservation of the Polish Acad of Sciences; contains the results of original research on nature protection and documentation of protected species and sites in Poland; monitors changes in the natural environment, identifies dangers and suggests appropriate courses of action for conservationists; in Polish and English.

Oecologia: Springer-Verlag, 1000 Berlin 33, Heidelberger Platz 3, Germany; tel (30) 82071; telex 185411; fax (30) 8207300; f 1968; monthly; describes advances in research into the interaction between plant and animal species and their environment; in English; circ 1,000.

Oikos, A Journal of Ecology: Munksgaard International Publishers Ltd, 35 Nørre Søgade, POB 2148, 1016 Copenhagen K, Denmark; tel 33-12-70-30; telex 19431; fax 33-12-93-87; Editor: Nils Malmer; f 1948; 9 a year in 3 vols; published for Oikos Society; in English; circ 1,550.

Oil Pollution Bulletin: Center for Short Lived Phenomena, POB 199, Harvard Sq Station, Cambridge, MA 02238, USA; tel (617) 492-3310; telex 3201628; fax (617) 492-3312; fortnightly; Co-Publr and Editor: Richard Golob.

Oil Spill Intelligence Report: Cutter Information Corpn, 37 Broadway, Arlington, MA 02174, USA; tel (617) 641-5125; telex 1009891; fax (617) 648-1950; Editor: Faith Yando; f 1978.

Okayama no Shizen (Nature in Okayama): Okayama no Shizen o Mamoru Kai, 1-21 Tsushima Minami 1-chome, Okayama-shi, Okayama-ken 700, Japan; every 2 months; published by the Study Group for Nature Protection in Okayama; in Japanese.

Okhrana i Uluchshenie Gorodskoi Sredy: Referativnyi Zhurnal, Vsesoyuznyi Institut Nauchno-Tekhnicheskoi Informatsii (VINITI), A-219 Moscow, Baltiiskaya ul 14, Russian Federation (Publr); Mezhdunarodnaya Kniga, 113095 Moscow, Dimitrova ul 39, Russian Federation (Subscriptions); f 1980; monthly; index optional.

Okhrana Prirody i Vosproizvodstvo Prirodnykh Resursov: Referativnyi Zhurnal, Vsesoyuznyi Institut Nauchno-Tekhnicheskoi Informatsii (VINITI), A-219 Moscow, Baltiiskaya ul 14, Russian Federation (Publr); Mezhdunarodnaya Kniga,

113095 Moscow, Dimitrova ul 39, Russian Federation (Subscriptions); f 1975; monthly; index optional.

Öko-Mitteilungen: Öko-Institut eV, 7800 Freiburg, Binzengrün 34A, Germany; tel (761) 473031; f 1977; quarterly; bulletin of Institute for Applied Ecology.

Öko-Test-Magazin: Öko-Test-Verlag GmbH & Co KG, 6000 Frankfurt am Main 1, Mainzer Landstr 147, Germany; tel (69) 75800540; monthly; ecology magazine intended for the general public; circ 90,100.

Ökologiepolitik: Ökologisch-Demokratische Partei, 5300 Bonn 1, Kaiserplatz 17, Germany; tel (228) 261633; fax (228) 211188; Editor: Edgar Guhde; f 1984; every 2 months; former title *Ökologie und Politik;* publication of the Ecological-Democratic Party; concerned with political ecology; provides information for party members and those interested in environmental politics; circ 5,200.

Ökowerkmagazin: Natur und Text Verlags GmbH, 1000 Berlin 37, Dahlenerweg 403, Germany; tel (30) 8175070; fax (30) 8177451; f 1987; monthly; concerned with nature conservation in Berlin and Brandenburg; circ 10,000.

Ontario Conservation News: Conservation Council of Ontario, Suite 506, 489 College St, Toronto, ON M6G 1A5, Canada; tel (416) 969-9637; fax (416) 960-8053; Editor: Chris Winter; f 1975; every 2 months; former title *Conservation News*; circ 1,000.

Ontario Recycling Update: Recycling Council of Ontario, 489 College St, Suite 504, Toronto, ON M6G 1A5, Canada; tel (416) 960-1025; fax (416) 960-8053; Editor: Pat McFarlane; f 1981; quarterly; members' bulletin dealing with reduction, re-use and recycling of waste; circ 1,400.

Options: International Institute for Applied Systems Analysis (IIASA), 2361 Laxenburg, Austria; quarterly; publishes IIASA's research results.

Organic Gardening magazine: Rodale Press, Inc, 33 East Minor St, Emmaus, PA 18098, USA; tel (215) 967-5171; fax (215) 967-8956; Pres Magazine Div: John Griffin; Publr: Barbara R Newton; Editor-in-Chief: Mike McGrath; f 1942; 9 a year.

Organic Grower: Organic Growers' Association of Western Australia, POB 213, Wembley, Perth, WA 6014, Australia; f 1977; quarterly; aims to demonstrate that practical alternatives exist for gardeners and farmers who are concerned by the increasing use of harmful chemicals; encourages soil restoration; circ 700.

Orion magazine: The Orion Society, Myrin Institute Inc, 136 E 64th St, New York, NY 10021, USA; tel (212) 758-6475; fax (212) 758-6784; Editor-in-Chief: George K Russell; f 1982; quarterly; literary journal devoted to the relationship between people and the earth.

Österreichische Wasserwirtschaft: Springer-Verlag Komm Ges, 1200 Vienna, Sachsenplatz 4–6, POB 89, Austria; tel (1) 330-24-15; telex 114506; fax (1) 330-24-26; Editors: A Kravogel, F Oberleitner, H Werner; f 1949; monthly; publication dedicated to water supply and management; circ 2,300.

Our Planet/Notre Planète/Nuestro Planeta: UNEP Information Service, POB 30552, Nairobi, Kenya; tel (2) 233930; telex 22068; fax (2) 520711; Editor: Shane Cave; f 1974; quarterly; encourages global awareness of all aspects of the current environmental crisis; published in English, French and Spanish editions; circ 17,000.

Ozone News: International Ozone Association, Pan-American Group, 31 Strawberry Hill Ave, Stamford, CT 06902-2608, USA; tel (203) 348-3542; fax (203) 967-4845; Editor: Dr Rip G Rice.

P & A: Paysage et Aménagement: SEDA Edition, 8 rue Léon-Serpollet, 87021 Limoges Cedex, France; tel 42-65-71-67; Editor: R Chaux; f 1984; quarterly; publication concerned with landscape and planning; in French, with summaries in English; circ 2,000.

Packaging News: Yaffa Publishing Group Pty Ltd, 17/21 Bellevue St, Surry Hills, NSW 2010, Australia; tel (2) 281-2333; fax (2) 281-2750; Publr: Martin Coleman; Advertising Manager: John Skinner; monthly; deals with all aspects of packaging, including relevant legislation, environmental considerations, recycling and new materials; circ 5,450.

Pamatky a Priroda: Obzor, Ceskoslobenskej Armady 35, 815 85 Bratislava, Slovakia; f 1976; every 2 months; published for Statni Ustav Pamatkove Pece a Ochrany Prirody (Slovak Institute for Care of Historical Monuments and Nature Protection); in Czech, with summaries in English, German, Russian and French.

Panda: World Wide Fund for Nature (WWF) Italy, Via Salaria 290, 00199 Rome, Italy; tel (6) 852492; Editor: Fulco Pratesi; f 1966; every 2 months; nature conservation magazine; circ 250,000.

Panda Junior: World Wide Fund for Nature (WWF) Italy, Via Salaria 290, 00199 Rome, Italy; tel (6) 852492; Editor: Fulco Pratesi; every 2 months.

Panos Features: Panos Institute, 9 White Lion St, London N1 9PD, UK; tel (71) 278-1111; telex 9419293; fax (71) 278-0345; monthly; environment and development news.

Panoscope—PS: Panos Institute, 9 White Lion St, London N1 9PD, UK; tel (71) 278-1111; telex 9419293; fax (71) 278-0345; Editorial Dir: Joe Hanlon; Publications Dir: Liz Carlile; Editor: Kelly Haggart; quarterly; environment and development magazine; includes *Focus* section containing several articles on a single, specific topic.

Park Watch: Victoria National Parks Association Inc, 247 Flinders Lane, Melbourne, Vic 3000, Australia (Publr); POB 785F, Melbourne, Vic 3001, Australia (Subscriptions); tel (3) 650-8296 (Publr); fax (3) 650-5684; Editor: James Calde; f 1952; quarterly; circ 3,400.

Parks: Science and Technology Letters, POB 81, Northwood, Middx HA6 3DN, UK; tel (9274) 23586; fax (9274) 25066; Editor: Paul Goriup; f 1990; 3 a year; international publication concerned with the world's protected areas; published for the Commission on National Parks and Protected Areas (CNPPA).

Perspectives: International Institute for Environment and Development (IIED), 3 Endsleigh St, London WC1H 0DD, UK; tel (71) 388-2117; telex 261681; fax (71) 388-2826; 2 a year; newsletter.

Perspectives in Energy: Moscow International Energy Club (MIEC), 127412 Moscow, Izhorskaya 13/19, Russian Federation (Editorial); Pion Ltd, 207 Brondesbury Park, London NW2 5JN, UK (UK Rep); tel (095) 485-96-63 (Editorial); telex 411959; fax (095) 484-17-29; Editor-in-Chief: A Sheindlin; Assoc Editors: G Aslanian, S Malyshenko; f 1991; quarterly; intends to inform scientists, politicians, business people, journalists and the general public of developments and problems in the field of energy; considers the environmental impact of energy development.

Physiological Zoology: The University of Chicago Press, Journals Division, 5720 S Woodlawn Ave, POB 37005, Chicago, IL 60637, USA; tel (312) 702-7600; fax (312) 753-0811; Editor: Warren W Burggren; f 1928; every 2 months; presents recent, original research in environmental, adaptational and comparative physiology and biochemistry.

Physiology and Ecology Japan: Kyoto Daigaku Rigakubu Dobutsugaku Kyoshitsu, Kita-Shirakawa, Sakyo-ku, Kyoto 606, Japan; telex 5422302; fax (75) 751-6149; Editor: Hiroya Kawanabe; f 1947; 2 a year; published by Department of

Zoology of Kyoto University; deals with all aspects of physiology and ecology; in English; circ 700.

The Pilot: UNEP, POB 30552, Nairobi, Kenya; f 1987; quarterly newsletter of Marine Mammal Action Plan.

Planet News: The Planet News Group, 87 Kirkstall Rd, London SW2 4HE, UK; tel (81) 671-7920; Co-Publrs: Ian Henshall, Janet Alty, Rowland Morgan; Advertising Manager: Nick Bradshaw; Editor: David Black; f 1991; quarterly; networking for 'Greens'; circ 20,000.

Planning and Design: Environment and Planning B: Pion Ltd, 207 Brondesbury Park, London NW2 5JN, UK; tel (81) 459-0066; fax (81) 451-6454; Editor: Michael Batty; Co-Editors: Michael Breheny, Helen Couclelis; Reviews Editor: Paul Longley; f 1974; every 2 months; international journal of theory and practice of urban and regional planning and design.

Plans and Programs: Department of the Environment and Natural Resources, Visayas Ave, Diliman, Quezon City, Philippines; tel (2) 976626; f 1976; annual; includes statistical tables.

Plant Genetic Resources Newsletter: International Board for Plant Genetic Resources (IBPGR), Via delle Sette Chiese 142, 00145 Rome, Italy; tel (6) 518921; fax (6) 5750309; f 1957; quarterly; contains research papers, articles and reports on work relating to plant genetic resources; includes reports of conferences, news items, abstracts, bibliographies and book reviews; published in English, French and Spanish editions; circ 5,000.

Polish Ecological Studies: Polska Akademia Nauk, Instytut Ekologii, 05-092 Lomianki, Dziekanów Lesny k-Warszawy, Poland; Editor: J Luczak; f 1975; quarterly; in English, with summaries in Polish; circ 430.

Pollution: Springfield Information Services, POB 31, Cross St, Peterborough, Cambs PE1 1XA, UK; tel (733) 52454; Editor: John Franks; f 1971; monthly; environmental news bulletin; contents include abstracts and book reviews.

Pollution Abstracts: Cambridge Scientific Abstracts, 7200 Wisconsin Ave, Bethesda, MD 20814-4823, USA; tel (301) 961-6750; fax (301) 961-6720; Publr: Ted Caris; Man Editor: Angela Hitti; contains summaries of articles on scientific research and government policy relating to all aspects of pollution and pollution control.

Pollution Atmosphérique: Association pour la Prévention de la Pollution Atmosphérique (APPA), 58 rue du Rocher, 75008 Paris, France; tel (1) 42-93-62-07; fax (1) 42-93-41-99; Editor: M Sommer; f 1959; quarterly; journal dealing with atmospheric pollution in cities, globally and within working and living environments; includes book reviews, bibliographies and statistics; in French, with summaries in English; circ approx 2,000.

Pollution Engineering: Cahners Publishing Co, 1350 E Touhy Ave, Des Plaines, IL 60018, USA; tel (708) 635-8800; fax (708) 390-2636; f 1969; 21 a year; technical journal; circ 60,100.

Pollution Equipment News: Rimbach Publishing Inc, 8650 Babcock Blvd, Pittsburgh, PA 15237-5821, USA; tel (412) 364-5366; fax (412) 369-9720; Pres: Norberta Rimbach; f 1968; every 2 months; circ 91,300.

Pollution Prevention Review: Executive Enterprises Publications Co, Inc, 22 W 21st St, 10th Floor, New York, NY 10010-6990, USA; tel (212) 645-7880; fax (212) 645-1160; Man Editor: Jean Stephenson; Editor: Ann Graham; f 1990; quarterly; articles intended for pollution prevention specialists, managers in industry, consultants, researchers and government agency representatives.

Pollution Research: Enviro Media, Shaniwar Peth, Patan Colony, POB 90, Karad 415 110, India; tel (2164) 44369; Man Editor: Sukhada R Trivedy; f 1986; quarterly; international journal published for the Indian Association for Pollution Chemists and Biologists; in English.

Polskie Archiwum Hydrobiologii: Polska Akademia Nauk, Instytut Ekologii, 05-092 Łomianki, Dziekanów Leśny k-Warszawy, Poland; tel (22) 513157; telex 817378; fax (22) 513100; Man Editor: Ewa Kamler; Editor-in-Chief: Romuald Z Klekowski; f 1953; reports of original research in every field of hydrobiology, theoretical papers, reviews and comprehensive bibliographies; in English, French and German.

Practical Alternatives: Victoria House, Bridge St, Rhayader LD6 5AG, UK; tel (597) 810929; Editor and Publr: David Stephens; f 1981; irregular; reports on the construction of a village of passive solar houses designed to reduce energy consumption greatly; circ 2,000.

Prairie Forum: Canadian Plains Research Center, University of Regina, Regina, SK S4S 0A2, Canada; tel (306) 585-4795; fax (306) 586-9862; Editor: Alvin Finkel; f 1976; 2 a year; journal of interdisciplinary research relating to the interaction of humans and nature in the Canadian plains region; circ 460.

Pro Natura Genova: Pro Natura Genova, c/o Museo di Storia Naturale, Via Brigata Liguria 9, 16121 Genova, Italy; Dir: Pierluigi Oneto; f 1976; quarterly; members' bulletin; contents include book reviews and statistics; circ 1,000.

Problemy Bol'shikh Gorodov (Problems of Large Metropolitan Areas): Moskovskii Gorodskoi Territorial'nyi Tsentr Nauchno-Tekhnicheskoi Informatsii i Propagandy, 101958 Moscow, Pr Serova 5, Russian Federation; tel (095) 921-67-05; Editor: Irina P Mikhailova; f 1971; fortnightly; general survey dealing with the development of urban environmental protection, urban traffic improvement, problems related to energy and water supply, and recreation; circ 800.

Problemy Kontrolya i Zashchita Atmosfery Ot Zagryazneniya: Izdatel'stvo Naukova Dumka, c/o Y A Khramov, Dir, 252601 Kiev, Ul Repina 3, Ukraine; Mezhdunarodnaya Kniga, Moscow G-200, Russian Federation (Subscriptions); Editor: A N Shcherban; f 1974; annual; published for the Institute of Engineering Thermal Physics of the Ukrainian Academy of Sciences; collection of scientific papers on problems related to pollution control and atmospheric preservation in Ukraine; in Russian.

Progress in Environmental Science and Technology: Divyajyoti Prakashan, 5 Bhagat-ki-kothi, Jodhpur 342 003, India; Editors: Alam Singh, U S Sharma; f 1984; annual; former title *Current Practices in Environmental Science and Engineering;* co-published by Geo-Environ Academia.

Project Appraisal: Beech Tree Publishing, 10 Watford Close, Guildford, Surrey GU1 2EP, UK; tel (483) 67497; fax (483) 67497; Publr: William Page; Editors: Dr Vary Coates, Dr John Weiss; f 1986; quarterly; international journal dealing with environmental impact assessment and related fields such as cost-benefit analysis and technology assessment; contains articles, book reviews and details of forthcoming conferences.

Protec—Il Progresso e L'Ambiente: Publi & Consult SpA, Via Tagliamento 29, 00198 Rome, Italy; tel (6) 8543267; telex 622368; fax (6) 8440697; Head of Advertising Dept: Sabrina Soldati; Editor: Paolo F Bancale; f 1986; 9 a year; publication dealing with the environment, ecology and civil defence; in Italian, with summaries in English; supplemented by *Annual Environment Guide & Directory;* circ 23,000.

Protection of Atmosphere Against Pollution: Lithuanian Academy of Sciences, Institute of Physics, K Pozelos g-ve 54, Vilnius 2600, Lithuania; tel (2) 641-672; Editor: R Baltramiejunas; f 1974; annual; describes methods for measuring the background concentration of chemical pollutants in air and rain, with particular attention to the Baltic region; in English and Russian, with summaries in English, Russian and Lithuanian; circ 350.

Pulse of the Planet: Orgone Biophysical Research Laboratory, POB 1395, El Cerrito, CA 94530, USA; tel (510) 526-5978; fax (510) 526-5978; Editor: James DeMeo; f 1989; annual; research report and journal containing original and reprinted articles on social, environmental and geophysical subjects.

Queensland Conservation Council Newsletter: POB 12046, Elizabeth St, Brisbane, Qld 4002, Australia; tel (7) 221-0188; fax (7) 229-7992; f 1973; quarterly; reports on all conservation issues in Queensland, covering initiatives by conservation groups, environmental legal disputes and the activities of government and industrial committees.

Radioactive Waste Management: US Department of Energy, Office of Scientific and Technical Information (OSTI), POB 62, Oak Ridge, TN 37831, USA; tel (615) 576-1188; fax (615) 576-2865; monthly; publishes most recent entries to relevant OSTI databases; contains abstracts of various types of sources.

Rain: POB 30097, Eugene, OR 97403-1097; Editors: Greg Bryant, Danielle Jones; quarterly; publishes articles which encourage environmental improvement.

Rapport annuel du Conseil consultatif de l'environnement et de la conservation du Québec: Ministère des Communications, Direction Générale des Publications Gouvernementales, 2ème étage, 1279 blvd Charest Ouest, Québec, PQ G1N 4K7, Canada; tel (413) 643-3895; annual report of the Advisory Council on the Environment and Conservation of the Province of Québec.

Reclamation Newsletter: Canadian Land Reclamation Association, POB 682, Guelph, ON N1H 6L3, Canada; tel (403) 427-4147; fax (403) 422-8233; Editor: Chris Powter; f 1977; 3 a year; co-published by American Society for Surface Mining and Reclamation; includes statistical tables and book reviews; in English and French; circ 950.

Recover: Recover Enterprises, 114 Dollery Court, N York, ON M2R 3P1, Canada; Editor: Nancy Phillips, Heather Sangster; f 1990; quarterly; environmental magazine.

Recyclage-Récupération (Recycling-Reclamation): SOCI-DOC, 142 rue Montmartre, 75080 Paris, France; tel (1) 42-33-00-61; fax (1) 42-33-28-42; f 1909; weekly; waste reclamation bulletin; circ 5,000.

Recycling: Handelsblatt GmbH, 4000 Düsseldorf 1, Kasernenstr 67, Postfach 102717, Germany; tel (211) 8870; Editor: Rolf Willeke; 3 a year; published for Bundesverband der Deutschen Schrottwirtschaft (National Association of German Scrap Metal Management); includes statistics and trade information; circ 1,000.

Reeflections: Great Barrier Reef Marine Park Authority (GBRMPA), POB 1379, Townsville, Qld 4810, Australia; tel (77) 81-8811; telex 47332; fax (77) 72-6093; Editors: Elaine Eager, Liz Hall; f 1982; 2 a year; contains news, articles, book reviews and descriptions of research relating to the Great Barrier Reef; circ 13,000.

Regulus: Letzeburger Natur & Vulleschutzliga ASBL, BP 709, 2017 Luxembourg, Luxembourg; tel 47-23-69; quarterly; published by the Luxembourg League for the Protection of Nature and Birds; in German, French and English; circ 16,000.

Reinwater Tijdschrift: Stichting Reinwater, Vossiusstraat 20, 1071 AD Amsterdam, Netherlands; tel (20) 6719322; fax (20) 6753806; Editor: Wim Verhoog; f 1980; quarterly; magazine of organization aiming to reduce pollution of main European surface waters; in Dutch; circ 1,500.

Remediation: The Journal of Environmental Cleanup Costs, Technologies and Techniques: Executive Enterprises Publications Co, Inc, 22 W 21st St, 10th Floor, New York, NY 10010-6990, USA; tel (212) 645-7880; fax (212) 645-1160; Man Editor: Deborah Wenger; Editor: Mark Dennison; f 1991; quarterly; examines benefits and costs of innovative environmental clean-up techniques; intended for environmental managers, engineers and consultants.

Remote Sensing of Environment: Science Press, Marketing and Sales Dept, 16 Donghuangchenggen Beijie, Beijing 100707, People's Republic of China; tel (1) 4012180; fax (1) 4019810; quarterly; in Chinese with English abstract.

Remote Sensing of Environment: Elsevier Science Publishing Co, 655 Ave of the Americas, New York, NY 10010, USA; tel (212) 633-3806; fax (212) 633-3880; Editor-in-Chief: Marvin E Bauer; Advertising Manager: Arthur J Carlucci; f 1969; monthly; concerned with ecology, the environment, management of natural resources and remote sensing; circ 12,000.

Renewable Energy Bulletin: Multi-Science Publishing Co Ltd, 107 High St, Brentwood, Essex CM14 4RX, UK; tel (277) 224632; fax (277) 224632; every 2 months; summaries of international research in all forms of renewable energy.

Report of Operations under the Canada Water Act: Department of Fisheries and Oceans, 200 Kent St, Ottawa, ON K1A 0E6, Canada; tel (613) 995-4031; f 1973; annual; co-published by Environment Canada; in English and French.

Report of the Secretary for Lands and Natural Resources: Government Printer, POB 8062, Causeway, Zimbabwe; f 1968; annual; includes statistical tables; circ 400.

Reports to the Nation on our Changing Planet: University Corporation for Atmospheric Research, Office for Interdisciplinary Earth Studies, POB 3000, Boulder, CO 80307, USA; fax (303) 497-1679; f 1991; 2–3 a year; intended for the general public and pupils in secondary education; aims to explain complex environmental issues with text and illustrations.

RERIC International Energy Journal: Regional Energy Resources Information Center (RERIC), Asian Institute of Technology, POB 2754, Bangkok 10501, Thailand; tel (2) 524-5866; telex 84276; fax (2) 524-5870; Information Specialist: On-Anong Suraniranat; information on energy conservation, energy management and renewable energy.

Research Bulletin of Western Australia Department of Conservation and Land Management: POB 104, Como, WA 6152, Australia; tel (9) 334-0333; fax (9) 334-0466; Editor: Marianne Lewis; f 1919; 2 a year; scientific journal covering biology, entomology, construction, conservation, environmental studies and forestry; circ 1,500.

Research on Environmental Disruption—Kogai Kenkyu: Iwanami Shoten Publishers, 2-5-5 Hitotsubashi, Chiyoda-ku, Tokyo 101-02, Japan; Japan Publications Trading Co Ltd, 1255 Howard St, San Francisco, CA 94103, USA (Int Distributor); fax (3) 3239-9618 (Publr); f 1971; quarterly; encourages interdisciplinary collaboration.

Research Papers of the Board of Environmental Studies: University of Newcastle, Newcastle, NSW, Australia; fax (49) 21-6905; f 1973; annual; includes book reviews, bibliographies and diagrams; circ 200.

RESILOG: Environmental Protection Publications, Technology Development Branch, Environment Canada, Ottawa, ON K1A 0H3, Canada; tel (819) 953-5750; fax (819) 953-7253; Editor: Paul Topping; quarterly; serves as a forum for the exchange of information and opinions on waste management in Canada; includes details of conferences and publications dealing with waste treatment and relevant legislation; in English and French.

Resource Recovery Report: 5313 38th St, NW, Washington, DC 20015, USA; tel (202) 362-6034; fax (202) 362-6632; Editor and Publr: Frank McManus; f 1976; monthly; newsletter covering alternatives to landfilling, such as recycling, composting and energy recovery; contains special reports on solid wastes and conferences on waste issues such as ash management, landfill mining, integrated waste management and mechanical processing.

Resource Recycling: Resource Recycling Inc, POB 10540, Portland, OR 97210, USA; 1206 NW 21st, Portland, OR 97209, USA; tel (503) 227-1319; fax (503) 227-6135; Principal Officers: Jerry Powell, Judy Roumpf; f 1980.

Resources, Conservation and Recycling: Elsevier Science Publishers BV, POB 181, 1000 AD Amsterdam, Netherlands; tel (20) 5803911; fax (20) 5803249; Editor-in-Chief: Walter M Shaub; f 1988; 8 a year in 2 vols; co-published by Pergamon Press plc (UK); scientific journal providing analyses and reviews of interdisciplinary aspects of the management of renewable and non-renewable resources, with particular attention to conservation issues; includes abstracts, statistics, indexes and book reviews; circ 1,050.

Resources newsletter: Resources for the Future, 1616 P St, NW, Washington, DC 20036, USA; tel (202) 328-5000; fax (202) 939-3460; concerned with natural resources and the quality of the environment.

Resources Policy: Butterworth-Heinemann Ltd, Linacre Hse, Jordan Hill, Oxford OX2 8DP, UK; tel (865) 310366; fax (865) 310898; Editor: Dr Rod Eggert; f 1975; international journal devoted to minerals policy and economics.

Resurgence Magazine: Ford House, Hartland, Bideford, Devon EX39 6EE, UK; tel (237) 441293; fax (237) 441203; Editor: Satish Kumar; f 1965; articles and information on ecology, the arts and spirituality.

Reviews on Environmental Health: Freund Publishing House Ltd, Suite 500, Chesham House, 150 Regent St, London W1R 5FA, UK; POB 35010, Tel-Aviv, Israel; tel (3) 615335 (Israel); Editor: S Samueloff; f 1972; quarterly; includes indexes and book reviews; in English.

Revue d'Ecologie (La Terre at la Vie): Société Nationale de Protection de la Nature, 57 rue Cuvier, 75005 Paris; BP 405, 75221 Paris Cedex 05, France; tel (1) 47-07-31-95; fax (1) 47-07-07-16; Editor-in-Chief: Prof François Bourlière; f 1931; quarterly; dedicated to the application of ecology to nature protection; describes the relationships between plant and animal species; contains essays by specialists, a bibliography and book reviews; international readership; circ 1,200.

Revue Juridique de l'Environnement: Centre International de Droit Comparé de l'Environnement (CIDCE), 4 place du Présidial, 87031 Limoges Cedex, France; Publications Périodiques Spécialisées, 11 rue d'Algérie, 69001 Lyon, France; Editor: M Prieur; quarterly; published for French Society for Environmental Law.

RICHMAC Magazine: Editrice Bias Sas, Viale Premuda 2, 20129 Milan, Italy; tel (2) 55181842; telex 352110; fax (2) 5400481; Editorial Dir: Ing Mario Gargantini; Advertising Man: Roberto Mura; f 1988; every 2 months; technical magazine for chemical and manufacturing industries with some environmental coverage; circ 14,000.

Riverlander Notes: Murray Darling Association Inc, POB 359, Albury, NSW 2640, Australia; tel (60) 21-3655; fax (60) 21-2025; Editor: G Scott; f 1946; quarterly; bulletin dealing with the environmental preservation of the Murray and Darling Rivers and the promotion of sustainable development of their valleys' resources; includes indexes and book reviews; circ 10,000.

Roessleria: Instituto de Pesquisas de Recursos Naturais Renovaveis 'Ataliba Paz', Rua Gonçalves Dias 570, 90.060 Porto Alegre, RS, Brazil; telex 513359; Editor: Helena Mondadori Mansan; f 1977; annual; published by research institute concerned with renewable natural resources; fields dealt with include botany, forestry, hydrology, ornithology, soil conservation and zoology; circ 500.

ROM: Vaktijdschrift voor Ruimtelijke Ordening en Milieubeheer: Delwel Uitgeverij BV, Postbus 19110, 2500 CC The Hague, Netherlands; tel (70) 3624800; fax (70) 3605606; Publr: N F van't Zet; Editor: H J Bakker; Deputy Editor: R C Ravestein; f 1983; monthly; professional environment and physical planning magazine; acts as a forum for new technological ideas and disseminates information about current research; in Dutch.

Rössing Magazine: Rössing Uranium Ltd, POB 22391, Windhoek 9000, Namibia; tel (61) 211721; fax (61) 228147; Editor: D Haarhoff; f 1979; 2 a year; aims to increase awareness of the people and environment of Namibia; each edition concentrates on a particular issue; in English; circ 4,000.

Die Rote Mappe: Niedersächsischer Heimatbund eV, 3000 Hannover 1, Goseriede 15, Germany; tel (511) 131565; fax (511) 17475; f 1960; annual; critical reports relating to conservation of the heritage of Lower Saxony; circ 7,000.

Safe Energy: Scottish Campaign to Resist the Atomic Menace (SCRAM), 11 Forth St, Edinburgh EH1 3LE; tel (31) 557-4283; fax (31) 557-4284; Editors: Mike Townsley, Graham Stein; every 2 months; journal dedicated to safe, renewable and efficient sources of energy.

Safety Mate: International Environmental Technology, 12 Alban Park, Hatfield Rd, St Albans, Herts AL4 0JJ, UK; tel (727) 855574; fax (727) 841694; Editor: Mike Patisson; f 1978; quarterly; environmental health and safety journal; circ 12,500.

Scan: Abstracts and International Highlights: DLO Winand Staring Centre for Integrated Land, Soil and Water Research, Marijkeweg 11, 6709 PE Wageningen, Netherlands; POB 125, 6700 AC Wageningen, Netherlands; tel (8370) 74200; telex 75230; fax (8370) 24812; f 1990; annual; description of research work and publications of the DLO Winand Staring Centre; in English.

Scandinavian Journal of Forest Research: Universitetsforlaget, POB 2959, Tøyen, 0608 Oslo, Norway (Publr); c/o Royal Swedish Academy of Agriculture and Forestry, POB 6806, 113 86 Stockholm, Sweden (Editorial); tel (2) 27-60-60 (Publr); Editor-in-Chief: Sven-Uno Skarp; Asst Editor: Gunilla Agerlid; quarterly; covers all aspects of forest research, in addition to forest planning, management and economy, and the utilization of forest resources.

SCENES (Scottish Environment News): Strome House, North Strome, Lochcarron IV54 8YJ, UK; tel (5202) 588; fax (5202) 660; Editors and Publrs: Michael Scott, Sue Scott; f 1988; monthly; independent digest of environmental news in Scotland, including conservation, farming, forestry, fisheries and fish-farming, the marine environment, energy and nuclear power, recreation and tourism, and news of organizations.

Schweizer Naturschutz/Protection de la nature: Schweizerischer Bund für Naturschutz (SBN), Postfach, 4020 Basel, Switzerland; tel (61) 3127442; fax (61) 3127447; f 1935; 8 a year; published by Swiss League for Nature Protection (SBN); ensures communication between SBN, its sections and members; in German and French; special issues on specific subjects.

Science et Changement Planétaires—Sécheresse: John Libbey Eurotext, 6 rue Blanche, 92120 Montrouge, France (Publr); Gauthier-Villars, Centrale des Revues, 11 rue Gossin, 92543 Montrouge Cedex, France (Distributor); tel (1) 47-35-85-52 (Publr), (1) 46-56-52-66 (Distributor); fax (1) 46-57-10-09 (Publr); quarterly; publication intended for those working to prevent drought and its consequences.

Science and Environment: M Abdula Munim, Wahid Manzil, Civil Lines, Aligarh 202 001, India; Editor: M Abdula Munim; 2 a year.

Science of the Total Environment: Elsevier Science Publishers BV, POB 211, 1000 AE Amsterdam, Netherlands; tel (20) 5803911; telex 18582; fax (20) 5803598; Editors: E I Hamilton, J O Nriagu; f 1972; 42 a year in 14 vols; international journal dedicated to scientific research on the environment and its interaction with humans, with particular attention to disturbances of the normal levels and distribution of chemical substances and their impact on the environment; deals with applied environmental chemistry in the areas of nutrition, pollution control, environmental health, policy and planning.

Scientia Limnologica Sinica: Institute of Hydrobiology, Chinese Academy of Sciences, Luojiashan, Wuhan 430072, Hubei Province, People's Republic of China; tel (27) 723481; fax (27) 725132; quarterly; Chinese journal of limnology; in Chinese, with an English abstract.

Scienza, Diritto e Economia dell'Ambiente: Edizioni Giuridico-Scientifiche, Via G Donizetti 37, 20122 Milan, Italy; tel (2) 794146; fax (2) 76009444; f 1987; every 2 months; concerned with environmental science, law and economics; circ approx 5,500.

Scoop: DLO Winand Staring Centre for Integrated Land, Soil and Water Research, Marijkeweg 11, 6709 PE Wageningen, Netherlands; POB 125, 6700 AC Wageningen, Netherlands; tel (8370) 74200; telex 75230; fax (8370) 24812; Editor: E C W M Ruyten; f 1989; 3 a year; bulletin containing abstracts of research results in agriculture, ecology, hydrology and related fields; in Dutch and English.

Scottish Forestry: Royal Scottish Forestry Society, Camsie House, Charlestown, Dunfermline, Fife KY11 3EE, UK; tel (383) 873014; fax (383) 872863; Editor: Jenny Johnson; quarterly; journal circulated world-wide; reports on Royal Scottish Forestry Society's activities; contains technical articles on forestry, items of more general interest, abstracts and reviews of publications dealing with forestry.

Section 46: Environmental Health and Pollution Control: Excerpta Medica, Elsevier Science Publishers, POB 548, 1000 AM Amsterdam, Netherlands; Molenwerf 1, 1014 AG Amsterdam, Netherlands; tel (20) 5803507; fax (20) 5803222; f 1971; 10 a year; deals with air, soil and water pollution, noise prevention, radioactivity and general environmental matters.

SEE Journal: Society for Environmental Education (SEE), 16 Trinity Rd, Enderby, Leicester, UK; Editor: R Edwards; f 1968; 2 a year; circ 200.

Seedling: Genetic Resources Action International (GRAIN), Jonqueres 16, 6° D, 08003 Barcelona, Spain; tel (3) 3105909; fax (3) 3105952; newsletter; contains current information on the seeds industry, biotechnology and international organizations, news from the GRAIN network and book reviews; in English; Spanish edition, *Semillas* (qv).

Seikatsu to Kankyo (Life and Environment): Nihon Kankyo Eisei Senta, 10-6 Yotsuya, Kami-cho, Kawasaki-ku, Kawasaki-shi 210, Japan; Editor: Kazumi Kuroda; f 1956; monthly; published by Japan Environmental Sanitation Centre; in Japanese; circ 10,000.

Semillas: CEIDER, Calle Pascual y Genís 21, pta 10a, 46002 Valencia, Spain; Contact: Paco García; Spanish edition of *Seedling* (qv).

SES Newsletter: Saskatchewan Environmental Society (SES), POB 1372, Saskatoon, SK S7K 3N9, Canada; tel (306) 665-1915; fax (306) 665-2128; Editor: Lynn Brown; f 1975; every 2 months; deals with environmental issues relevant to Saskatchewan province; circ 650.

SESI Journal: Tata Energy Research Institute (TERI), 103 Jor Bagh, New Delhi 110 003, India; tel (11) 699427; telex 3161593; fax (11) 4621770; Editor: Veena Joshi; f 1987; 2 a year; published for the Solar Energy Society of India (SESI); deals with all issues related to renewable sources of energy; in English; circ 700.

Shanghai Huanjing Kexue (Shanghai Environmental Sciences): Shanghai Huanjing Kexue Zazhishe, 508 Xinzhou Lu, Shanghai 200233, People's Republic of China; Guoji Shudian, China International Book Trading Corpn, POB 399, People's Republic of China (Int Distributor); Editor: Chen Jiangtao; monthly; published for Shanghai Huanjing Baohu Ju (Shanghai Environmental Protection Bureau); in Chinese, with list of contents in English.

Shengtai Xuebao (Acta Ecologica Sinica): Science Press, Marketing and Sales Dept, 16 Donghuangchenggen Beijie, Beijing 100707, People's Republic of China; tel (1) 4012180; fax (1) 4019810; f 1981; quarterly; published for Zhongguo Shengtaixue Xuehui (Ecological Society of China); contains original research papers on ecology, with particular emphasis on new aspects of ecology contributing to Chinese development; includes short items, debates and book reviews; in Chinese, with summaries in English; circ 6,000.

Shizen Hogo (Conservation of Nature): Nihon Shizen Hogo Kyokai, 8-1 Toranomon 2-chome, Minato-ku, Tokyo 105, Japan; f 1960; monthly; bulletin of the Nature Conservation Society of Japan; in Japanese.

Shizen Kansatsukai Kaiho: Shizen Kansatsukai, c/o Mr Okubo, 2-4 Yotsuya, Shinjuku-ku, Tokyo 160, Japan; quarterly; newsletter of Shizen Kansatsukai, an association for the observation of nature; in Japanese.

Sia/Skogsindustriarbetaren: POB 1138, 111 81 Stockholm, Sweden; tel (8) 23-04-25; fax (8) 11-27-42; Editor: Leif Nordström; f 1935; 22 a year; publication concerned with forestry and the timber industry; circ 147,400.

Sichuan Huanjing (Sichuan Environment): Sichuan Huanjing Bianjibu, 18 Renmin Nanlu 4 Duan, Chengdu, Sichuan 610041, People's Republic of China; tel (28) 580473; Editor: Zhan Yongdong; f 1979; quarterly; journal of Sichuan Research Institute of Environmental Protection; deals with local and national policies, laws and regulations relating to environmental protection and monitors the development of environmental science and technology around the world; in Chinese with English abstracts.

Siebenstern: Fichtelgebirgsverein eV, 8670 Hof, Auguststr 6, Germany; tel (9281) 2531; f 1927; every 2 months; membership journal dealing with heritage conservation, local history, walking and nature preservation; circ 13,000.

Sieppo: Luoto-Liitto, Tallberginkatu 1 D 113, 00180 Helsinki, Finland; tel (0) 6947899; fax (0) 6948367; published by Luoto-Liitto (the Nature League); nature conservation magazine for children.

Sierra: Sierra Club, 730 Polk St, San Francisco, CA 94109-7897, USA; tel (415) 923-5656; fax (415) 776-0350; Editor-in-Chief: Jonathan F King; Advertising Manager: Ilene Briggin; f 1893; every 2 months; former title *Sierra Club Bulletin*; covers conservation and related political and social affairs; circ 350,000.

Sierra Report: Sierra Club, Western Canada Chapter, 620 View St, Suite 314, Victoria, BC V8W 1J6, Canada; tel (604) 386-5255; fax (604) 386-4453; Editor: Geraldine Irby; f 1973; quarterly; deals with environmental issues affecting the provinces of Alberta, British Columbia, Manitoba and Saskatchewan and the Yukon Territory; covers transboundary issues with the Sierra Club Alaskan chapter; circ 5,000.

Silviculture: Maclean Hunter Ltd, Suite 900, 1130 W Pender St, Vancouver, BC V6E 4A4, Canada; tel (604) 683-8254; fax

(604) 683-8202; Editor: Deborra Schug; Publr: Rob Stanhope; f 1985; quarterly; emphasizes the importance of reafforestation in Canada; circ 7,500.

Simply Living: Otter Publications Pty Ltd, 46–54 Foster St, Surry Hills, NSW 2010, Australia; tel (2) 281-3177; fax (2) 281-4154; every 2 months; covers nature and the environment; intended for the general public; circ 40,000.

Sistemy, Pribory i Metody Kontrolya Kachestva Okruzhayushchei Sredy: Referativnyi Zhurnal, Vsesoyuznyi Institut Nauchno-Tekhnicheskoi Informatsii (VINITI), A-219 Moscow, Baltiiskaya ul 14, Russian Federation (Publr); Mezhdunarodnaya Kniga, 113095 Moscow, Dimitrova ul 39, Russian Federation (Subscriptions); f 1980; monthly; index optional.

SITE Notizie: Società Italiana di Ecologia (SITE), c/o Istituto di Ecologia, Viale delle Scienze, 43100 Parma, Italy; tel (521) 905609; fax (521) 905665; f 1980; annual bulletin of Italian Ecological Society.

Skogeieren (The Forest Owner): Norges Skogeierforbund, Stortingsgaten 30, POB 1438, Vika, 0115 Oslo 1, Norway; tel (2) 83-47-00; fax (2) 83-40-47; Editor: K Wedul; Advertising Manager: L Nordby; f 1914; 16 a year; published by the Norwegian Forest Owners Federation; deals with forestry and the timber trade; circ 57,000.

Skogen (Forestry): Sveriges Skogsvårdsförbund, POB 500, 182 15 Dandayd, Sweden; tel (18) 753-03-90; fax (18) 755-86-02; Editor: Bo Lindevall; Advertising Manager: Kerstin Edwards; f 1883; 11 a year; publication of the Swedish Forestry Association covering forestry and the timber industry; intended for forest owners and forestry professionals; circ 23,000.

Skoven (The Forest): Dansk Skovforening, Amalievej 20, 1875 Frederiksberg C, Denmark; tel 31-24-42-66; fax 31-24-02-42; monthly; journal of Danish Forestry Society; concerned with the conservation of the natural values of Danish forests; circ 4,000.

SLEJF Newsletter: Sri Lanka Environmental Journalists Forum (SLEJF), POB 30, 5 Fourth Lane, Nawala, Sri Jayawardenapura, Sri Lanka; tel (1) 566717; telex 22144; fax (1) 698315; quarterly; deals with communication and awareness-raising in the areas of environment and development; in English.

Sludge Newsletter: Business Publishers Inc, 951 Pershing Drive, Silver Spring, MD 20910-4464, USA; tel (301) 587-6300; fax (301) 587-1081; f 1976; fortnightly; covers the management and disposal of sludge.

Society and Natural Resources: Taylor & Francis, 1101 Vermont Ave, NW, Suite 200, Washington, DC 20005, USA; tel (202) 289-2174; fax (202) 289-3665; Editors: Donald R Field, Rabel J Burdge; Advertising Manager: Louis M Fancher; f 1888; quarterly; concerned with research in the social sciences and the environment; seeks to anticipate research requirements; intended for social scientists, biologists, natural resource managers, policy planners, and soil and water conservationists; circ 400.

Soft Technology: Alternative Technology Association, 247 Flinders Lane, Melbourne, Vic 3000, Australia; tel (3) 650-7883; fax (3) 650-4175; Editor: Ian Scales; f 1979; quarterly; concerned with renewable energy, energy conservation, energy policy and appropriate technology, with particular attention to their practical applications; promotes the adoption of environmentally safe technologies; circ 4,000.

Soil and Health: Soil and Health Association, POB 36-170, Northcote, Auckland 9 1309, New Zealand; tel (9) 480-6650; fax (9) 480-6650; quarterly; deals with organic agriculture and gardening, and related environmental and health issues; circ 2,000.

Soil Science: Department of Environmental Sciences, Rutgers University, New Brunswick, NJ 08903, USA (Editorial); Williams & Wilkins Ltd, 428 East Preston St, Baltimore, MD 21202-3993, USA (Publr); tel (908) 932-9810 (Editorial); fax (908) 932-8644 (Editorial); Editor-in-Chief: Robert L Tate, III; f 1912; monthly; contains papers describing research into soils and environmental soil–plant relationships.

Soil Use and Management: Blackwell Scientific Publications Ltd, Osney Mead, Oxford OX3 0BN, UK (Publr); Soil Science Dept, Rothamsted Experimental Station, Harpenden, Herts AL5 2JQ, UK (Editorial); tel (582) 763133 (Editorial); fax (582) 760981 (Editorial); Editor: Prof J A Catt; f 1986; journal of British Soil Science Society; contains scientific papers and book reviews relating to the use and management of soils in agriculture, forestry and recreation.

Soil and Water Conservation News: Soil Conservation Service, US Department of Agriculture, Washington, DC 20402, USA; US Government Printing Office, Superintendent of Documents, Washington, DC 20402-0001, USA (Distributor); tel (202) 447-4543; (202) 783-3238 (Distributor); monthly.

Soko no Sekai (World of the Creation): Shogakukan Publishing Co Ltd, 2-3-1 Hitotsubashi, Chiyoda-ku, Tokyo 101, Japan; tel (3) 3230-5211; telex 2322192; quarterly; environmental magazine intended for the general public.

Solar Energy Materials and Solar Cells: Elsevier Science Publishers BV, POB 181, 1000 AD Amsterdam, Netherlands; tel (20) 5803911; fax (20) 5803249; Editor: C M Lampert; f 1979; 12 a year in 3 vols; formerly two separate periodicals; international journal dealing with practical aspects of photochemical, photothermal and photovoltaic solar energy conversion; in English.

Solar Progress—Renewable Energy for Australasia: Australian and New Zealand Solar Energy Society, POB 124, Caulfield East, Vic 3145, Australia; tel (3) 571-7557; telex 154087; fax (3) 563-5173; Editor: Trevor R Lee; f 1962; quarterly; promotes the science, technology and application of solar energy; also covers wind and water power, and biofuels; circ 2,500.

Solaria: Keller und Co, Stapferstr 27, 5200 Brugg, Switzerland; Editor: Pierre Sabady; f 1978; every 2 months; journal devoted to solar energy and other alternative energy sources, and solar architecture.

Somos Ambientalistas: Movimiento Ambientalista Nicaragüense, Apdo A-99, Managua, Nicaragua; tel (2) 74835; fax (2) 784863; Editor-in-Chief: Carmen Herrera; f 1993; quarterly; publication of the Nicaraguan Environmental Movement.

Sonnenenergie/Energie Solaire: Verlag Sonnenenergie, Postfach, 8050 Zürich, Switzerland; tel (1) 3120909; fax (1) 3120540; every 2 months; solar energy journal; in German and French; circ 10,000.

Sonnenenergie und Wärmepumpe: Bielefelder Verlagsanstalt GmbH & Co KG, 4800 Bielefeld 1, Niederwall 53, Germany; Editor: A Urbanek; every 2 months; journal concerned with all renewable energy sources and with decentralized energy generation; circ 11,000.

South African Forestry Journal: POB 1022, Pretoria, South Africa; tel (12) 473479; Asst Editor: H A Van der Sÿde; f 1938; quarterly; publication of the South African Forestry Association; covers commercial plantations, forestry research, silviculture, conservation and the timber industry; in English and Afrikaans; circ 1,200.

South African Journal of Wildlife Research/Suid-Afrikaanse Tydskrif vir Natuurnavorsing: Bureau for Scientific Publications, POB 1758, Pretoria 0001, South Africa (Publr); Southern African Wildlife Management Association, POB 324, Bellville 7535, South Africa; tel (12) 3226422 (Publr), (2231) 70180 (SAWMA); fax (2231) 71606 (SAWMA); Editor: D T

Rowe-Rowe; f 1971; quarterly; research journal of Southern African Wildlife Management Association; in English; circ 800.

Species: IUCN—World Conservation Union, ave du Mont-Blanc, 1196 Gland, Switzerland; IUCN Species Survival Commission, c/o Chicago Zoological Society, Brookfield, IL 60513, USA (Subscriptions); tel (22) 649114 (Switzerland); telex 419605; fax (22) 644615; Editors: Diane Cavalieri, Timothy Sullivan; 2 a year; newsletter dedicated to endangered species and conservation of biodiversity; circ 3,000.

Stadtbau-Informationen: Deutscher Verband für Wohnungswesen, 5300 Bonn 1, Städtebau und Raumordnung, Ellerstr 58, Germany; tel (228) 657013; fax (228) 692271; Editors: D Menrath, N Burkard; f 1970; weekly; deals with issues related to housing, planning and the environment; circ 1,200.

State of the Environment—Report of the Executive Director: UNEP, Information Service, POB 30552, Nairobi, Kenya; tel (2) 230800; telex 22068; fax (2) 226886; Exec Dir: Elizabeth Dowdeswell; annual report; in English.

State of the World: Worldwatch Institute, 1776 Massachusetts Ave, NW, Washington, DC 20036-1904, USA; tel (202) 452-1999; fax (202) 296-7365; annual; report on progress towards a sustainable future; published in 27 languages.

Staub-Reinhaltung der Luft: Springer-Verlag GmbH & Co KG, 1000 Berlin 33, Heidelberger Platz 3, Germany; tel (30) 8207454; fax (30) 8207300; monthly; professional journal dedicated to air purification, with particular reference to dust particles; circ 2,350.

STOP Press: Society to Overcome Pollution (STOP), Inc, 716 St-Ferdinand, Montréal, PQ H4C 2T2, Canada; tel (514) 932-7267; Editor: D O'Leary; f 1976; quarterly environmental newsletter; in English and French; circ 2,500.

Stratégies Energétiques, Biosphère et Société (SEBES): c/o Département d'histoire du droit et des doctrines, Faculté de droit, Université de Genève, 1211 Geneva 4, Switzerland (Editorial); Editions Médecine et Hygiène, BP 456, 1211 Geneva 4, Switzerland (Publr); tel (22) 469355 (Publr); fax (22) 7350942 (Editorial), (22) 475610 (Publr); Editor: Prof Ivo Rens; f 1990; 1 or 2 a year; interdisciplinary magazine devoted to global sustainable energy strategies; in French, with summaries in English and German.

Streetwise: National Association for Urban Studies (NAUS), Lewis Cohen Urban Studies Centre, University of Brighton, 68 Grand Parade, Brighton BN2 2JY, UK; tel (273) 673416; fax (273) 673416; Editor: Richard Welsh; f 1971; quarterly; former title *BEE—Bulletin of Environmental Education*; journal of Nat Asscn for Urban Studies (NAUS); promotes education and action for sustainable cities; each issue focuses on a specific theme relating to the urban environment; has two supplements, *Streetwise Education Extra* and the *Sustainable City Newsletter* (qv); circ 1,000.

Streetwise Education Extra: National Association for Urban Studies (NAUS), Lewis Cohen Urban Studies Centre, University of Brighton, 68 Grand Parade, Brighton BN2 2JY, UK; tel (273) 673416; fax (273) 673416; supplement to *Streetwise*; contains information and practical ideas for using local resources in urban environmental education in schools.

Studies on Neotropical Fauna and Environment: Ecology and Systematics: Swets & Zeitlinger BV, Heereweg 347B, 2161 CA Lisse, Netherlands; tel (2521) 35111; telex 41325; fax (2521) 15888; Editor-in-Chief: E J Fittkau; f 1966; quarterly; former title *Beiträge zur neotropischen Fauna;* international journal containing original articles on the ecology, systematics and distribution of neotropical fauna in South America.

Suara Sam: Sahabat Alam Malaysia, 43 Jalan Salween, 10050 Pulan Pinang, Malaysia; tel (4) 375705; Editor: Chee Yoke Ling; f 1979; every 2 months; environmental periodical published by Friends of the Earth Malaysia; in English and Malay; circ 9,000.

Sudan Environment: Institute of Environmental Studies, University of Khartoum, POB 321, Khartoum, Sudan; Editor: M Tag el Seed; f 1981; 3 a year.

Sunceva Energije (Solar Energy): Hrvatsko Udruzenje za Suncevu Energiju, 51000 Rijeka, Narodnog Ustanka 58, Croatia; tel (51) 32210; fax (51) 515403; Editor: Bernard Frankovic; f 1979; 2 a year; dedicated to solar energy and other renewable sources of energy; in Croat, with summaries in English; circ 1,500.

Suomen Luonto: Suomen Luonnonsuojeluliitto, Runeberginkatu 15 A 23, 00100 Helsinki, Finland; tel (0) 642881; fax (0) 446914; nature conservation magazine.

Sustainable City Newsletter: National Association for Urban Studies (NAUS), Lewis Cohen Urban Studies Centre, University of Brighton, 68 Grand Parade, Brighton BN2 2JY, UK; tel and fax (273) 673416; supplement to *Streetwise* (qv); focuses on research, policy development and the communication of good practice by local authorities and voluntary organizations working towards sustainability in the urban environment.

Sveriges Natur: Svenska Naturskyddsføreningen, POB 4625, 116 91 Stockholm, Sweden; tel (8) 702-65-00; fax (8) 702-27-02; Editor: Lars Vaste; f 1910; every 2 months, with additional annual edition; published by Swedish Society for Nature Conservation; includes indexes and book reviews; in Swedish, with summaries in English; circ 200,000.

Svivot (Environments): Ben Gurion College, Co-operative of Sdeh-Boker College, Sde-Boger 84990, Israel; tel (57) 565720; Editor: Ezra Orion; f 1981; quarterly; deals with issues relating to environmental education; circ 400.

Swara: East African Wildlife Society, POB 20020 or 20110, Nairobi, Kenya; telex 22153; fax (2) 337423; Editor: Justine E Muchura; f 1978; every 2 months; magazine intended for the general public, dealing with wildlife and conservation in East Africa; in English; circ 20,000.

Taiki Osen Kenkyu (Journal of the Japan Society of Air Pollution): Taiki Osen Kenkyu Kyokai, 1-29-8 Shinjuku, Shinjuku-ku, Tokyo 160, Japan; tel (3) 3341-5632; fax (3) 3341-8224; Editor: Shun Araki; f 1966; every 2 months; in Japanese, with summaries in English; circ 2,000.

Taiyang Neng Xuebao—Acta Energiae Solaris Sinica: Guoji Shudian, Qikan Bu, POB 399, Beijing 100044, People's Republic of China; quarterly; published by Taiyang Neng Xuehui (Solar Energy Society); in Chinese.

Taiyangneng (Solar Energy): Guoji Shudian, Qikan Bu, Chegongzhuang Xilu 21, POB 399, Beijing, People's Republic of China; quarterly; in Chinese.

Taraxacum: International Youth Federation for Environmental Studies and Conservation (IYF), 36/626, Thiruvananthapuram 695 008, India; tel (471) 450877; fax (471) 446859; members' journal.

Tasmanian Conservationist: Tasmania Conservation Trust Inc, 102 Bathurst St, Hobart, Tas 7004, Australia; tel (02) 343-552; fax (02) 312-491; Editor: S Rundle; f 1969; 9 a year; deals with numerous local and international environmental issues affecting Tasmania; circ 500.

Tecnoambiente: Tecnología Internacional del Aguas, SA—TIASA, Calle Palmeras 23, San Sebastian de los Reyes, 28700 Madrid, Spain; tel (1) 6545011; fax (1) 6545021; Editors: Vicente Lara Ochoa, Rafael Fernandez-Rubio; f 1989; monthly; technical environmental journal concerned with landscaping and environmental engineering.

Tekhnologicheskie Aspekty Okhrany Okruzhayushchei Sredy: Referativnyi Zhurnal, Vsesoyuznyi Institut Nauchno-

Tekhnicheskoi Informatsii (VINITI), A-219 Moscow, Baltiiskaya ul 14, Russian Federation (Publr); Mezhdunarodnaya Kniga, 113095 Moscow, Dimitrova ul 39, Russian Federation (Subscriptions); f 1980; monthly; index optional.

Terra: Pátron, Via Badini 12, 40127 Quarto Inferiore, Bologna, Italy; tel (51) 767003; f 1987; quarterly; environmental science journal; circ approx 6,000.

Terra Grischuna—Graubünden: Terra Grischuna Verlag, Postfach, 4103 Bottmingen, Switzerland; tel (61) 4010011; fax (61) 4015005; f 1942; every 2 months; publication dealing with nature, culture, tourism and transport in the Grisons (Graubünden); in German, with summaries in Italian and Romansch; circ 24,000.

Teva Va'aretz (Nature and Land): Society for the Protection of Nature in Israel, 4 Hashfela St, Tel-Aviv 66183, Israel; tel (3) 375063; Editor: Seffi Ben Joseph; f 1957; monthly; in Hebrew.

Third Opinion: Movement Against Uranium Mining (NSW), POB K133, Haymarket, NSW 2000, Australia; tel (2) 212-4538; fax (2) 281-5216; f 1977; quarterly; contains reports on the nuclear cycle and gives particular attention to environmental issues, conservation, health, safety, economics, alternative energy sources and energy efficiency; circ 600.

TISGLOW—TERI Information Service on Global Warming: Tata Energy Research Institute (TERI), Habiart Complex, Lodi Rd, New Delhi 110 003, India; tel (11) 4625296; telex 61593; fax (11) 4621770; Editor: B Anil Kumar; Asst Editor: Beena Menon; f 1990; 2 a year (April and Oct); informs scientists and policy-makers of developments in the field of global warming and climate change, particularly in developing nations; contains specialist articles, abstracts, digests and book reviews; circ 1,000.

Today: POB 341, Mona Vale, NSW 2103, Australia; Editor: June McGovan; every 2 months; general environmental magazine; circ 40,000.

Toktokkie: POB 58, Cape Town 8000, South Africa; tel (21) 461-3213; fax (21) 45-2607; Editor: Beverley Reed; every 2 months; discovery magazine for young people; published by The Wildlife Society of Southern Africa.

Tokushima-ken Shizen Hogo Kyokai Chosa Hokoku: Tokushima-ken Shizen Hogo Kyokai, Tokushima Daigaku Kyoikugakubu, 1-1 Minami Josanjima-cho, Tokushima-shi, Tokushima-ken 770, Japan; f 1974; 3 a year; reports journal of Tokushima prefecture conservation society; in Japanese.

Tortoise: Nigerian Conservation Foundation, Plot 5, Moseley Rd, POB 74638, Ikoyi, Lagos, Nigeria; tel (1) 683563; quarterly; children's magazine dedicated to conservation issues.

Total Quality Environmental Management: Executive Enterprises Publications Co, Inc, 22 W 21st St, 10th Floor, New York, NY 10010-6990, USA; tel (212) 645-7880; fax (212) 645-1160; Man Editor: Jane Bensahel; Editor-in-Chief: Chris FitzGerald; Editor: John Willig; f 1992; quarterly; discusses the application of 'total quality management' to environmental issues and addresses the changes in industry's attitudes towards environmental regulation.

Toyama-ken Shizen Hogo Kyoukaiho: Toyama-ken Shizen Hogo Kyokai, Shinrin Suisan Kaikan, 4-19 Funabashi-kitamachi, Toyama-shi, Toyama-ken 930, Japan; fax (764) 41-5737; f 1962; every 2 months; newsletter of Toyama prefecture conservation society; contains general reports on nature conservation in Toyama prefecture; circ 600.

Trail and Landscape: Ottawa Field Naturalists' Club, POB 3264, Station C, Ottawa, ON K1Y 4J5, Canada; tel (613) 722-3050; Editor: F Brodo; f 1967; quarterly; indexed; circ 950.

TRANET newsletter: TRANET (Transnational Network of Appropriate Technology), POB 567, Pond St, Rangeley, ME 04970, USA; tel (207) 864-2252; every 2 months; newsletter and directory on all aspects of appropriate technology and philosophy; contains abstracts from more than 1,000 publications.

Transactions of the North American Wildlife and Natural Resources Conference: Wildlife Management Institute, 1101 14th St, NW, Suite 801, Washington, DC 20005, USA; tel (202) 371-1808; fax (202) 408-5059; annual conference proceedings.

Travaux Scientifiques du Parc Naturel Régional et des Réserves Naturelles de Corse: Parc Naturel Régional de Corse, rue Fiorella 4, BP 417, 20184 Ajaccio Cedex, France; tel 95-21-56-54; fax 95-21-88-17; f 1985; 5–6 a year; results of scientific research conducted in Corsican Regional Nature Reserves.

Trees and Natural Resources: Natural Resources Conservation League of Victoria, 593–615 Springvale Rd, Springvale S, Vic 3172, Australia; fax (3) 547-8791; Editor: Felicity Anderson; f 1959; quarterly; circ 6,500.

Trees—Structure & Function: Springer-Verlag, 1000 Berlin 33, Heidelberger Platz 3, Germany; tel (30) 82071; telex 185411; fax (30) 8207300; quarterly; contains original papers on the ecology, biochemistry, physiology and functional anatomy of trees; in English; circ 500.

Treespeak: Greening Australia (SA) Inc, POB 9868, Adelaide, SA 5001, Australia; tel (8) 207-8757; fax (8) 207-8755; f 1983; every 2 months; concerned with tree planting and conservation of the countryside; intended for the general public; circ 4,600.

Trends In Ecology and Evolution: Elsevier Trends Journals, 68 Hills Rd, Cambridge CB2 1LA, UK; tel (223) 315961; fax (223) 464430; Editor: Dr Andrew M Sugden; f 1986; monthly; news and reviews pertaining to ecology and evolutionary biology.

TRI News: Tropical Resources Institute (TRI), Yale University School of Forestry and Environmental Studies, 301 Prospect St, New Haven, CT 06511, USA; tel (203) 432-3660; fax (203) 432-3817; 2 a year; newsletter produced by students of the TRI; informs international readership of TRI activities, faculty and student research, and major activities of collaborators world-wide; circ approx 1,600.

Trias: Jeugdbond voor Natuur- en Milieustudie, Oudegracht 42, 3511 AR Utrecht, Netherlands; tel (30) 368925; fax (30) 343986; f 1973; quarterly; published by the Youth Federation for Natural and Environmental Studies; circ 2,500.

The Trumpeter: Journal of Ecosophy: Canadian Ecophilosophy Network, POB 5853, Station B, Victoria, BC V8R 6S8, Canada; tel (604) 598-7004; fax (604) 598-9901; Managing Editor: Victoria Stevens; Editor: Alan R Drengson; f 1983; dedicated to the exchange of ideas and valuations relating to ecophilosophy, a philosophy based on a commitment to respect nature and live according to ecocentric values.

Tuesday Letter: National Association of Conservation Districts (NACD), 509 Capitol Court, NE, Washington, DC 20002, USA; tel (202) 547-6223; fax (202) 547-6450; monthly; newsletter intended for all US district and state officials, and state and federal agency professionals; contains information on soil and water resource conservation.

Tydskrif Vir Skoonlug/Clean Air Journal: National Association for Clean Air, POB 5777, Johannesburg 2000, South Africa; tel (11) 7282418; Editor: G Kornelius; f 1971; 2 a year; discusses solutions to air pollution; in Afrikaans and English; circ 440.

Työ Terveys Turvallisuus: Institute of Occupational Health, Topeliuksenkatu 41 a A, 00250 Helsinki, Finland; tel (0)

47471; fax (0) 4747478; Editor-in-Chief: Matti Tapiainen; f 1971; 15 a year; specializes in occupational health and safety; includes articles by experts and reviews of current developments; annual edition in English, *Work Health Safety*; circ 88,523 (May 1991).

UK CEED Bulletin: UK Centre for Economic and Environmental Development (UK CEED), 3E King's Parade, Cambridge CB2 1SJ, UK; tel (223) 67799; fax (223) 67794; Editor: Justine Harbinson; f 1984; every 2 months; covers all aspects of environmental economics and policy; circ approx 6,000.

Umwelt: VDI-Verlag GmbH, 4000 Düsseldorf 1, Postfächer 8228/101054, Heinrichstr 24, Germany; tel (211) 61880; telex 8587743; fax (211) 6188112; Editor: R B Firnhaber; f 1971; 10 a year; published for the German Association of Engineers; deals with solid waste, prevention of water pollution and protection of the environment; includes book reviews, abstracts and statistics; circ 17,500.

Umwelt Aktuell: Oswald Möbius Verlag, 1060 Vienna, Amerlingstr 19, Austria; Editor: Erwin H Aglas; every 2 months.

Umwelt & Technik: Europa Fachpresse Verlag GmbH, 8000 Munich 83, Thomas-Dehler-Str 27, Germany; tel (89) 678040; fax (89) 6784108; f 1978; monthly; describes techniques for protection of the environment; circ 11,750.

Umwelt Technologie Aktuell: GIT Verlag GmbH, 6100 Darmstadt 11, Rösslerstr 90, Postfach 110564, Germany; tel (6151) 80900; fax (6151) 809045; Editor: E Giebeler; 2 a year; former title *Supplement Umweltanalytik—Umweltschutz*.

Umweltmagazin: Vogel Verlag und Druck KG, 8700 Würzburg 1, Postfach 6740, Max Planck Str 7/9, Germany; tel (931) 4180; fax (931) 4182640; Editor: J Jobst; f 1971; monthly; specialist journal dedicated to environmental technology in industry and the community; circ 18,000.

Umweltschutz: Bohmann Druck & Verlag GmbH & Co KG, 1110 Vienna, Leberstr 122, Austria; tel (1) 74-15-95; telex 132312; fax (1) 74-11-83; 11 a year; professional journal dedicated to environmental protection; circ 7,000.

Umwelttechnik: Cicero-Verlag AG, Spindelstr 2, Postfach, 8021 Zürich, Switzerland; tel (1) 4888400; fax (1) 4888300; Editor: P Schätzle; f 1966; monthly; published for Schweizerische Vereinigung für Gesundheitstechnik (Swiss Association for Health Technology); deals with the protection of air and water quality; circ 3,100.

Umwelttechnik Berlin: Flaskamp GmbH, Voltastr 5, 1000 Berlin 65, Germany (Publr); FGU, Kleiststr 23–26, 1000 Berlin 30, Germany (Subscriptions); tel (30) 21000311 (Publr); fax (30) 460065060 (Publr); f 1986; every 2 months; published for Fortbildungszentrum Gesundheits- und Umweltschutz (Centre for Further Education in Health and Environmental Protection); deals with research and technology from Berlin that is dedicated to environmental protection; circ 15,000.

Unasylva: FAO, Viale delle Terme di Caracalla, 00100 Rome, Italy; tel (6) 57971; telex 130181; fax (6) 57973152; quarterly publication dedicated to forestry.

Unsere Umwelt: Umwelt-Verlag, 1050 Vienna, Einsiedlerplatz 4, Austria; tel (222) 545-75-16; fax (222) 545-72-45; Editor: Peter H Nemeskal; f 1986; every 2 months; environmental periodical distributed throughout Europe; circ 24,000.

Va-Nytt: BYGGDOK, Swedish Institute of Building Documentation, Environmental Dept, Haelsingegatan 47, 113 31 Stockholm, Sweden; tel (8) 34-01-70; telex 12563; fax (8) 32-48-59; Editor: Gerard Lingre; f 1966; every 2 months; reviews material on environmental management; circ 1,000.

VAM Mededelingen: NV Vuilafvoer Maatschappij, Stadhouderskade 40, Postbus 75380, 1070 AJ Amsterdam, Netherlands; tel (20) 673-03-07; fax (20) 679-11-12; Editor: H M Swinkels; f 1946; quarterly; covers the removal, treatment and recycling of waste, soil pollution and the development of new products and technologies; circ 6,000.

Vand og Miljø (Water and Environment): Forlaget John Vabø A-S, Hartmannsvej 47-49, 2920 Charlottenlund, Denmark; tel 39-40-80-00; fax 39-40-82-80; Man Dir: John Vabø; Sales Man: Claes Kronholm; Man Editor, Dir: Dorte Kardel; f 1984; 8 a year; published for Selskabet for Vand og Miljø (Soc of Water and Environment), Tech Univ of Denmark; circ 1,200.

Vandteknik: Dansk Vandteknisk Forening, Vilh Becks Vej 60, 8260 Viby J, Denmark; tel 86-11-23-33; fax 86-11-79-39; published by Danish Water Supply Association.

Varstvo Narave (Nature Conservation): Zavod Slovenije za Varstvo Naravne in Kulturne Dediscine, 61001 Ljubljana, Plecnikov trg 2, POB 176, Slovenia; fax (61) 213120; Editor: Jelka Habjan; f 1962; annual; includes bibliographies and book reviews; in Slovene, with summaries in English; circ 800.

The Vasculum: The Northern Naturalists' Union, Sunderland Museum and Art Gallery, Borough Rd, Sunderland, Tyne and Wear SR1 1PP, UK; tel (91) 514-1235; fax (91) 510-0675; Editor: L Jessop.

Veröffentlichungen für Naturschutz und Landschaftspflege in Baden-Württemberg: Landesanstalt für Umweltschutz, 7500 Karlsruhe 21, Postfach 210752, Germany; tel (721) 8406424; fax (721) 8406456; f 1925; annual; published by Baden-Württemberg Institute for Environmental Protection; concerned with nature conservation and landscape management; circ 2,000.

Vesitalous: Tontunmäentie 33 D, 02200 Espoo, Finland; tel (0) 425530; fax (0) 425207; Editor-in-Chief: Aimo Maasilta; Deputy Editor: Marja-Leena Järvi; f 1960; every 2 months; journal specializing in water economy, water technology, hydraulic and agricultural engineering and environmental problems; circ 3,000.

Vida Silvestre: Ministerio de Agricultura, Pesca y Alimentación, Instituto Nacional para la Conservación de la Naturaleza, Centro de Publicaciones, Paseo de la Infanta Isabel 1, 28071 Madrid, Spain; tel (1) 2273939; Editor: Ramón Hernandez; f 1971; 2 a year; wildlife magazine published by National Institute for the Conservation of Nature; includes bibliographies; circ 6,000.

Vivre Autrement: ENDA Publications, Environmental Development Action in the Third World (ENDA), 54 rue Carnot, BP 3370, Dakar, Senegal; tel 22-98-90; telex 51456; fax 22-26-95; 5 a year; consumers' magazine for Francophone Africa; provides advice and suggests alternative products.

Voice of Nature: Pt Suara Alam, Jalan Palmerah Barat 31, Jakarta 10270, Indonesia; tel (21) 5495358; fax (21) 5495360; Editor: Linus Simanjuntak; f 1987; monthly; concerned with conservation, the environment, nature and ethnic cultures; circ 12,000.

VQ: The Vegetarian Society UK Ltd, Parkdale, Dunham Rd, Altrincham, Cheshire WA14 4QG, UK; tel (61) 928-0793; fax (61) 926-9182; Principal Officers: Sean McElherron, Carol Timperley; magazine for adult members of Vegetarian Society; campaigns on issues including the environment, animal welfare, developing nations and health.

Warmer Bulletin: The Warmer Campaign, 83 Mount Ephraim, Tunbridge Wells, Kent TN4 8BS, UK; tel (892) 524626; fax (892) 527287; Editor: Maggie Thurgood; information on minimization, re-use and recycling of waste and energy from waste.

Washington Environmental Protection Report: POB 1173, McLean, VA 22101, USA; Callahan Publications, POB 3751, Washington, DC 20007, USA; tel (703) 356-1925; fax (703)

356-9614; Editor and Publr: Vincent F Callahan, Jr; f 1975; newsletter on environmental legislation, regulations and developments.

Wasser/Abwasser GWF: R Oldenbourg Verlag GmbH, 8000 Munich 80, Postfach 801360, Rosenheimer Str 145, Germany; tel (89) 41120; telex 529296; fax (89) 4112207; Advertising Manager: K Haslbeck; f 1858; monthly; concerned with water supply and the treatment of waste water; circ 4,000.

Wasser-Boden-Luft-Umweltschutz: Verlag Laupper AG, Emil Frey-Str 163, 4142 Münchenstein 1, Switzerland; tel (61) 463646; fax (61) 469020; f 1963; 10 a year; publication concerned with protection of the environment; circ 7,000.

Wasser Energie Luft: IVA Ltd, Postfach, Mühlebachstr 43, 8032 Zürich, Switzerland; tel (1) 2512450; telex 816438; fax (1) 2512741; f 1908; 7 a year; former title *Wasser- und Energiewirtschaft*; concerned with water, air, energy and the environment; circ 3,000.

Wasser, Luft und Boden (WLB): Vereinigte Fachverlage Krausskopf-Ingenieur Digest GmbH, 6500 Mainz, Lessingstr 12, Postfach 2760, Germany; tel (6131) 99201; telex 187752; fax (6131) 992100; Editor: Hans Joachim Schmitz; f 1957; 9 a year, with annual supplement *WLB—Handbuch-Umwelttechnik;* former title *Wasser, Luft und Betrieb;* independent journal dedicated to water supply, air purification, waste recycling and environmental science; includes abstracts, indexes and book reviews; circ 15,000.

WasserAbwasserPraxis: Bertelsmann Fachzeitschriften GmbH, 33311 Gütersloh, Carl-Bertelsmann-Str 270, Postfach 120, Germany; tel (5241) 802241; telex 933868; fax (5241) 73055; Editor-in-Chief: Dipl-Ing Ulrich Knorra; Editor: Rainer Sürig; f 1992; every 2 months; environmental trade journal dealing with all aspects of waste water management; circ 7,000.

Wasserwirtschaft: Franckh'sche Verlagshandlung, 7000 Stuttgart 1, Pfizerstr 5–7, Germany; tel (711) 2191296; telex 721669; monthly; publication concerned with water economy; circ 2,350.

Waste Business West: CHMM Inc, 401 Richmond St W, Suite 139, Toronto, ON M5V 1X3, Canada; tel (416) 348-9922; fax (416) 348-9744; Editor: Guy Crittenden; f 1990; every 2 months; former title *Waste Business International;* provides information on new technology and environmental regulations for the control of pollution; intended for businesses producing waste in western Canada; circ 15,000.

Waste Disposal and Water Management in Australia: Editorial and Publishing Consultants Pty Ltd, 29 First Ave, Klemzig, SA 5087, Australia; tel (8) 261-5837; fax (8) 261-2697; Editor: Frank H Schmidt; f 1973; every 2 months; aims to provide a national forum for the analysis and solution of problems in the areas of solid and liquid waste management and disposal, and water management, by government, industry and individuals in Australia; circ approx 2,100.

Waste & Environment Today: AEA Technology, Bldg 7.12 Harwell, Didcot, Oxon OX11 0RA, UK; tel (235) 433484; telex 83135; fax (235) 432854; Editor: Peter J Doyle; fmrly *Waste Management Today*; publication of AEA Technology (UK Atomic Energy Authority).

Waste Information Digests (WID): Environmental Studies Institute, International Academy at Santa Barbara, 800 Garden St, Suite D, Santa Barbara, CA 93101, USA; tel (805) 965-5010; fax (805) 965-6071; Publr: Joanne St John; Man Editor: Mary Ann Short; f 1989; newsletter containing summaries of recent significant articles, books, reports and news releases on all aspects of waste management and reduction, pollution and global warming; also lists computer software, databases and videos relevant to the environment.

Waste Management and Research: Academic Press Ltd, 24–28 Oval Rd, London NW1 7DX, UK (Publr); International Solid Wastes and Public Cleansing Association (ISWA), Bremerholm 1, 1069 Copenhagen K, Denmark; tel 33-91-44-91 (ISWA), (71) 267-4466 (Publr); fax 33-91-91-88 (ISWA), (71) 482-2293 (Publr); Editor: P Rushbrook; f 1983; every 2 months; articles on all aspects of solid waste management, with particular attention to industrial and municipal wastes.

The Waste Manager: National Association of Waste Disposal Contractors (NAWDC), Mountbarrow House, 6–20 Elizabeth St, London SW1W 9RB, UK; tel (71) 824-8882; fax (71) 824-8753; journal providing updated information on the commercial waste management industry in the United Kingdom.

Water, Air and Soil Pollution: Kluwer Academic Publishers, Postbus 17, 3300 AA Dordrecht, Netherlands; tel (78) 33-49-11; telex 29245; fax (78) 334254; Editor: Billy M McCormac; f 1971; 24 a year; international journal; includes indexes and reviews.

Water Conditioning and Purification: Publicom Inc, 4651 N First Ave, Suite 101, Tucson, AZ 85718-5677, USA; tel (602) 293-5446; fax (602) 887-2383; Publrs: Sharon Peterson, Jerry Peterson; f 1959; magazine for the domestic and commercial water treatment industry.

Water Pollution Control Federation Journal: Water Pollution Control Federation, 601 Wythe St, Alexandria, VA 22314-1994, USA; tel (703) 684-2400; f 1928; monthly; scientific and technical journal; circ 29,050.

Water Pollution Research Journal of Canada: Canadian Association on Water Pollution Research and Control, 867 Lakeshore Rd, POB 5068, Burlington, ON L7R 4L7, Canada; tel (416) 336-4698; fax (416) 336-4858; Editors: J M Barica, D T Chapman; f 1966; quarterly; in English and French; circ 440.

Water Quality International: International Association on Water Quality (IAWQ), 1 Queen Anne's Gate, London SW1H 9BT, UK; tel (71) 222-3848; telex 918518; fax (71) 233-1197; Editor: W A Horobin; f 1987; quarterly; former title *IAWPRC Newsletter;* contains news and reports on legislation and new technology; circ 5,500.

Water Research: International Association on Water Quality (IAWQ), 1 Queen Anne's Gate, London SW1H 9BT, UK; tel (71) 222-3848; telex 918518; fax (71) 233-1197; scientific journal containing original papers.

Water Science and Technology: International Association on Water Quality (IAWQ), 1 Queen Anne's Gate, London SW1H 9BT, UK; tel (71) 222-3848; telex 918518; fax (71) 233-1197; scientific journal including conference proceedings.

Water, Sewage and Effluent: Brooke Pattrick Publications (Pty) Ltd, POB 422, Bedfordview, Johannesburg 2008, South Africa; tel (11) 6224666; fax (11) 6167196; Man Dir: John Pattrick; Financial Dir: Eileen Manley; Editor: Helen Gow; quarterly; journal covering all stages of water supply and management, including the development of ground-water reserves, treatment and distribution of water for domestic and industrial consumption, control of water pollution, and the treatment of sewage and industrial effluent before disposal or recycling; circ 2,800.

Water & Wastes: NZ Water and Wastes Association, POB 60217, Auckland, New Zealand (Publr); 120 Scenic Drive, Titirangi, Waitakere City, New Zealand (Editorial); tel (9) 817-5478 (Publr); fax (9) 817-5478 (Publr); Editor: David Ogilvie; f 1958; every 2 months; members' journal; covers issues related to water, wastes and the environment; intended for government officials, relevant professionals, scientists, engineers, lawyers and other interested parties; circ 1,000.

Water and Wastewater Treatment Technology: Metra Martech Ltd, Glenthorne House, Hammersmith Grove, London W6 0LG, UK; tel (81) 563-0666; fax (81) 563-0040; Man Dir: Peter Gorle; Dir: David Cotton; guide to European centres of expertise in water quality and treatment.

Wattenmeer International: WWF Wattenmeerstelle Schleswig-Holstein, 25813 Husum, Norderstr 3, Germany; tel (4841) 62073; fax (4841) 4736; f 1983; quarterly; published by the WWF Schleswig-Holstein Mud-flat Station; concerned with the conservation of the habitats of coastal migratory birds.

Weather and Climate Report: Nautilus Press Inc, 1054 National Press Bldg, Washington, DC 20045, USA; tel (202) 347-6643; Editor: John R Botzum, Sr.

West Coast Environmental Law Research Foundation Newsletter: West Coast Environmental Law Research Foundation (WCELRF), 1001-207 W Hastings, Vancouver, BC V6B 1H7, Canada; tel (604) 684-7378; fax (604) 684-1312; Editor: Bill Andrews; f 1977; quarterly; circ 300.

Western Canada Outdoors: McIntosh Publishing Co Ltd, POB 430, N Battleford, SK S9R 2Y5, Canada; tel (306) 445-4401; fax (306) 445-1977; Editor: Stan Nowakowski; f 1977; monthly; members' conservation magazine; circ 38,750.

What's On The Planet: POB 161, Brighton-Le-Sands, NSW 2216, Australia; tel (2) 281-6179; fax (2) 281-2075; Publr and Editor: Leon Nascon; f 1989; 10 a year; issues covered include the environment, recycling, nutrition, natural beauty and health, global music, travel, leisure and personal development; circ 45,000.

Wilderness: The Wilderness Society, 900 17th St, NW, Washington, DC 20005, USA; tel (202) 833-2300; f 1935; quarterly; concerned with wilderness conservation and management of public land; circ 170,350.

Wilderness Alberta: Alberta Wilderness Association, POB 6398, Station D, Calgary, AB T2P 2E1, Canada; tel (403) 283-2025; fax (403) 270-2743; Editors: Brooke Campbell, Cyndi Smith; f 1968; quarterly; includes statistics and book reviews; circ 4,000.

Wilderness News: Wilderness Society, 130 Davey St, Hobart, Tas 7000, Australia; tel (04) 349-366; fax (02) 233-651; Editor: Alan Thomas; f 1978; every 2 months; former title *Wilderness;* deals with topical national and global environmental issues; circ 16,000.

Wildfields: 15 Dayaram Gidumac Rd, Karachi 3, Pakistan; Editor: Mansoorul Hasan; monthly; general conservation publication; in English.

Wildland News: Wildlands League, 160 Bloor St E, Suite 1335, Toronto, ON M4W 1B9, Canada; tel (416) 324-9760; fax (416) 972-0760; Editor: Caroline Schultz; f 1968; quarterly; members' bulletin; circ 1,500.

Wildlife Australia: Wildlife Preservation Society of Queensland, Level 4, 160 Edward St, Brisbane, Qld 4000, Australia; tel (7) 221-0194; fax (7) 221-0701; Editor: Saren Starbridge; f 1963; quarterly; environmental education, with particular emphasis on native Australian wildlife and its habitats; circ 16,500.

Wildlife News: WWF Australia, 8–12 Bridge St, Sydney, NSW 2000, Australia; tel (2) 247-6300; fax (2) 247-8778; quarterly; wildlife conservation magazine; circ 15,000.

Wildlife Research: CSIRO, 314 Albert St, POB 89, East Melbourne, Vic 3002, Australia; tel (3) 418-7333; fax (3) 419-4096; Editor: A Grant; f 1974; every 2 months; concerned with the biology and management of wild vertebrate species in Australia; indexed; circ 1,350.

Wildlife Research News: David G Stead Memorial Wildlife Research Foundation of Australia, POB 4840, Sydney, NSW 2001, Australia; Editors: C Burns, D Clark; f 1964; quarterly.

Wildlife Safari Magazine: Berouw 128, 9000 Ghent, Belgium; tel (91) 23-41-60; fax (91) 24-43-77; quarterly; conservation magazine; in Flemish; circ 10,000.

Wind-Energie Jahrbuch: Wind-Energie Sekretariat, 5300 Bonn, Gielsdorfer Str 16, Germany; Editor: Walter Schonball; f 1970; annual; circ 1,200.

Wind Engineering: Multi-Science Publishing Co Ltd, 107 High St, Brentwood, Essex CM14 4RX, UK; tel (277) 224632; fax (277) 244632; f 1977; every 2 months; devoted to technology of wind energy; contains articles, reviews and summaries of papers.

Windkraft Journal: Verlag Natürliche Energie GmbH, 2372 Brekendorf, Germany; tel (4353) 551; fax (4353) 796; Editor: Luise Junge; f 1980; quarterly; dedicated to wind energy; circ 2,500.

Windpower Monthly: Torgny Møller, Vrinners Hoved, 8420 Knebel, Denmark; tel 86-36-59-00; telex 64728; fax 86-36-56-26; Editors: Lyn Harrison, Ros Davidson; f 1985; monthly; international news and opinion concerning wind energy; deals with environmental impact, technological developments, market trends and public opinion; includes statistics, book reviews and news of forthcoming events; in English; circ 3,000.

The Workbook: Southwest Research & Information Center, POB 4524, 105 Stanford Drive, SE, Albuquerque, NM 87106, USA; tel (505) 262-1862; fax (505) 262-1864; quarterly; magazine containing original documents on issues related to the environment, consumer and citizen action, and social change.

World Wastes (Waste Management): Communication Channels Inc, 6151 Powers Ferry Rd, NW, Atlanta, GA 30339-2941, USA; tel (404) 955-2500; telex 4611075; fax (404) 955-0400; Publr and Editor: William Wolpin; f 1958; monthly; journal intended for professionals involved in the management of solid, liquid and hazardous wastes; circ 32,850.

World Wastes International: Communication Channels Inc, 6151 Powers Ferry Rd, NW, Atlanta, GA 30339-2941, USA; tel (404) 955-2500; fax (404) 955-0400; Editor: William Wolpin; f 1993; every 2 months; international journal; covers all issues related to the collection, management and recycling of wastes, and landfill; circ 15,000.

World Watch: Worldwatch Institute, 1776 Massachusetts Ave, NW, Washington, DC 20036-1904, USA; tel (202) 452-1999; fax (202) 296-7365; f 1988; every 2 months; short articles on global environmental issues intended for a general readership; in English, German, Italian, Japanese and Russian.

WPSQ News: Wildlife Preservation Society of Queensland (WPSQ), Level 4, 160 Edward St, Brisbane, Qld 4000, Australia; tel (7) 221-0194; fax (7) 221-0701; Publication Officer: Saren Starbridge; f 1966; quarterly; members' newsletter; promotes and supports the work of the Society.

WWF Annual Review: Maggie Sutton, Information Officer, WWF International, ave du Mont-Blanc, 1196 Gland, Switzerland; tel (22) 3649556; fax (22) 3645358; Editor: Ben Barkow; Production Editor: Zandra McGillivray; in English, French and Spanish.

WWF News: Maggie Sutton, Information Officer, WWF International, ave du Mont-Blanc, 1196 Gland, Switzerland; tel (22) 3649556; fax (22) 3645358; Editor: Lesa Griffith; every 2 months; world conservation news; in English and Spanish.

WWOOFnews: Working Weekends on Organic Farms (WWOOF), 19 Bradford Rd, Lewes, E Sussex BN7 1RB, UK; tel (273) 476286; every 2 months; members' newsletter listing organic farms in the UK requiring voluntary help at weekends; includes details of events, developments, training and job opportunities in the organic movement, members' contributions and advertisements.

Yearbook of International Environmental Law: Graham and Trotman Ltd, Sterling House, 66 Wilton Rd, London SW1V 1DE, UK; tel (71) 821-1123; telex 298878; fax (71) 630-5229; Editor-in-Chief: Günther Handl; Assoc Editors: Daniel Bodansky, Lothar Gündling, Linda A Malone, Charles O Okidi, Philippe Sands; f 1990; review of international legal developments; contains short reports by experts, theoretical articles, documents, reviews and a bibliography.

Yidion: Nature Reserves Authority, Rehov Yermeyahn 78, Jerusalem 94 467, Israel; tel (2) 536271; Editor: O Greenstein; f 1975; quarterly; publication providing information on nature reserves in Israel; includes book reviews; circ 500.

Zahlentafeln der Physikalisch-Chemischen Untersuchungen des Rheinwassers/Tableaux Numériques des Analyses Physico-Chimiques des Eaux du Rhin: Internationale Kommission zum Schutze des Rheins gegen Verunreinigung, 5400 Koblenz, Hohenzollernstr 18, Postfach 309, Germany; fax (261) 36572; f 1963; annual; published by International Commission for the Protection of the Rhine Against Pollution; results of physical and chemical analyses of the River Rhine; in German and French; circ 1,200.

Zalioji Lietuva (Green Lithuania): Lithuanian Green Movement, Jaksto 9-127, Vilnius 2300, Lithuania; tel (2) 766-737; fax (2) 766-737; Editor: Daina Karlonaite; f 1989; weekly; concerned with ecology and conservation of nature; circ 10,000.

Zaschita Rastenii (Plant Protection): 107807 Moscow, Sadovaya-Spasskaya ul 18, Russian Federation; tel (095) 207-21-30; fax (095) 207-28-70; Editor-in-Chief: V Savzdarg; f 1932; articles on pest and weed control, progressive technology, ecological plant protection and preservation of the environment.

Zeitschrift für Energiewirtschaft: Vieweg Publishing, 6200 Wiesbaden 1, Postfach 5829, Germany; tel (611) 160216; telex 4186928; fax (611) 160229; quarterly; concerned with sources of energy, and energy supply and policy; circ 1,450.

Zeitschrift für Lärmbekämpfung: Springer-Verlag GmbH & Co KG, 1000 Berlin 33, Heidelberger Platz 3, Germany; tel (30) 8207454; fax (30) 8207300; every 2 months; professional journal for noise prevention; circ 2,000.

Zeitschrift für Ökologie und Naturschutz: Gustav Fischer Verlag Jena, 6900 Jena, Villengang 2, Postfach 176, Germany; tel (3641) 27332; telex 588676; fax (3641) 22638; f 1992; quarterly; journal dedicated to ecology and nature conservation.

Zeitschrift für Umweltpolitik: Deutscher Fachverlag GmbH, 6000 Frankfurt am Main, Schumann-Str 27, Germany; tel (69) 7433530; telex 411862; fax (69) 7433333; Editor: Prof Dr Klaus Zimmermann; Advertising Manager: Klaus Michael Heinz; f 1978; quarterly; published for the International Institute for Environment and Society; focuses on legal, economic and sociological aspects of environmental policy and research; circ 400.

Zeitschrift für Wasser- und Abwasser-Forschung: VCH Verlagsgesellschaft mbH, 6940 Weinheim, Postfach 101161, Pappelallee 3, Germany; tel (6201) 6020; telex 467155; fax (6201) 602328; Editor: K E Quentin; f 1968; every 2 months; journal dedicated to research in the areas of water and waste water; intended for scientists, engineers and ecologists; circ 1,000.

Zelenyj Kalendar (Green Calendar): 252054 Kiev-54, Presa Ukrainy Publishing House, 11 Pavlivska St, Ukraine; tel (044) 216-40-25; Editor-in-Chief: M Khalymononko; fortnightly environmental bulletin; circ 70,000.

Zielone Brygady: 30-060 Kraków, Wydział Chemii UJ, Ingardena 3/100, Poland; tel (12) 336377; telex 322297; fax (12) 340515; Editors: Piotr Rymarowicz, Katy Wimble, Andrzej Zwawa; f 1989; monthly; ecologists' magazine in Polish.

Zimbabwe Wildlife: Wildlife Society of Zimbabwe, POB 66070, Kopje, Harare, Zimbabwe; POB 3497, Harare, Zimbabwe; tel (4) 738722; telex 26334; fax (4) 707130; Editor: Dick Pitman; f 1973; quarterly; former title *Wild Rhodesia;* deals with tourism and the conservation of wildlife; in English; circ 1,400.

Ziran Ziyuan (Journal of Natural Resources): Science Press, Marketing and Sales Dept, 16 Donghuangchenggen Beijie, Beijing 100707, People's Republic of China; tel (1) 4012180; fax (1) 4019810; f 1977; quarterly; contains articles on the land, water, biological and atmospheric resources of China and research papers dealing with the development, use and conservation of these natural resources; covers the application of new technologies to such research; in Chinese, with English abstracts; circ 11,000.

Životné Prostredie, revue pre teóriu a tvorbu životného prostredia (The Environment, review of theory and shaping of the environment): Štefánikova ul 3, POB 254, 814 99 Bratislava, Slovakia (Editorial); Veda Publishing House, Klemensova 19, 814 30 Bratislava, Slovakia (Publr); tel (7) 335-377 (Editorial); fax (7) 332-560; Editor-in-Chief: Prof Dr Milan Ružička; Editor: Marta Ciranová; f 1967; every 2 months; edited by Inst of Landscape Ecology, Slovak Acad of Science; contains articles on current problems and solutions in the field of environmental protection; each issue focuses on a single topic; intended for the general public, managers in industry and environmental activists; in Slovak, with summaries in English, German and Russian.

Ziyuan Yu Kaifa (Resource Development and Conservation): Sichuansheng Ziran Ziyuan Yanjiusuo, 24 Yihuan Lu Nan Erduan, Chengdu, Sichuan 610015, People's Republic of China; tel (28) 553272; telex 60219; fax (28) 589983; Editors: Lu Yongli, Hu Dieze, Jiang Yunzhi; f 1985; quarterly; published by Sichuan Institute of Natural Resources; includes results of scientific research into development, utilization and conservation of natural resources, topography, applications of aerial remote sensing and surveying; in Chinese.

WHO'S WHO
in the
ENVIRONMENT

A

ABDEL NOUR, Prof Hassan Osman, PH D; Sudanese forester and entomologist; b 1 Jan 1944, Abu Usher; s of Osman Abdel Nour Adam and Zeinab Mohamed Ahmed Abdalla; m Shane Moscati 1965; 1s 2d. *Education:* Aberdeen Univ, UK, Khartoum Univ. *Career:* Asst Forest Conservator 1967–72; Lecturer, Forest Rangers Coll, Soba 1972–81 and 1983–85; Sr Forest Conservator, Blue Nile Province 1981–83; Dean, Agric Studies Coll, Sudan Science and Tech Univ 1985–89. *Present Position:* Gen Man, National Forest Corpn 1989–. *Address:* POB 658, Khartoum, Sudan. *Telephone:* 451575. *Telex:* 24307.

ABRAHAM, Martin, PH D; Malaysian marine biologist; b 4 June 1953; m Rajeswari Kanniah 1985; 1d. *Career:* Visiting scientist and consultant, Nat Inst of Cholera and Enteric Diseases, India; visiting scientist, Maryland Univ and N Wales Univ; research co-ordinator, FoE, Malaysia; Head of Environment and Information Research, Int Org of Consumer Unions; bd mem, Environmental Liaison Centre Int. *Present Position:* Project Dir, Asian Environment Programme, WWF Int. *Awards and Honours:* UNEP Global 500 Award for Environmental Achievement in the area of Consumers and Environment. *Address:* POB 911, Jalan Sultan, 46990 Petaling Jaya, Malaysia. *Telephone:* (3) 7579192. *Fax:* (3) 7565594.

AHAS, Rein; Estonian environmentalist; b 1966, Tartu. *Education:* Tartu Univ. *Career:* Pres, Tartu Student Nature Protection Club 1987–90. *Present Position:* Estonia Co-ordinator, Taiga Rescue Network. *Address:* Lembitu St 1, Tartu EE2400, Estonia. *Telephone:* (1434) 28-180. *Fax:* (1434) 62-084.

AL-WADAEE, Abdul-Elah, PH D; Bahraini industrial hygienist; b 1955, Manama; s of Jawad Al-Wadaee; m Naaem Al-Asfour 1984; 1s 1d. *Education:* Baghdad Univ, Iraq; Paris Univ and Aix-Marseilles Univ, France. *Career:* Occupational Hygienist, Ministry of Health 1979–90. *Present Position:* Chemist, Environmental Protection Cttee 1990–. *Address:* POB 26909, Adliya, Bahrain. *Telephone:* 293693. *Fax:* 293694.

ALAWNEH, Ziyad, B SC; Jordanian environmentalist; b 1 Sept 1955, Zarqa; m Huda Hiyasat 1985; 1s 1d. *Education:* Beirut American Univ. *Career:* Environmental health researcher, Ministry of Environment 1982–84; Dir, Preventive Medicine Dept, Armed Forces Hospital, Saudi Arabia 1984–91. *Present Position:* Man, Nat Environmental Information and Educ Programme, sr ed, environmental newsletter. *Address:* POB 340636, Amman 11134, Jordan. *Telephone:* (6) 699844. *Fax:* (6) 695857.

ALBIGNAC, Roland; environmentalist. *Present Position:* UNESCO and UNDP, Madagascar. *Address:* BP 1346, Antananarivo, Madagascar.

ALI, Mohamed A; Sudanese environmentalist. *Present Position:* Sec-Gen, Higher Council for Environment and Natural Resources. *Address:* Higher Council for Environment and Natural Resources, POB 10488, Khartoum, Sudan.

ALOISI DE LARDEREL, Jacqueline Marthe, MS, MBA; French international civil servant; b 8 July 1942, Orléans; d of Victor and Anne (Voyant) Delzenne; m Corso Aloisi de Larderel 1971; 1s 1d. *Education:* Paris Univ; Yale Univ. *Career:* Head, Planning and Devt Programme, OTAM consultants; Asst Head then Head, Waste Div, Ministry of Environment 1972–84, Asst to Dir of Pollution Prevention 1985–87. *Present Position:* Dir, Industry and Environment Programme Activity Centre, UNEP 1987–. *Awards and Honours:* Chevalier, Ordre Nat du Mérite 1986. *Address:* 39–43 quai André-Citroën, Tour Mirabeau, 75739 Paris Cedex 15, France. *Telephone:* (1) 44-37-14-50. *Telex:* 204997. *Fax:* (1) 44-37-14-74.

ALON, Azaria; Israeli teacher, writer and editor; b 15 Nov 1918, Vollodarsk, Ukraine; s of Haim and Sarah (Kipnis) Kozirovsky; m Ruth Diamant 1952; 2s 1d. *Education:* Tel Aviv Univ. *Career:* Youth leader and agricultural worker 1936–49; biology teacher 1951–67; co-founder, Soc for Protection of Nature 1953; mem, Nature Reserves Authority and Council 1964–76; Gen Sec, Soc for Protection of Nature 1969–77, Chair of Bd 1977–90. *Present Position:* Chair, Council, Soc for Protection of Nature. *Awards and Honours:* Radio award 1967; government prizes for environmental activities 1978, 1984; Israel Prize 1980; UNEP Global 500 1987. *Publications:* Encyclopedia of Plants and Animals of Israel (Ed) 1991. *Address:* Kibbutz Bet Hashita, 18910 Bet Hashitta, Israel. *Telephone:* (6) 536867. *Fax:* (6) 536577.

ALVAREZ DEL TORO, Miguel; Mexican zoologist, naturalist and conservationist; b 23 Aug 1917; s of Carlos Alvarez and Marcela Del Toro (both deceased); m 1st Aurora Rincon 1944 (deceased); m 2nd Clementina Perez 1975; 1s 3d. *Career:* Field Scientist, Philadelphia Acad of Natural Sciences 1938; taxidermist then Deputy Dir, Museo de la Flora y Fauna, Mexico City 1939–42; Viveros Tropicales y Museo de Historia Natural 1942–44. *Present Position:* Dir, Inst de Historia Natural, Chiapas 1944–. *Awards and Honours:* J Paul Getty Wildlife Conservation Prize 1989; UNEP Global 500 1992. *Publications:* Los Animales Silvestres de Chiapas 1952; Los Reptiles de Chiapas 1960; Las Aves de Chiapas 1971; Los Cocodrilos de México 1974; Los Mamíferos de Chiapas 1977; Asi era Chiapas 1985; Las Aranas de Chiapas 1992. *Address:* Apdo Postal 60, Tuxtla Gutierrez, Chiapas, Mexico. *Telephone:* (322) 23754. *Fax:* (322) 29943.

ANADA, Tiéga; Niger environmentalist. *Present Position:* Dir, IUCN Nat Office, Niger. *Address:* BP 10933, Niamey, Niger.

ANADU, Pius Amaeze, PH D; Nigerian consultant on biodiversity, conservation and environmental studies; b 11 Nov 1940, Nnewi; s of Joseph Afamefuna Anadu and Uduka Anadu; m Christiana Akunwa Anadu 1970; 2s 2d. *Education:* Ibadan Univ. *Career:* Asst Lecturer, Zoology Dept, Ibadan Univ 1973–74, Lecturer 1974–79; Sr Lecturer, Zoology Dept, Benin Univ 1980–84, Acting Head, Forestry and Wildlife Dept 1984–88; Exec Dir, Nigerian Conservation Foundation 1988–93; Vice-Pres, Nigerian Field Soc and mem, Programme Cttee, WWF Int 1990–93; mem, Editorial Bd, *International Journal of Primatology* 1990–92. *Present Position:* Chair and CEO, Bio-Resources (Nigeria) Ltd 1992–. *Publications:* Wildlife Conservation in Nigeria, Problems and Strategies 1987; Prospects for Primate Conservation in Nigeria 1987; Progress in the Conservation of Nigeria's Wildlife 1987; Antelopes (Co-Author) 1990. *Address:* 2 Oweh St, Yaba, PMB 80009, Victoria Island, Lagos, Nigeria. *Telephone:* (1) 866968. *Telex:* 27469. *Fax:* (1) 860732.

ARAUJO, Joaquin; Spanish writer and documentary maker. *Career:* Lecturer; founder several Spanish environmental socs; Pres first State Centre for Environmental Education in Spain. *Awards and Honours:* UNEP Global 500 Roll of Honour 1991. *Publications:* Publications include 15 books and more than 300 articles. *Address:* Avda America 8, 28028 Madrid, Spain.

ARCHIBALD, George William, PH D; Canadian conservationist; b 13 July 1946, New Glasgow; s of Donald and Lettie Archibald; m 1981. *Education:* Cornell Univ; Dalhousie Univ. *Career:* Co-founder, Int Crane Foundation 1973. *Present Position:* Dir, Int Crane Foundation 1973–. *Awards and Honours:* Order of Golden Ark 1983; MacArthur Award 1984; WWF Gold Medal 1985; UNEP Global 500 1987; Distinguished Conservationist Award, Nat Audubon Soc 1989. *Address:* E-11376 Shady Lane Rd, Baraboo, WI 53913-9778, USA. *Telephone:* (608) 356-9462. *Telex:* 297778. *Fax:* (608) 356-9465.

ATMONTE, Carlos L; Brazilian environmentalist. *Present Position:* Mem, Asociación Latinoamericana Multidisciplinaria de Planeación Estrategica del Medio Ambiente. *Address:* Avda da Ingreja 346, 2°, 95590 Tramandai, RS, Brazil.

ATTENBOROUGH, Sir David Frederick, CVO, CBE, MA, FRS; British broadcaster and writer; b 8 May 1926, London; s of Frederick and Mary Attenborough; m Jane Elizabeth Ebsworth Oriel 1950; 1s 1d. *Education:* Clare Coll, Cambridge. *Career:* Royal Navy 1947–49; editorial asst, publishing house 1949–52; BBC TV 1952–73, Trainee Producer 1952–54, Producer, zoological, archaeological, travel and political programmes 1954–64, Controller, BBC2 1964–68, Dir of Programmes 1969–73, mem, Man Bd 1969–73; mem, Nature Conservancy Council 1975–82; Int Trustee, WWF 1979–86; Fellow, Soc of Film and TV Arts 1980; Trustee, British Museum 1980–, Science Museum 1984–87. *Present Position:* Freelance broadcaster and writer 1973–. *Awards and Honours:* Special Award, Guild of TV Producers 1961; Silver Medal, Royal TV Soc 1966; Silver Medal, Zoological Soc of London 1966; Soc of Film and TV Arts Award 1970; UNESCO Kalinga Prize 1982; Philadelphia Acad of Natural Sciences Medal 1982; Founders Gold Medal, Royal Geographical Soc 1985; Encyclopedia Britannica Award 1987. *Publications:* Zoo Quest to Guiana 1956; Zoo Quest for a Dragon 1957; Zoo Quest in Paraguay 1959; Quest in Paradise 1960; Zoo Quest to Madagascar 1961; Quest under Capricorn 1963; The Tribal Eye 1976; Life on Earth 1979; The Zoo Quest Expeditions 1982; The Living Planet 1984; The First Eden, The Mediterranean World and Man 1987; Atlas of the Living World (Co-Writer) 1989; The Trials of Life 1990. *Address:* 5 Park Rd, Richmond TW10 6NS, UK.

AVEROUS, C P; French environmentalist. *Career:* Contrib, UNEP Environmental Data Report 1992. *Present Position:* OECD. *Address:* c/o 2 rue André-Pascal, 75775 Paris Cedex 16, France. *Telephone:* (1) 45-24-82-00. *Telex:* 620160. *Fax:* (1) 45-24-85-00.

AZNAR, Guy; French environmentalist. *Present Position:* Pres, Amis de la Terre. *Address:* 62 bis rue des Peupliers, 92100 Boulogne, France.

B

BAASANDORJ, Davaagiyn; Mongolian environmentalist and politician. *Career:* Mem, Alliance of Greens; Chair, Mongolian Greens Party 1992. *Present Position:* Mem, Mongolian Greens Party. *Address:* CPOB 1089, Ulan Bator, Mongolia. *Telephone:* 323871. *Telex:* 79236.

BABAEV, Agadjan G; Turkmen scientist and physical geographer; b 10 May 1929; s of Geldy and Ogulbek Babaev; m D Babaeva 1950; 2s 6d. *Education:* Turkmen Univ, Ashkhabad. *Career:* Teacher then Chair of Geography, Turkmen Univ 1952–60; Pres, Acad of Sciences 1975–93. *Present Position:* Dir, Desert Inst, Acad of Sciences of Turkmenistan 1960–. *Awards and Honours:* USSR State Prize Laureate 1981. *Publications:* The Karakum Desert 1963; Deserts in the USSR 1977; Deserts of the World 1986. *Address:* 744000 Ashkhabad, ul Gogolya 15, Turkmenistan. *Telephone:* (3632) 29-54-27. *Fax:* (3632) 25-37-16.

BAČÍK, Martin; Slovak hydraulic engineer; b 13 June 1953, Bratislava; m Anna Bačík 1977; 1d. *Education:* Slovak Tech Univ, Bratislava. *Career:* State Navigation Bureau, Bratislava 1976–78; mem, Czechoslovak Cttee for Hydrology 1986–92; Moscow State Univ 1988. *Present Position:* Deputy Dir for Hydrology and Hydraulics, Water Research Inst. *Publications:* Mathematical Modelling of Total Load Sedimentation in Reservoirs 1986; Mathematical Modelling of Morphological Flow Development 1989; The Influence of the Gabčíkovo Hydrostructure on the Environment 1992. *Address:* nábr Gen L Svobodu 5, 812 49 Bratislava, Slovakia. *Telephone:* (7) 343-311. *Fax:* (7) 315-743.

BAGCHI, Ajoy; Indian environmentalist. *Present Position:* Exec Dir, People's Comm on Environment and Devt. *Address:* 15 Institutional Area, Lodi Rd, New Delhi 110 003, India.

BAHOUTH, Peter; American environmentalist. *Present Position:* Exec Dir, Greenpeace USA Inc. *Address:* 1436 U St, NW, Washington, DC 20009, USA. *Telephone:* (202) 462-1177.

BAJSAROWICZ, Janusz, M SC; American int environmental man consultant; b 1936, Poland; s of Jozef and Bogumila Bajsarowicz; m Lola Joan Bajsarowicz 1960. *Education:* California Univ, Berkeley, Colorado State Univ, Kraków Tech Univ. *Career:* Eng Man, Bechtel Corpn, San Francisco 1973–81; Pres and CEO, Aqua Resources Inc, Berkeley, CA 1983–92; Founder and Pres, US Cttee on Poland's Environment 1989; Speaker, World Affairs Council, UN Asscn, San Francisco 1992, 1993; Hon Mem, Polish Hygiene Soc 1992; Contrib, *Los Angeles Business Journal* and *Wall Street Journal* 1993. *Present Position:* Consultant, World Bank, Science Applications Int Corpn and Ministry of Environmental Protection, Poland 1989–. *Publications:* The Greening of Eastern Europe; Pollution and Solutions in Poland 1989; Environmental Reform: The Polish Model 1993. *Address:* 21 Del Mar Ave, Berkeley, CA 94708, USA. *Telephone:* (510) 987-8499. *Fax:* (510) 893-1321.

BAKE, Gernot Friedrich Wilhelm, PH D; German geographer; b 3 Oct 1950; s of Wilhelm and Minna Bake; m Elizabeth Migongo 1983; 1d. *Education:* Hamburg Univ; Hanover Tech Univ. *Present Position:* Nat Resources Man, Marsabit Devt Programme 1992–. *Publications:* Desertification in South-West Sicily 1979; Climatology of Marsabit District 1983; Water Resources of Ten Districts in Kenya. *Address:* POB 47051, Nairobi, Kenya. *Telephone:* (2) 520434. *Fax:* (2) 562670.

BALČIAUSKAS, Linas, PH D; Lithuanian biologist; b 30 April 1957, Biržai; m Laima Paulauskaitė 1978; 3d. *Education:* Vilnius State Univ. *Career:* Jr research assoc 1981–89. *Present Position:* Sr Researcher, Inst of Ecology, Lithuanian Acad of Sciences 1989–. *Publications:* Hoofed Animals of Lithuania (Co-Author) 1987. *Address:* Ecological Diversity of Lithuania,

POB 2147, 2017 Vilnius, Lithuania. *Telephone:* (3702) 359-278. *Fax:* (3702) 359-257.

BARACATT SABAT, Gabriel; Bolivian lawyer; b 26 Nov 1966, Tarija. *Education:* Tarija Univ. *Career:* Founder, Protección del Medio Ambiente Tarija (PROMETA) 1990; Contrib, *Linea Verde* magazine 1991–92; video documentaries 1991–. *Present Position:* Exec Dir, Protección del Medio Ambiente Tarija (PROMETA) 1990–; Gen Co-ordinator, UNITAR Network 1993–. *Address:* Calle O'Connor 556, Tarija, Bolivia. *Telephone:* (66) 45865. *Fax:* (66) 45865.

BARAHONA-ISRAEL, Dr Rodrigo; Costa Rican professor of law and consultant; b 28 Nov 1945; s of Gabriel Barahona; m Maria Mata 1983; 3c. *Education:* Costa Rica Univ, Florence Univ. *Career:* Prof of Law, Costa Rica Univ 1974–93; Dir, Cen Bank 1983; Under-Sec of Int Trade 1986–87; Founder and Pres, Centro de Derecho Ambiental y de los Recursos Naturales (CEDARENA) 1989–93. *Present Position:* Legal Counsel, Earth Council 1993–. *Publications:* Derecho Agrario 1971; Recursos Naturales 1992. *Address:* Apdo 134-2050, San Pedro, Costa Rica. *Telephone:* 23-8463. *Fax:* 33-8256.

BARDOT, Brigitte; French actress and environmentalist; b 28 Sept 1934, Paris; d of Louis and Anne-Marie (Mücel) Bardot; m 1st Roger Vadim; m 2nd Jacques Charrier; m 3rd Gunther Sachs 1966 (dissolved 1969); m 4th Bernard d'Ormale 1992; 1s. *Career:* Stage and film actress 1952–. *Present Position:* Founder and Pres, Brigitte Bardot Foundation. *Awards and Honours:* Étoile de Cristal, Cinema Acad 1966; Chevalier, Légion d'honneur 1985. *Publications:* Films: Manina: la fille sans voile; Le fils de Caroline chérie; Futures vedettes; Les grandes manoeuvres; La lumière d'en face; Cette sacrée gamine; La mariée est trop belle; Et Dieu créa la femme; En effeuillant la marguerite; Une parisienne; Les bijoutiers du clair de lune; En cas de malheur; La femme et le pantin; Babette s'en va-t-en guerre; Voulez-vous danser avec moi?; La vérité; Please Not Now?; Le mépris; Le repos du guerrier; Une ravissante idiote; Viva Maria; A coeur joie 1967; Two Weeks in September 1967; Shalako 1968; Les femmes 1969; Les novices 1970; Boulevard du rhum 1971; Les pétroleuses 1971; Don Juan 1973; L'Histoire très bonne et très joyeuse de Colinot trousse-chemise 1973; Noonah, le petit phoque blanc. *Address:* 4 rue Franklin, 75116 Paris, France.

BARISICH, Alessandro; Italian int civil servant. *Present Position:* Head, Civil Protection Div, Directorate A (Nuclear Safety, Waste Man, Prevention and Control of Pollution), Directorate-Gen XI (Environment, Consumer Protection and Nuclear Safety), Comm of EC. *Address:* 200 rue de la Loi, 1049 Brussels, Belgium. *Telephone:* (2) 235-11-11. *Telex:* 21877. *Fax:* (2) 235-01-22.

BARNES, Christopher Richard, PH D, FRSC; Canadian professor of earth and ocean sciences; b 20 April 1940, Nottingham, UK; m Susan Barnes 1961; 3d. *Education:* Birmingham Univ, UK; Ottawa Univ. *Career:* NATO Post-Doctoral Fellow, Wales Univ 1964–65; Asst, Assoc Prof, Prof then Chair, Dept of Earth Sciences, Waterloo Univ, Ontario 1965–81; Head, Dept of Earth Sciences, Memorial Univ, Newfoundland 1981–87; Dir-Gen, Geological Survey of Canada 1987–89; Dir, School of Earth and Ocean Sciences, Victoria Univ 1991–. *Present Position:* Dir, Centre for Earth and Ocean Research, Victoria Univ 1989–. *Awards and Honours:* Past President's Medal 1977; Bancroft Award 1982; Willis Ambrose Medal 1991. *Address:* POB 1700, Victoria, BC V8W 2Y2, Canada. *Telephone:* (604) 721-8847. *Fax:* (604) 721-6200.

BARNES, James Neil, BA, JD; American int environmental lawyer and organization official; b 28 June 1944, Tulsa, OK; m Anne L Fuhrman; 2d. *Education:* Northwestern Univ; Michigan Univ. *Career:* Law clerk, US Dist Court, Washington, DC; Staff Attorney, Center for Law and Social Policy, Washington, DC 1971–72, Sr Staff Attorney 1977–81; Staff Attorney Public Defender Service 1972–74; Assoc, Hudson and Co, Port Vila, Vanuatu 1974–75, Wilmer, Cutler and Pickering 1976–77; Consultant, Council for Public Interest Law, Washington, DC 1975–76; UN Rep, Greenpeace Int, Dir Greenpeace-US 1985–86; E Coast Dir, Threshold Int Center for Environmental Renewal, est Small Grants Program 1981–87; Consulting Attorney, Environmental Defense Fund, Washington, DC 1986–87. *Present Position:* Head, Int Dept and Sr Attorney, FoE, Washington, DC 1987–, FoE rep at World Bank, UNCED, IMF, Asian Devt Bank and numerous other int asscns. *Awards and Honours:* International Environmentalist of the Year, National Wildlife Federation 1991. *Publications:* Let's Save Antartica (jtly) 1982; An Introduction to Southern Ocean Conservation Issues 1981; Russian Roulette: Nuclear Power Reactors in Eastern Europe and the Former Soviet Union (jtly) 1992; numerous contribs to professional journals. *Address:* 2633 15th St, NW, Apt 3, Washington, DC 20009, USA. *Telephone:* (202) 265-3603.

BARTELMUS, Peter L P, PH D; German UN official; b 31 Jan 1942, Brunn; m Liza Morales 1990; 2c. *Education:* Heidelberg Univ. *Career:* Mem, scientific staff, Fed Statistical Office, Wiesbaden 1968–71; Admin, Environment Div, Fed Ministry of Interior 1971–72; Econ and Statistical Adviser, E African Community, Nairobi 1972–75; Programme Officer, UNEP 1975–81; Chief, Environment Statistics Section, Statistical Office of the UN 1981–91; Chair, Task Force on Environment and Devt, Dept of Int Econ and Social Affairs 1987–92; mem, Int Advisory Bd, *Journal of Environmental Statistics* 1993–; mem, Dept of Econ and Social Information and Policy Analysis 1993–; mem, German Statistical Soc, Int Statistical Inst. *Present Position:* Officer-in-Charge, Environment and Energy Statistics Branch, UN Statistical Div 1991–. *Publications:* Heterogenitätsprobleme in industriestatistischen Systemen 1972; Economic Development and the Human Environment: A Study of Impacts and Repercussions with Particular Reference to Kenya 1980; Environment and Development 1986; Integrated Environmental and Economic Accounting: A Case Study for Papua New Guinea 1992. *Address:* 4C, 205 E 22nd St, New York, NY 10010, USA. *Telephone:* (212) 963-4581. *Fax:* (212) 963-4116.

BASKYTE, Ruta; Lithuanian geographer; b 30 Nov 1958; 2s. *Education:* Vilnius Univ. *Career:* Nat parks man schemes, regional planning policy 1981–90. *Present Position:* Sr Expert and Chief Engineer, Landscape Bd, Environmental Protection Dept. *Address:* Juozapaviciaus str 9, 2600 Vilnius, Lithuania. *Telephone:* (3702) 352-572. *Telex:* 261191. *Fax:* (3702) 358-020.

BASU, Biplab Bhushan; Indian academic. *Career:* Founder Young Scientists Society, Calcutta, 1964, now School of Fundamental Research, dedicated to spreading scientific and environmental awareness to the grass-roots level by developing environmental awareness programmes. *Present Position:* Head School of Fundamental Research. *Awards and Honours:* UNEP's Global 500 Roll of Honour 1991. *Address:* School of Fundamental Research, 49/B Tolly, Gunge Rd, Calcutta 700026, India.

BATAREVSKIS, Olegs; Latvian lawyer; b 21 March 1951, Riga; s of Vladimirs Batarevskis and Maiga Batarevska; m Aija Batarevska 1978; 1d. *Education:* Latvian Univ, Riga. *Career:* Vice-Pres, Latvian Environment Protection Club 1989; MP, mem, Presidium, Supreme Council and Chair, Environmental Protection Comm 1990–93. *Present Position:* Co-Chair, Latvian Green Party (LZP) 1990–. *Address:* Kalku iela 8-1A, Riga, LV 1050, Latvia. *Telephone:* (3712) 613-692. *Fax:* (3712) 228-892.

BATISSE, Michel R J, DR SC; French international civil servant; b 3 April 1923, Chateauroux; m 1949; 2d. *Education:* Paris Univ. *Career:* Scientific Officer for Middle E, UNESCO 1951–56, Co-ordinator, Arid Lands Project 1957–62, Dir, Natural Resources Research Div 1963–73, Organizer, Int Hydrological Decade 1965–74 and Man, Biosphere Programme 1968–83, Deputy Asst Dir-Gen for Environment 1975–83, Asst Dir-Gen for Science 1983–85. *Present Position:* Pres, Mediterranean Blue Plan and Sr Environmental Adviser, UNESCO 1985–. *Awards and Honours:* John Phillips Medal for Int Conservation; UNEP Global 500 1988. *Publications:* Futures for the Mediterranean: The Blue Plan (Co-Ed) 1989. *Address:* 9 ave Leonce-Bucquet, 92380 Garches, France. *Telephone:* (1) 45-68-40-51. *Fax:* (1) 40-65-98-97.

BATOV, Prof Dr Simeon; Bulgarian university professor; b 1922, Kustendil; s of Georgi and Ljuba Batov; m Julia Batov 1949; 1s 1d. *Career:* Engineer, Energoprojekt 1946–48; Lecturer, Sofia Tech Univ 1948, later Assoc Prof, Prof; Head, Power and Nuclear Stations Dept 1964–85. *Present Position:* Energetics Branch Co-ordinator, Nat Ecological Club. *Address:* 1421 Sofia, Milin Kamak Str 1, Bulgaria. *Telephone:* (2) 66-17-30. *Fax:* (2) 59-10-92.

BEHREND, Hans Nikolaus (Juan); Danish environmentalist. *Career:* Fmr Deputy Sec-Gen, Green Group, EP. *Present Position:* Sec-Gen, Green Group, EP. *Address:* Bâtiment Remorqueur, 79 rue Belliard, 1040 Brussels, Belgium. *Telephone:* (2) 284-30-45. *Telex:* 26999. *Fax:* (2) 284-78-37.

BELLAMY, David James, PH D, F I BIOL, FLS; British botanist, writer and broadcaster; b 18 Jan 1933, London; s of Thomas and Winifred (neé Green) Bellamy; m Rosemary Froy 1959; 2s 3d. *Education:* Chelsea Coll of Science and Technology and Bedford Coll, London Univ. *Career:* Lecturer, then Sr Lecturer Dept of Botany, Univ of Durham 1960–80, Hon Prof of Adult and Continuing Education 1980–; Visiting Prof Massey Univ, NZ 1988–89; Special Prof of Botany Nottingham Univ; TV and radio presenter and scriptwriter; TV series include Life in Our Sea 1970, Bellamy on Botany 1973, Bellamy's Britain 1975, Bellamy's Europe 1977, Botanic Man 1978, Up a Gum Tree 1980, Backyard Safari 1981, The Great Seasons 1982, Bellamy's New World 1983, End of the Rainbow Show, SWALLOW, Turning the Tide 1986, Bellamy's Bugle 1986, 1987, 1988, Bellamy on Top of the World, Bellamy's Journey to the Centre of the World 1987, Bellamy's Bird's Eye View, Wheat Today What Tomorrow? 1989, Moa's Ark 1990, Bellamy Rides Again 1992, Blooming Bellamy 1993; Pres Youth Hostels Asscn 1983. *Present Position:* Founder Dir Conservation Foundation 1982–; Pres WATCH 1982, Population Concern 1987– Nat Inst of Environmental Education 1989–. *Awards and Honours:* Dr hc CNAA 1990; Dutch Order of the Golden Ark 1989; UNEP Global 500 1990. *Publications:* Peatlands 1974, Life Giving Sea 1977, Half of Paradise 1979, The Great Seasons 1981, Discovering the Countryside with David Bellamy (Vols I, II) 1982, (Vols III, IV) 1983, The Mouse Book 1983, The Queen's Hidden Garden 1984, Bellamy's Ireland 1986, Bellamy's Changing Countryside (4 vols) 1988, England's Last Wilderness 1989, Wetlands, Wilderness Britain, Moa's Ark 1990, How Green Are You? 1991, Tomorrow's Earth, World Medicine 1992. *Address:* c/o The Conservation Foundation, 1 Kensington Gore, London SW7 2AR, UK. *Telephone:* (71) 823-8842.

BELLION, Uta, M SC; German environmentalist; b 3 Jan 1956; d of Kurt and Kläre von Strünck; m Michael Bellion 1986; 1s. *Education:* Bielefeld Univ. *Career:* Research Asst, Aachen Univ 1982–1983; Research Scientist, Hanover Tech Univ 1983–1986; Mem, Programme Cttee, Die Grünen, Lower Saxony 1985–86; Int Project Co-ordinator, European Rivers Campaign, Greenpeace 1986–87, Int Campaign Co-ordinator, Toxics Campaign 1987–89; Mem, Bd of Dirs, Greenpeace Int and Greenpeace Sweden 1990–; Chair, Bd, Greenpeace Austria 1990–92, Voting Rights Cttee, Greenpeace Int 1990–91, Licensing Cttee 1991–. *Present Position:* Chair, Bd, Greenpeace Int. *Address:* Temple House, 25-26 High St, Lewes BN7 2LU, UK. *Telephone:* (323) 811025. *Fax:* (323) 811026.

BENDAHMANE, H; environmentalist. *Career:* Contrib, UNEP Environmental Data Report 1992. *Present Position:* environmentalist. *Address:* c/o UNEP, POB 30552, Nairobi, Kenya. *Telephone:* (2) 230800. *Telex:* 22068. *Fax:* (2) 226890.

BENJAMIN, Shirley; Israeli environmentalist. *Present Position:* Dir, Ecological Network (Econet Israel). *Address:* POB 581, Karkur 37105, Israel. *Telephone:* (6) 377072. *Fax:* (6) 377072.

BENNETT, Burton G, PH D; American environmental scientist and diplomatist; b 7 July 1939; m Meta Bennett 1965; 2s. *Education:* Oregon State Univ, Washington Univ, New York Univ. *Career:* Environmental Scientist, Environmental Measurements Lab, New York 1964–78; Deputy Dir, Monitoring and Assessment Research Centre, London 1979–87. *Present Position:* Sec, UNSCEAR 1988–. *Publications:* UNEP Environmental Data Report (Ed) 1987; UNSCEAR Report 1988 and 1993. *Address:* Vienna Int Centre, POB 500, 1400 Vienna, Austria. *Telephone:* (1) 21131-4330. *Telex:* 135612. *Fax:* (1) 237608.

BENNETT, Edward; British int civil servant; b 11 Sept 1931. *Education:* London Univ. *Career:* Surgeon-Lieut, RN 1957–60; Medical Officer, Wellcome Foundation 1960–64; Lecturer then Sr Lecturer, St Thomas's Hosp Medical School, London Univ 1964–70, Prof and Head, Clinical Epidemiology and Social Medicine Dept, St George's Hosp Medical School 1974–81; Tutor in Occupational Health, London School of Hygiene and Tropical Medicine 1966–68; Regius Prof of Medicine, Oxford Univ 1970–77; Sec, Epidemiology and Preventive Medicine Section, Royal Soc of Medicine 1972–76, Soc for Social Medicine 1972–77, Panel for Epidemiology and Social Medicine of European Communities 1975–79 and Asscn of Profs of Community and Social Medicine 1977–79; mem, Regional Nurse Training Cttee, Oxford Regional Health Bd 1974–76 and Council, Asscn of Clinical Profs 1977–79; Scientific Adviser, S Thames Cancer Registry 1976–81 and to Chief Scientist, Dept of Health and Social Security 1977–81; Ed-in-Chief, *International Journal of Epidemiology* and Pres, Int Epidemiological Asscn 1977–81; Dir of Health and Safety, Directorate-Gen V, Comm of European Communities 1981–87. *Present Position:* Head, Directorate A (Nuclear Safety, Waste Man, Prevention and Control of Pollution), Directorate-Gen XI (Environment, Consumer Protection and Nuclear Safety), Comm of European Communities. *Address:* 200 rue de la Loi, 1049 Brussels, Belgium. *Telephone:* (2) 235-11-11. *Telex:* 21877. *Fax:* (2) 235-01-22.

BERGMANS, Wim; Dutch environmentalist. *Present Position:* Head, IUCN Netherlands. *Address:* Plantage Middenlaan 2B, 1018 DD Amsterdam, Netherlands. *Telephone:* (20) 6261732. *Fax:* (20) 6279349.

BERKOW, Charles; Swedish environmentalist. *Present Position:* Chair, Jordens Vänner (FoE). *Address:* Fjällgatan 23A, 11628 Stockholm, Sweden. *Telephone:* (8) 702-20-17. *Fax:* (8) 642-62-61.

BICHEL, Prof Svend; Danish conservationist. *Present Position:* Pres, Danish Soc for Conservation of Nature. *Address:* Noerregade 2, 1165 Copenhagen K, Denmark.

BLAJAN, L; French UN official. *Career:* Int Office of Epizootics, Paris. *Present Position:* Adviser, UNEP 1992–. *Publi-*

cations: UNEP Environmental Data Report (Co-Author) 1992. *Address:* POB 30552, Nairobi, Kenya. *Telephone:* (2) 230800. *Telex:* 22068. *Fax:* (2) 226890.

BLANCO, José, M SC; Costa Rican biomass resource manager; b San José; m Vivienne Solís 1989; 1d. *Education:* Costa Rica Univ, Pennsylvania Univ. *Career:* Chief of Energy and Transportation, Ministry of Planning 1980; Tech Adviser, Ministry of Natural Resources 1988; Researcher, Centre for Tech Research (CISAT) 1990. *Present Position:* Regional Dir, Biomass Users Network (BUN) 1991–. *Publications:* Cooking in Rural Areas in Costa Rica 1988; The Importance of Bio-Energy 1990; Sustainable Use of Energy 1991. *Address:* POB 1800-2100, Guadalupe, Costa Rica. *Telephone:* 40-8997. *Fax:* 40-8998.

BLOOMFIELD, Michael I, M SC; Canadian wildlife biologist and environmental educator; b 1 Feb 1950, Cleveland, OH; s of Daniel and Frances Bloomfield; m Chris French 1971. *Education:* Ohio State Univ, Alberta Univ. *Career:* Regional wildlife biologist and provincial caribou specialist, Alberta Fish and Wildlife 1978–83; founder, Summer Inst for Environmental Values Educ, ECOmmunity and Green Works. *Present Position:* Founder and Exec Dir, Harmony Foundation of Canada 1985–. *Awards and Honours:* UNEP Global 500 1992; Environmental Leadership Award (Green Print). *Publications:* A Canadian Response to the Challenge of Sustainable Development (Ed) 1988; Home and Family Guide (Co-Author) 1989; The Workplace Guide to Practical Action for the Environment 1992. *Address:* 225–501 Metcalfe St, Ottawa, ON KZP 1P9, Canada.

BLUZMA, Dr Petras; Lithuanian zoologist, ecologist and scientist; b 1940, Pasvalys Dist; m Irena-Marijona Mauručaotė 1971; 1d. *Education:* Vilnius State Univ. *Career:* Scientist 1965–. *Present Position:* Head, Theriology Dept, Ecology Inst, Acad of Sciences. *Publications:* Ungulates of Lithuania (Co-Author) 1987; Fauna of Lithuania: Mammals (Co-Author) 1988. *Address:* 2600 Vilnius, Akademijos str 2, Lithuania. *Telephone:* (3702) 359-248. *Telex:* 303266. *Fax:* (3702) 359-257.

BODUROV, Dimo; Bulgarian environmentalist; b 1951, Kardjali; m 1977; 2d. *Career:* Traffic Police, Kardjali 1977–85; human rights org 1985–88; Chair, United Front for Ecological Salvation of Kardjali. *Present Position:* Head, Municipal Market, Kardjali 1988–. *Address:* 6600 Kardjali, 30 Republikanska Str, Bulgaria. *Telephone:* (361) 30007.

BOJKOV, Prof Rumen Dimiter, PH D, D SC; Canadian meteorologist; b 1 May 1931, Vellingrade, Bulgaria; m Anneliese Scheffler 1965; 1s. *Education:* Sofia Univ, Moscow Univ, Rostock Univ. *Career:* Sr Weather Forecaster, Sofia 1955–60; Assoc Prof, Sofia Univ 1961–65; Visiting Scientist, Boulder, CO 1966–68; Prof, New York State Univ, Albany 1968–69; Chief, Atmospheric Science Div, WMO 1971–83; Chief Scientist, Ozone Research Programme, Atmospheric Environment Service, Canada 1987–; Contributor, UNEP Environmental Data Report 1992. *Present Position:* Chief, Environment Div, WMO 1988–. *Address:* CP 2300, 41 ave Giuseppe Motta, 1211 Geneva 20, Switzerland. *Telephone:* (22) 7308455. *Telex:* 23260. *Fax:* (22) 7342326.

BOLIN, Prof Bert; Swedish meteorologist. *Present Position:* Climatologist, World Climate Change Study. *Address:* Meteorology Dept, 18451 Osterkar, Sweden. *Telephone:* (8) 15-77-31. *Fax:* (8) 15-71-85.

BONINGTON, Christian John Storey, CBE; British mountaineer, writer and photographer; b 6 Aug 1934; s of Charles and Helen Anne (Storey) Bonington; m Muriel Wendy Marchant 1962; 3s (1 deceased). *Education:* Royal Military Acad. *Career:* Royal Tank Regt 1956–61; Man Trainee, Unilever 1961–62; climbs include: Annapurna S Face Expedition 1970, British Everest Expedition 1972, Brammah, Himalayas 1973, Changabang, Himalayas 1974, British Everest Expedition (1st ascent, SW Face) 1975, Mt Vinson, Antarctica 1983, Everest summit 1985; Vice-Pres, British Mountaineering Council 1976–79, 1985–88, Pres 1988–91; Vice-Pres, Army Mountaineering Asscn 1980–; Pres, British Orienteering Fed 1985–, Council for Nat Parks 1992–; Fellow, Univ of Cen Lancashire. *Present Position:* Writer and photographer 1962–. *Awards and Honours:* Royal Scottish Geographical Soc Medal 1991. *Publications:* I Chose to Climb 1966; The Next Horizon 1973; Quest for Adventure 1981; The Everest Years 1986; Mountaineer 1989; The Climbers 1992. *Address:* Badger Hill, Nether Row, Hesket Newmarket, Wigton, Cumbria, UK. *Telephone:* (6998) 286.

BORN, Rubens Harry, M SC; Brazilian environmentalist; b 20 June 1955, São Paulo; s of Theodor and Ilse Graupe Born; m Gemima Cirilo Cabral Born 1982. *Education:* São Paulo Univ. *Career:* Environmental activist 1977–; Founder Mem and Co-ordinator, CEACON, Sobrevivencia and Ondazul NGOs 1978–; Founder, Brazilian NGOs Forum for UNCED 92, Int Rep to Int Facilitating Cttee on CFCs and UN 1990. *Present Position:* Co-ordinator, Vitae Civile devt, environment and peace inst 1989–; mem, Admin Bd, Fundação Ondazul 1992–. *Address:* CP 20635, São Paulo, SP 01498-970, Brazil. *Telephone:* (11) 869-9774. *Fax:* (11) 869-9774.

BOULAND, Patrick; Senegalese environmentalist. *Present Position:* Regional Rep, Regional Office for Western Africa, IUCN. *Address:* BP 3215, Dakar, Senegal. *Telephone:* 32-05-45. *Telex:* 3103. *Fax:* 32-92-46.

BOURDEAU, Philippe François Joseph, PH D; Belgian consultant and univ lecturer; b 25 Nov 1926, Rabat, Morocco; s of Michel and Lucienne (Imbrechts) Bourdeau; m Flora Gorirossi 1954; 3d. *Education:* Duke Univ, USA. *Career:* Teacher and researcher, N Carolina State Univ and Yale Univ 1954–62; Dir of Environmental and Energy Research, EURATOM and EC Comm 1962–91. *Present Position:* Special adviser, EC Comm and Head, Task Force for European Environmental Agency 1991–; Lecturer, Brussels Free Univ 1972–; Sec-Gen, ICSU (SCOPE). *Address:* 26 ave des Fleurs, 1150 Brussels, Belgium. *Telephone:* (2) 296-88-14. *Telex:* 21877. *Fax:* (2) 296-95-62.

BOUTA-ANGER, Anna; Swedish civil servant. *Present Position:* Mem, Swedish Environmental Protection Agency. *Address:* 171 85 Solna, Sweden.

BOWERS, Mary; Irish environmentalist. *Career:* Co-ordinator, Comhaontas Glas (Green Party) 1992. *Present Position:* Mem, Comhaontas Glas (Green Party). *Address:* 5A Upper Fownes St, Dublin 2, Ireland. *Telephone:* (1) 6771436. *Fax:* (1) 6771436.

BRANDENBURG, Jim; photographer. *Career:* Photographer for National Geographic; Dir Wolf Ridge, an environmental centre; founder Concerts for the Environment; created a photographic portfolio for the Nature Conservancy to inaugurate their prairie conservation programme. *Present Position:* Freelance photographer. *Awards and Honours:* has won numerous photography awards. *Address:* POB 787, Ely, Minnesota, MN 55731, USA.

BREVIK, T J; environmentalist. *Career:* Contrib, UNEP Environmental Data Report 1992. *Address:* c/o UNEP, POB 30552, Nairobi, Kenya. *Telephone:* (2) 230800. *Telex:* 22068. *Fax:* (2) 226890.

BRINKHORST, Laurens Jan. *Present Position:* Dir-Gen, Directorate-Gen XI (Environment, Nuclear Safety and Civil Protection), EC. *Address:* EC Environment, Civil Protection and Nuclear Safety Commission, 34 ave Belliard, 1049 Brussels, Belgium. *Telephone:* (2) 235-11-11. *Telex:* 21877. *Fax:* (2) 235-01-44.

BROWER, David Ross, D SC, PH D; American conservationist; b 1 July 1912, Berkeley, CA; s of Ross J and Mary Grace (Barlow) Brower; m Anne Hus 1943; 3s 1d. *Education:* California Univ, Maryland Univ, San Francisco Univ. *Career:* 1st ascents, Sierra Nevada 1933–41, Shiprock, NM 1939; conservation lecturer 1939–; Editorial Bd, Sierra Club 1935–69, Dir 1941–43, 1946–53; Ed, California Univ Press 1941–52; 1st Lieut, 10th Mountain Div, Infantry 1943–45; activist, conservation campaign, Dinosaur Nat Monument 1952–56, Grand Canyon 1952–68, N Cascades Nat Park 1955–68, Redwood Nat Park 1963–68, nuclear proliferation 1969–; Exec Dir, Sierra Club 1952–69, Vice-Pres 1971–; Founder, Trustees for Conservation 1954, Sec 1960–61, 1964–65, Vice-Pres and Trustee; Dir, Citizens Comm for Natural Resources 1955–78; mem, Natural Resources Council of America, Chair 1955–57; Dir, N Cascades Conservation Council 1957– and Rachel Carson Trust for Living Environment 1966–72, Conservation Expert 1973–; Founder, Sierra Club Foundation 1960; Vice-Pres, John Muir Inst for Environmental Studies 1968–72, Dir 1969–71; Founder, Steering Cttee, League of Conservation Voters 1969–, Amis de la Terre 1970; Founder Pres, FoE 1969–79 then Chair, Guarantor, FoE UK 1970–, Founder Pres, FoE Foundation 1972–79, then Chair; Chair, Earth Island Ltd, London 1971–74; mem, Bd, Environment Liasion Center, Nairobi 1975–, Nat Parks Asscn and Mountaineers. *Present Position:* Mem, Bd, FoE Foundation. *Awards and Honours:* California Conservation Council Award 1953; Nat Parks Asscn Award 1956; Audubon Naturalist Soc of Cen Atlantic States Award 1967; Sierra Club Award 1977. *Publications:* Only a Little Planet 1972; Song of the Earth Spirit 1973; Of All Things Most Yielding 1973; Celebrating the Earth (Ed) 1973; Guale: The Golden Coast of Georgia 1974; Micronesia: Island Wilderness 1975; Headlands 1976; At Home in the Wild: New England: White Mountains 1978; Wake of the World 1979; The Earth's Wild Places (Ed) 1979. *Address:* 530 7th St, SE, Washington, DC 20003, USA.

BROWN, Lester R, BS, MA, MPA; American founder and pres, non-profit research inst; b 28 March 1934; s of Calvin C and Delia (Smith) Brown; m Shirley Ann Woolington 1960 (divorced); 1s 1d. *Education:* Rutgers Univ, Maryland Univ, Harvard Univ. *Career:* Foreign agricultural policy adviser to Sec of Agric 1964–66; Admin, Int Agricultural Devt Service, Dept of Agric 1966–69; Co-Founder, Overseas Devt Council 1969; Founder, Worldwatch Inst 1974; bd mem, Green Seal; mem, Advisory Cttee, Inst for Int Economics and Advisory Council, Cttee for Nat Insts for Environment. *Present Position:* Pres, Worldwatch Inst. *Awards and Honours:* Nat Wildlife Fed Special Conservation Award 1982; UNEP Environmental Leadership Award 1982; UNEP Prize, WWF Gold Medal 1989; Environmentalist of the Year, Tokyo Jaycees 1992; Certificate of Special Recognition, Asscn of American Geographers 1993. *Publications:* Man, Land and Food 1963; World Without Borders 1972; By Bread Alone (Co-Author) 1974; Building a Sustainable Society 1981; The World Watch Reader (Ed) 1991; Saving the Planet: How to Shape an Environmentally Sustainable Global Economy (Co-Author) 1991. *Address:* Suite 701, 1776 Massachussetts Ave, NW, Washington, DC 20036, USA. *Telephone:* (202) 452-1999. *Fax:* (202) 296-7365.

BROWN, Noel J; int civil servant and environmentalist. *Present Position:* Dir, UNEP. *Address:* POB 30552, Nairobi, Kenya. *Telephone:* (2) 230800. *Telex:* 22068. *Fax:* (2) 226890.

BRUCE, James. *Present Position:* Pres, Organizing Cttee, World Climate Programme Summit, Geneva 1993. *Address:* c/o WMO, CP 2300, 41 ave Giuseppe Motta, 1211 Geneva 2, Switzerland. *Telephone:* (22) 7308111. *Telex:* 414199. *Fax:* (22) 7342326.

BRUNDTLAND, Gro Harlem, MD; Norwegian physician and politician; b 20 April 1939, Oslo; d of Gudmund and Inga Harlem; m Arne Olav Brundtland 1960; 3s (1 deceased) 1d. *Education:* Oslo Univ, Harvard Univ. *Career:* Consultant, Ministry of Health and Social Affairs 1965–67; Medical Officer, Oslo City Health Dept 1968–69; Deputy Dir, School Health Services, Oslo 1969; Minister of Environment 1974–79; Deputy Leader, Lab Party 1975–81, Leader, Parl Group 1981–92; mem, Parl Standing Cttee on Foreign Affairs, fmr mem, Parl Standing Cttee on Finance; Chair, UN World Comm on Environment and Devt; fmr Vice-Chair, Sr Secondary Schools Socialist Asscn, Students Asscn of Lab Party. *Present Position:* Prime Minister 1981, 1986–89, 1990–. *Awards and Honours:* Third World Prize for Work on Environmental Issues 1989. *Address:* Akersgt 42, POB 8001 Dep, 0030 Oslo, Norway. *Telephone:* (2) 22-34-90-90. *Telex:* 21565. *Fax:* (2) 22-34-95-00.

BRUSASCO-MACKENZIE, Margaret; int civil servant. *Present Position:* Head, Int Affairs Div, Directorate C (Environmental Instruments and Int Affairs), Directorate-Gen XI (Environment, Consumer Protection and Nuclear Safety), Comm of EC. *Address:* 200 rue de la Loi, 1049 Brussels, Belgium. *Telephone:* (2) 235-11-11. *Telex:* 21877. *Fax:* (2) 235-01-22.

BRYSON, Prof Reid A; American climatologist; b 7 June 1920, Detroit, MI; m Frances Edith Williamson 1942; 3s 1d. *Education:* Denison Univ, Chicago Univ. *Career:* Asst Prof of Geography and Meteorology, Wisconsin Univ 1946–48, Asst Prof of Meteorology 1948–50, Founder, Dept of Meteorology 1948, Assoc Prof of Meteorology 1950–56; Prof of Meteorology, Arizona Univ 1956–57; Prof of Meteorology, Wisconsin Univ 1957–86, Prof of Meteorology and Geography 1968–86, Emer Prof of Meteorology and Geography 1986–; Prof, Inst for Environmental Studies, Wisconsin Univ 1968, Dir 1970–85, Emeritus Prof 1986–. *Present Position:* Sr Scientist, Center for Climatic Research, Inst for Environmental Studies, Univ of Wisconsin 1986–. *Awards and Honours:* UNEP Global 500 1990; Foundation for Environmental Conservation Best Paper of the Year in Environmental Conservation 1990. *Address:* 1225 W Dayton St, Madison, WI 53706, USA. *Telephone:* (608) 262-2839. *Fax:* (608) 262-5964.

BUBLINEC, Eduard; Slovak soil scientist, ecologist and forester; b 1937, Komárno; m Anna Macová 1962; 2s 2d. *Education:* Zvolen Tech Univ. *Career:* Lecturer, School of Forestry 1960–63; Research Worker, Forestry Research Inst 1963–84; Head, Dept of Soil Ecology 1984–90; Asst Prof 1991. *Present Position:* Dir, Inst of Forest Ecology 1990– and Head, Dept of Natural Environment, Zvolen Tech Univ 1993–. *Awards and Honours:* Slovak Acad of Sciences Prize 1987. *Publications:* Industrial Emissions and Soils 1990. *Address:* Šturova 2, 960 53 Zvolen, Slovakia. *Telephone:* (855) 209-14. *Fax:* (855) 274-85.

BUCHMAN, Andrée; French environmentalist. *Present Position:* Spokesperson, Les Verts (Green Party). *Address:* 50 rue Benoît Malon, 94250 Gentilly, France. *Telephone:* (1) 49-08-91-31. *Fax:* (1) 49-08-97-11.

BUCHNER, Josef; Austrian politician and environmentalist. *Present Position:* Chair, Vereinte Grüne Österreichs (United Green Party of Austria). *Address:* 4020 Linz, Göthestr 9, Austria. *Telephone:* (732) 66-83-91. *Fax:* (732) 5-06-68.

BUCKINGHAM, David; environmentalist. *Present Position:* Mem, Commonwealth Environmental Protection Agency, Commonwealth Secr. *Address:* Marlborough House, Pall Mall, London SW1Y 5HX, UK. *Telephone:* (71) 839-3411. *Telex:* 27678. *Fax:* (71) 930-0827.

BULLOCK, P; British soil scientist. *Present Position:* Mem, UK Soils Survey Unit. *Address:* Silsoe Campus, Beds MK45 4DT, UK. *Telephone:* (525) 60428. *Fax:* (525) 61147.

BURHENNE-GUILMIN, F; environmentalist. *Career:* Contrib, UNEP Environmental Data Report 1992. *Address:* c/o IUCN, 28 rue Mauverney, 1196 Gland, Switzerland. *Telephone:* (22) 9990001. *Telex:* 419624. *Fax:* (22) 9990002.

BURKE, David Thomas (Tom); British environmental adviser; b 5 Jan 1947; s of J V and Mary (Bradley) Burke. *Education:* Liverpool Univ. *Career:* Dir, Green Alliance 1982–93; mem, Exec Cttee, Nat Council of Voluntary Orgs 1984–89, European Environmental Bureau 1985–91; mem, Council, Royal Soc of Arts 1990–92, Advisory Cttee, Business in Environment Task Team 1991–, Overseas Advisory Cttee, Save the Children Fund 1992–; founding mem, Council, Inst of Environmental Man. *Present Position:* Special adviser to Sec of State for Environment. *Awards and Honours:* UNEP Global 500 1993. *Publications:* Europe: Environment (Co-Author) 1981; Pressure Groups in the Global System (Co-Author) 1982; Ecology 2000 (Co-Author) 1984; The Green Capitalists (Co-Author) 1987; Green Pages (Co-Author) 1988; Ethics, Environment and the Company (Co-Author) 1990. *Address:* Room N14/13, 2 Marsham St, London SW1P 3EB, UK. *Telephone:* (71) 276-4299. *Fax:* (71) 276-3269.

BURLEY, Jeffrey, CBE, PH D; British forest geneticist and university lecturer; b 16 Oct 1936, Portsmouth; s of Jack and Eliza N V (Creese) Burley; m Jean Shirley Palmer 1961; 2s. *Education:* New Coll, Oxford, Yale Univ. *Career:* Officer-in-Charge, Forest Genetics Research Lab, Agric Research Council of Cen Africa 1965–69; Forest Geneticist, Unit of Tropical Silviculture, Oxford Univ 1969–76, Head, Dept of Forestry 1983–85. *Present Position:* Dir, Oxford Forestry Inst 1985–. *Address:* Dept of Plant Sciences, Oxford Univ, S Parks Rd, Oxford OX1 3RB, UK. *Telephone:* (865) 275050. *Telex:* 831. *Fax:* (865) 275074.

BURLEY, Lisa; environmentalist. *Present Position:* Mem, Regional Office for S America, IUCN. *Address:* c/o Casilla 17-17626, Avda de los Shyris 1240, Edif Albatros, Of 601, Quito, Ecuador. *Telephone:* (2) 430-974. *Fax:* (2) 430-974.

BURNETT, Sir John Harrison, D PHIL; British botanist and academic; b 21 Jan 1922, Ripon; s of Rev T H and Florence Hilda (Waite) Burnett; m Enid Margaret Bishop 1945; 2s. *Education:* Merton Coll, Oxford. *Career:* Fellow, Magdalen Coll, Oxford and Lecturer in Botany, Oxford Univ 1949–54; Prof of Botany, St Andrews Univ 1955–60, Newcastle-upon-Tyne Univ 1961–69; mem, Scottish Cttee, Nature Conservancy 1955–60, English Cttee 1961–66; Regius Prof of Botany, Glasgow Univ 1969–71; Sibthorpian Prof of Rural Economy and Fellow, St John's Coll, Oxford 1971–79; Principal and Vice-Chancellor, Edinburgh Univ 1979–87; mem, Scottish Cttee, Nature Conservancy Council 1982–87, Acting Chair 1987–90. *Present Position:* Chair, Co-ordination Comm for Biological Recording 1990–, Int Org for Plant Information (IOPI) 1991–; Head, Dept of Plant Sciences, Oxford Univ. *Publications:* The Vegetation of Scotland 1964; Fundamentals of Mycology 1968, 1976; Maintenance of the Biosphere (Co-Author) 1990; Surviving with the Biosphere (Co-Author) 1993. *Address:* 13 Field House Drive, Oxford OX2 7NT, UK. *Telephone:* (865) 513324. *Fax:* (865) 513324.

BURTON, Tony, M PHIL; British environmental campaigner; b 14 Feb 1964, Hobart. *Education:* St John's Coll, Cambridge, Univ Coll, London. *Career:* Campaigner on land use planning and transport issues. *Present Position:* Sr Planner, Council for Protection of Rural England 1988–. *Address:* Warwick House, 25 Buckingham Palace Rd, London SW1W 0PP, UK. *Telephone:* (71) 976-6433. *Fax:* (71) 976-6363.

BUSTAMENTE, Maria Ines; Ecuadorean environmentalist. *Present Position:* Mem, Ecuador Office, UNDP. *Address:* Foch 265 y Avda 6 de Diciembre, Casilla 17-03-4731, Sucursal 3, Quito, Ecuador.

BUSTAMANTE, Teodoro; Ecuadorean anthropologist; b 27 Sept 1953, Quito; m M Rocio Rosero 1973; 3s 1d. *Education:* Catholic Univ, Quito, Facultad Latinoamericana de Ciencias Sociales (FLACSO), Quito. *Career:* Social consultant 1975–80; Adviser on Indian Affairs, Ministry of Social Welfare 1980–85; Co-ordinator, Asociación Latinoamericana para los Derechos Humanos (ALDHU) 1982–85; Consultant, Rural Devt Programmes, Pichincha Provincial Council 1985–87; Dir, Pesticides Projects and Amazonian Issues, Fundación Natura (FNE) 1986–89; Co-ordinator of Interdisciplinary Studies, Facultad Latinoamericana de Ciencias Sociales (FLACSO) 1987–92. *Present Position:* Adviser, Fundación Natura (FNE) 1992–, Teacher, Facultad Latinoamericana de Ciencias Sociales (FLACSO), Quito and Facultad de Ciencias Humanas, Catholic Univ, Quito. *Publications:* La Larga Lucha del Kakaram contra el Sucre 1977. *Address:* Avda America 5653 y Vozandes, Casilla 17-01-253, Quito, Ecuador. *Telephone:* (2) 447-341. *Fax:* (2) 434-449.

BUTTLE, Eileen, PH D; British environmental admin; b 19 Oct 1937; d of George Ernest and Mary Stewart Linford; m Hugh Langley Buttle 1970. *Education:* Southampton Univ. *Career:* Research Fellow, Southampton Univ 1963–65; Research Scientist, Nat Inst of Research in Dairying 1965–71; Cattle Breeding Centre, Ministry of Agric and Fisheries 1971–76, Policy Admin 1976–89. *Present Position:* Chief Exec, Natural Environment Research Council. *Address:* Polaris House, N Star Ave, Swindon SN2 1EU, UK. *Telephone:* (793) 411500.

C

ČABOUN, Dr Vladimír; Slovak forestry engineer; b 9 Aug 1952, Zvolen; s of Kvetoslav Čaboun and Anna Čabounova; m Marta Kalmánekovǻ 1977; 1s. *Education:* Zvolen Tech Univ. *Career:* Man of Regional Environmental Protection 1975–77; Researcher, Dept of Ecology and Ecophysiology of Forest Tree Species, Zvolen Forest Research Inst 1977–. *Present Position:* Lecturer, Ecology Faculty, Zvolen Tech Univ 1989–. *Awards and Honours:* Man and Environment Prize for Young Researchers, Czechoslovak Acad of Sciences 1987. *Publications:* Environment and Forest Management 1986; Climatic and Therapeutic Influence and Characteristics of Forests 1989; Allelopathy in Forest Ecosystems 1990. *Address:* T G Masaryka 22, 960 92 Zvolen, Slovakia. *Telephone:* (855) 273-11. *Fax:* (855) 233-97.

CACCIA, Charles, MA; Canadian politician; b 28 April 1930, Milan, Italy. *Education:* Vienna Univ. *Career:* Mem Faculty of Forestry, Toronto Univ and Trade Analyst, Italian Trade Comm 1955–58; Founder, Caccia and Assocs 1958; Co-Founder and Pres, COSTI (Centro Organizzativo Scuoli Tecniche Italiane); mem, City of Toronto and Metro Council 1964–68; MP (Lib) for Davenport 1968–; Chair, ON Lib Caucus 1977–78, Canadian Parl Helsinki Group 1979 and External Affairs Sub-Cttee on CSCE; Minister of Lab 1981–83; Minister of Environment 1983–84; Opposition environment spokesman 1984–89; Founder, Parl Centre for Environmentally Sustainable Devt 1989. *Present Position:* Chair, Cttee on Sustainable Devt and mem, Priority and Planning Cttee, Nat Lib Caucus. *Awards and Honours:* UNEP Global 500 1985. *Address:* House of Commons, Room 353S, Centre Block, Ottawa, ON K1A 0A6, Canada. *Telephone:* (613) 992-2576. *Fax:* (613) 995-8202.

ČALA, Pavle; Croatian environmentalist. *Present Position:* Pres, Green Action Ecological Movt. *Address:* 59000 Šibenik, Roberta Vissiania 3, Croatia. *Telephone:* (59) 29097. *Fax:* (59) 29097.

CALDER, Nigel David Ritchie, MA; British science writer; b 2 Dec 1931; m Elisabeth Palmer 1954; 2s 3d. *Education:* Sidney Sussex Coll, Cambridge. *Career:* Physicist, Mullard Research Labs 1954–56; mem, Editorial Staff, *New Scientist* 1956–66; Science Corresp, *New Statesman* 1959–62, 1966–71, Science Ed 1960–62, Ed 1962–66; Chair, Asscn of British Science Writers 1962–64; Vice-Pres, Cruising Asscn 1981–84; mem, Initiative Group, Foundation Scientific Europe 1987–90. *Present Position:* Science writer, TV broadcaster. *Awards and Honours:* UNESCO Kalinga Prize for Popularization of Science 1972. *Publications:* Electricity Grows Up 1958; Robots 1958; Radio Astronomy 1958; The World in 1984 (Ed) 1965; The Environment Game 1967; Unless Peace Comes (Ed) 1968; Technopolis: Social Control of the Uses of Science 1969; The Violent Universe 1969; Living Tomorrow 1970; The Mind of Man 1970; The Restless Earth 1972; Nature in the Round: A Guide to Environmental Science (Ed) 1973; The Life Game 1973; The Weather Machine 1974; The Human Conspiracy 1976; The Key to the Universe 1977; Spaceships of the Mind 1978; Einstein's Universe 1979; Nuclear Nightmares 1979; The Comet is Coming! 1980; Timescale 1983; 1984 and After 1983; The English Channel 1986; The Green Machines 1986; Future Earth (Ed) 1988; Scientific Europe (Ed) 1990; Spaceship Earth 1991. *Address:* 8 The Chase, Furnace Green, Crawley RH10 6HW, UK. *Telephone:* (293) 526693.

CALDWELL, Prof Lynton Keith, PH D, LL D; American academic and environmental and science policy researcher; b 21 Nov 1913, Montezuma, IA; s of Lee Lynton and Alberta (Mace) Caldwell; m Helen W Walcher 1940; 1s 1d. *Education:* Chicago Univ, Harvard Univ, Western Michigan Univ. *Career:* Worked at Chicago Univ, Oklahoma Univ, Syracuse Univ and California Univ, Berkeley; served in Senate, Commerce Dept, Defense Dept, Interior Dept, Office of Tech Assessment, Nat Inst of Health, UN and UNESCO; participated in Nat Research Council, Nat Comm on Materials Policy, Sea Grant Advisory Bd, Science Advisory Bd, Int Joint Comm and Pacific Science Congress; mem, Bd of Governors, Nature Conservancy 1959–69; drafted Nat Environmental Policy Act 1969. *Awards and Honours:* UNEP Global 500 1991. *Address:* 4898D Heritage Woods Rd, Bloomington, IN 47401, USA. *Telephone:* (812) 336-5901.

CALOW, Prof Peter, PH D, D SC; British professor of zoology; b 23 May 1947; 1s 2d. *Education:* Leeds Univ. *Career:* Lecturer and Reader, Glasgow Univ 1972–84; Ed, *Functional Ecology* 1986–, *Integrated Environmental Management*; Dir, Inst of Environmental Sciences and Tech; Chair, Advisory Cttee on Hazardous Substances (UK Govt Statutory Cttee) 1991–. *Present Position:* Prof of Zoology, Sheffield Univ 1984–. *Address:* Dept of Animal and Plant Sciences, POB 601, Sheffield S10 2UQ, UK. *Telephone:* (742) 824627. *Fax:* (742) 780694.

CAMBOT, Guy; French politician and environmentalist. *Present Position:* Nat Sec, Les Verts (Green Party). *Address:* 50 rue Benoît Malon, 94250 Gentilly, France. *Telephone:* (1) 49-08-91-31. *Fax:* (1) 49-08-97-44.

CAMPEAU, Arthur H, QC, BA, BCL; Canadian lawyer and diplomatist. *Education:* McGill Univ, Montréal, PQ. *Career:* Special Adviser on Int Affairs to Minister of Environment 1989–; Head, Canadian del to negotiations on Convention on Biological Diversity 1992. *Present Position:* Amb and Personal Rep (Sherpa) of Prime Minister to UNCED 1992–. *Address:* c/o Dept of External Affairs, Lester B Pearson Bldg, 125 Sussex Drive, Ottawa, ON K1A 0G2, Canada. *Telephone:* (613) 995-1851. *Telex:* 0533745. *Fax:* (613) 996-9288.

CAMPREDON, Dr Pierre; French environmentalist; b 26 March 1951, Fès, Morocco; 2d. *Education:* Montpellier Univ. *Career:* Researcher into duck hibernation, Camargue 1977–83; Scientific Adviser, Banc d'Arguin Nat Park, Mauritania and researcher into terns 1984–89, Sec-Gen, Banc d'Arguin Int Foundation 1989–. *Present Position:* Rep, IUCN Guinea-Bissau 1989–. *Address:* Apdo 23, Bissau Codex 1031, Guinea-Bissau. *Telephone:* 201230. *Fax:* 201168.

CANADAY, Christopher, BS, MA; American conservation biologist; b 13 Aug 1961, CA. *Education:* Humboldt State Univ, Florida Univ. *Career:* Extensive experience with Amazonian fauna, protected areas and native peoples. *Present Position:* Consultant in conservation projects in NE Ecuador, especially Cuyabeno Reserve. *Publications:* La Avifauna Amazónica frente a la Actividad Humana. *Address:* Consejo Ecuatoriano para la Preservación de las Aves Cecia (CECIA), Casilla 17-17-906, Quito, Ecuador. *Telephone:* (2) 244-734. *Fax:* (2) 244-734.

CANNELL, Melvin, PH D, D SC, FRSE; British environmental researcher; b 12 Aug 1944; m 1966; 2d. *Education:* Reading Univ. *Career:* Research Officer, Coffee Research Station, Kenya 1966–71; mem, Natural Environment Research Council 1971–. *Present Position:* Head of Station, Inst of Terrestrial Ecology 1987–. *Address:* Edinburgh Research Station, Bush Estate, Penicuik, Midlothian EH26 0QB, UK. *Telephone:* (31) 445-4343. *Telex:* 72579. *Fax:* (31) 445-3943.

CANNON, Clare Virginia, M SC; Australian dir of environmental org and social researcher; b 14 Sept 1960; d of L Gordon and Margaret Darling; m Andrew Cannon 1989; 1s. *Education:* Melbourne Univ. *Career:* Dir, Earthwatch Australia 1984–88. *Present Position:* Chair, Earthwatch Australia 1991– and Dir, Victorian Community Foundation 1992–. *Address:* 28 Albany Rd, Toorak, Vic 3142, Australia. *Telephone:* (3) 822-2348. *Fax:* (3) 824-7915.

CARDENAS, Mario; Ecuadorean environmentalist. *Present Position:* Mem, Ecuador Office, FAO. *Address:* Avda Amazonas y Eloy Alfaro, Quito, Ecuador.

CARDY, W Franklin G, MA; Canadian geologist, environment man specialist; b 14 Aug 1937, Belfast, Northern Ireland; s of William M and Monica Cardy; m Kate Costelloe 1968; 1s 1d. *Education:* Keble Coll, Oxford. *Career:* Geologist, Geological Survey of Ghana 1960–62; Eng Geologist, Lower Swansea Valley Project 1963; Sr Hydrogeologist and Sr Engineer, Howard Humphreys UK 1965–71; Hydrogeologist, NB Water Authority, Canada 1971–72; Dir, Water Resources Branch, NB Environment 1973–87; Chair, Nat Research Council, Cttee on Hydrogeology, Canada 1983–86; Chief Tech Adviser, UNDP Water Project, Cape Verde 1986–88; Environment Man Specialist, World Bank, Washington 1988–92. *Present Position:* Deputy Asst Exec Dir, Environment Man and Dir, Desertification Control Programme Activity Centre, UNEP, Nairobi 1992–. *Publications:* Groundwater: The Overlooked Resource 1975; Land Use Policy: A Positive Approach 1983; Water Use Management Reform 1984; Africa's Urban Environment: Crisis in the Making 1991. *Address:* UNEP, DC/PAC, POB 30552, Nairobi, Kenya. *Telephone:* (2) 230800. *Fax:* (2) 215615.

CARRERA DE LA TORRE, Luis; Ecuadorean consultant and builder; b 30 Oct 1934, Quito; s of Luis Carrera Pazmiño and Imelda de la Torre Chávez; m 1st Gladys Montalvo 1962; m 2nd Francisca Pérez 1985; 1s 5d. *Education:* Cen Univ, Quito. *Career:* Univ Prof 1961–86; CEO, public and pvt insts 1968–; Consultant, OAS, World Bank, Netherlands 1970–; Pres, Latin American Asscn of Regional Devt Corpns 1977–79. *Present Position:* Exec Sec, Tratado de Cooperación Amazónica (TCA) 1990–; Pres, Ecuadorean Environmental Comm 1993–. *Publications:* Las Obras Hidraúlicas y la Supervivencia del Ecuador 1972; El Proyecto Binacional Puyango-Tumbes 1990; La Gestión Ambiental en el Ecuador 1993. *Address:* Edif Metrocar, 10 de Agosto 3560 y Ma de Jesus, 4°, Quito, Ecuador. *Telephone:* (2) 437-243. *Fax:* (2) 565-809.

CARSON, Dr David; British meteorologist, climatic modeller. *Present Position:* Dir, Hadley Centre, Meteorological Office. *Address:* London Rd, Bracknell, RG12 2SY, UK. *Telephone:* (344) 856653. *Fax:* (344) 854898.

CARTER, Jimmy (James Earl, Jr), BS; American farmer and politician; b 1 Oct 1924, Plains, GA; s of James Earl Carter Sr and Lillian Gordy; m Rosalynn Smith 1946; 3s 1d. *Education:* Georgia Southwestern Coll, Georgia Inst of Tech, US Naval Acad, Annapolis, MD. *Career:* US Navy 1946–53; peanut farmer and warehouseman, Carter Farms and Carter Warehouses, GA 1953–77; mem, School Bd, Sumter Co, GA 1955–62, Chair 1960–62; mem, Americus and Sumter Co Hosp Authority 1956–70, Sumter Co Library Bd 1961; Dir, GA Crop Improvement Asscn 1957–63, Pres 1961; Senator (Democrat), GA 1962–66, Gov 1971–74; Pres, Plains Devt Corpn 1963; Chair, W Cen GA Area Planning and Devt Comm 1964; Georgia Planning Asscn 1968; Chair, Congressional Campaign Cttee and Democratic Nat Cttee 1974; Pres, USA 1977–81; Prof, Emory Univ 1982–; Founder, Carter Presidential Center 1982; leader, int observer team, Panama 1989, Nicaragua, Dominican Repub and Haiti 1990; host, peace negotiations, Ethiopia 1989. *Present Position:* Chair, Bd of Trustees, Carter Center Inc 1986–. *Awards and Honours:* World Methodist Peace Award 1984; Albert Schweitzer Prize for Humanitarianism 1987; Matsunaga Medal of Peace 1993. *Publications:* Why Not the Best? 1975; A Government as Good as Its People 1977; Keeping Faith: Memoirs of a President 1982; The Blood of Abraham: Insights into the Middle East 1985; Everything to Gain: Making the Most of the Rest of Your Life 1987; An Outdoor Journal 1988; Turning Point: A Candidate, a State and a Nation Come of Age 1992. *Address:* 1 Copenhill Ave, NE, Atlanta, GA 30307, USA.

CASTELLO, Luis Manuel; Ecuadorean engineer and environmentalist. *Present Position:* S America Rep, IUCN. *Address:* Casilla 17-17-626, Avda de los Shyris 1240, Edif Albatros, Of 601, Quito, Ecuador. *Telephone:* (2) 430-974. *Fax:* (2) 430-974.

ČEŘOVSKÝ, Jan, PH D; Czech conservationist, ecologist and botanist; b 2 Feb 1930, Prague; s of Jan Čeřovský and Bozena B Brabcova; m Jarmila B Misikova 1976; 2s. *Education:* Charles Univ, Prague. *Career:* Ed-in-Chief, *ABC* scientific monthly for young people 1956–59; specialist, State Inst for Protection of Monuments and Nature Conservation (renamed Czech Inst for Nature Conservation 1990) 1959–69, Scientist 1973–; Educ Exec Officer, IUCN HQ, Switzerland 1969–73. *Present Position:* Sr Scientist, Czech Inst for Nature Conservation 1990–; Councillor, IUCN 1988–, Vice-Pres 1991–; mem, bd of dirs, Czech and Slovak ecological foundation, Pres 1992–. *Publications:* Fundamentals of Nature Conservation 1966; A Handbook for the Nature Conservationist (3 vols) (Co-Author) 1988; Environment and Culture (Co-Author) 1989; Trails to Nature (Co-Author) 1989. *Address:* IUCN, Slezská 9, 129 29 Prague 2, Czech Republic. *Telephone:* (2) 4172609. *Fax:* (2) 254555.

CHADWICK, Prof Michael John, MA, PH D; British prof of biology and dir of environmental inst; b 13 Sept 1934; s of Henry Chadwick (deceased) and Hilda Corman; m Josephine Worrall; 1s 2d. *Education:* Wales Univ, Downing Coll, Cambridge. *Career:* Lecturer, Khartoum Univ 1959–62; Lecturer, Sr Lecturer, Reader and Prof of Biology, York Univ 1966–91. *Present Position:* Dir, Stockholm Environment Inst 1991–. *Address:* POB 2142, 103 14 Stockholm, Sweden. *Telephone:* (8) 723-02-60. *Telex:* 19580. *Fax:* (8) 723-03-48.

CHARLES, Gabriel Lewis, MBE; St Lucian conservationist, forestry specialist; b 10 Feb 1936; s of James Luc Charles and Rosalind Lewis; m Margaret Lewis Charles 1962; 1s 3d. *Education:* Toronto Univ, Aberdeen Univ. *Career:* Forestry Asst, Forestry Dept 1961–73, Chief Forest Officer 1975–90; Chair and Founder, Environmental Comm 1977–; mem, Nat Trust Council 1975–; founder mem, Parks and Beaches Comm 1986–89, Caribbean Foresters Org 1981–; Nat Co-ordinator, Tropical Forest Action Plan 1990–. *Present Position:* Asst Dir of Conservation Educ, RARE Center for Tropical Conservation. *Awards and Honours:* UNEP Global 500 1988. *Publications:* St Lucia Environmental Profile 1988. *Address:* POB 1537, Castries, St Lucia. *Telephone:* 452-0864. *Fax:* 452-0864.

CHARTERS, Prof William W S (Bill); Australian solar energy researcher. *Present Position:* Dean, Eng Faculty, Melbourne Univ. *Address:* Parkville, Vic 3052, Australia. *Telephone:* (3) 344-6619. *Fax:* (3) 344-5031.

CHATURVEDI, Amar Nath, M SC; Indian energy researcher; b 5 Feb 1930; s of Hira Lal and A Chaturvedi; m Manorama Chaturvedi 1953; 1s 2d. *Education:* Agra Univ. *Career:* Uttar Pradesh Forest Dept 1955–67, seconded to Forest Research Inst and Coll, Dehradun 1967–73, seconded to Bhutan govt 1976–79; Consultant, FAO, Philippines 1987. *Present Position:* Sr Fellow, Tata Energy Research Inst 1986–. *Awards and*

Honours: Hari Om Ashram Award, Indian Council of Agricultural Research 1986. *Publications:* Forest Mensuration; Eucalyptus for Farming; Bamboos for Farming; Popular Farming in Uttar Pradesh. *Address:* 1st Floor, 101 Jor Bagh, New Delhi 110 003, India. *Telephone:* (11) 4622189. *Telex:* 3161593. *Fax:* (11) 4621770.

CHEN PEIXUN, Prof; Chinese scientist; b 5 July 1927; m 1956; 1d. *Education:* Hua Zhong Univ, Wuhan. *Career:* Asst, Agricultural Coll, Wuhan 1951–52; Lecturer and Assoc Prof, Inst of Hydrobiology, Chinese Acad of Sciences 1953–78, Prof 1978–. *Present Position:* Head, River Dolphin Research Dept 1978–; mem, IUCN 1987–. *Publications:* Fishes of the Yangtze River 1976; Baiji, A Rare Treasure 1992. *Address:* Wuhan 430072, Hubei, People's Repub of China. *Telephone:* (27) 723481. *Fax:* (27) 725132.

CHERKASOVA, Mariya; Russian ecologist. *Present Position:* Chair, Social-Ecological Union. *Address:* Moscow, POB 273, Russian Federation. *Telephone:* (95) 945-40-94. *Fax:* (95) 292-65-11.

CHURCH, Chris; British environmental biologist and independent environmental activist. *Career:* Environmental activist 1977–; worked for FoE; UK Co-ordinator, Earth Summit Tree of Life Project, Contributor, 'Earth Summit 92' report; Adviser, *One World* season, BBC TV; Founder, Environmental Law Foundation. *Present Position:* Devt Dir, Global Action Plan UK. *Publications:* Living in a Time of Change 1992. *Address:* c/o Wildlife Programme Dept, BBC TV, Wood Lane, London W12, UK. *Telephone:* (81) 743-8000.

CLARK, Howard; American civil servant. *Present Position:* Environmental Affairs Assessor, US Agency for Int Devt (USAID). *Address:* Avda Colombia 1573, Casilla 538, Quito, Ecuador.

COLLINS, Kenneth Darlingston (Ken), M SC; British politician; b 12 Aug 1939; s of Nicholas Collins and Ellen Williamson; m Georgina Frances Pollard 1966; 1s 1d. *Education:* Glasgow Univ, Strathclyde Univ. *Career:* Deputy Leader Lab Group, EP 1979–84, Socialist Spokesperson on Environment, Public Health and Consumer Protection 1984–89; mem, Socialist Int Environment Cttee, Advisory Cttee, EP Public Policy Inst, Warwick Univ, FoE (Scotland); Dir, Inst for European Environment Policy; Hon Vice-Pres, Royal Environmental Health Inst of Scotland, Inst of Environmental Health Officers, Int Fed on Environmental Health, European Food Law Asscn, European Consumer Safety Asscn, Socialist Environment and Resources Asscn, Energy Action Scotland, Nat Soc for Clean Air, Town and Country Planning Asscn. *Present Position:* Chair, Cttee on Environment, Public Health and Consumer Protection, EP 1989–. *Address:* 11 Stuarton Park, E Kilbride G74 4LA, UK. *Telephone:* (3552) 37282. *Fax:* (3552) 49670.

CORVALÁN, Osvaldo Bernabé; Argentinian environmentalist; b 6 April 1943, Santiago del Estero; s of Bernabé Corvalán and Eloísa Saganías; 2c. *Education:* Santiago del Estero Nat Univ. *Present Position:* Pres, Amigos de la Tierra Argentina 1992–. *Address:* Zapata 343, 1426 Buenos Aires, Argentina. *Telephone:* (1) 773-5947. *Fax:* (1) 331-4323.

COUSTEAU, Capt Jacques-Yves; French explorer, inventor, film director, writer and environmentalist; b 11 June 1910, St André de Cubzac, Gironde; s of Daniel and Elizabeth (Duranthon) Cousteau; m 1st Simone Melchior 1937 (deceased); m 2nd Francine Triplet 1991; 3s (1 deceased) 1d. *Career:* Served in Navy; acquired retd minesweeper 'Calypso' 1950, adapted for use as oceanographic vessel; Founder, Cousteau Soc 1973; Chair, Council for Rights of Future Generations, Paris 1993; mem, High-Level Advisory Bd on Sustainable Devt, UN, Geneva. *Present Position:* Pres, Cousteau Soc, USA 1973– and Équipe Cousteau, France 1981–. *Awards and Honours:* Palme d'Or, Cannes Film Festival 1956; Acad Award, Hollywood 1957, 1965; UN Int Prize for Environment 1977; UNEP Global 500 1988. *Publications:* Le monde des océans 1974; Calypso (Co-Author) 1978; Almanach Cousteau de l'environnement (Co-Author) 1981; La planète des baleines (Co-Author) 1989; Films: Silent World 1955; World without Sun 1963; Voyage to the Edge of the World 1975. *Address:* 233 rue du Faubourg St Honoré, 75405 Paris Cedex 08, France. *Telephone:* (1) 40-53-63-00. *Telex:* 641586. *Fax:* (1) 40-53-63-03.

CRANBROOK, Gathorne Gathorne-Hardy, Lord, PH D, FLLS, FZS, FRGS; British environmental admin; b 20 June 1933; s of Lord and Countess Cranbrook; m Caroline Jarvis 1967; 2s 1d. *Education:* Corpus Christi Coll, Cambridge, Birmingham Univ. *Career:* Asst, Sarawak Museum 1956–58; Fellow, Yayasan Siswa Lokantara, Indonesia 1960–61; Sr Lecturer in Zoology, Malaya Univ 1961–70; Ed, *Ibis* 1973–80; mem, Suffolk Coastal Dist Council 1974–83, Environment Sub-Cttee, House of Lords Select Cttee on EC 1979–85, 1987–90, Chair 1980–83, 1987–90, Royal Comm on Environmental Pollution 1981–, Nat Environmental Research Council 1982–88, Broads Authority 1988–, Harwich Haven Authority 1989–; Pres, Suffolk Trust for Nature Conservation 1979–; Non-Exec Dir, Anglian Water 1987–. *Present Position:* Chair, English Nature 1990–. *Publications:* Birds of the Malay Peninsula 1976; Mammals of Borneo 1977; Mammals of Malaya (Co-Author) 1978; Riches of the Wild: Mammals of South East Asia 1987; Key Environments: Malaysia (Ed) 1988. *Address:* Northminster House, Peterborough, Cambs, UK. *Telephone:* (733) 340345.

CRICKHOWELL, Roger Nicholas Edwards, Rt Hon Lord, PC, MA; British civil servant; b 25 Feb 1934, London; s of Ralph Edwards and Marjorie Ingham Brooke; m Ankaret Healing 1963; 1s 2d. *Education:* Trinity Coll, Cambridge. *Career:* MP (Cons) for Pembroke 1970–87; Sec of State for Wales 1979–87; Dir, Associated British Ports Holdings PLC and HTV Group PLC; Pres, Cardiff Coll, Wales Univ; mem, Cttee, Automobile Asscn. *Present Position:* Chair, Nat Rivers Authority. *Address:* 30 Albert Embankment, London SE1 7TL, UK. *Telephone:* (71) 820-0101. *Fax:* (71) 582-7840.

CRUTZEN, Prof Paul J, PH D; German meteorologist; b 3 Dec 1933, Amsterdam; s of Josef C Crutzen and Anna Gurk; m Terttu Soininen 1958; 2d. *Education:* Stockholm Univ. *Career:* Asst, Stockholm Univ 1973–74; Scientific Dir, NCAR, Boulder, CO, USA 1974–77, Research Dir 1977–80; mem, Max Planck Foundation. *Present Position:* Dir, Max Planck Inst 1980–. *Awards and Honours:* Scientist of the Year 1984; Tyler Prize 1989. *Publications:* Environmental Consequences of Nuclear War 1985; Schwarzer Himmel 1986. *Address:* Saarstr 23, POB 3060, 6500 Mainz, Germany.

D

DADE, Drita; Albanian interpreter; b 4 Aug 1968; d of Drita Zefloka and Prena Loka; m Shyqyri Dade 1989; 1d. *Education:* Tirana Univ. *Present Position:* Head of Information and Public Awareness Cttee of Environmental Protection. *Address:* c/o Ministry of Health and Environmental Protection, Ministria e Shendetesisë, Tirana, Albania. *Telephone:* (42) 27904. *Fax:* (42) 27907.

DANCHEV, Jordan, PH D; Bulgarian ecologist; b 9 April 1939, Assenovgrad; s of Kiril Danchev and Anastasija Dancheva; m Tiha Dancheva 1968; 2s. *Education:* Veterinary Univ, Sofia. *Career:* Veterinary surgeon 1964–69; work at Veterinary Univ 1969–72, Asst 1972–75; scientist, Ecology Inst, Acad of Sciences 1975–79, Asst 1979. *Present Position:* Assoc Prof, Ecology Inst, Acad of Sciences 1988– and Pres, Soc for Conservation of Rhodope Mountains 1990–. *Publications:* Bulgarian Domestic Animals 1984. *Address:* 2 Gagarin St, 1113 Sofia, Bulgaria. *Telephone:* (2) 70-51-78. *Fax:* (2) 70-54-98.

DAVIS, R M; American scientist. *Career:* Contrib, Environmental Data Report, UNEP 1992. *Present Position:* Mem, Center for Disease Control. *Address:* Rockville, MD, USA.

DE AMORIM, Maria; Mexican int civil servant and environmentalist. *Present Position:* Dir, Latin America and Caribbean Regional Office and Regional Rep, UNEP. *Address:* Pres Mazaryk 29, 5°, 50 Piso Col Chaputlepec Morales, 11570 México, DF, Mexico. *Telephone:* (5) 203-4579. *Telex:* 1771055. *Fax:* (5) 203-4465.

DE HAEN, Dr H; int civil servant. *Career:* Mem, programme on climate change and agric, FAO. *Present Position:* Asst Dir-Gen, Agric Dept, FAO. *Address:* Via delle Terme di Caracalla, 00100 Rome, Italy. *Telephone:* (6) 57971. *Telex:* 610181. *Fax:* (6) 57973152.

DE HAES, Charles; Belgian environmentalist; b 5 April 1938, Antwerp; m Elizabeth Keindl 1982; 1s. *Education:* Capetown Univ, S Africa, Northwestern Univ, USA. *Career:* Rupert Int Group 1963; Chief Exec for E Africa, Rothman's Int Group 1964; Personal Asst to Pres, WWF 1971, Jt Dir-Gen 1975. *Present Position:* Dir-Gen, WWF 1977–. *Address:* World Conservation Centre, ave du Mont-Blanc, 1196 Gland, Switzerland. *Telephone:* (22) 649111. *Telex:* 419618. *Fax:* (22) 644238.

DE JANOSI, Peter; Austrian scientist and environmentalist. *Present Position:* Dir, Int Inst for Applied Systems Analysis (IIASA). *Address:* 2361 Laxenburg, Austria. *Telephone:* (2236) 71521599. *Fax:* (2236) 71313.

DECROLY, Vincent; Belgian politician and environmentalist; b 1963, Charleroi. *Education:* Liège Univ. *Present Position:* Fed Sec, Ecolo (French-speaking green party) 1992–. *Address:* 8 rue du Séminaire, 5000 Namur, Belgium. *Telephone:* (81) 22-78-71. *Fax:* (81) 23-06-03.

DEL VALLE, Guillermo; Argentinian environmentalist. *Present Position:* Pres, Asociación para la Difusión de los Programas de las Naciones Unidas para el Medio Ambiente (ADNUMA). *Address:* Avda 27 de Abril 864, Dirección de Visita 835, 5000 Córdoba, Argentina.

DELGADO, José Gregorio; Venezuelan environmentalist. *Present Position:* Mem, Amigos del Ambiente (Friends of Environment). *Address:* Apdo 47272, Of Ip Los Chaguaramos, Calle 66-A con Avda las Delicias, Edif Alcarava VII, Apt 7B, Los Alcaravenez, Mara, Caracas, Venezuela.

DELL'ALBA, Gianfranco; Italian politician and environmentalist. *Present Position:* Deputy Sec-Gen, Green Group, EP. *Address:* Bâtiment Remorqueur, 79 rue Belliard, 1040 Brussels, Belgium. *Telephone:* (2) 284-30-45. *Telex:* 26999. *Fax:* (2) 284-78-37.

DI CARPEGNA, Ranieri; Italian int civil servant. *Present Position:* Dir, Directorate C (Environmental Instruments and Int Affairs), Directorate-Gen XI (Environment, Consumer Protection and Nuclear Safety), Comm of EC. *Address:* 200 rue de la Loi, 1049 Brussels, Belgium. *Telephone:* (2) 235-11-11. *Telex:* 21877. *Fax:* (2) 235-01-22.

DI CASTRI, Francesco; int civil servant and environmentalist. *Present Position:* Mem, UNESCO. *Address:* 7 place de Fontenoy, 75700 Paris, France. *Telephone:* (1) 45-68-10-00. *Telex:* 204461. *Fax:* (1) 45-67-16-90.

DIEZ de RIVERA, Carmen; Spanish politician; b 29 Aug 1942. *Education:* Complutense Univ, Madrid. *Career:* Head of Int Relations, Spanish TV; Dir, Cabinet of Pres 1976; mem, Council, Unión Europea de Radiodifusión; Founder, Movimiento por el Desarrollo, la Paz y la Libertad and Entente Ecológica Europea; mem, PSOE. *Present Position:* MEP for Balearic Islands and Menorca 1987–, mem, Socialist Group, Inter-Parl Del for Relations with Israel, Cttee on Environment, Public Health and Consumer Protection. *Awards and Honours:* Federación Amigos de la Tierra Prize 1988. *Publications:* Nuevos Aires para Europa 1990. *Address:* 97-113 rue Belliard, 1040 Brussels, Belgium. *Telephone:* (2) 284-58-95. *Fax:* (2) 284-98-95.

DIOUHA, Jana; Czech environmentalist. *Present Position:* Mem, Green Circle. *Address:* Prague, Czech Republic.

DOMOTO, Akiko, BA; Japanese politician and business exec; b 31 July 1932, CA, USA. *Career:* Reporter, Camera Operator, Newscaster and Dir, Tokyo Broadcasting System 1959–89; Bd Mem, Family Planning Fed; Vice-Pres, Global Legislators Org for a Balanced Environment (GLOBE) Japan and Chair, Biodiversity Working Group 1991–; Co-Founder, Japan Women's Global Environment Network Int (GENKI) 1992–. *Present Position:* Mem, House of Councillors, Japanese Diet 1989–. *Publications:* TV Documentaries: Baby Hotel 1981; Tibetan New Year 1984; The Phantom Kingdom 1984; Monitoring Japan's ODA 1987; The Age of Child Slavery 1989. *Address:* 2-1-1 Nagata-cho, Chiyoda-ku, Tokyo 100, Japan. *Telephone:* (3) 3508-8422. *Fax:* (3) 3506-8085.

DONOVAN, Richard Zell, MS; American forestry specialist; b 6 Aug 1952; s of Lawrence and Berniece Donovan; m Karen Ann Alfonsi 1982; 1s 1d. *Education:* S Florida Univ. *Career:* Peace Corps volunteer, Paraguay 1975–77; logger, photographer and training consultant 1978–81; natural resource consultant 1982–86; Sr Fellow and Project Dir, WWF, Costa Rica 1987–91. *Present Position:* Dir, Timber Project and Smart Wood, Rainforest Alliance 1992–. *Address:* Suite 512, 270 Lafayette St, New York, NY 10012, USA. *Telephone:* (802) 899-1383. *Fax:* (802) 899-2018.

DORJI, Tandi; Bhutanese civil servant. *Present Position:* Communications Officer, Nat Environment Comm. *Address:* POB 466, Thimphu, Bhutan. *Telephone:* 23384. *Fax:* 23385.

DORJI, Paljor; Bhutanese diplomatist and environmentalist; b 24 Aug 1943, Kalimpong, India; s of Jigme Palden Dorji (deceased) and HRH Ashi Tessla Dorji; m 1st Ashi Louise Dorji 1966; m 2nd Ashi Dechen Chhoden; 4s (3 from 1st marriage) 1d. *Career:* Founder, Royal Soc for Protection of Nature, Pres 1989–91. *Present Position:* Amb and Perm Rep to UN, Geneva 1991–. *Awards and Honours:* UNEP Global 500 1988. *Publications:* Film: On the Wings of Prayer. *Address:* 17-19 chemin Champ d'Anier, 1209 Geneva, Switzerland. *Telephone:* (22) 7987971. *Telex:* 415447. *Fax:* (22) 7882593.

DOTTO, Lydia Carol; Canadian science writer and ed; b 29 May 1949; d of August and Assunta (Paron) Dotto. *Education:* Carleton Univ, Ottawa. *Career:* Science writer, *Toronto Globe and Mail* 1972–78; Pres, Canadian Science Writers Asscn 1979–80; Exec Ed, Canadian Science News Service 1982–92. *Present Position:* Freelance science writer and ed. *Publications:* Ozone War 1978; Planet Earth in Jeopardy 1986; Canada in Space 1987; Thinking the Unthinkable: Social Consequences of Rapid Climate Change 1988; Asleep in the Fast Lane 1990; Blue Planet 1991; Ethical Choices and Global Climate Warming 1993; The Astronauts: Canada's Voyagers in Space 1993; Yucca Mountain Socioeconomic Study 1993. *Address:* Unit 303, 2650 Marsdale Drive, Peterborough, ON L9L 1Y1, Canada. *Telephone:* (705) 741-1476. *Fax:* (705) 741-1476.

DOW, Douglas Morrison; British civil servant; b 1 July 1935; s of George Torrance Dow and Grace Morrison MacFarlane; m Felicity Margaret Mona Napier 1959; 2s. *Career:* Served in Navy 1952–92. *Present Position:* Dir, Nat Trust for Scotland 1992–. *Address:* The National Trust for Scotland, 5 Charlotte Sq, Edinburgh EH2 4DU, UK. *Telephone:* (31) 226-5922. *Fax:* (31) 220-6266.

DOWDESWELL, Elizabeth; Canadian int civil servant. *Career:* Deputy Environment Minister 1992. *Present Position:* Exec Dir, UNEP 1993–. *Address:* POB 30552, Nairobi, Kenya. *Telephone:* (2) 230800. *Telex:* 22068. *Fax:* (2) 226890.

DRAMMEH, Halifa; int civil servant. *Present Position:* Sr Environmental Affairs Officer, Regional Office for Africa, UNEP. *Address:* POB 30552, Nairobi, Kenya. *Telephone:* (2) 230800. *Telex:* 22068. *Fax:* (2) 226890.

DRDOŠ, Ján, D SC; Slovak environmentalist and geographer; b 6 Jan 1934, Víglaš; s of Jozef Drdoš and Amália Drdošová; m Oldřiška Drdošová 1963. *Education:* Comenius Univ and 17 Nov Univ, Bratislava. *Career:* Research Asst, Inst of Nature Protection 1958–63; Sr Researcher, Inst of Landscape Ecology 1963–76, Inst of Geography 1976–; mem, IUCN Comm on Environmental Strategy 1972–; Pres, Slovak Geographical Soc 1978–88; Sec, IGU Working Group on Landscape Synthesis 1980–88. *Present Position:* Vice-Chair, Nat Cttee of Geography and Assoc Prof of Landscape Ecology, Zvolen Tech Univ. *Awards and Honours:* Slovak Acad of Sciences Silver Medal 1984. *Publications:* Landscape Synthesis (Ed) 1984; Sagarmatha National Park (Co-Author) 1987. *Address:* Inst of Geography, Slovak Acad of Sciences, Stefánikova 49, 814 73 Bratislava, Slovakia. *Telephone:* (7) 492-751. *Fax:* (7) 491-340.

DRURY, Dr David; British scientist. *Present Position:* Dir, British Antarctic Survey. *Address:* High Cross, Madingley Rd, Cambridge CB3 0ET, UK. *Telephone:* (233) 61188.

DRUZHININ, Igor P, PH D; Russian hydrology and hydropower specialist; b 10 Feb 1929; m Larisa Ya Druzhinina; 1d. *Education:* Moscow Energy Inst. *Career:* Researcher, Dept of Hydrological Problems, Acad of Sciences 1952–57; Lab Head, Inst of Energy and Water Economy, Kirghiz Acad of Sciences 1957–61; Deputy Dir, Siberian Energy Inst, Acad of Sciences 1961–87. *Present Position:* Dir, Inst of Water and Ecological Problems, Acad of Sciences 1987– and Chair, Khabarovsk Science Centre 1991–. *Publications:* Long-Term Prognosis and Information 1987; Life Input and Ecology 1990; Water and Land Resources of a Country: Optimization of Use 1992. *Address:* Inst of Water and Ecological Problems, Far E Branch, Acad of Sciences, 680063 Khabarovsk, 65 Kim Yu Chen St, Russian Federation. *Telephone:* (7421) 33-38-49.

DUNKEL, Arthur, B SC; Swiss int civil servant; b 28 Aug 1932, Lisbon, Portugal; m Christiane Muller-Serda 1957; 1s 1d. *Education:* Lausanne Univ. *Career:* Fed Office for Foreign Econ Affairs 1956, Head, OECD Section 1960–64, Developing Countries Section 1964–71, World Trade Policy Section 1971–73; Vice-Chair, UNCTAD Inter-Governmental Group on Supplementary Financing 1968, Rapporteur, UNCTAD Bd 1969; Chair, GATT Cttee on Balance of Payments Restrictions 1972–75; Perm Rep to GATT 1973–76; Asst Lecturer, Journalism Inst, Fribourg Univ 1974–; Fed Council Del for Trade Agreements and Amb 1976–80; Chair, UN Conf on New Wheat Agreements 1978; Sr Lecturer, Econ Social Sciences Faculty, Geneva Univ 1983–. *Present Position:* Dir-Gen, GATT 1980–. *Address:* Centre William Rappard, 154 rue de Lausanne, 1211 Geneva 21, Switzerland. *Telephone:* (22) 7395111. *Telex:* 412324. *Fax:* (22) 7395458.

DURRELL, Gerald Malcolm, OBE, FRGS, FZS; British zoologist and writer; b 7 Jan 1925, Jamshedpur, India; s of Lawrence Samuel and Louisa Florence (Dixie) Durrell; m 1st Jacqueline Sonia Rasen 1951 (divorced); m 2nd Lee Wilson McGeorge 1979. *Career:* Student Keeper, Whipsnade Park 1945–46; zoological collecting expeditions: British Cameroons 1947, 1948, 1956, British Guiana 1949, Argentina and Paraguay 1953, Trans-Argentine 1958, NZ, Australia and Malaysia 1961, Sierra Leone 1964, Mexico 1968, 1979, Mauritius 1976, 1977, Assam, India 1978, Madagascar 1978, Mauritius and Madagascar 1981; Founder, zoo, Jersey 1959, Wildlife Preservation Int 1973, Wildlife Preservation Trust Canada 1985; Fellow, Int Inst of Arts and Letters and Inst of Biology; mem, British Ornithologists Union. *Present Position:* Founder, Jersey Wildlife Preservation Trust 1964–. *Publications:* The Overloaded Ark 1952; Three Singles to Adventure 1953; The Bafut Beagles 1953; The Drunken Forest 1955; My Family and Other Animals 1956; Encounters with Animals 1959; A Zoo in My Luggage 1960; The Whispering Land 1962; Menagerie Manor 1964; Two in the Bush 1966; Rosy is My Relative 1968; Birds, Beasts and Relatives 1969; Fillets of Plaice 1971; Catch Me a Colobus 1972; Beasts in My Belfry 1973; The Stationary Ark 1976; Golden Bats and Pink Pigeons 1977; The Garden of the Gods 1978; The Picnic and Suchlike Pandemonium 1979; The Mockery Bird 1981; Amateur Naturalist 1982; How to Shoot an Amateur Naturalist 1984; Durrell in Russia 1986; Marrying Off Mother and Other Stories 1991; The Aye-Aye and I 1992; Children's Books: The New Noah 1956; Island Zoo 1961; Look at Zoos 1961; My Favourite Animal Stories 1963; The Donkey Rustlers 1968; The Talking Parcel 1974; The Fantastic Flying Journey 1987; The Fantastic Dinosaur Adventure 1989; Best Dog Stories (Ed) 1990. *Address:* Les Augres Manor, Trinity, Jersey JE3 5BF, UK. *Telephone:* (534) 864666.

DYANKOVA, Elena Detecheva, MS; Bulgarian forester; b 30 June 1957, Isperih; d of Detcho Dyankov and Ivanka Dyankova; m Ivo Simeonov Nedyalkov; 1d. *Education:* Higher Institute of Forestry and Forest Industries (HIFFI), Sofia. *Career:* *New Ludogorie* newspaper 1980–85. *Present Position:* Researcher, Forest Research Institute 1985–. *Publications:* Biological Cycling of Minerals in Temperate Deciduous Forest Ecosystems in Northeastern Bulgaria; Throughfall and Stem-Flow Chemistry in a Spruce Forest on Rila Mountain. *Address:* Forest Research Institute of the Bulgarian Academy of Sciences, 1756 Sofia, 132 Kliment Ohridski Blvd, Bulgaria. *Telephone:* (2) 62-29-61. *Fax:* (2) 62-29-65.

E

EDINBURGH, HRH Prince Philip, Duke of, KG, KT, OM, GBE; British royal and environmentalist; b 10 June 1921, Corfu, Greece; s of HRH Prince Andrew of Greece and Denmark and HRH Princess (Victoria) Alice Elizabeth Julia Marie; m HRH Princess Elizabeth (now HM Queen Elizabeth II); 3s 1d. *Education:* Royal Naval Coll, Dartmouth. *Career:* Mediterranean Fleet and British Pacific Fleet, SE Asia and Pacific 1939–45; renounced right of succession to Throne of Greece and Denmark, naturalized British subject, adopting surname Mountbatten 1947; Personal ADC to HM King George VI 1948–52; Chancellor, Wales Univ 1948–76, Edinburgh Univ 1952–, Cambridge Univ 1977–, Salford Univ; PC 1951–; Pres English-Speaking Union of Commonwealth and RSA 1952–, Commonwealth Games Fed 1955–, Royal Agricultural Soc of Commonwealth 1958–, BMA 1961–81, Council for Nat Academic Awards 1965–75, Scottish Icelandic Asscn 1965–, Maritime Trust 1969–79, Nat Council of Social Service 1970–73, Australian Conservation Foundation 1971–76, David Davies Memorial Inst of Int Studies 1979–; Admiral of Fleet, Field Marshal, RAF 1953–, RNZAF and NZ Army 1978; Patron and Chair of Trustees, Duke of Edinburgh's Award Scheme 1956–; PC, Canada 1957–; Visitor, RCA 1967–. *Present Position:* Pres, WWF Int 1981–. *Publications:* Birds from Britannia 1962; Wildlife Crisis (Co-Author) 1970; The Environmental Revolution: Speeches on Conservation 1962–77 1978; Men, Machines and Sacred Cows 1984; Down to Earth 1988; Living off the Land 1989. *Address:* Buckingham Palace, London SW1, UK.

EDWARDS, Gordon Douglas, MS, MA, PH D; Canadian mathematics teacher and consultant; b 18 July 1940, Landsdowne, ON; s of Alan Edwards and Mary Coghlan; m Karen Rogers 1966; 3s. *Education:* Toronto Univ, Chicago Univ, IL, USA, Queen's Univ, Kingston, ON. *Career:* Ed, *Survival* int ecology magazine 1970–74; Prof of Mathematics, Vanier Coll, Montréal 1974–; Dir, Green Energy Conf, Montréal 1989. *Present Position:* Founder and Pres, Canadian Coalition for Nuclear Responsibility 1975–. *Awards and Honours:* White Owl Conservation Award 1976. *Publications:* Nuclear Waste: What, Me Worry? 1978; The Myth of the Peaceful Atom: Canada's Nuclear Dilemma 1983. *Address:* 53 Dufferin Rd, Hampstead, PQ H3X 2X8, Canada. *Telephone:* (514) 489-2665. *Fax:* (514) 489-5118.

EHRLICH, Paul Ralph, PH D; American population biologist; b 29 May 1932, Philadelphia, PA; s of William and Ruth (Rosenberg) Ehrlich; m Anne Fitzhugh Howland 1954; 1d. *Education:* Pennsylvania Univ, Kansas Univ. *Career:* Assoc Investigator, USAF Research Project, Alaska and Kansas Univ 1956–57; Research Assoc, Acad of Sciences, Chicago and Entomology Dept, Kansas Univ 1957–59; Stanford Univ 1959–, Prof of Biology 1966–; Pres, Centre for Conservation Biology 1988–; Correspondent, NBC News 1989–; Fellow, American Acad of Arts and Sciences and NAS. *Present Position:* Prof of Population Studies, Biological Sciences Dept, Stanford Univ 1976–. *Publications:* How to Know the Butterflies 1961; Population Resources; The Population Bomb 1971; How to be a Survivor (Co-Author) 1971; Man and the Ecosphere (Co-Ed) 1971; Global Ecology (Co-Author) 1971; Environment (Co-Author) 1972; Human Ecology (Co-Author) 1973; Ark II (Co-Author) 1974; The Process of Evolution (Co-Author) 1974; The End of Affluence (Co-Author) 1974; Biology and Society (Co-Author) 1976; The Race Bomb (Co-Author) 1977; Ecoscience: Population, Resources, Environment (Co-Author) 1977; Introduction to Insect Biology and Diversity (Co-Author) 1978; The Golden Door: International Migration, Mexico and the US (Co-Author) 1979; Extinction: The Causes and Consequences of the Disappearance of Species (Co-Author) 1981; Machinery of Nature 1986; Earth (Co-Author) 1987; The Birder's Handbook (Co-Author) 1988; New World, New Mind (Co-Author) 1989; The Population Explosion (Co-Author) 1990; Healing the Planet (Co-Author) 1991; Birds in Jeopardy 1992. *Address:* Stanford, CA 94305-2493, USA. *Telephone:* (415) 723-3171.

EL-ASHRY, Mohamed, PH D; American geologist and environmental scientist; b 21 Jan 1940, Cairo; s of Taha el-Ashry and Faika Fadda; m Patricia R Murphy 1962; 2d. *Education:* Cairo Univ, Illinois Univ. *Career:* Asst Prof of Geology, Cairo Univ 1966–69, Prof of Environmental Sciences 1969–79; Sr Scientist, Environmental Defense Fund 1975–79; Dir of Environment, TN Valley Authority 1979–83; Sr Vice-Pres, World Resources Inst 1983–91. *Present Position:* Chief Environmental Adviser to Pres and Dir of Environment, World Bank 1991–. *Awards and Honours:* Air Conservationist of the Year 1983; UNCED Honour for Outstanding Contribution to Earth Summit 1992. *Publications:* Air Photography and Coastal Problems 1977; Water and Arid Lands of the Western United States 1988; Air Pollution's Toll on Forests and Crops 1989. *Address:* 1818 H St, Washington, DC 20433, USA. *Telephone:* (202) 473-3202. *Telex:* 248423. *Fax:* (202) 477-0565.

ELIÁŠ, Pavol, PH D; Slovak ecologist and environmental botanist; b 19 Oct 1949, Velčice; m 1972; 3s. *Education:* Comenius Univ, Bratislava. *Career:* Fellow, Botany Inst, Slovak Acad of Sciences 1972–74, Inst of Experimental Biology and Ecology 1975–79; presented *Ecology* exhibition 1984. *Present Position:* Chair, Nat Cttee, SCOPE, Scientific Sec, Cttee for Environment, Presidium, Slovak Acad of Sciences and Sr Research Scientist, Botany Inst 1990–. *Awards and Honours:* Agrokomplex Nitra Award 1984. *Publications:* A Dictionary of Ecology and the Environment (Co-Author) 1993; Principles of Ecology. *Address:* Dúbravská 14, 842 23 Bratislava, Slovakia. *Telephone:* (7) 373-294. *Fax:* (7) 371-948.

ELKINGTON, John Brett, M PHIL, FID, FRSA; British environmental consultant and writer; b 23 June 1949; 2d. *Education:* Essex Univ, Univ Coll, London. *Career:* Regular contrib on environmental, energy and devt issues, *New Scientist* 1975–78; columnist, *Guardian* 1989–92; Sr Planner and Assoc, Transport and Environment Studies; Co-Founder, Environmental Data Services Ltd (ENDS) 1978, Man Dir 1981–83, Founder Ed *ENDS Report*; independent environmental consultant 1983–88; Non-Exec Dir, Asscn of Environmental Consultancies; mem, Council, Inst of Environmental Managers, Royal Town Planning Inst, Strategic Planning Soc, Council, Environmental Law Foundation; Patron, New Economics Foundation; Ed, *Biotechnology Bulletin*. *Present Position:* Co-Founder and Dir, SustainAbility 1987–. *Awards and Honours:* Hon Fellow, Centre for Social and Environmental Accounting Research (CSEAR), Dundee Univ; UNEP Global 500 1989. *Publications:* The Green Consumer Guide; The Green Capitalists; The Green Business Guide; The Corporate Environmentalists; Holidays that Don't Cost the Earth; Coming Clean. *Address:* People's Hall, 91-97 Freston Rd, London W11 4BD, UK. *Telephone:* (71) 243-1277. *Fax:* (71) 243-0364.

ELOY, Antonio; Portuguese environmentalist. *Present Position:* Dir, Amigos da Terra (FoE). *Address:* Travessa da Laranjeira 1A, 1200 Lisbon, Portugal. *Telephone:* (1) 470788. *Fax:* (1) 473586.

ELTON, Benjamin Charles (Ben), BA; British comedian and writer; b 3 May 1959; s of Prof Lewis Richard Benjamin Elton. *Education:* S Warwicks Coll of Higher Educ, Manchester Univ. *Career:* First professional appearance at the Comic Strip Club 1981; fmr Host TV shows, Friday Night Live, Saturday Night Live. *Present Position:* Comedian, writer and TV dir. *Publications:* Novels: Stark 1989, Gridlock 1991, This Other Eden 1993; TV scripts: The Young Ones (jtly), Blackadder II 1987 (jtly), Blackadder the Third 1988 (jtly), Blackadder goes Forth 1989 (jtly), Filthy Rich, Catflap, Happy Families; Plays: Gasping 1990, Silly Cow 1991; Video: Ben Elton Live 1993. *Address:* c/o Phil McIntyre, 15 Riversway, Navigation Way, Preston, Lancs PR2 2YP, UK.

EMMERIJ, Prof Louis Johan, PH D; Dutch int civil servant; b 25 Dec 1934, Rotterdam; m 2nd Vera Emmerij-Marinova 1977; 1d. *Education:* Paris Univ, France, Johns Hopkins Univ and Columbia Univ, USA. *Career:* Directorate of Scientific Affairs, OECD 1962–70; Dir, World Employment Programme, ILO 1971–76; Rector, Social Studies Inst, Hague 1976–85; mem, Governing Council, Int Devt Soc 1978–84, 1988–, Chair, Programme Cttee and mem, Exec Cttee 1982–84, 1988–; mem, N-S Round Table 1980–, Governing Body, Devt Studies Inst, Sussex Univ 1987–, Scientific Advisory Cttee, Int Food Policy Research Inst and Research Advisory Cttee, Philippine Inst for Devt Studies 1989–. *Present Position:* Pres, Devt Centre, OECD 1986–. *Publications:* Development Policies and the Crisis of the 1980s (Ed) 1987; One World or Several? (Ed) 1989; Science, Technology and Science Education in the Development of the South (Co-Author) 1989. *Address:* 94 rue Chardon-Lagache, 75016 Paris, France. *Telephone:* (1) 45-24-82-90. *Telex:* 620160. *Fax:* (1) 45-24-79-43.

EPSTEIN, Howard Michael, LL B; Canadian lawyer; b 17 Jan 1949; s of Ray Epstein and Leah Z Epstein (deceased); m Ilga Leja 1984; 2c. *Education:* Carleton Univ, Dalhousie Univ. *Career:* Lawyer, Justice Dept 1974–76; Lecturer, Osgoode Hall Law School 1977–78; Exec Dir, ON Confed of Univ Faculty Asscns 1984–88. *Present Position:* Exec Dir, Ecology Action Centre 1991–. *Address:* 3rd Floor, 3115 Veith St, Halifax, NS B3K 3G9, Canada. *Telephone:* (902) 454-7828. *Fax:* (902) 454-4766.

ESPINOZA, J O; environmentalist. *Career:* Contrib, UNEP Environmental Data Report 1992. *Present Position:* Mem, Regional Office for Europe, WHO. *Address:* c/o 8 Scherfigsvei, 2100 Copenhagen Ø, Denmark. *Telephone:* (1) 39-17-17-17. *Telex:* 15348. *Fax:* (1) 31-18-11-20.

EHRENFELD, David; writer and conservationist; *Present Position:* Prof of Biology, Rutgers Univ, NJ, USA; writer on environmental issues. *Publications:* Conserving Life on Earth 1972; Beginning Again 1993. *Address:* Dept of Biology, Rutgers University, POB 1201, New Brunswick, NJ, USA. *Telephone:* (908) 932-1766.

F

FAIRCLOUGH, Anthony John, CMG, MA; British environmental consultant; b 30 Aug 1924; s of the late Wilfrid and Lillian Anne (Townsend) Fairclough; m Patricia Monks 1957; 2s. *Education:* St Catharine's Coll, Cambridge. *Career:* Head of New Towns Div, Dept of Environment 1970–72, Under-Sec 1973–74, Dir, Cen Unit on Environmental Pollution 1974–78; Chair, Environment Cttee, OECD 1976–79; Del, Governing Council, UNEP and Cttee on Challenges of Modern Soc, NATO; mem, Nat Cttee on Problems of Environment, Royal Soc; Dir of Int Transport, Dept of Transport 1978–81; Dir for Environment, Directorate-Gen XI, EC Comm 1981–85, Acting Dir-Gen for Environment, Consumer Protection and Nuclear Safety 1985–86, Deputy Dir-Gen for Devt, Directorate-Gen VIII 1986–89. *Present Position:* Special Adviser, EC Comm, Dir, Environmental Resources Man and Groundwork Foundation. *Address:* 6 Cumberland Rd, Kew, Richmond TW9 3HQ, UK. *Telephone:* (81) 940-6999. *Fax:* (81) 940-3758.

FAITONDJIEV, Lyudmil, PH D; Bulgarian soil chemist and environmentalist; b 11 Aug 1943, Pernik; s of Petar Faitondjiev and Nadya Faitondjieva; m Kichka Faitondjieva 1972; 1s 1d. *Education:* Sofia Higher Inst of Chemistry and Tech. *Career:* Scientist, N Poushkarov Inst of Soil Science and Agroecology 1975–85. *Present Position:* Sr Scientist and Head, Dept of Soil Chemistry, N Poushkarov Inst of Soil Science and Agroecology 1985–. *Publications:* Problems of Soil Pollution 1984; Problems of Land Use and the Control of Heavy Metal Pollution of Soils 1987; Air Pollution and its Effect on Soils and Vegetation 1989. *Address:* Sofia 1080, 7 chaussée Bankya, Bulgaria. *Telephone:* (2) 2-52-71. *Telex:* 22701. *Fax:* (2) 24-89-37.

FALKENMARK, Malin, PH D; academic. *Career:* Consultant on water-scarcity issues to the FAO; Chair Int Water Resources Asscn's Cttee on Water Strategies for the 21st Century, Working Group on the Int Geosphere-Biosphere Programme for the Int Asscn for Hydrological Sciences. *Present Position:* Prof of Int Hydrology at the Natural Science Research Council. *Awards and Honours:* UNEP Global 500 Roll of Honour 1991. *Publications:* has written more than 150 water science publications. *Address:* National Science Research Council, POB 6711, S-113 85, Stockholm, Sweden.

FARMAN, Dr Joe; British scientist. *Career:* Discoverer of hole in ozone layer over Antarctic. *Present Position:* Mem, British Antarctic Survey. *Address:* High Cross, Madingley Rd, Cambridge CB3 0ET, UK. *Telephone:* (223) 61188.

FASELLA, Prof Paolo Maria, MD, PH D; Italian int civil servant; b 16 Dec 1930, Rome; s of Felice Fasella and Margherita Parazzoli; m Sheila Hauck Dionisi 1957; 4d. *Education:* Rome Univ. *Career:* Asst Prof of Biological Chemistry, Rome Univ 1959–63, Assoc Prof 1963–65, Prof 1971–81; Research Assoc, MIT 1961–62; Assoc Prof of Biochemistry, Parma Univ 1965–68, Prof 1968–71; Visiting Prof, Chemistry Dept, Cornell Univ, NY 1966. *Present Position:* Dir-Gen, Directorate-Gen XII (Science, Research and Devt), Comm of EC 1981–. *Address:* 200 rue de la Loi, 1049 Brussels, Belgium. *Telephone:* (2) 235-35-70. *Telex:* 21877. *Fax:* (2) 236-20-07.

FEARNSIDE, Philip Martin, PH D; academic. *Career:* Carried out study of human carrying-capacity on the Osa Peninsula in Costa Rica; helped in the creation of the Ouro Preto do Este Ecological Research Reserve; has made study of the environmental impacts of major World Bank projects 1988–89. *Present Position:* Research Prof, Dept of Ecology, Nat Inst for Research in the Amazon (INPA), Brazil. *Awards and Honours:* UNEP Global 500 Roll of Honour 1991. *Address:* National

Institute for Research in the Amazon, CP 487, Manaus-Amazonas, Amazon, Brazil.

FELDMANN, Fabio José; Brazilian politician and environmentalist; b 14 May 1955; 3c. *Education:* São Paulo Univ. *Career:* Elected to Congress 1986, 1990; Co-ordinator, Chapter on Environment, Nat Constitution 1988; bd mem, Conservation Int and WRI. *Awards and Honours:* UNEP Global 500 1990. *Address:* Av Brigadeiro Luiz Antonio, 4442 CEP 01402-002, São Paulo, Brazil. *Telephone:* (11) 887-8228. *Fax:* (11) 884-2795.

FERNANDEZ-GALIANO, Eliado, PH D; Spanish int civil servant and environmentalist; b 27 June 1953; s of Dimas Fernandez-Galiano and Maria Victoria Ruiz Perez; m Leticia Asensio 1983; 2s. *Education:* Complutense Univ, Madrid. *Career:* Researcher and Asst Prof, Ecology Dept, Univ Autónoma, Madrid 1977–85. *Present Position:* Civil servant, Council of Europe, Strasbourg 1985–, Sec, European Wildlife Convention (Berne Convention). *Address:* BP 431, 67006 Strasbourg Cedex, France. *Telephone:* 88-41-22-59. *Telex:* 870943. *Fax:* 88-41-27-81.

FÉRON, Eric Maurice, M SC; French wildlife man; b 1963, Kimbo-Fria, Guinea. *Career:* Co-Founder and Dir, Vetaid, UK and Mozambique 1988–90; Asst Project Man, Communal Areas Goat Research, Zimbabwe 1989–90; Consultant, Campfire and French Ministry of Foreign Affairs, Zimbabwe 1990–91; Country Rep, IUCN, Zambia 1991–92. *Present Position:* Wildlife and Livestock Project Man, CIRAD-EMUT, Zimbabwe 1992–. *Address:* POB 1378, Harare, Zimbabwe. *Telephone:* (4) 705885. *Fax:* (4) 701357.

FERRETTI, Janine Helene, BA; Canadian environmentalist; b 26 June 1958, Washington, DC, USA; d of James Joseph and Irma Priebe; m Gary T Gallon 1983; 2d. *Education:* California Univ, Santa Cruz, York Univ, Downsview, ON. *Career:* Research Asst, IUCN 1977; Researcher, Environment Liaison Centre, Nairobi 1980–81; Dir of Int Programmes, Pollution Probe Foundation; mem and Vice-Chair, ON Round Table on Environment and Economy. *Present Position:* Exec Dir, Pollution Probe Foundation 1989–. *Awards and Honours:* Environmentalist of the Year, City TV 1991. *Address:* 12 Madison Ave, Toronto, ON M5R 2S1, Canada. *Telephone:* (416) 926-1907. *Fax:* (416) 926-1601.

FETZ, Anita; Swiss historian, activist in women's movement and environmentalist; b 19 March 1957, Basel. *Career:* Councillor, Basel Canton 1984–88; MP 1985–89; Founder, FEMMEDIA 1985–; mem, Advisory Council, Alternative Bank Schweiz (ABS) 1990–. *Present Position:* Mem, Steering Cttee, Greenpeace Switzerland 1992–. *Publications:* Strukturwandel der Gesellschaft und Veränderung der Frauenrolle (Co-Author) 1988; Hoffnungspotential Frauen in der Wirtschaft 1989; Mut zur Karriere, Laufbahnplanung für Frauen (Co-Author) 1992; Rahmenkonzept für die Aus- und Weiterbildung in den öffentlichen Verwaltungen des Kantons Zürich, der Stadt Winterthur und der Stadt Zürich (Co-Author) 1992. *Address:* Mühlenberg 12, 4052 Basel, Switzerland. *Telephone:* (61) 2720323. *Fax:* (61) 2720383.

FISHER, Jonathan; British economist. *Present Position:* Econ Adviser on Environmental Protection, Dept of Environment. *Address:* Room A105, Romney House, 43 Marsham St, London SW1P 3PY, UK. *Telephone:* (71) 276-8496. *Fax:* (71) 276-8501.

FISK, David John, PH D; British civil servant; b 9 Jan 1947; s of the late John Howard Fisk and of Rebecca Elizabeth (Haynes) Fisk; m A Anne Thoday 1972; 1s 1d. *Education:* St John's Coll, Cambridge, Manchester Univ. *Career:* Higher Scientific Officer, Bldg Research Establishment 1972–73, Sr Scientific Officer (Energy Conservation Research) 1973–75, Principal Scientific Officer 1975–78, Sr Principal Scientific Officer, Mechanical and Electrical Eng Div 1978–84; Asst Sec, Cen Directorate of Environmental Protection, Dept of Environment 1984–87, Deputy Chief Scientist 1987–88; Visiting Prof, Liverpool Univ 1988–. *Present Position:* Under-Sec, Dept of Environment 1987–, Chief Scientist 1988– and Dir, Air, Climate and Toxic Substances Dept 1990–. *Publications:* Thermal Control of Buildings 1981. *Address:* Romney House, 43 Marsham St, London SW1P 3EB, UK. *Telephone:* (71) 276-3000. *Telex:* 22221. *Fax:* (71) 276-0818.

FLEISCHER, L A; environmentalist. *Career:* Contrib, UNEP Environmental Data Report 1992. *Present Position:* Chair, Int Whaling Comm. *Address:* c/o Red House, Station Rd, Histon, Cambridge CB4 4NP, UK.

FLOSS, Frank; Austrian environmentalist. *Present Position:* Chair, Grüne Alternative (Green Alternative). *Address:* 1070 Vienna, Stiftgasse 6, Austria. *Telephone:* (222) 52-125-0. *Fax:* (222) 52-125-40.

FONTANA, Walter Atilio; Argentinian journalist and teacher; b 1968, Córdoba. *Career:* Founder, Juvenile Action and Ecological Fight Centre (CALEJ) 1984; Pres, Fundación de Acción Ecológica 1989–, Organizing Cttee, 'World Year of Children and Environment' 1990; radio writer and dir. *Present Position:* Dir, *Foro Ecologista* magazine. *Awards and Honours:* UNEP Global 500 1991. *Address:* Casilla 519, Código Postal 5000, Córdoba, Argentina. *Telephone:* (51) 512631. *Telex:* 51894. *Fax:* (51) 894353.

FORBES, Leslie; Canadian politician and environmentalist. *Career:* Fmr Exec Sec, Green Party of Canada/Canadian Greens. *Present Position:* Mem, Green Party of Canada/Canadian Greens. *Address:* 831 Commercial Drive, Vancouver, BC V5L 3W6, Canada. *Telephone:* (604) 254-8165. *Fax:* (604) 254-8166.

FRISCH, Dieter; German int civil servant; b 25 March 1931, Bad Homburg; m José Reichling 1959; 2s 1d. *Education:* Heidelberg Univ, Bonn Univ. *Career:* Interpreter 1954–58; Directorate-Gen for Personnel and Admin, Comm of EC 1958–60, for Competition 1960–63, Asst to Dir-Gen, Directorate-Gen for Devt then Head of Div 1964–77, Dir, Directorate-Gen for Budgets 1977–82. *Present Position:* Dir-Gen, Directorate-Gen VII (Devt), Comm of EC 1982–. *Address:* 200 rue de la Loi, 1049 Brussels, Belgium. *Telephone:* (2) 235-10-94. *Telex:* 21877. *Fax:* (2) 235-01-22.

FULLER, Kathryn S; American environmentalist. *Career:* Mem, WWF US. *Present Position:* Pres, WWF. *Address:* World Conservation Centre, ave du Mont-Blanc, 1196 Gland, Switzerland. *Telephone:* (22) 3649111. *Telex:* 419618. *Fax:* (22) 3643239.

FUTRELL, J William, BA, LL B; American lawyer; b 6 July 1935; s of J W and Sarah Ruth (Hitesman) Futrell; m Iva MacDonald 1966; 1s 1d. *Education:* Columbia Univ, Tulane Univ, Berlin Free Univ. *Career:* Prof of Law, Alabama Univ 1971–74, Georgia Univ 1974–80. *Present Position:* Pres, Environmental Law Inst 1980–. *Publications:* Sustainable Environmental Law 1993. *Address:* Suite 200, 1616 P St, NW, Washington, DC 20036, USA. *Telephone:* (202) 328-5150. *Fax:* (202) 328-5002.

G

GÁBOR, Rácz; Hungarian nature film director and producer; b 24 March 1940, Nyiregyháza; s of Rácz Gábor and Josefin Lázár; m Balogh Emöke 1960; 1d. *Present Position:* Head, Naturfilm. *Awards and Honours:* Pro Natura Award 1987; UNEP Global 500 1988. *Publications:* A Course in Hungarian 1973; Last Shelter for the Animals of the World 1983; Films: Vipers; The Animal World of the Gobi; Twelve Months in the Forest; The Injured Planet; Landscape. *Address:* 1221 Budapest, Leányka u 3, Hungary. *Telephone:* (1) 226-5662. *Fax:* (1) 226-5662.

GAKAHU, Christopher Gatama, B SC, PH D; Kenyan conservation biologist; b 15 Sept 1951, Kenya; m Alice Nyambura 1977; 2s 2d. *Education:* Nairobi Univ. *Career:* Monitoring feeding habits and dispersal of large mammals, Amboseli Nat Park 1977–80; teacher and researcher, Nairobi Univ 1981–84; Founder, Wildlife Programme, Moi Univ 1985–88. *Present Position:* Conservation Biologist and Consultant, Wildlife Conservation Int 1988–. *Publications:* Tourist Attitudes and Use Impacts in Maasai Mara National Reserve 1992. *Address:* POB 62844, Nairobi, Kenya. *Telephone:* (2) 221699. *Fax:* (2) 215969.

GALAL-GORCHEV, Hend, PH D; American scientist; b 19 Nov 1932, Egypt. *Education:* Harvard Univ. *Career:* Office of Research and Devt, US Environmental Protection Agency 1967–78. *Present Position:* Scientist, Programme on Chemical Safety, WHO 1978–. *Address:* ave Appia, 1211 Geneva 27, Switzerland. *Telephone:* (22) 7913537. *Telex:* 415416. *Fax:* (22) 7881949.

GALAZY, Dr Gregory I; Ukrainian botanist, ecologist and limnologist; b 5 March 1922, Mechebilovo, Charkovskaya; s of Jogan Semenov and Praskovia Laurentia Galazy; m Alevtina Vasilievna Galazy 1945; 2s. *Education:* Irkutsk Univ. *Career:* Dir, Limnological Inst, Acad of Sciences 1954–87, Baikal Ecology Museum 1988–93. *Present Position:* Head, Pribaikalia Ecological Research Dept. *Awards and Honours:* UNEP Global 500 1987. *Publications:* Baikal and the Problem of Pure Water in Siberia 1968; Problems of Baikal 1973; The History of Botanic Research on Baikal 1980. *Address:* 664000 Irkutsk, Rossijskaya St 8-6, Russian Federation. *Telephone:* (3952) 33-44-83.

GALLOPIN, Dr Gilberto C; Canadian ecologist. *Present Position:* Int Inst for Sustainable Devt. *Address:* 212 McDermot Ave, Winnipeg, MB R3B OS3, Canada. *Telephone:* (204) 958-7700. *Fax:* (204) 958-7710.

GAO XING-YI, Prof; Chinese research scientist and zoologist; b 18 March 1935, Chongqing, Sichuan; m Xu Ke-Fen 1961; 1s 1d. *Education:* Sichuan Univ, Chengdu. *Career:* Technician, Xinjiang Inst of Biology, Pedology and Desert Research, Acad of Sciences 1961, Asst Engineer 1976, Engineer 1980, Assoc Prof 1986, Prof 1992; wildlife man, Xinjiang 1961–64, hunt man and man in charge of introduction of species 1965–77, research and classification of birds and mammals 1978–87. *Present Position:* Researcher on endangered species, Xinjiang 1988–. *Publications:* Animal Resources in the Inland Arid Area in Xinjiang 1987; The Distribution and Status of Equidae in China 1989; The Brown Bear of Xinjiang 1991; Newly Recorded Birds in Xinjiang 1992; Survival, Status and Breeding of Red Deer in Xinjiang 1992; The Influence of Human Activity on the Survival of Mammals in the Arid Area of Xinjiang 1992; The Fauna and Zoogeographical Division of Birds in Xinjiang 1992. *Address:* 40 S Beijing Rd, Urumqi, Xinjiang 830011, People's Repub of China. *Telephone:* (991) 337358. *Fax:* (991) 335459.

GÁPER, Ján, PH D; Slovak teacher; b 18 Sept 1957, Fačkov; m Svetlana Gáperová 1991; 1c. *Education:* Comenius Univ, Bratislava. *Career:* Scientist, Dendrobiology Inst 1981–90. *Present Position:* Deputy Dir, Forest Ecology Inst 1991–; Teacher, Ecology Faculty, Zvolen Univ 1993–. *Address:* Štúrova 4, 960 53 Zvolen, Slovakia. *Telephone:* (855) 6-35. *Telex:* 72267. *Fax:* (855) 22-654.

GARBA, Adamou; Niger journalist; b 1954, Bara; m Fati Ibrahim 1988. *Education:* Lomé Univ, Poitiers Univ, Paris Univ. *Present Position:* Dir of Communications, Ministry of Communications 1993–; Pres, Rassemblement pour un Sahel Vert. *Publications:* La Télévision au Niger 1983. *Address:* BP 12515, Niamey, Niger. *Telephone:* 74-11-25. *Fax:* 72-23-36.

GARCÍA NUÑEZ, Blas Antonio; Venezuelan politician and environmentalist. *Present Position:* Pres, Partido Ecológico Venezolano (PEV). *Address:* Apdo 51694, Caracas 1050, Venezuela.

GARDIOL, Irène; Swiss politician and environmentalist. *Career:* Fmr Pres, Grüne Partei der Schweiz/Parti écologiste suisse. *Present Position:* Mem, Grüne Partei der Schweiz/Parti Écologiste Suisse. *Address:* Marienstr 11, 3005 Berne, Switzerland. *Telephone:* (31) 441441.

GARRETT, Peter, LL B; singer and environmentalist. *Career:* Mem Midnight Oil pop group 1977–, toured USA and Europe 1989, performed protest concert outside New York offices of Exxon oil co 1990; Nuclear Disarmament Party Parl Candidate 1987; campaigns for Koori (Australian Aborigines) rights. *Present Position:* Mem Midnight Oil, Pres Australian Conservation Foundation. *Publications:* Albums: Midnight Oil 1978, Head Injuries 1979, Place without a Postcard 1980, 10,9,8,7,6,5,4,3,2,1 1982, Red Sails in the Sunset 1985, Diesel and Dust 1987, Blue Sky Mining 1990, Scream in Blue–Live 1992, Earth and Sun and Moon 1993. *Address:* Australian Conservation Foundation, 340 Gore St, Fitzroy, Vic 3065, Australia. *Telephone:* (3) 416-1455. *Fax:* (3) 416-0767.

GARRIDO, Francisco; Spanish politician and environmentalist. *Present Position:* Spokesperson, Los Verdes (Spanish Green Party). *Address:* Apdo 565, 38400 Puerto de la Cruz, Tenerife, Spain.

GAŠPARÍKOVÁ, Otília, PH D; Slovak plant physiologist; b 1935, Tvrdošín; d of Jozef Kubáni and Žofia Kubáni-Zmrazek; m 1963; 1d. *Education:* Comenius Univ, Bratislava. *Career:* Fellow, Botany Inst, Acad of Sciences 1961–64, Scientist 1965–, Head, Plant Physiology Dept 1977–90; Deputy Dir, Inst of Experimental Biology and Ecology 1982–90; mem and Chair, Scientific Bd for Biological and Ecological Sciences 1985–; Lecturer, Comenius Univ, Bratislava 1988–; Vice-Pres, Int Soc for Root Research. *Present Position:* Dir, Botany Inst, Acad of Sciences 1990–. *Awards and Honours:* Acad of Sciences Gold Medal 1989. *Publications:* Structural and Functional Aspects of Transport in Roots 1989; Root Metabolism 1992. *Address:* Dúbravská cesta 14, 842 23 Bratislava, Slovakia. *Telephone:* (7) 373-507. *Fax:* (7) 371-948.

GATINONI, Melvin; Argentinian environmentalist. *Present Position:* Dir, Greenpeace Argentina. *Address:* Junin 45, 3°, 1026 Buenos Aires, Argentina.

GIOVANNINI, Bernard; Swiss environmentalist. *Present Position:* Mem, Int Acad of Environment and Grüne Partei der Schweiz/Parti Écologiste Suisse. *Address:* Marienstr 11, 3005 Berne, Switzerland. *Telephone:* (31) 441441.

GIRARDET, Herbert, B SC; German writer and film-maker; b 28 May 1943; s of Dr Herbert Girardet and Ingrid Hoffman; m Barbara Hallifax 1967; 2s. *Education:* Eberhard-Karls Univ, Tübingen, Berlin Free Univ, LSE, UK. *Present Position:* Environmentalist, writer and film-maker 1974–. *Awards and Honours:* UNEP Global 500 1992. *Publications:* Land for the People (Ed) 1976; Far from Paradise (Co-Author) 1986; Blueprint for a Green Planet (Co-Author) 1987; Earthrise 1992; The Gaia Atlas of Cities 1993; Films: Jungle Pharmacy 1989; Halting the Fires 1990. *Address:* 93 Cambridge Gardens, London W10 6JE, UK. *Telephone:* (81) 969-6375. *Fax:* (81) 960-2202.

GOLDSMITH, Edward René David, MA; British author and environmentalist; b 8 Nov 1928; s of Frank B H and Marcelle (Mouiller) Goldsmith; m 1st Gillian Marion Pretty 1953; m 2nd Katherine Victoria James 1981; 3s 2d. *Education:* Magdalen Coll, Oxford. *Career:* Parl Cand, Eye, Ecology Party 1974, EP Cand, Cornwall and Plymouth 1979, Founder, Green Party; Adjunct Assoc Prof, Michigan Univ 1975; Visiting Prof, Sangamon State Univ 1984. *Present Position:* Publisher and Ed, *Ecologist*, Ecosystems Ltd 1970–. *Publications:* Can Britain Survive? (Ed) 1971; A Blueprint for Survival (Co-Author) 1972; The Future of an Affluent Society: The Case of Canada 1976; The Stable Society 1977; La Médecine à la Question (Co-Ed) 1981; The Social and Environmental Effects of Large Dams (Co-Author) 1986; Green Britain or Industrial Wasteland? (Co-Ed) 1986; The Earth Report (Co-Ed) 1988; The Great U-Turn 1988; 5,000 Days to Save the Planet (Co-Author) 1990. *Address:* Agriculture House, Bath Rd, Sturminster Newton DT10 1DU, UK. *Telephone:* (258) 73476. *Fax:* (258) 73748.

GOLITSYN, Prof Georgui S, D SC; Russian atmospheric scientist; b 23 Jan 1935, Moscow; s of Sergei Golitsyn and Klaudia Golitsyna; m Ludmila Lissitskaya 1956; 2d. *Education:* Moscow State Univ. *Career:* Atmospheric Physics Inst, Acad of Sciences 1958–, mem, Presidium 1992–. *Present Position:* Dir, Atmospheric Physics Inst, Acad of Sciences 1990–. *Awards and Honours:* Acad of Sciences Friedmann Prize in Meteorology 1990. *Publications:* An Introduction to the Dynamics of Planetary Atmospheres 1973; Climatic Catastrophes (Co-Author) 1987. *Address:* 109017 Moscow, Pyzhevsky 3, Russia. *Telephone:* (95) 231-55-65. *Fax:* (95) 233-16-52.

GOODALL, Jane, PH D; British primatologist, environmentalist and writer; b 3 April 1934, London; d of Mortimer Herbert and Vanne (Joseph) Morris-Goodall; m 1st Hugo Van Lawick 1964 (divorced 1974); m 2nd M Derek Bryceson 1975 (died 1980); 1s. *Education:* Cambridge Univ. *Career:* Sec, Oxford Univ; Asst Ed, Documentary Film Studio; waitress; Asst Sec, Olduvai Gorge then Gombe Stream Game Reserve 1964; Founder, Cttee for Conservation and Care of Chimpanzees 1986; Visiting Lecturer, Yale Univ. *Present Position:* Founder, Jane Goodall Inst for Wildlife Research, Educ and Conservation. *Awards and Honours:* Conservation Award, New York Zoological Soc. *Publications:* Shadow of Man, Chimpanzees of Goombe 1986; Through a Window 1990; The Chimpanzee: The Living Link between 'Man' and 'Beast' 1992. *Address:* POB 41720, Tucson, AZ 85717, USA.

GOODMAN, Gordon; British scientist and environmentalist. *Career:* Fmr Dir, Stockholm Environmental Research Inst, UN; int activist. *Present Position:* Environmental researcher. *Address:* 20 Dunstal Field, Cottenham, Cambridge CB4 4UH, UK. *Telephone:* (954) 52259. *Fax:* (954) 50425.

GORBACHEV, Mikhail Sergeyevich; Russian politician and environmentalist; b 2 March 1931, Privolnoye, Krasnogvardeisky Dist; m Raisa Titirenko 1956; 1d. *Education:* Moscow State Univ. *Career:* Machine operator 1946; joined CPSU 1952; 1st Sec, Stavropol Komsomol City Cttee 1955–58, then Deputy Head, Propaganda Dept, then 2nd then 1st Sec, Komsol Territorial Cttee; Party Organizer, Stavropol Territorial Production Bd of Collective and State Farms 1962–66; Head, Party Bodies Dept, Territorial Cttee, CPSU 1962–66; 1st Sec, Stavropol City Party Cttee 1966–68; 2nd Sec, Stavropol Territorial Cttee, CPSU 1968–70, 1st Sec 1968–70; mem, Cen Cttee, CPSU 1971, Agric Sec 1978–85, alt mem, Political Bureau 1979–80, mem 1980–91, Gen Sec 1985–91; Deputy, Supreme Soviet 1970–91, Chair, Foreign Affairs Comm, Soviet of Union 1984–85, mem, Presidium 1985–88, Chair 1988–90; Chair, Supreme Soviet, RSFSR 1979–90; mem, Congress of People's Deputies 1989, Chair 1989–91; Pres, USSR 1990–91; Founder, Social and Political Research Foundation (Gorbachev Foundation) 1992–. *Present Position:* Pres, Green Cross/Green Crescent Int 1992–. *Publications:* A Time for Peace 1985; The Coming Century of Peace 1986; Speeches and Writings 1986; Peace Has No Alternative 1986; Moratorium 1986; Perestroika: New Thinking for Our Country and the World 1987. *Address:* 160 route de Florissant, 1231 Geneva, Switzerland. *Telephone:* (22) 7891662. *Fax:* (22) 78991695.

GORDON, John Keith, MA; British diplomatist and dir of environmental research centre; b 6 July 1940; s of Prof James and Theodora (Sinker) Gordon; m Elizabeth Shanks 1965; 2s. *Education:* Cambridge Univ, Yale Univ, USA and LSE. *Career:* Diplomatic Service 1966–90, Perm Del to UNESCO 1984–85; Head, Nuclear Energy Dept, FCO 1986–88; Visitor, Centre for Environmental Tech, Imperial Coll, London 1988–90. *Present Position:* Deputy and Policy Dir, Global Environmental Research Centre, Imperial Coll, London 1990–. *Publications:* Institutions and Sustainable Development: Meeting the Challenge 1991. *Address:* 56 Queen's Gate, London SW7 5JR, UK. *Telephone:* (71) 225-1818. *Fax:* (71) 225-0627.

GOULANDRIS, Niki, PH D; Greek ecologist and man of natural history museum; d of Minas and Eleni Kefala; m Angelos Goulandris. *Education:* Goethe Univ, Frankfurt, Germany. *Career:* Co-Founder, Goulandris Natural History Museum 1964; exec, women's org. *Present Position:* Vice-Pres and Man, Goulandris Natural History Museum. *Publications:* Wild Flowers of Greece 1968; Peonies of Greece. *Address:* 13 Levidou St, 145 62 Kifissa, Greece. *Telephone:* (1) 8015870. *Fax:* (1) 8080674.

GRANDJEAN, E C; environmentalist. *Career:* Contrib, UNEP Environmental Data Report 1992. *Present Position:* Mem, Regional Office for Europe, WHO. *Address:* c/o 8 Scherfigsvej, 2100 Copenhagen Ø, Denmark. *Telephone:* (1) 39-17-17-17. *Telex:* 15348. *Fax:* (1) 31-18-11-20.

GRANT, James Pineo, D JUR; American int civil servant; b 12 May 1922, Beijing; s of John B Grant and Charlotte Hill; m 1st Ethel Henck; m 2nd Ellan Young 1989; 2s. *Education:* California Univ, Harvard Univ. *Career:* UN Relief and Rehabilitation Admin, People's Repub of China 1946–47; Deputy Asst Sec of State for Near E and S Asian Affairs; Dir, USAID Programme, Turkey 1964–67, Asst Admin, USAID 1967–79; Pres, Overseas Devt Council 1969–80. *Present Position:* Exec Dir, UNICEF 1980–. *Address:* 3 UN Plaza, New York, NY 10007, USA. *Telephone:* (212) 326-7000. *Telex:* 7607848. *Fax:* (212) 888-7465.

GREEN, Reg, B SC; British health, safety and environment officer; b 12 March 1952, Liverpool; s of James Reginald and Lilian (Tulloch) Green; m Marianne Wenning 1987; 1d. *Education:* LSE. *Career:* Asst, Legal Dept, Nat Union of Agricultural and Allied Workers 1980–85, Head, Legal and Health and Safety Depts 1985–87; Health, Safety and Environment Officer, ICFTU 1987–92; Trade Union Co-ordinator, UNCED 1992, Contrib, *Earth Summit '92* report. *Present Position:* Health, Safety and Environment Officer, Int Fed of Chemical, Energy and Gen Workers' Unions (ICEF) 1992–. *Publications:* Priorities for the Future 1992. *Address:* 109 ave Émile de Béco, 1050 Brussels, Belgium. *Telephone:* (2) 647-02-35. *Telex:* 20847. *Fax:* (2) 648-43-16.

GREGUS, Ctibor, D SC; Slovak forestry engineer and ecologist; b 6 Oct 1923, Kremnica; s of Jan Gregus and Julia Dobra; m Maria Pajevská 1951; 1s 2d. *Education:* Slovak Tech Coll, Bratislava, Agricultural Coll, Brno. *Career:* Head, Forest Man Inst, Zilina 1951–64; Co-ordinator, Forestry Research Inst, Zvolen 1964–85; Head, Working Party, Int Union of Forestry Research Orgs and mem, Forestry Man Union, Fed Repub of Germany 1972–80. *Present Position:* Scientific Consultant, Acad of Sciences 1985–. *Awards and Honours:* Slovak Environmental Union Prize 1984; Acad of Sciences Gold Medal 1992. *Publications:* Management of the Stelterwood Forest 1976; The Origins of Forestry Ecology Research 1985; Forestry Strategy in the Slovak Republic 1987. *Address:* Stúrova 2, 960 53 Zvolen, Slovakia. *Telephone:* (855) 223-12. *Fax:* (855) 274-85.

GRINBERG, Miguel, M SC; Argentinian professor of social ecology; b 18 Aug 1937, Buenos Aires; s of Samuel Grinberg and Esther Liba Zynger; m Flavia N Canellas Da Costa 1983; 3s 1d. *Education:* Buenos Aires Univ, New York Univ. *Career:* Lecturer, Georgetown Univ and Maryland Univ 1964; Prof of Social Ecology and Holistic Philosophy 1980–; Bd Mem, Environment Liaison Centre, Nairobi 1982–86; Founder, Buenos Aires Multiversidad 1982–87, Nat Network for Eco-Action 1986, Citizens' Perm Ass on Environment 1989; Co-Founder, Latin American Pact for Eco-Action, Chile 1989. *Present Position:* Dir, Promundo Int 1979– and Multiversidad de Campinas, Brazil 1992–. *Awards and Honours:* UNEP Global 500 1988. *Publications:* Ecología Vivencial 1988; Introducción a la Ecología Social 1992; Ecología Cotidiana 1993. *Address:* Casilla Cen 1933, 1000 Buenos Aires, Argentina. *Telephone:* (1) 343-4377.

GROSSMAN, Dr Wolf-Dieter; Austrian environmentalist. *Present Position:* Dir, Research Unit for Ecosystems and Environmental Studies. *Address:* 1010 Vienna, Kegelgasse 27, Austria.

GROVE-WHITE, Robin Bernard, BA; British scientist; b 17 Feb 1941; s of Charles William Grove-White and Cecile Mary Rabbidge; m 1st Virginia Harriet Ironside 1970 (dissolved); m 2nd Helen Elizabeth Smith 1979; 3s 1d. *Education:* Worcester Coll, Oxford. *Career:* Freelance writer for TV, radio, press and advertising, UK, Canada and USA 1963–70; McCann-Erickson Ltd, London 1970; Asst Sec, Comm for Protection of Rural England 1972–80, Dir 1981–87; Research Fellow, Centre for Environmental Tech, Imperial Coll, London 1987–89. *Present Position:* Sr Research Fellow in Environmental Research Policy, Lancaster Univ 1989–. *Publications:* Politics of Physical Resources (Co-Author) 1975; Future Landscapes (Co-Author) 1976; Nuclear Prospects (Co-Author) 1976. *Address:* Lancaster LA1 4YN, UK. *Telephone:* (542) 63806. *Fax:* (542) 843934.

GROZEV, Ognyan, PH D; Bulgarian forester; b 6 April 1955, Plovdiv; s of Grozu Grozev and Elena Simeonova; m Mariana Penkova Grozeva 1979; 1s 1d. *Education:* Higher Inst of Forestry and Forestry Eng, Sofia. *Career:* Agrolesproject 1980–85; Forest Research Inst, Bulgarian Acad of Sciences 1985–. *Present Position:* Researcher, Ecology and Environmental Conservation Dept, Forest Research Inst, Bulgarian Acad of Sciences 1987–. *Publications:* Genesis of Growth Variety of Pinus Sylvestris L in the Region of the West Rhodopes 1989; Results of the Introduction of Cedrus Atlantica Manetti in South-West Bulgaria 1993. *Address:* 132 Kliment Ohridski Blvd, 1756 Sofia, Bulgaria. *Telephone:* (2) 62-29-61. *Fax:* (2) 62-29-65.

GRUBB, Michael; British environmentalist and energy researcher. *Present Position:* Adviser, UNEP; mem, Energy and Environmental Programme, Royal Inst of Int Affairs. *Address:* Chatham House, St James Sq, London SW1Y 4LE, UK. *Telephone:* (71) 957-5700. *Fax:* (71) 957-5710.

GUDAL, Hassan; int civil servant. *Present Position:* Assoc Dir, Regional Office for Africa, UNEP. *Address:* POB 30552, Nairobi, Kenya. *Telephone:* (2) 230800. *Telex:* 22068. *Fax:* (2) 226890.

GUDYNAS, Eduardo; Uruguayan social ecologist; b 4 Jan 1960, Montevideo. *Education:* Uruguay Univ, Montevideo, Franciscan Multiversity, Montevideo. *Career:* Co-ordinator for Environment and Devt, CIPFE 1984–, 1st Latin American Congress on Ecology 1989; Duggan Fellow, Natural Resources Defense Council 1990; Faculty Assoc, Atlantic Coll, USA 1991–92; mem, Council, Latin American Network of Social Ecology 1991. *Present Position:* Co-ordinator for Environment and Devt, Latin American Center for Social Ecology 1989. *Publications:* Ethics, Environment and Development in Latin America 1989; Praxis for Life: Social Ecology Methodologies 1991; Environment and Trade in Latin America 1993. *Address:* Casilla 13000, 11700 Montevideo, Uruguay. *Fax:* (2) 793132.

GUO, Paul, PH D; Canadian sanitary engineer; b Taiwan; m Helen M H Guo 1964; 2s. *Education:* Nat Univ, Taiwan, Hanover Tech Univ, Germany. *Career:* Lecturer, Nat Univ, Taiwan 1964–67; Visiting Research Scientist, Hanover Tech Univ 1967–72; Sr Project Devt Engineer, Wastewater Tech Centre, Dept of Environment, Burlington, Canada 1972–78; Water Pollution Control Adviser, WHO, Seoul 1978–84, Water Quality Man Adviser, W Pacific Regional EHC 1985–88, Acting Dir and Water Quality Man Adviser 1988–89. *Present Position:* Dir, W Pacific Regional EHC, WHO 1989–. *Address:* POB 12550, 50782 Kuala Lumpur, Malaysia. *Telephone:* (3) 9480311. *Fax:* (3) 9482349.

GUPPY, Nicholas; British environmentalist. *Present Position:* Founder, Survival Int. *Address:* 310 Edgware Rd, London W2 1BY, UK. *Telephone:* (71) 723-5535.

H

HAIGH, Nigel, OBE, MA; British dir, environmental policy inst; b 23 Feb 1938, Tokyo; s of Anthony and Pippa (Dodd) Haigh; m Carola Pickering 1971; 2d. *Education:* King's Coll, Cambridge. *Career:* Civic Trust 1973–80; Vice-Pres, European Environmental Bureau 1975–79; Dir, London Office, Inst for European Environmental Policy 1980–89. *Present Position:* Dir, Inst for European Environmental Policy. *Publications:* EEC Environmental Policy and Britain 1987; A Manual of Environmental Policy: The EC and Britain 1992. *Address:* 158 Buckingham Palace Rd, London SW1W 9TR, UK. *Telephone:* (71) 824-8787. *Fax:* (71) 824-8145.

HAILES, Julia Persephone; British environmental consultant and author; b 23 Sept 1961, Templecombe; m 1991. *Career:* Consultant, British Airways, EC, ICI, Procter & Gamble; mem, UK Eco-Labelling Bd, UK Cttee, UNEP, UK Packaging Standards Council and Cttee on Educ and Training for Industry and Environment, Council for Environmental Educ; participated in Global Forum, Earth Summit, Rio de Janeiro 1992, European Co-ordinator, Inaugural Council. *Present Position:* Jt Founder, Dir and Sec, SustainAbility. *Awards and Honours:* UN Global 500 1989. *Publications:* Green Consumer Guide (Co-Author); Green Pages: The Business of Saving the World; The Green Consumer's Supermarket Shopping Guide; The Young Green Consumer Guide; The Green Business Guide; Holidays that Don't Cost the Earth; The Life-Cycle Assessment Sourcebook. *Address:* People's Hall, 91-97 Freston Rd, London W11 4BD, UK. *Telephone:* (71) 243-1277. *Fax:* (71) 243-0364.

HAIR, Jay D, M SC, PH D; American executive; b 30 Nov 1945, Miami, FL. *Education:* Clemson Univ, Univ of Alberta. *Career:* Mem Investment Policy Advisory Cttee (INPAC) to the US Trade Rep; lecturer; writes weekly syndicated column; mem Bd of Dirs Clean Sites Inc, The Windstar Foundation, Kids for Saving Earth, Earth Day 1990; Chair of the Bd, Pres IUCN-US (World Conservation Union); mem National Wetlands Policy Forum 1987-88, US Environmental Protection Agency Biotechnology Science Advisory Cttee 1987–89. *Present Position:* Pres (CEO) National Wildlife Fed 1981–. *Awards and Honours:* Awards include Environmental Educator of the Year, Ball State Univ 1988, Edward J Cleary Award, Nat Acad of Environmental Engineers 1989, Theodore and Conrad Wirth Environmental Award, National Park Foundation 1990. *Publications:* has written numerous articles on the environment for periodicals. *Address:* 1400 16th St, NW, Washington, DC, 20036-2266, USA. *Telephone:* (202) 797-6800.

HALL, Prof David Oakley, PH D; British biologist; b 14 Nov 1935; s of Charles Hall and Ethel Oakley; m Peta Smyth 1981; 2d. *Education:* Natal Univ, S Africa, California Univ, Berkeley, USA. *Career:* Fellow, Johns Hopkins Medical School, Baltimore, MD 1963–64; Lecturer, King's Coll, London 1964–68, Reader 1968–74; Visiting Sr Research Scientist, Princeton Univ 1990–91; Contrib, Environmental Data Report, UNEP 1992. *Present Position:* Prof of Biology, King's Coll, London 1974–. *Publications:* Photosynthesis 1972; Plants as Solar Collectors 1983; Biomass 1987. *Address:* Campden Hill Rd, London W8 7AH, UK. *Telephone:* (71) 333-4317. *Fax:* (71) 937-7783.

HANLEY, Paul J; Canadian writer and researcher; b 9 Aug 1952; m V O'Grady 1991; 2s. *Career:* Founding mem, Saskatchewan Earthcare Group 1975, New Roots Community Land Trust 1976, Agriculture Seminar, Asscn for Baháí Studies 1986, Saskwatch 1993; mem, Saskatchewan Eco-Network. *Present Position:* Environmental Columnist, *Saskatoon Star-Phoenix*. *Publications:* Earthcare: Ecological Agriculture in Saskatchewan 1980. *Address:* 921 6th Ave, N, Saskatoon, SK S7L 2T3, Canada. *Telephone:* (306) 244-9520. *Fax:* (306) 664-0437.

HANSEN, Dr James; American space scientist and environmentalist. *Present Position:* Mem, Inst for Space Studies, NASA. *Address:* 400 Maryland Ave, SW, Washington, DC 20546, USA.

HANSSEN, Alain; Belgian environmentalist. *Present Position:* Int Sec, Amis de la Terre (FoE). *Address:* place de la Vingeanne, 5100 Dave, Belgium.

HARDOY, Jorge E; Argentinian environmentalist. *Present Position:* Pres, Int Inst for Environment and Devt (IIED) Latin America. *Address:* Avda Corrientes 2835, Cuerpo A, 6°, 1193 Buenos Aires, Argentina.

HARRISON, James S; British environmentalist. *Present Position:* Deputy Chair, Council, Nat Soc for Clean Air and Environmental Protection. *Address:* 136 N St, Brighton BN1 1RG, UK.

HARRISON, Paul Anthony, MA, M SC; British writer and devt consultant; b 10 July 1945; 2s. *Education:* Cambridge Univ, LSE. *Career:* Social policy Ed, *New Society* 1972–75; Assoc Ed, *People* 1982–84; Contributing Ed, *People and Planet* 1992–; Consultant, World Bank, FAO, UNFPA, UN and nat devt agencies. *Present Position:* Self-employed writer and devt consultant. *Awards and Honours:* UNEP Global 500 1988; Global Media Award, Population Inst 1992. *Publications:* Inside The Third World 1979; The Third World Tomorrow 1980; Inside The Inner City 1983; The Greening of Africa 1987; The Third Revolution 1993. *Address:* 29A Nassington Rd, London NW3 2TX, UK. *Telephone:* (71) 794-6921. *Fax:* (71) 794-6921.

HARROP, Peter John, KCB, MA; British civil servant; b 18 March 1926, Morecambe; s of Gilbert and Frances May Harrop; m Margaret Elliott-Binns 1975; 2s. *Education:* Peterhouse, Cambridge. *Career:* New Towns Dept, Ministry of Town and Country Planning 1949–51; Ministry of Housing and Local Govt 1951–70, Head, Clean Air Dept 1956–60, SE Regional Planning Dept 1966–69; Regional Dir for Yorks and Humberside, Dept of Environment 1971–73; Treasury 1973–76; Head, Inner Cities Div, Dept of Environment 1977, Water and Environmental Protection Directorate 1977–79, 2nd Perm Sec for Environment, Water and Housing 1981–86; Chair, UK Cttee for European Year of Environment 1986–88, UK Cttee for UNEP 1988–93, Advisory Cttee, Waste Watch 1988–91; Trustee, British Museum. *Present Position:* Dir, Thames Water Authority and Thames Water plc 1986–, Chair, Bd Environmental Audit Cttee. *Address:* 19 Berwyn Rd, Richmond, Surrey, TW10 5BP, UK. *Telephone:* (81) 876-0752. *Fax:* (81) 876-0752.

HASSAN, Parvez, LL M, SJD; Pakistani lawyer; b 30 Sept 1941; s of S Ahmad Hassan; m 1967 (divorced 1992); 1s 1d. *Education:* Harvard Univ, Yale Univ, Punjabi Univ. *Career:* Mem, UNEP Mission to Bangladesh 1978, Pakistan 1981; Consultant, Workshop on Environmental Co-ordination of Devt Agencies, World Bank 1979, ESCAP 1985-86; Chair, Scientific Cttee, WWF Pakistan 1987. *Present Position:* Mem, Int Council of Environmental Law 1979–; Pakistan Environmental Protection Council 1989–; Chair, Council of Environmental Law, IUCN 1991–. *Awards and Honours:* UNEP Global 500 1991. *Address:* 2nd Floor, Paaf Bldg, 7D Kashmir Egerton Rd, Lahore, Pakistan. *Telephone:* (42) 6360880. *Telex:* 44805. *Fax:* (42) 6360811.

HASSELL, Prof Michael Patrick, D PHIL, D SC, FRS; British biologist; b 2 Aug 1942, Tel Aviv; s of Albert Marmaduke Hassell and Gertrude Loeser; m 1st Glynis Mary Everett 1966; m 2nd Victoria Anne Hassell 1982; 3s 1d. *Education:* Clare Coll, Cambridge, Oriel Coll, Oxford. *Career:* Research Fellow, Entomology Dept, Oxford Univ 1968–70; Lecturer, Biology Dept, Imperial Coll, London 1970–75, Reader 1975–79, Prof 1979–, Deputy Head 1984–93. *Present Position:* Head, Biology Dept, Imperial Coll, London 1993–. *Awards and Honours:* Scientific Medal, Zoological Soc, London 1981. *Publications:* Insect Population Ecology (Co-Author) 1973; The Dynamics of Competition and Predation 1975; The Dynamics of Arthropod Predator-Prey Systems 1978. *Address:* Silwood Lodge, Silwood Park, Ascot SL5 7PZ, UK. *Telephone:* (344) 294207. *Fax:* (344) 874957.

HASSI, Satu, PH D; Finnish politician; b 1957, Helsinki; s of Osmo and Maija-Liisa Hassi; 2d. *Education:* Helsinki Tech Univ, Tampere Tech Univ. *Career:* Design and Research Engineer, Oy Tampella Ab 1979–81; teacher, Tampere Tech Univ 1981–85; freelance writer 1985–91; Chair, Parl Group, Vihreä Liitto (Green Union). *Present Position:* Mem, Tampere City Council 1985– and MP 1991–. *Publications:* Magdalene, Ashamed No More (poems) 1984; A Stationary Model of an MHD Generator as a Part of an Electrical Network 1985; Female Screw (novel) 1986; The Serpent and the Tree of Knowledge (report on women and tech) 1987; A Wall Which Does Not Exist 1988; The New Clothes of King Adam (TV play) 1989. *Address:* Päivölänkatu 10D, 33250 Tampere, Finland. *Telephone:* (931) 149443. *Fax:* (90) 4322722.

HELLMAN, Richard A; American environmentalist. *Present Position:* Pres, US Cttee for UNEP. *Address:* 2013 Que St, NW, Washington, DC 20009, USA. *Telephone:* (202) 234-3600.

HENNINGSEN, Jorgen; Danish int civil servant; b 7 June 1941, Copenhagen. *Career:* Asst Prof, Tech Univ 1965–73; Head of Unit, Ministry of Environment 1973–76; Deputy Dir-Gen, Nat Agency of Environmental Protection 1976–80, 1983–87; Vice-Pres, Danish Oil and Natural Gas Corpn 1980–83. *Present Position:* Dir, Directorate B (Protection of Water and Air and Conservation), Directorate-Gen XI (Environment, Consumer Protection and Nuclear Safety), Comm of EC. *Address:* 200 rue de la Loi, 1049 Brussels, Belgium. *Telephone:* (2) 236-23-16. *Telex:* 21877. *Fax:* (2) 235-01-44.

HENRY, Donald James; Australian environmentalist. *Present Position:* Mem, WWF Australia. *Address:* Level 17, St Martin's Tower, Market St, Sydney, NSW 2001, Australia.

HENS, Dr Luc; Belgian professor of ecology; b 5 Dec 1951, Mechelen; s of Gaston Hens and Louisa Bruyminckx; m Erna De Cock 1982; 2s. *Education:* Free Univ of Brussels. *Career:* Asst, Dept of Human Ecology, Free Univ of Brussels 1974–80, Assoc Prof 1980–90. *Present Position:* Prof and Head, Dept of Human Ecology, Free Univ of Brussels 1990–. *Awards and Honours:* Laureate Belgian Acad of Sciences. *Publications:* numerous contribs to Belgian and int journals. *Address:* Laarbeeklaan 103, B1090 Brussels, Belgium. *Telephone:* (2) 477-42-81. *Fax:* (2) 477-49-64.

HESSELINK, Frederik Joost; Dutch environmentalist; b 25 Aug 1945, Amersfoort; s of Sander Hesselink and Sofia Cornelia Klaertje Leusvelt; m Namgyal Lhamo 1979. *Education:* Utrecht State Univ. *Career:* Co-Founder, SME 1975, Man Dir 1983. *Present Position:* Dir, Inst of Environmental Communication, IUCN 1990–. *Address:* POB 13030, 3507 LA Utrecht, Netherlands. *Telephone:* (30) 802444. *Fax:* (30) 801345.

HEYDEMANN, Prof Dr Berndt; German politician and professor of ecology; b 27 Feb 1930, Kiel. *Education:* Kiel Univ. *Career:* Prof of Ecology, Kiel Univ 1970–, Dir, Biology Centre, Founder, Dept of Applied Ecology and Coastal Research and Research Station for Ecosystems Research and Eco-Tech; Research and Tech Centre, Büsum; Co-ordinator, Centre for Ecology and Project Research; Vice-Pres, Int Advisory Cttee for Wadden Sea; Pres, nat nature conservation union; Adviser, Fed Environment and Agric Ministry, Fed Research Ministry and German Research Community; Fed Pres, School Science Asscn. *Present Position:* Minister of Nature and Environment, Schleswig-Holstein 1988–. *Awards and Honours:* Schubert Prize for Nature Conservation. *Publications:* Biologischer Atlas Schleswig-Holstein (Co-Author) 1980; Elementare Kunst in der Natur (Co-Author) 1989. *Address:* Grenzstr 1-5, 24149 Kiel, Germany. *Telephone:* (431) 219200. *Telex:* 292290. *Fax:* (431) 219209.

HEYERDAHL, Thor, PH D; Norwegian anthropologist and explorer; b 6 Oct 1914, Larvik; s of Thor and Alison (Lyng) Heyerdahl; m 1st Liv Coucheron Torp 1936; m 2nd Yvonne Dedekam-Simonsen 1948; 2s 3d. *Education:* Oslo Univ. *Career:* Field research, Marquesas Islands, Polynesia 1937–38, British Columbia (theory of American Indian movts into Pacific) 1939–40; free Norwegian Mil Forces 1941–45; research, Europe and USA 1945–47; Kon-Tiki Expedition from Peru to Polynesia (4,300 miles) 1947; Co-Founder, Kon-Tiki Museum, Oslo 1949; research and lectures, Europe and USA 1948–52; Leader, Norwegian Archaeological Expedition to Galapagos (evidence of visits by S American Indians) and field research, Bolivia, Peru and Colombia 1954; Leader, Norwegian Archaeological Expedition to Easter Island and E Pacific 1955–56; Vice-Pres, World Asscn of World Federalists 1966–, Worldview Int Foundation 1982–; attempted crossing of Atlantic in papyrus boat Ra I 1969 (2,800 miles); sailed from Morocco to Barbados in papyrus boat Ra II (attempt to prove ancient Mediterranean civilization could have sailed reed boat to America) 1970; Int Patron, United World Colls 1973–; Leader, expedition with Sumerian-type reed boat Tigris from Asia to Africa 1977–78; archaeological excavations, Maldive Islands 1982–83; Leader, archaeological expeditions to Easter Island 1986, 1987; Organizer, Norwegian-Chilean Archaeological Expedition to Easter Island 1986–88. *Present Position:* Int Trustee, WWF 1979–, Leader archaeological project, Peru 1988–. *Awards and Honours:* Academy Award 1951; UN Pahlavi Environmental Prize 1978. *Publications:* På Jakt Efter Paradiset 1938; Kon-Tiki Ekspedisjonen 1948; American Indians in the Pacific: The Theory behind the Kon-Tiki Expedition 1952; Archaeological Evidence of Pre-Spanish Visits to the Galapagos Islands 1956; Aku-Aku: Påskeøoyas Hemmelighet 1957; Reports of the Norwegian Archaeological Expedition to Easter Island and the East Pacific (Co-Author) 1965; Indianer und Alt-Asiaten im Pazifik 1966; Sea Routes to Polynesia 1968; The Ra Expeditions 1971; Fatuhiva: Back to Nature 1974; Zwischen den Kontinenten 1975; The Art of Easter Island 1975; Early Man and the Ocean 1978; The Tigris Expedition 1980; The Maldive Mystery 1986; Easter Island: The Mystery Solved 1989; The Kon-Tiki Man (Co-Author) 1990. *Address:* Colla Micheri, 17020 Laigueglia, Italy.

HILL, Julie Elizabeth, M SC; British organization dir; b 22 March 1961, Middlesex. *Education:* Leeds Univ, LSE, London Univ. *Career:* Mem staff, Deloitte, Haskins, Sells 1982–83, World Food Assembly 1984–85; Admin, Green Alliance 1985–87, Parl Officer 1987–91, Dir of Policy 1991–92, Dir 1992–. *Awards and Honours:* Fellow, Royal Soc of Arts, Manufacturing and Commerce. *Publications:* Ethics, Environment and the Company (Co-Author) 1990, Towards Good

Environmental Practice, A Book of Case Studies (Co-Author) 1992. *Telephone:* (71) 836-0341. *Fax:* (71) 240-9205.

HILLARY, Sir Edmund Percival; New Zealander explorer and diplomatist; b 20 July 1919, Auckland; s of Percival Augustus and Gertrude Hillary; m 1st Louise Mary Rose 1953 (died 1975); m 2nd June Mulgrew 1989; 1s 2d (1 deceased). *Education:* Auckland Univ. *Career:* RNZAF, Pacific 1944–45; NZ Garwhal Expedition to Himalayas 1951; British Expedition to Cho Oyu 1952; British Mt Everest Expedition 1953; Leader, NZ Alpine Club Expedition to Barun Valley 1954; NZ Antarctic Expedition 1956–58, S Pole 1957; Leader, Himalayan Expeditions 1961, 1963, 1964; Pres, Volunteer Service Abroad 1963–64; built hosp for Sherpa tribesmen, Nepal 1966; Leader, expedition to Mt Herschel, Antarctica 1967; River Ganges Expedition 1977; High Commr to India 1984; Consultant, Sears Roebuck & Co, Chicago. *Present Position:* Bee farmer. *Publications:* High Adventure 1955; The Crossing of Antarctica (Co-Author) 1958; No Latitude for Error 1961; High in the Thin Cold Air (Co-Author) 1963; Schoolhouse in the Clouds 1965; Nothing Venture, Nothing Win 1975; From the Ocean to the Sky: Jet-Boating up the Ganges 1979; Two Generations (Co-Author) 1983. *Address:* 278A Remuera Rd, Auckland SE2, New Zealand.

HILLER, Wolfgang; Austrian economic and environmental consultant; b 8 Nov 1960; s of Ernst and Edeltrude Hiller. *Education:* Vienna Econs Univ. *Career:* Project Man, Voest-Alpine AG, Linz 1985–86; Econ Co-ordinator, Fed Chancellery 1986–90; Foreign Dept, Girozentrale and Bank of Austrian Loan and Savings Asscns 1990–91. *Present Position:* Admin, Cttee on Environment, Public Health and Consumer Protection, Socialist Group, EP. *Address:* 79-113 rue Belliard, 1040 Brussels, Belgium. *Telephone:* (2) 284-31-72. *Fax:* (2) 230-66-64.

HINDÁK, František, D SC; Slovak biologist; b 25 March 1937, Trnava; s of Jozef Hindák and Vilma Bočková; m Dr Magdaléna Hindáková 1961; 1s 1d. *Education:* Comenius Univ, Prague. *Career:* Microbiology Inst, Acad of Sciences, Třeboň 1962–68, Botany Inst, Bratislava 1968–; Sciences Faculty, Comenius Univ 1968–; Construction Faculty, Slovak Tech Univ, Bratislava 1987–; Pres, Slovak Botanical Soc 1990–93; Ed, *Preslia* and *Biologia*; Chair, Cttee for Environment, Presidium, Acad of Sciences. *Present Position:* Head, Dept of Lower Plants, Botany Inst, Acad of Sciences and Asst Prof, Sciences Faculty, Comenius Univ, Bratislava. *Publications:* The Key to the Determination of Algae 1975; Freshwater Algae 1978; Studies on the Chlorococcal Algae 1990. *Address:* Dúbravská cesta 14, 842 23 Bratislava, Slovakia. *Telephone:* (7) 373-294. *Fax:* (7) 371-948.

HISLAIRE, Peter; environmentalist. *Present Position:* Co-ordinator, Western Africa/Sahel Programme, IUCN. *Address:* 28 rue de Mauvernay, 1196 Gland, Switzerland. *Telephone:* (22) 9990001. *Telex:* 419624. *Fax:* (22) 9990002.

HOEKSTRA, Hayo H; environmentalist. *Present Position:* Head, Centre Naturopa, Council of Europe. *Address:* BP 431 R6, 67006 Strasbourg, France. *Telephone:* 88-41-20-00. *Telex:* 870943. *Fax:* 88-41-27-84.

HOLCÍK, Vladimír; Slovak water researcher; b 8 Oct 1944, Bratislava; m Dr Marta Holčíková 1975. *Education:* Slovak Tech Univ, Bratislava and City Univ, Bratislava. *Career:* Hydroprojekt, Bratislava 1967–69; Slovak Hydrometeorological Inst 1969–72; Danube Basin Admin 1972–91; URBACO, Constantine, Algeria 1980–83, Eastern Hydroproject 1986–87. *Present Position:* Dir, Water Research Inst 1991–. *Address:* nábr arm gen L Svobodu 5, 812 49 Bratislava, Slovakia. *Telephone:* (7) 343-336. *Fax:* (7) 315-743.

HOLDGATE, Martin Wyatt, CB, PH D; British biologist; b 14 Jan 1931, Horsham; s of Francis W and Lois M (Bebbington) Holdgate; m Elizabeth M Weil (née Dickason) 1963; 2s. *Education:* Queens' Coll, Cambridge. *Career:* Research Fellow, Queens' Coll, Cambridge 1953–56; Jt Leader, Gough Island Scientific Survey 1955–56; Lecturer in Zoology, Manchester Univ 1956–57, Durham Univ 1957–60; Leader, S Chile Expedition, Royal Soc 1958–59; Asst Dir of Research, Scott Polar Research Inst, Cambridge 1960–63; Chief Biologist, British Antarctic Survey 1963–66; Deputy Dir (Research), Nature Conservancy 1966–70; Dir, Cen Unit on Environmental Pollution, Dept of Environment 1970–74, Dir-Gen of Research 1976–81; Chief Scientist and Deputy Sec (Environment Protection) and Chief Scientific Adviser, Dept of Transport 1981–88; Dir, Inst of Terrestrial Ecology 1974–76. *Present Position:* Dir-Gen, IUCN 1988–. *Awards and Honours:* UNEP Silver Medal; UNEP Global 500 1988; Commdr, Order of Golden Ark 1991; Conservationist of Year Award, Nat Wildlife Fed, USA 1993. *Publications:* Mountains in the Sea: The Story of the Gough Island Expedition 1958; A Perspective of Environmental Pollution 1979. *Address:* ave du Mont-Blanc, 1196 Gland, Switzerland. *Telephone:* (22) 9990001. *Fax:* (22) 9990029.

HORVAT, Manfred; Austrian academic and environmentalist. *Present Position:* Dir, Univ Extension Centre, Vienna Tech Univ. *Address:* 1040 Vienna, Gusshausstr 28, Austria. *Telephone:* (1) 58-80-1. *Telex:* 131000. *Fax:* (1) 505-49-61.

HOTI, Namik V; Albanian politician and environmentalist. *Career:* Fmr Chair, Partija Ekologjike (Ecology Party). *Present Position:* Mem, Partija Ekologjike (Ecology Party). *Address:* Tirana, Albania.

HOUGHTON, Sir John, CBE, D PHIL, FRS; British physicist; b 30 Dec 1931, Dyserth, Clwyd; s of Sidney and Miriam (Yarwood) Houghton; m 1st Margaret E Broughton 1962 (died 1986); m 2nd Sheila Thompson 1988; 1s 1d. *Education:* Jesus Coll, Oxford. *Career:* Research Fellow, Royal Aircraft Establishment, Farnborough 1954–57; Lecturer in Atmospheric Physics, Oxford Univ 1958–62, Fellow, Jesus Coll 1960–, Reader 1962–76, Prof 1976–83; Pres, Royal Meteorological Soc 1976–78; Dir, Rutherford Appleton Lab, Science and Eng Research Council 1979–83; Chair, Earth Observation Advisory Cttee, European Space Agency 1980–, Jt Scientific Cttee, World Climate Research Programme 1981–83, Royal Comm on Environmental Pollution 1992–; Dir-Gen, Meteorological Office 1983–90, Chief Exec 1990–91; mem, Exec Cttee, WMO 1983–91, Vice-Pres 1987–91. *Present Position:* Scientific Chair, Scientific Assessment Working Group, UN Intergovernmental Panel on Climate Change (IPCC), WMO 1988–. *Publications:* Infra Red Physics (Co-Author) 1966; Remote Sounding of Atmospheres (Co-Author) 1984; The Physics of Atmospheres 1986; Does God Play Dice? 1988. *Address:* Space Science Dept, Rutherford Appleton Lab, Chilton, Didcot OX11 0QX, UK. *Telephone:* (235) 446624. *Fax:* (235) 446434.

HOWE, Elspeth Rosamund Morton, Lady, JP, B SC; British businesswoman; b 8 Feb 1932; d of Philip Morton Shand and Sybil Mary Sissons; m Rt Hon Sir Geoffrey Howe 1953; 1s 2d. *Education:* LSE. *Career:* JP, Inner London Juvenile Court Panel 1964–; mem, Lord Chancellor's Advisory Cttee on Appointment of Magistrates for Inner London Area 1965–75, on Legal Aid 1971–75; Vice-Chair, Cons London Area Women's Advisory Cttee 1966–67, Pres, Contact Group 1973–77, mem, Cons Women's Nat Advisory Cttee 1966–71; Gov, Cumberlow Lodge Remand Home 1967–70; mem, Inner London Educ Authority 1967–70, Briggs Cttee on Nursing Profession 1970–72; Gov, Froebel Educ Inst 1968–75, LSE 1985–, James Allen's Girls School 1988–; Chair, Inner London Juvenile Court 1970–90; Parole Bd 1972–75; Deputy Chair, Equal

Opportunities Comm 1975–79; Pres, Peckham Settlement 1976–, Women's Gas Fed 1979–, Fed of Recruitment and Employment Services 1980–; Vice-Pres, Pre-School Playgroups Asscn 1979–83. *Present Position:* Dir, Kingfisher plc 1986–, United Biscuits (Holdings) plc 1988–, Legal & Gen Group 1989–, Chair, Hansard Soc Comm on Women at Top 1989–90, Women's Econ Devt Target Team, Business in Community and BOC Foundation for Environment and Community 1990–. *Publications:* Under Five 1966. *Address:* c/o Barclays Bank, 4 Vere St, London W1, UK.

HROMAS, Dr Jaroslav; Czech nature conservationist; b 2 June 1943; m Eva Ruprechtová 1962; 1s 2d. *Education:* Charles Univ, Prague. *Career:* Admin, Koněprusy Cave 1962–65; Head, Geological Section, Beroun Dist Museum 1965–67; State Inst for Care of Historical Monuments and Nature Conservation 1967–90; Founder, Czech Asscn for Nature Conservation 1981–. *Present Position:* Dir, Czech Inst for Nature Conservation, Ministry of Environment 1990–. *Address:* Slezská 9, 120 00 Prague 2, Czech Republic. *Telephone:* (2) 774855.

HTUN, Nay; Thai int civil servant. *Present Position:* Dir, Asia and Pacific Regional Office, UNEP. *Address:* UN Bldg, 10th Floor, Rajadamnern Ave, Bangkok 10200, Thailand. *Telephone:* (2) 282-9161. *Telex:* 82392. *Fax:* (2) 280-3829.

HUBA, Mikuláš, PH D; Slovak geographer and scientist; b 24 March 1954, Bratislava; m Olga Martinčeková 1985; 2s 1d. *Education:* Comenius Univ, Bratislava. *Career:* Sr Researcher, Geography Inst, Acad of Sciences 1987–; Chair, SZOPK 1989–93; MP and Chair, Parl Environment Cttee 1990–92; mem, IUCN, SCOPE, Slovak Asscn, Club of Rome and Slovak Cttee, UNESCO. *Present Position:* Chair, Sustainable Living Soc. *Awards and Honours:* Slovak Nature and Landscape Union Medal 1989. *Publications:* Bratislava (Co-Author) 1987; Slovak Forests SOS 1990; The Danube Story 1990; Slovak Agriculture SOS 1991; Slovakia's Message for Rio 1992; Slovakia's Message for Lucerne 1993; Film: Settlement and Landscape 1987. *Address:* Stefánikova 49, 814 73 Bratislava, Slovakia. *Telephone:* (7) 314-635. *Fax:* (7) 313-968.

HUNT, Prof Julian Charles Roland, PH D, FRS; British meteorologist and environmentalist; b 5 Sept 1941; m Marylla Ellen Shephard 1965; 1s 2d. *Education:* Trinity Coll, Cambridge, Warwick Univ. *Career:* Fellow, Trinity Coll, Cambridge 1966–; Researcher, Cornell Univ 1967, Cen Electricity Research Labs 1968–70; Lecturer in Applied Mathematics and Eng, Cambridge Univ 1970–78, Reader in Fluid Mechanics 1978–90, Prof 1990–; mem, Cambridge Co Council 1971–74, Leader, Lab Group 1972; Visiting Prof, Colorado State Univ 1975, N Carolina State Univ and Environmental Protection Agency 1977, Colorado Univ 1980, Nat Center for Atmospheric Research, Boulder, CO 1983; Sec, Int Meteorological Asscn 1984–89, Vice-Pres 1989–; Founder Dir, Cambridge Environmental Research Consultants Ltd 1986–; Gen Sec, European Research Community for Flow Turbulence and Combustions 1988–. *Present Position:* Dir, Meteorological Office 1992–. *Address:* London Rd, Bracknell RG12 2SZ, UK. *Telephone:* (344) 854463.

HURNI, Hans, PH D; Swiss geographer and teacher; b 21 Dec 1950, Erlenbach; m Marlies Alt 1978; 2s. *Education:* Berne Univ. *Career:* Park Warden, WWF Int, Ethiopia 1975–77; Founder and Dir, Ethiopia Soil Conservation Research Project, Berne Univ 1981–87, Lecturer 1989–91, Privat-Docent 1991–; Asst Prof, Addis Ababa Univ 1986. *Present Position:* Dir, Group for Devt and Environment, Geography Inst, Berne Univ 1988– and Pres, World Asscn of Soil and Water Conservation, Ankeny, IA, USA 1991–. *Publications:* Climatic and Geomorphic Studies in the Simen High Mountains 1982; Soil Conservation in Ethiopia 1986; Simen Mountains Management Plan 1986; Soil Erosion in Agricultural Environments 1990. *Address:* Hallerstr 12, 3012 Berne, Switzerland. *Telephone:* (31) 658876. *Fax:* (31) 428843.

HURTADO, Mario; Ecuadorean maritime and coastal conservationist; b 13 Dec 1953; s of Segundo Hurtado and María Gualán; m Laura Domínguez 1978; 2s 1d. *Education:* Guayaquil Univ. *Career:* Research on sea turtles, Galapagos Islands 1978–82, co-ordinator, biological research on fish 1983–84, Deputy Dir, Charles Darwin Scientific Station 1984–86, Research Dir, Charles Darwin Foundation for Galapagos Islands 1987, Administrative and Technical Officer. *Present Position:* Scientific and Technical Co-ordinator, Charles Darwin Foundation for Galapagos Islands. *Publications:* Tortugas Marinas en Galápagos 1984; Aves Marinas en el Ecuador 1986; Mamíferos Marinos en el Ecuador 1990. *Address:* Francisco de P Icaza 115 y Pichincha, 8°, Casilla 09-01-10355, Guayaquil, Ecuador. *Fax:* (4) 565-049.

HUSBAND, Victoria (Vicky), BA; Canadian environmentalist; b 1940, Victoria, BC; m 1973. *Education:* British Columbia Univ, Vancouver. *Career:* Film dir and producer 1980–82; public speaker; Pres, Friends of Ecological Reserves, BC 1985–89; campaign for S Moresby/Gwaii Haanas Nat Park Reserve 1987; Conservation Chair, Sierra Club of Western Canada 1989–90; campaign for Khutzeymateen, first Canadian grizzly bear sanctuaries 1992; campaign for Clayoquot Sound temperate rainforest 1993. *Present Position:* Chair, Sierra Club of Western Canada 1990–. *Awards and Honours:* UNEP Global 500 1988. *Publications:* Ancient Rainforests at Risk 1993. *Address:* POB 10 RR5, Durrance Rd, Victoria, BC V8X 4M6, Canada. *Telephone:* (604) 478-0388. *Fax:* (604) 386-4453.

HUTTUNEN, Satu Orvokki, PH D; Finnish botanist; b 27 Sept 1945, Sotkamo; d of Sulo O Pulkkinen and Sirkka O Achren-Pulkkinen; m Jouko Huttunen 1965; 1d. *Education:* Oulu Univ. *Career:* Jr Fellow then Sr Fellow, Finland Acad 1976–88; Acad Docent in Environmental Hygiene, Kuopio Univ 1977; Acad Docent in Botany, Oulu Univ 1982. *Present Position:* Prof of Botany, Oulu Univ 1989–. *Awards and Honours:* Finnish Woman of the Year 1986; UNEP Global 500 1987. *Publications:* Pelastakaa metsät 1988. *Address:* POB 400, 90571 Oulu, Finland. *Telephone:* (81) 5531527. *Telex:* 32375. *Fax:* (81) 5531500.

I

ICKE, David; British environmentalist and writer. *Career:* Goalkeeper, Hereford United Football Club; Sports Presenter, BBC TV; Spokesman, Green Party. *Present Position:* Writer. *Address:* c/o Cen Books Ltd, 99 Wallis Rd, London E9 5LN, UK. *Telephone:* (81) 986-4854.

IDRIS, Mohamad; Malaysian environmentalist. *Present Position:* Pres, FoE Malaysia. *Address:* 43 Jalan Salween, 10050 Pulau Pinang, Malaysia.

IGNATAVIČIENĖ, Irena; Lithuanian meteorologist; b 18 Sept 1933, Lithuania; d of Antanas Sabaliauskas and Jadvyga Sabaliauskienė; m Algimantas Ignatavičius 1956; 2d. *Education:* Vilnius Univ. *Career:* Sr Researcher, Forestry Man Dept, Agric Acad 1983–91; mem, Lithuanian Green Party 1989–. *Present Position:* Chair, Lithuanian Green Party. *Publications:* Weather and Man 1989. *Address:* B Radvilaites 1, Vilnius 232000, Lithuania. *Telephone:* (2) 770-870. *Telex:* 261191. *Fax:* (2) 358-020.

IRA, Dr Vladimír, C SC; Czech geographer and researcher; b 25 Oct 1952, Písek; m Ľubica Raposová 1975; 1s 1d. *Education:* Comenius Univ, Bratislava. *Career:* Asst Researcher, Czechoslovak Environment Centre 1980–81; Asst Researcher, Geography Inst, Slovak Acad of Sciences 1981–85, Sr Researcher 1985–90, Prin Researcher 1990–; mem, Nat Cttee, Man and Biosphere 1990–; Dir, Postgrad Environmental Studies Programme, Acad of Sciences 1991–; Lecturer in Geography and Environment, Comenius Univ, Bratislava 1991–. *Present Position:* Vice-Pres, Sustainable Living Soc 1993–. *Publications:* Bratislava (Co-Author) 1987. *Address:* POB 92, 840 02 Bratislava, Slovakia. *Telephone:* (7) 820-597. *Fax:* (7) 491-340.

ISSA, Aboubacar; Niger civil servant; b 1954, Namounou; m Haoua Hima 1980; 3s. *Education:* Niamey Univ. *Career:* Regional Dir of Water and Forests, Dosso 1982–83; Dir of Forestry Research 1983–85. *Present Position:* Co-ordinator, Cellule de Gestion des Ressources Naturelles (CGRN) 1992–. *Publications:* Rôle des combretacées dans l'approvisionnement en bois de chauffage en région soudano-sahelienne 1988; Aménagement de terroirs d'Azourou-Dessa 1991; Plan d'action forestier tropical 1991. *Address:* BP 12 946, Niamey, Niger. *Telephone:* 73-37-17.

J

JACOBI, Abbes; Luxembourgeois politician and environmentalist. *Present Position:* Sec, Déi Gréng Alternativ (Green Alternative Party). *Address:* BP 454, 2014 Luxembourg, Luxembourg. *Telephone:* 46-37-40. *Fax:* 46-37-43.

JECMÍNEK, Jan; Czech environmentalist and politician. *Present Position:* Chair, Czech Green Party. *Address:* nám Republiky 7, 111 49 Prague 1, Czech Republic.

JEFTIC, Ljubomir, PH D; Croatian UN official; b 25 Feb 1936, Novi Becej; s of Mile and Slavica (Elblinger) Jeftic; m Ivanka Karacic 1961; 2c. *Education:* Zagreb Univ, Warsaw Univ. *Career:* Marine Scientist, Rudjer Boskovic Inst, Centre for Marine Research, Zagreb 1961–81; Research Assoc, Kansas Univ 1967–69, Brookhaven Nat Lab, NY 1969–70; Prof of Environmental Protection, Zagreb Univ 1979–85; Dir, Planning and Environmental Protection Dept, Croatian Cttee for Bldg, Housing and Environmental Protection 1981–85. *Present Position:* Sr Marine Scientist and Prin Officer, Mediterranean Action Plan Co-ordinating Unit, UNEP, Athens 1985–. *Publications:* Electrochemical Oxidation Pathways of Benzo (a)pyrene (Co-Author) 1970; Chromium (II) Catalyzed Aquation of Hexacyanochromate (III) to Pentacyanomonohydroxychromate (III) (Co-Author) 1971; Climatic Change and the Mediterranean (Co-Ed) 1992. *Address:* Nevrokopiou St 8A, 156 69 Athens, Greece. *Telephone:* (1) 6517886. *Fax:* (1) 6517886.

JENDROSKA, Jerzy, LL D; Polish lawyer; b 4 Jan 1956, Wrocław; s of Jan and Krystyna (Jandy) Jendroska; m Ewa Jendroska 1979; 2s. *Education:* Wrocław Univ. *Career:* Asst Prof, Environmental Law Group, Law Inst, Acad of Sciences 1978; Sec, Bd of Dirs, Polish Environmental Law Asscn 1987–; mem, Polish Environmental Law Reform Cttee 1989–92; Visiting Fellow, Univ Coll, London 1990–91. *Present Position:* Adjunct Prof, Environmental Law Group, Law Inst, Acad of Sciences 1986–. *Publications:* Access to Environmental Information 1991; Environmental Impact Assessment in Polish Law 1993. *Address:* 50-138 Wrocław, Kuzwioza 46-47, Poland. *Telephone:* (71) 444747. *Fax:* (71) 444747.

JIVAN SHAH, Kantilal, FRGS; Seychelles businessman and historian; b 1922, Bombay, India; m Urmila Hathibhai 1957; 1s. *Education:* Seychelles Coll. *Career:* Mem, Int Council for Bird Protection, Seychelles Nat Archive and Museum, Town and Country Planning Authority, Chamber of Commerce and Industry, Nature Conservation and Nat Parks Comm, Seychelles Island Foundation, Nat Research and Planning Council; Vice-Chair, Victoria Dist Council; Hon Sec, Seychelles Soc. *Present Position:* Chair, Tourism Consultative Cttee and Tourism Marketing Monitoring Cttee. *Awards and Honours:* UNEP Global 500 1990. *Address:* POB 699, Mahé, Seychelles. *Telephone:* 322310. *Fax:* 324137.

JONAITIS, Dr Vytautas; Lithuanian entomologist; b 10 Nov 1938; m Vladislava Jonaitienė 1960; 1s. *Education:* Agric Acad, Kaunas, Zoology and Parasitology Inst, Acad of Sciences, Vilnius. *Career:* Forester, Širvintos 1962–65; Researcher, Ecology Inst, Acad of Sciences 1965–. *Present Position:* Head, Entomology Lab, Ecology Inst, Acad of Sciences 1991–. *Publications:* Key to Insects of the European Region of the USSR 1981; Ichneumonids of Lithuania 1983; Resources, Formation and Functioning of Host-Parasite Entomocomplexes in Ecosystems 1990. *Address:* Akademijos str 9, 2600 Vilnius, Lithuania. *Telephone:* (2) 359-280. *Fax:* (2) 359-257.

JOVANOVIĆ, Djura, PH D; Yugoslavian (Serb) teacher; b 31 July 1934, Novi Sad; s of Slavko and Dragica Jovanović; m Antonia Jovanović; 1s. *Education:* Belgrade Univ, Imperial Coll, London. *Career:* Yugoslav Nuclear Programme; mem, Yugoslav Soc for Water Protection. *Present Position:* Chief, Water Quality Testing Lab, Hydrometeorological Service 1977–. *Publications:* The International Commission on Large Dams (Co-Author) 1988. *Address:* POB 100, 11030 Belgrade,

Yugoslavia. *Telephone:* (11) 543186. *Telex:* 11748. *Fax:* (11) 545378.

JÜSSI, Fred; Estonian freelance writer and photographer; b 29 Jan 1935, Aruba; s of Johannes and Elise-Anita Jüssi; m Helju Jüssi 1960; 2s 1d. *Education:* Tartu Univ. *Career:* Teacher, Hiiumaa 1958–60; Insp of Nature Protection 1961–75; radio broadcaster 1961–91; freelance writer and nature campaigner 1976–93. *Present Position:* Pres, Eestimaa Looduse Fund (ELF, Estonian Fund for Nature) 1992–. *Publications:* Record: Frogs and Toads of Estonia 1986. *Address:* 13 Tina str 4, Tallinn EE0010, Estonia. *Telephone:* (142) 421747.

K

KAASIK, Arne; Estonian nature conservationist; b 2 Oct 1947, Rutja; m Mare Kaasik 1968; 1s 1d. *Education:* Agric Univ, Tartu. *Career:* Ranger then Deputy Dir, Lahemaa Nat Park and Section Chair, Estonian Nature Conservation Union 1972–; mem, IUCN Comm of Nat Parks and Protected Areas 1991–; Dist Gov, Lions Clubs Int 1993–. *Present Position:* Dir, Lahemaa Nat Park 1988– and Chair, Eesti Kaitsealade Liit (Union of Protected Areas of Estonia) 1991–. *Awards and Honours:* Fred M Packard Parks Merit Award, IUCN 1993. *Publications:* Lahemaa National Park 1983. *Address:* Lääne-Virumaa, EE2128 Viitna, Estonia. *Telephone:* (142) 324-5759. *Fax:* (142) 324-5759.

KAMUKALA, Godfrey Lweyemamu, MA; Tanzanian physical planner and environmentalist; b 30 May 1943; m Esther Kamukala 1975; 2s 2d. *Education:* Tech Univ, Germany. *Career:* Acting headmaster 1973–74; Lecturer, Ardhi Inst 1976–78; Regional Urban Planning Officer 1978–79, Acting Dir of Urban Planning 1980–83; Head, Environmental Section, Nat Environment Man Council 1983–86. *Present Position:* Dir-Gen, Nat Environment Man Council 1986–. *Publications:* Implications of Climate Change on the Marine Environment 1992; Implications of Climate Change on Coastal Settlements in the Eastern African Region 1992; Environment and Development: Balancing the Imperatives 1993. *Address:* POB 63154, Dar es Salaam, Tanzania. *Telephone:* (51) 34603.

KARAKACHANOV, Aleksandur; Bulgarian politician and environmentalist. *Present Position:* Pres, Green Party. *Address:* 1000 Sofia, Yanko Sakazov St 30, Bulgaria.

KARAS, Jacqueline; British climatic researcher. *Present Position:* Mem, Climatic Research Unit, E Anglia Univ. *Address:* Norwich NR4 7TJ, UK. *Telephone:* (603) 56161. *Telex:* 97154. *Fax:* (603) 58553.

KAROLES, Kalle, PH D; Estonian forestry expert and nature conservationist; b 1 June 1953, Tartu; s of Aleksei and Ilme Karoles; m Malle Karoles 1978; 1s 2d. *Education:* Agric Univ, Tartu, Forest Acad, St Petersburg. *Career:* Kaarepere Forest Research Station 1971–76; Estonian Inst of Forestry and Nature Conservation 1980–. *Present Position:* Dir, Estonian Inst of Forestry and Nature Conservation. *Address:* Rôômu tee 2, EE2400 Tartu, Estonia. *Telephone:* (142) 36-381. *Fax:* (142) 36-375.

KASSAS, Prof Mohamed, PH D; Egyptian professor of botany; b 6 July 1921; m Freda K Hosny 1957; 1s 1d. *Education:* Cairo Univ, Cambridge Univ. *Career:* Science Faculty, Cairo Univ 1950–; Ecological Cartographer, Mediterranean Basin, UNESCO Arid Lands Research Programme 1950–60; Pres, IUCN 1978–84; Sr Adviser, UNEP 1973–92; UN Conf on Desertification 1977; mem, Shoura Council 1981–, Egyptian Acad of Science, Egypt Inst, World Acad of Art and Science and Club of Rome. *Present Position:* Emer Prof of Botany, Cairo Univ 1981–. *Awards and Honours:* First Order of Science and Arts 1959; UN Prize for Environment 1978; ALECSO Gold Medal 1978; State Science Prize 1982. *Publications:* The World Environment 1972–1982 (Co-Ed) 1982; Desertification (Co-Ed) 1987; Financial Support for the Biosphere (Co-Ed) 1987. *Address:* Giza 12613, Egypt. *Telephone:* (2) 727202. *Telex:* 94372. *Fax:* (2) 628884.

KATSOUPAS, Vassilis; Greek environmentalist; b 1 March 1964, Athens; m 1993. *Education:* York Univ, Toronto. *Career:* Activist, Tools for Peace and Popular Summit Facilitation Cttee (PSFC) NGOs 1988, N Watch NGO, ON 1989; Host, environmental radio show, CHRY Radio, York Univ, Toronto 1988–89; Researcher and Campaign Asst, Greenpeace Canada 1990–91. *Present Position:* Campaigns Officer, WWF Greece 1992–. *Address:* c/o World Conservation Centre, ave de Mont-Blanc, 1196 Gland, Switzerland. *Telephone:* (22) 3649111. *Telex:* 419618. *Fax:* (22) 3643239.

KAVALIAUSKAS, Prof Dr Paulius; Lithuanian geographer; b 1945, Anykščiai; s of Povilas Kavaliauskas and Monika Kavaliauskienė; 1s 1d. *Education:* Vilnius Univ. *Career:* Regional Planning Man, Nat Parks Protected Areas, Inst of Architecture and Bldg 1967–76; Founder, Nature Frame (Lithuanian ecological network). *Present Position:* Prof and Head, Landscape Section, Nature Faculty, Vilnius Univ 1976–. *Publications:* Perspectives of Protected Areas in Lithuania 1988; The Integrated Environmental Protection Scheme of Lithuania for the Period to the Year 2000 1988; Nature Framework Theory Questions 1990; Methodological Fundamentals of Land Management 1992. *Address:* Čiurlionio 21-27, Vilnius 2031, Lithuania. *Telephone:* (2) 631-417.

KEAR, Dr David, CMG, FRSNZ; British and New Zealander scientific administrator and geologist; b 29 Oct 1923, London; s of Harold and Constance (Betteridge) Kear; m Joan Kathleen Bridges 1948; 2s 1d. *Education:* Imperial Coll, London. *Career:* Geologist, NZ Geological Survey 1948–67, Dir 1967–74; UN Groundwater Consultant, Western Samoa 1969–78; Asst, NZ Scientific and Industrial Research Dept 1974–80, Dir-Gen 1980–83; mem, Commonwealth Science Council 1980–83, Ross Dependency Research Cttee and NZ Medical Research Council 1980–83 and UN Advisory Cttee on Science and Tech for Devt 1988–90; Geothermal Energy Consultant, Kenya and Indonesia 1985–87. *Present Position:* Consultant in geology, energy and science administration. *Publications:* Geology and Hydrology of Western Samoa (Co-Author) 1959; Jurassic Sequence at Kawhia Harbour (Co-Author) 1960; Geology of Northland: Accretion, Allochtons and Arcs (Co-Author) 1989. *Address:* 34 W End, Ohope, Whakatane, New Zealand. *Telephone:* (7) 312-4635.

KEC, Jiří; Czech forestry engineer; b 1956, Brno; m Maria Kec 1980; 3s. *Education:* Forestry Tech Univ, Brno. *Career:* Army Forest Man 1980–91. *Present Position:* Dir, Šumava Nat Park Admin 1991–. *Address:* Zámek 385 01, Vimperk, Czech Republic. *Telephone:* (339) 21334. *Fax:* (339) 22947.

KELANY, Prof Ibrahim, PH D; Egyptian environmentalist; b 1944, Minia; m Amany Abd el-Maksoud 1976. *Education:* Zagazig Univ, Giessen Univ, Germany. *Career:* Lecturer 1979–84, Assoc Prof 1984–88, Prof 1988–. *Present Position:* Sec-Gen, Tree Lovers Asscn 1991–; Vice-Pres, Egyptian Green Party 1993–. *Address:* POB 262, el-Maadi, 11728 Cairo, Egypt. *Telephone:* (2) 3530621. *Fax:* (2) 712937.

KENMOGNE, Rev Jean-Blaise; Cameroonian pastor and environmentalist; b 10 June 1952, Bahouan; m Suzanne Djouinndje 1977; 1c. *Education:* Montpellier Univ, Paris Univ. *Career:* Del-Gen, N-S Movt for Environment and Devt (MONSED). *Present Position:* Pastor, Cameroon Evangelical Church and Dir-Gen, CIPCRE environmental org. *Address:* BP 1256, Bafoussam, Cameroon. *Telephone:* 44-62-67. *Fax:* 44-51-91.

KENT, Bruce, LL B; British campaigner for nuclear disarmament; b 22 June 1929; s of Kenneth and Rosemary (Marion) Kent; m Valerie Flessati 1988. *Education:* Oxford Univ. *Career:* Curate, Kensington N and S 1958–63; Sec, Archbishop's House, Westminster 1963–64; Chair, Diocesan Schools Comm 1964–66; Catholic Chaplain, London Univ 1966–74; Chaplain, Pax Christi 1974–77; Parish Priest, Somers Town, London 1977–80; Gen Sec, Campaign for Nuclear Disarmament (CND) 1980–85, Vice-Chair 1985–87, Vice-Pres 1985, Chair 1987–90; resigned from ministry 1987; Lab Party Parl Cand, Oxford W and Abingdon 1992. *Present Position:* Pres, Int Peace Bureau 1985–. *Publications:* Undiscovered Ends 1992. *Address:* 11 Venetia Rd, London N4 1EJ, UK.

KERN, Eha; Swedish environmentalist. *Present Position:* Mem, Barnens Regnskog. *Awards and Honours:* Goldman Environmental Foundation Prize 1991. *Address:* c/o Ministry of Educ and Science, Drottninggt 16, 103 33 Stockholm, Sweden. *Telephone:* (8) 763-10-00. *Telex:* 13284. *Fax:* (8) 723-11-92.

KHOSHOO, Triloki Nath, PH D; Indian scientist and researcher on biodiversity; b 7 April 1927, Srinagar, Kashmir; s of S C and Vanamala Khoshoo; m Mohini Khoshoo 1945; 2s. *Education:* Punjab Univ, Lahore, Punjab Univ, Chandigar. *Career:* Sr Lecturer, Punjab Univ, Chandigar 1948–62; Head, Postgrad Botany Dept, Srinagar Univ 1962–64; Asst Dir, Nat Botanic Garden, Lucknow 1964–74, Deputy Dir 1974–76; Dir, Nat Botanical Research Inst, Lucknow 1976–82; Vice-Chair, Governing Council, UNEP 1982; Sec, Environment Dept 1982–85; Distinguished Scientist, Council of Scientific and Industrial Research 1985–90; mem, Plant Advisory Group, IUCN 1988– and Advisory Cttee to Dir-Gen, UNESCO 1993–. *Present Position:* Fellow, Tata Energy Research Inst 1991–. *Publications:* Plant Systematics and Evolution (Co-Ed) 1978; Silvae Genetica (Co-Ed) 1984. *Address:* Tata Energy Research Inst, 9 Jor Bagh, New Delhi 110 003, India. *Telephone:* (11) 4623935. *Telex:* 3161593. *Fax:* (11) 4621770.

KINNERSLEY, David John, MA; British water consultant; b 28 May 1926, Essex; s of Arthur Kinnersley and Doris Angier; m Barbara Fair 1950; 1s 2d. *Education:* Corpus Christi Coll, Cambridge. *Career:* Energy industry 1951–64; water industry 1964–83; Fellow, Nuffield Coll, Oxford then Mansfield Coll, Oxford 1983–86; Consultant, World Bank 1983–93; Special Consultant to Sec of State for Environment 1987–88; mem, Bd, Nat Rivers Authority 1989–91. *Present Position:* Mem, Council, WATERAID 1983– and Monopolies and Mergers Comm. *Publications:* Troubled Water: Rivers, Politics and Pollution 1988. *Address:* 111 Church St, Chesham HP5 1JD, UK. *Telephone:* (494) 778118. *Fax:* (494) 791627.

KISBY, Steve; Canadian environmentalist. *Present Position:* Exec Sec, Green Party of Canada/Canadian Greens. *Address:* c/o 831 Commercial Drive, Vancouver, BC V5L 3W6, Canada. *Telephone:* (604) 254-8165. *Fax:* (604) 254-8166.

KNEEPKENS, Manuel; Dutch politician and environmentalist. *Career:* Fmr Chair, De Groenen (Green Party). *Present Position:* Mem, De Groenen (Green Party). *Address:* POB 3244, 1001 AA Amsterdam, Netherlands. *Telephone:* (20) 6387004. *Fax:* (20) 341545.

KNILL, Prof John Lawrence; British engineering geologist; b 22 Nov 1934, Wolverhampton; s of William Cuthbert and Mary (Dempsey) Knill; m Diane Constance Judge 1957; 1s 1d. *Education:* Imperial Coll, London. *Career:* Geologist, Sir Alexander Gibb & Partners 1957; Asst Lecturer, Lecturer, Reader then Prof of Eng Geology, Imperial Coll, London 1957–93; Chair, Natural Environment Research Council 1988–93. *Present Position:* Emer Prof of Eng Geology, Imperial Coll, London 1993–. *Publications:* Industrial Geology 1977. *Address:* Highwood Farm, Shaw-cum-Donnington, Newbury RG16 9LB, UK. *Telephone:* (635) 40542.

KOCH, David H; American environmentalist. *Present Position:* Dir and Adviser, Earthwatch. *Address:* c/o 680 Mt Auburn St, POB 403, Watertown, MA 02272-9104, USA.

KODRÍK, Milan, PH D; Slovak research scientist; b 16 May 1957, Zvolen; m 1987; 1s. *Education:* Zvolen Tech Univ. *Present Position:* Chair, Soil Ecology Dept, Forest Ecology Inst, Acad of Sciences 1989–. *Publications:* Forest Ecosystems in Slovakia (Ed) 1991; Root Biomass Distribution of Norway Spruce under Different Air Pollution Impacts 1992. *Address:* Štúrova 2, 960 53 Zvolen, Slovakia. *Telephone:* (855) 20-914. *Fax:* (855) 27-485.

KOHANOFF, Roberto; Argentinian politician and environmentalist. *Present Position:* Mem, Los Verdes (Green Party). *Address:* Ciudad de la Paz 2944, 1429 Buenos Aires, Argentina.

KOITJÄRV, Teet; Estonian environmentalist; b 17 March 1963, Tapa; m 1990; 1s. *Education:* History and Archives Inst, Moscow. *Career:* Archivist, Tallinn City Archives 1986–87; Historian and Guide, Lahemaa Nat Park 1987–88. *Present Position:* Deputy Dir, Lahemaa Nat Park 1988– and Sec, Asscn of Baltic Nat Parks (ABNP) 1991–. *Publications:* TV Documentary: Even one Vargamäe (Co-Writer) 1989. *Address:* EE2128 Viitna, Lääne-Virumaa, Estonia. *Telephone:* (142) 45-759. *Fax:* (142) 45-759.

KOLLÁR, Miroslav; Slovak chemical engineer and water technology specialist; b 21 Nov 1961; s of Emil Kollár and Helena Kollárová; m Eleonóra Kollárová 1987; 1c. *Education:* Slovak Tech Univ, Bratislava, City Univ, Bratislava. *Present Position:* Scientific Sec, Výskumný ústav vodného hospodárstva (Water Research Inst) 1992–. *Address:* Nábr arm Gen L Svobodu 5, 812 49 Bratislava, Slovakia. *Telephone:* (7) 343-345. *Fax:* (7) 315-743.

KONDO, Jiro, D ENG; Japanese engineer; b 23 Jan 1917; s of Shuzo and Fujie Kondo; m Kei Mizuki 1944; 1s 1d. *Education:* Kyoto Univ, Tokyo Univ. *Career:* Prof, Aeronautics Dept, Tokyo Univ 1958, Dean, Eng School 1975. *Present Position:* Pres, Nihon Gakujutsu Kaigi (Science Council of Japan) and Dir, Nat Inst of Environmental Studies (RITE) 1980–. *Publications:* Integral Equations 1991; Supercomputing 1991. *Address:* 7-chome, Roppongi, Minato-ku, Tokyo 106, Japan. *Telephone:* (3) 3403-6291. *Fax:* (3) 3403-6224.

KOROBKO, Mykola; Ukrainian geologist; b 17 Jan 1937, Sochi, Russia; s of Ivan and Anastasiya Korobko; m Liudmila Korobko 1965; 3s 1d. *Education:* Dniepropetrovsk Univ. *Career:* Geologist, Ministry of Geology 1959–68; postgraduate student, Simferopol 1968–70; Teacher, Mining Coll, Krivoy

Rog 1970–74, Research Scientist 1985–90; Chief Expert, Ministry of Ferrous Metals Industry 1974–85. *Present Position:* People's Deputy, Supreme Soviet of Ukraine and Chair, Zeleny Svit ecological asscn. *Publications:* The Development of Railway Transport and Building Materials Production in the Krivoy Rog Basin 1991. *Address:* 254213 Kiev, Piririchna St 37, ap 201, Ukraine. *Telephone:* (44) 412-33-90.

KORYZMA, Andrés R; Chilean politician and environmentalist. *Present Position:* Pres, Los Verdes (Green Party). *Address:* Apdo, Santiago, Chile.

KOSTENKO, Dr Yuriy I; Ukrainian engineer and politician; b 12 June 1951, Vinnitsia Region; s of Ivan Kostenko and the late Lyudmila Kostenko; m Alina Kostenko; 1s. *Education:* Machinery Inst, Zaporizjia, Electric Welding Inst, Ukraine Acad of Science. *Career:* Engineer, Electric Welding inst, Ukraine Acad of Science 1973–82, student 1982–85, mem staff 1985–90; elected to Ukrainian Parl 1990, Deputy Chair of Parl Ecological Comm 1990–92. *Present Position:* Minister for Environmental Protection 1992–. *Awards and Honours:* Honourable Diploma (for participation in rescue expedition in Armenia) 1989. *Publications:* several articles in newspapers and periodicals. *Address:* 127 Kiev, Ap 102, 29 Lomonosov St, Ukraine. *Telephone:* (44) 228 06 44. *Fax:* (44) 229 83 83.

KÖSTER, (Kajus) Friedemann, DR RER NAT; German filmmaker and environmentalist; b 14 Jan 1944, Breslau; m Heide Köster-Stoewesand; 2s. *Education:* Rheinische Friedrich Wilhelms Univ, Bonn, Hamburg Univ, Panama Univ, Colombia Nat Univ. *Career:* Zoologist and Deputy Dir, Inst Columbo-Alemán de Investigaciones Cientificas 'Punta de Betin', Santa Marta, Colombia 1974–80; Expert, German Agency for Tech Co-operation (GTZ) 1976–80; Dir, Charles Darwin Research Station, Galapagos Islands, Ecuador 1980–84. *Present Position:* Scientific Adviser, Zoologist and Wildlife Photographer, Anglia TV/Survival Anglia Ltd 1984–86 and 1987–. *Publications:* TV Documentaries: Vampires on Wolf Island 1986; How They Got There 1986; Cold on the Equator 1986; The Ocean Travellers 1986; Man Came To Eden 1986; Perspectives of Paradise 1986; The Year of El Niño 1986; An Ocean Nursery 1986; Quest in the Flooded Forest 1989; The Green Leaf Code 1990; The Last Flight of the Condor 1993. *Address:* Casilla 17-16-1014, Quito, Ecuador. *Telephone:* (2) 356-086. *Fax:* (2) 443-188.

KOTHARI, Ashish, MA; Indian researcher, trainer and environmental activist; b 15 Aug 1961, Vadodara; m 1992. *Education:* Delhi School of Econs. *Career:* Research Assoc, Indian Inst of Public Admin (IIPA) 1985–92, Lecturer, 1992–; Project Dir, Status Report on Biodiversity in India 1992–93. *Present Position:* Founder, Kalpavriksh environmental action group 1979–. *Publications:* Narmada Valley Project: A Critique 1988; Management of National Parks and Sanctuaries in India 1989; The Directory of National Parks and Sanctuaries in India (Ed) 1991; Conservation of Biological Diversity in India 1993; Birds in India: Status and Conservation 1993. *Address:* C17/A Munirka, New Delhi 110 067, India. *Telephone:* (11) 3317309.

KOUDA, Dr Michel; Burkinabé environmentalist; b 1949; s of N Kouda and K Dianda; m Marie Bonafos 1981; 1s 2d. *Education:* Paul Sabatier Univ, Toulouse. *Career:* Ministry of Rural Devt 1979; Deputy Dir, CRTO 1980; mem, UN Group of Experts on Remote Sensing 1983; Dir-Gen of Agriculture 1984; Dir of Research and Planning, Ministry of Water 1985, Minister of Water 1986; Head, Canada-Burkina Faso Devt Project 1988. *Present Position:* Head, IUCN Mission to Burkina Faso 1993–. *Publications:* Analyse synchronique et diachronique de l'évolution de la végétation par télédétection multispectrale en zone semi-aride de Haute Volta 1981. *Address:* 01 BP 3133, Ouagadougou 01, Burkina Faso. *Telephone:* 30-16-51. *Fax:* 30-13-51.

KOZALIEV, George; Bulgarian teacher; b 5 Sept 1927, Smoljan, Rajkovo; m Elena Kazalieva 1952; 2d. *Education:* Sofia Univ. *Career:* Teacher, Smoljan 1952–63, instructor 1963–. *Present Position:* Co-ordinator, Ecological Club. *Address:* 4700 Smolyan, Rajkovo dist, Zh K Izvor, etr B, Flat 4, Bulgaria. *Telephone:* (2) 12-28-80.

KOZHOVA, Prof Dr Olga Mikhailovna; Russian hydrobiologist; b 3 March 1931, Irkutsk; d of Prof Dr M M Kozhov and R V Kozhova; m 1st Ravil Gasin (deceased); m 2nd Boris Pavlov; 1d. *Education:* Irkutsk State Univ. *Career:* Scientist, Limnological Inst, Acad of Sciences 1956–69; Chair of Invertebrate Zoology and Dir, Biological Inst, Irkutsk State Univ 1969–82; Head of Lab, Ecological Toxicology Inst 1982–90. *Present Position:* Dir, Biology Research Inst, Irkutsk State Univ 1990–. *Publications:* Phytoplankton of Lake Baikal: Structural and Functional Characteristics 1987. *Address:* 664003 Irkutsk, Lenin St 3, POB 24, Russian Federation. *Telephone:* (3952) 24-30-79.

KOZOVÁ, Mária, PH D; Slovak landscape ecologist and environmentalist; b 6 Dec 1948, Čierny Brod; m Peter Koza 1971; 1d. *Education:* Comenius Univ, Bratislava. *Career:* Asst Researcher, Landscape Biology Inst 1972–74; Researcher, Experimental Biology and Ecology Inst 1975–89; Sr Researcher and Deputy Dir, Landscape Ecology Inst and mem, Council of Experts, Federal Ministry of Environment 1990–91; Asst Prof, Faculty of Natural Sciences, Comenius Univ, Bratislava 1992–. *Present Position:* Founder and Head, Landscape Ecology Dept, Faculty of Natural Sciences, Comenius Univ, Bratislava and Head, Environmental Section 1992–. *Publications:* Use of Measures of Network Connectivity in the Evaluation of Landscape Stability (Co-Author) 1986; Umweltbericht Wien-Bratislava (Co-Author) 1993; The EIA Process: The Present State of Methodology, Legislation, Implementation and Education in the Slovak Republic 1993. *Address:* Mlynská dolina B-2, 842 15 Bratislava, Slovakia. *Telephone:* (7) 720-462. *Fax:* (7) 729-064.

KRAJČOVIČ, Roman; Slovak regional development and environment man; b 25 April 1958, Bratislava; s of Dr Štefan Krajčovič and Adriena Krajčoviča; 2d. *Education:* Comenius Univ, Bratislava, Univ Coll, London. *Career:* Scientist, State Nature Conservation 1982–88; Sr Researcher, Landscape Ecology Inst, Acad of Sciences 1988–91. *Present Position:* Dir, Slovak Environmental Agency 1991–. *Awards and Honours:* Slovak Union for Nature and Landscape Protection Bronze Medal 1986. *Address:* Ďurková 19, 949 01 Nitra, Slovakia. *Telephone:* (89) 24-189. *Fax:* (89) 24-189.

KRAMER, Paul Jackson, PH D; American plant physiologist; b Brookville, IN; s of LeRoy Kramer and Minnie Jackson; m Edith Vance 1931; 2c. *Education:* Miami Univ, Ohio State Univ. *Career:* Teacher and Researcher, Botany Dept, Duke Univ, NC; Dir, Sarah P Duke Gardens 1945–74; Program Dir in Regulatory Biology, Nat Science Foundation 1960–61. *Present Position:* Emer Prof of Botany, Duke Univ, NC 1974–. *Publications:* Physiology of Trees (Co-Author) 1960; The Physiological Ecology of Woody Plants 1991; Water Relations of Plants 1993. *Address:* POB 90338, Durham, NC 27708-0338, USA. *Telephone:* (919) 942-4480. *Telex:* 802829. *Fax:* (919) 684-5412.

KRASSILOV, Prof Valentin A; Russian ecologist and palaeontologist; b 1 Dec 1937; m 1970; 1d. *Education:* Kharkov State Univ, Ukraine. *Career:* Head, Evolutionary Dept, Far Eastern Acad of Sciences, USSR 1961–89; Dir, All-Russian Research Inst on Nature Conservation and Reserves, Moscow 1990–92. *Present Position:* Leading Scientist, Geography Inst,

Russian Acad of Sciences, Moscow 1992–. *Awards and Honours:* Int Org of Palaeontology Medal 1992. *Publications:* Palaeontology of Terrestrial Plants 1975; Nature Conservation: Principles, Problems, Priorities 1993. *Address:* c/o 113628 Moscow, PO Vilar, Znamenskoe-Sadki, VNII Priroda, Russian Federation. *Telephone:* (95) 423-03-22. *Fax:* (95) 423-23-22.

KRAUSKOPF, Konrad Bates, PH D; American geologist; b 30 Nov 1910, Madison, WI; s of Francis Craig and Maude Luvan Bates Krauskopf; m Kathryn Isabel McCune 1936; 1s 3d. *Education:* Wisconsin Univ, California Univ, Berkeley, Stanford Univ. *Career:* Chemistry Instructor, California Univ 1934–35; Physical Science Instructor, Stanford Univ 1935–39, Asst Prof, Assoc Prof then Prof 1939–76; Geologist, US Geological Survey 1943–; Head, G2 Geographic Section, US Army, Tokyo 1947–49. *Present Position:* Emer Prof of Geochemistry, Stanford Univ 1976–. *Publications:* Introduction to Geochemistry 1967; Radioactive Waste and Geology 1988; The Physical Universe (Co-Author) 1992. *Address:* Stanford, CA 94305, USA. *Telephone:* (415) 723-3325. *Fax:* (415) 725-2199.

KREBS, Prof John Richard, D PHIL, FRS; British research professor; b 11 April 1945, Sheffield; s of Sir Hans Adolf Krebs and Margaret Cicely Fieldhouse; m Katherine Anne Fullerton 1968; 2d. *Education:* Pembroke Coll, Oxford. *Career:* Zoology Lecturer, Pembroke Coll, Oxford 1969–70, Zoology Fellow 1981–88; Asst Prof, Animal Resource Ecology Inst, British Columbia Univ, Vancouver 1970–73; Zoology Lecturer, N Wales Univ Coll, Bangor 1973–74; Researcher, Animal Behaviour Research Group, Zoology Dept, Oxford Univ 1975–76; Zoology Lecturer, Edward Grey Field Ornithology Inst 1976–88; Pres, Int Soc of Behavioural Ecology 1988–90; mem, Council, Zoological Soc, London and Chair, NERC Review of Evolution, Taxonomy and Biodiversity 1991–92, BBC Science Consultative Group 1992–93; Pres, Asscn for Study of Animal Behaviour 1993–. *Present Position:* Research Prof, Oxford Univ, Fellow, Pembroke Coll, mem, Agricultural and Food Research Council (AFRC) 1988–, Dir, AFRC Ecology and Behaviour Unit and NERC Behavioural Ecology Unit 1989–, Sr Scientific Consultant and Chair, Animals Research Cttee, AFRC 1991–. *Publications:* Foraging Theory 1987; Foraging Behaviour 1987; Behavioural and Neural Aspects of Learning and Memory (Co-Ed) 1991; Behavioural Ecology: An Evolutionary Approach (Co-Ed) 1991; An Introduction to Behavioural Ecology (Co-Author) 1993. *Address:* S Parks Rd, Oxford OX1 3PS, UK. *Telephone:* (865) 271166. *Fax:* (865) 310447.

KREISEL, Wilfried, PH D; German physical chemist; b 20 July 1942, Mährisch-Ostrau; m 1971. *Education:* Heidelberg Univ. *Career:* Asst Prof, Dortmund Univ 1972–76; Regional Adviser on Environmental Quality Man, WHO 1977–86. *Present Position:* Dir, Environmental Health Div, WHO 1986–. *Address:* ave Appia, 1211 Geneva 27, Switzerland. *Telephone:* (22) 7913583. *Telex:* 415416. *Fax:* (22) 7884218.

KRIKIS, Dr Andris; Latvian forester and chemical engineer; b 16 Feb 1936, Livani; s of Julijs and Nora-Emma Krikis; m Dagnija Snikere 1964; 1d. *Education:* Agric Acad, Jelgava, Riga Polytechnic Inst. *Career:* Sr Engineer 1959–68; Head, Research Group, Organic Synthesis Inst, Acad of Sciences 1968–78, Deputy Head of Lab 1980–90; Polymer Researcher, Science Inst, Jelgava and Head, Nature Conservation Lab 1978–80. *Present Position:* Head, Environmental Protection Co-ordination Dept, Environmental Protection Cttee. *Publications:* Structure and Functions of Low Molecular Peptides 1980; Unnatural Amino-Acids 1987. *Address:* Peldu iela 25, Riga LV 1494, Latvia. *Telephone:* (2) 226-930. *Telex:* 161175. *Fax:* (2) 228-159.

KRIŠTÍN, Dr Anton; Slovak microbiologist; b 1956, Bratislava; s of Anton and Eva Krištín; m Maria Krištín 1977; 1s 1d. *Education:* Comenius Univ, Bratislava, Bayreuth Univ, Munich. *Present Position:* Mem, Forest Ecology Inst. *Publications:* Trophic Ecology of the Tree Sparrow 1984; Stictocephala bisonia 1988; Trophic Relationships of Birds and Vertebrates 1992. *Address:* Štúrova 2, 960 53 Zvolen, Slovakia. *Telephone:* (855) 20-914. *Fax:* (855) 27-485.

KUČERA, Dr Bohumil; Czech scientist and nature conservationist; b 15 Oct 1939, Prague; s of Bohumil Kučera and Bozena Kučerová; m Alena Šittnerová 1969; 2s 1d. *Education:* Charles Univ, Prague. *Career:* Alpine research expeditions 1961, 1962, 1963, 1968; Specialist, State Inst for Care of Historical Monuments and Nature Conservation 1963; research in geology, Mongolia 1970–71; study expedition, India 1974, Cen Asia 1988, Lake Baikal 1989 and Vladivostock region 1990; geomorphological mapping, Zagros Mountains, Iraq 1976. *Present Position:* Deputy Dir, Nature Conservation Inst 1990–. *Publications:* Development of State Conservancy 1986; Principles of Karsology and Speleology (Co-Author) 1992. *Address:* Slezská 9, 120 00 Prague 2, Czech Republic. *Telephone:* (2) 24172986. *Fax:* (2) 254555.

KULFAN, Ján, PH D; Slovak entomologist; b 10 June 1957; s of Viliam Kulfan and Olğa Kulfanová; m Katarína Kulfanová 1984; 2s. *Education:* Comenius Univ. *Career:* Researcher, Forest Ecology Inst, Zvolen 1984–93, Chair, Scientific Bd and Co-ordinator, UNESCO Zoological Investigations in Forest Ecosystems in Polanov Biosphere Reserve 1991–93; mem, Cttee, Slovak Entomological Soc. *Present Position:* Head, Zoology Dept, Forest Ecology Inst 1993–. *Publications:* Structure of Heliophilous Butterflies' Taxocoenoses in Several Biotopes of Western Slovakia 1990; Changes in the Distribution of Thermophilous Butterflies in Slovakia (Co-Author) 1992; The Structure and Seasonal Dynamics of Caterpillar Communities on Oak Trees 1992. *Address:* Štúrova 2, 960 53 Zvolen, Slovakia. *Telephone:* (855) 22-312. *Fax:* (855) 27-485.

KULL, Prof Kalevi, PH D; Estonian biologist; b 12 Aug 1952, Tartu; m Tiiu Veski 1978; 2s 2d. *Education:* Tartu Univ. *Career:* Research Asst, Zoology and Botany Inst, Tartu 1977–83, Head, Dept of Ecophysiology 1984–, Prof of Ecophysiology 1992–. *Present Position:* Pres, Estonian Naturalists' Soc 1991–. *Awards and Honours:* Estonian Life Science and Culture Award 1984. *Publications:* Lectures in Theoretical Biology (Co-Ed) 1993. *Address:* Vanemuise 21, EE2400 Tartu, Estonia. *Telephone:* (142) 28-619. *Fax:* (142) 33-472.

KÜLVIK, Mart, B SC; Estonian ecologist; b 20 Dec 1960, Tallinn; m Maaris Kiis 1988; 1s 1d. *Education:* Tartu Univ. *Career:* Research Fellow, Tallinn Botanical Gardens, Acad of Sciences 1984–85; Ecological Adviser, Tartu City Council 1985–91; mem, Congress 1990–92, Tartu City Council and Bd, Estonian Fund for Nature 1990–, Advisory Group, E European Programme, IUCN 1992–; Head, Nature Conservation Dept, Ministry of Environment 1991–92. *Present Position:* Dir-Admin, Nature Conservation Research Centre, Ministry of Environment 1993–. *Awards and Honours:* International Man of the Year Award 1993. *Publications:* IUCN Environmental Status Report: The Baltic Countries—Estonia (Ed) 1993. *Address:* POB 222, EE2400 Tartu, Estonia. *Telephone:* (142) 36-385. *Telex:* 173238. *Fax:* (142) 62-084.

L

LAFFERTY, William M, PH D; Norwegian/Irish political scientist; b 2 May 1939, Cleveland, OH; m Gro Helgesen 1978; 2d. *Education:* King's Point Acad, Long Island, NY, Florida Univ, Gainesville, FL. *Career:* Research Dir, Social Research Inst, Oslo 1970–77; Prof of Political Science, Oslo Univ 1977–. *Present Position:* Project Dir, Alternativ Framtid (Alternative Future) 1992–. *Publications:* Economic Development and the Response of Labour in Scandinavia 1971; Industrialization, Community Structure and Socialism 1973; Participation and Democracy in Norway 1981. *Address:* Sognsvn 70, 0855 Oslo, Norway. *Telephone:* 22-18-11-70. *Fax:* 22-18-20-77.

LALONDE, Olivier Brice Achille; French politician and environmentalist; b 10 Feb 1946, Neuilly-sur-Seine; m Patricia Raynaud 1986. *Education:* Paris Univ. *Career:* Man, textile co 1969–70; Pres, Amis de la terre asscn and journalist, *Sauvage* 1971–77; mem, Nat Planning Comm, Ministry of Planning and Land Man; Admin, European Environment Bureau 1982–85; Dir, Paris Office, Inst for European Environment Policy 1986–88; Sec of State for Environment 1988. *Present Position:* Pres, Génération Écologie. *Address:* 57 blvd de la Villette, 75010 Paris, France. *Telephone:* (1) 44-52-32-00.

LAMBERT, Gérard O J G; Belgian environmentalist; b 29 April 1952, Namur; m Agnès Lefevre 1976; 2d. *Education:* Louvain Catholic Univ. *Career:* Univ researcher 1976–78; trade unionist 1978–84; publisher 1984–92. *Present Position:* Fed Sec, Ecolo (French-speaking green party of Belgium) 1992–. *Address:* 8 rue du Séminaire, 5000 Namur, Belgium. *Telephone:* (81) 22-78-71. *Fax:* (81) 23-06-03.

LANDBORN, Gunnar, B SC; Swedish environmentalist; b 15 July 1948, Limhamn; m 1989; 1s 1d. *Education:* Stockholm Univ. *Career:* Teacher 1974–84, Deputy Headmaster then Headmaster 1984–88; 1st Prin Officer, Nat Bd of Educ 1988–90, Dir of Educ 1991. *Present Position:* Sec-Gen, Naturskydds föreningen (Swedish Soc for Nature Conservation) 1992–. *Publications:* The School 1988; To Give and Take 1990. *Address:* Åsögatan 115, POB 4625, 11691 Stockholm, Sweden. *Telephone:* (8) 53032947. *Fax:* (8) 53035536.

LANG, Dr Winfried; Austrian diplomatist and univ prof; b 16 Nov 1941, Vienna; m Jutta Lang 1969; 2s 1d. *Education:* Vienna Univ. *Career:* Diplomatic service, France, Japan and Geneva 1966–; Prof of Int Law, Vienna Univ 1982–; Pres, Ozone Conf, Vienna 1985, Montréal 1987, Biological Weapons Conf, Geneva 1986. *Present Position:* Amb and Perm Rep of Austria to UN, EFTA and GATT, Geneva 1991–. *Awards and Honours:* UNEP Global 500 1990. *Publications:* Internationaler Regionalismus 1982; Integration und Kooperation in Nord und Süd 1988; Internationaler Umweltschutz 1989; Environmental Protection and International Law 1991. *Address:* 9-11 rue de Varembé, 1211 Geneva 20, Switzerland. *Telephone:* (22) 7337750. *Fax:* (22) 7344591.

LANGRAND, Oliver; environmentalist. *Present Position:* Madagascar Rep, WWF. *Address:* BP 738, Antananarivo, Madagascar.

LANGSLOW, Derek Robert, PH D; British environmentalist; b 7 Feb 1945, London; s of Alec and Beatrice (Wright) Langslow; m Helen Katherine Langslow 1969; 1s 1d. *Education:* Queens' Coll, Cambridge. *Career:* Research Fellow, Cambridge Univ 1969–72; Lecturer in Biochemistry, Edinburgh Univ 1972–78; Sr Ornithologist, Nature Conservancy Council 1978–84, Asst Chief Scientist 1984–87, Policy and Planning Dir 1987–90. *Present Position:* Chief Exec, English Nature 1990–. *Address:* Northminster House, Peterborough PE1 1UA, UK. *Telephone:* (733) 318364. *Fax:* (733) 898290.

LANGSTON, John; British environmentalist. *Present Position:* Vice-Pres, Nat Soc for Clean Air and Environmental Protection. *Address:* 136 N St, Brighton BN1 1RG, UK.

LARSEN, Kai, M SC, BA; Danish botanist; b 15 Nov 1926, Hilleröd; s of Axel G and Elisabeth (Hansen) Larsen; m Supee Saksuwan 1971. *Education:* Copenhagen Univ. *Career:* Scientific Asst, Botanical Lab, Copenhagen Univ 1952–55; Asst Prof, Royal Danish School of Pharmacy 1955–62, Assoc Prof 1963; Leader, botanical expeditions to Thailand 1958–59, 1961–62, 1963, 1966, 1968, 1970, 1972, 1974, 1990, 1991, 1992, 1993; Visiting Prof, Royal Forestry Dept and Kasetsart Univ, Bangkok 1974; mem, Royal Danish Acad of Sciences and Letters, Royal Norwegian Acad of Sciences, Linnean Soc, London, Presidium, WWF Denmark and Exec Bd, Flora Malesiana Foundation, Leiden, Netherlands; corresp mem, Musée Nat d'Histoire Naturelle, Paris; hon mem, Botanical Garden Asscn, Thailand; Pres, Int Asscn of Botanical Gardens 1981–87; Ed-in-Chief, *Nordic Journal of Botany* and *Opera Botanica*; Danish Ed, *Flora Nordica*; Ed, *Flora of Thailand*; Adviser, *Flora of China*. *Present Position:* Prof, Aarhus Univ 1963–and Dir, Systematic Botany Dept. *Address:* Nordlandsvej 68, 8240 Risskov, Denmark. *Telephone:* 86-21-06-77. *Fax:* 86-21-18-91.

LAVÉRIE, Michel; French environmentalist; b 14 Nov 1945, Versailles. *Present Position:* Dir-Gen, Inst Nat de l'Environnement Industriel et des Risques (INERIS) 1993–. *Address:* Parc Tech ALATA, BP 2, 60550 Verneuil-en-Halatte, France. *Telephone:* 44-55-63-01. *Telex:* 140094. *Fax:* 44-55-66-99.

LAY, Vladimir; Croatian environmentalist. *Present Position:* Mem, Zelana Akeija. *Address:* c/o Ministry of Environmental Protection and Construction, 41000 Zagreb, Ave Vukovar 78, Croatia. *Telephone:* (41) 633444. *Fax:* (41) 612929.

LEAKEY, Richard Erskine Frere, FRAI; Kenyan palaeontologist; b 19 Dec 1944, Nairobi; s of Louis and Mary Leakey; m Maeve Gillian Epps 1970; 3d. *Career:* Leader, expeditions to W Natron, Tanzania 1963, 1964, Baringo, Kenya 1966, Omo River, Ethiopia 1967 and E Rudolf, Kenya 1968–70; Researcher on origin of man, Lake Turkana 1971–76, Bulok 1980, Rusinga Island 1983–84, W Turkana 1984–85; Admin Dir, Nat Museums of Kenya 1968–74, Dir 1974–89, mem, Bd of Govs; Chair, Wildlife Clubs of Kenya 1969–84, Kenya Exploration Soc 1969–72, Foundation for Research into Origin of Man; Chair, E African Wildlife Soc 1985–; Dir, Wildlife and Conservation Man Service 1989–90; Trustee, Nat Fund for Disabled and Foundation for Social Habilitation. *Present Position:* Dir and Exec Chair, Kenya Wildlife Service 1990–. *Awards and Honours:* Golden Ark Medal for Conservation 1989. *Publications:* Origins (Co-Author); People of the Lake: Man, His Origins, Nature and Future (Co-Author) 1979; The Making of Mankind 1981; Human Origins 1982; One Life 1984; Origins Reconsidered (Co-Author) 1992; TV Series: The Making of Mankind 1981. *Address:* POB 40658, Nairobi, Kenya. *Telephone:* (2) 742161.

LEES, Andrew John, B SC; British campaigner; b 8 June 1949, Great Yarmouth. *Education:* Wales Univ, Cardiff. *Career:* Ecological surveys, Crymlyn Bog and Pant-y-Sais Fen, Wales 1979, mobilized local community direct action against waste dumping 1980, obtained judicial review of govt decision not to notify part of Crymlyn Bog as a Site of Special Interest 1981; mobilized local community direct action against drainage and ploughing, Havergate Marshes, Norfolk 1983; exposed ground water and river pollution by May & Baker chemical factory. *Present Position:* FoE, England, Wales and Northern Ireland 1986–, Campaigns Dir 1990–. *Address:* 26-28 Underwood St, London N1 7JQ, UK. *Telephone:* (71) 490-2664. *Fax:* (71) 566-1689.

LEGGETT, Jeremy; British scientist and environmentalist. *Present Position:* Dir, Science, Atmosphere and Energy Div, Greenpeace Int. *Address:* 30 Islington Green, London N1 2PN, UK. *Telephone:* (71) 354-5100. *Fax:* (71) 359-4062.

LEVY, Jean-Pierre, PH D; French int civil servant; b 29 Oct 1935, Mulhouse; m Maya Mansukhani 1987. *Education:* Strasbourg Univ. *Career:* Joined UN 1966, Deputy Chief, Ocean Economy and Tech Branch 1967–73, Head 1974–86; Sec, Group of Experts on Marine Science Tech 1967, Deputy Sec, Sea Bed Cttee and Sec, Sub-Cttee I 1968–73, Sec, 3rd Cttee, Conf on Human Environment 1972, 1st Cttee, 3rd UN Conf on Law of Sea 1974–82. *Present Position:* Dir and Deputy to Special Rep of Sec-Gen, Preparatory Comm, UN Int Sea-Bed Authority, Div for Ocean Affairs and Law of Sea, Office of Legal Affairs, UN 1987–. *Publications:* La conférence des Nations Unies sur le droit de la mer: histoire d'une négociation singulière 1983. *Address:* Room DC-2 0450, UN Plaza, New York, NY 10017, USA. *Telephone:* (212) 963-3975. *Fax:* (212) 963-4116.

LIKENS, Gene Elden, PH D; American ecologist; b 6 Jan 1935, Pierceton, IN; s of Col Benjamin and Josephine Estella Garner Likens; m Phyllis Irene Craig Likens 1983; 1s 2d. *Education:* Wisconsin-Madison Univ. *Career:* Instructor, Asst Prof then Assoc Prof, Dartmouth Coll 1961–69; Assoc Prof, Prof then Chair, Ecology and Systematics Section, Cornell Univ 1969–83; mem, NAS; Co-Dir, Hubbard Brook Ecosystem Study. *Present Position:* Dir, Ecosystem Studies Inst and Vice-Pres, New York Botanical Garden 1983–, Adjunct Prof, Ecology and Systematics Section, Cornell Univ, Prof, Grad Ecology and Evolution Program, Rutgers Univ and Prof of Biology, Yale Univ. *Awards and Honours:* USDA Forest Service 75th Anniversary Award 1980; American Soc of Limnology and Oceanography Award 1982; Tyler World Prize for Environmental Achievement 1993. *Publications:* Long-Term Trends in Precipitation Chemistry at Hubbard Brook, New Hampshire (Co-Author) 1984; An Ecosystem Approach to Aquatic Ecology: Mirror Lake and its Environment (Ed) 1985; Evidence for Sulfate-Controlled Phosphorus Release from Sediments of Aquatic Systems (Co-Author) 1989; Changes in the Chemistry of Surface Waters: 25-year Results at the Hubbard Brook Experimental Forest (Co-Author) 1989; Some Aspects of Air Pollution on Terrestrial Ecosystems and Prospects for the Future 1989; Dry Deposition of Sulfur: A 23-year Record for Hubbard Brook Forest Ecosystem (Co-Author) 1990; The Impact of Changing Regional Emissions on Precipitation Chemistry in the Eastern United States (Co-Author) 1991; Limnological Analyses (Co-Author) 1991; The Ecosystem Approach: Its Use and Abuse 1992. *Address:* Box AB, Millbrook, NY 12545-0129, USA. *Telephone:* (914) 677-5343. *Fax:* (914) 677-5976.

LINET VERLINDEN, Christiane; Belgian environmentalist; b 29 April 1934; m Marc Linet 1957. *Education:* Brussels Univ. *Career:* Publisher, *Panda* and *Pandapress* magazines 1967–; mem, Species Survival Comm, IUCN; contrib, World Conservation Strategy, WWF/IUCN/UNEP 1980 and Caring for the Earth, IUCN/UNEP/WWF 1991–. *Present Position:* Pres, WWF Belgium 1992–. *Address:* chaussée de Waterloo 608, 1060 Brussels, Belgium. *Telephone:* (2) 347-01-11. *Fax:* (2) 344-05-11.

LISICKÝ, Mikuláš Juraj, PH D; Slovak ecologist and ecosociologist; b 9 May 1946, Trenčín; s of Mikuláš Lisický and Wanda Kucielová; m Eva Jelínková 1973. *Education:* Comenius Univ, Bratislava, Charles Univ, Prague. *Career:* Lecturer then Sr Lecturer in Zoology, Comenius Univ, Bratislava 1968–76, in Conservation Science 1976–85, 1990–93, Founder and Head, Ecosociology Dept 1990–93; Founder, Co-ordinator and Biomonitor, Danube in Slovakia 1989–91; Vice-Pres, Nature and Landscape Protection Union 1990–93; mem, Editorial Bd, *Ekologia*. *Present Position:* Founder and Head, Ecosociology Monitoring Dept, Acad of Sciences 1989–91 and 1993–. *Publications:* Living Symmetry 1980; Nature Conservation: An Introduction 1987; Molluscs of Slovakia 1991. *Address:* Mánesovo nám 2, 851 01 Bratislava, Slovakia. *Telephone:* (7) 840-678. *Fax:* (7) 840-363.

LÓPEZ DE URALDE, Juan; Spanish environmentalist. *Present Position:* Mem, Greenpeace España. *Address:* Rodriguez San Pedro 58, 4°, 28015 Madrid, Spain.

LÓPEZ IGLESIAS, Carlos A; Argentinian environmentalist. *Present Position:* Exec Dir, Greenpeace América Latina Cono Sur. *Address:* Bartolomé Mitre 26, 4°, 1032 Buenos Aires, Argentina.

LOVEJOY, Prof Derek; British environmental consultant. *Present Position:* Head, Derek Lovejoy Partnership. *Address:* 8 Denbigh Mews, London SW1V 2HQ, UK. *Telephone:* (71) 828-6392. *Fax:* (71) 630-6958.

LOVEJOY, Thomas E, PH D, FLS; American tropical and conservation biologist; b 22 Aug 1941, New York, NY; m Charlotte Seymour 1966 (divorced 1978); 3d. *Education:* Yale Univ. *Career:* Zoological Asst, Yale Peabody Museum 1963–64; Yale Carnegie Teaching Fellow 1964–65; Research Asst, Natural History Museum, Smithsonian Inst 1970, Exec Asst to Scientific Dir and Asst to Vice-Pres for Resource and Planning 1972–73; Program Dir, WWF USA 1973–78, Vice-Pres for Science 1978–85; Exec Vice-Pres, WWF 1985–87; mem, Pres's Council of Advisers in Science and Tech 1989–92, Advisory Bd, New York Botanical Garden, Acad of Natural Sciences, Philadelphia, Resources for the Future, World Resources Inst, Environmental Defense Fund, WWF, Center for Plant Conservation, Rainforest Alliance and Rachel Carson Council, Inc; Pres, American Inst of Biological Sciences; Hon Chair, Wildlife Preservation Trust Int; Chair, Advisory Bd, Earth Communications Office, Biosphere II Scientific Advisory Cttee and US Man and Biosphere Program. *Present Position:* Asst Sec for External Affairs, Smithsonian Inst 1987–, Science Adviser to Sec of Interior 1993–. *Awards and Honours:* UNEP Global 500 1992. *Publications:* Nearctic Avian Migrants in the Neotropics (Co-Author) 1983; Key Environments: Amazonia; Conservation of Tropical Forest Birds (Co-Ed) 1985; Global Warming and Biological Diversity (Co-Ed) 1992. *Address:* Castle Bldg 320, Washington, DC 20560, USA. *Telephone:* (202) 786-2263. *Telex:* 264-729. *Fax:* (202) 786-2304.

LOVELOCK, Prof James Ephraim, CBE, PH D, D SC, FRS; British scientist; b 26 July 1919; s of Tom Arthur and Nellie Ann Elizabeth (March) Lovelock; m 1st Helen Mary Hyslop 1942 (died 1989); m 2nd Sandra Jean Orchard 1991; 2s 2d. *Education:* Manchester Univ, London Univ. *Career:* Staff Scientist, Nat Inst for Medical Research 1941–61; Fellow, Harvard Univ 1954–55, Yale Univ 1958–59; Prof of Chemistry, Baylor Univ Coll of Medicine, TX 1961–64; Visiting Prof, Reading Univ 1967–90; Pres, Marine Biology Assn 1986–90. *Present Position:* Independent scientist 1964–. *Publications:* Gaia 1979; The Great Extinction (Co-Author) 1983; The Ages of Gaia 1988. *Address:* Coombe Hill, St Giles on Heath, Launceston PL15 9RY, UK.

LOZA-SAWIRIS, Yousriya, MA; Egyptian financial adviser and environmentalist; b 1953, Assuit; d of Nassif and Rosa Loza; m Onsa Naguib Sawiris 1953; 3s. *Education:* Cairo Univ. *Career:* Voluntary work 1973; Founder, Asscn for Protection of Environment 1977. *Present Position:* Pres, Asscn for Protection of Environment 1983–. *Awards and Honours:* Ministry of Agriculture Golden Prize 1987. *Publications:* Organic Compost Production 1990; Paper Recycling 1990; Rag Rugs 1993. *Address:* 4th Floor, Apt 41, 160 26 July St, Agouza, Cairo, Egypt. *Telephone:* (2) 3026930. *Telex:* 92768. *Fax:* (2) 3440201.

LUCAS, Percy Hylton Craig (Bing); New Zealander consultant; b 9 June 1925, Christchurch; s of Percy Charles Lucas and Ethel Lena Craig; m Kura Joyce Pitcher 1948; 1s 1d. *Education:* Canterbury Univ, Christchurch. *Career:* Dir, Nat Parks and Reserves 1969–75; Asst Deputy Dir-Gen of Lands 1975–81, Dir-Gen of Lands 1981–86; consultant 1986–90. *Present Position:* Chair, Comm on Nat Parks and Protected Areas (CNPPA), IUCN 1990–. *Publications:* Conserving New Zealand's Heritage 1970; Protected Landscapes: A Guide for Policy-Makers and Planners 1992. *Address:* 1/268 Main Rd, Tawa, Wellington 6203, New Zealand. *Telephone:* (4) 232-5581. *Fax:* (4) 232-9129.

LUPASCU, Prof Dr Mihail F; Moldovan plant biologist; b 27 Aug 1928, Kuizovka Rezina; s of Feodosie and Eudokia Lupascu; m Zinaida Armash 1953; 1s 1d. *Education:* Agric Inst, Kishinau. *Career:* Dir-Gen and Dir of Research, Plant-Growing Inst, Beltsy 1962–78; Vice-Pres, Acad of Sciences, Kishinau 1978–80, Ministry of Agric 1980–85, Head of Lab 1986–90. *Present Position:* Chair, Man and Biosphere Nat Cttee, UNESCO 1979–, Vice-Pres, Acad of Sciences 1990–. *Publications:* Environmental Protection and Intensive Agricultural Production 1984; Ecology and Intensification of Forage Production 1989; Alfalfa 1992. *Address:* Stefan cel Mare 1, Chişinău 277612, Moldova. *Telephone:* (0422) 26-44-91. *Fax:* (0422) 26-20-91.

LYNGSTAD, Anni-Frid; Swedish singer and environmentalist; m Benny Andersson (divorced). *Career:* Entered Eurovision Song Contest as solo artist 1971; mem ABBA pop group 1973–82, entered Eurovision Song Contest with ABBA 1974. *Present Position:* Solo singing career 1982–; environmental campaigner. *Awards and Honours:* Winner Eurovision Song Contest (with ABBA) 1974. *Publications:* Recordings include: Albums (with ABBA): Waterloo 1974, ABBA 1975, Arrival 1976, The Album 1978, Voulez-vous 1979, Super Trouper 1980, Visitors 1981, Gracias por la Musica 1981, Thank You For The Music 1983, ABBA Live 1986, ABBA Live in Concert 1993; Single: Something's Going On 1982.

M

MAATHAI, Wangari, PH D; Kenyan environmentalist; b 1 April 1940, Nyeri; m 1969; 2s 1d. *Education:* Mt St Scholastica Coll, KS, and Univ of Pittsburg, PA, USA, Univ of Nairobi. *Career:* Visiting Research Asst, Univ of Munich, Germany 1967–69; lecturer Univ of Nairobi 1969–71, Sr Lecturer 1973, Assoc Prof 1978, Chair, Dept of Anatomy 1973–81; Dir Kenya Red Cross Soc 1973–77; mem, then Chair of Bd, Environment Liaison Centre Int 1974–84; mem Nat Council Women of Kenya 1977–87, Club of Rome 1987–; Judge Sasakawa Environment Prize. *Present Position:* Founder, Co-ordinator The Green Belt Movement 1977–, public speaker on human rights, women's issues, devt and the environment. *Awards and Honours:* numerous awards include Woman of the Year Award 1983, Better World Soc Award 1986, UNEP Global 500 Roll of Honour (first woman) 1987, Woman of the World Award 1989, Green Century Environmental Award for Courage 1990, Goldman Environmental Prize 1991. *Publications:* The Green Belt Movement 1985; The Green Belt Movement: Sharing the Approach and the Experience 1988; numerous scientific papers in journals. *Address:* The Green Belt Movement, POB 67545, Nairobi, Kenya. *Telephone:* (2) 504264. *Fax:* (2) 504264.

MACDONALD, Gordon James Fraser, PH D; American professor of international relations and geophysicist; b 30 July 1929, Mexico City; s of Gordon James Fraser and Josephine Bennett MacDonald; m 1st Marcelline Kuglen (deceased); m 2nd Betty Ann Kipniss 1969; m 3rd Margaret Stone Jennings 1991; 3s 1d. *Education:* Harvard Univ. *Career:* Asst Prof of Geology and Geophysics, MIT 1954–55, Assoc Prof 1955–58; Prof of Geophysics, California Univ, Los Angeles 1958–68; Vice-Chancellor, California Univ, Santa Barbara 1968–70; mem, Council on Environmental Quality, Exec Office of Pres 1970–72; Prof and Dir of Environmental Studies, Dartmouth Coll 1972–79; Vice-Pres and Chief Scientist, Mitre Corpn 1979–90; Founder, *Journal of Environment and Development*, California Univ, San Diego 1992–. *Present Position:* Prof of Int Relations and Research Dir, Environmental Studies Dept, California Univ, San Diego. *Awards and Honours:* American Academy of Sciences Monograph Prize 1959; Man of Science 1966. *Publications:* The Rotation of the Earth (Co-Author) 1959; Global Climate and Ecosystem Change 1990. *Address:* 1221 Robinson (0518), California Univ, San Diego, 9500 Gilman Drive, La Jolla, CA 92093-0518, USA. *Telephone:* (619) 534-3254. *Fax:* (619) 534-7655.

McEWAN, Angus David, PH D, FAA; British/Australian science administrator; b 20 July 1937, Scotland; s of D N R McEwan and Annie Marion Rowan; m Julie Britten 1962 (divorced 1982); 2d. *Education:* Melbourne Univ, Cambridge Univ. *Career:* Research Scientist, Australian Aeronautical Research Labs 1966–69; Queen Elizabeth Fellow 1969–71; Leader, Geophysics and Fluid Dynamics Lab, CSIRO, Australia 1971–81; Fellow, Woods Hole Oceanographic Inst 1975; Chair, Cttee on Climatic Changes and Ocean 1987–90. *Present Position:* Head, Oceanography Div, CSIRO 1981–, mem, Jt Scientific and Technical Cttee for Global Climate Observing System 1990–. *Address:* POB 1538, Hobart, Tasmania 7001, Australia. *Telephone:* (02) 325212. *Fax:* (02) 325125.

McKENNA, Virginia; British actress and conservationist; b 1931, London; d of Terence McKenna and Anne Oakeley; m Bill Travers 1957; 3s 1d. *Career:* Film, theatre and TV actress 1949–. *Present Position:* Founder, Born Free Foundation 1984–. *Awards and Honours:* British Film Acad Award 1954; Variety Club Award 1965; Soc of W End Producers Award 1979. *Publications:* Films: A Town like Alice 1954; Carve Her Name with Pride 1957; Born Free 1964; Ring of Bright Water 1968; Books: Some of My Friends Have Tails 1971; Beyond the Bars 1987; Into the Blue 1992. *Address:* Coldharbour, Surrey RH5 6HA, UK. *Telephone:* (306) 712091. *Fax:* (306) 713350.

McQUAID, James, PH D; British engineer; b 5 Nov 1939; s of James and Brigid McQuaid; m Catherine Anne Hargan 1968; 2s 1d. *Education:* Univ Coll, Dublin, Jesus Coll, Cambridge. *Career:* Grad Eng Apprentice, British Nylon Spinners 1961–63; Sr Research Fellow, Safety in Mines Research Establishment 1966–68, Sr Scientific Officer 1968–72, Prin Scientific Officer 1972–78; Safety Adviser, Petrochemicals Div, ICI 1976–77; Deputy Dir, Safety Eng Lab 1978–80, Dir 1980–85; Chair, Electrical Equipment Certification Management Bd and mem, Safety in Mines Research Advisory Bd and Court, Sheffield Univ 1985–, Council, Midland Inst of Mining Engineers 1987–; Pres, Sheffield Trades History Soc and mem, Council, S Yorks Trades History Trust 1989–. *Present Position:* Research Dir, Health and Safety Exec 1985–. *Address:* Broad Lane, Sheffield S3 7HQ, UK. *Telephone:* (742) 752571. *Fax:* (742) 780525.

MAHOUD, Omar Osman, B SC; Sudanese agronomist and plant conservationist; b 1 Jan 1955, Taiba; 2s 2d. *Education:* Alexandria Univ. *Career:* Plant Protection Dept, Agric Ministry 1979–81; Head of Agric Devt, Islamic Relief 1981–86; Founder and Dir, NAFIR environmental NGO 1986–88; NGO consultant 1988–89; Deputy Exec Dir, Council of NGOs 1989–91. *Present Position:* Sec-Gen, Sudanese Red Crescent 1992–. *Publications:* The Role of NGOs and the International Community in Crisis Management in Sudan (Co-Author) 1990. *Address:* POB 235, Khartoum, Sudan. *Telephone:* 72011. *Telex:* 23006. *Fax:* 72877.

MALMBERG, Prof Torsten, PH D; Swedish human ecologist; b 5 July 1923, Helsingborg; s of Sigfrid Vilhelm Malmberg and Lydia Henrietta Bergsten; m Aimée Birgit Magnussen 1948; 1s 1d. *Education:* Lund Univ. *Career:* Asst, Museum Keeper and Librarian, Zoology Dept, Lund Univ 1946–55, Asst Prof, Geography Dept 1978–83. *Present Position:* Prof and Head, Human Ecology Div, Lund Univ. *Awards and Honours:* Swedish Ornithological Soc Bronze Medal; Swedish Nature Protection Soc Silver Medal; UNEP Global 500 1992. *Publications:* Population Fluctuations and the Influence of Pesticide in the Rook, Corvus frugilegus L, in Scandinavia 1970; Human Territoriality 1980. *Address:* POB 2015, 220 02 Lund 2, Sweden. *Telephone:* (46) 10-96-85. *Fax:* (46) 10-42-07.

MALUKA, Nevruz; Albanian politician and environmentalist; b 23 March 1931, Erseke; s of Rustem and Shefikat Maluka; m Violeta Shkembi 1961; 2s 1d. *Education:* Tirana Univ. *Career:* Hydrometeorology Inst 1955–60; Prof, Tech School, Tirana 1960–80; mem staff plastics and rubber factory 1980–88. *Present Position:* Adviser, Industrial Science Inst, Tirana 1988– and Chair, Partia e Blerte Shqiptare (Green Party of Albania) 1991–. *Publications:* Inorganic Chemistry; Chemical Technology. *Address:* POB 749, Tirana, Albania. *Telephone:* (42) 33309. *Fax:* (42) 33309.

MANSER, Bruno; German environmental campaigner. *Present Position:* Campaigner for Amazon rainforests. *Address:* c/o Die Grünen (Greens W), 5303 Bornheim-Roisdorf, Im Ehrental 2, Bundesgeschäftsstelle, Germany. *Telephone:* (2222) 7008-0. *Fax:* (2222) 7008-99.

MARGUBA, Lawan Bukar, M SC; Nigerian ecologist and environmentalist; b 4 July 1951, Marguba; m Saadatu Kilishi Bukar 1980; 3s 1d. *Education:* New Mexico State Univ, USA, Coll of African Wildlife Man, Mweka, Tanzania, Salford Univ, UK. *Career:* Range man asst 1970–72; livestock devt asst 1973–74; wildlife ecologist 1979–85; Chief Conservator of Forests, Borno State 1985–89, Dir of Forestry 1989–90; Head, Afforestation Programme Co-ordinating Unit 1990–91. *Present Position:* Dir and CEO, Nigeria Nat Parks 1991–. *Publications:* Wild Animal Capture and Management in Captivity 1981; Desertification and its Impact on the Sahel Region of West Africa 1984. *Address:* PMB 0258, Garki, Abuja, Nigeria. *Telephone:* (9) 5230429. *Fax:* (9) 5230409.

MARTÍN, Isidoro Vicente; Spanish environmentalist. *Present Position:* Dir, Inst de Promoción y Apoyo al Desarollo. *Address:* Fuencarral 139 1D, 28010 Madrid, Spain.

MARTÍNEZ de PISÓN, Dr Eduardo; Spanish physical geographer; b 1 Jan 1937; m Margaret Fernandez-Ahuja 1976. *Education:* Autonomous Univ, Madrid. *Career:* Prof of Geography and Physical Geography, Canary Islands 1976; Autonomous Univ, Madrid 1977, Researcher in Physical Geography 1982–. *Present Position:* Prof of Physical Geography, Autonomous Univ, Madrid 1982–. *Awards and Honours:* Premio Nacional Medio Ambiente 1991. *Publications:* El Teide, Estudio Geográfico 1981; Geomorfología del Mount Everest, Tibet 1988; La Antartica y las Regiones Glaciares 1992. *Address:* Carretera de Colmenar, Canto Blanco, 28049 Madrid, Spain. *Telephone:* (91) 3974577.

MASTROVIĆ, Margita, M SC; Croatian environmental adviser; b 2 July 1939, Rijeka; d of Ivo and Andra Ivanišević; m Jurica Mastrović 1967; 2s. *Education:* Zagreb Univ. *Career:* Ecology Team Leader, INA oil refinery, Rijeka 1962–79; Adviser, Physical Planning and Environmental Protection Inst, Ass of Communes, Rijeka 1979–90. *Present Position:* Sr Adviser, Ministry of Civil Eng and Environmental Protection 1990–. *Address:* 51000 Rijeka, Užarska 2, Croatia. *Telephone:* (51) 213499. *Fax:* (51) 214324.

MATHEW, Thomas; environmentalist. *Present Position:* Mem, WWF India. *Address:* 172B Lodi Rd, New Delhi 110 003, India.

MATOLA, Sharon, BA; Belize environmentalist and zoo director; b 3 June 1957; d of Edward and Janice Matola. *Career:* Circus exotic dancer and lion tamer, Mexico; natural history film co; Founder, Belize Zoo 1983. *Present Position:* Dir, Belize Zoo and Tropical Educ Center and Chair, Tapir Specialist Group, IUCN. *Publications:* Hoodwink the Owl 1988; The Further Adventures of Hoodwink the Owl 1993. *Address:* POB 1787, Belize City, Belize.

MAYER, Susan Jean, PH D, MRCVS; British veterinary surgeon and environmental scientist; b 14 April 1955, Louth, Ireland. *Education:* Bristol Univ. *Career:* House Surgeon, Veterinary Surgery Dept, Glasgow Univ 1980–81; Veterinary Surgeon, Wilts 1981–84; Research Fellow, Bristol Univ 1984–87, Lecturer in Veterinary Therapeutics and Toxicology 1987–89; Dir, RSPCA Seal Assessment Unit, Norfolk 1989–90. *Present Position:* Science Dir, Greenpeace UK 1990–. *Address:* Canonbury Villas, London N1 2PN, UK. *Telephone:* (71) 354-5150. *Fax:* (71) 696-0014.

MEADOWS, Dennis L; American environmentalist; m Prof Donella H Meadows. *Present Position:* Dir, Inst of Methodological and Social Research, New Hampshire Univ. *Address:* Durham, NH 03824, USA. *Telephone:* (603) 862-1234. *Fax:* (603) 862-2030.

MEADOWS, Prof Donella H; American environmentalist; m Dennis L Meadows. *Present Position:* Mem, Inst of Methodological and Social Research, New Hampshire Univ. *Publications:* Limits to Growth 1972; Beyond the Limits 1992. *Address:* Durham, NH 03824, USA. *Telephone:* (603) 862-1234. *Fax:* (603) 862-2030.

MECI, Izini; Albanian politician and environmentalist. *Present Position:* Sec-Gen, Partija Ekologjike (Ecology Party). *Address:* Tirana, Albania.

MEHANDJIEV, Marin Roussev, PH D; Bulgarian chemical engineer and ecologist; b 8 March 1927, Sofia; m Krassimira Mehandjiev 1961; 1s. *Education:* Sofia Polytechnic, Chemical Tech High School, Bourgas. *Career:* Tech Dir, Itus chemical factory, Sofia 1950; Head, Lab and Sampling Div, Pan Mines 1953; man, hydrometal plant, Eliseyna 1959; Assoc Prof of Non-Ferrous Metallurgy, Plovdiv 1970; Ecologist, Chimcomplect co, Sofia 1978; Assoc Prof in Non-Equilibrium Thermodynamics, Varna 1988; Assoc Prof in Chemical Eng, Razgrad 1989. *Present Position:* Scientific Sec and Head, Ecological Modelling and Combined Assessment Lab, Bulgarian Ecologists Asscn (ABECOL). *Publications:* Ecology and Ecological Assessment 1981; Ecological Estimation and Ecological Modelling 1982; Thermodynamics and Ecology 1989; Combined Ecological and Economic Assessment 1992. *Address:* Secr, POB 52, 1231 Sofia, Bulgaria. *Telephone:* (2) 37-23-50. *Fax:* (2) 66-14-55.

MELLANBY, Kenneth, CBE, PH D; British environmental consultant, editor; b 26 March 1908, Barrhead, Scotland; s of Prof A L Mellanby; m 1st Helen Nielson Dow 1933 (dissolved); m 2nd Jean Louie Copeland 1949; 1s 1d. *Education:* King's Coll, Cambridge, London Univ. *Career:* Researcher, London School of Hygiene and Tropical Medicine 1930–36, Reader in Medical Entomology 1945–47; Lecturer, Sheffield Univ 1936; Maj, RAMC 1942–45; Prin, Ibadan Univ Coll, Nigeria 1947–53; researcher 1953–55; Head, Entomology Dept, Rothamsted Experimental Station 1955–61; Dir, Monks Wood Experimental Station, Huntingdon 1961–74; Ed, *Environmental Pollution*. *Present Position:* Chair, Council for Environmental Science and Eng 1981–. *Publications:* Scabies 1943; Human Guinea Pigs 1945; The Birth of Nigeria's University 1958; Pesticides and Pollution 1967; The Mole 1971; The Biology of Pollution 1972; Can Britain Feed Itself? 1975; Talpa, The Story of a Mole 1976; Farming and Wildlife 1981; Air Pollution, Acid Rain and the Environment (Ed) 1988; Waste and Pollution 1992; The DDT Story 1992. *Address:* 13 Wonford Rd, Exeter EX2 4LH, UK. *Telephone:* (392) 78715.

MELLOR, Diana Margaret, B SC; New Zealander microbiologist; b 15 Sept 1948; d of Stewart and Margaret Chave; divorced; 1s 2d. *Education:* Otago Polytechnic, Liverpool Polytechnic, Liverpool Univ. *Career:* Mem, Kampuchean Refugee Cttee 1980–82; Research Fellow in Experimental Medicine, Otago Univ 1984–93; Tutor in Zoonoses, Otago Polytechnic 1985–93; mem, Save Manapouri Campaign 1991; Gen Sec, Green Party of Aotearoa/NZ 1991–92. *Present Position:* Co-Convenor, Green Party of Aotearoa/NZ 1993–. *Address:* POB 11-652, Wellington, New Zealand. *Telephone:* (3) 473-9823. *Fax:* (3) 473-9823.

MENA VÁSCONEZ, Patricio A, MA; Ecuadorean conservation biologist; b 5 June 1959, Quito. *Education:* Catholic Univ, Quito, City Univ, New York. *Career:* New York Botanical Garden 1987; Prof of Botany and Evolution, Catholic Univ, Quito 1987–; Prof of Botany and Conservation Biology 1992–. *Present Position:* Pres, Fundación Ecuatoriana de Estudios Ecológicos (Ecociencia) 1989–. *Awards and Honours:* Nat Journalism Prize on Environment 1990. *Publications:* A Revision of the Genus Arcytophyllum 1991; La Nueva Faz del Ambiente (Co-Author) 1992; La Investigación para la Conservación de la Diversidad Biológica en el Ecuador (Co-Ed) 1993. *Address:* Av 12 de Octubre 959 y Roca, Edif Mariana, Of 701, Casilla 17-12-257, Quito, Ecuador. *Telephone:* (2) 526-802. *Fax:* (2) 526-802.

MICHAJLOW, Prof Dr Wlodimierz; Polish environmentalist; b 24 March 1905, Kiev, Ukraine; s of Constantin and Vera Michajlow; 3d. *Education:* Warsaw Univ. *Career:* Secondary school teacher 1933–39; Prof, Acad of Sciences 1956–75, Pres, Man and Environment Comm 1969–82, Prof of Parasitology 1975; mem, Presidium, UNESCO 1973–75; Del, UNEP/UNESCO Ecology Conf, Tbilisi 1977, Moscow 1982. *Present Position:* Mem, Acad of Sciences 1961–. *Awards and Honours:* UNEP Global 500 1987. *Publications:* Protection of Man's Natural Environment 1973. *Address:* Noakowskiego St 14, Apt 11, 00-666 Warsaw, Poland. *Telephone:* (22) 255557.

MIKLOS, Dr Ladislav; Slovak landscape ecologist; b 24 Jan 1949, Tornala; m Viktória Miklos 1973; 2d. *Education:* Comenius Univ, Bratislava, J E Purkyne Univ, Brno. *Career:* Scientist, Ecological Inst, Acad of Sciences 1973–90; Ecological Co-ordinator, Stavoprojekt 1979–89, URBION 1989; Slovak Environment Comm 1989–93; Co-Founder, 3rd Millenium Movt; Deputy Minister of Environment 1990–92; Head, Czecho-Slovak Del, UNCED 1992; mem, Trustee Bd, Regional Environmental Centre, Budapest. *Present Position:* Chair, Scientific Bd, Landscape Ecology Inst, Acad of Sciences, mem, Presidium and Chair, Environmental Section, Slovak UNESCO Comm 1973–. *Publications:* Systematization and Automization of Decision-Making 1986; The General Ecological Model of the Slovak Socialist Republic: Methodology and Content 1989; Basic Premises and Methods in Landscape-Ecological Planning and Optimization 1990; Protecting the Environment during Regional Development 1991. *Address:* POB 254, 814 99 Bratislava, Slovakia. *Telephone:* (7) 330-167. *Fax:* (7) 332-560.

MIRANDA, Rui; Guinea-Bissau civil servant; b 1950, Binar; s of Manuel Silva and Maria Socorro Nogueira Miranda; m Maria Isabel N Miranda 1974; 1s. *Education:* Panama Univ. *Career:* Head, Zootech Research Dept 1982–88; consultant, Addis Ababa 1988–89; Head, Wildlife Dept 1989; Dir, Nature Protection Service and Dir, Coastal Planning Project 1989. *Present Position:* Nat Dir of Coastal Planning 1989–. *Address:* Apdo 23, 1031 Codex Bissau, Guinea-Bissau. *Telephone:* 211065. *Fax:* 201168.

MISHRA, Veer Bhadra, PH D; Indian professor of civil eng and environmentalist; b 1939, Varanasi; s of Bankey Ram Mishra and Annapurna Devi; m Seva Devi 1959; 3s 2d. *Education:* Banaras Hindu Univ, Varanasi. *Career:* Lecturer, Civil Eng Dept, Banaras Hindu Univ, Varanasi 1961, Reader 1968, Prof 1979, Head of Dept 1984, Co-ordinator, Water Resource Man and Training Centre 1989–. *Present Position:* Pres, Sankat Mochan Foundation 1982–, mem, Divisional Task Force, Cen GANGA Authority, Varanasi 1986–. *Awards and Honours:* UNEP Global 500 1992. *Address:* Tulsi Mandir, Tulsi Ghat, Varanasi 221001, Uttar Pradesh, India. *Telephone:* (542) 311458. *Fax:* (542) 311658.

MITCHELL, Andrew William, B SC; British zoologist and author; b 15 Feb 1953, Jersey; s of William Hamish and Anne Dawn Mitchell; m Laura Jane Grumitt 1988; 2s. *Education:* Bristol Univ. *Career:* Scientific Co-ordinator, Scientific Exploration Soc 1978–82; Research Dir, Operation Drake expedition 1980–82; Bd Mem, Operation Raleigh expedition 1983–87; Producer, Natural History Unit, BBC TV 1986–88; Adviser, *Greystoke, The Legend of Tarzan* film, Warner Bros. *Present Position:* Deputy Dir, Earthwatch Europe 1990–. *Publications:* Voyage of Discovery 1981; The Enchanted Canopy 1986; A Fragile Paradise: Nature and Man in the Pacific 1989. *Address:* Belsyre Court, 57 Woodstock Rd, Oxford OX2 6HU, UK. *Telephone:* (865) 311600. *Fax:* (865) 311383.

MITROV, Prof Ivan, D SC; Bulgarian silviculturist; b 15 July 1934, Borilovetz; s of Marin Mitrov Mateev and Gergina Nikova Mateeva; m Velitchka Raditchkova Mitrova 1960; 1s. *Education:* Higher Forestry Inst, Sofia. *Career:* Head, regional forest admin 1959–61; Dir, forestry inst 1962–73; Researcher, Forest Research Inst, Acad of Sciences 1974–. *Present Position:* Head, Dept of Forest Genetics, Physiology and Man-Made

Plantations, Forest Research Inst 1986–. *Publications:* A Method for the Intensive Production of Ash Wood 1982. *Address:* 132 Kliment Ohridski Blvd, 1756 Sofia, Bulgaria. *Telephone:* (2) 62-29-61. *Fax:* (2) 62-29-65.

MKANDA, Francis Xavier, M SC; Malawian wildlife biologist; b 3 Nov 1957, Likoma Island; s of Amos Frederick Mkanda and Margaret Evelyn Kapunda; 2s 1d. *Education:* Edinburgh Univ, Malawi Univ. *Career:* Secondary school teacher 1983. *Present Position:* Research Officer, Nat Parks and Wildlife Dept 1984–. *Address:* POB 30131, Lilongwe 3, Malawi. *Telephone:* 723566. *Fax:* 723089.

MOHAMED, Yagoub A, PH D; Sudanese professor of environmental studies; b 1 Jan 1941, El Fasher; m 1971; 3s 3d. *Education:* Khartoum Univ, Liverpool Univ. *Career:* Research Asst, Khartoum Univ 1965, Lecturer 1975, Assoc Prof 1983–, Dir, Environmental Studies Inst 1986, Training Co-ordinator 1989. *Present Position:* Dir, Environmentalists Soc. *Publications:* North Kodofan Rural Water Supplies 1983; Planning for Environmental Education in Sudan 1983; The Environmental Impact of Agricultural Development in the Arab World 1990. *Address:* POB 321, Khartoum, Sudan. *Telephone:* 80993.

MOLDAN, Dr Bedřich; Czech scientist, teacher, writer and politician; b 15 Aug 1935, Prague; s of Dr František Moldan and Lidmila Moldanová; m Dobrava Moldanová 1957; 2s 1d. *Education:* Charles Univ, Prague. *Career:* Geological Inst, Prague 1958–89; activist, Czech Union of Nature Conservationists; Minister 1989–91; MP 1990–92. *Present Position:* Dir, Environmental Scholarship Centre and Assoc Prof (Docent) of Geochemistry, Charles Univ, Prague 1992–, Vice-Chair, Sustainable Devt Comm, UN 1993–. *Publications:* Geochemistry of the Atmosphere 1977; Atmospheric Deposition 1991; Ecology, Democracy and the Market 1992; Biogeochemistry of Small Catchments 1993. *Address:* Petrská 3, 110 00 Prague 1, Czech Republic. *Telephone:* (2) 2315334. *Fax:* (2) 2315324.

MOLLER, Jeanne; Danish environmentalist. *Present Position:* Gen Sec, Int Solid Wastes and Public Cleansing Asscn. *Address:* Vodroffsvej 22A, 2000 Frederiksberg C, Denmark.

MONTEITH, Prof John Lennox, PH D, DIC, FRS, FRSE, FINSTP; British agricultural scientist; b 3 Sept 1929; s of Rev John and Margaret Monteith; m Elsa Marion Wotherspoon 1955; 4s 1d. *Education:* Edinburgh Univ, Imperial Coll, London. *Career:* Physics Dept, Rothamsted Experimental Station 1954–67; Prof of Environmental Physics, Nottingham Univ 1967–86, Dean, Agricultural Science Faculty 1985–86, Emeritus Prof 1989–; Gov, Grassland Research Inst 1976–83; Vice-Pres, Ecological Soc 1977–79; Pres, Royal Meteorological Soc 1978–80; mem, Nat Environmental Research Council 1980–84, Nat Cttee for World Climate Programme 1980–86, Lawes Agricultural Trust Cttee 1983–86; Visiting Scientist, Resource Man Programme, Int Crops Research Inst for Semi-Arid Tropics 1984, Dir 1987–91; Nat Research Council Sr Research Assoc, Goddard Space Flight Center, MD, USA 1985; Lecturer, Leeds Univ 1986, Florida Univ 1991; Adjunct Prof, Depts of Agricultural Eng and Agronomy, Florida Univ 1991–. *Present Position:* Inst of Terrestrial Ecology. *Awards and Honours:* Royal Meteorological Soc Prize 1962; Int Soc of Biometeorology Award 1983; Rank Fund Nutrition Prize 1989. *Publications:* Instruments for Micrometeorology (Ed) 1972; Heat Loss from Animals and Man (Co-Ed) 1974; Vegetation and the Atmosphere (Ed) 1975; Soil, Water and Nitrogen (Co-Ed) 1981; Principles of Environmental Physics (Co-Author) 1990. *Address:* Bush Estate, Penicuik EH26 0QB, UK. *Telephone:* (31) 445-4343. *Fax:* (31) 445-3943.

MOORE, Sir Norman Winfrid, PH D; British biologist; b 24 Feb 1923; m Janet Singer 1950; 1s 2d. *Education:* Trinity Coll, Cambridge, Bristol Univ. *Career:* War service, Germany and Holland 1942–45; Prin Scientific Officer, Nature Conservancy Council 1958–65, Sr Prin Scientific Officer 1965–83; Visiting Prof of Environmental Studies, Wye Coll, London Univ 1979–83. *Present Position:* Environmental scientist. *Publications:* The Bird of Time 1987. *Address:* Farm House, Swavesey, Cambridge, UK. *Telephone:* (954) 30233.

MORRIN, Noel; British businessman and environmentalist. *Present Position:* Project Dir, Business in Environment. *Address:* 5 Cleveland Place, London SW1Y 6JJ, UK. *Telephone:* (71) 321-6432. *Fax:* (71) 321-6410.

MORTON, Brian, PH D; British professor of zoology; b 10 Aug 1942; m Janice Elizabeth Burn 1967; 2s 2d. *Education:* London Univ. *Career:* Lecturer, Zoology Dept, Hong Kong Univ 1970–75, Reader in Marine Biology 1975–82; Chair, Texas Christian Univ, Fort Worth 1981. *Present Position:* Prof of Zoology, Hong Kong Univ 1982–, Dir, Swire Marine Lab 1990–. *Awards and Honours:* UNEP Global 500 1989. *Publications:* The Future of the Hong Kong Sea Shore 1979; Sea Shore Ecology of Hong Kong (Co-Author) 1983; Partnerships in the Sea 1988; Shore Ecology of the Gulf of Mexico (Co-Author) 1989. *Address:* Cape d'Aguilar, Hong Kong. *Telephone:* 8092179. *Fax:* 8092197.

MOUDUD, Hasna J, MA; Bangladeshi conservationist and writer; b 11 Nov 1946, Faridpur; m Ahmed Moudud 1972; 1s 1d. *Education:* Dayton Univ, OH, Iowa Univ. *Career:* UNDP and US Cttee, UNICEF 1971–72; Lecturer, Dhaka Univ 1972–73; Alternate Rep, Gen Ass, UN 1983; MP 1986–90, Chair, Special Parl Cttee for Coastal Devt and Environment 1989–90; Int Elections Observer, Philippines 1988. *Present Position:* Pres, Nat Asscn for Resource Improvement 1986–, Coastal Area Resource Devt and Man Asscn 1987–, Vice-Pres, Global 500 Forum 1992–. *Awards and Honours:* UNEP Global 500 1992. *Publications:* Selected Poems of Jasimuddin (Ed) 1975; Women in China 1980; Bengali Mystic Poetry 1992. *Address:* 159 Gulshan Ave, Dhaka, Bangladesh. *Telephone:* (2) 600076. *Fax:* (2) 865547.

MOUSEL, Michel; French environmentalist. *Present Position:* Agence de l'Environnement et de la Maîtrise de l'Énergie. *Address:* Paris, France.

MUNN, Dr R E; Canadian environmental scientist. *Present Position:* Mem, Inst for Environmental Studies, Toronto Univ. *Address:* 215 Huron St, Toronto, ON M5S 1A1, Canada. *Telephone:* (416) 978-8202. *Telex:* 06218915. *Fax:* (416) 978-3884.

MUÑOZ-PEDREROS, Andres, M SC; Chilean environmentalist, univ prof and researcher; b 1955; m 1981; 2d. *Education:* Austral Univ, Valdivia. *Career:* Founder, Wildlife Soc of Chile, 1976–, Chair, Research Dept; Research Asst, Illinois Univ and Austral Univ, Valdivia 1983–85; Prof, Catholic Univ, Chile 1986–90, Temuco 1990–; Founder, Agric Study Research Centre (CEA) 1986–. *Present Position:* Chair, Natural Sciences Dept, Catholic Univ, Temuco 1991–. *Publications:* Environmental Education 1992; Mammals of Chile (Ed) 1993. *Address:* Casilla 164, Valdivia, Chile. *Telephone:* (63) 21-5846. *Fax:* (63) 21-3149.

MURILLO, Manuel M; Costa Rican oceanographer. *Present Position:* Chair, Intergovernmental Oceanographic Comm (IOC), UNESCO. *Address:* c/o 7 place de Fontenoy, 75700 Paris, France. *Telephone:* (1) 45-68-39-83. *Telex:* 204461. *Fax:* (1) 40-56-93-16.

MUYANDA-MUTEBI, Peter, PH D; Ugandan environmentalist and univ teacher; b 1941; m 1981; 1s 5d. *Education:* Makerere Univ, Kampala, Columbia Univ, NY, California Univ, Los Angeles. *Career:* Lecturer then Sr Lecturer, Makerere Univ, Kampala 1972–76; Sr Lecturer, Kenyatta Univ 1976–83. *Present Position:* Exec Dir, African Social and Environmental Studies Programme (ASESP) 1983–; Consultant, UNESCO, UNICEF and UNEP 1989–; Chair, Educ for All Network of Eastern and Southern Africa (EFANESA) 1992–, mem, Bd, Centre for Int Studies. *Publications:* Working Document for the First African Regional Conference on the Environment and Sustainable Development 1989; Educational Research for Development in Africa (Co-Ed) 1993. *Address:* POB 44777, Nairobi, Kenya. *Telephone:* (2) 747960. *Fax:* (2) 747960.

MYERS, Norman, PH D; British scientist; b 24 Aug 1934, Whitewell, Yorks; s of John and Gladys (Haworth) Myers; m Dorothy Halliman 1966; 2d. *Education:* Oxford Univ, California Univ, Berkeley. *Career:* Scientific consultant, UN, World Comm on Environment and Devt 1985–86, OECD, European Comm, US Dept of State, US Dept of Energy, NASA, NAS, US Nat Research Council, MacArthur Foundation, Intergovernmental Panel on Climate Change, World Resources Inst, Shell Int, Mobil UK, British Telecom, Hitachi, Int Distillers and Vintners Ltd, World Business Council; Fellow, Green Coll, Oxford; Prof-at-Large, Cornell Univ; Visiting Prof, Kent Univ and Utrecht Univ; Adviser, Global Environment Facility, World Bank; Sr Fellow, WWF US; Sr Adviser, UNCED 1992. *Present Position:* Scientific consultant 1970–. *Awards and Honours:* UNEP Global 500; Distinguished Achievement Award, Soc for Conservation Biology; Gold Medal and Order of Golden Ark, WWF Int; Special Achievement Award, Sierra Club; Gaia Int Prize; Volvo Environment Prize 1992. *Publications:* The Long African Day 1972; The Sinking Ark 1979; Conversion of Tropical Moist Forests 1980; A Wealth of Wild Species 1983; The Primary Source: Tropical Forests and Our Future 1984; The Gaia Atlas of Planet Management 1984; Future Worlds: Challenge and Opportunity in an Age of Chance 1990; Population, Resources and the Environment: The Critical Challenges 1991; Tropical Forests and Climate (Ed) 1992; Ultimate Security: The Environmental Basis of Political Stability 1993; Scarcity of Abundance: A Debate on the Environment 1993. *Address:* Upper Meadow, Old Rd, Headington, Oxford OX3 9SZ, UK. *Telephone:* (865) 750387. *Fax:* (865) 741538.

N

NADER, Ralph; American lawyer, author and consumer advocate; b 27 Feb 1934, Winsted, CT; s of Nadra Nader and Rose Bouziane. *Education:* Princeton Univ, Harvard Univ. *Career:* US Army 1959; law practice, Hartford, CT 1959–; Lecturer in History and Govt, Hartford Univ 1961–63; Head, Public Citizen Inc; Lecturer, Princeton Univ 1967–68; syndicated columnist 1972–; Contributing Ed, *Ladies Home Journal* 1973–. *Present Position:* Co-Founder, Princeton Project 55 1989–. *Publications:* Unsafe at Any Speed 1965; Who Runs Congress? 1972; The Consumer and Corporate Accountability (Ed) 1974; Taming the Giant Corporation (Co-Author) 1976; The Menace of Atomic Energy (Co-Author) 1979; The Lemon Book 1980; Who's Poisoning America? 1981; The Big Boys 1986; Winning the Insurance Game (Co-Author) 1990. *Address:* POB 19367, 2000 P St, NW, Washington, DC 20036, USA.

NANAJKAR, Shekhar Devidas, B SC; Indian electronics dealer, writer and environmentalist. *Career:* Chair, Inst of Environment and Wildlife Studies. *Present Position:* Founder Pres, Wild, Pres, Civic Environment Protection Cttee and Wild Adventure Inst, Ed, *Wild Vrutta*, monthly magazine. *Address:* 21A Samita, Swanand Soc, Sahakar Nagar 2, 411 009 Pune, India. *Telephone:* (212) 430877.

NESTEROV, Prof Dr Peter I; Romanian zoologist; b 1 Aug 1924, Bolotina, Moldova; m Marija Nesterov 1926; 2d. *Education:* Tiraspol Pedagogical Inst. *Career:* Romanian Army 1944; Soviet Army, Poland and Germany 1944–45; teacher 1947–62; Acad of Sciences, Moldova 1962–. *Present Position:* Pres, Soc for Protection of Animals and Scientist, Zoological Inst, Acad of Sciences. *Publications:* Phytoparasites and Nematodes of the South-West USSR 1979; The Systematic Ecology and Practical Importance of Nematodes 1988; Honey Resources of Moldova 1988; The Black Stork 1990; The Forest Stream 1990. *Address:* 277028 Chişinău, Str Academiei 1, Moldova. *Telephone:* (0422) 73-98-21.

NETHERLANDS, HRH Prince Bernhard Leopold Frederik Everhard Julius Coert Karel Godfried Pieter of, GBE; Dutch royal and environmentalist; b 29 June 1911, Germany; s of HSH Prince Bernhard zur Lippe and Princess Armgard, Baroness von Sierstorpff-Cramm; m Juliana Louise Emma Marie Wilhelmina, Queen of Netherlands 1937–80; 4d. *Education:* Lausanne Univ, Munich Univ, Berlin Univ. *Career:* Chief Netherlands Liaison Officer, British Forces, Col then Maj-Gen and Chief, Netherlands Mission to War Office; Lieut-Gen and Supreme Commdr, Netherlands Armed Forces 1944–45; mem, Bd, Netherlands Trade and Industries; Founder-Pres WWF Int; Pres, Rhino Rescue Trust. *Present Position:* Pres, WWF Netherlands 1976–. *Awards and Honours:* UNEP Global 500 1991. *Address:* Soestdijk Palace, Baarn, Netherlands.

NICHOLSON, (Edward) Max, CB, CVO, FZS; British environmentalist; b 1904; m 1st Eleanor Mary Crawford 1932 (dissolved 1964); m 2nd Marie Antoinette Mauerhofer; 3s. *Education:* Hertford Coll, Oxford. *Career:* Head, Allocation of Tonnage Div, Ministry of War Transport 1942–45; Sec, Office of Lord Pres of Council 1945–52; mem, Advisory Council on Scientific Policy 1948–64; Dir-Gen, Nature Conservancy 1952–66; Convenor, Conservation Section, Int Biological Programme 1963–74; Sec, Duke of Edinburgh's Study Conf on Countryside in 1970 1963; Lecturer, California Univ 1964; Chair, Land Use Consultants 1966–89; mem, Council, Int Inst of Environment and Devt 1972–88, Int Council, WWF 1983–86; Chair, Environmental Cttee, London Celebrations for the Queen's Silver Jubilee 1976–77, London Looks Forward Conf 1977, Trust for Urban Ecology 1977–87, UK Standing Cttee for World Conservation Strategy Programme 1981–83, Common Ground Int 1981–; Pres 1987–88; Vice-Pres, RSA 1978–82, Wildfowl and Wetlands Trust, WWF UK; Dir and Man Ed, Environmental Data Services Ltd 1978–80; Pres, RSPB 1980–85, Simon Population Trust 1985–; Trustee, Earthwatch Europe 1985–. *Present Position:* Prin, Land Use Consultants. *Awards and Honours:* IUCN Medal 1963; Europa Preis für Landespflege 1972; Commdr, Order of Golden Ark, Netherlands 1973. *Publications:* Birds in England 1926; How Birds Live 1927; Birds and Men 1951; Britain's Nature Reserves 1958; The System 1967; The Environmental Revolution 1970; The Big Change 1973; The New Environmental Age 1987; The Birds of the Western

Palearctic (Co-Ed) 1991. *Address:* 13 Upper Cheyne Row, London SW3 5JW, UK.

NIMIR, Mutasim Bashir, PH D; Sudanese wildlife biologist; b 19 Nov 1945, Khartoum; m Gawahir Sidahmed 1981; 1s 3d. *Education:* Khartoum Univ, Montana State Univ, Colorado State Univ. *Career:* Wildlife research officer 1968–75; Dir of Wildlife Research, Agric Research Corpn 1976–79, 1983–90. *Present Position:* Exec Dir, Sudanese Environment Conservation Soc 1991–. *Publications:* Wildlife Values and Management in Northern Sudan 1983. *Address:* POB 44266, Khartoum Cen, Sudan. *Telephone:* 77017. *Telex:* 22064.

NOGUEIRA NETO, Paulo. *Career:* Environmental Adviser, Ministry of Culture; First Brazilian Secretary of Environment. *Present Position:* Visiting Professor Univ of São Paulo. *Awards and Honours:* John Paul Getty Prize 1982. *Address:* c/o Univ of São Paulo, Cidade Universitária, CP 8191, 05508 São Paulo SP, Brazil.

NORIEGA, Beatriz; Mexican environmentalist. *Present Position:* Consultant, Greenpeace Mexico. *Address:* c/o Grupo de Consultores en Mexico, Apdo 76210, Cerro del Cubilete 240, int 1, Col Campestre Churubusco, 04200 México, DF, Mexico.

NORMAN, Sir Arthur Gordon (Gerry), KBE, DFC; British business exec; b 18 Feb 1917, N Pertherton, Somerset; s of Christopher William Norman and Mary Christine Milton; m Margaret Doreen Harrington 1944 (died 1982); 3s 2d. *Career:* Thomas de la Rue & Co 1934–87, Asst Gen Man 1947, Dir 1951, Man Dir 1953, Chair, de la Rue Co plc 1964–77, Non-Exec Chair 1977–87; RAF 1941–46; Pres, CBI 1968–70; Dir, SKF (UK) Ltd 1970–87; Chair of Trustees, WWF (UK) 1977–84; mem, Nature Conservancy Council 1980–86; Vice-Chair, Sun Life Assurance Soc 1984–87; Dir, Kleinwort, Benson, Lonsdale plc 1985–88; Treas, Int Inst for Environment and Devt. *Present Position:* Chair, UK Centre for Econ and Environmental Devt 1984–. *Address:* Fir Tree Cottage, Hammoon, Sturminster Newton DT10 2DB, UK. *Telephone:* (258) 860296.

NORRIS, John; British politician and environmentalist. *Present Position:* Exec Chair, Green Party. *Address:* 10 Station Parade, Balham High Rd, London SW12 9AZ, UK. *Telephone:* (81) 673-0045.

NORTH, Klaus, PH D; German prof of int man; b 29 May 1954, Giessen. *Education:* Darmstadt Tech Univ. *Career:* Researcher and Lecturer, Ergonomics Inst, Darmstadt Tech Univ 1978–82; Assoc Expert, ILO Regional Office for Latin America and Caribbean, Lima 1982–84; Man Consultant, Intra, Düsseldorf 1984–85; Head, Int Dept, REFA (Industrial Eng and Productivity Asscn), Darmstadt 1985–90; Sr Research and Programme Devt Officer, Entrepreneurship and Man Devt Branch, ILO 1990–93. *Present Position:* Prof of Int Business Admin, Wiesbaden Polytechnic 1993–. *Publications:* Environmental Business Management 1992; Environmental Concerns in Management Development Institutions: A Survey 1992; Developing Human Resources for Environmental Business Management Proceedings 1993. *Address:* Jahnstr 119, 64285 Darmstadt, Germany. *Telephone:* (6151) 42-2039.

NOWACKI, Konrad Jerzy, LL D; Polish lawyer; b 23 April 1946; m Eva Seminowicz 1970; 1s. *Education:* Wrocław Univ, Cologne Univ. *Career:* Political Sciences Inst, Wrocław 1970–72; Asst and Research Assoc, Law Admin Faculty, Wrocław Univ 1972–80; Research Fellow, Humboldt Foundation, Cologne Univ and Vienna Univ 1983–84, 1989; Pres, Silesian Branch, Asscn for Protection of Monuments 1985–92; Visiting Research Asst, Bremen Univ 1987, 1989. *Present Position:* Sr Lecturer, Admin Sciences Inst, Wrocław Univ 1981–, mem, Polish Environmental Law Asscn and Int Environmental Law Network, Darmstadt. *Awards and Honours:* Ministry of Culture Award for Protection of Landscape and Monuments 1990. *Publications:* Free Access to Environmental Files in Poland (Co-Author) 1990; Öffentlichkeit von Umweltinformationen (Co-Author) 1990. *Address:* 50-145 Wrocław, ul Uniwersytecka 22/26, Poland. *Telephone:* (71) 489542. *Fax:* (71) 402784.

NTAHUGA, Laurent Semuryama, DR RER NAT; Burundi zoologist; b 10 Jan 1951, Gisigi; m Marie Thérèse Gateretse; 1s 3d. *Career:* Asst Lecturer, Burundi Univ 1978–80, Lecturer in Biology 1984–90, Dean, Sciences Faculty 1989–91, Assoc Prof 1990–. *Present Position:* Dir-Gen, Inst Nat pour l'Environnement et la Conservation de la Nature (INECN) 1991–. *Publications:* La protection de la nature au Burundi 1989; The Fauna of Lake Tanganyika Plain's Protected Sites and Surrounding Areas 1991; Environmental and Nature Conservation in Burundi 1993. *Address:* BP 1855, Bujumbura, Burundi. *Telephone:* (40) 2071. *Telex:* 3000. *Fax:* (40) 2075.

NTAKIMAZI, Prof Gaspard; Burundi hydrologist; b 15 Dec 1951, Ngozi; m Gahwa Asha 1987; 2s. *Career:* Asst, Burundi Univ 1975–78, Head of Course 1985–90, Head, Biology Dept 1986–89, Research Dir 1989–92; Asst, Luxembourg Univ Foundation 1978–85. *Present Position:* Assoc Prof, Burundi Univ 1990–, Dir, Regional Centre for Research into Applied Hydrobiology 1993–. *Publications:* Hydrobiology of Lakes Cohoha and Rweru 1985. *Address:* BP 2700, Bujumbura, Burundi. *Telephone:* (2) 25556. *Telex:* 5161. *Fax:* (2) 22500.

O

OBASI, Prof Godwin Olu Patrick, D SC; Nigerian int civil servant; b 24 Dec 1933, Ogori; s of Albert B Patrick Obasi and Rhoda A Akande; m Winifred O Akande 1967; 6c. *Education:* McGill Univ, Canada, MIT, USA. *Career:* Expert and Sr Lecturer, WMO and UNDP, Nairobi Univ 1967–74, Acting Head, Meteorology Dept 1972–73, Dean, Science Faculty and Prof, Meteorology Dept 1974–76; Meteorology Adviser, Nigerian Govt and Head, Nigerian Inst for Meteorological Research and Training 1976–78; Dir, Educ and Training Dept, WMO 1978–83. *Present Position:* Sec-Gen, WMO 1984–. *Address:* CP 2300, 41 ave Giuseppe Motta, 1211 Geneva 2, Switzerland. *Telephone:* (22) 7308200. *Telex:* 414199. *Fax:* (22) 7342326.

OBEIDAT, HE Ahmad Abdul-Majeed; Jordanian lawyer and politician; b 18 Nov 1938, Jordan; s of Abdul Majeed M and Maryam Obeidat; m Amal Obeidat 1964; 3s 2d. *Education:* Baghdad Univ. *Career:* Minister of Interior 1982–84, Prime Minister and Minister of Defence 1984–85; Attorney-at-Law and Senator 1985–; Chair, Royal Comm for Nat Charter and Environment Cttee, Senate. *Present Position:* Pres, Jordanian Soc for Control of Environmental Pollution (JSCEP). *Address:* POB 2480, Amman, Jordan. *Telephone:* (6) 672222. *Telex:* 24277. *Fax:* (6) 699222.

O'GORMAN, Fergus, M SC; Irish environmental educationalist; b 26 Nov 1934, Dublin; s of Martin and Kathleen (Swan) O'Gorman; m 1st Sheila Scott 1962 (deceased); m 2nd Frances Hardiman 1976; 3s 5d. *Education:* Univ Coll, Dublin, Univ Coll, Cork, London Univ. *Career:* Biologist, British Antarctic Survey 1957–60; Researcher, Charing Cross Hosp Medical School, London 1960–62; Asst Lecturer, Univ Coll, Cork 1962–65; Scientific Adviser, Ministry for Lands and Forestry 1962–75, Head of Wildlife Research 1973–75; Chair, Irish Nat Trust 1964; Founder Mem, Irish Wild Bird Conservancy 1966, Irish Wildlife Fed 1979, Chair 1979, Pres 1986; Sr Wildlife Officer, FAO 1971–73; Dir, Int Programme, Tour du Valat, Camargue, France 1975–78, Int Environmental Consultants, Dublin and London 1978–82; Dir of Environmental Educ, Nat Environment Inst 1982–86, Environmental Awareness Bureau 1982–85; Controller, Govt Supplies Agency and Project Devt Dir, Nat Conservation Educ Centre 1986–89; Irish Mem, Exec, European Environmental Bureau, Brussels 1986–90, Rep, UNCED 1992. *Present Position:* Dir, Nat Conservation Educ Centre 1992–. *Publications:* Irish Wildlife: A Blueprint for the Future (Ed) 1974; The World Around Us 1976; The Irish Wildlife Book (Ed) 1982; The International Wildlife Congress (Ed) 1984. *Address:* Knocksink Wood, Enniskerry, Co Wicklow, Ireland. *Telephone:* (1) 2866609. *Fax:* (1) 2866610.

OHISHI, Buichi; Japanese environmentalist. *Present Position:* Pres, Defence of Green Earth Foundation. *Address:* 7A Howa Mansion, 3-25-5 Asakusabashi Taito, Tokyo 111, Japan.

OJA, Tonu; Estonian environmentalist. *Present Position:* Chair, Eesti Roheline Liikumine (Estonian Green Movement) and Tartu Ecocentre. *Address:* POB 318 (Veski 35), Tartu EE2400, Estonia. *Telephone:* (1434) 32-986. *Telex:* 173243. *Fax:* (1434) 35-440.

OKE, Prof Timothy Richard, PH D, FRSC; Canadian/British geographer and urban climate expert; b 22 Nov 1941, Kingsbridge, Devon, UK; s of Leslie Oke and Kathleen Smerdon; m Margaret Lowe 1967; 1s 1d. *Education:* Bristol Univ, McMaster Univ, Hamilton, ON. *Career:* Lecturer, McMaster Univ, Hamilton, ON 1966–67; Asst Prof, McGill Univ 1967–70; Asst Prof, British Columbia Univ 1970–71, Assoc Prof 1971–78; Ed, *Atmosphere Ocean* 1977–80; Assoc Ed, *Atmospheric Environment* 1987–; Visiting Fellow, Keble Coll, Oxford 1990. *Present Position:* Prof, British Columbia Univ 1978–, Head, Geography Dept 1991–. *Publications:* Boundary Layer Climates 1987; Vancouver and It's Region 1992. *Address:* 1984 W Mall, Vancouver, BC V6T 1Z2, Canada. *Telephone:* (604) 822-2900. *Fax:* (604) 822-6150.

OKIGBO, Bede Nwoye, PH D, D SC; Nigerian agronomist; b 26 Sept 1929, Ojoto-Uno, Idemili, Anambra; s of Chief Nnaebue and Nwangkoli Okigbo; m Philomena Nkiruka Mbanwusi 1961; 4s 2d. *Education:* Washington State Univ, Cornell Univ, NY. *Career:* Researcher, W Nigeria 1959–60; Research Fellow, Ibadan Univ 1960–61; Sr Lecturer, Nigeria Univ, Nsukka 1961–64, Assoc Prof 1965, Prof and Head, Plant and Soil Science Dept 1965–72, Dean, Agric Faculty 1965–73; Dir, Farming Systems Programme, IITA, Ibadan 1973–77, Deputy Dir-Gen 1977–88. *Present Position:* Dir, African Natural Resources Programme, UNU 1990–. *Publications:* Plants and Food in Igbo Culture and Civilization 1981; Development of Sustainable Agricultural Production Systems in Africa 1991. *Address:* c/o UNESCO, POB 30592, Nairobi, Kenya. *Telephone:* (2) 621234. *Telex:* 22275. *Fax:* (2) 520043.

OKUN, Daniel A, D SC; American environmental engineer; b 19 June 1917, New York; m Beth Griffin 1946; 1s 1d. *Education:* Harvard Univ. *Career:* Engineer, Public Health Service 1940–42; 1st Lieut then Maj, US Army 1942–46; Teaching Fellow, Harvard Univ 1946–48; Malcolm Pirnie Inc Consulting Engineers, New York 1948– 52; Prof of Environmental Eng, N Carolina Univ 1952–82; consulting engineer, UN, World Bank 1952–. *Present Position:* Emer Prof of Environmental Eng, N Carolina Univ, Chair, Water Science and Tech Bd, Nat Research Council. *Publications:* Water and Waste Water Engineering (Co-Author) 1968; Regionalization of Water Management: A Revolution in England and Wales 1977; Surface Water Treatment for Consumption in Developing Countries 1984. *Address:* 900 Linden Rd, Chapel Hill, NC 27514, USA. *Telephone:* (919) 966-4898. *Fax:* (919) 966-7646.

OLANIYA, Mahendra Singh, PH D; Indian scientist; b 10 Jan 1937, Jaipur; s of S S and Chandrawati Olaniya; m Vimla Olaniya 1962; 1s 2d. *Education:* Osmania Univ, Hyderabad. *Career:* Researcher, water resources 1962–72, pollution and wastes 1972–80, environmental planning 1980–. *Present Position:* Asst Dir, Solid Wastes Div, Nat Environmental Eng Research Inst. *Address:* Nagpur 440020, India. *Telephone:* (712) 526071. *Telex:* 0715233. *Fax:* (712) 523893.

OLSSON, Lars E, PH D; Swedish int civil servant; b 31 Jan 1937, Stockholm; 2s. *Education:* Royal Tech Inst, Stockholm, Michigan Univ. *Career:* Asst Prof, Oregon State Univ 1969–73, Stockholm Univ 1974–75; Dir of Climatology, Meteorological and Hydrological Inst (SMHI), Norrköping 1975–86. *Present Position:* Chief, World Climate Programme Div, WMO 1986–. *Address:* 41 ave Giuseppe Motta, CP 2300, 1211 Geneva 2, Switzerland. *Telephone:* (22) 7308277. *Telex:* 414199. *Fax:* (22) 7342326.

OMENN, Gilbert S, PH D; American professor of community medicine; b 30 Aug 1941, Chester, PA; m Martha Darling 1984; 2s. *Education:* Princeton Univ, NJ, Harvard Univ, Washington Univ, Seattle. *Career:* Sr Fellow, Washington Univ, Seattle 1969–; White House Fellow, Atomic Energy Comm 1973–74; Deputy Science Adviser and Assoc Dir, Man and Budget Office and Chair, Environmental Studies and

Toxicology Bd 1988–91; mem, Nat Environment Comm 1991–93. *Present Position:* Prof of Medicine and Environmental Health and Community Medicine 1961–, and Dean, Public Health and Community Medicine School, Washington Univ, Seattle 1982–. *Address:* SC-30, WA 98195, USA. *Telephone:* (206) 543-1144. *Fax:* (206) 543-3813.

O'RIORDAN, Timothy, MA, MS, PH D, FRSA; British professor of environmental sciences; s of Kevin Denis and Norah Joyce (neé Lucas) O'Riordan; m Ann Morison Philip; 2d. *Education:* Univ of Edinburgh, Cornell Univ, NY, Univ of Cambridge. *Career:* Asst Lecturer and Assoc Prof Dept of Geography, Simon Fraser Univ, Canada 1967–74; Reader, School of Environmental Sciences, Univ of East Anglia 1974–80. *Present Position:* Prof of Environmental Sciences, Univ of East Anglia. *Awards and Honours:* Gill Memorial Award, Royal Geographical Soc 1982. *Publications:* Environmentalism 1976 (2nd edn 1981), Countryside Conflicts (Co-Author) 1986, Sizewell B: An Anatomy of the Enquiry (Co-Author) 1988. *Address:* School of Environmental Studies, Univ of East Anglia, Norwich NR14 7TJ, UK. *Telephone:* (603) 810534.

OSBORN, Derek, CB, BA, B PHIL; British civil servant; b 14 Jan 1941; s of Rev George and Betty Osborn; m Caroline Osborn 1971; 1d 1s. *Education:* Oxford Univ. *Career:* Ministry of Housing and Local Govt 1965; Dept of Environment 1970–. *Present Position:* Dir-Gen of Environmental Protection, Dept of Environment 1989–. *Address:* Romney House, 2 Marsham St, London SW1P 3PY, UK. *Telephone:* (71) 276-3000. *Telex:* 22221. *Fax:* (71) 276-0818.

O'SULLIVAN, Prof Patrick; British energy and bldg expert. *Present Position:* Head, Bartlett School of Architecture and Planning. *Address:* Wates House, 22 Gordon St, London WC1H 0QB, UK.

OSZLÁNYI, Július, PH D; Slovak scientist; b 19 Sept 1944, Trnava; m Anna Hanáková 1979; 1d. *Education:* Zvolen Tech Univ, Comenius Univ, Bratislava. *Career:* Engineer, Forestry Faculty, Zvolen Tech Univ 1962–67, Researcher in Production Ecology 1969–74; Researcher then Sr Researcher, Forestry Research Inst 1974–84. *Present Position:* Sec, Slovak Ecological Soc, Chair, Nat Cttee, MAB, Sr Researcher, Sr Scientist and Deputy Dir, Ecobiology Inst, Acad of Sciences 1984–. *Publications:* Vertical Distribution of Overground Biomass 1977; Energy Value of Overground Biomass in Oaks 1979; Analysis of the Surface Area of Tree Species 1980; Production and Productivity of Beech Forests 1988. *Address:* Štefánikova 3, POB 259, 814 34 Bratislava, Slovakia. *Telephone:* (7) 332-016. *Fax:* (7) 334-968.

OTORBAEV, Prof Kaip, PH D; Kyrgyz geographer and economist; b 22 April 1922, Tegermenti, Kemin Dist; s of Otorbai Kulambaev and Danapija Baiterekova; m Maria Nanaeva 1954; 2s. *Education:* Kyrgyz State Univ, Frunze, Moscow Geography Inst. *Career:* Dir, Economy Inst, Acad of Sciences 1962–66, Chief Scientific Sec 1965–69; Dir, Economy Inst, Kyrgyzstan Planning Cttee 1978–79; Rector, Kyrgyz State Univ, Frunze 1979–88. *Present Position:* Chair, Productive Forces Study and Natural Resources Comm, Acad of Sciences 1988–. *Awards and Honours:* Order of Red Star 1943; Order of Red Banner of Labour 1981; Order of Patriotic War 1985. *Publications:* The Jalal-Abad Region of the Kyrgyz SSR 1957; Nature in Kyrgyzstan 1962; Mining and Industrial Development in Kyrgyzstan 1987. *Address:* 720001 Bishkek, Chui Prospekt 261-12, Kyrgyzstan. *Telephone:* (3312) 25-31-86. *Telex:* 251245.

OUEDRAOGO, Ram; Burkinabé environmentalist. *Present Position:* Sec-Gen, Union des Verts pour le Développement du Burkina (UVDB). *Address:* Ouagadougou, Burkina Faso.

OUIMET, Jean; Canadian politician and environmentalist. *Present Position:* Leader, Parti Vert du Québec/Green Party of Quebec. *Address:* 2093 rue St-Denis, Montreal, PQ H2X 3K8, Canada. *Telephone:* (514) 287-7392. *Fax:* (514) 288-7337.

P

PACHAURI, Prof R K, PH D; Indian energy researcher; b 20 Aug 1940. *Education:* N Carolina State Univ. *Career:* Visiting Prof of Resource Econs, Mineral and Energy Resources Coll, W Virginia Univ 1981–82. *Present Position:* Dir, Tata Energy Research Inst. *Address:* 9 Jor Bagh, New Delhi 110003, India. *Telephone:* (11) 4627651. *Telex:* 3161539. *Fax:* (11) 4621770.

PACYNA, Prof Jozef Marian, PH D; Norwegian environmental health scientist; b 12 Jan 1949, Kepno, Poland; m Elisabeth G Pacyna 1972; 1s 1d. *Education:* Wrocław Tech Univ, Poland. *Career:* Assoc Prof, Wrocław Tech Univ, Poland 1972–81; Visiting Scientist, London Univ 1982, Nat Center for Atmospheric Research, Boulder, CO, USA 1989. *Present Position:* Sr Research Scientist, Norsk Institutt for Luftforskning (Norwegian Air Research Inst), Lillestrøm 1983–; Visiting Prof, School of Public Health, Michigan Univ, USA. *Awards and Honours:* Polish Ministry of Higher Educ and Tech Award for Environmental Research 1979; Nordic Aerosol Asscn Aerosol Scientist of the Year 1991. *Publications:* Quantitative Assessment of Worldwide Contamination of Air, Water and Soils by Trace Elements 1988; Control of Atmospheric Heavy Metals 1989. *Address:* POB 64, 2001 Lillestrøm, Norway. *Telephone:* (63) 81-41-70. *Fax:* (63) 81-92-47.

PACZUSKI, Prof Ryszard; Polish lawyer and teacher; b 1929, Zakopane; s of Seweryn Paczuski and Barbara Buczkowska; m Halina Paczuski 1957; 1s. *Education:* Poznań Univ. *Career:* Secondary school teacher 1953–58; Univ Teacher 1974–, Prof 1992–. *Present Position:* Prof and Head, Environmental Law Dept, Nicolaus Copernicus Univ, Torun. *Publications:* Environmental Law as a Subject in High Schools 1986; Legal-Administrative Issues in the Protection of the Natural Environment against Extreme Emergencies 1987; The Water Boards: History, Legal Foundations and New Development Trends 1989; Problems of Health and Natural Resources in Poland 1990. *Address:* 85-644 Bydgoszcz, ul Cisowa 4 m 33, Poland. *Telephone:* (52) 410426.

PADUA, José Augusto, PH D; Brazilian political scientist and ecologist; b 24 March 1959, Rio de Janeiro; s of João Lima and Lucia Valladares Padua; m Ligia Saramago Padua 1982; 2d. *Education:* Rio de Janeiro Catholic Univ. *Career:* Mem, Brazilian environmental movt 1976–; Sec, Amazon Defence Cttee 1979; Prof, History Dept, Rio de Janeiro Catholic Univ 1984–90; mem, Nat Bd, Brazilian Green Party 1986–88; Co-ordinator, Amazon Ecological Disorder Project, Int Social Science Council 1988–90; Environmental Studies Co-ordinator, Brazilian Inst for Social and Econ Studies (IBASE) 1990. *Present Position:* Latin America Co-ordinator, Forests Campaign, Greenpeace Brazil 1991–. *Publications:* What Is Ecology? 1984; Ecology and Politics in Brazil 1988; The Birth of Green Politics in Brazil 1992. *Address:* Rua México 21, 13°, 20031-144 Rio de Janeiro, Brazil. *Telephone:* (21) 240-4476. *Fax:* (21) 240-1690.

PALZ, Dr Wolfgang; German solar energy researcher. *Present Position:* Head, Solar Energy Programme, Directorate-Gen XII, Comm of EC. *Address:* sq de Meeus 8, 1040 Brussels, Belgium. *Telephone:* (2) 235-69-22. *Fax:* (2) 236-30-24.

PARKIN, Sara Lamb; British politician; b 9 April 1946, Aberdeen; d of George Lamb and Marie Munro (Rankin) McEwan; m Donald Maxwell Parkin 1969; 2s. *Education:* Michigan Univ. *Career:* Staff Nurse then Ward Sister, Edinburgh Royal Infirmary 1970–74; Research Asst, Nursing Research Unit, Edinburgh Univ; sex and relationship counsellor 1976–80; Press Officer and Newsletter Ed, Yorkshire and Humberside Branch, Ecology Party 1978–80, mem, Nat Council 1980–81, Int Liaison Sec, Green Party 1983–90, Speaker on Int Affairs 1989–91, Speaker, European Election Campaign 1989, Chair, Nat Exec 1992; Co-Sec of Co-ordination, European Greens 1985–91, Ed, *Newsletter of the European Greens* 1986–89; Fellow, Schumacher Soc. *Present Position:* Writer and speaker on green politics 1981–, mem, Advisory Bd, Earth Centre and Environmental Working Group, Eng Council 1991–. *Publications:* The European Greens Joint Programme (Co-Author) 1984; The UK Green Party General Election Manifesto (Co-Author) 1987; The European Greens Election Programme 1989; The European Greens Common Statement (Co-Author); Green Parties: An International Guide 1989; Green Futures: Agenda for the 21st Century; Green Light on Europe (Ed); Green Politics: An International Perspective 1990. *Address:* 18 blvd Pinel, 69003 Lyon, France. *Telephone:* 72-33-65-97. *Fax:* 72-34-13-31.

PARRY, Prof Martin L; British environmentalist. *Career:* Head, Atmospheric Impacts Research Group and Prof of Environmental Man, Birmingham Univ; contrib, study of impacts of climate change on world agric, UN. *Present Position:* Dir, Environmental Change Unit, Oxford Univ 1991–; Chair, Climate Change Review Group. *Address:* 1A Mansfield Rd, Oxford OX1 3TB, UK. *Telephone:* (865) 28220. *Fax:* (865) 282281.

PASCHEN, Elizabeth Emily, M ED; Canadian electrical contractor and environmentalist; b 27 Feb 1926, Condor, AB; m Gerhard F Paschen 1960; 1s 2d. *Education:* Alberta Univ. *Career:* School teacher 1948–62, 1978–83, 1985–88; raised children 1962–78; civil servant, AB 1984–85. *Present Position:* Leader and Spokesperson, Alberta Green Party. *Publications:* Moosehide Jackets 1991. *Address:* POB 133, Station M, Calgary, AB T2P 2H6, Canada. *Telephone:* (403) 436-4913. *Fax:* (403) 439-9677.

PASKALEV, Prof Zdravko Doukov; Bulgarian physicist, biophysicist, radiobiologist and ecologist; b 25 June 1935, Sofia; s of Douka Paskalev and Olga Paskaleva; m Lili Neikova 1963 (died 1992); 1d. *Education:* Sofia State Univ, St Petersburg State Univ. *Career:* Electronics factory 1958–59; Specialist, Hygiene Research Inst 1959–63; Prof, Nat Radiobiology Centre 1963–. *Present Position:* Head, Radioecology Lab 1992–; Deputy Head, Lab for Protection against Ionizing Radiation 1993–. *Address:* 1756 Sofia, Climent Ochridsky Bul 132, Bulgaria. *Telephone:* (2) 62-11-23. *Fax:* (2) 62-10-59.

PATOČKA, Prof Ján, D SC; Czech entomologist; b 25 Sept 1925; s of Ján Patočka and Matylda Patočkova; m Vlasta Patočkova 1952; 1s 2d. *Education:* Charles Univ, Prague. *Career:* Assoc Prof, Zvolen Tech Univ 1964. *Present Position:* Consultant, Forest Ecology Inst, Acad of Sciences, Zvolen. *Awards and Honours:* Acad of Sciences Gold Medal 1992. *Publications:* Lepidoptera 1960; Larvae and Pupae in Oak Trees 1980; Lepidoptera of Central Europe 1980. *Address:* Stúrova 2, 960 53 Zvolen, Slovakia. *Telephone:* (855) 22-312. *Fax:* (855) 27-485.

PATTERSON, Dr Walter C; British writer and environmentalist. *Present Position:* Writer on energy and environmental issues. *Address:* c/o Logic, 10 Chesham Rd, Amersham HP6 5ES, UK. *Telephone:* (494) 726748.

PEARCE, Prof David William, MA; British economist; b 11 Oct 1941, Harrow; s of William and Gladys Pearce; m Susan M Reynolds 1965; 2s. *Education:* Lincoln Coll, Oxford. *Career:* Sr Lecturer in Econs, Southampton Univ 1967–74; Dir, Public Sector Econs Research Centre, Leicester Univ 1974–77; Prof of Econs, Aberdeen Univ 1977–83; Adviser to Sec of State for Environment 1989–92. *Present Position:* Prof of Econs, Univ Coll, London 1983–, Dir, Centre for Social and Econ Research on Global Environment (CSERGE) 1991–. *Awards and Honours:* UNEP Global 500. *Publications:* Blueprint for a Green Economy 1989; Economics of Natural Resources and the Environment 1990; Sustainable Development 1990; Economic Valuation and the Natural World 1993. *Address:* Gower St, London WC1E 6BT, UK. *Telephone:* (71) 380-7874. *Fax:* (71) 916-2772.

PELIKAN, Wolfgang W; Austrian environmentalist; b 1947, Wels; m Anna M Spitzer 1976; 3s 2d. *Education:* Vienna Univ. *Career:* Parl journalist 1974; local govt officer; local councillor, Hornstein 1982–. *Present Position:* Gen Sec, Vereinte Grünen Österreichs (United Green Party of Austria) 1982–. *Address:* 7053 Hornstein, Rosengasse 4, Austria. *Telephone:* (2689) 24-16. *Fax:* (2689) 24-16.

PELLEW, Robin; British conservationist. *Career:* Biodiversity expert. *Present Position:* Dir, World Conservation Monitoring Centre, UN. *Address:* 219 Huntingdon Rd, Cambridge CB3 0DL, UK. *Telephone:* (223) 277314. *Fax:* (223) 277365.

PEREIRA, Sir Herbert Charles, D SC, FRS; British agricultural scientist and consultant on tropical land use and research; b 12 May 1913, London; s of H J Pereira and Maud Edith Machin; m Irene Beatrice Sloan 1941; 3s 1d. *Education:* London Univ. *Career:* War service, Royal Engineers 1941–46; Coffee Research Team, Colonial Agric Service, Kenya 1946–52; Founder, Physics Div, E Africa Agric and Forestry Research Org 1952–61; watershed hydrology studies, Kenya, Uganda, Tanganyika (now Tanzania) 1955–61; Dir, Rhodesia and Nyasaland Agric Research Centre and Cen Africa Agric Research Centre 1961–67, E Malling Research Centre, Kent 1969–72; Chief Scientist and Deputy Sec, Ministry of Agric and Fisheries and mem, Agric Research Council and Nat Environmental Research Council 1972–77. *Present Position:* Independent consultant. *Awards and Honours:* Haile Selassie Prize for Research in Africa 1966. *Publications:* Land Use and Water Resources 1973; Policy and Practice in the Management of Tropical Watersheds 1989. *Address:* Peartrees, Nestor Court, Teston, Maidstone ME15 5AD, UK. *Telephone:* (622) 813333.

PÉREZ, Efraín; Ecuadorean consultant; b 16 Oct 1944, Guayaquil; m Carolina Mendoza 1981; 1s 2d. *Education:* Catholic Univ, Guayaquil. *Career:* Prof, Catholic Univ, Guayaquil 1974–76, Catholic Univ, Quito 1989–93; Consultant and Researcher on Public Law and Legislation on Environment and Natural Resources 1981–93; mem, Environmental Law Comm, IUCN 1993. *Present Position:* Exec Dir, Corpn Estade (Estudios de Estructura y Administración del Estado 1989–. *Awards and Honours:* Govt Lab Award 1981. *Publications:* Government Contracts Digest 1983; Legal Institutions of the Galapagos 1984; Legislation on Coastal Resources Management 1987; A Handbook on Natural Resources and Environmental Legislation 1993. *Address:* Rusia y Shyris, Edif Shyris

I, 4°, Of 401, Casilla 17-17-8, Quito, Ecuador. *Telephone:* (2) 242-080. *Fax:* (2) 467-830.

PERRY, Dan; Israeli environmentalist; b srael. *Career:* Paratroops Unit 1956–59; Malkia Kibbutz 1964–69; nature reserves, Galilee and Golan 1964–68; Dir-Gen, Mt Hermon Authority 1969–73; Co-ordinator, Haleh Reserve Rehabilitation Project and Environment Unit, Nature Reserves Authority 1973–74, Asst Dir-Gen 1975–90. *Present Position:* Dir-Gen, Nature Reserves Authority 1990–. *Address:* 78 Yirmeyahu St, Jerusalem 94467, Israel. *Telephone:* (2) 387471. *Fax:* (2) 374887.

PETROVA, Dimitrina Gueorguieva, PH D; Bulgarian philosophy teacher and politician; b 11 Jan 1957, Burgas; d of Georgui Petrov and Trandafilka Sotirova; m 1st Todor Georgiev 1978 (divorced 1984); m 2nd Krassimir Kanev; 2d. *Education:* Sofia Univ. *Career:* Philosophy teacher and human rights activist 1982–; Founder, Ecoglasnost environmental org 1988–89; MP 1990–91. *Present Position:* Dir, Social and Environmental Research Inst 1991– and Spokesperson, Bulgarian Helsinki Cttee 1992–. *Address:* 1000 Sofia, Narodno Sabranie Square 12A, 4th Floor, Bulgaria. *Telephone:* (2) 88-34-06. *Fax:* (2) 88-51-94.

PETROVA, Rossitsa Tsvetkova, MS; Bulgarian forester; b 14 April 1954, Sofia; d of Tsvetko Petrov and Assenka Petrova; m Petar Dimitrov Danin 1986; 1s. *Education:* Higher Inst of Forestry and Forest Eng, Sofia. *Career:* Minproekt Mining Research and Design and Construction Inst 1979–86. *Present Position:* Researcher, Ecology and Environmental Conservation Dept, Forest Research Inst, Bulgarian Acad of Sciences. *Publications:* Biological Recultivation with Pinus Silvestris L 1992. *Address:* 132 Kliment Ohridski Blvd, 1756 Sofia, Bulgaria. *Telephone:* (2) 62-29-61. *Fax:* (2) 62-29-65.

PIDDINGTON, Kenneth; int civil servant. *Present Position:* Dir, Environment Dept, World Bank. *Address:* 1818 H St, NW, Washington, DC 20433, USA. *Telephone:* (202) 477-1234. *Telex:* 248423. *Fax:* (202) 477-6391.

PLATT OF WRITTLE, Beryl Catherine, Baroness, CBE, MA; British engineer; b 18 April 1923; d of Ernest and Dorothy Myatt; m Stewart Sydney Platt 1949; 1s 1d. *Education:* Girton Coll, Cambridge. *Career:* Technical Asst, Experimental Flight Test Dept, Hawker Aircraft; Research and Devt Section, British European Airways; mem, Chelmsford Rural Dist Council 1959, Essex Co Council 1965, Vice-Chair, Educ Cttee 1969–71, Chair 1971–80, Council Vice-Chair and Chair, Co-ordinating and Finance Cttee 1980–83; Mem, House of Lords 1981–, mem, Select Cttee for Science and Tech; Chair, Equal Opportunities Comm 1983–88; Fellow, Girton Coll, Cambridge 1988; Non-Exec Dir, British Gas plc 1988–; Chancellor, Middlesex Univ 1993–. *Present Position:* Pres, Nat Soc for Clean Air and Environmental Protection 1991–. *Address:* 136 N St, Brighton BN1 1RG, UK. *Telephone:* (273) 326313.

PLUT, Dr Dušan; Slovenian politician and environmentalist. *Present Position:* Pres, Zelena Slovenije (Greens of Slovenia). *Address:* 61000 Ljubljana, Komenskega 11, Slovenia. *Telephone:* (61) 311629. *Fax:* (61) 313065.

POLICANSKY, David J. *Present Position:* Assoc Dir and Program Dir for Applied Ecology and Natural Resources, Bd on Environmental Studies and Toxicology, Nat Research Council, USA. *Address:* c/o National Research Council, 2101 Constitution Ave, Washington, DC 20418, USA.

POLUNIN, Prof Nicholas, CBE, D SC, D PHIL, FLS, FRGS; British environmentalist and plant ecologist; m Helen Eugenia Campbell 1948; 2s 1d. *Education:* Christ Church, Oxford, New Coll, Oxford, Yale Univ, Harvard Univ. *Career:* Arctic, sub-Arctic and high Alpine explorations; Sr Research Fellow, New Coll, Oxford; Fellow, Yale Univ, Assoc Fellow; Founder, Biosphere Day (21 Sept) Fund and Prizes 1991–; Ed, *Biological Conservation*. *Present Position:* Ed, *Environmental Conservation* 1974–; Pres and Admin, Foundation for Environmental Conservation and World Council for Biosphere 1975–. *Awards and Honours:* UNEP Global 500 1991. *Publications:* Botany of the Canadian Eastern Arctic 1948; Circumpolar Arctic Flora 1959; An Introduction to Plant Geography and Related Sciences 1960; Éléments de Géographie Botanique 1967; The Environmental Future (Ed) 1972; Growth without Ecodisasters? 1980; Ecosystem Theory and Application 1986; Maintenance of the Biosphere (Co-Author) 1990; Surviving with the Biosphere 1993; Environmental Challenges (Co-Ed) 1993. *Address:* 7 chemin Taverney, 7th and 8th Floors, 1218 Grand-Saconnex, Geneva, Switzerland. *Telephone:* (22) 7982313. *Fax:* (22) 7982344.

PORRITT, Hon Jonathon; British environmental consultant; b 6 July 1950; m Sarah Staniforth 1986; 1d. *Education:* Magdalen Coll, Oxford. *Career:* Schoolteacher, London 1975–84, Head, English Dept, Burlington Danes School, London 1980–84; Chair, Ecology Party 1979–80, 1982–84, Parl Cand 1979, 1983; Dir, FoE 1984–90. *Present Position:* Environmental consultant 1991–. *Publications:* Seeing Green: The Politics of Ecology 1984; The Friends of the Earth Handbook 1987; The Coming of the Greens 1988; Save the Earth (Ed) 1990; Where on Earth Are We Going? 1991; Captain Eco 1991. *Address:* 9 Lypiatt Terrace, Cheltenham GL50 2SX, UK.

PRANCE, Prof Ghillean Tolmie, D PHIL, FRS; British botanist; b 13 July 1937, Brandeston; s of Basil Camden and Margaret Hope (Tolmie) Prance; m Anne Elizabeth Hay 1961; 2d. *Education:* Keble Coll, Oxford. *Career:* Research Asst, New York Botanical Garden 1963–66, Assoc Curator 1966–68, Curator 1968–75, Dir of Research 1975–81; Dir of Grad Training, Inst Nacional de Pesquisas da Amazonia, Manaus, Brazil 1973–75; Sr Vice-Pres and Dir, Econ Botany Inst 1981–88; Visiting Prof, Reading Univ 1988–. *Present Position:* Dir, Royal Botanic Gardens, Kew 1988–. *Awards and Honours:* Henry Shaw Medal, Missouri Botanical Garden 1988; Linnean Medal 1990. *Address:* Richmond TW9 3AB, UK. *Telephone:* (81) 332-5111. *Telex:* 296674. *Fax:* (81) 948-4237.

PRESTT, Ian, CBE, M SC; British biologist, ecologist and conservationist; b 26 June 1929, Wigan, Lancs; s of Arthur and Jessie Prestt; m Jennifer Ann Wagstaffe 1956; 1s (deceased) 2d. *Education:* Liverpool Univ. *Career:* 2nd Lieut, Royal Artillery 1947–49; Ornithological Officer then Toxic Chemicals and Wildlife Section, Nature Conservancy 1956–74; Deputy Dir, Cen Environmental Pollution Unit, Cabinet Office and Dept of Environment 1970–74; Deputy Dir, Nature Conservancy Council 1974–75; Dir-Gen, Royal Soc for Protection of Birds (RSPB) 1975–89; Chair, Exec Cttee, Int Council for Bird Preservation 1979–90. *Present Position:* Pres, Royal Soc for Protection of Birds (RSPB) 1989–, Cambs Branch, Council for Protection of Rural England 1990–, Vice-Pres, Birdlife Int 1991–; Trustee, Herpetological Conservation Trust 1992–; Chair, Sensory Trust 1993–. *Publications:* Effects of Organochlorine Insecticides on European Birdlife; Investigations into the Possible Effects of Organochlorine Insecticides on Wild Predatory Birds 1967; The Breeding Cycle of the Grey Heron 1969; An Ecological Study of Vipera Berus in Southern Britain 1971. *Address:* The Lodge, Sandy SG19 2DL, UK. *Telephone:* (767) 680551. *Telex:* 82469. *Fax:* (767) 292365.

PUGH, David; British oceanographer. *Present Position:* Head of Information and Scientific Services, Inst of Oceanographic Sciences. *Address:* Brook Rd, Wormley GU8 5UB, UK. *Telephone:* (428) 684141. *Fax:* (428) 683066.

PULVER, Bernhard; Swiss politician and environmentalist; b 1965, Berne. *Career:* Sec, Swiss Peace Council 1985, Centre Suisse d'information sur la Sauvegarde des Biens Culturels 1986–87. *Present Position:* Gen Sec, Grüne Partei der Schweiz—Parti écologiste suisse (Swiss Green Party) 1987–. *Address:* Marienstr 11, 3005 Berne, Switzerland. *Telephone:* (31) 441441.

PYLE, Dr John; British chemist. *Present Position:* Head, CFC and Ozone Research Programme, Chemistry Dept, Cambridge Univ. *Address:* Lensfield Rd, Cambridge CB2 1EW, UK. *Telephone:* (223) 337733.

Q

QUEIROZ DE HENEEUS, Lina; Brazilian politician and environmentalist. *Present Position:* Mem, Los Verdes (Brazilian Green Party). *Address:* 715 Co, Avda Comandante Julio de Moura 2, Barra de Tijuca, Rio de Janeiro, RJ, Brazil.

QUIÑONES, Prof Ana María Cabello, M SC; Chilean environmentalist; b 5 March 1956; m Manuel Herrera Acevedo; 2s. *Education:* Concepción Univ; Santiago Univ. *Career:* Univ Docent 1977–. *Present Position:* Co-ordinator of Environmental Educ, Maule Professional Inst, Talca; Pres, Asociación para la Difusión de los programas de la Naciones Unidas para el Medio Ambiente (ADUMA Chile). *Publications:* La Antartica; Nuestro Medio Ambiente. *Address:* Casilla 102, Talca, Chile. *Telephone:* (71) 227046. *Fax:* (71) 227046.

R

RÁDAI, Ödön, PH D; Hungarian hydrologist; b 17 June 1927, Budatétény; s of Béla Rádai and Paula Glasz; m Maria Weidinger 1960; 2d. *Education:* Budapest Science Univ. *Career:* Vituki Research Centre for Water Resources Devt 1949–90, UNESCO Lecturer in Advanced Hydrology 1982–; Sub-Surface Water Expert, FAO, Laos 1979–81. *Present Position:* Head of Dept, Nat Authority for Nature Conservation, Ministry of Environment and Regional Policy 1991–. *Publications:* Aerial Photo Interpretation in Karstwater Exploration 1969; Aerial Photography in Water Management 1978; The Status of Hungarian Nature and Nature Conservation 1993. *Address:* 1121 Budapest, Költö u 21, Hungary. *Telephone:* (1) 175-1093. *Fax:* (1) 175-7457.

RADECKI, Wojciech, D SC; Polish lawyer; b 13 July 1943, Sosnowiec; s of Andrzej Radecki and Jadwiga Rosnowska-Radecka; m Jadwiga Popiel-Radecka 1965; 1s. *Education:* Wrocław Univ. *Career:* Prosecuting Attorney, Jelenia Góra 1965–73; Lecturer, Wrocław Univ 1973–76; Lecturer then Assoc Prof, Law Inst, Acad of Sciences 1976–91. *Present Position:* Prof, Law Inst, Acad of Sciences 1992–. *Publications:* Environmental Protection under the Criminal Law 1982; Responsibility in Environmental Law 1987. *Address:* 54-242 Wrocław, ul Jelenia 42/5, Poland. *Telephone:* (71) 551209.

RAGEB, Hassan; Egyptian politician and environmentalist. *Present Position:* Chair, Green Party. *Address:* Cairo, Egypt.

RAI, Laghu Dhan, MA; Nepalese environmental researcher; b 25 July 1958, Nepal; s of Bir Singh and Maheshwori Rai; m Nanda Kumari Rai 1982; 3d 1s. *Education:* Tribhuvan Univ, Kathmandu. *Career:* Social worker 1975; mem, Nepali Congress 1978, Parl Cand 1986. *Present Position:* Founder and Chair, Mt Everest Environment Conservation Foundation (EECF) 1992–. *Address:* POB 3508, Kathmandu, Nepal. *Telephone:* 417757. *Fax:* 414324.

RAKOVSKY, Prof Slavtcho, PH D; Bulgarian chemist; b 1947; m Rositza Rakovsky 1968; 1s 1d. *Education:* Sofia Inst of Chemical Tech. *Career:* Chemist, Organic Chemistry Inst 1970; Asst Prof, Kinesis and Catalysis Inst, Sofia 1983, Assoc Prof 1986. *Present Position:* Deputy Dir, Kinesis and Catalysis Inst, Sofia; Co-ordinator, Science and Educ Branch, Nat Ecological Club. *Publications:* Ozone Reactions with Organic Compounds 1983. *Address:* Sofia 1000, POB 1653, Bulgaria. *Telephone:* (2) 713-25-23. *Fax:* (2) 75-61-16.

RALL, David; American environmental health scientist. *Present Position:* Dir, Nat Inst of Environmental Health Sciences. *Address:* c/o Environmental Health Inst, Cincinnati Univ, 3223 Eden Ave, Cincinnati, OH 45267-0056, USA. *Telephone:* (513) 556-6000.

RAMANCHANDRAN, Dr Arcot; Indian int civil servant. *Present Position:* Exec Dir, UN Centre for Human Settlements (UNCHS/Habitat). *Address:* POB 30030, Nairobi, Kenya. *Telephone:* (2) 520600. *Telex:* 22996. *Fax:* (2) 226473.

RAMPHAL, Sir Shridath Surendranath (Sonny), GCMG, OM, ONZ, AC, QC, SC, LL M, FRSA; Guyanese barrister, politician and int official; b 3 Oct 1928, New Amsterdam; s of James I and Grace (Abdool) Ramphal; m Lois Winifred King 1951; 2s 2d. *Education:* King's Coll, London, Harvard Univ. *Career:* Crown Counsel, British Guiana 1953–54; Asst to Attorney-Gen 1954–56; Legal Draftsman 1956–58, W Indies 1958–59; Solicitor-Gen 1959–61; Attorney-Gen, Guyana 1965–73; mem, Nat Ass 1965–75; Minister of State for External Affairs 1967–72; Minister of Foreign Affairs 1972–75; Minister of Justice 1973–75; Sec-Gen, Commonwealth 1975–90; Chancellor, Guyana Univ 1988–92, Warwick Univ and W Indies Univ 1989–, mem, Council for Legal Educ; mem, World Comm on Environment and Devt, S Comm, Brandt Conf Working Group, Bd of Govs, World Maritime Univ, Commonwealth of Learning, Int Advisory Council, TV Trust for Environment, Int Bd, United World Colls, Exec Cttee, Int Inst for Environment and Devt, and Council, Int Negotiation Network, Carter Center, GA, USA 1991–; Patron, One World Broadcasting Trust; Chair, W Indian Comm 1990–92, Int Steering Cttee, Leadership for Environment and Devt Programme, Rockefeller Foundation 1991– and Selection Cttee, 3rd World Prize; Co-Chair, Comm on Global Governance

1992–; Vice-Chair, Centre for Research on New Int Econ Order, Oxford; Exec Pres, Willy Brandt Int Foundation; Vice-Pres, Royal Overseas League; Adviser to Sec-Gen, UNCED 1992. *Present Position:* Pres, IUCN. *Publications:* One World to Share: Selected Speeches of the Commonwealth Secretary-General 1975–79; Nkrumah and the Eighties 1980; Sovereignty or Solidarity 1981; Some in Light and Some in Darkness: The Long Shadow of Slavery 1983; The Message not the Messenger 1985; The Trampling of the Grass 1985; For the South, A Time to Think 1986; Making Human Society A Civilised State 1987; Inseparable Humanity: An Anthology of Reflections of Shridath Ramphal 1988; An End to Otherness 1990; Our Country, The Planet 1992. *Address:* ave du Mont-Blanc, 1196 Gland, Switzerland. *Telephone:* (22) 3649111. *Telex:* 419618. *Fax:* (22) 3643239.

RANDLA, Tiit; Estonian biologist; b 11 April 1940, Tallinn; s of Erich and Eve Randla; m Silje Randla 1980; 1s 1d. *Education:* Tartu Univ. *Career:* Biology teacher, secondary school 1963–68; Vice-Chair, Estonian Hunters Soc 1969–91; Ministry of Environment 1991–. *Present Position:* Head, Nature Conservation and Wildlife Man Dept, Ministry of Environment 1992–. *Awards and Honours:* Nature Conservation Award 1985. *Publications:* Arctic Wildlife 1990; Film: The Estonian White-Tailed Sea-Eagle 1989. *Address:* 24 Toompuiestee, EE0100 Tallinn, Estonia. *Telephone:* (142) 451559. *Fax:* (142) 453310.

REDFORD, Robert; American actor, film director, film producer and environmentalist; b 18 Aug 1937, Santa Monica, CA. *Education:* Colorado Univ. *Present Position:* Head, Sundance. *Awards and Honours:* Acad Award 1981; Audubon Medal 1989. *Publications:* Films: War Hunt 1961; Situation Hopeless But Not Serious 1965; Inside Daisy Clover 1965; The Chase 1965; This Property is Condemned 1966; Barefoot in the Park 1967; Tell Them Willie Boy is Here 1969; Butch Cassidy and the Sundance Kid 1969; Downhill Racer 1969; Little Fauss and Big Halsy 1970; Jeremiah Johnson 1972; The Candidate 1972; How to Steal a Diamond in Four Uneasy Lessons 1972; The Way We Were 1973; The Sting 1973; The Great Gatsby 1974; The Great Waldo Pepper 1974; Three Days of the Condor 1975; All the President's Men 1976; A Bridge Too Far 1977; Ordinary People (Dir) 1980; The Electric Horseman 1980; Brubaker 1980; The Natural 1984; Out of Africa 1985; Legal Eagles 1986; Milagro Beanfield War (Dir, Producer) 1988; Promised Land (Dir, Exec Producer) 1988; Havana 1991; Sneakers (Dir) 1992; A River Runs Through It (Dir) 1992. *Address:* Rural Route 3, Box 624-A, Sundance, UT 84604, USA.

REDMAN, Nigel James; British ornithologist; b 1952, Taplow; s of L J and S M W Redman; m Irina Redman 1987 (separated); 1d. *Career:* Mem, Council and Treas, Ornithological Soc of Middle E (OSME) 1980–87; Tour Guide, Birdquest Ltd 1984–; Consultant Ed, Helm Information Ltd 1991–. *Present Position:* Chair, Oriental Bird Club 1991–. *Publications:* Birdwatching in Britain: A Site-by-Site Guide 1987. *Address:* 74 Chatburn Rd, Clitheroe BB7 2AT, UK. *Telephone:* (200) 442755.

REIJNDERS, Dr Lucas; Dutch environmental scientist; b 1946, Amsterdam; m G M van Asperen 1974; 1s. *Education:* Amsterdam Univ. *Career:* Biochemistry Dept, Amsterdam Univ 1969–73; Molecular Biology Dept, Wageningen Agric Univ 1973–74; Environmental Studies Dept, Groningen State Univ 1974–80. *Present Position:* Mem, Foundation for Nature and Environment 1980–; Prof, Environmental Science Dept, Amsterdam Univ 1988–. *Publications:* Pleidooi voor een duurzame relatie met het milieu 1984. *Address:* Anna van den Vondelstraat 10, 1054 62 Amsterdam, Netherlands. *Telephone:* (20) 5256206. *Fax:* (20) 5256272.

REILLY, William K, BA, MS, JD; American govt official, lawyer and conservationist; b 26 Jan 1940, Decatur, IL. *Education:* Harvard and Columbia Univs. *Career:* Lawyer, Ross and Hardies, Chicago, IL 1965; Assoc Dir, Urban Policy Center, Urban American Inc, Nat Urban Coalition, Washington, DC 1969–70; sr staff mem Pres's Council on Environmental Quality 1970–72; Exec Dir, Task Force on Land Use and Urban Growth 1972–73; Pres, WWF 1985–89; Chair, Partners for Livable Place; Dir, Concern Inc, Winrock Int, Inst of Agric Devt; Trustee, German Marshall Fund, American Farmland Trust. *Present Position:* Pres Conservation Foundation 1973–; Admin Environmental Protection Agency 1989–. *Address:* Environmental Protection Agency, 401 M St SW, Washington, DC 20460, USA.

REINIGER, Róbert; Hungarian mechanical engineer and economist; b 1950, Budapest; m Mária Albrecht 1975; 1s 1d. *Education:* Budapest Tech Univ. *Career:* Chief Design Engineer, Civil Eng Planning Office 1968–86; Sr Engineer, Cen Planning Bureau 1987–88; Head of Dept, Ministry for Environment and Water Man 1988–90; Deputy Head, Nat Authority for Environment 1991–93. *Present Position:* Head, Nat Authority for Environment 1993–. *Publications:* Environmental Protection: Hazardous Waste 1991; A Manual of Waste Treatment 1993. *Address:* 1054 Budapest, Alkotmány utca 29, Hungary. *Telephone:* (1) 132-2787. *Fax:* (1) 132-7196.

RENKIN, Franz; Austrian politician and environmentalist. *Present Position:* Chair, Die Grüne Alternative (Green Alternative). *Address:* 1070 Vienna, Stiftgasse 6, Austria. *Telephone:* (222) 52-125-0. *Fax:* (222) 52-125-40.

REYNOLDS, Fiona Claire, MA, M PHIL; British environmentalist; b 29 March 1958, Alston, Cumbria; d of Jeffrey Alan Reynolds and Margaret Mary Watson; m Robert William Merrill 1981; 2d. *Education:* Newnham Coll, Cambridge. *Career:* Sec, Council for Nat Parks 1980–87; Asst Dir, Council for Protection of Rural England (CPRE) 1987–91. *Present Position:* Dir, Council for Protection of Rural England (CPRE) 1992–. *Awards and Honours:* UNEP Global 500 1990. *Address:* Warwick House, 25 Buckingham Palace Rd, London SW1W 0PP, UK. *Telephone:* (71) 976-6433. *Fax:* (71) 976-6373.

RIOJA BALLIVIÁN, Guillermo; Bolivian environmental anthropologist; La Paz; m Cynthia Luizaga Paz 1979; 1s 1d. *Education:* San Andres Univ, Puebla Autonomous Univ, Mexico. *Career:* Prof of Anthropology and Archaeology, San Andres Univ 1988–; Anthropologist, Nat Ethnography and Folklore Museum 1988–90. *Present Position:* Dir, Bolivia Programme, Conservation Int 1990–. *Publications:* Bolivia: Cuestión Étnica y Lucha de Clases 1989; Conservación y Pueblos Indigénas 1990. *Address:* Avda Villazón 1958, Edif Villazón, 10°, Of 10A, Casilla 5633, La Paz, Bolivia. *Telephone:* (2) 32-5839. *Fax:* (2) 35-6529.

RIPA DI MEANA, Carlo; Italian politician and journalist; b 15 Aug 1929, Marina di Pietrasanta, Lucca; m Marina Punturieri. *Career:* Journalist, *Il Lavoro* weekly and *L'Unità* daily 1950–53; Co-Founder, *Nuova Generazione* weekly 1953–56; Co-Founder and Ed, *Passato e Presente* magazine 1957; resigned from Communist Party 1957; joined Socialist Party 1958; Ed, Feltrinelli & Rizzoli 1958–66; Regional Councillor, Lombardy, Leader, Socialist Party Group 1970–82; mem, Bd, La Scala Theatre, Milan 1970–74; Chair, Venice Biennale 1974–79; Head of Int Relations, Socialist Party 1979–80; mem, EP 1979, Comm of EC for Citizen's Europe, Information, Culture and Tourism 1985–89, for Communication 1986–89, for Environment, Nuclear Safety and Civil Protection 1989–93; Chair, Istituto per la Cooperazione economica internazionale e i problemi dello sviluppo 1983–; Minister for Environment 1992–93; left Socialist Party. *Present Position:* Leader, Green

Party 1993–. *Publications:* Un viaggio in Viet-Nam 1956; Dedicato a Raymond Roussel e alle sue impressioni d'Africa 1965; Il governo audiovisivo 1973. *Address:* Camera dei Deputati, Rome, Italy. *Telephone:* (6) 4957383.

ROBERTS, Lewis Edward John, CBE, D PHIL, FRSC, FRS; British research scientist and univ teacher; b 31 Jan 1922, Cardiff; s of Rev W E and L L Roberts; m Eleanor Mary Luscombe 1949; 1s. *Education:* Oxford Univ, California Univ, Berkeley. *Career:* Chemical research in nuclear power production, Canada and UK 1946–54; Fellow, California Univ 1954–55; Research Admin, Atomic Energy Research Establishment, Harwell 1955–75, Dir 1975–86; Prof of Environmental Risk Assessment, E Anglia Univ, Norwich 1986–91. *Present Position:* Emer Prof, Environmental Sciences School, E Anglia Univ, Norwich 1990–. *Publications:* Nuclear Power and Public Responsibility 1984; Power Generation and the Environment (Co-Author) 1990. *Address:* Penfold Wick, Chilton, Didcot OX11 0SH, UK. *Telephone:* (235) 834309.

ROBERTS, Prof Thomas Michael; British ecologist; b 12 May 1948, Wales; m 1968; 1s 2d. *Career:* Researcher in applied ecology and scientific man; univ and industrial research; research on impact of air pollution on vegetation, derelict land reclamation and land use change; Prof, York Univ and London Univ; Fellow, Inst of Biology. *Present Position:* Dir, Terrestrial Ecology Inst, Natural Environment Research Council 1989–. *Publications:* Planning and Ecology (Co-Author) 1984. *Address:* Monks Wood Experimental Station, Abbots Ripton, Huntingdon PE17 2LS, UK. *Telephone:* (4873) 381. *Telex:* 32416. *Fax:* (4873) 590.

ROBINSON, Steve, M SC; British environmentalist; b 7 March 1956, Altrincham, Cheshire; m Katharine Robinson 1992. *Education:* Durham Univ, Newcastle upon Tyne Univ. *Career:* Fisheries Man, Malawi 1979–81; dir, light eng co 1981–83; Commonwealth Inst and Conservation Foundation 1984–88; mem, Cttee on Educ and Training for Industry, Council for Environmental Educ, Comm on Environmental Strategies and Planning and Comm on Environmental Educ and Communication, IUCN; Gov, Capel Manor Horticultural and Environmental Centre. *Present Position:* Chief Exec, Environment Council 1988–. *Publications:* Healthier Profits, Business Success and the Green Factor 1986. *Address:* 21 Elizabeth St, London SW1W 9RP, UK. *Telephone:* (71) 824-8411. *Fax:* (71) 730-9941.

RODDA, Dr John C; scientist. *Present Position:* Hydrology and Water Resources Dept, WMO. *Address:* POB 5, 41 ave Giuseppe Motta, 1211 Geneva 20, Switzerland. *Telephone:* (22) 7348245. *Telex:* 414199. *Fax:* (22) 7342326.

RODDICK, Anita Lucia, OBE; British business exec; b 23 Oct 1942, Littlehampton; d of Henry Perilli and Gilda de Vita; m Thomas Gordon Roddick 1970; 2d. *Career:* Teacher of English and History; Cuttings Dept, UN, Geneva; owner and man, restaurant and hotel 1962–76; opened 1st Body Shop, Brighton 1976; Body Shop Int floated on Unlisted Securities Market 1984. *Present Position:* Group Man Dir, Body Shop Int plc. *Awards and Honours:* Veuve Clicquot Business Woman of the Year 1985; Co NatWest Retailer of the Year Award 1989; UNEP Global 500 1989. *Publications:* Body and Soul 1991. *Address:* Watersmead, Littlehampton BN17 6LS, UK. *Telephone:* (903) 717107.

RODRIGUEZ, Arsenio, PH D; American environmentalist and coastal zone manager; b 18 Dec 1943, Puerto Rico; m Julie C Rodriguez 1975; 2s 2d. *Education:* Puerto Rico Univ, Mexico Nat Univ, Miami Univ. *Career:* Research Asst, Medicine School, Puerto Rico Univ 1964–66, Sr Research Asst 1968–69; Research Fellow, Mexico Univ 1966–68; Scientific Adviser, Environmental Quality Bd, Puerto Rico 1972–73, to Sec of Nat Resources 1973–74; Dir, Puerto Rico Coastal Zone Man Programme 1974–77; Sr Scientist, ECLAC/UNEP Jt Programme for Environmental Man of Caribbean 1977–80; Sr Programme Officer, Regional Sea Programme, UNEP, Geneva 1980–84, Regional Adviser for Latin America and Caribbean, Mexico 1984–89, Deputy Dir, Latin America and Caribbean Office, Mexico 1990–92. *Present Position:* Dir and Regional Rep, Latin America and Caribbean Office, UNEP 1992–. *Publications:* Ambio (Ed). *Address:* Apdo Postal 10793, Bd de los Virreyes 155, Col Lomas Virreyes, 11000 México, DF, Mexico. *Telephone:* (5) 282-5371. *Telex:* 01771055. *Fax:* (5) 202-0950.

ROELANTS DU VIVIER, François, PH D; Belgian politician and environmentalist; b 5 Nov 1947, Brussels; m Chantal de Coninck 1972; 1s 1d. *Education:* Catholic Univ, Louvain. *Career:* Gen Sec, Inter Environnement Wallonie 1975–83; mem, Exec Cttee, European Environmental Bureau (EEB) 1979–83; mem, EP 1984–89, Vice-Chair, Cttee on Environment, Public Health and Consumer Protection, Vice-Pres, Parl Office for Scientific and Technological Options Assessment, Substitute, Agric Cttee; Special Adviser to European Commr for Environmental Matters 1990; Chair, Green Party ERE (Europe, Regions, Environment); mem and Chair, Environment Cttee, Brussels State Parl; Lecturer, Fondation Univ Luxembourgeoise; Chair, Bd, European Citizen Action Service (ECAS). *Present Position:* Founder, Global Legislators for a Balanced Environment (GLOBE) 1989–, Dir 1990–. *Publications:* Agriculture européenne et environnement: un avenir fertile 1987; Les vaisseaux du poison, la route des déchets toxiques 1988. *Address:* 50 rue du Taciturne, 1040 Brussels, Belgium. *Telephone:* (2) 375-07-65. *Fax:* (2) 375-88-66.

ROLSTON, III, Holmes, PH D; American theologian, teacher and philosopher; b 19 Nov 1932, Staunton, VA; s of Holmes and Mary Winifred (Long) Rolston; m Jane Irving Wilson 1956; 2s. *Education:* Pittsburgh Univ, Edinburgh Univ. *Career:* Pastor, Walnut Grove Presbyterian Church, Bristol, VA 1959–67; Asst Prof of Philosophy, Colorado State Univ 1968–71, Assoc Prof 1971–76, Prof 1976–; Visiting Scholar, Center for Study of World Religions, Harvard Univ 1974–75; Lecturer, Yale Univ and Vanderbilt Univ; Assoc Ed, *Environmental Ethics* 1979–; mem, Editorial Bd, *Oxford Series in Environment, Philosophy and Public Policy* and *Zygon: Journal of Religion and Science*. *Present Position:* Pres, Int Soc for Environmental Ethics. *Publications:* The Cosmic Christ 1966; John Calvin versus the Westminster Confession 1972; Religious Inquiry: Participation and Detachment 1985; Philosophy Gone Wild 1986; Science and Religion: A Critical Survey 1987; Environmental Ethics 1988. *Address:* c/o Philosophy Dept, Colorado State Univ, Fort Collins, CO 80523, USA. *Telephone:* (303) 491-6315. *Fax:* (303) 491-4900.

ROMERO DIAZ, Aldemaro, PH D; Venezuelan biologist; b 11 Sept 1951; m Ana Mayayo 1977; 2d. *Education:* Barcelona Univ, Miami Univ. *Career:* Adjunct Assoc Prof, Miami Univ 1990–. *Present Position:* Exec Dir, Bioma (Fundación Venezolana para la Conservación de la Diversidad Biológica) 1986–. *Address:* Edif Camara de Comercio de Caracas, 4°, Avda Este 2, Los Caobos, Apdo 1968, 1010-A Caracas, Venezuela. *Telephone:* (2) 571-3120. *Fax:* (2) 571-1412.

ROMERO MORENO, Claudia; Colombian biologist; b 19 March 1959, Bogotá; m; 1d. *Education:* Andes Univ, Bogotá. *Career:* Dir, Gorgona Island Nat Park 1986–88; Environmental Project Co-ordinator, INDERENA 1988–90; Parks Co-ordinator, Peilr 1990–92. *Present Position:* Project Co-ordinator, Fundación Natura 1992–. *Publications:* Del Orinoco al Siglo XXI: El Hombre, la Fauna y Su Medio 1989. *Address:* Avda 13 No 87/43, Apdo Aereo 55402, Santa Fe de Bogotá, Colombia. *Telephone:* (1) 616-9262. *Fax:* (1) 236-3077.

ROSIVAL, Prof Ladislav, MD, D SC; Slovak doctor of medicine; b 19 Aug 1924, V Polana; s of J and E Rosival; m Ann Rosival 1966; 1d. *Education:* Bratislava Univ. *Career:* Asst Prof 1961; Research Dir, Hygiene Inst, Bratislava 1968–78, Occupational Medicine Inst 1971–73, Dir, Hygiene Centre, Preventive Medicine Inst 1979–91; mem, Int Acad for Protection of Environment 1976. *Present Position:* Head, Chemicals Registration and Information Centre, Preventive and Clinical Medicine Inst 1991–. *Publications:* Toxicology and Pharmacobiodynamics of OPI Compounds 1959; Foreign Substances in Food 1983; Health 1984; Preventive Medicine 1992; Hygiene 1992. *Address:* Limbová 14, 833 01 Bratislava, Slovakia. *Telephone:* (7) 373-782. *Telex:* 92279. *Fax:* (7) 373-906.

ROSZAK, Theodore; American academic, writer and environmentalist. *Career:* Writer on environmental affairs, Bantam Press, UK. *Present Position:* Mem staff, California State Univ. *Address:* Hayward, CA 94542, USA. *Telephone:* (415) 881-3000. *Fax:* (415) 881-3808.

ROTHSCHILD, Emma; British scientist; b 16 May 1948; d of Baron and Lady Rothschild; m Prof Amartya Kumar Sen 1991. *Education:* Somerville Coll, Oxford, MIT, USA. *Career:* Mem, OECD Group of Experts on Science and Tech in the New Socio-Econ Context 1976–80; Assoc Prof of Humanities, MIT 1978–80, of Science, Tech and Soc 1979–88; Research Dir, École des Hautes Études en Sciences Sociales, Paris 1981–82; mem, Bd of Dirs, *Bulletin of Atomic Scientists* 1983–89, Governing Bd, Stockholm Int Peace Research Inst 1983–, Bd, Olof Palme Memorial Fund, Stockholm and Royal Comm on Environmental Pollution 1986–, Governing Bd, Stockholm Environment Inst 1989–; OECD Scientific Examiner, Australia 1984–85; Trustee, Inst for Public Policy Research 1988–; Research Adviser, World Inst for Devt Econs Research 1989–. *Present Position:* Sr Research Fellow, King's Coll, Cambridge 1988–. *Publications:* Paradise Lost: The Decline of the Auto-Industrial Age 1973. *Address:* Cambridge CB2 1ST, UK. *Telephone:* (223) 350411.

ROTKO, Jerzy, LL D; Polish lawyer; b 1956, Wrocław; s of Michal Rotko and Stefania Władyka; 1d. *Education:* Wrocław Univ. *Career:* Asst Prof, Law Inst, Acad of Sciences 1979; mem, Polish Environmental Law Asscn 1987–, Polish Environmental Law Reform Cttee 1989–92. *Present Position:* Adjunct Prof, Environmental Law Group, Law Inst, Acad of Sciences 1986–. *Publications:* Entwicklung des Natur- und Umweltschutzrechts in Mittel- und Osteuropa 1991. *Address:* 50-138 Wrocław, Kuźnicza 46/47, Poland. *Telephone:* (71) 444747. *Fax:* (71) 444747.

ROWLAND, Frank Sherwood, PH D; American chemist; b 28 June 1927, DE; s of Sidney A and Margaret Lois Drake Rowland; m Joan Lundberg 1952; 1s 1d. *Education:* Chicago Wesleyan Univ, Ohio Wesleyan Univ. *Career:* Instructor in Chemistry, Princeton Univ 1952–56; Asst to Prof, Kansas Univ 1956–64; Guggenheim Fellow 1962, 1974; mem, American Acad of Arts and Sciences 1977–, NAS 1978–. *Present Position:* Prof, California Univ, Irvine 1964–. *Awards and Honours:* Tyler Prize. *Address:* Physical Sciences 571, Irvine, CA 92717, USA. *Telephone:* (714) 856-6016.

ROY, Dunnu; Indian environmentalist. *Present Position:* Mem, WWF India. *Address:* 172B Lodi Rd, New Delhi 110 003, India.

RUŽIČKA, Prof Milan, D SC; Slovak scientist; b 29 May 1929, Bratislava; s of František Ružička and Ludmila Ružičkova; m Helena Ružičkova 1967; 2d. *Education:* Comenius Univ, Bratislava, Zvolen Tech Univ. *Career:* Head, Biology and Landscape Man Dept, Acad of Sciences 1962–64, Dir 1965–74, Head, Landscape Ecology Dept, Experimental Biology and Ecology Inst 1975–90; Chief Co-ordinator, Landscape Ecology project 1970–90. *Present Position:* Chair, Slovak Ecological Soc, Dir, Landscape Ecology Inst, Acad of Sciences and Lecturer, Ecology Faculty, Zvolen Tech Univ. *Publications:* Principles of the Biological Landscape Plan as the Basis for Housing Estate Planning in the Model Area of Bratislava-Lamač (Co-Author) 1974; Landscape Components, Elements and Structure in Biological Landscape Planning (Co-Author) 1978; Ecological Landscape Planning (Co-Author) 1982; Methods for Ecological Evaluation of Territory for Zone and Settlement Formation: Rimavská Sobota (Co-Author) 1982; Basic Premises and Methods in Landscape-Ecological Planning and Optimization (Co-Author) 1990. *Address:* Štefánikova 3, 814 99 Bratislava, Slovakia. *Telephone:* (7) 334-311. *Fax:* (7) 332-560.

RUBAN, Yuriy G, PH D; Ukrainian computer engineer; b 25 Oct 1958, Poltava; s of Grigoriy and Anastasija Ruban; m Ann Ruban 1979; 1d. *Education:* Kiev Polytechnic Inst. *Career:* Sr Research Asst 1981–88; Asst Prof 1988–92; Consultant, Parl Econ Reform Comm 1992–. *Present Position:* 1st Deputy Minister of Environmental Protection. *Address:* 252001 Kiev, 5 Khreschatyk St, Ukraine. *Telephone:* (44) 226-25-01. *Fax:* (44) 228-81-81.

RÜDEL, Walter; Jordanian environmentalist. *Present Position:* Mem, Friedrich Naumann Foundation (FNF). *Address:* Amman, Jordan.

RÜHLE, Heide; German politician and environmentalist. *Present Position:* Gen Sec, Die Grünen (Greens W). *Address:* 5303 Bornheim-Roisdorf, Im Ehrental 2, Bundesgeschäftsstelle, Germany. *Telephone:* (2222) 7008-0. *Fax:* (2222) 7008-99.

RYDELL-JANSON, Elisabet; Swedish politician and environmentalist. *Present Position:* Sec-Gen, Miljöpartiet de Gröna (Green Party). *Address:* Int Secr, POB 25014, 100 23 Stockholm, Sweden. *Telephone:* (8) 695-02-02. *Fax:* (8) 656-00-11.

S

SABA, Abdulrasheed Kolawole, M SC; Nigerian civil servant and ecologist; b 25 Oct 1943, Lagos State; s of Alhaji and Alhaja M R Saba; m Alhaja B R Saba 1976; 4c. *Education:* Ahmadu Bello Univ, Zaria, Ibadan Univ, Nairobi Univ, Michigan Univ. *Career:* Sec, Nat Resources Conservation Council. *Present Position:* Dir-Gen, Fed Environmental Protection Agency 1993–. *Publications:* Ecology of Carabids and Cicindello Beetles of River Gulma and Kubani Area, Zaria 1969; The National Conservation Strategy for Nigeria 1988; Mobilization of Finance for Natural Resource Conservation. *Address:* The Presidency, Plot 444, Maitama Dist, PMB 176, Garki, Abuja, Nigeria. *Telephone:* (9) 5232263. *Fax:* (9) 5232556.

SABO, Peter, PH D; Slovak environmental manager; b 19 March 1954, Nitra; m Ľudmila Hüttlová 1980; 2s. *Education:* Slovak Tech Univ. *Career:* Nature conservation, SZOPK 1978–; MP (Green Party) 1990–92; Exec Officer, Ecopoint 1992–; Head, Bratislava Project Co-ordination Unit, IUCN 1993–; writer on technical cybernetics, computer applications in nature sciences, practical nature conservation and botany and green alternatives and politics. *Present Position:* Pres, Green Alternative Foundation 1992–; mem, Exec Cttee, Slovak Union of Nature and Landscape Protection 1993–. *Address:* Vajanského 3, 921 01 Piešťany, Slovakia. *Telephone:* (838) 23-927.

SALIH, El-Tigani Mohamed, PH D; Sudanese soil scientist; b 1945, Omdurman; s of Mohamed and Sumu Salih; m Monah Salih 1977; 4s 2d. *Education:* Friendship Univ, Moscow. *Career:* Soil Scientist, Soil Conservation Admin 1973, Head, Soil Survey Section 1975, Asst Dir 1983, Head, Remote Sensing Section 1987; Regional Dir of Soil Conservation and Land Use, Water Programme Admin, Khartoum Prov 1985; Asst Nat Co-ordinator for Desertification and Drought Control, Desertification and Drought Control and Monitoring Unit (NDDU) 1989, Lecturer on Desertification 1991. *Present Position:* Nat Co-ordinator and Dir, Desertification and Drought Control and Monitoring Unit (NDDU) 1990–. *Publications:* Particulars of Compact Soils of Central Sudan 1981; Main Types of Soils in Central and Eastern Sudan, Their Genesis, Classification and Agricultural Evaluation 1982; Soil Genesis and Geography in Eastern Sudan (Co-Author) 1983; Desertification Assessment and Mapping Methodology in Sudan (Co-Author) 1992. *Address:* POB 1942, Khartoum, Sudan. *Telephone:* 72023.

SALMON, Prof John Tenison, CBE, D SC, FRSNZ, FRPS; New Zealander writer and zoologist; b 28 June 1910, Wellington; s of Charles Tenison Salmon and Marie Patterson; m Pamela Naomi Wilton 1948; 4s. *Education:* Victoria Univ, Wellington. *Career:* Entomologist and Photographer, Dominion Museum 1934–48; Sr Lecturer in Biology, Victoria Univ, Wellington 1948–58, Assoc Prof of Zoology 1959–64, Acting Head, Zoology Dept 1964–65, Prof and Head 1965–75, Dean of Science 1971–72. *Present Position:* Emer Prof, Victoria Univ, Wellington 1976–. *Awards and Honours:* Nature Conservation Council Citation 1982. *Publications:* Heritage Destroyed 1960; New Zealand's Flowers and Plants in Colour 1963; A Field Guide to the Alpine Plants of New Zealand 1968; The Native Trees of New Zealand 1980; Collins Guide to the Alpine Plants of New Zealand 1985; A Field Guide to the Native Trees of New Zealand 1986; Native New Zealand Flowering Plants 1991; The Stick Insects of New Zealand 1991; A Field Guide to the Alpine Plants of New Zealand 1992. *Address:* Rongonui, 60 Harvey St, Taupo, New Zealand. *Telephone:* (7) 378-5187.

SAMIMI, Ergjin; Albanian cartographer; b 9 Aug 1940, Tirana; s of Vasfi and Saxhide Samimi. *Education:* Tirana Univ. *Career:* Topo-geodetic work 1961–63; school and thematic cartography 1964–87. *Present Position:* Dir and Chief of Cartography, Geographic Studies Centre, Acad of Sciences 1988–. *Publications:* Cartography and Topography 1984; Mathematical Cartography 1985; Compiling and Printing Maps 1987; Cartometry 1990. *Address:* Quendra e Studimeve Gjeografike, Tirana, Albania. *Telephone:* (42) 22901. *Fax:* (42) 23693.

SANCHEZ, Silvia, M SC; Peruvian biologist and conservationist; b 6 July 1955. *Education:* Lima Agric Univ. *Career:* Visiting Researcher, Washington Zoo Park, USA 1984; Asst, Agric Dept 1985; WWF, Washington 1987; work on buffer zones of protected areas 1988; Officer, Conservation and Man of Lake Titicaca Natural Reserve, Peruvian Asscn for Conservation of Nature (APECO) 1988–91. *Present Position:* Pres, Asociación para la Conservación de la Naturaleza (Peruvian Asscn for Conservation of Nature—APECO) 1990–. *Publications:* Operation Plans for Lake Titicaca Natural Reserve 1990. *Address:* Parque José de Acosta 187, Altos Magdalena del Mar, Lima 17, Peru. *Telephone:* (14) 625410. *Fax:* (14) 633048.

SANDBROOK, Richard; British environmentalist. *Career:* Adviser, UN. *Present Position:* Mem, Int Inst for Environment and Devt (IIED). *Address:* 3 Endsleigh St, London WC1H 0DS, UK. *Telephone:* (71) 388-2117.

SANTOS, Maria Amélia Mota; Portuguese politician; b 25 Aug 1952, Lisbon. *Career:* Univ lecturer; mem, Exec Cttee, Os Verdes (Green Party) 1983–, Leader; MP and Chair, Green Parl Group 1985–89. *Present Position:* Mem, EP and Co-Chair, Green Parl Group 1989–, mem, Cttee on Devt and Co-operation. *Address:* 97 rue Belliard, 1040 Brussels, Belgium. *Telephone:* (2) 234-21-11. *Telex:* 26999. *Fax:* (2) 230-69-33.

SAOUMA, Édouard; Lebanese int civil servant; b 6 Nov 1926, Beirut; m Inès Forero; 1s 2d. *Education:* St Joseph's Univ, Beirut. *Career:* Sec-Gen, Nat Fed of Agronomists 1955; Dir-Gen, Nat Inst for Agricultural Research 1957–62; mem, Governing Bd, Nat Grains Office 1960–62; Del, FAO 1955–62, Deputy Regional Rep for Asia and Far E 1962–65, Secr 1963–, Dir, Land and Water Div 1965–75; Minister of Agric, Fisheries and Forestry 1970. *Present Position:* Dir-Gen, FAO 1976–. *Address:* Via delle Terme di Caracalla, 00100 Rome, Italy. *Telephone:* (6) 57971. *Telex:* 610181. *Fax:* (6) 5797-3152.

SARATHY, M A Partha; Indian environmentalist. *Present Position:* Chair, IUCN Comm on Educ and Communication. *Address:* 1 12th Cross, Rajmahal Vilas Ext, 560 080 Bangalore, India. *Telephone:* (812) 345595. *Telex:* 0818452334. *Fax:* (812) 341674.

SATOO, Prof Dr Taisitiroo; Japanese environmentalist; b 18 July 1923, Chonju, Korea. *Education:* Tokyo Imperial Univ, Tokyo Univ. *Career:* Research Scientist, Forestry Experiment Station, Korea 1941; military service 1942–46; Instructor, Tokyo Univ 1946, Assoc Prof 1950, Prof 1962, Emer Prof 1979. *Present Position:* Pres, Japan Wildlife Research Centre 1978– and Nagao Natural Environment Foundation 1989–. *Awards and Honours:* Japanese Forestry Soc Scientific

Achievement Award 1955; Ministry of Environment Distinguished Contribution to Conservation Award 1991; UNEP Global 500 1992. *Publications:* Silviculture 1982; Biomass Production of Forests 1987. *Address:* Yusima 2-29-3, Bunkyouku, Tokyo 113, Japan. *Telephone:* (3) 813-8506. *Fax:* (3) 813-8958.

SAURI, Pekka Markus, PH D; Finnish politician and environmentalist; b 31 May 1954, Helsinki; s of Markus and Elnikki Sauri; m Gandul Saris 1979; 3s. *Education:* Helsinki Univ, Brunel Univ, London. *Present Position:* Chair, Vihreä Liitto (Finnish Green Party) 1991–. *Address:* Eerikinkatu 24, 00100 Helsinki, Finland. *Telephone:* (0) 1351692. *Fax:* (0) 6933799.

SAWYER, John Stanley, MA, FRS; British meteorologist; b 19 June 1916, Wembley, Middx; s of Arthur Stanley and Emily Florence Sawyer; m Betty Vera Beeching (née Tooke) 1951; 1d. *Education:* Jesus Coll, Cambridge. *Career:* Meteorological Office 1937, Researcher 1946–76; Pres, Royal Meteorological Soc 1963–65; Pres, Atmospheric Sciences Comm, WMO 1968–73; mem, Natural Environment Research Council 1975–81. *Present Position:* Meteorological expert. *Awards and Honours:* IMO Prize, WMO 1973. *Address:* Ivy Corner, Corfe, Taunton TA3 7AN, UK. *Telephone:* (823) 421612.

SAWYER, Steve; environmentalist. *Present Position:* Exec Dir, Greenpeace Int. *Address:* Keizersgracht 176, 1016 DW Amsterdam, Netherlands. *Telephone:* (20) 5236555. *Telex:* 18775. *Fax:* (20) 5236500.

SCHNEIDER, Stephen Henry, PH D; American climatologist; b 11 Feb 1945, New York; s of Samuel and Doris C (Swarte) Schneider; m Cheryl Kay Hatter 1978. *Education:* Columbia Univ. *Career:* Research Assoc, NASA 1971–72; Fellow, Nat Center for Atmospheric Research 1972–73, Scientist and Deputy Head, Climate Project 1973–78, Acting Leader, Climate Sensitivity Group 1978–80, Head, Visitors Program and Deputy Dir, Advanced Study Program 1980–87; Affiliate Prof of Atmospheric Research, Lamont-Doherty Geological Observatory, Columbia Univ; mem, Carter–Mondale Science Policy Task Force 1976; Ed, *Climatic Change* 1976–. *Present Position:* Head, Interdisciplinary Climate Systems Section, Nat Center for Atmospheric Research 1987–. *Publications:* The Genesis Strategy: Climate and Global Survival (Co-Author) 1976; The Primordial Bond: Exploring Connections between Man and Nature through Humanities and Science (Co-Author) 1981; Interactions of Food and Climate (Co-Author) 1981; Social Science Research and Climate Change: An Interdisciplinary Appraisal (Co-Author) 1983; The Coevolution of Climate and Life (Co-Author) 1984; Forecasting in the Social and Natural Sciences (Co-Author) 1987. *Address:* POB 3000, Boulder, CO 80307, USA. *Telephone:* (202) 225-6161.

SCHREIBER, Dr Heinz; Austrian civil servant and expert on renewable energy sources and environmental planning; b 21 June 1939, Vienna; s of Georg and Anna Schreiber; m Christiane Vogt 1965; 1s 1d. *Education:* Vienna Tech Univ, Vienna Univ. *Career:* Asst Prof and Lecturer in Econs, Vienna Tech Univ 1970–75; Fed Ministry for Science and Research 1975–, Dir 1980–. *Present Position:* Dir-Gen, Fed Ministry of Environment, Youth and Family 1987–. *Address:* 1030 Vienna, Radetzkystr 2, Austria. *Telephone:* (1) 71158-4100. *Fax:* (1) 71158-4226.

SCHULTES, Richard Evans, PH D; American botanist, explorer and conservationist; b 12 Jan 1915, Boston, MA; s of O Richard and Maude B (Bagley) Schultes; 2s 1d. *Education:* Harvard Univ. *Career:* Ethnobotanical researcher, NE Oaxaca, Mexico 1939–40; Nat Research Council Fellow, Colombian Amazon 1941–42, Researcher on Hevea Rubber, Agric Dept 1943–53; Research Fellow, Harvard Botanical Museum 1941–53, Curator, Orchid Herbarium 1953–58, Curator of Econ Botany 1958–85, Dir 1967–85; mem, Latin American Selection Cttee, Guggenheim Foundation 1964–85; Prof of Biology, Harvard Univ 1970–72, 1980–85, of Natural Sciences 1973–80; Adjunct Prof of Pharmacognosy, Pharmacy Coll, Illinois Univ, Chicago 1975–. *Present Position:* Emer Prof of Biology, Harvard Univ 1985–. *Awards and Honours:* WWF Annual Gold Medal 1984; Tyler Ecology-Energy Prize 1987; Linnean Gold Medal 1992. *Address:* Harvard Botanical Museum, 26 Oxford St, Cambridge, MA 02138, USA. *Telephone:* (617) 495-2326.

SCHWALBA-HOTH, Frank; German environmentalist; b 12 Dec 1952, Hamburg. *Education:* Philipps Univ, Marburg. *Career:* Co-Founder, German Green Party 1980; mem, State Parl, Hesse 1982–83, EP 1984–87, Pres, Green Group 1986–87; Head, EC Unit, Greenpeace 1991–93. *Present Position:* Dir, Worldwatch Inst (Europe) 1993–. *Address:* rue Biddaer 17, 1070 Brussels, Belgium. *Telephone:* (2) 523-21-85. *Fax:* (2) 523-21-85.

SCURLOCK, Jonathan Michael Oliver, PH D; British consultant on energy, environment and plant physiology; b 16 July 1959, Oxford; s of Prof Ralph G and Maureen M Scurlock. *Education:* Magdalen Coll, Oxford, King's Coll, London, Imperial Coll, London. *Career:* Asst Co-ordinator, Photosynthesis Project, UNEP 1983–, Contrib, Environmental Data Report 1992; Bulletin Ed, Parl Alternative Energy Group 1987–; UK Corresp, *Windpower* monthly magazine 1988–92; Consultant, FoE, Greenpeace, WWF (UK) 1988–. *Present Position:* Research Asst, King's Coll, London 1988–. *Publications:* Techniques in Bioproductivity and Photosynthesis (Co-Ed) 1985; Photosynthesis and Production in a Changing Environment (Co-Ed) 1993. *Address:* Campden Hill Rd, London W8 7AH, UK. *Telephone:* (71) 333-4315. *Fax:* (71) 937-7783.

SEGLINŠ, Dr Valdis; Latvian geologist; b 26 May 1958, Ligatne; s of Edgars Seglinš and Dzintra Dreimane; m Elita Seglina 1985; 1s. *Education:* Latvian Univ, Riga. *Career:* Field Geologist, Head of Group then Chief Geologist, Geology Dept 1981–89; Head of State Ecological Expertise, Nature Protection Cttee 1989–90. *Present Position:* 1st Vice-Chair, Environmental Protection Cttee. *Publications:* The Environmental Situation and Project Identification in the Republic of Latvia 1991; A Pre-Feasibility Study of the Gulf of Riga and the Daugava River Basin 1992; The State of the Environment in Latvia 1993. *Address:* Peldu 25, LV-1282 Riga, Latvia. *Telephone:* (2) 229540. *Telex:* 161180. *Fax:* (2) 228159.

SELFO, Dr Lirim; Albanian chemist; b 28 July 1942, Saranda; m Marita Afezolli 1975; 1s 1d. *Education:* Tirana Univ. *Career:* Researcher, Industrial Projects Inst 1963–75; Engineer, Elbasan Cement Factory and Metallurgical Combine 1975–82; Researcher, Chemical Tech Inst 1982–92. *Present Position:* Chair, Cttee for Environmental Preservation and Protection (CEPP) 1992–. *Address:* Tirana, Albania. *Telephone:* (42) 27907. *Fax:* (42) 27907.

SETH, Anand Kumar; Indian economist; b 24 Oct 1945, India; m Savita Seth 1972; 1s 1d. *Education:* Indian Tech Inst, New Delhi, Pittsburgh Univ, PA. *Present Position:* Chief, Environment Div for Middle E, N Africa, Europe and Cen Asia, World Bank. *Address:* 1818 H St, NW, Washington, DC 20433, USA. *Telephone:* (202) 473-2502. *Fax:* (202) 477-0711.

SHIVA, Dr Vandana; Indian environmentalist and writer. *Present Position:* Mem, Int Advisory Council, Worldwide Network and Research Foundation for Science, Tech and Resource Policy. *Address:* 105 Rajpur Rd, Dehra Dun 248 001, India.

SIDIKOU, Fatoumata, LL M; Niger int civil servant; b 2 June 1949, Say; m 1971; 3s 2d. *Education:* Abidjan Univ, Côte d'Ivoire. *Career:* Credit Services Dept, Int Bank of W Africa 1973–77; UNDP 1978–83. *Present Position:* Programme Head, UNFPA, Dakar 1983–. *Address:* 19 rue Parchappe, BP 154, Dakar, Senegal. *Telephone:* 239168. *Telex:* 216935. *Fax:* 235500.

SILINA, Mara; Latvian organic chemist; b 26 April 1949, Riga; d of Voldemars Fridermanis and Dzidra Fridermane; m Aldis Silins 1977 (divorced 1992); 1d. *Education:* Latvia State Univ, Riga. *Career:* Scientific Researcher, Applied Biochemistry Inst 1971–83; Experimental Plant, Organic Synthesis Inst 1983–89; Latvian Environmental Protection Club 1989–90. *Present Position:* Programme Co-ordinator, FoE Int 1990–. *Address:* POB 163, 814 99 Bratislava, Slovakia. *Telephone:* (7) 827-213. *Fax:* (7) 827-213.

SINGH, Karan, PH D; Indian statesman and author; b 1931, Cannes, France; s of Maharaja Hari Singh and Maharani Tara Devi of Jammu and Kashmir; m Princess Yasho Rajya Lakshmi of Nepal 1950; 2s 1d. *Education:* Delhi Univ. *Career:* Regent, Sadr-e-Riyasat (Head of State) and Gov, Jammu and Kashmir State 1946–66; Minister of Civil Aviation and Tourism 1967–73; Minister of Health and Family Planning 1973–77; Amb to USA 1989–90. *Present Position:* Pres, People's Comm on Environment and Devt; mem, Steering Cttee, Global Forum of Parliamentarians and Spiritual Leaders on Human Survival. *Publications:* Prophet of Indian Nationalism 1963; Population, Poverty and the Future of India 1975; Humanity at the Crossroads 1988; Essays on Hinduism 1990. *Address:* 15 Institutional Area, Lodhi Rd, New Delhi 110003, India. *Telephone:* (11) 4627102. *Fax:* (11) 4620102.

SMITH, David G; British environmental consultant. *Career:* Consultant, Comm of EC. *Present Position:* Dir of Environment Services, PA Consulting Group. *Address:* Cambridge Lab, Melbourn, Royston SG8 6DP, UK. *Telephone:* (763) 261222. *Fax:* (763) 260023.

SMITH, Dr John; scientist and environmentalist. *Present Position:* Mem, European Science Foundation. *Address:* 1 quai Lezay-Marnesia, 67080 Strasbourg Cedex, France. *Telephone:* 88-76-71-31. *Fax:* 88-37-05-32.

SOCHACZEWSKI, Paul Spencer; American nature conservationist and author; b 1 Aug 1947, New York. *Education:* George Washington Univ, Washington, DC. *Career:* US Peace Corps, Sarawak, Malaysia; film dir; journalist; nature conservationist. *Present Position:* Head of Creative Services, WWF Int. *Publications:* Soul of the Tiger (Co-Author) 1988; Eco-Bluff Your Way to Instant Environmental Credibility (Co-Author) 1991. *Address:* World Conservation Centre, ave du Mont-Blanc, 1196 Gland, Switzerland. *Telephone:* (22) 3692227. *Telex:* 419618. *Fax:* (22) 3643239.

SOMERVILLE, Hugh, PH D; British scientist; b 13 Sept 1939, Castle Douglas, Scotland; m 1964; 1s 1d. *Education:* Edinburgh Univ, Sheffield Univ. *Career:* California Univ, Los Angeles and Berkeley 1965–68; Researcher, Shell 1965–68, Houston 1978–80, Head, Environmental Affairs Dept, Exploration and Production Section, Aberdeen 1983–87; Technical Sec, CONCAWE, Hague 1980–83; Sr Adviser on Environment and Occupational Health, Occidental Petroleum, Aberdeen 1987–89. *Present Position:* Head, Environment Dept, British Airways 1989–. *Address:* S285, Speedbird House, POB 10, Heathrow Airport, Hounslow TW6 2JA, UK. *Telephone:* (71) 562-5699. *Telex:* 8813983. *Fax:* (71) 562-8604.

STAHR, Elvis J, Jr, MA; American conservationist, lawyer and univ official; b 9 March 1916, Hickman, KY; s of Elvis J and Mary A (McDaniel) Stahr; m Dorothy Howland Berkfield 1946; 2s 1d. *Education:* Kentucky Univ, Oxford Univ, Yale Univ. *Career:* Law firm, New York; US Army, 2nd World War; Assoc Prof of Law, Kentucky Univ 1947–48, Prof and Dean, Law Coll 1948–54, Provost 1954–56; Army Dept, Korean War; Exec Dir, Pres's Cttee on Educ beyond Secondary School Level 1956–57; Vice-Chancellor, Pittsburgh Univ 1957–59; Pres, W Virginia Univ 1959–61; Army Sec 1961–62; Pres, Indiana Univ 1962–68, Nat Audubon Soc 1968–79, Sr Counsellor 1979–81; mem, Presidential Aviation Advisory Comm 1970–73, Bd of Advisers, Nat Exec Service Corps, Population Inst, Fed for American Immigration Reform; mem, Del to Tripartite Environmental Pollution Conf (UK, Canada, USA), Ditchley Park 1970, to UN Conf on Human Environment, Stockholm 1972 and to Int Whaling Comm, London 1975, 1978; mem, Jt US–USSR Cttee on Co-operation for Protection of Environment, Washington 1973, Nat Cttee for World Population Year 1974, Nat Petroleum Council 1974–79; Pres, Univ Assocs Inc 1981–; Special Partner, Chickering & Gregory law firm 1982–; Exec Vice-Pres, Public Resource Foundation 1982–; Sr Assoc, Cassidy & Assocs Inc 1984–; Chair, Washington Conservation Roundtable 1986–87; Dir, Environmental and Energy Study Inst, Alliance to Save Energy, Nat Water Alliance, Acacia Bank; fmr Pres, US Army Asscn; fmr Nat Chair, United Service Orgs; fmr Deputy Chair, Bd, Fed Reserve Bank of Chicago. *Present Position:* Pres Emer, Nat Audubon Soc 1981–. *Awards and Honours:* Nat Wildlife Fed Conservation Achievement Award 1978; Dept of Interior Conservation Service Award 1979; Natural Resources Council Award of Honor 1988. *Publications:* The Economics of Pollution (Co-Author) 1971. *Address:* c/o Chickering & Gregory, 1815 H St, NW, Washington, DC 20006, USA. *Telephone:* (202) 463-7456.

STEINHÜBEL, Prof Gejza, D SC; Slovak scientist and tree ecophysiologist; b 6 Oct 1922, Banská Bystrica; s of Gejza Steinhübel Sr and Mária Stolmannová; m Vlasta Zoulová 1954; 1s. *Education:* Comenius Univ, Bratislava. *Career:* Asst and Lecturer, Comenius Univ, Bratislava 1947–51; Scientist, Tree Ecophysiologist and Assoc Prof, Nitra Agric Univ and Mlynany Arboretum, Acad of Sciences 1951–69; Chief Scientist, Forestry Research Inst, Zvolen 1969–86. *Present Position:* Chief Scientist and Consultant, Forest Ecology Inst, Acad of Sciences, Zvolen 1987–. *Awards and Honours:* Biology Golden Award of Merit 1992; Zvolen Tech Univ Ecology Prize 1992; Forestry Research Inst Merit Award 1992. *Publications:* Inhibiting Factors in the Reproduction of Liriodendron Seeds in Slovakia 1961; An Introduction to the Ecophysiology of Evergreens 1967. *Address:* Štúrova 2, 960 53 Zvolen, Slovakia. *Telephone:* (855) 20-914. *Fax:* (855) 27-485.

STEINNES, Eiliv, PH D; Norwegian environmental scientist; b 21 Sept 1938, Elverum; m Randi Surdal 1962; 3d. *Education:* Oslo Univ. *Career:* Research Scientist, Atomic Energy Inst 1963–68, Head, Radioanalytical Section 1969–77, Analytical Div 1978–79; Assoc Dean, Science Faculty, Trondheim Univ 1981–84, Rector, Arts and Science Coll 1984–90. *Present Position:* Prof in Environmental Science, Trondheim Univ 1980–. *Address:* Dept of Chemistry, AVH, 7055 Dragvoll, Norway. *Telephone:* (7) 59-62-37. *Fax:* (7) 59-62-55.

STOREY, James M; American environmentalist. *Present Position:* Adviser and Dir, Earthwatch. *Address:* 680 Mt Auburn St, POB 403, Watertown, MA 02272-9104, USA.

STOYANOVA, Nadezhda Georgieva, MS; Bulgarian forester; b 20 Oct 1948, Sofia; d of Georgi Georgiev and Kera Georgieva; m Dimitar Tzenkov Stoyanov 1975; 1s 1d. *Education:* Higher Inst of Forestry and Forest Eng, Sofia. *Career:* Forest Research Inst, Bulgarian Acad of Sciences 1974–. *Present Position:* Researcher, Ecology and Environmental Conservation Dept, Forest Research Inst, Bulgarian Acad of Sciences.

Publications: Ecological and Typological Investigations into Low Mountain Forest Ecosystems of the Natural Range of Fagus Sylvatica L 1992; The Ecosystems of Common Beech (Fagus Sylvatica L) 1992. *Address:* 132 Kliment Ohridski Blvd, 1756 Sofia, Bulgaria. *Telephone:* (2) 62-29-61. *Fax:* (2) 62-29-65.

STRONG, Maurice F, OC, FRSA, FRSC; Canadian environmentalist; b 29 April 1929, Oak Lake, MB; s of Frederick Milton and Mary Fyfe Strong; m 1st Pauline Olivette Williams 1950 (divorced 1980); m 2nd Hanne Marstrand 1981; 2s 2d 1 foster d. *Career:* UN Secr 1947; pres and dir, Canadian and int corpns 1954–66; head, devt and int affairs orgs; Dir-Gen, External Aid Office 1966; Chair, Canadian Int Devt Bd; Alt Gov, World Bank, ADB and Caribbean Devt Bank; Under-Sec-Gen for Environmental Affairs, UN 1970–72, Chief Exec, Human Environment Conf, Stockholm 1972, Under-Sec-Gen 1985–87, Sec-Gen, UNCED 1990–; Exec Dir, UNEP 1973–75; Chair, Petro Canada and Pres, Stronat Investments Ltd 1976–80; Chair, Bd of Govs, Int Devt Research Centre 1977–78; Chair, Strouest Holdings Inc, Procor Inc 1978–79, AZL Resources Inc 1978–83, Int Energy Devt Corpn 1980–83, N–S Round Table Soc for Int Devt and Canadian Devt Investment Corpn 1982–84; Chair and Pres, American Water Devt Inc, Denver 1986–89; Pres, World Fed, UNA 1987– and Baca Corpn; Dir, Better World Soc 1988–; Head, Ontario Hydro 1992. *Present Position:* Under-Sec-Gen, UN 1989–. *Awards and Honours:* Onassis Int Award 1993. *Address:* c/o UN HQ, Palais des Nations, 1211 Geneva 10, Switzerland. *Telephone:* (22) 330730.

STUTZIN, Godofredo; Chilean lawyer; b 21 Sept 1917, Darmstadt, Germany; m Nora Schottlander 1950; 2s. *Education:* Chile Univ, Santiago. *Career:* Founder, Unión de Amigos de los Animales 1957, Comité Nacional pro Defensa de la Fauna y Flora (CODEFF) 1967; newspaper correspondent on nature conservation 1959–; Dir of Publications, IPPF Int Conf, Santiago 1967; mem, Exec Bd, IUCN 1972–78. *Present Position:* Pres, Comité Nacional pro Defensa de la Fauna y Flora (CODEFF) and mem, Legislation and Species Survival Comm, IUCN. *Awards and Honours:* UNEP Global 500 1990. *Publications:* Die Natur der Rechte und die Rechte der Natur 1980; Un Imperativo Ecológico: Reconocer los Derechos de la Naturaleza 1984; A Task for the Law: To Re-Establish the Ecological Balance 1993. *Address:* Camino El Alto 17220, Parcela 14, El Arrayan, Santiago 52, Chile. *Telephone:* (2) 217-1643. *Fax:* (2) 737-7290.

SUMNER, Gordon Matthew (Sting); British singer, musician, songwriter, actor and environmentalist; b 2 Oct 1951, Northumberland; m 1st Frances Tomelty (divorced 1984); m 2nd Trudie Styler 1992; 5c. *Career:* Teacher, primary school, Cramlington, Newcastle; mem, Police pop group 1977–; tours, UK, Europe, USA; stage acting debut, The Threepenny Opera 1989. *Present Position:* Pop singer, actor and environmentalist. *Awards and Honours:* Ivor Novello Award 1989. *Publications:* Jungle Stories: The Fight for the Amazon 1989; Records (with Police): Outlandos d'Amour 1977; Regatta de Blanc 1979; Zenyatta Mondatta 1980; Ghosts in the Machine 1981; Synchronicity 1983; Records (Solo): The Dream of the Blue Turtles 1985; Bring on the Night; Nothing like the Sun 1987; The Soul Cages 1991; Ten Summoner's Tales 1993; Films: Quadrophenia 1979; Brimstone and Treacle; The Bride of Frankenstein; Plenty; Dune 1985; The Adventures of Baron von Munchausen 1989; Stormy Monday 1989; Rosencrantz and Guildenstern are Dead. *Address:* IRL, Bugle House, Noel St, London W1, UK.

SUTTON, Rodney; British businessman and environmentalist. *Present Position:* Dir, ICC UK. *Address:* 14 Belgrave Sq, London SW1X 8PS, UK. *Telephone:* (71) 823-2811. *Fax:* (71) 235-5447.

SUZUKI, David T; Canadian zoologist. *Present Position:* Mem, Zoology Dept, British Columbia Univ. *Address:* 6270 Univ Blvd, Vancouver, BC V6T 1Z4, Canada. *Telephone:* (604) 822-2211. *Fax:* (604) 822-5785.

T

TABOR, Prof Harry Zvi, PH D; Israeli applied physicist; b 7 March 1917, London, UK; s of Charles and Rebecca Tabor; m Vivienne Landau 1947; 2d. *Education:* London Univ, Hebrew Univ, Jerusalem. *Career:* Industrial research physicist 1937–49; Head, Physical Sciences Div, Israeli Research Council 1949–55; Dir, Nat Physical Lab 1950–74; Scientific Dir, Scientific Research Foundation, Jerusalem 1971–87; renewable energy specialist and Pres, Int Solar Energy Soc (ISES) 1981–83. *Present Position:* Consultant, Ministry of Science and Tech, Ministry of Energy and Infrastructure, UNESCO. *Awards and Honours:* Royal Soc Energy Award 1975; Int Solar Energy Soc (ISES) Award 1981; Krupp Foundation Energy Award, Essen, Germany 1981. *Address:* POB 3745, Jerusalem 91036, Israel. *Telephone:* (2) 435785. *Fax:* (2) 437470.

TAESLER, Prof Roger; Swedish climatologist. *Career:* Urban climatology expert; Adviser, UN. *Present Position:* Mem, Swedish Meteorological and Hydrological Inst. *Address:* POB 923, 601 76 Norrköping, Sweden.

TARAND, Dr Andres; Estonian climatologist and ecologist; b 11 Jan 1940, Tallinn; s of Helmut and Leida Tarand; m Mari Viiding 1963; 2s. *Education:* Tartu Univ. *Career:* Soviet Army 1963–65; Antarctic expedition 1968–70; Researcher-Dir, Tallinn Botanical Gardens 1970–89; Chair, Environment Cttee, Supreme Soviet 1990–92. *Present Position:* Minister of Environment 1992–. *Publications:* Cassiopeia 1993. *Address:* 24 Toompuiestee, EE0001 Tallinn, Estonia. *Telephone:* (142) 452-507. *Fax:* (142) 453-301.

TEIWAKI, Roniti, M PHIL; Kiribati politician; b 20 Dec 1949, Kiribati; m Taberauea Yee-Ting 1972; 2s 3d. *Education:* Canterbury Univ, S Pacific Univ, Wales Univ. *Career:* Minister 1974–82, of Natural Resources Devt 1979–82; Dir, Atoll Research Unit 1982–84, Kiribati Centre, S Pacific Univ 1982–90. *Present Position:* MP (S Tarawa) 1990–, Leader of Opposition 1993–; Dir, Kiribati Centre, S Pacific Univ. *Awards and Honours:* S Pacific Univ Gold Medal 1978. *Publications:* Management of Marine Resources in Kiribati 1989. *Address:* POB 300, Abarao Village, Bikenibeu, Tarawa, Kiribati. *Telephone:* 28136. *Fax:* 21004.

TEJADHAMMO, Phra Ajahn Pongsak; Thai contemplative monk, conservationist and teacher; b 20 July 1932, Tamnop Village, Nakorn Sawan. *Career:* Mahatat Temple, Bangkok 1951–53; Suan Mokh Monastery 1959–64; solitary meditational retreats in forests and caves, attaining deep insight into nature of man and environment, Tudong, N Thailand;

Founder, Forest Meditation Centre, Mae Soi Valley 1980; Abbot, Palad Monastery 1983; Founder, Mae Soi Villagers Asscn for Protection of Watersheds 1987. *Present Position:* Pres, Dhammanaat Foundation for Conservation and Rural Devt and Dir, Mae Soi Valley Project 1987–; Pres, Monks Asscn for Protection of Life and Environment 1990–. *Awards and Honours:* Exemplary Citizen's Award, Thai Rath Foundation 1989; Royal Forest Dept Award 1989; UNEP Global 500 1990. *Publications:* First Stages in Meditation 1985; Buddhism and Conservation 1990; Films: The Chaophya River 1989; The Monk, the Princess and the Forest 1990; The Monk, the Trees and the Concrete Jungle 1992; Save the Earth 1992. *Address:* Wat Palad, Doi Suthep, Chiangmai 50000, Thailand. *Telephone:* (53) 248209. *Fax:* (53) 223062.

THOMPSON, Hilary J; British businesswoman and environmentalist. *Career:* Personal Asst to Lord Alexander. *Present Position:* Mem, Environmental Man Unit, Nat Westminster Bank plc. *Address:* 1st Floor, 1 Broadgate, London EC2M 2AD, UK. *Telephone:* (71) 714-4455. *Fax:* (71) 714-4464.

THORNTON, Prof Iain; British environmental scientist. *Career:* Del, UNCED, Rio de Janeiro 1992; Adviser, UN. *Present Position:* Dir, Global Environmental Research Centre, Imperial Coll of Science and Tech, London. *Address:* 56 Queen's Gate, London SW7 5JR, UK. *Telephone:* (71) 225-1818. *Fax:* (71) 225-0627.

TICKELL, Sir Crispin Charles Cervantes, GCMG, KCVO, FRGS, FZS; British academic and diplomatist; b 25 Aug 1930; s of Jerrard Tickell and Renée Haynes; m 1st Chloë Gunn 1954 (divorced 1976); m 2nd Penelope Thorne 1977; 2s 1d. *Education:* Christ Church, Oxford. *Career:* HM Diplomatic Service 1954, Netherlands 1955–58, Mexico 1958–61, France 1964–70, Head, Western Orgs Dept, FCO 1972–75, Amb to Mexico 1981–83, Deputy Under-Sec of State, FCO 1983–84, Perm Sec, Overseas Devt Admin 1984–87, Perm Rep, UN 1987–90; Pvt Sec to Chancellor of Duchy of Lancaster 1970–72; Fellow, Center for Int Affairs, Harvard Univ 1975–76; Chef de Cabinet to Pres, Comm of EC 1977–81; Visiting Fellow, All Souls Coll, Oxford 1981; Special Adviser to Prime Minister on Environmental Matters and Chair, Int Inst for Environment and Devt, Earthwatch Europe, Climate Inst, Washington, DC and Marine Biological Asscn 1990–; Pres, Royal Geographical Soc; Dir, IBM (UK) and BOC Foundation 1990– and Mexican Horizons Investment 1991–; Trustee, Natural History Museum 1991–. *Present Position:* Warden, Green Coll, Oxford 1990–. *Publications:* Climatic Change and World Affairs 1977. *Address:* Oxford OX2 6HG, UK. *Telephone:* (865) 274770.

TIENSUU, Roland; Swedish environmentalist. *Present Position:* Mem, Barnens Regnskog. *Awards and Honours:* Goldman Environmental Foundation Prize 1991. *Address:* c/o Ministry of Educ and Science, Drottninggt 16, 103 33 Stockholm, Sweden. *Telephone:* (8) 763-10-00. *Telex:* 13284. *Fax:* (8) 723-11-92.

TIMBERLAKE, Lloyd; British environmentalist and writer. *Career:* Only One Earth TV series; Consultant, World Comm on Environment and Devt, Business Council for Sustainable Devt. *Present Position:* Dir of External Affairs, Int Inst for Environment and Devt (IIED). *Awards and Honours:* World Hunger Media Award. *Publications:* Africa in Crisis; Only One Earth; Our Common Future (Co-Author); Changing Course (Co-Author). *Address:* 3 Endsleigh St, London WC1H 0DD, UK. *Telephone:* (71) 388-2117.

TIMOSHENKO, Alexandre, PH D; Russian lawyer; b 18 Sept 1939, Moscow; s of Sergei and Olga Timoshenko; m Elena Timoshenko; 1s 1d. *Education:* Moscow State Inst of Int Relations, State and Law Inst, Acad of Sciences. *Career:* Econ Analyst, Ministry for Environment and Trade 1961–65, USSR Embassy, Somalia 1965–69; Sr Legal Researcher 1973–86; Dir, Environmental Law Dept, Acad of Sciences 1987–91. *Present Position:* Head, Int Legal Instruments Dept, Environmental Law and Insts Programme Activity Centre, UNEP 1991–. *Publications:* International Environmental Co-operation within the UN System 1986; Formation and Development of International Environmental Law 1986. *Address:* POB 47074, Nairobi, Kenya. *Telephone:* (2) 230800. *Telex:* 22068. *Fax:* (2) 230198.

TINKER, Jon, BA; British journalist; b 19 Jan 1940, Huddersfield; 4d 1 stepson. *Education:* Cambridge Univ. *Career:* Sec to Bertrand Russell, British philosopher 1961–62; Press Officer, Council for Nature 1964–66; Ed, *Wildlife and the Countryside* 1964–70; Environment and Development Ed, *New Scientist* 1969–76; mem, Royal Comm on Environmental Pollution 1973–74; Founder and Dir, Earthscan 1974–86. *Present Position:* Founder and Pres, Panos Int 1986–; Sr Assoc, Sustainable Devt Research Inst, British Columbia Univ 1993–. *Awards and Honours:* UNEP Global 500 1992. *Address:* B5-2203 Main Mall, Vancouver, BC V6T 1Z4, Canada. *Telephone:* (604) 822-8198. *Fax:* (604) 822-9191.

TOBIAS, Phillip Vallentine, PH D, D SC, FLS, FRAI; South African prof of anatomy; b 14 Oct 1925, Durban; s of Joseph Newman and Fanny (Rosendorff) Tobias. *Education:* Witwatersrand Univ, Emmanuel Coll, Cambridge, UK. *Career:* Anatomy Lecturer, Witwatersrand Univ 1951–52, Sr Lecturer 1953, Prof and Head, Anatomy Dept 1959–90, Dean, Medicine Faculty 1980–82; Visiting Prof, Cambridge Univ; Founder Pres, Inst for Study of Man in Africa, Anatomical Soc of Southern Africa and S African Soc for Quaternary Research; Vice-Pres and Acting Exec Pres, S African Asscn for Advancement of Science; Pres, Royal Soc of S Africa 1970–72; mem, Perm Council, Int Union of Prehistoric and Protohistoric Sciences, Pan-African Congress of Prehistory and Quaternary Studies and Int Asscn of Human Biologists. *Present Position:* Prof, Dept of Anatomy and Human Biology, Witwatersrand Univ Medical School. *Publications:* Chromosomes, Sex-Cells and Evolution 1956; Man's Anatomy (Co-Author) 1968; The Brain in Hominid Evolution 1971; The Meaning of Race 1972; The Bushmen 1978; Dart, Taung and the Missing Link 1984; Hominid Evolution: Past, Present and Future 1985; Olduvai Gorge 1991. *Address:* York Rd, Parktown, 2193 Johannesburg, S Africa. *Telephone:* (11) 647-2516. *Telex:* 427125. *Fax:* (11) 716-8030.

TÖLGYESSY, Juraj, PH D, D SC; Slovak chemist and environmentalist; b 27 Jan 1931, Dunajská Streda; m Eva Cichá 1957; 1s 2d. *Education:* Bratislava Tech Univ, Chemical Tech Univ, Prague, Lomonosov Univ, Moscow. *Career:* Asst, Bratislava Tech Univ 1952–55, Asst Prof 1955–62, Assoc Prof 1962–72, Prof 1978–, Deputy Dean, Chemical Tech Faculty; Founder and Ed-in-Chief, *Journal of Radioanalysis and Nuclear Chemistry*. *Present Position:* Head, Environmental Science Dept, Bratislava Tech Univ. *Publications:* Chemical Aspects of the Environment 1981; Nuclear Environmental Chemical Analysis 1987; Environmental Chemistry 1989; Radionuclide X-Ray Fluorescence Analysis with Environmental Applications 1990; Nuclear Environmental Applications 1990; Chemistry and Biology of Water, Air and Soil: Environmental Aspects 1993. *Address:* Radlinského 9, 812 37 Bratislava, Slovakia. *Telephone:* (7) 495-243. *Fax:* (7) 493-198.

TOORNSTRA, Franke Hendrik, M SC; Dutch environmentalist; b 19 Sept 1959; m Heidi de Bruyn 1986. *Education:* Wageningen Agric Univ. *Career:* Research Asst, Wageningen Agric Univ 1983–84; Research Fellow, Environmental Studies Centre, Leiden Univ 1984–86, staff mem 1986–. *Present Position:* Dir, Environment and Devt Study Centre (CEDC), Cameroon 1989–. *Address:* c/o CML, POB 9518, 2300 RA

Leiden, Netherlands. *Telephone:* (71) 277482. *Telex:* 39427. *Fax:* (71) 277496.

TOPPING, John C; American climatologist. *Present Position:* Pres, Climate Inst. *Address:* 316 Pennsylvania Ave, SE 403, Washington, DC 20003, USA. *Telephone:* (202) 547-0104.

TÓTH, Dezider, PH D, D SC; Slovak scientist; b 10 April 1941, Strážske; s of Ján Tóth and Gizela Tóthová; m Libuša Tóthová 1987; 1s 2d. *Education:* Bratislava Tech Univ, Comenius Univ, Bratislava. *Career:* Head, Analytical Lab, E Slovakia Steelworks, Košice 1964–69; Research Assoc, Acad of Sciences 1964–75; Visiting Teacher of Microbiology, Bratislava Tech Univ 1979–84; Assoc Prof of Microbiology and Biochemistry, Maiduguri Univ, Nigeria 1984–86. *Present Position:* Dir, Ecobiology Inst, Acad of Sciences 1992–. *Publications:* Changes in the Cell Components of Escherichia Coli and Pseudomonas Fluorescens in Deficient Medium (Co-Author) 1983; Microbial Interactions with Chemical Water Pollution (Co-Author) 1989. *Address:* Stefanikova 3, 814 34 Bratislava, Slovakia. *Telephone:* (7) 334-595. *Fax:* (7) 334-967.

TOWNSEND, Michael Hugh Crofton; Zimbabwean farmer; b 1939, Harare; s of R O C Townsend and G A P Shand; m Catherine Mary Glanfield 1967; 2s 1d. *Career:* Chair, Harare W Farmers Asscn 1975–79, Cen Mashonaland Branch, Commercial Farmers Union 1979–85, Wildlife Producers Asscn 1986–92, Sussex Cattle Breeders Soc 1985–88, Cattle Cttee, Agric Show Soc 1985–90. *Present Position:* Pres, Wildlife Soc of Zimbabwe 1991–; Chair, Rhino Custodians Cttee 1992–, Bushcrafters and Bowhunters Asscn 1993–. *Address:* Lowdale Farm, POB A392, Avondale, Harare, Zimbabwe. *Telephone:* (4) 393038.

TRAVERS, Bill; British actor and conservationist; m Virginia McKenna 1957; 3s 1d. *Career:* Film actor. *Present Position:* Trustee, Born Free Foundation. *Publications:* Films: Born Free 1964; Ring of Bright Water 1968. *Address:* Coldharbour RH5 6HA, UK. *Telephone:* (306) 712091. *Fax:* (306) 713350.

TROTZ, U O'D; scientist. *Present Position:* Science Adviser to Sec-Gen and Science Council, Commonwealth. *Address:* Marlborough House, Pall Mall, London SW1Y 5HX, UK. *Telephone:* (71) 839-3411. *Telex:* 27678. *Fax:* (71) 930-0827.

TRUBÍNOVÁ, Lubica; Slovak environmentalist. *Present Position:* Head, Greenpeace Slovakia. *Address:* POB 58, 815 99 Bratislava, Slovakia.

TRUDA PALAZZO, José; Brazilian naturalist and writer; b 28 July 1963, Porto Alegre; s of José Truda Palazzo and Ilka Ferreira Palazzo; m Maria do Carmo Both 1989; 1c. *Career:* Co-ordinator, Whale Research and Conservation Project 1982– and Int Wildlife Coalition 1986–; mem, Council, Brazilian Conservation Foundation 1987–. *Present Position:* Pres, Brazilian Branch, Int Wildlife Coalition. *Awards and Honours:* Brazilian Nat Defence League Civic Merit Award 1985; WWF Award 1990. *Address:* POB 5087, 88040-970 Florianópolis, SC, Brazil. *Telephone:* (482) 821150. *Fax:* (482) 341580.

TSENKOV, Lyudmil Boyanov; Bulgarian forest engineer; b 12 March 1951, Valtchedrum; m Vesselina Ivanova Tsenkova 1978; 1c. *Education:* Sofia Forestry Inst. *Career:* Forester 1977–78. *Present Position:* Researcher, Erosion Control Dept, Forest Research Inst, Acad of Sciences, Kurdjaly 1978–. *Publications:* Fortifying the Tidal Marshes of the East Rhodopean Region 1988; The Deformations of a Torrential Bed after Biotechnical Corrections 1992. *Address:* 132 Kliment Ohridski Blvd, 1756 Sofia, Bulgaria.

TUKAHIRWA, Joy Margaret, M SC; Ugandan environmentalist and scientist; b 11 April 1955, Kabale; m Dr Eim Tukahirwa 1984; 1d 2s. *Education:* Makerere Univ, Kampala. *Career:* Lecturer, Makerere Univ, Kampala 1982–83, Sr Lecturer 1991; Visitor, E Anglia Univ, UK 1990; Visiting Scholar, Ohio State Univ, USA 1992. *Present Position:* Researcher, Geography Dept, Makerere Univ, Kampala 1991–. *Awards and Honours:* FAO Research Award 1990; Rockefeller Research Award 1991. *Publications:* Prospects and Sensitivity of Soil Resources in the Highlands of Uganda 1986; Decision-Making in Soil Conservation among Small-scale Farmers in Uganda 1990; Soil Erosion Control 1991. *Address:* POB 7062, Kampala, Uganda. *Telephone:* (41) 531261. *Fax:* (41) 245597.

TWIDELL, Dr John; British scientist. *Present Position:* Head, Energy Studies Unit, Strathclyde Univ. *Address:* 16 Richmond St, Glasgow G1 1XQ, UK. *Telephone:* (41) 522-4400. *Telex:* 77472. *Fax:* (41) 552-0775.

U

UNSWORTH, Prof Mike; British scientist. *Career:* Researcher on climate change and agric. *Present Position:* Dept of Physiology and Environment, Nottingham Univ. *Address:* Univ Park, Nottingham NG7 2RD, UK. *Telephone:* (602) 484848. *Fax:* (509) 63917.

UNTAWALE, Arvind Gajanan, PH D; Indian scientist; b 26 June 1940, Deori; s of Dr G B and Shantabai Untawale; m Kshma Untawale 1972; 1s 1d. *Education:* Nagpur Univ. *Career:* Lecturer 1966; Govt Research Fellow 1967–70; Research Assoc, Nat Science Acad 1970–72; Scientist, Nat Inst of Oceanography, Goa 1972–; Exec Sec, Mangrove Soc 1986–93. *Present Position:* Sr Asst Dir, Nat Inst of Oceanography, Goa 1989–. *Address:* Biological Oceanography Div, Dona Paula, Goa 403 004, India. *Telephone:* (832) 46253. *Telex:* 0194216. *Fax:* (832) 53360.

UPPENBRINK, Martin; int civil servant and environmentalist. *Present Position:* Dir, European Regional Office, UNEP. *Address:* Palais des Nations, 8 ave de la Paix, 1211 Geneva 10, Switzerland. *Telephone:* (22) 7988400. *Telex:* 415465. *Fax:* (22) 7983945.

URBAN, František; Czech forester; b 15 Nov 1937, Olomouc; s of Josef Urban and Bohumila Palánová; m Irena Brunová 1961; 1s. *Education:* Prague Tech Univ, Kostelec Agric Univ, Charles Univ, Prague. *Career:* Forest Man Service 1961–66; Regional Nature Conservation Centre, České Budějovice 1966–90. *Present Position:* Dir, Nature Conservation Div, Ministry of Environment 1990–. *Address:* Vršovická 65, 100 10 Prague 10, Czech Republic. *Telephone:* (2) 7121111. *Fax:* (2) 731357.

URIARTE, Filemon, Jr, PH D; Filipino chemical engineer; b 15 Oct 1945, Philippines; s of Filemon Uriarte Sr and Paz Abuel; m Genevieve Pacis 1968; 4s 1d. *Education:* Philippines Univ, Carnegie-Mellon Univ, Pittsburgh, PA and Denver Univ, CO, USA. *Career:* Dir, Nat Inst of Science and Tech 1983–86; Dir for Science and Tech, ASEAN Secr 1986–89; Under-Sec for Science and Tech 1989; Regional Adviser on

Environmental Man, ESCAP 1989–91; Pres, World Integrated Tech Systems Inc and Foundation for Environment and Devt Studies 1991–93. *Present Position:* Dir, Functional Co-operation Bureau, ASEAN Secr 1993–. *Awards and Honours:* Earth Award 1976. *Publications:* Pollution Control in Electro-Plating Plants and Battery Manufacturing 1982; Energy, Environment and Technology Transfer 1987. *Address:* 70-A Jalan Sisingamangaraja, POB 2072, Jakarta 12110, Indonesia. *Telephone:* (21) 712272. *Fax:* (21) 7398234.

V

VAN ERMEM, Raymond; environmentalist. *Present Position:* Sec-Gen, European Environmental Bureau. *Address:* 22 rue de la Victoire, 1060 Brussels, Belgium. *Telephone:* (2) 514-12-50. *Telex:* 62720. *Fax:* (2) 514-09-37.

VAN KOOLWYK, Dr Theo; Dutch environmentalist; b 6 April 1956, Eindhoven; m Marja Van Tienhoven 1987; 1s 1d. *Education:* Amsterdam Univ. *Career:* Research in taxonomy and ecology 1978–82; research in marine ecology, Europe and India 1978–79; research in freshwater ecology, Cen Africa 1980–82; Policy Adviser, Ministry of Agric 1984–85; research in trade and awareness-raising, WWF and IUCN, Netherlands 1985–86; Founder, Environment and Devt Service for NGOs (Both ENDS) 1987–. *Present Position:* Exec Dir, Environment and Devt Service for NGOs (Both ENDS) 1991–. *Address:* Damrak 28, 1012 LJ Amsterdam, Netherlands. *Telephone:* (20) 6230823. *Fax:* (20) 6208049.

VARTÍKOVÁ, Elena, PH D; Slovak botanist and environmental teacher; b 13 March 1945, Kisač, Vojvodina, Yugoslavia; m 1969 (divorced 1993); 1s 1d. *Education:* Comenius Univ, Bratislava. *Career:* Lecturer, Comenius Univ, Bratislava 1967–79; Specialist, Nature Protection Dept, Slovak Ministry of Culture 1979–80; Specialist in Environmental Educ and Int Affairs, Slovak Union of Nature and Landscape Protectors, Bratislava 1981–93. *Present Position:* Int Co-ordinator, Greenway 1993–. *Address:* POB 163, 814 99 Bratislava, Slovakia. *Telephone:* (7) 827-213. *Fax:* (7) 827-213.

VENIZELOS, Lily Theresa; Greek conservationist; b 1933, Athens; m Lefteri Venizelos 1954; 1d. *Career:* Marine turtle conservationist 1983–; mem, Council, Hellenic Soc for Protection of Nature (HSPN) 1986–, Greek Animal Welfare Fund (GAWF), UK 1987–, Inaugural Council, European Region, Global 500 Forum 1992–. *Present Position:* Pres, Mediterranean Asscn to Save Sea Turtles (MEDASSET) 1988–. *Awards and Honours:* UNEP Global 500 1987. *Address:* 1c Licavitou St, 106 72 Athens, Greece. *Telephone:* (1) 3613572. *Fax:* (1) 7243007.

VENTO, Dr Domenico; Italian civil servant. *Present Position:* Dir, Meteorological and Ecological Centre, Ministry of Agric and Forestry. *Address:* Via del Caravita 7A, 00186 Rome, Italy. *Telephone:* (6) 6793376. *Telex:* 610148. *Fax:* (6) 4742314.

VERHEIRSTRAETEN, Herman; politician and environmentalist. *Present Position:* Sec-Gen, Rainbow Group (ARC), EP. *Address:* Bâtiment Ardennes, 97 rue Belliard, 1040 Brussels, Belgium. *Telephone:* (2) 234-11-11. *Telex:* 26999. *Fax:* (2) 230-69-33.

VERSCHUREN, Jacques C F, D ÈS SC; Belgian bioecologist and specialist in ecology and conservation of habitats and mammals; b 1926, Brussels. *Education:* Louvain Univ. *Career:* Biologist, Albert Nat Park, Congo 1957–62; conservation and environmental consultant 1956–92; Dir-Gen of Nat Parks and Environment, Zaire 1969–75. *Present Position:* Biologist, Inst Royal des Sciences Naturelles Belgique (IRSNB) 1948–. *Awards and Honours:* WWF Int Gold Medal 1970. *Publications:* The Ecology and Biology of Garamba Park, Zaire 1958; Le Parc Albert 1960; Mourir pour les Éléphants 1970; Évolution des Habitats en 70 Ans: Parc National des Virunga 1993. *Address:* 29 rue Vautier, 1040 Brussels, Belgium. *Telephone:* (2) 770-86-36. *Fax:* (2) 646-44-33.

VICENZIO, Raul; scientist and environmentalist. *Present Position:* Deputy Sec, Science Council, Commonwealth. *Address:* Marlborough House, Pall Mall, London SW1Y 5HX, UK. *Telephone:* (71) 839-3411. *Telex:* 27678. *Fax:* (71) 930-0827.

VIGNAROLI, Prof Vivian Beatriz Lavalle de; Argentinian teacher and environmentalist; b 18 Nov 1948, New Orleans, LA, USA; d of Hector Lavalle and Lidia Castiglioni de Lavalle; m Ernesto O Vignaroli 1973; 3s 3d. *Career:* Co-ordinator and Dir, Educ Dept, Vida Silvestre Foundation 1977–87; environmental educ workshops 1983–; Asst, 1st Pan-American Conservation Congress 1990. *Present Position:* Dir, Grupo de Educadores Ambientalistas (Environmental Educators Group). *Address:* Laprida 1373, San Isidro, 1642 Buenos Aires, Argentina. *Telephone:* (1) 743-9467. *Fax:* (1) 812-1350.

VIRBICKAS, Prof Juozas; Lithuanian scientist; b 8 Oct 1939, Anykščiai; s of Balys Virbickas and Zofija Virbickienė; m Aldona Virbickienė 1960; 1s 1d. *Education:* Vilnius Univ. *Career:* Ecology Inst, Acad of Sciences 1962–, Head, Ecology and Physiology Lab 1964–; mem, Editorial Bd, *Acta hydrobiologica Lithuanica*, *Thermal Power Generation and Environment* and *Voprosy ichthyologii*, Russian Acad of Sciences. *Present Position:* Dir, Ecology Inst, Acad of Sciences 1989–. *Awards and Honours:* Honoured Scientist in the Field of Nature Protection, Supreme Council of Lithuania 1989. *Publications:* Biology of the Pike and Perch in Lithuanian Waters 1974; Fish of Lithuania 1986; The Ecosystem of the Cooling Water Reservoir of the Ignalina Nuclear Power Plant at the Initial Period of its Exploitation 1993. *Address:* 2600 Vilnius, Akademijos 2, Lithuania. *Telephone:* (2) 359-275. *Telex:* 303266. *Fax:* (2) 359-257.

VO QUY, Prof Dr; Vietnamese biologist and environmental scientist; b 31 Dec 1929, Ha Tinh; m Le Thanh 1956; 2s. *Education:* Vietnam Univ, Moscow Univ, Russia. *Career:* Biology and chemistry teacher, secondary schools 1949–55; Lecturer then Prof, Hanoi Univ 1956–, Head, Zoology Dept 1967–75, Educ Dept 1975–80, Dean, Biology Faculty 1980–90; research on long-term effects of herbicides used in Vietnam War 1972–85; Chair, State Programme on Environmental Studies 1980–90; Vice-Chair, Man and Biosphere Cttee 1986–, Biologists Asscn 1986–and Asscn for Protection of Nature and Environment 1989–; mem, Steering Cttee, Hanoi Environmental Protection Comm 1990–. *Present Position:* Dir, Centre for Natural Resource Man and Environmental Studies (CRES), Hanoi Univ 1985–. *Awards and Honours:* WWF Gold Medal 1988; UNEP Global 500 1992. *Publications:* The Life of Birds 1978; Birds of Vietnam 1981; Vietnam's National Conservation Strategy (Co-Author) 1985; Birds of Hanoi (Co-Author) 1986; The Red Data Book of Vietnam (Co-Author) 1993; Films: Vietnam after the Fire 1988; Vietnam: A Country, Not a War.

Address: 19 Le Thanh Tong St, Hanoi, Vietnam. *Telephone:* 4253506. *Fax:* 4262932.

VOGGENHUBER, Johannes; Austrian politician and environmentalist. *Present Position:* Leader, Parl Group, Die Grüne Alternative (Green Alternative). *Address:* 1070 Vienna, Stiftgasse 6, Austria. *Telephone:* (222) 52-125-0. *Fax:* (222) 52-125-40.

VOLLENWEIDER, Richard Albert, PH D, FRSC; Canadian scientist, limnologist, environmentalist and teacher; b 27 June 1922, Zurich, Switzerland. *Education:* Zurich Univ, Geneva Univ. *Career:* Research Asst then Research Fellow, Hydrobiological Inst, Kastanienbaum, mem, Scientific Cttee 1946–54; Teacher, Lucerne Secondary School 1949–54; Fellow, Istituto Italiano di Idrobiologia, Pallanza 1954–55, Research Fellow 1959–66; Fellow and Research Asst, Limnological Inst, Uppsala Univ, Sweden 1955–57; Inst of Oceanography and Fisheries Research, Alexandria, Egypt 1957–59; Consultant on Inland Water Eutrophication, OECD 1966–68, Chair, Int Co-operative Programme on Eutrophication 1971–80; Fisheries Research Bd 1968; Prof of Biology, McMaster Univ, Hamilton 1979–. *Present Position:* Sr Scientist, Nat Water Research Inst, Centre for Inland Waters, Burlington, ON 1973–. *Awards and Honours:* Tyler Prize 1986; Int Limnological Soc Medal 1987; UNEP Global 500 1988. *Publications:* Methods for Measuring Primary Production in the Aquatic Environment (Co-Author); Scientific Fundamentals of the Eutrophication of Lakes and Flowing Waters, with Particular Reference to Nitrogen and Phosphorus as Factors in Eutrophication 1968; Eutrophication of Inland Waters: Monitoring, Assessment and Control 1982. *Address:* POB 5050, Burlington, ON L7R 4A6, Canada. *Fax:* (416) 336-4989.

VOYNET, Dominique, D EN MED; French politician; b 1958, Montbéliard; 1d. *Education:* Besançon Univ. *Career:* Co-Founder, Les Verts (French Green Party) 1984; Sec-Gen, Green Group, EP 1989–91, mem 1991. *Present Position:* Mem, Dole Municipal Council 1989–, Regional Councillor, Franche-Comté 1992–, Leader, Les Verts 1993–. *Address:* 6 rue Bauzonnet, 39100 Dole, France. *Telephone:* 84-82-28-05. *Fax:* 84-82-41-78.

W

WAECHTER, Dr Antoine; French ecologist and engineer; b 11 Feb 1949, Mulhouse; m Martine Charbonnel 1990; 1d. *Education:* Strasbourg Univ. *Career:* Co-founder, Écologie et Survie 1973; environmental consultant 1978–; Regional Fed of Nature and Environment Protection Asscns Rep, Social and Econ Cttee, Alsace 1980, Vice-Pres 1983; Regional Councillor, Alsace 1986–; Spokesperson, Les Verts (French Green Party) 1986–89, 1991–93, Cand, Presidential Election and Cantonal Election, Mulhouse E 1988, Municipal Councillor, Mulhouse and Cand, European Election 1989; mem, EP, Pres, Cttee on Regional Policy and Land Man, Substitute, Cttee on External Econ Relations 1989–91. *Present Position:* Vice-Pres, Regional Council, Alsace 1992–. *Publications:* Dessine-Moi une Planète 1990. *Address:* 26 rue de l'Est, 68100 Mulhouse, France. *Telephone:* 89-66-22-60. *Fax:* 89-56-31-91.

WALKER, Brian Wilson, MA; British environmentalist; b 31 Oct 1930, Chipping Norton; s of Arthur Harrison and Eleanor Charlotte Mary Walker; m Nancy Margaret Gawith 1954; 1s 5d. *Education:* Manchester Univ. *Career:* Business 1952–74; Founder and Chair, New Ulster Movt 1969–74; Dir-Gen, Oxfam 1974–83; Dir, Independent Comm on Int Humanitarian Issues, Geneva 1983–84; Pres, Int Inst for Environment and Devt 1985–89; Int Comm on Food and Peace 1989–. *Present Position:* Exec Dir, Earthwatch Europe 1989–. *Address:* 14 Upland Park Rd, Oxford OX2 7RU, UK. *Telephone:* (865) 515473. *Fax:* (865) 311383.

WALLACE, Catherine Ceinwen, BA; New Zealander economist and environmentalist; b 17 Jan 1952, Hamilton; d of Dr Lindsay and Dr Charlotte Wallace. *Education:* Victoria Univ, Wellington. *Career:* Fisheries Economist, Ministry of Agric and Fisheries 1980–81, 1983–85; Chair, Environment and Conservation Orgs of NZ 1985–86, 1989–, Vice-Chair 1987–88; Convenor, Antarctic and Southern Ocean Coalition 1985–. *Present Position:* Lecturer in Econs and Public Policy, Victoria Univ, Wellington 1986–. *Awards and Honours:* Goldman Environmental Foundation Prize 1991. *Address:* 3 Finnimore Terrace, Vogeltown, Wellington 6002, New Zealand. *Telephone:* (4) 389-1696. *Fax:* (4) 384-6971.

WALLEN, C; climatologist. *Career:* Organizer, Stockholm Conf; Adviser on Climatology, UNEP. *Present Position:* Climatologist, WMO. *Address:* 41 ave Giuseppe Motta, CP 5, 1211 Geneva 20, Switzerland. *Telephone:* (22) 7308111. *Telex:* 414199. *Fax:* (22) 7342366.

WARRICK, Dr R A; British climatologist. *Career:* Climatologist, Climate Research Unit, E Anglia Univ. *Present Position:* Mem, CEARS, Waikato Univ 1993–. *Address:* Private Bag 3105, Hamilton, New Zealand. *Telephone:* (7) 856-2889. *Fax:* (7) 856-0135.

WATERS, Joe; American ozone researcher. *Present Position:* Chief Researcher into Ozone Layers in Northern Hemisphere, NASA. *Address:* Fed Office Bldg 6, 400 Maryland Ave, SW, Washington, DC 20546, USA. *Telephone:* (202) 453-1000.

WEBER, Otto; Romanian politician and environmentalist. *Present Position:* Chair, Partidul Ecologist Român (Romanian Ecological Party). *Address:* Bucharest, Blvd George Casbuc 1, Block P58, Romania. *Telephone:* (1) 6312350.

WEEKS, John; British dir of devt inst and environmentalist. *Present Position:* Dir, Centre for Devt Studies, School of Oriental and African Studies (SOAS), London. *Address:* Thornhaugh St, Russell Sq, London WC1H 0XG, UK. *Telephone:* (71) 580-4812. *Fax:* (71) 436-3844.

WEIHE, Prof Wolf H; Swiss biologist and environmentalist. *Career:* Leading medical expert on health and environment. *Present Position:* Mem, Cen Biological Lab, Univ Hosp, Zurich Univ. *Address:* 8091 Zurich, Switzerland. *Telephone:* (22) 2571111.

WEREKO-BROBBY, Charles Y; Ghanaian energy consultant. *Career:* Mem, Commonwealth Science Council. *Present Position:* Energy Policy Adviser, Ministry of Fuel and Power, Nat Energy. *Address:* POB 40, Stadium PO, Accra, Ghana. *Telephone:* (21) 667151.

WERIKHE, Michael Mayeku; Kenyan conservationist; b 25 May 1956; m 1987; 2d. *Career:* Wildlife Conservation and Man

Dept 1976–77; walked from Mombasa to Nairobi (490 km) to raise awareness and funds for rhino projects in Africa 1982, across E Africa (2,300 km) 1985, Europe (2,200 km) 1988, N America 1991; Advisory Dir, World Soc for Protection of Animals 1987; Chair, E African Wildlife Soc 1988. *Present Position:* Founder, Michael Werikhe Conservation Centre 1993–. *Awards and Honours:* Guinness Stout Award 1983; E African Wildlife Soc Conservation Medal 1987; UNEP Global 500 1989; Goldman Environment Prize 1990; Heroes of the Earth Prize 1991. *Address:* POB 80310, Mombasa, Kenya. *Telephone:* (11) 433311. *Fax:* (11) 434461.

WESTING, Arthur H, PH D; American ecologist and environmental consultant; b 18 July 1928, USA; s of S W and Paula R Westing; m Carol E Westing 1956; 1d 1s. *Education:* Columbia Univ, Yale Univ. *Career:* Sr Researcher, Stockholm Int Peace Research Inst 1976–78, 1983–87; Prof of Ecology and Dean of Natural Science, Hampshire Coll, Amherst 1978–83; Sr Researcher, Int Peace Research Inst, Oslo 1988–90. *Present Position:* CEO, Westing Assocs in Environment, Security and Educ 1990–. *Awards and Honours:* UNEP Global 500 1990. *Publications:* Ecological Consequences of the Second Indochina War 1976; Warfare in a Fragile World: The Military Impact on the Human Environment 1980; Cultural Norms, War and the Environment 1988. *Address:* RFD 1, Box 919, Putney, VT 05346, USA. *Telephone:* (802) 387-5450. *Fax:* (802) 387-4001.

WIGLEY, Prof Tom M L; British climatologist. *Career:* Leading modeller of climatic change. *Present Position:* Climatologist, School of Environmental Sciences, E Anglia Univ. *Address:* Norwich NR4 7TJ, UK. *Telephone:* (603) 592840. *Telex:* 97154. *Fax:* (603) 507719.

WILCOX, Bruce; American environmentalist. *Present Position:* Pres, Inst for Sustainable Devt. *Address:* Palo Alto, CA, USA.

WILKNISS, Peter E, PH D; American polar scientist; b 28 Sept 1934, Berlin, Germany; s of Fritz Wilkniss and Else Stueber; m Edith Koester 1963; 1s 1d. *Education:* Munich Tech Univ. *Career:* Research Asst, Munich Tech Univ 1959–61; Research Chemist and Radiological Protection Officer, US Naval Ordnance Station 1961–64, Head, Nuclear Chemistry Branch 1964–66; Research Oceanographer, US Naval Research Lab 1966–70, Head, Chemical Oceanography Branch 1970–75, Chair, Radiological Cttee 1974–75; Man, Nat Center for Atmospheric Research Program, Nat Science Foundation 1975–76, Int Phase, Ocean Drilling/Ocean Sediment Coring Program 1976–80, Ocean Drilling Project Team 1980, Dir, Div of Ocean Drilling Programs 1980–81, Sr Science Assoc, Office of Dir 1981–82, Deputy Asst Dir, Science, Technological and Int Affairs Directorate (STIA) 1982–84, Dir, Div of Polar Programs 1984–93. *Present Position:* Mem, US Sr Exec Service; Sr Science Assoc, Geosciences Directorate, Nat Science Foundation 1993–. *Awards and Honours:* Designation of Wilkniss Mountains, Antarctic 1992. *Address:* 1800 G St, NW, Rm 520, Washington, DC 20550, USA. *Telephone:* (202) 357-7357.

WILLIAMS, Jonathan; British scientist and environmentalist. *Present Position:* Sr Project Leader, Centre for Exploitation of Science and Tech. *Address:* 5 Berners Rd, London N1 0PW, UK. *Telephone:* (71) 354-9942. *Fax:* (71) 354-4301.

WILLIAMS, Phil; American environmentalist. *Present Position:* Pres, Int Rivers Network. *Address:* Suite B, 301 Broadway, San Francisco, CA 94133, USA. *Telephone:* (415) 986-4694.

WINFIELD, Mark, PH D; Canadian researcher and environmentalist; b 1 Nov 1963, Toronto. *Education:* Toronto Univ. *Career:* Instructor, Toronto Univ 1991–. *Present Position:* Dir of Research, Canadian Inst for Environmental Law and Policy and mem, Exec Cttee, Conservation Council, ON 1992–. *Publications:* The Ultimate Horizontal Issue: Environmental Policy and Politics in Ontario and Alberta 1971–1992 1992. *Address:* Suite 400, 517 College St, Toronto, ON M6G 4A2, Canada. *Telephone:* (416) 923-3529. *Fax:* (416) 923-5949.

WOOLF, Timothy; British environmentalist. *Career:* Fmr Research Man, Asscn for Conservation of Energy, London. *Present Position:* Mem, Tellus Inst. *Address:* Boston, MA, USA.

WRIGHT, Wayne; Canadian environmentalist. *Present Position:* Chair, Nature Conservancy of Canada. *Address:* 794A Broadview Ave, Toronto, ON M4K 2P7, Canada.

WUORI, Matti; Finnish environmentalist. *Present Position:* Chair, Greenpeace Int. *Address:* Keizersgracht 176, 1016 DW Amsterdam, Netherlands. *Telephone:* (20) 5236555. *Telex:* 18775. *Fax:* (20) 5236500.

Y

YABLOKOV, Alexey; Russian biologist and politician; b 3 Oct 1933, Moscow; s of Vladimir S Yablokov and Tatiana S Sarchyeva; m Dil'bar N Klado 1990; 1s. *Education:* Moscow State Univ. *Career:* Head, Post-Natal Onthogenesis Lab, Devt Biology Inst, Acad of Sciences 1969–91; Chair, Fish Studies Cttee, Ministry of Fisheries 1988–91; People's Deputy, Supreme Soviet and Deputy Chair, Ecology Cttee 1989–91; Chair, Presidential Ecological Policy Council 1992–93, Interdepartmental Comm on Ecological Security, Nat Security Council 1993. *Present Position:* Presidential Adviser on Ecology and Health 1991–. *Awards and Honours:* IUCN Roll of Honour 1991; Green House Int Award 1993. *Publications:* Whales and Dolphins 1974; Variability of Mammals 1974; Levels of Protection of Living Nature 1985; Conservation of Living Nature and Resources: Problems, Trends and Prospects 1987. *Address:* Kremlin, Moscow, Russia. *Telephone:* (95) 925-23-27. *Fax:* (95) 206-09-41.

YAKOVENKO, Vasily Timofeevich; Belarusian geologist, journalist and writer; b 5 May 1936, Gomelskaya Region; s of Timofey and Vera Yakovenko; m Larisa Furmanova 1972; 2d. *Education:* Belarus State Univ, Minsk. *Career:* Engineer and geologist 1966–72; Ed, Essay Dept, *Polyma* literary magazine 1972–77; writer 1977–; Chief Ed, *Nabat* newspaper 1990–. *Present Position:* Pres, 'Chernobyl' Socio-Ecological Union of Belarus 1990–. *Publications:* The Earth Opened to Us 1971; Test Stone 1974; While a Sun is at the Zenith 1978; Getting to a Harmony 1985; Rural Public Debates 1987. *Address:* 220116 Minsk, Izvestiya ave, Home 17, Flat 72, Belarus. *Telephone:* (172) 20-39-04. *Fax:* (172) 23-90-14.

YARRANTON, George Anthony, PH D; Canadian plant ecologist; London, UK; m Madeleine Yarranton 1962; 2s. *Education:* Univ Coll, London, Exeter Univ. *Career:* Prof of Botany, Toronto Univ 1965–76; Dir, Arctic Land Use

Research Programme, Canada 1976–79; Environmental Man, Northern Pipeline Agency 1979–83; environmental consultant 1983–91; Vice-Chair, Alberta Natural Resources Conservation Bd 1991–93; Contributor, *Journal of Ecology*, *Ecology*, *New Phytologist*, *Vegetatio*, *Canadian Journal of Botany*, *Pollen and Spore*; Adviser, Natural Resources Conservation Bd, Canada. *Present Position:* Dir, Environmental Research Centre, Calgary Univ. *Address:* 2500 University Drive, NW, Calgary, AB T2N 1N4, Canada. *Telephone:* (403) 220-5271. *Fax:* (403) 282-1287.

YOUMBI, Augustin; Cameroonian environmentalist; b 9 June 1952, Mbalmayo; s of Marcel Youmbi and Marie Motoum; m Matagne Henliatte 1980; 3s 1d. *Education:* Yaoundé Univ, Brussels Free Univ. *Career:* Researcher, Nat Cttee for Man and Biosphere 1980–81, Technical Adviser 1981–86, Head of Research 1986–89; environmental consultant 1990–. *Present Position:* Sec-Gen, Int Asscn for Protection of Environment in Africa (Enviro-Project) 1990–. *Publications:* Dégradation de la végétation et des sols au Cameroun 1979; Les problèmes de l'environnement au Cameroun 1990. *Address:* BP 3110, Yaoundé, Cameroon. *Telephone:* 23-54-35. *Fax:* 22-18-73.

Z

ZACH, Peter, C SC; Slovak forest entomologist; b 15 Sept 1962, Zvolen; s of Dr Oto Zach and Veronika Zachová; m Gabriela Zachová 1990; 2s. *Education:* Zvolen Tech Univ. *Present Position:* Forest Entomologist, Forest Ecology Inst 1985–. *Address:* Štúrova 2, 960 53 Zvolen, Slovakia. *Telephone:* (855) 27-485. *Fax:* (855) 27-485.

ZÁVESKY, Aleš; Czech teacher; b 1926, Mošovce, Slovakia; m Ludmila Blumová 1950; 2s. *Education:* Charles Univ, Prague. *Career:* Dir, nature schools, Prague 1945; teacher of biology and nature conservation, Prachatice Dist 1958; founder, nature school reserve 1967, Ecological Educ Centre for Young Conservationists 1972, Dřípatka NGO 1990; radio and TV broadcaster; writer. *Present Position:* Nature trails consultant. *Publications:* Bear's Trail Guidebook 1967; Nature Trails 1989. *Address:* Rumpalová 402, 383 01 Prachatice, Czech Republic. *Telephone:* (338) 22806.

ZÁVODSKÝ, Dušan, PH D; Slovak physicist and environmentalist; b 8 June 1941, Bratislava; s of Andrej and Emma Závodský; m Eva Závodský 1965; 1s 1d. *Education:* Comenius Univ, Bratislava. *Career:* Asst Prof, Nitra Agric Univ 1963–64; Head, Deputy Dir then Dir, Radioactivity Lab, Air Pollution Dept, Hydrometeorology Inst, Bratislava 1964–; Lecturer in Air Pollution Meteorology, Comenius Univ, Bratislava 1972–; Pres, Slovak Meteorological Soc 1992–. *Present Position:* Sr Scientist, Hydrometeorology Inst, Bratislava 1993–. *Publications:* Air Pollution Analysis (Co-Author) 1981; Monitoring Trans-Boundary Movement of Air Pollutants (Co-Author) 1987; Atmospheric Ozone 1992. *Address:* Jeséniova 17, 833 15 Bratislava, Slovakia. *Telephone:* (7) 373-870. *Telex:* 93265. *Fax:* (7) 375-670.

ZHANG ZHILIANG, Prof; People's Repub of China human ecologist and teacher; b 9 Jan 1934, Wuxi, Jiangsu; m Wei Xiuzhen 1960; 2s 1d. *Education:* NW Teachers Univ, Lanzhou, Gansu. *Career:* Lecturer, Geography Dept, Lanzhou Univ 1960–80, Assoc Prof 1980–85; mem, Nat Population Asscn, Nat Sociology Soc on Population and Environment; Chair, Population Ecology Cttee. *Present Position:* Prof of Human Ecology, Lanzhou Univ 1985–, Dir Population Studies Dept. *Publications:* Relieving Poverty by Development 1993; Capacity and Migration 1993. *Address:* 78 Tianshui Rd, 730000 Lanzhou, Gansu Prov, People's Repub of China. *Telephone:* (931) 412907. *Telex:* 72144. *Fax:* (931) 485076.

ZHOU KAIYA, Prof, BS; People's Repub of China teacher; b 8 Oct 1932, Fenghua, Zhejiang Prov; s of Zhou Jingting and Wang Cuilian; m Xu Luoshan 1959; 1s 1d. *Education:* Jiangsu Coll, Suzhou. *Career:* Teaching Asst, Jiangsu Coll 1953–55, Nanjing Coll 1955–60, Lecturer 1960–78, Assoc Prof and Dir, Cetacean Research Lab 1973–83; Dean, Biology Dept, Nanjing Univ 1983–90. *Present Position:* Prof, Nanjing Univ 1983–. *Awards and Honours:* IUCN Marine Turtle Specialist Group Certificate of Merit 1989. *Publications:* On the Conservation of the Baiji 1982; A Project to Translocate the Baiji, Lipotes vexillifer, from the Mainstream of the Yangtze River to Tongling Baiji Nature Reserve 1986; The Baiji and its Preservation 1989; Baiji: The Yangtze River Dolphin and Other Endangered Animals 1991. *Address:* 122 Ninghai Rd, Nanjing 210024, Jiangsu, People's Repub of China. *Telephone:* (25) 3303666. *Fax:* (25) 3307448.

ZHVANIA, Zurab B; Georgian biologist, politician and environmentalist; b 1963, Tbilisi. *Education:* Tbilisi State Univ. *Career:* Mem, Georgia Green Movt. *Present Position:* MP; Head, Cen Council, Mtsvanta Partia (Georgia Greens). *Address:* 380012 Tbilisi, David Aghmashenebeli Ave 182, Mushtaid Park, Georgia. *Telephone:* (8832) 34-80-68. *Telex:* 212289. *Fax:* (8832) 35-16-74.

ZURITA PEREZ, Luis Gabriel; Ecuadorean economist; b 11 July 1955, Conocoto; m Laura Lagos Riederer 1977; 1s 1d. *Education:* Cen Univ, Ecuador. *Career:* Work in countryside, Youth Dept 1972–73; man, textile business 1977–89; technical asst to indigenous communities 1978–92; sustainable devt projects 1991–93; research in ecological conservation 1992–93, in water supply 1992–93. *Present Position:* Exec Dir, Alternative Social Investigation Centre (CISA) 1991–; econ and environmental consultant. *Publications:* Estudio de la Educación Laíca en el Ecuador 1973; Estudio y Caracterización de las Areas de Amortiguamiento del Parque Nacional Machalilla (Co-Author) 1993. *Address:* Calle Vargas 760 y Matovelle, 3°, Casilla 17-15-167C, Quito, Ecuador. *Telephone:* (2) 218-460. *Fax:* (2) 509-202.

ZYLICZ, Tomasz, PH D; Polish environmental economist; b 22 Sept 1951, Lublin; s of Marek Zylicz and Teresa Zmigrodzka; m Barbara Stankiewicz 1976; 2d. *Education:* Warsaw Univ, Wisconsin-Madison Univ. *Career:* Teacher, Econs Dept and Mathematics Dept, Warsaw Univ 1974–; environmental researcher and project co-ordinator 1981–; Researcher, Environment and Behaviour Project, Colorado Univ 1988–89; Head, Econs Dept, Ministry of Environment 1989–91, Adviser 1991–; mem, Scientific, Managing and Editorial Bd, *Ecological Economics* 1989–, *Environmental and Resource Economics* and Beijer Int Inst of Ecological Ergonomics, Royal Swedish Acad of Sciences 1991–. *Present Position:* Assoc Prof of Econs, Warsaw Univ 1989–. *Awards and Honours:* Pew Scholar in Conservation and Environment, USA 1993. *Publications:* Economics and Environmental Issues 1989. *Address:* 44 Długa St, 00-241 Warsaw, Poland. *Telephone:* (22) 313201. *Fax:* (22) 312846.